WORLD RAILWAY ACCIDENTS 1946-1960

Francis Voisey

ISBN: 978-1-916201-19-4

This book is dedicated to signalman Brian Stone, who first awakened my lifelong interest in railways way back in 1966.

Thanks to the staff of the Exeter Reference Library and the University of Exeter Library for their assistance while I was researching this book, to my friend June Nightingale for encouraging me to go ahead and get it finished when my enthusiasm was waning a little, and to my publishing team, without whose guidance and assistance it is doubtful if I would have been able to navigate the mysteries of getting my first book into print.

CONTENTS

INTRODUCTION

This book is the sixth of what it is hoped will be a series detailing with railway accidents worldwide. A good many books have been written over the years about railway accidents, some concentrating on single disasters, others covering accidents in a given country or geographical area. Accidents in countries such as Great Britain, Germany, the Netherlands, the United States, Australia, New Zealand and so on have been documented comprehensively in one or more books but no attempt has been made previously to detail every significant accident worldwide. This is clearly an ambitious project, but five volumes dealing with the earlier years having been completed; this book will attempt to detail the accidents which occurred between 1946 and 1960. Previous volumes have covered the periods 1811-80, 1881-1900, 1901-15,1916-30 and 1931-45. Further volumes are planned to cover the remainder of the twentieth century and the opening years of the twenty-first century. As before, an attempt will be made to include all accidents involving five or more fatalities to train passengers. A number of less serious accidents which have been considered of interest have also been included. The accidents have been grouped into chapters according to their primary cause and are dealt with as far as possible chronologically within each chapter.

The gathering of information about accidents in distant parts of the world has not been easy, though it is much easier nowadays than when this project began in the days before the internet. Outside of Great Britain and Ireland few countries seem to have had a formal body charged with investigating all serious railway accidents until recent years, though one or two major disasters such as the Bussey Bridge failure in the United States and the Baptiste Creek collision in Canada were the subject of full formal investigations even in the early days. Otherwise reliance has largely had to be placed on contemporary newspaper reports and previously published works on the subject, though the author has tried as far as possible to keep to primary sources. If the book appears to be biased towards accidents which occurred in Great Britain and the United States this is probably partly because these are the countries about which information on accidents has been easiest to obtain, though these were also in any case the countries where many of the most serious accidents did occur.

Contemporary spellings of place names have been used where appropriate and for the many American accidents, terms such as railroad, engineer, caboose and so on have been used instead of railway, driver, brake van etc. By this date the 24-hour clock was in widespread use, except in Great Britain and the United States. The 12-hour clock and 24-hour clocks will both be used depending upon the custom in the country concerned. Distances likewise will be given in miles and yards/feet or kilometres and metres depending on the practice in the country concerned.

By 1946 railways had spread to almost all parts of the world, though some countries only had very small systems or perhaps just one line.

The names and the boundaries of countries have undergone many changes over the years. The names of a number of countries and of various places have also changed. The names of locations which were in use at the time an accident occurred have been given, together with present names where appropriate, for example Ürsküb (now Skopje).

CHAPTER 1

DERAILMENTS: TRACK DEFECTS

The tracks over which trains run are obviously a vital part of the railway and any problems with the track are likely to lead to the train being derailed. By 1946 the technology involved in laying railway track had much advanced since the early days, but train speeds and weights had also much increased, while accidents occurred from time to time due to failures to keep the track in a good state of maintenance, to insufficient allowance being made for hot weather expanding the rails, to damage caused to the track from floodwaters and so on. A lack of track maintenance due to the pressures of the Second World War was also a factor in a number of countries at the beginning of the 1945-60 period. Also included in this chapter, though they were not strictly due to defects in the track, are accidents due to trains having become derailed on open points. Broken rails were another hazard which caused mishaps and worse from time to time.

Eastbound Southern Pacific Railroad first-class passenger train No. 58, *The Owl*, was derailed at 2.45 a.m. on Friday, 17th January 1947 at Lerdo, California, United States. It consisted of engine No. 4346, three baggage cars, five coaches, two sleeping cars, a dining car and four sleeping cars, all of steel construction, in charge of Conductor George Cochrane with Engineer H.P. Lawrence. It had passed Famoso, the last open office, slightly less than 8 miles west of Lerdo, at 2.36 p.m., and was travelling at an estimated 65 miles an hour when all but the engine and three leading baggage cars were wholly or partly derailed into a potato field. The fifth to twelfth cars were badly damaged. Eight passengers died and 86 were injured. The casualties were taken in ambulances to Kern General Hospital. [1]

The cause of the derailment was quickly identified as a broken rail. The rail concerned was 33 feet long, weighed 110lbs, had been manufactured by the Tennessee Coal, Iron and Railroad Company in December 1923 and had been laid in February 1925. It had shattered into many pieces; 17 were recovered.

A rail detector car had passed over the line on 7th October 1946 without finding any problem; neither had an inspection 'by the mirror method' a week before the accident disclosed anything. A westbound train had passed the spot half an hour before the

[1] The Bakersfield Californian, Bakersfield, California, United States, 17th January 1947

derailment. The Interstate Commerce Commission investigated the accident. [2]

Five coaches of passenger train P 747 left the track as it was entering the station at Welver, Germany, on Sunday 1st June 1947. Seven people died and 39 others were injured. Excessive stresses in the track on a hot day had caused the rails to buckle. [3]

On the morning of Monday 21st July 1947 the 8.30 a.m. Down London Midland and Scottish Railway express from London Euston to Liverpool Lime Street was crowded, with 800 passengers, 150 of whom had to stand. All went well until it was approaching the station at Polesworth, England, when its engine and 14 of its 16 vehicles became derailed on the outside of a left-handed curve. The locomotive and two leading coaches overturned and these coaches were badly damaged, while the ends of six following coaches were crushed together, though there was no telescoping. Four passengers were killed outright and one died subsequently in hospital. Another 19 passengers were admitted to hospital and 45 others, including dining car staff, suffered slight injuries or shock. The driver and fireman were unhurt and the fireman went forward at once to protect the opposite running lines, while another fireman travelling as a passenger went back and was able to stop a goods train on the Down Slow line. Meanwhile the Polesworth signalman, unhappy about the non-appearance of the train, had stopped the 8.30 a.m. Up express from Manchester to London. [4]

Several passengers told reporters of their experiences. 'Mr S Sugar, of Ambrose Road, Hackney, a passenger in the second coach, said: - "The first indication I had of anything amiss was of the coach swaying from side to side. When I came to I was lying face downwards on the permanent way with a seat on my back and people were walking over me."

'Mr Louis Martinelli, of West Derby, Liverpool, told a reporter: I was in the corridor between two coaches talking to a man when the train suddenly began to rock alarmingly and a gap appeared between the two coaches. The man I had been chatting to fell into the gap and was crushed to death."

'Mr Sidney Collingwood, of Quebec Road, Ilford, who arrived in London last night, said: "I was in the second coach and felt the train rocking and then heard a cracking noise as the coach left the rails. For one moment I saw the white faces of my fellow passengers and then I felt my arm being dragged along the sleepers. Piles of luggage fell on me and when the coach finally came to a halt on its side my face was pressed into the ground. Others helped me to escape and we set about releasing as many as we could."

'Mr Albert Peebles, a New Zealand seaman, who was travelling to Liverpool to join a ship with other New Zealanders and Australians, said: "My escape was a miracle. I was in the front compartment behind the engine. The coach began to list and then toppled over. I could feel the shingle of the track grazing my back. I can remember nothing after that, but eventually found myself on the side of the embankment with only a cut leg and a bruise on my head."' [5]

---

[2] Interstate Commerce Commission report

[3] *Katastrophen der deutschen Bahnen Teil II* by H.J. Ritzau, pg. 121

[4] Ministry of Transport report

[5] *The Times*, London, England, 22nd July 1947

The express consisted of 4-6-2 steam locomotive No. 6244 *King George VI* and 16 bogie vehicles, in charge of an unnamed guard with Driver Samuel Owen and Fireman Frank Parr.

The accident was investigated by Lieut. Colonel G.R.S. Wilson on behalf of the Ministry of Transport.

Driver Owen told the accident inquiry that he had had a good run until he had neared Polesworth and that his engine had been in first-class condition and was a steady runner. He put the speed at the time of the derailment at 60 miles an hour; the inspecting officer thought it had probably been somewhat higher – 65 to 70 miles an hour – after three miles of downhill running. Driver Owen said that just after he had passed overbridge No. 67 (200 yards short of the first mark of derailment) his engine had begun to roll alarmingly from side to side. He had applied the brakes lightly and after half a dozen rolls the engine had steadied itself momentarily, then it had left the rails, and he had made a full brake application. The engine had rattled over the sleepers and overturned. He had never experienced any problem previously on the curve in question.

Fireman Parr agreed with his driver's evidence.

A detailed examination of the locomotive was made at Crewe after the accident and nothing could be found wrong with it. With the condition of the engine clearly having nothing to do with the cause of the derailment, with no evidence of any obstruction on the line and with the speed of the train being within the maximum permitted, the focus of the inquiry shifted to the condition of the track.

Permanent Way Inspector Reynolds stated that relaying proposals were normally submitted to the District Engineer two years in advance and that for the section of the Down Fast line involved in the derailment they had been submitted early in 1946. He made every effort to get the last bit of life out of the sleepers in the interest of economy. He considered the sleepers and rails of the track in question (laid in 1929) to be "fairly well worn out," but even so thought that it would last with a good margin of safety for two more years. On 27th June Ganger A. Scaysbrook had informed him of movement of sleepers on chairs and a widening of the gauge by nearly ½ in. He had instructed the ganger to refasten the sleepers and had then been satisfied on a further inspection that the track was safe for the passage of trains at 70 miles an hour. He had been 'amazed' at the accident only 11 days later and added that the sleepers might not have been in as good a condition as they had appeared and that "once a thing starts to go, it goes very rapidly."

Ganger Scaysbrook stated that he had examined the Down Fast line on Sunday 20th and had found all in order. He had been on holiday the next day, leaving Sub-Ganger J. Lees in charge. The maintenance of gauge of the Down Fast Line, the ganger continued, had given continual trouble during recent months. No repair work had been possible during the severe frosts of February or March. Early in April he had re-gauged some sleepers in the Down Fast line but he had not considered the condition of the track to be so bad as to need to ask Permanent Way Inspector Reynolds for extra help and, anyway, "everyone was shorthanded, we were all alike, and there was a lot of frost work to be dealt with." In May he had checked the gauge and had refastened a few sleepers at the Atherstone end of the length. Further refastening had been done at the end of June as described by his permanent way inspector. The formation was well drained and although he was aware that the high rails of the curve were badly side cut, he had not thought it required bringing to his superior's attention. He had been generally satisfied with the condition of the track and "did not feel that the job was getting on top of him."

Sub-Ganger Lees stated that he had examined the Down Fast line at 9.30 a.m. on the morning of the accident and that he would have noticed had there been anything substantially wrong with the gauge at that time. He had witnessed the derailment.

Chief Permanent Way Inspector S.A. Saunders said he had only moved to the area eight weeks earlier. He was unable to suggest how the track could have got into such a condition that its fastenings had failed to hold under the passage of a train.

District Engineer F. Everitt said that he had very briefly inspected the track six days before the accident, together with Permanent Way Assistant to the Chief Engineer H.B. Everard, in connection with the acceptance of the proposal for relaying the stretch in 1948. He said that had the condition of the rails and sleepers been brought to his attention he would have arranged for the relaying of the line at the first opportunity, in advance of the programme.

Mr Everard informed the inspecting officer that following the unavoidable stoppage of track maintenance during the prolonged frost, District Engineers had been instructed to give priority to track maintenance, especially where the track was nearing the end of its life, while they were given a free hand to impose speed restrictions pending track renewal. Regarding the derailment he considered that the side cutting of the rails had been a material factor in the derailment, not only by facilitating the climbing action of the flange of the engine wheels but also by virtually widening the gauge, so that the thrust on the rail fastenings was accentuated by the additional lateral freedom of the axles. As a general rule he thought that rails should be turned when the side cutting had progressed downwards to within a ¼ in of the underside of the rail head.

Concluding his report, Lt. Col. Wilson noted that the track was found to have been reaching the end of its life in 1946 and had been scheduled for renewal in 1948, but he considered that the extent to which the margin of safety had decreased in the meantime had been seriously misjudged in spite of warning signs. Eventually the point had been reached where the holding power of the screw fastenings in 18-year-old sleepers had been insufficient to withstand the passage of a heavy engine at moderately high speed, while the fastenings had been unfairly stressed by the effect of the side cut rails, which in their turn had facilitated the derailment of the leading bogie wheels of the engine. The rail wear and the condition of the fastenings were thus interdependent factors, each of which had reacted on the other, while the regular wave form variations of the cant in the track had also contributed.

Responsibility for the continuing fitness of the track must, the inspecting officer stated, rest at all times with the District Engineer, but it appeared that Mr Everitt was out of touch with the position, notwithstanding the two special reminders which had been issued by the Chief Engineer, though Mr Everitt had not had the support to which he was entitled from Permanent Way Inspector Reynolds. The latter had not appreciated that the side wear of the high rails had been approaching the danger point for some time, and so the rails had not been turned. There had also been a misjudgement as to the strength of the fastenings, despite the warnings provided by the increasing difficulty in maintaining the gauge. Mr Reynolds should have brought the matter to the attention of the District Engineer, rather than relying on refastening individual sleepers.

Ganger Scaysbrook also appeared to have misjudged the position. He had done his best to maintain the gauge with the means and time at his disposal and the track had been solidly packed with clean and tidy ballast, but he had not paid sufficient attention to the cross level, thus adding to his difficulties.

No responsibility rested with Chief Permanent Way Inspector Saunders who had only recently come to the district.

The inspecting officer remarked that although the recent war had not directly contributed to the accident, accumulated arrears of maintenance had led to an acceptance of a standard of track maintenance which was inferior to that before the war. Close supervision and control of track conditions thus became even more important and since the accident the Chief Engineer had sent out an instruction requiring District Engineers to carry out special and personal inspections of main lines by engine riding fortnightly until further notice, and they had been given full discretion to impose speed restrictions where required. A separate letter instructed the District Engineer to immediately begin a programme of turning side worn rails on Sundays. A special gang had turned the rails on the Polesworth curve and replaced doubtful sleepers after the accident.

Two further points which Lt. Col. Wilson drew attention to were the regular wave form of variations in cant on the curve, which he described as being 'particularly objectionable,' and confusion as to the permissible extent of side wear before the rail needed turning, which the Chief Engineer confirmed should not exceed ¼ in. before the rail was turned, as suggested by Mr Everard. [6]

The inquest on the five victims returned a verdict of "Accidental death." Not having the benefit of having read the Ministry of Transport report, which would not have been published so quickly, the Coroner commented that the expert evidence had not established the cause of the accident. [7]

The 11.45 a.m. British Railways, London Midland Region, express from London St Pancras to Bradford Forster Square was fortunately not crowded by the time that it reached Wath Road Junction, England, on the afternoon of Tuesday 18th May 1948, many passengers having travelled on a relief train which had preceded it 15 minutes earlier, while others had left the train at Sheffield. As the train reached the junction on the Down Passenger line, travelling at 60 to 65 miles an hour, it became derailed on a 30 ft. embankment at 3.52 p.m.. The pilot engine broke away from the train, the train engine overturned, the leading coach was derailed, the four following coaches were wrecked, and the next three coaches were derailed. Six passengers were killed outright and the driver of the train engine and another passenger were fatally injured. Thirty-three passengers and a railwayman were detained in hospital because of their injuries and 21 passengers and a railwayman were less badly hurt.

The derailment occurred close to Manvers Main Colliery and was witnessed by colliery staff, so help was immediately available, including an ambulance and doctors from the colliery, and the first casualty was admitted to Mexborough Montague Hospital, three miles away, only 11 minutes after the accident. [8] Mr John Greenwood, of Mexborough, who worked in the power house of the colliery, described the rescue work. "First we got out a baby girl and a boy. Then we got out three women and a stretcher case. Then we started to try and get out a chef of the train staff, who was trapped, and there was a dead woman beside him. Two doctors helped to get the chef out. It was about three and a half hours before we could do it. We were going to try to get the woman out,

---

6 Ministry of Transport report

7 *The Times*, London, England, 28th August 1947

8 Ministry of Transport report

but they stopped us because they thought the train was going to start sliding down the embankment." [9]

The express consisted of Class 5XP 4-6-0 steam locomotives Nos. 5605 and 5609 and 12 coaches, in charge of Passenger Guard F.R.W. Chase with Driver W. Thornhill and Fireman F.W. Orford on the pilot engine and Driver Herbert Wilshire and Fireman E.A. Hubbard on the train engine.

Driver Thornhill was one of the principal witnesses at the Ministry of Transport inquiry into the accident. He stated that he had passed through the Wath Road curve at about 60 miles an hour and had seen the Manvers Main Colliery distant 'off' beneath the Wath Road Junction starter, had attended to the engine injector, and looking out again had seen that the track ahead was distorted. He went on, "Both rails were badly distorted to the left or near side: the rails were still level but the distortion was so bad I could hardly believe my own eyes. I should say the right hand rail was very nearly in the centre of the track and both rails were symmetrical – a perfect kink in the road – I applied the brake in full immediately I saw it. I am not sure whether I closed the regulator; if it was open, it was only open slightly, as it is not necessary to work the engine heavily. The next thing I knew we were up in the air. After we got over the first uplift we came down again and I should say we went about 30 yards before we came to a dead stand."

Driver Thornhill agreed that in fact he must have travelled considerably further than 30 yards before his engine stopped but said that everything had happened so quickly that he had been confused. He remembered looking back and seeing the train engine check and topple over in a cloud of dust, with the coaches piling up behind it. There had been no severe lurching or rolling short of the point of derailment, though he pointed out that the track between Sheffield and Leeds was not as good as other sections due to long stretches of ash ballast and colliery subsidences, so that drivers drove more slowly over it.

The two firemen could add little of value.

Passenger Guard Chase and Travelling Ticket Collector C.P. Henderson both spoke of having felt a violent brake application followed very quickly by a severe jolting and the train rapidly coming to a halt. Neither had noticed anything unusual prior to the brakes being applied.

Passed Fireman H.R. Groom and Fireman D.W. Green, on an engine of a train on the Up Goods line, stated that they had seen the express approach and the pilot engine become derailed before clouds of dust obscured their view.

Passed Fireman J. Swire, in charge of the engine of the relief train which had preceded the express, said that he had not noticed any problem as his train had passed the point of derailment. On the other hand Guard Reginald Adey, riding in the leading end of the last coach, said he had experienced a severe lurch between Wath Road Junction and Manvers Main which made him stagger and shook down a parcel. He picked up the parcel before looking out; when he did so there was nothing to see but a dust cloud. The severe lurch had not, however, alarmed him, and he had not reported it.

Three passengers who had been riding in the same coach as Guard Adey independently described their experiences. One of them said, "When travelling along a straight stretch of line at an estimated speed of 60 to 65 m.p.h. the rear coach suddenly commenced to sway violently from side to side to such an extent that pieces of baggage

---

[9] *The Times*, London, England, 19th May 1948

were dislodged from the racks. The rear bogie appeared to leave the lines completely and for a moment it seemed that the carriage must be derailed. This violent rocking probably lasted for three or four seconds in all."

None of the drivers of the previous three trains had noticed any problem, though Driver A.E. Llewellyn Marshall, who had passed the spot on a light engine at 30 to 35 miles an hour at 1.23 p.m., had noticed a slight roll and a bump. Upon braking and looking back, the track had appeared normal, so he had given the matter no more thought.

In charge of the length of track was Ganger E. Kitchen, who said that he had done some minor repairs to the Down Passenger line about ten days before the accident, lifting some low spots with jacks and slewing the line an inch at a couple of points. The ash-ballasted track at that point required much more maintenance than the stone-ballasted track at the northern end of his stretch, he pointed out. He said that the fishplates had been oiled early in March though not all of them had been removed, as per his instructions, only loosened before oiling, because, he said, removing them all was 'a big job.'

Ganger Kitchen had only a sub-ganger and three lengthmen under him instead of the normal five and his most experienced lengthman had less than three years' service, while the other men were constantly changing, so that he was feeling the strain of constantly having to train new staff. He had walked the track the day before the accident (Whit Monday) but on the day of the accident he had instructed Sub-Ganger T.W. Galloway to walk the length in the morning and as it was a very hot day, to inspect the passenger lines again in the afternoon.

The inspecting officer, Brigadier C.A. Langley, commented that he regarded some of the evidence of Sub-Ganger Galloway as unreliable. The latter said that he had walked the track on the morning of the accident, starting at Wath-on-Dearne at 7.30 a.m., going north along the Up Passenger line and back along the Down Passenger line, taking three hours to walk the 1¾ miles. His afternoon inspection had begun at 12.45 p.m., when he had walked along the ashes in the six feet between the passenger lines to the south end of the length, reaching there at 3 p.m., having taken 2¼ hours to cover less than a mile. He had then returned to Wath-on-Dearne and had been in the ganger's hut there when he had heard of the derailment. During the morning, he said, the joints had all been in good order, with gaps varying between ¼ and ½ in., and although it had been much hotter in the afternoon the joints had still been wide open, though he had not slackened them.

Permanent Way Inspector B. Johnson said he had last examined the track on 29th April when he had thought that the general condition of the track had been good and the joints wide. He knew the ash ballast required much attention and that Ganger Kitchen had difficulty because of constantly changing lengthmen. Johnson was on leave at the time of the accident.

Permanent Way Sub-Inspector A.E. Mycock, who had relieved Johnson, stated that he had inspected the Down Passenger line on 14th May and had found it to be good and satisfactory.

Chief Permanent Way Inspector S. Howell considered that the Down Passenger line was of "good strength," but the top and alignment were only fair and the cross levels at the north end of the Wath Road curve were poor. Track on ash ballast could not be maintained to the same standard as that on stone ballast, plus the ash on the embankment was soft and he much preferred plenty of clinker in it. Despite this he had been satisfied that the line was safe for speeds up to 75 miles an hour. He agreed that the fishplates which had been removed for inspection after the accident had not been properly oiled and considered that owing to expansion the rails had been abnormally tight on the day of the accident.

Brigadier Langley stated that there was doubt that the derailment had been caused by a severe distortion of the track. The stretch on the embankment had been lifted and packed ten days before the accident and with the subsequent warm and dry weather it was probable that the fine ash loosened by the repair work had not fully consolidated, and its resistance to lateral movement had been much lower than normal.

The evidence of Sub-Ganger Galloway that the expansion joints had been wide was unreliable. If, as he stated, the joints had been as wide in the afternoon as in the morning then the fishplates must have been so tight that the rails could not move. This was probably the case in some places due to tight bolts and insufficient oiling of fishplates. Other joints were found on the day after the accident to be only ⅛ or ¼ in. and thus insufficient to take up the full theoretical expansion of ⅜ in. per 60 ft rail caused by the rail temperature rise of 80° F on the day of the accident. The rail expansion had thus been restricted partly by tight fishplates and partly by butting joints.

On well-maintained track buckling should not occur even in the hottest weather, and responsibility for the accident must therefore, the inspecting officer remarked, rest on the members of the engineering department responsible for track maintenance.

On the day of the accident Ganger Kitchen had not walked the track but had left the task to his sub-ganger, who had apparently made only a cursory inspection of the line that afternoon. Had he made a more careful examination he night have noticed the tight fishplates and butting joints and realised the dangerous condition of the track. He could then have cautioned trains until the problem had been rectified. Ganger Kitchen had realised the dangers of track distortion but would have been more prudent to satisfy himself personally as to the safety of the track rather than leaving it to a junior member of his gang. He was, however, handicapped by constant staff changes, and had concentrated his efforts on training new men.

In general Ganger Kitchen's length was not up to the standards required of a first-class main line. There was a shortage of ballast on the embankment, the cross levels at the southern end of the length were unsatisfactory and the track just north of the point of derailment was also below standard. The ganger admitted that not all the fishplates had been removed when they were oiled and some of them were too tight.

Neither Permanent Way Inspector Johnson, Sub-Inspector Mycock, nor Chief Permanent Way Inspector Howell appeared to have fully realised the weakness of the dry soft ash ballast. Divisional Engineer Dickinson had not had any anxiety about the section of line either, though all agreed that ash-ballasted track was more difficult to maintain than stone-ballasted and did not stand up well to traffic. Brigadier Langley stated that in view of the deficiency of ballast on the embankment and the general condition of the track a speed restriction should have been imposed and that he considered that the Divisional Engineer and his Inspectors must share some responsibility for the accident.

Had Guard Adey of the relief train looked back *immediately* after he had felt his coach lurch he might have seen the track distortion, stopped his train at Manvers Main Colliery Sidings box, warned the signalman there, and the accident would have been averted. The guard could not, however, be criticised, the inspecting officer said, for not taking such rapid action as the lurch had not unduly alarmed him and he had looked back a little later, when his view had unfortunately been obscured. [10]

Guard Adey became another victim of the accident. Blaming himself for not having stopped his train he became "run down and ill," and committed suicide on 30th July. At his inquest the Coroner commented that there had been undue delay in clearing up the facts of the accident, saying, "I do not think it is sufficiently realised that delay causes certainly worry and often grief and anxiety to people concerned." [11]

Regarding recommendations, the inspecting officer stressed the need for further research into the strength of different types and conditions of ballast to resist the lateral forces set up by highly stressed rails whose expansion had been supressed by tight or butting joints. The need for loosening fishplates during hot weather should not need emphasising, he continued, but clearly had not been fully appreciated.

It was doubtful whether the general maximum permitted speed of 75 m.p.h. for the region should be permitted over ash-ballasted track unless it was fully consolidated and maintained in first-class condition with sleepers fully boxed and ample shoulders. The substitution of stone ballast was highly desirable where the formation was stable and there was no expectation of colliery subsidence.

The staff difficulties experienced by Ganger Kitchen appeared to be general but it was hoped that they would soon be overcome. [12]

Tragedy struck on the Canadian Pacific Railway at 12.45 a.m. on Thursday 27th May 1948 at a culvert over the flooded Rock Creek two miles west of Lundbreck, Alberta, Canada. Three and a half hours earlier a westbound work train had passed over the culvert without difficulty but when a freight train, consisting of empty waggons bound from Lethbridge to collieries in the Crow's Nest Pass, reached the spot it was derailed and the locomotive and ten coal cars plunged into the water. Engineer William Dommett, Fireman Edward James and Head End Trainman Gordon Cripps were trapped in the cab of the engine and drowned. Passenger train No. 11 was due to cross the culvert at about 1 a.m. but by chance was 40 minutes late and the empty waggon train was allowed to precede it westwards.

The waters of Rock Creek, a tributary of the Crow Nest River, which it joined a short distance downstream, were carried under the railway in a concrete culvert. On the day of the accident either the flow of water had been simply too great and had overwhelmed the capacity of the culvert, or else it had been blocked by debris brought down from upstream by the floodwater. As a result the water had accumulated on the north side of the railway until it had formed a lake about 20 feet deep. The grade had been washed away over some distance and the track left completely unsupported. Engine No. 5218 remained upright in the river and the cars piled up on top of it, some of them telescoping one another. Passenger train No. 11 was stopped by the conductor and other railwaymen at the rear of the freight train and the alarm was raised. Mail and express carried on train No. 11 was forwarded by truck and the train returned eastwards as eastbound train No. 12. It was expected that it would take 48 hours to reopen the line. [13]

---

[10] Ministry of Transport report

[11] *The Times*, London, England, 2nd August 1948

[12] Ministry of Transport report

[13] *The Lethbridge Herald*, Lethbridge. Alberta, Canada, 27th May 1948

A passenger train from Chuncheon to Seoul was derailed near Namyangju, South Korea, 20 kilometres from the capital, on Friday 6th August 1948. Four coaches of the train fell down a 100 metre embankment. Initially it was stated that seven passengers, all Koreans, had been killed and about 100 injured, but American newspapers were more concerned with 49-year-old Veronica Welsko, a civilian nurse attached to the military government's health department, who was an American from White Haven, Pennsylvania. She had suffered a fractured pelvis and was taken to the Army hospital in Seoul, where her condition was later stated to be satisfactory. Two of the injured passengers later died and the number of injured was re-estimated at 129. At first there was a suspicion that the derailment had been the result of an act of sabotage [14] but later it was subscribed to track subsidence following heavy rain. [15]

Rather more than a year after the Wath Road Junction derailment it was the turn of the French to suffer an accident from the same cause at Emberménil, in the Est region of France, on Monday 4th July 1949. The accident was not, a French newspaper declared, the fault of the train driver, or of a signalman, or of defective material.

Express train No. 12, which had left Strasbourg for Paris at 11:27 consisted of a locomotive and eight vehicles, including a postal van and a restaurant car. As it was 200 metres from Emberménil station at 12:48 and approaching a curve at 110 kilometres an hour, its driver felt an abnormal resistance. He applied the brakes and his engine continued for a further two kilometres, with the two leading vehicles attached, before it stopped. The fourth vehicle, a steel third-class coach, left the track and its bogies were caught by the sleepers and torn off. Its momentum carried it 100 metres along the track, then it overturned on its left-hand side – the corridor side – and literally planed (as a reporter described it) along the opposite track. Behind it the couplings snapped. The next vehicle, the restaurant car, was bumped by the following coach and overturned at 45 degrees. Twenty-seven passengers were enjoying lunch in the car at the time. Only one of the diners was hurt, together with a dishwasher who was struck in the back by a cooking pot. The last three coaches also overturned but only a few passengers in them were hurt.

Most of the deaths and injuries occurred in the fourth coach. The train was busy, though not as crowded as it would have been on 14th July (Bastille Day). As it was a hot day though the passengers had opened the windows and had been standing with their elbows resting on the window bars. When the coach overturned a number had been crushed between the coach and the ground and seven of them were killed or fatally injured. [16] About 40 other passengers were injured, a dozen of them seriously.

Société nationale des chemins de fer français (SNCF) technicians immediately proceeded to the accident site and quickly established that the derailment had been due to the track having been buckled by the heat of the sun. They also noticed that there were a number of sleepers which were in poor condition and which were marked with a white cross indicating that they required replacing. It was clear that the steel carriages had stood up well to the derailment and had prevented worse consequences. [17]

---

[14] *Oakland Tribune*, Oakland, California, United States, 6th August 1948
[15] *Pacific Stars and Stripes*, Tokyo, Japan, 9th August 1948
[16] *L'Aurore*, Paris, France, 5th July 1949
[17] *La Croix*, Paris, France, 6th July 1949

The train known as *El Entrerriano* (named after a tango written in 1897) left Federico Lacroze station in Buenos Aires, Argentina, at 12:15 on Friday 30th December 1949, bound for Posadas. It was carrying many holiday passengers, also parcels and general goods. All went well until it was crossing an arroyo 200 metres short of the River Luján, but here a number of its vehicles left the track. The accident occurred between Empalme and Pilar, about four kilometres from the latter place and 51 kilometres from its starting point. The train had covered the distance in about an hour.

There was only a small amount of water in the arroyo, despite the recent heavy rain, which made the rescue of the victims of the accident, according to one report, a little easier. The carriages were, however, badly damaged.

The passengers who survived the accident said that there had been a moment of great panic when their train had become derailed. They explained that their train had been running normally when the carriages had suddenly begun to jerk violently, which was what had frightened them. Moments later several of the coaches left the track. Some of them had overturned and others had been left leaning over dangerously. Three of them were precipitated into the arroyo and completely wrecked.

An initial police report stated that the bodies of three passengers had been retrieved from the wreckage. They were those of Martin Bastos, a barman, and two children; Lucy Raquel Ross, aged 13, and Marta Ross, aged 10. The rescue operation was reported to be proceeding as vigorously as possible, but the situation of the coaches which had fallen into the arroyo made the task difficult, because as well as being shattered, they were half covered by the water. Breakdown cranes sent by the railway were not able to immediately reach the coaches in the arroyo because the line was blocked by other derailed vehicles. There were known early on to be eight dead, but it was impossible to determine if there were more than that until the carriages had been hoisted back onto the line. Four of those who had lost their lives were soon identified as the mother, father and two children of the Ross family. Several national deputies were passengers on the train; they were all said to have been unhurt. [18]

The railway company stated that the derailment had resulted from the recent heavy rains having damaged the track. [19]

A passenger train on the Occidente line of the Ferrocarril del Pacifico was derailed near Buenaventura, Colombia, on Wednesday 22nd March 1950. According to a statement from Doctor Daniel Lorza Ayora on behalf of the company, four people lost their lives in the accident and another 28 were injured. Those passengers killed were Gerardo Montoya Hernández, aged 23; Alvaro Torres, aged 10; and a third person who could not immediately be identified. The fourth fatality was the conductor of the train, 36-year-old Hernán Maldonardo Oliveros. Ironically the conductor, who had been with the company 15 years, had tendered his resignation and was working what would have been the last month of his service. The injured were taken to the Santa Elena hospital in Buenaventura. A report from there the day after the accident stated that the lives of another passenger, Luis Climaco Vegara, and of Brakeman Galvis of the train were despaired of; the latter had suffered a head injury and had lost his mind.

The cause of the derailment was unclear. Some of the passengers alleged that the train

[18] *El Orden*, Santa Fe, Argentina, 31st December 1949
[19] *The Times*, London, England, 31st December 1949

had been going too fast, causing suitcases to fall from luggage racks on the curves, but more informed reports blamed it on track subsidence. Driver José López and 'other members of the train crew' were detained pending an investigation. A newspaper pointed out that several accidents had occurred previously in the same area as the latest one, at a curve between kilometres 34 and 35, and there was a need for a study to discover how to eliminate the dangerous curve at that point. [20]

On Thursday 30th March 1950 mail train No. 982 from Madrid to Córdoba and Málaga became derailed on a set of facing points as it was entering the station at Bobadilla, Spain, the engine, a mail van and the following mail coach all leaving the track. A third-class carriage, which was occupied almost entirely by conscripts from Lérida bound for Melilla, was pushed under the mail coach and the rescue of its passengers was very difficult. The initial rescue work was performed by railwaymen, members of the Civil Guard and members of the air force stationed at Bobadilla. Eight people, seven soldiers and a woman bound for Málaga, were killed, and 30 others were injured. Rescuers from Antequera and Málaga left for the scene immediately upon hearing of the accident. The first injured arrived at Antequera at 18:15 and were admitted to the hospital there and to a health clinic.

Once all the injured passengers had been cared for, attention turned to clearing the line so that the train could continue to Málaga. Travelling on the train had been 200 members of the Frente de Juventudes (the youth section of the FET, the only political party permitted in Spain during the Franco dictatorship) from different Spanish provinces who were to participate in the Campeaonato de España (the annual national knockout football competition) between 1st and 3rd April. Fortunately, they all escaped injury. [21]

The train, which had left Madrid at 22:15 the previous night, had been due to call at Bobadilla at 15:25, thence to run non-stop to Málaga, where it was due at 16:45. The accident apparently occurred at about 17:00, so the train must have been considerably behind time. [22]

An overcrowded passenger train of the Estrada de Ferro Leopoldina from the Barão de Mauá terminus in Rio de Janeiro to Campos dos Goytacazes and Vitoria, Brazil, was involved in a major catastrophe in the first half of 1950. Leaving Rio at 20:45 the previous evening the train had arrived at Itaborai at about midnight. Here it was combined with a train from Niterói. The united train, which now carried about 700 passengers, continued its journey through torrential rain. At about 01:00 on Thursday 6th April 1950 it reached Tanguá station, where there was a large sugar mill, and shortly afterwards it was crossing a bridge over the Indios River, between Tanguá and Rio Bonito. It was here that disaster overtook it. According to initial reports the bridge, whose foundations had been fatally weakened by an exceptionally large volume of water in the river, collapsed beneath the train, though from the engineers' reports (see below) it was in fact one of the approach embankments which had failed, derailing the train. The engine, luggage van and six coaches fell down the embankment and one coach was carried into the middle of the river, the others remaining near the bank. The train had been composed

[20] *El Tiempo*, Bogotá, Colombia, 24th March 1950
[21] *La Vanguardia*, Barcelona, Spain, 31st March 1950
[22] Webpage: Federación Castellano Manchega de Amigos del Ferrocarril (in Spanish)

of 28 vehicles, five of which were later taken on to Campos and Vitoria by another engine.

There was reportedly an indescribable panic among the passengers when the accident happened and the scene was described as Dantonesqe. There was endless confusion in the darkness and the incessant rain, with the cries of the injured and the moans of children, while the uninjured and slightly injured passengers worked to pull their less fortunate fellow passengers out of the wreckage and the muddy water.

A special train was sent to the scene from Rio de Janeiro with doctors and equipment as soon as news of the disaster was received there. It left Niterói at 04:00.

Reports from the accident scene described the situation as the most despairing and confusing possible. Upon learning of the disaster the governor of Rio de Janeiro province decreed three days of mourning on the 6th, 7th and 8th and announced an exceptional grant of 300,000 cruzeiros to cover the expenses of the rescue operation and the re-opening of the Norte main line. [23]

By 17:00 the next day the bodies of 40 victims had been recovered, 70 or 80 others were still missing, and there were 40 injured, 15 or 20 seriously. Driver Abilio Miranda de Jesus was seriously injured and Firemen Alvaro da Silva and [24] Joelsan Ataida Correla (name illegible in newspaper) were both killed. The driver was unable to explain what had happened other than that his engine had suddenly left the track and fallen down the embankment. The train had been travelling at 15 or 18 kilometres an hour at the time. Upon being told of the death of his colleagues the driver hid his grief, thanking God for saving his life. [25]

On 9th April the exact number of those who had died was still not known. There were 57 bodies in the mortuary at Niterói. It was believed that many of the panicking passengers had thrown themselves through the windows of the carriages of the train. Señor Gesu Lima, assistant administrator of the Leopoldina company, stated that by this time all the coaches of the train had been removed from the river apart from the first Pullman car. The removal of this vehicle had proved extremely difficult and had been temporarily abandoned. It was almost completely submerged, with only its roof visible, and buried in that part of the river where the water was deepest. Concerned that there might be more bodies in this coach he had asked for a diver to examine the vehicle, but he was unable to verify if there were any victims in it. The Pullman had a capacity of 32 seated passengers, and it would be impossible to be certain if there further victims until the coach had been raised.

Engineer Alcides Lins, technical director of the company, outlined the mechanism of the disaster. 'The real cause of it was the floodwater. The waters had washed away part of the approach embankment, causing it to subside and the train to derail as it reached the bridge, with such regrettable, tragic, and inevitable consequences.' What had made matters worse was that it was such a large train, with 25 coaches. As for the overcrowding of the train, that was understandable as there had been hundreds of people wishing to travel to the interior to spend the Easter holiday with their families. This was always the case at such times on the Leopoldina as on other lines. At Barão de Mauá, according to the train conductor and the stationmaster, the crowds of passengers wishing

---

23 *O Estado de São Paulo*, São Paulo, Brazil, 7th April 1950
24 *Jornal do Brasil*, Rio de Janeiro, Brazil, 7th April 1950
25 Ibid, 10th April 1950

to board the train had been so great that the police had had to be called in to control them.

The engineer added, mysteriously, that this was not the worst railway disaster to have occurred in Brazil as there had been one on the Central do Brasil in 1919 in which a train had fallen from a bridge and caught fire, killing 'dozens and dozens' of passengers and injuring hundreds more. Despite an extensive search it has not proved possible to trace this disaster.

Engineer Feliciano de Sousa Agular, the commercial director of the company, added, "What happened was really an unpredictable catastrophe. Impossible to prevent. During the night a very severe storm caused a torrent which diverted the course of the river in such a way as to cause damage which is still visible. The embankment adjoining the Indios bridge was undermined. Because the water had risen to rail level at the bridge it was impossible for the driver to spot the danger. As soon as the engine left the bridge the embankment gave way and the engine was derailed, dragging seven coaches with it. Matters would have been worse had not a passenger boarded the train at Tanguá station with his wife and three children, who had been killed. Had the train not stopped at that station for them to join it its speed would have been higher and the consequences more serious still. The sleeping car passengers escaped, apart from suffering bruises and shock."

Asked if the excessive number of passengers could not have contributed to the disaster, he replied, "Indeed, since as far as we know about 130 passengers were travelling without tickets, in addition to those who were packed in the vestibules between the coaches. It should be noted that we had to request police assistance to deal with the crowds at the stations."

He went on to say that work of repairing the embankment was already well under way. About 15 metres had been damaged and a number of workmen were busy filling the gap with stones and sand so that it was hoped to be able to reopen the line the following day.

Driver Miranda de Jesus, interviewed again in hospital, said that due to the stop at Tangua station his train had not been going at more than 15 kilometres an hour as it had crossed the bridge. The journey had been normal and he had instructed his firemen to maintain the boiler pressure as it was or to try and increase it a little if possible. It was raining in torrents as his engine crossed the bridge. It had already reached the far bank when he felt it dropping violently. Realising what was happening he had struggled to save himself, and with great difficulty he managed to get out of his cab, managed to cling on to a piece of iron and had made it to the riverbank. [26]

The total number of passengers who died in the accident is uncertain, though reports in some American newspapers a few days later of 200 deaths are almost certainly exaggerated. At least 87 and possibly 110 died. [27]

An announcement from Radio Belgrade stated that five passengers had died and four were injured when a train became derailed as it was entering the station at Ripanj, Yugoslavia, 15 miles from the capital on Friday 7th April 1950. The accident was due to a mistake by a signalman, who had reversed a set of facing points as the last carriage of the train was still passing over them. [28] [29]

---

[26] *O Estado de São Paulo*, São Paulo, Brazil, 9th April 1950

[27] *Rushville Republican*, Rushville, Indianan, United States, 11th April 1950

[28] *The Sunday Star*, Washington, D.C., United States, 9th April 1950

Bridge Junction, Doncaster, England, on British Railways, Eastern Region, was the location of the next serious derailment, on Friday 16th March 1951. The 10.06 a.m. Up express from Doncaster to London King's Cross had only just departed from its starting point when it became derailed at a scissors crossing leading from the Up Slow to the Up Main line at a speed of 20 to 25 miles an hour. Fourteen passengers lost their lives, 12 were seriously injured and 17 were slightly hurt or suffered from shock.

Mr Leslie Shadbolt, of Hull, travelling with Mr Albert Rogers, also of Hull, said: "We knew there was going to be an accident when the train started to vibrate and sway. Another passenger told us to pull our leg up, and we all hung up to the sides. Then there was a long rending crash, a tumult of noise, and to our horror we saw the carriage behind coming past us by the side of our windows."

'Mr Rogers said: "The whole of the corridor caved in on us. The carriage seemed to rise in the air and come apart, and the next minute it was all a mass of wreckage. There was a woman standing in the midst of the debris with timber all around her. I broke a window and got out and saw that every passenger in the next compartment was trapped. I broke a compartment window and, with a porter, helped them to safety."

'The Rev. J.H. Wilson, whose home is in Cornwall, said: "We were just out of the station and going slowly when it happened. Our carriage went over on its side. I was thrown on the floor. There was very little panic – just a few children screaming. One of the carriages was split right in two."

'Rescuers seeking the dead and injured were faced with a great heap of wreckage packed against the pier {of Balby Bridge} and reaching to the underside of the bridge. A policeman who was amongst the earliest helpers described how quickly the victims were extricated from what looked an impenetrable tangle of metal. Two were found pinned underneath the front bogies of the fourth coach. Eight bodies were ultimately recovered from this part of the wreckage – one that of a baby. The rescuers were equipped with jacks, acetylene-burners, and axes, and it was 1.45 p.m. before the last body was recovered. Elsewhere among the derailed coaches passengers jumped or were lifted or dropped down to the ground. In a short time members of the W.V.S. {the Women's Voluntary Service for Civil Defence – formed in 1938 for air raid defence} were helping housewives from the neighbourhood to provide tea for the injured, who were taken to Doncaster Royal Infirmary.' [30]

From No. 4 platform, from which the express had started, southbound trains could reach the Up Main line either via a crossover at the end of the platform or along the Up Slow line and thence to the Up Main via a scissors crossover 600 yards south of the platform. The express was routed over the second crossover. At the scissors crossover the Up Main line was on a curve with 4 inches of super-elevation and the cant through the trailing lead from the Slow line had to be run up very steeply between the V crossing and the switches. There was a 10 m.p.h. speed limit along the Up Slow line from the platform to Bridge Junction box, south of the crossover.

The train consisted of A2/2 class 4-6-2 steam locomotive No. 60501 *Cock O' the North*, six coaches which had formed the 8.45 a.m. train from Hull and eight coaches and a horse box which had formed the 9.15 a.m. train from York. Manning the engine were

---

[29] *The Independent*, Long Beach, California, United States, 9th April 1950

[30] *The Times*, London, England, 17th March 1951

Driver D. Wadsworth and Fireman R.B. Darley.

The engine and leading two carriages negotiated the crossover safely but the trailing bogie of either the third or fourth carriage was derailed to the left just as it cleared the trailing V crossing of the Up Main connection and the following six coaches were also derailed to the left at the same point. The first pair of derailed wheels kept fairly close to the track alignment until they reached another crossover 60 yards further on where they were diverted further to the left. The rear of the third coach and front of the fourth were carried further and further to the left as the front of the train continued to move forward until the third coach struck a pier of Balby Bridge. The coach ended up virtually wrapped around the pier, crushed against the brickwork, and demolished. All the fatalities were in this coach.

Stationmaster J.E. Fisher of Doncaster told the subsequent Ministry of Transport inquiry that the express had made a normal departure from Platform 4, assisted for a short distance by the 0-6-0 shunting engine which had attached the York coaches to the rear of the train. The train had been travelling at about 10 m.p.h. as it had cleared the platform.

Signalman C.C. Cutforth in Doncaster South box explained that he sent the express forward along the Up Slow line because he had accepted another train on the Up Main. He had obtained *Line Clear* for the train from Bridge Junction at 10.07 a.m. and cleared his platform Starting signal (No. 32), a shunting signal (No. 108) and his advanced starting signal (No. 40). The train had been travelling at between 10 and 15 m.p.h. as it passed him.

Signalman P. Cooper in Bridge Junction box stated that having accepted the train from Doncaster South he had obtained *Line Clear* from Balby Junction and cleared his Home signal (No. 38). He had watched the train passing through the scissors crossover at what he considered an "average" speed and it passed his box, accelerating, at 15 or 20 m.p.h. He had seen plenty of trains negotiating the crossover at higher speeds. As the train passed he had felt his box shaking and seen coaches rolling about, then there had been a loud crash as the train hit the bridge and he had sent the *Obstruction Danger* signal in both directions.

Signal and Telecommunications Labourer N. Peach, who was working near Balby Bridge, said that he saw the express the express approaching. Its engine passed him and he was about to resume work when he saw sparks coming from, probably, the third coach and then the coaches left the rails and were coming directly towards him. He barely had time to jump into a recess before the coach hit the brickwork. There was nothing abnormal in the speed of the train as it negotiated the crossover but its driver had seemed to "open up" between the crossover and the bridge.

Driver Wadsworth said that the platform starting signal had been yellow, the shunting signal off and the advanced starting signal green. He had started normally with the regulator open "to half of the first valve." After slipping once the wheels had gripped. He had not wanted to accelerate too quickly because of the restriction through the scissors crossover; he estimated the speed of his train through the crossover as 15 m.p.h. At the north end of Balby Bridge he had opened the regulator fully and about halfway through the bridge he had seen the needle of the train pipe vacuum gauge drop to zero. Assuming that the guard had applied the brakes he had closed the regulator and the engine had stopped just beyond the bridge. He had not noticed any lurching as the train passed through the crossover, nor any sudden jerk when the couplings between the second and third coaches had separated. As soon as his train had stopped he sent his fireman forward to protect the down line with detonators and to advise the Bridge Junction signalman. Driver Wadsworth admitted that he had previously driven through the crossover at speeds up to 20 miles an hour, though he knew that there was a speed restriction of 10 miles an hour through it. The day after the accident he stated that he had come through the crossover at 25 miles an hour on the day of the disaster, but at the inquiry he explained that he had been overwrought then and had overestimated his speed.

Fireman Darley corroborated his driver's statement.

Sub-Ganger T.F. Molloy was Acting Ganger for the section involved. He stated that he examined the length daily but had not reached the scissors crossover on the day of the accident before it had happened. He had not noticed that a bolt was missing from the right-hand trailing V crossing or that a bolt was loose in the trailing V crossing opposite on his inspection the previous day; the day before that the right-hand bolt had been present and the opposite bolt tight. That day he had noticed sleepers and a worn out packing piece under a chair but had not considered the defects such as to need immediate attention. He had seen one or two trains passing through the crossover at speeds that he put as high as 25 m.p.h. but had taken no particular notice of them. He was unaware of the 10 m.p.h. restriction.

Permanent Way Inspector G. Taylor said that he had examined the track after the accident; it was he who had found the missing and loose bolts. He did not consider that a missing bolt would have unduly weakened the crossing and would have permitted traffic over the crossover at 5 m.p.h. even if all the bolts had been out. He could not account for the derailment and did not think that an engine, even going considerably faster than permitted, would have broken the bolts, and burst open the crossing. He had seen one or two trains going through the crossover at "a nice speed – 10 to 15 m.p.h." He could not recall any going faster than that, though he said that he found it hard to judge speed.

A number of speed tests were conducted to try to determine the speed of the express through the crossover, using the same engine and a train of identical composition and the speed of other Up trains through the crossover was noted.

Lieut. Colonel G.R.S. Wilson and Brigadier C.A. Langley jointly investigated the accident. They stated that it had been unclear whether the initial derailment to the left had been caused by a failure of the wing rail supports or whether the exceptionally steep cant gradient in relation to the speed had forced the wing rail aside. A length of track was therefore laid to represent the scissors crossover and further tests were conducted with coach underframes. The results of the investigations into 13 previous derailments of passenger trains at trailing V crossings since 1901 were also examined.

As regards the speed of the express the inspecting officers agreed that the residual speed when the train became divided must have been 15 m.p.h. and the speed at the moment of derailment considerably higher. They noted that shortly after the derailment trains were still negotiating the scissors crossover at speeds of 20 m.p.h., twice as high as the maximum permissible speed.

Examination of the engine and carriages of the express showed no defects.

An unusual feature of the accident was the exceptionally steep cant gradient through a section of the scissors crossover. At a point 5 feet behind the trailing crossover the cant was only ⅛ inch, while 33 feet further on it had risen to 3⅝ inches, while at the point of derailment the cant was 1 in 81. The voids under the sleepers and the decayed packing piece might have accentuated the gradient. The curvature of the crossover might also have accentuated the oscillation of the carriage bogies, allowing a flange to mount the rail and initiate the derailment.

In conclusion, the inspecting officers stated that although the steep cant gradient and sharp curvature could, in their opinion, have given rise to dangerous conditions at speeds of 20 to 25 m.p.h., the damage to the right-hand crossing could not be reconciled to an initial derailment to the left. The behaviour of a train during a derailment was unpredictable but the evidence pointed, they considered, more strongly to the failure of the right-hand wing rail having been the cause rather than the result of the derailment. The disaster had thus been initiated by the bursting of the crossing. The speed of the train was considerably higher than that permitted but no serious results would have followed had it not been for the proximity of the bridge.

Although the speed of the express was not, it appeared, a major contributory cause of the derailment it could not be left out of consideration and the derailment either would not have occurred or would have been less serious at 10 m.p.h. Driver Wadsworth was at fault for taking the crossover at too high a speed but his engine was not fitted with a speedometer and the speed might seem deceptively slow on a large engine. It had also been commonplace for some time for the speed restriction to be ignored without any action having been taken to ensure that it was obeyed.

Sub-Ganger Molloy had not been paying sufficient attention to the maintenance of the scissors crossover, given the pumping sleepers, worn out packing piece, missing bolt, and loose bolt. Neither he nor Permanent Way Inspector Taylor appeared to have appreciated the standard of maintenance required for heavy passenger traffic.

With the unavoidable break in the continuity of the rails the crossing was an unavoidable weak point, with all its components continually subjected to severe shock loading and vibration from the traffic. The strength of the crossing depended largely on the integrity of the transverse bolts. Here, one was missing and the other two had fatigue fractures which probably led to their failure. It was desirable that the behaviour of crossing bolts in service and whether they should be periodically renewed depending on the character and strength of traffic should be investigated.

The scissors crossover had existed since 1910 though the 10 m.p.h. restriction had only been imposed four years earlier. At some points steep cant gradients were unavoidable but at the request of the inspecting officers the layout at Bridge Junction had been examined and it was proposed to replace the scissors with a simple crossover with flatter crossings and improved cant gradient so that it could be negotiated safely at 20 m.p.h.

In contrast with practice in other parts of the world British locomotives were not fitted with speed recorders. It was recommended that these should be provided on the new standard-class locomotives and in due time on all engines used on important passenger services.[31]

At the end of May an inquest jury returned a verdict of "Accidental Death" on the victims of the accident, adding a rider that 'more supervision should be given to the speed limit over crossings.'[32]

---

[31] Ministry of Transport report

[32] *The Times*, London, England, 30th May 1951

Heavy rain caused flooding which washed away a culvert and caused the derailment of a Calcutta-bound express at a location about 220 miles from Gauhati, Assam, India, on Sunday 26th August 1951. The engine and three bogies (coaches) were derailed, the remainder of the train remaining on the track. An official statement by the Assam Railway was that there were seven dead but passengers related that they had counted 11 bodies and several others might have been washed away. [33]

Ten people – five men, three women and two children – reportedly lost their lives in the derailment of the *Shahdara Express* on the narrow (2 ft. 6 in.) gauge Shahdara-Saharanpur Light Railway at a point 11 miles from Delhi, India, on Monday 3rd September 1951. About 30 others were injured and were brought to Shahdura on a relief train. Five vehicles, including two passenger coaches, derailed, and overturned. The exact cause of the accident was not immediately clear, but a contributory, if not the main cause, was stated to have been the ramshackle condition of the track.

The Railway Board had been considering taking over the small private company running the line for several years but the heavy financial investment involved had resulted in a postponement of the takeover and the recent extension of the company's lease for another five years. [34]

The line was eventually taken over by Indian Railways and was converted to broad gauge in 1970. [35]

A major catastrophe in Brazil on Tuesday 4th March 1952 had its origin in the derailment of a freight train due to a broken rail. The Estrada de Ferro Central do Brasil freight train involved, No. S3, was en route from Dom Pedro II station in Rio de Janeiro to Santos Dumont, in Minas Gerais state. It was in charge of Conductor Cirne Kopke with Driver Januario Felisberto (surname illegible in newspaper). It had not travelled very far and was crossing a bridge over the muddy Rio Anchieta stream near Anchieta station when it ran over a broken rail at 08:40. (Most reports on the disaster speak of the larger Rio Pavuna but a Brazilian newspaper specifies the smaller Rio Anchieta stream.) Several tank waggons loaded with petrol were derailed and fouled the opposite running line just as an electric suburban passenger train from Nova Iguaçu was passing. The four rear cars of the passenger train, No. UM-3, bound for Rio de Janeiro, were derailed in turn. As was usual the train was grossly overcrowded, to the extent that passengers were hanging onto the outside of the cars, riding in the vestibules between them and so on, which must have increased the number of casualties. In all no less than 119 passengers were said to have lost their lives and another 250 or so were injured.

One tragic case recorded in a newspaper was that of a mother travelling on the passenger train with her children. She was sat on the side of the train adjoining the opposite running line and was killed instantly when the derailed waggons struck the side of her coach. Because there had been no room for her children to sit beside her they had been sat on the opposite side of the coach and had escaped. Afterwards they were found crying, vainly trying to revive their dead mother. Other young children were found wandering about the tracks after the accident until kind hands took them to a place of

---

[33] *The Railway Gazette*, London, England, 7th September 1951
[34] *The Indian Express*, Madras, India, 4th September 1951
[35] Wikipedia – Shadara-Saharanpur Light Railway

safety or to their homes.

Assistance was soon at hand after the accident and the injured were taken to Carlos Chagas and other hospitals, while the bodies of the many dead were taken to the mortuary at the Instituto Medico Legal, where there were insufficient tables to accommodate them, so that many had to be laid on the floor of the hall to await identification by relatives or friends. Blood supplies ran low due to the many operations needed and there was a request for blood donors, which was well responded to.

Dr João Roberta Guimarãco, who had been a passenger on the train, described his experiences. He said that the brakes of the train had come on very suddenly, so that everyone had fallen on top of one another. With the violent impact the walls of the coach had collapsed inwards and trapped passengers. They had struggled to free themselves and the more that they moved the more that they had bled from their injuries. In his words men who had had the appearance of madmen, with bulging eyes, had torn out their hair in despair. He, himself, had tripped over bodies as he made his way out of the car and his shoes had been soaked in blood. He concluded, "I have seen such terrible scenes, I hope to God that I never see such a sight again."

Dr Silvio Miranda Freitas, the deputy director of the railway company, left for the scene of the accident. Meanwhile it was reported that one or perhaps both of the drivers had fled the scene because of an obscure law which stated that if they were arrested at the scene of the accident they could be held indefinitely without bail, whereas if they remained at large for 48 hours they would remain at liberty until and if convicted of being responsible for the accident. Whatever the truth of this both drivers were present to make preliminary statements regarding the disaster by the following day. Driver Braz Guerino da Silva, of the passenger train, stated that he had first seen the freight train as it had passed Anchieta station, travelling at speed. Everything had gone normally until the two trains had been alongside one another, just after a bridge over the Rio Merni. Then, suddenly, he had heard a terrific crash and it had felt as if the cars of his train were being pulled backwards. He had immediately applied his brakes in emergency and stopped his train. Getting down from his cab he had gone back to find what was wrong, when he had been confronted with a pile of twisted metal and dozens of corpses.

Driver Felisberto began, "I swear by the saints that I am not to blame for the many deaths caused by this disaster." He continued, "I felt that something unusual was happening as the trains passed one another. It seemed as if my train had tightened up and then there was a violent shock. My engine shot ahead because of the reduced load. I quickly applied the brakes. When I alighted from my engine it was near Olinda station." [36]

Colonel Souza Gomes, the director of the Estrada de Ferro Central do Brasil, confirmed that the derailment had originated in a broken rail, adding that such fractures were 'very natural' (common?). The problem was that accidents on the Brazilian railways, including those involving fatalities, were very frequent at this time. In 1950 alone there had been 1,442 derailments in the country. Part, at least, of the difficulty, was that the tracks were carrying more frequent and heavier trains than those that they had been designed to cope with. As the death toll from the disaster mounted there was a public outcry against the 'slaughter on the railways.' The government-owned Central do Brasil was accused of negligence and extra police were drafted in at Dom Pedro II station for fear of possible demonstrations.

[36] *Jornal do Brasil*, Rio de Janeiro, 5th and 6th March 1952

The inquiry into the disaster confirmed that there had been a broken rail but found also that the railway in general was in a 'deplorable and dangerous condition.' The Brazilian president, Getúlio Vargas, in response, ordered the renewal of 200 kilometres of track known to be in a poor condition, together with the purchase of 200 new electric locomotives, as well as new rolling stock, to be financed by foreign loans. [37]

An express from Lübeck was approaching Hamburg Hbf, West Germany, when it became derailed on the Steinhauerbrücke due to a broken rail on Thursday 18th September 1952. The force of the impact was such that a crowded coach wrapped itself around a concrete bridge support. Eight passengers were killed and about 50 were injured, 30 of them seriously. Rescuers were still combing the wreckage for victims three hours after the accident. The train was running late at the time of the accident. [38] [39] [40] [41]

A bad mistake by permanent way staff led to an accident to a Milano – Venezia express near Verona, Italy, on Tuesday 23rd December 1952. It was reported that the maintenance work which they were carrying out entailed lifting a section of track, perhaps replacing a rail. They waited until they thought that the express had passed then began the lifting operation, apparently without having arranged adequate protection before the job was begun. It turned out that what they had assumed was the express was in fact an unscheduled extra working, probably a relief train put on at short notice because of the volume of pre-Christmas traffic. No sooner had the track been disturbed than the regular express appeared and its derailment could not be prevented. The engine, leading van and following first- and second-class composite coach were derailed and the two vehicles were reduced to 'a mass of twisted steel and splintered wood.' Fortunately the coach had been only half full.

Assistance soon came from Verona in the form of ambulances, nurses, police, firemen and volunteers so that rescue work was begun within a few minutes of the derailment. One report stated that five people had died in the derailment, [42] another stated that two bodies had been recovered by the early afternoon and that it was feared that six more were buried beneath the debris. [43] Many other passengers were injured. Fire apparently also broke out in the wreckage after the accident. [44]

According to an initial report, express train No. 5045 from Madrid to Coruña was derailed at 07:30 on Friday 28th December 1956 at kilometre 284·390 between Quereño and Sobrádelo, Spain. The engine, mail van, a second-class coach, a first-class coach, and a dining car left the track. The second-class coach went down an embankment, the first-class coach overturned onto its left-hand side and the dining car was left leaning over. The derailed vehicles were all destined for Coruña. Immediately news of the accident was

---

[37] Wikipedia – Anchieta rail disaster
[38] *Katastrophen der deutschen Bahnen Teil II* by J.R. Ritzau, pg. 122 (in German)
[39] *The Times*, London, England, 19th September 1952
[40] *Escanaba Daily Press*, Escanaba, Michigan, United States, 19th September 1952
[41] *Eau Claire Leader*, Eau Claire, Wisconsin, United States, 19th September 1952
[42] *Sarastota Herald Tribune*, Sarastota, Florida, United States, 24th December 1952
[43] *The Times*, London, England, 24th December 1952
[44] *Blytheville Courier News*, Blytheville, Arkansas, United States, 24th December 1952

received a rescue train was arranged and three ambulances and some lorries also went to the scene from Ponferrada. By midday about 30 injured had evacuated to the latter place. Torrential rain made the work of getting the injured from the wreckage more than usually difficult. It was expected that the line would be clear by about 18:00.

A further report altered the time of the accident to 08:15 and added that the train had been running three hours late. The derailment had occurred not far from the exit from Quereño station, close to the disk for it and 300 metres from Quereño Tunnel. A section of rail 50 metres long had broken and the engine and five vehicles had left the track. Two of the coaches had gone down an embankment, one on either side, the other three had remained on the embankment. The engine had ended up against the right-hand side of a cutting at kilometre 274·200 (thus ten kilometres short of the distance originally quoted). The five vehicles had been a van, a mail van, a first-class coach and a first- and second-class composite coach. The engine, van, mail vans and restaurant cars were wrecked. The first-class coach was left overhanging the bank of the River Sil and after its occupants had been rescued it was knocked to the bottom of the embankment to clear the line.

At least five people lost their lives in the accident, probably a sixth and possibly a seventh. About 17 others were seriously injured. [45] [46]

An express train from Buenos Aires to the winter resort of San Carlos de Bariloche (commonly called simply Bariloche) was derailed near Bahia Blanca, Argentina, on Tuesday 20th August 1957, eight coaches leaving the track. Initially it had been reported that nine people had died but the following day the Transport Ministry stated that there had been six deaths and 23 cases of injury. [47] The derailment was apparently the result of a broken rail. [48]

Disaster struck a Lagos – Kano passenger train in the early hours of Sunday 29th September 1957 after several days of heavy rain. As the train was crossing a six-foot embankment between Lalupon and Odo-Oba, Nigeria, 20 miles north of Ibadan, the track gave way, derailing the engine and first seven coaches of the 16-coach train. An initial report spoke of 20 dead and 122 injured, but it was feared that there were many more casualties. The following day the number of known dead had risen to 35 but it was believed that there were more dead in two coaches still in the water. The passengers on the train had included 87 students of the Zaria College of Arts and Technology returning from their holidays, who had been travelling in a separate coach. Only 20 of them had been accounted for. [49] By 4th October 70 bodies had been recovered and 147 were known to be injured. Many of the 370 passengers on the train were still unaccounted for, but it was suspected that many of them had just left the scene after the accident. [50]

Reports of the unsafe condition of the track had been received before the passenger train started out and it had been running at reduced speed as a result. A goods train had passed over the same stretch two hours earlier without its crew noticing anything wrong.

---

[45] *La Vanguardia*, Barcelona, Spain, 29th December 1956

[46] Webpage: Federación Castellano Manchega de Amigos del Ferrocarril (in Spanish)

[47] *The Billings Gazette*, Billings, Montana, United States, 22nd August 1957

[48] *Wisconsin State Journal*, Madison, Wisconsin, United States, 21st August 1957

[49] *The Times*, London, England, 30th September and 1st October 1957

[50] *Daily News Record*, Harrisburg, Virginia, United States, 5th October 1957

In the interim, though, a culvert through the embankment had become blocked and as a result the water had risen over the permanent way, washing away the track formation over a distance of 60 yards.

Following the investigation into the accident it was ordered that sections of the main line prone to flooding should be patrolled four times a night. [51] [52]

The rear two coaches of a Société nationale des chemins de fer belges electric passenger train from Antwerp to Brussels were derailed at Vilvoorde station, Belgium, on Friday 18th July 1958. According to eyewitnesses the penultimate coach was shattered and overturned, with its wreckage and twisted rails thrown over a wide area, and it was in this coach that five passengers were killed and 20 injured, five seriously. The Belgian Minister of Communications, M. Paul Segers, stated that according to an investigation by railway officials the derailment had resulted from the track buckling due to the heat on a hot day. [53]

---

[51] *The Railway Gazette*, London, England, Vol 107, pg. 529
[52] https://trid.trb.org/view/6704
[53] *The Times*, London, England, 19th July 1958

CHAPTER 2

DERAILMENTS - BROKEN TYRES, BROKEN AXLES AND OTHER VEHICLE DEFECTS

One of the frequent causes of derailments of trains other than problems with the track, was defects in the trains themselves, such as broken tyres, wheels, and axles. Such accidents were not as frequent by the twentieth century as in the early days of railways, but they still happened quite often and were to a fair degree unavoidable.

Mixed train No. 1003 from Rossio to Porto was derailed a few metres from the station at Vila Franca de Xira, Portugal, at 21:18 on Saturday 16th August 1947. The axle of a goods waggon had broken. This waggon left the track and in turn caused the derailment and piling up of other waggons and two third-class passenger carriages.

Stationmaster Joaquim dos Reis Bravo reacted calmly to the accident and rail traffic was soon moving again on a deviation line.

A large number of sleepers were marked and some were broken, damaged by the wheel which had become detached due to the breakage of the axle.

A squad of police kept the curious away from the wreckage so they did not interfere with the clearance operations and to prevent looting. This was a distinct possibility, for a 21-year-old girl named Maria Emilia, who had been on her way to her godfather's house in Lisbon, and who had suffered a leg injury, had had 900 escudos stolen from her bag shortly after the accident,

Numerous workmen recovered crates, suitcases, mail bags, bicycles, rolls of lead and wire and all kinds of luggage and goods, including a large quantity of fish which the President of Camara, Major Cunha Nery, ordered washed and distributed amongst the poor.

A reporter described the scene at the station, 'Almost in front of the station could be seen third-class carriage OD-TYF-798, 12·80 metres long. Another, old carriage, CT, with a capacity of 62 passengers, had, when the derailment had occurred, telescoped two thirds of it. Suffice it to say that of its 82 seats only 23 escaped. The rest were all swept away, with the passengers sitting and standing in them, and everything was jumbled up in a mixture of cadavers, mutilated bodies, injured, ironwork, splintered wood, and luggage. As for the CT coach, only its undercarriage remained after the accident, to be lifted by a heavy crane and placed at the side of the right of way. Brigades of workers strove constantly to remove the massive amount of derailed and destroyed material and clear the track, which they hope to do by tomorrow. Caminhos de Ferro Portugueses engineers and officials accompanied them. Wreckage, baggage, and merchandise was despatched everywhere with great speed, and in the middle of it all a sewing machine, broken, a tricycle, a stroller and a woman's shoe. Next to the undercarriage of coach CT were three intact closed goods waggons, then a group of 13 waggons, a large part of which were damaged, and in front an American J waggon, which was re-railed.

Next to the telescoped third-class carriage there was a poor woman in her 50s, Joaquina des Neves, sister of railwayman José das Neves, of Albergaria Dos Doze. She did not know how she had escaped. She said, "I felt a great shock, a massive noise, the carriage all bunched up, people were falling on top of one another. I just thought of Nossa Senhora das Neves {Our Lady of the Snows – The Blessed Virgin Mary} and asked her to save me. Had it not been for that, I don't know..." She suffered only small bruises on her nose and arm.

The day after the accident it was reported that there were 17 dead and 72 hurt. Some of the injured were in a grave condition and it was feared that the number of dead might well increase.[54]

The passenger train which left Bogotá, Columbia, at 10:00 on Sunday 10th October 1948, was crowded, for in Colombia, as in other South American countries, Tuesday 12th was a public holiday, Dia de la Raza or Columbus Day, and many people were travelling from Bogotá to spend the holiday at Apulo or Girardot. At 14:10 the train was travelling normally and close to the station at Anapoima, at kilometre 90, when its third-class passengers suddenly felt that their coaches were shaking violently. A moment later the train was derailed. Three third-class coaches were overturned on the left-hand side of the track and wrecked, while the restaurant and luggage cars, in which passengers were also riding due to the lack of room elsewhere, were also derailed, as was a first-class coach.

The first assistance arrived in a few minutes from the town of Anapoima. Later doctors who had been on the train were supplemented by others from neighbouring towns, a contingent of soldiers was sent from Girardot and chemists with drugs arrived by bus from La Mesa. The injured were taken to Girardot hospital and to the railway hospital at Dolma, while breakdown cranes were sent from Girardot and Bogotá to clear the line.

By Tuesday 12th there were known to be 20 dead and about the same number of injured, seven of them in a serious condition. The great majority of the victims were poor peasants. Four of the injured were still unidentified, including a boy aged about ten, despite a description of the clothes that he had been wearing being published in the newspapers.

Initially the cause of the derailment was unclear. Possible causes suggested were the excessive speed of the train on a curve, its heavy loading, and, at first the most favoured theory, a gap between the sleepers, that is, track defects. After further investigation excessive speed was ruled out, for the train had been reducing speed to stop at Anapoima station. Nor were the third-class coaches which had left the rails apparently overloaded, though a number of their passengers, who just boarded the train at the previous station (San Joaquin), had preferred to ride on the platforms of the cars for their short journey, and it was their presence on the platforms which had led to the high death toll when the

[54] *Diário de Lisbòa*, Lisbòa, Portugal, 17th August 1947

coaches had overturned. Track defects were now also excluded as a cause and from the evidence of Driver Eduardo Gonzáles Briceño, in charge of engine No. 42, the cause of the accident was now presumed to have been a broken axle on coach No. 5.[55]

A train carrying 4,500 supporters of the People's Party to the capital to celebrate the 25th anniversary of the foundation of the Republic was derailed between Irmak and Ankara, Turkey, on Friday 29th October 1948. Forty-nine people were said to have died and many were injured when 25 coaches were telescoped. A broken axle had caused the derailment.[56] [57]

A Nepal – Joynagar passenger train was derailed a mile from the station at Mahinathpur, Bihar, India, on Saturday 18th June 1949, due to the breaking of the axle of a coach. Four passengers were killed outright and 52 were seriously injured, of whom two had succumbed to their injures in Madhuban hospital by the next day.[58]

Twelve people died and 40 or 50 were injured in the derailment of a Cerbère – Avignon ADX prototype X 5001 railcar at Barbentane, France, on Tuesday 25th October 1949. The accident happened at 09:55, six kilometres from Avignon. The railcar was pulling three passenger carriages, one of them a wooden one, when it left the track in front of the station. After leaving the rails the momentum of the train carried it along for some distance until it crashed into the platform of a warehouse adjoining the line and the driving car smashed into the following coach, the wooden one. The other two coaches were derailed but little damaged.

The investigation undertaken by the Société nationale des chemins de fer français technicians encountered great difficulty in determining the cause of the accident due to the destruction of the train, but it was believed to have been due to the breakage of the leading axle of the driving car.

Surviving passengers stated that the train had ridden roughly from Tarascon and the conductor had had it stop exceptionally for some time at Graveson-Maillane, the previous station, while an examination was carried out to try and determine the cause of the rough running.[59] [60]

A major catastrophe at Sirhind, India, early on Sunday 29th January 1950, began with an accident to a goods train. As a Down goods train was leaving the station at 3.15 a.m. a brake block from a waggon fell into a set of points, derailing two waggons loaded with cotton. Before it could be stopped the engine of No. 27 Up *Kashmir Mail*, bound for Pathankot, collided with the derailed trucks and was itself derailed and overturned. In addition three coaches of the mail train, one of them reserved for military personnel, were overturned and 'smashed to smithereens.' The driver and two firemen of the mail escaped

---

55 *El Tiempe*, Bogotá, Colombia, 11th and 12th October 1948
56 *The Times*, London, England, 30th October 1948
57 *Edwardsville Intelligencer*, Edwardsville, Illinois, United States, 9th November 1948
58 *The Indian Express*, Madras, India, 21st June 1949
59 *La Croix du Nord*, Lille, France, 26th October 1949
60 *The Railway Gazette*, London, England, 4th November 1949

death, though the driver was trapped under his locomotive until released by rescuers.

After the accident relief and medical trains were rushed to the scene from Ludhiana and Ambala and ambulances and civil, military and railway doctors were standing by at Ambala City and Cantonment stations to receive the casualties. The line was blocked and trains were diverted to Raipura, Dhuri and Ludhiana. The undamaged portion of the train arrived back at Ambala Cantonment station at 12.30 p.m. with 26 injured, eight of them soldiers, and the driver and two firemen of the mail.

An initial report stated that there were 19 dead and 82 injured. [61]

By the following day the number of known dead had increased to 50 and injured to 112. Thirty of the dead were reportedly military personnel, another 30 of whom were injured. The work of clearing the tracks went on all night and the down line was cleared by 9 a.m. the next morning.

The number of dead had increased to 63 by the next day and then to 65, 34 of them military personnel. [62]

Mail train No. 1521, which had left Madrid at 20:30 the previous evening, and was due at Gijon at 11:30 was derailed at Villayana, Spain, near Pola de Lena, at 09:20 on Thursday 6th April 1950. Two wooden carriages, a first-class and a second-class, went down a seven-metre-high embankment and were crushed. The engine and following mail van, sleeping car and restaurant car remained on the top of the embankment. The derailment occurred at a point where work was being carried out on the track. Nineteen people died and 65 were injured.

Asked as to the cause of the derailment the Minister of Public Works, Fernández Ladreda, stated that it was too early to be certain, but it was probably something to do with the condition of the track, which was, as already stated, under repair at the time. He surmised that possibly it was the derailment of the restaurant car which had jolted the following vehicles and forced them off the line. He repeated, however, that it was too early to come to a definite conclusion. The Madrid correspondent of an English newspaper, on the other hand, wrote that it had been established that the derailment had resulted from a broken axle. [63]

Questioned as to the use of wooden coaches in the train, the minister stated that 100 steel carriages had just left the factory but pointed out that the total replacement of the wooden rolling stock would cost millions of pesetas and was being done as quickly as circumstances permitted. [64] [65]

Two named express trains of the Atchison Topeka and Santa Fe Railway were involved in an accident at Monica, Illinois, United States at about 4.40 a.m. on Thursday 6th July 1950. The trains, both eastbound, were No. 22, the *El Capitan*, an ‘extra fare all coach service,’ from Los Angeles to Chicago, and No. 10, the *Kansas City Chief*, a ‘combined Pullman and coach express,’ from Kansas City to Chicago. There were only

---

[61] *The Indian Express*, Madras, India, 30th January 1950
[62] Ibid, 31st January and 1st and 2nd February 1950
[63] *The Times*, London, England, 8th April 1950
[64] *La Vanguardia*, Barcelona, Spain, 8th and 9th April 1950
[65] Webpage: Federación Castellano Manchega de Amigos del Ferrocarril (in Spanish)

two main tracks through Monica but in American, as opposed to British practice, each could be used by trains in either direction during normal working. There were no lineside signals except at interlockings, instead there was automatic cab signalling with an automatic train control system.

First-class train No. 22 consisted of diesel electric units 29, 29A, 29B, and 29C, coupled in multiple unit control, a mail baggage car, a mail car, a baggage car, a baggage dormitory car, three chair cars, a dining car, two chair cars, a lounge car, two chair cars, a dining car, a chair car, and a chair observation car. It left Shopton, Iowa, about 86¼ miles from Monica, on the eastbound track at 3.07 a.m., passed Williamsfield, 10 miles from Monica and the last open office, at 4.32 a.m., 18 minutes late, and as it was overtaking train No. 10 just west of Monica, and travelling at 90 miles an hour, the rear wheels of the leading car derailed. General derailment of the vehicles of the train followed at once, one or more cars fouled the westbound track and the chair car tenth in the train was struck by train No. 10.

First-class train No. 10 consisted of diesel electric units 6 and 50A, coupled in multiple unit control, two baggage cars, four chair cars, and lounge car and six sleeping cars. It had left Shopton at 2.50 a.m., 15 minutes late, on the eastbound track, passed Williamsfield at 4.27 a.m., 24 minutes late, and there was diverted to the westbound track. It was travelling at 55 miles an hour when it struck the chair car of train No. 22. As a result of the collision the two diesel units and first three cars of train No. 10 were also derailed but the most damaged cars were the fourth to tenth ones of train No. 22. Nine (according to the Interstate Commerce Commission report, newspaper reports said ten) passengers died and 134 were injured. All the fatalities and most, at least, of the injured, were in train No. 22. [66]

Lock Adams, a dining car attendant on train No. 22, told his story to a reporter. "The shock of the collision threw me out of my bunk onto the floor of the dormitory car for employees aboard the El Capitan. It bounced me down pretty hard, but I wasn't hurt. I could hear yelling and screaming behind me. I grabbed a sledge hammer from the emergency kit and jumped out of a door. I ran to where the head end of the Chief's locomotive was stuck into one of the coaches of El Capitan. I swung up on to the locomotive and used the sledge to smash a window into the coach. That window went into the men's washroom and when I climbed in the first thing I saw were four dead men. They looked as though they had been killed while they were shaving to go to the dining car.

"Just then a man stumbled up to me with blood running from his head. Bill Durn, El Capitan bartender, helped me get the injured man out the window. After that, we found five or six others and helped them out. Everybody in that car seemed to have been killed or hurt. A lot of them were in bad shape.

"Further back in the El Capitan, things were calmer. But there were babies, luggage and clothing scattered all over as the people picked themselves up and tried to get off the train. Pretty soon our conductor, Mr. A.B. Billows, of Streator, Ill., came in with the two stewardesses. They're both registered nurses. Those two girls went to work as calmly as though they were in a hospital.

"Billows ran off and got to a telephone to call for ambulances and doctors. He was in the 15th car, the next to last one. He said that back there all he felt was a jerk and then came a sudden stop. Things weren't so bad in the back but he said that as he walked

[66] Interstate Commerce Commission report

forward it became worse and worse. It seemed to me like the fourth car in our train must have derailed first and the two after that swung across the other tracks. At least the diesel engine, two baggage cars, and our dormitory car stayed on the track.

"This was my first trip on the El Capitan since I began working for the Santa Fe." [67]

The Interstate Commerce Commission (ICC) report noted that Engineer C.E. Croull of train No. 22 and his unnamed fireman had inspected their trains at curves en route without noticing any problems. Nor had the operator at Williamsfield, who had inspected the train as it passed, given any signal to indicate defective equipment. The crew of train No. 10 did not notice anything wrong either as train No. 22 overtook them, but their view of it had suddenly become obscured by dust. The engineer of train No. 10 had at once made an emergency brake application, reducing the speed of his train from 70 to 55 miles an hour before it struck the chair car of train No. 22.

Examination of the track at the point of derailment showed it to be well maintained and there was no sign of any obstruction having been on it. The first indication of the cause of the derailment of the mail car was a damaged bond wire on the north rail 4,145 feet west of the point where it had left the track. Eastwards from that point to a crossover east of Monica station the track showed indications of dragging equipment at various points. A flange mark on the gauge side of the south rail was the first mark of the actual derailment.

Examination of train No. 22 showed that rear brake-frame assembly hanger box of the rear truck (bogie) of the first car had become detached from the transom. The brake-frame assembly had then been forced downwards and had come into contact with the track. Chafing marks on the transom indicated that the hanger-box connection had been loose before the accident occurred. The left horizontal hanger-box cap bolt was missing and the threads in the transom indicated that the bolt had been pulled out. The right horizontal hanger-box cap bolt was broken, and its missing part was found 2,580 feet west of the point of derailment. The left vertical hanger-box cap bolt was broken with a progressive fracture covering 70 per cent of the cross-sectional area of the bolt. Failure of the right vertical hanger-box cap bolt had been caused by a twisting action. The part of the hanger-box normally bearing on the transom was broken and was not recovered. All the brake shoes of the brake-frame assembly were broken. The left brake disc was destroyed, as were the inside face, ventilating fins, and part of the outside face of the right brake disc. The broken portions of the brake shoes and discs which were recovered did not show any indication of overheating or excessive wear, and all the brakes were new. What damage had occurred before the accident to cause the failure of the cap bolts could not be determined.

Since the accident the railway company had issued instructions for the hanger box to be welded to the transom and as a temporary measure to apply safety chains on future applications of this type of brake assembly. As a permanent measure safety supports made of metal straps were to be installed as soon as possible. All trucks equipped with this type of brake assembly were also to be inspected annually. [68]

The 8.20 a.m. British Railways, London Midland Region Up express passenger train from Liverpool Lime Street to London, Euston, met with an accident at Weedon,

---

[67] *The Vidette Messenger*, Valparaiso, Indiana, United States, 6th July 1950

[68] Interstate Commerce Commission report

England, not far from the scene of the 1915 disaster, at about 11.15 a.m. on Friday 21st September 1951. It was travelling at 60 to 65 miles an hour when its engine and 13 of its 15 coaches were derailed, the engine and leading eight coaches going down a shallow embankment and piling up on one another. Seven passengers and a member of the dining car staff were killed outright and seven more passengers died in hospital. Twenty-six other passengers, nine dining car staff and the driver were admitted to hospital.

The derailment was fortunately witnessed by Signalman Frank Denny in Heyford box in time to stop the Down *Royal Scot* at his Outer Home signal. Communications northwards had been severed by the accident but he was able to contact Control via Blisworth to the south. Four Metropolitan Police sergeants, travelling along the nearby Towcester – Weedon road, also witnessed the accident and raised the alarm. Local doctors arrived on the scene within seven minutes. Police and firemen were also soon on the scene, plus detachments from the Royal Army Ordnance Corps and the Royal Army Medical Corps. Miss Raynor, a 70-year-old retired nurse, walked two miles to the scene with a case of medical supplies. The injured were all on their way to hospital within an hour. [69]

Others also helped. 'Mr Grey, of Tamborough Farm, Weedon, said he heard the train coming at a fair speed, and he saw the engine rear up and crash down the embankment, ploughing into the field on its side. The leading coaches followed it, piling up in confusion. The noise of their crash was drowned by the hiss of escaping steam. Mr Grey, with his son and some farm workers, was harvesting in a wood nearby. One of the farm workers, Mr Johnston, ran to help passengers who were climbing through broken windows and out of the wreckage. The rest of the family prepared bandages, hot water, and brandy, and as the uninjured passengers came off the train they went into the small farmhouse to rest.'

Some passengers were lucky. 'One woman with a five-month-old baby and two other children had a remarkable escape. She was travelling with her back to the engine in the leading coach. Although the coach was wrecked, she and her children escaped without injury. She had crossed to Liverpool from Ireland the previous night.

'Mr G. Brett, of Maghull, Lancashire, was in the restaurant car. "A few minutes before the crash the restaurant car was crowded," he said. "It would have been much worse if the crash had come then. The restaurant car swayed, then turned over on its side."' [70]

The express consisted of 4-6-2 Pacific-class steam locomotive No. 46207 *Princess Arthur of Connaught* and 15 bogie coaches, in charge of Guard P.J. Burns with Driver W. Tomlin with Fireman Alfred Wallis.

Driver Tomlin told the Ministry of Transport inquiry into the accident that he had checked his engine at Edge Hill, Liverpool, before taking out the express and had found all in order. The journey had likewise been uneventful until his engine had emerged from Stowe Hill Tunnel, south of Weeden station, when the front end of the locomotive had begun "shaking." He had applied the brakes but the derailment had occurred almost at once. He had been buried in coal and had a lucky escape from serious injury or death.

Fireman Wallis likewise stated that he had noticed nothing wrong until his driver had remarked that there was a problem with the front engine bogie, a few seconds before the derailment. Afterwards, he had made his way to Heyford box, where Signalman Denny

---

[69] Ministry of Transport report

[70] *The Times*, London, England, 22nd September 1951

said he had appeared 'in a very shocked condition.'

Guard Burns said that he had felt nothing unusual until the shock of the derailment had thrown him to the floor. Nor had Travelling Ticket Collector B. Madeley or Driver W. Robertson (riding as a passenger) noted anything unusual.

Assistant Guard R.W. Webb, Assistant Cook B.J. Sorrell, Kitchen Porter H. Lever, and Travelling Carriage Cleaner D. Purcell had, however, noted more or less rough riding as the train had entered Stowe Hill Tunnel, as had a couple of the passengers, but none of them had thought it severe enough to pull the communication cord.

Several railwaymen gave evidence as to the condition of the track. Permanent Way Inspector J. Hughes stated that he considered the condition of the Up line at the derailment site to be "fairly good," though the clogged and dirty ballast required riddling.

Ganger Stone stated that he always had plenty of work to do as the cutting near where the derailment had begun was inclined to be wet. He had repacked seven or eight lengths of the Up line just over a week before the accident.

The derailment having begun with the engine it was subjected to a detailed examination after the accident. This found a significant problem. The accident report explained 'both leading axle boxes of its bogie were much too tight in the fore and aft direction, and there were unusually bright patches of considerable area on all of the four white metalled faces, which were clear evidence of excessive friction. The two trailing axle boxes of the bogie were a very slack fit, with more than three and three-thirty-second inches fore and aft play on the right hand side.'

It had been the practice for some time to change engine bogies from front to back or from left to right, to even out tyre wear so as to increase the amount of time before tyres needed to be turned. On the 13th September an examining fitter at Edge Hill Motive Power Depot had reported that the flange of the left leading bogie wheel of the engine was beginning to wear sharp, so Foreman Fitter W. H. Backhouse had noted on the repair card that the bogie axles needed changing. Fitter H. Bennett had duly stripped the engine on 19th September and Fitter W. Taylor had changed over the bogies.

Concluding his report into the accident Lieut. Colonel G.R.S. Wilson, the inspecting officer, stated that there could be no doubt that the derailment of the leading wheels of the engine bogie and hence of most of the remainder of the train had been caused by the lack of freedom of the axle boxes in the horns.

Apart from clogged and dirty ballast, which had prevented drainage, the general maintenance of the track had been good, and slight variations in cross-level due to wetness in the formation had not been such as to give rise to any serious disturbance at speed with a well-designed engine in roadworthy condition. Nor was there any suggestion that the engine had been unusually sensitive to track conditions.

The engine had been making its first trip since its bogie axles had been changed over and in its faulty condition had been at continuous risk of derailment since leaving Liverpool. The accident was clearly the result of the mismanagement of the changing of the axles by Fitter Taylor, although as he was a skilled man of long experience the task should have been well within his capacity. His work should, however, have been more directly supervised by Foreman Fitter Backhouse.

It was understandable that the rough riding had not been noticed sooner by the enginemen as experience showed that the derailment of a leading pair of wheels was often not felt in the engine cab. It was, however, felt further back in the train, and it was hard to believe that it had not been noticed by Guard Burns, especially as Mrs Purcell, the carriage cleaner, who had been riding with him, had remarked upon it to him. However, the riding through Stowe Hill Tunnel had been smoother, and the most that could have been expected was the stopping of the train at the next signal box to get the track examined.

Turning to recommendations, the inspecting officer commented that although axle boxes had been changed for many years without problems, it was apparent that such work done in running sheds needed closer supervision. Since the accident the Railway Executive was considering whether the procedure of changing bogie axles should be standardised or simply abandoned as not worthwhile.

The accident demonstrated that flat bottom track was less vulnerable to damage by derailed wheels than bull head track. [71]

The mixed train which left Meknès, French Morocco, at 13:00 on Saturday 22nd January 1955 for Rabat had only covered 10 kilometres when a coach and three vans left the track. Seven passengers and the conductor of the train were killed and 15 other passengers were seriously injured. The derailment was apparently due to a broken axle, though an inquiry also implicated a faulty set of points. [72]

A Soviet military train carrying 280 soldiers and a number of tanks was derailed near Plauen, East Germany, on Saturday 26th February 1955. News of the accident was apparently censored by the East German authorities and books on German railway accidents make no mention of it, but it was reported in a number of American newspapers a week or so after it had occurred. Seventy soldiers and the train driver were said to have lost their lives and a large number of soldiers were injured. Two carriages of the train, each containing 40 passengers, were wrecked, and most of their occupants killed.

News of the accident was received in the West following the flight there of the fireman of the train, who explained that he had feared being arrested for sabotage. He said that his driver had protested against taking out the train because the wheels of the locomotive had been 'worn out and wobbling,' implying that the derailment had been caused by an engine defect. [73] [74]

Twelve passengers were reportedly killed and 37 others injured in the derailment of an express train from Buenos Aires to Zapala near Fortin Uno, Argentina, halfway between Bahia Blanca and Neuquen on or about Wednesday 23rd October 1957. The train included at least two tank waggons and the axle of one of them broke, leading to the derailment of the waggon together with a mail van, a luggage van, two passenger coaches and a restaurant car. [75]

Readers opening a morning London paper on Friday 22nd January 1960 read that, 'Four men and a woman lost their lives when a Glasgow to London express was in

---

[71] Ministry of Transport report

[72] *La Bourgogne Républicaine*, Paris, France, 24th January 1955

[73] *Valley Morning News*, Harlingen, Texas, United States, 5th March 1955

[74] *The Daily Herald*, Gulfport, Mississippi, United States, 4th March 1955

[75] *The Times*, London, England, 25th October 1957

collision with an express goods train from Leeds to Carlisle at 1.45 a.m. today {Thursday 21st January}. The accident occurred about half a mile north of Settle station, about 200 yards from a viaduct over the Keighley to Kendal road.

'The dead are James Smith Allen, of Tamworth Street, Glasgow, Patrick Ian Wilson, of King's Hedge Road, Cambridge, Mr E.W. Mahany, of Ranfurley Road, Bridge of Weir, Renfrewshire, Mrs Ross, of Shaw Place, Dalry, Ayrshire, and John Miller, aged 23, of Pennelds Road, Glasgow. Miller was on his way to join the R.A.F., died from his injuries in hospital.' [76]

Eight other passengers and the guard of the freight train were injured.

The accident was naturally the subject of a full Ministry of Transport report which stated that the passenger train involved had been the 9.05 p.m. Up express from Glasgow St Enoch to London St Pancras, consisting of *Britannia* class 4-6-2 steam locomotive No. 70052, a brake second, two second-class coaches, a first and second-class composite coach, a second-class sleeper, two first-class sleepers and a brake van, in charge of Guard W.C. Bird with Driver F. Waites and Fireman J. Chester.

The 10.40 p.m. Down H class freight train from Leeds to Carlisle consisted of a Class 5 MT 2-6-0 steam locomotive, 20 waggons and a brake van in charge of Goods Guard F.G. Atkinson with Driver G. Stewart and Fireman J.M. Graham.

Driver Waites of the express was the first person to give evidence at the inquiry into the accident. He stated that he had taken over the engine at St Enoch station when it had arrived there from Corkerhill depot. He had not examined it there but had checked the axle and motion bearings and the corks in the lubricating holes on arrival at Dumfries. All had appeared in order. Carlisle had been left two minutes late but with a light load the train had been on time when it had reached Ais Gill summit, at the end of the climb from Carlisle, at about 1.10 a.m. It was after passing the summit, at about 35 miles an hour, that he first became aware of a "knock," which he thought came from the big end. He had closed the regulator, when the knocking had got worse, so he had opened it slightly, and slowed the train with the brake. He had decided to stop and examine his engine at Garsdale (formerly Hawes Junction) three miles from Ais Gill.

It had been typical high fells weather at Garsdale; there were gale-force winds, it was snowing, and it was pitch dark. Using a small electric torch Driver Waites had tried to examine his engine to find the cause of the problem. He had examined first one side and then the other, feeling the axle boxes and bearings and looking at the various parts as best he could in the minimal light. The bearings had all been cool, with no sign of overheating, and he could find nothing wrong. The best thing to do, he had decided, was to continue at reduced speed to Hellifield, 25 miles further on, where there was a small motive power depot and where it would be easier to find the problem. He maintained that he had kept the speed of his train down to 20 miles an hour but the Ministry of Transport inspecting officer, Brigadier C.A. Langley, pointed out that according to the signal box timings the train had averaged 42 miles an hour between Garsdale and Settle. Driver Waites finally agreed that running down the long falling gradients the speed of his train must have risen to a greater extent than he had realised.

Passing the Settle Station Up distant signal, Driver Waites said, he and his fireman had noticed sparks flying up from the right-hand side of the engine and ballast was flying

---

[76] Ibid, 22nd January 1960

up against the cab. He had then made a full brake application, a Down freight train had passed and his engine had shuddered. When his train stopped he had alighted from his engine and found the right-hand connecting rod with piston rod and cross head attached lying in the ballast.

Fireman Chester more or less agreed with his driver's evidence.

Guard Bird thought that his train had stopped for three minutes at Garsdale and that its speed from there had not exceeded 20 miles an hour.

Driver Stewart of the freight train stated that his engine had been working hard on the rising gradient when they passed Settle. Shortly afterwards his fireman had shouted, "I think we have hit something," and his engine had lurched, rolled, shuddered, and stopped.

Fireman Graham said that he had seen the express approaching. Seeing sparks directly in line with their engine he had then thrown himself across the cab and called to his driver to look out. At the same moment there had been a big bang, the engine had lurched and the cab had filled with steam. As soon as his train stopped he had alighted and seen that the engine and several vehicles were derailed. Going back he had found Guard Atkinson sitting on the track, bleeding badly. As the guard was able to walk he had told him to go to Settle cabin for first aid and he (the fireman) had run on ahead to advise the signalman of the accident, to block both lines and to call for doctors and ambulances, as the express appeared to be badly damaged. He had then gone forward as far as the Down distant to protect his train.

Goods Guard Atkinson was thrown about his cab by the sudden stoppage of his train and badly hurt. He got out of his cab to go back and protect his train but Fireman Stewart took over the task.

Signalman E. Ashton at Garsdale stated that he had received the *Train Entering Section* signal from Ais Gill at 1.11 a.m. for the Up express. The train had reached his cabin at 1.16 a.m. and had stopped 90 yards from his box, though it was not booked to stop there. Through the falling snow he had seen a man with a torch walking around the engine. The train had then restarted and as it passed his cabin he had seen the driver and fireman looking out, clearly trying to locate a violent knocking which he (the signalman) could also hear. He had called out, "She is knocking badly," but received no response from the enginemen, though he had thought that they might have asked him to advise Hellifield that they needed assistance. The train left Garsdale at 1.20 a.m. and had taken six minutes rather than the usual three to clear the next box at Dent.

None of the other signalmen between Garsdale and Settle had noticed anything wrong with the express. Signalman J.W. Wilson at Stainforth Sidings said that he had received the *Train Entering Section* signal from Helwith Bridge for the express at 1.41 a.m. and it had passed him at 1.45 a.m., travelling at about 30 miles an hour. Sparks coming from its wheels the length of the train showed that its brakes were applied. Because of the weather his box windows had been closed, which might have prevented him from hearing any knocking. The Down freight train he had accepted from Settle Station at 1.41 a.m. and he had received the *Train Entering Section* signal for it at 1.45 a.m. A couple of minutes later the Settle signalman had telephoned him to say the freight train had stopped in the section and he had enquired as to the running of the Up express. Settle had sent the *Obstruction Danger* signal at 1.56 a.m., when he had placed detonators on the Up line.

Signalman A. Bullock at Settle Station said that he had received the *Train Entering Section* signal for the Down freight from Settle Junction at 1.42 a.m. and it had passed him at 1.46 a.m., travelling at 30 miles an hour or more. The Up express, Stainforth Bridge had sent on at 1.45 a.m., and three minutes later he had seen its headlights and realised that it had stopped down the line. At the same time the tail lights of the Down freight train had been still in sight. Unable to work out what was wrong he had telephoned Signalman Wilson at Stainforth, who could only tell him that the express had been going more slowly than usual when it had passed there. Then, at 1.56 a.m., Fireman Graham had arrived at his cabin, had told him that the goods had collided with the express, and to stop all trains. He had then sent the *Obstruction Danger* signal both ways, advised Control and telephoned Hellifield South cabin to call the emergency services. Assistance had come very quickly, including a doctor who had attended to Guard Atkinson in his box.

Shed Master M. Cowley of Hellifield Motive Power Depot said that he had arrived at Settle at 4 a.m. He described seeing the detached parts of the express engine lying in the six foot and said that the Down line had been badly damaged, and it appeared that some object, probably the connecting rod of the *Britannia* engine, had struck the sleepers and pushed the track out of gauge. There was a deep hole at one point where the track had been pushed inwards.

Examining Fitter G.A. Neil, of Polmadie Motive Power Depot, Scotland, stated that he said that he had examined the express engine on the night of 17th/18th and had tested the bolts for tightness by tapping them with a hammer. He had found the right-hand inside front slide bar bolt to be loose and had made out a repair card.

Fitter J. Aitkenhead of Polmadie said that he had used a thin spanner to tighten the inside bolt on the morning of the 18th, though he had not replaced the split pin, which had been in good condition.

Examining Fitter C. Warriner of Holbeck Motive Power Depot, Leeds, stated that he had examined the engine when it had arrived there on the night of 19th/20th. He had found that the right-hand inside front slide bar bolt was loose and there was a slight gap between the packing piece and the bottom bar. The outside front bolt and both rear bolts had appeared to be tight. He had made out a repair card. He said that the slide bar bolts of *Britannia* engines sometimes worked loose. It was almost always the front bolts, which were difficult to reach.

Fitter E.J. Sadler, of Holbeck, said that he had attended to the engine on the morning of the 20th. The right-hand inside front slide bar bolt had been loose and the bottom bar had dropped about one 64th of an inch. He had tightened the bolt with a thin spanner and used a 4 lbs hammer to tap it. He had put the spanner on three times and moved the nut a third of a turn. He had not renewed the split pin and agreed that as he had tightened the nut by one sixty-fourth of an inch the split pin could no longer be bearing against the head of the nut. He also agreed that the inside front bolt was difficult to reach and therefore to tighten. He had done nothing to the rear bolts.

Examining Fitter H.A. Wallach of Corkerhill said that he had examined the engine on the evening of the 20th and found nothing wrong with it.

Brigadier Langley remarked that the circumstances of the accident were remarkable. It had been initiated by the failure of the motion of a modern express engine which had eventually led to part of the motion assembly causing the derailment of the engine and leading vehicles of a freight train on an adjoining line. The freight train engine had then struck and ripped open the sides of the coaches of the express, causing serious casualties.

The failure had begun with the falling off of a slide bar of the *Britannia* engine 30 miles short of the point of derailment, followed by a second one eight miles further on. At this point the driver had been alerted to the problem and had examined the engine at Garsdale but failed to locate the trouble and elected to continue at reduced speed. The engine had run for 20 miles in this dangerous condition before the right-hand piston rod, cross head and connecting rod had come adrift. The displaced motion assembly had been driven deep into the formation and the displaced cross head had shifted one rail of the Down line sufficient to cause the leading wheels of the freight engine to drop into the four foot, whereupon the track had burst open, derailing the front vehicles of the freight train.

The failure of the right-hand motion assembly had undoubtedly been initiated by the loosening of the right-hand inside front slide bar bolt due to insufficient attention having been paid to the tightening of the front slide bar bolts, which were not easy to reach and difficult to tighten.

The failure of Driver Waites to notice the loss of the slide bars at Garsdale was understandable in view of the fact that his examination of the engine had been made under conditions which could hardly have been worse. In view of the inclement weather, the late hour, and the exposed position of the train the decision of the driver to continue was also reasonable, but he should have advised the shedmaster of the problem and continued very cautiously. He should also have gone forward much more carefully than he had, and therefore bore some measure of responsibility for the accident. He had clearly not appreciated how quickly a powerful engine could gain speed with a light train on a steep falling gradient.

Loose slide bars had been a frequent occurrence on the *Britannia* engine involved in the accident. Loose bolts had been reported on nine occasions in less than four months. This type of defect was common on *Britannia* class engines and there had been four cases in the previous two years of slide bars becoming displaced or falling off. The problems arose from the failure of maintenance staff to properly tighten the bolts and to ensure that the split pins were always bearing correctly against the top of the nuts. A modification of the slide bar fastenings of *Britannia* class engines to make them more accessible had therefore been prepared but only four engines had been modified, when the engines were due for periodic overhaul and that involved in the accident was not one of them. All 55 engines of the *Britannia* class had subsequently been modified.

Brigadier Langley commented that the standard of maintenance of the motion of *Britannia* engines at a number of depots had not been satisfactory and recommended that the facts of the accident should be brought to the notice of all motive power staff 'and that special steps be taken to ensure that proper attention be paid to the security of fastenings on which so much of the safety of the engine depends.'

In closing his report the inspecting officer praised the speed and efficiency with which the emergency services had come into action in the early hours in adverse weather conditions. The train crew had promptly advised the signalman of the accident and despite snow-covered roads and a raging blizzard the first ambulance had reached the scene in less than ten minutes; the first doctor had arrived five minutes later and the police and fire service had been not far behind. The injured passengers were on their way to hospital in less than an hour and the uninjured ones were given every assistance by the Settle station staff.[77]

[77] Ministry of Transport report

CHAPTER 3

OTHER DERAILMENTS

Other chapters in this book cover derailments caused by track defects, defects with the trains themselves and those caused deliberately by acts of vandalism or sabotage. This chapter will deal with derailments due to other causes such as excessive speed, collisions with animals and other accidental obstructions, traction shocks and so on, together with those for which no cause was given.

Seventeen people died and 97 were injured in a derailment on Thursday 14th March 1946 'on the Rumoi line' of the Japanese National Railways. [78] The accident probably occurred at Mashike. [79]

The news resonated painfully through this capital, a São Paulo, Brazil, newspaper wrote, of a grave accident that had occurred on Wednesday 20th March 1946 to a passenger train in Sergipe state. According to news from Aracaju a train which had left there had met with an accident between Laranjeiras and Riachuelo. It had become derailed, resulting in the death of a large number of people and injury to more than 200. The Surgical Hospital in Aracaju was the scene of intense activity and had a police cordon to prevent it being invaded by the families of injured passengers. There had never been a disaster of like magnitude in the state of Sergipe previously and rarely one in the country as a whole.

Speedy measures were taken to rescue the injured in the disaster which had occurred at kilometre 458 of the Viação Férrea Federal Leste Brasileiro, the newspaper report on the disaster continued, as well as in removing the bodies of those who had died. Vehicles belonging to the 28 Batalhão de Caçadores (an army battalion), the city of Aracaju, and firefighters were all used to bring victims of the disaster to the hospital. Every available medical person was involved in the rescue work. So far 70 dead had been recovered and taken to the hospital, which was overwhelmed with hundreds of passengers who had been on the fateful train. Worthy of praise was the tireless work of the doctors and nurses attending the victims, many of whom were in a critical condition and expected to die at any moment. Businesses and public offices in the city had been shut down and its inhabitants were in a state of consternation, with groups of people everywhere talking

---

[78] http://www.ikjeld.com/japannews/00000147.php
[79] Wikipedia Commons – Rumoi Main Line

about the terrible calamity.

Driver João Claro dos Santos, in charge of the engine of the train, was found in a state of severe shock, and could not explain to the Laranjeiras authorities how the accident had happened. He was clearly suffering from shock and unable to give a rational explanation.

Many doctors and a large amount of medical equipment were sent from Aracaju to the scene of the accident. Engineers of the railway company also went to the scene to investigate the cause. The fate of the conductor of the train, Edgar Simas, remained at the time of the first reports unknown.

Driver João Claro, who, the newspaper supposed, might have been responsible for the disaster, was threatened by surviving passengers, and leaving the scene had made his way to Laranjeiras, where he had surrendered to the police, clearly severely distressed.

A later report stated that there were possibly 200 dead and 300 injured, and that many of the injured had died of their wounds. As well as those taken to Aracaju victims were taken to medical establishments at Laranjeiras, Riachuelo, Maoim and Rosario. The post and telegraph offices, as well as the telephone company, were crowded with people seeking news of relatives and friends who had been on the train.

Aracaju station was full of people awaiting the arrival of trains from the accident scene, a reporter wrote. Some trains had already arrived, bringing injured and numerous bodies. There were emotional scenes when parents and other relatives recognised the bodies of one of the dead passengers. Some of the corpses were badly mutilated, making the work of identification difficult. Many of the passengers who died had been crushed between the carriages and the walls of the cutting through which the train was running at the time, and the presence of the cutting had increased the magnitude of the disaster. [80]

The train had been bound for the town of Capela and was carrying about 1,000 passengers when it had become derailed while descending an incline, the engine, baggage car and three coaches leaving the line. A report from Rio de Janeiro the day after the disaster put the number of dead at 185, making it the worst accident in the history of the country. [81] [82]

The cause of the disaster was not stated. Possibly the driver let the speed of the train get too high on the descent and then lost control of it.

Train No. 10, the Down Surma Mail, was derailed between Talsahar and Brahmanbaria stations, India, on the Bengal and Assam Railway, on the morning of Monday 9th September 1946. The engine and three coaches capsized and three other coaches were derailed. Seven people died according to the first report and 32 others were injured. It was hoped that the line would be cleared within 24 hours. The cause of the derailment was under investigation. [83]

Twenty-seven passengers were killed and about 70 others were injured in the derailment of a Kathiawar – Lucknow express on the Oudh and Tirhut State Railway at 1.15 a.m. on Monday 23rd September 1946 between Dighwara and Baragopal stations,

---

[80] *O Estado de São Paulo*, São Paulo, Brazil, 21st March 1946

[81] *The Kingston Daily Freeman*, Kingston, New York, United States, 21st March 1946

[82] *The Amarillo Daily News*, Amarillo, Texas, United States, 22nd March 1946

[83] *The Bombay Chronicle*, Bombay, India, 10th September 1946

India, about 185 miles from Gorakhpur. The engine, a brake van, a third-class and an inter-class bogie capsized. The engine driver escaped injury.

At 5.30 a.m. a relief train left Gorakhpur for the scene of the accident, accompanied by a number of officials, including the Resident Engineer and the Chief Engineer. A second relief train was sent from Sonepur with medical supplies and two doctors from Dighwara and the Civil Surgeon, Chupra, also attended the accident. A passenger train took 11 injured passengers to Smastipur Railway Hospital and a special train took 19 more to Chupra Civil Hospital. It was hoped that the line would be clear by the evening. The cause of the derailment was not immediately clear. [84]

A second report stated that the number of dead in the accident had risen to 30, though it was feared that other bodies might remain in the wreckage. Over 100 were injured. The engine crew had escaped but two ticket examiners had been injured, one of whom had succumbed. The engine of the train had been completely thrown from the track and had embedded itself squarely into the ground on the side of the embankment, while the four leading vehicles were derailed. The heaviest casualties were in the women's inter-class compartment of the leading coach, which had been telescoped and completely smashed. The other two derailed coaches were not badly damaged. A whole length of rail had been torn up over a considerable distance. At the beginning there had been some looting of luggage and armed police subsequently cordoned off the area. Squads of volunteers including Khaksar and Muslim League workers were busy with relief work. [85]

Six people (five passengers and a chair car attendant) died and 132 (123 passengers, six dining car employees and three railwaymen) were injured in the derailment of the Union Pacific Railroad Transcon passenger train at Oro Grande, near Victorville, California, United States, at 7.02 a.m. on Thursday 26th September 1946.

Marty Jewett, publisher of the *Victorville News-Herald*, described the accident scene. "The engine and tender were off the track. The baggage car was spread across the rails, and the first coach lay on its side, parallel to the track. The next coach was telescoped into the car ahead, and the following one was also derailed."

W.E. Helm, a maintenance engineer at a power station near the accident site, added, "It looked like the engine might have exploded. I don't see how the fireman escaped alive. He was pinned in the cab of the engine, and a piece of rail was jutting right up through it." [86]

Westbound first-class Union Pacific train No. 223 consisted of 4-8-4 steam locomotive No. 835, a baggage car, three coaches, two dining cars, and five Pullman sleeping cars in charge of Conductor Joseph Murphy with Engineer Perl Parker and Fireman R.J. Zeltamier. It had left Barstow, California, at 6.28 a.m., eight minutes late, passed Helendale, rather less than 10½ miles east of Oro Grande and the last open office, at 6.51 a.m., six minutes late, had passed Oro Grande and was travelling at a speed in excess of 75 miles an hour when the engine and leading six cars were derailed. [87]

The Pullman car passengers had still been asleep in their berths at the time of the derailment. Oil from the wrecked engine flowed over the scene and the Victorville fire

[84] Ibid, 24th September 1946
[85] Ibid, 25th September 1946
[86] *Berkeley Daily Gazette*, Berkeley, California, United States, 26th September 1946
[87] Interstate Commerce Commission report

department and a squad from an army air base stood by to prevent any fire. Ambulances were unable to get close to the wreck in the rugged mountainous country and the injured victims were put into a boxcar and taken a mile along the track to a point where the ambulances could gain access. [88]

The curve on which the derailment had occurred had a speed restriction of 40 miles an hour, indicated by a rectangular sign with the figures 40-30 in black on a yellow background. Engineer Parker asserted at the Interstate Commerce Commission (ICC) inquiry into the accident that his train had been travelling at 50 miles an hour as it had approached the sign. He said that he had then moved the throttle lever to the drifting position and made an 8 or 10 pound brake pipe reduction to reduce the speed of the train to 40 miles an hour for the curve. As the engine had entered the curve he moved the automatic brake valve to the running position. His first warning of anything being wrong had been when his engine surged to the left, when he had immediately moved the brake valve to the emergency position. He thought that he had caught hold of the throttle lever as his engine had overturned to prevent his falling from the seat box and in doing so might have opened the throttle.

Fireman Zeltamier was pinned under the engine and too badly hurt to be questioned during the investigation, while Conductor Murphy and his flagman were unable to add any useful information.

There was no indication of any problem with the engine before the derailment, of dragging equipment, track defects or any obstruction on the line. After the accident the throttle lever of the engine was found to be fully open and nothing was found which could have prevented the proper application of the brakes. The theoretical overturning speed of the engine was 76.8 miles an hour and of its tender 78.1 miles an hour. The ICC concluded that the train must have been moving at overturning speed when the engine had overturned to the left side of the curve without marking the rails and had slid on its left side until it stopped. [89]

An inquest jury at the beginning of October, not having the benefit of seeing the yet to be published ICC report, was unable to determine the cause of the accident. [90]

Eleven people died and 26 were injured, seven seriously, when a train ran into a snowbank near Taguchi station, Japan, on the Shinetsu line of Japanese National Railways on Thursday 19th December 1946. Two engines and two coaches overturned. [91] [92]

Labour Day is celebrated in the state of Queensland, Australia, on the first Monday in May, which in 1947 was the 5th. On that day a number of special trains were chartered by different groups to convey groups of people to various picnic locations in the country north-west of Brisbane. Among the trains planned to run along the quiet Dayboro line was a train chartered by employees of the Brisbane Customs Department Social Club bound for a picnic venue at Closeburn.

---

[88] *San Mateo Times*, San Mateo, California, United States, 26th September 1946

[89] Interstate Commerce Commission report

[90] *The Bakersfield Californian*, Bakersfield, California, United States, 2nd October 1946

[91] *Sarasota Herald-Tribune*, Sarasota, Florida, United States, 19th December 1946

[92] http://www.ikjeld.com/japannews/00000147.php

Train No. E91 consisted of 4-8-0 C17 class steam engine No. 824 and tender, a water gin (a tank waggon carrying water for the engine) and six wooden carriages, in charge of Guard George Evans, with Driver Clyde Hind and Fireman Augustus Knight. It left Brisbane Roma Street station at 8.50 a.m. and Central station at 8.59 a.m. with about 230 passengers, many of them children.

The Dayboro line was steeply graded, with gradients between 1 in 50 and 1 in 66, and sharply curved; there was a speed restriction of 25 miles an hour even on the straight portions, reduced to 20 miles an hour on the curves. At 9.50 a.m. the train reached Ferny Grove station on the branch, where a number of passengers alighted and tablets were exchanged, the branch being single track. Leaving Ferny Grove the train reached Camp Mountain Knob, the summit of the Samford Range, and began its descent to Samford station, two miles on. Almost at once the passengers realised that the train was going too fast, as the carriages began to rock dangerously and luggage fell from the overhead racks. Women and children began to scream.

Halfway down the descent the train was travelling at an estimated 40 miles an hour as it entered a six chain left-hand curve. This proved to be too fast and the engine overturned on its right-hand side. Driver Hind was trapped in his cab and could not be released until 3.30 p.m. when a hole had been cut in the roof; Fireman Knight was caught between the engine firebox and tender and killed instantly. The tender was thrown up into the air and overturned and the water gin telescoped the leading carriage and part of the second carriage. The front of the third carriage was also damaged, but the rear three carriages suffered very little. Fourteen passengers died and 38 were injured. Driver Hind was unconscious by the time he was released and passed away in hospital the following day, becoming the 16th fatality. [93]

One of the passengers, a Mr Edward Hart, was the first person to go for help. 'With blood streaming from a cut over his eye, he clambered out of a carriage and ran on down the line towards Samford.' There an emergency rescue squad was formed in a few minutes from local shopkeepers and farmers who climbed into trucks to get to the scene of the disaster, where they found the uninjured passengers doing what they could to help the victims. The first outsiders to reach the scene were Sergeant J. Kunkel and Constable L. Fitchie of Mitchelton, who had received a phone call informing them of the disaster from a nearby homestead a quarter hour after the derailment. By 11.05 a.m. 14 ambulances, including two from the Labour Day processions, had reached the scene.

Mr Hart, 41, of Albion Street, Brisbane, said later that he had been in one of the rear carriages. "The impact threw me against a window, the glass shattered and a lump of it cut me across the eye. I realised that something terrible had happened. All the women in the carriage were screaming. I climbed through the window and saw other people clambering from the wreckage." [94]

Guard Evans had been thrown across his van by the sudden stoppage of the train. He had gone to the Westinghouse brake but found that there was no air. Getting out of his van and seeing the destruction he had returned to apply the hand brake, given a first-aid box to someone and then protected the rear and front of his train with red flags and detonators. [95]

---

[93] *Australian Railway Disasters* by Kenn Pearce, pgs. 121-6
[94] *Warwick Daily News*, Warwick, Queensland, Australia, 6th May 1947
[95] *Australian Railway Disasters* by Kenn Pearce, pgs. 121-6

Newspapers carried detailed accounts of the rescue operations: 'Dr E. Marks prepared a hypodermic morphia syringe, which ambulance men, crawling through the wreckage, administered to still conscious sufferers. The engine driver, O. Hinds, had to administer the syringe to himself, as the ambulance man was unable to reach far enough. When the morphia injections were completed passengers, ambulance men, and emergency breakdown gangs made a determined onslaught on the wreckage. Beneath the wheels of the overturned engine they found the terribly crushed body of the fireman, A. Knight. Inside the engine, bent almost double, the engine driver was jammed by the control lever and steam pipe.

'Later oxy-acetylene torches were brought into action, with the arrival of more railway gangs, in an effort to release the dead and injured from the tangled wreckage. Dr Marks then moved to the second carriage. Visible from the waist up in the first two compartments were Miss Linda Glenny and Mr and Mrs T. McLean. McLean was wedged against the right side of the carriage. His wife was lying half across him, with Miss Glenny wedged tightly against her. Forced across Miss Glenny's lap was the unidentified body of a small boy. Between the jammed bodies could be seen the head of a dead man. Protruding from the wreckage was the arm of a woman, with a heavy gold bracelet round her wrist.

'Dr Marks gave injections to those passengers who were conscious, although suffering considerable pain and severe shock. With axes and saws rescue gangs cut away the top of the first carriage, which was barring their way in their fight to reach the driver. At the same time men started to cut away the side of the second carriage, so they could reach the trapped people. Constant morphia injections were given to the trapped survivors. On his own in the shattered cab the engine driver, Hind, actually assisted with a hacksaw and urged on his rescuers, two of whom were overcome with heat and had to be assisted into the open. In the carriage both the women and McLean smoked cigarettes and joked with ambulance men who clambered inside and were supporting their heads.

'As the work progressed and the strain of the wreckage shifted more weight fell on the limbs of these three. Several times Miss Ginny screamed with pain. At 3.30 p.m. the engine driver, Hind, was lifted through a hole cut in the roof. Then unconscious, he was rushed to hospital. At 3.40, the bodies of three children, including a boy who had been jammed against Miss Ginny's legs, were removed. Five minutes later rescue gangs simultaneously lifted Miss Ginny and a dead man, believed to be Frank Delaney, from the compartment. Redoubling their efforts, they had both Mr and Mrs McLean clear in the next ten minutes. They had been trapped for more than six hours.

'Mrs McLean's first request was to ask the ambulance to get in touch with her mother, Mrs J.R. List, of Torbanlen, near Maryborough, who was looking after her two children, Dorothy (14) and Thomas (12). When Mr and Mrs McLean had been released work was continued to free bodies still in the wreckage.' [96]

The Court of Inquiry into the accident last 14 days and examined 50 witnesses. As part of the inquiry a special train carrying members of the inquiry board was run to the scene of the accident.

Early on in the inquiry it was stated that exanimation of the engine after the accident had revealed no defects with it such as could have caused the accident. Evidence was given that Driver Hinds was not very familiar with the Dayboro line, not having been

---

[96] *Warwick Daily News*, Warwick, Queensland, Australia, 6th May 1947

over it for a year, and that he was therefore given Fireman Knight as a pilot. Guard Evans, on the other hand, was rather more familiar with the line; on the day of the accident he had been making his third trip over it in charge of a train that year. [97]

Chief Mechanical Engineer Vincent Hall told the inquiry on 27th May that he was convinced that excessive speed was the cause of the derailment. He said it was clear that the brakes had not been applied at the time of the derailment. [98] He said that the driver had not applied the hand brake, had not reversed his engine, nor had he whistled for the guard's brakes. Coming down the bank the driver should have made three light brake applications; the evidence indicated he had made none, and running without brakes the train would have reached an estimated speed of 44 miles an hour. [99]

The next day of the inquiry Traffic Inspector S.W. Chippindall stated that he had questioned Guard Evans at the scene of the accident and had asked him what had happened. "I don't know," the guard had replied. Asked what speed his train had been travelling at, the guard had said, "The train was not going over the regulation speed, not more than 25 mph." Pressed as to exactly how fast the train had been going the guard had replied that he did not know. Trains Clerk Oliver Fenwick stated that Guard Evans had said in a report to the General Manager that he had noticed nothing unusual on the trip until, without warning, the guard's van had given a sudden lurch. The guard could give no reason for the derailment. [100]

In contrast to the assertion by the guard that the train had not been going extra fast was the evidence of passengers given on 30th May. Leslie McCormick, who had been riding in the second carriage with his mother, father, fiancée, and two sisters, said that he believed that the train had "got away" on the down grade. "The train was swaying like the switchback railway in Lune Park, in Sydney," he said. 'McCormick said that on the run down the range the train's speed continued to increase until the crash. There was no slackening. "The train wheels were screeching as we went round the two curves before the fatal curve," he said, "and people started to sing out. We were all thrown about." McCormick said that the train was swaying the same at the second curve, and they were thrown off balance again. After that they seemed to be going faster on the straight stretch. "The train was still swaying," he said. "Then came the crash. It made a terrific sound. Our carriage just turned straight on its side and there was the sound of rending woodwork. We were all piled in a heap on the floor. The front of the carriage was smashed in, but not the third compartment," said McCormick. His mother was hurt and was bleeding profusely. He said that he had not felt any "braking." To Mr Kissick he said that he would say that they were travelling more than 40 mph.'

Likewise, 'George Nathaniel Smoothey, who was to have been sports starter at the picnic, said: "The further we went the faster we went." Accompanied by his grandson, he was in the first carriage, in which many were killed. Asked by Mr Justice Mansfield (chairman) if he had felt the brakes being applied, Smoothey said: "My candid opinion is that no brakes were applied from the top of the range until the crash."

Also giving evidence the same day was Patrick Campbell, a former fireman himself who knew both Driver Hind and Fireman Knight. 'Campbell said that he had called out

---

[97] *The Telegraph*, Brisbane, Queensland, Australia, 26th May 1947
[98] *The Central Queensland Herald*, Rockhampton, Queensland, Australia, 29th May 1947
[99] *The Telegraph*, Brisbane, Queensland, Australia, 28th May 1947
[100] *Cairns Post*, Cairns, Queensland, Australia, 30th May 1947

amid the engine wreckage, "Are you there driver?" He heard a voice answer, "Yes," and asked, "How are you doing?" The driver answered, "Not too good, I can't do anything with this hacksaw." Campbell said that with others he pulled away the loose timbers of the engine cab top. He said to the driver: "What happened? Did your pump stop?" {of the Westinghouse brakes} "No, nothing like that, I didn't know the road, and neither did my mate."' [101]

On 2nd June, James Murray, a tax agent, and another passenger on the train, said that it had been climbing all the time to the summit and had accelerated very quickly once it had passed the summit. "I don't think the brakes were applied. I remarked that it was 'dangerous driving,' and that we should not have been travelling at such a speed. Then it seemed as if the brakes were applied. The train shuddered and the crash followed. I think the brakes were applied and then the engine turned over. After seeing the train I came to the conclusion that the shudder was caused by the engine ploughing through the bank. [102]

Passenger Ernest Wood, of Annerley, said: "My little fellow decided that we were chasing bushrangers and said to me, 'We are gaining on them, Dad,' but I was worried and told him to be quiet." [103]

A surprising piece of evidence was given by Constable Kenneth Johnson, who had visited Driver Hind in hospital the morning after the disaster and had asked him to make a statement. 'Hind replied, "Not at present, mate. I think I should not say anything until I have made a report to them." He told Hind he would return the following day. When he was leaving, Hind said, "I know I am in the clear, all right, mate." Hind died that afternoon.'

Defending Driver Hind, Mr T. Kissick, representing the Federated Loco Enginemen's Union, maintained that every passenger train exceeded the regulation and that they had to do so in order to keep their schedules. He said that if drivers kept to all the speed limits they would be charged with conducting a "go slow" strike and that the Railway Department were well aware that speed restrictions were disregarded. He attributed the accident to Driver Hind not knowing the road.

Guard Evans agreed that he knew the line. He said that the train had gathered speed past the summit and that he had been writing at his desk at the time. [104] 'Evans said that there were no unusual noises before the accident. Just the wheels rattling against the rails on the range. "I climbed halfway up the cutting," Evans said, "I thought we had a head-on collision. My watch showed 9.50. It was just as though we ran into a brick wall. The van stopped dead." He claimed there was no unusual swaying and he felt two applications of the brakes. "I was expecting a check of the brakes at any time when I was thrown off my feet right across the van. I rushed to the end and turned on the emergency brake tap, but there was no air. By the time I turned the tap the van was at a standstill."

'Evans said that when he heard the next train coming, he got a red flag and detonators and ran back about 500 yards. He then returned to the scene of the accident and was helping to lever a seat off a jammed passenger when he heard the other train coming down. He ran back and found it had stopped at the red flag. When the engine of the other train was hooked on to the last three carriages of the picnic train air came through the

---

[101] *The Courier Mail*, Brisbane, Queensland, Australia, 31st May 1947

[102] *Townsville Daily Bulletin*, Townsville, Queensland, Australia, 3rd June 1947

[103] *The Telegraph*, Brisbane, Queensland, Australia, 3rd June 1947

[104] *Townsville Daily Bulletin*, Townsville, Queensland, Australia, 5th June 1947

brake line satisfactorily.' [105]

At the close of the inquiry on 9th June Mr J. Seymour, assisting the board, submitted that given the surviving passengers' overwhelming evidence of speed it was impossible to believe the guard's evidence. Mr J. Ruddy, for the Australian Railways Union, on the other hand, suggested that excessive speed had not been the primary cause of the derailment and that there might have been an obstruction on the line or a defect in the engine or tender. [106]

The results of the inquiry were published at the beginning of July. It found that the derailment had begun with the tender of the engine leaving the rails, due to excessive speed, and that Driver Hind, Fireman Knight and Guard Evans were all guilty of a breach of duty. There had been no defect in the track, no foreign body on it, and no defect in the rolling stock, though the condition of the leading bogie castings of the tender had contributed to its overturning. The Westinghouse brake had been in good order and functioning properly.

The inquiry board stated that the fact that Driver Hind was unfamiliar with the line and was being piloted by his fireman had not been a contributing cause as the driver must have known that he was exceeding the maximum permissible speed. Fireman Knight should have ensured that his driver observed the speed limits, as should Guard Evans, who should have applied the Westinghouse brake from his van when he found that his driver was going too fast. The Board considered that greater safety would be provided if drivers learning the road were tutored by a qualified driver familiar with the road rather than a fireman and that said driver should be in control of the locomotive. The board conceded that Fireman Knight had had other duties to perform and that his warnings to the driver might not have been heard or have been misunderstood. As for guards their training for emergencies could be improved.

The speed of the train at derailment, estimated by the Board to have been between 35 and 50 miles an hour, was slightly below the theoretical overturning speed of the tender, indicating that other factors must have played a part. There were slight imperfections in the permanent way which might have caused undue swaying of the engine and tender and combined with the movement of the driving wheels, crankshafts, and pistons to reduce the overturning speed.

The Board was not in favour of speedometers on engines, saying that were extremely sensitive instruments and likely to get out of order or give incorrect speeds, and that even without them experienced drivers could estimate speeds to within five miles an hour.

Finally, the Board held that the driver, fireman and guard had all been competent railwaymen and it praised the guard for protecting his train after the accident and giving assistance despite suffering from severe nervous shock. [107]

Train No. 8 Up, the night mail, which had left Kuala Lumpur at 9.30 p.m. and was due in Penang at 9.50 a.m. was derailed at Bukit Tengah, Malaya, five miles from Prai, at 9.38 a.m. on Wednesday 21st May 1947. The engine and four carriages left the rails, overturned, and caught fire. The train was travelling at 15 to 20 miles an hour. Eight

---

[105] *The Northern Miner*, Charters Towers, Queensland, Australia, 6th June 1947
[106] *Morning Bulletin*, Rockhampton, Queensland, Australia, 10th June 1947
[107] *Cairns Post*, Cairns, Queensland, Australia, 2nd July 1947

passengers were reportedly killed and between 30 and 50 injured, 13 of whom were taken to hospital. At the time of the report other passengers were said to be still trapped under the wreckage. The dead were all Indian and Chinese third-class passengers. Police and officials stated that there was no sign of sabotage and it appeared to be purely an accident. [108]

Eyewitnesses of the accident stated that the train had been rounding the curve into Bukit Tengah Pass when "the engine and coaches seemed to sway and then rolled over." Ambulances took the injured to Bukit Metajam and Butterworth and uninjured passengers were brought on to Penang by a special train from Prai and by launch. Through communication was restored by the afternoon.

Driver J.W. McIntyre stated that as the train had slowed down about 50 yards from Bukit Tengah station he had felt a bump. He had applied the brakes, felt a wheel leaving the rails and was thrown from his seat. He was badly bruised and scratched. Of his two firemen, Fireman Govindasamy was thrown against him and apparently only slightly hurt while Fireman D.W. Stewart was thrown from the cab and injured his feet. The three of them had helped uninjured passengers into a special train after which Fireman Govindasamy, who perhaps had been more seriously hurt than it appeared, had collapsed. (Another newspaper states that it was Fireman Stewart who was thrown against the driver, Fireman Govindasamy who was thrown from the engine and Fireman Stewart who collapsed – clearly one or other of the reporters got his names mixed up.) The accident occurred just as the train was passing a signal box at a point where a disused branch constructed by the Japanese went off into an estate. [109]

Mr S. N. King, Resident Commissioner, Penang, visited the scene during the afternoon accompanied by Mrs King. Mr. J.O. Sanders, General Manager of the Malayan Railway, commented, "We are very proud of our record. This is most distressing."

The newspaper report stated that there had been some very serious derailments in the country during the Japanese occupation, mostly due to guerrilla sabotage, one of the worst being near Johore, but the Japanese had published no details of them. Otherwise there had only been four relatively minor derailments since mid-1932:

On 5th August 1932 a passenger train was derailed at Tanjong Malim. Two firemen were killed and several passengers injured.

On 6th September 1937 a passenger train left the rails at Tenang, south of Johore, injuring two passengers.

On 26th November 1937 a passenger train was derailed at Kulai, north of Johore, killing one Tamil trespasser on the line.

On 28th May 1939 a driver died and one person was hurt in a passenger train derailment at Gemas. [110]

A passenger in the fifth coach of the train described the derailment of the four coaches in front of him. He said that he had had his head out of the window and had seen the third coach break horizontally in half. "The steel chassis remained derailed on the track while the wooden coachwork, with the passengers inside, was flung into the lalang {coarse, weedy Malaysian grass} beside the line."

---

[108] *Indian Daily Mail*, Singapore, 22nd May 1947
[109] *The Malaya Tribune*, Singapore, 22nd May 1947
[110] *The Straits Times*, Singapore, 22nd May 1947

By the following day the number of dead had risen to 11. [111]

An animal on the line, specifically a cow, caused a major disaster in Argentina on Sunday 15th June 1947. A passenger train departed from Fererico Lacroze station, Buenos Aires, on the Ferrocarril Central Buenos Aires, at 14:15 that day, and made its way as usual to Zárate. From the latter place there was a ferry across the Paraná River, thence the journey continued from Ibicuy on the Ferrocarril Nordeste Argentino through Basavijbaso and Concordia to the train's destination at Posadas. The train was between Concordia and Pasados and two and a half kilometres from the town of La Cruz when, at 19:30, it hit the cow at kilometre 333·900. The engine, a van, a first-class and two second-class coaches were thrown from the track. Sixteen passengers and Driver Bibiano Ortiz were killed and 38 people were taken to local hospitals and clinics. Two conscripts, who must have been badly hurt, were flown to hospital in Buenos Aires. [112]

A doubled-headed Japanese National Railways passenger train was derailed between Kudamatsu and Hikari stations on the Sanyo line in Yamaguchi Prefecture, Japan, on Tuesday 1st July 1947. The locomotives and three carriages overturned. Initial reports stated that between 60 and 150 people had been killed or injured. It was later stated that 15 people had died and 72 were injured. [113] [114] [115]

An express train from Bratislava to Praha was derailed near the station at Kuty, Czechoslovakia, in Moravia, on Tuesday 5th August 1947. The accident apparently resulted from an error by either the train driver, the permanent way staff, or perhaps both. A section of track was under repair and the express was to be diverted to the opposite track. Instead the engine reportedly ran onto the track which had already been lifted, was derailed, and overturned. The express had supposedly failed to slow down despite signals to do so. There was some telescoping of the express coaches but it was stated that a steel coach had probably saved the remainder of the train. The bogies of this coach were smashed but it remained upright.

Sixteen passengers, all Czechs, were killed and 37 were injured, 16 seriously. A party of Swedish tennis players who had originally planned to take the train chose to fly instead and thus escaped. [116] The driver, fireman and a third railwayman were arrested. [117] [118]

A train was being brought into the platform at Jullundur Cantonment, India, on Thursday 9th October 1947 and waiting passengers scrambled to load their luggage onto it. In the rush one of their trunks fell onto the line and under the wheels of the train, causing the derailment of a bogie (carriage). Nine people died and seven were seriously injured. [119]

---

[111] Ibid, 23rd May 1947

[112] *El Orden*, Tucumán, Argentina, 17th May 1947

[113] *Doylestown Daily Intelligencer*, Doylestown, Pennsylvania, United States, 1st July 1947

[114] *The Marion Star*, Marion, Ohio, United States, 1st July 1947

[115] http://www.ikjeld.com/japannews/00000147.php

[116] *The Times*, London, England, 7th August 1947

[117] *Le Figaro*, Paris, France, 7th August 1947

[118] *The Singapore Free Press*, Singapore, 8th August 1947

[119] *The Indian Express*, Madras, India, 11th October 1947

A very brief report from India stated that the Khandwa – Bombay passenger train carrying refugees towards Hyderabad had been 'derailed causing casualties' on about Sunday 19th October 1947. No further details were given. [120]

Only two days after the disastrous collision on the Southern Railway near South Croydon (see Rear Collisions chapter), England suffered a second disaster, this time in the north-east, at Goswick, on the London and North Eastern Railway, on Sunday 26th October 1947. The second disaster was a derailment.

The train involved in the Goswick accident was the 11:15 a.m. Up express from Edinburgh Waverley to London King's Cross. It consisted of 4-6-2 A3 class steam locomotive No. 66 *Merry Hampton* and 15 bogie carriages, in charge of Guard W. Blaikie with Driver T. Begbie and Fireman W.R.E. Baird. Also riding on the engine, though he had no right to be there, was a naval rating, Leading Stoker T.A. Redden. At Goswick the train was to be diverted from the Up Main to the Up Goods Independent line because of engineering work on underbridges on the Up Main line. The Company Engineer stated that a safe speed through the crossover from the Up Main to the Up Goods Independent would be 15 or 20 miles an hour. The express attempted to negotiate the turnout at a much higher speed, with the result that the engine and eight leading carriages of the train were derailed. There had been 420 passengers on the train. Twenty-seven of them and one train attendant were killed, 59 passengers, six railwaymen and the naval rating were admitted to hospital and 25 passengers and three railwaymen were slightly hurt, treated, and sent home. [121]

'Rescue workers informed Press representatives that although the accident happened in a sparsely populated area doctors, ambulance men, farmers, and others arrived from miles around in nearly 100 cars. They went to the assistance of the uninjured passengers, who were helping the injured and trying to free others who were trapped in the wrecked coaches. Luggage and wreckage lay scattered about the fields. A small Presbyterian chapel near the track was also converted into a casualty clearing station and mortuary. Members of the congregation who were arriving at the chapel for a service were among the first to arrive at the wrecked train and helped to carry victims to the chapel.

'Berwick Infirmary, where casualties were taken, was so busy that a telephone call by a Press representative was answered by one of the casualties. He said the hospital was fully occupied in dealing with the seriously injured. A number of casualties, after receiving treatment, were sent by ambulances to hospitals at Newcastle, Ashington and Edinburgh. Ambulances were fully deployed throughout the afternoon, and rescue work was still going on when darkness fell.

The bodies of two of the kitchen staff were found when the dining car was lifted by crane. Two passengers described how they helped the others out of their carriage. Cries for help were heard coming from a heap of wreckage where one carriage had fallen on top of the other, after toppling down the embankment, and they found a woman with head injuries trapped under wreckage which they were unable to lift. A man pinned by a steel girder had only his head clear. Other passengers joined them in trying to lift the wreckage, but it was not until a crane arrived that the injured were released. The rescuers

[120] *Indian Daily Mail*, Singapore, 21st October 1947

[121] Ministry of Transport report

next concentrated on the buffet car and helped to free the stewards. All were injured. A woman passenger was taken out dead. An Australian warrant officer was next freed and carried on a stretcher to Goswick station. Passengers stopped the first ambulance that came along and he was taken to Berwick Hospital but died on the way.

Other passengers described the accident as having happened so suddenly that they were in a heap in the carriage before they realised what had happened. One commented that he had thought that it was raining, but in fact it had been water pouring from the engine.

'Mr K.J. Mulvaney, of Notting Hill Gate, London, and Mr D.D. Roberts, of Ladbroke Square, London, were in the first carriage of those that did not leave the rails. Mr Mulvaney said he felt two jolts as if the brakes had been suddenly and firmly applied. Then the carriage began to heel over, and finally came to rest at an angle. All the people in the compartment were thrown in a heap by the floor farthest from the corridor. They had to climb up into the corridor and out of a window to reach the ground. They found a pile of wreckage and saw injured people lying all along the line.

'Mr G. Kay, of Bonnyrigg, Midlothian, who was uninjured, said that in his compartment were thrown in a heap on the floor. One of the first to recover from the shock was a doctor, who said he must help, and was assisted through the window. Many helpers soon arrived, and the first batches of injured were taken in buses to hospital. Those who could not be put into buses were put into open trucks. Another passenger described finding the fireman lying injured with his feet near the fire-box. After trying to get the fireman out by himself he obtained the assistance of two soldiers, and they managed to free the injured man. The train carrying survivors arrived at King's Cross just before 2.15 a.m. About 100 people were waiting to meet it.' [122]

The job of renewing the three underbridges had originally been planned to take place on Sunday 19th, then postponed for a week. The arrangements at Goswick had been discussed by Stationmaster G.R. Danskin and Relief Signalman T. White the day before. Signalman White came on duty at 6 a.m. on the Sunday morning and as soon as he had been advised that the engineering work was ready to begin he had reset the road from the Up Main to the Up Goods Independent, bolted the points and put a collar on the points lever. With the road thus set the Up Main distant (No. 24) and Up Main starting signals (No. 18) were necessarily locked at danger.

Signalman White told the Ministry of Transport inquiry that for each Up train he had kept the home signal (No. 22) at danger until the train was almost at a stand, then lowered it to bring the train up to the starting signals and then lowered the Up Main to Goods Independent Starter (No. 19). By thus only clearing the home signal at the last moment he had been, on his own initiative, applying the provisions of Rule No. 39 (a) which was more than he needed to do according to the Rules and Regulations. He said that he had received the *Train Entering Section* signal for the Edinburgh express from Scremeston box at 12.45 p.m., it had been accepted by Beal box at the same time on the Up Goods Independent line. He had shortly seen steam issuing from the engine of the approaching express, something which he had noticed with other trains, and assuming that it was slowly nearing the home signal he had pulled the signal off. The train was then about 300 yards from the signal. Then, suddenly, he realised that the train was not pulling up, and shouting to Lengthman McIntyre, who was in the box as flagman, he had thrown the

[122] *The Times*, London, England, 27th October 1947

home signal back to danger. The Up Goods Independent starting signal, No. 19, he had not cleared at all. Pulling the emergency detonator placer he had then gone to the box window to motion to the driver of the express as it steamed past him at 60 miles an hour, exploding the detonators. The train then entered the crossover and was derailed. Signalman White sent the *Obstruction Danger* signal to Scremeston and Beal, stopped a Down goods train, told the lengthman to put down detonators on the Up line and inform the stationmaster and telephoned Signalman Burnett at Scremeston to call out the emergency services.

Lengthman McIntyre stated that he had been sent to Goswick box by the Permanent Way inspector to protect the engineering work and had placed a red flag and three detonators on the Up main. He confirmed Signalman White's account of what had happened. He said that he had been eating his dinner in the box when the Edinburgh express had arrived and the signalman had shouted, "Good God, he won't pull up!"

Stationmaster Danskin said that he had been on duty for a couple of hours on the Sunday morning because of the engineering work and was in Goswick box between 8 a.m. and 9.30 a.m. He confirmed that Signalman White had dealt with several Up trains as the latter had stated. After visiting Beal box he had gone off duty at 11 a.m. until he was told of the accident by a cyclist. Back in the box at 1.05 p.m. he found that Signalman White had gone to help in the rescue work. The Up distant signal lever was back in the frame, the repeater for the signal showed it to be at *Danger* and the rear coaches of the express were stopped on the facing points, showing that they could not have been reset after the derailment.

Driver Begbie said that he knew Naval Stoker Redden, whose brother was a Haymarket (Edinburgh) fireman. Redden told him that he was shortly to begin as an engine cleaner and had asked if he could ride with Begbie and his fireman on the engine to Newcastle, a request to which he (Begbie) had eventually reluctantly agreed. Begbie said that he had signed on at Haymarket at 10 a.m. on the Sunday morning and had studied the notices, though he had failed to see the late posting referring to the Goswick diversion. This notice read:

TO ALL CONCERNED

SIGNALLING PROGRAMME No. 22

Please note the following alterations: -

SUNDAY 26TH OCTOBER

PAGE 528 – INSERT: -

BETWEEN BEAL AND GOSWICK SIGNAL BOXES

7.30a.m. to about 4.30 p.m. reconstruction of superstructures of bridges Nos. 170 and 172 at 58m. 69ch and 59m. 33ch.

7.30 a.m. to about 1.0 p.m. possession of Up and Down Main lines. All traffic to be diverted to Up and Down Goods Independents.

During time traffic is being diverted over the Goods Independent absolute block working to be maintained and any points not fitted with facing point gear to be securely clamped or wedged.

1.0 p.m. to about 4.30 p.m. possession of Up Independent.

The run from Edinburgh, Driver Begbie said, had been normal, and he had obeyed speed restrictions at Berwick and over the Royal Border Bridge. As his train neared Goswick he had failed to see the Up distant signal because of smoke and steam from the chimney of his engine. He had shut off steam but had not applied the brakes as, he maintained, a few seconds later he had seen the Goswick Up home signal at *Clear* and assumed that he had a clear road past there. The home signal, he stated, had remained *Off* until after he had passed it and it was only then that he had seen one of the starting signals at clear, though it had been thrown to danger as he was close to it. Signalman White, on the other hand, stoutly denied that he had ever cleared the starting signal. The driver could not explain why he had not crossed over to the fireman's side to see the distant signal and had not seen the starting signal earlier. He denied having had any conversation with the naval stoker.

Fireman Baird explained that he had been a little late signing on at Haymarket and so had not had time to read the notices. As they were nearing Goswick he had been firing until he had seen the home signal at *Clear* at short range. His driver had then shut the regulator and applied the brakes. He did not know who Redden was and had barely spoken to him on the journey.

Leading Stoker Reddin stated that he had been a porter before joining the Royal Navy in 1941. Now out of the Navy, he had been anxious to rejoin the railway and had applied to the Locomotive Department. He confirmed that he had not spoken to the driver during the journey and only very briefly to the fireman.

Like the enginemen, Guard Blaikie said that he had not seen the late notice relating to the Goswick derailment. He had not been aware that there was a third person on the footplate from Haymarket. The journey from Edinburgh had been normal; the train had been ten minutes late at Berwick due to single line working at Drem. He put the speed of his train at 60 miles an hour at Goswick. He had not been observing the signals and did not notice any brake application before the derailment.

The inquiry into the accident had been begun by Colonel A.C. Trench but he had taken ill and it was completed by Lieut. Colonel G.R.S. Wilson. In his conclusions, the latter inspecting officer stated that it was clear that there had been no defect in the track or locomotive and the derailment had been solely due to the train passing through the turnout at excessive speed.

Signalman White admitted that he had misjudged the speed of the express when he had cleared the home signal but his denial that he had cleared the starting signal was impressive.

Driver Begbie admitted having failed to see the Goswick Up distant signal but he had not reduced speed. Seeing the home signal lowered he had then assumed that the road was clear. Even if the starting signal had been lowered, which the signalman denied, it could only have been the one for the diverging route and it was unlikely that he had closed the regulator as early as he stated. Although there was no definite evidence it was probable that his grave breach of discipline in having an unauthorised person on his engine was a factor in his failure to exercise reasonable caution. He must bear the main responsibility for the accident.

Contributory to the accident were the failures of any of the train crew to observe the notice at Haymarket relating to the diversion.

It was hard to criticise Signalman White for the premature lowering of the home signal given that according to the regulations he could have lowered it earlier anyway.

The inspectors recommended that consideration be given to the general application of the provisions of Rule No. 39 (a) for movements involving diversions over low-speed junctions. (Rule 39 (a) stated that, 'When a stop signal is at Danger the stop signal next in the rear of it and worked from the same box must not be lowered for an approaching train until it is close to such signal and has been brought quite or nearly to a stand'.)

Improvements in the arrangements for the display of late notices were suggested so that they were less likely to be missed.

The accident would (or should) have been prevented had Automatic Train Control been installed at the distant signal and high priority should be given to a general programme for bringing it into use. [123]

A mixed train to Sourabaya, Dutch East Indies, was derailed at an unstated location on New Year's Eve, Wednesday 31st December 1947. The engine and passenger coach at the front of the train remained on the rails but 18 waggons at the rear of the train were derailed and overturned and several hundred metres of track were torn up. Seven people died and 17 were hurt. Three of those who lost their lives were brakesmen, the other dead and the injured were unauthorised passengers who had been riding on the running boards and in the goods van in the middle of the train. [124]

Thirty-five people died when a crowded two-car electric train of the Nagoya Railway derailed and overturned on a curve at Moriyama, Japan, on the Seto line, on Monday 5th January 1948, possibly due to excessive speed. Another 154 of the estimated 450 passengers on the train were injured. [125 126]

Less than a year after the Camp Mountain accident in Australia a derailment with similar features occurred on the South Island of New Zealand, between Seddon and Blind River. On Wednesday 25th February 1948 a Picton to Christchurch express passenger train left the track on a curve. Six passengers died and 61 passengers, the driver and fireman were injured.

The train which was derailed consisted of Ab class 4-6-2 steam locomotive No. 696 and tender, five second-class and two first-class carriages, a guard's van and a Z waggon, in charge of Guard J.R. Parmenter with Driver Joseph Gubb and a unnamed fireman. As at Camp Mountain, the line over which the train was running had quite severe curvature, and when the train about 2½ miles south of Seddon and not too far from Blind River it entered a 10-chain curve at excessive speed. The engine had almost exited the curve when it overturned, rather than becoming derailed, the first vehicle stopped behind the tender, the next three vehicles went past the engine, the next three vehicles were also derailed but the guard's van and Z waggon remained on the track. The first, second and third carriages were wrecked. [127] The leading carriage was almost empty and it was in the

[123] Ministry of Transport report
[124] *The Straits Times*, Singapore, 2nd January 1948
[125] http://www.ikjeld.com/japannews/00000147.php
[126] *The Lethbridge Herald*, Lethbridge, Alberta, Canada, 5th January 1948
[127] *Tragedy on the Track* by Geoff Conly and Graham Stewart, pgs. 109-13

next two that most of the casualties occurred.

Guard Parmenter sent his assistant to a telephone a mile away to ring Blenheim for doctors and ambulances while local settlers were among the first outsiders at the scene.[128] Later two lorryloads of men from the Royal New Zealand Air Force station at Woodbourne rendered assistance. Stretchers were improvised from pieces of timber torn from the wreckage and some of the injured were given morphia injections to control their pain before ambulances, lorries, school buses and private cars took them to the hospital in Blenheim.[129]

'Many passengers had amazing escapes. One man was hurled through a gaping hole in the side of the second carriage after it had gone past the engine. He landed on the side of the track away from the wreckage. A young woman who was in the second carriage was lifted bodily from her seat and was hurled through what was left of the window onto the gravel at the side of the line.'[130]

An inquiry was conducted into the accident. The direct cause of the derailment was obviously excess speed, the estimated overturning speed for the curve being 52 miles an hour. The board of inquiry found that Driver Gurr was guilty of a serious dereliction of duty in driving the engine at a speed in excess of that prescribed and that he could offer no explanation for the accident. It commented that it was bound to say that they were unable to understand how a driver of his experience and good record could fail to observe well-known regulations and exercise ordinary vigilance and judgement within sight of clearly defined and easily observed warning aids. The actions of the locomotive foreman at Christchurch and the senior driver at Kaikoura in giving Gurr the job of driving the express the board found to have been justified.

It transpired that Driver Gurr was only what in Great Britain would be called a passed fireman. He had less than 18 months' experience driving trains and the day of the accident had been the first time that he had driven an express. Driver Gurr candidly admitted that he had been 'over confident' and had made 'a terrific error' in entering the curve at the speed that he had. He had been over the line a number of times as a fireman but only three times previously as a driver, working goods trains. On an express he was of course under a lot more pressure to keep time. He said that his train had lost five minutes at Blenheim and more time at Dashwood, just before Seddon. He had seen the warning board indicating that he was approaching the curve and had made a light brake application – the standard procedure for a goods train to bunch the trucks – and the next thing that he knew there had been a surge of his train and he had been flying through the air.

The reason that Gurr had been given the job was that there was a shortage of drivers at Kaikoura, more experienced Christchurch drivers being reluctant to take up a position there because of the rural location and lack of facilities.

Driver Gurr subsequently stood trial for manslaughter but was acquitted by a Blenheim jury, who added a rider that the cause of the accident was 'the inefficient administration of the Railways Department.'[131]

The passenger train which left Cali for Popayán, Colombia, at 13:00 on the afternoon

---

[128] *Cairns Post*, Cairns, Queensland, Australia, 26th February 1948
[129] *Tragedy on the Track* by Geoff Conly and Graham Stewart, pgs. 109-13
[130] Ibid.
[131] Ibid.

of Sunday 1st August 1948 was derailed just after 15:00 at kilometre 228, between San Francisco and Asmazú stations. The engine, No. 76, an equipment van and three third-class passenger coaches left the track. First information was that there were many victims. Four bodies of passengers were soon recovered, which it was not possible immediately to identify, together with those of two railwaymen, identified as Brakemen Manuel J. Garcia and Manuel Varela. It was feared that there were a good many more dead and injured.

As soon as news of the accident became known the railway management arranged for rescuers to go to the scene, together with engineers to examine the wreckage. News of the derailment had caused great upset in Cali.

At 22:00, Driver R. Cifuentes, Assistant Driver J.I. Restrepo and Fireman Obdulio Nieto, all badly hurt, were brought to Cali, together with eight seriously injured passengers, whose names were unknown, and were taken to the San Juan de Dios Hospital and clinics. Another 20 injured had been taken to the hospital at Asnazú. A breakdown crane was at the site. It had lifted at least one of the derailed carriages and six more bodies had reportedly been discovered. There were believed to be further bodies beneath the carriages which had yet to be rerailed but until then it was impossible to tell.

A further report from Cali stated that there were eight dead and more than 15 injured. The accident had happened, it was said, at 13:30, not 15:00. Immediately news of the accident was received an autoferro expresso (railmotor?) with rescuers and equipment had been despatched to the scene. The autoferro had returned with injured, some of them in a serious state, and the bodies of those who had lost their lives. The bodies of the dead were taken to the mortuary pending identification. It was the most serious railway accident in western Columbia in recent years. Some of the injured were treated at the mineral company hospital in Asnazú, others were transferred to hospitals in Popayán and Cali. The victims belonged mostly to poor families residing in various regions of the Cauca department, who took advantage of Sundays to go shopping in the market at Timba. The people lived on the two banks of the river of that name, which serves as the boundary of the departments of Valle and Cauca. On receiving news of the accident the head of the Ferrocarril de Pacifico ordered all possible aid to be sent to the area.

Alberto Arango Montova, one of the inspectors sent by the railway company to the scene, to investigate the causes and magnitude of the disaster, returned to Cali on the 3rd. He told creepy and Dantonesqe accounts of the disaster which had affected many poor households of the departments. This official would furnish a report on the magnitude and consequences of this unprecedented accident. He was to return to the scene of the accident to continue the investigation, then send his report to the senior railway administrators.

The cause of the derailment remained unclear, though all parties agreed that it had not been due to excessive speed. If it had been, a newspaper report continued, there would have been almost total loss of life, as the accident site was on one of the most dangerous places on the line through the mountains, with deep gorges. [132]

Fourteen people died in a derailment between Furkating and Kamarbandha Ali stations, India, on the Lunding – Tilsukia section of the Bengal and Assam Railway, at about 09:30 on the morning of Monday 15th November 1948. Twelve people were killed

[132] *El Tiempo*, Bogota, Colombia, 2nd and 3rd August 1948

outright, including two railwaymen, when the engine and five leading bogies derailed and overturned and a number of people were injured, two of whom died later. Doctors were rushed to the scene from Mariani, Chakiting and Furkating and a relief train left Lunding for the accident site at 10:55. Twenty-one injured were taken to Jorhat Civil Hospital and five to Mariani Railway Hospital. Through traffic was not interrupted as trains were diverted from Furkating via Mariana. [133]

As a result of either a derailment or possibly a buffer stop collision a freight train collided with the station platform at Shimo Goto, Japan, on Friday 14th January 1949. Seven people died and 50 were injured. [134]

The first through train from Nanking to Canton, which had left Nanking on the 17th, was derailed at Chuchow, China, 350 miles north of Canton, on Friday 22nd April 1949. Eleven people died and ten were injured. [135]

Seven people were reportedly killed and 40 injured when a coach of a Guayaquil – Quito passenger train was derailed on a bridge at Bucay, Ecuador, on Wednesday 22nd June 1949. [136] [137]

A train carrying 700 children from Belgrade and Zagreb to a summer camp on the Adriatic coast was derailed near Pula, Yugoslavia, on the evening of Monday 8th August 1949. Fourteen of the children lost their lives and 60 were injured. [138] [139]

'Unofficial but reliable sources' stated that over 200 people had died in the derailment of a Gdańsk – Warszawa express on a curve at Nowy Dwór Mazowiecki, Poland, on Saturday 22nd October 1949. Another 200 or more passengers were reportedly injured when the locomotive and several carriages left the rails on a curve. It seems probable that the derailment was due to excessive speed. By Monday the track had been repaired and trains were running normally past the scene. [140] [141] Official sources gave no further information on the disaster and it has been suggested that it never actually happened. [142]

Eleven Africans died and many others were hurt when a train was derailed and fell down a steep embankment near Dar-es-Salaam, Tanganyika, on Tuesday 24th January 1950. [143]

---

[133] *The Bombay Chronicle*, Bombay, India, 16th and 17th November 1948
[134] *Sunday Tribune*, Singapore, 16th January 1949
[135] *The Post Standard*, Syracuse, New York, United States, 24th April 1949
[136] *The Charleston Gazette*, Charleston, Virginia, United States, 23rd June 1949
[137] *The Logansport Press*, Logansport, Indiana, United States, 23rd June 1949
[138] *The Canberra Times*, Canberra, Australian Capital Territory, Australia, 10th August 1949
[139] *Gazette de Lausanne*, Lausanne, Switzerland, 9th August 1949
[140] *The Independent Record*, Helena, Montana, United States, 24th October 1949
[141] *Manitowoc Herald-Times*, Manitowoc, Wisconsin, United States, 24th October 1949
[142] https://de.wikipedia.org/wiki/Eisenbahnunfall_von_Nowy_Dw%C3%B3r
[143] *The Railway Gazette*, London England, 3rd February 1950

Seven people died in the derailment of a Chemins de fer de la Côte d'Ivoire mixed train at an unknown location between Abidjan and Dioulasso, Côte d'Ivoire, on Wednesday 8th February 1950. A salon coach and a flat waggon left the track. [144]

Fourteen men, women and children lost their lives and 16 others were gravely injured in an accident a few miles from the town of General Macarthur, on the island of Samar, the Philippines, late on Saturday 8th April 1950 (possibly early on Sunday 9th). They had been riding into town on top of iron ore on a train from the Elizalde mines in order to take part in the Easter sunrise observances and were buried under large quantities of ore when the train was derailed. At the time of the report rescue work was continuing as it was not known how many people had been riding on the eight loaded cars. [145] [146]

In what was probably a derailment on the Chemins de fer de la Côte d'Ivoire on the evening of Sunday 30th July 1950, ten people died and 90 were injured at an unstated location in Abidjan and Dioulasso, Côte d'Ivoire. [147]

An electric train was derailed as it was descending a steep incline at speed in Sicily, Italy, on Sunday 6th August 1950. The train was rounding a curve near Castelvetrano when the accident occurred. Peasants rushed to the scene and removed the badly mangled bodies of six passengers from the twisted wreckage. Thirty-five injured passengers were taken to hospitals in Castelvetrano and Marsala. [148]

On Sunday 27th August 1950 a passenger train from Ath was derailed at Havinnes station, Belgium, on the line between Leuze and Tournai. The train consisted of an engine running tender first and six wooden coaches. Normally it was formed of a Michelin railcar but a larger train was provided to cope with the anticipated extra traffic in connection with a fair (La Grande Kermesse) at Ath. The derailment occurred on a secondary line as the train was negotiating a set of points.

According to the driver and fireman, who escaped unhurt, there was a sudden shock, whereupon the driver had immediately applied the brakes. The violence of the sudden stop caused the second carriage to climb onto and telescope the first and the middle of the third carriage was also damaged by the wreckage of those in front.

After a brief moment of panic a rescue operation was quickly organised and the bodies of five dead were removed to the mortuary at Beclers while ten seriously injured were hospital in Tournai. Four of the dead were identified as three women and a man from Leuze. The fifth victim, another woman, could not be immediately identified.

The cause of the derailment was unclear, though track subsidence was thought a possibility. [149]

---

[144] *Le Monde*, Paris, France, 11th February 1950

[145] *Lebanon Daily News*, Lebanon, Pennsylvania, United States, 10th April 1950

[146] *The Taylor Daily Press*, Taylor, Texas, United States, 10th April 1950

[147] *The Charleston Gazette*, Charleston, West Virginia, United States, 2nd August 1950

[148] *Ce Soir*, Paris, France, 8th August 1950

[149] *Nord-Matin-Sports*, Paris, France, 28th August 1950

Electric train N-2 of the Estrada de Ferro Sorocabana, on its way from Ourinhos to São Paulo, was derailed at 05:30 on Wednesday 6th September 1950 at Pantojo, Brazil. The train consisted of engine No. 3045, with Driver Aristides Mariano and Assistant Driver João Moreira Prado, and an unspecified number of coaches. As it was passing through Pantojo three first-class and two second-class coaches left the track, collided with a stationary freight train in the station yard and were badly wrecked. The coupling between the first and second vehicles of the train gave way and the engine and leading coach continued for about a kilometre down the line before they stopped.

As soon as news of the accident became known ambulances were despatched from Santa Casa Assistencia Municipal and Santo Antonio and São Severino hospitals in Sorocaba, together with doctors, firemen and police. Fifteen passengers lost their lives in the accident and 49 were injured, of whom 12 were just slightly hurt. Of the remainder, nine were hospitalised at Sorocaba and the other 28 were transferred to São Paulo and hospitalised there.

The cause of the derailment was not immediately apparent; a broken axle or track defects were put forward as possibilities. Engineers Augusto Nonan and Vicente Leito Sampaio, heads of the traffic and operating departments of the company respectively, were tasked with carrying out an investigation, including an examination of the tachograph of the locomotive, but their work was hampered by the disappearance of Driver Mariano soon after the accident (probably because he feared indefinite detention under Brazilian law – see Anchieta accident above). [150]

Express train No. 102, which had left Nancy at 06:22 and was due in Paris at 11:27, was derailed at about 08:00 on Monday 25th September 1950. Six passengers died and 26 were injured, seven seriously.

Between Lérouville and Bar-le-Duc there was a long descent. Over a considerable length of this stretch work on the track was in progress on the line on which the express usually ran, and single-line working was therefore in operation over the opposite running line from Lexoville, France. On the approach to the mains crossover a *signal de ralentissement* warned drivers to reduce speed to 30 kilometres an hour. In a repetition of the Goswick disaster in England three years earlier the driver of the express failed to reduce the speed and the train was still travelling at 60 kilometres an hour when it reached the crossover. At the last moment the driver realised his error and applied the brakes, but it was too late, in fact the sudden, violent, brake application might have contributed to the derailment.

At the point of the derailment the line ran on a five-metre-high embankment. The engine overturned on the permanent way while the two leading carriages, after tearing up over 100 metres of track, fell down the embankment. The third carriage ended up with one end suspended in mid-air, held by the couplings to the remainder of the train. Although the leading carriages were of steel construction they were not strong enough to completely resist the shock of the fall down the embankment.

Most of the casualties in the accident were young men who were apprentices working in the Wendel steel works in Lorraine and who had been on their way to a holiday camp

---

[150] *O Estado de São Paulo*, São Paulo, Brazil, 7th September 1950

in Tours. One of the apprentices, who escaped with only relatively minor injuries, told his story to a newspaper reporter while a Red Cross nurse was attending to him: "The young people in the coach felt that they were already on holiday. They were playing cards or singing. Suddenly a violent brake application threw them one upon the other. The luggage fell down from the racks onto us. The coach rolled over on the slope of the embankment, into the meadow, then, all of a sudden, it lay down. Without knowing what had happened I found myself flat on my stomach on the grass. Then the injured were crying out for help and the passengers from the rear of the train were running to their aid."

The rescue operation was quickly organised. The injured were taken to hospital in Bar-le-Duc and a *chapelle ardente* was set up in the village hall where families in tears were to come to identify the dead by the end of the afternoon.

At the end of the morning the Bar-le-Duc parquet had opened an investigation into the accident, while a Société nationale des chemins de fer français technician was called in to examine the Flaman speed recorder on the locomotive.

As in the Goswick disaster the driver should have been issued with a bulletin advising him of the engineering work, the single-line working and the need to reduce speed at Lexoville. [151]

In a follow-up report on the accident the next day the two enginemen were named as Driver Genet and Fireman Claude.

The fireman told his interviewers, "I saw the *signal de ralentissement* and shouted to Genet, 'Attention!'"

Driver Genet declared, "I applied the brakes when the engine reached the points. But it was too late. I had seen the service bulletin regarding the reduction of speed to 30 kilometres an hour and specifying that (exceptionally) that the *signal de ralentissement* would be on the right-hand side."

The Flaman recorder had now been analysed and showed that the speed of the train had been 60 kilometres an hour at the moment it reached the crossover.

The three most seriously injured victims, who had been unconscious when admitted, had all now improved and were able to mutter a few words. [152]

The Estrada de Ferro Leopoldina in Brazil suffered a major derailment on Friday 23rd December 1950 near Engano station. Engineer Felicania de Souza Agiar announced in a statement on Saturday, Christmas Eve, that the general manager of the railway company deeply regretted to inform the public that a serious accident had occurred the previous night on the southern line in Esperito Santa state to the passenger train which had left the Barão de Mauá terminus in Rio de Janeiro at 20:45 on Thursday night. The train had become derailed and as a result 17 passengers and railway employees had lost their lives and about 50 had been injured. Everything had been done for the care of the victims and the injured had been taken to Vitória, the state capital and the train's destination. It was hoped to reopen the line at 14:00 (on Saturday). The railway company would meet all the costs of the funerals of those who had died. A rigorous investigation had been opened to determine the cause of the disaster and those responsible would be severely punished.

---

[151] *L'Aurore*, Paris, France, 26th September 1950
[152] Ibid, 27th September 1950

The accident had happened at about 20:00 on Friday at kilometre 539, about 300 metres from Engano station, which was described as being located at a place absolutely devoid of resources. It was not stated how long it had taken for news of the disaster to reach Vitória but as soon as it did numerous doctors, medical supplies and firemen set off for the scene. The relief train from Vitória was delayed en route for an hour or more at Pedra de Vento due to the derailment of three waggons and it was Saturday morning before the injured began to arrive at Vitória, followed by a train carrying the bodies of 13 of the dead on Saturday afternoon. Two more bodies were taken south to Campos and the other two to the municipality of Alfredo Sales. Of the injured, 37 were taken to the Santa Casa hospital in Vitória, one of whom was reportedly comatose, six were taken to the Casa de Saúde São Sebastão hospital there, and four were treated in the emergency room and released.

A reporter managed to interview one of the passengers from the train, a Dr. Teofilo Costa, who had been returning from Campos with a party of students from the Vitória Technical School. He and his party had been riding in one of the two rear coaches of the 11 which made up the train, these being the only two coaches which were not derailed. With his colleague, Deputy Nilton Barros, the doctor had provided the initial medical aid for the victims. He said that a large number of them had been bleeding badly and in the absence of anything else to use, the victims' own shirts had had to be ripped up for use as bandages.

A luggage van and two carriages were described as having been reduced to a heap of twisted metal in the derailment and three more carriages were badly damaged. Initially, at any rate, the accident was blamed on excessive speed, several passengers apparently having commented that the train was going too fast. As in other Brazilian accidents at this time the driver, Nery Tinoco, 49, fled the scene, but his whereabouts was reported to the police by a newspaper and he was arrested shortly afterwards, as were his two firemen, 34-year-old João Francisco Barcelos and 25-year-old Noronha Coutinho de Souza. [153]

Driver Tinoco made a long statement denying responsibility for the accident. He said that on the morning of 23rd December he had taken charge of the engine at Campos station from the driver who had brought the train from Barão de Mauá. Before departure from Campos he and his firemen had carried out a thorough check of the locomotive to be sure that all was in order. Campos had left at 09:00, considerably behind time, as the train had arrived there late. From Campos the train had consisted of the engine and 15 vehicles. One coach was detached at Murudum, and on the orders of control three more vehicles were left at Mimoso do Sul, including the restaurant car, the train continuing with just 11 vehicles. Upon arrival at Cachoeiro de Itapemirim three more coaches were detached from the train and it had then waited there for the three coaches with had been left at Mimoso do Sul. When these had arrived at Cachoeiro de Itapemirim they were added back to the train, and it had continued its journey towards Vitória, now later than ever, at 17:00. (The reason for the working of three coaches separately between Mimoso do Sul and Cachoeiro de Itapemirim was not stated, possibly a 14-coach train would have been too heavy for the engine to manage if there were severe gradients between the two stations.) A further stop was then made at Kilometre 497 'to take on supplies,' presumably water and perhaps also coal. From there the train had proceeded at a normal speed, no attempt being made to regain time because of its great weight. When they had reached Engano, Driver Tinoco said he had suddenly felt a bump behind him and had

[153] *O Estado de São Paulo*, São Paulo, Brazil, 27th December 1950

attempted to apply the brakes but the train had already become separated from his engine so it was no use. After the accident he had been going to assist but two soldiers had appeared and had ordered him to remain with his engine, so he had done so. The accident had occurred at exactly 20:00. He denied that the train had been going too fast. He added that he had over 20 years' service with the company, had been a driver 14 years, and had never previously had an accident.

The two firemen backed up their driver's story and pointed out that he had lost a brother in the disaster. [154]

The Oak Grove Times, a local New Jersey, United States, newspaper, had a very big story to put on its front page on Friday 9th February 1951, for a catastrophic railway accident had happened in the state on Wednesday evening, 7th February. The paper's account began, 'Death rode the rails Tuesday night, when a Pennsylvania railroad train, homeward bound with shore commuters, was derailed in Woodbridge at 5.43 p.m. and toppled down a 30-foot embankment. The death roll has reached 84 persons, three of whom were Oak Grove residents. Over 400 were injured, including 17 from here, and many are still hospitalized.

'It was the nation's worst train disaster in 33 years that killed Miss Katherine M. Frank, 55, of 122 Mt Hermon Way, and two brothers, Edwin A. VanExter, 24, and Walter VanExter, 26, sons of Mr and Mrs VanExter, of 85, Franklin avenue.

'The 11-car train left Jersey City at 5.10 bound for Bay Head Junction and packed to capacity. The tragedy occurred as the high-speed express crossed a newly opened temporary trestle 1,000 feet south of the Woodbridge, over the New Jersey Turnpike, now under construction.

'Travelling 50 mph in an emergency 25 mph zone, the steam locomotive, tender, and seven passenger cars left the track while on the trestle, with four coaches plunging down the embankment. The right side of one car was sheared off by the impact and another coach was crushed and twisted.' [155]

Westbound first-class passenger train No. 733, *The Broker*, consisted of 4-6-2 steam locomotive No. 2445, ten coaches and a club car, in charge of Conductor John Bishop with Engineer Joseph Fitzsimmons and Fireman Albert Dunn.

Before leaving Jersey City the conductor and engineer had received copies of General Order No. 1806, the relevant part of which read:

Effective 1.01 p.m. Tuesday February 6, 1951

Applies in Zone C

(1) PERTH AMBOY AND WOODBRIDGE BRANCH WOODBRIDGE-GEMASCO

No. 2 main track and catenary moved 60 feet northward between a point 1,000 feet west of Woodbridge Station and a point 4,000 feet west of Woodbridge Station. Trains and engines must not exceed a speed of 25 miles an hour between these points.

---

[154] *Jornal do Brasil*, Rio de Janeiro, Brazil, 27th December 1950

[155] *The Ocean Grove Times*, Ocean Grove, New Jersey, United States, 9th February 1951

As stated in the newspaper the train had left Jersey City at 5.10 p.m., on time, passed Union, the last open office, at 5.28 p.m., one minute late, passed signal A-229, showing Proceed, and was travelling at 50 miles an hour when the engine, tender, first seven cars and the front bogie of the eighth car were derailed on or near the trestle.

Eighty-three passengers and Fireman Dunn lost their lives and according to the Interstate Commerce Commission 345 passengers, Engineer Fitzsimmons, Conductor Bishop, and Brakemen E.M. Craige, W.D. Jones and Vincent D. Lowery were injured. [156]

Broker Leslie Cavanagh, of Seagirt, New Jersey, told a reporter, "It happened too quickly for anyone to scream. We were going at a pretty good speed when our car lurched suddenly at an exaggerated angle. We could feel we were going off the track. Then the car, the fourth or fifth from the front, flipped over on its right side. The train was crowded because many of the passengers normally took the Jersey Central, which was strikebound. The aisles were filled with people standing. Those standing probably got it worse than the others. A great many were severely injured and practically everyone got cut with flying glass. I guess most of the riders got cut in some way. I was pretty badly bruised and got a few cuts, but nothing like some of the others. There were five or six persons wedged in between the seats. God knows how badly they were hurt. It was a disaster, all right. I was able to climb up a ladder that was put into the car and then down the side of the car and then down the embankment." [157]

Another reporter wrote, 'Immediately following the ear-splitting crash, neighbourhood residents rushed to aid survivors who were crawling to safety amid clouds of steam and dust. Ambulances and rescue equipment raced to the scene, removing the injured to nearby hospitals where all doctors and nurses were on emergency duty. Thousands of spectators jammed the area, hindering rescue, but State Police and 200 soldiers from Camp Kilmer eventually cleared the area and barred the crowds.

'Many of the survivors who were not reported on the injured list suffered minor cuts and bruises and applied their own remedies or called doctors when they finally arrived home. Glenn Trout, 103 Stockton Avenue, escaped by breaking a window in his coach and crawling thru. Others from here {Ocean Grove} had similar experiences and were shaken up by the crash. H.A. Marshall, 24, Ocean Pathway, stayed in his seat until the silence that followed pandemonium. His car was tilted on a 45-degree angle but he crawled thru a window and was taken down an aerial ladder to safety. He was treated on the scene by a nurse and related afterwards that the interior of his car was a mass of struggling humanity in the inky blackness.

'Like any disaster, fate played a part in protecting the lives of many. Walter Nischman, 4 Main Avenue, had boarded the train at Newark, found it too crowded, and jumped off to await the next train. Richard Stirling, 12 Pitman Avenue, shortly after boarding the train, had left his seat, and discovered after the disaster that he was safe in a rear coach still on the tracks while death struck his first coach choice. Harold Timms, 146 Cookman Avenue, rushed to the station only to miss the train. Many had worked late that night or driven home with friends. Stories like these flooded the shore area the day following the crash.

'The Pennsylvania railroad company revealed that The Broker, the tragic train, was the seventh one to cross the trestle since its opening at 1.01 p.m. on the day of the crash.

---

[156] Interstate Commerce Commission report

[157] *Wisconsin State Journal*, Madison, Wisconsin, United States, 7th February 1951

A speed limit of 25 m.p.h. was placed in effect there. William C. Scott, of Elberon Park and Ocean Grove, who rode an earlier train home from his West Side high school, Newark, teaching duties, reported that his carrier had crept over the same trestle at a snail speed of 15 m.p.h.

'Governor Driscoll ordered an immediate inquiry and the Public Utilities Commission, the railroad, the Turnpike Authority and Middlesex County prosecutor are investigating.' [158]

Coming after the Rockville Centre and Richmond Hill collisions on the Long Island Railroad the accident was the third railway disaster to occur to a commuter train in the New York area in slightly less than a year.

The Interstate Commerce Commission (ICC) report on the catastrophe explained that in connection with a highway construction project the two main tracks west of Woodbridge were to be taken out of use for several months. Instead trains were to be operated over two temporary tracks north of and approximately parallel to the permanent tracks. The northerly temporary track had been brought into use on the afternoon of the accident, the southerly temporary track was to be brought into use on 8th February. Eight westbound trains had used the temporary track before train No. 733.

Examination of the track showed no indication of dragging equipment. The track structure on the first curve of the temporary track was displaced to the south a maximum of 6½ inches. At a point 72 feet east of the east end of the trestle, the north rail was pulled apart at a joint. Beyond that point the track was destroyed over a distance of 275 feet. [159]

Engineer Fitzsimmons, speaking from his hospital bed, said, "The moment my engine passed over the trestle and lurched sharply I felt the rest of the cars would never make it. I hit the trestle at about 25 miles an hour and the speed of the train certainly couldn't be blamed for the crash. When I started to sway it was apparently too late." However, 'Passengers and at least one railroad detective – who did not give his name – disagreed with the engineer on the speed of the train. The detective said it was going at top speed when it hit the trestle.' [160] The 57-year-old engineer was described as a veteran with 33 years' experience. [161]

The ICC stated that 'the engineer stated that the engine seemed to slip to one side in the vicinity of the trestle, and it was his opinion that the tender was the first unit to become derailed. The damage to the equipment of No. 733 and the locations of the units after the derailment indicate that the speed was about 50 miles per hour when the accident occurred. Both the conductor and the flagman testified that they did not notice any reduction in speed immediately before the derailment occurred.'

A possible factor in the derailment was that the line over which the train ran was governed by the operating rules of the Pennsylvania Railroad over the 24½ miles between Jersey City and WC (Woodbridge Junction) and the rules of the New York and Long Branch Railroad over the 37¾ miles between WC and Bay Head Junction. On the Pennsylvania section as long as the general order could be issued in good time no train order was issued reminding the engineer of the restriction, nor were warning signs erected at the spot. On the New York and Long Branch section, on the other hand,

---

[158] *The Ocean Grove Times*, Ocean Grove, New Jersey, United States, 9th February 1951

[159] Interstate Commerce Commission report

[160] *Abilene Reporter News*, Abilene, Texas, United States, 7th February 1951

[161] *The Titusville Herald*, Titusville, Pennsylvania, United States, 7th February 1951

warning signals were provided in addition. Engineer Fitzsimmons testified that he had overlooked the fact that because the restriction was on the Pennsylvania section no warning signs would be provided. The ICC noted that had such signs been present they would have provided an additional safeguard against a failure to reduce speed.

About 60 trains daily passed over the line, 70% of which were passenger trains. Automatic block signal and cab signal systems were provided, but these were provided to indicate track occupancy and did not display restrictive indications to warn an engineer of speed restrictions.

The ICC found that the accident had been caused by excessive speed over a section of temporary track. It recommended that a system of automatic train control be installed by the Pennsylvania company, beginning with the New York Division, which would automatically and continuously enforce a speed restriction of 20 miles an hour through a block occupied by a preceding or opposing train and in like manner enforce a safe speed where temporary restrictions were in force, as at Woodbridge. Unless such automatic protection was provided the use of speed restriction warning signs was also recommended. [162]

A fortnight after the accident a Middlesex county grand jury returned 84 separate indictments for manslaughter against the Pennsylvania company, charging that it did "feloniously kill and slay the 84 people who had died. As no individual defendants were specified in the indictment no jail sentences were possible but the company reportedly faced a maximum fine of $84,000 on each indictment. [163] The case dragged on. In April 1952 the New Jersey Supreme Court ruled that rather than being fined $84,000 per indictment as originally announced the Pennsylvania company could only be fined $1,000 in total. [164] Finally on Friday 19th December 1952 the indictments were dismissed. Middlesex County Prosecutor Alex Eber remarked that had the case been pursued it would have punished the county of Middlesex rather than the railroad company as it would have cost over $25,000 to try the case for a fine of $1,000. [165]

Four women died and at least 40 people were injured when what was described as a 'speeding diesel hauled train' from Capilla del Mont to Cordoba was derailed near Thea, Argentina, on Sunday 18th March 1951. About 70 others suffered minor cuts and bruises when the engine, tender and four coaches derailed.

On the evening of Tuesday 21st August 1951 a passenger train from Belgrade, Yugoslavia, jumped the tracks at an unstated location near the border with Hungary. The engine and five coaches were derailed, killing ten people, and injuring 30. According to the Belgrade newspaper *Borba* the derailment was due to excessive speed. [166]

A passenger train from Grato to Fortaleza in Ceara state, Brazil, was derailed near Piquet Carneiro on Monday 17th December 1951. The train was reportedly crowded with

---

[162] Interstate Commerce Commission report

[163] *Burlington Hawk-Eye Gazette*, Burlington, Iowa, United States, 20th February 1951

[164] *Independent Leader*, Woodbridge, New Jersey, United States, 10th April 1952

[165] *Red Bank Register*, Red Bank, New Jersey, United States, 23rd December 1952

[166] *The Cumberland News*, Cumberland, Maryland, United States, 22nd August 1951

passengers travelling before the Christmas holidays and at least 34, and according to one report 53 people died. [167] Nine of the ten coaches of the train left the rails and four overturned. Most reports state that the cause of derailment was unknown, but it was possibly due to excessive speed. The accident occurred in a remote location with no doctors or medical resources nearby and the limbs of some victims had to be amputated with a carpenter's saw, according to despatches. In addition to those who died about 150 passengers were injured. Rightly or wrongly the driver of the train, João da Cruz, was blamed for the disaster, and afterwards he was manhandled and came close to being lynched. [168]

Thirty-two people were killed, according to updated reports, in a railway accident which occurred on Saturday 17th May 1952 between Ulukişka and Adana, Turkey. The initial reports spoke of a derailment of a mixed train; follow-up reports spoke of a collision. The wreckage of 51 vehicles of the train(s) was described as having been strewn over a mountainous area of several square miles. [169] [170]

When news first reached the West of the wreck of a Soviet troop train in Poland on Wednesday 9th July 1952, resulting in perhaps 160 deaths, 'responsible sources were inclined to doubt its veracity,' despite the statements of survivors who had arrived in Berlin with eye-witness accounts of the catastrophe. [171] Matters were not helped by the refusal of the authorities to acknowledge that any such accident had happened, but confirmation came a few days later from members of the West Berlin Railway Workers Union, some of whose members worked in the Soviet zone. The train involved was the luxury *Blue Express*, which ran between Moscow and East Berlin, 'a favourite with Soviet officials.' [172]

The cause of the catastrophe, according to reports, was the obstinacy of a Soviet general, who insisted in having his automobile loaded onto a flat car and marshalled in the middle of the train, despite the protests of the railway officials that running with the light waggon between heavy coaches would make the train liable to derailment on curves and that the flat car should be at the end of the train. Heading for Moscow with Soviet troops going on leave, the predictions of the railwaymen came true on a curve as the train was running on the east side of the Oder River between Boczow and Rzepin, and at least part of the train fell into a lake. There was no suspicion of sabotage. The stubborn general was amongst those who died. [173]

A Polish resident, who recalled the crash from memory, apparently stated that it had occurred on 9th February 1952 but the July date seems more likely. [174]

---

[167] *Railroad Wrecks* by Edgar A Haine, pg. 146
[168] *The Evening Star*, Washington, D.C., United States, 18th December 1951
[169] *The Bradford Era*, Bradford, Pennsylvania, United States, 19th May 1952
[170] *Traverse City Record-Eagle*, Traverse City, Michigan, United States, 19th May 1952
[171] *Syracuse Herald-Journal*, Syracuse, New York, United States, 12th July 1952
[172] *Montana Standard*, Butte, Montana, United States, 19th July 1952
[173] *Sarasota Herald-Tribune*, Sarasota, Florida, United States, 19th July 1952
[174] https://www-waszymzdaniem-zielonogorska-pl.translate.goog/2009/12/zatajona-katastrofa-kolejowa-z-czasow.html?_x_tr_sch=http&_x_tr_sl=pl&_x_tr_tl=en&_x_tr_hl=en&_x_tr_pto=sc

An express train travelling at about 100 kilometres an hour was derailed on Tuesday 26th August 1952 at Suchdol nad Odrou, Czechoslovakia, on the Přerov – Bohumín line. Twelve people died and 108 were injured. The derailment was blamed on the fact that only one of two green lights of a warning signal (of a temporary speed restriction?) had been lit. [175]

Early on Monday 20th October 1952 the Durban – Johannesburg night mail had just passed through Heavitree station, South Africa, about ten miles from Estcourt and not far from Frere, on the Natal main line, when its first ten coaches left the track on a wide curve, though the electric locomotive remained on the rails. One coach slewed around and ended up on top of another and there was considerable damage to the overhead line equipment and support masts.

'The catering staff,' a Durban correspondent wrote, 'worked with great gallantry in pitch darkness amid the screams of the injured and trapped until help arrived from neighbouring farms and ambulance units from Estcourt and Ladysmith.' [176] Further assistance came from Colenso and Mooi River and the South African Red Cross and St John's Ambulance were involved.

Both lines were blocked by wreckage and the downed overhead line equipment so that traffic was completely dislocated. First reports spoke of 21 dead and the final figure was 23 dead, all 'non-Europeans,' while 33 'non-Europeans' were seriously injured and three of the 55 European passengers were slightly hurt. A board of inquiry was constituted to investigate the cause of the disaster. [177]

Passenger train P-4 of the Estrada de Ferro Mogiana, en route from Ribeirão to Campinas, was derailed at kilometre 99, near Urtuba station, Brazil, at 14:30 on Christmas Day, 25th December 1952. Five people died and 12 were injured, two seriously. Dr Augusto Cesar de Andrade told a newspaper reporter that it was one of the most serious disasters to have happened in the Mogiana region in recent years, and had it not been for the fact that there were only a few passengers on the train, because it was Christmas Day, the number of dead would have been considerably higher.

The railway company issued a statement to the effect that luggage van E-66 of the train had been the first vehicle to derail and that it had left the track at a curve and where the line was on a falling gradient approaching a bridge over the Orisanga River. The latter was also referred to as a corrego or stream, so was clearly not a very substantial watercourse. The luggage van had fallen into the river following the derailment and had dragged second-class coach C-58 with it. Luggage van E-18, first-class coach B-43 and restaurant-car R-4 were all derailed but did not go into the water.

Those who died were named as Conductor João Gabriel Silva, Assistant Conductor Francisco Damasceno and passengers Aguinaldo Moraes Paiva, from Igarapava, Alberto Machado, and a girl, Elma Machado, presumably his sister or daughter, both from Aguai.

As soon as they became aware of the accident villagers from Mogi-Guaçú and Mogi-Mirim went to the scene to begin the rescue work and were joined by the Mogi-Guaçú

---

175 Wikipedia – Liste von Eisenbahnunfällen in Tschechien
176 *The Times*, London, England, 21st October 1952
177 *South African Railways and Harbours Magazine*, November 1952, pg. 700

police. The superintendent of the railway company and his assistants arranged medical treatment for the injured, for the removal of the bodies of the dead and for the clearance of the line, which was reopened at 04:30 on Boxing Day. Five of the injured were taken to the Santa Casa (holy house) in Mogi-Mirim, the others were taken to Campinas.

The cause of the derailment was being investigated, though it was quickly established that neither a broken rail nor a defective vehicle were to blame. [178]

Benevento station, Italy, was the scene of a serious derailment on Sunday 15th February 1953, involving Bari – Napoli direct train No. 816, which consisted of an electric locomotive E428.212 and nine first-, second- and third-class coaches. At 16:37 the train entered the station at a speed of 106 kilometres an hour and left the track on a set of points at the entrance to the platforms. The engine and leading coaches overturned to the left, the mail van penetrated a building, injuring several workmen, and other carriages slid along the ground. Fortunately there were few people on the station platform. A severe rainstorm was raging at the time of the accident. The Benevento stationmaster immediately informed Bari and Napoli of the accident to stop the movement of trains along the line and the station staff and local people at once began the rescue work, with police and firemen soon on the scene. The dead were lined up on the platform and 21 seriously injured, five of whom were critically hurt, were taken to the civil hospital. In all, there were more than 80 hurt.

There had been about 500 passengers on the train, many of them football fans bound for Napoli to watch the game between the Inter and Neapolitan teams. Most of those killed, however, were emigrants bound from poverty-stricken and overpopulated southern Italy to Venezuela, who were sailing thence from Napoli. They had been riding in the leading third-class carriage.

An investigation into the disaster was launched at the instigation of the Ministry of Transport under the direction of engineer Di Raimondo, general manager of Fellovie dello Stato (FS), the state railway company. At the same time a separate criminal investigation was opened by Ambrogio Romano, the public prosecutor. The fact that the train had been travelling at 106 kilometres an hour was established by an examination of the tachograph of the wrecked locomotive by the chief of the Napoli railway department, Engineer Lissone. A technical examination of the brakes of the train by the FS showed no abnormalities, implying that the responsibility for the derailment rested entirely upon the driver of the train, Roberta Pitera, and Assistant Driver Nicola Ricciardi. Two university professors appointed by the public prosecutor's office, on the other hand, stated that they could not exclude a problem with the brakes.

At the end of May 1954 Driver Piera and his assistant stood trial charged with manslaughter and causing an accident through negligence. At the trial the driver stated that he had begun braking 1,600 metres from the station, where his train had been travelling at the permitted speed of 90 kilometres an hour, but he had found that the brakes would not respond. Having commenced, the trial of the two men was postponed until the end of the following month. [179]

The newspaper report on the conclusion of the trial is available as an internet archive but is unfortunately very hard to decipher. It would appear that Driver Piera was sentenced to

---

[178] *Correio paulistiano*, São Paulo, Brazil, 27th December 1952

[179] Wikipedia – Incidente ferroviario di Benevento

five years and six months' imprisonment while his assistant was acquitted, the prosecutor having demanded sentences of 15 and 14 years imprisonment respectively. [180]

Diesel railcar No. 52, forming train No. 130, was carrying passengers home from the popular Atlantic Coast resort of Playa de Rada Tilly in Patagonia when it was derailed on a curve with a speed restriction of 80 kilometres an hour at Punta Piedras, near Comodoro Rivadavia, Argentina, and fell down a 40-metre embankment, on Sunday 15th February 1953, the same day as the Benevento disaster in Italy. [181] Twenty-four people were killed and 57 were injured according to early reports. The train was reportedly travelling at high speed when it became derailed and ended up embedded in marshy ground. Ambulances and private cars rushed to the scene to extricate the dead and injured. [182] A special train with doctors, nurses and medical equipment also went to the scene. Some doctors came from Buenos Aires. The steepness of the embankment made the rescue work very difficult. [183] The injured were taken to the Municipal Hospital in Comodoro Rivadavia, to Astra Hospital and to the Napolitano Sanatorium. After the disaster the Mardis Gras carnivals which had been planned were suspended and three days of mourning were declared.

Ramós Jesús Miguez, who witnessed the accident, told a newspaper reporter, "It was terrible. The car was lying with its wheels up in the air. There were people trapped between seats, struggling amidst the screams and shouts of others, which paralysed any action." [184]

The motorman of the train was among those who were hurt and was taken to hospital, where he was placed under arrest but had yet to make any statement at the time of the initial reports. [185] There were allegations that the motorman had been drunk and that when a passenger had complained that the train was going too fast he had increased speed rather than braking. The railcar was also seriously overcrowded as it was carrying at least 75 passengers, but only had a seating capacity of 48. [186] One source states that the train was carrying about 100 passengers when it left Rada Tilly at 17:00 and that there were 36 dead and a further 65 injured. The line was also narrow gauge (75cm) which probably contributed to the instability of the overloaded railcar.

The investigation into the disaster confirmed that excessive speed had caused the derailment and that the motorman, who had a history of driving too fast on roads, had been drunk. He was subsequently tried, convicted, and jailed. [187]

Two express passenger trains and two freight trains were involved in a disaster on the New York Central Railroad on the quadruple-track main line near Conneaut, Ohio, United States, late on the evening of Friday 27th March 1953. Sixteen passengers, an express messenger, an assistant conductor, a brakeman and two other employees lost their

---

180 *La Stampa*, Torino, Italy, 29th June 1954

181 *The Montana Standard*, Butte Anaconda, Montana, United States, 16th February 1953

182 *The Charleston Gazette*, Charleston, West Virginia, United States, 17th February 1953

183 *La Vanguardia*, Barcelona, Spain, 17th February 1953

184 https://wwwcronicaferroviaria.blogspot.com/search/label/Ferrocarril%20Patag%C3%B3nico?m=0

185 *Daily Sikeston Standard*, Sikeston, Missouri, United States, 17th February 1953

186 http://ferrocarrilesenelconosur.co.uk/07Sncomrivport.html

187 https://www.elpatagonico.com/la-tragedia-la-chanchita-n773667#fotogaleria-id-881016

lives and 30 passengers, a Pullman Company employee, a person carried under contract, two engineers, two firemen, a conductor, an assistant conductor, two baggagemen and a flagman were injured.

Freight train No. 1871 East had started on the Pittsburgh and Lake Erie Railroad at East Youngstown, Ohio, at 5.34 p.m. It consisted of a diesel-electric engine, 103 cars and a caboose. It left OD interlocking (Ashtabula) on track No. 4 at 9.18 p.m. and was travelling at 36 miles an hour as it approached Conneaut. The Conneaut operator gave its conductor a proceed signal to indicate that he had observed no defective condition with the train. Immediately beyond Conneaut station train Extra No. 1736 West was passed on track No. 3 and the flagman of train No. 1871 East saw that stop signals were being given with a white light from the caboose of the former train. The flagman acknowledged the signals, informed his conductor, the brakes of train No. 1871 East were applied by the conductor opening the valve in the caboose, and the train stopped near Springfield, Pennsylvania, 3¼ miles east of Conneaut. While the flagman went west to protect his train the conductor inspected it. When he reached the 73rd car, Baltimore and Ohio Railroad car No. 254645, loaded with 58 pieces of pipe, he found that its side stakes were broken on the north side of the car and that some of the pipes had fallen off. The ends of the high-tension bands used to secure the load were trailing over the sides of the car. Counting the pipes he found that there were only 49, so that nine were missing. He contacted the dispatcher to tell him what had happened and was told of the accident. The car was detached and left on an auxiliary track.

Train Extra No. 1736 West consisted of diesel-electric units 1736 and 1735, 120 cars and a caboose. It passed Dock Junction, just over 22¼ miles east of Conneaut and the last open office at 9.16 p.m., passed signal No. 113.3, indicating *Proceed*, and was moving at 31 miles an hour on track No. 3 when its 76th car was struck by the locomotive and vehicles of westbound train No. 5, derailing its 75th to 90th cars.

First-class westbound passenger train No. 5 from Buffalo to Chicago (*The Mohawk*) consisted of diesel-electric locomotives 4018 and 4003, a mail car, two express cars, a baggage car, two coaches, a sleeping car, a club-sleeping car, two sleeping cars and an observation sleeping car. It left Erie, Pennsylvania, 25¼ miles east of Conneaut, at 9.36 p.m., passed Dock Junction at 9.41 p.m., passed signal No. 113.1, indicating *Proceed*, and was moving on Track No. 1 at 76 miles an hour, when its engine and ten cars were derailed and collided with Extra 1736 West. The derailed vehicles of train No. 5 fouled track No. 2 and were struck by train No. 12.

First-class eastbound passenger train No. 12 from St Louis to New York (*The Southwest Limited*) consisted of diesel-electric locomotives 4020 and 4111, a baggage dormitory car, two coaches, a dining car, two sleeping cars, a coach, four sleeping cars, and an observation-sleeping car. It passed ER interlocking at 9.11 p.m., on time, passed OD interlocking (Ashtabula), the last open office, at 9.46 p.m., passed signal No. 114.2, indicating *Proceed*, and was moving on track No. 2 at 71 miles an hour. The first six cars of train No. 5 and the first two cars of train No. 12 were destroyed in the accident. [188]

Passenger Earl B. Mitchell, of Garwood, New Jersey, said, "We were just riding along, and then I felt a jolt. The jolt didn't seem too hard at the time, but the next thing I knew I was laying on the floor and my whole left side felt numb. Just as I was going down I saw my bag, which was filled with machinery, fly off the rack and go soaring

[188] Interstate Commerce Commission report

about 10 feet up. It just missed a woman. Some of the people hit the backs of their seats so hard that a lot of the seats were all lopsided. But everything was very quiet. It's funny, but instead of saying much everybody started looking for their luggage."

Andy Aranyons, a service station worker, was one of the first rescuers to reach the scene. He declared, "I helped carry out at least 10 dead…There was no panic among the passengers when I got there. I started walking around and saw a Catholic priest. Then the two of us started going through the cars. There were so many injured we couldn't get any help at first from the passengers. In one car, a coach, I found two dead. Then I walked around the car in the mud and I heard some people crying in another car. In another coach I found one fellow dead and about five injured. We walked up the tracks and saw a conductor. Sirens were blowing and the Pennsylvania state police were starting to evacuate the wounded. The doctors were giving drugs to the badly injured people. Then some more people started to help out with the injured."

'Hundreds of people stood in the damp cold early today outside Erie's Union station as a mercy train arrived from the three-train wreck 27 miles of here. The street thoroughfare was crowded with automobiles, ambulances, and hearses – ready to rush some 152 survivors to hospitals and hotels. A dozen Erie physicians had medical kits and emergency equipment in the station's baggage room. The Red Cross had set up a reception and registration center.

'Civil Defence officials, police and firemen were on hand to lend aid – and they were needed as the first passengers were helped carried from the elevator leading from the train level. It was a wonderful and terrible sight – wonderful to see people walking, terrible to watch the pain-wracked faces of the critically injured as they were borne on stretchers and placed in ambulances and hearses pressed into service. Twenty-three stretchers passed through the baggage room, 14 women and nine men. Most were unconscious. Others looked up at the long row of faces with dazed expressions. Still others limped through the room aided by the ever more fortunate who had escaped serious injury.

'Two boys, 2 and 4 years of age, dragged along behind their mother, bewildered by this arena of confusion. Another little boy, Gregory Quinn, jr., of Erie, was wrapped in a blanket and carried in the arms of a fireman. The total shock will come when he learns that his mother, Mrs Alpha Quinn, was left lying along the railroad tracks covered with a blanket – forever gone. Mrs Quinn was the first victim to be identified.

'Then there was a 2-year-old girl, Barbara Marks, of St Louis, Mo., who was going home. She reached out for the waiting arms of her grandfather, Martin Rieman, of Orchard Park, N.Y., who had driven more than a 100 miles to Erie after hearing of the wreck.

'There were shrieks, screams, and fears as those awaiting at the station identified some of the passengers being helped from the mercy train.

'That was Saturday morning in Erie's Union station.' [189]

The Interstate Commerce Commission (ICC) noted that all four tracks at Conneaut had been destroyed for about 200 feet. Six pieces of pipe were found amongst the wreckage and the other three pieces about two miles further east. An examination of the car by loading inspectors of the American Association of Railroads determined that it had

---

[189] *The Evening Independent*, Massillon, Ohio, United States, 28th March 1953

been loaded in accordance with loading specifications in all respects, except that the high-tension bands securing the load had not been properly sealed. None of the seals which remained on the car and were accessible for inspection had been crimped properly to provide 85 per cent of the load strength of the band as required by the loading rules. In their opinion the faulty application of the seals had resulted in the loosening of the high-tension bands during the movement of the car such that the load had shifted sufficiently to exert excessive pressure on the side stakes. These had then broken on the north side of the car and some of the pipes had fallen off.

The ICC deduced that when the pipes had fallen from the car one of them had become wedged between track No. 1 and the next car of the train, Canadian National Railway No. 487579, and the thrust submitted to the track had forced it northwards about 18 inches. The Canadian National car, a box car loaded with wire nails, showed signs of damage from contact with the pipe.

The Baltimore and Ohio car had been loaded at the Jones and Laughlin Steel Company at Aliquippa, Pennsylvania, on the Aliquippa and Southern Railroad and was bound for New York, New York via the Pittsburgh and Lake Erie, New York Central and Delaware Lackawanna and Western Railroads. It was one of several cars loaded with pipes which had been examined by interchange inspectors between 6.55 p.m. and 8.15 p.m. on Tuesday 24th March on receipt at West Blacks Run Yard. It was their practice to climb onto the car if they saw that the load had shifted or the side stakes or tie hands were not in place. They could not recall if they had done so with this car and had observed nothing wrong with it. The car arrived at East Youngstown at 11.55 p.m. on Thursday 26th March. Here the inspector did climb onto the car but he did not examine the high-tension bands or seals. He found nothing wrong. Finally, the car was inspected on the evening of Friday 27th March before it left East Youngstown at 5.34 p.m. Again no defective condition was found.

The ICC concluded that the Baltimore and Ohio car had not been properly loaded and that the inspections of it en route had been inadequate. [190]

A major disaster occurred in West Pakistan on the morning of Thursday 21st January 1954. When the Lahore – Karachi mail train on the North Western Railway was between Jhampir and Braudabad it came into collision with the derailed waggon of a freight train passing in the opposite direction at 05:30. The freight train was carrying diesel-oil, paraffin and petrol, and the waggon which was struck was loaded with petrol. Fire broke out following the collision and involved the leading coaches of the diesel-hauled passenger, especially the first two, and other waggons of the freight train. Initially it was believed that over 100 passengers had died, then the figure was revised downwards to about 60 dead and 49 injured, 11 seriously. All the deaths were thought to have occurred in the leading two lower-class carriages.

One of the passengers on the train was Sir Mohammed Zafrullah Khan, the Pakistani Foreign Minister. His saloon carriage was further back along the train and he escaped with a severe shaking. [191]

The accident happened in an isolated part of the Sind Desert. The pilot of an aircraft which flew over the accident scene spoke of seeing injured lying near the wreckage.

---

[190] Interstate Commerce Commission report

[191] *The Times*, London, England, 22nd January 1954

Their plight was serious due to a lack of food and water. The dining car was smashed and the water tanks were destroyed. There was no road access to the accident site but a special train with doctors, nurses and medical equipment was despatched from Karachi, 70 miles away. [192] The relief train returned to Karachi with the injured about 12 hours after the accident.

A passenger train from Campeche to Merida in Yucatan Province, Mexico, became derailed near the small town of Pomucho on Saturday 3rd April 1954. Eleven people died and 43 were injured. [193]

Twenty-four people died in an unspecified train crash, probably a derailment, on Tuesday 3rd August 1954, neat Humbia, Portuguese West Africa. The 23 African and one European, presumably Portuguese, victims were employed in the construction of a new section of the Caminhos de Ferro Moçâmedes. [194] [195]

At 01:52 on Monday 6th September 1954 an overnight passenger train from Malmö to Stockholm was derailed at Rörvik, Sweden, not far from Savsjo. Nine of the 14 carriages of the train left the rails and it was reported that there were six fatalities and 53 injured, a number seriously. Some of the injured were trapped in the wreckage and could only be released by the use of cutting equipment. Sailors who had been on their way to Stockholm to join their ships helped with the rescue work, which lasted all night.

The derailment was blamed on excessive speed, it being stated that the train had exceeded the permanent restriction of 40 kilometres an hour through the station area. Statens Järnvägar, the state railway company, was already seriously short of rolling stock before the accident, and it was feared that with the further loss of coaches in the accident, older wooden rolling stock would have to be brought back into use. [196]

A major disaster occurred in Portugal on Monday 13th September 1954, involving passenger train No. 8012 which had left Vila Real de Santo António at 13:16 and was due at Barreiro at 20:36. The train was derailed about midway between Sabola and Pereiro stations. It consisted of steam engine No. 559, a van, three third-class carriages, a restaurant car and two first-class carriages, in charge of Conductor Manuel Rosa Marques with Chief Driver Manuel António Maximo, Driver José Guilherme de Oliviero and Fireman Joaquim José Zeferino. The steel carriages of the train had been bought from the Budd company of the United States. The accident occurred on a curve and in a cutting. In the derailment the locomotive climbed the rail then fell into the ditch adjoining the line, dragging its tender, the van and the leading third-class carriage with it and the sudden stoppage resulted in the body of the second third-class carriage cutting into half of the carriage in front, while the following third-class carriage was derailed.

---

[192] *The Bakersfield Californian*, Bakersfield, California, United States, 21st January 1954
[193] *Albuquerque Journal*, Albuquerque, New Mexico, United States, 4th April 1954
[194] *The Straits Times*, Singapore, 5th August 1954
[195] Wikipedia – Moçâmedes Railway
[196] https://blidan54-blogspot-com.translate.goog/2019/09/tisdagen-den-7-sept.html?_x_tr_sch=http&_x_tr_sl=sv&_x_tr_tl=en&_x_tr_hl=en&_x_tr_pto=sc

The accident occurred in an isolated, unpopulated area, difficult to access by road, but a number of cars and buses nevertheless managed to reach the scene, and relief trains were sent from Beja, Funcheira and Faro. Most of the injured were taken to hospital in Beja and the bodies of those who had died were taken to a temporary mortuary at Santa Clara-a-Velha. Initially it was reported that 27 people had died and the final official figure was 34 dead, with the same number of injured, though there was a suspicion that the true number of dead was somewhat higher, perhaps even 50.

A newspaper reporter interviewed Chief Driver Maximo at his home. Still clearly suffering from the effects of the accident, the latter said that he and Fireman Zeferino owed their lives to Driver de Oliviero, who had been the first to sense that the locomotive was about to leave the track. Immediately afterwards the engine had overturned against the side of the cutting, throwing up clouds of dust on both sides, while a quantity of water had escaped from the tender. Somehow de Oliviero had got him and the fireman out of the wreckage, as screams and cries of pain sounded all around.

Asked as to the cause of the disaster the chief driver said that he could not be certain, though he had seen a broken rail. Questioned as to suggestions that the train had been running at high speed he denied it, saying that the train could not run at more than 60 kilometres an hour at the location of the accident due to the curves and reverse curves and pointing out that had the train been going too fast all of its carriages would have left the line.

Fireman Zeferino was interviewed lying in his hospital bed in Beja, still with dust and dirt on his face and in his hair and with a badly injured arm, a head injury, and bruises all over his body.

"How did the accident happen?"

"I don't know, because it all happened so quickly, so confusingly…"

"You didn't know what was going to happen?"

"No, the front driver, when he felt the train 'run away from him,' moved his hand to the {brake} handle, but there was nothing that he could do, despite the fact that we were going at somewhat more than 45 kilometres an hour, according to my estimation. I heard a tremendous noise, there was a confusion of twisted iron and broken wood and an indescribable roar. It seemed as if a bomb had gone off. I thought that the boiler, which had a pressure of 13 kilos, was about to explode. All I know is that my head was struck against the side of the cutting, to the left of the track, where the engine had overturned. There was a cloud of escaping steam and screams were already coming from the train. I was close to losing consciousness and it was the front driver who helped us out of our critical situation. I remember being taken to an adjoining field and watching dozens of terrified passengers fleeing from the wreckage. Later I was taken to Beja."

"Did the accident happen on a curve?"

"Yes, it was on a curve and I believe that the train was derailed due to the poor condition of the track."

The fireman's wife, the reporter added, was sat beside him on his bed, clutching her husband's hands and crying convulsively.

In the adjoining bed to the fireman was 50-year-old António Augusta Lavandeiro, who had just returned to Portugal after spending 30 years in Angola, where he had owned a car repair shop – Autoreparadora do Congo – in Uige province. He had been going to Lisbòa with his wife, 43-year-old Maria da Conceição Lavandeiro, and their 16-year-old son Fernando José, a good student who had completed his fifth year in high school in Luanda and was to complete his studies by taking an engineering course in Lisbòa. Now Sr Lavandeiro was crying uncontrollably while a young priest sat next to him reading from a prayer book. Tears running down his face the middle-aged settler repeats endlessly, "My son! My son! My son! How wretched I am!" Then, recovering himself a little, he told his story to the reporter. "I had been looking at the country that I had not set eyes upon for 30 years. As we approached a bend I felt everything shake. I felt the train leave the track and hit the side of the cutting. We were in the second carriage, which, I saw later, had been overtaken and cut in two by the one behind. Seeing that an accident was about to occur I pulled my son to me as a shower of suitcases and wreckage fell upon us. There was an indescribable panic when the crash came. I lost consciousness. When I came to I saw a nightmare scene. Dripping blood I searched the wreckage for my family. There were destroyed, mutilated bodies and screams from all sides. I found the point where we had been and under a seat was the corpse of a woman whom we had been talking to during the journey, her body horribly broken. For hours I searched for the bodies of my wife and child but could not find them. Now, I am waiting here for someone to tell me where they are." Twisting his hands he ended, "I have lost my wife and child, everything that I had in the world, now there is just me left to rot in pain." [197]

Passengers who lost their lives in the accident included Albano da Silva Pestana, responsible for the Photographic Archive of the National Museum of Ancient Art, and footballer Francisco Coelho. [198]

Companhia dos Caminhos de Ferro Portugueses (CP) engineers immediately began an investigation into the causes of the disaster. The results of their inquiry were made public in an official note from the Ministry of Communications at the end of October. It was a very thorough investigation. The track and train were described in considerable detail. The inquiry concluded that the speed of the train had not exceeded 60 kilometres an hour. The course of the derailment, as evidenced by an examination of the track and wreckage, was detailed. Various theories as to the cause of the derailment were then examined in turn:

(a) The possibility of a broken rail was ruled out because the track was intact at the initial point of the derailment.

(b) The sleepers at the accident site were in good or at least moderately good condition and the gauge of the track was correct, it not having been widened even by the violent shock to which it had been subjected when the train was derailed.

(c) The curve was not perfect, having some irregularities, but these did not appear sufficient to have caused a derailment.

(d) Engine defects or malfunctions. Experience with locomotives of the class involved in the accident showed that their leading bogies exerted a strong pressure on the outer rails on curves, especially those with small radii. A lack of lubrication of the bogie would contribute to the problem, making it difficult to enter the curve and increasing the angle of attack of the wheel flange against the high rail.

In the case of a rail in good condition, none of these features would be likely to pose a danger to traffic, but the same was not the case in a rail with such marked lateral wear as that at the accident site.

---

197 *Diario de Lisbòa*, Lisbòa, Portugal, 14th September 1954

198 Wikipedia – Desastre do Rápido do Algarve

There was no damage to the engine which could have contributed to the accident.

(e) The profile of the front locomotive bogie wheel was of interest because it was this wheel which had climbed the high rail of the curve. The profile of the wheel flange was considerably worn, almost to the permissible limit. It was not likely that a wheel in this condition, running on a rail which was new or only slightly worn, would lead to an accident. However, in the case of a rail with very severe wear, such as that at the site of the derailment, there might be a danger under certain conditions.

From diagrams it can be seen that the inclined face of the rail and the surface of the wheel flange can combine under favourable conditions of friction and lateral thrust to create a contact zone which facilitates wheel climbing.

(f) The speed of the train. As noted earlier, the speed of the train was about 60 kilometres an hour and was not likely to have been a factor in causing the derailment. Combined with other unfavourable conditions, however, it could have contributed to the derailment.

(g) It was impossible to be certain of the locomotive wheel loads were normal. A very large drop in the load on the derailed wheel would make it more liable to climb the rail, but this is an unlikely scenario as the engine weights are checked on a weighbridge.

It appeared, therefore, the report continued, that the derailment could only be explained by a combination of the aforementioned factors and others which had created an optimum condition for the bogie wheel to climb the rail. The wear of the rail and the flange were considered to be the main factors leading to the derailment and the combination of the two, together with other factors, created favourable conditions for a derailment.

The contributory factors were:

(a) The irregularities of the curve, which were one of the main causes of vehicles oscillating.

(b) The speed of the train, which had been close to 60 kilometres an hour on a steep descent.

(c) The strong pressure exerted by the front bogie of the locomotive on the high rail of the curve for the reasons indicated above and the increased angle of attack of the wheel flange against the worn side face of the rail.

It was also possible that abnormal oscillations of the rolling stock could have contributed to the derailment by creating strong lateral forces against the rail and the rail-flange interface, facilitating the wheel climbing.

Interestingly, the last railway companies congress discussed the side wear of rails in curves and concluded that accidents caused by such defects were infrequent. Obviously these rails were ones whose wear was within the limits permitted by the companies. Our opinion that a combination of circumstances led to the disaster was thus confirmed.

The conclusion of the report was that the derailment had been due to the rail head chamfer having reached a dangerous limit, the right front wheel rim flange of the locomotive bogie being very severely worn, and these two circumstances being simultaneously aggravated by other aforementioned factors.

The inquiry made eleven recommendations:

(a) Remove rails on curves which have accentuated lateral wear and study the maximum limit for the angle of the head side chamfer, which in the opinion of the inquiry should not exceed 28 degrees.

(b) Replace sleepers found to be in poor condition.

(c) Modify the plan for relaying the line between Funcheira and Tunes where necessary, laying 30 sleepers per rail length.

(d) Eliminate the exaggerated mismatching of rail joints.

(e) Correct the irregularities of the curves.

(f) Make track maintenance a priority.

(g) Replace or correct wheel rims whose flanges show very marked wear close to or beyond permitted limits.

(h) Reduce the maximum permitted speed of trains on dangerous sections of line, such as those with severe curvature or steep descents, to 50 kilometres an hour.

(i) Always keep the bogies of locomotives well lubricated and carry out a general revision of 500 class engines.

(j) Conduct a study into the conditions leading to the transverse displacement of the bogies of locomotives on sharp curves, with a view to improving their running around such curves.

(k) To improve the attachment of the Budd carriages to avoid a repetition of what had happened in the accident, when the body of the second carriage separated from its undercarriage, contributing to the severity of the disaster. [199]

An international football match had been played between England and West Germany on Wednesday 1st December 1954 (England won 3-1). The next morning the special train of 12 Deutsche Bundesbahn coaches carrying 700 returning German football fans from Ostende was involved in a major disaster in Belgium, at Wilsele, near Louvain. Its engine, tender, and four carriages left the rails at a spot where the track temporarily became single because of engineering work on a bridge over the River Dijle, the derailment occurring at a set of points 200 metres short of the bridge. Most of the casualties were in the second coach, which was crushed between those in front of and behind it and almost completely destroyed. The Belgian driver and fireman were amongst those who lost their lives and the retrieval of their burned bodies was not immediately possible.

Many of the passengers were thrown from the train and found lying on the railway embankment and in adjoining fields. The first rescuers, who were Dominican monks, found what was said to have been an indescribable scene, with steel plates and beams torn apart and passengers begging for help. There were some Deutsche Bundesbahn railwaymen travelling on the train. The police and firemen arrived about half an hour after the accident. Cutting equipment had to be used to free some of the trapped passengers. Ambulances took the injured to Leuven, 15 kilometres away, where the medical services of the two hospitals and the university were mobilised. At the scene a Roman Catholic priest gave the last rites to the fatally injured passengers.

A second football train – there were nine trains in all – was following via the same

[199] *Gazeta dos Caminhos de Ferro*, Lisbòa, Portugal, No. 1605

route but had, of course, to be diverted after the accident. [200]

Many of the passengers lost all their possessions in the accident and they wandered through Leuven after the crash. In the afternoon about 600 of them were able to resume their journeys to Germany on another train, while the injured remained in hospitals in Leuven.

None of the West German team had been on the train. [201] Initially it was reported that there were 19 dead and about 80 injured; a second report stated that the number of fatalities had risen to 21 and there were 20 injured in hospital in Leuven. [202]

Wilhelm Fuchs, a supporter whose hand was crushed and who also had a foot injury, spoke to a newspaper reporter from his bed in a Leuven hospital. “I was asleep when the accident happened. I was thrown backwards over my seat. My hand was forced through the window and someone fell right on top of me. I lay there half stunned for about ten minutes, listening to the groans and screams of the victims. Suddenly I realised that I had to get out. I crawled through an open door and began to pull other passengers out.”

A derailment had occurred at the same place three days earlier, without serious consequences. A newspaper reporter theorised that the points involved had been damaged in the earlier derailment without this being noticed. According to experts, however, the derailment was probably due to excessive speed over the points. The engine speedometer (tachograph?) registered a speed of 78 kilometres an hour. The sound of the derailment could be heard kilometres away. [203]

Several coaches of a well-filled passenger train which left Aracaju in the early morning of Sunday 9th January 1955 for Salvador, were derailed at Rampo do Cristo, near São Cristãvel station, Brazil. An early report spoke of 40 dead and 100 injured. As soon as news of the accident was received the prefect of Aracaju appealed to the local population to donate blood as it was presumed that there would be many victims requiring transfusions. The hospital soon reported receiving 100 casualties. However, a report from Salvador on the day after the accident, quoting a director of the Viação Férrea Federal Leste Brasileiro, the railway company on whose line the accident had happened, as saying that although the exact number of dead was still unknown, at present it appeared that only nine people had died and a reporter who was early on the scene confirmed that the initial report of the magnitude of the accident had been exaggerated. This reporter stated that the exact location of the accident was a place named São Gonçalo, and that it had occurred at 06:00.

The reporter added that two second-class carriages, a restaurant car, a goods waggon, and a van had violently left the track, some vehicles being wrecked. At 17:00 only nine bodies had been recovered. Three of them were identified as Senhora Alda Santana, Senhor João Francelino and Maria do Socorro, a child. Over 50 injured were taken to Aracaju, others were treated at São Gonçalo. There were still other bodies buried in the wreckage, but their number was not that great. A large number of people rushed to the scene of the accident and did all they could to help in the removal of the injured. São Cristãvel and São Gonçalo factories collaborated to send staff and vehicles to the

200 *Overijsselsch Dagblad*, Zwolle, the Netherlands, 3rd December 1954

201 *The Times*, London, England, 3rd December 1954

202 Ibid, 4th December 1954

203 *Trouw*, Amsterdam, the Netherlands, 3rd December 1954

accident site to help in the transport of the injured and the work of clearing the line. Nurses from the two textile factories, friars, and other religious individuals, as well as people of all classes, 'worked hard in an unequivocal demonstration of human solidarity.' Meanwhile cars bringing relatives and friends of passengers who had been on the train arrived from Aracaju and elsewhere all the time, seeking news of what had happened to their loved ones.

The cause of the accident remained unclear. The conductor of the train, who was badly hurt, told a reporter that one of the carriages had had a broken spring and this had unbalanced the whole train on the incline where the accident occurred. A brakeman and a *garimpeiro* (an independent prospector for minerals) were among those killed.

The railway company later issued a statement. "Passenger train R.N.-2, which left Aracaju yesterday at 05:00 for Salvador, suffered an accident near the town of São Cristãvel. As a result of the accident a freight waggon, a baggage and mail van, two second-class carriages and a restaurant car fell down an embankment. As a result nine passengers have died and about 100 have been injured. The first-class passengers were unhurt. As soon as news of the accident was received the company chartered an aircraft to fly engineers and doctors from Salvador to Aracaju, from whence they could proceed to the accident location. A train left Aracaju today {Monday 10th January} at the same time as train R.N.-2 and brought the passengers from the accident train on to the capital."

Engineer Francisco Burity, speaking on behalf of the company director, informed the newspaper that the identification of the victims of the accident would be left the Sergipe (local) police, adding that this could be a difficult job because most of the passengers did not carry documentation 'due to their modest social positions,' that is, they were poor people.[204]

Reports of the accident in United States newspapers suggested that it had been due to the excessive speed of the train rather than a defective vehicle as suggested by its conductor.[205]

1955 was not a good year for English railways, with three major disasters and three further accidents involving passenger fatalities. In contrast no passengers lost their lives in train accidents in either 1954 or 1956.

The disasters began with a derailment at Sutton Coldfield on at 4.13 p.m. on the afternoon of Sunday 23rd January. Sutton Coldfield lay on the secondary main line of British Railways, London Midland Region, running between Wichnor Junction and Birmingham via Lichfield. The line was only maintained to Class C standards, with a speed limit of 60 miles an hour, steep gradients, and considerable curvature, in contrast to the roughly parallel main line via Tamworth, which was maintained to Class A standards, had a speed limit of 75 miles an hour and easy gradients and curvature.

On weekdays the Sutton Coldfield line was used by about 15 local trains, all of which called at Sutton Coldfield station, plus a handful of freight trains and light engines. There was no Sunday service during the winter months but the line was used for diversions on Sundays when engineering work was in progress on the Tamworth line. From Lichfield City station the line climbed at 1 in 100, fell at 1 in 100 for rather less than a mile, shortly

[204] *O Estada de São Paulo*, São Paulo, Brazil, 11th January 1955
[205] *The Oakland Tribune*, Oakland, California, United States, 10th January 1955

climbed for 3½ through Shenstone, mostly at 1 in 100, fell at 1 in 100 for 2 miles through Four Oaks, then climbed at 1 in 101 from the Lichfield end of the short Sutton Coldfield tunnel and through Sutton Coldfield station itself. There was a 40 miles an hour speed restriction between Four Oaks and Sutton Coldfield and a permanent speed restriction of 30 miles an hour over the 15 chain left-hand curve through Sutton Coldfield station.

The train involved in the accident was the 12.15 p.m. express from York to Bristol via Sheffield, Derby, and Birmingham, diverted via Sutton Coldfield due to track renewals on the Tamworth line. It consisted of 4-6-0 5MT Class steam engine No. 45274 and ten corridor coaches, in charge of Guard F.J. Harrison with Driver James Martin and Fireman John Howell. The leading carriage was an all wooden vehicle, the other nine carriages had steel underframes. Driver Martin was not familiar with the Sutton Coldfield and Driver Harold Allen joined the engine at Burton-on-Trent to act as a conductor.

Approaching Sutton Coldfield the speed of the train was not reduced as it should have been and the train entered the left-hand curve at 55 or 60 miles an hour and was derailed. The engine and tender overturned, all but the rear coach were derailed, and the rolling stock was badly damaged, the leading, fourth and fifth coaches being destroyed. Fifteen of the 300-odd passengers on the train were killed (one of them was Driver E. Blount, who was travelling as a passenger) as were Driver Allen and Fireman Howell. Forty passengers, Driver Martin and a goods guard travelling as a passenger were taken to hospital and 24 passengers were slightly hurt or suffered shock.

Both lines were blocked and Sutton Coldfield box was switched out. The 1.20 p.m. express from Bristol to York was already in the extended Erdington – Four Oaks block section but Travelling Ticket Collector G.A. Attenborough and Fireman D.H. Smith broke into the box and replaced the Down signals, stopping it short of the wreckage. In addition two lineside residents, Mr and Mrs Fairey, who had seen the accident, ran along the line and would doubtless have stopped the express had the signals not been replaced.

The first notice of the accident probably came via a 999 emergency call from a passenger, Marine Harold Swann, who had broken out of the closed station and found a public call box. Meanwhile Superintendent G.C. Gardner of the Warwickshire County Police had witnessed the accident and telephoned his headquarters and Ticket Collector Attenborough had informed Railway Control from Sutton Coldfield box. By 4.30 p.m. the first ambulance had reached the scene and it was followed by others, doctors and nurses, the police, fire brigade, Civil Defence Units, a Royal Air Force detachment, the Home Guard, and the Woman's Voluntary Service. A special train eventually took the uninjured passengers on to Bristol. The lines were blocked until Wednesday morning. [206]

Newspaper coverage of the accident included stories from passengers. 'A passenger in the fifth coach, Mr W Forrest, of South Shields, who was on his way to Coventry, said: "The train had been swaying from side to side. Suddenly there was a terrific rending noise and we were all thrown in a heap. Some of us were thrown up and struck the roof."

'A member of a Marine Commando, Marine Harold Swann, aged 19, of Long Eaton, Derbyshire, who was travelling in the third coach, said that he had been asleep in his compartment and awoke to find himself lying on the track. The roof of his coach had been ripped off. He began to help to extricate his fellow passengers. Two women were trapped by the legs but had their arms free and were able to lift out a small girl to him.

[206] Ministry of Transport and Civil Aviation report

'Pilot Officer John Wolstenholme, of Buxton, who was returning to the R.A.F. station at Hullavington, Wiltshire, after a 48-hour leave, said that he was thrown through the roof of his carriage. "I heard screams of a child coming from the next compartment to mine. I was told by rescue workers that it was a seven-year-old boy who was apparently travelling alone. I tried to get at him but it was no good and eventually the screams died down." Pilot Officer Wolstenholme suffered only a cut finger.

'Vincent Staham, aged 19, of Cromwell Street, Lancashire, who was on his way back to an R.A.F. camp at Locking, Somerset, said: "I should think the speed was 65 to 70 m.p.h." There was a terrific impact, he said and then his coach was being dragged along the sleepers.' [207]

Guard Harrison was the first witness to give evidence at the Ministry of Transport inquiry into the accident. He stated that he had been riding in the rear coach with Ticket Collector Attenborough. The journey had been perfectly normal as far as Four Oaks, where the speed had been 40 miles an hour. He had not noticed any braking at the Sutton Coldfield Up Distant signal and he and the ticket collector had commented to one another that the train was going too fast for the sharp curve at Sutton Coldfield station. He had then got up and lifted the brake valve for about three seconds, without any appreciable reduction in speed. Then he had looked out of the right-hand window as the engine had entered the tunnel and there had been a sharp retardation, followed by some violent surges and his van had come to rest between the tunnel and the station platform. Asked why he had not kept the brake valve open longer to make a strong brake application Guard Harrison pleaded that he had hoped to attract the driver's attention, adding that he had hesitated to take the control of the train out of the driver's hands.

Ticket Collector Attenborough agreed that the train running had been normal until Four Oaks but after there it had been going too fast, as he and the guard had remarked to one another. He said that he had not been alarmed by the speed but had 'felt that something ought to be done.' The guard had then applied the brake, though only briefly, not fully. A few seconds later there had been some severe bumps. The speed then had been 55 to 60 miles an hour.

It was almost two months before Driver Martin was well enough to give evidence. He said that he had left the engine when it had stopped to take water at Lichfield (Trent Valley) to ride in the leading coach, in contravention of Rule No. 127 (iii). He excused his doing so by saying that the rough riding of the engine had made him uncomfortable, especially as he had had to stand behind Driver Allen and he had had a previous leg injury. Regarding his obligation to study an unfamiliar route he said that he had not supposed that he would ever need to go via Sutton Coldfield again, and so he had not been interested.

The late Driver Allen had been stationed at Burton (Horninglow) motive power depot for 37 years and had driven local passenger and freight trains for 17 years. He was said to have been reliable, temperate and of the highest possible character. His health had been good and he had passed his most recent sight test.

Ganger J.H. Gilbert, whose length was between Four Oaks and Sutton Coldfield, said that there had been no special problems with his section. The speed restrictions had been in existence during the whole of his 42 years' service. He had been in his garden at Four Oaks when the York train had passed and had seen that it was going fast but had felt no

[207] *The Times*, London, England, 24th January 1955

anxiety about it.

Permanent Way Inspector E. Nunnerley agreed that there had been no problems with the stretch of line on which the accident had occurred. He said that diverted trains sometimes exceeded the 30 miles an hour restriction but never before to such a degree as to be worth reporting.

Station Inspector G.W. Overton, who lived in the Sutton Coldfield station house, said that the vibration and noise caused by the York train had been so noticeably abnormal that he had remarked to his daughter, "If he keeps up that speed he'll be in trouble." She had gone to the door leading to the Down platform to find that the train had been wrecked.

The Ministry of Transport inspecting officer, Lieut. Colonel G.R.S. Wilson, made a couple of trial runs over the line with different drivers and locomotives to collect more information for his report.

Concluding his report, the inspecting officer stated that there had been no defects in the engine, rolling stock or track which could have contributed to the derailment. From the evidence of Signalman Gilbert at Four Oaks and Mr A. Reason, a keen railway photographer, it was clear that the engine of the express had still been under steam at that point, while there was no evidence of any effective breaking before the derailment. It was likely that the express had been travelling at at least 52 miles an hour at Four Oaks, and at least 57 miles an hour when it was derailed.

The excessive speed was too high to be explained simply by a misjudgement and it appeared that Driver Allen had made no attempt to observe the 40 or 30 miles an hour restrictions. It was very difficult to explain such an extraordinary lapse on the part of a driver with intimate knowledge of the line and in apparent good health. There was no evidence of any sudden engine defect such as a blow back or an injector failure. The train had been running late, though the lateness had been reduced from 30 minutes on arrival at Derby to 10 minutes at Lichfield (Trent Valley). The only explanation for Driver Allen's failure seemed to be that in his anxiety to regain more of the lost time he had entirely forgotten about the speed restrictions.

As for Driver Martin abandoning the engine at Lichfield, the inspecting officer remarked, 'In my opinion no driver who was physically fit to take the rough with the smooth in all the vicissitudes of main line work, as Driver Martin appeared to have been from his medical history, should have found the vibration unbearable, or even disconcerting; it was clear from his evidence that he was thinking only of his own personal comfort, and I do not accept the excuse which he put forward for leaving his post of duty.'

Regarding Guard Harrison, 'He could have prevented the accident or at least lessened its consequences if he had made a full brake application, and it is much to be regretted that he did not have the courage of his first convictions. I think he was honest in saying that he was anxious not to take the control of the train out of the hands of the driver, but no such consideration should have affected him in view of the short time and distance available in which to reduce the speed.'

Ticket Inspector Attenborough and Fireman Smith were praised for their actions in breaking into the signal box and stopping the Down train.

Turning to recommendations, the inspecting officer considered that lineside speed restrictions or signs, had they existed, might well have reminded Driver Allen of the need to reduce speed. Such signs were not in general use on British railways, though the former Great Western Railway had used illuminated signs at selected locations and "cut out" signs had been in general use on the former London and North Eastern Railway. The matter had been discussed by the Railway Executive in 1949 and it had been decided to provide signs where there were no landmarks or where speeds were repeatedly exceeded. Few such signs had, however, actually been provided. In general, reliance had been placed on drivers having a thorough knowledge of the road and this had worked well, with no previous serious derailment of this type since that at Canal Junction, Carlisle, in 1931, though before that there had been similar accidents at Aylesbury in 1904, Salisbury in 1906 and Shrewsbury in 1907.

Conditions were, however, changing, with the introduction of diesel and electric locomotives fitted with speedometers, capable of higher speeds. In addition driver training was changing, with drivers having less years of experience and making longer runs over more routes. It was recommended that, in view of these and other factors, the British Transport Commission should reconsider the matter, with a view to the systemwide introduction of lineside signs of the ex-London and North Eastern Railway type. The replacement of simple speedometers with speed recorders, which could also provide a permanent record. was also suggested.

The start to pass timing of 15 minutes for the 9¾ miles between Lichfield (Trent Valley) and Sutton Coldfield was unrealistic and should be modified, as should any other unrealistic timings elsewhere.

Although the nominal radius of the curve at Sutton Coldfield was 15 chains, at one point between the tunnel mouth and the platform it sharpened to a radius of 8¼ chains. This irregularity could not be considered responsible for the derailment, but the track should be re-aligned to remove it. [208]

The inquest on the victims of the accident returned a verdict of "Accidental Death," Coroner C.W. Iliffe commenting that there appeared to have been an error of judgement on the part of Driver Allen, and that had he survived it was unlikely that the jury would have considered the question of criminal negligence and manslaughter against the driver. [209]

Nine vehicles of the *Ouro Verde* (Green Gold) passenger train of the Estrada de Ferro Sorocabana left the track at kilometre 40, between Itapevi and Amador Buena stations, Brazil, at 10:45 on Tuesday 1st March 1955. The accident occurred close to the spot where 14 people had been hurt in a collision between two trains a day earlier. The Ouro Verde train, prefix NS, with Driver Benedito de Oliveria, had been coming from Presidente Epitácio and was bound for São Paulo.

The derailment apparently originated with the overturned of an unstable refrigerated waggon, which, because of the high speed of the train, led in turn to the overturning of other vehicles of the train on the opposite side of the line. In addition to the refrigerator car the train consisted of the engine, five ordinary passenger carriages, three sleeping cars, a restaurant car and a luggage van. Of these, seven vehicles overturned and two more were derailed. Eleven people died and dozens were injured, 15 of them seriously. Posts supporting the overhead line equipment of the electrified line were torn down,

---

[208] Ministry of Transport and Civil Aviation report
[209] *The Times*, London, England, 24th February 1955

causing a violent short circuit. Notice of the accident was immediately sent to the Cotia police station, which in turn notified the central police station, who requested the services of the PSM and fire service, all available ambulances and men of the Força Publica security forces and the Civil Guard, in order to police the accident site as well as carry out the rescue work.

The injured had to be taken 600 metres to a point which ambulances could reach. Rescue workers remarked that they had been expecting such an accident for a long time. Fifty injured were taken to hospitals at Itapevi and Coita and to the Hospital das Clinicas e Beneficencia Portuguesa in São Paulo. The firemen worked until 19:00 to rescue victims and clear the track. They returned in the early hours of the next day, when cranes were to be used to lift the overturned vehicles to check that there were no further bodies beneath them.

Engineer João Caetano Alvares, Transport and Public Works Secretary, issued a statement concerning the accident: "The Transport and Public Works Secretary has the painful duty of informing the public that at 10:45 today an accident occurred to an overnight train from Presidente Epitácio of the Estrada de Ferro Sorocabana between Itapevi and Amador Buena stations. As a result of the accident 11 people were killed and 46 others were injured, 15 of them seriously. The train was driven by J.B. Oliveira and consisted of 13 vehicles, of which seven were overturned and two derailed.

"The reason for the derailment is not yet known. A full inquiry has been opened in order to determine the responsibilities for the accident.

"The Estrada de Ferro Sorocabana has provided all possible assistance to the victims of the accident and made available to the families of the deceased the means to remove their bodies." [210]

A major disaster, supposedly the worst to have occurred in Indonesia up to that time, occurred to the Djokja – Bandoeng express when it was derailed between Tjiamis and Manondjaja, at about 11 a.m. on Thursday 12th May 1955. The bodies of 30 victims were initially located in the wreckage, including those of the driver and two firemen, and 39 passengers were more or less injured. The dead and injured were taken to hospitals at Tasik and Tjiamas.

At first it was supposed that the derailment was the result of an act of sabotage but a goods train bound for Tasik had passed the spot safely half an hour earlier. Also D.K.A. (Djawatan Kereta Api – the post-independence railway company) employees were working on the embankment at the spot and there was an army post in the vicinity.

Speaking to a newspaper reporter before he left for the scene of the disaster on an Air Force plane, Effendi Saleh, the head of the DKA, stated that it seemed very unlikely that terrorist action had caused the disaster. What the actual cause was had still to be determined.

The disaster occurred as the train had descended the steep incline from Tjirakong and was increasing its speed to climb the incline on the far side of the railway bridge. The track workers, who had stopped their labours while the train passed, had watched in astonishment as the engine left the rails in front of them, pulling the leading luggage van and the following three third-class carriages with it. Following the derailment, a reporter

---

[210] *O Estada de São Paulo*, São Paulo, Brazil, 2nd March 1955

wrote, 'the sound of tearing metal, breaking glass, and escaping steam mingled with the cries of the passengers in the wreckage, many of whom were unable to free themselves. Other passengers, in the rear coaches, who had escaped unscathed, climbed out of the windows in a panic and fled right and left along the line, apparently fearful that it was a terrorist attack and that they would be fired on. Even without any shooting the results were bad enough, for the military and police who were soon on the scene counted 30 dead in the wreckage of the telescoped coaches. In addition many of the injured were in a bad condition and work to release them from the wreckage began as quickly as possible.'

"I have the greatest admiration," Mr Saleh told reporters, "for the work of the police, military, permanent way staff and others in the minutes after the disaster, without which the consequences of the accident would have been even more serious." Mr Saleh also praised the officials who arrived at the scene shortly afterwards from Tasik and Tjiamas and the medical staff, who had worked for many hours caring for the victims.

A work train left Bandoeng at 05:00 the next morning for the scene. The extent of the damage was so great that it was not possible immediately to say when it would be possible to reopen the line. Nor was it known if any of the injured were in a critical condition.[211]

By the following day the number of known dead had risen to 39, comprising the driver, both firemen and 36 passengers. The bodies of the driver, two firemen, a schoolboy from Djokja and an unknown woman were still imprisoned in the wreckage. There were 60 injured in the two hospitals. During the whole of the previous day the telephones in Tasik, Tjiamas, Bandjar and Garoet hospitals, plus the DKA office in Bandoeng, had been ringing constantly with inquiries from concerned people regarding relatives or friends whom they believed had been travelling on the unlucky train.

At the accident site at kilometre 284, only a few hundred metres from the viaduct over the Tjitanduj, there ruled, a reporter wrote, an oppressive silence, despite the presence of hundreds of people, with only whispers spoken as the railwaymen, ambulance workers and police did their work. Although most of the victims had been rescued by Thursday afternoon, several appeared to be trapped, and their bodies could only be released from the broken beams and smashed panels with the aid of cutting equipment, and in some cases even then the work of freeing them proved very difficult.

'Even more poignant than the sight of the coaches tossed and overturned amidst the green hills, however, were the scenes which unfolded in the little hospitals at Tasik, Tjiamas, Bandjar and Garoet, where under white sheets lay the motionless bodies of those who had lost their lives in this most disastrous train accident in Indonesian history. Nearby, possessions such as a cigarette lighter or a plastic bag waited to be identified by relatives, friends or acquaintances, a horrendous task for the men, women, and children, who, full of fearful premonitions, moved between the white biers. There were heart breaking moments when someone recognised a body on one of the biers as that of a loved one, a man, woman, or child. Then lamentations filled the narrow space and words could not describe the grief which welled up in the little hospital wards, which were permeated with chloroform and ether.'

It was noteworthy that almost all the victims, dead and injured, had been passengers in third-class coach CL 9113, which had been marshalled second in the train, behind the luggage van. The coach, which had been occupied by about 100 passengers, had been

---

[211] *Algemeen Indisch dagblad de Preangerbode*, Bandung, Indonesia, 13th May 1955

thrown across the track by the impact and was then pushed by the following coach against the derailed engine (Krupp locomotive No. 52080). The coach was thus telescoped at each end and hardly any passenger in it escaped death or injury. Those in the front and rear compartments were killed and most of those in the middle compartments were injured. That only a small number of passengers in the other remaining coaches of the train were injured was thanks to the sturdy construction of the modern carriages which, according to DKA head Saleh, were especially strong in a longitudinal direction. Had fate not decreed that the first passenger coach was thrown crossways over the line the consequences of the accident would not have been anything like as severe.

As regards the cause of the disaster, Mr Saleh appeared to attribute it to excessive speed, stating that due to the high speed with which the train had descended the fairly steep slope leading to the bridge, the engine had been thrown off the rails when it reached the next bend. It was suspected that the driver had been trying to make up time, the train having left Tjiamas 18 minutes late, though it was equally possible that he just wished to gain sufficient speed to easily surmount the bank on the far side of the viaduct. At any rate the possibility of sabotage as a cause of the disaster had been firmly ruled out.

It was noted that the same engine had been derailed by a terrorist attack at Malangbong in 1953 (July 1952?), that time with much less serious consequences. [212]

That work was being done on the track at the point of the derailment does not appear to have been considered as a factor.

Fifty people reportedly lost their lives in a derailment at or near Sidi Yahia du Gharb, French Morocco, on or about Sunday 22nd May 1955. [213]

A British Railways Western Region Sunday excursion trip from South Wales to London ended in tragedy at Milton, England, on 20th November 1955. The train had left Treherbert at 8.30 a.m. and was due in London Paddington at 2.30 p.m. Because of engineering work on the Up Main line it was to be diverted onto the Up Goods line at Milton box but as at Goswick eight years earlier (see above) its driver failed to reduce speed for the low speed diversion through the Up Main to Goods crossover, the train was badly derailed and its engine and leading three carriages fell down an 18ft embankment. Of the 294 passengers on the train 11 lost their lives and 157 were injured, of whom 62 were detained in hospital. The driver and a travelling ticket collector were seriously hurt and the guard and three other railwaymen were slightly injured. [214]

The accident occurred outside of the Royal Air Force depot at Milton and help was quickly available from there, soon supplemented by soldiers from the Vauxhall Barracks at nearby Didcot, police, firemen and ten doctors. Fifty injured were taken by ambulance to the Radcliffe Infirmary at Oxford and when they could take no more, to Battle Hospital at Reading and other local hospitals.

A newspaper reporter described the scene and interviewed several passengers. 'At intervals during the next four hours stretchers were lifted down the steep slope into the

---

[212] Ibid, 14th May 1955

[213] Author's personal records, source mislaid

[214] Ministry of Transport and Civil Aviation report

meadow, bearing bodies wrapped in grey blankets. Among them was that of a girl aged 14.

'The cafeteria coach, derailed but upright, stood with its lights burning and behind it were two other derailed coaches. The two coaches which remained on the track were filled with uninjured passengers, and a light engine took them back to Steventon station. From there some went on to Paddington and others back to their homes.

'One rescue worker, Police-constable James Griffiths, had to be taken from the wreckage after he had worked for two hours. He recovered and went back to the task. Another who worked unceasingly was a passenger, Donald Harris, aged 24, a polisher, of Fairholme Road, West Kensington. He had hurt his knee but refused aid. He said, "I was asleep in one of the front coaches. I found myself in the smashed coach in the field, but another man and I forced our way out into the open, and we began removing the dead and injured. The train was not going very fast. I don't think there was any panic. Everyone was very good indeed."

'Mrs Gladys Seller, of Cardiff, who was with her husband in the front portion of the train, said: "It was a gentle sort of crash. I thought the brakes were going on, and then we all went sideways. We managed to clamber through the top of the compartment."

Another front-coach passenger who escaped without injury was Mr Arthur Rogers, of Castle Street, Cwmparc, Glamorgan, who said: "The crash came without any warning at all."

'Mr W.P. Henderson, of Coldsmead Road, Redhill, Surrey, who joined the train at Cardiff, was in one of the coaches that fell into the meadow, but he was unhurt. He said: "In my carriage there were two men, a small boy, and a woman. The boy's father jumped on top of him to protect him when the crash came and I tried to protect the woman. When it was all over we just had to lift up a piece of the door and we were out. Then I gave a hand with the injured."' [215]

The excursion train consisted of British Railways Standard No. 7 mixed traffic class steam locomotive No. 70026 and ten wooden coaches with steel panels and underframes, all ex-Great Western Railway except the sixth, the cafeteria car, which was ex-London Midland and Scottish Railway. It was in charge of Temporary Passenger Guard L.W. Wall with Driver W.T. Wheeler and Fireman A.G.P. Marsh. Cardiff had been left 7 minutes late. A speed restriction for a diversion onto the Up Goods line at Lliswerry had been correctly observed, as had a 5 miles an hour restriction at Over Junction, near Gloucester. Smart station working saved four minutes at Gloucester but this was lost by signal checks outside of Swindon. Beyond there a 15 miles an hour restriction over a bridge between Wantage Road and Lockinge was strictly observed and the train had passed Steventon, a mile and a half short of Milton, at 1.10 p.m., 10 minutes late.

Driver Wheeler had over 40 years' service on the railway. He had driven 12 Up passenger trains and four Up goods trains past Milton in the previous six months, including on two occasions a *Britannia* class engine pulling the *Red Dragon* express. He told the Ministry of Transport inquiry that he had signed on at Cardiff Canton shed at 8.53 a.m. on the morning of the accident and had studied the book of Weekly Notices. He admitted that although he had seen the notice regarding the Lliswerry diversion he had failed to see the entry relating to the Milton diversion on another page and added that he had never imagined that his train would be put onto the Up Goods line there. The

[215] *The Times*, London, England, 21st November 1955

automatic train control (A.T.C.) siren had worked correctly on the trip to Gloucester, though the bell (given at a clear distant signal) had been a little irregular and sometimes faint. At Lliswerry the Up Distant had been at caution for the diversion and the A.T.C. siren had sounded correctly. At Over Junction the Distant signal had likewise been at caution, Fireman Marsh had reset the A.T.C. equipment and he (Driver Wheeler) had reprimanded him for doing so. There had been no problems but leaving Swindon the engine had been very dirty and at speed the noise had been terrific. He had carefully observed the Wantage Road restriction.

Approaching the Milton Up Distant signal he had seen that his fireman was having trouble with the exhaust injector. This had distracted him and he had failed to see the signal; he did hear the A.T.C. siren and had not reset it. He had failed to see the Milton Home signals for the same reason and had not realised that his train was being diverted until he had seen the points set for the loop. At that point he had shut the regulator and made an emergency brake application. His engine had been travelling at 40 m.p.h. when it was derailed a moment or so later.

Driver Wheeler said that the view of signals from a *Britannia* class engine was poor. Often he had to lean right outside of the windscreen so that his eyes got full of dirt and on one occasion a cinder had badly burnt his eye. The Milton signals were especially hard to see from the left-hand side of the cab. His regular fireman usually helped him in calling out signal aspects and he had expected Fireman Marsh, with whom he had not worked previously on a *Britannia* class engine, to do the same, as indeed he had at other places on the journey, and doubtless would have done at Milton had it not been for the injector trouble.

Fireman Marsh said that he had been busy firing after leaving Cardiff and had not noticed the A.T.C. siren or bell except when nearing Gloucester, when he had reset it. He did not recall his driver commenting on his having done so. He had been firing between Swindon and Steventon, where his driver had remarked that they were 10 minutes late. He had then noticed that the right-hand exhaust steam injector was wasting water. Unable to correct the problem he had switched to the live steam injector. He had not heard the A.T.C. siren nor reset the A.T.C. equipment, nor had he observed the Milton signals, through being occupied with the injectors. After the derailment he had extricated his driver from beneath a pile of coal, then, unable to find any detonators, set off to the signal box for some. Going back to protect his train he found that the guard had already done so, and had shut off the injectors and steam heater, then assisted with the rescue of the passengers.

Guard Wall testified that he had tested the vacuum brake at Cardiff. He had been aware of the Milton diversion, though he had not mentioned it to the driver. He had been waiting for his driver to reduce speed when there had been a sudden brake application, three severe jolts and the train stopped. He admitted that he had been somewhat worried at the failure of his driver to reduce speed sooner but feared that if he applied the brakes at the same moment as the driver "the vacuum would have parted the train." After the accident he had gone back to protect his train as far as Steventon, where he reported to the stationmaster.

Travelling Ticket Collector W.H. Cheeseman and Driver R.G. Cheeseman, riding as a passenger, confirmed that the Wantage Road restriction had been observed and put the speed of the train at derailment at 35 or 40 miles an hour.

Driver S.J. Wooster, who had been in charge of the last Up train to use the Milton crossover before the excursion train – a freight train at 11.45 a.m. – stated that the Milton

Up Distant signal had been at *Caution*, the A.T.C. siren on his engine had sounded, and he had reduced speed from 25 to 5 miles an hour for the crossover.

Signalman G.C. Birt said that Milton box had been opened specially on Sunday morning for the diversion. District Inspector W.F. Checkley had been in charge of the arrangements and told him (Signalman Birt) that he (Inspector Checkley) had not appointed a handsignalman at the Up Main to Goods crossover as it was not necessary. No handsignalman had been appointed on previous occasions when trains were diverted either but the practice was to keep No. 32, the Up Main to Up Goods Home signal, at danger until the train was almost at a stand. The Home signal was 350 yards from his box and there was a berth track circuit to its rear 400 or 500 yards long. Signalman Birt said that he had judged the speed of Up trains by watching his track circuit indicator, though it was difficult to judge their speed exactly. Foxhall Junction box had blocked back at 6.57 a.m., when he bolted the Up Main to Up Goods points (No. 22) for the loop and placed collars on the Up Main Home and Distant signal levers (Nos. 34 and 35). Three passenger trains had passed over the crossover at about 10 miles an hour. He had accepted the excursion train from Steventon at 1.04 p.m. and immediately got it accepted from Foxhall Junction on the Up Goods line under absolute block regulations. He had received the *Train Entering Section* signal from Steventon at 1.09 p.m. When the track circuit indicator had showed *Occupied*, he waited for a time which judged by a later re-enactment was 8½ seconds, then cleared the Up Main to Up Goods Home (No. 32). As the train had come closer he had thought that it was going rather too fast, the engine had lurched as it reached the crossover and the train had gone down the embankment. He had at once sent the *Obstruction Danger* signal both ways and also telephoned the Foxhall and Steventon signalmen, recording the time in his train register book as 1.10 p.m.

Signalman A. Bennett at Steventon stated that he had accepted the excursion train from the box in the rear at 1.02 p.m. and it had been accepted by Milton when he had received the *Train Approaching* signal at 1.06 p.m. The excursion train had passed him at about 40 miles an hour, the same speed as the previous three Up passenger trains that morning. Milton box had sent him the *Obstruction Danger* signal at a time he had recorded as 1.15 p.m. Upon hearing from Signalman Birt that the train had gone down the embankment he operated the Up detonator placer and informed the stationmaster, the district inspector, and Swindon Control.

Brigadier C.A. Langley, the Ministry of Transport inspecting officer, commented on the five-minute difference in the times recorded by the Milton and Steventon signalmen and stated that he thought that the derailment had probably occurred at 1.13 p.m.

District Inspector Checkley explained that he had not posted a handsignalman at the Milton crossover because he had considered that the regulations only required one for an emergency diversion and not a booked diversion shown in the Weekly Notices. He had worked that way for the previous nine years and understood that the practice elsewhere in the region was the same. He had visited Milton box at 12.10 p.m., thence gone to Moreton Cutting and back to Didcot, where he had heard of the accident and bicycled back to Milton. Everything had been in order at Milton box on his first visit and again upon his return, except that the Up Main to Goods signal had still been off due to wreckage on the signal wire.

Motive Power Inspectors Jenkins and Pettifor agreed that the *Britannia* engines were very noisy at speed, making it hard to hear the A.T.C. bell and siren, though the brakes were not affected.

Ganger F. Greenaway stated that the Milton crossover had been in good order on the morning of the accident.

Signal Inspector W.S. Stark stated that he had checked the signalling at Milton box at 1.35 p.m. and it had been in order. He also inspected the A.T.C. ramp at the Milton Up distant signal and found it to be in order.

A.T.C. Inspector C.V. Pragnell examined the A.T.C. equipment on the derailed engine where it lay in the field the morning after the accident, and found it undamaged. After the engine had been rerailed and taken back to Swindon it was examined personally by the inspecting officer. It worked correctly and applied the brakes, though due to dirt contamination the siren made a spluttering sound rather than a steady whine. A running test was also made between Swindon and Milton with a train of similar composition to the excursion train and a *Britannia* class engine. With the Milton Up Distant signal at *Caution* the A.T.C. equipment applied the brakes and stopped the train a yard beyond the Up home signal.

Brigadier Langley also rode on a *Britannia* locomotive to ascertain how easily the signals could be observed by a driver. He found that the view of the Milton Up Distant and Home signals from the left-hand side of the engine (the normal driving position) was seriously interfered with by the hand rails and smoke deflector plates, and even more so on engines fitted with automatic ejector isolating elements, except by leaning right outside of the cab. The signals were much easier to observe from the cab of an ex-Great Western Railway Castle class engine.

In conclusion the inspecting officer stated that there could be no doubt that excessive speed had caused the derailment. Driver Wheeler frankly admitted that he had failed to observe the Milton Up Distant and Home signals and alleged that the A.T.C. apparatus had failed to work. It had worked correctly earlier on the journey and there was nothing in its condition afterwards to suggest a sudden failure. It was more likely that Driver Wheeler had re-set it unthinkingly and then forgotten that he had done so, or possibly the fireman had re-set it. Possibly Driver Wheeler had failed to close the large ejector properly after the Wantage Road check and the brake application had thus been neutralised, though the warning siren would still have sounded. Driver Wheeler must therefore accept full responsibility for the accident.

Driver Wheeler had had ample time to study the Weekly Notices but had overlooked the Milton diversion. In addition the view of the Milton signals from the cab of *Britannia* class engine was far from good and it was possible that the driver had mistaken the Up Main to Goods Home signal when it was lowered by Signalman Birt for the Up Main Home signal. The signals were easily visible from the right-hand (fireman's) side of the engine and Fireman Marsh could easily have seen them had he looked out at the critical time. Both men were, however, used to driving Western Region engines with right-hand driver and Driver Wheeler had not arranged with Fireman Marsh for the latter to look out for the Milton signals. As the fireman was also not familiar with the route and had had trouble with the injector he could not be blamed for the accident.

It was unfortunate that Guard Wall had not discussed the Milton diversion with his driver. He had realised that his train was going too fast approaching Milton and his fear of dividing the train was quite unfounded. The inspecting officer conceded that, 'a guard is placed in a difficult position because he is naturally loth to take control out of the driver's hands unless danger is imminent,' but, 'This was clearly a case for action, and Guard Wall's judgement was clearly at fault.'

In the absence of a handsignalman Signalman Birt had rightly decided to wait until the berth track circuit in the rear of signal No. 32 was occupied before clearing the signal. However, the track circuit was over 500 yards long and the box was almost 400 yards from the signal. It was thus difficult for him to judge the speed and position of an approaching train and likely that the excursion train was still 300 or 400 yards away when the signal was lowered.

District Inspector Checkley had been under a serious misapprehension in thinking that a handsignalman was not required at the Up Main to Goods home signal except during emergency diversions. Had there been one present he might have been able to attract the attention of Driver Wheeler early enough to avert the derailment.

Turning to recommendations the inspecting officer stated that there had been only one previous serious accident due to unthinking cancellation of the A.T.C. warning (at Norton Fitzwarren in 1940, described in the previous volume of this series). How many accidents had been prevented by its action it was, of course, impossible to say, but 69 of 664 accidents investigated since 1912 might have been prevented by it. The British Transport Commission Modernisation Plan called for a gradual extension of an improved version of the equipment as well as other signalling improvements.

The excursion train engine had not been fitted with the vacuum ejector isolating equipment and the A.T.C. brake application might not, therefore, have been fully effective, but should have prevented the derailment had it not been cancelled or neutralised by the large ejector. The fitting of the ejector isolating equipment to British Railways standard class Western Region engines was now almost complete.

The cleaning of the A.T.C. sirens of *Britannia* class engines was now to be done every fortnight rather than every 28 days.

It was recommended that the Western Region A.T.C. equipment be provided with the visual indicators found on the new British Railways apparatus.

Drivers should be reminded that their firemen were not to interfere with the A.T.C. equipment.

The design of *Britannia* class engines should be modified to improve the driver's view, beginning with engines on the Western Region where, until recently, the driver had stood on the right-hand side. Eventually the introduction of main-line diesel locomotives would give drivers an unobstructed view.

The Milton Up signals were excellently sited and their position should not be changed.

There was no special significance to the accident having occurred on a Sunday other than the fact that many of the accidents on that day were more or less associated with engineering works.

The Weekly Notices were clearly set out but diversions of passenger trains could be added to Section A, Temporary Speed Restrictions.

A rule might be added that passenger trains diverted over what were normally goods lines were only to do so after they had received verbal authority at the diverging point.

Although it was not related to the accident the fitting of speedometers to all steam locomotives was under consideration.[216]

---

[216] Ministry of Transport and Civil Aviation report

Excessive speed was quickly established as the cause of the derailment of an Atchison Topeka and Santa Fe Railway passenger train at Redondo Junction, Los Angeles, California, United States, on the evening of Sunday 22nd January 1956. Thirty passengers lost their lives in the accident and 117 passengers, four railwaymen and a chair-car attendant were injured.

After the accident, 'Emergency calls went out for doctors, nurses, clergy and blood. Scores of ambulances lined up at the scene. Clergymen of all faiths lined up among the dead and injured, administering final rites or giving comfort.

'The first ambulance attendant to reach the scene, Daniel J. Cesarotti, said: "We gave them morphine until we didn't have any more – and still there were many people screaming out in pain. It was like a nightmare. When we got there we found the two cars tipped over. We climbed up on top of them and broke windows so we could climb inside. There were screams and moans all about us. It was horrible. Bodies and parts of bodies were everywhere. Many of the passengers had been crushed in their seats. Others had been thrown through the windows of the railroad cars and crushed under them. We saw the bodies of two children. This is the worst we've ever seen."

'Firemen used axes and sledges to smash into the cars for victims inside. Heavy cranes later righted the cars. When it brought up one, the assembled crowd gasped. A body hung out of a jagged window. At the rear door were parts of two or three other bodies. Other victims were strewn along the right of way in the wake of the sliding cars. More than 40 physicians were there. Others were dispatched to hospital. The Red Cross made 1,500 pints of blood available.

'A detachment of soldiers joined hands forming a human chain to keep back the throngs of morbidly curious who jammed the area near Washington Boulevard and Soho Street in the southeastern part of the city.

'Marine Capt. Roger Barnard, 26, of Triangle, Va., en route from Los Angeles to Camp Pendleton, near San Diego, said: "The train tilted. I saw the suitcases start coming down but I got down on the seat. I felt like a die in a dice game. The lights went out and then came on again. I got out through the window on the left side even though the car had fallen on the left side. There was still some room to crawl through the window and roll ourselves out. I got only a small bruise."

'Many servicemen struggled to avert a panic among survivors. "People were numb with fear," said a woman marine, Reatha Corbett. "All around us servicemen were telling the passengers to 'lie down and keep quiet."'' [217]

Eastbound first-class train No. 82, the *San Diegan*, consisted of diesel passenger units DC-191 and DC-192, in charge of Conductor George Spickard with Engineer Frank Parrish and Secondman Homer Smith (referred to a fireman in the Interstate Commerce Commission (ICC) report but secondman seems more appropriate). It had left Los Angeles Union Passenger Terminal at 5.30 p.m. and was derailed on a 6 degree 12 minutes curve to the left only 3¾ miles from its starting point, at about 5.42 p.m. The maximum authorised line speed was 79 miles an hour but there was a speed restriction of 15 miles an hour around the curve where the derailment had occurred. A sign 9½ inches wide and 24 inches high, with the numerals '15' in 8 inch black figures on a yellow

[217] *The Bakersfield Californian*, Bakersfield, California, United States, 23rd January 1956

reflectorized background, warned engineers of the restriction. The train had been travelling at an estimated 68 or 69 miles an hour when it left the track. The two units remained coupled together and overturned to the north. They stopped on their sides, considerably damaged.

As the train approached the curve Engineer Parrish and Secondman Smith had been in the positions at the front of the leading unit and Conductor Spickard and a flagman had been in the second unit.

The train brakes had been tested before departure and its headlight was burning brightly. Engineer Parrish told the ICC investigators that he remembered passing under a street viaduct a mile and a half short of the point of derailment and seeing an engine in a yard at that point. After that he remembered nothing, no landmarks or anything, until he recalled the fireman standing beside him telling him to apply the brakes in emergency. He did not know if he had done so, but he remembered the train leaving the track.

Fireman Smith stated that he had called out all the signals and also the speed restriction sign. His engineer had acknowledged this last call by raising his hand, had reduced the throttle to No. 1 position and made a service brake application. This application had failed to reduce the speed of the train. He had then moved to the engineer's side and warned him to apply the brakes in emergency. He said that he had not taken any action to stop the train himself because the engineer had appeared normal and had at once moved the brake to the emergency position, but the derailment had occurred before there was any drop in the train's speed. The service brake application had been 1,700 feet short of the curve and the emergency application over 1,000 short of the curve, according to the fireman.

Conductor Spickard and the flagman said that Engineer Parrish appeared normal when they had seen him at the terminal before departure. The conductor said that immediately before the derailment he had been busy dealing with tickets and had not noticed that they were approaching the curve. There had been 161 passengers on the train, all of whom had been seated.

The flagman said that he had been concerned about the speed of the train but had not worked out exactly where they were before the derailment had occurred. The conductor and flagman agreed that the brakes had not been applied prior to the derailment.

The ICC noted that there was no sign of dragging equipment or of any obstruction on the line.

The bodies of the units had not been broken open in the accident such as to cause any casualties. However, with the exception of two in the leading unit all the windows on the left-hand side had been broken out and it was evident that the large number of casualties had been caused by passengers falling or being pulled through the openings where the 25½ by 55¾ inch windows had been displaced and coming into contact with the track, as the units were sliding along on their sides.

Two days after the accident the brake equipment of the leading unit was tested and found to be functioning normally. That of the second unit was too badly damaged to be tested.

The equipment of train No. 82 had arrived in Los Angeles as a westbound train at 4.30 p.m. The engineer of that train had found the brakes to be functioning properly and a car inspector who had examined the train after its arrival found no problems. The brakes had also worked properly at several points between Los Angeles Passenger Terminal and

Redondo Junction. The speed recorders on the two units indicated speeds of 66 (front unit) and 72 (rear unit) m.p.h. at the time of derailment. Corrected for errors the speed of the train was estimated as having been between 68 and 69 m.p.h. at the moment of the accident. There was no indication on the recorder tapes of any brake application at the curve.

Engineer Parrish was 61 years of age. He had started as a fireman in 1918 and had been an engineer since October 1939. He had passed a general health examination in August 1955. His physical condition after the accident was still found to be generally good and in the opinion of physicians had suffered an epileptic fugue attack (a state of altered consciousness) and had not been conscious of his actions as the train approached the curve. Two electroencephalograms made during the examinations showed abnormal brain waves in specific areas, consistent with such a diagnosis. This condition was not one which could have been discovered during a routine medical examination.

The ICC report made no comment as to Fireman Smith's failure to intervene in time to prevent the accident. [218]

At the inquest on the victims of the accident Engineer Parrish was said to have been chain smoking and to have trembled and buried his head in his hands when the roll call of the victims was read out. He said that he blamed himself for the dead and orphans he had made and took full responsibility for the accident. Deputy District Attorney Adolph Alexander told the jury that they could find one of three verdicts:

(a) Criminal negligence, involving Engineer Parrish, Fireman Smith and Conductor Spickard.

(b) Negligence.

(c) Unavoidable accident.

The railway company stated at the beginning of the inquest that a medical examination had confirmed the engineer's story that he had suffered a blackout before the accident. It also transpired that the engineer was far from being a healthy man, having suffered from tuberculosis, a perforated peptic ulcer and peritonitis in the previous ten years.

The inquest jury returned a verdict that the deaths had been accidental and that no person had been criminally responsible. [219]

As a mixed train was leaving the station at Orejo, Spain, at 22:00 on Thursday 9th February 1956 one of its leading vehicles, which was a trifle wider than the engine hauling the train, collided with a waggon which had become derailed on an adjoining siding. This led to the derailment in turn of the following vehicles of the train, which piled up in a heap against the base of a road overbridge.

The rescue operation which followed the accident was conducted in the bitter cold and it was snowing. Several hours later the bodies of only two victims had been recovered but it was eventually discovered that five people had lost their lives and some others were injured, two seriously. Those who died were Brakeman Luis Hoyo and passengers Antonio Costa, a textile manager from La Cavada; his son, also named Antonio Costa, who worked in the same factory; Julia Barquin Maza, a fishmonger from Seseña, and

---

[218] Interstate Commerce Commission report

[219] *The Bakersfield Californian*, Bakersfield, California, United States, 3rd and 7th February 1956

Clemente Miguel, a teacher at the La Cavada national school.

The track was cleared at about 05:00 the next morning and the bodies of the dead were taken to a *chapelle ardente* at Orejo church. [220] [221]

Train No. 340 Down from Rajkot to Okha was derailed at mile 182.10 between Hadmatiya and Jamwanehali stations, India, at about 4 p.m. on Saturday 19th May 1956 on the Western Railway. Seven passengers were killed and 28 were injured, five seriously, when nine carriages left the line. A relief train was sent to the scene from Jamnagar and the seriously injured were admitted to the civil hospital there. It was hoped to reopen the line by the end of the following afternoon. [222] [223]

Travelling at the maximum authorised speed of about 100 kilometres an hour, a Paris – Luxembourg express was derailed on a set of facing points about a hundred metres from the station at Fismes, France, on the morning of Thursday 14th June 1956. The engine, tender and first four vehicles of the 12 making up the train passed over the points safely but the fifth vehicle, a bar car, jumped the track at the points, as was described by two railwaymen who witnessed the accident. Help was summoned by the sounding of the Fismes air raid siren and every available ambulance was sent from Reims, 24 kilometres away, plus fire engines from there and elsewhere. The bodies of nine victims were recovered from the wreckage and of those passengers taken to hospital two subsequently passed away. Another 38 people were detained in hospital at Reims, mostly with broken bones, and the condition of an 11-year-old boy, who had suffered a fractured skull, was early on reported to be giving concern.

A technical investigation into the accident was opened by M. Charles Boyaux of the Société nationale des chemins de fer français and a second inquiry was launched by juge d'instruction Roussel and Reims procurer Flammarion. The cause of the derailment was not immediately apparent but it was stated that no blame for it attached to Driver Armand Didier or Fireman Roland Chailler, the two enginemen of the train. The likely causes were a defect in the fifth vehicle of the train or a defect in the points themselves, which had recently been repaired. [224]

It was expected that the line would be blocked for 36 hours. [225]

On the night of Saturday 30th June 1956 a train from Mexico City was derailed on a curve 74 miles northwest of Oaxaca, the capital of the state of the same name, and 197 miles from Mexico City. Two coaches of the train left the track and fell down a ravine. Fifteen people lost their lives and 57 were injured, some very seriously. The accident to the Ferrocarril Nacional de México train was initially blamed on it having run into a landslide after heavy rain [226] but further investigation resulted in the arrest of Driver Jesus de la Torre, a railwayman with 25 years' experience, who was charged with being

[220] *La Vanguardia*, Barcelona, Spain, 11th February 1956
[221] Webpage: Federación Castellano Manchega de Amigos del Ferrocarril (in Spanish)
[222] *The Bombay Chronicle*, Bombay, India, 20th May 1956
[223] https://rsdebate.nic.in/bitstream/123456789/574848/2/IQ_17_22051957_S101_p1051_p1053.pdf
[224] *Le Monde*, Paris, France, 15th June 1956
[225] *The Times*, London, England, 15th June 1956
[226] Ibid, 3rd July 1956

responsible for the derailment by having attempted to negotiate curves at excessive speed. He was also, himself, one of those injured in the accident, and was taken to hospital in Oaxaca along with the injured passengers. It was reported that, 'Military authorities were forced to throw a cordon around the hospital to prevent hundreds of hysterical persons from breaking in seeking relatives missing in the wreck.' [227]

Japanese National Railways (JNR) Down Limited Express No. 243 from Nagoya to Toba on the Sangu line, which consisted of two Class C51 steam locomotives and nine carriages, normally crossed an Up train at Matsuzaka but on 15th October 1956 the crossing of the two trains was altered to Rokken, the preceding station. The driver of the express failed to observe that the signal on the approach to the station was at *Caution*, entered the station at a speed of 58 kilometres an hour, passed the starting signal protecting the beginning of the single line at *Danger* and his train became derailed, while still moving at 27 kilometres an hour, either after colliding with a set of buffer stops or on a set of trap points. Less than a minute later Up Limited Express No. 256 arrived. This train, bound for Nagoya, consisted of a Class 57 steam locomotive and 11 carriages, and was travelling at 55 kilometres an hour when it collided with the derailed engines and three overturned carriages of the Down train. Escaping steam from the damaged boilers of the two engines added to the number of casualties and in all 42 passengers died and 94 were injured. Many of the victims were senior high school students who had been on a field trip from Tokyo and the accident caused a great shock throughout the country.

An investigation into the accident was reportedly inconclusive and a court blamed both the driver of the Down express and the station staff, though the driver was held to be principally to blame. The braking ability of the trains also came into question.

After the disaster the JNR pressed ahead with a cab signalling system capable of giving a warning to a train driver and with the automation of the signalling. [228] [229] [230]

A freight train carrying construction materials for a dam 50 kilometres from Tokyo, Japan, was derailed in the suburbs of the capital on Sunday 4th November 1956 and plunged into a river. Six people died. [231] Railway officials stated that a landslide had caused the accident, explaining that the train had rounded a curve on a falling gradient to find tons of rock and soil blocking the line. [232]

Train No. 38, *El Capillense*, was carrying at least 650 and possibly 800 tourists back to Buenos Aires from the mountain resort of Capilla del Monte when it was derailed at Alta Córdoba, Argentina, on Monday 25th February 1957. The train consisted of an engine and 10 carriages in charge of Guard Zenón Barrionuevo with Driver German Arguello and Fireman José Bazan. Running down a falling gradient it left the rails on a curve and seven of the coaches overturned and piled up in a heap of wreckage, some of

[227] *The Brownsville Herald*, Brownsville, Texas, United States, 3rd July 1956

[228] Webpage: Breakthroughs in Japanese Railways 2 Learning from Past Railway Accidents – Progress of Train Control by Masayuki Matsumoto

[229] Webpage – Japanese Railway Safety and the Technology of the Day by Masao Saito

[230] Wikipedia – Rokken Rail Accident

[231] *Daily News Record*, Harrisonburg, Virginia, United States, 5th November 1956

[232] *Ukiah Daily Journal*, Ukiah, California, United States, 5th November 1956

the vehicles being thrown 25 or 30 metres by the force of the derailment. Nine passengers lost their lives in the accident and a large number were injured. The first help came from neighbouring Villa Paraisos and soon the injured were being transferred to the San Roque hospital and health centres in Jeeps, trucks, taxis, and private cars. The number of victims overwhelmed the emergency department of the hospital and the army set up facilities in an adjoining school for the care of some of the victims. [233] Radio stations made requests for blood donations and soon more blood than was needed had been donated, while 17 operations were reportedly performed in an hour.

It seemed clear that the principal cause of the derailment was excessive speed as a number of passengers reported that the train had been travelling at high speed, perhaps 115 kilometres an hour, when it had left the rails. Passengers Enrique Rosental, of Valentin Alsina, and Fernando Caprani, of Bánfield, agreed as to the high speed of the train and said that both of them had complained to an inspector, pointing out to him that bags and suitcases were repeatedly falling from the luggage racks due to the speed of the train and its very unsteady motion. The inspector had replied that he would speak to the 53-year-old driver (presumably at the next stop). It was suggested that Driver Arguello had been running late, had been anxious to make up time, and that he had regained three minutes in the short distance between Demesnil and Rodríguez del Busto, and three more minutes between the latter place and the point of derailment. Driver Arguello denied that he had been travelling at anything above normal speed.

Federal Judge Dr Antonio Cornejo opened an investigation into the causes of the disaster and the driver, fireman, guard, and inspector were all arrested and held incommunicado pending the results of the inquiry. There were suggestions that in addition to the excessive speed of the train the poor condition of the rolling stock, its unsuitability for the mountainous route, and the condition of the track might have contributed to the accident. The possibility of sabotage was also raised, as various other acts of sabotage had been committed by supporters of exiled former President Juan Perón. [234] [235]

Train No. 326, an express from Tourcoing to Paris Nord, had left Arras and had reached a speed of about 80 kilometres an hour when it was derailed near Boisleux-au-Mont station, France, at 14:30 on the afternoon of Tuesday 18th June 1957. The rear postal van was the first to leave the track and it pulled the 11th and 12th coaches, which immediately preceded it, off the line in turn before the train could be brought to a stand 200 metres further on. The 11th coach overturned and the 12th smashed into it. Eight passengers were killed outright and 20 were injured; three seriously injured passengers died later. The number of casualties was increased because it being a hot day many of the passengers were standing near open windows to get some fresh air and they were crushed when the carriages overturned. Two postal sorters in the van were cushioned by the mail bags and escaped harm.

A rescue operation was quickly organised with soldiers from a local infantry battalion. The injured were taken to hospital in Arras and a building in the station was transformed into a *chapelle ardente* for the bodies of the dead. Two of the four running lines were

---

[233] *La Voz*, Córdoba, Argentina, 17th January 2010

[234] *El Litoral*, Buenos Aires, Argentina, 26th February 1957

[235] *The Victoria Advocate*, Victoria, Texas, United States, 26th February 1957

blocked: one was reopened at 05:00 the next day and normal working was resumed at 17:00 the same day. The mail bags from the derailed van were retrieved and postal deliveries were not affected.

As is sometimes the case the reason for the derailment was not easy to determine. Three different mechanisms were suggested:

(a) The postal van had suffered a broken axle

(b) Heat distortion of the track

(c) Subsidence of the track because an excessive amount of ballast had been removed. Work was in progress at the time in preparation for the forthcoming electrification of the Paris – Lille main line. [236] [237]

Rede Ferroviária Federal (formerly Estrada de Ferro Leopoldina) mixed train No. 3, which left the Barão de Mauá station in Rio de Janeiro at 05:25 on Thursday 20th June 1957 for Campos was derailed at about kilometre 156 between Indaiaçu and Poço D'Antas stations, Brazil at about 10:00. Some of the coaches of the train overturned. A statement from the railway company indicated that only one passenger had been killed while a number were injured, one seriously. Help quickly arrived from Macaé, the injured were taken care of and the line was quickly cleared. [238]

According to reports in a couple of American newspapers all but five of the 23 vehicles of the train had left the track, six passengers were confirmed as having lost their lives and 97 were injured.

Whether the railway company was trying to minimise the seriousness of the accident or whether the reports in the American Press were exaggerated is not clear. [239]

France suffered several bad accidents in 1957, including one early on Friday 19th July at Bolléne la Croisière (commonly called simply Bolléne).

Train No. 20 from Ventimiglia via Nice to Paris consisted of a 241P class steam locomotive, luxury sleeping cars and a restaurant car, in charge of Conductor Julien Parregno with Driver Paul Mazi and a fireman referred to only as 'F.' It left Marseille St Charles station at 23:00 on Thursday 18th, on a dark and wet night, and the run to Avignon was made safely. A few kilometres north of Avignon was the station at Orange and Bolléne was 18 kilometres further north again. Train No. 20 was being followed by a relief train, No. 10080. The latter train was running 13 minutes late and Control wished to advance northbound freight train No. 4572 from Orange in the short interval between the two fast trains, which would mean shunting it out of the way of train No. 20 at Bolléne. Confusion arose here, related to the fact that Bolléne received its instructions from Control at Valence, further north, while Orange took its orders from Control at Marseille to the south. The official at Bolléne understood that the freight train was preceding train No. 20 and set his facing points for a siding, but Marseille Control probably decided that the interval was too short and the freight train was held at Orange until train No. 20 and the following train No. 4 (*Le Train Bleu* – the Mediterranean – Calais express) had

236 *Le Monde*, Paris, France, 20th June 1957

237 Wikipedia – Boisleux Station

238 *O Estada de São Paulo*, São Paulo, Brazil, 21st June 1957

239 *Albuquerque Journal*, Albuquerque, New Mexico, United States, 21st June 1957

passed. The trap had been set.

Signal protection for such permanent speed restrictions was normally provided by a colourlight signal displaying two horizontal lights, which was followed by a second signal close to the crossover showing two vertical lights. At Bolléne, however, the first signal displayed a single yellow light, which could mean either that the route was set for the siding, or that the main line ahead of the next signal was occupied by another train. As train No. 20 was never routed into the siding at Bolléne Driver Mazi had clearly assumed that the latter was the case, and that he was overhauling a slower train which would be shunted out of his way at Pierrelatte, the next station. The drivers of prestige trains such as No. 20 were in the habit of not immediately braking when they received a yellow signal, on the assumption that the signal would clear to green. Although it had stopped raining it was also somewhat foggy, which might have been a factor. At any rate the train arrived at the facing points still travelling at 90 kilometres an hour, much too fast to negotiate the crossover, through which trains were restricted to 30 kilometres an hour, despite Driver Mazi making an emergency brake application at the last moment when he found the second signal was not showing another yellow, as he must have assumed, but a double yellow for the diverging route. The engine overturned on the points and the following vehicles piled up behind it, one of them climbing onto the locomotive. By an unlucky chance the boiler of the locomotive was penetrated in the derailment and scalding steam at over 400°C enveloped the sleeping cars and passed through their open windows, causing most of the deaths and injuries. Seventeen people died more or less at once and three more passed away in the days that followed. Seventy-four people were reportedly injured. Conductor Parregno and Driver Mazi were among those who lost their lives while the fireman survived but was seriously scalded. In a dying statement the driver said, "The bastards, they sent me on to the siding." [240] There were a number of English passengers on the train, and the dead included a honeymooning couple, Mr and Mrs Dennis Skellington, from Nottingham, who had been wed only a fortnight earlier. A French passenger who survived the accident unhurt was reportedly on the telephone to his wife in Cannes to tell her that he was safe when he collapsed and died from a stroke. [241]

Assistance was quickly sent to the scene after the accident from Bolléne itself, Orange, Avignon, Valréas, Le Pontet, Sorgues, Pont Saint-Esprit, Pierrelatte and Montélimar. Most of the injured were taken to hospitals in Orange and Avignon, though the most seriously injured were flown by helicopter to Marseille.

Work on clearing the line began the following morning, with breakdown cranes from Montélimar and Avignon. Three lines were badly blocked but one was cleared to enable trains to pass at reduced speed with single line working. The locomotive was rerailed, sent to the workshops and quickly repaired.

M. Louis Armand, the head of the Société nationale des chemins de fer français (SNCF), went to the scene of the disaster and afterwards made a statement to the Press. He began by pointing out that the magnitude of the accident was due to the chance that the poor passengers had been scalded by escaping steam from the boiler of the locomotive, something which had never happened before in the history of the French railways. This was not controversial, but he then went on to say that the Flaman train

---

[240] https://rail-en-vaucluse.blog4ever.com/la-catastrophe-de-bollene

[241] *The Times*, London, England, 20th July 1957

recorder had been examined and it clearly showed that the accident was entirely the fault of Driver Mazi, who had both disregarded the acoustic warning of the 'crocodile' warning device and the visual warning of the signal, and had failed to brake for the diverging route.

Asked what lessons the railway company drew from the disaster he replied that first of all it showed the need to educate drivers even more to obey signals.

Secondly, safety should be further automated, with the automatic warning system being converted into a system of automatic train control, whereby the brakes were automatically applied should a train pass a signal at caution. Experiments in this direction were in progress.

Finally it was anticipated that the line would be electrified within two years, which would improve not only the passengers' comfort but also their safety.

Asked about the mix-up which had led to the route being set for the siding at Bolléne, in anticipation that train 4572 was coming first, he stated that the matter was still being investigated, but repeated that had the driver obeyed signals the accident would not have occurred in any case.

M. Armand's comments, which pre-empting the investigation into the accident cast all the blame for it on a railwayman who was no longer alive to defend himself, provoked the anger of the railway unions. In protest they called upon the departures of all trains in the Lyon region to be delayed for 5 minutes on the day after the accident, though in the end only about 20 trains were affected.

A number of other factors were shown to have combined to cause the accident. One of the most important was the statement of Fireman F after the accident that the 'crocodile' had not sounded its warning at the signal approaching Bolléne. This proved to be true. Another driver had reported the failure of the 'crocodile' to work at Avignon on 17th July and a message was duly sent to Marseille Blanchard depot at 17:10 that day. However, for reasons of economy there was no mechanic stationed at the depot at night capable of carrying out the repair, and so it was not done. Engine 241 P 22 worked train No. 20 that night with the 'crocodile' inoperative, returned from Lyon to Marseille on the 18 with the *Mistral*, and left again with train No. 20 again with the 'crocodile' not working.

In addition the regulation time of 55 minutes for the preparation of the engine was cut to a mere 20 minutes on the 18th.

It also appeared that Driver Mazi and Fireman F had only finished their previous duties at Marseille Blanchard at 13:30 on the 18th and had had to return to work train No. 20 at 21:25 after only 7 hours and 55 minutes' rest.

The fact that Control were also pressurised by management to be as efficient as possible, as for example by taking advantage of the short gap between trains Nos. 20 and 10080 to advance train No. 4572 a few kilometres was also criticised, perhaps unfairly in the author's view.

The inquiry into the disaster subsequently placed the blame for the accident on the SNCF for the various deficiencies which had combined to cause it, rather than on Driver Mazi, who was a 'top-link' driver with a very good record, and who had been within a few months of retirement. He left a widow and two children aged 14 and 10.

With the death of further victims from their injuries the final number of fatalities was probably 31.

Less than two months after the Bolléne disaster France suffered another major derailment, again due to excessive speed. Société nationale des chemins de fer français (SNCF) overnight passenger train No. 1115 was timetabled to leave Paris Lyon at 23:13 on Friday 6th September 1957 for Nîmes. It was very crowded although it had been strengthened with two extra carriages, with at least 500 passengers, many of them standing in the corridors and even in the toilets. Many passengers were returning holidaymakers, others going to the harvest. All went well until the train reached Nozières-Brignon, not too far from its destination, but here it was derailed at 12:35 on Saturday 7th September 1957, with considerable loss of life. Twenty-seven people died and 134 were injured, 30 seriously.

Train No. 1115 consisted when the accident occurred of a Class 141 F steam locomotive (141 F 95), a wooden mail van, a wooden luggage van, a first- and second-class composite OCEM sleeping car, a first and second-class OCEM coach and five second-class OCEM coaches. (OCEM vehicles were all-steel coaches built between 1924 and 1940 for various French railway companies to replace older wooden coaches by the office central d'études de material de chemins de fer, hence the abbreviation OCEM.)

The train was due to arrive at Clermont Ferrand at 05:59 on Saturday morning. Here 15 minutes were allowed for its engines to be changed, but the exchange of engines was postponed to Langeac, where the train was due to arrive at 08:12. At Langeac engine 141 F 95 was added to the train, and Driver Elie Gagnes and Fireman Joseph Cubizolles took over from their colleagues. The train was booked to leave Langeac at 08:19 and according to a newspaper reporter it left on time, though its actual departure time is uncertain. Possibly time was lost at Langeac, for it is likely that the train did not arrive at Alès station until 12:09, 17 minutes late.

Down the line at Nozières-Brignon *voie unique temporaire* (temporary single-line working) over the opposite running line, Track 2, had been in operation since Thursday between Nozières-Brignon and Fons Saint Mamert for relaying work, concrete sleepers being provided in place of the timber ones on Track 1. The facing crossover from Track 1 to Track 2 required a reduction in speed to 30 kilometres an hour. The restriction was marked by a diamond-shaped sign, consisting of black figures on a white background (illuminated at night), situated at kilometre 695.884, over 1,000 metres in the rear of the crossover. The train passed the warning sign without any reduction in speed and according to the Flaman train recorder was still travelling at 92 kilometres an hour when it reached the crossover. At the points the engine leapt into the air and fell sideways across the track, forming a 150 tonne obstacle for the following vehicles. The mail van was demolished, the luggage van was thrown against the embankment, the sleeping car was thrown across the line, the next three carriages were driven into one another and the seventh vehicle of the train was derailed. The rear two vehicles escaped. Driver Gagnes and Fireman Cubizolles survived the accident, though both were injured.

Crossing PN 135 was close to the scene of the derailment and Crossing Keeper Teral witnessed the accident. She said later, "I hardly had time to think. The train was going too fast! Then I saw the engine and two carriages tip over and settle down on the ballast. There wasn't much noise. Then a huge jet of steam, a huge cloud of dust rose up into the air. Then horrible cries dominated by children's voices filled the air, following an even more disturbing silence…"

Jeanne Deleuze, aged 17, riding on the train with her mother and two brothers, told a newspaper reporter, "First there was a great silence, then came the screams of panicking people…Later the driver appeared. His hands were burnt, his skin was in tatters. He kept repeating, 'What have I done? What have I done?'"

After the accident help was quickly organised and the first ambulances arrived at the scene between 13:05 and 13:10 while four doctors had also arrived at 13:05 and there was one on the train. By then some of the injured had already been sent to hospital by car and van. Assistance came from Alès, Nîmes and Avignon. Twenty-six bodies were recovered from the wreckage and one of the injured later died in hospital. Those who lost their lives were taken to a *chapelle ardente* established in the Boucoiran town hall. The uninjured passengers were forwarded by motor coach during the afternoon. A 50 tonne breakdown crane arrived from Avignon at 17:20 and another from Nevers at 02:00 the following morning. The work of cutting into the steel carriages was a long job and the rescue operation lasted for 48 hours. With all lines blocked passenger trains were diverted via Quissac and Sommières and freight and parcels trains were cancelled. The line could be reopened at 03:00 on Tuesday 10th September.

It was immediately clear that the excessive speed of the train was the cause of the derailment and Driver Gagnes was interrogated as to his actions. It emerged that he had failed to sign a register when he had come on duty at Langeac which would have advised him of the temporary diversion at Nozières-Brignon. He had, in fact, omitted to sign the register since 22nd July and his superior at the depot had failed to note this omission or do anything about it.

When the train had been stopped at Alès there had been a loudspeaker announcement to the passengers advising them of the engineering work so they did not look out of the windows but these had not been noticed by Driver Gagnes from his position on the engine.

Asked by a police officer about the VUT *(voie unique temporaire)* sign warning him of the speed restriction Driver Gagnes could only answer, "I didn't see it." It was pointed out that the Flaman tape recording showed that the 'crocodile' automatic warning equipment had sounded and he had cancelled it, to which he replied, "I saw it too late," and, "I took a long time to stop because I was confused by hearing the 'crocodile' alarm and seeing the VUT sign at the same moment." It was pointed out that there had been almost 1,200 metres between the VUT sign and the crossover itself, plenty of time for him to have applied the brakes and brought the speed of his train down to 30 kilometres an hour. He could only reply, repeatedly, "I didn't have time." The distance between the VUT sign and a fixed reminder signal was 1,152 metres, and between the latter signal and the beginning of the facing crossover 44 metres. The speed of the train could have been reduced from 92 to 30 kilometres an hour in only 400 metres, leaving an ample margin of safety. Later Driver Gagnes said that he had not seen the VUT sign but could not explain why he had just silenced the 'crocodile' alarm without applying the brakes. He added that he had seen the normal *semaphore de cantonment* (block signal) which was in the open (*Clear*) position but bore a white cross showing that it was out of use, and had "hesitated, and thought for a moment." It seems that not having been previously aware of the diversion and speed restriction he had been taken completely surprise when the 'crocodile' had sounded at an unexpected place and had failed to act. The regulations stated that should a 'crocodile' alarm sound unexpectedly the driver must immediately reduce speed to a level at which he could stop short of any train, signal, or obstruction.

On 20th November Driver Gagnes was charged with manslaughter, involuntarily causing injury and violation of railway regulations.

Before the Court at Alès on 4th June 1958 the driver was sentenced to six months' imprisonment and fined 108,000 francs. On appeal his term of imprisonment was amended to a two years' suspended sentence and the fine was reduced to 100,000 francs. [242] [243]

At least 15 people reportedly died and 60 others were injured when a passenger train from Buenos Aires Retiro was derailed at 01:00 on Sunday 10th January 1958 at suburban Palermo, Argentina. The derailed train fouled the adjoining running line and a collision with an incoming train followed the initial derailment. One carriage of one or another of the trains also caught fire.

Newspaper reports of the accident added that, 'There was no immediate indication that the derailment was an incident of the rail sabotage that has been a common occurrence here in recent months.' [244]

A later report in an Argentine newspaper stated that there had been only six deaths and 47 cases of injury. [245]

Train No. 35, the Up *Kirti Express* from Viragam to Rajkot, was derailed at 01:20 on Wednesday 21st May 1958 near the station at Chamraj, India, on the Western Railway. The engine and three leading carriages were reportedly derailed and the next five carriages overturned. The initial report stated that at least 21 passengers were killed and 44 injured, 18 seriously. A second report stated that nine of the injured had died, raising the number of dead to 30. [246]

An 11-coach South African Railways passenger train carrying 400 black migrant workers from Johannesburg back to their native townships was derailed near New Canada Junction, South Africa, about 9 miles from Johannesburg, on the evening of Thursday 26th June 1958. Initially it was reported that up to 30 people might have lost their lives but later it was stated that there had been eight deaths – six black workers and two white railwaymen, one of whom was the driver and the other possibly the fireman. According to different estimates from 60 to 130 people were injured, at least 35 seriously. Five carriages and probably the locomotive left the track and were wrecked. Police radioed for all available doctors to go to the scene of the accident, on the Langlaate side of New Canada Junction, while there was an appeal for Johannesburg residents to donate blood. [247] [248] [249]

A 12-coach Rede Ferroviária Federal train coming from Fortaleza to Crato in the south of Ceará state was derailed on Saturday morning, 13th December 1958, near Diogo

---

242 https://noziersbrignon1957.e-monsite.com/

243 https://www.geocaching.com/geocache/GC391HP_memoire-aux-victimes-du-paris-nimes-07-sep-1957?guid=b57041c6-e6a7-42cd-b475-ef2dae09da79

244 *The Lowell Sun*, Lowell, Massachusetts, United States, 10th January 1958

245 *El Litoral*, Santa Fe, Argentina, 9th November 1958

246 *The Times*, London, England, 22nd and 24th May 1958

247 *The Lincoln Star*, Lincoln, Nebraska, United States, 27th June 1958

248 *Kingsport News*, Kingsport, Tennessee, United States, 28th June 1958

249 *Provinciale Dretsche en Asser Courant*, Drentsche, the Netherlands, 27th June 1958

Siqueira, Brazil. Five coaches crowded with passengers overturned and fell down an embankment. According to a report the next day in a Brazilian newspaper 12 people were killed outright and about 50 were injured. Upon news of the accident being received at Fortaleza rescuers and ambulances were despatched from there. [250] The north-east region of Brazil was suffering from a severe drought at the time and the train was reportedly carrying some of the refugees who were being taken to other regions. [251] A United Press report stated that the coaches had piled up in a mass of twisted steel and broken wood at the bottom of the embankment and that at least 20 people had died with others possibly still buried in the wreckage. A preliminary investigation blamed the derailment on excessive speed. [252]

A train carrying a good number of holidaymakers travelling from Sarajevo to Zagreb was derailed on the afternoon of Thursday 27th August 1959 near Banja Luka, Yugoslavia. After leaving the rails the engine of the train struck a bridge abutment and the leading two vehicles, a mail van, and a second-class carriage, plunged down an embankment into a river. Fourteen people were reported killed and at least 42 injured but at the time of the report it was feared that there might be further victims still in the wreckage. The cause of the derailment was not stated. [253]

Six of the eight coaches of a crowded passenger train left the track on a curve near Rotifunk and 58 miles from Freetown, Sierra Leone, on Christmas Eve, Thursday 24th December 1959. Thirteen people died and 20 were injured, four seriously. Railwaymen and police tended the injured, aided by doctors from a nearby hospital. [254] [255]

A doctor who lived near the scene of the accident and was called to the scene by one of the train crew later came across an unconscious and severely injured girl with her mother outside the hospital where he worked. He recognised the child, Marie Jones, as a victim he had come across when he had gone to the accident scene. She had a severe neck injury and assuming that she must be fatally hurt, he had moved on to other victims. The child was operated on and survived. 'Morning time saw Marie battered and broken, swathed in bandages…but still very much alive. Her mother, who had lost Marie's sister, uncle and two cousins in the accident was still in a state of shock and could not appreciate, just yet, that Marie was going to live…Marie, in simple trust and with great courage, progressed from day to day. She must have had quite a deal of pain when her dressings were done, but she made few complaints. Eating must have been quite a problem, but she sucked away at oranges and progressed from a liquid diet to the rice chop she loved…She acted much older than her years, perhaps she had lost some of her babyhood as she had passed through this great ordeal – for she was only five.' The child was eventually well enough to be discharged and the writer met Marie again some years later, when she was a student at a mission school. [256]

---

[250] *O Estada de São Paulo*, São Paulo, Brazil, 14th December 1958

[251] *The Arizona Republic*, Phoenix, Arizona, United States, 14th December 1958

[252] *Independent Press-Telegram*, Long Beach, California, United States, 14th December 1958

[253] *The Times*, London, England, 28th August 1959

[254] Ibid, 26th December 1959

[255] *The Winnipeg Tribune*, Winnipeg, Manitoba, Canada, 26th December 1959

[256] *Sierra Leone Remembered* by Esther J. Megell, pgs. 67-8

A disaster with similarities to that at Woodbridge, New Jersey, United States, in 1951 (see above) occurred near Monza, Italy, on Tuesday 5th January 1960. About two kilometres from Monza station on the Sondrio line a tramway running along the Viale Libertà passed under the railway via a narrow underpass. Possibly there was also a level crossing for road traffic at this point, the newspaper report is a little unclear. In 1959 the authorities decided to improve matters by constructing a more spacious underpass, work on which began in October 1959. In connection with this work a provisional bridge was provided for the railway, using steel scaffolding, taking three pairs of tracks, and situated some metres away from the permanent bridge. As at Woodbridge a pair of sharp curves were needed to connect the permanent tracks with those over the temporary bridge. This temporary deviation had been in use almost a month at the time of the disaster, that is, since 7th December. The usual speed of passenger trains over the line was 80 to 90 kilometres an hour, rising to perhaps 100 kilometres an hour on favourable stretches, but over the deviated tracks at the accident site a severe speed restriction of only 10 kilometres an hour was in force. Warning of the restriction was given to train drivers by two signs placed 1,200 and 1,000 metres from the bridge. It was the failure of the train to abide by this speed restriction which led to the disaster. The train involved was *diretta* No. 341 from Sondrio via Lecco and Monza to Milano, consisting of an engine and 11 coaches, carrying about a thousand workers, commuters, and students.

The warning signs for the speed restriction were of course illuminated at night, and the responsibility for lighting and extinguishing the night-time signals rested with the signalmen in a box near the 5 kilometre post, about midway between Monza and Arcore. Signalman Giuseppe Garzaro, 48, who was on duty at the time of the accident and was the only witness to it, related his story to a newspaper reporter. He began, 'I was saved by a miracle. Instead of jumping to the right I leapt to the left.' He had been on his way to extinguish the lights of the warning signs at 08:00. He continued, 'I left the box at 07:40 and cycled on my bike by the path which runs alongside the tracks. I reached the first sign, 1,200 metres from the bridge. It was already daylight, there was a little fog, but the visibility was generally good. I dealt with that sign and continued to the one closer to the bridge. Putting my bicycle down I began to cross the tracks to reach the second sign. I had just heard the sound of the factory hooters so it must have been a couple of minutes after 8. *Diretto* No. 341 from Sondrio, due in Monza at 07:58, must therefore have been a little late. At that moment I heard it coming. It was running very fast, about 90 kilometres an hour, despite the 10 kilometre restriction at the bridge, and I immediately anticipated disaster. I had to jump out of the way of it or the train would have run me down.'

At this point in the interview the signalman covered his eyes and was silent for a while. 'After I jumped,' he continued at length, 'I looked around as the train emerged from the fog. Very shortly it reached the bridge, and the world ended. There was a terrible, endless crash, hissing, the scraping of metal, a great shower of sparks, coaches everywhere, up in the air, tangled up. For a moment there was silence, then there were screams, groans, cries for help.'

He was silent again and covered his face with his hands. After a while he went on, 'I was dazed, paralyzed by fright, by shock. Then I pulled myself together. I ran to the next level crossing where there was a telephone and rang Monza station, told them to call the emergency services, that there had been an accident and there were dozens and dozens of injured. I heard the passengers screaming and moaning louder as I returned to the bridge.

Many workers from nearby factories were already on the scene. I helped some of the injured to safety. Then I went to the engine, which had run off the line and collided with a pole carrying the overhead line equipment, which had demolished its cab. Through the cab window I saw Driver Pierino Vacchini {the driver's Christian name is also given as Pietro on the same page of the same newspaper}. He was sat motionless, with his head bowed, his hands folded on his chest. His body below his waist was fearfully mangled. He must have been killed instantly. Next to him Assistant Driver Andrea Giuliano was lying on the cab floor, injured and moaning. Then doctors, nurses and firemen arrived.' [257]

A passenger on the train, Lucaiano Ferrari, also told his story. 'I was going to work on the regular Sondrio – Monza – Milano commuter train this morning with hundreds of others who ride it every day. Without warning the terrifying crash happened. I was in a compartment of the fifth coach with three young girls. The girls were chatting. The students were reading newspaper accounts of the funeral of Fausto Coppi, Italy's great bicycle racer, who had died on Saturday.

'Suddenly, as the coach inclined on a curve, the girls were hurled into our laps. Then, for what seemed several long seconds, there was a violent crashing and grinding sound, a gigantic scraping of steel on steel. This seemed to continue for several hundred metres. Now the coach was on its side and we were all piled together. The coach continued to grind along.

'The girls screamed. Only one of them had a slight bruise. We crawled out of shattered windows on our stomachs, along the ground to get clear of the wreckage. We had no idea of the state of the crash. Fog was so thick you could scarcely see the coach ahead. Then I saw it. It had apparently also scraped along on its side and then bounced back on its wheels. I could scarcely believe what I saw. A rail stuck out of the coach, pointing skyward. I looked and saw that the rail had pierced and killed three passengers in that coach. One was a Roman Catholic priest. It was just as though a child had stuck a wire through a thin paper toy.

'I couldn't see the third coach, it just seemed to have disappeared. Then I saw it. It was on the tracks. To get there it had to hurtle over five railway tracks and up into the air about six metres. I don't know how it did that. I could now hear screams coming from the coach. There must have been at least 100 persons in that coach. It was packed with labourers, on the way to their jobs.

'I could still not see the last coaches of the train because of thick fog. There must have been 1,000 persons in all aboard the train.

'One girl with her hands under her coat asked me to hand her purse to her. I did. She put out her hands. They were bleeding. She could not move her fingers. "I guess they're broken," she said and turned away and walked towards town.

'I counted 17 dead in the wreckage, but there must have been more bodies trapped in the tangled cars. Many injured were also trapped, and it was difficult for rescuers to find them in the dense fog.' [258]

The rescue operation was described as impressive. All the ambulances in Monza and Milano were mobilised, and together with buses and private cars they took the injured to hospital. Firemen then got to work to release the bodies of the dead from the wreckage,

---

[257] *La Stampa*, Torino, Italy, 6th January 1960

[258] *Adirondack Daily Enterprise*, Saranac Lake, New York, United States, 5th January 1960

using oxy-acetylene torches to cut their way through the twisted metal. In addition to Driver Vacchini, 14 passengers died in the accident, 76 were badly enough hurt to be taken to hospital and another 48 were slightly injured. Forty doctors and consultants took part in the care of the victims. The medical director later paid thanks to the people from Monza, Sesto San Giovanni, Milano, Lissone, Desio, Rho, Pavia, Villasanta, Cinisello, Balzamo, Merate and Brughera, who had donated blood. In all, there had been 104 blood donations and up to 15:00 there had been 38 transfusions of blood and 12 of plasma.

It was immediately clear that the accident was due to the excessive speed of the train. The director general of the Fellovie dello Stato, Engineer Severo Rissone, flew to Monza from Roma the same afternoon together with Undersecretary for Transport Finelli. The director general told reporters, "The train was running nine minutes late. After leaving Usmate station the train ran into a dense fog and its driver may have missed seeing the signals and failed to realise that he was entering a dangerous location until too late. He did apply the brakes in emergency at the last moment but the weight of the train and its high speed made the accident inevitable, with the consequences you have seen."

As well as the testimony of Signalman Garzaro and the degree of destruction caused by the accident, evidence of the train's excessive speed was provided by the tachograph fitted to the locomotive of the electric train, which showed that its speed had been 90 kilometres an hour as it had neared the bridge, falling at the last moment to 75 kilometres an hour. The evidence provided by the tachograph was carefully examined on the afternoon of the disaster by Director General Rossone, who carried out the examination in the undamaged rear carriages of the train, accompanied by the public prosecutor. Dr. Volterra; the deputy prosecutor, Dr. Gatti; Mayor Alfredo Caseraghi; the commander of the carabinieri, Captain De Siena, and other officials. Director General Rissone and Public Prosecutor Volterra also visited the hospital and conducted a brief interview with Assistant Driver Giuliano but he was in a state of shock and his examination was postponed.

Examination of the accident site showed that the temporary trestle had stood up to the shock of the derailment and its supporting beams had remained intact, but the overturning engine had torn up the track.

The line was reopened with single-line working the same evening, though the Monza – Montena – Oggiono secondary line, which paralleled the others, remained out of use. It was expected to be three months before normal working could be fully resumed. Damages, not including compensation to the victims, were estimated at 200 million lira. [259]

Train No. 312 of the New York Central Railroad from St Louis, Missouri, to Cleveland, Ohio, otherwise *The Southwestern Limited*, left the track just after 8 p.m. on the evening of Sunday 10th January 1960 at Wellington, Ohio, United States. The accident occurred at the crossing of the New York Central and New York Chicago and St Louis tracks and the train demolished the signal tower there. Operator R.E. Weiss had a lucky escape; he had been about to deliver orders to the train. Seven of the ten cars of the train were derailed. The wreckage and the wooden Wellington signal tower caught fire.

Initial reports were that four people had died in the accident and 58 had been injured. Those who lost their lives were Fireman Phillip Lehman, Kenneth Hale, of Rosedale,

---

[259] *La Stampa*, Torino, Italy, 6th January 1960

Indiana, and George Rummel, of Indianapolis, Indiana, both Pennsylvania Railroad conductors on their way to a convention in Cleveland, and Audrey Cox, of St Paris, Ohio.

The cause of the accident was clearly excessive speed. The tachograph on the train showed that it had been travelling at 90 miles an hour approaching Wellington and at 74 miles an hour near the point of derailment. The maximum permissible speed at the point of derailment was only 15 miles an hour. Engineer Henry Kirstein, who was injured, told a rescuer, "I have no brakes, no air, no horn, no speedometer and no nothing."

A newspaper reporter interviewed an eyewitness to the accident. '"When I first saw it, it was about half a mile away and I knew it was coming too fast but I thought it would slow down. As soon as I saw the headlight dip straight down, though, I knew it was a wreck." These are the words of a 20-year-old Wellington youth who pulled out two of the dead and helped rescue more than 50 passengers who were trapped in overturned, twisted, tangled masses of metal following the derailment of a crack New York Central train last night.

'Marion Tudor, 136, Wheeling Ave., Wellington, was walking along the tracks with his fiancée, Nancy Bingham, a 17-year-old Clarksville girl, who attends New London High School, when the couple first saw the high-speed train hurtling towards them. Marion and Nancy had been on a "Coke date" and were returning to his house, only a block from the wreck scene.

'When Tudor saw the train end its out-of-control, crazy journey, he asked her to go call an ambulance. He then ran to the engine, where he found the engineer, apparently thrown clear of the wreckage, with blood splattered over his clothing from large, raw, facial lacerations. The engineer asked Tudor to get the fireman out of the engine room. "I tried, but all I could see was his foot sticking out. It wasn't moving at all. I crawled on top of the engine, which was turned on its side, and looked in. He was covered in dirt that had been forced through the window when the engine fell on its side. I couldn't have got to him, and even if I could have it wouldn't have done any good because he was covered in too much dirt," he explained.

'While Tudor was looking for a way to help the fireman, the engineer told him that the train didn't have any speedometer, brakes or horn, Tudor said.

'"After I left the engine I went to the next car back, where there was a conductor trying to force the door open. We tried together, him pushing and me pulling, but couldn't get it. Finally he handed me a sledge-hammer and I was able to force it open enough to allow people to get out. The first person that came out was a Cleveland girl. She ran straight toward Nancy. Nancy told me she was badly cut," he said, lighting a cigarette and looking at Nancy. After the girl came out, Tudor entered the car and began carrying injured passengers out. He estimated that he helped 15 injured passengers out of the car.

'At the second car, the first door he tried was covered with broken power lines. The second door of the car he was able to force open with the sledge-hammer. After escorting about 20 passengers from this car, he returned for one last look and found a soldier trapped under a seat with one leg amputated just below the knee. Another man in the car, whom he came to know only as "Fred," and Tudor tried to free the man unsuccessfully. Finally they broke open a suitcase and used ladies' lingerie as a tourniquet to stop the flow of blood. Fred stayed with the trapped man while Tudor went to the next car. There he found about ten people in the car all helping one another to get out the exit. After the passengers had left, he entered the car and found two babies with their clothing on fire and one man attempting to smother the flames with his coat jacket. Tudor went to the back of the car and helped the unidentified man save the babies' lives, then stayed in the car to fight the fire after the man had carried the babies out of the car.

'The last car he went to was the mail car, also aflame. There, on the floor, he found the bodies of two men he assumed were dead. Both bodies were sprawled on the floor motionless. He dragged them to the door and then placed them on the ground. Before being driven from the car by the intense heat of the fire, he threw out many bags of mail.

'Tudor is employed by the Thow Shovel Co. of Lorain as a shipping clerk. His father, Henry, also a hero in the rescue work, is also an employee of the Thow Co. Lorain.' [260]

In mid-February the railroad company announced that Engineer Kirstein had been dismissed. It was stated that no mechanical defects had been found which would have prevented the engineer from reducing speed and that the wreck had been caused by the "engineer's failure to control the speed of the train in accordance with signal indications." [261]

The Interstate Commerce Commission report on the accident appeared in May. It stated that the derailment had been due to excessive speed, which had been 59 miles an hour where the maximum permitted was 15. The speedometer on the leading diesel engine had been inoperative and it had been foggy, both of which factors might have affected the judgement of the engineer, but it dismissed his claim that he had had no brakes. In the words of the report the speed tape device indicated "that no action was taken to reduce the speed until the train was closely approaching the point of the accident. And then, because of the excessive speed, there was insufficient distance to reduce the speed of the train...It is possible that the fog in this vicinity and the inoperative speed indicator materially affected the engineer's speed of the train...However, no defective condition was found in the brake equipment of the train and it is evident that if braking had been initiated in accordance with the requirements of the signal indications, this accident would have been averted." [262]

A Sunday excursion train carrying 250 employees of the Braden Copper Co. and their families from Sewell, Chile, site of the El Teniente mine, to Rancagua, 20 miles to the west, was derailed on a mountain curve and tumbled down a 30ft bank. The train had only covered 2½ miles when seven of its coaches fell into a ravine at 07:25 on Sunday 7th February 1960. According to one report 33 people died and 107 were injured. Other reports put the number of fatalities at 25 or 31. 'So far as is known, none of the casualties was American.' [263] [264] The accident was said to have been the first fatal mishap in 30 years on the narrow-gauge line. According to one report the train was travelling at excessive speed when one coach left the rails and pulled six more after it. [265] An estimated 30,000 people were said to have attended the funeral of the victims. [266]

---

[260] *The Chronicle Telegram*, Elyria, Ohio, United States, 11th January 1960
[261] *Reflector Herald*, Norwalk, Ohio, United States, 20th February 1960
[262] *The Coshocton Tribune*, Coshocton, Ohio, United States, 12th May 1960
[263] *The Capital Times*, Madison, Wisconsin, United States, 8th February 1960
[264] *Times-News*, Twin Falls, Idaho, United States, 8th February 1960
[265] *Pasadena Independent*, Pasadena, California, United States, 8th February 1960
[266] http://hitosdechile.blogspot.com/2013/04/sewell-la-tragedia-ferroviaria-de-agua.html

CHAPTER 4

REAR COLLISIONS

The risks associated with one train running into the back of another had been low in the formative years of the railways when trains were few and train speeds not very high, but as the years progressed train speeds and traffic intensity often grew faster than advances in brake power, rolling stock construction and signalling and as a result by the early twentieth century serious rear collisions had become common. This chapter is therefore a long one, even though a great number of less serious accidents have not been included.

On Thursday 24th January 1946 workmen's train P 2631 from Kaltennordheim was following 20 minutes behind workmen's train P 2502 from Vacha, both being bound for Bad Salzungen. Train P 2502 was stopped longer than usual in the station at Merkers, Germany, while its driver carried out minor repairs to its engine. Distracted by passengers the *fahrdienstleiter* (signalman) failed to notice that the train was still standing in the unlit station area despite his having cleared his signals for it to depart and as a result of a misunderstood telephone call he cleared it back and accepted train P 2631. Five people died and 47 were injured in the collision which followed. The *fahrdienstleiter* was sentenced to death by firing squad by a Soviet court. [267]

A serious rear collision at Eisleben in Germany on Thursday 18th April 1946 (some sources say 16th April) occurred in the abnormal conditions prevailing in the early post-war period. Difficult driving conditions in the form of a thick fog also played a part. The passenger train involved in the accident was seriously overcrowded with returning soldiers and city dwellers who had gone into the country in search of food in this time of great scarcity. Up until the end of the war the Halle – Eisleben line on which the accident occurred had been double track but the Soviet authorities had requisitioned one track for war reparations. The other train involved in the accident was a work train consisting of engine, No. 44081 and a G waggon carrying a Soviet escort.

The singling of the railway predictably led to operating problems. On the morning of the accident the *fahrdienstleiter* at Helfta had four trains to juggle with. Westbound passenger train No. 592 had arrived at Helfta from Halle and there met an eastbound freight train. The work train was approaching Helfta from Röblingen am See behind the

---

[267] *Deutsche Eisenbahn-Katastrophen* by Hans Joachim Ritzau, Jürgen Hörstel and Thomas Wolski, pg. 212 (in German)

passenger train while another eastbound freight train was ready to follow the first one from the Eisleben direction. In order to save a path over the single line the *fahrdienstleiter* wished to couple the work train to the rear of the passenger train. This did not happen because the Soviet officer in charge of the work train objected to the proposal as he wished it to make several stops in section to supervise the work of lifting the other running line, which was still in progress. Instead, in defiance of the regulations, the work train had followed the passenger train through the section, initially keeping close behind it, but because of the stops in the section the gap between the two trains soon increased considerably.

When the passenger train reached Eisleben, *Einfahrsignal* A (the home signal) was at *Danger* because the train had arrived on what would have been, before the singling of the railway, the wrong line, and a written order was required before it could proceed. Before it had restarted the work train arrived and in the fog its driver failed to see the passenger train in time. A violent collision ensued at 10:11. The partly wooden carriages of the passenger train suffered badly. The rear carriage was demolished and the penultimate one telescoped that in front. The front coaches largely escaped, the passengers in them suffering only bruises from falling luggage. On the other hand, the occupants of the rear coaches fared much worse; 17 of them were killed outright and six more died later (another source states that there were 24 dead rather than 23). Forty others were injured.

The first help for the injured came from the passengers in the front of the train, who were advised by a Soviet officer of the severity of the accident and called upon to assist. It was some hours before outside help arrived.

A Soviet military court found that the driver of the work train was mainly to blame for the disaster and the conductor of the passenger train, an inspector at Röblingen am See and three other railwaymen were also found guilty to a lesser degree. They were sentenced to terms of up to nine years' imprisonment in one of the camps (gulags) which had been established in the Soviet occupation zone. Herr Oelkers, the Deutsche Reichsbahn President in Halle, bravely commented that in the difficult circumstances of the accident it did not consider that his employees had been much at fault. He was reprimanded by the public prosecutor for daring to suggest that the Soviet military was partly to blame, and the sentences on the railwaymen remained unaltered. News of the accident was hushed up by the Soviet occupying forces and later by the East German government, but eventually emerged. [268] [269] [270] [271]

An early report in an Illinois, United States, evening newspaper on Thursday 25th April 1946 spoke of a collision between two Chicago Burlington and Quincy Railroad passenger trains at Naperville. It said that about 75 injured passengers had been brought to the St Charles hospital in nearby Aurora and the railroad reported that an "undetermined number" of persons were killed or injured. [272] The full magnitude of the

---

268 *Eisenbahnunfälle bei der Deutschen Bahn: Ursachen – Hintergründe – Konsequenzen* by Erich Preuβ pg. 161 (in German)

269 https://harz-saale.de/das-eisenbahnungluck-von-eisleben-vom-april-1946/

270 https://www-mz-de.translate.goog/lokal/eisleben/schweres-zugungluck-vor-70-jahren-23-menschen-sterben-bei-crash-1145244?_x_tr_sl=de&_x_tr_tl=en&_x_tr_hl=en&_x_tr_pto=sc

271 https://www.gerdboehmer-berlinereisenbahnarchiv.de/Statistiken/BBU-DR-DB.html

272 *Centralia Sentinel*, Centralia, Illinois, United States, 25th April 1946

accident would soon become evident. By the following day the newspapers were reporting that there were 44 people dead and 125 injured. The final figures according to the Interstate Commerce Commission (ICC) were 45 dead and 69 injured.

The trains involved in the accident were No. 11, *The Advance Flyer*, from Chicago to Lincoln and Omaha, Nebraska, and No. 39, *The Exposition Flyer*, from Chicago to San Francisco, California.

Westbound first-class passenger train No. 11 consisted of diesel electric units 9920A and 9920B, a baggage car, a storage mail car, two baggage cars, a mail car, a refrigerator express car, two baggage cars, two coaches, a dining car, a parlor-lounge car, and a coach, all of steel construction, in charge of Conductor Jack Aue with an unnamed engineer and fireman. It had started from Chicago Union station, 28½, miles from Naperville, on track 2, at 12.35 p.m., passed Downers Grove, 7¼ miles east of Naperville, at 12.57 p.m., one minute late, and stopped at Naperville for inspection at 1.03 p.m. because an object had been seen flying from beneath one of the cars. About two minutes later it was run into by train No. 39.

Westbound first-class passenger train No. 39 consisted of diesel electric units 9910A and 9910B, three coaches, a dining car, two tourist sleeping cars and three Pullman sleeping cars, all of steel construction, in charge of Conductor George Hill with Engineer William Blaine and Fireman G.H. Grayton. It had also left Chicago Union station at 12.35 p.m., on track 1, crossed to track 2 at Kedzie Avenue, 23½ miles east of Naperville, passed Downers Grove at 1 p.m., two minutes late, passed signal No. 227.1, displaying *Approach next signal prepared to stop*, passed signal 228.1 displaying *Stop, then proceed*, passed Flagman James Tangney of train No. 11, and was still travelling at an estimated speed of at least 45 miles an hour when it ran into the rear of train No. 11.

As a result of the collision the ninth car of the forward train was derailed, both ends of its tenth car were buckled, the eleventh car was turned around, bent into a U-shape and three quarters of it were demolished. The twelfth car, the parlor-lounge car, was derailed across track 1 and both its trucks were badly damaged. As might be expected the thirteenth and rearmost car suffered the most. It was penetrated for about three-quarters of its length by unit 9910A. In train No. 39 the two diesel electric units were derailed and damaged, its first to fifth cars were wholly or partly derailed and some telescoping occurred between the second and third cars. [273]

After the collision, 'At first there was complete confusion. Huge, shining passenger coaches were strewn across tangled tracks, some in tangled wreckage. The cries of the dying came mostly from the rear coach of the Advance Flyer, where passengers were trapped. Others groped in bewilderment for escape from the mass of steel wreckage...Through the night, hours after the accident, search continued for additional bodies. This was discontinued at dawn, however, when searchers were convinced all casualties were accounted for.' [274]

A newspaper reporter managed to find Mrs Ann Miller, of Naperville, who had witnessed the accident. She told him, "I saw the railroad wreck yesterday. It happened as I was just starting to wash the lunch dishes. I had just finished reading a letter from the National Safety Council when I started my work. No sooner had I put my hands into the dishwasher when I saw the Advance Flyer stop. I thought it peculiar because the train

[273] Interstate Commerce Commission report
[274] *The Evening Star*, Washington D.C., United States, 26th April 1946

usually goes right through Naperville. I went to the back door to wave at the passengers on the train. Sometimes there are soldiers on the trains, and I like to wave to them. My house is only 75 feet from the tracks, so I can see the passengers well.

"Then I heard the other train coming. It was on a curve, coming lickety split. The brakes screeched terribly, and then, right before my eyes it crashed into the other train. Oh, it was terrible. It made a horrible noise. I rushed to the telephone and called my nephew to tell him about the accident. Then I called the telephone operator and told her to call the police and all the doctors. I knew many people were killed and injured. I didn't know how many, but I knew that it was a major disaster.

"The engineer of the Exposition Flyer was brought into my kitchen. I gave him coffee and a doctor bandaged his face, which was black. He didn't say anything. He just drank coffee. I did what I could to help. Every available pair of hands was needed. Throughout the afternoon I gave coffee to survivors of the wreck. When the crisis had passed, I returned to my dishes.

"I looked out of the window again. The lilacs blooming in my yard framed the horrible wreckage. The blue cross letter fluttered in the evening breeze. I hope nothing like this ever happens in Naperville again." [275]

A couple of passengers spoke to reporters. 'Mrs Irene W. Cook, 20, Cohoes, N.Y., was in the rear car, "I was facing the rear of the train," she said, "then everything came down on me. My head feels funny." She was taken to St Charles hospital at Aurora.

'Sol Greenbaum, 27, St Louis, Mo., also was in the last car and suffered an ankle injury. "When the crash came I was thrown to the top of the car, turned a somersault and came down," he said. "A pile of people fell on me. I kicked out a window and climbed out. I think a woman behind me was killed."

'William McBride, 26, Chicago Pullman waiter, was in the diner one car forward. "The whole kitchen was messed up like scrambled eggs," he said. "If it hadn't been for the steel car we would all have been goners. When we crawled out and looked around we couldn't even see anything that looked like a dining car."

'The Rev F.R. Stenger, Catholic priest, who administered last rites to the dying alongside the wreck, made his way into the last car and found "decapitated bodies, dismembered, mutilated, and crushed beyond human shape."

'A passenger saw a soldier carrying a baby in his arms. When the crash came, the baby was hurled from its father's grasp into the arms of a sailor, who carried it from the train, unharmed. The soldier father, however, lost a leg.

'Master Sgt. John A. Sramovsky, 29, Ft. Robinson, Neb., was eating lunch in the diner with his wife, Sophie, 29, Plymouth, Pa. Both were injured, and Mrs Sramovsky died. Sramovsky, suffering painful leg injuries, refused morphine as he was carried from the train. "Other people need it worse than I do," he told a rescuer.' [276]

Engineer Blaine suffered a fractured skull in the accident and was admitted to hospital. The following day State Attorney Lee Daniels announced that a manslaughter warrant had been issued against Blaine but he would not be arranged for at least two weeks or when he was released from hospital, where he was placed under guard. [277]

---

[275] *The Daily Register*, Hattiesburg, Illinois, United States, 26th April 1946
[276] *The Daily Independent*, Murphysboro, Illinois, United States, 26th April 1946
[277] *The Evening Star*, Washington D.C., United States, 26th April 1946

Miss Leona Sayler, 28, a government employee, of Brunswick, Nebraska, died in St Charles hospital three days after the accident to become the 45th fatality. [278]

Describing the event which led to the stoppage of train No. 11 the ICC stated that Brakeman Sherman Grant of that train had been making a running inspection from the right rear vestibule of the tenth car when he had seen an unidentified object fly up from beneath the train, which had then been approaching Naperville at 80 miles an hour. He had informed Conductor Aue and the latter had signalled to his engineer to stop the train. [279]

Flagman Tangney had been stationed at the front of the twelfth car, his flagging equipment had been on the platform at the rear of the rear car. He said that he had made his way to the rear as soon as he had felt the brakes being applied. According to newspaper reports he had remarked to passengers, "I'm going to stop that train behind us." [280] He had not been in the rear car because with the equipment used it was impossible to inspect the train from that point. He had set off back as soon as the train had stopped and had covered 300 feet when train No. 39 passed him as he waved a red flag at it. He had not thrown down any 10-minute fusees because from his experience they did not remain alight when thrown down from trains travelling at over 40 miles an hour. Tests made after the accident confirmed this. He had not turned on the oscillating red light at the rear of his train. With the line being on a curve and the sun shining it was debatable if the engineer would have seen the light in time anyway.

Train No. 39 had also been travelling at 80 miles an hour as it approached Naperville. Engineer Blaine and Fireman Grayton had been in the control compartment at the front of the leading unit. The engineer was too badly injured to make any statement to the ICC. In a statement to Attorney Daniels he said that he had been travelling at 85 miles an hour when he had seen the yellow signal and had applied the brakes at once but "we were going too fast," and they had passed the red signal still "trying to bring her down." He had remained at the controls and thought that if he had been pulling more cars he could have stopped. [281] Immediately before the collision Fireman Grayton apparently shouted to his engineer, "Bill, you're going to hit!" The fireman had then jumped from the unit. He was killed, the only railwayman on duty to lose his life in the accident, though one railwayman off duty and four dining car employees were among the dead. Apart from Engineer Blaine the injured comprised 48 passengers, 19 dining car employees and a porter.

None of the other members of the train crew were aware of any problem before the collision. Some had noticed a light brake application near Naperville. The brakes of the train were tested after the accident and found to be working normally.

Unable to question Engineer Blaine, the ICC could not determine why he had not applied the brakes. The weather had been clear, the signals had been visible for 5,000 feet, the brakes had been working properly and so on. The wording of the rules "approach next signal prepared to stop," was criticised as it did not stipulate a maximum permitted speed. Also criticised was the close headway between the two trains; after the accident their departure times were amended to 12.30 p.m. and 12.45 p.m. What was really required though, the ICC stressed, was an adequate automatic train stop or train control

278 *The Montana Standard*, Butte, Montana, United States, 29th April 1946
279 *The Galva News*, Galva, Illinois, United States, 2nd May 1946
280 *The Evening Star*, Washington D.C., United States, 26th April 1946
281 *The Daily Independent*, Murphysboro, Illinois, United States, 26th April 1946

system. In this context its report stated:

There is now pending before the Commission docket No. 29543, which is an investigation instituted May 20, 1946, on its own motion, to determine whether it is necessary, in the public interest, to require any common carrier by railroad to install block signal system, interlocking, automatic train stop, train control and/or cab signal devices, and/or other similar appliances, methods and systems intended to promote the safety of railroad operation, upon the whole or any part of the railroad on which any train is operated at a speed of 50 miles or more miles per hour. Hearing therein will be held in the near future.

In addition the ICC noted that the dining car, eleventh in train No. 11, had been of lightweight construction, less able to resist buffering forces than the heavier cars on either side of it. A recommendation made after a 1938 head-on collision had suggested that lightweight cars should not be marshalled in trains between heavier ones, but:

Notwithstanding this recommendation, and also the subsequent action of the Association of American Railroads, establishing a minimum requirement of resistance to end buffing forces for cars in unrestricted service, cars which do not conform to this standard are continued in operation in association with cars of substantially heavier construction and which meet this minimum requirement. The number of casualties which resulted in this case may have been attributable in part to this condition. Only the three lightweight cars were equipped with tightlock couplers, and had all cars conformed to the standard for end buffing resistance, it is probable that the disastrous consequences of this accident would have been greatly reduced.

The ICC recommended that the Burlington company cease mixing lightweight cars with heavier cars in its trains. It made no recommendations regarding train control as the matter was being dealt with in its aforementioned docket. [282]

Towards the end of June an inquest jury confirmed that Engineer Blaine was guilty of manslaughter but on 4th October a grand jury refused to indict him on the grounds of insufficient evidence. The grand jury recommended the installation of radar and two-way radios on all high-speed trains. [283]

Described as an 'old timer', Engineer Blaine was 68 years of age. No mention was made in the ICC report of his having received any medical examination. It is tempting, in view of his age, to wonder if his failure to reduce speed had resulted from some medical condition such as a transient ischaemic attack (minor stroke).

At about 23:00 on Sunday 28th July 1946 the Allahabad passenger train from Gorakhpur ran into the rear of the Katihar passenger train, which was standing in Bhatni Junction station, India, 40 miles from Gorakhpur on the Oudh and Tirhut State Railway. The guards van, an upper-class carriage and a staff carriage were described as having been 'reduced to matchwood.'

---

[282] Interstate Commerce Commission report

[283] *Centralia Sentinel*, Centralia, Illinois, United States, 4th October 1946

A relief train was sent to the scene from Gorakhpur and returned there the following morning with 23 injured passengers who were admitted to local hospitals. The same number of bodies were recovered from the wreckage. The total number of injured was unknown. [284]

Seven people died and 14 were injured when a passenger train collided with the rear of a freight train in the station at Kokura, Japan, on Monday 12th August 1946. [285]

Passenger train P 4201 collided with *Umsiedlerzug* (a resettler train – presumably Germans displaced from that part of their country allocated to Poland at the end of the war) 5019 between Reichenbach and Linde block post, Germany, on the Leipzig – Hof line, on the night of Wednesday 18th September 1946. Train 5019 had been at a stand in the section for seven minutes due to lack of steam. The Linde pointsman was aware that train 5019 had not reached him and did not clear it back but when the passenger train was offered to Reichenbach by Neumark the Reichenbach *zugmelder* (dispatcher), assuming that train 5019 must have cleared Linde, told his *fahrdienstleiter* (signalman) to let train P 4201 proceed, without consulting the Linde pointsman. Six people died and 49 were injured in the ensuing collision. [286]

A court subsequently sentenced the *fahrdienstleiter* to ten years' imprisonment and the *zugmelder* to eight years. The Linde pointsman was acquitted. The Soviet military administration also ruled that the driver of train P 4201 should be reduced to fireman for a year on the grounds that he should have observed the train in front of him sooner. It seems likely that the tail lamp of the resettler train was also unlit because of a lack of oil. [287]

A Strasbourg – Lyon Michelin railcar which left Héricourt, France, at 10:30 on Sunday 22nd September 1946 had reached full speed and was rounding a curve when its driver saw the rear waggons of a freight train in front of him and a collision ensued. The freight train had come to a stand in the section due to brake problems. It had been protected in the rear by detonators put down 900 metres behind it but the railcar had been moving at 90 kilometres an hour and a brake application by Motorman Gunther for some reason failed to prevent a severe collision. The whole of the front part of the leading car of the Michelin was smashed and it was here that the casualties occurred. The passengers in the rear half of the front car and in the second car mostly escaped injury and after the accident went to the aid of their less fortunate fellow passengers, assisted by residents from the neighbouring village of Bethoncourt. The freight train had been loaded with the massive trunks of fir trees and jacks had to be used to reach the victims trapped beneath them. Seven people died and 13 were seriously injured. Motorman Gunther not only seems to have survived the collision but apparently was not hurt; possibly he jumped from his cab.

The Montbéliard parquet (examining magistrate) strove to determine the circumstances of the accident. Motorman Gunther was arrested while the young Société nationale des chemins de fer français apprentice at Héricourt, 19-year-old Marcel Chalot,

[284] *The Bombay Chronicle*, Bombay, India, 30th July 1946
[285] *Malaya Tribune*, Singapore, 12th August 1946
[286] https://www.sachsenschiene.net/bahn/sys/unf.htm?134
[287] *Eisenbahnunfälle bei der Deutschen Bahn* by Erich Preuß, pgs. 152-3 (in German)

who had allowed the railcar to leave the station, 'and whose responsibility appears to be great,' was also arrested. [288] [289] Motorman Imbert, driving the second car of the train, claimed not to have heard the alarm signal given by Motorman Gunther. [290] M. Joseph Zimmerman was the conductor of the train; he was among the injured.

How the railcar was permitted to leave Héricourt with the freight train still in the section and why, when he had exploded detonators 900 metres in the rear of the freight train, the motorman of the railcar had still been unable to prevent a serious collision, were, of course, the key points which an investigation had to clear up.

A probable rear collision occurred at Łódź Kaliska station, Poland, at 05:40 on Saturday 28th September 1946. Passenger train No. 524 was wrongly stopped by a station official (signalman?) and express passenger train No. 502 from Wroclaw collided with it. A van (?) and two coaches of the first-named train were destroyed and the steam locomotive pulling the express was derailed and presumably damaged. According to a newspaper 17 people were killed more or less outright, six more died in hospital and 150 were injured. The Ministry of Communications stated that 14 people had died at the scene, seven had died later and 42 were seriously injured. It has been alleged that the real death toll was much higher but the true figures were censored. [291]

Another probable rear collision occurred in Poland at 00:25 on Friday 4th October 1946, in the station at Szczecin Turzyn, Poland. Stationary passenger train No. 4291 was run into by passenger train No. 171 and three carriages were wrecked. Seven people died and 15 were injured. The signalman at Wzgórze Hetmańskie, and the driver and assistant driver of train No. 171 were arrested and found to have been responsible for the accident. [292]

Passenger train No. 122 from Bar-le-Duc to Châlons-sur-Marne was stopped in the station at Revigny, France, at 06:30 on the morning of Tuesday 12th November 1946 when it was run into from behind by freight train No. 1138. All six carriages of the passenger train were wrecked and the death toll was heavy. An official announcement on the evening of the disaster stated that there were 31 dead and another 30 injured; other reports stated that there were 32 dead and 55 hurt. Almost all of the victims came from Revigny or its neighbourhood and about half were schoolchildren aged under 17, on their way to schools at Vitry-le-Francois and Châlons-sur-Marne. The driver, fireman and conductor of the freight train were among those who lost their lives. Several of the victims had been stood on the station platform at the moment of the collision. The driver and fireman of the passenger train were unhurt.

A rescue operation was quickly organised and the injured were taken to hospital in Bar-le-Duc. About a dozen of them were in a serious condition.

There had been a thick fog at the time of the accident but according to an official

[288] *Le France-tireur*, Paris, France, 24th September 1946

[289] *Ce Soir*, Paris, France, 24th September 1946

[290] *L'Aurore*, Paris, France, 24th September 1946

[291] Wikipedia – Train disaster at Łódź Kaliska station

[292] Wikipedia – Railway disaster at the Szczecin Turzyn station

statement the colourlight caution and stop signals which the freight train had overrun were fitted with powerful electric beams which should have penetrated the fog (but see St Johns, Lewisham, 1957, below). The freight train also exploded some detonators, probably co-acting ones at the stop signal, and the driver of the passenger train attempted to move away at the last moment. The freight train was said to have still been moving at about 55 kilometres an hour when it struck the passenger train. Also significant was probably the fact that the locomotive of the freight train was a new American one, which was not fitted with either a speedometer or the automatic warning system fitted to French locomotives. There were reports that the boiler of the engine gave way after the collision, spraying the victims with steam. [293] [294]

Georges Prevost, 18, had gone to Revigny station with his brother, Pierre, who was to take the train to school at Châlons-sur-Marne after the Armistice weekend holiday. He said, "My brother had just boarded the last coach of the train when I heard the noise of the second train and the detonators exploding. I called to him to jump, but it was too late. The big locomotive of the freight train ground right over him." Georges was also a victim of the accident as flying wreckage amputated his left leg just above the knee. [295]

On the evening of Wednesday 11th December 1946 a goods train came to a stand between Borgsdorf and Lehnitz, Germany, due to lack of steam. After a while sufficient steam was raised for the train to begin moving forward again at 10 kilometres an hour but as it was doing so it was run into from behind by a following S-Bahn passenger train. The motorman of the railcar and eight passengers were killed, its conductor and 14 passengers were seriously hurt and 19 passengers were slightly hurt. Although darkness had fallen the tail lamp at the rear of the freight train was unlit.

Following the wartime reduction of the Berlin – Stralsund line to single track the *Streckenblock* system of operation which ensued the automatic spacing of trains was taken out of use. Furthermore, the line was used by both long-distance and S-Bahn suburban trains. Nor was there any telegraph apparatus at Borgsdorf, and the *blockwärter* (pointsman) there had to communicate with the *fahrdienstleitern* (signalmen) at Lehnitz and Birkenwerder by telephone. When the S-Bahn train was offered to the pointsman at Borgsdorf from Birkenwerder he forgot that Lehnitz had not cleared back the goods train and set his signals at *Fahrt (Clear)* for the railcar. He was sentenced to 10 years' forced labour for his error. [296]

On Thursday 2nd January 1947 there was a rear collision on the London and North Eastern Railway (LNER) main line from London Liverpool Street to Colchester and Ipswich at Gidea Park, England. Darkness and a dense fog (doubtless in fact a smog) were features of the accident. The trains involved were the 10.28 p.m. suburban passenger service from Liverpool Street to Southend Victoria and the 10.25 p.m. mail and passenger train from Liverpool Street to Peterborough, which had started behind the Southend train.

The Southend train consisted of 4-6-0 tender steam engine No. 1565 and eight non-

---

[293] *The Times*, London, England, 13th November 1946
[294] *Journal de Genève*, Geneva, Switzerland, 13th November 1946
[295] *The Montana Standard*, Butte, Montana, United States, 13th November 1946
[296] *Eisenbahnunfälle bei der Deutschen Bahn* by Erich Preuß, pgs. 153-4 (in German)

corridor bogie carriages with wooden bodies on steel underframes. It was in charge of Guard A.E. Bailes with Driver G.W. Morris and Fireman S.A. Wood. It started from Liverpool Street punctually but cautious running due to the fog and temporary speed restrictions near Stratford meant that it lost 16 minutes on the way to Gidea Park. It was being caught up by the Peterborough train and would have been switched from the Down Through to the Down Local line at the country end of Gidea Park station; it could not have been so diverted before because of other trains on the local line. It had stopped at Gidea Park station and was just pulling away from the platform when it was run into by the Peterborough train.

The Peterborough train consisted of 4-6-0 tender steam engine No. 1602, four bogie carriages, two bogie mail vans, two bogie carriages, and a three bogie twin composite carriage, all with wooden bodies on steel undercarriages, in charge of Guard G. Whitmore with Driver Harold Baybut and Fireman Phillip Cole. Dislocation due to the fog meant that it left Liverpool Street 25 minutes late at 10.50 p.m. but it then had a more or less clear run to Gidea Park and did not lose more than three minutes on its booked time.

In the collision the three rear coaches of the Southend train were demolished and the trailing end of the fifth coach was smashed. The engine of the Peterborough train was damaged, the leading mail van was demolished, the second mail van was damaged and the three leading coaches were damaged to some degree.[297]

Five passengers were killed outright in the accident and two more died later of their injuries. Forty-five were badly enough hurt to need to stay in hospital, others were treated and sent home. Six of the passengers who lost their lives were Mrs and Mrs Percy Dimond of Wickford, on their way home after seeing the Ice Revue at the Stoll Theatre, Kingsway; Mr Duncan Clark, of Brentwood, another theatre goer; Mr William King, of Brentwood, a Post Office worker on his way home from work; Mr Reginald Neale, of Brentwood and Mabel Payton, of Harold Wood, who died in hospital. The name of the second passenger who died in hospital was not given. Mr Dimond was an LNER platelayer who had taken a day off work to see the show. His 12-year-old son, Dennis, suffered two broken legs and was said to have been critically hurt; he survived.[298]

The Gidea Park station staff carried out the initial first-aid work and summoned ambulances, doctors firemen, and the police and warned local hospitals. The intensity of the continuing fog meant that some of the ambulances had to be piloted by walkers carrying torches or white handkerchiefs but the first casualty was received by the Victoria Hospital, Romford, before midnight and the last was removed from the wreckage at 1.40 a.m. (According to a statement issued by the LNER the accident had occurred at 11.23 p.m. but the Ministry of Transport report states that it happened at 11.17 p.m.) All four lines were blocked by the accident but two were cleared within a couple of hours and the other two by noon the next day.

There were four tracks through Gidea Park, which from north to south were the Down Local, Up Local, Down Through and Up Through. Between Liverpool Street and Romford, a mile on the London side of Gidea Park, signalling was by Sykes Lock and Block with semaphores while on the country side of Gidea Park and between there and Shenfield there was full track circuiting with automatic colourlight signalling. Between Romford and Gidea Park the lines were also fully track circuited but the semaphore

---

[297] Ministry of Transport report

[298] *The Times*, London, England, 3rd and 4th January 1947

signalling was retained; instead of block working between the two boxes train describers were in use. In the absence of block working between Romford and Gidea Park the normal working during fog, where in the absence of a fog signalman the Romford signalman would not be permitted by the signalling regulations to accept a Down train from Crowlands, the box in the rear, until it had been accepted by Gidea Park box, had to be modified. A special local regulation therefore existed, which read in part:

In foggy weather or during falling snow, until the fogmen arrive at their ports at the Down Distant and Outer Home signals, down trains may be accepted when the line is clear to the starting signal.

This provided an overrun of 1,150 yards beyond the outer home signal.

Fogsignalmen had been called out by the Crowlands signalman at about 8.50 p.m. and there was a man at the Crowlands Down distants from about 9.45 p.m. The man at the Crowlands Down Starting signals, below which were the Romford Down Distants, was not home when called and did not report at Romford box until about 10.20 p.m. The fogmen at the Romford Outer Homes and onwards to Gidea Park had not arrived before the accident.

Driver Morris of the Southend train told the accident inquiry that he had just been able to pick out Crowlands signal box and the Romford Distant signal. Both were clear and seeing no green light on the ground he had assumed the fogsignalman had not reached his post yet. Neither Fireman Wood nor Guard Bailes could make out the signal but the guard saw what he assumed was the handlamp of the fogsignalman making his way towards it. At Romford Driver Morris saw the Outer Home, Fireman Wood saw the Inner Home, Porter A.W. Day saw that the Starter was clear before giving Driver Morris the right-away from the platform, and Driver Morris got a fleeting view of the arm of the Advanced Starter and the Gidea Park Distant below it from the light from the engine firebox. The fog was a little less dense at Gidea Park and both driver and fireman saw the Outer Home and the banner repeater of the Inner Home at short range. At Gidea Park Guard Bailes had a narrow escape; he was just about to re-enter his van as the train began to pull away from the platform when the Peterborough train appeared out of the fog, and he remained on the platform.

Signalman H. Looker at Romford stated that he had cleared back the Southend train when it left his station and had almost at once accepted the Peterborough train from Crowlands, which he could not have done unless his Down Distant, Outer, and Inner Homes had all been replaced to Danger, because of the block controls. Crowlands shortly gave the *Train entering Section* signal for the Peterborough train but he had not cleared any of his signals at that point as the Southend train was on the controlling track circuits. When the Peterborough train occupied the berth track circuit in the rear of his Outer Home signal he had cleared just that signal in accordance with Rule No. 39 (a). He had then seen that the train was not stopping at the Inner Home and had shouted and waved a lamp at it to no effect. When he had seen that the train had also run past his starter he had got on the telephone to his colleague at Gidea Park. He admitted at the inquiry that sending the *Train Running Away* signal on the block bell would have been quicker. He had not had time to turn his hand lamp to red but thought that he would have been able to put a detonator on the line under the train, if not the engine, had the box been provided with lever-worked detonator placers. He was positive that he had not cleared any of his Down signals other than the Outer Home.

Signalman S. Pilgrim at Gidea Park said that he had replaced his Down Through

Outer Home signal as soon as he had seen from his illuminated diagram that the Southend train had reached the station platform. His Inner Home and Starting signals had been clear for that train. His box was nearly 400 yards on the country side of the station platforms and all that he could do when he had heard from Romford that the Peterborough train was running through signals was to try and ring the station staff to get the Southend train moving, but he could get no reply.

Driver Baybut of the Peterborough train stated that the last Down signal that he had seen had been the Crowlands Down Distant beneath the Chadwell Heath Yard Down Starter. Neither he nor Fireman Cole had heard a detonator explode at the Crowlands starter, below which were the Romford splitting distants, nor was one heard by Guard Whitmore in his van, Inspector A.J. Ferguson, riding in the front of the Peterborough train, nor Signalman S.A. Witherington in Crowlands box. Driver Baybut maintained that he had seen a green light near the foot of the Crowlands starter, which he had assumed was that of the fogsignalman, and that the Romford Distant must therefore be off and that he had a clear run through to Gidea Park, and he said that he had remarked to this effect to his fireman. The fogman's post was, however, not in that position, but on the far side of the Down Local line. At Romford Fireman Cole had seen the Outer Home at *Clear* at close range. Driver Baybut stated that looking out for the Romford Inner Home he had seen another green light at low level, reinforcing his belief that the Inner Home was clear. No explanation for this light could be found, there was certainly no fogsignalman there. Driver Baybut admitted that he had not been able to see Romford box, he had not heard Signalman Looker shouting at him, and he had seen no signals between Romford and Gidea Park. He put his speed at 25 to 30 miles an hour but Porter Day thought that the Peterborough train had been travelling at "the usual speed for an express" and that Driver Baybut "must be running on his bangs." Driver Baybut stated that he had still been looking for the Romford Advanced Starter with the Gidea Park Distant when he had run into the Southend train.

Lengthman G.F. Veralls, who fogged the Romford Down Distant signals, said that he had been called out by the late turn Romford signalman at 8.50 p.m. but had not been at home. He had signed the Romford train register at 10.20 p.m., had collected some detonators and made his way to the Distant signals, taking about half an hour to cover the three-quarters of a mile. His evidence as to his further actions the Ministry of Transport inspecting officer, Lieut. Colonel E. Woodhouse, described as 'far from satisfactory.' The fogsignalman stated that he had found a partly filled magazine in the Clayton fogging machine for the Down Through line. Two trains had then passed on that line (that is, the two involved in the accident). The Distant signal had been cleared for the first so he had removed the detonator from the rail. The second (Peterborough) train had exploded a detonator and he had shown it a yellow light. Going to Crowlands box for a fresh supply of detonators a little later Signalman Witherington had told him of the accident. He maintained that he could not have shown a green signal with his hand lamp as the green glass had been broken some days earlier; he had not attempted to swop it for a different lamp. After the accident he had obtained a green flag from Crowlands box to put over the white light of the lamp.

The inspecting officer noted that:

(a) Guard Bailes stated that he had seen the fogsignalman apparently still on his way to the distant signals after the Southend train had passed them.

(b) Sub-Ganger W.J. Norton, who had been the last person to fog the Romford Distants before Veralls was called out, stated that when he had left duty at 6.45 p.m. on 1st January he had returned the unused detonators from the fogging machine to Crowlands box, where they were always kept when not in use. Signalman Witherington agreed this was the practice.

(c) No sound of a detonator exploding under the Peterborough train had been heard by five men in the vicinity.

After the accident three cases of exploded detonators were found near the Clayton machine. Two were rusty and had clearly been there some time; the third, which was quite clean, had clearly been exploded by the breakdown train after the collision.

Coming to his conclusions, Lieut.-Colonel Woodhouse stated that the direct cause of the accident must be considered to be Driver Baybut's serious mistake in running through the Romford – Gidea Park section at fairly high speed while unsure of his location in ignorance of the signal indications. Rule No. 127 (xxii) stated that in such a situation a driver must assume that the Distant signals were at *Caution* and the Stop signals at *Danger*, unless a fogsignalman indicated otherwise. The driver's statement, which he had adhered to, that he had still been looking out for the Gidea Park Outer Distant signal when the collision occurred, was remarkable, given that the signal was only 386 yards beyond Romford station, the platform lights of which he had seen, and was 1,412 yards short of the point of collision. It seemed likely that he had confused the Gidea Park Outer Distant signal with the Inner Distant signal, 560 yards short of the point of collision. It also seemed likely that the two green lights Driver Baybut alleged that he had seen were ones which he had seen earlier on the journey. In any case after passing Romford he had shown a regrettable lack of caution in proceeding at a speed which was manifestly too high given the thickness of the fog and the absence of fogsignalmen. He should have continued at a crawl until certain of his location. Given his serious error of judgement he must be considered primarily responsible for the accident.

Fogsignalman Veralls' account of his actions was not, the inspecting officer considered, convincing. It seemed clear that there had been no detonators in the Clayton fogging machine when he eventually had arrived at his post and that by the time he had collected a magazine from Crowlands box the Peterborough train had passed. There had, however, been nothing to prevent him fixing an ordinary detonator (these were not suitable for the fogging machine) to the rail before going to Crowlands box for the magazine and before the Peterborough train had passed. His failure to do so and his excuses about his defective hand lamp implied that he had been carrying out his duties 'in a casual and unintelligent manner.' He might be considered indirectly responsible for the accident.

Signalman Looker did not escape criticism either. Although his action in lowering the Outer Home when the Peterborough train had occupied the 100 yard berth track circuit to its rear was in accordance with the rules the inspecting officer felt that given the poor visibility Signalman Looker would have acted more prudently if he had made sure that those in charge of the train had seen the signal at *Danger* before lowering it.

Turning to remarks and recommendations the inspecting officer doubted if Signalman Looker would have had time to operate a detonator placer under the Peterborough train after seeing that it was not stopping. Nevertheless, had one been provided it could have been used in accordance with Rule No. 94, which read:

When the Signalman requires to stop an approaching train during fog or falling snow and a Fogsignalman is not on duty, he must keep his signals at Danger, and, when practicable, place one detonator on the rail of the line to which the signals apply.

At signal boxes where emergency detonator placers are provided, the provisions of the Rule must always be carried out during fog or falling snow whether a Fogsignalman is on duty or not.

It was noted that the provision of detonator placers had been mentioned in the report on the January 1944 collision at Ilford on the same stretch of line (see 1931-45 Accidents volume) and the inspecting officer recommended that in view of the fact that the provision of detonator placers in signal boxes was neither costly nor complicated, they should be provided at boxes in the area lacking them before the next winter, rather than the matter being delayed until the line was resignalled. This was especially important in view of the density of traffic, the prevalence of fogs, delays in fogsignalmen reaching their posts and the difficulty of placing detonators by hand at a box controlling four running lines.

Had colourlight signals been installed the accident would probably have been prevented. Their installation had been delayed by the War. Even colourlight signals could be passed in fog, though, as the Ilford accident had shown, and the only absolute safeguard was some form of absolute train control. Experiments and trials with various systems were in hand and it was hoped that one or another system would soon be provided by the LNER, at least on busy stretches of line where fogs were common. [299]

At the close of the inquest on the victims of the accident in February 1947 the jury returned a verdict of "Accidental death arising out of a serious error of judgement on the part of Driver Baybut." They added a rider that a strong recommendation should be made to the railway company that their methods of calling out fogmen should be completely revised. [300]

According to reports in several American newspapers five people died and 53 were injured, 13 seriously, when one electric passenger train on the Tohoku Line ran into the rear of another near Tabata station, Japan, on the outskirts of Tokyo, on Tuesday 22nd April 1947. 'No Allied personnel were involved.' [301] According to a presumably more reliable Japanese source the casualties were a good deal higher than stated in the American reports, as it states that there were 118 casualties. The accident appears to have been due to a driver overrunning signals. There had been experiments with an automatic train stop system on the busy commuter line in the early 1930s but the project had been abandoned in favour of quadrupling the tracks. [302]

A mistake by a *fahrdienstleiter* resulted in a rear collision at Husby station, Germany, on the Flensburg – Kiel line, close to the Danish border, on Monday 5th May 1947. Mixed train GmP 2192 was stopped in the station and he inadvertently let passenger train P 1207

---

[299] Ministry of Transport report

[300] *The Times*, London, England, 5th February 1947

[301] *Miami Daily News-Record*, Miami, Oklahoma, United States, 22nd April 1947

[302] Webpage: Breakthroughs in Japanese Railways 2 Learning from Past Railway Accidents – Progress of Train Control by Masayuki Matsumoto

in behind it. Five people died and 27 were injured. [303]

Two London and North Eastern Railway expresses from London to Leeds were involved in a major disaster just south of the station at Doncaster, England, on the late afternoon of Saturday 9th August 1947. By Monday, with the death of three of the seriously hurt in hospital, the number of dead had risen to 18. Another 118 passengers were injured or suffered from shock, 51 of whom were kept in hospital, and a driver and a train attendant were also hurt. [304]

Happening when and where it did, the accident probably did not get as much coverage as it might otherwise have done in the national newspapers, though it was described in some detail on the following Monday. After detailing what had happened, the casualties and the damage caused, a newspaper continued, 'A breakdown gang was quickly at work, and the police summoned every ambulance in Doncaster, several from colliery villages, many of the town's taxicabs, and some buses, lorries, and private cars to take the injured to Doncaster Royal Infirmary. Doctors hurried to the accident and to the infirmary, where 12 of the surgeons worked through the night. In about two hours the last of the victims was released. Ambulance men crawled under the wreckage to give morphia injections.

'People from houses nearby tore up sheets and household linen for bandages, and for some hours provided tea for the injured and rescue workers. Breakdown gangs worked throughout the night cutting and removing the wreckage, and by noon today the rails had been replaced. The down main line was cleared at 4 p.m. today {Sunday 10th} and the up main line an hour later. Normal working was resumed, although wreckage is still to be taken away and a few repairs to be made.

'Most of the passengers were returning from holidays in the south to Yorkshire, though among the casualties were people from London and the south. The saddest case was that of the Carter family, of Williams Avenue, Rothwell, near Leeds. As Margaret, aged seven, waited at the infirmary her father and brother, aged four, were dead, and her mother was being admitted with severe injuries. The child was placed in the care of a Doncaster family.

'Experts are of the opinion that the steel frames of the coaches prevented the casualties from being higher. Although the four rear coaches were splintered, the impetus of the crash was largely borne by the frames. Many of the hundreds of passengers in both trains complained of nothing more than a shaking.' [305]

The first train was No. 956, the 1.10 p.m. Down express passenger from London King's Cross to Leeds Central, consisting of 4-6-2 A3 Class steam engine No. 50, and 14 coaches, all but one of which, the all-steel Pullman which was the third vehicle in the train, had wooden bodies on steel frames. It was in charge of Guard G.T. Wells, with Driver H. Mason and an unnamed fireman.

The second train was No. 958, the 1.25 p.m. Down passenger express, also from London King's Cross to Leeds Central, consisted of 2-6-2 V2 Class steam engine No. 956 and 12 coaches, all with wooden bodies on steel frames except the all-steel articulated set third and fourth in the train. It was in charge of Guard J. Gilmore with

---

303 *Die Katastrophenzene der Gegenwart* by Hans Joachim Ritzau and Jürgen Hörstel, pg. 107 (in German)

304 Ministry of Transport report

305 *The Times*, London, England, 11th August 1947

Driver R. Foster with Fireman R.B. Huffingley.

The signal boxes involved in the accident, from north to south, were South Yorkshire Junction, Bridge Junction, Balby Junction and Red Bank. The relevant signals at Bridge Junction were the Down Home, No. 67, and its Distant, formerly worked by lever No. 68. At Balby Junction, on the other side of the line, the Down Home was No. 2 with its associated Distant, formerly worked by lever No. 1. A partial modernisation of the signalling had taken place shortly before the disaster and the previously semaphore Bridge Junction Down Home signal was now a four-aspect colourlight, while the Down Distant signals of the two boxes had been converted to automatic operation, so that they cleared when their respective Home signals were cleared. All the other signals remained as semaphores. Because of the short distances between the boxes indicator working was provided, so that the Balby Junction signalman was prohibited from lowering his Down Home and Distant signals until the Distant signal at Bridge Junction had been cleared, or until the train had passed his (the Balby Junction) Distant signal. An annunciator bell was provided to indicate when he could do so. The situation as it existed at the time of the accident was that the Bridge Junction Down Distant signal would clear whenever the Bridge Junction colourlight Home signal was cleared, when there was still a train in the rear of the said Home signal, and at this point the annunciator bell would ring in Balby Junction box to indicate that it had done so. There were no controls over the Balby Junction Down Home signal or the Down line block instrument there.

Driver Mason told the Ministry of Transport inquiry into the accident that the Balby Junction Down Distant signals had both been *On*, the Home signal had been *Off*, and the Bridge Junction Distant below it had been *On*, as had the colourlight Home signal (No. 67), at which he had come to a stand. After two or three minutes the latter signal had cleared to *Yellow* and then to *Double Yellow*, but it had taken some time to restart his heavy train up a rising gradient and a sharp curve. Just as he had finally begun to move forward, at the third attempt, his train had been run into.

Guard Wells agreed with his driver's evidence.

Driver Foster testified that the Red Bank Down Home and Distant signals, the Balby Junction Home and Distant signals and the Bridge Junction Distant signal had all been *Off*, so he had been expecting a clear run into Doncaster. He had begun to brake before passing Red Bank box to stop at the station. As he had been passing Balby Junction his fireman had shouted for him to stop and he had made a full brake application.

Fireman Huffingley spoke of seeing his driver closing the regulator as they passed Decoy No. 1 box (700 yards south of Red Bank), and the Balby Junction Inner Distant and Bridge Junction distant at *Clear*. He had then seen the train in front of them and shouted a warning to his driver.

Guard Gilmore could only say that he had felt the brakes going on before the collision.

Signalman W. Sampson at Red Bank stated that the Down train preceding the 1.10 p.m. had passed him at 4.31 p.m., when he had given the *Out of Section* for it to Decoy No. 1 and accepted the 1.10 p.m.. At 4.32 p.m. he had offered the latter train to Signalman J.W. McKone at Balby Junction and it had been accepted at once. The 1.10 p.m. had passed him at 4.34 p.m. and been cleared back by Balby Junction at 4.36 p.m. He had accepted the 1.25 p.m. train from Decoy No. 1 at 4.35 p.m. and offered it to Balby Junction as soon as the 1.10 p.m. train had been cleared back to him at 4.36 p.m. The 1.25 p.m. had passed him at 4.40 p.m.; he had not noticed the position of the Balby

Junction Distants.

Signalman H.S. Taylor at Bridge Junction stated that he had been offered the 1.10 p.m. by Balby Junction at 4.33 p.m. but had not been able to accept it until 4.34 p.m. It had been offered to South Yorkshire Junction box three times before it had been accepted at 4.38 p.m. He had then cleared his Home signal (No. 67) after the train had been stopped at it for two minutes. This was a common occurrence on a Saturday, as was the train having difficulty in restarting. The 1.25 p.m. had not been offered to him and his first notice of anything being wrong had been receiving the *Obstruction Danger* from Balby Junction at 4.41 p.m. McKone had then told him over the telephone that the coaches of the 1.10 p.m. were obstructing the Up line but had said nothing about the 1.25 p.m. train.

Signalman F.E. Buckett at South Yorkshire Junction explained that the acceptance of the 1.10 p.m. train had been delayed because he had been crossing the 4.35 p.m. Up train to Sheffield from Doncaster Up platform across the Down Main to the South Yorkshire branch. He had then had the 1.10 p.m. train accepted from Doncaster South box, the Doncaster South Distant signal indicator had cleared to *Slot Off*, and he had cleared his Home signal and Distant signal indicator, which in turn had cleared the Bridge Junction Home from *Yellow* to *Double Yellow*, caused the Distant indicator at Balby Junction to change to *Slot Off* and the annunciator bell there to ring.

Trains were not booked at Balby Junction but Signalman McKone stated that he remembered accepting the 1.10 p.m. train from Red Bank, clearing it back after it had passed and accepting the 1.25 p.m. train. He had not noticed that the 1.10 p.m. train had stopped ahead of his box. Shortly afterwards, he said, he had been called to the telephone as the annunciator bell had rung. Turning round he had then pulled lever No. 2, clearing at one stroke his Home signal, the Bridge Junction Distant signal beneath it, and his Inner and Outer distant signals, although he had not had the 1.10 p.m. cleared back from Bridge Junction or offered the 1.25 p.m. train forward. He alleged that the 1.25 p.m. train must have been close to his Home signal but other evidence, such as that of Driver Foster, indicated that it must still have been outside of his Distant signals.

Signalman McKone stated that upon hearing the sound of the collision he had thought that the 1.25 p.m. train had become derailed; the 1.10 p.m. train "had gone clean out of my mind." He said that it was only a quarter of an hour later that he had remembered the 1.10 p.m., though this was not the impression he left when he had spoken on the phone to Signalman Taylor. He stated that the ringing of the annunciator bell had confused him. He admitted that he had not looked at his block instrument before clearing his signal, not had he obtained Line Clear for the 1.25 p.m. train. He alleged that although he had forgotten to do so on this occasion, he normally reset the road from the Down Main to the Down Goods after each train had passed so as to obtain the required clearing point and enable him to accept another train at once, even though this would have entailed a total of 10 extra lever movements 30 or 40 times a shift. He could not explain why he had failed to observe the stationary 1.10 p.m. train.

Signalman K.T. Bird, who also worked Balby Junction box, testified that it was not an easy box to work as with no booking, trains had to be committed to memory. Trains were often held at Bridge Junction and he was sometimes unclear if he had offered a train forward and had to confirm via the telephone with the Bridge Junction signalman. He had used to rely on the ringing of the annunciator bell to clear lever No. 2 but now it rang even when a train was standing at the Bridge Junction Home signal, and he made a practice of placing a lever collar on lever No. 2 when a train was stopped in the section ahead. He did not normally reset the road from the Down Main to the Down Goods for acceptance purposes.

Yard Master E.S. Gellatly stated that he could not trace a telephone call to Balby Junction at the time stated by Signalman McKone but the signalman might have mistaken a call to another box.

Retired District Inspector W.G. Northfield agreed that the Balby Junction signalmen were inclined to take the ringing of the annunciator bell as an indication that the line was clear but was emphatic on the need for correct block working. He ridiculed Signalman McKone's statement that he regularly reset the road from the Down Main to the Down Goods for acceptance purposes.

Lieut. Colonel A.H.L. Mount and Brigadier C.A. Langley jointly investigated the disaster. They stated that the sequence of events leading up to the collision was clear. The 1.10 p.m. train had been stopped at the Bridge Junction Down Home due to train movements ahead. That signal had then cleared, causing the annunciator bell to ring in Balby Junction box, whereupon Signalman McKone, who had already irregularly accepted the 1.25 p.m. train without having the required clearing point ahead of his Home signal, had pulled No. 2 lever, clearing that signal, the Bridge Junction Distant controlled by and beneath it, and the Balby Junction Inner and Outer distants, while the 1.25 p.m. train had still been 2 miles from Red Bank box, and had thus implied that it had a clear run into Doncaster station. The collision had occurred two or three minutes later, the 1.10 p.m. train having been slow to restart.

Very serious responsibility rested on Signalman McKone for having cleared his signals for the 1.25 p.m. train before he had had it accepted by Bridge Junction or the latter box had even cleared back the 1.10 p.m. train. McKone had frankly admitted his failure, but his explanations in mitigation were evasive and inconsistent. He had failed to observe the rear coaches of the 1.10 p.m. train and had wrongly accepted the 1.15 p.m. train even though he had been working under no great pressure. His excuse that he normally reset the road for Down Goods was clearly an afterthought. If he had really repeatedly done so there would have been complaints about interference with traffic on the goods lines, none of which had been received. Nor had it been possible to trace a telephone call to Balby Junction at the crucial time, though he might have mistaken a call to another box. His suggestion as to confusion caused by the ringing of the indicator bell might have some substance but could not excuse a fundamental breakdown in block working. The inspecting officers concluded that given his lapse and unreliable evidence Signalman McKone had shown himself no longer fit for a position as a main-line signalman.

The inspecting officers noted that the accident could not have occurred before the partial modernisation of the signalling and would not have done so if the modernisation had been completed, or if the Balby Junction Down Home signal had been locked by the position of the block instrument, so that it could not be cleared until the instrument displayed *Line Clear*. Investigation showed that no similar situation occurred elsewhere. Following the accident a *Line Clear* release was provided for the Balby Junction Home signal, the Balby Junction Distant signals were reconnected to lever No. 1, and track circuit H to the rear of the home signal at Bridge Junction when occupied was made to hold the Bridge Junction Distant signal at *Caution*, thus preventing the annunciator bell at Balby Junction from ringing. Then on 14th September 1947 the Balby Junction Home signal, No. 2, was also converted to a four-aspect colourlight. In view of these actions the inspecting officers had no recommendations to make, but he emphasised that the accident showed the need for the closest co-operation between departments regarding alterations and controls during the changeover from mechanical to electrical signalling. [306]

Twenty-three passengers died and 118 were injured when No. 18 Up passenger train ran into the stationary No. 14 Up passenger train in the station at Kolaghat, India, on the Bengal-Nagpur Railway on the night of Wednesday 27th August 1947. [307] [308]

Two major train disasters occurred in England towards the end of October 1947. The second, the derailment at Goswick on Sunday 26th, has been dealt with in a previous chapter. The other accident was a rear collision at South Croydon on the London – Brighton main line of the Southern Railway during the morning rush hour on Friday 24th, when an electric commuter train from Haywards Heath was run into by a similar electric commuter train from Tattenham Corner.

The 7.33 a.m. service from Haywards Heath to London Bridge consisted of two 2-car sets from Haywards Heath to which a 4-car set had been attached at Redhill to form an 8-coach train. It was in charge of Guard W.C. Waters with Motorman A. Clack and was carrying about 800 passengers (seating capacity 536). It had been checked at the South Croydon Junction Up Main Distant signal and was passing the Up Main Home Signal at low speed when it was run into from behind by the Tattenham Corner train.

The 8.4 a.m. service from Tattenham Corner to London Bridge consisted of two 3-car sets from Tattenham Corner to which another 3-car set from Caterham had been attached at Purley to form a nine coach train. It was in charge of Acting Guard A.F. Sims with Motorman H. Goldsmith and was carrying about 1,000 passengers (seating capacity 750). It collided with the Haywards Heath train at about 8.37 a.m. during a thick fog. Thirty passengers and Motorman Goldsmith were killed outright and one injured passenger died later in hospital. 183 passengers were injured or suffered shock, of whom 58 were taken to hospital and 41 were detained. Most of the casualties were in the leading carriage of the Tattenham Corner train. [309]

Post-war paper rationing meant that the disaster probably did not get as much coverage as it might otherwise have done in the newspapers of the time given its magnitude but the immediate aftermath of the disaster was described in some detail by a reporter.

'Help was quickly forthcoming for the injured. Residents of Sussex Road who ran from their houses described the impact as being similar to a bomb explosion. One man said he saw a blue flame pierce the fog. A spiked iron fence had to be broken before the injured could be reached from the road, and householders threw open their doors to let people through. They provided blankets, sheets for bandages, and hot water and tea. Nearly all the casualties came from the wrecked coaches. A man in one compartment was trapped for nearly half an hour before he could be extricated. South Croydon station sent an emergency message for ambulances, doctors, and blankets. Within a few minutes police cars which had been notified by wireless were at the scene. Accident squads were

---

[306] Ministry of Transport report

[307] *San Mateo Times*, San Mateo, California, United States, 28th August 1947

[308] *The Railway Gazette*, London, England, 15th April 1949, pg. 400

[309] Ministry of Transport report

also sent from Scotland Yard, and a mile of streets was cleared to give passage for the 55 ambulances which answered the call. More than 100 firemen equipped with oxy-acetylene cutters were sent to extricate those trapped.

'Everyone in the neighbourhood who could do so gave a hand to free the injured and remove the dead from the wreckage. Shrouded in fog, the scene after the accident was both eerie and distressing. The moans of the injured were heard in the wrecked coaches and from those who had been thrown clear. All, however, remained calm. The rescuers placed the injured on the embankment above the road to await the ambulances. Part of the iron fence was removed, and the stretchers were then carried through and the occupants taken to four local hospitals. Here extra beds had been assembled to meet the emergency. Within two hours of the accident all the dead and the injured had been removed.'[310]

The main line at the point of the accident had four tracks running from south to north which from east to west were designated Down Main, Up Main, Down Local and Up Local. From the south (Brighton) end were found Coulsdon North signal box, from where trains were described to Purley North and South Croydon Junction boxes; Purley station, where the Tattenham Corner and Caterham branches joined the main line; Purley North box (at the London end of Purley station), Purley Oaks box, Purley Oaks station, South Croydon Junction box, where the Oxted line joined the main line; and South Croydon station. The line was worked by Sykes Lock and Block with semaphore signals. Up Main passenger trains which were described forward from Coulsdon North box were signalled 3 pause 1 but Tattenham Corner and Caterham trains, which of course had not been described from Coulsdon, were described as 3 pause 3 pause 1. Fogmen (fogsignalmen) had been called out before dawn on the day of the accident and were in position at all signals by 8 a.m.

Motorman Clack told the Ministry of Transport inquiry that the Purley Oaks Distant signal had been at Caution and he had brought his train to a stand at the Up Starter (No. 19). After seven or eight minutes it had cleared and he had moved off. Because the line was worked by Sykes Lock and Block there was no need for anyone to go to the box in accordance with Rule No. 55. The South Croydon Up Main Distant had been at *Caution* and his train had exploded a detonator there. He had then found the South Croydon Junction Up Main Home with the South Croydon Station distant beneath it both *Clear* and had just released the brakes and opened the controller when his train was run into.

Guard Waters generally confirmed his motorman's account. His compartment was shattered by the collision and he was very lucky to escape with minor injuries.

Acting Guard Sims of the Tattenham Corner train stated that a large number of passengers had joined it at Purley. He thought that the brakes had been applied immediately before the collision.

Lengthman J. Langridge, fogging the South Croydon Junction Up Main and Local Distants, put the visibility at 20 to 30 yards. The Up Main Distant had been at *Caution* for the Haywards Heath train; he had shown it a yellow hand signal and it had exploded a detonator. A minute and a half or two minutes later the Distant signal had come off and the Tattenham Corner train had passed him at a considerably higher speed than the Haywards Heath one, perhaps 40 miles an hour.

Lengthman L. Tipper, the fogsignalman at the South Croydon Station distants, put the

---

[310] *The Times*, London, England, 25th October 1947

visibility at 20 yards. Immediately after the front of the Haywards Heath train had passed the Distants at *Clear* at about 15 miles an hour he had heard the sound of the collision.

Signalman B. Neary at Purley North box stated that he had offered the Haywards Heath train to Porter Signalman Horace Hillier at Purley Oaks at 8.23 a.m. and given the *Train Entering Section* signal for it at 8.27 a.m. The Tattenham Corner and Caterham trains had then arrived at Purley and had been combined while Coulsdon South box had described an Ore train at 8.33 a.m. By then Purley Oaks should have cleared back the Haywards Heath train and he had to decide whether to give the Tattenham Corner or Ore trains priority on the Up Main. He had therefore telephoned Purley Oaks and asked Hillier how he was looking on the Up Main. Hillier had replied simply, "Gor blimey," put the phone down, and a few seconds later had given the *Train out of Section* signal on the Up Main. At this he (Neary) had decided to give precedence to the Tattenham Corner train. Signalman Hillier had accepted this train as a 3-3-1, the signals at Purley had been cleared and the train had departed at 8,34.

Lieut. Colonel A.H.L. Mount, the inspecting officer, described Signalman Neary as a good witness. Neary said that he had spoken to Hillier in a friendly manner; he knew that the latter was a porter signalman who had just been discharged from the Army. He wished to avoid making Hillier "feel uncomfortable by saying anything about delaying the *Out of Section* signal as this would have been inclined to put the man off of his balance." Regarding Hillier "he had never worked with a better porter signalman." He considered, though, that Hillier should not have been put straight into a busy main line box, "because it did seem in a way unfair to put a porter signalman with a few weeks' experience out on the Main Line."

Purely Oaks box had 56 levers, but 43 of them were spares and it was only a block post. Porter Signalman Hillier had opened the box at 5.58 a.m. It had become increasingly foggy, a fogsignalman had reached his Up Distant signals at 7.20 a.m. and by 8.30 a.m. he could not see the station platforms, 170 yards away. He remembered accepting the Haywards Heath train. He could not get it accepted at once by South Croydon Junction so he had checked it at his Up Home signal (No. 18) before letting it into Purley Oaks station and up to his starting signal (No. 19). It had by then passed out of sight and by the time Signalman Neary had phoned him he had forgotten about it. Seeing that the upper tablet of the Lock and Block instrument for his home signal showed *Locked* and the lower tablet was blank he had jumped to the conclusion that the instrument had failed and had used the release key to reset it. He said, "I just turned round (from the telephone) and looked at the instruments and thought I had stopped something, and immediately freed the instrument…at this time No. 19 Starting signal was section locked." He had then given the *Out of Section* signal for the Haywards Heath train and accepted the Tattenham Corner train, a 3-3-1. Shortly afterwards South Croydon Junction had accepted the Haywards Heath train with a 3-1 and plunged, releasing No. 19 Starting signal. Although this was a different bell code than that for the Tattenham Corner train which he had just accepted from Purley North and although he had not offered the Tattenham Corner train to South Croydon Junction, Hillier, who had clearly become completely confused at this point, had pulled off all of his Up Main signals for the Tattenham Corner train. As the Haywards Heath train had moved off and over the treadle connected with signal No. 19 it had freed the Sykes instrument for the latter signal, which changed from *Locked* to *Free*. Thinking that this must just be another malfunction Hillier had just reset the indication to *Locked*. Nor did the fact that he saw from its headcode that the second train was a Tattenham Corner one, and hence a 3-3-1, and not the 3-1 which the South Croydon man had just accepted, alert him that anything was wrong. He had just telephoned South Croydon and advised Signalman E. Walder there that it was a Tattenham Corner train coming.

In accounting for Porter Signalman Hillier's mistake the report mentioned the intensity of the traffic. It stated that in an hour (during the morning rush hour) there were 13 trains to deal with on the Up Main, 12 on the Up Local and 13 more in total on the Down Main and Local lines. These required 38 plunges, 90 to 100 bell signals and the pulling and replacing of the signals for each train. In addition there was the handicap of the fog. Hillier said that "in the normal way, I was at home in that box," but, with the fog, "I could not see further than the other side of the tracks, looking directly out." Asked if he had placed the switch hook over the plunger on receiving the *Train Entering Section* signal for the Haywards Heath train Hillier replied, "I do not know whether I actually did this time. It is quite possible that in offering the train to South Croydon, I may have been at the other end of the frame and ringing on bells far away from the instrument…I am afraid I must have overlooked it." Use of the switch hook would have prevented use of the plunger and raised the block indicator at Purley North.

Signalman E. Walder was another experienced signalman. He testified that at 8.24 a.m. Selsdon box on the Oxted line had offered him a Tunbridge Wells train. The Haywards Heath train had been offered by Hillier at 8.26½ a.m. but he had decided to give the Tunbridge Wells train priority, so he had refused the Haywards Heath train at that point. At 8.35 a.m. South Croydon Station had cleared back the Tunbridge Wells train and he had accepted the Haywards Heath train from Purley Oaks. A minute later Hillier sent the *Train Entering Section* signal, following it up with information that it was a Tattenham Corner train rather than a Haywards Heath one. This did not surprise him as the order of trains was often altered. A minute later he heard the sound of the collision and blocked all lines. Asking Hillier if the Tattenham Corner train had been overrunning signals Hillier had replied, "No, it was all my fault."

Inspector Akehurst, a promoted signalman who had recommended Hillier for the post, stated that he had considered him "perfectly suitable," above average and fully competent to take charge but then, of course, as Mandy Rice-Davies was famously to say a few years later at the Profumo trial, "Well, he would say that, wouldn't he?" The inspector would not accept that the appointment of Hillier had been a mistake. He was "quite prepared to accept him as a signalman to work any turn in the box at any time," adding that, "It is left to the man to decide when he is ready to take over." The fog and telephone messages should not have put Hillier, "off his balance." Instead the inspector considered that Hilliard had been "prompted by the signalman at Purley that he was delaying traffic, which caused him to take the action he did on the spur of the moment." The inspector pointed out that he had stressed upon Hillier the care to be exercised in using the release key and "to treat it as a red hot poker and handle it was such."

Porter Signalman Hiller was 29 years old. He joined the railway before the war, then had been called up and had taken part in the abortive Norway campaign, then fought at Tobruk in North Africa, and spent three years as a prisoner of war. Rejoining the railway in May 1946 he had worked as a shunter and a ticket collector before becoming a porter signalman. He had worked the box for nine weeks before the accident, but never during a fog.

During the period when Hillier had been working Purley Oaks box there had been three failures of the Sykes Lock and Block, one of which had been a wrong-side failure where the instrument had gone to *Free* when it should have remained at *Locked.* Inspector Akehurst and Assistant Signal and Telegraph Engineer Boucher agreed that such a run of failures was probably coincidental and exceptional.

Concluding his report, Lieut. Colonel Mount commented that the primary cause of the disaster was clearly Porter Signalman Hillier entirely forgetting about the Haywards Heath train and then wrongly concluding that the Sykes Lock and Block instrument had failed and using the release key, a very serious error, especially in a fog. Subsequent warnings such as the dropping of the Sykes instrument from *Locked* to *Free* and the fact that the Tattenham Corner train had been a 3-3-1 whereas South Croydon had accepted a 3-1 had failed to alert him to his error. He had frankly admitted his mistake at once and expressed deep regret.

The inspecting officer considered that Hillier must have failed to appreciate the danger of irresponsible use of the release key, despite an instruction that irregular use of it might result in instant dismissal. Interestingly, the report states that Hillier was not, in fact, dismissed after the disaster but was 'found other work,' perhaps as a clerk or something. Lieut. Colonel Mount believed that Hillier's actions might have been influenced by the previous failures of the Lock and Block apparatus plus his difficulties in working in a fog for the first time. While accepting that signalmen of long service had misused the release key, the inspecting officer considered that Hillier had been allowed to assume too much responsibility too rapidly and recommended that the training and examination of signalmen should be reviewed. As regards the failures of the Sykes Lock and Block treadles, another of which occurred while he was inspecting Purley Oaks box, it must be emphasised that all such incidents must be reported. There appeared, however, to have been no deterioration in the general standard of reliability of the equipment.

As was the case elsewhere, resignalling with colourlight signals and full-track circuiting had been delayed by the War. The present signalling had been in use for over 50 years and 'it is probably true to say that no other mechanical system could have coped so efficiently with such heavy main line and suburban electric services.'

The overcrowding of the trains had doubtless increased the number of casualties. It had been due to insufficient rolling stock, peak-hour traffic and insufficient line capacity. Improved allocation of resources was the answer to the first two factors, resignalling should improve the last named. [311]

At the inquest on the accident victims the Coroner, in summing up, stated that he did not think Hillier had deliberately neglected his duty. That he was negligent was frankly admitted. The jury returned a verdict of "Death by misadventure, contributed to by the forgetfulness of Porter Signalman Horace Hillier". They added a rider that they deprecated the practice of putting an inexperienced signalman alone in a main line signal box. [312]

No. 16 Down express ran into the rear of a goods train stopped at a signal at Kanchrapara, India, on the East Indian Railway on Monday 8th December 1947. Seven

---

[311] Ministry of Transport report

[312] *The Times*, London, England, 25th November 1947

people lost their lives and 45 were hurt. [313]

In what appears to have been a repetition of the 1947 Husby accident, an error by a *fahrdienstleiter* at Flensburg, Germany, resulted in passenger train P 1353 from Husum running into the rear of mixed train Gmp 9345 standing at the *Einfahrsignal* at Flensburg on Friday 12th December 1947. [314]

New Year Day, 1st January 1948, saw a rear collision near Syracuse, Missouri, United States, on the Missouri Pacific Railroad, involving two sections of *The Missourian* passenger train. The first section was running at reduced speed due to adverse signals when it was struck by the second section. Twelve passengers and two Pullman car employees died in the accident and 32 passengers, three railwaymen, a dining car employee and a porter were injured.

Westbound train First No. 9, *The Missourian*, a first-class passenger train, consisted of engine No. 5326, three mail cars, a baggage car, a passenger baggage car, two coaches, a dining lounge car and four sleeping cars in charge of Conductor N.A. Bradley with Engineer Elmer Lynam and an unnamed fireman. At Tipton, 5¼ miles east of Syracuse, it received an order which in part read:

There are no opposing trains in block between Tipton and MKT Crossing.

MKT Crossing was 19½ miles west of Syracuse.

The train left Tipton at 7.20 a.m., three hours and 45 hours minutes late. Because of restrictive indications at five signals short of the point of collision it had stopped at each of them and it was approaching signal No. 1703, which was displaying *Stop then proceed at low speed* at about 15 miles an hour when it was run into by train Second No. 9.

Westbound train Second No. 9, a first-class mail express train, consisted of engine No. 2213, two box express cars, two express refrigerator cars, three box express cars and a coach, in charge of Conductor J.B. Rodgers, with Engineer Fred Butler and Fireman Ulysses Parks. At California, 17¾ miles east of Syracuse, it had received a train order which read in part:

There are no opposing trains in block between Clarksburg and MKT Crossing.

Clarksburg was 12½ miles east of Syracuse. The train left California at 6.43 a.m., 3 hours and 23 minutes late. Subsequently it passed 23 signals without stopping, nine of which displayed *Stop* and the remainder *Stop then proceed at low speed.* It passed signal No. 1687, which was displaying *Stop then proceed at low speed* and moving at 40 miles an hour collided with the rear of the first section rather more than a mile and a half further on. As a result of the impact of the collision, the first section of the train became separated at three different places and the rear, twelfth car was telescoped at its front end by the eleventh car and at its rear end by the engine of the second portion. All the fatalities occurred in this car. The engine of the second portion was damaged, the second car of this train was demolished and the first and third to sixth cars were badly damaged. [315]

A newspaper reporter described the experiences of the passengers who had been in the

---

[313] *The Railway Gazette*, London, 15th April 1949, pg. 400

[314] *Katastrophen der deutschen Bahnen Teil II* by J.R. Ritzau, pg. 121 (in German)

[315] Interstate Commerce Commission report

front coaches of the first portion of the train in the aftermath of the accident:

'A hundred chilled and frightened passengers clambered down unhurt from forward coaches of Missouri Pacific's No. 9 this afternoon surprised to learn that all the passengers in the rear Pullman of their train had been killed more than seven hours earlier. The uninjured passengers said they felt a jolt about 7.40 a.m. when No. 9's second section rammed into their slowly moving train but they were asked to remain in their seats and most of them didn't know what had happened. Rumors flew up and down the aisle during the long wait.

'The collision burst the steam line in the forward train and the coaches cooled rapidly in below freezing weather. A bitter wind wiped snow around the cars and the passengers got cold before the steam line could be repaired. At 3.30 this afternoon the uninjured filed out of the cars, lugging their bags and parcels down a snowy country lane to the highway. Four coaches waited to take them on their way over dangerously-iced highways.

'There were exclamations of horror and surprise as the travellers trudged along the snowy right-of-way past the shell of the rear Pullman and the waiting ambulances. Workmen recruited from the crowd helped the passengers carry their luggage over the debris and icy crossing. At one side a bonfire kept coffee warm for the crews of state patrolmen and railroad laborers.' [316]

The Interstate Commerce Commission (ICC) report on the accident revealed that about eight hours before it had happened a sleet storm had damaged the power transmission lines and circuit control wires at numerous points in the 62½ miles between Jefferson City and MKT Crossing and had disrupted communications between Tipton and MKT Crossing. As a result the absolute signals had all displayed *Stop* and the intermediate and approach signals had all displayed *Stop then proceed at low speed* over a distance of 25¼ miles east of the point of collision. Power to the colourlight signals was supplied by batteries and so they were unaffected by the storm.

According to the rules of the company, where automatic signals were provided and communications had failed trains were only authorised to pass signals displaying *Stop* under the protection of a flagman preceding the train on foot and should a signal displaying a *Stop then proceed at low speed* indication follow a signal displaying *Stop*, the flagman must continue to precede the train until a signal displaying a less restricting aspect was reached.

Engineer Lynam of the first section of train No. 9 told the ICC inquiry that as he had understood from the dispatcher's train order that there was no opposing train in the territory involved, and in view of the icy conditions, he had not required a flagman to precede his train. He had, however, stopped at each automatic signal showing its most restrictive indication and then proceeded at low speed.

The flagman of the first section of train No. 9 stated that he had dropped a lighted 10-minute fusee each time his train had stopped at a stop signal. The last fusee he had put down about 12 minutes before the collision and 2½ miles back along the line. His supply of 18 fusees had by then dwindled to four; he had intended to get a new supply at Sedalia, 18½ miles further on. There had been no time to put down torpedoes during the brief signal stops of his train.

Engineer Butler of the second section of train No. 9 admitted that he had

---

[316] *The Sedalia Weekly Democrat*, Sedalia, Missouri, United States, 2nd January 1948

misunderstood the dispatcher's order as meaning that he needed neither a flagman to precede his train, nor to comply with the indications of automatic signals but instead could proceed at medium speed so as to be able to stop on seeing a lighted fusee. Although the dispatcher's order only mentioned opposing trains for some reason he had got it into his head that the first section had cleared the territory involved.

Fireman Parks had not made any previous trips over this stretch of line and had not been sufficiently knowledgeable of the rules to find fault with his engineer, while Conductor Rodgers and his flagman had depended on the engineer complying with the rules.

The ICC concluded that the accident would not have occurred if proper flag protection had been provided for the first section of the train or if the engineer of the second section had obeyed signal indications. It noted that although the rear Pullman car of the first section, the *Golden Cloud,* which had been of lightweight low-alloy steel construction, had met the specifications of the Association of American Railroads, it had not been strong enough to resist the heavy buffing stresses of the collision.

Engineer Butler was subsequently dismissed by the railway company, Fireman Parks was suspended and Engineer Lynam, who would not seem to have been greatly at fault, was given 30 demerits (90 demerits resulted in dismissal). [317]

Early on Wednesday 14th January 1948 two Canadian National Railways passenger trains were involved in a rear collision near Wykes, Quebec, Canada, a small flag station on the transcontinental main line. Local train No. 11, bound for Cochrane, Ontario, had stalled on a curve due to engine trouble when it was run into from behind by express train No. 21 at 8.05 a.m. Eastern Standard Time. Both trains were running late as the local was due at Parent, about 5 miles further on, at 6.20 a.m. and the express was due there at 7.55 a.m. [318] As a result of the collision the steel sleeping car at the rear of the local train telescoped the wooden coach in front of it and it was in this carriage that the casualties occurred. Initial reports spoke of five dead and 43 injured and the final figures were nine dead and 55 injured. Eight of those who lost their lives were passengers, the ninth was Baggageman J.A. Vaillancourt. 'The crash occurred in 35 degrees below zero weather, and the uninjured passengers clambered over smashed coaches and debris in the bitter cold to aid the others.' Rescue trains were sent from Parent and Fitzpatrick with doctors and nurses and a third special seems to have taken the injured passengers to the nearest hospital at La Tuque. [319]

A passenger spoke to reporters. 'Mrs John Cleary of Noranda, who suffered a twisted neck and back, said, "We sat in the cars for about two hours without any heat or water. It was so cold that people hesitated to get out. In the other cars water was running ankle deep. We didn't care about the fact there was no light or heat because we were glad to be alive," she said. [320]

The line was apparently not worked on the block system, so it was vital that any train stopped unexpectedly be protected in the rear. Thus it was that an inquest jury at La Tuque returned a verdict of "Accidental death" on the victims of the collision but found

---

[317] *Daily Sikeston Standard*, Sikeston, Missouri, United States, 19th January 1948
[318] *The Winnipeg Tribune*, Winnipeg, Manitoba, Canada, 14th January 1948
[319] *The Western Star*, Corner Brook, Newfoundland, 16th January 1948
[320] *The Lethbridge Herald*, Lethbridge, Alberta, Canada, 15th January 1948

that Brakeman E.G. Rousseau of the local train had shown a 'lack of foresight,' though this did not amount to criminal negligence. There had been an interval of 16 minutes between the stopping of the local train and the arrival of the express. He said that he had been riding in the last coach when he heard the whistle of the express and had run back. A red flag which he had shown to the driver of the express had not been seen while it had been too icy to put down torpedoes. The jury remarked that he could have gone back much further in the time available and also commented that in future the Canadian National should operate its express trains in front of its local services. [321]

At the beginning of February the Transport Commissioners recommended that the Canadian railway companies dispense with their remaining wooden coaches as soon as possible and ordered that in the meanwhile such coaches should always be marshalled at the rear of trains to prevent their being telescoped by heavier coaches behind. [322]

Late editions of London newspapers on Saturday 17th April 1948 carried brief reports of an accident involving an express from Scotland and a postal train near Crewe. Many casualties were feared, and police, troops, fire brigades and civilian helpers from a radius of 20 miles around the scene were helping to free people from the wreckage. 'First news of the accident was given by a passenger on the express, who ran from the wreckage across fields to a telephone to give the alarm. Rescuers were hampered by lack of adequate lighting, and emergency apparatus was rushed to the scene of the smash.' [323]

The Monday morning newspapers had considerably more on the accident, and by then it was known that 24 people had lost their lives when the 5.40 p.m. Up passenger train from Glasgow Central to London Euston was run into by the 6.25 p.m. Up postal train, also from Glasgow to London. The collision had occurred on British Railways, London Midland Region at Winsford, England, about 8 miles north of Crewe.

The passenger train consisted of a 4-6-2 Pacific class engine No. 6207 *Princess Arthur of Connaught,* a brake van, four passenger coaches, two brake vans and three coaches, in charge of Guard G.H. Iveson with Driver T.D. Jones and Fireman J.E. Price. Apart from the leading van, which was of all-steel construction, the coaches all had timber bodies with steel underframes. The Up postal was booked to overtake the passenger train at Lancaster but it was running late so the passenger was preceding it to Crewe. The passenger train passed Winsford Junction signal box at 12.09 a.m. and stopped between there and Winsford Station signal box at about 12.10 a.m. because the communication cord had been pulled by a passenger. It emerged later that this had been a young soldier who lived near the line and had pulled the cord so that he could alight from the train and make his way homeward across the fields. The train was still stationary 17 minutes later when it was run into by the postal train.

The postal train consisted of 4-6-2 Pacific class engine No. 6251 *City of Nottingham,* a milk van, a newspaper van, a postal brake van, eight postal vans and two brake vans, in charge of Guard T. Horne with Driver J. Howie with Fireman W. Miller. It entered the Winsford Junction – Winsford Station block section under clear signals and was only 400 yards from the passenger train when Driver Howie saw the red light exhibited by Guard Iveson and applied the brakes as detonators also exploded, bringing the speed of his train

---

[321] Ibid, 21st January 1948

[322] *Winnipeg Free Press*, Winnipeg, Manitoba, Canada, 2nd February 1948

[323] *The Times*, London, England, 17th April 1948

down to 45 or 50 miles an hour before the impact. The rear coach of the passenger train and half of its penultimate coach were demolished and there was severe telescoping between the second, third and fourth vans of the postal train. In addition to the 24 people killed 14 passengers and four postal workers were badly enough hurt to be kept in hospital and 12 passengers and two guards were slightly hurt. [324] The passengers who died included a number of servicemen and also a man named Wermer Hans Kolln, described as a German prisoner of war, and who very likely had been finally on his way home. [325]

Driver Jones told the Ministry of Transport inquiry that he had passed Winsford Junction at 60 miles an hour and shortly afterwards the brakes had come on, he had seen that the vacuum had fallen from 21 ins. to 10 ins. and had realised that the communication cord had probably been pulled. Fireman Price had at once gone back to meet the guard and after a few minutes he (the driver) had also gone back and had met his fireman returning to the engine. The fireman had told him that he and the guard had been unable to find where the disc had been pulled and that the guard had gone back to protect the train. He and the fireman had eventually found a turned disc on the third coach, reset it, and the fireman had attempted to find which passenger had pulled it. Then he had heard the sound of detonators going off and the collision had occurred.

Fireman Price agreed with his driver's account and said that after the collision he had gone to Winsford Station box, putting down detonators on the Down line on the way. The signalman had at first seemed unable to grasp what had happened and told him that he had thought the passenger train had gone by at the same time as a Down goods train.

Lieut. Colonel G.R.S. Wilson, the inspecting officer, commented that Guard Iveson had clearly been much affected by the accident and his account of what had happened had been confused. The guard said that he had gone forward after his train stopped and met Fireman Price, neither of them having by then found where the cord had been pulled. He (the guard) had realised that according to Rule No. 179 (a) the train should be protected in the rear and that in the circumstances it was the fireman's duty to do so, while he dealt with the pulled cord in accordance with Rule 181 No. (f). However, as he (the guard) was the more experienced man he had decided that it would be better if he went back and left the fireman to deal with the cord. Accordingly, he had returned to his van, picked up some detonators and set off back. After about a minute he had tripped over a sleeper, fallen over and his handlamp had gone out. He had relit the lamp, put down a detonator and started back again when he heard the postal train coming. He had then put down a second detonator and exhibited a red light to the oncoming train. He had subsequently gone back to Winsford Junction box where the signalman had told him that the postal train had been accepted ahead and the signals had been off for it.

Driver Howie told the inquiry that he passed Winsford Junction under clear signals at 50 to 60 miles an hour. Just after passing beneath overbridge No. 21 and near the Up Home signal of the intermediate Winsford Goods Yard box (which was switched out) he had seen Guard Iveson's red light about the Up cess. He had at once shut the regulator and applied the brake fully. A detonator had exploded, he had seen the tail lamp of the passenger train, another detonator had exploded and his engine had struck the forward train before he had been able to reverse his engine.

---

[324] Ministry of Transport report

[325] The Times, London, England, 19th April 1948

Fireman Miller confirmed his driver's account.

Guard Horne stated that he had gone back to protect his train after the collision and reached Winsford Junction box at 12.45 a.m. Despite severe chest bruises he had then worked for a long time along the injured passengers.

Signalman F. Chamberlain had worked Winsford Junction for 33 years. He testified that he had accepted the passenger train from Verdins Siding's box at 12.06 a.m. and had it accepted by Winsford Station box at the same time. He had sent it on (given the *Train Entering Section* signal) to Winsford Station at 12.09 a.m. Up expresses usually only took one or two minutes to clear the forward section so when it had not been cleared back by 12.14 he had rung Signalman William Hulse there to enquire what had become of it. He had been told by the latter that the Down Winsford goods had been passing at the same time so that the Winsford signalman missed seeing the tail lamp of the passenger train. He had not thought this sounded very convincing as the goods train must have passed Winsford Station box well before the passenger train had reached there, and he said so to Hulse. When he had got the *Train out of Section* signal, though, at 12.17 a.m., he had been entirely satisfied, assuming that Hulse had received the *Train Out of Section* signal from the next box south at Minshull Vernon. He had thought no more about the matter and offered the postal the postal train to Winsford Station at 12.22 a.m. and had it accepted. At 12.26 a.m. the train had passed, moving at about 55 miles an hour, and he had sent it on to Signalman Hulse. A minute or so later he had heard a dull thud but had not realised that there had been a collision until Guard Iveson arrived at his box at about 12.34 a.m., when as a precaution he had refused the offer of an Up Liverpool train from Verdin's Sidings. What happened when the guard reached the box would in other circumstances have been mildly amusing, for Signalman Chamberlain had been able to make nothing of what the guard, doubtless suffering from shock and giving a confused account in a broad Scottish accent, had been trying to tell him. It was only when Guard Horne had arrived at 12.45 a.m. that all had become clear, and Signalman Chamberlain had been able to explain to Control what had happened.

The train register book at Winsford Junction was being kept by Signalman T. Sandwich, a fully qualified signalman who was working as a booking boy there only because he had been made redundant at another box. He confirmed Signalman Chamberlain's account.

Signalman Hulse had been at present Winsford Station box for 21 years, and previous to that in the old box for 12 years. He said that he had accepted the Glasgow train from Winsford Junction at 12.05 a.m. (the Winsford Station times were generally a minute earlier than those at Winsford Junction) and it had been accepted by Minshull Vernon at 12.07 a.m. when a Windemere express had cleared there. His account of what had happened then was, the inspecting officer said, confused and contradictory. He admitted that the entry in his train register book that the passenger train had passed at 12.11 a.m. could not be correct and could not explain why he had altered the time from 12.10 a.m. He said that he had sent the passenger train on to Minshull Vernon early as he had been anxious to report some Down trains to Control before starting his supper, plus his kettle had been boiling over. He maintained that having been preoccupied with speaking to Control and getting his meal out of the oven he had completely forgotten about the passenger train, despite his Up signal levers being over in the frame and a track circuit indicator not having gone to *Occupied* and back to *Clear,* until the enquiry from Signalman Chamberlain. Control, however, had no record of a conversation with

Winsford Station box at this time. After speaking to Winsford Junction he had dealt with a Down express, which passed his box at 12.16 a.m., then had spoken on the telephone to Signalman F.J. Morris at Minshull Vernon, who was asking where the passenger train was, as it had not reached there. Before or after speaking to Minshull Vernon he had gone down to the foot of the box steps and had looked both ways but he had been unable to see anything. He had then given the *Train Out of Section* signal for the passenger train to Winsford Junction at 12.17 a.m. and replaced his Up signals. As he had been starting his meal Winsford Station then offered him the Up postal, which he had accepted, and Signalman Chamberlain sent that train on three minutes later. At 12.31 a.m. Minshull Vernon had offered him the Down postal but at this point some uncertainty over the situation had clearly entered his mind and he had not offered this train forward to Winsford Junction.

Signalman Morris at Minshull Vernon stated that Winsford Station had sent the Glasgow passenger train on at 12.11 a.m. and he had it accepted by Coppenhall Junction at the same time. Trains usually took 4 minutes to reach him and after 6 or 7 minutes he had telephoned Winsford Station and asked if the passenger train had been "doing all right", thinking that it might have been checked. "Yes," Hulse had replied. When the passenger train had still not appeared he had stopped and cautioned the Down postal train, as per the Block Regulations. He first heard about the accident from someone on the omnibus telephone circuit; afterwards he had spoken to Signalman Hulse about it but had not asked how it had come about. He said that he had been relieved to learn that the Down postal had not been involved.

Lieut. Colonel Wilson remarked that despite reminders from his colleagues on either side, Signalman Hulse continued to assume that the passenger train had passed his box in the usual way. Signalman Hulse admitted his responsibility for the accident and expressed his deep regret for it. Signalman Chamberlain, who had worked with him for 30 years, said that Hulse was always prompt to answer the block bells and telephone and he had never previously had doubts as to his block working. He considered, however, that Hulse had not been in as good health since a hernia operation three years earlier. Hernias were an occupational hazard of manual signalmen, as the author, who was an unofficial signalman for 20 years, can testify, having also suffered one.

Signalman Hulse was a teetotaller and moderate smoker. He said that he had not been tired on the night of the accident, though he had been suffering from the after effects of a bad cold and had been somewhat worried about a possible severe illness of his wife.

Beginning the conclusions of his report, the inspecting officer pointed out that although the passenger train had been stopped due to the irresponsible action of a passenger, it could have stopped for a number of other reasons (such as engine failure, collision with animals on the line, a derailment due to a broken rail and so on) and that it should have been fully protected by the block system. It had occurred because of a grave breach of the regulations by Signalman Hulse, who must bear responsibility for it.

There could be no suggestion that Signalman Hulse had been over-pressed by traffic and as an experienced signalman he should have dealt with the situation without difficulty and himself enquired of Winsford Junction as to the non-arrival of the train, as the Minshull Vernon signalman had enquired of him. His suggestions that he had missed seeing the passenger train because of the passing of the Down Winsford goods or a telephone call to Control were clearly without foundation and it seemed more likely that he had been distracted by the preparation of his meal. The premature sending on of the passenger train to Minshull Vernon was probably the start of his confusion, 'and the circumstances cannot help suggesting that despite his statement to the contrary, he had fallen into the habit of attending to block signals as a mere matter of routine, depending on every train passing normally without hitch, and was thus taken unawares by the stoppage of the passenger train before it reached him. A state of mind in which he was paying so little regard to the integrity of his block working may also explain his subsequent confusion and lack of honesty with himself, as is illustrated by his readiness to assume, in spite of reminders from his colleagues on either side, that the train had passed in the ordinary course, and that he had neglected for some reason or another, to clear the block to Winsford Junction and to restore his Up line signals to danger.'

Signalman Chamberlain had been on the alert. His suspicion of Hulse's reply had proved to have been justified but it would not be fair to criticise him for accepting the *Train Out of Section* when he had received it at 12.17 a.m.

Signalman Morris had also acted correctly in enquiring as to the non-appearance of the passenger train and in cautioning the Down postal.

The rules for the protection of trains stopped in mid-section were designed as a secondary line of defence. According to the rules it had been Fireman Price's duty to go back, as his driver had appreciated. Instead both the fireman and the guard had wasted time searching for the turned disc before the guard had elected to go back, contrary to the rules. Both men had no doubt been trying to do their best and they had been handicapped by the absence of the passenger who had pulled the cord, but if each had concentrated on his own task it was probable that the fireman could have got back considerably further, perhaps even to Winsford Junction box.

Driver Howie of the postal train had clearly been on the alert, though he had probably been running rather faster than he thought.

Driver Jones could not be criticised for stopping where he had, though had he run 650 yards further he would have occupied the berth track circuit in the rear of Winsford Station Up home signal, No. 19, preventing the Winsford Station signalman from clearing his block back to Winsford Junction and the signalman at the latter box clearing his starting signal, No. 56, for the postal train.

Regarding remarks and recommendations, the inspecting officer noted that Class "C" block controls had been installed on practically all of the main lines from London Euston and London St Pancras to Carlisle but this could not prevent the irregular clearance of the block if a train had not reached the track circuit. (Class C block controls involved a number of measures: combined block instruments, the block indicators not to be used for routing purposes, annunciators to be provided where berth track circuits were provided in the rear of the home signals, separate track circuit indicators, proving of the arm as well as the lever of home signals, a warning light to indicate when the distant signal was not in the "on" position when the lever was normal, locking of the home signal lever when the starting signal was at danger until the berth track circuit was occupied, and the starting signal of the box in the rear being released for one pull only when the needle of the block instrument was placed to "Line Clear".) [326] Continuous track circuiting could not be justified other than on exceptionally busy lines but an extension of the so-called "Welwyn" control or the similar Rotary Interlocking Block used on former Midland

[326] London Midland and Scottish Railway instruction re-Standardisation of Block dated 9th December 1930 – with thanks to Mike Turner via Facebook

Railway lines was recommended, with the controls arranged so that the co-operation of the signalmen at each end of the section was required. The inspecting officer would probably not be happy to learn that on former Great Western main lines in the Westcountry at any rate, and presumably elsewhere, no such block controls were found in all the boxes as late as the mid-1980s.

Noting that not all block signals were recorded in the train registers in the Winsford area (on the London Midland Region in general?) the inspecting officer recommended that all block signals should be recorded, as they were in other regions.

It was not likely that a booking boy was needed at Winsford Station box, as had been suggested.

The question of flares to protect trains had been considered after the 1928 Dinwoodie collision and rejected but the 1946 Browney accident and advances in pyrotechnics had led to its being considered again; trials were still in progress. [327]

At the inquest on the victims of the accident in May the jury returned a verdict "Accidental Death", saying that they thought the soldier who pulled the communication cord had acted selfishly. Coroner G.W. Wain, in his summing up, said that he did not think the jury could say more than that the soldier had started the chain of events. 'The jury recommended to the British Railways Executive that a system of warning the driver and guard by communication cord in case of emergency should be adopted, thus taking out of the hands of passengers the power to pull the train up.' [328]

Gunner Harry Bishop, 19, pleaded "Guilty" at Winsford magistrates court on 5th June to unauthorised use of the communication cord and was fined £5 with £6 5s costs. He was given time to pay. [329]

An accident in South Korea in 1948 was well covered in American newspapers because almost all of its victims were American soldiers. Up until 1945 Korea had been a Japanese colony but with the defeat of Japan in August of that year the northern part of the country had been occupied by the Russians. The Americans, fearful that the Russians would gain control over the whole peninsula, had occupied the south for three years. In September 1945 the country was divided along the 38th parallel into Russian and American zones and in September 1948 these divisions became permanent. The United States handed over power in the south on 15th August 1948 and Kim Il-sung took over power in the north on 9th September of the same year. [330] The accident occurred on Tuesday 14th September 1948.

One of the trains involved in the accident was an American troop train from Pusan to Seoul, named *The Liberator,* carrying soldiers and others who were on the first stage of their way home to the United States, plus others who were returning to duty after furlough in Japan. Problems with its engine had resulted in the train running four hours late by the time it stopped at Napnin, South Korea, for repairs. About 110 American soldiers (GIs) and military government employees were riding in the last three coaches of the train. While the train was stationary it was run into from behind by a fast passenger train from Mokpo to Seoul. Early reports stated that 35 Americans and one or two

---

327 Ministry of Transport report
328 *The Times*, London, England, 7th May 1948
329 Ibid, 8th June 1948
330 Wikipedia – Division of Korea

Koreans had been killed and 80 Americans and 45 Koreans had been injured in the accident, 35 and four respectively seriously. The driver of the colliding train was unhurt. *The Liberator* had been diverted from its normal route because of flooding.

One of the surviving American passengers, Melborn Penry, said that the casualty list would have been higher had not a number of passengers left the train to stretch their legs, he himself being one of them. He said that he had heard another train whistle and then seen the Mokpo train approaching on the same line. "I realised that the Mokpo train was on the same track, but I thought there was some kind of a block system. The Mokpo train ploughed into the back of *The Liberator* and telescoped the last three cars. I estimated that about half the 110 Americans riding on the train had been killed or injured. We laid them out on the tracks as soon as we could extricate them from the wreckage. An Irish Catholic priest and a Protestant chaplain who were aboard did a fine job of helping extricate the seriously injured and dying and give them comfort. I saw no evidence of sabotage. In my opinion there was no particular person to blame for the accident, only the system. There were no lights on either of the engines, no red or green lights on the last coach and only dim lights in the coaches."[331]

Another report stated that the collision had occurred just outside of a tunnel and that the driver of the Mokpo train had apparently not seen the train in front until he was about 30 yards from it. A passenger named M.E. Penry of Flushing, New York, presumably the aforementioned Melborn Penry named in another newspaper, said that he had shouted a warning when he had seen the oncoming train and the coach he had been in had emptied quickly. 'He said he ran alongside the train, waving a small pocket flashlight in an effort to warn the soldiers in the two coaches ahead. Seconds late and just before the crash, he jumped from the right-of-way into a rice paddy. The crash, he said, sent the rear car "right through the one ahead – riding through at the level of the seat tops."[332] He added, "I've got to hand it to an M.P. {military policeman} who was giving a warning from the other side of the train. He got onto one of the soldier's cars and got caught in the vestibule when the crash came."[333]

Charles McCafferty, of Chicago, 'a labour relations adviser to the military government in the American zone, said, "We saw the train come out of the tunnel about 1,000 yards away. We yelled to the passengers {in the rear coach} and all got off except one man in the washroom who was not hurt seriously."[334]

The wreckage reportedly caught fire following the collision and a passenger likened the scene to a Dante's Inferno. Rescuers workers worked all night releasing the dead and injured, using acetylene torches to cut into the wreckage. 'Army officials said relatives of the dead were being notified but it may be necessary to bring personnel records from outlying detachments to complete identifications. About daybreak Wednesday a hospital train arrived in Seoul with 20 bodies and 16 of the injured.'[335]

Syngman Rhee, President of the Republic, offered his condolences to Major General John B. Coulter, commander of the United States occupation forces, and to General Douglas McArthur. Major General Coulter personally met the second relief train to reach

---

331 *San Mateo Times*, San Mateo, California, United States, 15th September 1948

332 *The Escanaba Daily Press*, Escanaba, Michigan, United States, 15th September 1948

333 *The Bismark Tribune*, Bismark, North Dakota, United States, 15th September 1948

334 *The Camden News*, Camden, Arkansas, United States, 15th September 1948

335 *The Bismark Tribune*, Bismark, North Dakota, United States, 15th September 1948

Seoul from the site of the accident. [336]

An injured soldier died in hospital two days after the accident, increasing the number of American soldiers killed in the accident to 36. [337] Two Korean children also died. The seriousness of the accident was reportedly not realised in Seoul until the arrival there of the first hospital train. [338]

Up train No. 103, the Chandpur – Sylhet Express, ran into the rear of a goods train standing in Gangajagar station, East Pakistan, on the Lakshmi – Akhaura section of the Eastern Bengal Railway, at 3.15 a.m. on Sunday 19th September 1948. Nine people died and 27 were injured. A relief train was rushed to the scene and the injured were taken to Comilla hospital. [339]

Thick fog covered most of England on Tuesday 30th November 1948 for the fourth day in a row, and the dislocation of road and rail traffic in London and the Home Counties was described as the worst ever known. Further north, in Stockport, the same conditions led to a serious railway accident on British Railways, London Midland Region. [340]

The dense fog led to great delays on the railway and to reduce the number of traffic paths various passenger train services were combined. Thus the 5.11 p.m. and 5.17 p.m. trains from Manchester London Road to Crewe and Disley departed as one train, as did the 5.40 p.m. and 5.50 p.m. services from Manchester to Buxton. In the winter darkness and fog the two trains crawled along the line towards Stockport. The platforms there were all occupied and the Crewe and Disley train was stopped for 7 or 8 minutes at the Up Fast home signals waiting for one to become clear. While it was still standing there it was run into from behind by the Buxton train at about 7.39 p.m. As a result four passengers on the Crewe train were killed outright, one was fatally hurt, five were seriously injured, 21 others and a guard were detained in hospital and nine were slightly injured.

The Crewe and Disley train consisted of 2-6-4 tank engine No. 2354 piloting 4-6-0 tender engine No. 5143 and twelve bogie coaches.

The Buxton train consisted of 4-6-0 tank engines Nos. 42370 and 2367 and nine coaches in charge of Guard J.F. Walker with Driver F.H. Williamson and Fireman W.R. Clarke on the pilot engine and Driver G.E. Wilson and Fireman R.E. Woodward on the train engine.

The severity of the fog may be gauged by the fact that the Crewe and Disley train took half an hour to cover the five miles from London Road station to Heaton Norris No. 2 box. Signalman W.H. Hayward was on duty there, assisted by a second signalman doing the booking and with a fogsignalman on the ground to check for tail lamps. According to the train register the Crewe train was accepted by Signalman B. Swindells in Heaton Morris No. 1 at 7.19 p.m. and passed No. 2 box at 7.20 p.m. It was held at the Heaton Norris No. 2 Up Fast Home signal for two or three minutes with its rear coaches fouling the junction at

[336] *Lubbock Morning Avalanche*, Lubbock, Texas, United States, 16th September 1948
[337] *The Tipton Daily Tribune*, Tipton, Indiana, United States, 17th September 1948
[338] *The Straits Times*, Singapore, 16th September 1948
[339] *The Indian Express*, Madras, India, 21st September 1948
[340] *The Times*, London, England, 1st December 1948

No. 2 box so that Signalman Hayward was unable to give the '*Train out of Section*' to No. 3 box until 7.25 p.m.

By then the Crewe train had been accepted by Signalman J.H. Ford at Stockport No. 2 box and it then restarted from Heaton Norris No. 1 Home signal and moved slowly through the station. Signalman Swindells gave the train a green light, which was seen by the engine crew. The train almost stopped at the Up Fast Starting signal at the south end of the platform before its engine crew could see that it was clear. Heaton Norris No. 1 box gave the *Train out of Section* signal to No. 2 box at 7.28 p.m. The Crewe train moved forward for 500 yards to the Stockport No. 2 Up home signals, where it stopped at about 7.32 p.m. The fireman of the train engine was forced to climb the signal gantry to verify that the signals were at danger.

Signalman Swindells thought the fog the worst that he had ever experienced. He told the Ministry of Transport inquiry into the accident that having been assured by Senior Porter H. Wilkinson that the Crewe train had passed his Up Starter complete with tail lamp he had cleared the train back to No. 2 box. The Buxton train was immediately offered to him and accepted, as the fogsignalmen were at their posts.

The Buxton combined train had left London Road station at 7.05 p.m. Signalman Hayward at Heaton Norris No. 2 box had accepted it on the Up Fast line at 7.29 p.m. as soon as he had let the 6.12 p.m. Down passenger train from Stockport to Stalybridge across the junction onto the Down Branch. He had already had it accepted by No. 1 box a minute earlier. This train passed No. 2 box at 7.30 p.m., came almost to a stand at the Up Fast Home signal for No. 1 box and then drew forward to the Up Starter for No. 1 box. Signalman Hayward cleared it back to No. 3 box at 7.33 p.m.

Signalman Swindells stated that having checked with Porter Wilkinson that his Up starter had returned to danger he felt that he had complied with Rule 65 (requiring that a signalman verify that a signal had responded to its lever) and that in accordance with local instructions he had been permitted to let the Buxton train up to the Starting signal, as he had been instructed that it was at *Danger*, though the local instructions really only applied to trains booked to stop at the station, which the Buxton train was not. He had not cleared his Home signal until the Buxton train had exploded the fogsignalman's detonator there. As the engines passed his box he had shouted, "I have no road for you," and showed a white light from the box door.

As the Buxton train came to a stand at the Up Fast Starter its coaches were alongside the platform and several passengers alighted, although it was not meant to stop there – did they perhaps think they were at Stockport? Guard Walker said that he had closed some doors, had seen a porter with a white light and had been about to make his way to No. 1 box to carry out Rule 55 when he heard a repeated shout of, "Right Away Driver," and his train had begun to move off.

Driver Williamson of the pilot engine of the Buxton train spoke of himself and his fireman having heard the explosion of a detonator at the Heaton Norris No. 1 Up Fast Home signal (with Stockport No. 2 Outer Distant signal beneath it). Neither of them had been able to see the signal but the fogsignalman had called out, "One off." This was confirmed by Fogsignalman J. Burns, who had been halfway up the signal post. Neither driver nor fireman had seen the white light shown by Signalman Swindells, nor heard his shout. At the Heaton Norris No. 1 Starter Fireman Clarke had been able to make out the yellow light of the Stockport No. 2 Up Fast Inner Distant signal but neither driver nor fireman could see the aspect of the Heaton Norris No. 1 Starting signal above it on the post.

Driver Williamson testified that after two or three minutes he and his fireman had heard a shout of, "Right Away Driver," which he had assumed must have come from the guard and he thought that the shout had been repeated. A moment later the train engine had "popped" its whistle and further reassured, he had opened the regulator and started off. Steam had been shut off after a short distance and he had been coasting when his engine ran into the back of the Crewe train. He admitted that he had forgotten that his train had been stopped at the platform where a member of the station staff might have called out, and said it had not occurred to him that the guard would have had insufficient time to get to the box, find that line was clear and return to the train.

Driver Wilson and Fireman Woodward on the train engine said that they had also heard a repeated "Right Away" call from a shadowy figure in the fog whom they had assumed must be their guard, while Driver Wilson alleged that he had also been given a green light. Neither Wilson nor Woodward had been able to see the Starting signal though they had assumed that Driver Williamson had.

It transpired that the shouted message heard by the engine crews had been given by Junior Porter A Hill, who was only 16 years old with six months' service. The latter said that after the doors of the Buxton train had been closed he had gone to the Stockport end of the platform and said to the drivers, "O.K. When you are ready." He had only said this once and was sure that he had not shouted, "Right Away, Driver," as this was an expression which he never used. He also maintained throughout that the hand lamp which he had been carrying had not at any time shown anything other than a white light and said that he had been instructed by Senior Porter Wilkinson to never on any account show a green light to a driver as he had no authority to do so. Porter Hill remembered speaking to Guard Walker as the train had pulled away (which Guard Walker confirmed) and added that he had then made his way towards No. 1 box, where Signalman Swindells told him that the Starting signal had not been cleared for the Buxton train.

Signalling Inspector T.G. Clarkson said that he considered Swindells to be a very good signalman. After the collision he had asked the signalman if his Starting signal had been at danger and the reply had been, "Yes, the senior porter told me."

By chance Divisional Signal and Telegraph Engineer C.G. Derbyshire had been a passenger on the Crewe train, riding in the front part of it. Realising what had happened he had gone after the collision first to Stockport No. 2 box and then to Heaton Norris No. 1 box to check the block controls. Signalman Ford at Stockport told him that the Buxton train had never been accepted and the Up Fast line block instrument there was correctly locked at *Train on Line* by the occupation of Track Circuit No. 2057 in the rear of the Up Fast Home signals by the Crewe train. Likewise, at Heaton Norris No. 1, the Up Fast Starting signal was locked normal by the block instrument at *Train on Line*. He had also climbed the Heaton Norris No. 1 Up Fast Starting signal to check that its arm was actually at danger and that its light was burning, which they had been.

Lieut. Colonel G.R.S. Wilson, the Ministry of Transport inspecting officer, remarked at the beginning of the conclusions to his report that the accident would not have happened but for the persistent dense fog, which had put no little strain on the operating staff.

There was no doubt, the inspecting officer said, that the Buxton train had not been accepted by Stockport No. 2 box and that it had entered the occupied section against the Heaton Norris No. 1 Starting signal at danger. Primary responsibility for the accident therefore rested with Driver Williamson on the pilot engine of the Buxton train, and to a lesser degree upon Driver Wilson on the train engine. Neither driver had made a serious effort to see the Starting signal but they had relied upon the assumption that a shouted message they had heard through the fog was from the guard. The exact wording of the message was immaterial. It seemed that each driver had relied upon the other. Porter Hill's denial of having shown a green light was convincing and even if he had done it was unlikely that it could have been seen through the fog by Driver Nelson.

Guard Walker was entitled to assume that his drivers had seen the Starter at clear when the train had started away.

Porter Hill had not been authorised to speak to the drivers but it was clear that he had only meant to indicate that the station work was complete.

The accident would not have happened had the Buxton train been held at the Heaton Morris No. 1 Up Fast Home signal but it would be unfair to criticise Signalman Swindells for letting the train down to the Starting signal in the circumstances.

Turning to remarks, the inspecting officer noted that there were different interpretations of Rule No. 39 and that the special local instructions had since been withdrawn.

The accident was a reminder of the difficulty in working in fogs with oil-lit semaphores and it was a tribute to the vigilance and detailed knowledge of enginemen that there were not more accidents. The substitution of colourlight signals would greatly improve matters and it was good to learn that there was a scheme for their installation on the busy line between Manchester and Stockport where fog was also prevalent during the winter months. [341]

What was described as the worst railway accident in the history of the Union of South Africa occurred on a high embankment at a location given in newspaper reports as Orlando in Natic township, 10 miles outside of Johannesburg, on the Midway – Langlaagte line. Three well-filled electric trains were involved in the disaster, which happened on Thursday 28th April 1949. The first was stopped at a signal, the second drew up behind it and the third then crashed into the rear of the second. As a result, the two rear coaches of the second train were telescoped and in addition the second coach of the front train was telescoped into the leading power car. The coaches were 'smashed to matchwood' and there were 40 deaths in the second car of the front train alone.

During the rescue operation firemen with oxy-acetylene torches cut their way into the wreckage while passengers who were able to climbed out of windows and fled into the surrounding countryside. Thousands of 'natives' from Orlando rushed to the scene fearing that their relatives were aboard the train. Eventually the police were forced to rope off the area to allow the rescue work to go on without impedance. A trapped woman passenger reportedly had her leg amputated by a fireman using a penknife 'sterilised' in the flame of matches. It was necessary to lash the injured to stretchers so that they could be carried to the top of the steep cutting in which the accident had occurred. 'Native' schoolchildren riding in the third train had a lucky escape, though one was thrown through a window. [342]

The number of dead was variously given in different newspaper accounts as 70, 72 and 74. Two of those who died were reportedly Europeans, these being the conductor of

---

[341] Ministry of Transport report

[342] *The Singapore Free Press*, Singapore, 29th April 1949

the second train and the motorman of the third. Of the others 14 were said to have been Indians and the remainder were described as 'natives.' [343] [344] A professional railway magazine gives the presumably more accurate figures of 68 fatalities, comprising the conductor, the motorman and 66 'Non-Europeans,' adding that 166 'Non-Europeans had been injured.' This magazine gives the location of the collision as between Mlamwankunzi Halt and New Canada station and states that the first train had been stopped at the New Canada Junction Intermediate Home signal, with the second train drawn up two coach lengths behind it. Despite unfavourable weather the lines were cleared in 37 hours. [345]

An investigation into the disaster determined that it had occurred due to the block system being faulty.

A relief fund was set up to support the victims of the accident by the mayor of Johannesburg, towards which the city council voted to donate £2,500. [346]

A signalling error led to a rear collision on the Malmbanan in northern Sweden on Sunday 4th September 1949. Iron ore train No. 95 was on its way to the port of Luleå and was being followed by passenger train No. 21 from Narvik in Norway. The iron ore train stalled on a steep gradient between Gällivare and Harrträsk. It was being set back in order to get a run at the gradient when it was run into from behind by the passenger train. Four people died, six were seriously injured and others were slightly hurt. The Gällivare *tågklareraren* (signalman) had mistakenly allowed the passenger train to depart despite the ore train not having been reported as having arrived at Ripats, 7 kilometres beyond Harrträsk and 20 kilometres from Gällivare. There was presumably no *tågklareraren* on duty at Harrträsk station because it was a Sunday. [347]

At 19:45 on the evening of Monday 7th November 1949 a serious rear collision occurred between Las Matas and Las Rosas stations, Spain. The two electric trains services involved were train No. 7009, the 19:25 departure from Madrid Principe Pio to Segovia, and train No. 6005 from the same station to Ávila, which departed at 19:30. The Segovia train was stopped at a red signal at kilometre 21.950 because shunting operations were in progress in the station ahead, and before it could proceed it was rammed by the Ávila train. There was a suggestion that there had been no tail lamp at the rear of the Segovia train so that the motorman of the Ávila train had only become aware of its presence at the last moment when he had applied the brakes in emergency. Because of this the damage was limited to the rear car of the front train and the front car of the colliding one, but even so 14 people died, including the motorman of the Ávila train, and 41 more were injured. [348]

The initial first aid was provided by members of the Civil Guard and a group of youths from the Madrid Hogar del Empleado, both of which parties had been travelling on the Segovia train. (The Hogar del Empleado was a charitable organisation which had

---

343 *The Times*, London, England, 29th April 1949
344 *Bluefield Daily Telegraph*, Bluefield, West Virginia, United States, 29th April 1949
345 *The Railway Gazette*, London, England, 3rd June 1949
346 *The Times*, London, England, 29th April 1949
347 https://www.ekeving.se/hi/SJdok/SakTj/1969.pdf
348 Webpage: Federación Castellano Manchega de Amigos del Ferrocarril (in Spanish)

been set up in 1946 to provide aid for young people who had moved to the big cities in search of employment; the youths on the train had been going on holiday to Segovia.) Three priests who had been riding on the trains provided spiritual aid to the victims while two women who had been passing the scene in a car on the adjoining highway also lent their help. The first relief train arrived from Madrid two hours after the accident, and further trains followed; the first train returned to the capital with twenty seriously injured passengers and the bodies of three of the dead.

The work of clearing the line continued throughout the night. Neither of the trains, which had each consisted of a motor unit and a trailer, was derailed and the track was little damaged. Single-line working was brought into operation soon after the accident and normal was resumed at 08:47 the following morning.

Red Nacional de los Ferrocarriles Españoles (RENFE), the state railway company, issued a statement regarding the accident in which it strove to correct incorrect statements which had appeared in the Press. It stated that the line on which the accident occurred was provided with automatic colourlight signals, which had been working normally. The interior lights of the Segovia train had also all been lit and the collision had occurred on a level and straight portion of the line, so the signals could be easily seen. Conductor José Antonio Alvarez Merallo of the Ávila train, who was fortunate to survive the accident, had stated that his motorman had stopped the train at the previous red signal, but only halted very briefly before setting off again. Realising the danger the conductor had then hurriedly made his way to the motorman's cab, where he had found the latter in a semi-conscious state, as if he were suffering from the effects of a nervous shock. At the last moment the motorman had released the 'dead man's handle,' thus activating the brakes of the train, but too late to prevent the collision.

The motorman was an experienced employee with 15 years' service and an excellent record who had operated the most modern and complicated trains. It seemed clear therefore that the accident had resulted not from his negligence but from a medical condition which had left him unable to control the train. [349] Probably he had suffered either a heart attack or a severe case of angina with crushing chest pain. Such a condition should have been revealed at a post mortem, were one carried out.

The evening express train from Cairo to Alexandria ran into the back of another passenger train between the two cities at Hagar Nawatieh, Egypt, four miles outside of Alexandria, on Wednesday 7th December 1949. The first train had stopped in the section because of engine trouble. Six or seven people died and between 30 and 45 were injured, a number of them seriously. [350] [351]

The driver of the colliding train stated that the signals had been clear for him and he had only seen the train in front when he had been 600 yards from it. He had then applied his brakes in emergency but had still been travelling at 40 miles an hour when the collision had occurred. [352]

Seven people died and 37 were injured when an express train ran into the rear of a

---

[349] *La Vanguardia*, Barcelona, Spain, 9th November 1949

[350] *The Times*, London, England, 8th December 1949

[351] *The Independent Record*, Helena, Montana, United States, 8th December 1949

[352] *The Western Australian*, Perth, Western Australia, Australia, 9th December 1949

freight train on Friday 9th December 1949, probably between Lipnik and Hranice, Moravia, Czechoslovakia. [353] [354]

Korea and an American troop train were involved in the next accident to be considered, which happened on 11th September 1950, though unlike the one two years earlier this collision occurred in the United States. The Korean War had broken out on 25th June 1950 and the troop train involved in the accident was carrying members of the Pennsylvania National Guard from their home state to Camp Atterbury, Indiana, where they were to undergo training before departing for Korea. The train came to an unintended halt east of Coshocton, Ohio, on the Pennsylvania Railroad, and before it could be restarted it was run into by the *Spirit of St Louis* express passenger train. Thirty-three passengers (all troops) were killed and 258 passengers, the conductor of the troop train, the conductor, engineer, fireman, five dining car employees, five Pullman employees, three mail clerks, a coach attendant and a passenger representative on the express were injured.

The troop train, Passenger Extra 5444 West, consisted of steam engines Nos. 5444 and 5526, two express cars, five coaches, a kitchen car, five coaches, a kitchen car and six coaches, in charge of Conductor E.D. Doak with Engineers P.W. Longstreth and J. Shustich and two firemen, one of whom was named G.T. Vorres. It had left Carbondale, Pennsylvania, at 2 p.m. the previous day and Wilkes-Barre at 4.30 p.m. Departure from Pittsburgh had been at 1.40 a.m., the last open office at Newcomerstown, Ohio, rather more than 6¾ miles east of the scene of the accident, was passed at 4.28 a.m., and the train was stopped by an emergency brake application at 4.38 a.m., 7½ miles east of Coshocton and 185 feet west of signal No. 1149.

The *Spirit of St Louis,* westbound first-class passenger train No. 31 from New York to St Louis, Missouri, consisted of diesel electric locomotives Nos. 5774A and 5776A, coupled in multiple, a storage mail car, a mail car, a baggage car, four sleeping cars, two dining cars, two sleeping cars, five coaches and an observation car, in charge of Conductor M. McCafferty with Engineer William Eller and Fireman Edwin Kearns. It had left Pittsburgh at 2.09 a.m., 29 minutes late, passed Newcomerstown at 4.35 a.m., 40 minutes late, passed signal No. 1129, indicating *Approach*, passed the flagman of the troop train, passed signal No. 1149, indicating *Stop and Proceed*, and was still moving at almost 50 miles an hour when it struck the troop train at about 4.42 a.m. [355]

Standing at the rear of the troop train when the collision occurred had been Captain F.R. Brannan of the 109th Infantry regiment, who saw the *Spirit of St Louis* approaching. He told a newspaper reporter, "The red flares were out on the track at the time and we couldn't figure out what was going on. I dashed through the train yelling to other officers and men to fall to the floor, I pulled a lot of them down with me. But it happened so fast we couldn't do much."

The newspaper account continued, 'Many of the victims were asleep when the disaster occurred and survivors said most of them were unaware that the train even had stopped.

'Fog was so dense that some ambulances and doctors had trouble locating the scene of the disaster, which was just at the rear of the Harry F. Shurtz residence, east of West

353 *Nevada State Journal*, Reno, Nevada, United States, 11th December 1949
354 *Sarasota Herald-Tribune*, Sarasota, United States, 11th December 1949
355 Interstate Commerce Commission report

Lafayette. A narrow farm road in the rear of the Shurtz farm was jammed with ambulances, trucks, and autos within a half hour after the tragedy. Curiosity seekers were asked to keep away as salvage work proceeded today. West Lafayette volunteer firemen, American Legion members and Red Cross workers were at the scene early. Rev. Glenn F. Lockard, pastor of the Main St. Methodist church at West Lafayette, and other ministers were on hand to offer assistance. The state highway patrol {were} also on duty, and sandwiches and coffee were served at the scene by Boy Scouts and by the American Legion Auxiliary.

'Lt-Col. Frank Townsend, Wilkes-Barre, Pa., was in charge of the troop train, one of four taking guardsmen to Camp Atterbury yesterday, today, and tomorrow. Two other troop trains preceded the one which was wrecked.

'Sgt. Joseph Warbola, of Simpson, Pa., related that he was standing in a car about eight coaches from the rear of the troop train when the crash occurred. "I was getting a drink of water," Sgt. Warbola said. "There were two jolts. The first knocked me against a wall and the second threw me to the floor." [356]

'One man, Sgt. Lester Kuhn, of Wilkes-Barre, Pa., died of injuries in the local {Coshocton Memorial} hospital last night. An ironical twist of the disaster is that Sgt. Kuhn and his son, Cpl. Daniel L. Kuhn, also of Wilkes-Barre, both were members of the Pennsylvania National Guard, which was moving to Camp Attlebury, Ind., for training when the accident occurred. Cpl. Kuhn, in a forward car, was not seriously hurt, and assisted in the rescue work. In the darkness and fog he helped carry his mortally wounded father to a stretcher. "Is he badly hurt?" the youth inquired anxiously. "I'm afraid so," someone replied, "why do you ask?" "He's my father," Cpl Kuhn replied, and returned to the rescue work.

'The corporal was with his father when the latter died in the local hospital at 9.15 last night. He was then granted a furlough and when his mother and sister, Miss Clara Kuhn, arrived here last night they stayed at the Verle Rucker home, 1141 Kenilworth av. They planned to return home today for the planned funeral.' [357]

Flagman W.G. Lancaster of the troop train told the Interstate Commerce Commission (ICC) inquiry that near Dennison, Ohio, 24½ miles from the site of the accident, he had become aware of something dragging near the rear of the train. He had opened the conductor's valve to stop the train and an inspection had shown that the steam connector at the front of the last car had become detached and was striking the track. He and the conductor made a temporary repair and the train had continued to Dennison, where both engines had taken water and he had gone eastwards to provide flag protection. The *Spirit of St Louis* had arrived a few minutes later, had acknowledged his signals and had stopped at a *Stop and Proceed* signal behind the troop train. The troop train had then proceeded, passing Uhrichsville, 23¼ miles from the scene of the collision, at 4.13 a.m. When the brakes had come on suddenly in emergency he dropped a five-minute lighted red fusee. The conductor and brakeman found that the steam connector had again struck the track and this time had also damaged the air hose, causing the brakes to apply. Meanwhile the flagman said that he had again gone back to protect his train, equipped with white and red lights, torpedoes and fusees. He had seen the headlight of the express three-quarters of a mile away, signal No. 49, which was approach lit, not yet being

[356] The Coshocton Tribune, Coshocton, Ohio, United States, 11th September 1950
[357] Ibid, 12th September 1950

illuminated because the express had not come onto the approach track circuit. Continuing eastwards, the flagman testified, he had shown a red light to the express, then had lit another red fusee. His signals had not been answered. As the express passed him he had thrown his fusee at it. By then signal No. 1149 had been lit and was displaying *Stop and Proceed*. He saw that the first fusee he had dropped was still burning.

Fireman Kearns of the express told the inquiry that their train had been stopped at a *Stop and Proceed* signal at Dennison where he had telephoned the operator at Uhrichsville, who informed him that the troop train had stopped for water but was now on its way. A speed restriction at Newcomerstown had been complied with and the signals had all displayed *Proceed* up to No. 1129, which had displayed *Approach*. He had called the indication to Engineer Eller and the cab signal display on the locomotive had changed from *Proceed* to *Approach*. A warning whistle had also sounded, which Engineer Eller cancelled by pressing a foot pedal, and a service brake application had been made. The speed of the train had dropped at first but had then increased again to 70 miles an hour on the falling gradients. Engineer Eller and Fireman Kearns agreed that it had been somewhat foggy. When first the stop signals being shown by Flagman Lancaster and then the *Stop and Proceed* indication of signal No. 1149 had come into view another service brake application had been made, immediately followed by an emergency application when the rear of the troop train was seen. The speed of the train had only fallen to 48 miles an hour when it had struck the troop train.

The ICC noted that as a result of the collision the leading diesel unit of the express had crushed the superstructure of the rear car for about 10 feet at each end. The diesel unit had then been deflected upwards and penetrated the penultimate car above floor level. It was in this latter car that most of the casualties had been found. The all-steel cars of the troop train dated from 1909 but had conformed to the minimum standards of strength adopted by the Association of American Railroads in 1945.

Engineer Eller alleged that he had been travelling at 50 miles an hour when he had seen the signals of the flagman but according to the regulations he should have reduced speed to 30 miles an hour after passing a signal displaying *Approach* and at such a speed as to be able to stop at the next signal.

The ICC concluded that the accident had been due to the failure of the engineer of train No. 31 to operate his train in accordance with signal indications. [358]

In January 1951 County Prosecutor Robert Carton informed a grand jury that despite the conclusions of the ICC there was no Ohio law under which Engineer Eller could be charged with any offence. [359]

Just over two months after the Coshocton disaster a considerably more serious rear collision in terms of fatalities occurred on the Long Island Railroad in the New York commuter area, at a location commonly known as Richmond Hill, though the Interstate Commerce Commission (ICC) gives the location as 'near Jamaica.' Two eastbound commuter trains were involved in the accident on Wednesday 22nd November 1950. Train No. 780 came to a stand because of brake problems and was run into by the following train, No. 174. The accident happened at the end of the evening rush hour and both trains were crowded. Seventy-seven passengers and the motorman of the colliding train lost

---

[358] Interstate Commerce Commission report

[359] *The Coshocton Tribune*, Coshocton, Ohio, United States, 9th January 1951

their lives and 358 passengers, the flagman and two ticket collectors of train No. 780, the conductor, front brakeman, flagman and two ticket collectors of train No. 174, and three other employees were injured.

Newspaper reporters described the scene of the disaster. 'The crash occurred at 128th street and Metropolitan avenue, on the outskirts of Kew Gardens and not far from the Jamaica station. One train made an unscheduled stop in front of another and was rammed in the rear. Eyewitnesses reported a great roar and a flash of fire as the trains hit.

'"One passenger was thrown on top of another," said accountant Jeff Kaufman, a passenger on his way home to Bellmore, N.Y. "Everyone was screaming, hollering and moaning." Kaufman said he heard the conductor of the train ask the motorman to stop at the Jamaica station to let off two women who had boarded the train by mistake at the Pennsylvania station. The accountant said the train was not due to stop until it reached Rockville Centre station, scene of a disastrous Long Island wreck last Feb. 17 when 31 persons were killed and 100 injured {covered in the next chapter of this book}.

'Tonight's crash occurred on Thanksgiving eve as trainloads of persons headed out of the city for the holiday. Every available ambulance was summoned from Queens Borough, the third largest in New York City. Additional aid was sought from neighbouring Brooklyn. At Queens General Hospital a spokesman said the casualties coming in there "are very badly injured people."

'Dave George, a publicity representative for the railroad, was aboard one train. "There was a dead silence for seconds after the accident," he said, followed by screams and cries of 'Help, Help.'"

'Kaufman said he was on the platform near the front of the lead train. He said: "Suddenly the lights went off. Everything shook. I smelled something like dust – like in the last war when I was wounded. I saw steel shooting through the car. The car was sheared right in half – like a slice of bread. I was knocked down and shaken up but not hurt. A man said to me, 'You were lucky to get out alive.'"

'Kaufman said he heard the two women – one a well-dressed redhead – tell the conductor they boarded the train by mistake. They had intended to take another train to Port Washington. The accountant said the conductor walked forward and within his hearing asked the motorman to stop at Jamaica to let the women off. Kaufman said the train had slowed down almost to a stop on its approach to Jamaica when the crash came. After the crash, Kaufman said, "bodies were thrown on top of one another – one piled right on top of another. I saw a woman's body beside the tracks where she had been thrown," Kaufman went on. "She was dead. It looked to me like blood was pouring all over. People were screaming and hollering but there was no panic. I tried to quiet some of them down and helped some injured off the train."

'The figure on the number believed dead came from police on the 102nd precinct, in which the wreck occurred. Doctors and firemen tried desperately to enter the top portion of the telescoped cars, in which, apparently, some of the victims had been jammed to the roof. Many victims were removed on stretchers. At least 100 passengers must have been riding in the two cars. Priests came to the scene to give the last rites. Firemen placed ladders against the jammed cars.' [360]

Another newspaper carried an article describing the impact of the disaster on

---

[360] *Bluefield Daily Telegraph*, Bluefield, West Virginia, United States, 23rd November 1950

individual families. 'The telephone rang in a small suburban home in Garden City. A 17-year-old boy picked up the receiver. "Oh," he said, not wanting to believe. Then, slowly, "M-m-my father always took the 6.09."

'The telephone rang again at Floral Park. A father answered. "Is it certain. Are you sure he is really dead?" "Yes." "Oh, my God…Why, that's our last boy…The other was in a B-17."{The Boeing B-17 Flying Fortress was a heavy bomber aircraft, the man's other son had doubtless been killed during the war.}

'The telephone rang again – a shrill, impatient death knell making the round of the commuter communities that dot Long Island. A woman answered at East Rockaway. She screamed.

'All night long the telephone rang in scores of homes that will not celebrate Thanksgiving today. The calls brought the tragic news of a major disaster on the Long Island Railroad – the worst train wreck in the history of New York State – and one of the worst in the history of the Nation.

'A warm and happy crowd had pushed and jostled its way through the loading gates at Manhattan's Pennsylvania Station only a few minutes before. The day was over, tomorrow was Thanksgiving, and the big city's commuters were hurrying home to wives and families. About 1,200 of them boarded No. 780 of the bankrupt Long Island Railroad. Another thousand crowded onto No. 174. It was the 6.13, headed down the same track for Babylon through Rockville Center. It was dark early, and a cold, chill moon, almost full, gleamed on the track ahead. Inside the cars, hot and stuff, the extra crowds stood in the aisles, swaying as the 12-car trains plunged eastward. It had been nine months since the disastrous head-on collision at Rockville Center killed 31. Probably nobody was thinking of that wreck when the 6.09 made an unscheduled stop.' [361]

Eastbound first-class passenger train No. 780 consisted of a 12-car multiple unit set of all-steel construction, in charge of an unnamed conductor with Motorman William Murphy. It had passed Win, the last open office, 4¼ miles west of the scene of the accident, on time at 6.19 p.m., passed signal 'C,' indicating *Approach*, and stopped on track No. 2, 3,516 feet east of the signal and 1¼ miles west of Jamaica, at about 6.26 p.m. Roughly three minutes later it was struck by train No. 174.

Eastbound first-class passenger train No. 174 likewise consisted of a 12-car multiple unit set of all-steel construction, in charge of Conductor Howard Austin with Motorman Benjamin Pokorney. It passed Win on time at 6.23 p.m., passed signal No. 66, indicating *Approach*, and stopped at signal 'C', which indicated *Stop and Proceed*. It then set off again and was still travelling at 30 miles an hour when it hit train No. 780. As a result of the collision the rear car of train No. 780 was destroyed and the eleventh car somewhat damaged, while in train No. 174 the upper portion of its leading car was sheared off by the rear car of train No. 780, which had been deflected upwards.

Engineer Murphy of train No. 780 told the ICC investigation that on seeing signal 'C' at *Approach* he had made a light service brake application. Releasing the brakes when the speed of his train had been reduced to 30 miles an hour he had continued without applying power. The next signal, No. 114R, had indicated *Restricting* and he had made a further service application. When the speed of his train had dropped to 15 miles he released his brakes again but this time the speed of his train continued to drop; he

[361] *The Evening Star*, Washington, D.C., United States, 23rd November 1950

released that the brakes had failed to release properly and his train had come to a halt. He had been still trying to release the brakes when his train had been run into.

Flagman Bertrand Biggam stated that when train No. 780 had stopped he had gone to the rear vestibule of the twelfth car with a red lantern and had alighted to provide flag protection. He had only gone back a few feet when he heard power being applied so he had reboarded the train and given a proceed signal. When the train did not move he had repeated the signal. He had then seen train No. 174 approaching and about 1,000 feet away. There had only been time to give a stop signal with his flashlight before the collision had occurred. No lighted fusee had been displayed at any time after his train had stopped. The marker lights of his train had been in position and lit when they had left New York; he had not observed them after that.

Members of the crew of train No. 174 agreed that their train had stopped at signal 'C' and then proceeded. Conductor Austin testified that at first speed had been low but when the second car of his train, in which he had been riding, had been at Kew Gardens, 1,960 feet from the scene of the accident, the speed of the train noticeably increased until it had reached about 35 miles an hour, when there had suddenly been an emergency brake application, a few seconds before the collision.

The ICC remarked that with the death of the motorman it was impossible to determine why he had not operated his train at restricted speed through the section. After his train had passed signal 'C' the indication of signal 116R had changed from *Restricting* to *Approach* as a preceding train had cleared the block section ahead. It seems likely that the motorman of train No. 174 had 'read ahead' to this signal, unaware until he was close to it that train No. 780 was stopped behind it.

Automatic train stop devices of the tripper type, the commission continued, were provided on the Atlantic Beach branch between Dunton Tower, Jamaica, and Flatbush Avenue and at two drawbridges on the Rockaway Beach branch. These, however, would probably not have prevented the accident as they could be reset by the motorman and the train then continue. The same was true as regards other train stop devices in use on other railroads. Automatic block signals were in use on 160 miles of the Long Island system and an automatic cab signalling system was in use on 44 miles. These provided a continuous display in the cab of trains of the track conditions ahead. Had this system been in use on the line where the accident occurred the motorman of train No. 174 could have received a continuous *Restricting* indication after passing signal 'C' and the accident would probably have been prevented. The trains were already provided with the equipment needed and the necessary track equipment should be provided immediately to extend the area over which an automatic cab signalling system was in use and thus increase protection.

Even an automatic cab signalling system combined with automatic block signalling did not, the ICC noted, enforce obedience to signal indications, as had been shown by the February Rockville Centre accident. On an average weekday 370 trains passed the point of the accident, 99% of which were passenger trains. Given the volume of traffic an automatic train control system which automatically and continuously enforced signal compliance should be installed. Should notice not be received within 30 days that these recommendations were being complied with, the ICC warned, consideration would be given to instituting further proceedings under Section 25 of the Interstate Commerce Act.

362

The section referred to stated in part that, 'The Commission may, after investigation, if found necessary in the public interest, order any carrier within a time specified in the order, to install the block signal system, interlocking, automatic train stop, train control, and/or cab-signal devices and/or other similar appliances, methods, and systems intended to promote the safety of railroad operation which comply with specifications and requirements prescribed by the Commission, upon the whole or any part of its railroad such order to be issued and published a reasonable time (as determined by the Commission) in advance of the date for its fulfilment.' [363]

A rear collision occurred near Bryn Mawr, Pennsylvania, United States, on the Pennsylvania Railroad, at 6.39 a.m. on Friday 18th May 1951. Seven passengers and a Pullman employee were killed and 95 passengers, 14 Pullman car employees, eight dining car employees and six railwaymen (the conductor, engineer and baggageman of the front train and the engineer, fireman and brakeman of the colliding train) were injured. [364]

A newspaper report wrote, 'The *Red Arrow* rammed into a night express train from Pittsburgh to Philadelphia which had stopped one quarter mile from the Bryn Mawr station in response to a stop signal. Witnesses said the *Red Arrow* tore through the rear sleeping car "Poplar Vale," from Cleveland. It cut it in two and rose with it into the air in a flash of flame. Then the locomotive crashed nose first into the ground. So completely was the sleeping car demolished that its parts could not be separated from the locomotive. A berth was caught against the front of the locomotive and hung there swinging.

'Clothing, suitcases, train equipment and jagged bits of steel littered the four main line tracks, blocking the east-west route. Three hours after the wreck officials were not sure how many were dead. Giant cranes had to lift the locomotive to pry the bodies from the mass of twisted steel. The locomotive dragged an empty crew dormitory car off the tracks when it telescoped the sleeping car of the Pittsburgh to Philadelphia train. A postal car just behind the engine of the halted train also was derailed.

'A Pennsylvania Railroad official said that the train from Pittsburgh had stopped west of Bryn Mawr when it set off a trouble signal which automatically threw the track signals ahead into {the} stop position. In railroad talk the train "accentuated" {sic} the drag detector, a device between the two rails to catch anything hanging down from the cars. This is a safety measure to warn crew the brake rigging is faulty. While crew members were trying to find out what was wrong with the night express, the Red Arrow ran into it with its whistle screaming, witnesses said. A railroad spokesman said the Red Arrow was travelling at "only about half its normal speed" at the time or the casualties would have been higher. He said the Red Arrow engineer "apparently had gotten some sort of signal or he wouldn't have slowed down"

'One huge piece of the telescoped car pointed into the air; its car ripped open as if a can opener had cut it. The drawn out whistle before the crash, the terrific noise, and the flash of flame when the locomotive and rear sleeper rose together into the air, aroused residents of the main line community outside of Philadelphia. There were approximately

---

362 Interstate Commerce Commission report

363 https://ideas.dickinsonlaw.psu.edu/cgi/viewcontent.cgi?article=2526&context=dlra

364 Interstate Commerce Commission report

200 passengers in each of the trains. The Red Arrow was made up of 10 cars and the Pittsburgh express of 19 cars.

'The Pennsylvania Railroad said it expects to have two of its four main line tracks at the Bryn Mawr wreck scene open for traffic by "early afternoon." Trains on the main east-west PRR route are being detoured over the Trenton cut off between Downingtown, Pa., to Trenton, N.J.

'George Webb, a passenger on the Pittsburgh to Philadelphia train, said he saw three decapitated bodies from the rear car of his train and one crew member "apparently killed" in the second train. Webb, of Glenmore, near Pottstown, Pa., said the engine of the Red Arrow "ploughed right the rear car of the standing night express. "It was awful," he said. "It was as if you took both sides and the roof of the rear car of our train, lifted them up in the air and piled them on top of the Red Arrow engine. The rear car was demolished and the second car from the rear was knocked over. The engine of the Red Arrow also toppled over across the adjacent tracks." Webb was pinned beneath two seats but "I was lucky because I was sitting in the front seat of the first car. People were screaming and were hurt as far back as the 10 car," he said. "There were about 15 cars in our train." Webb said the train had been stopped for about 15 minutes "apparently by crossed signals." Webb said that after the crash, the engineer of the Red Arrow "was dangling out of the engine window. I guess he was dead. He looked all limp like a dummy." Webb said the Pittsburgh train was running 30 minutes late when it pulled up near the Bryn Mawr station. What I can't understand," he said, "was why we didn't get the drag signal that an express was coming through."

'Mrs Lucille Reardon, 29, Detroit, took her three children to the lounge car of the Red Arrow express train today, "to stretch our legs before we got to Philadelphia." When they returned to the roomette, "It was a complete shambles," said Mrs Reardon. "I'm certain the children would have been injured if we had been in the room when our train hit the one in front," she said. We were further away in the lounge car. Mrs Reardon was taking Paul, 3; Maureen, 2; and David, 10 months, to visit her parents, Mr and Mrs C.H. Cobbs Jr., in Philadelphia. A fourth child, Mary Beth, 4, came ahead of them on the same train last week.'[365]

As noted in the newspaper report the line on which the accident occurred was quadruple track. From north to south the tracks were: No. 4 westbound passenger, No. 3 westbound freight, No. 2 eastbound freight and No. 1 eastbound passenger. The lines were fully track circuited and colourlight position light signals were provided. Those applying to track No. 2, the one on which the accident occurred, were Nos. 10R, 118 and 132, situated respectively 1,554 feet east of and 1 mile 1109 feet and 2 miles 3485 feet west of the accident point.

Eastbound first-class passenger train No. 36, the Pittsburgh express, consisted of electric locomotive No. 4868, five storage mail cars, two express cars, a combination passenger baggage car, a coach, and ten sleeping cars. It had left Paoli at 6.24 a.m., 41 minutes late on track No. 1, had been diverted onto track 2 immediately afterwards, had passed signal No. 118, indicating *Approach,* and had stopped at signal No. 10R, displaying *Stop*, at 6.32 a.m. Seven minutes later it had been run into from behind.

Eastbound first-class passenger train No. 68, otherwise, the *Red Arrow*, consisted of electric locomotive No. 4839, a dormitory sleeping car, a baggage car, two coaches, two

[365] *The Daily News*, Huntingdon and Mount Union, Pennsylvania, United States, 18th May 1951

sleeping cars, a dining car and three sleeping cars. It had left Paoli on track No. 4 at 6.24 a.m., five minutes late, and was switched back to track No. 2 at the east end of the station. It passed signal No. 132, indicating *Approach*; stopped at signal No. 118, which was displaying *Stop and Proceed*, restarted and was travelling at an estimated 40 miles an hour when it ran into the back of train No. 36.

The Interstate Commerce Commission (ICC) report on the accident noted that when train No. 36 had stopped at signal No. 10R its fireman had gone to the telephone adjoining the signal and spoken to the Bryn Mawr operator, who had advised him that a dragging equipment detector near signal No. 132 had detected a problem and put signal No. 10R to danger. The engineer, fireman and conductor had then proceeded to inspect the train, looking for the problem, while the flagman had gone back with a red flag, torpedoes and a fusee to protect the train. Seeing train No. 68 approaching when he was about 600 feet from the rear of his train the flagman said that he had waved his flag. The oncoming train had whistled and its brakes had been applied as it passed him.

The brakes of train No. 68 had been tested before it left Harrisburg. As it had approached signal No. 132 the fireman had left the engine cab to adjust the steam heating apparatus. The engineer had called out that the signal was showing *Approach* and the fireman had acknowledged this. Signal No. 118 had displayed *Stop and Proceed* and the train stopped 75 feet short of it. The train then proceeded, its cab signal displaying *Approach medium* as it passed the signal and accelerated to 40 miles an hour. Engineer F.B. Yentzer (who contrary to the impression given by the newspaper report had been injured but not killed) told the ICC that as soon as his engine had passed Signal No. 118 the cab signal changed to *Restricting*. He had repeated the *Approach medium* and *Restricting* aspects to his fireman, who was now back in the cab. Two hundred yards east of signal No. 118 the engineer said that the cab signal had changed to *Approach* and the speed had been increased to 30 miles an hour, while about 1,000 feet east of signal No. 118 the aspect had changed to *Approach medium* and the speed had been increased again to 45 miles an hour. The fireman stated that he had seen the flagman of train No. 36 in the vicinity of Rosemont station, 1,613 west of the point of the accident. He had shouted a warning to the engineer, who had then, himself, seen the flagman and applied the brakes in emergency, to little effect, before the collision.

The engineer's account of the various cab signal indications which he had received after passing Signal No. 118 were not supported by his fireman. As might be expected very extensive tests were carried out after the accident on the signals and on the cab signal equipment of the engine and nothing was found which could have caused false signal indications. The ICC concluded that the lineside and cab signals had been working properly. It noted that the engine cab signals provided a continuous display of lineside signal aspects but did not enforce compliance with them, as an automatic train control device would have done. The Pennsylvania company stated that it was progressing with the installation of speed controlling devices as rapidly as possible following the recommendations of the ICC after the Woodbridge derailment (see above). [366]

Forty people probably lost their lives on the evening of Sunday 17th June 1951 when a special freight train loaded with wheat ran into the rear of a stationary passenger train at Ghotki, West Pakistan, on what since 1947 had been the Pakistan Western Railway

---

[366] Interstate Commerce Commission report

(previously known as the North Western Railway). Reports reaching Karachi from Sukkur, 50 miles south of Ghotki, 'where a relief station has been set up,' said that the exact toll was not yet known but that the number of injured was 'very heavy.' Rescuers worked at the scene through the night. [367] Five coaches of the Quetta – Lahore passenger train were reportedly destroyed and it took rescuers over six hours to reach all the dead and injured. Relief trains took the injured to Sukkur. [368]

An express commuter train from Plaza Constitución to Temperley ran into the rear of a local train to San Vicente standing in Hipórito Yrigoyen station, on the Ferrocarril General Roca line, in the Buenos Aires suburbs of Argentina, on Saturday 23rd June 1951. Some reports give the location of the accident as Barracas, the suburb where the accident occurred, and the station had formerly been called Barracas al Norte. Two coaches were telescoped and another was thrown into the air to land on that in front. Six or seven people reportedly died and 45 were injured, some critically. [369] A dense early morning fog was believed to have contributed to the accident. [370] [371]

Ford station, England, on the West Coast line of the former London Brighton and South Coast Railway, now part of British Railways, Southern Region, was the scene of a serious accident on Sunday 5th August 1951. Two electric passenger trains were involved. A train from Brighton to Portsmouth ran into the rear of one from Three Bridges to Bognor at 11.58 a.m. Seven passengers and the motorman of the Brighton train were killed outright, an eighth passenger was fatally injured and died in hospital, and 47 more passengers were hurt, 40 of whom were admitted to hospital.

After the accident, 'In answer to a general call for all available help, doctors, police and ambulances from Bognor, Littlehampton, Arundel, and Chichester went to the scene and helped to extricate the injured. Doctors gave first-aid as the passengers, many in holiday clothes, lay beside the damaged trains. Six fire appliances were sent by West Sussex fire brigade. There was no fire, but firemen undertook the rescue of trapped passengers. They were helped by members of the Royal Naval Air Service station at Ford.

'Seventeen passengers were admitted to St Richard's Hospital, Chichester, and 12 to the Royal West Sussex Hospital, Chichester. Thirteen were taken to Worthing Hospital, of whom seven were detained.

'Mrs Hoad, wife of the licensee of the Arundel Arms, outside the station, sent out the calls at the request of Mr A. Miles, Stationmaster, and many injured were carried into the inn. "I thought two jet aircraft had crashed in the garden," said Mrs Hoad. "My daughter tore up sheets for bandages, and my niece, Joan Cornford, ran across the line with blankets, tea, and brandy for the injured."

'Assistant Divisional Officer L.E. Walwin said: "Those in the front coach who escaped death were mostly facing the front. They were catapulted forward clear of the wreckage. The others were trapped in the wreckage when seats were forced upwards in a heap, and the bodies crumpled between them. One woman was trapped with two broken

[367] *Pacific Stars and Stripes*, Tokyo, Japan, 20th June 1951

[368] *The Daily Inter Lake*, Kalispell, Montana, United States, 18th June 1951

[369] *Lowell Sunday Sun*, Lowell, Massachusetts, United States, 24th June 1951

[370] *Syracuse Herald-American*, Syracuse, New York, United States, 24th June 1951

[371] *The Railway Gazette*, London, England, 27th July 1951

thighs. It took three-quarters of an hour, first to cut her free and then to cut out a section of the roof to lift her clear. Her 12-year-old son was trapped by his ankles. He ignored his own pain and kept asking, "How's Mother?"

'Police under Superintendent Catt organized a wireless post outside the station to assist rescue operations.' [372]

The 10.47 a.m. Three Bridges to Bognor train was an eight-coach multiple unit, in charge of Guard S.E.C. Eames with Motorman H.J. West. It was routed via Littlehampton and thence via the west leg of the Arundel Junction triangle to the Down Loop platform at Ford, where it was standing when it was struck by the Brighton train.

The 11.17 a.m. Brighton to Portsmouth train consisted of a six-coach electric multiple unit set, in charge of Guard W.S. Knight with Motorman Horace Sherwood. Both trains had steel underframes and composite steel and wooden bodies dating from 1937. When it collided with the Three Bridges train the rear car of the latter and the leading car of the Brighton train were telescoped into one another, while four other coaches were damaged to various degrees.

The Bognor and Portsmouth trains were both offered to Signalman R.W. Arnell in Ford box at 11.51 a.m. The Bognor train was running two minutes late but he decided to give it priority as per the timetable. It was booked to stand for five minutes at the Down Loop platform while it was overtaken by the Portsmouth train using the Down Main platform. Signalman Arnell had set the points for the loop platform, closed the gates of a level crossing at the east end of the station, and cleared his signals for the Bognor train to run into the Down Loop platform. A complication arose here, as eight-car trains, such as ran on Sundays, were too long for the loop platform. As soon as the train had stopped he therefore reversed a set of catch points and cleared a shunt signal, so that the Bognor train could move forward a little into a siding once the station work had been completed. He also accepted the Portsmouth train from Arundel Junction box and received the *Train Entering Section* signal for it. Until the Bognor train had moved forward he could not reverse the Down Main to Down Loop crossover because the rear of the train was standing on the controlling track circuit. Outer and Inner Home signals were provided at Ford and when he saw that the Portsmouth train had come to a stand at the Outer Home he had cleared that signal to let the train draw forward to the Inner Home, then shouted to Guard Eames to hurry up and get his train to move out of the way. The Portsmouth train had appeared to be stopping at the Inner Home until he had heard it rattling through the facing points. He had then rushed to the box window but the front of the train had already passed and the collision occurred almost at once. Signalman Arnell had then promptly sent the *Obstruction Danger* signal to the boxes on either side and telephoned Havant control for the current to be cut off.

Sitting in his car at the level crossing had been Lieutenant L.E. Hall of the Royal Navy, who told the Ministry of Transport inquiry into the accident that he had seen the Bognor train enter the loop platform with its rear coach projecting over the crossing. A couple of minutes later another train had appeared which he had expected to pass on the main line. Instead, it had suddenly swerved to the left. The motorman of this second train, who had been alone in his cab, had a look of astonishment on his face and had been pulling at a lever. The lieutenant had a narrow escape; his car was covered with glass and splinters. After the accident he had gone to help but his children in the car had become hysterical, so

[372] *The Times*, London, England, 6th August 1951

he had returned to his car and driven to the Naval Air Station to raise the alarm.

Motorman West of the Bognor train said that he had stopped at the platform and had been waiting for a tip from the guard. He had seen the shunt signal cleared and had been preparing to pull forward but had still been waiting for notice from his guard when his train was struck.

Porter D. Harrigan on the platform said that he had been waiting the guard to signal to the driver while Guard Eames said on previous occasions the porter had waved to the motorman. There had also been a delay trying to find the owner of a headscarf which had been handed to the porter.

Guard Knight of the Portsmouth train remembered it stopping at the Ford Outer Home signal. He said that he had seen that signal being cleared through his periscope, then he had been busy booking some delays in his journal and so failed to notice the aspect of Inner Home signal. The speed of his train had been the normal speed for a train running into the main platform. Motorman Sherwood had seemed to be in good health and spirits and the brakes of his train had worked normally during the journey.

Tests showed that the Portsmouth train must have been travelling at about 18 miles an hour as it passed over the level crossing.

The Ministry of Transport inspecting officer, Brigadier C.A. Langley, commented that with the death of Motorman Sherwood the reason for his overrunning the Inner Home signal could not be determined. The mechanical and electrical interlocking proved that the Inner Home signal must have been at danger when the Portsmouth train had passed it, and that the Home signals must both have been at danger when the Portsmouth train had been accepted. This acceptance by Signalman Arnell had been in order and he had properly stopped the train at the Outer Home signal in accordance with Rule 39(a). The Inner Home signals were also clearly visible against the sky. The only possibility seemed to be that when the Outer Home had been cleared and Motorman Sherwood had seen that the down main platform was clear he had assumed that the way was clear and that the Inner Home was off. A post mortem had shown no evidence of any medical condition which might have contributed to his mistake.

Rule No. 148 required a guard to carefully watch the running of the train and take any necessary action when approaching important junctions and stations. While it was appreciated that guards had other duties and could not keep a continual look-out, Guard Knight should not have allowed his attention to be diverted when his train had been approaching the Inner Home signal at Ford, especially as it had already been stopped at the Outer Home.

The misunderstanding between Guard Eames and Porter Harrigan was unfortunate. Had either of them acted promptly the Bognor train could have been moved forward, the track circuit cleared and the facing points reversed to allow the Portsmouth train to take the Down Main line.

Brigadier Langley remarked that the accident was another example of the failure of the human element. The need to observe signals required no emphasis.

Work to lengthen the Down platforms at Ford to take eight coach trains was to be undertaken before the following summer.

The damage to the coaches was not abnormal for a collision speed of 18 miles an hour. The latest coaches would have provided greater resistance but the 1937 rolling stock had many more years of useful life, and premature scrapping could not be justified. [373]

The jury at the inquest at Littlehampton on the victims of the accident returned a verdict of 'Accidental Death.' The Coroner told the jury that as far as could be ascertained Motorman Sherwood had been responsible for the accident through an error of omission but he doubted that, even if Sherwood had survived the accident, the jury would have returned a verdict of manslaughter against him, as to do so would have required a very high degree of recklessness on the motorman's part, such as had not been present here. [374]

A correspondent to a London newspaper, describing the failure of Hungary to meet the targets of a five-year plan, mentioned in passing a catastrophic railway accident which had occurred on Saturday 18th August 1951 at Székesfehérvár, probably a rear collision. A newly trained signalwoman had let an express train into the station 10 minutes too early and it had collided with another passenger train, killing at least 150 people. It has not been possible to obtain any more information on this accident. [375]

At 04:00 on the morning of Friday 24th August 1951 a Société nationale des chemins de fer français Frankfurt – Paris express was run into by a Basle – Calais express at Sanry-sur-Nied, France, east of Metz. The luggage van at the rear of the Frankfurt express was said to have practically disintegrated and only its wheels and two doors could be found. The penultimate coach of this train, in which most of the passengers were asleep, was thrown onto the roof of the coach in front. Twenty-three passengers lost their lives in the accident, including six American servicemen, and 40 were injured, half of them French servicemen. The servicemen had been stationed at Metz and Verdun and were returning from leave. The bodies of the dead were taken to a room transformed into a *chapelle ardente* at Metz. The line between Metz and Béning-lès-Saint-Avold was blocked and a powerful breakdown crane was called to the scene from Strasbourg. Two C-82 transport aircraft flew from Mayence at 11:30 to a point near the scene to take the injured American soldiers to hospital in Munich, Germany.

'Detachments of French and American troops and the Metz fire brigade carried out rescue work. The military authorities at Fontainebleau sent Military Police to Sanry-sur-Nied by air to inquire into the state of the injured, and to find out what had happened to supplies for American units in the region which were in the luggage vans. Reports that they were pilfered are officially denied.'

Meanwhile, 'Nearly 300 British holidaymakers who were passengers in the Basle-Calais express which was in collision with the Frankfurt-Paris train reached Folkestone to-night nine hours late. The *SS Maid of Orleans*, which brought them from Boulogne, to which port the Basle train was diverted, reached Folkestone at 9 p.m. The passengers should have arrived at 11 a.m. Most of them were suffering from bruises.' [376]

The driver of the Frankfurt express had considered a signal as doubtful and thought that he had exploded a detonator. He had therefore stopped his train. The conductor of this train, knowing that the Basle express was likely to be close behind, had gone back to

[373] Ministry of Transport report
[374] *The Times*, London, England, 23rd August 1951
[375] Ibid, 16th November 1951
[376] Ibid, 25th August 1951

protect his train with detonators but had only covered 250 metres before the latter train, travelling at 100 kilometres an hour, had borne down upon him. [377]

On the evening of the same day a train carrying a detachment of French police to the border with the Saarland collided with another train in the station at Conflans-en-Jarny. Eleven people were injured in this second accident and were taken to the military hospital in Metz. [378]

Four days after the accident *Sémaphoriste* (Signalman) René Didelon was arrested on a charge of manslaughter. It was alleged that he had first inadvertently stopped the Frankfurt train and then had failed to keep his signals at danger against the following Basle train, though a failure of the signalling apparatus was also a factor. [379] [380] The signalman was duly found guilty, as well, it would appear, as one or more other railwaymen. They then appealed, with the result that their sentences were confirmed and increased by the Colmar Court of Appeal.

On 13th December 1952 the mechanical signalling at Sanry-sur-Nied was replaced by an automatic system. [381]

On the evening of Wednesday 19th September 1951 a serious collision occurred in Brazil. An official note from the railway company, the Estrada de Ferro Central do Brasil, which appeared in the newspapers the following days, stated, "Yesterday, at 21:00, between the stations Antonio Carlos and João Ayres, train R-4 collided with the rear of train R-138, at Kilometre 352. As a result of the collision, two second-class carriages were telescoped. The accident resulted in 11 dead and 87 injured.

"As soon as the Administration became aware of the accident all possible resources were brought into use to care for the injured passengers, and we could count on the effective collaboration of the civil and military authorities of Barbacena and of the medical staff of that city and of Santos Dumont. The Barbacena hospitals were immediately prepared to receive the victims, who were taken there on a special train.

A special commission has been appointed to inquire into the causes and responsibility for the accident, composed of engineers Vitor Figueira de Freitas, Luiz Gonzaga de Lima, and Raimundo Alceu de Souza. Although the line suffered great damage in the accident, traffic is already back in operation."

A further report on the accident on the Norte – Sul line stated that there were 16 dead and 90 injured, 30 seriously, while another report the same day spoke of 21 dead and 100 injured, of whom 40 were in a serious condition. Assistance was provided by the military via the 9 Batalhão de Calçadores the School of Air Cadets in Barbacena and the state Military Police. The injured were taken to the Santa Casa hospital, the San José Misericordia hospital and the Policlinica in Barbacena. Colonel Eurico de Souza Gomez Filho, the director of the railway company, immediately made his way to the scene of the accident upon being informed of it.

---

[377] https://www.republicain-lorrain.fr/edition-de-metz-ville/2017/05/23/crash-entre-deux-trains-l-horreur-a-20-km-de-metz-ibys

[378] *Gazette de Lausanne*, Lausanne, Switzerland, 25th August 1951

[379] *The Times*, London, England, 29th August 1951

[380] *Gazette de Lausanne*, Lausanne, Switzerland, 29th August 1951

[381] https://cdn.website-editor.net/470adc47427d4720bad6ebb69ad0858c/files/uploaded/courcelles_histoire.pdf

The passenger train involved in the accident, R-4., was known as *Rapido dos Balansos.* Its passengers were mainly internal migrants coming from Bahia and Pernambuco states to search for work in São Paulo, part of a mass internal migration to the cities in Brazil in the 1950s and 1960s which involved almost 40 million people. Most of those who lost their lives had not been carrying any papers, making their identification difficult. The other train, R-138, was described as a mineral train. Exactly what mineral was being transported is not stated, very possibly it was iron ore.

Responsibility for the accident was soon pinned on Driver Francisco dos Santos of the passenger train, who it was stated, had overrun signals. He was only slightly hurt and at 08:00 the following was reportedly refusing to leave the scene; presumably he was helping with the rescue work. He stated later that he had been unable to see the signals because of a thick fog, and that when he had seen the freight train in front of him at a short distance it had been too late to stop.

The director of the railway company pointed out that the line on which the accident occurred had modern signalling, with centralised traffic control (though not, presumably, automatic train control to guard against drivers overrunning signals). The vehicles of the passenger train, on the other hand, had been 40-year-old wooden coaches. Had they been steel coaches there would probably have been no deaths and fewer injuries. Measures were in hand to finance the acquisition of more modern rolling stock. [382] [383]

There were conflicting reports regarding the nature of the two trains involved in a rear collision just after midnight on Tuesday 6th November 1951 at Vinkovci, Yugoslavia, on the Belgrade – Zagreb main line. The accounts agreed that the front train had been stopped at a signal. Some, however, stated that a passenger train had been run into by a goods train, others that a local passenger train had been run into by the Simplon Orient express from Istanbul and Athens to Paris. The reports all stated that there had been 17 deaths and 32 had been injured, 19 seriously. The passengers at the front of the train that was run into escaped injury. It was officially announced that a mistake by a signalman had caused the accident. [384] [385]

Monday 12th November 1951 saw a collision between two streamliner express trains of the Union Pacific Railroad at Wyuta, Wyoming, United States (the location of the accident is often given as Evanston). During what was described as a blinding blizzard *The City of San Francisco* collided with the rear of *The City of Los Angeles.* Eleven passengers, three Pullman employees, the flagman of the Los Angeles train and the engineer and a diesel-electric maintainer of the San Francisco train were killed and 142 passengers, two Pullman employees, the front brakeman of the Los Angeles train, the conductor, fireman and flagman of the San Francisco train, a mail clerk and ten dining car employees were injured. [386]

'Thomas E. Sharp, owner of radio station KFSD in San Diego, was sitting in a compartment of the City of San Francisco. He was drinking coffee. "The coffee flew out

[382] *O Estrada de São Paulo*, São Paulo, Brazil, 20th and 21st September 1951
[383] *Jornal do Brasil*, Rio de Janeiro, Brazil, 21st September 1951
[384] *Gazette de Lausanne*, Lausanne, Switzerland, 7th November 1951
[385] *Ironwood Daly Globe*, Ironwood, Michigan, United States, 7th November 1951
[386] Interstate Commerce Commission report

of my hand and the conductor flew into the room," Sharp reported. He explained that the conductor was standing in the aisle outside his doorway.

Alex Henetz, of Cheyenne, Wyo., was riding in a coach three cars back of the San Francisco units. Everything was running as usual when suddenly there was a terrific roar and a sound like someone screaming, he said. "I guess that was the steel being torn apart." Henetz said he pushed outside and helped a woman from the top of the shattered wreckage. "How she was thrown out, I'll never know," he said.

'A Bible also was thrown into the snow beside the carnage. A woman picked it up. Nearby lay a "zoom-ray gun," the toy of some child.

'The snow let up as rescue operations began. It was a foot deep near the wreck, drifts nearby were deeper. The snow began at nightfall. Workmen rigged floodlights. They continued their work in sub-freezing temperatures.

'A.F. Denham, of Detroit, was riding with his wife in a compartment of the City of San Francisco. Denham's wife suffered a bruised leg. Denham was not hurt. He said, "There was not a lot of screaming. It was rather quiet for such a terrible wreck."'[387]

The City of Los Angeles, eastbound first-class passenger train No. 104, consisted of diesel-electric units Nos. 998, 988B and 985B, coupled in multiple control, a baggage mail car, a dormitory car, two coaches, two dining cars, a lounge car, four sleeping cars and an observation-dining car, in charge of Conductor J.A. Barnes with Engineer D. Kellher.

The City of San Francisco, eastbound first-class passenger train No. 102, consisted of Chicago and North Western diesel-electric unit No. 5007B and Union Pacific diesel-electric units Nos. 928B and 987B, coupled in multiple unit control, two mail cars, a dormitory car, two coaches, two dining cars, a lounge car and five sleeping cars, in charge of an unnamed conductor with Engineer Rees Paul, Fireman John Branstiter and Diesel Electric Maintainer Norman Evans.

Also involved indirectly in the accident was a freight train, Extra No. 1475 East, which consisted of diesel-electric units Nos. 1475, 1484B and 1484C, coupled in multiple unit control, 89 cars and a caboose, in charge of Conductor T.C. Richens with Engineer Matthew Ortega. This train was standing in the middle siding at Wyuta at the time of the collision.

Train No. 104 left Ogden, Utah, at 9.45 a.m. and passed Wahsatch, the last open office, at 11.09 a.m., nine minutes late. Subsequently it stopped successively at colourlight signals Nos. 9242, 9224 and 9214 because the lenses of the signals were covered with snow and ice and their aspects could not be determined. Immediately after it restarted from the last signal it was struck by train No. 102.

Train No. 102 left Ogden at 10.07 a.m., 12 minutes late, passed Wahsatch at 11.21 a.m., 10 minutes late, passed signals Nos. 9242 and 9224 and was still travelling at 77 miles an hour when it struck train No. 104 at 11.27 a.m.

All the cars of train No. 104 were derailed as a result of the collision. The rear three cars were demolished, the eighth and ninth cars were badly damaged, the fourth to seventh cars were somewhat damaged and the first and second slightly damaged. All but the rear car of train No. 102 were likewise wholly or partly derailed. The three diesel units and the first and fifth to ninth cars were badly damaged, the second, third, fourth,

[387] *The Idaho Free Press*, Nampa, Idaho, United States, 13th November 1951

tenth and eleventh cars were somewhat damaged. The thirty-sixth to forty-third cars of train Extra No. 1475 East were struck by derailed cars of the express trains and derailed.

The crew of train Extra No. 1475 East told the Interstate Commerce Commission (ICC) that rain, sleet, and snow had been falling intermittently between Ogden and Echo, 40 miles from Ogden. Beyond Echo heavy wet snow had fallen all the way to Wyuta. This train had left Ogden at 5.45 a.m. and Wahsatch at 10.55 a.m. The green aspects of the signals had been hard to make out from the locomotive and speed had had to be restricted to 20 to 25 miles an hour. Conductor Richens and Flagman Ray Wilson said that the red aspects of some of the signals between Wahsatch and Wyuta had been visible from the caboose, but not others. The train had gone into the siding at Wyuta to allow the two streamliners to pass. When the train had stopped to enter the siding the flagman had thrown off a 5-minute fusee, following it with a second fusee as the train entered the siding. The freight train had been completely in the siding at 11.20 a.m. The flagman said he had closed the switch and soon afterwards he had seen train No. 104 stopped at signal No. 9224. Train No. 104 had then passed him, he had seen Flagman Hyram Preece of that train standing in the vestibule of the rear car and the marker lights and oscillating signal light at the rear of the car had been working. Not long afterwards train No. 102 passed signal No. 9224 and he had assumed that train No. 104 must have passed signal No. 9214.

Conductor Richens stated that after his train had stopped he had been making his way along the north side of it. He heard train No. 104 restart from signal No. 9224 but had not realised that it had stopped at signal No. 9214. After he had walked another 800 feet train No. 102 had passed; from the sound of its engines they were working under load and he did not think that they had been shut down before the collision.

The engine crew of train No. 104 stated that after leaving Wahsatch they had been able to make out signal No. 9268, displaying *Proceed*, and signal No. 9256, displaying *Approach*. When they reached signal No. 9242 its post had been visible but they could not see its indication even after stopping at it. They had then proceeded at caution, had seen the freight train going into the siding and its flagman closing the switch, but signal No. 9214, at the entrance to the siding, had also been completely obscured. The train had stopped again, restarted, had run alongside of the freight train to signal No. 9224, which was likewise invisible, stopped again and had just restarted when it was run into by train No. 102.

The conductor and front brakeman, riding between the third and fourth cars, said that they had not been able to see any of the last three signals either. The actions of Flagman Preece between Wahsatch and Wyuta were unknown as he was killed in the collision.

Fireman Branstiter was the only survivor of the engine crew of train No. 102. He was seriously injured. He told the ICC inquiry that the signals had all been visible as far as Wahsatch but beyond there the snow had become heavier and the power windscreen wiper had failed to keep the left-hand side of the locomotive windscreen clear of slush. He had been able to make out signal No. 9256 as indicating *Proceed* from a mile distant; closer to it the aspect could not be distinguished. Signals Nos. 9242 and 9224 he had not been able to distinguish at all but Engineer Paul and Maintainer Evans had called out both as *Proceed*. He had assumed that they could see them better from their side of the cab. No fusees had been seen or torpedoes exploded after Wahsatch. The track curvature combined with the presence of the freight train meant that the red oscillating signal at the rear of train No. 104 had only become visible at short range. As soon as it had been seen Engineer Paul had applied the brakes in emergency but the collision occurred almost at once before speed had been reduced.

The other members of the crew of train No. 102 had noticed nothing wrong with the operation of their train.

The speed recorder on the engine of train No. 102 indicated a speed of 48 miles an hour approaching Wahsatch, reduced to 30 miles an hour and then increased to 77 miles an hour, which had been maintained until it struck train No. 104.

Going back after the accident the flagman of train No. 102 found the lenses of signal No. 9224 covered in snow but could just see that it was showing *Stop and Proceed.* Continuing back as the storm abated he had found signal No. 9256 indicating *Approach.*

An hour and he half after the collision a signal maintainer arrived at Wyuta. He found the lenses of signal No. 9224 covered by three or four inches of ice and snow, which he had to remove before the aspects were visible. Moving on to signals Nos. 9256 and then 9214 he had found them to be in the same condition.

The ICC noted that Flagman Preece should have thrown down lighted fusees each time his train had stopped at a signal beyond Wahsatch but none had been seen by the fireman of train No. 102 or the flagman of the freight train. Conductor Barnes had relied on the flagman, as an experienced man, doing so without instruction.

The aspects of signals Nos. 9242 and 9224 had been obscured by snow and ice. Under the rules Engineer Paul should have stopped at each signal and then proceeded at reduced speed, as Engineer Kellher had done. In view of his death it could not be determined why he had not done so.

The automatic signals on the line were provided with snow shields but these were not effective when wet snow was blown directly against the signal lens, as on the day of the disaster. When the aspect of a signal could not be determined the rules decreed that a train must treat it as showing its most restrictive indication and in the case of these signals should stop and then proceed at restricted speed. The rules also stated that when a train was likely to be overtaken by a following train the flagman must provide full protection. On the day of the accident train No. 102 had not proceeded at restricted speed and train No. 104 had not been provided with flag protection.

At other points on the Union Pacific Railroad the automatic block signals were supplemented by an automatic cab signal system whereby the signal aspects were continuously displayed in the cab of the locomotive. Had such a system been in place at the accident location the disaster would probably have been averted, and the ICC recommended that the Union Pacific install such a system over the remainder of its line between Ogden, Utah, and Omaha, Nebraska.[388]

In April 1952 Fireman Branstiter featured in a court case in which he claimed $40,000 damages from the Union Pacific Railroad for injuries received in the accident. Speaking of it in evidence he said that approaching Wyuta he had remarked to his engineer, "I hope you can see those signals, because I can't." He had been assured that they were green.

Then, immediately before the collision he had seen a slash of red and Engineer Paul had cried, "Oh, God, what's happened."

"Just then we hit," the fireman continued. "Glass began to fly and I threw my arms over my eyes and ducked below the dashboard. The engine seemed to climb and turn. It

[388] Interstate Commerce Commission report

stopped suddenly and came down with a bang." [389]

The fireman, who had been in hospital for two months after the accident, was awarded $17,500. [390]

At 6.50 a.m. on the morning of Wednesday 7th May 1952 two crowded suburban electric trains were in collision at Berala station, in the suburbs of Sydney, New South Wales, Australia. The 6.28 a.m. from Liverpool to Sydney Central had stopped in Berala station and the 6.42 a.m. train from Bankstown to Sydney Central was stopped at a signal a quarter mile in the rear. The Bankstown train then restarted and travelling at 25 miles an hour it ran into the rear of the Liverpool train as the latter was about to restart from Berala station. The two rear coaches of the Liverpool train and a further two coaches towards the front of the train were telescoped. The police said that they could not understand how the Bankstown train could have gained so much speed in such a short distance; they obviously did not appreciate how rapidly electric trains were able to accelerate.

At the time of the accident there was a thick fog, with visibility reportedly down to 50 or less yards.

Guard Maurice Wills was in charge of the Liverpool train, with Motorman Leslie Watts. In charge of the Bankstown train was Guard M.T. Millgate with Motorman Osric Martin.

Passengers on the trains and others told their stories to reporters. 'Mrs R. Burgess, of Wilga Street, Regent's Park, said: "I was in the last carriage of the Liverpool train. We all got knocked over, and then everybody scrambled out as quickly as they could. Women were screaming, and I could not look at the people who were hurt." Mrs Burgess suffered slight head and leg injuries.

Mr Watson, of Hunter Street, Berala, his wife, and Mr W. Cameron, of Bankstown, were sitting just behind the driver's cabin in the Bankstown train. All except Mr Watson escaped with only bruises and shock. Mr Watson suffered a probable fracture of the leg. "My lunch case is full of glass, but I haven't suffered a cut," said Mr Cameron, at Auburn District Hospital. Mr Watson said: "Ours was a steel carriage and it took the shock of the crash. Only one man in it was seriously hurt. Most of the people injured on our train were in the wooden carriage, third from the front.

'Mr Stanley West, of Epsom Road, Moorebank, said he tore off a woman's frock to use as bandages. "I was sitting in the Liverpool train when I heard someone yell, 'Hang on, here's a crash,'" he said. Next moment we were thrown everywhere. I had a seat thrown across my back. People were crushed up everywhere. A lot were trapped. We levered the seats up and pulled some of them out. It was a mass of twisted steel, broken glass, mangled bodies. Some of the bodies were jammed right up against the roof of the carriage. We ran for blankets, towels, and bandages. A woman on the platform let me tear off her dress for bandages.

'Harold Sharp, 29, labourer, of the Hargrave Park housing settlement, had a remarkable escape from death. He was sitting in the rearmost seat of the rear carriage of the stationary train. In Lewisham Hospital last night Sharp said, "I felt a terrific impact

---

[389] *The Salt Lake Tribune*, Salt Lake City, Utah, United States, 23rd April 1952

[390] Ibid, 25th April 1952

and knew nothing until I found myself lying with glass and wreckage around me. I was dazed and didn't know what had happened. But I remember a lot of people racing around, people shouting and women crying and screaming. My mate, Ken Wise, took me out of the carriage. Outside I sat down. There were people with blood all over them lying around me. It was like a nightmare. An ambulance fellow wanted to take me to hospital, but I said I wanted to go to work. Then my mate put me in a taxi and brought me here." Sharp, with his scalp stitched, was suffering from concussion.

'Mr N.K. Tame, of Oakland Street, Wentworthville, said, "I was at work at the Commonwealth Engineering Company when news of the accident came through. "My supervisor loaded us into a utility truck with all kinds of rescue equipment, including oxy-acetylene torches, and sent us to the train crash. I worked with the acetylene torches for a while, then helped to carry out the dead and give first-aid. A thing I will never forget was a man's head which had been severed and was resting on a window sill in the second last carriage of the Liverpool train. It had been severed clean by a piece of twisted wire."

'Mr D.A. Mobbs, of Hyde Park Road, Berala, said, "I arrived just after the accident happened, and took the first load of five injured in my utility truck to St Joseph's Hospital. One man had the whole of his face completely smashed. A Balt girl I got to the hospital went mad as she was being put into a bed."'

Initially it was reported that seven passengers had lost their lives in the accident; with the death of three of the injured, the last several days after the accident, the final number of fatalities was 10, with another 140 injured.

In explanation of the accident it was stated that the line was divided into 400-yard sections, with automatic colourlight signals. The signal immediately in the rear of an occupied section displayed two red lights and the preceding signal a single red. Tripping devices were provided at red signals which applied the Westinghouse brakes when a train arrived at a signal at danger. Motormen were, however, permitted to 'trip' trains past signals at danger after the brakes had recharged, proceeding 'with the utmost caution' and at a speed such as to be able to stop short of any obstruction.

Defending the *Stop and Proceed* system Mr R. Winsor, Director of Transport and Highways, stated that the trip system was the safest in the world. [391]

The inquest on the victims of the accident took place in July. Amongst other witnesses was Motorman Watts, who testified that he had tripped past the Berala signal on other occasions but had not done so on the day of the accident because of the fog. He added, however, that a driver was free to trip past a signal in such circumstances if he thought fit to do so.

Motorman Harold Nicholls, who had driven a train through Berala in the opposite direction shortly before the accident, was asked by Mr Cahill, the lawyer acting for Motorman Martin, if he considered it safe for a driver to 'trip' his train past a stop signal in the foggy conditions of the day of the accident. He replied, "Yes, if the driver had such control that he was able to halt the train at short notice."

Motorman Martin did not go into the witness box at the inquest on the grounds that evidence he gave might incriminate him, but statements that he had previously made to the police were read out. He allegedly said, "I thought I had the train under control, There

---

[391] *The Sydney Morning Herald*, Sydney, New South Wales, Australia, 8th May 1952

was a terrible fog. Only for the fog I can say I did have control of the train. I could have stopped the train but for the fog. The fog was fluctuating. One moment it was clear, and the next moment it was like a blank wall." The police officer pointed out that it was obvious that he had not got the train under control or there would have been no collision, to which he had allegedly replied, "I suppose so. That's the thing that has got me beat. I hit the train all right but I do not think I caused much damage. I reckon I was doing five to ten miles an hour when I hit the train. There was about 300 tons behind me and I suppose that caused the damage." Asked if he considered it safe to drive through fog at 25 miles an hour Martin is said to have replied, "Well, the department expects you to keep to a timetable. I am always careful. I have been in the job for nearly 30 years." When it was suggested that he had forgotten that he had passed a signal at danger the apparent reply was, "I could not tell you. I am always stopping and starting. You pass so many signals at times you just forget." [392]

At the close of the inquest Coroner E.T.A. Smythe committed Martin to trial. He was initially charged with the manslaughter of all ten passengers who had died but eventually stood trial for causing the death of just one passenger, Irene Ward, of Kiama Avenue, Villawood. [393]

At the beginning of the trial of the motorman in December Judge Berne commented that a regulation which allowed train drivers to pass signals at danger shocked and alarmed him and that measures should be taken to stop the practice. Told that the practice was worldwide the judge replied, "I don't know anything about railways, but I think it is alarming." [394]

Later in the inquest it emerged that the Liverpool train had still been stationary at Berala station, where the automatic signal at the end of the platform had been red, because of some confusion between Guard Wills and Motorman Watts as to who was to start the train, the motorman saying that he had been waiting for a bell from the guard and the guard leaving it to the motorman to decide if he wished to 'trip' the train past the signal.

At the close of the trial the jury acquitted Motorman Martin, a verdict with which Judge Berne said he agreed, and he praised the motorman for telling the truth and not attempting to evade responsibility. [395]

An empty ambulance train was stationary in the station at Seadong, South Korea, two stations north of Taegu, on the line between Seoul and Pusan, on the evening of Tuesday 7th October 1952, when it was run into from behind by a second empty ambulance train. Five people died, two of them American soldiers. [396]

Early on the morning of Sunday 9th August 1953 a passenger train came to a stand with an overheated axle box ('hot-box') in the station at Samsong, South Korea, 60 miles north of Pusan. Due, apparently, to a signalman's error a freight train ran into its rear, telescoping three coaches. Fifteen South Korean civilian passengers were killed,

---

[392] Ibid, 17th July 1952

[393] Ibid, 18th July 1952

[394] Ibid, 2nd December 1952

[395] Ibid, 6th December 1952

[396] *Singapore Standard*, Singapore, 9th October 1952

including four women and a child, and six more were seriously injured. The Korean National Railway officials were carrying out an investigation. [397] [398]

A serious collision happened on the narrow-gauge (1,000mm) electric line of the Ferrocarriles de Peñarroya y Puertollano at kilometre 105.8 between Minas del Horcajo and La Garganta, Spain, at about 15:30 on Monday 28th September 1953. A work train was stopped in the section but due, presumably, to human error on someone's part, passenger train No. 24 from Puertollano was allowed to enter the occupied line. The engine of the passenger train destroyed four compartments of the van at the rear of the work train, and it was in that train that all the casualties occurred. Four workmen were killed on the spot, two died in the rescue train taking them to Puertollano and two more passed away in the hospital there. Five other workmen were injured. It was expected that the line would be blocked for 30 hours. The railway company donated 3,000 pesetas to the widows of each of the men who had been killed. [399] [400]

A Vienne – Lyon omnibus train stopped in the station at Sérézin-du-Rhône, France, a few kilometres south of Lyon, was run into from behind by a light engine at 06:30 on Friday 16th October 1953. The violence of the collision forced the rear three coaches of the passenger train up into the air. The packed wooden carriages – a newspaper referred to them as 'the criminal wooden carriages' – were smashed. Portions of the coaches were thrown onto the station platform while cries arose from all parts of the train.

Immediately after the accident there were known to be six dead and 40 injured, 15 seriously, and three more of the injured had succumbed before reaching hospital. Fifteen of the injured were taken to the Edouard-Herriot Hospital in Lyon and 30 to Vienne. Three more passengers died later.

Passengers described their experiences. M. Gouzoubachan, from Hasse, said: "The train was full as usual, and everyone was chatting cheerfully because we were a group of friends. We were all waiting for the train to start from Sérézin when suddenly there was a terrible shock. We were one thrown upon the other. I was in the third coach from the end, and I realised at once that its roof had gone. We jumped out immediately because we feared that the carriage might catch fire, and we then rescued some of the passengers from the last two coaches. We were able to save some, but it was horrible."

Mme Dal-Gobbo was on her way to a factory at Saint Fons and was riding in the second carriage from the front of the train. She related: "The shock was so violent that I was knocked to the other side of the compartment, against the wall. I didn't know what was happening to me. Immediately we heard frightful screams and knew that there had been an accident."

Her husband, who was among the first rescuers, added: "I thought everything fell on our heads. Pieces of wood and iron fell on us. I was busy reading a newspaper and obviously didn't see the engine coming. Everyone was screaming; I jumped out onto the platform and was shocked by what I saw."

---

[397] *Pacific Stars and Stripes*, Tokyo, Japan, 12th August 1953
[398] *The Singapore Free Press*, Singapore, 12th August 1953
[399] *La Vanguardia*, Barcelona, Spain, 29th September and 10th October 1953
[400] Webpage: Federación Castellano Manchega de Amigos del Ferrocarril (in Spanish)

Discussing the accident, a newspaper wrote that the train had arrived at Sérézin at the usual time. It had been crowded with workmen bound for factories on the outskirts of Lyon and there had been many other passengers waiting on the station platform to join it. While it was in the station the engine had suddenly appeared, 300 metres on the Vienne side of the station. The locomotive had been in charge of Driver Gustave Fournier, with Fireman Roger Fauget. The signals stood at danger to protect the train standing in the station but witnesses saw the engine approach at the normal speed of 60 kilometres an hour. About 150 metres from the stationary train the driver and firemen apparently saw the signal at danger – its aspects were also repeated in the cab of the locomotive – and applied the brakes, but it was then too late.

An investigation began immediately and Driver Fournier and his fireman were detained at the station for questioning and later arrested. According to their initial statements they had not been able to see the signals soon enough to stop the engine. There had been a thick fog at the time of the collision. [401]

M. Jacques Chastellain, the Minister of Public Works, was subsequently questioned in Parliament regarding the continued use of wooden coaches in workmen's trains. He agreed that the Sérézin accident would have been less serious had all steel rolling stock been in use but stated that the Société nationale des chemins de fer français was unfortunately far from having sufficient all-steel vehicles. New rolling stock was being provided as quickly as the budget would permit but to replace all the wooden coaches would cost 145,000,000,000 francs. [402]

'Suitcases filled with kiddies' toys, clothing, gayly covered packages and bloodstained parcels were strewn along the track after Saturday's train crash at Sydenham. These told of shoppers homeward bound, counting the days to the Christmas convivialities and the greatest day of the year for the children. In the twinkling of an eye, the fun of living and giving was gone, homes were bereft of mothers, and untold suffering became the lot of scores whose warning of danger was too sudden to be availed of.'

Thus began an account in an Australian newspaper of a rear collision on Saturday 19th December 1953 on New South Wales Government Railways at Sydenham, New South Wales. Two suburban electric trains were involved in the accident. The first had left Sydney St James station at 12.15 p.m. for East Hills. The second train had left Wynard station for Lidcombe via Bankstown at 12.16 p.m., the trains using the same line as far as Sydenham. Each train was crowded with about 1,200 passengers, a mixture of Christmas shoppers and people returning home from work. Motorman Harold Joseph O'Neill was driving the Lidcombe train. The East Hills train was stopped at the Sydenham home signal and, of course, the Lidcombe train should have been stopped by signals behind it. Instead it kept going and collided with the rear of the East Hills train, telescoping the two rear cars of the front train and the two front cars of the rear one. The accident happened at about 12.30 p.m. Two passengers were killed outright and three more succumbed to their injuries in the hours and days following, while no less than 748 were more or less injured.

The same newspaper whose leading paragraph has been reproduced above contained the accounts of several passengers: 'Mr A. Wolraven, of Marrickville, said at Royal

---

401 *Journal de Genève*, Geneva, Switzerland, 18th October 1953

402 http://4e.republique.jo-an.fr/page/1954_p575.pdf

Prince Albert Hospital, "At St Peters, the Bankstown train overshot the station by about three or four carriages. The passengers near me were alarmed and after we left St Peters we began to get very worried. "I remember a terrific crash and the next thing I knew I was at the hospital."

'Keith Burden, 28, of Staples Street, Kingsgrove, who escaped with minor injuries, said: "I was in a wooden carriage standing up when the crash occurred. I was thrown off my feet. One of my legs became jammed under my feet. I could hear women and children screaming. I could not move my leg. Two men dragged the seat away from me. The crash happened in a split second. I lost my glasses and some Christmas presents for the children, but I am thankful that I did not lose my life."

'Charles Stippard, 17, of Punchbowl, said: "I was in the front of the first carriage of the train. After we left St Peters I thought something was wrong because the train was not on the line indicated at Central. As we approached Sydenham I felt a terrific jolt and must have blacked out. The next thing I knew someone was giving me a smoke on a stretcher."

'Dozens of people hurried to the wreckage to ask railway officials to help them trace relatives and friends. A father collapsed when told his son had been sent to hospital with minor injuries. The father, George Charty, of McMillan Street, North Bankstown, had been walking along the rail track asking, "Has anyone seen Glenn." They had been travelling home in one of the trains, but were separated after the crash. A railway workman told Charty his son had been taken to hospital with leg injuries. Charty fainted.'[403]

The cause of the collision was a mystery for several days, not helped by the fact that Motorman O'Neill was too badly hurt to be interviewed. By the following Tuesday it appeared that a wrong-side signal failure might be to blame and two signal technicians had been suspended and were being questioned by detectives. Guard Herbert Graeme Walker of the East Hills train was also being questioned.[404]

The inquest on the victims of the accident opened in February 1954. It was told that visibility had been good and that the brakes of the Lidcombe train had been working correctly. It was stated that the Lidcombe train had overrun the platform at Erskineville by three carriage lengths and had had to be reversed, and that Motorman O'Neill had at first admitted, and later denied, that he had had a couple of drinks on the morning of the accident. It was alleged that he had drunk the equivalent of 11 middles (10 oz. glasses) of beer or five whiskies. Government Analyst Thomas McDonald told the inquest on 12th February that a specimen taken from the motorman on 19th December had contained 0.194% alcohol and had added that he had never known an individual to consume this quantity of alcohol without being "appreciably affected" by it.[405]

However, the immediate cause of the accident was a wrong-side signal failure. On 3rd March Detective Sergeant R. Hamer told the inquest that Linemen Keith Venn and Eric Clissolm had been working on the signals before the accident, dealing with a track circuit failure. Venn admitted that he had placed a screwdriver in the signal relay to create a false current and put the automatic signal to *Clear*, 'so as to avoid delaying the business trains.' He had then been called away to the telephone and had forgotten to remove the screwdriver, so that the signal had remained at *Clear* for the following Lidcombe train

---

[403] *The Propeller*, Hurstville, New South Wales, Australia, 24th December 1953

[404] *The Northern Star*, Lismore, New South Wales, Australia, 22nd December 1953

[405] *Cairns Post*, Cairns, Queensland, Australia, 13th February 1954

even though the East Hills train was still in the section. [406]

At the close of the inquest the Coroner, Mr E.J. Forrest, committed both Lineman Vann and Motorman O'Neill for trial on charges of manslaughter. [407]

The trial of Motorman O'Neill at Sydney Central Criminal Court in November 1954 ended with the jury finding him 'Not Guilty' without them even bothering to hear the evidence for the defence. [408]

Nothing can be found of Lineman Venn standing trial other than reports that the two men were to be tried separately. Possibly after the acquittal of the motorman the case against the linesman was dropped.

Following the accident the regulations were strengthened with a rule stating that *in the event of a track circuit failure, no attempt shall be made to clear any of the signals controlled by the track circuit by manipulation of the track circuit relay.* [409]

According to the Czechoslovak Ministry of the Interior 103 people died and a further 83 were injured at 01:00 on Christmas Eve, Thursday 24th December 1953, in a collision at Šakvice, Czechoslovakia. The *Balkan Express* from Prague via Brno to Bratislava ran into the rear of a local train, demolishing its four rear carriages. Rescue work went on throughout the night with railwaymen, firemen, doctors, troops, staff from the Ministries of Transport and the Interior all involved [410] and by noon on Christmas Day the line had been cleared and the line reopened. Thirty more of the injured reportedly succumbed to their injuries in hospital, increasing the number of dead to 133. There were allegations that the collision had been caused by gross negligence by a number of railwaymen, who had been arrested. [411] Some accounts stated that members of the engine crew had drunk a number of bottles of wine. [412] An American writer, on the other hand, suggests that lack of maintenance of rolling stock and track, together with excessive working hours had been factors in the disaster. [413] The engine crew had apparently fallen asleep, their train ran past three stop signals and was still travelling at 90 kilometres an hour when it hit the stationary train. [414]

What, at the date that it occurred, was the most serious railway accident ever in Chile, happened at 08:55 on the morning of Sunday 17th July 1955 at San Bernardo station on the Linea Central Sur of the Empresa de los Ferrocarriles de Estado, the state railway company. Ordinary passenger train No. 3 from Santiago Alameda to San Rosendo had departed 30 minutes late. At its rear were three third-class wooden carriages. This train arrived in due course at San Bernardo and was stopped here for longer than usual, reportedly because of track repairs ahead; possibly overnight engineering work which had not been completed on time.

---

406 *The Morning Bulletin*, Rockhampton, Queensland, Australia, 4th March 1954
407 *Goulburn Evening Post*, Goulburn, New South Wales, Australia, 4th March 1954
408 *The Sun*, Sydney, New South Wales, Australia, 11th November 1954
409 Wikipedia – Sydenham rail disaster
410 *Journal de Genève*, Geneva, Switzerland, 29th December 1953
411 *The Times*, London, England, 29th December 1953
412 Wikipedia – Šakvice train collision
413 *Railroad Wrecks* by Edgar A. Haine, pgs. 149-150
414 *Eisenbahnunfälle in Europa* by Erich Preuβ, pgs. 93-4 (in German)

Behind train No. 3 was train No. 11, another ordinary passenger train from Santiago Alameda to Pichilemu, hauled by Class 80 *Mikado* type steam engine No. 842. As this second train approached San Bernardo a thick fog prevailed, which prevented Driver José René Peña González from seeing the first train in front of him, while he claimed to have seen the home signal at clear. His engine rammed and demolished the rear two coaches of the San Rosendo train. Thirty-five people were killed outright and another ten were fatally injured, only three of the occupants of one of the coaches escaping death. The death toll would have been even higher had not a number of the passengers alighted onto the station platform during the lengthy stoppage of the train.

Firefighters, police, members of the Chilean Red Cross and soldiers from an Infantry School co-operated in the rescue work and by chance there was a hospital nearby, though the large number of injured meant that some were sent to other, more distant hospitals. The President of the country, Carlos Ibáñez del Campo, arrived at the scene at 11:00 and ordered an immediate investigation of the accident. [415]

After the accident the railway company began to withdraw the wooden third-class carriages and replace them with steel ones, but they were to feature in another disaster at Cerrillos only seven months later (see below). [416]

Thick fog was a factor in another South American accident in Argentina on Tuesday 23rd August 1955. A local train was stopped in the station at Ciudadela, in the suburbs of Buenos Aires. According to an initial report from the Transport Ministry there were 14 deaths and 35 injured but observers at the scene feared that the number of dead might rise to 30 and of injured to 100. Driver Eduardo Garcia of the colliding fast electric train, which was carrying commuters and schoolchildren, was among those who lost their lives. Many of the injured were said to be in a grave condition. Five coaches were smashed and parts of them were thrown into nearby Rivadavia Avenue. Telescoping occurred between the rear coach of the local train and the leading coach of the express, which were reported crumpled into less than the length of one coach. Passengers on the station platform were likely also among the casualties. Some reports suggested that the express should have bypassed the platform line but was diverted onto it in error. [417]

The last of several major disasters on English railways during 1955 occurred at about 11.28 p.m. on Friday 2nd December on British Railways, Southern Region at Barnes, in South-West London. The 10.55 p.m. freight train from Battersea to Brent was run into from behind by the 11.12 p.m. passenger train from London Waterloo to Windsor and Chertsey. Fire broke out following the collision and 11 passengers, the guard of the freight train and the motorman of the passenger train were killed, while 41 passengers were injured, 20 of whom were detained in hospital. [418]

'Mr Alfred Lester, aged 56, of Staines Road, Twickenham, a passenger in the front coach, said: "The train was fairly full and most of us in my carriage were dozing when we were suddenly flung about by a tremendous crash. The coach turned over on its side and was plunged into darkness. As we tried to drag ourselves we heard a voice shouting:

---

[415] http://accidentesanbernardo.blogspot.com/

[416] Wikipedia – Accidente Ferroviario de San Bernardo

[417] *The Medicine Hat News*, Medicine Hat, Alberta, Canada, 24th August 1955

[418] Ministry of Transport report

'Stay where you are,' and then, about five minutes after the crash, there was an explosion and we saw flames. The electricity seemed to be flashing all around us. I could hear people screaming, and then the next thing I remember was somebody dragged open the door above us and hauled us up. Two girls and four other men in the carriage with me were dragged up to safety."

'The other three coaches of the train all remained upright after the collision. A man who was travelling in one of the back coaches said the train was running fairly fast at the time of the accident. He and the other passengers were flung violently from their seats by the impact.

'Miss Shirley Harris, 20, of Gresham Road, Staines, in the last of the coaches, said: "We were flung to the floor of the carriage and then a few minutes later we heard shouts and screams and saw that the front of the train appeared to be on fire. We could see the flames leaping up. Shortly afterwards a railway official helped us out along the track. None of us in our coach appeared to be injured."

'Mr F. Bonor, of Twickenham Road, Hanworth, Middlesex, also in the rear of the train, said that he was asleep when the crash came and was thrown with other passengers in a heap on the floor. "We looked out of the window and saw the front coach on its side. Minutes later part of it seemed to burst into flames. People were shouting for help and screaming. It was terrible."

'Police-constable Thomas Oliver, aged 45, who lives at York Way, Hanworth, told a Press representative that he had rescued the driver of the passenger train. "With my sergeant I dashed across the tracks. The rails were still live. I could see the front coach was burning fast," he said. "We got to the driver's compartment and I pushed the wreckage aside with my bare hands to get a man whom I could see. I dragged him clear, but beyond him I could another man writhing in the flames. He was trapped as far as I could see and we just couldn't get at him." Many of the injured are suffering from burns, received when they tried to get clear of the burning train.

'Mrs B. Alderson, living in a house in Gipsy Lane, immediately overlooking the rail track by the bridge, said: "I was trying to get to sleep and could hear the goods train moving along slowly. Suddenly I heard the crash and went to the window. Within minutes flames were roaring high over the bridge. I could see people struggling to get out of the coach in the flames. They were screaming and shouting for help. It was a fearful sight."

"Mr G. Bevan, who lives in a house overlooking the scene, said: "Within minutes of the crash people were at our door asking for help; but there was little that we could do. The front coach was like a raging furnace. We got down on to the track. The top of the first coach had been torn away from its wheels and was on fire. All we could hear was screams coming from people in the first coach. We tried to get to them but were beaten back by the flames. I think there were people trapped underneath it."

'Hundreds of people from the residential district bordering the tracks watched the rescue efforts by firemen who fought the flames from a bridge over the track, and as the injured were taken from the wreckage of the derailed front coach they were rushed by a fleet of ambulances to Putney, Barnes and Hammersmith hospitals.' [419]

The four tracks of the Windsor lines from London Waterloo diverged from the Main

---

[419] *The Times*, London, England, 3rd December 1955

lines at Clapham Junction and ran as far as Barnes. Close to Barnes Junction signal box, at the west end of Barnes station, the four lines divided into two double lines to Richmond (left) and Chiswick (right). All four lines were electrified on the standard third rail system (650 volts D.C.). From north to south the tracks were Up Local, Up Through, Down Through and Down Local. Between London Waterloo and Clapham Junction the lines were continuously track circuited with multiple aspect colourlight signals and there was no manual block working, but westwards from Clapham Junction E box the lines were worked by Sykes Lock and Block, although continuous track circuiting and colourlight signals were provided as far as Point Pleasant Junction, the next box to Clapham Junction E. Onwards from there past Putney and Barnes East boxes to Barnes Junction box upper quadrant semaphore signals were provided. Putney and Barnes East boxes had been switched out at the time of the accident giving an extended block section from Point Pleasant to Barnes Junction. As at South Croydon in 1947 (see above) the collision was determined to have resulted from a misuse of the release key of the Sykes Lock and Block by the Barnes Junction signalman.

The freight train was a transfer working from the Southern Region to the London Midland Region. It consisted of a 2-8-0 Class 8F steam locomotive, 13 loaded waggons, 29 empty waggons and a brake van, in charge of Guard E.A. Hinman with Driver A.R. Nield and Fireman A.W. Feltham. It had joined the Windsor lines from Battersea at Clapham Junction A box and was routed onto the Down Local line rather than the usual Down Through line to keep it out of the way of a following train, with a 15 minute margin on the Local line in front of the Windsor and Chertsey train. It had run from Clapham Junction to Barnes Junction and came to a stand at the Down Local Home signal there at about 11.18 p.m. It was just moving forward again, when that signal was cleared, when it was run into by the Windsor and Chertsey train.

The electric passenger train consisted of a pair of two car units, the front two for Chertsey and the rear two for Windsor. It was in charge of Guard W.E. Jeffery with Motorman W.G. Flanders. It stopped at Clapham Junction at 11.23 p.m., four minutes late, then had a clear run to Barnes, where it ran into the rear of the freight train.

Driver Nield was the first person to give evidence at the Ministry of Transport inquiry. He said that he had had a clear run between Clapham Junction and Barnes Junction, where the signals had been against him, and he had come to a stand at the Barnes Junction Down Local Home, signal No. 49, at *Danger*. After what he estimated was a short interval the Home signal had cleared and he had begun to move forward. They had only moved a few engine lengths when the collision had occurred. Looking back, his fireman had told him that their train was on fire, so he had gone to the signal box and told the signalman that they had been run into and were on fire. The signalman, he related, had appeared almost "dumbfounded."

Fireman Feltham thought that they had only been stopped at the signal for two minutes. He remembered his driver remarking that it was unusual to be held at that signal.

Guard Jeffery of the electric train stated that he had seen signal No. 49 at green; his train, not being booked to call at Barnes station, had been coasting at 30 to 35 miles an hour at that point. He had noticed no whistling or brake application before the collision, which had thrown his across the compartment. Recovering himself, he had gone back to protect his train and had spoken to "somebody" from Barnes East box over the telephone, requesting assistance. He had then returned to his train, where the fire was by now too intense to help anyone in the front of the train.

Motorman G. Peters, booked to work the rear unit of the passenger train from Windsor to Slough, said he had been riding in the third coach with Station Foreman S.N. Rogers. He recalled a brake application at the last moment. With Rogers he had tried to free a trapped passenger when severe arcing had occurred from the third rail. He had attempted to place a short circuiting bar between the Down Through conductor rail and the running rail. It had burnt through but another attempt with a second bar had succeeded, opening the Down Through circuit breaker at Clapham Junction sub-station at 11.34 p.m.

Station Foreman Rogers spoke of checking that the Barnes Junction signalman had called the emergency services and of helping a policeman release a passenger.

Special Class Relief Signalman A.T. Blundy admitted that Putney box should not have been switched out until 11.45 p.m., with a latitude of a quarter hour due to the need for all sections to be clear of trains, but he had closed on the Through lines already at 11.07 p.m. and on the Local lines at 11.09. p.m., after refusing the offer of the freight train from Point Pleasant. (Barnes East box had already closed, at the correct time of 9.45 p.m.). The switching out operation had, however, been done correctly. He excused his early closing by saying that he usually worked in power boxes, so was not used to the harder work of the Point Pleasant manual lever frame, while no harm, he considered, had been done by closing early. In fact two trains had been severely checked at Point Pleasant due to the longer block section.

Special Class A Signalman J.W.G. Davis at Point Pleasant box stated that immediately after Putney box had closed he had offered the freight train to Barnes Junction, it had been accepted, the Sykes instrument connected with lever No. 42 (his Down Local Starter, controlling the entrance to the block section to Barnes Junction) had dropped to *Free* and he had cleared the signal. He assumed that he must have seen the semaphore arm of his instrument go up when he had sent the *Train Entering Section* signal to Barnes. When Clapham Junction E had offered him the Windsor and Chertsey train he had offered it forward to Barnes in turn, it had been accepted, the Sykes instrument for signal No. 42 had again gone from *Locked* to *Free* and he had cleared the signal for the electric train. He admitted that he might have offered the electric train to Barnes before he had received the *Out of Section* signal for the freight train, in contravention of the Block Regulations, and said that normally, after a certain lapse of time, he would offer another train and rely on the signalman in advance not accepting it if he was not in a position to do so. The problem was that on the Western Section of the Southern Region (the former London and South Western lines) the block indicator of the Sykes instrument, when raised, meant either that no train was in the section or that one was and had not been cleared out. The only thing that the signalman in the rear could rely on to show him that a train had cleared the box ahead was his having received the *Out of Section* signal on the block bell. (To the author, who has worked many manual boxes but never Sykes Lock and Block, this seems a faulty arrangement, despite the other safeguards of the system.) Signalman Davis pleaded that in a busy box, with many bells ringing, it was difficult to always be able to pick out the ringing of the *Out of Section* signal.

At 11.30 p.m., Davis continued, he had received the *Obstruction Danger* signal from Special Class A Signalman B.C. Parish at Barnes, when they had spoken on the telephone:

Parish: What is down the Local line?

Davis: You have got a Windsor going down the Local.

Parish: No I have not, I have got a goods.

Davis: You have got a Windsor.

Parish had then hung up. Davis was unsure whether this conversation had happened before or after receiving the *Obstruction Danger* signal. He said there had previously been a brief phone call in which Parish had said that he had "heard a bang and was going out to look." Davis added that he had had every confidence in the Sykes instrument at Point Pleasant.

Signalman Parish was working alone at Barnes Junction, though during the day it was worked by two men and a booking boy. He stated that up until the accident the trains had been running fairly normally, though out of order. Putney box had switched out normally, after which he had exchanged the 2-1 *Out of Section* signal with Point Pleasant, instead of the proper 16 to test the working of the bells. He had then had to use the release key to restore the Down Through and Down Local Sykes plunging instruments from *Train On* to blank, as per the normal procedure. Point Pleasant had then offered him the freight train on the Down Local line, at the same time as the 10.50 p.m. Waterloo – Teddington passenger train on the Down Through line. He had then been busy with Up trains, had cleared his signals for the Teddington train (taking the Richmond line), then had cleared signal No. 49 to let the freight train into the station. He had not heard the freight train whistle at the signal (tests showed that it would have been difficult to hear such a whistle in the box). The Teddington train had been followed on the Down Through by the 10.56 p.m. Waterloo – Hounslow passenger train (Chiswick line) and as soon as that train had cleared the next box at Grove Park he had offered and had accepted the freight train but could not let the latter train across the junction until the passage of the 11.03 p.m. Waterloo – Kingston train from the Down Through line to the Richmond line. Signalman A Ketchell at Grove Park said, however, that there had been an interval of 2½ to 3 minutes between his clearing back the Hounslow train and accepting the freight train, to which Parish said that he would not dispute it.

Parish denied having been offered the Windsor train by Point Pleasant at all and said that his first warning that anything was wrong had been Driver Nield shouting from the ground, "I have had a bang in the back." He had not at that point realised that it was anything serious but then the driver had come into his box, saying that he had been knocked across his cab and that a porter at the station had shouted that there was a fire. Realising that this was serious he (Parish) had telephoned Point Pleasant to say that he was going down to see what was wrong. Leaving his box he had met a passenger who had told him to call an ambulance, so he had dialled 999 from a public call box on the Down platform, requesting an ambulance. He had then seen "a mass of flames" so had made a second 999 call requesting ambulances, the fire brigade, and the police (these calls were recorded as having been made at 11.33 p.m. and 11.34 p.m.). Returning to his box he had phoned Point Pleasant again and told Davis that something had hit the freight train. Davis had replied, "You have got the Windsor down there somewhere," and he (Parish) had answered, "Well, I did not clear you for the Windsor."

Parish said that he had sent the *Obstruction Danger* signal to Point Pleasant box immediately he had seen that there was a fire (recorded as at 11.30 p.m.) and to White Hart Lane box on the Richmond line after calling the emergency services. It was not sent to Grove Park. He had then tried to contact Barnes substation to get the current cut off and had eventually managed to get a message through to the crossing box on the Windsor lines at 11.35 p.m.

Told that the freight train had been stopped for several minutes at signal No. 49 Parish could only suggest that the signal had failed to clear until "jerked off" by another lever movement. He denied having forgotten the freight train or having used the release key to accept the Windsor train. Given the evidence of Signalman Davis he could only suggest a failure of the Sykes instrument there had wrongly freed No. 42 signal.

The inspecting officer, Lieut. Colonel G.R.S. Wilson, suggested that Parish might have momentarily overlooked the freight train under pressure of traffic and that when the Windsor train had been offered to him by Point Pleasant and he had seen that the Sykes instrument was showing *Train On* had imagined that he had forgotten to 'key out' after Putney box had closed and had used the release key to accept the Windsor train. Parish would only go as far as admitting that "this was a possibility," while stating that he was sure that he had not done so, adding that had he received the *Train Entering Section* signal for the Windsor train he would have had time to run back along the line to stop it.

Area Inspector G. Brant spoke highly of both Signalman Davis and Signalman Parish; he said he did not know Signalman Blundy so well. It was hard to detect boxes closing early if false entries were made in the books. He was satisfied with the working of the Sykes equipment in the 30 boxes that he supervised.

Inspector S.A. Attle, Sub-inspector J.W.D. Burden and Chief Lineman R. McGovern described in detail the tests that they had carried out on the Sykes equipment in Point Pleasant and Barnes Junction boxes after the accident, under the general direction of Signal Engineer L.J. Boucher. No faults had been found but in view of Signalman Parish's complaints of the unreliability of the Sykes instruments a list was requested by the inspecting officer of every failure of signalling equipment during the last 25 months in Point Pleasant, Putney, Barnes East, and Barnes Junction boxes. There had been 238 such failures, 67 of which had been of the Sykes equipment. Most of these had been failures of plungers to release the front locks of levers or of the treadles to release the back locks. Six, however, which had been specially investigated, had been wrong side failures. They were:

	Signalman reported	*Signal Department record*
BARNES EAST		
21 12 1954	Backlock of Up Through Advanced Starter wrongly freed with lever reversed	Fault verified. Water in treadle. Drain hole choked.
PUTNEY		
29 10 1954	Backlock of Down Local Advanced Starter treadle wrongly freed with lever reversed	Fault not repeated on test, but line insulation bad. Line renewed
29 10 1954	Backlock of Down Through Starter wrongly freed with lever reversed.	Fault verified. Broken arm in treadle. Made good.
31 10 1954	Instrument of Down Through	Fault verified. A broken cotter

	Advanced Starter "dropped to free."	pin had allowed a fork screw to work out, disconnecting the vertical rod between the lever and the instrument
PT PLEASANT JCT		
26 10 1954	Instrument of Down Local Starter (No 42) "dropped to free" on slamming the lever hard back.	Failure not repeated on test. Considered possible that armature had failed to "latch up." Armature slightly slack and armature bedplate slightly worn. instrument changed as a precaution
26 10 1954	Backlock of Down Through home (No 33) Failed to hold up on "pulling off" for 7.8 p.m. Waterloo-Windsor train	Fault not repeated on test. Fitter could find no fault with instrument, but Changed as a precaution.

The inspecting officer remarked that the first three faults had been due to false energisation of the treadle line, the fourth had been a mechanical failure which should have been obvious to the signalman, and the last two could not be reproduced but had possibly been due to a slight mechanical slackness. None of them had been at Barnes Junction.

Signal Engineer Boucher would not say that he was satisfied with the record of failures but he did not consider it excessive given the intensity of traffic, especially as he considered that the Sykes equipment was inclined to be more troublesome than modern electrical signalling because of its mechanical complexity. Nor had the Down Local Home signal at Barnes Junction failed to respond to the lever during tests.

The report went on to discuss at considerable length the delays in the cutting off of the electric current to the third rails and in calling of all the emergency services.

Concluding his report, Lt-Col Wilson stated that the statement that the Point Pleasant Starting signal (No. 42) had been off for the Windsor train was confirmed by the evidence of the driver of the 11.08 p.m. London Waterloo – Hounslow train, which had been stopped alongside at the Down Through Starter (No. 32), and of a lengthman. The Windsor train had therefore entered the section to Barnes Junction under clear signals and the Down Local Home signal at Barnes Junction (No. 49) had also been clear for the Windsor train. With the darkness and the curvature of the line Motorman Flanders would have been taken by surprise by the freight train in front of him and he was not responsible for the accident. The fault must therefore lie with the signalling.

It was clear that signal No. 42 at Point Pleasant must have been restored to danger behind the freight train. The aforementioned lengthman was sure that it had been and had it not, the locking would have prevented Signalman Davis from clearing his Down Local Home signal, No. 43, for the Windsor train. The accident could there only have occurred in one of two ways:

(a) If Signalman Davis had disregarded the elementary principles of block working by allowing the Windsor train to go forward without it having been accepted by Barnes Junction *and* there had been a coincidental wrong-side failure of the Sykes equipment freeing the lever of signal No. 42, or,

(b) If Davis had offered the Windsor train to Barnes Junction and it had been accepted by Parish, who had used the release key to enable him to plunge while the freight train was still in the section.

Possibility (a), the inspecting officer considered, was most unlikely. There had been nothing to disturb Davis's balance. Putney box had been closed correctly, if too early, there was no junction working at Point Pleasant, his actions had been simple and his account of them had been straightforward. There had been no reason for him to disturb the regular sequence of block working and had he done so there would have had to be a coincidental sudden wrong side failure of the Sykes equipment, which could not be reproduced.

The only conclusion must therefore be that the collision must have been due to possibility (b). In contrast to the situation at Point Pleasant, Signalman Parish at Barnes Junction had been working under some pressure with junction movements and trains running out of course (due to an earlier power failure). His account of events leading up to the collision had been disconnected and it was clear that he had had no idea of the position of the freight train after he accepted it, while he had not cleared signal No. 49 for it as early as he alleged. His contention that the signal had failed to clear when he pulled its lever was most improbable. There could be little doubt that he had forgotten about the freight train and that when he had seen the Down Local line Sykes instrument at *Train On* he assumed that he had failed to clear it after Putney box had switched out (as noted above). The inspecting officer did not accept the alleged time of a conversation with Signalman D.T. Maher at White Hart Lane crossing as evidence that Parish had not forgotten the freight train and believed that Parish had only been reminded of it when the collision occurred at about 11.28 p.m., at which time he had offered it to Grove Park, according to the evidence of Signalman Ketchell there.

Signalman Parish must therefore take full responsibility for the accident, and it was to be regretted that he had taken such pains to place the blame elsewhere. Parish would not, however, have made his mistake had not Davis offered the Windsor train before receiving the *Out of Section* signal for the freight train. The inspecting officer did not criticise Davis too severely for this error, given that his block indicator gave him no indication if the forward section was occupied. Davis had a very good record; it was a pity that Signalman Blundy had persuaded Davis to record false times for the closing of Putney box, to conceal the fact that he was closing it too early.

Signalman Blundy could not be held responsible for the accident, but his standard of self-discipline and lack of honesty were not such as might be expected of a senior signalman.

Driver Nield must have been stopped at signal No. 49 longer than he stated but was free of responsibility for the collision.

Summarising his main conclusions, these were, Lt-Col Wilson stated:

(i) The Windsor train had entered the section at Point Pleasant under clear signals. No responsibility rested upon its motorman, who had been killed at his post.

(ii) The Windsor train had been offered to and wrongly accepted by Barnes Junction while the freight train was still in the section.

(iii) There was no fault with the Sykes Lock and Block, and in order to accept the Windsor train the Barnes Junction signalman had made irregular use of the release key. He was therefore responsible for the collision.

(iv) The Windsor train had been offered to Barnes Junction irregularly before the Point Pleasant signalman had received the *Out of Section* signal for the freight train but he had had no visual indication of the state of the forward section.

(v) The premature switching out of Putney signal box had no more than an indirect bearing on the course of events, but the action of the signalman had been undisciplined.

(vi) The electrical arcing which had persisted for about 5 minutes had been sufficient to start an intense fire in the wooden body of the overturned coach under the conditions which had been present.

(vii) The arcing had been due solely to the continuation of the feed to the Down Through line from Clapham Junction substation, 3 miles from the site, until the circuit breaker was tripped by the application of a short circuiting bar by a motorman of the Windsor train. This action was most creditable.

(viii) The old type of slow speed circuit breakers at the distant Clapham Junction substation did not trip out, because the current passed by the short circuits due to the resistance of the conductor and running rails did not rise to the trip setting, except perhaps momentarily.

(ix) The electrical staff had no responsibility for the continuance of the feed, but the Barnes Junction signalman should have telephoned for the current to be cut off earlier than he did.

(x) No responsibility rested with the fire service for the late attendance from Barnes fire station. They had not received a fire call until 12.40 p.m., 12 minutes after the accident. There had then been no delay in turning out the appliances.

(xi) The fire station should have received a fire call 7 minutes earlier. The Barnes Junction signalman had asked for all three services when he had made the first 999 call at 11.33 p.m. and had been promptly connected by the exchange to the fire station. In his agitation, the signalman had then failed to make it clear that a serious accident had occurred and only asked for an ambulance, which had been sent. The mistake might have been corrected had the telephonist remained in circuit to keep the situation under control.

Turning to remarks and recommendations, the inspecting officer noted that the Sykes Lock and Block, though old fashioned in design, had a long record of reliable service in dense traffic, though accidents due to misuse of the release key had occurred from time to time. There were, however, limits to the density and complexity of traffic that such a system could handle, plus a difficulty in recruiting suitable staff. Modernisation of the signalling on the line with continuous track circuiting and colourlight signals should be given a high priority. Meanwhile the provision of berth track circuits at the Barnes Junction Home signals had been authorised and they should be installed at other similar locations.

It was most undesirable that a second train should be offered before a previous train had cleared the section. The Western section arrangement regarding the Sykes block indicators should be altered to that on Eastern and Central sections, where the block indicator was normally lowered, raised when the train was accepted, and lowered when the *Out of Section* signal was given.

The current on the Down Local line had been cut off within a few seconds of the collision but it had remained on to the Down Through line with short circuiting by metal parts of the overturned coach causing the arcing and fire, the circuit breakers having failed to clear the fault before serious damage had been caused. More modern equipment now being installed would reduce the risk but there must always be a dependence on staff at the scene to get the current cut off as a matter of urgency in the event of an accident. The risk of a short circuit was much less with overhead electrification.

A modern steel coach would not have caught fire in the same manner. Wooden coaches in electric trains were expected to be eliminated in 1957, the replacement of all the wooden rolling stock on British Railways would take rather longer.

The General Post Office was considering steps to avoid the delay which had occurred to the calling out of the fire service. [420]

Only seven months after the San Bernardo accident (see above) Chile suffered another disastrous rear collision. This happened between the halt at Cerrillos and Maipú station, on the line from Santiago Alameda to the Pacific coast at Cartagena, at 07:55 on Tuesday 14th February 1956. Two excursion trains from Santiago to Cartagena were involved. The first train, carrying about 200 passengers in wooden carriages, had only covered just over 7 kilometres when it came to a stand just beyond Cerrillos due to an unspecified mechanical problem. Trains were apparently being worked on the time interval system, with a 12-minute interval between them. If this was so, it was clearly imperative, in the event of a train stopping out of course, that immediate steps be taken to protect it in the rear, but this does not seem to have been done, and the second excursion train ran into the first one. The colliding engine embedded itself in the rear two third-class coaches of the first train and it was here that the deaths and injuries occurred, few of the occupants of these two carriages escaping unhurt. In total 23 passengers died and 158 were more or less injured.

The driver of the second excursion, fearing for his safety, fled the scene and hid at Maipú station. People living along the line performed the first rescue work and called the emergency services and firemen, doctors, nurses, police, and Civil Defence workers hurried to the scene. Ambulances took the injured to San Juan de Dios, Posta Central and Posta 3 hospitals. Carlos Ibáñez del Campo, the President of the Republic, ordered an investigation into the cause of the disaster. [421] [422] [423]

Snow, rather than fog, was the condition which caused a disaster at Swampscott Junction, Massachusetts, United States, on Tuesday 28th February 1956. Two Boston and Maine Railroad passenger trains were involved; a Portsmouth, New Hampshire, to Boston train was run into from behind by a Danvers to Boston train. Eleven passengers and the engineer and fireman of the Danvers train were killed and 260 passengers, the engineer, fireman, conductor, two assistant conductors and flagman of the Portsmouth train and the conductor, assistant conductor, baggageman and flagman of the Danvers train were injured. [424]

---

[420] Ministry of Transport report

[421] http://chile-catastrofes-tragedias.blogspot.com/2010/06/tragedia-ferroviaria-en-queronque.html

[422] http://hitosdechile.blogspot.com/2012/06/accidente-ferroviario-en-los-cerrillos.html

[423] Wikipedia – Accidente ferroviario de Cerrillos de 1956

[424] Interstate Commerce Commission report

Howard Ross, of West Newbury, Massachusetts, was on his way to Lynn where he worked as a stereotyper. He related, "I was a passenger in the fourth car of the train from Portsmouth. I didn't see too much of it, I was reading a paper and felt this terrific crash. My head seemed to go one way while I went the other. The car I was in stayed on the rails and I guess most of us were unhurt, except for one woman. What happened was that the Danvers train crashed into the last car of ours and derailed it. The first car of the Danvers train split in half right down the middle. When I saw it a piece of metal was up ahead of our last car. There was nothing left of the first car of the Danvers train except the platform. Our last car was derailed and jacknifed. The crash occurred about a half mile north of Swampscott. I got out and came along to work. I don't like to see death. I walked into Swampscott and took a taxi from there into Lynn." [425]

The Portsmouth train was No. 214, a westbound first-class passenger train, consisted of diesel-electric unit No. 1516, five coaches and a combination baggage and smoking car, in charge of Conductor George Buckley with Engineer Everett Storey. It had left Portsmouth at 6.40 a.m., passed Salem Tower, the last reporting station, at 8.10 a.m., and passed signal No. P 146 at low speed because its roundel was covered with snow and only distinguishable at low speed. This signal, like the next two, was of the automatic searchlight colourlight type. The roundels of the next two signals, No. P 138 and RA2, were completely obscured and the train passed the first at low speed and stopped at the second. Two or three minutes later, at about 8.18 a.m., it was struck by the Danvers train.

The Danvers train, westbound first-class passenger train No. 2406, consisted of three diesel-powered passenger units and a diesel-powered passenger-baggage unit, in charge of Conductor Thomas Gill with Engineer Ernest Tourtellotte and Fireman Raymond Jones. It had been held at Salem tower for train No. 214, then left there at 8.12 a.m. It passed signal No. P 146, which should have indicated *Prepare to stop at next signal*, passed signal No. 138, which should have indicated *Stop, then proceed at reduced speed*, and was still moving at 40 to 55 miles an hour when it struck the Portsmouth train. As a result of the collision the underframe of the rear car of train No. 214 overrode the leading car of train No. 2406 and sheared it off. The rear car was itself considerably damaged, the fifth car of the front train was somewhat damaged and the first, second and fourth cars were slightly damaged. The second car of the rear train was badly damaged at its front end and the third and fourth cars were slightly damaged.

Heavy, wet snow was falling at the time of the accident. The Portsmouth train was about 25 minutes late and the Danvers train was about 11 minutes late when they collided.

Engineer Storey of train No. 214 told the Interstate Commerce Commission (ICC) inquiry that after leaving Newburyport, about 19½ miles west of Portsmouth the supercharger of the diesel engine of his train had developed a fault, and from then on his train had lost time as a result. It was snowing and there was a strong wind. At Beverly, 19 miles further on, the roundels of the signal at the interlocking had been obscured by snow so that he had had to ask the operator for permission to proceed. Beyond Salem, 40½ miles west of Portsmouth, blowing snow had restricted his vision. He had been able to see signal No. P 154, a lower quadrant semaphore, showing *Proceed*. Having reduced speed to 5 miles an hour he had just been able to distinguish a faint green light showing from signal No. P 146. The next signal, No. P 138, he had not been able to make out at

[425] The Lowell Sun, Lowell, Massachusetts, United States, 28th February 1956

all, but because of the defective supercharger he had been reluctant to stop at it and instead continued at 2 or 3 miles an hour. Signal No. RA 2 had also been invisible and the train had been stopped and Conductor Buckley had alighted and gone to the signal post telephone to ring the operator at Lynn tower. He (Engineer Storey) had then seen train No. 2406 coming and had heard its horn sounding.

Flagman M.J. Rival said that he had dropped off 5-minute lighted fusees in the vicinities of signals Nos. P 146 and P 138. He did not know if they had remained alight.[426] He said that as soon as his train had stopped at signal No. RA 2 he started back with a red flag to protect his train. He continued, "All at once I could hear the click of wheels as if a train was coming. I couldn't imagine a train coming on our track so I looked on the other track and there wasn't one there. I was about 150 to 200 feet from my train then. I started to get panicky and started to run up the track. Something appeared under the {railroad} bridge {nearby}. I started to wave the flag frantically. I could see the fireman and engineer standing up. I don't want to cast reflections on the deceased engineer. I knew him. I worked with him. But I could not recognise him at that speed."[427] The flagman said he had not had time to light a fusee before the train passed.

Conductor Gill testified that the train had been making a return trip to Boston after a journey from that place to Danvers. The brakes of his train had been tested before leaving both Boston and Danvers. After leaving Danvers three station stops had been made before his train had been held 14 minutes at Salem tower for train No. 214, after which it had stopped at Salem station. The brakes had worked normally at each stop. He had spoken to the engineer and fireman several times during the morning and they seemed to be in normal mental and physical condition. Leaving Salem his engineer had remarked that train No. 214 was close ahead. Visibility had been about 200 feet. Neither Conductor Gill, Assistant Conductor Clifford Homan nor the baggageman noticed any brake application or reduction of speed after leaving Salem and they estimated the speed of the train at the moment of collision at between 40 and 50 miles an hour. Conductor Gill stated that immediately before the collision Engineer Tourtellotte had stepped into the passenger compartment and warned him that they were about to strike train No. 24.

Following train No. 2406 was train No. 2506, which was being driven by its fireman and had passed Salem tower at 8.20 a.m. Its engineer said that signal No. P 154 had indicated *Proceed* and his fireman had called signal No. P 146 as also indicating *Proceed* but then apparently had his doubts and he had applied the brakes. Neither the engineer nor the fireman had been able to distinguish the indication of signal No. P138, at which they had been stopped by Flagman Joseph Joyce of train No. 2406.

Some of the uninjured passengers involved in the accident boarded a Marblehead branch train at Swampscott to continue their journeys to Boston (the junction with the Marblehead branch was at Swampscott, south of the accident site, and so trains from the branch were not affected by the blockage of the main line). They then found themselves in a second collision in identical circumstances only an hour later at Revere, seven miles nearer Boston. This collision involved trains Nos. 2206 and 2208, which had left Marblehead at 7.57 a.m. and 8.30 a.m. Train No. 2206 was stopped at signal No. 4, which was obscured by snow and ice, and its fireman was cleaning the signal when it was run into by train No. 2208. This time it was only a low-speed collision and ten railwaymen,

---

[426] Interstate Commerce Commission report

[427] *The North Adams Transit*, North Adams, Massachusetts, United States, 9th March 1956

but no passengers, were injured and there were no deaths.

Engineer Henry McCarthy of train No. 2208 testified that had the brakes of his train been effective he could have pulled up short of train No. 2206 and that on previous occasions he had had trouble making station stops with trains equipped with disc brakes of the type on train No. 2208 when rail conditions were unfavourable and the anti-wheel-slide device was activated. Another engineer agreed.

A representative of the manufacturer of the brake equipment stated that it was designed to produce maximum braking under normal conditions. The anti-wheel-slide device was designed to prevent wheels from sliding when rail adhesion was reduced. When a wheel began to slide the brake pressure was reduced for about three seconds to permit the wheel to regain rail speed, sand being applied during this time.

The ICC carried out a series of brake tests at the end of March. It noted that should the sanding device fail to deliver any or sufficient sand, repeated operation of the anti-wheel-slide devices under unfavourable conditions was likely to impair braking, also that the freezing of wet snow and moisture on brake discs and shoes would make them ineffective until it melted. The Boston and Maine management was aware of these problems and had issued instructions for engineers to apply the brakes earlier during snowy and icy conditions.

The ICC joint report on the two collisions came to eight conclusions:

(1) The signalling equipment had worked properly at the time of the accidents.

(2) The signal roundels had been obscured by snow to the extent that according to the rules the trains should not have been operated at more than 15 miles an hour.

(3) Both accidents had resulted from trains failing to obey signal indications.

(4) Both trains had been operated at excessive speed.

(5) The engineer at Swampscott had sounded the pneumatic horn and left his seat before passing the flagman of train No. 214 but the train had not appreciably reduced speed.

(6) The engineer at Revere had applied his brakes but the stopping distance of the train might have been increased due to the operation of the anti-wheel-slide devices.

(7) The efficiency and reliability of the disc brakes on the trains had proved satisfactory under normal operating conditions and the brakes had worked very well in the tests of 30th March 1956.

(8) The aforementioned tests had not been conclusive as to the braking obtained under adverse weather conditions.

The ICC recommended that the Association of American Railroads carry out a series of comprehensive tests of the efficiency of the disc brakes involved in unfavourable weather conditions. [428]

Possible problems with the combination of disc brakes and anti-wheel-slide devices in bad weather conditions notwithstanding, the Boston and Maine dismissed Engineer McCarthy of train No. 2208 at the end of March. He protested, "I do not believe I was at fault and I'll fight to prove I was not at fault." [429]

---

[428] Interstate Commerce Commission report

[429] *Biddeford Daily Journal*, Biddeford, Maine, United States, 29th March 1956

Late on Saturday 23rd June 1956 a rear collision occurred in the suburbs of Madrid, Spain. The Andalusia passenger and mail train, No. 482, had departed from Madrid Atocha station at 23:05 with 600 passengers, five minutes late, shortly followed by a parcels train, No. 6404, also bound for Andalusia. The mail train had only travelled a few kilometres, and had reached the station of Santa Catalina, when it was run into from behind by the parcels train at 23:15.

Train No. 482 consisted of an engine, a first- and third-class composite coach for Almería, two coaches for Granada, a composite coach for Jaén, a first- and third-class composite sleeping car for Córdoba, two coaches for Sevilla, a coach for Cádiz, two mail vans and a guards van. The order of vehicles in the train was not reported, other than the fact the Granada coach was the rearmost and the Almería coach was the penultimate; it was these two carriages which suffered the most damage in the accident, especially the Almería one. These two carriages overturned against the poles supporting the overhead line equipment, but the latter were not much damaged. Police and Civil Guards on the mail train immediately alerted Atocha station of the accident and then began the rescue operation. A van hauled by a diesel locomotive soon arrived at the scene, followed by a second rescue train pulled by a steam locomotive and numerous Red Cross ambulances also attended.

Eight passengers and Driver Eufrasio Rigau Vargas of train No. 6404 were killed outright and 76 passengers were injured, one of whom had died in hospital by the following day.

The line was reopened for normal working at 08:45 the next morning.

It was stated that the collision was undoubtedly due to a failure of the electrical signalling caused by a severe thunderstorm which had begun at 20:00 and which had in some manner put the signals to danger against the mail train while clearing them for the parcels train.[430]

On the morning of Wednesday 1st August 1956 a disastrous collision occurred at Barro do Piraí station, Brazil, on the line of the Estrada de Ferro Central do Brasil, due to a freight train bound for Japaranã disregarding signals and colliding with passenger train R-4, the so-called *Trem dos bainos* (Bahian train) which ran between Belo Horizonte and São Paulo. The rearmost two coaches of the passenger train were wrecked and 14 passengers were killed and 50 injured, five seriously. The passenger train had reportedly featured in several recent collisions. Most of the victims were again immigrants from the interior of the country who were on their way to look for work in the cities. The injured were cared for at the Santa Rita nursing home at the expense of the railway company and the cost of the funerals of the dead was likewise met by the company. There were a large number of children amongst the victims, who came mostly from Bahia, Pernambuco, and Sergipe states. The drivers of both trains disappeared after the accident, perhaps fearing reprisals from surviving passengers or else arrest.

The railway company issued an official statement regarding the disaster: 'At 05:35 today freight train prefix P-2, when approaching Barro do Piraí, ran past a danger signal and collided with train R-4 from Belo Horizonte bound for the Paulista capital. This accident resulted in the death of 14 passengers, gravely injured three and slightly injured

---

430 *La Vanguardia*, Barcelona, Spain, 26th June 1956

many more. A rescue operation was immediately begun and the injured people were treated on the spot by local railwaymen, who were soon assisted by medical staff, then promptly removed to hospital at the expense of the company.

'Normal working was resumed at 10:00. The delays to trains caused by the accident were as follows: train D-4 from Rio was 2 hours 45 minutes late and train M-2 from there was 4 hours and 50 minutes late.

'In the opposite direction train R-1 to Rio was 2 hours and 25 minutes late, train D-1 was 52 minutes late and train S-1 was 2 hours and 56 minutes late. Train M-1 was cancelled and the train involved in the disaster continued its journey 10 hours and 20 minutes late.

'The serious consequences of the accident were due largely to the fact that the passenger train was made up of wooden carriages and their replacement by steel vehicles has been urgently requested by the railway management.

'Driver A.C. Lemos of the freight train has been suspended with immediate effect and the company has opened an immediate investigation into the accident.

'The railway company greatly regrets the disaster. It intends to take steps to withdraw from traffic, as soon as possible, all obsolete wooden rolling stock, whose lack of strength is unacceptable with today's heavy traffic.' [431]

Nine people reportedly died and 29 were injured in a collision between an express passenger train and a goods train near Přerov, Moravia, Czechoslovakia, on or about Tuesday 7th August 1956. The express, which according to one report had come from Banská Bystrica in eastern Slovakia, collided with the goods train which had stopped at a level crossing. [432] Another source states that the express had come from Bohumin and had been bound for Praha, gave figures of 7 dead and 56 injured, and stated that the accident had been due to the express overrunning signals. [433]

A major accident occurred in Chicago, Illinois, United States on Monday 5th November 1956, at the Wilson Avenue station of the Chicago Transit Authority (CTA), on a stretch of elevated railway over which the Chicago North Shore and Milwaukee Railroad interurban had trackage rights (running powers). A train of each company was involved in the accident. The six-car North Shore train, with over 300 passengers, had stopped at the station at 5.28 p.m. and was just about to restart when it was run into by the following crowded eight-car CTA train at about 5.30 p.m., thus at the height of the evening rush hour. The CTA train was travelling at an estimated 15 miles an hour.

'Screams of panic ran through the CTA as lights failed after the impact. Passengers were thrown to the floors and piled atop each other. Mrs Margot Witcoff, a passenger, said that when the lights went out "people started yelling, pushing, and trampling over others who had fallen in the aisle."

'More than a dozen people were trapped in the forward section of the first elevated car, including the motorman, Michael O'Driscoll, 52. He was the last to be removed from the wreckage, nearly five hours after the pileup. He had been given blood plasma and

[431] *O Estada de São Paulo*, São Paulo, Brazil, 2nd August 1956
[432] *Gazette de Lausanne*, Lausanne, Switzerland, 8th August 1956
[433] Wikipedia – Liste von Eisenbahnunfällen in Tschechien

hypodermic injections of morphine. Dr J.R. Guttman said O'Driscoll told him he tried to apply the brakes as he neared the station but they had failed.

'CTA spokesmen said brakes of the last seven cars of the train were in operating order. They said condition of the brakes in the lead car could not be determined because of damage caused by the impact. Some 100 firemen worked on the rescue. Most of the passengers trapped were rescued only after an entire section of the front car was pulled away with chain hoists.' [434]

Initial reports were that seven people had died and 160 had been injured. The 6000 series cars of the CTA train were considerably lighter than the cars of the North Shore train and almost all the casualties occurred in the CTA train.

An inquest on the accident was opened the following day, by which time an eighth passenger had died and the number of known injured had reached 200. A witness spoke of the lead car of the CTA train being crumpled "like a cracker box." Joseph Block, brother of Mrs Frances Strauss, one of the passengers who died, wanted to know, "Why only two of the four tracks at Wilson avenue are in use, why the CTA train was running so close to the North Shore, and why the six-car North Shore train was not badly damaged whereas the CTA lead coach was?" Motorman O'Driscoll did not answer any questions. He had been badly hurt; his left leg was amputated below the knee and his right leg was broken in three places. The inquest was adjourned for further investigations. [435]

A couple of weeks later the motorman, sitting in a wheelchair, was interviewed by reporters at a news conference. He stated that he had not been drinking and alleged that the brakes of his train had slipped on several previous occasions during the fateful journey but he had not been especially concerned as this had happened before. He said that his train's brakes had four positions, the fourth being reserved for emergencies. Describing the moments before the collision he said, "I put on the No. 1 brake about 150 feet south of the Wilson platform. It held. Then about 50 to 60 feet south of the platform I applied No. 2 brake position. It held. Then at the beginning of the platform, about 100 to 150 feet south of the North Shore train, I gave her No. 3. No. 3 didn't hold. It slipped." [436]

At the conclusion of the inquest the jury found that the collision was entirely due to the failure of Motorman O'Driscoll to see the North Shore train ahead of him.

Twenty years after the Granville collision of 24th November 1936 (dealt with in the previous volume of this series) and despite the intensity of traffic on the line trains were still run "on sight." The jury recommended that the CTA install automatic block signals but it would be another 20 years before the signalling was modernised. [437]

Because it occurred on an urban transit line the accident was not investigated by the Interstate Commerce Commission.

At about 04:30 on 7th April 1957 Rede Ferroviária Federal (formerly Estrada de Ferro Central do Brasil) freight train No. CP 110, consisting of an engine driven by José Joaquim Gomes and 32 waggons of coal bound for Volte Redondo, Brazil, passed through Magno. It then stalled on the incline between Magno and Turiaçu stations and its

---

434 *Moline Daily Dispatch*, Moline, Illinois, United States, 6th November 1956

435 *Freeport Journal-Standard*, Freeport, United States, 6th November 1956

436 *Daily Journal-Gazette*, Mattoon, United States, 24th November 1956

437 https://www.chicago-l.org/mishaps/wilson.html

driver was forced to set back towards Magno in order to get a run at the bank. The signalman at Magno, Nilson Candido Pinto, who had seen the freight train pass at apparently normal speed, was reportedly advised by his colleague at Turiaçu, Targino José da Silva, that electric passenger train UA 10, which was waiting at Magno, could proceed, and let the passenger train go, unaware that the freight train had stalled and was setting back, and presumably in the mistaken impression that the it had been reported as having arrived at Turiaçu.

Driver Julio da Silva of the electric train duly set off and was only a few hundred metres from Turiaçu station when he saw the rear of the freight train ahead of him. He frantically applied the brakes and sounded his horn loudly. Nothing helped. The electric train tore into the rear of the freight train 'like a meteor,' destroying half of its rear waggon. Driver da Silva was killed instantly. The trains collided at 04:47. The force of the impact threw all the passengers in the front car of the electric train, where their bodies were piled up in a confused heap among the twisted wreckage. In a graphic description of the aftermath of the accident a reporter spoke of headless bodies, arms and legs dangling amongst the remains of the carriage, and piercing screams from the survivors, many of whom were sandwiched between the bloody bodies of the dead. Indescribable scenes took place as passengers from the undamaged part of the train tried to help the victims. Men, women, and children were pulled out through windows as ambulances arrived from several hospitals, together with firemen who worked to free the trapped victims. Among those who were badly hurt and caught in the wreckage was Conductor Vilarinho Sampalo of the electric train. Initial reports stated that 19 people had died in the accident and about a dozen had been injured, most of whom were taken to Carlos Chagas Hospital.

Commissioner Moura Brasil initiated an immediate inquiry into how the disaster had occurred. He first interviewed Traffic Inspector Joaquim Floriana da Fonseca, who explained in detail what had happened. The commissioner next questioned Signalman José Amancio Ferreira, who stated that the Magno signalman had accepted the electric train from him, and then Signalman Targino José da Silva at Turiaçu, who placed the blame entirely on Signalman Nilson Candido Pinto at Magno. Signalmen Ferreira and da Silva agreed that there had been no problems with the signalling equipment. The Magno signalman disappeared after the accident.

On the other hand the police authorities and the railway company place some of the responsibility for the accident on Driver Gomes, who having almost reached Turiaçu, had then set back his train towards Magno without advising either of the stations of the reverse movement. It was also clarified that Driver da Silva would have had no means of averting the collision once the freight train had come into view, especially in the prevailing somewhat foggy conditions and restricted visibility. [438] [439]

Two Indian suburban electric trains, both bound for Bombay Victoria terminus (now Mumbai Chhatrapati Shivaji Maharaj Terminus or Mumbai CST) were in a collision at 08:00 on Sunday 2nd June 1957, between Sewri and Vadala Road stations on the Harbour Branch of the Central Railway. Up Bandra-Bombay Local train No. B-12 was stopped at signals when it was struck by Up Mankhurd Local train No. M-22.

As a result of the collision the roof of the leading first-class coach of the colliding

---

[438] *A Noite*, Rio de Janeiro, Brazil, 8th April 1957

[439] *Morris Daily Herald*, Morris, Illinois, United States, 8th April 1957

train caved in, its sides were partially ripped open and its floor was covered with splintered glass, fans, cushion seats and dislodged luggage racks. The steel undercarriage was twisted out of shape and splinters of wood, glass and other fittings were scattered over a wide area on either side of the track. The damage to the rear vehicle of the front train was not detailed but presumably was as great if not greater, and the next two coaches of the colliding train were also damaged. Uninjured and slightly injured passengers who were able to jumped out of the trains while others were trapped in the wreckage. Churchgoers and 'burly Pathans,' alerted by the noise of the collision, immediately began the rescue work. An announcement by the Central Railway Public Relations Department stated that there were 18 dead and 66 injured. Driver Augustus of the Mankhurd train had time to jump from his cab immediately before the collision, while the guard of the Bandra train also survived, though he was injured. Among the dead were a woman and child but most of the victims were workers bound for the oil installations in the docks close to the terminus. It was pointed out that had the accident not happened on a Sunday the number of casualties would have been greater.

How the accident could happen in broad daylight was not immediately clear, as it had occurred on a straight stretch of track and the stationary train had been visible from a furlong (220 yards) away.

Railwaymen were able to clear the line by late afternoon. [440]

By the following day two of the injured had succumbed, increasing the number of fatalities to 20 [441] and the final number of dead was 23 and of injured 64, while damage to rolling stock and the permanent way amounted to 47,770 rupees. The accident was blamed on the failure of Driver Augustus to comply with the regulations regarding the passing of automatic signals at danger. [442]

A collision occurred at Stuttgart Bad Cannstatt, West Germany, on Tuesday 1st October 1957. The *fahrdienstleiter* assumed or was wrongly informed that passenger train P 1414 to Schorndorf had left the station and admitted passenger train N 3108. Train P 1414 was in fact still there, hidden by thick fog, and a collision ensued at a closing speed of 40 kilometres an hour, in which 10 people lost their lives and 56 were injured. The rear coach of the Schorndorf train was so badly damaged that it had to be cut up. [443] There had been 600 or 700 passengers in each train, one of which was carrying workmen and the other mostly schoolchildren. The responsible *fahrdienstleiter* was sentenced to a term of imprisonment. [444] [445]

In what was probably a rear collision a Split – Zagreb express collided with a freight train in the station at Zagreb, Yugoslavia, on the evening of Thursday 5th December 1957. Three tank waggons of petrol exploded and set the wreckage on fire. Three or five people

---

440 *The Indian Express*, Bombay, India, 3rd June 1957

441 Ibid, 4th June 1957

442 Report on Indian Railways 1957-1958, Vol. 1

443 *The Times*, London, England, 2nd October 1957

444 *Katastrophen der deutschen Bahnen Teil II* by Hans Joachim Ritzau, pg. 125 (in German)

445 https://www-hna-de.translate.goog/lokales/melsungen/melsungen-ort45520/melsunger-fand-alte-zeitungen-aehnlichem-zugunglueck-bayern-6112639.html?_x_tr_sl=de&_x_tr_tl=en&_x_tr_hl=en&_x_tr_pto=sc

died according to different reports and about 20 were injured. [446] [447]

On New Year's Eve, Tuesday 31st December 1957, an electric suburban passenger train ran into the rear of a long-distance express near Otwock, Poland, not far from Warsaw. Six people died and a good number were injured. [448] [449]

Less than two months after the catastrophe at St Johns, Lewisham (see Train Accidents Involving Bridges chapter) another rear collision in fog occurred in England at Dagenham East, on the British Railways, Eastern Region main line between London Fenchurch Street and Shoeburyness. The accident occurred at about 7.34 p.m. on Thursday 30th January 1958 and involved the 6.20 p.m. Down passenger train from London Fenchurch Street to Thorpe Bay and the 6.35 p.m. Down passenger train from London Fenchurch Street to Shoeburyness. The first train was run into from behind by the second. There were about 500 passengers on each train and 10 passengers lost their lives and 89 passengers and four railwaymen were injured.

The Thorpe Bay train consisted of a Class 4 P 2-6-4 tank engine and 11 coaches, in charge of an unnamed guard with Driver R. Wilson and Fireman T.E. Spicer. It had started from Fenchurch Street 10 minutes late and due to delays caused by the fog was 31 minutes late at Barking East. Another 17 minutes were occupied in running the three and a half miles between Barking East and Dagenham East, outside of which the train was stopped because the 6.15 p.m. departure from Fenchurch Street was in Dagenham East station.

The Shoeburyness train consisted of a Class 4 tank engine and another 11 coaches, in charge of Guard W.T. Pryde with Passed Fireman Reginald Barnes and Fireman S.G. Fryer. It left Fenchurch Street five minutes late and was 29 minutes late at Barking East. It then took only eight minutes between Barking East and the point where it collided with the Thorpe Bay train.

As a result of the collision the engine of the Shoeburyness train ploughed through the rear coach of the Thorpe Bay train, while the front of that coach telescoped part of the coach in front. The engine of the colliding train, which was running tender first, was also badly damaged. [450]

Following the accident, 'Every available ambulance that could get through the thick fog was ordered to Blackborne Road, Dagenham, and police car crews broadcast over their radios for every possible assistance to be sent to the scene of the crash. Nurses and doctors were taken through the fog in police cars from Oldchurch Hospital, Romford, and the King George Hospital, Ilford, to attend to the injured on the track. Police cars on patrols in Essex and the Metropolitan areas were ordered to make their way to the scene to assist with rescue work and those from districts as far apart as Woodford and Loughton were told to pick up flares and searchlights and help the rescue teams. All available firemen from the Essex, East Ham and West Ham brigades drove in slow possession through the fog carrying oxy-acetylene equipment.

---

446 *Reno Evening Gazette*, Reno, Nevada, United States, 6th December 1957

447 *Gazette de Lausanne*, Lausanne, Switzerland, 7th-8th December 1957

448 *The Corsicana Daily Sun*, Corsicana, Texas, United States, 1st January 1958

449 *The Times*, London, England, 2nd January 1958

450 Ministry of Transport report

'Telephone operators at Rainham exchange volunteered to stay on after their tour of duty to help to cope with the rush of calls. Rescue workers, who included civil defence staff, collected blankets from homes nearby for the injured awaiting ambulances. The current was switched off on the electric lines alongside the debris.

'The rescuers worked by searchlight, flares, and hundreds of candles held by helpers. Residents of a council estate nearby brought out tea, whisky, and brandy, while extra police cleared the lines and held back hundreds of sightseers. The rear coach of the 6.20 p.m. train was forced over at an angle by the crash and had to be propped up with cranks and iron bars to stop it toppling.

'First on the scene were Mr Charles Hendy, Mr George Milson, and Mr Joe Heron, all of Pavett Close, which adjoins the track. Mr Hendy said: "I was in bed and heard one train stop. When I heard another train coming along I sensed that something was going to happen. I ran down to the scene and through the fog could see the tender had ploughed into the rear coach of the first train and crushed it against the next coach. We got crowbars and pulled part of the coach away and forced open some of the doors. Some of the passengers were standing in the compartment. I saw the seat section of one compartment had been forced into the next compartment."

'Mr Milson said: "I found a woman still sitting down on a seat which had been forced up to the roof. She was hurt but she escaped with her life."

'Mr J.P. Martin, of Newmarket Way, Hornchurch, a passenger in one of the middle carriages of the Southend train, and there was a "terrific crash" and the lights went out. Many passengers were unaware that there had been an accident and sat in the darkness of their compartments for about 10 minutes before a railwayman ran along the track and told them what had happened. Mr Martin said that after walking along the track, he spoke to the engine driver of the Southend train, who appeared to be shaken, and asked him how he was. He replied that he was "all right."

Mr A. Wiseman, of Charlotte Road, Dagenham, said: "We climbed into several of the coaches and dragged out many of the passengers, all of them dazed, and many injured. The rear coach of the first train was buckled high into the air and all the people inside were packed tight, even up to the roof. We could not get into that coach as all the doors had been jammed and inside was complete chaos. By the time we had got 14 people out firemen and ambulances arrived and began rescue work."' [451]

Between Barking and Dagenham the double-track main line from Fenchurch Street was paralleled by a pair of London Transport lines, the four lines from north to south being the Down Local, Up Local, Down Main and Up Main. The two Local Lines were used by all London Transport trains but steam trains could be diverted onto them at Barking and Dagenham East. The Through Lines were not electrified and could only be used by steam trains. The Local lines were equipped with two- and three-aspect colourlight signals with train stops at each signal. The Through lines retained semaphore signals with Hudd type automatic train control (A.T.C.) at the distant signals and were worked on the absolute block system with Class C block controls (for description of these see Winsford, above). Between Barking East and Dagenham East signal boxes there were boxes at Upney and Becontree which were only block posts and were switched out of use at quiet periods.

[451] The Times, London, England, 1st February 1958

On the evening of the accident a variable fog was affecting the area where the accident occurred with visibility down to probably 10 or 20 yards. This was affecting the running of trains so that services were leaving London Fenchurch Street between 5 and 10 minutes late and with cautious running were about half an hour late by the time that they reached Barking. Upney box normally closed at 7.15 p.m. but was being kept open later because of the late running of trains, in accordance with normal practice. Becontree box was normally only open from 7 a.m. until 11.15 a.m. and 4 p.m. to 6.50 p.m. to cover the morning and evening rush hour periods. The other signalman having left for another job there was only one signalman there, doing a 12-hour shift, in the middle part of which, when the box was switched out, he attended to the signal lamps and did portering work. With the late running Down trains he was unable to switch out until 7.24 p.m.

Signalman Alfred Tinson at Upney told the Ministry of Transport inquiry into the accident that visibility in the fog had been 20 to 25 yards, deteriorating further after dark, so that he could barely make out electric trains on the Up Local line 12 yards away. He had been going out to check that the Down Home signal had returned to danger after each train had passed. The 6.15 p.m. Down train had passed him under clear signals, except perhaps for a check at his Distant signal, and as soon as it had cleared the track circuit (which extended from the rear of his Home signal to his Starting signal) he had accepted the 6.20 p.m. train at 7.11 p.m. He had received the *Train Entering Section* signal from Barking for the 6.20 p.m. train at 7.14 p.m. and had offered it to Becontree, the latter box having cleared back the 6.15 p.m. train. The Becontree signalman had at first refused it, because, Tinson assumed, he wished to switch out and go home, having been on duty for over 12 hours, but had then accepted it. He had given the *Train Entering Section* signal for the 6.20 p.m. to Becontree at 7.17 p.m. as soon as the train reached the track circuit. It had been going slowly and had not cleared the track circuit until 7.21 p.m., when he accepted the 6.35 p.m. train from Barking East. He had received the *Train Entering Section* signal for this train from Barking East at the same time and two minutes later it had come to a stand at his Down Home signal, which he had not cleared because he had not yet received the *Out of Section* signal for the 6.20 p.m. train (Rule 39a). Before or immediately after the 6.35 p.m. train had stopped at his Down Home Becontree box had switched out and he had spoken on the telephone to the Dagenham East signalman to ensure that everything had been done correctly. He was positive that the Down block instrument to Becontree/Dagenham East had remained at *Train on Line* the whole time.

At this time the block regulations of the former London Midland and Scottish Railway (LMS) were still in force and these permitted a signalman to switch out his box with a train in the section, provided that the block instruments at the boxes in the rear and in advance of the box being closed both displayed either *Line Clear* or *Train on Line.* On other British Railways, Eastern Region lines, which were not ex-LMS, in contrast, boxes could only be switched out when there was no train in the section, a safer procedure. Thus if the Dagenham East block instrument had still been at *Line Clear* at Becontree for the 6.20 p.m. train and the Upney block instrument at Becontree had been at *Train on Line* for the same train, when Becontree box was closed the *Line Clear* indication from Dagenham East would appear at Upney, although the 6.20 p.m. was still in the section.

Signalman Tinson stated that after holding the 6.35 p.m. train at his Down Home for a minute or so he had let it down to his Starting signal. He had then accepted the 6.10 p.m. ex-Shoeburyness Up train from Dagenham East and as the 6.35 p.m. Down train was passing its guard had called out, “Are all the boxes in?” to which he had replied, “No,

Becontree is switched." Dagenham East had sent on the Up train, he had it accepted by Barking East, cleared his Up signals, then had gone outside to relieve himself. When he returned he had been surprised and shocked to find that his Down track circuit had been showing clear. Instead of at once sending the *Train Running Away* signal, 4-5-5, he telephoned Dagenham East, saying something like, "I have good reason to believe this Down train has passed my starter." He had then rushed out of his box in the forlorn hope of seeing the tail lamp of the train, before returning to his box and sending the 4-5-5, to which Dagenham East replied with the *Obstruction Danger* signal.

Signalman Tinson was grilled for several hours by Brigadier C.A. Langley, the inspecting officer, but stuck to his story.

Porter Signalman M. Sass at Becontree put the visibility at only 10 yards after dark. He said that he had been unable to close his box at 6.50 p.m. because he had a train at his Down Intermediate Block Signal (Becontree or Heathway I.B.S) and had accepted another train from Upney. He had also got to the bottom of the page of his train register and already written the next day's date on the following page so he had written the times of the signals from Upney for the 6.10, 6.15 and 6.20 p.m. trains on a scrap of paper, being in a hurry to get away. The times to Dagenham he had not recorded at all. He had accepted the 6.15 p.m. train from Upney at 7.02 p.m., received the Train Entering Section signal for it at 7.05 p.m. and offered it to Dagenham East as soon as the 6.10 p.m. train had been cleared back. The 6.20 p.m. train he had accepted from Upney at 7.15 p.m. and he stated that he had phoned Signalman Tinson and told him that he would not give the *Out of Section* signal for it as he wished to switch out. Signalman Tinson denied this conversation. Sass said that he had also spoken to Signalman H.F. Andrews at Dagenham East after the 6.15 p.m. train had passed and had told him, "If you give me out of section in time before the 6.20 reaches my box, I will close on this," and Andrews had replied, "Yes, that will be all right." He had then waited anxiously for Dagenham East to clear the 6.15 p.m. train. This happened at 7.24 p.m. when he offered and had had accepted by Dagenham East the 6.20 p.m. train, cleared his Down I.B.S. Home, had sent the *Train Entering Section* signal for the 6.20 p.m. train to Dagenham East and had then switched out his box. He was closely questioned as to whether he might have omitted to give the *Train Entering Section* signal to Dagenham East so that the block instrument at Upney box would have wrongly gone to *Line Clear* but could not be shaken in his statement.

Signalman Andrews at Dagenham East, a Class I signalman with 34 years' service, also put the visibility at no more than 10 yards. He said that he had been unable to clear the 6.15 p.m. train back to Becontree until 7.24 p.m. because he had had to hold it at his Down Home signal until the 6.10 p.m. train cleared the River Rom I.B.S. Home track circuit. Meanwhile Signalman Sass had telephoned him and said, "I've got another train coming to me now. Will it be all right to switch out with train on line?" He had replied, "That will be alright, lad. When I call you back for this train, you give me the 'train on' and I'll accept it, and then give me 'train entering section.'"

Describing Sass, who had less than a year's service, and had only been a signalman five weeks, Andrews added that Sass was, "an inexperienced lad and he kind of leant on my shoulder a bit. He had confidence in me and I had confidence in him." He had helped Sass because, "It was foggy, and to make sure that the lad made no error." He remembered Sass giving the *Train Entering Section* signal for the 6.20 p.m. immediately before he had switched out Becontree box. He had not been able to clear his Down signals for the 6.20 p.m. train until about 7.30 p.m. and shortly afterwards Tinson had telephoned him and said, "I believe that the Down train has passed my starter at danger. He (Andrews) had asked, "What gave you that impression?" and Tinson had replied, "I can't see it. I will go outside and make sure." He had received the *Train Running Away on Right Line* signal at 7.36 p.m. and had then sent the *Obstruction Danger* signal both ways. Shortly afterwards he had been told of the accident. The fireman of the 6.35 p.m. train had reached his box at 7.40 p.m. and later returned with the driver. He had given them both a cup of tea and then asked the driver, "What happened, did you pass the Upney signals at danger?" The driver had responded, according to Andrews, "No, of course not. He pulled us up at the starter and pulled it off – you know, the starter alongside the little cabin by the bridge."

Driver S.A. Church of the 6.15 p.m. train said that the fog had been very thick and he had only been able to see the signals from a few yards away. The Barking signals had been off, the Upney Distant had been at caution, the Becontree signals had been off, the Dagenham East Distant at caution and he had been stopped at the Dagenham East Home before it had cleared and he had proceeded.

Driver Wilson of the 6.20 p.m. train agreed that visibility had been bad and said that he had only seen the semaphore signals when he had almost reached them. They had left Fenchurch Street at 6.28 p.m. and gone very slowly to Barking. Leaving there the Upney Distant had been at caution and he had crawled to the Home. The Upney Starter had been clear, as had the Becontree signals. The Becontree I.B.S. Distant had been at *Caution*, the Home had been *Off* but the Dagenham East Distant below it had been at *Caution.* He had been coasting towards the latter signal when he had felt a bump in his rear and lost his vacuum. He had sent his fireman back to investigate and while the latter was away the fireman of the 6.35 p.m. train had appeared, had told him of the collision and had gone on to the box. Regarding the A.T.C. Wilson said it was a great help, "It lets you know just where your distant signals are – it is as good as a third man. It helps the driver to judge the position of the stop signals."

Fireman Spicer of the 6.20 p.m. train generally corroborated his driver's evidence.

Passed Fireman Barnes of the 6.35 p.m. train was still suffering from shock when he gave evidence at the inquiry. He stated that he had driven over the Down line in question 13 times since passing as a driver, though not, he thought, previously in a fog. He made several statements at different times, the gist of which was that he had been stopped at the Upney Down Home, that signal had cleared and he had moved forward to the Upney Starter. That signal had been clear and he had continued at a steady speed past the Becontree signals and the Becontree or Heathway I.B.S. signals, all of which had been *Clear.* The red tail light of the 6.20 p.m. train had been his first warning of danger. He did not remember any conversation with Signalman Andrews regarding the signals.

Fireman S.G. Fryer, who was described by the inspecting officer as 'a keen young man with a great interest in railway working,' correctly described the signals between Fenchurch Street and Barking, their having been stopped at the Upney Home, and having seen the Upney signalman in his red jersey as they passed the box. He stated that the Upney Starter had been green. "It was on a single post, one arm fairly high." He denied that he could have been referring to a Distant signal at green, but at one point stated, "I meant that it was the first time that we had a green right away, sort of thing. More so when we got the Becontree Distant off." Although cut and shaken in the accident he had gone back and to protect his train in the rear before going to the signal box.

Guard Pryde of the 6.40 p.m. train recalled it stopping at the Upney Down Home, then moving forward, asking the Upney signalman what was the matter and being told that the trains were running "block to block." He had not seen the Upney starter, after which his train had continued at about 15 miles an hour. He had gone back to protect his train after the collision.

Concluding his report, Brigadier Langley estimated that at the time of the collision the 6.20 p.m. train had been moving at 5 miles an hour and the 6.35 p.m. must have been moving at 25 miles an hour. The signalling and A.T.C. equipment had been working properly. The chief question must be the position of the Upney Down Starter when the 6.35 p.m. train had passed it. It was highly improbable that it had stuck off because it was working freely afterwards and the evidence of Mr Richardson, a local resident, indicated that it had returned properly to *Danger* after the passage of the 6.15 p.m. train. The signal must therefore have been at *Danger* when the 6.35 p.m. train had approached the Home signal and it could not then have been released except by a *Line Clear* from Becontree box, or from Dagenham East box if Becontree was switched out. There was no question of an irregular *Line Clear* having been given (as per, for example, at Winsford, see above) but there was a possibility that an accidental clearance had been given during the closure of Becontree box.

One possibility was that Signalman Sass had only obtained *Line Clear* for the 6.20 p.m. train before switching out. However, Sass and Andrews at Dagenham East had discussed Becontree switching out and Andrews knew that the 6.20 p.m. train was on its way to Becontree from Upney and that he must therefore receive the *Train Entering Section* signal for it before Becontree switched out. The arrival time of the 6.35 p.m. train had been altered by Tinson in the Upney train register but the inspecting officer did not attach much significance to this. The current rules permitted Tinson to let the 6.35 p.m. train down to his Starting signal even during fog. If Tinson had then received a wrong *Line Clear* during the switching out of Becontree box he might have then cleared his Starting signal, then panicked when he had realised his mistake. It was clear that he had, indeed, lost his head somewhat when the 6.35 p.m. had gone past his Starter, but he was adamant that at no time had the indication of his block instrument changed from *Train on Line*.

Brigadier Langley wrote, 'Thus, although some of the circumstances suggest that a wrong *Line Clear* might have been given to Upney, this presupposes collaboration and untruthfulness on the part of all three signalmen, which seems hardly likely, as, until my demonstration on 4th February, they were apparently unaware that a wrong *Line Clear* could be transmitted. The three men were very closely questioned for many hours, and, although some minor inconsistencies arose, they gave their evidence clearly and I believe honestly. This evidence shows that a wrong *Line Clear* was not transmitted to Upney during the closing of the Becontree box, and that the Upney Starter was never off for the 6.35 p.m. train, but before reaching a final decision the evidence of the driver and fireman of the 6.35 p.m. train must be examined.'

Passed Fireman Barnes, the inspecting officer continued, was a young and relatively inexperienced driver. He had travelled constantly over the line for 4½ years as fireman but never before in fog as a driver and it was doubtful if his knowledge of the line was sufficient for him not to miss a signal in such fog.

Fireman Fryer's confident remark that he had seen the Upney Starter at green was mistaken, Brigadier Langley decided, and what the fireman had really seen had been the Becontree Down Distant signal.

In conclusion, taking all the evidence into consideration, the inspecting officer concluded that Passed Fireman Barnes and his fireman had missed seeing the Upney Starter signal, whose small red glow would only have been visible for two or three seconds. It was likely that the train had not accelerated very much until Barnes had received a clear A.T.C. indication at the Becontree Distant signal. The signals had then all been clear until he had struck the 6.20 p.m. train, as the Dagenham East Distant signal had not yet been replaced behind the latter train.

Signalman Tinson should have sent the *Train Running Away* signal earlier to Dagenham East.

The collision would probably not have occurred had Becontree box still been in switch but the regulations permitted the box to be switched out with a train in the section and Signalman Sass had closed the box correctly.

Turning to remarks and recommendations, the inspecting officer noted that a train stop at the Upney Starter would have prevented the collision but he did not recommend them for the same reason that he had not after the St Johns, Lewisham accident two months earlier (that they would be expensive and complicated to employ on a line with differing train speeds).

Rule 39 (b) should be amended to prohibit trains being permitted to be forwarded to Starter signals during fog or falling snow unless the starting signals were colourlights or the signals were equipped with co-acting detonator placers. Colourlight signalling would be installed on the line in conjunction with the forthcoming electrification.

Although the switching out of Becontree box had been in order, the danger of switching out boxes with trains in the section existed and the prohibition of boxes being closed in such circumstances should be made universal in all regions.

A more intense tail lamp at the rear of the 6.20 p.m. train would probably have made little difference given the density of the fog, though improved types of lamp were being investigated.

The rolling stock involved in the collision had been built pre-war with wooden bodies on steel underframes. Stronger all-welded stock would be provided upon electrification.

At the inquest on the accident the Coroner, Lieutenant Colonel H.H. Kenshole, in summing up, pointed out that Passed Fireman Barnes had claimed that the Upney starter had been green, "but even if he were mistaken, can you say that he was being reckless of his duty in missing a signal?" [452]

The inquest jury returned a verdict of "Accidental Death." [453]

A major catastrophe occurred in Brazil on the former Estrada de Ferro Central do Brasil section of the Rede Ferroviária Federal on Friday 7th March 1958 at Paciência.

The circumstances of the accident were explained in an official statement issued by the railway company the following day: 'The accident occurred at about 19:00 yesterday. At approximately 18:00 a tremendous storm broke out along the lines of the Central do Brasil, especially affecting the Santa Cruz branch, and all communications, including those necessary for the normal signalling of trains, were disrupted. In addition there were problems with the overhead line equipment at Banga and Santa Cruz. As a result

---

[452] Ministry of Transport report

[453] *The Times*, London, England, 7th March 1958

emergency measures had to be introduced for train operation. These involved the issuing to each train driver of a ticket on which was written a clear warning in red that there was a failure of the telegraph system and that he must proceed with great caution due to the possibility of a previous train being in the same section ahead of him. A further effect of the storm was the flooding of the railway at the Santa Cruz side of Paciência station, together with the washing away of the embankment at various other points along the line.

'Train US-65 passed through Paciência before the full effects of the storm were felt but train US-67, which followed it at 17:54, received a caution ticket from the agent at Inhoaíba {the next station on the Rio de Janeiro side of Paciência} and stopped in Paciência station {possibly because of the flooding of the line beyond the station}. At 18:02 train SS-55 followed from Inhoaíba with a caution ticket and travelling slowly stopped in turn behind train US-67. At 18:29 train US-69 received a caution ticket at Inhoaíba and also reached Paciência safely. Train SS-7, called the COFAP train, was given a caution ticket at Inhoaíba at 18:45. It collided slightly with the rear of train US-69, but without any serious consequences. According to the Inhoaíba station agent the next train, US-73, received a caution ticket in the same manner as the previous trains, but eyewitnesses stated that it nevertheless ran through the section at high speed, with the result that it collided violently with the rear of train SS-7, the last two cars of which were wrecked and thrown to one side. The motorman of train US-73 was crushed to death in the collision and the motorman of train SS-7 has not been found and it is presumed was also killed.

'Immediate assistance was provided upon news of the accident being received and the work of rescue continued through the night, with the commendable assistance of the Army, the Fire Brigade, and the Police. The chiefs of the Department of Inquiry and Research, the chief of the Department of Transport, the chief of the Division on which the accident occurred, the engineer of the line, the plant engineer, the electrical engineer, and other officials also attended with their teams. All necessary measures were taken to provide assistance to the victims and to restore traffic, staff remaining on duty all night at Dom Pedro II station. Teams from the rolling stock, overhead line equipment, track and signalling departments are striving to remove the victims and to restore normal working as soon as possible.

'The initial findings suggest that the motorman of train US-73 was responsible for the accident as he failed to drive through the section with enough care, as per the requirements of the caution ticket issued to him,' [454]

By Tuesday 11th March the bodies of 55 victims had been recovered, these including that of Motorman Octaviano Gomez Coelho, of train US-73, and there were about 100 injured. Motorman Francisco de Oliveira of train SS-7, at first thought to have been among the dead, had however survived. The work of recovering the bodies of the victims, conducted during heavy rain, was very difficult, and the last victim was apparently not discovered until Monday evening, 10th March.

With the death of Motorman Coelho the reason for his lack of caution (his train had apparently been travelling through the section at 70 to 80 kilometres an hour) could, of course, not be determined. [455]

A rear collision occurred on the Ferrocarril General Bartolomé Mitre line of the

---

[454] *O Estado de São Paulo*, São Paulo, Brazil, 9th March 1958

[455] *Jornal do Brasil*, Rio de Janeiro, Brazil, 11th March 1958

Empresa Ferrocarriles del Estado Argentino in the Palermo area on the outskirts of Buenos Aires, Argentina, at 13:30 on the afternoon of Saturday 8th November 1958. Electric train No. 309 from Buenos Aires Retiro to Tigre was stationary when it was run into by train No. 47 from Buenos Aires Retiro to Pergamino, hauled by steam locomotive No. 1101. In the violent collision the steam engine penetrated to halfway along the crowded rear car of the electric train, smashing wood and metalwork, while to add to the horror steam reportedly escaped from the damaged engine and the wooden parts of the electric train caught fire.

Initial reports stated that 20 passengers had died and about 100 were injured. Staff from the adjoining Aeroparque Airport assisted firemen, police, and railwaymen in the rescue operation. Most of the injured were taken to Fernandez Hospital but some went to other establishments. [456]

The cause of the accident was not immediately announced but Driver Humberto Vergara [457] of the steam train, before he was detained and held incommunicado, declared that it had been due to a failure of the automatic electrical signalling. He added that he had not been able to see the stationary train very quickly because of a curve in the line. [458] Motorman Nelson Monasterio of the electric train, who was likewise detained, stated that his train had stopped due to a mechanical problem.

The line was completely blocked until 21:50 when two of the four tracks were brought back into use, with train proceeding with great caution through the accident site, due to concerns over the reliability of the signalling.

By the following day the number of dead had risen to 22 and it was feared that it would rise further.

The detention of the two drivers and of one or more signalmen, met with an angry response from the railway unions, who threatened to immediately begin an indefinite strike, and to avert this the men were released again the following day.

A newspaper report theorised that the cause of the accident was related to the severing of the cables connected with the signalling system during a previous strike. Repairs to the signalling installation had apparently not been completed at the time of the accident. The same newspaper pointed out that there had been previous bad accidents in the same area in October 1949 and in January of the current year. [459]

A passenger train from Puebla to Mexico City stopped near Xomeltla, Mexico, because it had developed a 'hot-box,' late on Thursday 5th March 1959. After the overheated axle-box had been examined the decision was made for the train to continue at reduced speed. As it was doing so it was run into from behind by a freight train proceeding from Apizaco to Mexico City. The collision wrecked several first-class carriages of the passenger train. Seeing that a collision was about to occur some of the passengers jumped out of the train, although at the same time the driver of the passenger train tried to accelerate out of the way as the driver of the freight train attempted to pull up. Some of the passengers who died were reportedly killed in the process of getting off the train. According to different accounts 10 or 11 passengers died and 40 or over 70

---

[456] *Oakland Tribune*, Oakland, California, United States, 9th November 1958

[457] *The Arizona Republic*, Phoenix, Arizona, United States, 9th November 1958

[458] *Clarín*, Buenos Aires, Argentina, 9th November 1958

[459] *El Litoral*, Santa Fe, Argentina, 9th November 1958

were injured. Red Cross ambulances took the injured to hospital in Mexico City, others were brought there in buses and lorries. [460 461 462]

The driver of the freight train disappeared after the collision and was being sought by the police. However, investigator Javier Garcia Travesi stated that the crews of both trains were apparently at fault. The driver of the freight train had failed to give the passenger train a 10-minute 'head start,' indicating that the line was not worked on any sort of block system, and he supposedly was driving at high speed. On the other hand the crew of the passenger train had failed to take any measures to protect its rear and so might be regarded as being equally to blame. [463]

A signalling error led to wrong clearing of a signal and a heavily laden goods running into the rear of a stationary passenger train near Minino, USSR, on Tuesday 2nd June 1959. The collision occurred near an oil tank, a fire broke out and it spread to the passenger train. Fifty-six children died. [464]

A disastrous collision between Engenheiro Goulart and Engenheiro Trinidade in Brazil in early June 1959 (described in Single Line Collisions chapter below) was followed by two further accidents in the vicinity of the capital, Rio de Janeiro, on Tuesday 16th June 1959. When the president of the republic became aware of the first he personally telephoned and harshly recriminated the director of Rede Ferroviária Federal (the state railway company formed in March 1957), Engineer Jorge de Abreu Schilling. Eleven significant accidents had occurred on former Estrada de Ferro Central do Brasil lines in the previous 9 years, resulting in 412 deaths and 1,169 injuries.

The first and more serious of the two accidents occurred at 01:33. Train UA-1, with Driver Francisco Sena Villasboas, ran past the starting signal at danger at Vieira Fazenda station and subsequently collided with the rear of freight train CP-229 at Maria da Graça station. A statement from the railway read, "The directors of the Central do Brasil lines regret to state that in consequence of the collision four people have died and many others are injured. The movement of trains on the Auxiliary line is partially halted as the line is still blocked." According to a newspaper report five passengers died in the accident, possibly one of the injured later succumbed.

Although the railway company immediately blamed the driver for overrunning signals, Engineer Aluisio Coelho dos Santos, the regional railway inspector, stated that the signalling of the line was inadequate to cope with the automatic brake system, which consisted of train stops which applied the brakes should a train pass a signal at danger. Such automatic train stops were provided at the two stations involved but were not working. The engineer decided that Driver Villasboas had been taken by surprise by a signal at danger and had not had time to apply the brakes.

There were also contradictions between the two stations as to the signalling of the two

---

[460] *Correio da Manhã*, Sábado, Portugal, 7th March 1959
[461] *El-Paso Herald Post*, El Paso, Texas, United States, 6th March 1959
[462] *The Arizona Republic*, Phoenix, Arizona, United States, 6th March 1959
[463] *Austin Daily Herald*, Austin, Minnesota, United States, 6th March 1959
[464] https://de.wikipedia.org/wiki/Liste_schwerer_Unf%C3%A4lle_im_Schienenverkehr#1941%E2%80%931950

trains. Signalman Valter Pereira at Vieira Fazenda stated that he had informed his colleague at Maria da Graça, Signalman Carlos Costa, of the departure of the freight train. As the freight train had not arrived at his station, Signalman Costa should have communicated with the next station at Del Castilho and blocked the lines, but he had not done so.

At the scene of the accident Engineer Schilling declared that such accidents were "practically unavoidable." Human errors could not be prevented. A system of centralised traffic control was being introduced which would provide greater safety but even this would not be entirely proof against mistakes. The inquiry into the accident would examine the private life of Driver Villasboas, who was initially believed to have lost his life in the accident, to find if there was anything which could have caused him to make his mistake. Until recently the driver had been employed in Belo Horizonte; he had been transferred to Rio to drive electric trains on the Auxiliar lines.

The second accident occurred on the former Estrada de Ferro Leopoldina line at 10:32 between Penha Circular and Brás de Pina, when train RZ-6, bound for Inhumirim, near Raiz de Serra, and Barão de Mauá. At that moment it reached the junction with the express lines, having until then travelled on the suburban lines. When negotiating the junction one of the bogies of the rear vehicle became derailed, overturned, and was dragged 50 metres. The penultimate coach also overturned. Both coaches were crowded; two passengers died and 14 were injured. The Leopoldina company stated that Driver Antonio Lafaiete, in charge of diesel engine No. 2803, was not to blame for the accident. The damaged coaches were examined during the afternoon and Engineer Knudsen, director of the Leopoldina company, stated that the accident was the result of a broken axle and hence unpredictable and unavoidable. [465]

Driver Villasboas had in fact, contrary to initial reports, survived the accident, and on Thursday 18th appeared with his lawyer to give a statement regarding it. He stated that he was not to blame for the accident, maintaining that he had only received a yellow *Caution* signal at Vieira Fazenda and not a red *Stop* signal. As his train had rounded the curve at the exit from the station he had seen a figure in black, but not knowing that anything was wrong he had kept going. When he had discovered that there was a train in front of him he applied the brakes but the collision had then been unavoidable. Fortunately there had been a low waggon at the rear of the freight train and the force of the impact had thrown him up into the air. After that he lost consciousness and only regained it at the Meyer Posto de Assistencia. He had received medical treatment there, then gone home.

In a statement to the Press the driver declared that a rigorous investigation would prove that he was not responsible for the accident. "I shall investigate," he said, "with my colleagues, to determine the real state of the signalling, especially on the Auxiliar line, and then point it out to those responsible for accidents such as that which happened to me." On the other hand Señora Ieda Villasboas, the driver's wife, speaking to a newspaper reporter, said that on the morning of the accident her husband had nervously confessed to her that he had run past a signal at Vieira Fazenda.

Engineer Schilling gave a detailed report on the state of the railway, in an attempt to explain the series of disasters. He stated that the company had the largest suburban service of any railway in the world, which caused it enormous deficits and much wear to rolling stock. He admitted that the signalling on the Auxiliar lines was not as up to date as

---

465 *O Estado de São Paulo*, São Paulo, Brazil, 17th June 1959

it might be and said that a more modern signalling system had been purchased but not yet been installed due to lack of resources. He added that sabotage was also a factor in some accidents, for example two days earlier a driver had had to suddenly stop his train because criminals had damaged the track.

Engineer Brito Pereira for the Leopoldina stated that all the rolling stock of the railway was worn out and they could not afford to renew it. All the bogies were old and needed repair or replacement. Their axles were calculated to support a load of 15 tons but due to contiguous transport were actually supporting a 20-ton load. [466]

In what was presumably a rear collision a freight train collided with a stationary passenger train near Devon, South Africa, on the night of Monday 1st February 1960. The accident happened on the line between Johannesburg and Bethel in the Transvaal. Fourteen or fifteen 'Africans' were killed and two 'whites,' (the driver and fireman of the colliding train?) were seriously injured. [467] [468]

Two passenger trains carrying members of Premier Fidel Castro's militia force and peasants were involved in a rear collision near Camaguey, Cuba, on Thursday 28th July 1960. The occupants of the trains were returning from celebrations at Las Mercedes marking the seventh anniversary of an attack on 26th July 1953 of army barracks in Santiago de Cuba in an attempt to overthrow dictator Fulgencio Batista. The 'coaches' of the two trains were freight cars normally used to carry sugar cane, used presumably due to a shortage of passenger rolling stock. The peasants were riding in the first train and members of the militia in the second. At least seven and possibly as many as 24 passengers died in the accident and about 200 more were injured. The colliding train was reportedly unable to stop due to brake trouble. A breakdown train was hurried to the scene to raise an overturned coach and a Cuban Army helicopter flew in emergency supplies of plasma and helped in the evacuation of injured passengers to Camaguey hospitals. [469] [470] [471]

---

466 Ibid, 19th June 1959

467 *Gazette de Lausanne*, Lausanne, Switzerland, 2nd February 1960

468 *The Capital Times*, Madison, Wisconsin, United States, 2nd February 1960

469 *Independent*, Long Beach, California, United States, 28th July 1960

470 *The Emporia Gazette*, Emporia, Kansas, United States, 29th July 1960

471 *Corpus Christi Times*, Corpus Christi, Texas, United States, 29th July 1960

CHAPTER 5

SIDELONG, HEAD-ON, DOUBLE AND FOULING COLLISIONS

Trains can come into collision with one another in a variety of ways. Rear collisions, that is ones involving one train running into the back of another, trains meeting one another head-on on a single track and trains colliding on rail/rail level crossings are detailed in other chapters. This chapter will deal with other types of collision such as those involving trains running into one another at junctions, meeting head-on on multiple tracks and colliding with vehicles left in sidings but which end up foul of running lines.

After a serious accident at Northwood on New Year's Eve 1945, English railways went on to suffer two major disasters in the first five days of 1946. The first disaster occurred a day after the Northwood accident, on Tuesday 1st January 1946, at Lichfield (Trent Valley), on the London Midland and Scottish Railway. On the evening of that day an Up local passenger train was stopped in the station platform when it was run into from behind by an Up fish train which had been wrongly diverted from the Up Fast line to the Up Slow line. The four coaches of the passenger train were wrecked and 20 passengers were killed outright or fatally injured. Twenty-one passengers, the driver, fireman and guard of the fish train and a porter standing on the platform were hurt, some of the passengers seriously.

Local doctors and nurses, the National Fire Service, police, residents and a Royal Air Force (RAF) detachment from a nearby airfield lent assistance after the accident. Ambulances reached the station only ten minutes after the accident and the first casualties reached hospital only ten minutes later. Breakage of a gas main in the accident had plunged the station into darkness but the RAF were able to provide emergency lighting.[472] Motorists took some of the injured to hospital, one woman driver made three trips. Mr A.N. Ballard, Lichfield Town Clerk, commented that, "This is the most serious train accident that has ever occurred at Trent Valley station. It is amazing that anyone on the local train is alive."

A woman who visited the scene likewise stated, "It was just a mass of wheels and steelwork and splintered wood. There were only four coaches on the local train. The last

---

[472] Ministry of Transport report

three were piled one on top of the other. I don't see how anyone who was in those coaches could possibly have lived through it." [473]

The Ministry of Transport report on the accident described the layout at Lichfield (Trent Valley). The main line north and south of the station was only double track but through the station and from a point 600 yards to the north of it there were four tracks, with the two Fast lines in the centre and two Slow lines serving the station platforms on either side. Immediately north of the station a chord went off to the east from the Up Slow line to connect with the Walsall – Burton line. The Up Slow line could be reached from Up Fast line via facing crossovers at both Lichfield No. 2 and Lichfield No. 1 signal boxes, the passenger and subsequently the fish train had both reached the Slow line via the crossover at No. 1 box, through points Nos. 33 and 34 reversed.

The 6.08 p.m. Up local passenger train from Stafford to Nuneaton consisted of 4-6-0 *Prince of Wales* Class steam engine No. 25802 and four electrically lit bogie coaches with wooden bodies on steel or composite underframes, in charge of an unnamed guard with Driver Edward Major and Fireman D. Owen. It was stopped at the Up platform with its brakes applied when it was struck.

The 2.50 p.m. Fleetwood to London Broad Street fish train consisted of 4-6-0 Class 5 steam engine No. 5495, seven fish vans and covered goods waggons, and a brake van, in charge of Guard F.K. Freeman with Driver Frederick Reed and Fireman G. Beckett. It had approached the station travelling at about 35 miles an hour when it had been accidentally diverted onto the Up Slow line and so into the rear of the passenger train. Despite having its brakes on the passenger train was driven forward 100 yards and as already stated its coaches were wrecked. The vehicles of the fish train escaped damage.

Points Nos. 33 and 34 had stood normal for the Up Fast at Lichfield No. 1 box for an hour before they were reversed by Signalman Edward Williams to admit the passenger train to the platform. The train was more usually diverted onto the Up Slow line at No. 2 box but could not be on this occasion because a freight train was standing on the Up Slow line between the two boxes waiting to proceed towards Burton via the chord line. He then had to place levers Nos. 32 (the facing point lock), 33 and 34 to normal and pull lever No. 32 again before clearing out the passenger train to No. 2 box and accepting the fish train at 6.52 p.m. He had noticed no unusual difficulty in working the levers. He had then cleared the Up Fast Starting Home and Distant signals (Nos. 5, 4, and 3). As the fish train was passing he sent the *Train entering Section* signal to Hademore Crossing box at 6.58 p.m. A few seconds later he had heard the noise of the collision. Thinking that part of the fish train had struck his box he had sent the *Stop and Examine Train* signal to Hademore Crossing. Only when Guard Freeman reached his box several minutes later had he realised what had really happened. He had then sent the *Obstruction Danger* signal to Hademore and a down freight train was stopped there. The guard had been followed to the box by Fireman Beckett, who Signalman Williams had taken to be the driver of the fish train, and he told the fireman that his signals had all been off for the Up Fast.

After the accident the levers in the signal box were left as they had been until they were examined by officers of the company from Crewe, Nuneaton and Rugby. It was found that points No. 33 were reversed and locked in that position despite the lever in the signal box being normal in the frame. Further examination disclosed that an accumulation of frozen ballast in the points had prevented them from being reset, though the signalman

[473] *The Times*, London, England, 2nd January 1946

had been able to replace the lever in the frame by (unknown to him) buckling the down-rod beneath the signal box floor. With the lever replaced and the facing point lock lever reversed again the interlocking between the levers had been freed and he had been able to clear his signals for the Up Fast despite the wrong position of the points, or rather he had been able to clear the Starting and Distant signals. A further safeguard was provided in the form of what was known as 'detection,' a mechanical device at the points which prevented the arm of the protecting signal (No. 4) from clearing if the points were not properly closed, regardless of the lever for the signal being pulled in the box. The detection was carefully examined after the accident and was found to be in perfect order. It seemed clear, therefore, that the Home signal at any rate had remained at danger.

Driver Read of the fish train, though not badly hurt, suffered from severe shock in the accident and it was over six weeks before he could give evidence at the Ministry of Transport inquiry. He said that the fish train had suffered several signal checks south of Stafford and been stopped by signals once. At Lichfield the colour light distant for No. 2 box had been green and the inner distant for No. 1 box beneath the Home signal for No. 2 box had also been off. He insisted that the No. 1 box Up Fast Home had also been clear though his memory of the configuration of the signals was faulty. He did not recall seeing either the Starting signal, No. 5, or the tail lamp of the passenger train.

Fireman Beckett and Guard Freeman confirmed that the colour light distant had been green. Shortly after the accident the fireman heard from Signalman Williams that all the Up Fast signals had been off and told his driver and guard this.

Driver G. Kendall of the Burton freight train had been waiting for his signal (No. 12) to clear and testified that he had seen signal No. 4 clear for the fish train and had remarked to his fireman, W. Hartley, "They've pulled off for the main again, they're busy tonight," or words to that effect. Guard J.W. Moors of the freight train also believed that he had seen signal No. 4 at clear. Finally, Signalman J.T. Shone in No. 2 box said that he had seen signal No. 4 at clear.

These various statements as to signal No. 4 having been clear proved a puzzle for Lieut. Colonel E Woodhouse, the inspecting officer, and a very thorough set of tests was made to determine if there was any manner in which the signal could have been off despite the detection. No way in which this could have happened could be found.

Concluding his report the inspecting officer wrote that three factors had contributed to the accident:

(a) The failure of the facing point bolt to disengage from the stretcher bar of No. 33 points due to frozen ballast, which had prevented their restoration to normal.

(b) The buckling of the down-rod in the locking room of No. 1 box which had permitted Signalman Williams to replace the lever of No. 33 points despite their being in the wrong position and had thus freed the interlocking between the levers allowing the Up Fast signals to be cleared.

(c) Either the failure of the detection to prevent signal No. 4 from clearing or else the failure to observe that signal at danger.

The inspecting officer accepted that evidence of Ganger W.F. Wright, who had stated that there had been nothing wrong with the points when he had examined them three hours before the accident. It was quite possible that the vibration and slight movement of the track caused by the frequent passage of fast trains had caused a small amount of ballast to fall into and obstruct the rocking shaft lever. This first of the chain of events leading to the accident was, the inspecting officer said, due to extreme mischance, though the arrangement of the rocking shaft lever was not satisfactory.

It was not surprising that Signalman Williams had failed to notice the unusual resistance in reversing lever No. 33 as no little effort was required to work points 200 yards from the box (even a set of points quite near to a box can require considerable effort, as the author can testify). The signalman was of medium height and weighed 9½ stones; he normally worked his levers very smartly. The real fault lay in the arrangement of the down-rod. It was, however, not entirely satisfactory that the signalman could not remember if he had seen signal No. 4 clear and it seemed likely that he had merely assumed that it had done so.

The Up Fast Home signal, No. 4, was visible for a distance of 755 yards, which was quite a sufficient distance for the fish train to have been pulled up in. The evidence of Driver Read, corroborated by that of Driver Kendall, Guard Moors and Signalman Shone that the signal had been clear was incompatible with the tangible evidence that it had been at danger. Having very carefully reviewed all the evidence Lt-Col Woodhouse was forced to conclude that these four witnesses were mistaken. He considered that Driver Read, having finally had clear Distant signals at Lichfield after a series of signal checks, had assumed that he had finally got a clear road and missed seeing the Home signal.

Signalman Williams, mystified after the accident as to how it had happened, had told Fireman Beckett that all his up signals had been clear and either the fireman or, in the case of Signalman Shone, Signalman Williams, had told the other witnesses the same thing. By a process of auto-suggestion the witnesses had then all convinced themselves that this had been the case. This, the inspecting officer considered, was the most likely explanation for what was undoubtedly a perplexing feature of the case.

Lt-Col Woodhouse ended his report with recommendations as to alterations to the rocking shafts at the points and the down-rods in the signal box to prevent a recurrence of the accident.[474]

At the close of the inquest on the victims a verdict of 'Accidental Death' was returned, and the Coroner commented that, 'One must bear in mind the terrific strain under which the railway companies of this country had been working during the past few years. They had done wonderful work and it was a relief to know that further men were being released from the services to work on the railway.'[475]

Labour Home Secretary James Chuter Ede was a passenger in the London – Newcastle express which was involved in the second disaster in England in four days, when it collided with some derailed waggons at Browney cabin, on the London and North Eastern Railway, at 5.40 a.m. on Saturday 5th January 1946. He was riding in a sleeping car towards the rear of the train and was not hurt. After the accident a police car took him to Newcastle en route to his constituency at South Shields.[476] Others were not so lucky; ten passengers lost their lives in the accident and 25 were injured, 18 of whom were taken to hospital. Four railwaymen were also hurt.

The passenger train involved in the accident was the 11.15 p.m. Down express from

---

474 Ministry of Transport report

475 *The Times*, London, England, 1st February 1946

476 *The Times*, London, England, 7th January 1946

London King's Cross to Newcastle, consisting of Class V2 2-6-2 steam locomotive No. 4865 and 15 coaches, in charge of an unnamed guard with Driver George Furlanger and Fireman William Evans. When it collided with the derailed waggons of an Up goods train its engine and tender were thrown to the right across the Up line while the leading vehicle, a brake van, was thrown to the left. The second, third and fourth coaches were piled up in a heap and wrecked, the fifth coach ended up on top of them and considerably damaged, the sixth, seventh and eighth coaches were derailed and less damaged and the remainder of the train more or less escaped.

The mostly unfitted goods train was the 3.40 a.m. Low Fell to Doncaster, consisting of Class B 16 4-6-0 steam locomotive No. 842, 44 four-wheeled waggons and a brake van, in charge of Guard Wear, with Driver J. Slinger and Fireman W. Clarke. Before its waggons were struck by the Down express the train had become divided and its two portions colliding at a considerable speed, two waggons had been thrown partly down an embankment, and about 30 were piled up on one side of the Up line, some of them forming the obstruction into which the Down express had run. The rearmost 11 waggons and guard's van had remained on the line.

Driver Slinger told the Ministry of Transport inquiry into the accident that his train had had a clear run through Durham, beyond which the Up Distant signal at Relly Mill Junction had been at *Caution*, as had the Up Distant signal of the next cabin, Bridge House Junction. The Bridge House Up Home signal had been at *Danger* and there had been a man standing in the six-foot (the space between the Up and Down lines). This had proved to be Relief Signalman Gavin, who had climbed onto the slowly moving engine and explained that Control had authorised him to travel on the train to take up duty at Ferryhill No. 3 cabin, as he had done the previous two days. The Home signal had then been cleared before the train had completely stopped and Driver Slinger said that he had then released the brake but had not opened the regulator, until he had felt that all the waggon couplings were tight. There had been no appreciable pluck. Guard Wear in the rear van said that he had not applied the hand brake and on the falling gradient at that point it was likely, the inspecting officer, Colonel A.C. Trench, considered, that most if not all of the train must have closed up at this point behind the two vacuum-fitted vans next to the engine. Driver Slinger said that he had had no suspicion at this point that the train had become divided. Continuing at 25 miles an hour past Littleburn cabin (out of switch) he had found the Up Distant signal at Browney, the next cabin, about two miles from Bridge House, at *Caution* and its Up Home signal at *Danger*. The Home signal had been lowered at the last moment and the signalman had been exhibiting a red hand signal from his cabin.

Fireman Clarke and Guard Wear agreed that there had been no appreciable jerk at Bridge House.

Signalman Gavin, on the other hand, said that he had felt a jolt as he mounted the footplate of the engine and that he had remarked to the driver that he hoped he had not pulled the train in two, though he doubted if Driver Slinger had heard him. He thought that Fireman Clarke had looked back between Littleburn and Browney cabins but was not sure if he had seen the lights of the brake van.

Relief Signalman Handysides at Bridge House described having kept his signals at Danger to enable Signalman Gavin to join the train; when he had had a torch signal from Gavin to indicate that he was aboard the engine he had cleared his Home and Starting signals. As the train had passed his cabin he had seen a gap of a few yards in the train, six or seven waggons back from the engine, and had sent the *Train Divided* signal to Relief Signalman W. Johnson at Browney. He had then telephoned Johnson to tell him that the second portion of the train was following. Only after this phone call had he tried to attract the attention of Guard Wear, but by then the brake van had passed his cabin and it was too late to use his detonator placer.

Guard Wear remained had unaware that his train had become divided as the rear portion with his van ran down the gradient at 25 to 30 miles an hour. It was only when the two portions collided, throwing him down in his van, that he realised what had happened.

As the goods train had come to a stand at Browney cabin Relief Signalman Johnson had shouted to Driver Slinger that his train had become divided. "Where's the rear portion?" the driver had asked. Relief Signalman Johnson had replied that he would find out but at this moment the question was answered as driver, fireman and signalman saw the rear portion bearing down upon them. Driver Slinger had opened his regulator to try and pull away but the front portion of the train had only just begun to move when the collision occurred. Relief Signalman Johnson had then belatedly called his mate in the next cabin at Hett Mill on the telephone to tell him that he had had the *Train Divided* signal but Signalman Travers there did not answer the phone at once as he was watching the express go by. When he did answer and found what was up he had dropped the telephone and rushed to his levers, just too late to stop the express.

Fireman Clarke had just been leaving Browney cabin to protect the Down line when the collision happened and the pile-up of waggons had blocked both lines in front of the cabin. By the time he had found his way past the pile of wreckage the express was too close for him to be able to give it any useful warning by waving his torch. Worse, the debris had fallen on the signal wires, putting both the Down distant and the Down home signals to *Clear* without their levers having been pulled in the cabin. Driver Slinger might have been able to give the driver of the express some warning by opening the whistle of his engine but it did not occur to him to do so.

Colonel Trench found plenty to criticise in the actions of the railwaymen involved, beginning with Relief Signalman Handysides at Bridge House, who, he considered, had failed to rise to the occasion in not at once notifying Guard Wear that his train had split in two, the most urgent thing that should have been done after sending the *Train Divided* signal. Had he done so the guard could probably have stopped the runaway at that moment by applying the hand brake in his van.

Equally, if not more at fault, was Relief Signalman Johnson at Browney. He had already had the goods train accepted from Hett Mill, and according to the regulations should have shown a green hand signal to Driver Springer to make him aware of the situation and forwarded the *Train Divided* signal to Hett Mill. Signalman Johnson gave three reasons for acting as he had. He said that:

(a) He had been under the impression that there was a length of rising gradient between Bridge House and Browney sufficient to stop the second portion before it reached Browney. In fact the line fell continuously at varying gradients from Relly Mill Junction to a dip at Croxdale viaduct, half a mile beyond Browney cabin, before rising again towards Hett Hill.

(b) He had been anxious to avoid a derailment or other accident on Croxdale viaduct.

(c) If there was to be a collision between the two parts of the goods train he had wanted it to happen north of his Down Home signal which could be used to protect the express.

The inspecting officer suspected that reason (or excuse) (c) was an afterthought but accepted that Johnson might have been influenced by (a) and (b). However, having increased the risk of a collision between the portions of the goods train Johnson should have at once sent the *Obstruction Danger* signal to Hett Mill. Had he done so, or had he acted according to the rules and forwarded the *Train Divided* signal to Hett Mill, Signalman Travers there would have had ample time to stop the express. Colonel Trench commented that how a signalman with 16 years' experience, as Johnson had, could have acted as he did, was difficult to understand. The second line of defence, the Browney signals, had then failed because of the wreckage holding them off. The impact of some of the wreckage with Browney cabin itself had also demolished the repeater in the cabin for the Down Distant signal, preventing Relief Signalman Johnson from knowing that it was not at *Danger*.

Relief Signalman Johnson had been working Browney cabin for 5 weeks. The inspecting officer commented that it was obvious that he should have acquainted himself with the physical conditions, including gradients, on either side of his cabin. He must bear a considerable degree of responsibility for the accident.

That Driver Slinger had not whistled to warn the express was to be much regretted but the inspecting officer thought that he had probably been dazed by being thrown down in the initial collision.

It appeared that Signalman Travers at Hett Mill had acted promptly when he had been informed of the accident over the telephone and it was unfortunate that he had been just too late. Nor did the inspecting officer attach any blame to either Signalman Travers, Driver Furlanger of the express, or Fireman Clarke of the goods train, whose efforts to warn the express had been frustrated by the wreckage blocking his path.

Signalman Gavin's evidence, Colonel Trench considered, had been coloured by subsequent events. Had the signalman really thought when he had climbed up onto the engine at Bridge House that the train had become divided he would have been gravely remiss in not being sure that Driver Slinger was aware of this.

Given the condition of the drawbar which broke Driver Slinger could not be blamed for careless driving.

The drawbar concerned belonged to a London Midland and Scottish Railway waggon and dated from Lancashire and Yorkshire Railway (L & Y R) days, the waggon having been built at Newton Heath in 1914. The drawbar was of an old fashioned design and faulty construction. It should not have been passed for use when originally made or should have been discarded during a general overhaul, the most recent of which had been in 1941. The number of drawbar and coupling link failures annually had fallen substantially in the years before the war (no records were kept during the war) but they were still a serious problem. The inspection of waggon drawgear during periodical overhauls should therefore be tightened up even further than it had been after a previous collision at Dagenham Dock in 1931. Instructions had already been given for an inspection of all ex-L & Y R waggons when they came in for repairs.

Suggestions had again been made for the use of flares to warn other trains of an accident or obstruction. Exhaustive trials made after the Dinwoodie accident of 1928 had indicated that the disadvantages of flares outweighed their usefulness. Recent developments in pyrotechnics might modify this decision and it was possible, though no more, that they might have helped at Browney.

At the inquest on the victims of the accident it had been suggested that every signal box should have a diagram showing the gradients in the vicinity of the box. This Colonel Trench thought a good idea and that the study of such diagrams by new or relief signalmen should be a part of their routine learning of the box. This recommendation was adopted and a requirement was added in 1950 that gradient diagrams should be displayed in signal boxes. [477]

Two trains collided in Japan at an unstated location on the Fujisanroku – Dentetsu Railway on Monday 21st January 1946. Twenty-six people died and 82 were injured. [478]

The initial report on an accident on the East Indian Railway at 6 a.m. on Monday 4th March 1946 at Bhagauli station, Uttar Pradesh, India, stated that 31 persons had been killed and 55 were seriously injured. Train No. 10, the Down Dehra Dun express, had collided head-on with train No. 723, an Up through goods working. Some more bodies were believed to be lying beneath the wreckage. Most of the dead and injured were third-class passengers.

Immediate assistance was provided by the chief medical officer of the railway company, who chanced to be a passenger on the express, while a relief train was sent from Lucknow, 48 miles away, and returned there with 43 casualties. Eleven other injured were sent to Hardoi. The Government Inspector of Railways, Mr F.J. De Souza, was also en route to the scene. [479]

A second report the following day increased the number of dead to 'about 40' and said that further bodies might still be buried in the debris. Two upper-class bogies attached to the express engine were badly damaged while the adjoining third-class carriage was completely smashed. A passenger who had been riding in the third-class carriage said that he had been one of only four in that vehicle to have escaped death. [480]

Colonel Mahmud Hussain, Vice Chancellor of Dacca University, who had been a passenger on train No. 10, told a reporter that after the collision, in which he had been thrown violently to the ground, he had seen some people looting the train. Two men had come near him and ransacked the wreckage of the carriage. In the process they had broken open his attaché case, taking two watches and all his money. When he had asked the men to help him out of the wreckage, he added, they had demanded more money and had threatened to assault him. Telling them that he was a very high official he had picked up a piece of wreckage to defend himself and shouted for help. The men had then left without aiding him.

Meanwhile five more of the injured had succumbed, increasing the number of

---

[477] Note from Charles Eric Steele in reply to a question on Facebook.
[478] Webpage – A Brief Survey of Railway and Aircraft Accidents in Japan and Investigations Thereof, by Abe Seiji and Sato Takemune
[479] *The Indian Express*, Madras, India, 5th March 1946
[480] Ibid, 6th March 1946

fatalities to 48. [481]

According to a report in the *Railway Gazette* 60 people died and 84 were injured in the accident but the figures of 48 dead and 56 injured which appeared several days afterwards in an Indian newspaper seem more likely to be accurate. Both train drivers were apparently among the dead. [482]

A mistake by the stationmaster at Ponte Cagnano, Italy, was blamed for a serious collision there on Wednesday 19th June 1946. A passenger train from Reggio Calabria to Roma was diverted onto the wrong line and collided with a stationary freight train. A report from the ANSA press agency spoke of 13 dead and 52 injured but possibly later reports in several American newspapers indicated that there were 19 dead and 110 injured, of whom 54 were in a serious condition. [483] [484]

A major disaster occurred on the Madras and Southern Mahratta Railway at Ongole station, India, at about 2.20 a.m. on Friday 4th October 1946. Involved in the accident were train No. 1 Down Madras – Calcutta mail and passenger train (the *Deccan Mail*) and the engine of a goods train. The signals were reportedly clear for the mail train to enter the station on the main line but it was diverted onto a loop line on which was standing the engine of the goods, which had been detached from its train and was taking water. As a result of the collision both engines were badly damaged and the three leading vehicles of the mail train, a luggage van and two bogies, were telescoped.

Mr Rama Rao, a passenger on the mail, gave an account of his experiences. He said that he and his fellow passengers had felt a heavy jerk as the train was entering Ongole station and it had not taken them long to realise what had happened. All the lights had gone out. The engine of their train had capsized and the luggage van and two third-class compartments (coaches) had been smashed to smithereens. The casualties would have been greater had not the luggage van borne some of the impact. Other passengers confirmed his account and added that medical aid had been rushed to the scene and after first aid the injured had been taken to the Headquarters Hospital (elsewhere in the same article it was stated that they had been taken to the Ongole Government and Mission hospitals). The women's coach was not damaged and apart from one or two women who suffered shock none was hurt. The upper-class coaches were also not damaged and their passengers, who included the Maharaja of Parlakimedi, a former Premier of Orissa, were unhurt.

Initially it was stated that 28 people had been killed and 70 were injured; the figures were later revised to 36 killed and 81 injured. Railway, military, and missionary doctors gave first aid at the station and the police engaged a number of special buses and lorries to transport the injured to hospital. At first it was thought that the fireman and underguard of the mail train had been killed but later it was confirmed that they had survived. First Fireman Magey and Second Fireman Vovindu were subsequently listed among those injured. Driver B. Carr was less fortunate. He had shut off steam when his train had been diverted onto the loop line and then attempted to jump from his engine but was caught under a coach and expired before he could be released.

[481] Ibid, 7th March 1946
[482] *The Railway Gazette*, London England, 5th April 1946
[483] *The Times*, London, England, 20th June 1946
[484] *The Cedar Rapids Gazette*, Cedar Rapids, Iowa, United States, 20th June 1946

A reporter who visited the injured in hospital found three more passengers to interview. 'Natarajan (24), Arumugham (33) and Natmull (15) who narrowly escaped with injuries in the Ongole train disaster, were brought to Madras last night by the Waltair Passenger and admitted into the General Hospital.

'Natarajan is a Camp Clerk of the Joint Director of Industries, Madras, and he was accompanying his officer who was proceeding on an inspection tour by the ill-fated train. Interviewed in the hospital by an "Indian Express" representative, Natarajan said that he was travelling by the third coach which was smashed after the collision. At the time of the accident, he said he was just then returning after having had a drink, when all of a sudden he heard a loud crashing sound and the next moment he found his compartment breaking. Broken pieces of wood were flying about and he was himself struck by a falling door. He added that while a number of persons who were in that compartment were killed, he had a most providential escape with multiple injuries.

'Arumugham, an office attendant who was travelling with Natarajan, said that he was sleeping at the time of the accident and was awakened by the loud report of a crash when he found himself amidst a heap of broken pieces of wood. He has sustained a fracture of his collar bone.

'Natmull said he was on his way to Calcutta and had a narrow escape.'

Due to the accident the *Grand Trunk Express* and the *Waltair Passenger*, which should have left Madras Central at 8 a.m. and 9 a.m. respectively, did not start until 10.45 a.m. and 11.30 a.m., and the *Grand Trunk Express* was further held up for four hours near Ongloe. In the opposite direction the *Calcutta Mail*, due in Madras in the morning, was held for the night at Chinnaganjam, 5 miles short of Ongole, and did not reach Madras until the evening. A duplicate Calcutta-bound train left Ongole with the uninjured passengers at 11 a.m.

Large breakdown cranes were despatched from Madras to clear the line. [485]

A 37th victim of the disaster died in an Ongole the following morning. [486]

The disaster was the subject of a government inquiry, the results of which it has unfortunately not been possible to access.

At least 11 people reportedly died and 50 were injured in a collision between two passenger trains at La Mata, Mexico, on Thursday 31st October 1946. The accident happened near the Pacific port of Salina Cruz, 'on the Tehuantepec peninsula line.' [487]

Over two million ethnic Germans living in Czechoslovakia were forcibly expelled to Germany at the end of the Second World War, the expulsions continuing until 1948. [488] It was a one of a number of trains carrying these refugees which was involved in a disaster near the station at Tröglitz, Germany, in the Soviet-occupied zone. The train was bound for Thuringia and routed via the Alterberg – Zeitz line. When the train arrived at Tröglitz in the early hours of Friday 1st November 1946 it was diverted onto a dead-end line and collided violently with the buffer stops at the end of the track. The steam locomotive

485 *The Indian Express*, Madras, India, 6th October 1946

486 Ibid, 7th October 1946

487 *Pottstown Mercury*, Pottstown, Pennsylvania, United States, 1st November 1946

488 Wikipedia – Expulsion of Germans from Czechoslovakia

pulling the train was buried into the ground up to its axles and the two leading carriages of the train were telescoped against the tender.

After the accident first aid was given by local residents and farmers and the Deutsche Reichsbahn sent two relief trains from Halle. The most seriously injured passengers were taken to hospital in Zeitz. Thirty people died in the accident and 19 were injured. The main line was reopened after six hours.

Fourteen-year-old Gisela Grunert, cycling from her home at Langendorf to Zeitz for her first day of work at the Etzold and Baeβler battery factory, witnessed the accident and years later described hearing a deafening bang and seeing the engine and carriages becoming wedged together and passengers tumbling out of the coaches. A traumatic experience for her after the horrors of the war which had just ended.

The Soviet authorities immediately opened an investigation into the disaster and a *fahrdienstleiter* was arrested. Another railwayman believed to have shared responsibility for the accident had absconded. [489]

Two trains collided at Yeongdeungpo station, Seoul, Korea, in the American occupation zone, on Tuesday 12th November 1946. Contemporary American newspapers stated that there were 34 dead and 44 injured while a modern South Korean source gives the probably more reliable figures of 41 dead and 67 injured. [490] [491]

A collision at Savigny-sur-Orge, France, on 12th December 1946 in which 31 people died and 50 were injured is very briefly referred to in *Railway Disasters of the World* by Peter Semmens. It has proved impossible to find any other reference to this supposed disaster and it seems clear that no such accident occurred and that the reference is to the collision at Revigny (Revigny-sur-Ormain) on 12th November 1946 (described in the Rear Collisions chapter).

A Pennsylvania Railroad double collision in the United States on Friday 13th December 1946 killed 15 passengers and the engineer of the second freight train, the engineer and fireman of the train engine of the passenger train, and the brakeman of the same train, while 131 passengers, six railwaymen, a Pullman employee and a dining car employee were injured. Newspapers gave the location of the accident as Coulter, Ohio; the Interstate Commerce Commission (ICC) preferred 'near Guthrie.'

A local Mansfield, Ohio, newspaper reported the accident in great detail. In one paragraph it wrote, 'Rail officials said their preliminary investigation indicated one of the freight trains, eastbound, had stopped because of a broken air hose. The second freight train crashed into the rear of the first, piling up both freights and throwing freight cars into the path of the Chicago-bound passenger train.

'Wreckage of 13 cars, six of them passenger coaches, three locomotives, one caboose and rails were strewn for nearly 100 yards along the tracks, on both sides of the highly banked right of way. One passenger coach, the first car immediately behind the second

[489] https://www.mz.de/lokal/zeitz/schweres-zugungluck-von-1946-augenzeugin-erinnert-sich-an-die-schreckliche-tragodie-3152366

[490] *Lincoln Evening Journal*, Lincoln, Nebraska, United States, 15th November 1946

[491] https://en.yna.co.kr/view/AEN20181112000700320

locomotive, was bent around one of the locomotives. It was this car into which observers said live steam poured and in which hysterical passengers screamed and fought each other to escape.

'Soon after the crash ambulances, doctors and volunteer workers streamed into the scene from Mansfield, Ashland, Loudonville, Wooster, Shelby, Galion and other smaller communities. The disaster preparedness of the Richland county chapter of the American Red Cross, under the direction of Mrs Sam Campbell and Mrs W.B. Ford, arrived early with hot soup and coffee and rolls and doughnuts for the workers.

'Besides the Red Cross disaster unit, scores of volunteers from the Salvation Army, several veterans' organizations, the Volunteers of America and other groups rushed to the scene to offer their aid. Mansfield police and firemen aided in the rescue work and pressed into service many of the spectators who visited the wreck. Roads, by mid-morning, were blocked by onlookers' autos for miles around. Among the volunteers were Mayor Roy W. Vaughn, Service-Safety Director William W. Enlow, Fire Chief L.J. Steele, and Clarence G. Tugend, superintendent of the state highways department's county maintenance station.

'The city building telephone switchboard was put into operation at 5 a.m. to handle any emergency calls for municipal services but Miss Jean Kelly, receptionist, said she had surprisingly few calls.

'Pvt. Robert Radtke, of Hustiford, Wis., was one of the first servicemen to crawl free from the smashed cars. "The force of the crash knocked me to the floor. I was asleep at the time and when I woke up I was on the floor. Duffle bags were flying about my head. There was a lot of yelling and steam was hissing," he said. The car in which Radtke was riding crashed into a freight car and its front end was smashed in. "When I got my bearings I crawled through a hole in the side of a car and another soldier and I went for help. Our first impulse was to get as far away from there as possible. I was scared; everybody was scared. But we finally made our way to a nearby building and told a man there to call for help."

'Wrecking trains from Crestline, Canton, Toledo, and New Castle and Conway, Pa., were dispatched to the scene of the tragedy immediately. The one from Crestline arrived at 4.45 a.m. but its work was held up for nearly two hours while the dead and injured were being removed.

'Rail officials said traffic on the eastbound (or south) track should be open to traffic by 4 p.m. and on the westbound (north) by 10 p.m. today. Meanwhile trains were detoured over the B. and O. through Willard and over the Pennsylvania's own lines through Columbus.

'Mansfield firemen who helped remove still living soldiers from the wreck said one man was hanging head down, pinned to a section of the car only by his feet. He had to be cut down before blood plasma could be administered. One soldier was wedged between the side of the car and the ceiling. He was alive, but it took four hours to cut him free, firemen reported. Several other servicemen were killed when they were jammed against the ceiling of the forward end of the smashed and twisted car, firemen said.

'Rail officials said a special train, dispatched from Crestline, transported some 70 uninjured passengers to Chicago immediately after the wreck. They said there were no accurate figures on the number of passengers riding the *Golden Triangle*.'

A newspaper reporter also spoke to several people living near the accident site. 'In the eerie light of pre-dawn, scores of farm families milled about the scene of the Pennsylvania train wreck today. Most of the injured had already been removed to hospitals, but the folks who had dressed quickly and hurried to the scene to give what aid they could told of the tragedy.

'One of them was Mrs Harold Fox. She was among the first to reach the scene of the wreck. "It was terrible! You could hear them calling for help from inside the coaches. Some were screaming. Others moaned. One person called out, 'I'm dying,' she related. "It must have been around 2.15 when I heard the first crash. Then there was a big flash of light that lit up the bedroom. It was only a matter of seconds and the second crash came. Then there was an explosion." Mrs Fox, who lives just above Coulter, Pennsylvania telegraph station, ten miles south of Mansfield, was one of the first to help administer first aid to the injured.

'Another early-arriver was Mrs T.J. Miller, who resides directly across the tracks at the station. "Mrs Fox and I were among the first of the neighbors out here," she said. We set up our headquarters here in the Ohio Oil company's pumping station and tried to help as much as we could. Most of the people we saw were soldiers who said they were on their way home for Christmas before going overseas. One boy said he had obtained special leave to attend his father's funeral tomorrow in Chicago."

'Both women said none of the passengers to whom they administered were seriously injured. "Most of them were shaken up," Mrs Fox said. "A few had minor scratches. Only two out of probably 20 or 30 were burned with the boiling water from the exploded boiler."'

The first eastbound freight train, Extra 6191 East, consisted of engine No. 6191, 78 cars and a caboose. It had passed Lucas, 5 miles west of Guthrie, at 1.59 a.m. and stopped on track 1 at 2.05 a.m., 996 feet east of signal No. 164.4. Twenty minutes later it was run into by Extra 8263 East.

Extra 8263 East consisted of engine No. 8263, 50 cars and a caboose with Engineer Louis Petoskey and Fireman Robert Keith. It passed Lucas at 2.19 a.m., passed signal No. 166.4 without reducing speed, passed signal No. 164.4, displaying *Stop then proceed at reduced speed*, and was still moving at 25 miles an hour when it struck train Extra 6191 East. As a result of the collision the caboose and eight rear cars of the front train and the tender of the engine and first four cars of the colliding train were derailed. Parts of the derailed vehicles of both trains fouled track No. 2 and were immediately struck by the passenger train.

Westbound first-class passenger train No. 63, the *Golden Triangle*, consisted of engines Nos. 3848 and 3807, two coaches, a passenger-baggage car, a coach, a dining car, seven Pullman cars and an observation car with an unnamed conductor, Engineer H.E. Harbaugh and Fireman Steve Sye on the pilot engine and Engineer Edward Patterson and Fireman Charles Reed on the train engine, and Brakeman F.J. Hughes. It had passed Mohican, 6.1 miles from Guthrie, at 2.18 a.m., passed signal No. 162.3, displaying *Proceed*, and travelling at 45 miles an hour struck the obstructions from the first collision. Both engines and the first six coaches of the train were derailed and the leading engine went down the embankment. Both engines were badly damaged, their steam pipes were broken and steam from the second engine penetrated the leading coach of the train, in which most of the fatalities occurred.

Extra 6191 East had stopped because a brake pipe hose at the rear of the first car had ruptured, causing the brakes to apply. When the train came to a stand the conductor went to a lineside telephone, the flagman went back to protect the train, the engineer and front brakeman inspected the train and the fireman went forward to protect track 2. The engineer discovered the broken pipe, the front brakeman obtained a replacement hose, the fireman was recalled by the engineer, the Lucas operator was informed over the telephone of the delay and the conductor returned to the caboose. The flagman also returned to the train after having put down two torpedoes in the vicinity of signal No. 164.4. Not long afterwards the conductor and flagman saw the headlights of Extra 8263 East half a mile away, whereupon the flagman gave stop signals with a red fusee and the conductor did likewise with a red lantern.

Engineer Petoskey of train Extra 8263 East told the Interstate Commerce Commission (ICC) inquiry that signal No. 166.4 had been displaying *Proceed* until he had passed it and he had called that indication to his fireman, who had been tending the stoker conveyor. His train had continued eastwards at about 50 miles an hour. About 1,500 feet west of signal No. 164.4 Fireman Keith had shouted a warning and he (the engineer) had seen that the signal was displaying *Stop then proceed at reduced speed*, plus the red fusee and the red lantern being shown by the flagman and conductor. When the fireman had shouted he had applied the brakes in emergency and he and the front brakeman had jumped from the engine.

Train No. 63 had passed the front of Extra 6191 East before the initial collision occurred and Engineer Harbaught on the front locomotive had been unaware of the obstruction before his engine struck it, while Fireman Sye had been tending the fire.

The ICC noted that according to the rules flagging protection in automatic block section territory was only required sufficient to stop a train moving at restricted speed. Signal No. 166.4 should have been displaying *Proceed prepared to stop at next signal*, though the engineer of Extra 8263 East stated that it had displayed *Proceed.* Tests were made after the accident and the signal was kept under observation for 48 hours but nothing was found to indicate that it had been showing a false or phantom indication.

The ICC concluded that the accident had been due to the engineer of Extra 8263 East failing to obey signal indications and the wreckage fouling the opposite track in front of an approaching train.

Socialist Deputy Luigi Battisti was one of five people who lost their lives in a collision at Santa Fe, near Minturno, Italy, on the electrified Roma – Napoli main line, on the evening of Sunday 15th December 1946. A direct train from Roma collided with a freight train which had stopped due to a sudden loss of current and had just restarted. The collision was a very violent one. In the darkness a large fire was seen and the two engines were wedged into one another. The luggage van and two coaches of the direct train were wrecked and serious damage was also done to the waggons of the freight train. Shrill cries arose from the tangle of wreckage while the uninjured passengers did what they could to help the victims.

The first outside rescuers came from Sessa Aurunca and Minturno. They worked to free the dead and injured and five bodies were recovered. Three of them were identified as those of Luigi Battisti, investment manager Mario Menichini, and the commissioner of P.S. Eliseo Ventura. The other two were not immediately identifiable. Fifty more or less seriously hurt passengers were taken to hospitals at Formia and Napoli.

The judicial authorities, assisted by the railway company, opened an investigation into the accident. From the initial findings it seemed clear that the crew of the freight train were at fault. [492]

The body of Luigi Battisti was taken to Roma and placed in a room at the station which had been converted into a chapel of rest. The deputy was the son of Italian patriot Césare Battisti, who had been hanged by the Austrians during the First World War.

According to a Swiss newspaper the accident occurred because the driver of the freight train had temporarily left his cab when the traction current had failed. It had then been restored and the driverless train, the brakes of which had presumably not been properly applied, had then moved off and fouled the line on which the express was approaching. [493]

Madrid – Algeciras mail and passenger express No. 406 left the Spanish capital at 20:15 on Monday 16th December 1946. It consisted of an engine, a van, a second-class coach, a first-class coach, a sleeping car, a mail van and a van. The engine was in poor condition and the train was soon running late. By the time that the express had reached Aranjuez the problem had become serious enough for a second engine to be attached to avoid further delays. By the time that the train was passing Alcázar de San Juan at 01:00 the following morning it was two hours late but from here there was some easy running through the plains of Ciudad Real Province and the train was soon running at over 80 kilometres an hour.

At Cinco Casas station, meanwhile, goods train No. 1402 was busy carrying out shunting operations and its engine was taking water. The express was not booked to stop at Cinco Casa and when notice of its belated approach was received the order was given for the main line to be cleared for it. Measures were duly taken to shunt the goods train out of the way but during the process a coupling broke with the result that 18 of its 42 waggons were left on the main line. Either this went unnoticed in the darkness or there was no time to remove the waggons before the express arrived at 01:30, travelling at 80 or 90 kilometres an hour. What signal protection was provided was not stated in the newspaper reports, nor do the names of any of the railwaymen involved seem to have been recorded.

In the violent collision with the stray waggons the pilot engine of the express was badly damaged and its boiler possibly exploded, while the train engine, though overturned on its right-hand side, suffered a lot less. The express had vehicles for Malaga and Cadiz as well as Algeciras; these were detached at Cordoba and Bobadilla. The sleeping car was fully occupied. Of steel construction, it was also a good deal stronger than the three vehicles in front of it and in the collision and the sudden stoppage it telescoped them, the first-class coach, destined for Algeciras, being most damaged.

As soon as news of the accident was received a relief train was dispatched from Alcázar de San Juan while the dead and injured were forwarded there in the undamaged coaches of the express. First aid was initially provided by two doctors who had been riding in the sleeping car, who were able to make use of sheets and bedding from that car, but the bitter cold of the early morning and the isolated situation of the station added to the difficulties of the rescuers. Some of those who died reportedly did so from the effects of the cold. The first batch of injured, those who were most seriously hurt, set off for

[492] *La Nuova Stampa*, Torino, Italy, 17th December 1946
[493] *Journal de Genève*, Geneva, Switzerland, 16th December 1946

Alcázar de San Juan at 05:00 and all the injured had been evacuated by 08:00, some of them being taken to Madrid, arriving there at 12:30. The work of clearing the line then commenced and was hampered by a snowstorm but one line had been reopened by the early hours of Wednesday 18th after 24 hours. The first train through was the express mail from Seville.

Twenty-one passengers and the fireman of the pilot engine lost their lives in the accident and 82 people were injured. [494] [495]

Twenty-seven people reportedly died and 67 were injured in a collision at Segarcea, Romania, on the line between Craiova and Calasat, probably early on Saturday 8th February 1947, in a collision between an express and a mixed train during a snowstorm. Two coaches were overturned. The railway administration stated that there had been a signal failure (a points failure?) which had diverted the mixed train onto the line on which the express was running. [496] [497] [498]

Spain suffered another major disaster on Saturday 10th May 1947 at Emperador station. Mail and passenger train No. 1611, which was due to depart from Badajoz at 18:45 Friday evening and arrive at Madrid Delicias at 09:00 on Saturday morning, came into collision at 06:40 with courier (parcels?) train No. 6644, which had left Madrid Delicias at 23:05 on Friday night and was due in Badajoz at 23:58 on Saturday. The mail train should have left Ciudad Real at 04:35 and Emperador at 05:47 but was almost an hour late. The parcels train was likewise late; it was booked to leave Emperador at 04:41 and arrive at Ciudad Real at 06:20. When the mail train arrived at Emperador station the points were wrongly set and it was diverted into a head-on collision with the stationary parcels train which was waiting there to cross it. The two locomotives were embedded in one another, otherwise it was the leading van (of the parcels train?) and the first- and second-class composite coach of the mail train which suffered the most.

A relief train was sent from Ciudad Real and arrived back there with the injured at 12:30.

Twelve passengers lost their lives and 30 were injured, 14 seriously. [499] [500]

"I was standing outside my garage working on a car engine. It was about 10.45 p.m. (C.D.T.). I looked up and saw the Minaki train coming along the tracks about a hundred yards from my home. It was clipping right along. The Transcontinental was sitting motionless beside the station. I thought the other train was going to pass by it. There was no sign of brakes being applied. All of a sudden there was a terrific crash. Steam started to shoot in all directions from both engines.

"I leaped my backyard fence and started running towards the wreck. I scrambled

[494] *La Vanguardia*, Barcelona, Spain, 18th, 19th and 20th December 1946
[495] Webpage: Federación Castellano Manchega de Amigos del Ferrocarril (in Spanish)
[496] *The Straits Times*, Singapore, 10th February 1947
[497] *Riverine Herald*, Echuca, Victoria, Australia, 10th February 1947
[498] *The Malay Tribune*, Singapore, 10th February 1946
[499] *La Vanguardia*, Barcelona, Spain, 18th May 1946
[500] Webpage: Federación Castellano Manchega de Amigos del Ferrocarril (in Spanish)

through a deep ditch with water up to my knees. Then I heard two loud explosions. The presto-light tanks under one of the day coaches had exploded. Immediately flames started to leap along the cars. I ran past the locked engines, a baggage car and stopped at a day-coach. It was a mass of splintered wood. Flames were roaring in the car's wreckage. People were all but buried in the wreckage. They seemed dazed. There was no yelling or screaming. Just pleas of 'Help me, Mac! Help me!'

"I came across a little girl sitting in the wreckage. She didn't seem badly hurt. I picked her up and carried her down the track away from the flaming car. The flames were, by this time, starting to work their way along the entire length of the car. People seemed to be everywhere. Placing the little girl on the ground I ran back to the car. Sitting in the wreckage was a young woman, about 25 years old. She seemed to be suffering from shock more than anything else. I picked her up and took her to where I had left the little girl.

"Returning to the blazing car, I noticed a woman lying on her face in the wreckage. Her shoulder was gashed. I dragged her clear and then went in after a young fellow. He complained that his back was hurt. I placed him on a grain door. He was later taken to hospital. I saw a man pinned down by the legs. I pulled him out. Another man was lying at the back of the car. He seemed unconscious. I tried to reach him but the flames drove me back."[501]

The witness and heroic rescuer who told the above story to a reporter was Mr Gerald Shields, a Canadian National Railways (CNR) employee. His efforts left him with cut hands and arms blistered from the heat of the blaze. The disastrous head-on collision which he had witnessed occurred at Dugald, Manitoba, Canada, on Monday 1st September 1947.

It was the end of the summer. Dugald was a siding (crossing place) on the single-track CNR main line a few miles east of Winnipeg. The westbound *Minaki Special* was carrying families returning to Winnipeg from various resorts in the Lake of the Woods area. It consisted of steam engine No. 6001, two baggage cars, nine day coaches and two parlor cars. Significantly while the leading baggage car and rear parlor cars were of steel construction the remainder of the vehicles were old wooden ones, half of which were gas lit, even at this late date. It was in charge of Conductor Frederick Skogsberg with Engineer Gaylord Lewis and Fireman E.J. Papkie.

In the opposite, eastbound direction were Transcontinental expresses Nos. 2 and 4. Train No. 4, which was to figure in the disaster, consisted of steam engine No. 6046 and a train of steel, electrically lit coaches, in charge of Conductor F.O. Smith with Engineer J.R. Gibson and Fireman Hazen Lawrie.

Orders were issued by Dispatcher A.G. Smith at Winnipeg regarding the meeting of the *Minaki Special* with the two eastbound expresses. Oder No. 335 read:

To Psgr Extra 6001 West at Malachi. Psgr Extra 6001 West meet No. 2 engine 6000 at Nourse and No. 4 engine 6046 at Vivian.

According to the system of working in general use on North American railways as the *Minaki Special* was an extra it was inferior to regular trains of the same class, such as trains Nos. 2 and 4. As an inferior train it was named first in the train order and at meets with regular trains it always took the siding (loop) leaving the main track clear for the regular train, unless the train order specifically stated otherwise.

[501] *The Winnipeg Tribune*, Winnipeg, Manitoba, Canada, 2nd September 1947

After issuing the above order Dispatcher Smith learnt that train No. 4 was now running an hour and a half late, later than it had been when he had issued train order No. 335, and thus to avoid delay to the *Minaki Special* he issued order No. 338, advancing the crossing of the special and train No. 4 from Vivian to Dugald, a few miles further west:

To Psgr Extra 6001 West at Elma. That part of Order No. 335 reading and No. 4 engine 4046 at Vivian is annulled. Psgr Extra 6001 West has right over No. 4 engine 6046 Elma to Dugald.

This order only gave the special the right to the main track to the east switch at Dugald, where, as the inferior train, it had to take the siding.

The *Minaki Special* was also running behind time and according to some of its passengers who were timing it using the mileposts it was travelling at 72 to 75 miles an hour as it was approaching Dugald. Train No. 4 was standing on the main track there and Engineer Gibson had dimmed the headlight of his engine as per the regulations when he had seen the headlight of the *Minaki Special.* Train No. 4 had not been there very long and some passengers were disembarking from the day coach. The train order board was also at red as there were orders for the train. As Engineer Gibson was studying these orders Fireman Lawrie, alarmed at the speed of the approaching special, had shouted, "They're not stopping," at which point the engineer and fireman had jumped. Engineer Gibson testified at the inquest on the victims of the disaster that there had not been time for the crew of train No. 4 to reverse the east switch to put the special into the loop, and added that even had the switch been thrown the special "could not have taken it at the speed at which he was going." Trainman Thomas Nicholls of No. 4 agreed that the switch, 50 to 55 car lengths away from his train, had been too far away for him to have had any chance of reaching it when he had seen the special approaching at speed. [502]

As the special had come into view Station Agent Donald Teddie had shown it a green hand signal to indicate that it was clear to enter the yard area. It was suggested that this might have confused Engineer Lewis. It was also suggested that Engineer Lewis might have believed that train No. 4 had been further delayed and that he therefore had a clear road into Winnipeg, but the train order issued to him was clear and unambiguous.

The two large engines were jammed together in the violent collision and the Pintsch gas cylinders beneath the second baggage car of the *Minaki Special* immediately exploded and set it ablaze. The fire spread rapidly to the other wooden coaches of the train, to an adjoining grain elevator and finally to nearby oil storage tanks. The baggage car and three leading coaches were also thrown from the tracks. The second baggage car and nine day coaches of the special were completely burnt out, though the two rear parlor cars were pulled back clear of the inferno. Most of the 80 drums of gasoline (petrol) at the Imperial Oil Depot exploded. Some were thrown 40 ft into the air and the burning gasoline contributed to the fire. Two pumper waggons of Winnipeg Fire Brigade were soon on the scene and firemen poured weak streams of water taken from ditches onto the flames to little effect. Any attempts at rescue after the first few minutes were made impossible by the extreme heat.

A reporter described the scene after the fire had finally died down. 'Today, fires are still smouldering in the ashes of the wooden coaches. A sticker on one of the end coaches was somewhat intact. It read, "Pala Camp Party."

---

[502] *Winnipeg Free Press*, Winnipeg, Manitoba, Canada, 24th September 1947

'I saw coach seats, remnants of springs, wooden splinters, molten glass and metal scattered about the graveyard scene. One piece of wood 60 feet from the train was blood-splattered. Large coach air tanks, once fastened underneath the framework, were scattered along the track. Near one coach, I saw a pair of red socks, neatly enclosed in one another. Near another coach, I found a jar of berries, the pickings of an unknown vacationist.

'On the station platform, apparently salvaged from the metal baggage car that did not burn, were two new canoes, one red, the other green. Beside them was the charred remains of a child's sulky. A jar of peanut butter, still intact, reposed on a laundry bag of dirty linen.

'Not a sign of the baggage of the more than 200 holidayers could be seen in the wreckage. Rescue workers would occasionally withdraw a burned bit of clothing, sometime attached to a body.

'The first three coaches of the Minaki train are a tangled mass of twisted metal in the ditch beside the track. Seven others remain on the tracks, their framework sagging to the ground. Steel stairways on coaches in the middle of the train were crushed by the impact. The track beside the burned grain elevator is twisted from the intense heat. At noon today, emergency crews were still clearing remains of 6,000 bushels of wheat which toppled onto the track.

'A grim air of silence, broken only by grinding crowbars of emergency workers and the shunting wrecking engine, pervades the scene. There is sufficient unspoken comment in the remains of the bodies, taken from the wreckage to the waiting funeral car. Silent, too, are the curiosity-seekers and town citizens who lined the fence barring them from the tracks. As if by consent they spoke in whispers as they regarded the tragic scene. "I took one of the casualties to hospital in my car last night," one bystander said quietly.' [503]

Thirty-one people lost their lives in the disaster, including Engineer Lewis, Fireman Papkie, Conductor Skogsberg and Front Brakeman G. Rougeau. Initially it had been declared that there were 35 dead but several supposed victims subsequently proved to be alive; a family which had intended to catch the special missed its departure by a few minutes. Eighteen other people were injured.

The inquest on the victims of the accident concluded in mid-October. Its jury returned a verdict that, "We, the jury, consider the primary cause of the train wreck was due to the non-observance of the book of rules order No. 93, CNR, by those in charge of the Minaki train, and also, from the evidence given, we, the jury, are convinced there exists a certain confusion among train crews and signal operators of the railway as to the signal and block system in operation at Dugald and especially as it pertains to clearance of westbound trains into the block.

"We the jury earnestly recommend that in order to prevent further accidents of this kind that a signal system be installed to give adequate protection to all trains and that the railway take immediate steps to see that all operating rules are thoroughly understood by the running trades and strictly obeyed at all times.

"We, the jury, are positive that the use of wooden-constructed and gas-lighted coaches added greatly to the death toll resulting from the collision and strongly recommend discontinuance of the use of this type of coach on any main line or crowded holiday trains

---

[503] *The Winnipeg Tribune*, Winnipeg, Manitoba, Canada, 2nd September 1947

and to be discontinued completely when new steel equipment becomes available." [504]

The verdict of the Board of Transport Commissioners was announced only a few days after the verdict of the inquest jury. Like the jury it recommended accelerating the replacement of wooden coaches with steel ones, and stated that it would issue an order prohibiting the marshalling of wooden coaches between steel ones; they must be marshalled only at the rear of trains.

As to the cause of the disaster, "The board's finding is that the accident was due to a violation of the operating rules of the CNR by the crew of passenger extra 6001 west and the failure of the operating official who was on the train to detect such violation of the rules in time to prevent the accident. The evidence at the hearing was that all concerned did know or should have known that, according to the rules, passenger extra 6001 west had rights over train No. 4 only as far as the east switch of the passing track at Dugald and should have headed into the passing track at that point."

The rule that all other than first-class trains must approach yards limits prepared to stop unless the main line was known to be clear had also been violated, "Instead of complying with the rule in this respect, passenger extra 6001 continued past the east switch on the main track until it collided, at a speed of at least 30 miles per hour, with train No. 4 standing at the station. It is quite evident that the train was not prepared to stop within the meaning of this rule."

The board considered as significant a remark made by Conductor Skogsberg to his crew, "Watch the board at Dugald. They might take us in for No. 4. No. 4 might fall down some more" (that is, be further delayed).

Finally the board stated that it was evident that the rapid spread of the fire had been due to the breakage of the Pintsch gas cylinders under the coaches. It noted that the CNR had purchased no wooden cars since 1923 and was replacing them with steel equipment "as conditions permitted." Between 1930 and 1946 the number of gas-lit coaches had been reduced from 1,165 to 707 and of oil-lit coaches from 601 to 308. [505]

Through goods train Dg 6186 collided with passenger train P 1308 at Karlburg block post, Germany, on the Würzburg – Aschaffenburg line, on Thursday 11th September 1947. A driver had overrun signals in fog. Eight people died and 59 were injured. [506]

Thirty-two people lost their lives in a collision near Karnal, India, 75 miles north of Delhi, on the night of Friday 10th October 1947, in a collision between a train carrying refugees from India to Pakistan and a goods train. Another 73 were injured, 23 seriously. Four Muslims were killed and 25 injured on the refugee train. The other casualties were among unauthorised non-Muslim passengers (refugees fleeing Pakistan?) on the goods train. [507]

Two mail trains collided head-on at Tamaree station, Queensland, Australia, at 12.55 a.m. on Saturday 18th October 1947. The southbound Rockhampton mail train, carrying

---

[504] Ibid, 14th October 1947
[505] Ibid, 20th October 1947
[506] *Katastrophen der deutschen Bahnen Teil II* by H.J. Ritzau, pg. 121 (in German)
[507] *Morning Tribune*, Singapore, 13th October 1947

about 300 passengers, was standing in the station waiting to cross the northbound Townsville mail train, which had about 200 passengers. When the latter train arrived it was diverted onto the wrong line and into head-on collision with the stationary Rockhampton train. The two engines were locked together and derailed, the Rockhampton train, despite having its brakes applied, was pushed back over 100 feet, three carriages of the Townsville train were derailed and two were telescoped for half their length. Only the leading compartment of the first carriage of the Rockhampton train was damaged but the tender of its engine was telescoped into its cab. Driver Charles Edwards and Fireman John Heiniger of the Rockhampton train were killed outright, together with four passengers. Fifteen other passengers were injured.

After the accident Acting Driver Alexander Stevens and Fireman Henry Campbell of the Townsville train told police that they had been within 20 yards of the Rockhampton train before they had seen it. "I whipped on my brakes, but it was too late," Driver Stevens said, adding that his train had been travelling at about 15 miles an hour at the time of the collision.

A rescue operation was quickly begun. Dr J.T. Henry, of Gympie, was on the Rockhampton train returning from a holiday on Heron Island and he was able to administer morphine to the badly injured. Also able to deliver valuable assistance were a Rockhampton railway ambulance team returning from a State competition in which they had come second, while the Tamaree stationmaster, Mr Charles Jones, was roused from his bed. "I jumped out of bed and ran down to see what had happened," he said later. The injured were taken to hospital in Gympie and gangs from there and Maryborough worked through the night to clear the line. One line was reopened at 5.39 a.m. Sunday though the clearing of all the wreckage took considerably longer. [508]

A fifth passenger, Mrs Grace Wells, wife of the Yaroona stationmaster, died in hospital at Ipswich on Tuesday 21st. Her husband, who had suffered a fractured leg and shock, was taken from Gympie Hospital to her bedside when it was realised that her condition was deteriorating, and he was at her bedside when she passed away. Her sister, who flew from Sydney to Brisbane, only reached the hospital three-quarters of an hour after her death. Mr and Mrs Wells had been on their honeymoon. [509]

Meanwhile, Mervyn Gilliland, 20, night officer at Tamaree, had been arrested on Monday charged with the manslaughter of Roydon Smith, one of the passengers.

At the inquest on the victims of the accident in early November it emerged that Gilliland had only had about four hours' sleep on the morning preceding the accident and had been forced to rest in a tin shed in which various goods were stored because there was no other accommodation for him. Gilliland frankly admitted his responsibility for the accident. "It was an error on my part. Nobody else can be blamed." He said that he had admitted the Rockhampton mail to the Tamaree station platform about 12.50 a.m. and that the Townsville mail had arrived six minutes later. Asked to explain why he had not routed the Townsville train into the loop he said, "I forgot I had the Rockhampton mail at the platform at the time. You see, it is a practice, when possible, to bring all passenger trains into the platform. I got so used to doing it I forgot I had the Rockhampton mail standing at the platform. I was standing between the Rockhampton mail and the loop line when the collision occurred. I crossed between two of the carriages of the stationary train

---

[508] *Cairns Post*, Cairns, Queensland, Australia, 20th October 1947

[509] *The Northern Miner*, Charters Towers, Queensland, Australia, 22nd October 1947

to give the staff to the Townsville mail driver." He said he had been perfectly sober. He had only been working at Tamaree for 12 days. There were no safety devices to prevent the mistake which he had made. Stationmaster Jones, Guard William Sorrenson of the Rockhampton train and others also gave evidence and at the close of the inquest Gilliard was committed for trial. He was allowed bail. [510]

In the following weeks Gilliard received letters of support from all over Australia. He did not give evidence in the trial, which was held at Gympie towards the end of November. The evidence given then was basically the same as that given at the inquest. At the end of the proceedings it took the jury only 17 minutes to find him 'Not Guilty,' a verdict which was greeted with applause from the public gallery. As Gilliland left the dock he was embraced by his father and then kissed by his sister, Miss Daphne Gilliland, and his fiancée, Miss Donna Banks. [511] [512]

The partition of India in 1947 was marked by a mass movement of Muslims from what is now India to East and West Pakistan and of Hindus and Sikhs in the opposite direction. Many of them travelled in refugee trains and much violence accompanied the movements, with armed gangs of the opposite religion(s) boarding the trains and murdering passengers. There were also attempts to derail trains but the worst railway accident during this period was a collision at Shambhu, India, on Monday 24th November 1947. An Up refugee special train which had left Ambala Cantonment on the Eastern Punjab Railway for Pakistan had only covered 11 miles when it was diverted into a dead-end siding at Shambhu and collided with the buffers with catastrophic results at 2 p.m. One report spoke of 129 dead and about 200 injured, nine of whom died in hospital. Another, possibly later report, speaks of 171 dead and 300 injured. The train finally resumed its journey with the survivors at 2.30 a.m. the next morning. It was unclear if the collision was the result of gross negligence or a deliberate act, but the assistant stationmaster, a pointsman and a driver were arrested. [513] [514]

A major disaster occurred in what was at the time the French occupation zone of Germany on Monday 22nd December 1947, at a bridge over the River Wied between Neuwied and Fahr-Irlich stations. After having been damaged in November 1944 by an Allied bombing attack the double-track bridge was blown up in March 1945 by German troops to hinder the Allied advance. In September 1945 the destroyed bridge was replaced by a temporary structure, wide enough to accommodate just a single track, and instead double to single-line points being installed at each end of the bridge the north and southbound tracks were gauntleted over the bridge. The approach to the bridge was protected by *Haupt-* and *Vorsignals* (home and distant signals) on each side.

On the evening of the disaster Dortmund – München Hbf express D48 was approaching the bridge from the north and in the opposite direction Freiburg – Dortmund express D 269 was approaching from the south. According to the timetable train D 48 was booked to pass over the 400 metres of gauntleted track first and the signals had been

---

[510] *Truth*, Sydney, New South Wales, Australia, 9th November 1947

[511] *The Telegraph*, Brisbane, Queensland Australia, 19th November 1947

[512] *The Courier Mail*, Brisbane, Queensland Australia, 20th November 1947

[513] https://jan.ucc.nau.edu/~sj6/ravinderkaurlastjourney.pdf

[514] *The Railway Gazette*, London, England, 15th April 1949

cleared for it and stood against train D 269. By chance as the southbound train was nearing the bridge a passenger named Berthold Feuchtinger, sitting in a poorly lit compartment, wished to open a ventilator, because there was an unpleasant smell in the compartment. Instead, he inadvertently operated the communication cord, and the train came to a stand close to Fahr-Irlich station, despite the signals being clear.

Meanwhile Driver Johann Reinert of train D 269 had seen the *Vorsignal* at caution and he and his fireman were looking out for the *Hauptsignal* but they failed to see it in the darkness, probably because it was unlit. Signals were often unlit in the difficult conditions of the immediate post-war period due to a lack of spare parts, including wicks, and poor quality oil. It was not until the northbound train had actually run past the *Hauptsignal* and had entered the gauntletted section of track that the engine crew realised where they were and the driver applied the brakes in emergency.

The problem with the accidental stopping of train D 48 having been sorted, it had restarted and was moving at about 10 kilometres an hour when it collided with the northbound train, whose speed had been reduced to 40 or 50 kilometres an hour by the brake application, at the north end of the bridge, at 18:32. As a result of the collision engine 03 054 of train D 269 and the three leading vehicles of this train were thrown down the embankment. The second vehicle, a four-wheeled Prussian carriage, and the front end of the third carriage were badly damaged. Forty-two passengers were killed and 116 more injured, all of them in train D 269. The engine of train D 48 was badly damaged but its passengers escaped with a fright.

The Neuwied police were informed of the accident at 18:45 and a rescue operation was begun. Dr Baum from Fahr-Irlich assisted two doctors who had been travelling on one of the trains to carry out five amputations in the first few minutes, while Frau Baum, who was also a doctor, brought bandages and medicines to the scene on a bicycle. The French military provided searchlights, medicines and rations and bandages were provided by a Fahr-Irlich factory. All the injured were evacuated within three hours. Relief trains from Engers, Linz and Oberlahnstein arrived at the scene at 21:00.

One of the coaches of train D 269 was reserved for war-wounded passengers. One of them, an amputee, related, "I felt the train slowing down and it was as if it was running over the sleepers. I was thrown to the left and my eyes were filled with dirt. Afterwards I ended up lying head down under the luggage. Two women kicked me in the head, saying that I wouldn't get up, but I couldn't."

Driver Reinert subsequently stood trial for causing the deaths and injuries. He was stated to have an exemplary record during his 33 years' service. Experts also spoke of the problems with the lighting of the signals stemming from the war, for example some lighting parts had before the war been supplied exclusively from what was now the Soviet-occupied zone and were now unobtainable, with the result that signal lamps frequently suddenly went out.

The driver of a northbound freight train bound for Köln which had crossed the bridge at 17:45 stated that the signal lamp had then been lit, while a number of witnesses who had seen the *Hauptsignal* shortly after the accident all agreed that its lamp had then been extinguished. In any case, of course, it is a general rule on any railway that a driver must be wholly familiar with the line over which he is operating his train and with the position of all of the fixed signals on his route. Should he fail to see a signal where one should be exhibited he must assume that it is showing its most restrictive aspect and act accordingly, stopping his train if necessary.

Apart from unlit signals the life of a German train driver at this time was, however, by no means an easy one. According to the newspaper report of his trial Driver Reinert had come on duty at 18:00 on the previous evening and he would not have finished his shift until he had driven a train from Frankfurt to Wiesbaden on the evening of the accident. His only rest in this period would have been a six-hour break at Wiesbaden. His food during this time consisted of a few slices of bread and some potatoes with a little flour sauce.

The court sentenced Driver Reinert to a term of four months' imprisonment. He was probably luckier than he would have been had he been in the Soviet zone. [515] [516]

Early on the morning of Thursday 15th April 1948 Down passenger train No. 36 collided with a goods train which was being shunted out of its way at Jungshahi, Pakistan. [517] Fifteen people died and 30 were injured. [518]

Eight people died and 70 were injured, some seriously, in a collision on the morning of Tuesday 25th May 1948 in the station at Rákospalota, Hungary, in the suburbs of Budapest. A workman's train collided with a freight train and the leading coaches of the former were wrecked. A driver, a conductor and a signalman were arrested. [519]

Up to 60 people reportedly died in an accident, apparently a collision between a passenger train and a freight train, near Uroševac, Yugoslavia (then in southern Serbia and now Ferizaj in Kosovo), on Friday 9th July 1948. The Serbian ministry of communications confirmed that there had been an accident but refused to give any details in accordance with the usual policy in Yugoslavia regarding accidents and disasters. The passenger train had apparently been travelling from Skopje to Belgrade via a secondary route because of floods affecting the Nish (Niš) – Belgrade main line. Many of the passengers on the train were said to have been Yugoslavs who had been expelled from neighbouring Albania following the expulsion of Yugoslavia from the Cominform the previous month and worsening relations between the two countries. [520] [521]

Twelve people died and 21 were injured in a collision between a passenger train and a freight train at Samsun, Turkey, on Thursday 28th October 1948. [522]

Just after 06:00 on Friday 3rd December 1948 crowded passenger train P 213 from (Rostock and?) Schwerin collided, probably head-on, with local train Vt 1002 from Spandau West to Nauen in the station at Falkensee, Germany, in the Soviet-controlled part of the Berlin suburbs. Five coaches were derailed, two were smashed and 17 people

---

[515] *Die Katastrophenszene der Gegenwart* by Hans Joachim Ritzau and Jürgen Hörstel, pgs. 21-26 (in German)

[516] https://www.spiegel.de/politik/hein-da-gibts-was-zu-holen-a-ca1dc4e2-0002-0001-0000-000044415346

[517] *The Indian Express*, Madras, India, 16th April 1948

[518] Ibid, 17th April 1948

[519] *Journal de Genève*, Geneva, Switzerland, 26th May 1948

[520] *The Daily Herald*, Biloxi, Mississippi, United States, 9th July 1948

[521] *The Singapore Free Press*, Singapore, 10th July 1948

[522] *The Mail*, Adelaide, South Australia, 30th October 1948

were killed and about 20 were injured. The trains were reportedly so crowded that passengers were riding on the buffers of the carriages. [523 524 525]

On Sunday 5th December 1948 the *fahrdienstleiter* at Bissendorf, Germany, on the Hannover – Soltau line, mistakenly permitted passenger train P 1475 to make a shunting move which fouled the line on which passenger train P 1476 was entering the station. Six people died and 59 were injured in the ensuing collision. There was a thick fog at the time of the accident. [526]

A mistake made by a trainee driver led to a major disaster at Port d'Atelier, France, at 22:11 on Friday 18th February 1949. Trainee Driver William Faivre, 26, was in charge of a light engine en route from Belfort to Blairville under the supervision of Driver Guillaume Schaaf, 40. The engine he was driving overran signals and fouled the main line, where it was struck by a Metz – Lyon – Nancy express. Faivre pleaded afterwards that he had tried, too late, to brake his engine. The two locomotives overturned and fell down the embankment along which the line ran at that point. The express was mostly composed of steel rolling stock but the second and third vehicles were of wooden construction. Following the impact they were crushed between the leading steel luggage van and the steel mail van marshalled fourth in the train. As a result no less than 42 passengers were killed outright or fatally injured and over 100 were hurt, 26 seriously.

Driver Pierre Derhe of the express, half-blinded by scalding steam and suffering from a head injury, nevertheless made his way along the line to protect the obstruction with detonators and inform a crossing keeper of the disaster. He then collapsed and died soon afterwards. He was subsequently recommended for a posthumous medal. Student Driver Faivre and Driver Schaaf were arrested. [527 528 529]

Nine people died and 16 were injured on Friday 4th March 1949 when a passenger train and a freight train collided near Eisenach, Germany, in Soviet-occupied Thuringia Province. [530]

A very brief Reuter report spoke of a collision between two goods trains about 70 kilometres west of Phnom Penh, which would be in the vicinity of Romeas, Cambodia, on Wednesday 6th April 1949. Twenty Cambodians were killed and 16 injured. Why there were so many people riding on two goods trains was not explained, possibly they were unauthorised passengers. [531]

---

[523] http://www.14641-bredow.de/doku/zug.htm
[524] *The Times*, London, England, 4th December 1948
[525] *Dunkirk Evening Observer*, Dunkirk, New York, United States, 3rd December 1948
[526] *Katastrophen der deutschen Bahnen Teil II* by H.J. Ritzau, pg. 121 (in German)
[527] *The Times*, London, England, 19th, 21st and 22nd February 1949
[528] *The Sunday Star*, Washington, D.C., United States, 20th February 1949
[529] *Journal de Genève*, Geneva, Switzerland, 24th February 1949
[530] *Salt Lake Telegram*, Salt Lake City, Utah, United States, 4th March 1949
[531] *The Singapore Free Press*, Singapore, 7th April 1949

At 23:29 on the night of Tuesday 11th October 1949 a major disaster occurred in Argentina. Electric train No. 415 bound for Tigre had left Buenos Aires Presidente Perón* terminus seven minutes late. In the opposite direction freight train No. 1636 was proceeding from Evita (formerly Victoria) towards the terminus. About a kilometre outside of the terminus the freight train was crossing the track on which the passenger train was running when the latter, running at speed, came into sidelong collision with it. As a result of the collision the leading motor car of the electric train was thrown to one side and the second and third first-class cars embedded themselves in the waggons of the freight train and then overturned. Short-circuiting set the first three vehicles of the electric train on fire while its following coaches were derailed.

Firemen and ambulances were quickly on the scene of the accident. At first it was feared there were over 50 fatalities but the figure was later revised down first to 25 and finally to 18. About 80 injured were taken to various hospitals. The crews of both trains survived the accident, though the motorman of the electric train at any rate was seriously injured. The first three cars of the electric train were completely burnt out. It took 24 hours to clear the line.

An investigation into the cause of the accident failed to reveal any problems with the signals or the signalling and suspicion fell on the crew of the electric train. It appeared that Motorman Juan Rossi of the passenger train had been anxious to make up for the late start but had left the driving of the train to his inexperienced young assistant, Trainee Motorman Pedro Serrano, who, like his counterpart at Pont d'Atelier in February, had allowed his train to overrun signals. The two railwaymen were arrested and lost their jobs. [532] Driver Samuel Mirazon of the freight train stated that the signals had been clear for him and his engine and two or three waggons had passed through the connection when his train was struck by the passenger working. The signalmen, who had also been arrested, likewise declared that they had given priority to the freight train, so that the drivers of the passenger train were those at fault seems without doubt. [533]

*Before 1948 Buenos Aires Retiro. Reverted to its former name 1955.

Six people died and two were injured in a collision between an express from Cairo to the upper part of the country and an oil train at Dairut, Egypt, 170 miles south of the Egyptian capital, on or about Thursday 3rd November 1949. Both trains caught fire. [534]

Three major disasters occurred to commuter trains in the New York area of the United States in just over a year in 1950 and early 1951. Those at Woodbridge, New Jersey, and Jamaica (Richmond Hill) have been described in previous chapters. The other accident, which was actually the first of the three, happened near Rockville Centre, New York, at 10.35 p.m. on Friday 17th February 1950, on the Long Island Railroad.

Since late April of the previous year work had been in progress in the Rockville Centre area in connection with the elimination of 10 grade (level) crossings and on either side of Rockville Center station trains were being run over a stretch of temporary track to the south of what would be the new elevated tracks. Because of the limited clearance

---

[532] *The Railway Gazette*, London, England, 16th December 1949
[533] *El Orden*, San Juan, Argentina, 12th and 13th October 1949
[534] *The Times*, London, England, 4th November 1949

available due to the presence of lineside buildings over a short distance there was insufficient room to accommodate a double track and so, as at Neuwied, Germany, in 1945, the two tracks were gauntleted. Automatic signal T and semi-automatic signal 02L protected westbound movements over the gauntletted track and automatic signal S176 and semi-automatic signal 02R governed eastbound movements.

On the evening of the accident westbound first-class passenger train No. 175, was a 12-car electric multiple unit, with Motorman J.W. Markin. It left Babylon at 9.58 p.m., passed signal T, displaying *Approach Medium*, stopped at Rockville Centre station, left there at 10.34 p.m., three minutes late, passed signal 02L, indicating *Medium Clear*, and moving at 25 miles an hour collided with train No. 192 at a point 2,060 west of Rockville Centre.

Train No. 192, an eastbound first-class passenger train, consisted of a similar 12-car electric multiple unit, with Motorman Jacob Kiefer. It passed Valley, the last open office, 2¾ miles west of the accident site, at 10.31 p.m., passed signal S176, displaying *Approach*, passed signal 02R, indicating *Stop*, and travelling at 25 miles an hour collided with train No. 175 at a point about 11 feet east of the west end of the gauntlet. The first and second cars of train No. 175 and the first car of train No. 192 were badly damaged, and most of the other cars of both trains were damaged to a lesser degree. Twenty-nine passengers and two railwaymen not on duty were killed and 124 passengers, 18 railwaymen not on duty, Motorman Markin, the conductor, front brakeman, flagman, and two ticket collectors of train No. 175 and Motorman Kiefer, the conductor, front brakeman, flagman and six ticket collectors of train No. 192 were injured. [535]

'Railroad man R.W. Campbell was riding home in the third car. "We didn't have a warning," he said. "We didn't hear any brakes. It just came out of the blue. I skidded down the aisle. The lights went out. Then they went on. Then they went out again. There was a lot of confusion. I don't think anyone was hurt much in our car. The conductor said to keep our seats, but nobody paid much attention to him. Then he asked me to call for help. I could see a guy in a car of the other train. I knew his leg was broken, for it dangled."

'Then Campbell, 40, of Bayshore, N.Y., a shopman for the Long Island Railroad, went to work with a torch too. He burned out seven living and two dead before the night was out. "One girl was almost cut in half. We had to take her out in two pieces."' [536]

'One of the heroes of the accident was Joseph Keller, 39, Franklin Road, Massapequa. L.I. Keller, who was riding in a forward coach, said: "I heard the conductor call out, 'Rockville Centre.' Then the train shook and was filled with flying glass. The conductor's left arm was severed." Keller immediately began pulling injured from the cars and found a first-aid kit in one of the coaches. He dragged people to safety, commandeered cars and had the injured driven to Meadbrook Hospital. Working with him was Dr R.H. Wilks, a passenger.

'Another passenger, Warren Carter, who was aboard the east-bound train, said: "There was a dammed short stop. Passengers were falling all about with broken limbs. I helped drag a few out, but I can't remember how I got out myself."' [537]

Motorman Markin of train No. 175 told the Interstate Commerce Commission (ICC) that the headlight of his train had been burning brightly. He said that when he had realised that train No. 192 had passed signal 02R and was entering the gauntletted track

---

[535] Interstate Commerce Commission report

[536] *The Van Wert Times Bulletin*, Van Wert, Ohio, United States, 18th February 1950

[537] *The Bristol Courier*, Bristol, Pennsylvania, United States, 18th February 1950

from the opposite end, he had applied his brakes in emergency but the collision occurred before they had had time to take effect.

The crew of train No. 192 testified that its brakes had been tested before it left New York. The stop at Jamaica had been normal. Leaving there, Motorman Kiefer had been alone in the control compartment of the train. The conductor remembered the whistle being sounded for two crossings at Valley Stream, just under three miles short of the accident site, but not a whistle for the crossing about 450 feet west of signal 02R. None of the train crew other than the motorman saw the indications of signals S176 or 02R, nor was there any warning of anything wrong before an emergency brake application immediately before the collision.

Motorman Kiefer declined, through his attorney, to make any statement regarding the accident. He had worked as an engineer and motorman for 26 years and had been on duty about four hours at the time of the collision. He apparently had his train under control at signal S176 and had reduced speed to 25 miles an hour for the temporary track, then had failed to stop at signal 02R.

The signalling and train brakes were carefully examined after the accident and found to be working correctly.

Subsequent to the accident the maximum speed of trains through the gauntletted track was reduced to 15 miles an hour. The control circuits of signals T and S176 were arranged so that they indicated *Stop* when the route was set for eastbound and westbound trains respectively and manual train-stop devices, worked by the operator at Centre interlocking, were installed in connection with these two signals. It was expected that normal double-line working would be resumed on 10th May 1950. [538]

Motorman Kiefer was indicted by the grand jury, charged with the manslaughter of one of the passengers in March. A week later the same Nassau grand jury found fault with the New York State Public Service Commission, the State Department of Public Works, the ICC and the Long Island Railroad because they had failed "to give due and adequate consideration to safeguarding the public" and they had shown an "insufficient regard for safety" in using the gauntletted track in an area of heavy traffic, which was "needlessly dangerous." [539] Quite what the alternative was if there was insufficient room for two tracks the jury did not suggest.

Motorman Kiefer, aged 55, was acquitted of manslaughter at the end of his trial on 26th September 1950. 'The engineer collapsed in his chair as the jury reported its verdict. His wife broke into sobs. Kiefer arose and tried to address the jury, but his voice broke. As the jurors filed from the room he managed to day, "Thank you, gentlemen."

'Keifer's future – especially as to whether he would be taken back by the railroad, or even wanted to be – was uncertain today.' [540]

Twelve people died and another 12 were injured in a collision involving an express and a freight train between Žilina and Košice, Czechoslovakia early on Friday 26th May 1950. The drivers of both trains were arrested, the collision being blamed on their

[538] Interstate Commerce Commission report

[539] *The Kingston Daily Freeman*, Kingston, New York, United States, 31st March 1950

[540] *The Daily Messenger*, Canandaigua, New York, United States, 27th September 1950

negligence, or more probably the negligence of one of them. [541]

The second portion of the Up *Irish Mail*, which had left Holyhead for London Euston at 1.45 a.m. on Sunday 27th August 1950, was involved in a collision with a light engine at Penmaenmawr, Wales, at about 3 a.m. As a result of the accident four passengers and a sleeping car attendant were killed outright, one passenger was fatally injured and 31 passengers and four railwaymen were injured.

The *Irish Mail* consisted of Class 6P (Royal Scot) locomotive No. 46119, a brake van, a first-class sleeping car, a third-class sleeping car, a first-class brake composite coach, a first- and third-class composite coach, a tea car, four third-class coaches, a first- and third-class composite coach, a third-class brake composite coach, a third-class coach, a third-class brake composite coach, a first- and third-class composite coach, and a third-class brake composite coach, in charge of Guard A. Roberts with Driver William Williams and Fireman John Williams.

Penmaenmawr was a wayside station situated at the foot of Penmaenmawr Mountain, on the slopes of which there were extensive quarries, a source of granite for use as ballast on the railway. The ballast was despatched from sidings on the Up side of the station. An Up ballast train ran from Penmaenmawr as required at 3 a.m. It was the engine to work this train which was involved in the accident. On the day in question this was Class 5MT 4-6-0 locomotive No. 42885 with Driver E.G. Roberts and Passed Cleaner M.T. Davies, accompanied by Goods Guard W.H. Davies.

On the morning of the accident the *Irish Mail* was running 30 minutes late and Signalman G. Morgan accepted it from Aber box at 2.50 a.m. The light engine arrived from Llandudno Junction on the Down line two minutes later and stopped near the signal box. Guard Davies then alighted from the engine and crossing the line made his way to the goods yard, while Signalman Morgan crossed the engine over to the Up line. Although he had already accepted the mail train this move was in order as the Up Starting signal protecting the crossover was more than a quarter mile in advance of the Up Home signal. Once the engine had come to a stand on the Up line clear of the trailing points leading from the Up Main into the goods yard, he could then reverse these points and the engine could be moved back clear of the main lines. As this British Railways, London Midland Region station was on a former London and North Western Railway line there were no shunting signals provided and movements had to be made with the use of hand signals.

At about 2.54 or 2.55 a.m. the light engine came to a stand clear of the mains crossover and Engineer Roberts popped his whistle to indicate to the signalman that it was. Signalman Morgan told the subsequent Ministry of Transport inquiry into the accident that on hearing this whistle he had restored the crossover (levers 12 and 14) and reversed the points leading into the goods yard (levers 10 and 11). Going to the top of the box steps he had then shown a white light to the driver of the engine and had given two rings on a trembler bell to hurry up the reverse movement and so avoid delay to the *Irish Mail.* He had cleared the engine back to Llandudno Junction No. 2 box at 2.55 a.m. and had accepted a Down freight train, the 12.30 a.m. from Mold Junction, at the same time. At 2.56 a.m. Aber box had given him the *Train Entering Section* signal for the *Irish Mail* and he had obtained *Line Clear* for it from Llandudno Junction No. 2. At about 2.58 a.m. he had heard a second "pop" whistle from the engine and although he had neither seen

[541] *The Independent Record*, Helena, Montana, United States, 27th May 1950

nor heard it move back he had taken this as an assurance that the engine was in the yard clear of the main lines, had restored levers 10 and 11 and had cleared his Up Main signals for the express. He had then sat down in his chair to await the arrival of the *Irish Mail*, which usually took 4 or 5 minutes to reach him (it was 4¾ miles from Aber to Penmaenmawr). As he had got up again, thinking that the *Irish Mail* must be close at hand, he had caught sight of a red light being shown from the Llandudno Junction end of the Down platform. Realising that something must be wrong he had thrown back his signals against the *Irish Mail*, the headlights of which were now in sight, and sent the *Obstruction Danger* signal to Aber and Llandudno Junction boxes at 3.02 a.m. (recorded by them as 3 a.m.).

Driver Williams of the *Irish Mail* told the inquiry that the Penmaenmawr Up Distant had been showing a good green light and he had passed it at 60 miles an hour. The Up Home signal had also been clear but had been thrown back to red "just before" he had passed it. He had then made a full brake application, bringing the speed down to about 50 miles an hour when they struck the engine. He was injured in the collision and realising that the Mold Junction freight train must be due had told his fireman to go forward and try and stop it. Her had then himself gone back towards the signal box and had met Signalman Morgan on the platform, who confirmed that the signals had been off for him but had been put back in his face.

Fireman Williams said that he had also been somewhat hurt and had been unable to run but had managed to cover a quarter mile or so and put down some detonators in front of the Mold Junction train.

Guard Roberts stated that he had not seen any of the Penmaenmawr signals. He said that after his train had stopped he had gone back to protect the Up line and Signalman Morgan had assured him that all traffic was stopped.

The Mold Junction train consisted of a heavy freight engine, 58 waggons and a brake van in charge of an unnamed guard with Passed Fireman H.M. Brown and Passed Cleaner V.B. Williams. Passed Fireman Brown said that just after passing the Down Penmaenmawr Distant at clear he had seen the Down Home at danger. He had shut off steam and applied the brakes, then exploded two detonators, when he had applied the brakes fully, and he had been able to stop just short of the wrecked coaches of the *Irish Mail* though he had grazed the side of its engine. As he was pulling up he had run over more detonators and had seen a red light shown by Guard Davies.

Passed Cleaner Williams generally confirmed his driver's statement. Near the brake van of their train they had found Fireman Williams lying beside the line, with blood all over his face and in a state of collapse.

As a result of the collision the light engine was pushed forwards 240 yards while the engine of the express became detached from its train. The first-class sleeping car, marshalled second in the train, was almost completely telescoped into the leading brake van and was the location of five of the six fatalities. The next two coaches were little damaged but the fifth and sixth coaches were thrown onto their sides and their ends were damaged, as was the front of the seventh coach. The next three coaches were derailed but little damaged and the rear six coaches remained on the track, undamaged. [542]

There had been 520 passengers on the *Irish Mail*, mostly holidaymakers returning

[542] Ministry of Transport report

from Ireland.

A newspaper reported on the aftermath of the accident. 'Doctors from all parts of the county and ambulances were quickly on the spot. Volunteers helped the police, fire service, and St John's Ambulance Brigade men in rescue work. Residents, including the proprietors of a café, made tea for the survivors, who afterwards continued their journey in buses seven miles to Llandudno Junction, where they went on by train.

'Mr J. Elias Williams, chairman of Penmaenmawr urban council, whose home is on a hillside above the railway, organized a casualty clearing depot on the station, staffed by volunteers.

'Mr Joseph Jones, whose bungalow overlooks the railway, said: "I heard what I thought was a big gust of wind. Then the lady next door called me to say that the Irish Mail had crashed. I left, taking a hammer and a ladder with me. "The first person I got out was a lady aged about 70, who was with her son in the train. The carriage had turned on its side, but a space of two feet was made by breaking the window." Mr Jones said that two nurses in civilian clothing who were in the train were "wonderful" in rescue work.' [543]

The nurses were later identified as Elizabeth Cunningham and Margaret Dowling of the Connaught Hospital, Walthamstow. They were sent letters of commendation by British Railways, as were third-year student nurse Alice Mulrannen of the Kent and Canterbury Hospital, Canterbury,[544] and her sister Margaret from Ramsgate who had also been prominent in the rescue work. All four had been returning from holidays in Eire.

The statements of the crew of the light engine as to the events leading up to the collision varied from that of the signalman. Driver Roberts stated that after bringing his engine over the crossover onto the Up line he had waited several, but not as many as five, minutes for a hand signal from the signalman, but had seen none. Then he had seen an unmistakeable set back signal from Guard Davies. He had begun to set back only to find that he was going back along the Up Main line rather than into the goods yard. Stopping his engine and going forward again, he had seen that the Up Advanced Starting signal had been cleared. However, it had not occurred to him that the signalman had pulled off for an Up train, he had thought rather that his guard had found that there was no train for them and that they were to return light to Llandudno Junction and thence to Mold Junction. Guard Davies had then come up to the engine and told him to move beyond the connection to the goods yard and give a "crow" whistle. Shortly afterwards the guard had cried that the *Irish Mail* was coming. Driver Roberts strongly denied having given a second "pop" signal as stated by Signalman Morgan. He did not think that he had been stopped long enough to need to send his fireman to the box in accordance with Rule 55. The hand signal from Guard Davies he had assumed had been given with the signalman's authority.

Passed Cleaner Davies thought that the engine had been stopped two or three minutes before it moved back. He did not think his driver had given a second "pop" whistle but was not sure. From his side of the engine he had seen no hand signals from either the signalman or the guard. After the engine had moved forward again he had got down to change the red lamp of the engine from the front to the rear as instructed by his driver.

Guard Davies testified that on the engine arriving at Penmaenmawr he had gone straight to the goods yard. There he had heard the ringing of the trembler bell from which he had understood that the signalman wanted the Up line cleared, though he had not been

---

543 *The Times*, London, England, 28th August 1950

544 Ibid, 29th August 1950

told by the signalman that the *Irish Mail* was close at hand. At first he denied having given Driver Roberts a hand signal to set back but later said that he had wondered why his driver had not done so. Finding that the engine was coming back along the main line he had given his driver a red light to stop him and then shown a white light to the signalman for him to set the road into the goods yard. Almost at once he had become aware of the approach of the *Irish Mail* and had shown a red light towards the box and shouted to his driver to move forward. After the collision he had seen the signalman on the Up platform and told him that the engine had come back along the main line, to which the signalman had replied, "Good God." Finding that there was a Down train approaching he had then gone up the line to stop it. His statements as to what light had been carried on the rear of the engine and whether it had whistled once or twice were vague and confused.

Signalman Morgan continued to maintain that he had heard a second whistle from the engine, indicating to him that the engine had come back into the yard. He could see no tail lamp on the engine after it had moved forward through the crossover and when he had given its driver the hand signal to shunt back into the goods yard. Why the engine had not come back when he had signalled for it to do so he did not know. Such a thing had not happened before and he had heard no complaints that hand signals from the box were hard to see because of an intervening footbridge, though he had heard suggestions that ground signals and/or track circuits would be desirable.

Upon it being suggested that it would have been better to have brought the engine directly into the goods yard from the Down Main at the west end of the station, he said that this was not usually done because there were often waggons in the different sidings and he could never be sure of the positions of the hand points leading to them. He said that the red light from Guard Davies had been his first warning that the engine was still on the main line and he had not been anxious about where it was. He maintained that he had fully complied with Rule 69 (regarding signalmen not moving points until they had received an intimation that the movement had been completed, in the case of an unaccompanied light engine by a short whistle) as he had heard a "pop" from the engine, though he agreed that it would have been better if someone had come to the box and told him. The other two Penmaenmawr signalmen, A.E. Thomas, and G.O. Williams, stated that they would also have accepted a "pop" whistle, though had they not been perfectly satisfied they would have gone out onto the box steps or the platform to make sure.

Stationmaster R.G. Owen, Relief Stationmaster O.R. Jones, District Inspector J.H. Edwards and Assistant District Inspector J.H. Nichol all spoke well of Signalman Morgan and the stationmasters seemed a little uncertain as to whether a mere whistle could be sufficient to indicate that the engine was in clear, though the district inspectors agreed that it was not.

Concluding his report, the inspecting officer, Lieut. Colonel G.R.S. Wilson, stated that Signalman Morgan's evidence had been simple and straightforward, and he (the inspecting officer) thought it probable that Driver Roberts had given a second "pop" whistle, despite his denial of having done so.

In contrast to that of the signalman, the statements of the engine crew and guard were confused and contradictory and it was impossible to be sure as to the movements of the engine or what had passed between the driver and the guard. It was possible that the footbridge might have prevented the driver from seeing the signalman's hand signal and that having waited several minutes the driver had impatiently whistled again, then set back upon a hand signal from his guard. It was most unfortunate that Driver Roberts had assumed that the starting signal had been cleared for him to return to Llandudno Junction, and not for another train. Alternatively, as suggested by Signalman Morgan, the driver might have set back a short way into the yard, mistakenly thought he had come back on the main line and gone forward again.

Signalman Morgan had, nevertheless, taken insufficient pains to satisfy himself that the engine was in clear. It was surprising that he had not looked for the engine light or gone down the box steps, either to hand signal to the engine or to see where it was after the second "pop" whistle. It seemed likely that he had been over anxious to get the engine in the yard quickly and avoid further delay to the *Irish Mail* which was already running late. The fact that he had been looking eastwards so that he had at once seen the red light from the guard, rather than westwards towards the approaching train, suggested that at that point he had become concerned as to whether the engine really had come back into the yard, especially as he had not seen it there.

The final paragraph of Rule 69 stated, *When the shunting operations are clear the Signalman must see, or have intimation from the Guard, Shunter or Driver, that the running lines are clear.* The intention of this paragraph was clearly that the short whistle mentioned in a preceding paragraph to indicate that the engine was clear of the points was not to be regarded as an assurance that *the running line was clear* (italics in report) but the rule was somewhat ambiguous. The ambiguity did not, however, relieve the signalman of his duty to be sure that the engine was in clear, especially in view of the lack of ground signals and the intervening footbridge. He must therefore bear some responsibility for the accident. It would have been better if he had kept the engine on the Down line until the *Irish Mail* had passed.

Driver Roberts was also considerably to blame for failing to send his fireman to the box in accordance with Rule 55 to remind the signalman that he was on the main line, or at least to have sounded a series of "pop" whistles to attract the signalman's attention. His assumption that the main-line signal was cleared for him to return to Llandudno Junction was unwarranted and he should have clarified the situation with his guard.

Guard Davies emerged with little credit. Unlike his driver he knew that the engine had to come into the yard and when he had seen that it was coming back along the main line he should have immediately shown a red light towards the signal box. Instead he had done nothing until brought to his senses by the approaching train.

Driver W. Williams of the mail train, on the other hand, had been alert and immediately braked when the Home signal had been thrown back in his face. No responsibility was attached to him. The unselfish action of Fireman J.E.N. Williams in going forward to stop the Down freight train despite his injuries deserved high praise.

Turning to recommendations, Lt-Col Wilson spoke of the lax method of working which had grown up at Penmaenmawr and said that he did not consider that the training and supervision of young signalmen, such as Signalman Morgan, had been adequate. Rule 69 should also be revised to remove any ambiguity. At some places on the London Midland Region there was a specific whistle code to indicate that a train had been shunted clear of the running lines but as a general principle no whistle code should relieve a signalman of his responsibility to ensure that the line was clear.

In view of the frequent shunt moves into and out of the Up sidings, the fast main-line traffic and the restricted view from the box, a strong case could be made for a track circuit in the rear of the Up Advanced Starting signal, but the box was in any case to be renewed and re-sited, with track circuits and ground signals.

The general acceleration of the provision of track circuits and block controls had been planned by the Railway Executive in 1948 but delayed due to lack of money.

There should be a general review in all regions of the provision of ground signals, where there were none, and where shunting moves at some distance from a box were made using hand signals.

An examination of the wreckage showed the wisdom of the decision to employ central Buckeye-type couplings on new rolling stock. It was notable how well the steel-bodied coach placed fourth in the train had stood up to the collision compared to the destruction of the timber-bodied first-class sleeping car.[545]

At the inquest on the victims of the accident a verdict of 'Misadventure' was returned. The jury added a rider to their verdict: 'No single person can be saddled with a guilty act, but there was a very serious lack of co-ordination regarding exact signalling in the dark, which calls upon the railway authorities to make drastic reforms in the signal box system and introduce a better method of co-operation between the signal box and the drivers of engines in cross-over or shunting operations."

The Coroner, Mr J. Pritchard-Jones, praised the courage and prompt action of Driver William Williams and Fireman John Williams of the *Irish Mail.*

Two trains reportedly collided at Sarpenzele, Hungary, on Sunday 17th September 1950. An explosion and possibly a fire followed the collision and there were 100 fatalities.[546]

Seven or eight people died and about 30 were injured in a head-on collision between two express trains at Svitavy, Moravia, Czechoslovakia, on Sunday 1st October 1950.[547] [548]

Express train No. 4405 from Madrid to La Coruña, running an hour and a half late, collided with railcar No. 492 from La Coruña to El Ferrol del Caudillo 400 metres outside of Cecebre station, Spain, at 13:40 on Wednesday 6th December 1950. The railcar was completely crushed by the engine of the express. Both trains were travelling at some speed.

As soon as news of the accident was received in La Coruña assistance was sent from there and ambulances arrived at Cecebre within half an hour. There had been 40 people on the railcar and five of them were killed, including Motorman Marcial Rey Martínez. Five passengers were seriously injured and others were slightly hurt. On the express only the driver and a porter were hurt.[549]

This may well have been a single-line collision; the mechanism of the accident is not described in the newspaper report, which is largely a list of the dead and injured.

---

545 Ministry of Transport report
546 Author's own notes, source mislaid
547 *The Lethbridge Herald*, Lethbridge, Alberta, Canada, 3rd October 1950
548 *Malaya Tribune*, Singapore, 3rd October 1950
549 *La Vanguardia*, Barcelona, Spain, 7th December 1950

Nine people died and 11 were injured at Darshanga, India, on Friday 11th May 1951 when a passenger train collided with a switch engine. [550]

A group of Italian railwaymen from Venezia and Ancona, returning with their families from a visit to Wien, Austria, whence they had been invited by their fellow Austrian colleagues, were to become the victims of a disaster at Langenwang, Austria, on Tuesday 25th September 1951.

Train D 589, the 20:20 express from Wien Südbahnhof to Roma Termini consisted of engine No. 33.118, a luggage van, a mail van of all-steel construction, and ten coaches. The front coach, immediately behind the mail van, was a wooden ex-second-class Austrian carriage dating from 1907, and it was in this vehicle that the Italian railwaymen and their families were riding.

The express was delayed travelling through Austria by speed restrictions and further delayed by border controls between the Soviet and British zones of the still occupied country (the Allied occupation did not end until 1955). Meanwhile at Langenwang station express goods train Egz 1851 from Wien to Graz was conducting shunting operations, in the course of which it had to transfer some waggons from track 4 on the opposite side of the station to track 3, which involved crossing the southbound Südbahn main line (track 1) from Wien to Bruck an der Mur and Italy. The signalling relating to track I consisted of an *Einfahrsignal* (home signal) and its associated *Vorsignal* (distant signal) at the north end of the station and an *Ausfahrsignal* (starting signal) which had its own *Vorsignal* at the south end. The shunting train was protected by these signals but the signalling regulations prohibited a southbound train from being accepted from Mürzzuschlag, the previous station, while the line was blocked in the rear of the *Ausfahrsignal.*

The Langenwang *fahrdienstleiter*, anxious to avoid further delay to the express, chose, as other signalmen have done before, to 'stretch the rules,' or more accurately to disregard them, on the assumption that the driver of the express would obey his signals. When the express was offered by Mürzzuschlag he accepted it, in contravention of the regulations, and cleared his signals for it while the freight train was still being set back into track 3, the shunting having taken rather longer than anticipated. The *Einfahrsignal* and its associated *Vorsignal* then went to *Fahrt frei* and *Freie Fahrt erwarten*, and the shunting train was only protected by the *Ausfahrsignal* and its associated *Vorsignal.*

In accordance with the usual railway provision of more than one safeguard the accident should therefore not have occurred, but in addition to the breaking of the regulations by the *fahrdienstleiter* the driver of the express failed to see the *Vorsignal* for the *Ausfahrsignal* and ran through the station at full speed. Seeing what was about to happen the *fahrdienstleiter* desperately waved a red lamp at the driver, who belatedly also saw the *Ausfahrsignal* at *Danger*. He made an emergency brake application but his train was still travelling at 50 to 60 kilometres an hour when it struck the backing waggons. The engine of the express was derailed and slid along the track and the freight waggons were smashed. The driver was thrown from his locomotive, his fireman was trapped in the overturned engine and had to be cut free. Both survived. The leading baggage and mail vans were thrown up into the air and the mail van fell onto and crushed the wooden

[550] *The Kingston Daily Freeman*, Kingston, New York, United States, 12th May 1951

coach carrying the Italian railwaymen. Sixteen passengers were killed outright and five more were fatally injured. Another 50 people were injured, 11 seriously. The five mail clerks escaped with a shock. [551]

The *fahrdienstleiter*, the driver of the express and his fireman were sentenced to terms of imprisonment, which were reduced on appeal.

A further victim was the fireman of the freight train, who had not in any way been responsible for the disaster but who subsequently suffered from depression and committed suicide two months later. [552]

The first major accident on the West German Deutsche Bundesbahn occurred on the morning of Thursday 8th November 1951 at Walpertskirchen, on the single-track line between München and Simbach.

The events which led up to the accident began with the arrival of local freight train Ng 9721 from München at 05:12, 45 minutes late. It had to drop off two boxcars there, which was duly done, and was then booked to proceed to the nest station at Than-Matzbach to cross passenger train P 1102 from Mühldorf to München. Because it was running late this would have involved delaying the passenger train so Fahrdienstleiter Georg Wimmer agreed with his colleague at Than-Matzbach to alter the crossing of the trains to Walpertskirchen. A further problem then arose. The station buildings stood alongside the loop siding (track No. 1) across which passengers had to make their way to reach the platform serving the through line (track No. 2) When a freight train was standing on the loop siding the normal procedure was to separate it, so as to leave a gap through which the passengers could pass to reach the through line platform but on this day the train was too long to fit into the loop siding after this had been done. The only solution, therefore, was to place the freight train on the main track and bring the passenger train into the loop. Fahrdienstleiter Wimmer duly advised his counterpart at Than-Matzbach that this was what he intended to do so that the latter could issue the driver of the passenger train, Anton Schandl, with a caution ticket informing him of the diversion, and the need to reduce speed over the connect to the loop at Walpertskirchen. What Wimmer failed to do, as the freight train would be occupying the track for more than 10 minutes, was to apply a lever collar as a reminder that the freight train was there, because at this point he was interrupted by a woman passenger who wished to leave her bicycle in the luggage office. He took the bicycle, wrote out a receipt for it, then had to find the woman, who had disappeared, to give her the receipt. He then heard the passenger train approaching and hurrying to the signal box he reset the road for track No. 2, normally used by the passenger train, completely forgetting that the freight train was standing there.

Seven hundred metres from the station Driver Schandl saw that the *Einfahrsignal* (home signal) was displaying a green light rather than a green and a yellow light indicating a diverging route as he had expected. He assumed that the yellow lamp had gone out and reduced speed anyway to 65 kilometres an hour for the crossover. It was still too dark to make out the arm of the semaphore signal, which was off for the through line, but then he saw the headlights of the freight train engine directly in front of him and made an emergency brake application. Quite how soon he had applied the brakes was disputed.

---

[551] *Railway Accidents of Great Britain and Europe* by Ascanio Schneider and Armin Masé, pgs. 91-95

[552] Wikipedia – Eisenbahnunfall von Langenwang

As a result of the collision the leading vehicle of the passenger train, a Prussian compartment coach, was telescoped into the engine tender. An eyewitness who had been riding in the following coach said: "We were sat reading, smoking, or talking on the morning München – Simbach train. Shortly before entering Walpertskirchen station the slowed down – it might have been going at 30 or 40 kilometres an hour. Suddenly the brakes screeched, we heard a terrible crash, the window panes fell out of their frames, the lamps fell from the carriage ceiling. We fell off the train, what we saw was unbearable."

A newspaper report spoke of severely injured passengers inside the carriages calling out for help. Rescuers tried to free the injured with axes, hammers, and saws. Casualties were being carried to the station buildings, no one knowing if they were dead or alive. A farmer took an injured passenger to hospital in a horse and cart before the first ambulance arrived. News of the accident quickly spread. Almost all the victims came from Dorfener and Erdinger areas and relatives soon flocked to the scene. Just before 07:30 a München – Mühldorf passenger train reached the station. It reversed and took the uninjured passengers on to München, along, perhaps, with some of the injured. Meanwhile, amongst the desperate rescue operation, three workmen were said to have asked for documentation to indicate that they were arriving at work late through no fault of their own. Driver Schandl and his fireman were trapped in the cab of their engine. When they were released they were held by the police for questioning, while Fahrdienstleiter Wimmer was also arrested.

Sixteen people died in the accident and 41 were injured.

In July 1952 Georg Wimmer was found guilty by a Munich court of causing 16 deaths and 41 cases of injury through negligence and sentenced to 14 months' imprisonment. A possible factor in his error was tiredness – Wimmer had arranged with a colleague to come on early at 19:30 rather than 21:00 and had had only 4½ hours' sleep.[553] [554] [555]

A headline in a national English newspaper on the morning of Thursday 9th October 1952 read 'DEATH TOLL UP TO 85 IN RAIL CRASH', and the newspaper went on to describe what up to the present time (October 2022) has been the worst accident in English railway history. The final death toll was 112; 108 passengers and four railwaymen. Another 153 passengers and four railwaymen were taken to hospital and 183 passengers were treated on the scene for slight injuries or shock. The catastrophe occurred close to London on the West Coast Main Line of British Railways, London Midland Region at Harrow and Wealdstone station. An Up local train from Tring to London Euston was standing at the Up Fast platform there when it was run into from behind by a late-running Perth to London express, at about 8.19 a.m. Within a couple of seconds the wreckage of the first collision, which had fouled the opposite running line, was run into by a London Euston to Liverpool and Manchester express. An Up electric train from Watford was fortunately stopped well short of the wreckage because the short circuit caused by the wreckage falling on the line had cut off the current, as well as placing the Up Home signal to *Danger*. Of the passengers who died 64 had been in the local train, 23 had been in the Perth express, 7 in the Liverpool express and the remaining 14 had probably been standing on the Up platform. Three of the railwaymen who died

---

[553] https://www.drehscheibe-online.de/foren/read.php?017,8475370

[554] *Von Siegelsdorf nach Aitrang* by Hans Joachim Ritzau, pgs. 83-90 (in German)

[555] Wikipedia – Eisenbahnunfall von Walpertskirchen

were the driver and fireman of the Perth express and the driver of the pilot engine of the Liverpool express.

Immediately after the accident telephone calls were made by the station staff and there was an immediate response from the local emergency services and the Metropolitan Police. The first ambulance arrived at the station only three minutes after the collision, shortly followed by the fire brigade and the police. More ambulances, doctors, nurses, fire engines, police, and a medical unit of the United States Air Force soon followed, and later priests and ministers, members of the Salvation Army, the Women's Voluntary Service, and others. The first loaded ambulance left at 8.27 a.m. and the last two serious casualties left at 2.30 p.m., though the search for further survivors continued until 1.30 a.m. the next morning. [556]

A good many of the passengers on the local train were railway staff on their way to work. Thirty-six of them were amongst those killed and many others were injured. Two of them spoke to a newspaper reporter. '"There had been a good deal of fog and mist farther down the line," one of them said, "but I do not think we were more than a minute or so late at Harrow and Wealdstone; and there I made the comment that it was already a good deal clearer and the sun was coming through. I was about three coaches from the rear end of the local train and we were all seated." A second member of the party was in the front guard's van, standing with other people. The train was due at the station at 8.11 and would then have run fast to Euston. "Suddenly," he said, "we heard a train coming up behind us, and we were struck violently. We were thrown to the ground like so many ninepins; we were just picking ourselves up when we heard the Euston to Manchester train coming through the station in the opposite direction. We knew then we were bound to be hit."'

Other accounts from survivors and witnesses also featured in the newspaper. 'Passengers in the three trains, describing their experiences, related how they escaped serious injury when the compartments in which they were travelling were smashed by the impact.

'Mr G. Hattersley, of Thornton Way, Hampstead Garden Suburb, who was returning from Scotland on the night express, said that he was gathering his belongings together ready to leave the train at Euston when the carriage seemed to mount up beneath him, until he found himself suspended 20 feet in the air with other passengers clinging to the smashed windows and doors. "An airman sitting opposite me was thrown across the compartment and stunned. As I feared the whole ghastly structure might collapse in a few minutes I helped another man to lower him foot by foot down the sloping side of the carriage. I think my carriage had gone over the one in front, because I could hear screams and cries from directly beneath me."

'A clergyman, who preferred to be unnamed, said he helped to release two men and a woman from the corridor of a coach which had been thrown against the buildings on platform 4. He said that, although when he arrived about 15 minutes after the disaster there was still a great deal of distress on all sides, the passengers were remarkably calm and brave in the face of many signs of death and injury.

'Mr John Bond, of First Avenue, Garston, Watford, who was in a compartment of the local train, said: "I could see people waiting on platform 2 for the Underground train, reading their newspapers and talking to each other. The first we felt of the crash was the

[556] Ministry of Transport report

grinding, tearing noise of the Perth express hitting our rear. Coaches of our train were swung across the track, and a second or two later we felt the Manchester-bound train ripping into us. There was an indescribable noise of crashing, and I was flung from my compartment on to the platform."

'Mr John Bannister, of Walton Drive, Harrow, who was in a coach of the local train immediately under the footbridge, said: "It all happened in a second. There was a terrific crash and glass and debris showered on me. There were 15 people in the compartment with me, and we were all jammed together in a mass of broken wood. I blacked out for a moment, and when I came round I found I was lying on the line with debris on top of me. I managed to free myself and drag myself up on to the platform."' [557]

All lines through the station were blocked with much disruption to traffic. Some passenger trains terminated and started from Watford, other main-line passenger services were diverted via Nuneaton and Leamington to London St Pancras and London Paddington respectively. The Slow lines were reopened at 5.32 a.m. the next morning, the Electric lines at 4.30 a.m. on Saturday 11th and the Fast lines at 8 p.m. on Sunday 12th. Freight trains were cancelled or diverted.

The 7.21 a.m. local train from Tring to London Euston consisted of 2-6-4 tank engine No. 42389, a third-class brake composite coach, five third-class coaches, a third-class brake composite coach and two third-class coaches, some of wooden construction with steel panels and others which were all wooden. It was in charge of Guard W.H. Merritt, with Driver A.W. Payne and Fireman A.R. Hine.

The 8.15 p.m. Perth to London Euston express consisted of Class 8P 4-6-2 engine No. 46242, a milk van, a brake van, a third-class coach, a third-class brake composite, a first- and third-class composite, a third-class coach, two first- and third-class composite sleeping cars, a third-class sleeping car, a first-class sleeping car and a brake van, all of wooden construction with steel panels except the milk van and leading brake van, which were all-wooden vehicles. It was in charge of Guard J. Kent with Driver R.S. Jones and Fireman Colin Turnock.

The 8 a.m. London Euston to Liverpool and Manchester express consisted of 4-6-0 engine No. 45637 piloting 4-6-2 engine No. 46202, a third-class brake composite coach, a first- and third-class composite coach, a first-class coach, a third-class brake composite coach, a first- and third-class composite coach, a third-class brake composite coach, a third-class coach, a kitchen car, two first-class coaches, a third-class brake composite coach and four brake vans, the coaches were of wooden construction with steel panels or all-steel construction, the four brake vans were of all wooden construction. The train was in charge of Guard A. Smith, with Driver Albert Perkins and Fireman G. Cowper on the pilot engine and Driver W.H. Darton and Fireman G. Dowler on the train engine.

The Ministry of Transport inquiry into the disaster was naturally comprehensive and included interviews with a good number of witnesses. The evidence of the various signalmen appeared first in the published report on the accident, these being the men in Bushey and Hatch End boxes, about 4 and 2 miles on the country side of Harrow and Wealdstone station, Harrow No. 1 box at the station, and North Wembley box, about 2 miles on the Euston side.

Signalman J. Ashcroft at Bushey stated that his fog object (the object which when it

---

[557] *The Times*, London, England, 9th October 1952

became obscured in a fog necessitated the signalman calling out his fogsignalmen) had become obscured at about 7.15 a.m. and he had called out his fogsignalman, Lengthman H.J. Richards. In fact the 10.20 p.m. express from Glasgow to London Euston, the Perth express and the Tring local train, which had passed his box at 8.05 a.m. and 8.14 a.m. on the Up Fast line and 8.11 a.m. on the Up Local line respectively had all had clear distant signals anyway.

Signalman Ashcroft did not notice the enginemen of any of the trains but Lengthman Richards was sure that he had had a wave from the footplate of the Perth train, indicating that the engine crew must have been alert at that point. He said that with the drifting fog visibility had varied between 100 and 200 yards.

Signalman A.G. Horsfall at Hatch End put the visibility earlier at 300 yards and said that he had not considered it necessary to call out his fogsignalmen. By 8 a.m. the visibility had increased to 500 yards. The local and Perth trains had passed him with clear signals at 8.13 a.m. and 8.16 a.m. respectively. He testified that when the Perth express passed him he had assumed that the local train would either be crossing from the Up Local to the Up Fast at Harrow and Wealdstone or be already stopped at the Up Fast platform there. (Local trains were booked to use the Up Fast line from Harrow and Wealdstone to Euston so as to leave the Up Slow line clear for empty coaching stock movements and were given priority over late-running long-distance trains.) Signalman Horsfall said that he had received the *Obstruction Danger* signal from Harrow No. 1 at 8.19 a.m. He had not spoken to the Harrow signalman either before or after the accident. He considered his colleague at Harrow to be "a very good man to work to."

Relief Signalman A.G. Armitage at Harrow No. 1 testified that it had not been sufficiently misty to require fogsignalmen when he had come on duty at 6 a.m. but it had got more foggy at 6.35 a.m. He knew that there was no point in asking for a fogsignalman at that time as they would be on their way to work, so he had begun double block working instead. By 8.10 a.m. the fog had cleared considerably and he had reverted to normal working. The Glasgow express had passed on the Up Fast at 8.11 a.m. and as far as he could remember he had reset the crossover to bring the Tring local from the Up Slow to the Up Fast line immediately afterwards. Also at 8.11 a.m. he had accepted the Perth express on the Up Fast from Hatch End, as he was entitled to do during normal working, as he had an Up Fast Outer Home signal which gave him his quarter-mile clearance between it and the fouling point of the crossover. At 8.14 a.m. he received the *Train Entering Section* signal for the local train, North Wembley box cleared out the Glasgow train and had accepted the local train on the Up Fast line. He had waited until the track circuit in the rear of the Up Slow Home signal was occupied, then had cleared his signals for it to proceed from the Up Slow to the Up Fast. The local train had arrived at the Up Fast platform at 8.17 a.m., he had replaced the Up Slow Home signal behind it and at the same moment he had received the *Train Entering Section* signal for the Perth express from Hatch End on the Up Fast and for the Liverpool and Manchester express from North Wembley on the Down Fast.

Signalman Armitage said that a minute or so later he had been amazed to hear the Perth express approaching at speed and had then seen it coming out of the mist, passing his Outer Home signal at Danger, and making no attempt to pull up. He had immediately pulled his Up Fast line detonator placer and thrown back his Down Fast signals. In his opinion the Perth express had been under steam; it had been running much faster than the Glasgow express and he had seen no sign of braking. The collisions of the Perth express with the local train and of the Liverpool express with the wreckage had happened almost simultaneously. According to his train register book he had sent the *Obstruction Danger* signal in both directions at 8.18 a.m.

The Ministry of Transport inspecting officer, Lieut. Colonel G.R.S. Wilson, grilled Signalman Armitage as to whether he might not have at first decided to give the Perth express priority over the local train and then changed his mind, but Armitage declared that he was well aware that local trains had priority over late-running expresses and pointed out that had he done as the inspecting officer suggested he would have had to throw back his signals, reverse the crossover and, because of the block controls, obtain a cancellation of the train from North Wembley. There was no record of any such cancellation having been requested.

Signalman D. Pullen at North Wembley stated that his "fog points" had been visible at all times since 06:00 and he had not needed to consider calling out the fogsignalmen. According to his train register book he had cleared back the Glasgow express at 8.13 a.m. and had accepted another train. All Up trains which did not stop at Willesden were signalled to him as 4-4-4 so he could tell if it was the local train or the Perth express. He had only been offered one train and none had been cancelled. He had not spoken to Signalman Armitage that morning. His book recorded the *Obstruction Danger* signal as having been received at 8.20 a.m.

Driver F.B. Halliburton of the Glasgow express said that the weather had been clear as far south as Wigan; from there the fog had been practically continuous as far as Willesden. There had, however, been no difficulty in observing either colourlight or semaphore signals. His train had not been checked south of Bletchley. Harrow No. 1 Up Fast Distant (a three-aspect colourlight signal, with a main yellow aspect, an auxiliary yellow aspect which was lit should the main aspect fail, and a green aspect) had been *Green* and 'he could not possibly have missed it.'

Fireman J. Parkinson of the Glasgow train generally confirmed his driver's account. He said he had seen the Bushey semaphore distant at clear. He had not seen either the Hatch End or Harrow No. 1 Distant signals, which were on his driver's side, but his driver had raised his hand to indicate that these signals were off.

Driver Payne of the local train described the fog as having been "patchy," with thick fog at Tring, clearer weather at Watford, then thicker at Carpenter's Park (between Bushey and Hatch End) and Headstone Lane (between Hatch End and Harrow and Wealdstone). All the signals had been clear except for Watford Tunnel North End Up Slow Distant and Harrow No. 1 Up Slow Distant. The latter signal was always at *Caution* for his train as it had to cross there to the Up Fast line. He had seen the Harrow No. 1 Slow line Distant without difficulty but had not noticed the indication of the Up Fast line Distant. He described the conditions at Harrow as hazy rather than foggy. He thought that his train had stopped rather longer than usual at Harrow and Wealdstone due to the heavy load of passengers; then it had been run into and his train had been pushed forward an engine's length as the Liverpool train passed.

The evidence of Fireman Hine was described as vague. He thought that the sun had been trying to break through the fog at Hatch End. He saw the Harrow No. 1 Up Slow distant at 20 to 30 yards.

Guard Merritt of the local train said that he did not see the Harrow No. 1 Up Slow Distant but he saw the Up Slow to Fast Home signal off. The Harrow platform had been crowded so he had let some passengers into the brake compartment. As he had been closing the carriage doors he had seen the Perth express coming in fast and dived for shelter beneath a platform coping until after the second collision, when he had gone to Harrow No. 1 box and Signalman Armitage had assured him that the lines were all protected.

Guard Kent said that he had spoken to Driver Jones of the Perth express when he had joined the train at Crewe. His driver had seemed in good health; he had remarked about the fog but had not appeared at all anxious. They had left Crewe at 4.37 a.m., 32 minutes late, and continued to lose time, due, he thought, to the fog. The train brakes had acted normally for signal checks/stops at Baddesley Sidings, Atherstone and Watford Tunnel North End and for special stops at Nuneaton and Rugby. After Watford he had been busy tying up invoices into bundles for throwing into the net at Camden Goods Yard and his train had been "just getting into its swing again" when the collision occurred. There had been a severe brake application, followed after about five seconds by three violent lurches, then a violent rebound. His idea was that Driver Jones "had spotted something just too late." He had climbed down with detonators and a flag after the collisions to go back and protect the line but the signal box had been close so he had gone there instead and the signalman assured him that the lines were protected. As regards visibility he thought it had been "not dense but patchy" and not foggy enough to warrant fogsignalmen.

Travelling Ticket Collector J.E. Smith and Sleeping Car Attendant H.V. West of the Perth express agreed that the brakes had been applied at the last moment.

In the minutes before the collision the 4.32 a.m. Down freight train from Norwood to Northampton had passed on the Down Slow line. It consisted of a heavy goods engine, 58 empty waggons and a brake van in charge of Guard T. Starmer with Driver R.C. Brown and Fireman G.T. Lane.

Driver Brown said that his train had been stopped at Harrow and Wealdstone. When it had restarted he looked back and had been able to see about 30 waggon lengths at the most (about 200 yards). Asked if he thought that there should have been fogsignalmen out he said that it was hard to tell as the fog had been drifting. He had passed the Perth train somewhere north of Headstone Lane, where the fog had been thicker. Hatch End Down Slow colourlight Distant he had seen at yellow at about 50 yards. This signal was about 345 yards south of the Harrow No. 1 Up Distants.

Fireman Lane confirmed his driver's account. Neither man had noticed the indications of the Harrow No. 1 Up Distants; they had, of course, had no reason to do so.

Guard Starmer recollected the Up Local train passing after his train after they had left Harrow and Wealdstone. He also remembered seeing the Perth express go by and at the railway's internal inquiry he stated that the Up Fast Distant had been at *Caution* when it did so, but at the Ministry of Transport inquiry he was not quite so definite, though he said he was "pretty certain" that it had been. Regarding visibility, he said it had been fairly clear at Willesden, then had got misty as they proceeded northwards. At Harrow and Wealdstone visibility had been about 200 yards, at Headstone Lane less.

Fireman G. Cowper of the pilot engine of the Liverpool train, who had had a very lucky escape from death, remembered little of the accident other than that the Harrow No. 1 Down Fast Distant had been off and his train travelling at full speed.

Driver Darton of the Liverpool train said that it had started late but regained two minutes by Harrow and Wealdstone. He added, "We could not help picking up time with the power we had." There had been no trouble seeing signals, the sun had been coming through at Harrow and he had remarked to Fireman Dowler that it was going to be a nice day. He had seen the Harrow No. 1 Down Fast Distant and Home signals off and had not seen them replaced.

Fireman Dowler could remember little of the collision other than that his driver had been trying to make up time.

Guard Smith described the visibility after passing North Wembley as having been quite good.

Driver Jones was 43 years old. He had joined the railway as a cleaner in 1927, been appointed fireman in 1937 and driver in January 1948. He was familiar with the main line between Crewe and London and had made 41 trips over it since March 1950. He had come off duty at 2.40 p.m. on the day before the accident and had booked on at 12.45 a.m. on the 8th. There had been plenty of time for him to read his notices before he left Crewe with the Perth express. Crewe Running Shift Foreman J. Hallmark described Jones as steady, even-tempered and a good driver, adding that he had appeared perfectly normal before he had taken the train out. Firing Instructor W.S. Davies spoke of Jones as a “methodical young man” with a keen interest in his work who attended improvement classes in his spare time. It appeared that on returning home at 3 p.m. the previous afternoon Driver Jones had been busy repainting the interior of his house, then had gone to bed at 7.45 p.m.

A medical examination of Driver Jones in 1946 disclosed no problems and his colour vision was normal. A post-mortem examination showed no evidence of diabetes, alcohol abuse or carbon monoxide poisoning.

Fireman Turnock was 23 year of age. He had joined the railway as a bar boy in 1943, had become a cleaner in 1944 and been appointed fireman in 1945, aged only 16. He had fired express trains 44 times in 1951 and 1952. Firing Instructor Davies described him as “exceptionally good, and out to learn all he could and build himself up to be a good driver.”

The engine of the Perth express was very carefully examined after the accident. Nothing was found to be wrong with it other than damage which was obviously the result of the collision. The position of its controls indicated that it would have been coasting approaching Harrow, as would be normal. It was evident that Driver Jones had only applied the brakes in emergency at the last moment.

Signal and Telecommunications Engineer S. Williams had been a passenger on the local train and had gone to Harrow No. 1 box straight after the accident. He verified that the levers for the Up Fast Inner Home, Outer Home and Distant were locked in their normal positions and that the block instruments were all at ‘Train on Line.’ He described the visibility as “a half fog,” about 200 to 300 yards and clearing.

Telegraph Lineman C. Thorpe went back on foot and verified that both of the Harrow No. 1 Up distants were at *Caution*. By then, half an hour after the accident, the weather was clear. Telegraph Inspector H.A. Fishenden arrived in the box a little later, tested all the apparatus and found it to be working correctly, and also went back and verified that both distant signals were at caution.

Later a series of tests was conducted to determine if a false feed could have caused a wrong-side failure of the Distant signal but there was no evidence that this could have happened, and none of the signalling officers could recall any previous false clear signal aspects since the first installation of colourlight distant signals in 1936.

Stationmaster C.S. Rolinson at Harrow estimated the visibility at 8.10 a.m. as 200 yards, with the sun just appearing above the mist, and said that he would not have expected fogsignalmen to be out. There had been a large number of passengers on the platform. He had gone straight to Harrow No. 1 box after the collision. Signalman Armitage had been "deathly white" and very upset. Having checked that the *Obstruction Danger* signal had been sent on all lines he had helped the signalman out onto the box steps to get some fresh air, then had concerned himself with the rescue work.

Station Foreman H.E. Foskett stated that he had been counting passengers as the local train had arrived; 332 had come down the stairs to join the train and 20 had gone up. He put the visibility at about 50 yards at 6.45 a.m. and said that at 8 a.m. the fog had been clearing to the south and coming and going to the north.

Ganger A.H. Payne stated that he had only had one man who could have done fogsignalling that morning, Lengthman Honour. He could not explain why he had not sent the lengthman to Harrow No. 1 box at 7.30 a.m. to report for fogsignalling duties in accordance with the usual practice.

Ganger A. Turney, who was responsible for the next length of track to the north, put the visibility at Headstone Lane at 100 yards at 7.30 a.m. He said that men were joining and leaving his gang all the time and none except himself were trained in fogsignalling duties.

Permanent Way Inspector C.E. Penny and District Signalling Inspector E. Short agreed that there was a shortage of fogsignalmen due to the constant change of maintenance staff and the latter officer considered that a tail lamp man at the box (to inform the signalman that each train had passed complete) was more important than a fogsignalman at the Outer Home signal, a view supported by District Operating Superintendent Cox.

Concluding his report, Lt-Col Wilson stated that he was entirely satisfied that Driver Jones had failed to reduce speed at either the Harrow No. 1 Up Fast Distant or the Outer Home. It seemed likely from his last-minute brake application that he had seen either the Inner Home at *Danger* or the obstruction ahead. It was clear that no responsibility rested upon Signalman Armitage for the first collision and no one could have prevented the second.

Tests and calculations showed that had Signalman Armitage first cleared his signals for the Perth express and then changed his mind, replaced his Up Fast signals, reversed the Up Slow to Up Fast crossover, and cleared his signals for the local train the Perth express would still have been well outside of the point where the Harrow No. 1 *Distant* first became visible. Similarly there was no problem demonstrated with the signalling equipment and the possibility of a false clear distant signal could be dismissed.

The question remained, the inspecting officer continued, whether, under the prevailing conditions Signalman Armitage should have given the local train priority and whether, having decided to do so, he should have accepted the Perth express from Hatch End.

As to the first point Signalman Armitage was expected to give priority to local trains and to keep them to the booked route as laid down in the working timetable, which at Harrow was from the Up Slow to the Up Fast at Harrow No. 1. Driver Payne had expected to be crossed to the Up Fast, even though he was running a few minutes late.

As to the second point, Signalman Armitage was entitled to accept the Perth express in accordance with paragraph 25 (a) of the Block Regulations provided that the visibility *In the neighbourhood of his box* (italics in Ministry of Transport report) was sufficient to justify the express being accepted under full *Line Clear* in the absence of a fogsignalman at the Outer Home signal. No class of evidence, remarked the inspecting officer, could be less positive than recollections of visibility, but he did not consider that Signalman Armitage's statement that at 8.10 a.m. he could see the Up Slow Home signals 303 yards from his box was inconsistent with the evidence of other witnesses. The fact that the Harrow signalman had instituted fog working at 6.35 a.m. showed that he was fully alive to his responsibilities and his statement that the fog had begun to clear at 8 a.m. was accepted. Generally Signalman Armitage appeared to be an able relief signalman who was thoroughly conversant with the working at Harrow. There was no evidence that he had failed in any way to comply with the signalling regulations and the state of shock in which he had been found after the accident should not be held against him in view of his obviously sensitive temperament. He should be exonerated from all responsibility for the accident.

In view of the death of Driver Jones and Fireman Turnock it was only possible to speculate on the reason for their failure. Driver Jones had clearly been alert at Watford Tunnel North End and at Bushey, where he had waved to the fogsignalman. It was unlikely that he would have lost his location after that. By all accounts the visibility at the Harrow No. 1 distants was considerably less than at Hatch End, possibly only 100 yards, in which case with a train travelling at 50 miles an hour the distant would only have been visible for four seconds. This was not a long time, but the colourlights were well sited and two other drivers travelling at the same speed in the previous few minutes had had no difficulties. The post mortem appeared to rule out any sudden incapacitation due to illness, nor was there any indication of an engine problem or blowback. The only other possibility seemed to be that Driver Jones had briefly lost his concentration at the crucial moment and had not realised that he had gone past the Distant signal without seeing it. In still looking out for the Distant signal he might then have missed seeing the considerably higher Outer Home signal and had probably only realised his mistake when Harrow station had come into view. There was a better view of the Harrow signals from the driver's side and so Fireman Turnock had no special reason to be looking out for them. It was unlikely that a *Green* colourlight signal on the parallel electric lines could have been mistaken for the Up Fast Distant.

Turning to remarks and recommendations the inspecting officer remarked that 'no practicable form of carriage construction could be relied upon to prevent casualties when heavy trains collide at high speed, as the energy which has to be dissipated in a short time and space is so very great, and this has been shown by recent cases abroad where it is possible to operate steel coaches of very heavy construction. It does seem possible, however, that the wreckage at Harrow might have been less compact, and the killed and injured fewer, if a greater proportion of the rolling stock had been of the latest all-steel type and rigidly coupled.' There were 1,100 coaches of the new standard design in use on British Railways out of 37,000 steam-hauled passenger vehicles. Shortage of steel had prevented any new construction in 1952 but with an improved position it was hoped to construct 1,200 coaches a year until 1956 and 650 a year thereafter. However, coaches had a long life so it would be some years before standard coaches predominated, even on long-distance trains. Similarly it would be a considerable time before Buckeye couplings replaced screw couplings on the lines of the former London Midland and Scottish and Great Western Railways.

The two bogie vans marshalled at the front of the Perth train had undoubtedly reduced the damage to the following passenger carriages. The new standard coaches on the Liverpool train had been marshalled 4th, 6th, and 11th. It would be better if they could be marshalled together to take advantage of their Buckeye couplings.

Lt-Col Wilson found no problem with the signalling at Harrow and Wealdstone, noting that Outer Home signals were provided specifically to give a margin against the overrunning of signals and no margin could guard against the total disregard of signals, while the Rules and Regulations for working during fog were adequate. The means of guarding against such accidents as that at Harrow and Wealdstone lay not in making the regulations more restrictive but in reinforcing the vigilance of drivers by a positive link between the wayside signals and the footplate, in the form of Automatic Train Control, such as had been in use for many years on the Great Western Railway and was in use on the Tilbury line of the former London Midland and Scottish Railway.

Ten percent of the accidents which had occurred since 1912, the report went on, could have been prevented or mitigated by Automatic Train Control (ATC), and 28 per cent or 399 of the fatalities could have been prevented. Inspecting officers had been recommending the introduction of ATC but none of the companies except the Great Western had considered the installation worth the expense, though they had agreed to further trials, as a result of which a different system, the Hudd system, had been installed on the Tilbury line. After the 1937 Castlecary disaster (dealt with in the previous volume of this series) the London and North Eastern Railway had also begun installing the Hudd system between Edinburgh and Glasgow, but this installation had stopped due to the war. The Southern Railway, on the other hand, had preferred to install colourlight signalling and continuous track circuiting. The need for ATC had further been highlighted by the disasters at Bourne End in 1945 (see previous volume of this series) and Goswick in 1947 (see Other Derailments chapter above). In 1948 the Railway Executive had recommended the extension of either the Great Western or Hudd systems. It was eventually decided to develop a "compromise" apparatus incorporating the best features of each system. Delays and difficulties had then ensued but by March 1953, 54 locomotives had been fitted with the final prototype design. The British Transport Commission had announced a five-year plan to install ATC between London Euston and Birmingham, Liverpool, Manchester and Glasgow, London King's Cross, Edinburgh and Glasgow, London Liverpool Street and Norwich and London Waterloo and Exeter, Southampton, and Bournemouth. A second long-term plan involved the provision of ATC over a further 3,988 route miles. With the existing ex-Great Western and Tilbury systems this would mean that almost all the main lines carrying heavy traffic were equipped. [558]

The Quetta – Karachi mail train collided with a goods train which was being shunted at Jhatpat, West Pakistan, on the evening of Monday 13th October 1952. The engine and three carriages of the mail train were smashed. According to one report seven people were killed and 16 injured; other reports spoke of 12 dead and 40 injured. Most of the victims were women and children. [559] [560]

Thirteen people reportedly died in a collision between two trains at Dobrinje, Yugoslavia, on Sunday 21st December 1952. [561]

---

[558] Ministry of Transport report

[559] *The Times*, London, England, 15th October 1952

[560] *Tucson Daily Citizen*, Tucson, Arizona, United States, 14th October 1952

[561] Author's own notes, source mislaid

A major disaster occurred on the Estrada de Ferro Central do Brasil, at Clemente Falcão station, Brazil, in the suburbs of São Paulo, just after 21:30 on Monday 23rd March 1953. A freight train hauled by engine No. 742, in charge of Driver José Antonio Alves, and composed of waggons from the Estrada de Ferro Santos Jundai, was stopped at the station waiting for the previous train to clear the next station at Sebastião Gualberto. Behind it was suburban passenger train UP-70 from Presidente Roosevelt to Mogi das Cruzes, while approaching from the opposite direction was another passenger train from Mogi das Cruzes, driven by José Augusto. Train UP-70 ran into the rear of the freight train and the rear car of the train from Mogi das Cruzes was struck by the wreckage of the initial collision and derailed. A second-class coach in train UP-70 was badly damaged and thirteen or possibly 14 passengers in it lost their lives, while about 33 were injured. The passengers in the derailed coach of the train from Mogi das Cruzes escaped with a shaking. Within a day of the disaster the railway company had announced that it was the fault of Driver Arlindo Alves dos Santos of train UP-70, who had overrun a red signal.[562]

Two trains of the Ferrocarril General San Martin collided in a dense fog near Buenos Aires Presidente Perón station, Argentina, on the morning of Friday 14th August 1953. Initial reports spoke of five dead and 20 injured and added that rescuers were searching for additional victims. [563] [564]

A collision occurred on the North Eastern Railway zone of Indian Railways when No. 339 Down Passenger train bound for Raxaul, India, collided head-on with a goods train half a mile from there, at 05:30 on Saturday 2nd January 1954. [565] No. 899 Up goods had arrived at Raxaul at 04:35 and begun shunting operations. At 05:14 the Raxaul assistant stationmaster had given line clear for No. 334 Down passenger train but did not clear his Up signals in view of the continuing shunting operations. For some unexplained reason the driver of the goods train, going forward to detach a waggon, kept going past the Down home and Down outer signals, disregarding frantic hand signals from the station staff, and came into head-on collision with the passenger train. Nine people were killed outright, three were seriously hurt, one of whom died later in hospital, and five were slightly injured. Damages were estimated at 50,821 rupees. [566]

Eighteen people died and several were injured in a collision between a passenger train and a freight train at Pozanti, near Adana, Turkey, on Wednesday 20th January 1954. [567]

An accident at Sątopy Samulewo, Poland, on Friday 26th January 1954 is a little hard to classify as it possibly occurred on a single line and involved what was more or less a runaway train. It has been included here. The trains involved were passenger train No.

---

[562] *O Estado de S. Paulo*, São Paulo, Brazil, 25th March 1953' and 'Jornal do Brasil, Rio de Janeiro, Brazil, 25th March 1953

[563] *Lawrence Daily Journal-World*, Lawrence, Kansas, United States, 14th August 1953

[564] *The Charleston Daily Mail*, Charleston, West Virginia, United States, 14th August 1953

[565] *The Indian Express*, Bombay, India, 3rd and 4th January 1954

[566] https://archive.org/stream/dli.csl.1063/1063_djvu.txt

[567] *Beckley Post Herald*, Beckley, West Virginia, United States, 21st January 1954

0192, from Bialystock to Gdynia and freight train No. 401933 from Olsztyn to Korsze. The passenger train consisted of steam engine No. Pt47-45, a luggage van, a third-class coach, two second-class coaches, two third-class coaches, a refrigerator van, and a mail van. The freight train consisted of engine Ty23-154 and an unstated number of Wddo waggons loaded with 1,780 tons of coal.

There had been problems with the slow freight train and it had become divided at Górowo station and recoupled together again. When it reached Sątopy Samulewo its brakes failed, it overran the signal protecting the station, ran through the station and beyond it collided head-on with the passenger train at 22:19. Seventeen people died at the scene, another on the way to hospital, and 37 were injured. It was a bitterly cold day, with heavy snow. Only one ambulance was able to get to the scene and some of those who died probably did so from hypothermia.

Water in the brake pipe of the freight train had prevented it stopping but its crew had failed to carry out a brake test after it had been recoupled and had failed to replenish its supply of sand. [568]

Another accident, probably a collision, happened at Swiebodzice, Poland, on Wednesday 24th February 1954, when 10 people reportedly died. A Central Intelligence Agency report spoke of a third accident at nearby Walbrzych Glówny station on Friday 5th March 1954 in which 120 people supposedly died, but it has been suggested that this is the same as the February accident. [569]

The West Berlin *Telegraf* newspaper stated that 'hundreds' of people had been killed or injured in a collision between a workman's train and a freight train between Kottbus and Guben, East Germany, on Thursday 30th March 1954. No further details of the accident were given other than that the train crews had been arrested. [570]

A major disaster occurred in France on the evening of Wednesday 3rd July 1954 when a Lyons – Nîmes passenger train consisting of a railmotor and three trailers collided head-on with a work train at Châteaubourg, between Mauvres and Saint-Péray. Thirty-seven people died, including two who died of their injuries some time later, and almost 50 were injured. [571] The railmotor was almost completely destroyed in the collision and its leading trailer was partially telescoped but the rear trailer, in which were travelling 28 children on their way to a holiday camp, was little damaged and the children escaped with cuts and bruises. The motorman of the passenger train and the driver and fireman of the work train were initially reported as among the seriously injured while the conductor of the passenger train was missing and probably among the dead. Along the roadside, piled up by the rescuers, among the bloodstained stretchers, was a pile of heterogenous objects belonging to the victims.

A reporter at the site wrote, 'The work continues at the scene of the disaster. Rescuers

---

568 Wikipedia – Railway disaster in Sątopy – Samulewo
569 Wikipedia – Railway accidents in Poland
570 *Council Bluffs Nonpareil*, Council Bluffs, Iowa, United States, 1st April 1954
571 https://fr.wikipedia.org/wiki/Liste_des_accidents_ferroviaires_en_France_au_XXe_si%C3%A8cle#Ann%C3%A9es_1950

are working on the side of the shell of the red and yellow motor unit of the railcar, which is embedded in the locomotive of the work train, fearful that inside will be more corpses. A strong wind is blowing from the south and a storm is brewing above the Châteaubourg cliff, which borders the railway to the right, falling to the Rhône valley. A terrible smell floats above the chaos where can be distinguished pieces of bodywork, gutted seats, and pieces of axles, which lie miserably under the enormous cables of a giant crane. At the Hotel de Ville in Saint-Péray, hastily draped with black curtains to receive the corpses, heart-breaking scenes began as the first families of local victims arrived and began to recognise their dead.' [572]

The Minister of Transport, M. Chaban-Delmas, and the director general of the Société nationale des chemins de fer français, M. Armand, went to the scene of the accident the following morning. M. Armand later announced that the accident had resulted from "three or four coincidental professional errors" and that the direct cause of it appeared to be confusion in the giving up of the single-line working over the southbound line which had been in operation earlier in the day because of track repairs, with the result that the work train was proceeding northwards on the southbound line. A signalman had been arrested. [573]

At about 15:30 on Wednesday 7th July 1954 freight train No. 7066 from Segovia collided head-on with a multiple-unit passenger train, No. 3636, at Los Molinos station, Spain. The engine of the freight train embedded itself in the front of the passenger train and several freight waggons were derailed.

Assistance was immediately provided by the station staff, doctors attended from Los Molinos, Cercedilla and the local Navy Tuberculosis Sanatorium, relief trains were sent from Madrid and Segovia and road ambulances came from the same two places. The rescuers found that Conductor Matías Barrios, Motorman Rafael Vela, and Electrical Controller Bautista Chevaler of the passenger train were all dead, as were Conductor Antonio Cogollos, Driver Francisco Jiménez, and Station Agent Santiago Nieva of the freight train. The station agent had been off duty at the time. Assistant Driver Pablo de Pablos of the freight train was trapped in the wreckage, gravely injured, and desperately pleading to be released. It proved impossible to quickly extricate him and doctors gave him blood transfusions while the rescue work continued, combined no doubt with morphine for his pain and sedatives. At 21:00 he was still imprisoned and a local priest gave him the last rites. He died before he could be released. Three other railwaymen and 21 passengers were injured.

The accident was attributed to the driver of the freight train overrunning signals. [574] [575]

Early on Thursday 9th September 1954, Estrada de Ferro Central do Brasil passenger train No. UP-21 from Mozi das Cruzes to Sâo Paulo, crowded with workmen, collided with stationary freight train No. QP-3 at Engenheiro Clemente Falcâo station, Brazil. The collision, at 07:35, was a violent one and the vehicles of the passenger train piled up on the freight train. The force of the impact resulted in two of the carriages of the passenger

---

[572] *Gazette de Lausanne*, Lausanne, Switzerland, 5th July 1954

[573] *The Times*, London, England, 5th July 1954

[574] *La Vanguardia*, Barcelona, Spain, 8th July 1954

[575] Webpage: Federación Castellano Manchega de Amigos del Ferrocarril (in Spanish)

becoming telescoped into one another for a distance of four metres. Ten people reportedly died at the scene and two of the injured died soon after their arrival at hospital. Sixty-nine people were injured, 20 seriously. Shortly after the accident it was blamed by a Rio de Janeiro newspaper on Driver Joaquim Carvalho da Silva of the passenger train having overrun signals in the fog. He was said to have emerged from the collision unscathed and fled the scene.[576]

A more detailed account of the disaster appeared in a Såo Paulo newspaper. It stated that train No. UP-21, the *Especial da Vila Matilde*, being hauled by engine No. 459, with Driver Carvalho da Silva and Fireman José Benedito da Silva, had come from Engenheiro Gualberto station and was bound for the Roosevelt terminus in Såo Paulo. Because of the dense fog it had been delayed. It had whistled as it approached Clemente Falcåo station when the freight train, hauled by engine No. 714, had suddenly come into view. The driver of the passenger train had applied his brakes in emergency but the collision had been unavoidable. The freight train had been 150 metres from the station and outside of the signal, whereas it should have been inside the signal, which was 15 metres from the station platform. Had the freight train been within the signal the disaster would not have happened. The initial report in the Såo Paulo paper mentioned only nine dead with 110 injured and placed the blame for the accident on the unnamed driver of the freight train rather than that of the passenger train.[577]

What was at the time the most serious accident in the Netherlands for 14 years occurred at Elst station on Wednesday 13th October 1954. On that morning crowded Nijmegen – Arnhem – Utrecht – Amsterdam electric train No. 516 of the Nederlandse Spoorwegen was running a little late as it entered the station at about 15 kilometres an hour. As far as the signalman was concerned it was about to run into the platform as per normal. As his train entered the station Motorman E.J. Lobbers realised that it was being diverted to the opposite running line through a wrongly placed set of facing points, and that furthermore Enschede – Arnhem – Nijmegen – Roermond fast train No. 741 (or No. 3212 according to a different source) was approaching on that track at 100 kilometres an hour. Stopping his train as quickly as he could, he switched on its red tail lamp to warn the driver of the oncoming train and hurrying back into the passenger compartment of the leading third-class coach shouted to its occupants, "Get back, there's going to be a collision!" The trains met at 08:42.[578]

In the collision the diesel-electric fast train climbed over the stationary stopping train and was itself split open over a length of 20 metres. By 14:00 six dead and 11 injured had been recovered from the wreckage. All the casualties in the accident occurred in the fast train as the strong nose of the new, post-war stopping train had withstood the impact without collapsing. There had been only a few passengers in the leading car of the fast train and some of them owed their lives to the presence of mind of Motorman J.G. Spekman, who like his colleague had rushed back through the train when he had seen the imminent danger, dragging some of the passengers with him. He survived, though he sustained a head injury.

The first assistance came from the Elst station staff and from the nearby Taminiau

---

[576] *Jornal do Brasil*, Rio de Janeiro, Brazil, 10th September 1954

[577] *O Estada de São Paulo*, São Paulo, Brazil, 10th September 1954

[578] *Het Vrije Volk*, Rotterdam, the Netherlands, 13th October 1954

canning factory, while further help came from Arnhem and Nijmegen within half an hour. The rescue work went on for some hours. [579]

The wrong position of the facing points was determined to be due to the fact that the wire operating them from post T had broken where it ran in a pipe beneath a level crossing, stray electric currents having damaged the wire. The points had therefore failed to move when the signalman at post T operated the lever working them, making the accident analogous to that at Lichfield (Trent Valley), England, almost nine years earlier (see above). The accident still seems a little mysterious as one would have thought that some form of detection would be provided to prevent the signal in the rear of the points from being cleared when they were in the wrong position. So either no form of detection was provided, it had been provided but had failed to do its job, or the signal had remained at *Danger* and was overrun by the Amsterdam train, as apparently happened at Lichfield (Trent Valley). [580]

An accident occurred at Fürstenwalde, East Germany, on Sunday 24th October 1954. A train was diverted onto the wrong line and a collision occurred. To make matters worse a third train then came into collision with the first two. The casualties in this double collision are unknown. [581]

As railbus N 2903 was leaving Dortmund Hbf station, West Germany, on Saturday 18th December 1954, it was diverted via a wrongly placed set of facing points into the path of incoming children's special train 4224 and a major disaster resulted. Fifteen people died and 71 were injured. [582]

At Wallersdorf, West Germany, on Monday 20th December 1954, the *fahrdienstleiter* had the assistant of an 18-year-old trainee. Cattle train 5654 had been accepted by the *fahrdienstleiter,* who then temporarily left his post for some reason, perhaps to answer the call of nature. During his absence the trainee took it upon himself to clear his signals for the cattle train, but due to his inexperience he mistakenly cleared one of the *Ausfahrsignals* (starting signals). The *fahrdienstleiter* then returned and advised the trainee of his error, whereupon the latter cleared another *Ausfahrsignal* for track 4, on which for empty ballast train Dst 19 905 stood, instead. This second mistake the *fahrdienstleiter* failed to notice, the empty train departed and a collision was unavoidable. The railwaymen were trapped in their engines and six died and a seventh was injured. The trainee was convicted of causing the deaths through negligence, as was the *fahrdienstleiter* for failing to notice his second mistake in time. [583]

In the early morning darkness of Tuesday 26th April 1955 a serious collision occurred in the station at Deodora, Brazil, on the Estrada de Ferro Central do Brasil.

Dr Jair de Oliveira, the director general of the company, issued a statement to the

---

[579] *De Volkskrant*, Amsterdam, the Netherlands, 14th October 1954

[580] *Spoorwegongevallen in Nederland* by R.J. Jongerius, pg. 175 (in Dutch)

[581] Author's own notes, source mislaid

[582] *Katastrophen der deutschen Bahnen Teil II* by H.J. Ritzau, pg. 123 (in German)

[583] Ibid, pgs 123-4

press regarding the accident the next morning: 'Yesterday, at 03:44, in Deodora, electric locomotive No. 2002 was setting back on line No. 5 and propelling two broken down vehicles. At the same time passenger train UM-5 was leaving the station on line A bound for Japeri. A sidelong collision occurred between train UM-5 and the train of vehicles bound for the workshops, as a result, according to the initial investigation, of Driver Luiz Magdalena, in charge of engine No. 2002 running past a signal at danger. As a result of the collision seven people have lost their lives and 38 have been injured, 14 of whom were able to return to their homes after receiving treatment.

'The Central company immediately provided all assistance following the accident, including the services of the Fire Brigade. The injured were taken to the Carlos Chagas, Pronto Socorro, Getulio Vargas and Central do Exercito Hospitals.

'Driver Magdalena, who is on the run, was immediately suspended pending the completion of the inquiry.

'Rail traffic to Nova Iguaçu and Japeri was not affected, Santa Cruz trains were also still running but these suffered delays until line 5 was brought back into use. Traffic was back to normal from 12 noon.'

An newspaper report on the accident spoke of the brilliant work of the fire brigade and their Herculean efforts to rescue passengers trapped in the wreckage, also of the untiring efforts of the railwaymen to clear the lines by dawn and the arrival of large numbers of passengers at the station. Four of those who died were named as Braz Calman de Almeida, 27, single, operator; Sebastião Batista de Silva, 28; Domigos Fernandos Gomez, Portuguese, married, 42; and a child named Lucia, age not stated, daughter of Leonidio Ribeiro de Orvalha (word illegible). The bodies of the other three victims had borne no identity papers and they could not be immediately named. Among the injured was Driver Ciodoaido Siqueira de Souza, 41, of the Japeri train.

An editorial in the same issue of the newspaper as that which carried the report of the accident was highly critical of the Central company describing the cause of the accident as having been an "oversight" by a driver. The editorial asked if it was conceivable that the lives of hundreds of passengers could depend on the lack of attention and discipline of a single employee and stated that discipline on the railway must be imposed with absolute rigour, that the conductor must exercise absolute permanent surveillance of the driver, that the cause of the accident might be considered, frankly, criminal, and so on.[584]

A suburban train of the Estrada de Ferro Leopoldina, bound for Caixas, collided with a train going to Inhomirim which was stopped at a signal in the station at Vicario Geral, Brazil, at 18:00 on Thursday 16th June 1955. The violence of the collision telescoped the coaches of the front train at the exact moment that another train was passing in the opposite direction, and the magnitude of the disaster was thus increased. Five passengers died on the spot and a sixth succumbed after admission to the Pronto Socorro Hospital. Many other passengers were injured and were taken to various hospitals. Meanwhile other passengers, infuriated that the accident had occurred, invaded the station buildings, smashed things, and tried to set the place alight. The only railwayman who could be found was attacked and the stationmaster, Milien Lima (name illegible in newspaper) would also have been attacked had he not fled the scene. In the midst of the uproar one of

[584] *Jornal do Brasil*, Rio de Janeiro, Brazil, 27th April 1955

the passengers allegedly went mad and was picked up by the police.

To gain control of the situation it was necessary to call out the Special Police and the military. A detachment of firemen also arrived and worked to search the wreckage of the three trains for victims as well as dealing with the attempted arson of the station.

The police authorities opened an inquiry to determine the causes of the accident and the railway company sent men to clear the damaged trains from the line, while the local blood bank had used up the greater part of its blood plasma in treating the victims, so that it was forced to make an appeal for donors. [585]

Radio Warsaw reported that nine people had been killed and 48 were injured, 24 seriously, in a collision between a passenger train and a locomotive between Rawicz and Wroclaw, Poland, on Sunday 24th July 1955. [586]

A mixed train carrying passengers, American military aid, explosives, and petrol collided with a freight train 'south of Tehran,' Iran, on Tuesday 16th August 1955. Fourteen people died. [587]

Six people died and 50 were injured in a collision between a diesel railcar and a work train near Managua, Nicaragua, on Wednesday 31st August 1955. [588]

Several infringements of the operating regulations combined with poor visibility due to fog led to a major disaster on the Deutsche Reichsbahn at Bornitz, East Germany, on Saturday 25th February 1956.

The trains which were involved in the disaster were Dresden – Leipzig express D 94, crammed with people bound for the Leipzig Trade Fair, and double-headed through freight train Dg 7137. The mainline on which the collision occurred had been reduced to single track because one track had been confiscated by the Soviets at end of the Second World War, as elsewhere in Germany. (It was finally redoubled about 1968.) The trains normally crossed one another at Riesa but both trains were running late and the *fahrdienstleiter*, who had come on duty at 06:00, decided to alter the crossing to Bornitz. Train Dg 7137 arrived at Bornitz first and entered the loop there but stopped 400 metres in the rear of the *Ausfahrsignal* and foul of a level crossing. The Bornitz *dienststellenleiter* (stationmaster?) who had commenced duty at 06:30, therefore instructed Weichenwärter (Pointsman) Martin Schulz to get the freight driver to pull forward. According to the regulations the *dienststellenleiter* was apparently not authorised to issue orders to the *weichenwärter,* this being the duty of the *fahrdienstleiter*, though as the 'order' only involved moving the train along the siding this seems a little strange. In any case he subsequently informed the *fahrdienstleiter* of what he had done and the latter had raised no objection. What was worse was that before the shunting movement had been reported as completed, the *fahrdienstleiter* cleared the signals for the express to pass through on the main track and there was no protection

[585] *O Estada de São Paulo*, São Paulo, Brazil, 17th June 1955

[586] *Daily News-Record,* Harrisonburg, Virginia, United States, 26th July 1955

[587] *Long Beach Independent*, Long Beach, California, United States, 17th August 1955

[588] *Edwardsville Intelligencer*, Edwardsville, Illinois, United States, 1st September 1955

against a fouling movement from the loop line, such as the provision of a trap siding or points. Problems with its brakes meant that the freight train did not pull forward at once and the *fahrdienstleiter* may have assumed that its driver did not intend to do so at all. However, as train D 94 was approaching the freight train did begin to move slowly forward and either because the driver of the pilot engine mishandled the braking or because he failed to see signal K11 due to the fog and steam obscuring it, his train drifted out onto the main line at 5 to 10 kilometres an hour as the express was passing at 50 to 60 kilometres an hour. The collision occurred at 07:20.

The results of the collision were very serious. Forty-eight people died and 55 were injured, including the driver of the pilot engine of the freight train.

There were various reactions to the accident from railwaymen and the general public. There was naturally great sympathy for the victims of the disaster and calls for a thorough investigation into how it had happened, while railwaymen blamed the lack of a safety siding and the fact that railwaymen were overworked and the equipment was obsolete. There were also rumours that the accident had been the result of an act of sabotage and that the number of dead was really considerably higher than had been stated. [589]

At Leipzig Main Court in mid-1956 of four of the railwaymen involved were convicted of various offences relating to the disaster. Judged to be the main offender because he had issued instructions without the authority of the *fahrdienstleiter*, the *dienststellenleiter* was sentenced to five years' imprisonment. The *fahrdienstleiter*, who had failed to countermand the order, received four and a half years, the driver of the pilot engine received a year and a half and his fireman one year's imprisonment. The poor driver was not in court to receive his sentence as he was still a hospital patient due to the extent of his injuries. [590] [591]

A moment's thoughtlessness on the part of an experienced railwayman led to a major disaster on the Atchison Topeka and Santa Fe Railway (AT&SF) at Robinson, New Mexico, United States, in the early hours of Wednesday 5th September 1956. An 'open switch,' to use the American terminology, diverted *The Chief* passenger train into a head-on collision with a stationary passenger and mail train. The engineer of the mail train, the engineer, fireman and conductor of *The Chief,* a brakeman who was deadheading and 15 dining car employees sleeping in a dormitory car were killed and a passenger, two dining car employees and a railwayman were injured.

Robinson was a long siding (crossing place) on the single line just over 64 miles east of Las Vegas, Nevada. Eastbound train No. 8 was a first-class mail and express train consisting of diesel-electric units No. 41, 41A, 41B and 41C in multiple-unit control, 11 baggage cars, an express refrigerator car, a combination baggage car and coach, a smoking car, and a business car, in charge of Conductor S.A. Rutz, with Engineer L.J. Rush and Fireman Pete Caldarelli. At Las Vegas it had received a copy of train order No. 508, which read:

---

589 https://www-ddr--im--blick-de.translate.goog/jahrgaenge/jahrgang-1956/report/stimmung-zum-eisenbahnunglueck-am-25-februar-1956/?_x_tr_sl=de&_x_tr_tl=en&_x_tr_hl=en&_x_tr_pto=sc

590 *Die Katastrophenszene der Gegenwart* by Hans Joachim Ritzau and Jürgen Hörtstel. Pgs. 172-3 (in German)

591 https://www.lvz.de/Region/Oschatz/Vor-60-Jahren-48-Tote-bei-schwerem-Zugunglueck-in-Bornitz

No. 8 Eng. 41 meet No. 19 Eng. 30 at Robinson. No. 8 take siding.

The train had left Las Vegas at 1.23 a.m., 28 minutes late, and Wagon Mound, exactly 20 miles west of Robinson, at 2.21 a.m., 17 minutes late. It had entered the siding at Robinson and stopped with the leading engine 431 feet west of the east siding switch at 2.54 a.m. Seven minutes later it was struck by *The Chief.*

Westbound first-class passenger train No. 19, *The Chief,* consisted of diesel-electric units Nos. 30C, 30B, 30A and 30, in multiple-unit control, a baggage car, a baggage-dormitory car, two sleeping cars, a dining car, a dome chair-lounge car, a sleeping car, two chair cars, a lunch counter dining car, three chair cars and a chair-observation car, in charge of Conductor W.L. Schriever with Engineer E.O. Foster and Fireman W.H. Adams. It received a copy of train order No. 508 at Raton, New Mexico, and left there at 2.20 a.m., seven minutes late. Springer, just over six miles east of Robinson and the last open office, was passed at 2.55 a.m., one minute late. Travelling at 63 miles an hour it collided with train No. 8 at 3.1 a.m.

As a result of the collision train No. 8 was pushed back 40 feet. The leading three diesel units were derailed, the first was demolished and the second badly damaged. All four diesel units and the first seven cars of *The Chief* were derailed. The diesel units were all badly damaged, the first two cars were destroyed, the third and fourth cars were badly damaged and the fifth and sixth cars were considerably damaged. [592]

Various people told reporters of their experiences, among them Conductor Rutz of train No. 8, who related, "We were stacked up on the siding at 3 a.m., just waiting. It was dark as pitch when the crash came. It knocked my train back about a car length. As far as I could tell it (*The Chief*) wasn't coming very fast."

'John Venelte, a sailor from Pueblo, Colo., en route to San Diego, said he was "damn scared" when he was awakened and realized what had happened.

'V.P. Schmidt, a sailor from Milwaukee, was two cars back of the dormitory car. He said, "I looked out when things settled down and the front end of the car looked like a huge radiator." In the aisle of the car, Schmidt said, "a huge piece of steel stuck right up through the middle."

'T.H. McInnis, of Pinehurst, N.C., en route to a new station at Ft. McArthur, Calif., was knocked out of bed in his roomette and under the seat.

'One of the passengers on *The Chief* was Marjorie Boyd, of Peoria, Ill. She told newsmen: "I was asleep at the time. Our car started swaying and I was knocked out of the seat. All the seats in our car were twisted, but the car didn't turn over." She escaped with a bruised ankle, she said.

'Mrs Lillian Marks, of Chicago, said she was riding in the car next to the one in which the crew members were killed. "They were all lying outside our window dead," she said. "They had to break the window to get us out. I'm so shocked over it." Mrs Marks said she was asleep when the wreck happened. "The train just kept rocking and rocking, back and forth. I never did hear a crash," she said. "It was a horrible thing."

'A passenger in a rear coach on *The Chief* was A.W. Lindenmuth, of Fort Collins, Colo. He was riding in {the} Denver coach, in which no one was hurt. "It felt as if we were going over a couple of ties," he related.' [593]

---

[592] Interstate Commerce Commission report

[593] *The Albuquerque Tribune*, Albuquerque, New Mexico, United States, 5th September 1956

The Interstate Commerce Commission (ICC) noted that when train No. 8 had arrived at Robinson Fireman Caldarelli had opened the west switch to let it into the siding and after it had done so the flagman had alighted and closed the switch. The train had then proceeded through the siding towards its east end, had stopped, and Engineer Rush, Fireman Caldarelli told the ICC, had instructed him to attend to the east switch. He had proceeded towards that switch and the headlight of the front locomotive had been extinguished. He had positioned himself about 25 feet south of the switch. As train No. 19 had approached, at about a quarter mile away his own engineer had sounded the horn several times. Up to this point he had been positive that his own train was in the siding and that the switch had been properly set for train No. 19 to take the main track but the sounding of the horn had confused him and he thought that his engineer was indicating to him to set the switch for the siding. Crossing in front of the now close-at-hand train he had unlocked the switch and reversed it, diverting the oncoming train into the siding.

The other crew members of train No. 8 stated that the brakes of *The Chief* had been applied at the last moment. None of them had noticed the fireman operating the switch. They agreed that their engineer had sounded his horn. It was surmised by the ICC that this had been intended to warn the fireman to move away from the switch.

The ICC noted that the dormitory car in which all the fatalities had occurred, AT&SF No. 1382, was a vehicle of lightweight construction constructed in 1937. It had met the strength requirements for Post Office mail cars at that time and its underframe, end frame and collision posts had been reinforced in 1950 to meet the current specifications of the Association of American Railroads.

It was stated that the rules of the carrier provided that at meeting points the employee attending to the switch must not station himself nearer to it than the clearance point until the expected train had been met. At Robinson the clearance point was 295 feet west of the switch. Had the fireman stationed himself there he would not have had time to reach the switch and reverse it at the last moment.

Fireman Caldarelli had been appointed fireman in 1942 and though acting as such, at the time of the accident he had been promoted to engineer in 1947. [594]

The inquest on the victims of the disaster was held already the day after it occurred. Fireman Caldarelli was of course the principal witness. He told the court, "We pulled into Robinson switch about one minute ahead of schedule. *The Chief* was a couple of minutes late. I stood across the track from the switch about 25 feet. Just before *The Chief* reached the switch, it suddenly occurred to me that there must be something wrong with the switch. I don't know what made me think that. I ran across the track, opened the lock, and threw the switch just as *The Chief* came up. Then there was a terrible crash."

The jury returned a verdict simply that the deaths had been caused, "by reason of a head-on collision of passenger train and mail train AT&SF approximately six miles south of Springer on Sept. 5, 1956." [595]

The fireman thus escaped a possible manslaughter verdict.

The new *Westlander* luxury passenger train of Queensland Railways collided head-on with the *Western Mail* train at Wallumbilla, Queensland, Australia, at 6.14 a.m. on

---

[594] Interstate Commerce Commission report

[595] *Albuquerque Journal*, Albuquerque, New Mexico, United States, 7th September 1956

Saturday 1st December 1956. Five people died in the accident and 13 were badly enough hurt to be taken to hospital. Most of the injured were taken to the local Roma hospital but railwayman Reginald Oehlmann, who was very badly hurt, was taken by ambulance 190 miles to Toowoomba, accompanied by a nursing sister who administered morphine and oxygen to him.

The five who lost their lives were Carol Bowden, 15, of Summer Hope; Roland Ortwell Fehlberg, of Toowoomba; George Raatz, of Injune; Mary Sewell, 13, of Roma, and her grandmother, Mrs Rebecca Smith, of Guluguba. [596] Mary Sewell's mother, Mrs Olive Sewell, who had lost both her daughter and her mother or her mother-in-law, spoke to a reporter: "I was standing by the window of the *Western Mail* looking out. I heard the *Westlander* whistle, then there was a terrific crash. I can't seem to remember anything else until I woke up on Wallumbilla platform." Another passenger, Lyal Grant, said: "There was a smash and I felt myself crashing through a wall. A water tank fell on me and burst. Seats were flying everywhere. I tried to get clear of the wreckage and there was a burning sensation in my left leg. I knew it was broken. Somehow or other I finished up on a pile of wreckage with the other injured people trapped under me."

The two trains, No. 19 Down, the *Western Mail*, and No. 8S Up, the *Westlander*, had been due to cross at Wallumbilla. Train No. 19 had been stationary when train No. 8S had run past the Up home signal at Danger and was still travelling at 20 to 25 miles an hour when it had smashed into the Down train, pushing it back 50 yards. The *Westlander* was the heavier train and was composed of all-steel rolling stock; only a few of its 150 passengers were slightly hurt. All the fatalities and serious injuries were in the mail train.

The departmental inquiry into the accident heard evidence from a number of witnesses, including Matron Alma Reiken, of Wallumbilla Hospital, who said that she had spoken to a man who had introduced himself as Driver William McDougall of the *Westlander*. This man had told her, "It's all my fault, I was not well when I was at Yuleba and should have got off the train there." Yuleba was 15 miles short of Wallumbilla. The matron added that Driver McDougall had also told her that he must have dozed. Before she could speak to him further she had been called to attend to another patient.

Grazier and stock agent David Bassingthwaighte told the inquiry that the *Westlander* had entered the station faster than it had ever done so before. It had not whistled and the (home) signal had been "straight out" against it. [597]

Guard Aubrey Connor of the mail train testified that he had asked Driver McDougall if the signal had been off for him. The reply had been, "I don't know." Guard Conner said that he had also spoken to his colleague, Guard James Phillips, of the *Westlander*, who told him that the signal had been at danger when he had seen it and he had gone to the brakes when he realised that the train was not going into the loop line.

Stationmaster Walter May at Wallumbilla testified that he had run to the engine of the *Westlander* straight away after the collision and addressed Driver McDougall, "Bill, what are you doing here? You've passed signals in the stop position." The driver had replied, "Wal, I must have dozed off." The stationmaster stated that he had been amazed at this admission, though later in the inquiry English Electric Engineer William Wood had explained that locomotives such as diesel-electric engine No. 1200 could be "very

---

[596] *The Argus*, Melbourne, Victoria, Australia, 3rd December 1956

[597] Ibid, 5th December 1956

comfortable," leading to a tendency for drivers to "drop their heads."

By a majority verdict the inquiry found that the primary cause of the accident had been Driver McDougall overrunning the home signal at Danger before he had received a Caution hand signal to enter the loop, and the driver's failure to see the opposing train in time to pull up.

Guard Phillips was found partly to blame for having failed to take prompt action to stop his train when he saw that the engine had passed the home signal at Danger, while Fireman Andrews was to blame for failing to observe the home signal and warn his driver.

The two union representatives on the inquiry board were not happy that the board had declared their members guilty of an offence which might have caused the death of anyone and which might therefore subsequently be the subject of a criminal charge. They agreed, however, that the accident had been due to the *Westlander* passing the home signal at Danger and gave the opinion that something abnormal must have occurred in the cab of engine No. 1200. [598]

Five people reportedly lost their lives when a train collided with a crane in Ludwigshafen, West Germany, on Friday 21st December 1956, killing five people and injuring 20. No more information on this accident has come to light; possibly it was a level crossing collision. [599]

Two electric trains, one crammed with commuters, collided at Woodstock, a fashionable suburb of Cape Town, South Africa, on the evening of Monday 8th April 1957. The trains were both travelling at about 45 miles an hour. According to eyewitnesses the trains approached one another on different tracks but one was diverted into the path of the other when they were about 100 yards apart. According to some reports there were believed to be 26 dead and 70 injured, though the severe mutilation of the bodies made it difficult to determine the exact number of victims. [600] The trains involved were identified as the 4.34 p.m. departure from Cape Town and the 4.14 service from Belville, which collided near the old Esplanade station. [601] By 6 p.m. 19 bodies had been recovered from the wreckage. [602]

At least 20 people were reportedly killed and 30 injured when a freight train collided with a crowded passenger train near Cortazar, Mexico, on the evening of Wednesday 5th June 1957. Local Red Cross officials said that they feared that the death toll would reach 40. [603]

At the beginning of August 1957 the Spanish army conducted manoeuvres at Solsona in the Pyrenees. When these had been concluded three trains were chartered to take the

---

598 *Australian Railway Disasters* by Ken Pearce, pgs. 147-9

599 *The Times*, London, England, 23rd December 1956

600 *Humboldt Standard*, Eureka, California, United States, 9th April 1957

601 https://www.pinterest.com/pin/338684834450919712/

602 *The Times*, London, England, 9th April 1957

603 *Star News*, Pasadena, California, United States, 6th June 1957

soldiers back from Tàrrega to their bases. The first train returned to the Carabanchel camp near Madrid without a problem but the second train met with disaster, probably early on Sunday 4th August (though a contemporary Spanish newspaper states that the accident occurred early on Saturday 3rd).

The second train, carrying 800 soldiers, consisted of two engines, three passenger carriages, three flat waggons loaded with military material, another carrying Jeeps and an unspecified number of waggons loaded with lorries. At 05:10 on the day of the accident the train had almost reached its destination when it came into collision with a light engine performing shunting operations at the junction of the Madrid – Barcelona and Madrid – Extremadura lines at Villaverde. As a result of the collision all three engines were derailed and the three passengers carriages and two following flat waggons were piled up in a heap in a deep cutting spanned by an overbridge.

Occurring as it did close to Madrid, help was quickly on hand after the accident with Army, Red Cross and municipal ambulances, fire engines, the Civil Guard, miliary police and several priests. Once all the injured and the bodies of those who had died had been removed, Army and railway cranes worked to clear the line. [604] [605]

Seventeen soldiers died in the accident and 53 were injured. [606]

How the collision had occurred was not immediately apparent.

Early on the morning of Monday 12th August 1957 a broken rail was discovered on the line between Eindhoven and Best, the Netherlands. As a result the driver of train No. 54 from Eindhoven to Rotterdam was instructed to use the third track, a set of reserve rails which ran alongside the main tracks, as far as Woensel and then the opposing track from Woensel to Best. When train No. 54 reached Woensel it failed to wait at the end of the third track for a train bound for Eindhoven, No. 609, the 05.23 service from Amsterdam to Maastricht, travelling at this point at 110 kilometres an hour. The Nederlandse Spoorwegen trains collided at 07:25.

Driver Engster of train No. 609 jumped from his cab at the last moment. The collision was terrific and the fronts of the trains were crushed together. Fragments of twisted wood and metal were thrown through shattered windows. Eyewitnesses, such as a Mr Dekker, a passenger on train No. 54, said later that there was no panic amongst the passengers, but as soon as the train had come to a halt there was a tremendous jostling to get off. People returning home after a weekend break, workmen and holidaymakers fought to get to the exits, half stunned by being thrown against doors and seats and frightened by the cries and screams of the injured, some of whom were still trapped.

"Suddenly I heard cracking and screaming. I automatically grabbed a rod and suddenly saw ballast coming towards me through a window. It all seemed to happen very slowly. The whole coach seemed to have broken up. Moments later we couldn't see anything through the dust." This was the testimony of A.A. Disseldorp, a KNSM office clerk. (The KNSM was an Amsterdam shipping company.) He had been sat in the leading car of train No. 54. having spent the weekend with his fiancée in the village of Beck van Donk, 23 kilometres by bus from Eindhoven. At the last moment he had jumped on the

---

[604] *La Vanguardia*, Barcelona, Spain, 6th August 1957

[605] *The Times*, London, England, 5th August 1957

[606] Webpage: Federación Castellano Manchega de Amigos del Ferrocarril (in Spanish)

07:15 train from Eindhoven to Utrecht and Rotterdam. "It was not very busy in my coach," he continued. There was only one girl with me in the car. I had bought a newspaper, and had read a bit of it, then I believe I dozed off. Suddenly I heard cracking and yelling, especially from the girl sat opposite me, who started screaming horribly. She clutched me. Yet there was no panic. Someone shouted very loudly, "Keep calm, keep calm." When the dust had settled a little and we could see something, we tried to get out. I was afraid that we would be scalded by steam. When I looked around, I saw a big hole behind the car. That was what I had crawled through. It was a terrible sight. There were wounded everywhere. Help soon came from railwaymen and soldiers from a nearby barracks. The girl who had been sat opposite me had lost her glasses. 'What does it matter?' I told her. Another girl kept shouting that she had lost her scarf, and that while she was bleeding badly from a head injury. The funny thing was that I stayed very calm. I picked up my luggage and did not even forget my newspaper. My first reaction was to go and take some pictures of the accident but when I saw all those people lying about naturally I didn't, of course. I went to help with a crowbar. Then I did take some photographs anyway. The first pictures probably won't show anything. I was still shaking too much." [607]

The inquiry into the collision opened the next day and it was stated that it would concentrate on the order which had been issued by the Eindhoven signalman via the platform supervisor to Driver E. de Boer of train No. 54. The order authorised the driver to run on the wrong line, namely the western track, the left-hand one, from Woensel to Best. As noted above, for Driver de Boer, who had started his journey on track 3, an even more westerly line, this meant driving on the wrong line, track 2, after Woensel signal box, where track 3 ended. The intention was for train No. 54, to be passed at Woensel from track No. 3 to track 2, the line for Eindhoven-bound trains, and then proceed in the wrong direction along track 2 from Woensel to Best. Before train No. 54 could make this movement, however, it had to be held by signal at Woensel until train No. 609, bound for Eindhoven, had gone through on track 2. The accident occurred because Driver de Boer, misled by the order, somehow assumed that it gave him authority to ignore the protecting signal at Woensel, which stood at *Danger* against his train, and he at once took it out onto track 2 to continue towards Best. His train had not quite finished negotiating the connection onto track 2, right in front of Woensel signal box, when it was struck by train No. 609. Signalman F. Foederer, seeing that the trains were about to collide, jumped down his box steps and dived into the woods behind it, fearing that it would be demolished. When his box emerged unscathed he returned to it to be the first person to raise the alarm. [608]

It emerged a day or so later, as the inquiry progressed, that although the arrangement for train No. 54 to run wrong line had been agreed between Signalman R., at Eindhoven and his colleague at Best, Signalman Foederer at Woensel signal box had not been informed over the move. This omission was probably because while the movement was being arranged Signalman R. had ended his shift and been replaced by Signalman V. Signalman Foederer was, in fact, apparently unaware of the approach of train No. 54 until it was close at hand. Pending the conclusion of the inquiry the two Eindhoven signalmen and the platform supervisor were removed from safety-critical occupations and given

---

[607] Het Parool, Amsterdam, the Netherlands, 12th August 1957

[608] *De Volkskrant*, Amsterdam, the Netherlands, 13th August 1957

office jobs. [609]

Seven people were either killed outright in the accident or fatally injured; five passengers and two railwaymen, one of whom was probably Conductor J. Hamstra of train No. 609. The identity of the other railwayman to lose his life is not clear from the newspaper reports; Drivers de Boer and Engster appear to have both survived. Nineteen people, one of them a railwayman, were injured.

Following the accident there was a tightening up of operating procedures. [610]

An extra passenger train from Brno to Olomouc was diverted onto another track through a wrongly placed set of facing points and collided with a stationary passenger or freight train (accounts vary) at Vrbátky station, Czechoslovakia, on the Nezamyslice – Olomouc line, on Wednesday 11th September 1957. Fifteen people died and about 25 others were injured. [611] [612]

A major catastrophe occurred in West Pakistan late on Sunday 29th September 1957. An oil train was standing in Gambar station waiting for a Karachi-bound express to overtake it. Instead, the express, travelling at 45 miles an hour, was diverted onto the line on which the oil train was standing and ran into its rear with disastrous results. The wreckage immediately caught fire and only three passengers were reported to have escaped from the two leading carriages of the express. The diesel locomotive, a luggage van and several carriages were completely consumed by the fire and three other coaches were telescoped and wrecked. An initial report spoke of over 75 dead, the figure was revised upwards to 101 and by 1st October it was stated that 130 bodies had been removed from the wreckage and five of 250 seriously injured persons taken to Okara hospital had died. It was feared that in total there were at least 250 dead, possibly 300. Three relief trains went to the scene and two relief camps were opened.

Miam Jafar Shah, the Minister for Communications, went to the scene, and a signalman was arrested. [613]

No. 45 Up Delhi – Pathankot – Express collided with No. 2 DU Down Ambala – Delhi passenger train at about 4.16 a.m. on Wednesday 1st January 1958 at Mohri station, India, on the Delhi – Ambala section of the Northern Railway. Train No. 45 consisted of steam engine No. 7639 WP and 10 bogies (coaches). Train No. 2 DU consisted of steam engine No. 7773-HPS/2, eight bogies and a four-wheeled parcels van. Mohri was a Class B station with a single platform served by a loop line and a main line. According to the working timetable train No. 2 was due to leave Ambala at 3.40 a.m. and train No. 45 to leave Shahabad Markanda at 3.43 a.m., with train No. 45 taking the loop line at Mohri for train No. 2 to cross it. If, however, train No. 45 was late train No. 2 took the loop line and train No. 45 ran through. On the morning of the accident No. 1 Up train was late, delaying the departure of train No. 2 from Ambala and it did not arrive at Mohri until,

---

[609] De Telegraaf, Amsterdam, the Netherlands, 15th August 1957

[610] *Spoorwegongevallen in Nederland* by R.J. Jongerius, pg. 175 (in Dutch)

[611] *The Times*, London, England, 11th September 1957

[612] Wikipedia – Liste von Eisenbahnunfällen in Tschechien

[613] *The Times*, London, England, 1st and 2nd October 1957

apparently, 4.13 a.m. instead of 4.08 a.m. Train No. 45, meanwhile, was 21 minutes late, and did not leave Shahabad Markanda until 4.03 or 4.04 a.m. The Mohri Assistant Stationmaster consulted Control at Delhi and was instructed to put train No. 2 into the loop line to allow train No. 45 to pass it. Train No. 2 duly entered the loop. Three minutes later when train No. 45 arrived it, too, entered the loop line from the opposite end and collided head-on with train No. 2.

As a result of the collision the front ends of both engines were telescoped. In train No. 45 the first two coaches were reduced to matchwood, the third coach was badly damaged and four other coaches in the train were slightly damaged. In train No. 2 the leading coach was completely splintered and the following coach was so badly damaged that it had to be condemned.

Thirty-one people were killed outright, five died subsequently in hospital and 88 others were injured.

The accident was the subject of a very extensive inquiry by a Government Commission in order to determine whether it had resulted from a mistake made by the Mohri assistant stationmaster or by the driver of train No. 45. Mohri Assistant Stationmaster P.C. Gupta stated that after the arrival of train No. 2 he had been busy obtaining the line clear token for train No. 45 and before he had time to instruct his pointsman, Sham Lai, to lock the trap points for the loop line, set the facing points for the main line and clear the signals for train No. 45, the latter train had arrived and run into the loop. Driver Sadhu Ram and Fireman Ram Sunder of train No. 45, on the other hand, alleged that the Up Home and Up Main to Loop signals had been clear for them and that they had not realised that there was any danger until they had seen train No. 2 standing on the loop in front of them.

The inquiry decided that there was no evidence of sabotage. The evidence of Driver Ram Prasad of train No. 1 and other witnesses was that there was a slight mist but no fog and that the Mohri signals had been easily visible. There was no evidence that the rails had been excessively slippery or that there had been any problems with the track.

Counsel Shri Anthony for the driver and fireman outlined various problems with the Mohri signals, which were mostly failures of the signals to clear due to broken signal wires but there was one of a signal remaining in the *Off* position. The inquiry commissioners decided that the number of failures was not excessive. Evidence as to the possibility of the Up signals having failed at *Clear*, perhaps because the assistant stationmaster had cleared them for train No. 45, then replaced them and failed to notice that they had not returned to *Danger*, was pursued at length but nothing was revealed to show that this had happened. Guard Lachman Das Puri testified that the Up signals had been at *Danger* immediately after the collision.

Assistant Stationmaster Gupta had come on duty at 11 p.m. the night before. He had dealt with six trains before those involved in the accident. There was nothing to suggest that he had been ill, intoxicated, or asleep. Experiments performed also demonstrated that neither the Mohri assistant stationmaster nor the pointsman would have had time in the three minutes which elapsed between the arrival of trains Nos. 2 and 45 to have performed all the necessary operations to allow the Up signals to be cleared. There was also evidence that the trap points at the Shahabad Markanda end of the station had been locked in the open position and the Up facing points had been set for the loop line, precluding the lowering of either the Up Main or Up Loop signals.

From the amount of damage caused and the distance which train No. 2 had been pushed back it was evident that the speed of train No. 45 must have been at least 30 miles an hour at the moment of impact.

There was evidence from some student passengers on train No. 45 that the Mohri signals had been *Clear* but the Commissioners decided that it was contradictory and unreliable. As for the engine crew of train No. 45 Driver Sadhu Ram and Fireman Ram Sundar alleged that the Up Outer Home and Loop signals had both been *Clear*; the other fireman, Bachan Singh, said that he had not seen either. Other witnesses, however, denied that the signals had been cleared or that any signal levers had been pulled over. The commission decided that the evidence as to the signals having been lowered was limited to the interested evidence of the driver and fireman and the unimpressive statements of the student passengers. Against this was the lack of time for the signals to have been cleared and the position of the trap and facing points. There was no evidence that the signals had failed to return to *Danger* after the passage of train No. 1.

There was evidence that train No. 45 had been travelling at excessive speed when it entered the loop, as there was a speed restriction of 10 miles an hour over the connection, and that the brakes had only been applied at the very last moment.

The Commissioners concluded that Assistant Stationmaster Gupta had acted in accordance with the regulations and that Driver Sadhu Ram had overrun signals, while Fireman Ram Sander had failed to keep a proper lookout while not otherwise engaged.

The Commissioners suggested a good number of recommendations:

(a) Punishments for offences such as overrunning signals should fit the offence, should be uniform and should be issued as soon as possible.

(b) With regard to locomotive inspections they should be made more frequent and should involve a surprise element.

(c) There should be a greater follow through on circular orders.

(d) Rules should be expressed in simple language, be unambiguous and issued in regional languages as well as English. They should be kept under constant review. Subsidiary Rules framed by zonal railways should be brought to the attention of other zones.

(e) Standard III signalling should be introduced where traffic required it. {Standard III signalling involved having all points and signals interlocked with one another, mechanically or electrically, Distant (Warner), Outer and Inner Home, and Starter signals worked from one or two cabins with all points operated centrally or co-ordinated between the two cabins. Loop lines were isolated from the main lines by sand humps, trap sidings, trap points or derails. In the case of points too distant to be directly worked mechanically keys were to be provided interlocked with the block instruments.} [614]

(f) Sand drags should be installed instead of trap points or loops extended to permit the simultaneous acceptance of trains.

(g) The visibility of signals in bad weather should be improved, perhaps by the use of yellow reflecting beads.

(h) Research into the provision of an automatic warning system to alert the driver of adverse signals should be pursued.

---

[614] https://www.irfca.org/faq/faq-signal4.html

(i) Practical experiments should be carried out to determine the effect of weather on braking distances.

(j) Every passenger train should have a luggage van next to the engine.

(k) Anti-telescoping steel coaches should be marshalled next to the luggage vans where possible.

(l) Outer signals should be provided with a double wire and pulley arrangement to minimise the likelihood of wrong-side signal failures.

(m) Signals should be visible from station lever frames.

(n) All passenger engines should have speedometers, preferably manufactured in India.

(o) Drivers and guards needing glasses for both short and long distances should have glasses with bi-focal lenses.

(p) The headlights of all stationary trains should be kept on in the dimmed position.

(q) Signal glasses and lamps should be kept in good condition.

(r) The number of blankets, sheets and lanterns carried in relief trains should be increased.

(s) The number of breakdown cranes and their capacity was unsatisfactory and this matter should be rectified. [615]

As No. S241 Up local passenger train was leaving Sonanpur station, India, for Calcutta (now Kolkata) on Wednesday 26th February 1958 and was passing from No. 3 line across the Down Main line to the Up Main line, it was run into by No. S379 Down local passenger train coming from Calcutta. The engine of the Down train struck the fourth coach of the Up train and was derailed, while the third, fourth and fifth coaches of the up train were damaged. Five people lost their lives, 47 were injured, and about 44,100 rupees of damage were caused to the engine, rolling stock and permanent way. The accident was blamed on non-observance of the procedure for passing defective signals. [616]

Brazil suffered one of the worst railway accidents in its history on Thursday 8th May 1958, though an initial estimate of perhaps 300 dead proved to be too pessimistic. The accident was a head-on collision in the suburbs of Rio de Janeiro, between Mangueira and Triagem, involving two crowded electric commuter trains. Dr Guilherme Romano, the Rio municipal health director and the person who had theorised that there might be 300 fatalities, added when he reached the disaster scene, "There are corpses everywhere. I've never seen anything like it and I can't find the words to describe it."

The trains involved in the accident were incoming train No. UA-68 bound for the former Central do Brasil terminus, driven by Motorman Damiâo Inácio and outbound train No. UA-71 from the terminus bound for Belfort Roxo, driven by Motorman Ferminio Teixeira da Sousa. They collided at 18:17, at the height of the evening rush-hour. Train UA-71, travelling at 60 kilometres an hour, penetrated train UA-68 for a distance of 20 metres.

A reporter wrote that at midnight rescuers were using oxy-acetylene torches to cut into

---

615 https://archive.org/stream/in.gazette.e.1963.699/E-1868-1963-0149-84568_djvu.txt

616 Report on Indian Railways 1957-1958 Vol. 1

the wreckage of the two trains and remove more injured and corpses, and it was expected that it would take six more hours to release all the bodies. Thousands of people had flocked to the site 'where piles of bodies offered a spectacle of horror,' and President Juscelino Kubitschek and Cardinal Jaime de Barros Câmara, the Archbishop of Rio de Janeiro, also went to the scene and then to the injured in hospital. Meanwhile, responding to radio calls for blood donations, large numbers of men and women, teenagers and even children went to the city's blood bank, and finding its doors closed threatened to break in until officials arrived and opened them. 'In the early hours of the following morning a large part of the population remained awake following the unfolding of the tragedy in the hospitals, at the mortuaries or on the radio.' In the wreckage of train No. UA-71, 'Among the mutilated bodies that were piled up on all sides of the car and amid the frantic screams of dozens of men. Women and children crying for help came a calm, confident voice. It was the commanding voice of a 40-year-old man who was alone, half lying in the broken carriage, side by side with the screams of the wounded and the silence of the dead. Little by little the firemen who heard his voice approached, his calm voice directing them to where he was, cutting their way through the wreckage until finally he was released.'

Four ambulances had arrived from the Sousa Aguilar Hospital soon after the collision, to find dead bodies and wounded lying on the ground and in the cars, with blood running everywhere. They were shortly followed by firemen from various stations under the command of Colonel Sousa Aguilar (whose name the hospital presumably took). The colonel told the *Jornal do Brasil* newspaper that one car of the train had gone right inside the other like a box of matches entering its casing. President Kubitschek described the scene as 'a bit Dantesque.' Sr Renato Feio, Director General of the Rede Ferroviária Federal, the state railway company, told the president, misleadingly, 'Don't be scared, Sr President, there aren't many dead, the cars carry at most 12 people.' Leaving the scene the president told a newspaper reporter that it was a heart-breaking spectacle. The State would compensate the families of the victims and he would demand that the railway company take measures to avoid a repetition of the disaster.

Mayor Negrao de Lima told reporters that the city was in mourning. It was worse than the Paciencia disaster and Dr Miguel Vasconcelos, who had previously assisted in eight other accidents on the (former) Central do Brasil lines, said that this was worse than any of them.

The first rescue work was undertaken by residents of Morro de Mangueira and was impeded by darkness and steady rain. When the firemen arrived, with floodlights powered by generators, the rescue operation became easier and more organised. The first injured were taken to the Sousa Aguilar Hospital but it soon became clear that there were too many victims to all be taken there and all the hospitals in the area were mobilised. Later the army was mobilised to assist the medical staff, firemen and police. A number of surgical operations had to be carried out at the scene with the aid of small lanterns, including the amputation of the right arm of Motorman Inácio of train UA-68, who was trapped in his cab until 23:00, but remained lucid and calm. Motorman Teixeira da Sousa was among the dead.

A newspaper the following day contained a statement from the public relations department of the railway company regarding the cause of the disaster:

(1) Francisco Sá signal cabin, which controlled the signal from the station yard there, had advised cabin No. 2 at Derbe Clube, that he was going to send train No. UAW-1, an empty coaching stock, for a shunting move.

(2) Cabin No. 2 had agreed to the shunting move and cleared the signal for it.

(3) Instead of releasing train UAW-1, Francisco Sá signal cabin had released passenger train UA-71 for its normal run.

(4) Train UA-71, driven by Motorman Ferminio Teixeira da Sousa, had wrongly been driven past the shunting signal and had continued on line 2, the opposite running line, without having obtained the special licence required to do so, and as a result had collided with train UA-68, driven by Motorman Damiâo Inácio, between Mangueira and Triagem, the latter train having stopped at signal A-53, near the San Francisco Xavier bridge.

(5) The collision had occurred at 18:17. The high number of deaths and injuries was regrettable.

(6) This information was in accord with the data obtained in the first 24 hours and was subject to correction following later more accurate information. [617]

The following day Public Relations Chief José Joaquim Andrade Sobrinho announced that from the initial findings of the inquiry Motorman Ferminio Teixeira da Sousa of train UA-71 and Signalman Joåo Cosme dos Santos at Francisco Sá were jointly responsible for the accident. Assistant Signalman José Gonçalves at Francisco Sá and Signalman Henrique Rodrigues Alves, of Derbe Giube, has also been questioned at the inquiry.

The exact circumstances which had led up to the disaster were explained to a newspaper reporter:

Francisco Sá station, which served Rio Douro and Linha Auxiliar trains, was congested with trains, including UAW-1, which had broken down. As there was no free platform for it incoming train UA-68 was stopped at signal A-53. The signalmen at Francisco Sá and cabin No. 2 (Derbe Clube) had therefore agreed over the telephone for train UAW-1 to run from Francisco Sá to San Pedro II, where it could be repaired, via Auxiliar line 1 to shunting signal No. 4, where it would pass onto Auxiliar line 2. Train UAW-1 had then set off but due to a temporary problem with points No. 43 it could not proceed to San Pedro II. It had then returned to Francisco Sá via Auxiliar line 2. The problem with the points was soon fixed and cabin No. 2 had advised Francisco Sá that it was now possible to send train UAW-1 again. The signalman at Francisco Sá had said he would do so, but he would first send train US-27. This was done and train US-27 had proceeded onto Auxiliar line 1. Train UAW-1 should then have followed but, instead, the signalman at Francisco Sá had released train UA-71 from another platform. When train UA-71 had arrived at shunting signal No. 4 indicating that it was to cross to Auxiliar line 2 it should not have proceeded any further with authorisation from the cabin No. 2 signalman, involving the issue of a "TT5" form, but instead of stopping it had continued onto the opposite running line without permission and had increased speed until it collided with train UA-68 near the San Francisco Xavier bridge.

That Motorman Ferminio Teixeira da Sousa was responsible was disputed by his widow, Dona Dulcelina, who told a newspaper that her husband had been with the railway for 30 years, did not smoke or drink, and had received previous commendations in connection with a potential runaway at Barro do Pirai and for stopping his train short of an obstruction. She blamed the signalmen in cabin No. 2 who she alleged had caused the collision through being drunk on duty.

Signalman Joåo Cosme dos Santos was also interviewed by a reported at Francisco Sá

---

[617] *Jornal do Brasil*, Rio de Janeiro, Brazil, 9th May 1958

station, where he was being guarded by two members of the Military Police for his own protection, which was doubtless a wise precaution, as there had been rioting after the disaster and as the only surviving railwayman likely to be to blame he would have been an obvious target. He, too, blamed Chief Signalman Henrique Rodrigues Alves at cabin No. 2. He agreed that he had sent train UAW-1 to San Pedro II for repairs, as it had not been in a condition to run, having agreed with his colleague to do so. He had then let train UA-71 go and had not advised his colleague that it was preceding train UAW-1, as it was not customary to do so.

Engineer Jorge de Abreu Schilling, a former director of the Estrada de Central do Brasil, blamed the disaster on the fact that there was no automatic signalling installed on the section of line on which the collision had occurred. [618]

The continuing inquiry heard evidence in private from Signalman Joåo Cosme dos Santos and Assistant Signalman José Gonçalves at Francisco Sá, Signalman Henrique Rodrigues Alves, Signalman Teófilo Borges and Assistant Linesman Jadir Teixeira, who had corrected the fault with No. 43 points, at cabin No. 2. The inquiry was conducted by company engineers Oton Sousa Novais, Gilson Cruz and Dante Di Ulio. [619]

As a result of the disaster President Kubitschek dismissed three railway company officials: Acting Superintendent Alberto Whateley and directors Teixeira Lobao and Heitor Guimaraxes.

The exact final death toll in the disaster remains unclear. The last mention of the disaster in a Brazilian newspaper, on 14th May, states that 115 victims had been interred while four bodies remained at the mortuary because they had not been identified. Several American newspapers, however, state that there were 131 dead, plus 300 injured, 40 seriously. [620] [621]

Seven people died and 40 were seriously injured in a collision between two passenger trains near Cairo, Egypt, on about Friday 14th November 1958. [622]

A mail train collided head-on with a stationary goods train in the station at Laksham, East Pakistan, on Wednesday 19th November 1958. The engines of both trains were derailed and overturned, three coaches of the mail train were smashed to pieces and three waggons of the goods train were telescoped. Sixteen people were killed and 36 were injured. Relief train were sent to the scene and the army also sent medical aid. [623]

A serious accident occurred in Chile at an unstated date in March 1959 on the branch between Los Sauces and Lebu, at kilometre 38, between Purén and Nahuelbuta stations. The victims were platelayers pushing or riding on two trolleys, which were struck by engine No. 574, in charge of Driver Eduardo Alegrua, which had come from Purén. Six men were killed outright or died shortly afterwards, ten were seriously injured and 18

---

[618] Ibid, 10th May 1958
[619] Ibid, 13th May 1958
[620] Ibid, 14th May 1958
[621] *The Republican Courier*, Findlay, Ohio, United States, 10th May 1958
[622] *The Times*, London, England, 15th November 1958
[623] Ibid, 20th November 1958

were slightly hurt. A 13-year-old boy, who was not a member of the gang of workmen, was also somewhat hurt. The precise cause of the accident was not immediately clear but a newspaper report suggested that the Purén and Nahuelbuta stationmasters were responsible. [624]

A freight train collided with a stationary passenger train at a wayside station near Chittagong, East Pakistan, on Friday 24th July 1959. At least five people were killed and 15 were injured. [625]

Morés station, Spain, was the scene of a collision on Monday 24th August 1959. This was a crossing place on the single-track Madrid – Barcelona main line and freight train C I 2, which consisted of an engine and 27 waggons, was stopped in the station awaiting the arrival of train No. 813, a Barcelona – La Coruña express. The latter train consisted of an engine and eight coaches. As it arrived at Morés, on time, its driver, Venacio Sanz Arribas, and his fireman, José María Acero Millán, saw that the facing points were wrongly set. The brakes were immediately applied but a head-on collision with the waiting freight train could not be avoided. The braking reduced the force of the collision but even so the two engines, a third-class coach of the express and five waggons of the freight train were piled up in a heap of wreckage. Five people died and 38 were injured, two seriously. The driver and fireman of the express were among those who were slightly hurt, the crew of the freight train had been standing alongside their train awaiting the crossing and escaped injury.

Railwaymen and local people performed the initial rescue work and relief trains were sent from both Madrid and Barcelona. The ten most seriously hurt were taken to hospital at nearby Calatáyud. The accident happened at 08:38 and the express reportedly resumed its journey after only two hours, with a fresh locomotive and rolling stock. [626]

Seven people lost their lives and at least six were injured in a collision on Wednesday 6th January 1960 near Tlanalapan, Mexico. A passenger train bound for Mexico City reportedly failed to obey signals and collided with a special train. [627]

A very serious collision occurred at Leipzig Hbf, East Germany, on Sunday 15th May 1960. Fifty-four or possibly 59 lives were lost and another 206 were injured. At about 20:20 on the day of the disaster express freight train Sg 5556 from Halle arrived on track 9 and was routed across track 11 to track 15 in the station yard. At about 22:26 signal box W 7 received notice from temporary signal box B 3 of the approach of passenger train P 466 to Halle, due to leave on track 11 at 20:28. Meanwhile railbus E 237 from Halberstadt was due from the Halle direction on track 9 at 20:39, and like train Sg 5556 had to cross track 11 to reach track 12 in the station yard. At signal box W points No. 268 allowed a trailing movement from track 9 to track 11 and were interlocked with signal K 11 on track 11, but due to a fault signal K 11 was held at danger even though points No.

---

[624] http://chile-catastrofes-tragedias.blogspot.com/2014/06/blog-post_12.html

[625] *Wisconsin State Journal*, Madison, Wisconsin, United States, 25th July 1959

[626] *La Vanguardia*, Barcelona, Spain, 25th August 1959

[627] *Sandusky Register*, Sandusky, Ohio, United States, 7th January 1960

268 were correctly set, so train P 466 came to a stand at the signal. The driver of train P 466 was given an Ab order, authorising him to pass signal K 11 at danger, and he set off. However, just beyond the box points Nos. 262 and 263, for facing movements from track 9 to tracks 11, 12 or 15, had not been reset after the arrival of train Sg 5556 and train P 466 was diverted onto track 9 in the wrong direction, as it began to quickly accelerate. The *signalwärter* (signalman) saw what was happening but instead of warning his colleague ahead in box 10, 500 metres away, via the telephone, he told the man in box B3, which train P 466 had, of course, already passed. Presumably he had lost his head. Trains P 466 and E 237 collided head-on on track 9 at a curve near his home signal.

The apparent failure of the points which had locked signal K 11 and caused a shrill alarm to sound in the box was traced to a battery which had run down due to a blown fuse, something which could have been easily remedied. Having failed to trace the problem and assuring himself that points No. 268 were correctly placed, the W 7 signalman had assured the *fahrdienstleiter* (head signalman) in box B 3 that there was no reason for train P 466 not to pass signal K 11 at danger, forgetting that the signal was also locked because he had forgotten to reset the road after letting train Sg 5556 across track 11. The B3 *fahrdienstleiter* had not queried the matter any further and his assistant had issued the Ab order to train P 466. The accident was thus the fault of the men in both boxes and afterwards each attempted to lessen their responsibilities. In a further violation of the regulations the driver of train P 466 had also taken both copies of the Ab order, rather than one being given to the conductor, who had been in the van at the rear of the train.

The driver of train P 466 was, however, not charged with having caused the accident. The order that he had been given did not authorise him to run on the opposite running line, but possibly he had not studied it carefully and had assumed that it did. The other railwaymen, all signalmen, received stiff sentences in the subsequent trial of those involved: the W 7 signalman was sentenced to 15 years' imprisonment, the W 7 *weichenwärter* (pointsman) to 12, the B3 head signalman 10 years and his assistant 8 years. The sentences were founded on violations of the regulations and attempts by those involved to cover up such violations. These sentences were, however, reduced by a higher court to 5 years' imprisonment for the W7 signalman and pointsman, 3 years for the B3 head signalman and 2 years for his assistant. [628]

Thirteen passengers died and about 40 were injured in an accident to a passenger train at an unstated location north-east of Saigon, South Vietnam on Wednesday 20th July 1960. As the train was descending a steep incline it became divided. The front portion of the train came to a stand and the rear portion collided with it, two coaches overturning. [629] [630]

Six people died and 18 were injured in a collision between an express train and a goods train at Martin Brod, Bosnia, Yugoslavia, on Thursday 25th August 1960. [631]

---

[628] *Katastrophen der deutschen Bahnen Teil II* by H.J. Ritzau, pgs. 290-3 (in German)
[629] *Corpus Christi Times*, Corpus Christi, Texas, United States, 20th July 1960
[630] *The Straits Times*, Singapore, 21st July 1960
[631] *The Times*, London, England, 26th August 1960

CHAPTER 6

SINGLE LINE COLLISIONS

The operation of trains in opposite directions over a single line of rails carries the obvious danger of head-on collisions between them. By 1946 there were many hundreds of miles of single-track railways throughout the world, especially in places where the density of traffic did not justify the provision of a second line. Most of them were worked with the aid of the electric telegraph, with decisions as to the meetings between trains being either made locally or, as in the United States, by a central dispatcher. In Great Britain, an improved method of single-line control had been brought into use in 1880 after the invention by Edward Tyer of the electric train tablet system and by the mid-1940s most lines were worked by this method or its variants (electric train staff and key token), though these systems were not used extensively outside of the British Empire. The invention of a failsafe track circuit in 1872 by the American William Robinson [632] had led to the introduction in the United States of automatic block signals, by which train movements on a single line could be safeguarded by signals warning of the approach of an opposing train, but the cost of installing such signals was high and as late as 1906 the Interstate Commerce Commission was stating that out of 194,727 miles of railroad there were still only 41,916 miles protected by them. [633] With the protection of trains running over a single line left solely to the exchange of telegraph or telephone messages, rather than by the more secure electric train tablet or its later equivalents, and not supplemented by some form of automatic signalling. collisions due to human error in the transmission or interpretation of telegraph messages were all too common.

A collision at Ravenstein,* the Netherlands, at 01:00 on Saturday 26th January 1946 was apparently the fault of the driver of a British leave train, en route from Germany to England via Roosendal. The leave train, No. RF 39, was booked to cross freight train No. 4507 at Ravenstein but its driver ran through the station and back out onto the single line where it collided with the freight train head-on outside of the *inrijsein* (home signal). Three coaches of the leave train were wrecked, seven waggons of the goods train were crushed or derailed, five British soldiers were killed and ten were more or less injured. [634] [635]

---

632 Wikipedia – Track Circuit

633 Wikipedia – Automatic Block Signalling

634 *De Tijd*, Amsterdam, the Netherlands, 26th January 1946

635 *Spoorwegongevallen in Nederland* 1839-1993 by R.T. Jongerius (in Dutch)

*(This is the name given in contemporary newspapers. A Dutch book on railways accidents gives the name as Ravenstijn.)

At 21:25 on the evening of Saturday 28th September 1946 a Veynes – Briançon railcar and Briançon-Veynes mixed train MV 9724, hauled by a Class 242 DT steam locomotive, came into head-on collision between Veynes and Montmaur, France. Nine or ten people died according to different reports, including the motorman of the railcar and both train conductors, while 12 were seriously and 20 slightly injured. The accident was due to a misunderstanding on the part of the railcar motorman, who had interpreted a hand signal for him to carry out a shunting manoeuvre as a signal to proceed and had then left Veynes station before the mixed train arrived. [636] [637]

A single-line collision in England on the Great Western Railway near Hallen Marsh Junction on Tuesday 18th February 1947 was similar to that which had occurred near Great Holland signal box back in 1931 inasmuch as one train had taken the wrong line altogether at a junction.

Hallen Marsh was the point where the double-track line from Stoke Gifford to Avonmouth was joined by a single-track branch to Pilning, which provided a more direct route, though it involved a reversal, for trains from South Wales via the Severn Tunnel to Avonmouth.

On duty in Hallen Marsh Junction on the early morning of the accident was Signalman J. Wheeler. At 3.03 a.m. he had accepted the 1.15 a.m. Cardiff – Avonmouth freight train from Pilning Branch box on the branch and it had duly passed his box at about 3.45 a.m. After it had gone he had replaced his signals behind it but he omitted to replace No. 14 points leading from the Up Main to the Down Branch so that these lay for the main line. At 4.37 a.m. he accepted another Cardiff – Avonmouth freight train from Pilning Branch box, this one a special which had left Cardiff at 8.45 p.m. the previous evening. He received the *Train Entering Section* signal for it at 5.10 a.m. Then, at 5.18 a.m. Holesmouth Junction, the next box on the Avonmouth side, offered him a main line train, the late-running 9.20 p.m. Avonmouth – Salisbury freight train, which he obtained *Line Clear* for from Henbury on the Stoke Gifford line and then accepted from Holesmouth Junction. Signalman Wheeler then attempted to clear his signals for the Salisbury train to run through the junction before he let the Pilning train off the branch but found that, despite repeated attempts to do so, he was unable to clear his Up Starting signal (No. 64). This signal was controlled by a track circuit, the indicator of which plainly showed *Clear*, and perhaps because he was tired towards the end of his night shift he wrongly concluded that he could not clear the signal because of a track circuit failure. It did not occur to him that the real reason that he could not clear it was that he had failed to reset No. 14 points.

The Salisbury train was now nearing Hallen Marsh box. Signalman Wheeler lowered his Home signal, brought the train up to his box and instructed Driver N.A. Connett that the Starting signal had failed and that he was in order to pass it at *Danger*. He then lowered his advanced starting signal, exhibited a green light to the driver and sent the *Train Entering Section* signal to Henbury box. Watching the train restart, he had then

---

[636] Wikipedia – Liste des principaux accidents ferroviaires
[637] *Journal de Genève*, Geneva, Switzerland, 30th September 1946

gone to book the times in his train register. Having made some entries in his register he had looked up to see that the Salisbury train was going down the branch towards Pilning rather than up the main line. All that he could do at that point was telephone Signalman Elson at Pilning Branch to tell him that the Salisbury train was coming towards him. He then advised Avonmouth Control what was happening and whilst he was doing so he heard the noise of the collision.

Driver Connett's account of the events leading up to the accident, given at the Ministry of Transport inquiry, agreed with that of Signalman Wheeler. He admitted that he had not carried out the provisions of Rule 81 (l) and checked that the facing points (No. 14) were correctly set for the main line, nor had he asked his fireman to do so. Having seen that the Up Main Advanced Starter was off for him he had taken it for granted that he was on the main line and had not realised that he was going down the branch until he had seen the headlight of the Cardiff train at the last moment. He had not had time even to close the regulator and his train was still travelling, he estimated, at 25 miles an hour when it hit the Cardiff train.

Fireman K.E. Lovell confirmed his driver's evidence. He testified that after leaving Hallen March he had been busy firing in preparation for Henbury Bank and had not seen the Advanced Starting signal. A little later he had looked out for the Blaise Up distant signal (Blaise box was switched out) and had seen the headlight of the Cardiff train, which he had wrongly assumed was a train on the Down Main line. He had then resumed firing and the collision had shortly occurred, burying him in coal.

Guard F.J. May of the Salisbury train said that he had been busy making up his journal after leaving Avonmouth and he could throw no light on the accident.

Driver E.T. Colcomb of the Cardiff train stated that Signalman Elson at Pilning Branch had handed him the staff for the single line and he had left there at 5.15 a.m. Going slowly along the branch, his engine running tender first, he had passed Severn Beach station and box (also switched out) when he had seen the headlight of an approaching train. There had been time to close the regulator, but not to apply the brakes or sound the whistle before the collision.

Fireman E.V. Burrows of the Cardiff train testified that his first warning of an impending collision had been his driver shouting, "Look out." He was credited with having been largely responsible for rescuing the injured men of both engines and obtaining assistance.

Guard J.A. Parry estimated that the Cardiff train had been moving at 12 miles an hour when it had stopped abruptly. Afterwards he had gone back to Severn Beach box and contacted Signalman Elson at Pilning Branch from there to report the collision.

The Avonmouth – Salisbury train had consisted of a 6800 class 4-6-0 steam engine, 44 waggons, all but one loaded, and a brake van. The Cardiff – Avonmouth train had consisted of a 9000 class 4-4-0 steam engine, 48 empty waggons and a brake van. As a result of the collision both locomotives were derailed and damaged, 26 waggons were derailed and a number of them were smashed. All six railwaymen were injured; five of them were taken to hospital but were released the same day. The branch was blocked until Thursday evening.

Brigadier C.A. Langley, the inspecting officer, not surprisingly found Signalman Wheeler principally to blame for the accident for having accepted the Salisbury train with the facing points wrongly set and for having advised its driver to pass signal No. 64 at

Danger. The signalman was, he remarked, an intelligent, keen, young man of 22, with 3½ years' service as signalman, six months of which had been at Hallen Marsh. Admittedly the night had been very cold, but there had been a stove with plenty of coal in the box, which had been lit by two Tilley paraffin lamps. The inspecting officer considered that a more experienced signalman would probably not have made the same mistake.

Driver Connett must share some of the responsibility for having failed to notice the incorrect setting of the facing points and having failed to keep an adequate look out. Had he kept a proper look out for the Advanced Starting signal he would have realised that his train was going away from it rather than towards it.

It would have been prudent of Fireman Lovell to have kept a look out until they had cleared the junction, although he had naturally been anxious to make up a good fire for the climb of Henbury bank.

Guard May of the Salisbury train and the crew of the Cardiff train were not to blame.

Brigadier Langley made no recommendations, commenting that 'hasty judgement on the part of a young signalman, coupled with carelessness on the part of an experienced driver, brought about this accident, which could have been prevented if either of them had exercised ordinary attention.' [638]

The Chemin de fer du Sud Est de la Suisse or Südostbahn (SOB) suffered a serious collision just after 17:00 on Saturday 26th July 1947 between Biberbrücke and Einsiedeln, Switzerland. Two regular trains were involved: No. 84 from Einsiedeln to Wädenswil, consisting of a motor unit and a trailer and No. 183 from Biberbrücke, consisting of a motor unit, two passenger trailers and a van. Both trains were reportedly travelling at between 50 and 60 kilometres an hour when they collided on plain track. The two motormen's cabs were destroyed in the violent collision. The single trailer of train No. 84 escaped serious injury but the two trailers of train No. 183 were seriously damaged.

Ten people were killed outright in the accident, including three SOB employees: Motormen Kleiner and Stähli and Mechanic Eugène Spahr. Sixteen people were badly enough hurt to be taken to hospital and a number of others were slightly injured. After the accident a rescue operation was quickly organised from Biberbrücke station and doctors, firemen and police went to the scene, whilst the Chemins de fer fédéraux suisses (CFF) supplied a rescue train. The line was reopened to traffic on Sunday afternoon. [639]

The collision apparently occurred in the following manner. Einsiedeln station reported the departure of train No. 84 to Biberbrücke and the message was acknowledged by one of the railwaymen on duty there, Hans Eschmann. Shortly afterwards Eschmann forgot that he had done so and gave permission for train No. 183 to depart towards Einsiedeln. Becoming aware of what he had done he had then desperately tried to send a message to a platelayer who lived alongside of the railway to stop one of the trains, but his telephone call had come too late, as the train had already passed that point. He had also tried to use a car to prevent the accident but all his efforts had been in vain. Eschmann was arrested after the accident. A married man and the father of two children, he had 17 years' railway service and a clean record. The investigation into the accident was quickly closed and he

---

638 Ministry of Transport report

639 *Gazette de Lausanne*, Lausanne, Switzerland, 28th July 1947

was released again. [640]

On 5th August M. Zuber, traffic director of the company, gave his version of the cause of the accident to journalists. He stated that Agent Eschmann was one of three employees at Biberbrücke and was considered completely reliable.

Eschmann's act of forgetfulness had probably been related, the director said, to his having been distracted at the time that he had prematurely let train No. 183 go by the arrival at that moment of a goods train which had been reported to him, coupled with the need to arrange a connection at Einsiedeln with a delayed service. It had then taken him a minute or two to perceive his error by which time it had been too late to stop the two electric trains by arranging for the current to be isolated.

M. Zuber added that the two trains had been very solidly constructed and that the number of fatalities and injuries would not have been so high had not most of the passengers chosen to sit close behind the motormen so as to get a better view of the passing scenery. [641]

Two days after the surrender of the Japanese occupying forces at the end of the Second World War, in August 1945, the Dutch East Indies declared independence as Indonesia. However, like the French in Vietnam, the Dutch wished to re-possess their colony and administer it in the same way as they had done before the war. Four years of conflict then followed before the Dutch finally recognised an independent Indonesia at the end of 1949. On the main islands of Sumatra and Java the Dutch were able during the conflict to control the cities and major towns but not the villages and countryside.

On 20th July 1947 the Dutch launched Operation Product, a military offensive designed to destroy the Indonesian Republican forces and enable the Dutch to regain control of the natural resources of Sumatra and Java. [642] This offensive involved driving the Republican troops out of parts of Sumatra and East and West Java. It was during this advance, as Dutch troops swept into Pematangsiantar, Sumatra, 50 miles south of Medan, that a major railway catastrophe occurred on Tuesday 29th July 1947. Terror stricken passengers, desperate to escape the advancing Dutch, forced the driver of a train to proceed from Pematangsiantar along a single line, although warned that a train was expected from the opposite direction; their train had not progressed very far when the predictable collision happened. Dutch sources stated that over 400 people were killed and an unknown number had been injured. Several days later 20 wrecked carriages littered the right of way and many dead bodies remained unburied. [643] [644]

At 21:05 on Friday 12th December 1947 passenger train 42-13 bound for Gannat came into collision with a Clermont Ferrand – Montuçon railcar on the single line at Pontmort, France, three kilometres from Riom. Following the collision the railcar burst into flames. Firemen rushed to the scene from Clermont Ferrand and joined police and volunteers in a torchlight search for bodies. Seventeen people reportedly died in the collision and at least

[640] *Journal de Genève*, Genève, Switzerland, 28th July 1947
[641] Ibid, 5th and 6th August 1947
[642] Wikipedia – Indonesian National Revolution
[643] *Algemeen Dagblad*, Rotterdam, the Netherlands, 4th August 1947
[644] *The Montana Standard*, Butte, Montana, United States, 4th August 1947

20 were taken to hospitals in Clermont Ferrand and Riom. [645] [646] Both trains were running late. There had been 40 passengers on the railcar. [647] The collision was blamed on a signalling error.

France suffered another, more serious, single-line collision on Tuesday 17th February 1948, this time in the Nord region of the country on the line of the Compagnie générale de voies ferées d'intérêt local (CGL) between Douai and Pont-à-Marcq. Just after 17:00 on the day of the accident a passenger train arrived at Thumeries station, where it filled with workers from the adjoining Béghin sugar refinery, which employed 3,000 people, mainly young women. The station work having been completed, Stationmaster Dufour gave the train the departure signal at 17:13, and it set off towards Douai, without having waited for the arrival of a freight train coming from Douai, which was carrying 500 tonnes of coal for the refinery. Like his counterpart at Biberbrücke in Switzerland the previous year the stationmaster had almost immediately realised what he had done and he telephoned the next station at Moncheaux in the hope of stopping the freight train, but it had left. He also sent a cyclist after the passenger train in the forlorn hope of saving the situation.

The two trains collided on a curve a short time later at kilometre 10, half a kilometre from Thumeries. In the collision the wooden carriages of the passenger train were telescoped, though some of the passengers escaped because the walls of the carriages burst and they were thrown clear. Eighteen people were killed outright and another six passed away in the following days. About 80 were injured, ten seriously. Most of the dead were girls under the age of 20 from the sugar refinery, though the driver and fireman of one of the trains, probably the passenger, also died. [648] [649] [650]

Stationmaster Dufour was arrested after the disaster, accused of manslaughter through negligence. He had served as a stationmaster for ten years and was said to have been in a state of extreme prostration after the collision. Safety on the single line was supposedly assured by the regulations requiring the drivers of each train to be in possession of a staff (a *plaque pilote*) in the form of a copper plate stamped with the names of the stations between which applied, with the plates for different sections having different shapes. The stationmaster must therefore have failed to give the driver of the passenger train the *plaque pilote* and the engine crew of that train must also have been at fault for leaving Thumeries without it. [651] [652]

No. 171 Up Goods train and No. 314 Down Workman's train collided head-on on the East Indian Railway between Sultangange and Gangania stations, India, at about 20:20 on Friday 30th April 1948. Seven people died in the accident and 15 were injured, three

[645] *Journal de Genève*, Genève, Switzerland, 13th and 14th December 1947
[646] *The Times*, London, England, 13th December 1947
[647] *The Daily Herald*, Biloxi, Mississippi, United States, 13th December 1947
[648] https://lesamisdupp-online-fr.translate.goog/recit%20marquis.htm?_x_tr_sch=http&_x_tr_sl=fr&_x_tr_tl=en&_x_tr_hl=en&_x_tr_pto=sc
[649] https://douai.maville.com/actu/actudet_-Secouriste-a-l-usine-des-Asturies-en-1948-Rene-Semaesse-a-vecu-la-catastrophe-du-PP_loc-1489049_actu.Htm
[650] http://lesamisdupp.free.fr/presse/lvn%2020040214%20commemoration.htm
[651] Wikipedia – Compagnie du chemin de fer de Pont de Deûle-à-Pont-à-Marcq
[652] *Tribune de Lausanne*, Lausanne, Switzerland, 19th February 1948

seriously. A relief train with the Deputy Chief Mechanical Engineer, Jamalpur, the medical officer and other officials went to the scene. [653] Responsibility for the accident was shared by both drivers and the guard of the workman's train. [654]

In what was presumably a single-line collision train No. 8 Down, the Kathgodam – Lucknow Express, collided with train No. 49 Up, the Kashganj – Lucknow Fast Passenger of the Oudh and Tirhut State Railway between Puranpur and Shahgarh stations, Uttar Pradesh, India, at 2.28 a.m. on Tuesday 14th September 1948. A special train carrying a contingent of police from Pilibhit and two relief trains from Pilibhit and Sitapur were rushed to the scene while another train left Bareilly with a batch of medical students from Ayurvedic College to render medical aid. Seventeen people were reportedly to have died and 51 were seriously injured. [655]

A further report stated that two third-class bogies of the Up train had been smashed and two third-class bogies of the Down train had been telescoped. Fifteen to 20 persons had been extricated from the Up train. The number of casualties in the Down train was not stated. [656]

The assistant stationmaster at Shahgarh and the driver and guard of train No. 49 Up were found to have been responsible for the accident. [657]

By Sunday 19th the number of fatalities in the accident had risen to 28. [658]

Two trains collided in what was probably a single-line collision between Jamraniwah and Lodhran, West Pakistan, on Wednesday 22nd September 1948. Further details of this accident are unfortunately lacking. [659]

Strasbourg – Lauterbourg and Lauterbourg – Strasbourg omnibus trains collided head-on between the stations of Drusenheim and Herrlisheim, France, on the evening of Thursday 2nd December 1948. There was temporary single-line working over the normally double track between the two stations because of bridge repairs which were in progress. Four coaches overturned and fell down an embankment. By Saturday it was known that ten people had died, 12 were seriously hurt and 20 were slightly hurt. [660] [661]

In what was presumably a single-line collision No. 63 Up passenger train and a Down light engine collided head-on between Pipraich and Bodarwar, India, on the Savan Loop the Oudh and Tirhut State Railway at about 05:00 on Sunday 13th March 1949 (date unclear). About 19 people lost their lives in the accident. The assistant stationmaster at Pipraich, the stationmaster at Bodarwar and the driver of the light engine were all held to

---

653 *The Indian Express*, Madras, India, 2nd May 1948
654 https://eparlib.nic.in/bitstream/123456789/759956/1/ppd_08-04-1950.pdf
655 *The Indian Express*, Madras, India, 15th September 1948
656 Ibid, 16th September 1948
657 https://eparlib.nic.in/bitstream/123456789/759956/1/ppd_08-04-1950.pdf
658 *The Indian Express*, Madras, India, 21st September 1948
659 Author's own notes, source mislaid
660 *The Times*, London, England, 3rd and 4th December 1948
661 *Journal de Genève*, Genève, Switzerland, 3rd December 1948

be to blame. [662]

Southbound passenger train No. 1012 from Zagreb collided head-on with a northbound freight train two kilometres south of Plavno, Yugoslavia, early on Monday 15th August 1949. Twenty-one people lost their lives. The trains normally crossed elsewhere but because one of them was running late the meeting point was changed to Plavno. The stationmaster there, Luka Javorina, was duly advised by telephone of the altered arrangement. Unfortunately he and his fellow railwaymen had slaughtered and roasted a lamb and indulged in a heavy drinking session, consuming about 12 litres of wine. In his intoxicated state the stationmaster forgot all about the telephone call and failed to advise his colleagues of the altered meet. The signals at the station were left at clear and the passenger train passed through it unhindered, to meet the freight train just beyond.

After the accident the stationmaster fled and hid himself in a nearby cornfield, fearful he said of being lynched by the survivors. He and his colleagues subsequently stood trial and were found guilty, the court finding that, 'The accused Javorina came to a state of not only severe fatigue but also almost complete oblivion due to alcohol consumption.' The stationmaster was sentenced to death and executed by firing squad on 24th October. The other three Plavno railwaymen, who had not directly caused the accident but had also been drunk, were sentenced to terms of imprisonment. [663] [664]

A mistake by a stationmaster in prematurely allowing a train to depart before an opposing train had cleared the single line again led to a bad accident in France on the evening of Thursday 26th January 1950. Originally there had been two tracks but the Germans had removed one for use elsewhere during the war. The accident occurred at 19:50 at a level crossing two kilometres beyond Moncey station on the line between Vesoul and Besançon and involved two passenger trains. The driver of the train from Besançon to Vesoul saw the train from Vesoul to Besançon in time to jump but did not escape serious injury. The smokeboxes of both steam engines were smashed and tender of the engine of the train coming from Besançon wrecked five compartments of the leading carriage. In the other train the tender of its engine similarly damaged the leading vehicle, in this case a van. By 23:15 there were known to be nine fatalities and 15 injured; those who died included two daughters of a crossing keeper, [665] while both drivers and both firemen were among the injured. News of the accident quickly spread and doctors went to the scene from nearby towns. The injured were taken first to Mouncey station and then to Besançon. [666]

An investigation opened by Divisional Inspector Vairans of the Société nationale des chemins de fer français in co-operation with the police quickly established that the temporary stationmaster at Moncey had started the train waiting at his station for Vesoul before the arrival of the train from there, which was already en route from the next station at Rigney. The stationmaster was taken into custody. [667]

---

[662] https://eparlib.nic.in/bitstream/123456789/759956/1/ppd_08-04-1950.pdf
[663] https://www.executedtoday.com/tag/plavno/
[664] https://en.wikipedia.org/wiki/M604_railway_(Croatia)
[665] *The Railway Gazette*, London, England, 3rd February 1950
[666] Ibid, 27th January 1950
[667] *Le Monde*, Paris, France, 28th January 1950

Fog, rather than a mistake by a stationmaster, was apparently a factor in a head-on collision near between Lisle-sur-Tarn and Gaillac stations, France, on Saturday 11th February 1950. Two Michelin railcars, both reportedly crowded with workers, were involved. One was travelling from Toulouse towards Cap Denac and the other from Carmaux towards Toulouse. The trains normally crossed at Gaillac but that coming from Toulouse was apparently running late and the crossing was moved to Lisle-sur-Tarn, a little closer to Toulouse. The signals at Lisle-sur-Tarn were overrun, probably in the fog, by the outgoing train from Toulouse and the two Michelins collided at 06:55, with disastrous results. Twenty people died, including both motormen, and 38 were injured. Rescuers rushed to the scene from Toulouse and local farmhouses were turned into emergency first-aid posts. [668] [669]

A passenger who was fortunate to survive the crash because he was a smoker was artist Nicholas Greschny. When he boarded the train he entered the leading non-smoking car but he and a companion were moved by the conductor into the second car because of their lighted cigarettes, thus by chance probably saving their lives. [670]

At the end of June 1950 track and bridge repairs necessitated taking one line out of use between Hartenstein and Fährbrüke, East Germany. There were two methods of operating trains while one line was out of use, known as *zeitwise eingleisigen Betrieb* (temporary single-line working) and *Befahren des falschen Gleises* (running on the wrong track). It was determined to use the second method, which involved the *fahrdienstleiter* at each end of the section issuing so-called B orders to the crews of each train. All the messages sent and taken also had to be recorded in a register, adding to the signalmen's workload.

Between about 20:00 and 20:30 Aue – Zwickau local railcar N 3175 was timetabled to cross *Schichtwechselzugen* (shift change train) P 4930 at Hartenstein, with train P4932 behind it. The latter trains carried workers to the uranium mines at Aue and the Soviets required that these trains be given priority over ordinary passenger trains. On the day of the accident train N 3175 was a little late reaching Hartenstein and train P 4930 was also late, the communication cord having been pulled at some point. In order to avoid further delay to train N 3175, when train P 4930 finally arrived at Hartenstein the *fahrdienstleiter* there decided to let train N 3175 proceed. He therefore telephoned his *Stellwerkwärter* (pointsman) and instructed him to issue a B order to train N 3175 for it to proceed to Fährbrüke once train P 4930 had arrived. At 20:32 the Hartenstein *fahrdienstleiter* reported the arrival of train P 4930 back to Fährbrüke and offered train N 3175. His colleague at Fährbrüke replied, "No, wait," and offered train P 4932 instead. The Hartenstein *fahrdienstleiter* thereupon accepted train P 4932, and only after he had done so did he belatedly tell his pointsman to hold train N 3175. The pointsman seized his signal horn and blew it at the departing railcar, but the rear conductor was chatting with her colleague and a passenger and did not notice it. It was only when the conductor heard the shouting of a ganger that the brake was applied, and then only the handbrake and not

---

668 *The Times*, London, England, 13th February 1950

669 *The Racine Journal Times*, Racine, Wisconsin, United States, 11th February 1950

670 https://www.ladepeche.fr/2021/03/15/tarn-le-terrible-accident-ferroviaire-de-1950-a-lisle-sur-tarn-fait-20-morts-9429463.php

the emergency brake. Eighteen people died and 63 were injured in the collision. [671]

Two suburban passenger trains collided head-on at 07:05 on Saturday 12th August 1950 between the stations of Acari and Pavuna, Brazil, on the Rio Douro line of the Estrada de Ferro Central do Brasil. The trains were UX-7, with Driver Romulado Luis da Silva, bound for Belford Roxo, and train UX-104, in charge of Conductor Valdemar Carneiro de Barros o Acevedo, with Driver Belmiro da Silva Ribiero. Five passengers were killed outright in the collision and about 40 were injured. The conductor of train UX-104 died in hospital shortly after the accident while being operated on.

Speaking from his hospital bed the driver of train UX-104 declared that he had received authority to leave Pavuna and was not responsible for the accident and an official statement issued by the company agreed, blaming the collision on train UX-7 leaving Acari without permission. [672]

Another passenger died of his injuries two days after the accident. [673]

By 1950 the electric interurban railways which had once been so prevalent in the United States, especially in the Mid-West, were in terminal decline, unable to compete with the ever increasing ownership of automobiles. In an effort to preserve at least a small part of at least one of them, a wealthy railway enthusiast named Jay Maeder, from Chicago, purchased a 13-mile portion of the former Milwaukee Electric Railway, which he renamed the Milwaukee Rapid Transit and Speedrail. The line ran from Milwaukee along the former double-track main line as far as West Junction and then along a former single-track branch as far as Hales Corners, Wisconsin. In the autumn of 1950 the National Model Railroad Association held its annual convention at Milwaukee and in connection with the convention excursion trips were run for the modelling enthusiasts along the Speedrail line. It was two of these trains which were involved in a collision near Hales Corner on Saturday 2nd September 1950, on the Labor Day weekend.

At 8.15 a.m. on the morning of the accident a special train left Milwaukee for Hales Corners with Meader himself as motorman. An hour later it was followed by a considerably heavier train driven by Motorman LeRoy Equitz. The signalling on the single-line portion of portion of the railway consisted of red and white lights: a red light meant that the line was occupied by an opposing train, a white light that there was a preceding train running in the same direction and no lights that the line was clear. When the second train arrived at the beginning of the single line no lights were showing and Motorman Equitz proceeded, but a couple of miles further one, the two trains met one another. The heavier train demolished the lighter one, in which ten passengers lost their lives. Maeder, himself, jumped from his train and survived. [674] About 50 other passengers were injured. The Milwaukee-bound train was also reportedly carrying some mothers and their children bound for the city for pre-school shopping.

E.T. Stafford Jr., of Superior, Wisconsin, who had been riding in the leading car of the second train, spoke to a reporter. "I was sitting with Lowell Wood of Minneapolis.

---

671 *Katastrophen der deutschen Bahnen Teil II* by H.J. Ritzau, pgs. 284-6 (in German)

672 *Jornal do Brasil*, Rio de Janeiro, Brazil, 14th August 1950

673 Ibid, 15th August 1950

674 *A History of Railroad Accidents, Safety Precautions and Operating Practices* by Robert B. Shaw, pgs. 372-4

Suddenly we heard the air brakes screaming and someone ran through the car {probably Motorman Equitz}. We dived to the floor just before the crash. Bodies flew through the air. We climbed out of a window afterward and saw someone lying beside the tracks with one leg missing. We put on a tourniquet and then an ambulance took him away." [675]

Motorman Maeder said that when he had seen that a collision was inevitable he had jumped from the train and rolled down an embankment, being cut and scratched but not seriously hurt. He maintained, "I was at the controls when we passed the Oklahoma Avenue station. I was given the white permission to indicate that the next block signal was open." [676]

Maeder also told a reporter that he had dreamed repeatedly that there would be an accident. "I dreamed of a train coming around a curve at me – many, many times, maybe 50 times – in my sleep. Because of this dream, I determined to be close to the operation of the chartered trains on Saturday. I went along on one of the trains and took control so that nothing would happen because of my dream, but my dream came true."

About three weeks after the accident at the inquest a coroner's jury found Maeder guilty of gross negligence and he was charged with fourth-degree manslaughter. The charge was later dismissed. Ridership over the line dropped badly after the accident, it suffered insuperable financial difficulties, and closed at the end of June 1951. [677]

A misunderstanding by the motorman of a diesel railbus led to a single-line collision between Godegård and Mariedamm stations, Sweden, on Saturday 28th October 1950. The railbus, northbound train No. 756, was timetabled to meet two southbound trains, passenger train No. 611 and a local freight train, No. 7311. The two southbound trains were delayed and as a result the crossing of the railbus and the goods train only was moved northwards to Mariedamm. The driver of the railbus somehow misunderstood the situation and imagining that the crossing of both trains had been altered set off from Godegård with his 20 passengers.

The station staff at Godegård did what they could to avert the collision. They could not contact the railbus driver but the line was electrified, so that by isolating the current they could at least stop the southbound passenger train, which was electrically powered. An attempt was also made to pursue the railbus with a dressage (rail tricycle) but the railbus and the passenger train collided on a curve two kilometres south of Mariedamm. The railbus was still travelling at 70 or 80 kilometres an hour and in the collision the front of it was crushed to a depth of a metre and a half by the engine of the passenger train. The motorman of the railbus and four passengers were killed outright, four more people died later and 13 were injured.

News of the accident was overshadowed in Sweden by the death the next day of King Gustav V.

Two of the passengers on the railbus were Gotthard and Hildegaard Johansson, of Motala, who with their 5-year-old daughter had been on her way to Fjugesta to take over a fruit and vegetable shop. They were killed outright, while their daughter, Ann-Charlotte, was only very slightly hurt. Ambulance Driver Knut Bengt Berglund told a

---

675 *Manitowoc Herald-Times*, Manitowoc, Wisconsin, United States, 2nd September 1950

676 *Moline Daily Despatch*, Moline, Illinois, United States, 2nd September 1950

677 https://eu.jsonline.com/story/life/green-sheet/2016/09/06/our-back-pages-deadly-train-crash-helped-end-interurban/89867368/

newspaper reporter, "We had taken care of the injured, and I asked if there were any more who needed to come along. No, they said at first, but then I was called back, and they showed me the girl. 'Dad, Dad,' she shouted, throwing her arms around my neck. I have a girl myself of the same age and could hardly hold back my emotions when I felt her arms around my neck." The orphaned girl was taken briefly to hospital in Akersund, then spent the night with the Berglund family. The parents of the girl had not been very well off. She had an older sister, but she needed what money she had to train as a nurse. The newspaper, *Motala Tidning*, according launched a fundraising appeal, the money collected to be placed in a trust until the child reached adulthood. [678]

The late editions of some Canadian newspapers on Tuesday 21st November 1950 carried reports of a serious train crash on the Canadian National Railways main line near Canoe River, British Columbia, Canada, some 83 miles west of Jasper, Alberta. The trains involved were said to be an eastbound transcontinental express and a westbound troop train, carrying personnel of the 2nd Field Regiment, Royal Canadian Horse Artillery, from Camp Shilo, Manitoba, to Fort Lewis, Washington, United States, en route to Korea to fight as part of the United Nations forces. It was known that six and possibly ten people had been killed, all on the troop train, while the engine crews of both trains were "missing." The collision had occurred at 10.40 a.m. Mountain Standard Time. [679]

By the following day the number of dead was known to have risen to 20 (another person died later) and there were at least 61 injured, 22 seriously. Those who lost their lives in the disaster included Engineer J.J. Stimson and Fireman Harry Proskunik of the troop train and Engineer Harvey Church and Fireman Adam Oleschuk of the intercontinental.

A report from Jasper described the aftermath of the disaster. 'The mountain train wreck west of here yesterday was described as worse than any battle by a soldier on the hospital train which passed through here late last night. "I was near the men who were killed and it was worse than going into any sort of battle," the unidentified soldier said. "In this case nobody had a chance. Nobody knew what was coming."

'Another soldier said, "Everything was orderly. It is hard to believe that so many were killed and injured. There was so little confusion." That was about the only description made of the crash between the troop train and the transcontinental passenger train. Troops aboard the hospital train were ordered not to talk about the wreck. Many appeared willing to talk but weren't given the opportunity.

'Most of Jasper's 1,700 residents – 60 per cent of them railroad men – were out to meet the hospital train in one of the district's worst snowstorms of the season. A number of ambulances and makeshift ambulances with volunteer crews lined the station platform ready to take off the injured. The Jasper hospital was ready to take 20 injured and put up another 20 in beds set up in the school auditorium. But the hospital train continued on to Edmonton after the frozen steam line was thawed out. The train did not appear from the outside as though it had been in a wreck.' [680]

Another newspaper reported that the accident could have been still worse. It wrote, 'The difference of half a mile prevented Tuesday's death toll in the mountain train crash

---

[678] Wikipedia – Järnvägsolyckan I Mariedamm

[679] *The Lethbridge Herald*, Lethbridge, Alberta, Canada, 21st November 1950

[680] *The Medicine Hat News*, Medicine Hat, Alberta, Canada, 23rd November 1950

from reaching into the hundreds, men of Canada's Special Force said here {Edmonton} Wednesday. Their westbound troop train had crossed a bridge straddling a 500ft canyon and travelled another half a mile before it rammed head-on into a transcontinental passenger train on a mountain curve.

'"If we had met over that canyon our whole train would have plunged into the canyon and not very many of us would have survived," said Lieut. Paul Cullen of Montreal. Lieut. Cullen said he was "mighty lucky to be alive." He was sitting in the rear of the third passenger coach from the front of the troop train. The coach ahead telescoped and landed on top of the front section of the third coach. The trooper's engine shot into the air and landed on top of the second coach, "crushing it down to the rails."

'Almost all the dead and injured were in the second and third coaches. "Most of them never knew what hit them," said Lieut. Cullen. "Our coaches, all of wooden construction, just disintegrated. The wreckage was piled a good 50 feet high. It wasn't recognizable as anything but a jumble of twisted steel and splintered wood.'

Passenger Jack Thorson, of Winnipeg, who had been riding on the intercontinental train, also spoke to reporters. He said, "There was one bump and then two more. I was in the dining car when we hit and the shock didn't seem very severe. The diner was away near the back of train No. 2. I'm pretty sure our train was already at a stop when the collision occurred. We seemed to be just into or just past Canoe River when it happened. We were on a curve. We all climbed out of the diner and ran toward the front of the train. The engines were both all tangled up together and there were big clouds of steam over everything at the front of the trains. The baggage, mail and express cars on our train were off the track. The troop train was in far the worst mess. The first passenger car of the troop train was a twisted mass of steel and splintered wood. The second passenger coach was partially telescoped and almost right up or on top of the third car. It was in these three cars that most of the casualties occurred. There was no fire but many soldiers were scalded when steam heating pipes in the coaches broke. We all ran back into our own train when we saw there were no civilian passengers hurt and grabbed axes and hammers that are for use in case of emergency. Everybody came to pitch in. But we weren't really needed. An engine came from Red Pass Junction and supplied steam heat to the trains about an hour after the wreck. It was cloudy when the collision occurred but started to snow a few minutes after the crash. There was nothing resembling panic at any time. I'm so glad those soldiers were there. I'm afraid they will be a long time reaching the enginemen. It seems to me they will have to use acetylene torches to cut open the cabs. We were lucky that no cars rolled off the grade. There is a 75 foot drop on each side of the track where the wreck occurred."'

The first medical aid was provided by Dr. P.S. Kimmett, of Edson, Alberta, who chanced to have been a passenger on the intercontinental. His wife acted as a nurse and they were assisted by Jack Cook, Canadian National regional first aid officer, and Mrs Lennie of Edmonton. However, when Dr Straughan, of Jasper, arrived at the scene he was critical that there had been no medical staff riding on the troop train. He commented, "There was no army medical man aboard the troop train. That was a grave error. Something should be done about it. That is the only troop train I ever heard of without a doctor."

Meanwhile an official statement from Vice President J.R. McMillan of the Canadian National British Columbia division was that the collision had been due to the failure to deliver to the crew of the troop train an order to meet the passenger train at Cedarside.[681]

The events which had led up to the disaster had begun when Dispatcher A.E. Tinsdale at Kamloops, British Columbia, had set about arranging the crossing of eastbound intercontinental trains Nos. 2 and 4 with the westbound troop train. By this date orders were sent by telephone rather than telegraph and the dispatcher telephoned the order to operators Alfred Atherton at Red Pass Junction and Frank Parsons at Blue River. The order, as issued, read:

Passenger extra 3538 west to meet No. 2 engine 6004 at Cedarside and No. 4 engine 6067 at Gosnell.

From east to west the stations and sidings were Red Pass Junction, Cedarside, Canoe River, Gosnell, and Blue River.

According to Dispatcher Tinsdale both operators read back the order to him correctly but in a fatal error the order as delivered by Operator Atherton to Conductor John Mainprize of the troop train read:

Passenger extra 3538 west to meet No. 2 engine 6004 and No. 4 engine 6067 at Gosnell.

The two crucial words *at Cedarside* were thus missing, permitting the troop train to continue beyond Cedarside, where it had to meet train No. 2, as far as Gosnell. Operator Atherton maintained that he had repeated the incorrect order back to the dispatcher but Operator Parsons at Blue River had been on the line at the same time and backed up the dispatcher in saying that Atherton had repeated the order correctly back to him.

Having been identified as the person who had inadvertently caused the disaster Atherton was subsequently charged with the manslaughter of Fireman Proskunik and stood trial at Prince George, British Columbia, in May 1951. Much was made of at the trial of the use by the railway company of a mixture of steel and older, partly wooden cars, even though the use of this older rolling stock, while it had doubtless contributed to the number of deaths amongst the troops, had nothing to do with the death of the fireman. However, in trying to underline this point, the prosecuting council made the grossly tactless remark that "in this case we are not concerned about the death of a few privates going to Korea."

This hardly helped the prosecution's case, a juror being heard to remark, "Did you hear what that bastard said?" After a brief recess the jury found Atherton 'Not Guilty.'[682]

Like many other accidents previously in both Canada and the United States the disaster underlined the weaknesses of the train order system.

The death in the accident of the only person who could have explained how it had happened means that the exact cause of a collision between Kinstaby and Marmaverken, Sweden, on the evening of Monday 22nd January 1951 must remain a mystery, though it was clearly due to human error.

Northbound train No. 2548, a diesel railbus, left Söderhamn Central station 14 minutes late on the day of the accident at 18:10 for Kilafors. Normally it should have crossed train No. 2539, variously described as a passenger and a freight train, at Marmaverken, but because of the late running of the railbus, train No. 2539 was advanced southwards to Kinstaby. Stationmaster Robert Bäck at Kinstaby watched the

---

681 *Winnipeg Free Press*, Winnipeg, Manitoba, Canada, 22nd November 1950

682 *Wreck!* by Hugh A. Halliday, pgs. 194-201

railbus arrive and then went into his office, where a telephone call from Marmaverken announced the departure from that station of train No. 2539. As he left his office again he was alarmed to see the railbus pulling out again, though he had not given any departure signal to its motorman. He vainly ran towards the train, signalling for it to stop, then attempted to warn its motorman by manipulating the signal at the northern end of the station, and also tried to telephone the house of a trackman who lived by the lineside, all without result. In any case the railbus did not get far before it met the passenger train on a curve. The railbus was no match for the steam locomotive pulling train No. 2539, and it was smashed and pushed back several hundred metres in the collision. The motorman and ten passengers were killed or fatally injured and 15 other passengers were injured.

The rescue operation which followed took place in darkness, in snow 60cm deep and at a temperature of -23 degrees Celsius. It was impossible for ambulances to reach the scene of the accident and the injured had to be removed on horse-drawn sledges. The dead were taken to Kinstaby station and the injured to Söderhamn hospital. Fireman Esbjörn Gelhaar of train 2539 told of the horror of the scene and added that if the railbus had been ten seconds later the trains would have both been on a straight stretch of track, and the accident would have been averted.

As is often the case there were stories of passengers' lucky escapes and of other passengers who had been less fortunate. Börje Larsson and his friend Erik Hansson had joined the train at Söderhamn to go to the cinema at Bollnäs but at Kinsta, Larsson left the train, having had a sudden urge to go and visit his grandparents nearby. His friend Hansson stayed on the train and was one of those who died. Larsson later identified the body of his friend and as might be expected was deeply affected by what had happened. On the other hand another potential woman passenger survived because when she arrived at the station to catch the train she discovered that she did not have enough money for the fare and so had to abandon her plan. [683] [684]

The usual method of working single-line sections in Great Britain in the 1950s, with the use of electric tablet, staff, or token instruments, while relatively expensive was very safe, and since the Abermule disaster of January 1921 there had not been one single-line collision involving a passenger train. Then in the middle of 1951 there were two such collisions in three weeks.

The first accident occurred on Wednesday 11th July 1951 on British Railways, Western Region between Manorwen signal box and Fishguard and Goodwick station, England. The former Great Western Railway main line from London Paddington had two or more tracks throughout its length except for 2¼ miles near its far end, beginning at Manorwen box. This stretch of single track was worked by the electric train staff. Over this length the line fell steeply, with a mile and a half at a ruling gradient of 1 in 50, and there was a "Stop" board 375 yards beyond the beginning of the single line at Manorwen, where all Down freight trains had to stop and if necessary pin down brakes.

On the evening of 11th July Signalman W.M. Lee at Fishguard and Goodwick accepted the 12.35 p.m. Down freight train from Llandilo to Fishguard Harbour from Signalman F.W. Payne at Manorwen at 6.30 p.m. The train was half an hour late but

683 https://www.helahalsingland.se/2017-01-28/ohyggliga-scener-motte-de-som-var-forst-pa-plats--66-ar-sedan-elva-personer-miste-livet-i-tagolyckan-i-kinstaby

684 Wikipedia - Järnvägsolyckan I Kinstaby

having been told by Signalman Payne that the partially fitted freight train would not need to stop to pin down brakes there he decided that there would be time to get it into Fishguard before the departure of the 6.40 p.m. Up push-and-pull auto train from Fishguard Harbour to Clarbeston Road. At 6.38 p.m. he accepted the auto train from Fishguard Harbour box, at 6.39 p.m. Manorwen sent on the freight train and at 6.40 p.m. Fishguard Harbour box sent on the auto train. He then replaced facing point lock lever No. 24 at the Manorwen end of the station so that he could clear his Up home signal, No. 3, to let the auto train into the station, before locking the facing points again.

The auto train consisted of a 0-6-0 tank engine running chimney first and a single coach in charge of Guard B.P. Tudor with Driver D.H. Rees and Fireman K.J.A. Phillips. It arrived at Fishguard and Goodwick station at 6.42 p.m., stood at the platform there five or six minutes, then started away. As chance would have it a newly married couple had boarded the train and in accordance with the usual custom its departure was marked by repeated whistling from its engine and the explosion of a number of detonators. Signalman Lee saw what was happening and tried to stop the train by waving a red flag, shouting, and blowing a whistle but amidst all the other noise and commotion his warnings went unnoticed. The auto train proceeded past the Up Starting signal, No. 4, at *Danger*, burst the facing points and proceeded onto the single line. Only a minute or so later, seeing some smoke coming from a stationary engine at a point near an intermediate siding serving a brickworks, Lee assumed that the two trains had collided and he sent the *Obstruction Danger* signal to Manorwen and Fishguard Harbour at 6.48 p.m. He told the Ministry of Transport inquiry into the accident that he always gave the train staff to the driver personally and never via a member of the train staff.

Porter I.J. Griffiths and Clerk B. Hale at Fishguard and Goodwick said that a wedding party had arrived at the station a quarter hour before the train. They confirmed the signalman's story of the exploding detonators and whistles and that such a "send-off" was normal practice.

Guard Tudor said that he had been busy attending to passengers and had not noticed the position of the Starting signal. With the noises of the detonators and whistles he had not noticed the signalman's shouting and waving either. Soon after the train had left the station its brakes had been applied and looking out of the window he had seen the oncoming freight train, whereupon he had shouted to his passengers to, "Hold on," and fortunately none of them had been hurt. (Four railwaymen were injured but only Driver Rees was detained in hospital.)

Driver Rees testified that the Fishguard and Goodwick Up home signal had been off as he had approached the station and he had run into the platform. After a few minutes his guard had given the *Right Away* signal on the bell and he had opened the regulator. He had heard six detonators go off and his fireman had sounded the whistle two or three times but he had seen the wedding party on the platform and his fireman told him that he had put down four detonators so he had taken no notice of the explosions. He admitted that he had started off without the train staff and against the Starting signal. At the brickworks siding he had belatedly become aware that he did not have the train staff, had stopped as quickly as he could and had been about to set back when the collision had occurred. He explained that upon leaving Fishguard Harbour he had found that the glasses of his lubricator were blurred. At Fishguard and Goodwick he had been preoccupied with unscrewing, cleaning, and re-screwing the lubricator glasses and he thought that it had been this rather than the wedding party which had made him forget the train staff. Previously, when there had been a Down train on the single line he had been stopped at the Fishguard and Goodwick Up Home signal until the Down train had arrived, or else had been called in with a green flag.

Fireman Phillips confirmed his driver's evidence and that he had put down the detonators and blown the whistle. He added that leaving the station he had been helping his driver with the lubricator and then had been busy firing, hence he had not noticed either the absence of the train staff or the Starting signal at *Danger*. He and his driver had seen the freight train at the same moment and he had jumped just before the collision.

The Down freight train consisted of a 4-6-0 tender engine, six empty vacuum-fitted vans, 15 loose coupled waggons and a brake van, in charge of Goods Guard T.W. Rowe with Driver T.C. Phillips and an unnamed fireman.

Signalman Payne said that he had handed the Manorwen – Fishguard and Goodwick train staff to the fireman of the freight train. The train had slowed at the "Stop" board but he was not sure if it had actually stopped.

Driver Phillips stated that he had tested the brakes of his train at Letterston Junction. The Manorwen Down Home and Starting signals had been off and he had passed the box at 10 miles an hour, his fireman collecting the train staff. He had stopped at the "Stop" board but with the vacuum-fitted vans there had been no need to pin down any brakes so after exchanging signals with the guard he had restarted. Leaving a cutting near the brickworks siding his fireman had shouted that there was a train on the line ahead. He had immediately applied the brakes and put down sand while his fireman had applied the hand brake. Speed had been reduced to 15 miles an hour, then he and his fireman had jumped.

Goods Guard Rowe agreed that the brakes of the freight train had been tested at Letterston Junction and that there had been no need to pin down any brakes at Manorwen. He said that his train had been well under control and travelling at 20 to 25 miles an hour just before the collision.

Fishguard Harbour Station and Quay Superintendent J.H. Swann and Goodwick Locomotive Foreman W.H. Jones agreed that the use of detonators to send off wedding couples was common practice.

Brigadier C.A. Langley, the Ministry of Transport inspecting officer, concluded that Driver Rees was primarily responsible for the accident for leaving Fishguard and Goodwick without the train staff and against the Starting signal. Fireman Phillips had taken no steps to help his driver and had placed detonators on the line. 'It is clear that he was thinking more about the wedding party than his proper work,' the inspecting officer remarked. Guard Tudor could not be blamed for not checking the Starting signal before ringing his bell, which only indicated to his driver that the station work was complete, but had he looked out after his train had started the accident would have been averted.

Driver Phillips and his fireman had been keeping a good lookout and were in no way to blame.

Signalman Lee should not have unlocked facing points No. 22 in the face of the approaching freight train and should have checked the auto train at the Home signal given that the Starting signal was at danger.

The inspecting officer recommended the provision of either trap points or a sand drag if Up trains were to continue to be admitted to the platform while a Down train was in the section, also the provision of a subsidiary signal below the Up Home signal so that there was no need to unlock the facing points in front of a Down train or else to resort to hand signals.

Whilst 'everyone would wish to show goodwill' on such occasions as the sending off of wedding parties, the light-hearted use of detonators, which were intended to stop or warn trains in emergency and not to send them on their way, a practice which was not confined to Fishguard and Goodwick, had now been prohibited. [685]

The second British accident in three weeks occurred on British Railways, Scottish Region on the former Highland Railway main line from Perth to Aviemore and Inverness near Dalguise, Scotland, on Wednesday 1st August 1951. The 4.10 p.m. Down passenger train from Perth to Blair Atholl came into collision with an Up Officers' Special train. Eight passengers, three railway officers, the driver and guard of the special and the guard of the passenger train were injured.

The passenger train consisted of a Class 3 4-4-0 tender engine and three non-corridor coaches in charge of Guard A. Robertson with Driver J. McPherson and Fireman W. Riddell. The Officers' Special train consisted of another 4-4-0 tender engine and saloon coach No. 972002, with Passed Fireman D. Thompson and Fireman L. Ross on the engine.

The layout at Dalguise consisted of a crossing loop and some sidings at the Perth (south) end of the station. The signal box was also situated towards the south end of the station, with a three-lever ground frame working the double- to single-line connection at the Aviemore (north) end. The single line southwards to Dulkeld was worked by the electric train token and northwards to Ballinluig by the electric train tablet.

The Down passenger train arrived at Dalguise on time, Fireman Riddell gave the Dunkeld – Dalguise token to Signalman J. Doig but was not given the Dalguise – Ballinluig tablet. The engine then detached a horse-box which had to be left at Dalguise. Driver McPherson had not noticed the position of the Starting signal as he entered the station. The horse-box having been detached the train drew into the station platform, stopped briefly, then Guard Robertson gave his driver the *Right Away* signal to indicate that the station work was complete and Driver McPherson set off. McPherson did not look towards the starting signal and forgot that he did not have the Dalguise – Ballinluig tablet. He told the Ministry of Transport inquiry that he had been checking that a party of 70 schoolboys riding on the train had closed all the coach doors. As his engine was passing the Starting signal he saw that it was at *Danger*, closed the regulator, applied the vacuum and then the Westinghouse brakes and reversed his engine.

Fireman Riddell said that he had not told his driver that he had not got the tablet but thought the latter had realised and that he had received it at the station. He had looked back as they were leaving the station, then begun to fire. When his driver had applied the brakes in emergency he had looked out, seen the Officers' Special train and its fireman jumping from its engine, and had jumped himself.

Guard Robertson said that he had seen the Starting signal at *Danger* after part of his train had passed it and had gone to apply his brake but his driver had already done so.

Ganger R. Stewart and his two lengthmen had just finished work and were picking up their tools near the Up Home signal. He testified he had seen that the signal was off and the Officers' Special train approaching, then he saw the Down passenger train pulling out of the station. Shouting to one of the lengthmen to stop the passenger train he himself had

---

[685] Ministry of Transport report

run towards the Officers' Special, holding up his arms.

Passed Fireman Thomson said that he had noticed the ganger and the Home signal being thrown to *Danger* at the same moment and had made a full brake application and reversed his engine.

Fireman Rose of the special related that he had heard his driver make some exclamation, the brakes had been applied, he had seen the passenger train and realising that there would be a collision had jumped.

The Scottish Region Assistant to the Motive Power Superintendent, Mr J.M. Fyfe, saw the engine of the passenger train fouling the Up line and went to apply the brake in the saloon carriage but the passed fireman beat him to it. He put the speed of the special at 10 miles an hour at the moment of the collision, the passed fireman put it at 15 to 20 miles an hour.

Signalman Doig stated that he had left his box to give the Dunkeld – Dalguise token to the Officers' Special, then, seeing the passenger train leaving the station, he had run back to his box and replaced the Up Home signal.

The inspecting officer, Colonel D. McMullen, grilled Driver McPherson regarding his failure. The poor driver admitted that he had simply forgotten about the tablet and pleaded that it had been due to his anxiety about the schoolboys, who had been repeatedly changing compartments at stations. Guard Robertson confirmed this.

Colonel D. McMullen concluded that Driver McPherson was responsible for the collision and added that the driver was very distressed by his mistake.

The inspecting officer remarked that, 'These two serious lapses {Fishguard and Dalguise} which occurred within three weeks of one other, are in contrast with the high standard of discipline which is generally maintained in single line working and is proved by the freedom from accidents of this kind over a period of 30 years.'

Trap points at the Starting signal would have prevented the accident but were they to be provided at Dalguise they logically should be provided at a large number of other stations, and the necessary labour and materials could, the inspecting officer, be used to better advantage in other directions. [686]

The *Southern Belle* passenger train of the Louisiana and Arkansas Railroad collided head on with a troop train carrying 300 United States marines near Lettsworth, Louisiana, United States, on the tracks of the Texas and Pacific Railroad at 7.02 a.m. on Friday 10th August 1951. The collision occurred in what was described as a desolate swamp area, 60 miles north-west of Baton Rouge, Louisiana.

Southbound first-class passenger train No. 1 (the *Southern Belle*) from Kansas City to New Orleans consisted of diesel electric units Nos. 23 and 32A, coupled in multiple, a baggage-mail-dormitory car, two coaches, a dining car, a sleeping car, and a lounge-observation car, in charge of an unnamed conductor with Engineer L.L. Rainey and Fireman Walter Lucine. It had left Alexandria, Louisiana, at 6 a.m., eight minutes late, passed Simmesport, the last open office, at 6.48 a.m., on time, and was travelling at an estimated 55 miles an hour when it collided with the troop train.

Northbound Extra 51-A North, the troop train, from Camp Lejeune, North Carolina, to San Diego, California, consisted of diesel electric units 59D and 78C, an express

[686] Ibid.

refrigerator car, 11 sleeping cars, a baggage car, two kitchen cars and a sleeping car, in charge of Conductor A.W. Marcotte with Engineer C.A. Yates and Fireman James Reed. It had left West Yard, Louisiana, at 3.15 a.m., passed New Roads, the last open office, at 6.27 a.m. and passed Lettsworth about 7 a.m. This was six minutes after train No. 1 was due to leave Eden, almost 8½ miles further north, and the next place where the two trains could have crossed one another. Travelling at 40 miles an hour it had collided with train No. 1 a mile and a half beyond the north switch at Lettsworth.

All four engines, the first five cars of the troop train and the first four cars of train No. 1 were derailed. Two diesel units, the first three cars of the troop train and two cars of train No. 1 were destroyed by a post-accident fire. Six passengers, Engineer Rainey and Fireman Lucine of train No. 1, Conductor Marcotte, Engineer Yates, Fireman Reed, Front Brakeman C. Moore, and Trainmaster B.T. Moore of the troop train were killed and 69 passengers, six dining car employees, three sleeping car employees, the conductor, front brakeman and baggageman and the flagman of the troop train were injured. [687]

A newspaper wrote, 'A coroner's jury was called here {New Roads} today to investigate the deaths of nine victims of the pre-dawn collision between the streamliner "Southern Belle" with a troop train carrying 288 Korea bound marines. Coroner J.C. Roberts said, however, "I believe it will be possible {sic} to learn the cause of the collision since the train crewmen responsible are dead. In my opinion, however, orders were either misunderstood or ignored."

'A check of the hospitals in the area showed 45 persons injured in the crash and the oil-fed fire that swept the wreckage in a dense Mississippi river swamp near Lettsworth early yesterday.

'R.R. Suter, Kansas City Southern railway general superintendent, blamed the collision on the troop train's failure to wait on a siding for the New Orleans bound Southern Belle. The railway official said, "a complete investigation of the cause of the wreck will be made jointly by the KCS and the Interstate Commerce commission." It could not be determined at what speed the two trains were travelling as they hurtled into each other. "The maximum allowable speed on the line is 55 miles an hour," said Suter, "and I doubt that they were running any slower than that."

"Aubrey Stears Jr., nine, of Lettsworth, was an indirect victim of the wreck. An automobile struck him and his pony, killing both, as they galloped towards the scene of the wreck.

Diesel oil spewed from the two engines over the wrecked cars, feeding the flames that hampered rescue workers. Suter said they were forced to wait for the cars to cool, delaying the grim job of determining the number of dead. Flames trapped Lipa {Corporal Chester Lipa, of Detroit, Michigan, the first of the deceased marines to be identified} in a wrecked car. "We tried to get him out but he was stuck too tight and we couldn't make it," said another marine. "The car was beginning to burn from the oil."

'A marine lieutenant from Rhode Island, who would not identify himself, said he was trapped in a car and the burning oil was about to engulf him. "About 50 of my men picked up a section of rail that had been broken loose and enlarged a little hole in the side of the car until it was big enough for me to crawl through," he said. He said the marine who was killed was sitting directly across from him. He thought casualties among the

[687] Interstate Commerce Commission report

marines would have been worse, if they had not been at the back of the car eating breakfast.'[688]

The Interstate Commerce Commission (ICC) noted that train No. 1 was timetabled to leave Simmesport at 6.48 a.m., Eden at 6.54 a.m., Torras (not a crossing place) at 6.59 a.m., Lettsworth at 7.3 a.m., Batchelor at 7.11 a.m. and Morganza at 7.21 a.m. According to the rules the troop train was required to clear the schedule time of train No. 1 by at least 10 minutes or if unable to do so to be protected at that time in accordance with Rule No. 99.

The line on which the collision happened was provided with a train communication system allowing communication between wayside stations and the locomotive or caboose of a train over a distance of 25 to 30 miles in favourable conditions, though communication between trains was limited to about a mile.

Flagman M. Johnson, the only surviving member of the crew of the troop train, was the key witness at the ICC inquiry. He stated that the conductor and the trainmaster had checked times before leaving West Yard, where the brakes had also been tested. After passing North Baton Rouge the conductor had shown him several train orders and had remarked that they would go as far as they could, to meet train No. 1. Beyond New Roads the conductor had moved to the front of the train and from there the engineer, fireman, front brakeman, and trainmaster had been in the front unit, the conductor in the second car and he had remained at the rear of the train. No order had been received regarding train No. 1. When the train had passed Batchelor he had been aware that it was infringing on the time of train No. 1 but knowing that the trainmaster was riding on the train and had access to the communication equipment, he had assumed that arrangements had been made to advance the troop train beyond Batchelor. The train had complied with speed restrictions and had whistled for level crossings. The brakes had been applied in emergency immediately before the collision.

Surviving members of train No. 1 said that there had been no brake application on their train before the collision. They had received no orders regarding the troop train.

It emerged that earlier in the journey Trainmaster Moore had used the train communication system to advance the troop train against the schedule of southbound first-class train No. 9, which was due to leave Baton Rouge at 5.30 a.m., without train order authority. The trainmaster had called the operator at North Baton Rouge informing him that the troop train would cross train No. 9 there. When the troop train had then been delayed for some minutes from some undisclosed cause the trainmaster had attempted to communicate with the crew of train No. 9 directly but although they could hear him he could not hear them, due to a problem with the transmitting equipment on the engine of train No. 9. The trainmaster had then contacted the North Baton Rouge operator again and had been assured that train No. 9 was being held for the troop train.

The ICC questioned the operators at North Baton Rouge, Simmesport and Alexandria and the crew of Extra No. 56-A north, a northbound freight train which met train No. 1 at Bijou, 27½ miles north of Simmesport, to try and determine whether there had been any messages exchanged between the crews of the troop train and train No. 1 via the train communication system, regarding the movements of the two trains, but nothing concrete could be determined.

The ICC noted that an hour and a half earlier the troop train had passed Essen, south

[688] *The Sheboygan Press*, Sheboygan, Wisconsin, United States, 11th August 1951

of Baton Rouge, when it had not had time to clear the schedule of train No. 9 (this would have been when the trainmaster had been trying to contact the crew of that train). The troop train had then stopped two miles short of Baton Rouge and protection had belatedly been provided before the trainmaster had eventually learnt from the North Baton Rouge operator that train No. 9 was being held for it. Similarly the troop train had passed Lettsworth without time to clear the schedule of train No. 1. It could not be determined if this was because of information received via the train communication system but it had been established that the system had been used previously to arrange movements without train order authority, in clear defiance of the rules, which forbade such a practice. The accident had been due to the troop train occupying the main track without authority. [689]

On Wednesday 10th October 1951 the Southern Railway introduced a new traffic control system between Burstall and Tuscaloosa, Alabama, United States. All signals and switches (points) within the area were controlled from the train dispatcher's office at Birmingham, Alabama, where the positions of all trains were indicated on a panel. An automatic train stop system was also provided, whereby if a train should pass a signal showing a restrictive aspect, the brakes were automatically applied, although the brake application could be forestalled by the engineer. A very modern form of single-line control for its time. Unfortunately, the new signalling system failed to prevent a major disaster at Woodstock, Alabama, within the area in which it was in operation, only a month and a half later, on Sunday 25th November 1951.

Two passenger trains were involved in the accident. They were described in the newspapers as luxury streamliners and were the southbound *Crescent Limited* of the Louisville and Nashville Railroad and the northbound *Southerner* of the Southern Railway.

First-class passenger train No. 48, the *Southerner,* consisted of diesel-electric unit No. 2923, a passenger-baggage car, a coach, a lounge coach, three coaches, a dining car, and a tavern car, all of lightweight steel construction. It had left Tuscaloosa, the last open office, at 2.01 p.m., six minutes late, passed signal No. 1756, indicating *Approach*, passed signal No. 71L, at the south end of the siding (crossing place) at Woodstock, indicating *Restricting*, entered the siding, met southbound train First No. 47, passed signal No. 73LC, indicating *Stop*, ran through the points at the north end of the siding, entered the single line and stopped 101 feet north of the points. A minute later it was struck by southbound Second No. 47 at 2.35 p.m..

First-class passenger train Second No. 47, the *Crescent Limited,* consisted of diesel-electric units Nos. 2915 and 2801, coupled in multiple-unit control, a mail car, a baggage-dormitory car, three coaches, a dining car and seven sleeping cars. It left Birmingham at 1.55 p.m., 20 minutes late, passed 14th Street Tower, Birmingham, the last open office, at 2.05 p.m., 25 minutes late, passed signals 75R, 1673 and 1699, all indicating *Proceed*, and was travelling at an estimated 45 miles an hour when it collided with train No. 48.

As a result of the collision the front of the engine of train No. 48 was crushed inwards and it was pushed back 148 feet, the whole of the passenger-baggage car was telescoped by the second car and the latter was badly damaged. The front of the leading diesel unit of train No. 47 was likewise crushed inwards, its first four cars were demolished and the next four were badly damaged. Fifteen passengers, Engineer P.J. Powers of train No. 48 and Road Foreman of Engines Sam Pape of train No. 47 were killed, and 52 passengers,

[689] Interstate Commerce Commission report

10 dining car employees, a passenger traffic employee, the conductor of train No. 48, and Engineer Cecil Love, the conductor, the fireman, Flagman H.G. Rogers and the baggageman of train No. 47 were injured. All but one of the passengers who lost their lives had been negroes riding in the *Mississippi*, the baggage-dormitory car of train No. 48; this would have been the Jim Crow coach, where black passengers were forced to ride at this time.

An Anniston, Alabama, newspaper reporter managed to interview a local child, who had been a passenger on one of the trains. '"Long time ago when I used to think about train wrecks I never thought it could be as bad as the one yesterday," said ten-year-old Anne Emerson this morning. The fifth grade student at Woodstock School, daughter of Mr and Mrs Waldo Emerson, of 1111 Greenwood Terrace, was apparently the only Anniston passenger on the Southerner, which collided with The Crescent Limited passenger train at Woodstock yesterday afternoon.

'"I was coming home from Grandmother's. I had gotten on the train at Slidell, La., and Saturday night I had dreamed that the train wrecked. We called Grandmother last night, too, and she said she'd had a feeling something was going to happen. The spunky red haired youngster said she had never imagined being in a train wreck could be as horrible as the one yesterday. She said she was sitting in car S-5, "by a man who didn't talk much. I was trying to go to sleep because I had just finished my book when 'Bang!,' - it happened. I was thrown out of my seat onto the floor. I didn't know what had happened, but I couldn't find my pocket book...so I looked underneath the seat when a little girl started crying, real loud. I looked up and saw another girl about my age who had been at the drinking fountain...all cut up; the glass in the fountain must have broken. There was another man...you know that rail where you hang coats on the back of the seat, he must have hit his nose on it, because his nose was broken and you could see the bone on the nose bridge...Then everybody started screaming and crying. They said to look out of the window. I didn't want to but I did and I saw all these people, bodies, and everything. I didn't know what to do, so I just got back in my seat and sat there till some man told us to get out quick, leave all our bags and everything. He said the car might blow up. There weren't any steps or anything so we had to jump down to the ground. There was a little Negro boy down there, and his head had been cut off. I was scared, but one funny thing happened – the Negro waiter who had waited on me at the table said he saw the *Crescent* coming so he jumped, said he rolled down a gulley and landed in a stream. He told me he just had one neck and he wanted to save that one.

"After we got out of the train and saw all those people bleeding and crying and bodies and everything, I didn't know what to do. There was an older couple, a Mr and Mrs J.A. Gaston, of Jackson, Ga. They came up to me and led me away from the wreck. I never did cry. Finally this Mr and Mrs Lee who just happened to be riding around took us to their home in the country near Bessemer...before I left home Mother had taught me how to make a collect call in case anything had happened. "I called home but nobody answered. Then I called the train station, they didn't know anything. Later I called some friends in Anniston and they said Mother and Father were on their way to Birmingham."

'Meanwhile Mr and Mrs Emerson, who had heard about the wreck in Anniston, but not the extent of deaths, called some friends in Birmingham to see if they could find some information about their daughter. Subsequently they got in their car and rode to Birmingham, listening to the news releases on the radio on their way to the city. "We were numb," Mrs Emerson said, speaking of herself and husband. "We didn't say more than five words to each other all the way to Birmingham." She said the total of announced deaths mounted as they rode.

'The wreck had occurred at 2.30 o'clock that afternoon, young Anne said, and it wasn't until 7 o'clock that night that the Emersons were able to locate their daughter. She had been taken to the bus station along with several other of the passengers, and through friends in Birmingham the Emersons were able to locate Anne. "She's never looked as beautiful to me in all of her life," Mrs Emerson said this morning, explaining how she felt when she first saw her daughter after hours of anguished waiting.' [690]

Another newspaper wrote, 'The Southerner's eight cars were crowded with 214 passengers. Only 119 were on the Crescent. Fuel from the Crescent's engine caught fire after the impact but was quickly put out.

'Cecil Love, Birmingham, engineer on the Crescent, was seriously injured. He told the ambulance driver who brought him to Bessemer to a hospital he, "tried to jump out of the cab door, but I couldn't get it open in time." More than 40 ambulances from Birmingham, Bessemer and Tuscaloosa answered emergency calls broadcast by the state Highway Patrol. Many thousands of curious jammed the single narrow highway from Birmingham to Woodstock, delaying rescue work.

'Jim Buford, Tuscaloosa, boarded the Southerner at that city, its last stop before the wreck. "The crash was like a sledgehammer blow," he said, "I heard the porter yell. I must have blacked out. The next thing I remember, was the voice of a boy screaming, 'Mother, are you dead? Speak to me, Mother.' Another passenger was screaming for morphine. I went back to the club car and got some miniatures of whisky. The porter was hurt, but he said, 'Who's going to pay for those?' and I told him, 'Who cares?' and gave them to some of the injured."

'Woodstock lies in the middle of a coal and iron ore mining area. Most of the coal mines are now closed. Rescue crews reached the wreck scene up a narrow, dusty road from the little town. From it they had to carry stretchers and equipment 200 to 300 yards and then up the steep 40-feet embankment.

'H.G. Rogers, Birmingham, flagman on the Crescent, said he, "Looked around and saw the Southerner coming around a curve. Then I felt the crash." He was thrown clear of the train and was seriously hurt.

'Passengers from the northbound train were brought to Birmingham in buses and left that city for Atlanta at 9 p.m., some six hours late. Crescent passengers were taken to Tuscaloosa to await a train for New Orleans which was routed into Tuscaloosa over a branch of the L&N.

Schults {K.C. Schults, Alabama and Great Southern Railroad district superintendent} said he hoped to clear the main line tracks here during the day. Many of the injured were brought into Birmingham and Bessemer by a rescue train, hastily organised and sent to the scene. [691]

The Interstate Commerce Commission (ICC) stated that the Birmingham dispatcher had routed train No. 48 into the siding at Woodstock for First and Second No. 47. Although the rules did not require it in traffic control territory First No. 47 had been carrying markers to indicate that there was a second section and its engineer had given

---

690 *Anniston Star*, Anniston, Alabama, United States, 26th November 1951

691 *The Decatur Daily*, Decatur, Alabama, United States, 26th November 1951

the appropriate whistle signals to indicate the same thing. Engineer Powers of train No. 47 had known therefore that there was a second section, though not where he would meet it. His fireman told the ICC that there had been some discussion between them over whether they would meet the second section at Woodstock. As dwarf signal No. 73LA had come into view the engineer had remarked to his fireman that it was clear for them and the fireman said that he had looked and had agreed that it was. The engineer had not, as far as the fireman could recall, operated the forestalling device for the automatic train stop. As their engine had reached the points leading out of the loop the fireman said that he had seen that they were wrongly set and he had shouted a warning to his engineer, who had applied the brakes to the first service position. Realising that they were not going to stop short of the points the fireman said that he had called out again when his engineer had made a full service brake application. The brakes had not been applied in emergency at all, and the train had come to a stand with the engine and front of the leading car out on the single line. The fireman and engineer had at once alighted from their engine and the fireman said that he had run forward with an orange coloured cloth in his hand. He had only covered a short distance when train No. 47 had come into view, moving at such a speed that it was clearly not going to be able to pull up in time. He had given stop signals with his orange cloth, which Engineer Love had acknowledged. Engineer Powers had gone towards a lineside telephone booth, where he was struck and killed by derailed equipment.

Engineer Love testified that the signals for his train had all been clear. As they had neared Woodstock he and Road Foreman of Engines Pape had become aware that train No. 48 was fouling the main track and he had applied the brakes in emergency. At about the same time he had seen the fireman giving stop signals, which he acknowledged with his horn. He had managed to reduce the speed of his train from 65 to 45 miles an hour with the brake application. The road foreman had jumped from the engine before the collision.

The Birmingham dispatcher stated that train No. 47 had occupied the OS {OS was short for On Sheet, a station in a timetable and train order area or a control point in a centralised traffic control where the time of the passage of a train was noted on the train sheet [692]) section at the north siding switch at Woodstock at 2.35 p.m. He had not altered the route after setting it for train No. 47 to enter the siding. When the fireman of train No. 48 had reported the accident by telephone he told the dispatcher that he and his engineer had thought that the dwarf signal had been clear, but the switch had not been set for the main track.

As might be expected the signalling at Woodstock was subject to a very thorough and detailed inspection after the collision. Nothing was found which could have caused the dwarf signal to have shown a false clear signal. It was, however, determined that although the dwarf signal first became visible from over 2,000 feet away its red aspect was not perceptible from a distance of more than 800 feet. In contrast, the lights of the corresponding main-track signal, 73LA, situated 77 feet south of the dwarf signal, were easily visible from over 2,000 feet. It appeared from the statement of the fireman that he and his engineer had seen the dwarf signal at a point where its red aspect had not been

---

692 https://cs.trains.com/trn/f/111/p/30761/398610.aspx#:~:text=OS%20means%20%22On%20Sheet.%22,sheet%20in%20the%20proper%20place.

perceptible and had not then looked at it again. [693]

On Sunday 30th March 1952 Class M.131.1 diesel railcar collided head-on with a Class 313.4 light engine near Stvolinky, Czechoslovakia, killing 12 people and injuring another 34. A mix-up in the arrangements for crossing the two trains was blamed for the accident. [694]

At about 23:00 on Sunday 18th May 1952 a major disaster occurred in India. Down mixed train No. 24 from Merta Road to Bikaner collided head-on with Up goods train No. 221 from Bikaner to Merta Road between Palana and Bikaner stations on the Northern Railway. Both engines were badly damaged and the leading three third-class bogies (carriages) of the passenger train were completely telescoped and mounted by the first, second, Inter, and third-class composite bogie. Thirty-six people were killed outright and nine were fatally injured. Those who died included the fireman of the goods train. Sixty-seven people were hurt, of whom 47 were still in hospital almost a week later.

A spare guard brought news of the accident to Bikaner, 7 miles away, half an hour after midnight and a relief train left there at 02:00 with two senior doctors from Bikaner hospital, medical equipment, drinking water, a generator to provide electric lighting and empty coaches for the casualties, at 02:00, reaching the accident site about 02:30. Some of the injured had already been removed from the wrecked train by railwaymen and uninjured passengers, and the lighting provided by the relief train enabled the remaining casualties to be removed.

The Maharaja of Bikaner, who had been attending Parliament in Delhi, flew to the scene with railway officials. [695]

The lines were of course completely blocked, the two engines being so interlocked that it was difficult to separate them. Following the accident two Up passenger trains had to be cancelled and the passengers of four other trains had to be transhipped at the accident site. The line was reopened at 11:30 on Tuesday 20th. Damage to locomotives, rolling stock and the permanent way was estimated at 109,000 rupees. [696]

The collision was blamed on a wrong line clear having been given to one of the trains, the line being worked by paper line clear ticket. [697]

The Southern Railway single line between Mulacalacheruvu and Battulepuram, India, was worked by block instruments in conjunction with ball tokens but at about 00:25 on Saturday 13th June 1953 a total failure of the instruments and telephones occurred and there was no means of communication between the two stations. This was discovered by the assistant stationmaster at Mulacalacheruvu just before 03:30 when he attempted to obtain line clear for No. 1148 passenger train. He therefore resorted to the emergency

---

693 Interstate Commerce Commission report

694 https://de.wikipedia.org/wiki/Liste_von_Eisenbahnunf%C3%A4llen_in_Tschechien#1946%E2%80%931959

695 *The Times*, London, England, 20th May 1952

696 https://eparlib.nic.in/bitstream/123456789/55006/1/lsd_01_01_28-05-1952.pdf

697 https://www.rsdebate.nic.in/bitstream/123456789/590605/1/PD_01_21051952_5_p333_p336_5.pdf

procedure by which only timetabled trains were despatched between stations, advised the guard of the train of the situation, and gave him an *Advice of the Next Train* order indicating from which station the next train should start. At 14:25 this method of working was still in operation and passenger train No. 1145 entered the single-line section from Battulepuram, with an *Advice of the Next Train* order indicating that the next train would also start from Battulepuram. In accordance with this order the stationmaster then dispatched goods train No. 3303 from Battulepuram at 23:05. Meanwhile, in complete disregard of the order, the stationmaster at Mulacalacheruvu despatched mixed train No. 1144 from his end at 23:00. In the ensuing head-on collision 67 people were killed and 69 were more or less injured. Damage was put at 2,160,428 rupees. The accident would doubtless have been less serious had both drivers been proceeding at reduced speed, as they should have been while the emergency working was in operation. [698]

Only three days after the above accident, on Tuesday 16th June 1953, another single-line collision occurred in India, this time on the North Eastern Railway, and due to a mistake by a driver rather than a stationmaster. An Up train had been despatched from Katihar West for Semapur with a proper *Line Clear* order, to meet Down Mixed Passenger and Parcels train No. 442 at the latter station. Train No. 442 arrived at Semapur at 22:50 and left there again at 22:52, without a *Line Clear* order. It was not vacuum fitted so its guard was unable to stop it by applying the brakes, while efforts by the Semapur station staff to attract the attention of the driver were also in vain. The trains met in mid-section at about 23:00. Five people were killed outright, 15 were seriously injured, one of whom died in hospital, and 35 were slightly hurt. Damage was estimated at 1,009,000 rupees. [699]

In what appears to have been a single-line collision on Friday 17th December 1954 a diesel railcar consisting of a single coach and a troop train carrying an infantry battalion and also about 70,000 litres of petrol collided head-on between Le Kroub and Ouled Rahmoun, Algeria, at a point where there was a curve restricting visibility and the line ran in a cutting. A fierce blaze broke out following the collision and 15 people on the railcar died either from the impact or in the fire. The soldiers were able to rescue some of the railcar passengers but only from the rear of the vehicle, the flames keeping them away from the front. According to different reports either no soldiers were hurt or else four were, one seriously. The accident was attributed to 'negligence at a switching station.' [700] [701] [702]

Two freight trains met one another, probably in a single-line collision, near Riobamba, Ecuador, on Thursday 7th July 1955, killing three railwaymen and three soldiers guarding the trains. The railway authorities blamed the accident on a strike which had then been in progress for 11 days, because strikers had cut telegraph wires, resulting in a lack of

---

698 https://archive.org/stream/dli.csl.1063/1063_djvu.txt

699 Ibid.

700 *Somerset Daily American*, Somerset, Pennsylvania, United States, 17th December 1954

701 *The Advocate*, Burnie, Tasmania, Australia, 20th December 1954

702 *European Stars and Stripes*, Pfungstadt, Germany, 19th December 1954

communication between stations. [703] [704]

At 15:36 on the afternoon of Sunday 11th September 1955 passenger train P 4503 from St Veit an der Glan to Villach collided head-on with freight train G 5368 at km. 345.3, about midway between the stations of St Martin-Sittich and Mautbrücken, Austria. [705] The passenger train consisted of engine No. 52 2428 and seven coaches; the goods train consisted of engine No. 52.6810 and an unspecified number of waggons. Both engines overturned following the collision and fell down an embankment and the leading carriages of the passenger train were telescoped. Nine passengers and the driver of the freight train lost their lives and about 40 people were injured. After the accident rescuers worked by floodlight to recover the dead and the injured. [706] [707]

Sixty years later one of the passengers on the train, named Oskar Telsnig, who had been six years old at the time, spoke of his experiences. He related that he had taken the train to the Wiesenmarkt at St Veit with his father to buy a new pair of shoes for his first day of school the next day. They had also bought *kokosbusseri* (coconut biscuits – an Austrian speciality) and Turkish honey and had been on their way home to Feldkirchen when the accident had happened. Suddenly there had been a terrific bang and passengers, objects and parts of the carriage had flown through the air. He was trapped in the wreckage of the splintered wooden coaches. All around him trapped and injured passengers were screaming. Help soon arrived from locals who had been alerted by the noise of the collision but it was hours before he and his father could be released. There was only one fire engine at Feldkirchen and it took it some time to get to the scene. Finally he and his father were taken to hospital in Waiern and it was some weeks before he had recovered enough to start school, without his new shoes which had been lost in the accident. His father told him that it brought back memories of his wartime experiences fighting in Stalingrad. [708]

The accident was blamed on the pointsman at St Martin-Sittich, who had apparently been drinking and had, by shifting the wrong set of points, let the goods train out onto the single line, whence it had proceeded towards Mautbrücken. He was arrested.

A passenger train and a goods train again collided head-on on the single line between Katihar West and Semapur stations, Bihar, India, on the North Eastern Railway, on Saturday 7th January 1956. This accident was more serious than the previous one and 11 passengers lost their lives. [709]

As an economy measure to help reduce the cost of working poorly trafficked lines Statens Järnvägar, the Swedish railway company, introduced in 1955 Säkerhetsordningen (Safety Ordinance) No. 100, which permitted trains to cross one another at unstaffed

---

703 *The Daily Telegram*, Eau Claire, Wisconsin, United States, 7th July 1955
704 *The Independent*, Long Beach, California, United States, 7th July 1955
705 Wikipedia – Liste von Eisenbahnunfällen in Österreich
706 *The Times*, London, England, 12th September 1955
707 *Amarillo Daily News*, Amarillo, Texas, United States, 12th September 1955
708 https://www.meinbezirk.at/feldkirchen/c-lokales/zugsunglueck-jaehrte-sich-zum-60-mal_a1575442
709 *The Times*, London, England, 9th January 1956

stations, as had been the practice for many years in North America, but not, presumably, previously in Sweden.

One of the stations which became unstaffed as a result of this economy measure was that at Akkavare. On Wednesday 28th March 1956 train No. 1041 from Gällivare to Arvidsjaur, consisting of Class YCo6 diesel railbus No. 839, arrived at Akkavare, where it was due to cross local freight train No. 8524, consisting of Class E2 2-80 steam locomotive No. 1241 and an unspecified number of waggons. The station work at Akkavare having been completed Motorman Göte Holmström clean forgot about the timetabled meet with the freight train and set off towards Arvidsjaur. His railbus had covered just over 3 kilometres when it collided with the freight train on a curve north of the Byske River bridge. The much more massive freight train crushed the front of the railbus, which was also pushed back about 150 metres. Sixteen people died and 10 more were injured. The railbus was scrapped after the disaster but the steam engine was little damaged. It remained in service and was later preserved.[710] [711]

One of the passengers on the railbus was 27-year-old Anne-Marie Lundberg, who lived in the village of Ravenjaur, at this time inaccessible by road, with her forestry worker husband. She was on her way to Arvidsjaur to do some shopping. She remembered the railbus stopping at Akkavare, then setting off again and soon travelling at about 80 kilometres an hour on the gentle descent towards Arvidsjaur. Suddenly Motorman Holmström had come back into the passenger compartment and shouted for everyone to get down on the floor. A moment later, as she had been taking hold of the door handle to the rear compartment, the crash had come. Anne-Marie was seriously hurt in the crash and was in hospital 17 days but she was one of only two or three passengers from the front portion of the railbus to survive the collision. In the months that followed she had to make repeated trips to hospital for her follow-up care, travelling by the same train, which understandably was a traumatic experience.[712]

The practice of trains crossing one another at unstaffed stations was abolished the day after the disaster.

The Longmoor Military Railway was a British Army railway designed to train soldiers in railway construction and operation. At about 8.45 a.m. on Saturday 13th October 1956 it suffered a serious accident. The 8.40 a.m. ordinary passenger train from Liss to Longmoor collided with a works train which had left Longmoor for Liss at the same time. The passenger train consisted of Class 8F 2-8-0 steam engine No. WD 512, a brake van and a coach carrying passengers for Longmoor Camp. The works train consisted of 0-6-0 diesel shunting engine No. WD 877 *Bari*, a box van and four waggons. Control of the single line between Liss Forest Road and Weaversdown Halt signal box, England, was by telephone and ticket. The railway as a whole was worked by the Military Railways Rule Book, which was based on the British Railways rules and regulations.

On the morning of the accident, which was foggy, Blockman (the Army name for a signalman) Sapper Wilfred Wareing wrote out a ticket for the Liss Forest Road – Weaversdown Halt section before the work train had arrived and gave it to the fireman of

---

[710] https://banvakt.se/arvidsjaur-kabdalis/olycksplats-akkavareolyckan/

[711] Wikipedia - Järnvägsolyckan I Akkavare

[712] https://norran.se/nyheter/langlasning-16-doda-i-varsta-lokala-tagolyckan-nagonsin-anne-marie-var-ombord-556983

the passenger train in exchange for the Liss – Liss Forest Road one engine in steam train staff, thereby saving himself a walk up the station platform after the work train had arrived. Seeing that there were no passengers to pick up at Liss Forest Road station and his fireman having collected a ticket, Driver Sapper Albert Hole failed to stop at all at the station, as he was booked to, ran past the Starting signal at *Danger* in the fog, and entered the single line at 8.44 a.m. There were no trap points protecting the start of the single line. In the collision which followed the locomotive and vehicles of the passenger train were only slightly damaged but the leading box car of the works train was telescoped by the following waggons. Of the 19 soldiers of the Royal Engineers Corps who were riding in the box car three were killed outright, three were fatally injured and eight were hurt, one seriously.[713] [714]

After the accident soldiers from the passenger train went to the assistance of their comrades in the wrecked car, while one man ran a mile to Weaversdown Halt signal box to give the alarm. The location of the accident was inaccessible to ambulances and the injured were taken by train to Longmoor Camp, where they were given first aid before being taken to hospital.[715]

At the inquest on the victims of the accident at Whitehill the coroner remarked that the driver of the diesel train and the signalman at Longmoor (Weaversdown Halt) had acted with the utmost regularity, but it was "a different story when you turn your attention to what was going on at Liss Forest." The jury returned a verdict of Accidental Death with a rider that, "We consider in view of the fact that the line is worked by young, rather inexperienced national servicemen in the main, that the discipline thereon should be more strictly enforced."[716]

Finland suffered what was stated to be its most serious accident in peacetime on Friday 15th March 1957 between Kuurila and Iittala. Involved in the collision were overnight train P64 from Tornio to Helsinki, consisting of a Class Hr1 Pacific locomotive No. 1005 hauling sleeping cars, and express train MP 41, which consisted of Class Dm4 diesel railcar No. 1607. Both trains were reportedly running over an hour late. A misunderstanding between the staff at the two stations caused the accident. The trains were timetabled to cross at Kuurila but the employee there decided to advance train P64 to Iittala to cross railcar MP 41 there. The Kuurila railwayman was under the impression that his colleague at Iittala had agreed to this alteration, while the latter still thought that the meet would be at Kuurila. Both therefore let their trains depart from their respective stations. In the collision, which occurred during a blizzard, the heavy steam locomotive overwhelmed the lighter diesel railcar and it was in this railcar that most of the fatalities occurred. The steam engine overturned though and its driver died and his fireman was injured. The leading sleeping car was badly damaged. The collision occurred in a wooded area which was difficult to access, so horses and sleighs were employed, as well motor ambulances and cars. A full infantry company from Parola Camp assisted in the rescue.

---

[713] https://api.parliament.uk/historic-hansard/written-answers/1956/oct/30/longmoor-military-railway-accident#column_128w

[714] Wikipedia – Longmoor Military Railway

[715] *The Times*, London, England, 15th October 1956

[716] Ibid, 27th October 1956

Twenty-eight people died in the accident and 19 were injured. [717] [718]

Turkey suffered a major catastrophe on Sunday 20th October 1957. The trains involved were No. 8, the *Simplon – Balkan Express*, which left Istanbul Sirkeci Terminal as a ten-coach semi-local train with saloon, couchette, and sleeper cars for Paris France, hauled by 45171 class steam locomotive No. 45501: and train No. 3, a three-car diesel multiple unit from Edirne to Istanbul. Train No. 8 had started at 21:50, train No. 3 started at 16:00, and they collided at 22:45 between Ispartakule and Yarimburgaz stations.

That an accident was going to happen became evident when Stationmaster Baki Imözü, at Yarimburgaz, advised his colleague, Stationmaster Cahit Firat, at Ispartakule, that train No. 8 was on its way to him. Stationmaster Firat had already cleared train No. 3 to proceed towards Yarimburgaz and so, barring a miracle, a collision on the single line was inevitable. The two railwaymen did the only thing that they could and telephoned their superiors to advise them that there was likely to be a collision at a point about 34 or 35 kilometres from Istanbul, and to send the emergency services to that area. The trains met closer to kilometre 35 and again the lighter multiple-unit train was overwhelmed by the more massive engine and coaches of the express. Eight-five people were killed outright and another ten died later, while 150 were injured. All the deaths and injuries occurred in train No. 3, which had been crowded; five of the dead were railwaymen.

Fire engines were despatched to the scene from Istanbul as soon as news of a possible accident was known and a relief train was also sent at 23:30 and returned with the injured passengers. The noise of the collision was also heard by a guard at the nearby munitions depot of the 213th Infantry Regiment and members of the military helped in the rescue work. According to some accounts a fire broke out after the collision, [719] though another account states, however, that the fireman of train No. 8 removed the burning coals from his engine and prevented a fire. Oxy-acetylene torches were needed to cut through the wreckage and release sone of the victims. Ambulances took the injured from the relief train to ten different hospitals. The line was reopened at 20:00 the next day. The two stationmasters were arrested.

An investigation into the catastrophe revealed that apart from the mistake made by the stationmasters there were other factors which had contributed to its severity:

(a) The engine of train No. 8 had not had a headlight.

(b) The make-up of train No. 8 had made a quick brake application impossible.

(c) Train No. 3 had been composed of old, wooden carriages which had been poorly able to resist the forces of a collision.

In March 1959 Stationmasters Firat and Imözü were tried by the First Criminal Court in Istanbul and found 60 and 40 per cent guilty respectively of causing the deaths of the victims through negligence. They were sentenced to three and two and a half years' imprisonment and fined 2,250 and 2,000 Turkish lira respectively. In addition the Türkiye Cumhuriyeti Devlet Demiryollan (Turkish State Railways) were ordered to pay compensation to the families of the dead and to the injured. [720]

---

[717] Ibid, 16th March 1957

[718] https://de.wikipedia.org/wiki/Eisenbahnunfall_von_Kuurila

[719] *The Terre Haute Star*, Terre Haute, Indiana, United States, 22nd October 1957

[720] Wikipedia – Yarimburgaz train disaster

On Saturday 16th November 1957 the Société nationale des chemins de fer français (SNCF) suffered its fourth bad accident in five months when a Michelin railcar collided head-on with a freight train at level crossing PN No. 68, near Chantonnay, France.

The railcar, No. 812, had arrived at Chantonnay station at about 12:40 and was bound for Thouars. It consisted of a motor car and a trailer. Riding in the leading car were students from the Notre-Dame-de-la-Forêt agricultural school at La Mothe Achard, returning home while the school was closed due to an influenza outbreak, while the passengers in the trailer included an engaged couple, Jean Bourdet and Jocelyne Fuseau. Jean was on leave from military service in Algeria and they were on their way to announce their planned marriage to the bride's godmother.

Coming from the opposite direction, from Chavagnes-les-Redoux, and running some minutes late, was a freight train hauled by steam locomotive No. 30.817, which should have crossed the railcar at Chantonnay. The Chantonnay stationmaster, while the railcar was waiting at his station for the freight train, was busy registering luggage, having telegraphed to the next station at 12:48:

Chantonnay to Chavagnes. Although still waiting for train 30.817, I am reporting train 812.

Then at 12:55 Motorman Daniel Lenain revved up the motor of the railcar and Assistant Stationmaster Paul Brunet, who was out on the platform, without thinking about what he was doing and forgetting that the freight train had not yet arrived, gave the departure signal with his flag. Conductor Marcel Godet rang the bell to his motorman and the latter started away. The deputy stationmaster then turned to a colleague and made some casual remark, at which his colleague asked him about the freight train. This brought the assistant stationmaster to his senses and he realised what he had just done. He rushed to the signal frame to operate the *carré rouge* (Stop) signal but the railcar did not stop. He claimed that he had closed (placed at *Stop*) the signal before the railcar had passed it but Motorman Lemain alleged that it had still been open (at *Clear*) when he passed it. The truth could not be decided, even at the inquiry into the accident. Such stop signals were usually provided with detonator placers which placed fog signals on the line while the signal was at *Danger* but here these had been removed, reportedly following complaints from local residents about the noise of the explosions.

Having failed otherwise to stop the railcar the assistant stationmaster mounted his bicycle and set off down the track after it. To begin with he was not too far behind it and he waved frantically to the passengers in the trailer, who could see him, in the hope that one of them would pull the communication cord, but they seem to have supposed that he was merely a passenger who had missed his train and ignored him. Soon the railcar had picked up speed anyway and outdistanced him.

At crossing PN No. 68 Madame Jarny, the crossing keeper, had lowered the crossing barriers upon a bell ringing to alert her as to the approach of the freight train. She saw the indicator light showing the direction from which the freight train was approaching, then to her astonishment the other indicator light came on, meaning that there was a train coming from the opposite direction along the single line, and then she could hear both the horn of the railcar and the sound of the steam locomotive. Crossing keepers were sometimes supplied with flares to warn trains but because of the proximity of Chantonnay there were none at crossing PN No. 68. There was nothing that Madame

Jarny could do, even had there been time, and shortly afterwards the two trains collided right in front of her, at 12:58. Yolande Auneal, a cyclist on her way home from school, who had been waiting at the lowered barriers for the freight train, also witnessed the collision. The assistant stationmaster, when he reached the scene and saw the carnage, collapsed.

Motorman Lemain applied his brakes when he saw the freight train, as did Driver Robert Billy and Fireman Dugas on the steam engine when they saw the railcar, but the distance between the trains was too short, and the railcar was still moving at 60 kilometres an hour and the freight train at 40 kilometres an hour, when they met one another. As in other head-on collisions of steam engines with lighter trains (see for example the Akkavare and Yarimburgaz accidents described above in this chapter) the railcar was no match for its 'opponent' and it was ripped open and crushed by the much heavier steam engine. All but two of the passengers in the leading car were killed outright, though Motorman Lemain in his raised cab survived and Conductor Godet, who must have been in the trailer at the time, was only hurt. In all 29 passengers lost their lives and 22 were injured. Diesel from the damaged tank of the railcar was spilt and a small fire broke out but was quickly put out with an undamaged fire extinguisher from the car. Sitting in the trailer the previously mentioned engaged couple were both hurt but survived. They were wed in 1958, Jean Bourdet went on to spend all his life working for the SNCF and they were still married 50 years later. [721]

The sound of the collision was heard all over Chantonnay. The emergency siren, the modern replacement of the tocsin church bells, was sounded and a fire engine quickly set off for the scene, with the three firemen who were at the fire station at the time. Other rescuers followed, together with firefighters and doctors from Chantonnay, La Roche-sur-Yon and La Chaize-le-Vicomte. The rescue operation was somewhat helped by the proximity of the level crossing and the fact that there had previously been a second track, giving easy access to the wreckage. Ambulances and private cars took the injured to La Roche-sur-Yon and Fontenay-le-Comte. The dead were taken to Chantonnay town hall, part of which was converted into a *chapelle ardente*, staffed by Sœurs de la Sagesse (a religious charity). Families of the victims began arriving at the town hall at 16:00 and there were heart-breaking scenes as they identified loved ones among the bodies. [722]

During the evening an SNCF breakdown crane arrived from Thouars to begin the work of clearing the line, while firemen were still at the scene until 03:00. The Minister of Public Works, Edouard Bonnefous. M. Armand, Chairman of the SNCF, and other officials visited the scene on Sunday morning. Later M. Bonnefous made a statement in which he referred to public anxiety over the repeated accidents on French railways, "due to serious and inexcusable shortcomings on the part of railway employees responsible for safety," and promised that the results of the investigation into the disaster would be made public as soon as possible. The assistant stationmaster and the motorman of the railcar were taken into custody for examination by order of the *juge d'instruction* (examining magistrate). [723]

[721] https://larochesuryon.maville.com/actu/actudet_-1957-le-drame-ferroviaire-de-chantonnay-_15-467305_actu.Htm

[722] https://www.la-chouette-de-vendee.fr/blog/params/post/1518301/laccident-de-chemin-de-fer-de-chantonnay

[723] *The Times*, London, England, 18th November 1957

On Friday 16th May 1958 a passenger train and a freight train collided head-on, probably on a single line, between Afyon and Konya, Turkey. Twelve people, all of them apparently railwaymen, were killed and four people were badly hurt. [724]

Sweden suffered another single-line collision on Friday 5th September 1958 between Grycksbo and Sågmyra on the line from Falun to Rättvik. Train No. 848, consisting of railbuses YBo6 No. 1097, UBFo6 No. 1780, UBFo6 No. 1763 and YBo6 No. 1801, was given an order to meet extra train No. 4941 at Grycksbo. The extra train was a new diesel locomotive constructed by the ASJ company of Falun and which they hoped to sell to Statens Järnvägar, the state railway company. It was returning from Rättvik to Falun on a demonstration run, pulling a number of coaches. Having received the crossing order at Falun the motorman of train No. 848 probably stuffed it into the pocket of his jacket and did not attach it to the clamp in the cab of the leading railbus next to the working timetable. Meets of trains at Grycksbo were not very frequent; at some periods there was only one a week.

When train No. 848 arrived at Grycksbo the clerk was busy dealing with a passenger with a difficult ticket query so the stationmaster, probably without being asked, took over the despatching of the train. The station work was soon completed, whereupon the stationmaster gave the train the 'Right away' signal, without having checked the train register or clearance book and unaware therefore of the planned crossing with the extra train. The motorman, meanwhile, had obviously forgotten about the order that he had received at Falun and having been given the departure signal by the stationmaster, he set off without further thought.

As soon as the stationmaster returned to the office the clerk, who had seen and heard train No. 848 leave, enquired of him if the extra had arrived. Realising that the railbus should have been held the stationmaster then ran out with a red flag but the train was too far away and the motorman did not look back. No starting signals were provided at Grycksbo but some level crossing barriers at the north end of the station were cranked up in the vain hope of stopping the train. When this failed, calls were made for the emergency services, in the sure knowledge that an accident was going to happen.

Yet again the railbuses were no match for the diesel locomotive. The leading railbus was destroyed and the second was badly damaged. Ironically, the rear two railbuses, which were undamaged, had also been locked out of use. Seven passengers and two railwaymen, presumably the motorman and conductor, were killed and 11 passengers were injured. There had been about 30 passengers on the train in all. Most of the dead and injured were schoolchildren. The crew of the extra train were unhurt.

Badly affected by the accident, although he was not directly responsible for it, the station clerk was on sick leave for several months. The stationmaster accepted full responsibility for the collision. He was relegated to a non-safety critical position for the few years which remained before his retirement. Three years after the accident the signalling was modernised and starting signals were installed. [725] [726]

---

724 *The Times*, London, England, 20th May 1958
725 Wikipedia - Järnvägsolyckan I Grycksbo
726 https://www.jvmv2.se/forum/index.php?mode=thread&id=180551#p180620

In common with other railways in Germany the southern part of the formerly double-track Hürth – Ehrang line was singled at the end of the Second World War, the second track in this case being removed by the French for reparations rather than by the Russians. The line was still single in 1958 between Urft and Nettersheim, West Germany, with an intermediate crossing place at Gronrechtsmühle (also known as Posten 41). On Wednesday 24th September 1958 there was a serious head-on collision on this stretch, involving northbound local passenger train N 3515 from Jünkerath and Köln and a southbound Class 23 light engine from Krefeld on its way to Trier. The supervisor who was on duty in the signal box at Gronrechtsmühle cleared signal F for the passenger train to run into track 2 when he was advised it was leaving Nettersheim. Then at 05:57 he accepted the engine from Urft. His *Ausfahrsignal* E (starting signal and *Ausfahrvorsignal* (distant signal to the starter) were at *Danger*, protecting the route back onto the single line. According to his later statement the supervisor said that he had watched the engine approach and to his horror instead of it slowing down and stopping at *Ausfahrsignal* E it had run past the signals at unreduced speed, had burst the points and had continued onto the single line towards Nettersheim. The supervisor left his box and blew his signal horn at the receding engine, in vain. Returning to his box he replaced signal F to danger and shouted to his colleague at Posten 43 (1.4 kilometres from Gronrechtsmühle and 2 kilometres from Nettersheim) to stop the engine. Posten 43 had control of two level crossings. The man there stated later that when he had first received the message to stop the engine he had been stunned, then recovering himself he had seized a red hand lamp and a detonator and run towards the advancing engine, which he could see coming around a curve 600 metres away. The crew of the engine did not appear to see his red light. Then he saw the passenger train coming from the opposite direction and only 200 metres away, so that he had to jump clear to avoid being run down. There had been no time to put down the detonator and now it was too late anyway. Engine and train collided head on at 06:05 close to where he was standing. He immediately reported the accident to the stations on either side and requested assistance.

Both engine crews and seven of the 25 passengers were killed in the collision, two passengers were seriously hurt and five slightly injured. [727]

Following the accident blood tests on the driver of the light engine were carried out. These showed no carbon monoxide, eliminating poisoning as a cause of his failing to stop. The blood alcohol, however, was 0.2%, which would appear to suggest that the driver was to some degree at any rate, intoxicated, but a German author dismisses this and suggests that the driver simply had his mind elsewhere and was not concentrating. [728]

Two Rede Ferroviária Federal trains on a line of the former Estrada de Ferro Central do Brasil, one crowded with workers returning home at the end of the day, collided at about 18:27 between Engenheiro Goulart and Engenheiro Trindade stations, Såo Paulo, Brazil, on Friday 5th June 1959.

Electric train UP-240, had left Roosevelt station at 18:05 for Såo Miguel Paulista, driven by Motorman Agenor André Pereira, while electric train UP-237, had left Såo Miguel Paulista, also at 18:05, driven by Motorman João Rosas da Costa. In the collision

---

[727] https://de.wikipedia.org/wiki/Bahnstrecke_H%C3%BCrth-Kalscheuren%E2%80%93Ehrang

[728] *Von Siegelsdorf nach Aitrang* by Hans Joachim Ritzau, pgs. 92-4 (in German)

the leading car of train UP-240 was demolished and reduced to a heap of scrap metal, with train UP-237 mounting the wreckage. The passengers on train UP-240 were thrown on top of one another, lost their belongings, and started calling desperately for rescue. Fifty-one ambulances, 25 radio patrol vehicles and two fire engines went to the scene. An estimated 10,000 sightseers made the rescue work difficult.

At 22:00 the Hospital das Clinicas, where the injured had been taken, appealed to blood banks, health centres and private individuals for blood. The appeal was made on television, interrupting the programme 'O Céu é o Limite.' At the same time the Health Department requested all highway doctors to go to the Araçá mortuary to perform post mortems on the dead and the Hospital de Câncer offered blood to the accident victims.

Following the disaster angry people attempted to wreck Engenheiro Goulart and other stations and had to be subdued by soldiers armed with machine guns with fixed bayonets. The soldiers (who numbered nearly 500) had been called by the municipal authorities, who had found themselves powerless to stop the rioting.

Responsibility for the disaster was quickly attributed, for at 22:45 the same day a director of the railway company stated that it rested entirely on Motorman Rosa da Costa of train UP-237, who had failed to check the order given to him by the Engenheiro Goulart station agent. Secondly, shortly after the station agent had issued the order he realised that he had make a mistake and that, train UP-237 having left his station, a collision was likely to occur. However, nothing then could be done. On the other hand it was up to the motorman to check the order and if he had done so he would have seen the mistake and would not have started. Motorman Rosa da Costa had fled, aware of his culpability. As for Motorman Pereira, of train UP-240, who had lost his life in the collision, he had been travelling with a perfectly valid order and had not been in any way to blame. [729]

By the following day it was reported that 49 people had died and 110 had been injured.

About 3,000 workmen daily made the same journey over the line between their homes and work. Traffic to Rio de Janeiro was not interrupted because the accident happened on the bypass line. [730]

Two days after the disaster the number of fatalities had risen to 56 and there were many heart-breaking scenes at the mortuary as relatives, mostly women, came to identify the badly mutilated bodies of the victims. Meanwhile the number of injured at the Hospital das Clinicas was said to be 132, of whom 61 were said to be in a serious condition.

It was now (Monday 8th) stated that the three employees responsible for the disaster were Station Agent Agenor José Braga, Motorman Rosa da Costa of train UP-237, who had accepted the order to proceed, and Conductor José Nascimento of the same train, all of whom had reportedly absconded. [731]

By Tuesday 9th the three railwaymen had surrendered themselves to the authorities and each blamed the other two for causing the disaster. Station Agent Braga was accused of having prematurely ordered train UP-237 to depart, Motorman Rosa da Costa of departing with an invalid order and Conductor Nascimento of having handed over the

[729] *O Estada de São Paulo*, São Paulo, Brazil, 6th June 1959
[730] *Jornal do Brasil*, Rio de Janeiro, Brazil, 6th June 1959
[731] Ibid, 8th June 1959

wrong staff. [732]

On 12th June Motorman Rosa da Costa gave an account of the events leading up to the collision. He stated that had arrived at and stopped at Engenheiro Goulart, where Conductor José Nascimento had given him the order authorising him to continue his journey towards Roosevelt station. As the lighting of both the station and his train had been poor he had not attempted to verify that the order was in accordance with the regulations but had trusted the station agent and the conductor. Only after the accident had he realised that a mistake had been made. He admitted that had he verified the order before proceeding the accident would have been averted. He added that Conductor Nascimento had informed him that he would cross train UP-40 at the next station, Engenheiro Trindade. After the accident, he added, he had thrown away his cap (so as not to be identified as the motorman?) and then helped for hours in the rescue of the injured and the removal of the dead. [733]

In what was probably a single-line collision two freight trains collided head-on near Sivas, Turkey, on Monday 6th July 1959. Ten people reportedly died and eight were seriously injured. [734] [735]

Czechoslovakia suffered a major catastrophe on Monday 14th November 1960 near Stéblová. The trains involved were No. 608 from Liberec to Pardubice and No. 653 from Pardubice to Hradec Králové. The former train consisted of steam engine No. 354.7128, a mail van, a freight car, a passenger carriage, a service car and eight passenger carriages. The latter train was a railcar consisting of a motor unit, four passenger cars and another motor unit. Train No. 653 had left Pardubice a minute late at 17:22 and train No. 608 had left Hradec Králové seven minutes late at 17:25. The line was single track and the trains were timetabled to cross at Stéblová, where there were three tracks.

Train No. 608 arrived at Stéblová first at 1740 and ran into track No. 1, the signalman having set the route up for this train to use track No. 1, rather than track No. 2 as usual, and then left his office. The pointsman went away to operate No. 4 points for the departure of train No. 653 and then open the crossing gates after the arrival of train No. 608. Returning to his office the signalman reported the arrival of train No. 608 to the next station at Čeperka, obtained the road from there for train No. 653, and set the route for the latter train's arrival on track No. 2. Exactly what happened next was never explained. The two conductors of train No. 608 alighted from it. It was a very foggy evening, with visibility restricted to 50 metres. The senior conductor imagined that he had glimpsed a green light through the fog and shouted to the junior conductor that the signalman had given them the "Right Away" signal, though in fact the signalman was in his office at the time, rather than on the platform. The junior conductor had not seen the green light but passed the "Right Away" signal on to the engine crew. There was also a head guard, who did not see the signal either, but assumed that there had been one. Some of the train passengers said later that they thought that they had seen a green light, but they were not certain. Possibly the green light, if there had been one, was the junior guard's hand lamp,

---

[732] Ibid, 9th June 1959

[733] *O Estada de São Paulo*, São Paulo, Brazil, 12th June 1959

[734] *The Times*, London, England, 7th July 1959

[735] *Journal de Genève*, Genève, Switzerland, 7th July 1959

which could show a green aspect. Nor had the train driver seen the green light, but he took it for granted that there had been one and started his train away. There were also fixed signals at the station but neither the driver, his fireman, who was engaged with the firing and had just noticed that the water level in the boiler was low, nor the head guard, had noticed that the starting signal was at danger. The head guard confessed later that he had known that the two trains normally crossed at Stéblová, but assumed that the crossing had been altered.

Seeing train No. 608 departing the signalman had rushed from his office, waving a red lamp, and blowing a whistle. The senior conductor should have been keeping a lookout as the train left the station but had begun checking tickets instead. A pointsman jumped on his bicycle and attempted to pursue the train, while the signalman attempted to telephone his colleagues 2 kilometres down the line to stop train No. 653, but the staff were outside watching that train and only answered the phone after it had passed.

The two trains collided at 17:42 or 17:44 according to different reports; each was travelling at about 60 kilometres an hour. The freight car of train No. 608 demolished the leading mail van and the service car demolished (probably telescoped) the passenger carriage in front of it; the remainder of the train was little damaged. As might be expected the railcars fared worse. The leading motor car mounted the steam locomotive, the first passenger car crashed into the locomotive and the next three passenger cars were telescoped into the first. Only the rear motor unit escaped. To make matters worse the crew of the steam engine dropped its fire for fear of a boiler explosion and the hot coals ignited diesel fuel leaking from the damaged railcar, starting a fire.

No less than 110 or 118 people died and 106 or 110 were seriously hurt. The rescue operation involved medical staff, police, soldiers from a nearby military school and locals. Most of the survivors were released by 20:00 and the last, who died on the way to hospital, at 23:00. The injured were taken to hospitals in Hradec Králové and Pardubice, even a maternity home being brought into use. Breakdown trains were sent to the scene and the line was reopened at about 12:00 the next day. Damage was estimated at about 477,000 koruna (the average monthly wage at the time was 1,300 koruna).

The investigation into the accident found that it had been due to a flagrant violation of elementary railway regulations. Six members of the crew of train No. 608 subsequently stood trial at Hradec Králové, of whom five were convicted. The head guard was sentenced to 5½ years' imprisonment, the driver to 4½ years (he was released after 3), the senior conductor to 4 years, the fireman to 1½ years and the junior conductor to 1 year. All were also suspended from the service for several years. A train attendant was acquitted. [736]

---

[736] Wikipedia – Stéblová train disaster

CHAPTER 7

RUNAWAY TRAINS AND VEHICLES

A variety of situations can lead to a train or part of a train running away out of control. If there are railway staff in attendance the train or engine can only run away out of control if there is a failure of the brakes or if the brakes are incapable of holding it on a steep gradient. In the case of unattended vehicles, the action of gravity alone is sufficient to precipitate a runaway. In contrast to the situation on a road a train or vehicles can sometimes run for many miles without mishap, an accident occurring only if and when the runaway meets with another train, comes to the end of the line, or gets to a curve too sharp for the runaway to negotiate at the speed at which it is travelling.

Express train No. 294 from Shinjuku to Odawara developed a problem with its brakes between Hatana and Shibuzawa on Monday 28th January 1946, ran into a siding and was derailed at Tsurumaki station, Japan, at 10:30 on the Odakyu Railway. About 20 passengers were killed, 30 seriously and 50 slightly injured. The injured were treated at the Kokakuen Hotel and elsewhere. [737] Another source states that there were 30 killed and 165 injured. [738]

A train on the Onomichi Railway suffered a brake failure, ran away backwards down an incline, and became derailed at Kinosho, Japan, on Tuesday 13th August 1946. The Kyodo news agency reported 32 dead and 57 injured and the final figures were 37 dead and 101 injured. [739] [740]

Disaster struck the eastbound *Red Arrow* express of the Pennsylvania Railroad at 3.21 a.m. on Tuesday 18th February 1947 on Bennington curve, a miles from Gallitzin, Pennsylvania, United States. Fifteen passengers, six mail clerks, Fireman J.L. Parasock of the pilot engine and Engineer M.E. McCardle and Fireman R.K. Henry of the train engine died in the accident and 121 passengers, seven Pullman employees, four mail clerks, four

---

737 https://collections.dartmouth.edu/teitexts/Press_Translations_Japanese/diplomatic/social-1018-diplomatic.html
738 http://www.ikjeld.com/japannews/00000147.php
739 *The Ludington Daily News*, Ludington, Michigan, United States, 15th August 1946
740 http://www.ikjeld.com/japannews/00000147.php

dining car employees and two railwaymen were injured. [741]

A newspaper wrote, 'That the toll of lives was not greater than it was appears little less than miraculous. Upon their arrival in Altoona, most of the injured appeared dazed and unable to talk about the harrowing experience through which they had passed. Edward D. Russell, of Middlesborough, Yorkshire, England, said he was trapped in his Pullman for an hour and a half before the rescue workers pulled him through a broken window of the overturned car. He said the scene was a "shambles of twisted metal and battered cars." He suffered only a torn fingernail and said he considered himself, "a lucky passenger, indeed."

'The tragic scenes following a major railroad disaster were re-enacted for hours at the Altoona passenger station this morning as train after train from the west pulled slowly to a stop to discharge its cargo of dead and injured into the waiting ambulances and taxis. Hundreds clustered around the fences watched as the broken bodies of men, women, and children, still and dishevelled, were carried tenderly from the cars to the ambulances at both the Twelfth and Thirteenth street entrances of the stationway. Passengers in the trains looked on in awed silence at the tableau of suffering and death. Most tragic of all were the litters bearing a small form. On one large stretcher lay a small blonde girl, her eyes closed as if in sleep, her hair tousled and unkept from the effect of the horror through which she had passed in the night.

'On train 76, *The Trail Blazer*, which pulled into the station shortly after 7.30 o'clock, 10 adults and three children were removed on stretchers from the baggage car at the front of the train, while three dead were taken from the rear coach and carried temporarily to the baggage station. Grim-faced policemen and engine crews, in their work uniforms, aided in lifting the stretchers from the cars. They carried them to the gates to the accompaniment of screaming sirens as ambulances pulled away the gates and other ambulances rushed up to take their place.'

Another account of the disaster was given by a young trainee sailor, 'Chester J. Bialecki, 17-year-old apprentice seaman of 56, Roscoe street, Pittsburgh, who was en route to the Bainbridge, Md, naval training station at the time of the crash, described his experience in the disaster in terse and graphic sentences to an *Altoona Mirror* reporter this morning at the Altoona hospital. "I was sitting in the second car," the slender sailor said, "when it began acting kind of funny, rocking over to the side of the track and jerking. People started to scream and I braced myself against the window sill. The car buckled and stopped. I found myself clear, broke the window and helped a lady, her two children and her husband to get out of the car. Then I helped other people. Lord knows who all. There were eight midgets in the car, part of a show troupe, and they were terribly frightened and wouldn't let me come near them at first. Finally, after more air got in the car, they quieted down and were gotten out."

'Bilecki was still on his feet at the hospital, wearing part of his navy uniform. He was drawn and jumpy but outside of stiffness in his neck and back said he would be all right. "Put this in," he said, "my mother makes me put a St Christopher medal in my bag before I go home. I guess that's why I didn't get hurt, although I didn't think much about it at the time." [742]

Eastbound first-class passenger train No. 68, otherwise, *The Red Arrow,* consisted of

---

[741] Interstate Commerce Commission report

[742] *Altoona Mirror*, Altoona, Pennsylvania, United States, 18th February 1947

K2 Class 4-6-2 engines Nos. 422 and 3771, a baggage-mail car, a passenger-baggage car, a coach, two sleeping cars, a dining car, six sleeping cars, a coach, and a baggage-express car, in charge of Conductor G.R. Hershberger with Engineer M.S. Billig and Fireman Parasock on the train engine and Engineer McCardle and Fireman Henry on the other. It had left Pittsburgh, Pennsylvania, at 1.06 a.m., 56 minutes late, passed AR tower, the last open office, at 3.17 a.m., an hour and three minutes late, passed Gallitzin, passed signal No. 50R, displaying proceed, and was travelling at an estimated speed of 65 miles an hour on track No. 2 when it became derailed. Both engines and the first and sixth cars went down an embankment and the first eleven cars of the train were derailed. The first car was demolished and the second to tenth cars were all more or less damaged.

Speed in the area of the accident was limited to 35 miles an hour on the tangent track approaching the curve and 30 miles an hour on the curve itself.

Conductor Hershberger and Flagman Joel Bowers told the Interstate Commerce Commission (ICC) that the brakes of the train had been applied at the summit of the line, a mile and three-quarters short of the point of derailment, in accordance with the special instructions which required a running brake test to be made at that point. The brakes had been released, then re-applied in a service application and not again released. They put the speed of their train at 30 miles an hour when it left the rails.

A car inspector at AR tower saw no indication of any problem with the engine or vehicles of the train at that point.

The engineer of a westbound freight train on track 3 rather more than a quarter mile short of the point of derailment thought the speed of train No. 68 was at least 45 to 50 miles an hour.

Engineer Billig testified that the speed of the train had been about 25 miles an hour at the summit of the line. He said that after the running test there he had released the brakes and closed the throttle; 3,000 feet east of the summit the speed had been 30 miles an hour and an 8-pound brake application had been made. Further 8-pound reductions had followed and 2,500 feet from the start of the curve he estimated the speed at 22 miles an hour. He put the speed at about 31 miles an hour at signal No. 50R, where a 2-pound application had been made, which he had considered sufficient to reduce speed for the curve. As his engine had entered the curve his throttle lever had become unlatched and opened, but he closed it after a few seconds. He had not been aware of anything wrong until his engine had started to overturn.

The road foreman of engines stated that after the brake test at the summit the brakes were released and the throttle closed. A 10 or 15-pound brake application was then required to control the train on the 2.36 per cent (about 1 in 42) falling gradient, with further brake applications as required. This should reduce the speed of the train to 20 or 25 miles an hour upon reaching the curve. A check on the performance of Engineer Billig over the previous six months showed that he had usually negotiated the curve at a speed of between 15 and 28 miles an hour, rising to 30 miles an hour on only three occasions. There were no speedometers on the engines. An observation by the fireman of the freight train on track 3 that the brakes of the engines had not been applied at that point indicated that the brakes must have been released at that time, which was most unusual.

Post-accident examination of the engines and track showed no problems with either.

The division engineer stated that the theoretical overturning speed of engines Nos. 422 and 3771 at the curve was 65.1 miles an hour and the overturning speed of the tenders was 73.2 miles an hour. The ICC concluded that as the engines had overturned on the outside of the curve without marking the rails the train must have been travelling at the overturning speed at the time. [743]

A major disaster occurred in Java, Dutch East Indies, on Tuesday 30th September 1947. A heavily loaded passenger train left Mageland and passed through Temanggoeng, but two kilometres beyond there a steep incline proved too much for the train to surmount. The decision was therefore made to lighten the load by detaching the rear four vehicles, and this was done. The brakes of the detached coaches failed, however, to hold them on the steep gradient and they began to run back towards Temanggoeng. One of the coaches soon left the rails and overturned while the other three continued their journey, reaching an estimated speed of 70 kilometres an hour. It was not until the runaways had travelled about 5 kilometres and had reached Kranggan, where the line levelled off, that their brakes were able to bring them to a stand, one of the remaining coaches becoming derailed by the sudden stoppage. Twenty-five passengers in the coach which had overturned lost their lives and 70 were seriously hurt.

Indonesian and Chinese Red Cross staff gave first aid to the victims. The train service between Mageland and Parakan was temporarily suspended. [744]

A runaway accident occurred at Senohraby, Czechoslovakia, on Sunday 12th October 1947. The double track between Čerčany and Senohraby had been reduced to a single track since 29th September for track renewal. In the evening of the day of the accident coal train No. 7518, which because of its length had had to be divided, had been stopped for over an hour on track No. 2 at Senohraby. Sufficient hand brakes had presumably not been applied and when the air brakes leaked off the train began to run away, to collide with a passenger train at the junction of track No. 2 and the line on which the passenger train was running, at 21:34. Eight people died and 25 were injured (another source states that there were five deaths and 40 injured). The line was blocked until 10:00 the next day. [745]

A major runaway disaster occurred in Switzerland on the evening of Sunday 22nd February 1948 on the Chemin de fer du Sud Est de la Suisse or Südostbahn (SOB). Only one train was involved in the accident, so-called sports train No. 3770, which left Sattel at 16:49 for Zurich, was carrying returning skiers from the Hochstuckli district. The train passed through the station at Burghalden despite the signals there being at danger and stop signals being given by the station employee. Beyond that point the speed of the train increased on the falling gradient and it was travelling at 60 kilometres an hour when it reached Wädenswil, where it was diverted into a siding, collided with the buffers, and demolished the three-storey administrative building of the Fruit and Winegrowers Association behind the buffers. The first three coaches of the train were partly telescoped into one another and the fourth was badly damaged. The other coaches suffered less. Twenty-one people died in the accident and 50 were seriously hurt. There had been between 400 and 450 passengers on the train. One person the building was injured when the floor collapsed beneath him. His family had time to retreat to the rear of the building

---

[743] Interstate Commerce Commission report

[744] *Algemeen Indische Dagblad de Preangerbode*, Bandoeng, Dutch East Indies, 2nd October 1947

[745] https://www.koridory.cz/nehody-v-minulosti-senohraby-1947/

and escaped harm.

The train consisted of Class Ce 6/8 goods engine No. 14269 and nine old four- and six-wheeled coaches. From Sattel the route of the train was on rising gradients to Rothenturm and thence on moderate falling gradients to Biberbrücke. Beyond Biberbrücke, though, the line fell very steeply all the way to Wädenswil, at 1 in 42 between Biberbrücke and Schindellegi-Feusisberg, steepening to 1 in 23 between there and Samstagern and 1 in 20 between Samstagern and Burghalden, before the gradient eased marginally to 1 in 24 between the latter station and Wädenswil, apart from short lengths at 1 in 20. The maximum permitted speed over the last stretch into Wädenswil was 35 kilometres an hour.

An inquiry into the disaster was quickly opened. It determined that a brake continuity test had been carried out at Sattel and had been in order. The journey had been normal between Sattel and Biberbrücke, where a banking engine had been detached. Onwards from there the maximum permitted speed had been adhered to until just before Samstagern, where it had gradually increased, dropping a little only on short level stretches of the line. Driver Aebi, who survived the accident, told the inquiry that he had applied all the brakes at his disposal attempting to control the speed of his train, first the regenerative brake and the so-called regulator brake, then when these appeared to have no effect the Westinghouse brakes. As his train had passed through Burghalden he had indicated to the staff there that his train was out of control and for them to advise the next station at Wädenswil. Also travelling on the locomotive was SOB Traffic Manager Albert Sieber, who had been allocated as pilot to Chemins de fer fédéraus suisses (CFF) driver Aebi. When it had become apparent that the train was out of control the traffic manager had made his way back to the leading coach to apply the handbrake there. He had been caught between the engine and the coach in the collision and was one of those who died. Meanwhile Driver Aebi had begun sounding the whistle as the speed of his train increased.

At Wädenswil, which was a junction, SOB trains could only use tracks Nos. 1 and 2, both of which were occupied when the signalman there was advised at 17:26 by Burghalden that the sports train was running away. Before there was time to clear either of the tracks the runaway train had arrived at about 17:28 and entered the siding, for which the route was normally set when the platform lines were occupied.

Examination of the locomotive after the disaster showed no problem with the brakes of the locomotive or coaches and the brake blocks and tyres of the locomotive showed signs of severe braking. The regenerative braking controls, however, were set to apply power rather than to brake the train. It was apparent that Driver Aebi had lost his head and whilst the regulator and Westinghouse brakes had been applied and had braked the train properly, they had been working against the continual application of traction power. Surprisingly the traffic manager had not, apparently, realised what was happening either. [746] [747]

A major disaster occurred in Japan on the Nara – Kyoto line of Kinki Nippon Railways at 7.46 a.m. on Wednesday 31st March 1948 near Kawachi-Hananzono station.

---

[746] *Journal de Genève*, Genève, Switzerland, 23rd and 25th February 1948

[747] *Railway Accidents of Great Britain and Europe* by Ascanio Schneider and Armin Masé, pgs. 137-42

As at Wykes an express train ran into the rear of a stationary local train. The rear coach of the local train and the leading coach of the express were wrecked and another coach was badly damaged. As it was the morning rush hour both trains were crowded with passengers. Both trains were electric multiple units. The Kyodo news agency stated that there were 50 dead, the American Eighth Army raised this number to at least 74. A modern Japanese source states that there were 49 dead and 242 injured. It seems impossible to know which is the most accurate figure.

The motorman of the express survived. After the accident he alleged that his train had suffered a complete brake failure as it was passing through Ikomamayama Tunnel which had made it impossible for him to stop. He had overtaken two other trains stopped in platform loops at previous stations before the eventual collision. [748] [749] [750] [751]

What at the time that it happened was the most serious railway accident in peacetime in Norway occurred between Hjuksebø and Holtsås on the morning of Wednesday 15th November 1950. It involved train No. 72, the 07:45 *Sørlandsexpressen* from Kristiansand to Oslo, and four runaway waggons.

Train No. 72 was a three-car electric multiple unit set, a so-called *Lyntoget* (lightning train) consisting of a *motorvogn* (motor coach), *mellomvogn* (middle coach) and *styrevogn* (driving trailer). At the time of the accident it was being operated with the *styrevogn* leading. It was in charge of Head Conductor Johan Oscar Holm with Motorman Emil Grimsund, and had left its previous stop, Nordagutu, nine minutes late.

Hjuksebø was the junction for the Notodden branch and quite a busy station. On the fateful day it was in charge of Acting Stationmaster Ole Rolf Sommer with Telegraphist Alf Kårtvedt, Shunters Asbjørn Bjørnsen (in charge of the north end of the goods yard), Halvor Haslekås and (from 10:30) Anders Brennås (the latter two in charge of the south end of the yard).

Freight train No. 5343 from Drammen to Skien had arrived at Hjuksebø at 09:45 and had four waggons for Notodden. Shunter Haslekås uncoupled these from those behind, the engine pulled southwards with them and they were then pushed back into track No. 2 ready to be taken down the branch. He was not sure later if he had put their brakes down as he had been hurrying to set the road for freight train No. 5386 from Skien to Drammen, due at 09:55, though in fact it was delayed and did not arrive until 10:07. The leading three waggons of train No. 5386 were also for Notodden and Shunter Bjørnsen uncoupled them, the engine drew them forward to the north end of the station and they were taken back into track No. 2 from the opposite end and left there about a metre away from the previous four. The two lots of waggons were thus not coupled together and Bjørnsen explained later it was not his practice to do so. At 10:30 Shunter Brennås came on duty. Shunter Haslekås, seeing that the two sets of waggons were not coupled, was about to mention the fact to his colleague, Brennås, but at that moment he was preoccupied with ensuring that passengers did not cross the line in front of train No. 501 from Oslo to Kristiansand, which arrived at 10:45. Branch train No. 5443 from

---

[748] *New Castle News*, New Castle, Pennsylvania, United States, 31st March 1948

[749] *Delphos Daily Herald*, Delphos, Ohio, United States, 31st March 1948

[750] Webpage – A Brief Survey of Railway and Aircraft Accidents in Japan and Investigations Thereof by Seiji Abe and Takemune Sato

[751] https://blog.gaijinpot.com/groovy-osaka-a-sacred-mountain-with-unlucky-tunnels/

Notodden, with Driver Sigurd Ihlebæk and Fireman Johan Sandven, had previously arrived at 10:40. Driver Ihlebæk had been advised before he left Notodden that there would be waggons in track No. 2 at Hjuksebø for the branch and he had waggons from Notodden to drop off.

Aware that train No. 72 was nine minutes late Shunter Brennås decided that there was time before it arrived to take the waggons for Notodden from track No. 2 to the branch and Driver Ihlebæk agreed. Having detached his engine from the waggons he had brought from Notodden the latter pushed the seven waggons on track No. 2 southwards to the points at the Kristiansand end of the station prior to drawing them forward onto track No. 3. His engine was coupled onto the three waggons left by train No. 5386 and he was unaware that they were not coupled to the four which had been left by train No. 5343. Telegraphist Kårtvedt had informed Shunter Brennås that train No. 72 was a little delayed but had not told him to make the shunt before it arrived, while Stationmaster Sommer had said that the shunt must wait until after the *Sørlandsexpressen* had passed. The telegraphist telephoned Shunter Brennås, in the pointsman's hut at the south end of the goods yard, to tell him this, only to be informed that the shunt was already in progress. Telegraphist Kårtvedt had meanwhile already obtained clearance for train No. 72 from Øysteinstul, the next station towards Oslo. When the stationmaster came into the office for the green flag to show to the express to indicate that all was in order he was told by the telegraphist that the shunting was already in progress. "Ja, det gjørde de," (Yes, they have done it), he had replied, in a very annoyed tone.

Driver Ihlebæk testified later that, "I had given the waggons a shove and had just shut off steam again when I saw that the front four of them were not coupled to the others. Brennås noticed it at the same moment. According to my fireman Brennås jumped up on one of the waggons and put down a hand brake but the waggons kept going." The first waggon was loaded with empty barrels, the second with heavy wooden stakes, the third with paper and the fourth was an empty box van. The line fell steeply over the 4½ kilometres between Hjuksebø and Holtsås with gradients of 10‰ and 8.4‰ (1 in 100 and 1 in 119).

Shunter Brennås said, "I saw that the waggons were still moving and starting down the incline towards Holtsås. I sprang after them to apply the brakes and stop them by putting down the handbrakes. I got onto a waggon and was applying the brake when I lost my balance and fell off. I jumped up and again got onto another waggon, only to find that it had no handbrake and I could not get from it to the waggon in front. I jumped off and tried to get to the front waggon, but by now they were gaining speed and it was no use."

Driver Ihlebæk continued, "When I realised that the runaways could not be stopped I took the chance of following them with my engine, even though I knew that I was risking my life in doing so. I hoped that the brakes of the waggons might still stop them before they reached the Hjuksebø home signal." At the same time he sounded the whistle of his engine, both to inform the Hjuksebø station staff that the waggons had run away and in the hope of warning the crew of the *Sørlandsexpressen*.

Shunter Haslekås suddenly realised that both the waggons for Notodden and the engine of train No. 5443 had disappeared from the south end of the station. He climbed to the top of the cutting from where he could see for 500 metres down the line towards Holtsås, but there was nothing there. Then he heard the whistle of an engine at Haga Tunnel, rather more than a kilometre away.

Stationmaster Sommer was standing with the driver of train No. 501. They heard a sound which they thought at first was the horn of train No. 72, the *Sørlandsexpressen*, approaching, then realised that it was the whistle of a steam engine. The stationmaster dropped his green flag and ran down to the pointsman's hut at the south end, where he met Shunter Haslekås, then telephoned Holtsås to have the express stopped, to be told that it had already passed there.

The collision occurred at 10.50.27, 1,350 metres on the Hjuksebø side of Holtsås. The time can be given very precisely because part of the wreckage brought down the overhead line equipment, cutting the current. It was estimated that the *Sørlandsexpressen* had been travelling at 70 kilometres an hour at the moment of collision, despite its motorman having made an emergency brake application when his train rounded a curve and the runaways came into view at a distance of 500 or 600 metres; the waggons themselves were moving at about 60 kilometres an hour. The *styrevogn* of train No. 72 was almost completely destroyed in the collision and inevitably Motorman Grimsund was one of those who lost their lives. The other two vehicles of the train were almost undamaged but 13 passengers who had been riding in the *styrevogn* were also killed outright or fatally injured. Waitress Olaug Pettersen could have lost her life as well as she had just been about to enter the *styrevogn* when a passenger at the rear of the train called her back because he wanted more cream in his coffee. Conductor Holm was thrown down and injured but assisted in the rescue work until it became too much for him and he was taken to Notodden hospital. One of two prisoners who had been travelling on the train with a guard to Oslo to serve his sentence, Magnus Andreassen, was also prominent in the rescue work (the other prisoner was too badly hurt in the accident to be able to do much). He was later pardoned in view of his efforts that day.

The alarm was quickly raised by telephone calls from Hjuksebø station and from the residents of a property at Kleivrud, only 300 metres from the scene, who hearing a tremendous crash immediately realised that something must have happened on the railway. A Dr Karlsen from Gvarv was the first on the scene together with his sister, who was a nurse. By 11:15 news of the disaster had reached Notodden and Red Cross ambulances, doctors and police were immediately dispatched from there. In all 100 people took part in the rescue operation. The injured were mostly taken to hospital at Notodden, but also to Kongsberg and Skien.

Shunter Brennås was subsequently charged with the manslaughter through negligence of the 14 people who had died but the case against him was later dropped. Subsequently he was subjected to a disciplinary hearing by Norges Statsbaner, the state railway company, found guilty and lost his job as a shunter, though he was given another job in a non-safety-critical role in the goods shed at Notodden, doubtless on lower pay. He was said to have been a bitter and marked man for the rest of his life. Two Norwegian writers who have studied the accident in detail suggest that he was made a scapegoat. They point out a number of factors which contributed to the accident, such as the fact that at certain times of the day the staff at Hjuksebø were very busy, the lack of a clear regulation as to who should check that groups of waggons were coupled together, the risks of shunting trains at a station from which the line fell steeply, the fact that there was no facility at Hjuksebø for cutting off the current and so stopping the express, and so on. Contemporary newspapers also questioned why there had been no radio communication between stations and the trains, though this was really an idea somewhat ahead of its time in 1950. [752] [753]

A serious runaway accident occurred on a steeply graded line in the Slovakia part of Czechoslovakia on Sunday 22nd July 1951. A M 131.1 railcar pulling two trailers crowded with passengers stalled on the incline at Tatranská Lomnica station and could not be restated. The decision was then taken to detach the rear trailer to reduce the load, but before this was done the conductor tightened its handbrake as much as he could and placed some stones beneath its wheels. He also tried to get its passengers to disembark but most of them refused to. The trailer was then detached and it immediately began to run away down the incline, its speed rapidly increasing despite the application of the handbrake. It leaned over dangerously on curves and the conductor, who had remained with it, instructed his passengers to move to the inside of the curves to lessen the chance of it overturning.

The official at next station, at Studený Potok, was advised by telephone of the runaway. There was another trailer standing in his station so he tried to persuade passengers who were present there to manually push it back, presumably in the hope of being able to get it into a siding out of the way. Before the passengers had time to do so, though, they heard a warning shout that the runaway was approaching at high speed, so they had to abandon their task and seek safety. A violent collision ensued between the runaway trailer and the stationary one, they were locked together, and ran a further kilometre and a half before they could finally be stopped. Thirteen passengers died at the scene and six more succumbed to their injuries in hospital.

None of the railwaymen involved in the runaway was convicted of negligence as it could not be shown that they had broken any regulations. Measures taken to prevent a recurrence included a stipulation that every train pulling trailers must have a driving car at each end of it and a renewed stress upon the importance of always performing a brake continuity test. [754]

Express train No. 716 from Madrid to Santiago de Compostela was in collision with 12 waggons which had run away between the stations of Padrón and La Esclavitud, Spain on Tuesday 20th May 1952. The waggons, which had become detached from a freight train at La Esclavitud and had run back down the falling gradient, included two tank cars loaded with diesel oil and petrol. A farmworker going home from his work saw the danger and waved his arms to warn the fireman of the express. As a result the driver of the express had time to pull up his train but the collision could not be avoided. Following the impact there was an explosion and fire. The locomotive was completely destroyed and the luggage and mail vans at the head of the train mounted the leading passenger coach, in which there were known to have been 14 people. By 21:00 the mutilated and charred bodies of seven victims had been recovered, including those of a woman and a child, one of her two daughters. The woman was a Padrón teacher, Señora Clotilda Stoll, who had been returning to La Esclavitud at the end of her day's teaching duties. According to other passengers she had escaped from the coach but had returned to it to save her children, María Rosa, aged 11, and Elvirita, aged 8, but all three had perished.

---

[752] *Hjuksebø 1950* by Halvor Kleppen and Ingvar Skobba
[753] *På Sporet* (Journal of the Norsk Jernbaneklubb) No. 29
[754] https://www.rail.sk/arp/slovakia/history/h185-4.htm

The teacher was the wife of Stationmaster José Miguet Stoll at La Esclavitud. Other identified victims who died in the accident were Driver Generoso Carreras Segueiro, Fireman Angel Pazos Polanco, Conductor Domingo Garcia Gómez, Porter Angel Novea Vázquez, mail clerk Joseph Domingo Montes, Red Nacional de los Ferrocarriles Españoles (RENFE) agent Celso Pereira Caldero, who was travelling as a passenger (?) and Luis Apante Pérez, a passenger who died in hospital of his injuries. Two other victims could not quickly be identified, though by the following day one of them had been recognised as 28-year-old passenger Maunel Losada of Villagarcia. Several other passengers were injured. On the other hand members of the Galician Education and Leisure choir *Rosalía de Castro,* riding in the rear coach of the train, escaped injury. After the accident they managed to uncouple the train behind its third vehicle, thus preventing the fire spreading to the remainder of the train. Another passenger who escaped was a Señor Vázquez, who have lived for many years in Argentina and had been a prisoner of war of the Japanese in the Philippines during the war. He told a reporter that he had never been as frightened in the Philippines as he had been during the accident.

Relief trains were sent to the scene from Santiago and Vigo and later a breakdown train from Vigo. The civil governor of La Coruña, the interim governor of Pontevedra, the mayors of Santiago, Villagarcia and Padrón and other officials also went to the scene. An early estimate that the line would be cleared in eight hours proved too optimistic. By the following morning the work of removing the wreckage was still in progress, though it was hoped to finish the task by 18:00. [755]

A major disaster in Nigeria on Wednesday 15th October 1952 had its origins in an accident which had occurred about a week earlier. In the first accident a freight train bound for Ibadan and carrying drums of petrol had derailed and a fire had broken out in the wreckage. The fire having been brought under control a bypass line had been constructed around the wreckage, so that traffic could be resumed. [756] A breakdown crane towed by an engine was later brought to the scene to clear the derailed waggons. It was this crane which caused the second, much more serious accident, when it broke away from the locomotive and ran backwards down an incline for three miles, until it collided with a passenger train between Ogunshile and Illugun stations. [757]

After the accident a young man reportedly ran three miles to the nearest road. The first two cars which came along by chance contained the senior district officer and the senior medical officer, returning from a tour, and they immediately initiated rescue work. [758] The injured were taken to hospitals at Abeokuta and Ibadan.

As a result of the collision the engine cut right through the first coach of the train and telescoped the next two, the rear coaches of the train escaping with slight damage. Several days after the accident it was known that there were 34 dead, but other bodies were buried beneath the engine, which could not be removed until a breakdown crane had arrived from Enugu, 1,000 miles away. The only other crane which would have been capable of performing the task was that involved in the accident.

The Nigerian Minister of Transport, Mr Boded Thomas, sent a message of sympathy

---

755 *La Vanguardia*, Barcelona, Spain, 21st and 22nd May 1952
756 *Corpus Christi Times*, Corpus Christi, Texas, United States, 21st November 1952
757 *The Daily Gleaner*, Kingston, Jamaica, 17th October 1952
758 *The Times*, London, England, 17th October 1952

for the victims and in London, England, Mr Oliver Lyttleton, Secretary of State for the Colonies, sent a message to the Acting Governor of Nigeria: "Greatly shocked to learn of railway disaster and loss of life. Please convey expression of deep sympathy to relatives of those who have been killed and to the injured." [759]

The final death toll was probably about 50.

On the afternoon of Thursday 30th October 1952 a westbound goods train arrived at Fürstenstein station from Tittling on the line between Deggendorf and Kalteneck, West Germany, to do some shunting. On the main track were two waggons loaded with ballast, which were not coupled to one another, but were secured with scotchblocks. An eastbound railcar was due from Nammering at 17:12, so the main track must be vacated for it. Although the *fahrdienstleiter* had intended for the waggons to be put into the loading siding, the shunter suggested putting them onto the passing loop instead, and the *fahrdienstleiter* agreed to this. Meanwhile Nammering had offered the railcar to Fürstenstein and the *fahrdienstleiter* there accepted it, although the waggons were still standing on the main track, and the regulations prohibited a train being accepted unless the line was clear. The shunter then coupled the engine of the goods train to one of the ballast waggons, unaware that the second waggon was not attached to it and removed the scotchblocks. The line fell steeply from Fürstenstein towards Nammering and the second unattached ballast waggon immediately began to run away down the bank. An urgent call was made to Nammering station to stop the railcar but it had already departed. In the subsequent collision eight people were killed and another 20 were injured. [760]

Ten people died when several oil tank waggons broke away from a freight train, ran about eight kilometres down a steep incline and collided with a passenger train standing in the station at Barrganetal, Ecuador, on Friday 2nd January 1953. Two coaches were smashed. According to one report 39 other people were injured, 24 of whom were admitted to hospital at Guayaquil, 65 kilometres away. [761] The crippled train was moved to the next station at Milgrao, where a crowd of locals gathered around to inspect it. In the process two of them fell under the train and were killed. [762]

A spectacular accident occurred at Washington Union station, D.C., United States, on Thursday 15th January 1953. Southbound Pennsylvania Railroad overnight first-class passenger train No. 173 from Boston to Washington, *The Federal*, was entering the terminus at 8.38 a.m. when it failed to stop and collided violently with the bumping post (buffer stops), continued across the station concourse, struck the stationmaster's office, and demolished a newsstand, before the weight of the engine proved too much for the floor of the station concourse and it fell through it to the baggage and mail rooms in the basement below. Despite this no one was killed, but 62 passengers, 11 station employees, nine licensees of station properties, four Pullman car employees and the fireman of the train were injured.

The train had consisted of electric locomotive No. 4876, three coaches, a combination

---

[759] Ibid, 20th October 1952

[760] *Katastrophen der deutschen Bahnen Teil II* by H.J. Ritzau, pg. 123 (in German)

[761] *Albuquerque Journal*, Albuquerque, New Mexico, United States, 3rd January 1953

[762] *The Mexia Daily News*, Mexia, Texas, United States, 4th January 1953

baggage car and coach, and 12 sleeping cars, in charge of Conductor Thomas Murphy with Engineer Henry Brower and Fireman John Mayer. It had left Boston at 11 p.m. the previous night, left New York Pennsylvania station at 4.38 a.m. 36 minutes late, after a terminal brake test had been made and thence had stopped without difficulty at Philadelphia, Pennsylvania; Wilmington, Delaware; and Baltimore, Maryland. Engineer Brower told the Interstate Commerce Commission (ICC) inquiry that after leaving Baltimore he had had no occasion to use the brakes until he had approached automatic signal No. 1339, two miles short of the terminus, displaying *Approach*, while travelling at between 60 and 70 miles an hour. He had then applied the brakes but 3,600 feet beyond the signal he had noticed that the speed of the train was not being reduced as usual. He had then made an emergency brake application, still with no effect, and had shouted to his fireman that the brakes were not working. By then they had been near C tower and he had begun sounding the pneumatic horn as a warning that the train was out of control. The sanders had been applying sand to the track and he had attempted, unsuccessfully, to reverse the motors of the engine. Speed had been somewhat reduced, but the train had still been moving at 35 miles an hour when it had hit the buffer stops.

Conductor Murphy in the third car said that becoming aware of the excessive speed of the train he had opened his back-up valve, but nothing had happened. He had then warned the passengers in the car to be prepared for a sudden stop.

Front Brakeman Fred King in the fourth car related that he had tried to open the brake valve there but the violent lurching of the train had prevented him from doing so. Flagman John Meng in the rear car said he had become aware that the train was running away a little later and that a crowd of passengers in the aisle had then prevented him from reaching the brake valve.

Assistant Operator Harry Ball in C tower testified that he had seen that the train was going too fast and had at once warned Operator John Feeney in K tower. The latter had directed his leverman to route the train into track No. 16, then had telephoned Clerk Raymond Klopp at the station to warn him. [763]

Klopp acted quickly, and thus saved the day. A newspaper related, ‘A thousand tons of roaring metal aimed down your throat at 50 miles an hour – and 10 seconds to act. That was the hair-raising plight of Raymond Klopp, 49, a stocky mild-mannered clerk at Washington Terminal. The day was yesterday: the instant, when he got word that the runaway Federal Express was headed down track 16, straight at the stationmaster’s office and the center of the Union station concourse.

‘In that 10 seconds’ notice he had by telephone from a signal tower, a scant four city blocks from the station yard, Clerk Klopp did these things:

(1) Routed out four other employees in the stationmaster’s office.

(2) Warned away seven bystanders near gate 16 and the counterman at the central concourse newsstand.

(3) Got to a telephone close by to ambulances and fire engines.

‘The stationmaster’s office, the newsstand, and the gate where the train precipitously arrived were demolished. None of the people who had been there, however, suffered more than razed nerves.

‘Mr Klopp, a veteran of 12 years in the stationmaster’s office, has never had anything

[763] Interstate Commerce Commission report

comparable to this experience in his life. "It's the first time I've ever had to act fast," he said. In his private moments of quietude, away from the normally calm routine of clerking, Mr Klopp collects oriental curios in his bachelor apartment at 215 Constitution avenue, N.E. One item he points out, is a 1,300-year-old Chinese porcelain image of a horse.

'As the critical moment approached yesterday morning Mr Klopp was sitting at his desk checking schedules. Suddenly the phone rang. "It was John Feeney, out of K tower, Mr Klopp, his voice was excited. He said, "There's a runaway coming at you on track 16. Get the hell outa there! I didn't stop and look. I knew track 16 was aimed right at the office. I yelled at the others in the office. I ran into the telegraph room next door and shouted at them to get out. I ran out of the door and saw that train bearing down just a few car lengths away, and I yelled at the people standing at the gate. I ran to the phone booths near track 9, and just about as I got there the train hit.

'As the dust and steam billowed through the high-arched concourse, Mr Klopp said he put through his call to the station operator to "call the fire and rescue squads." Moments later, as doctors and nurses appeared on the scene, he followed their orders to get "all the morphine you can from the Union drugstore." And when things had settled down a bit, "one of the telegraphers came over and kissed me on the cheek for getting her out of the office in time."

'For himself, Mr Klopp added, he walked over to K tower and thanked Signalman Feeney, "as best I knew how," for calling the alarm.

'Mr Feeney, who lives at 10529 St Paul St, Kensington, Md., still spoke in an excited voice last night as he told of his alarm over the train's excessive speed and the emergency horn blast of the engine.' [764]

The accident had occurred towards the end of the peak of the Cold War with Russia and also happened five days before the inaugural ceremonies of the Republican Party, leading to suspicions of sabotage. It was investigated not only by the railroad itself and the ICC but also by the Senatorial Committee on Interstate and Foreign Commerce, where the senators were at a loss to understand the intricacies of the working of the Westinghouse brake system, the blockage of which had stopped the train at Kingston, then prevented its stopping at Washington.

The ICC was able to clearly determine the cause of the runaway. It noted that during the journey the train had been stopped by the conductor at Kingston, Rhode Island, because the brakes of the last two cars had been sticking. This problem was dealt with but the engineer could not then restart the train until he had found that the angle cock of the main brake pipe was closed at the rear of the third car and had opened it. After the accident the same angle cock was again found to be closed. Abrasions were found at the top of the angle cock handle matching abrasions on the bottom of the end frame of the car. The third and fourth cars, New York New Haven and Hartford Railroad cars Nos. 8665 and 6103, were off different types and with different vertical deflections. It was theorised that the repeated knocking of the air cock handle against the end frame of the car had closed the valve. When the air of the brakes was fully charged before the train started it slowly leaked off during the journey and was normally replenished from the locomotive. The closed brake valve had prevented this happening at Kingston causing the brakes to eventually apply. At Washington the same closed valve had prevented the air

[764] *The Evening Star*, Washington, D.C., United States, 16th January 1953

being exhausted from the engine and the air brakes being applied from there. Had either Brakeman King or Flagman Meng been able to get to the brake valve they could have stopped the train. [765]

The ICC recommended that all passenger cars with similar types of couplers should be examined to determine if they could come into contact with angle cocks in a similar manner.

A serious runaway accident occurred on the Cremallera de Montserrat (Montserrat Rack Railway) on the morning of Saturday 25th July 1953. This railway was a 5 kilometre 1,000mm-gauge line which ran from Monistrol de Montserrat up the side of a mountain to the Montserrat Monastery in Catalonia, Spain. It had a short level stretch at its beginning followed by an Abt system rack section with a ruling gradient of 15.6 per cent (about 1 in 7). The 25th of July was the day of the Feast of San Jaime (Saint James) and because this chanced to fall on a Saturday in 1953 there were many Catholic pilgrims from various places in the area wishing to visit the monastery. To cater for them no less than six trains were arranged to follow one another from Monistol up to the summit of the line in the morning, numbered 8, 6,5, 3,2 and 4, their numbers presumably corresponding to the numbers of the steam locomotives hauling them. The trains followed one another at a distance of about 300 metres, each consisting of an engine and two carriages and each carrying about 120 passengers. The first train arrived at the summit safely but the second, train No. 6, suffered a serious failure of its brakes at a place known as Recta del Pasteur. The gradient here was 12 per cent (about 1 in 8) and the train rapidly began to run back. It collided with train No. 5 and the two trains, running at considerable speed, then collided in turn with train No. 3. What was now three trains locked together continued to roll back down the incline and many of the passengers in train No. 2, seeing what was happening and hearing the continuous whistling of the engine of train No. 3, panicked and threw themselves off the train. The driver of train No. 2 stopped it, then began to reverse, and by skilful driving was able to cushion the impact from the combined runaway trains and bring the whole ensemble to a halt. An official from train No. 2 then ran back to stop train No. 4. The passengers from the latter were detrained at the Perro Guarián level crossing and the train returned to Monistrol de Montserrat to raise the alarm. The initial collision occurred at 08:55.

Most of the rack railway was in view of the town of Monistrol de Montserrat and some of its inhabitants seem to have witnessed the accident, though they did not realise the seriousness of it at first. When this became evident, many of the townspeople volunteered to help with the rescue work and the Civil Guard, police, Red Cross, and others were soon in action. About 25 slightly injured victims were treated at the infirmary attached to the monastery while the more seriously hurt were taken to San Andrés hospital and San José clinic in nearby Manresa. There were almost 200 people injured in all. Five bodies were recovered from the scene, three of whom were apparently passengers who had jumped from one or other of the trains and the other two who had been killed in the initial collision. Three more of the injured would die in the following days. Those injured included Driver Ginés Tudela and Fireman José Florez of train No. 3

[765] *A History of Railroad Accidents, Safety Precautions and Operating Practices* by Robert B. Shaw, pgs. 221-4

and Driver Diego Bernal Romero and Fireman Juan Sellés of train No. 6. (Details in the various newspaper reports are confused, for example one report states that both firemen were named Juan Florez, later corrected (?) to José Florez and Juan Sellés, and some reports imply that it was the third train which first ran away rather than the second. The above account is as accurate as the author can make it given these contradictions.)

The accident was believed to have been due to the overloading of the trains on a day of unusually heavy traffic. [766] [767]

A serious runaway accident occurred at Filattiera on the steeply graded electrified La Spezia – Parma main line in Italy in the early hours of Monday 4th October 1953. Two freight trains were involved and the accident resulted in the death of seven railwaymen.

The story of the accident was told by 43-year-old Night Porter Carlo Cassiani, who was on duty alone at Filattiera station each night between 22:15 and 04:15 and who was interviewed afterwards by Dr Rinaldi, the Public Prosecutor. Clearly traumatised by what he had witnessed he fainted at one point during his questioning. He related that he had worked for 12 years at Filattiera. Night-time trains normally just passed through his station without stopping; all he had to do was to attend to the signals, check trains as they passed and give them an *All Clear* signal with his hand lamp. He said that at 03:55 he had checked the time with the stationmaster at Villafranca, the next station on the south (La Spezia) side and the official there had informed him that northbound freight train No. 6652 was passing Villafranca. A couple of minutes later the telephone had rung and this time it had been the stationmaster at Scorcetoli, the adjoining station four kilometres to the north, who had shouted to him, "Carlo, put 5777 onto the siding." Train No. 5777 was a southbound freight train which should have crossed train No. 6652 at Pontremoli, the next station north of Scorcetoli and the southern end of a stretch of a double track, the line being only single through Scorcetoli, Filattiera and southwards towards La Spezia.

In answer to the Scorcetoli stationmaster's warning Cassiani said he had replied, "But I don't have the keys." The facing points to the siding were presumably padlocked. The Scorcetoli stationmaster had then told him, "The freight has passed me going at 120 kilometres an hour. There will be a disaster if you don't put it into the siding. If you don't have the keys to the points, smash the lock." Cassiani said he had then rushed out of his office and towards the points in the darkness, but they were nearly half a kilometre away and already he had been able to see the train coming, whistling furiously. Realising that there was no chance of reaching the points in time he had turned back, replaced the signals which had been cleared for train No. 6652, and going to the middle of the station platform had exhibited a red lamp in each direction. Shortly afterwards he had seen that train No. 6652 was slowing up, while train No. 5777 continued to advance at high speed. He had been able to see the contorted faces of the engine crew of train No. 6652 and hear their cries. He had then thrown himself back into his office as the trains had collided head-on with a tremendous roar.

Emerging from his office again he had found that the overhead line equipment had been torn down and there was a smell of 'molten iron, wood, fire and blood,' as a newspaper report described it. He had shouted out, calling for help, "Bring some lights!" A young man named Renato Capiferri, who lived near the station, had been the first

---

[766] *La Vanguardia*, Barcelona, Spain, 26th, 28th, 29th and 30th July 1953

[767] Webpage: Federación Castellano Manchega de Amigos del Ferrocarril (in Spanish)

person to arrive, and they had together searched the wreckage. The two engines had buried themselves into one another, some of the waggons had jumped over them, others had piled up against the side of the cutting. All the 45 waggons of one train and 30 waggons of the other were wrecked. About 20 metres from the wreckage they had found Assistant Driver Lionello Cecchinato and carried him to the waiting room, by which time he had succumbed. Next they had found Conductor Bruno Salati, who had finished his shift and was returning from Parma to Roma. He had been trapped by his arm. They had managed to free him, but he was dead.

The body of Driver Pietro Garzelli of train No. 6652 was also soon discovered and later the corpses of Driver Mauro Bulli, Assistant Driver Mauro Casadio and Conductor Giuseppe Aldrovandi of train No. 5777 were recovered from the wreckage. Perceiving that their train was going too fast for them to jump they had huddled together in the rear cab of the locomotive but in the collision this had been crushed by the leading three tank waggons of their train and the cab had become their coffin. The seventh railwayman to lose his life was Conductor Virgilio Portaluppi of train No. 6552. After seven hours of exhausting work he was finally released from the wreckage but his injuries proved too severe and he died a short time later in Pontremoli Hospital. Surviving the accident were the two brakemen of train No. 5777, Amos Gruzza and Giuseppe Pasquali. They had been riding in a van at the rear of the train; when they became aware that the train was running away and that there was nothing that they could do to stop it both had jumped in the vicinity of Pontremoli station. They escaped with minor injuries.

The cause of the runaway was clearly the failure of the air brakes of train No. 5777 between Guinardi and Pontremoli stations at a point where the line fell at 2·5% (1 in 40). [768] The reason for the failure was unclear and would have to be determined by an investigation. Once the runaway had passed Scorcetoli and train No. 6652 had left Villafranca disaster was inevitable. Even if Night Porter Cassiani had had time to divert the runaway into the siding it would have been going too fast to take the connection without becoming derailed. The only fortunate feature of the accident was that train No. 5777 usually carried explosive materials destined for the Stacchini powder factory in Roma but on the day of the accident it had consisted only of empty waggons. [769]

A coach became detached from a passenger train from Mexico City to Cuernavaca, Mexico, and its air brakes failing to hold it the vehicle ran back down an incline until it collided with a following engine on Friday 26th February 1954. Three passengers died and 25 were injured, 14 seriously. The crew of the engine were unhurt. [770]

The French ski resort of Superbagnès in the Pyrenees was served by a metre-gauge Strub system rack and pinion electric railway which ran up the mountainside from the town of Bagnères-de-Luchon. The 5.7 kilometre long line climbed 1,170 metres and so had an average gradient of about 1 in 5. On the evening of Sunday 28th February 1954 this line suffered a serious runaway accident. It was the high season for skiing. During the high season the railway company, the Société des chemins de fer et hôtel de montagne aux Pyrénées (CFHMP) employed seasonal workers to operate the points to

[768] Wikipedia – Parma – La Spezia railway
[769] *La Stampa*, Torino, 5th and 6th October 1953
[770] *Albuquerque Journal*, Albuquerque, New Mexico, United States, 27th February 1954

avoid delays. At 17:00 on the fateful day the day's skiing had finished. There were three trains already at Superbagnès and two trains at Luchon, at the other end of the railway. The Luchon stationmaster was unsure if both the trains at his end would also be needed to bring all the skiers back down and attempted to contact Superbagnès. Receiving no immediate reply he decided to send both his trains up. His colleague at Superbagnès then answered and said that only one train would be needed. By then both trains were on their way from Luchon so the stationmaster there contacted the staff at the intermediate loop at Mi-Sahage, advised them of the situation and asked then to send the second ascending train back to Luchon.

At about 17:20 the first descending train arrived at Mi-Sahage and went into the loop there to leave the main track free for the ascending trains. The two ascending trains then arrived and the brakeman of the second operated the points at the lower end of the loop to allow the first descending train to depart. A little later one of the temporary employees, unaware that the brakeman had already done so, also operated the points and failed to notice that he had reset them for the main line. The first descending train, hauled by tractor No. 4, then departed. It burst through the closed points, pitching to right and left, such that the cogwheels became disengaged and contact with the overhead wires caused a massive electrical arc, visible from Luchon, and a short circuit which cut the power. The train was not derailed though and was soon running downhill at a speed estimated at 80 kilometres an hour as opposed to the normal permitted maximum speed of 8 kilometres an hour. Six of the panicking passengers jumped from the train, five of whom came into contact with a rock wall, the posts supporting the overhead line equipment or the retaining wall bordering the track, ended up beneath the wheels of the train and were killed. A sixth was found seriously injured lying against a tree stump.

As chance would have it the tractor was not coupled to the two passenger cars, as it should have been according to the regulations, and when the runaway came to a slight curve the passenger cars left the rails, coming to a stand a few metres from the railway. The tractor continued its downward rush until it, too, was derailed at the sharper La Soulan curve. It fell into the Bagnartigue ravine, was badly damaged and the two railwaymen in it, M. Louis Caillau and M. René Viel, were killed. One was presumably the motorman and the other the conductor. In all six passengers and two railwaymen lost their lives in the accident and indirectly there was a ninth fatality as seven-year-old Anne Lemoine, seriously hurt in a sledding accident, could not be brought down to Luchon for possible life-saving treatment with the railway out of action. [771] [772]

The Red Nacional de los Ferrocarriles Españoles (RENFE) branch from Villabona to San Juan de Nieva, Spain, suffered a runaway accident at 06:11 on Saturday 16th April 1955 between Cancienes and Nubledo stations. Involved in the accident were freight train DU-3 from Villabona and omnibus passenger train No. 1532 from San Juan de Nieva to Villabona. The freight train consisted of an engine and 25 waggons loaded with coal and cement and weighing over 800 tonnes. The passenger train consisted of two engines, a van, a waggon and three second and third-class coaches. It was crowded with passengers, mostly RENFE employees who had just finished their night shifts at San Juan de Nieva or Avilés and were on their way home to Corbera de Asturias.

---

[771] https://histoire-superbagneres.fr/laccident-du-28-fevrier-1954/

[772] Wikipedia – Chemin de fer de Luchon à Superbagnères

The freight train got out of control on the falling gradient approaching Cancienes station and ran through there, despite the signals being at *Danger*, its crew unable to regain control of it. The violent collision occurred at kilometre 10.1. The three engines, 22 waggons of the freight train and the van and waggon of the passenger were wrecked.

First-aid was rendered by the Civil Guard and by locals who had witnessed the accident and motorised transport was provided by an Avilés steel factory. Five people died in the accident and six were injured. Those who lost their lives were Driver Ramón Vilalta Calzada of the pilot engine of the passenger train; his fireman, Herminio González; Driver Oscar Campos of the train engine; Conductor Francis Suárez of the passenger train and Driver Manuel Sánchez Garcia of the freight train. Fireman José Madreno and Brakemen Benigno Garcia and Jaime Caballero Gonzalo, all of the freight train, were among the injured. [773] [774]

Brazil suffered a major disaster on Christmas Eve, Saturday 24th December 1955. Estrada de Ferro Central do Brasil freight train TP-74 to Volta Redondo was ascending the incline between Guararema and São Silvestre when four coal waggons loaded with coal became detached due to a broken coupling. They ran back down the bank for four kilometres before colliding with passenger train RB-4 from São Paulo to Monte Azul, otherwise the *Expresso Baiano*, which was stopped at a signal at Guararema waiting for the freight train to clear São Silvestre. The accident happened at 17:40. Stationmaster José Romeiro Pantagua and Signalman Sebastião Firmino at Guararema watched the runaway waggons tear through their station without being able to do anything about it. Shortly afterwards they heard the sound of the collision and alerted the emergency services.

As a result of the impact the engine and tender of the passenger train crushed the leading mail van, which in turn climbed over a second-class coach and wrecked a third vehicle, while fire broke out in the wreckage. Driver Artur dos Santos and Fireman Mario Monteiro dos Santos escaped with slight injuries, while Conductor Otavio Silva Catarino, who also escaped injury, worked until 03:00 among the victims. Assistance came from the cities of Jacarei and Mozi das Cruzes and about 20 ambulances were sent from São Paulo with plasma. Because it was a hot December day many of the passengers were standing in the vestibules between the coaches and were crushed in the collision.

A statement from the railway company intimated that there were 14 dead and over 30 injured. By the following morning the number of dead had increased to 15. [775] [776]

A tragic accident occurred in Sweden on the evening of Friday 13th January 1956. A northbound Hilding Carlsson diesel railbus was carrying homeward bound schoolchildren from Kopparberg and was due to cross a southbound iron ore train from Grängesberg at Ställdalen. The ore train was being pulled by almost new Ma Class electric locomotive No. 405. The accident apparently arose through negligence, as the air brake connection between the engine and the leading waggon of the ore train had not been made when the engine was coupled onto its train, nor can a brake continuity test have been made. As a result braking power was available only on the locomotive itself and this was insufficient

---

773 *La Vanguardia*, Barcelona, Spain, 17th April 1955
774 Webpage: Federación Castellano Manchega de Amigos del Ferrocarril (in Spanish)
775 http://guararematem.com.br/o-acidente-trem-baiano/
776 *O Estado de São Paulo*, São Paulo, Brazil, 27th December 1955

to hold the heavy train on the falling gradients approaching Ställdalen, so that it ran away through the station. There was no time to divert the runaway into a siding at Ställdalen and a telephone warning to a lineside residence of a railwayman to stop the railbus came too late. An attempt to cut the electric current was also futile as the runaway was being powered by gravity and the railbus was of course diesel. Twenty people died in the accident, 12 of them schoolchildren. [777] [778] [779]

During a severe storm on the night of Monday 4th to Tuesday 5th June 1956, 20 tank waggons loaded with crude oil from the Dollard oilfield, seven miles west of Shaunavon, Saskatchewan, Canada, were blown out of a siding by the wind and ran away eastwards. After covering some distance they came to a stand at a dip in the line and there were struck by a Canadian Pacific Railway freight train. Three of the cars were set ablaze, as was the front of the locomotive of the freight train, which was badly mangled, while flaming oil set fire to the grass alongside the right of way for more than a quarter mile. No one was hurt. [780]

The British Railways, London Midland Region line from Buxton to Stockport initially climbed for two miles at 1 in 66 to Bibbington's Sidings signal box, then fell almost continuously for 17 miles to near Stockport, beginning with seven miles at 1 in 70/58, broken only by brief easier stretches through Dove Holes and Chapel-en-Frith stations. Unfitted goods trains were banked in the rear from Buxton to Bibbington's Sidings, where the train stopped for the guards to pin down the waggon brakes before beginning the long descent.

Late on the morning of Saturday 9th February 1957 several trains were making their way along the line. The first was the 8.45 a.m. goods from Rowsley to Stockport Edgeley, consisting of a 2-8-0 steam engine, 37 waggons and a brake van, in charge of Guard John Creamer with Driver G.L. Butler and Fireman D. Bradshaw. Next came a passenger train, then a second goods train, the 11.05 a.m. from Buxton to Warrington Arpley, consisting of another 2-8-0 steam engine, 33 waggons and a brake van, in charge of Guard Alfred Ball with Driver John Axon and Fireman Ronald Scanlon: and finally another passenger train. At Bibbington's box the Rowsley goods train was put into a Down Goods loop which ran between there and Dove Holes so that the first passenger train could overtake it.

Before the second goods train had left Buxton there had been a problem with steam escaping from the steam brake pipe. Fitter G. Bradwell had been inclined to fail the engine but Driver Axon had wanted to get going so the fitter had just tightened a nut with a spanner. Driver Axon meanwhile made out a repair card and the train departed on time. It had not travelled very far when the steam brake pipe joint began to blow again. Fireman Scanlon wrapped some rags around it but the escape of steam intensified, then

[777] Wikipedia - Järnvägsolyckan I Ställdalen
[778] https://www-tydal-nu.translate.goog/postvagnen/231213.htm?_x_tr_sl=sv&_x_tr_tl=en&_x_tr_hl=en&_x_tr_pto=sc
[779] https://www-geocaching-com.translate.goog/geocache/GC4WW08_tagolyckan-i-stalldalen?guid=7b5f4f09-995f-4d41-9be8-117685a4c0bb&_x_tr_sl=sv&_x_tr_tl=en&_x_tr_hl=en&_x_tr_pto=sc
[780] *The Medicine Hat News*, Medicine Hat, Alberta, Canada, 6th June 1956

the brake pipe was blown completely off, filling the engine cab with scalding steam. The driver and fireman were forced from the cab. They were able to screw down the tender brake but could not return to the cab long enough to locate and close the regulator. Efforts were made to knock the regulator shut with a long fire iron, without success, and worse, because of its design the regulator had to be fully open before it could be closed. Meanwhile they were near the summit of the line, with the banker still pushing them. Driver Axon then told his fireman to jump off, try and pin down some waggon brakes and tell the guard that they were running away.

The first passenger train having passed Dove Holes, the signalman there let the Rowsley goods out of the Down loop and accepted the Buxton goods on the loop line in turn, where it was to be held for the second passenger train from Buxton. Signalman D. Bowyer at Bibbington's had noticed that the Buxton goods train had not stopped to pin down brakes before entering the loop, that it had taken the connection to the loop line at considerable speed, and that a lot of steam had been coming from the cab. However, he had not been a signalman very long and apparently did not realise, despite these unusual circumstances, that anything was wrong. He did, however, see that the train had no tail lamp, possibly because Guard Ball had deliberately left it off to warn him, so he sent the *Train Passed Without Tail Lamp* signal to Dove Holes.

Signalman Eric Fox at Dove Holes saw the Buxton train coming towards him, with steam pouring from its cab and its driver motioning violently to him, doubtless trying to tell him to abandon his box in a hurry. He had a moment to make a decision and decided to let the train back out onto the main line, as he told the Ministry of Transport inquiry, 'to give the driver a chance to regain control of his engine,' and he was not criticised for doing this either by Brigadier C.A. Langley, the inspecting officer, [781] or by the Coroner at the inquest, where Mr F.R.S. Nesbitt commented that, "I for one would hate to have been faced with the decision he had to make." [782] A modern writer has commented, though, that 'this went against all the signalling regulations and indeed good sense,' given the fierce gradients, the runaway 775 ton train, and the fact that the Rowsley train was still in the block section, Signalman Fox should, instead, have abandoned his box and let the train run into the buffers at the end of the trap siding (Dove Holes box stood behind the buffers and would thus probably have been demolished). [783] The train passed Dove Holes at 11.17 a.m., travelling at 28 miles an hour. Signalman Fox did not send the *Train Running Away on Right Line* signal to Chapel-en-le-Frith but advised Signalman George Howe there over the telephone.

The 10.20 a.m. Up passenger train from Manchester London Road to Buxton was standing in the platform at Chapel-en-le-Frith and could go no further because a goods train was in the section to Dove Holes. Guard H. Dutton had just gone into the box to carry out Rule 55 when Signalman Howe advised him of the runaway, and that it was not the Rowsley train, which was then just entering the station, but another train behind it.

Driver A. Robinson, travelling on the Manchester train with his fireman, saw the runaway train approaching and the two of them just had time to get the passengers out of the train and away from the platform.

Driver L. Simnor of the multiple unit also saw the runaway and tried to warn the

[781] Ministry of Transport report

[782] *The Times*, London, 16th February 1957

[783] *Obstruction Danger* by Adrian Vaughan, pgs. 146-58

fireman of the Rowsley train but the collision occurred shortly afterwards.

The brake van and last three waggons of the Rowslay train were destroyed. The engine of the Buxton train was overturned, its tender demolished Chapel-en-le-Frith signal box and 30 of its waggons piled up in a heap 25 feet high. Guard Creamer and Driver Axon were killed outright; Signalman Howe, who was thrown from his box, and Guard Ball, had lucky escapes.

The Ministry of Transport inquiry found that the brake steam pipe which had failed had been reported as blowing ten times in the previous 15 months. The collar of the brake pipe had been badly brazed onto the connecting pipe from the combined injector steam valve to the driver's steam valve, probably at Horwich in April 1955. It had not been solidly brazed over the whole distance and some of the brazing had overflowed onto the inside face of the collar, making it thicker on one side than the other. There had thus been uneven pressure on the asbestos packing when the nut was screwed down, and the joint had not been steam tight unless the nut was screwed down very hard. The whole joint had thus been severely stressed until it had given way.

Following the accident work was begun to change all similar joints on the London Midland Region from the butted type to a coned type. This was expected to take three or four years and meanwhile all remaining butted joints were to be regularly inspected. Also all leaking steam pipes were to be removed and examined. The inspecting officer recommended that other regions follow suit. He also noted that had the train been fitted with continuous brakes the accident would not have happened. All freight trains were to be so fitted in connection with the British Transport Commission's modernisation plan. [784] Driver John Axon was subsequently posthumously awarded the George Cross for his bravery in sticking to his engine and attempting to stop the runaway train.

Carelessness amounting to negligence by a railwayman caused a serious accident at Bylnice, Czechoslovakia, on the line from there to Vsetin. On Sunday 28th April 1957 a waggon at Brumov was only secured on a falling gradient by a block of wood instead of by the proper wheel chock. As a result it ran away, until at Bylnice it collided with a Class M 131-1 diesel railcar. Ten people died and six were hurt. [785]

What was by far the worst railway accident to occur on the island of Jamaica happened on Sunday 1st September 1957. The train involved in the catastrophe was an excursion organised by St Anne's Catholic Church of Kingston, which was returning from Montego Bay to Kingston late in the evening. It was approaching the small station at Kendal at considerable speed, its coaches swaying, and passing through a narrow rock cutting, when there was a terrific jolting likened to an earthquake and eight carriages were wrecked. The train had consisted of two diesel engines and 12 wooden coaches crowded with an estimated 1,600 or more passengers. The two locomotives became detached from the train and ran on for some distance before coming to a stand. The local newspaper wrote, 'One of the coaches, its superstructure razed to the flooring, remained on the lines, and ran on for about a hundred yards. Of the remainder, five toppled into a

---

784 Ministry of Transport report

785 https://de.wikipedia.org/wiki/Liste_von_Eisenbahnunf%C3%A4llen_in_Tschechien#1946%E2%80%931959

gully beside the track, two remained in the cutting, one mounted its bank, the one behind was raised up so that its body came off the undercarriage and the front wheels were lifted off the line. Only the last two coaches remained undamaged.

'Most of the passengers in the four rear coaches escaped unharmed but for shock and minor injuries such as cuts and bruises. The remaining coaches were an utter shambles, with dead and injured inside and underneath them, survivors screaming, calling for help or crying the names of relatives or friends. Of the derailed coaches in the gully, one was smashed to slivers, with the wheels lying to one side; another was a mere grip of smashed boarding lying partly on the wheels, partly on the roof of another coach lying at an angle and on its side; the next had broken in unequal parts, the larger (forward) section lying on its side, the smaller still on the line and coupled, but stripped of roof and sides; the last had its side torn out; food, clothing and utensils, could be seen strewn over the floor under the semi-collapsed roof. The leading coach of the two in the gully was gutted and broken in half, the second was wrecked but less extensively.'

Eyewitnesses said that the engines had stopped briefly after the accident, then continued to Kendal to raise the alarm.

There were some lucky survivors. 'A miracle in the midst of the disaster was witnessed last night when a little girl of barely three years old was rescued alive from beneath a pile of bodies – reportedly the sole survivor from the coach in which she was travelling when the Montego Bay – Kingston train was derailed just outside Kendal station. The child, whose parents are believed to have been killed in the accident, was unhurt except for a scratch on the forehead. A doctor on the spot examined her for internal injuries but could find no sign of any. She has been taken to the Hanley Home (a recently opened and privately run home for children near Kendal), where she will be cared for and where any surviving relatives may claim her.

'There were two other miraculous escapes from death, one a woman who sheltered under a seat which saved her, and a man, ironically enough, an undertaker by trade. Both climbed out from the wreckage of a coach which had been reduced to matchwood.

'Great difficulty was at first experienced in getting out the injured, since although the wreck was only a couple of hundred yards from the main road from Porus to Balaclava, the intervening terrain was very rough and almost impassable in the darkness. However, a bulldozer was made available by Alumina Jamaica Ltd which set about digging a path from the main road to the scene. This eventually served to bring out many injured and some of the dead. A train of waggons was, however, made up at Kendal and hitched to one of the engines from the excursion train. This shuttled back and forth between the wreck and the station, first bringing injured persons for treatment and then uninjured survivors who had no reason to remain at the wreck. Eventually a train came up from Porus to take the latter and any lightly injured persons to Kingston.

'Meanwhile ghastly scenes were being enacted at the site of the wreck. There was little light but for a few lanterns and fires stoked from the wreckage, round which survivors huddled, their faces pallid and drawn, often streaked with blood and dust. A few chatted, but for the most part they stood, sat, and lay silent, dazed with the tragedy of it.

'Father Eberle {the organiser of the excursion} and another priest, who had been going among the crowd, comforting the injured and blessing the dead, sat exhausted on one of the benches that had been pulled out from the wrecked coaches. They said that Father Brennan had come from May Pen – and three Passionist priests from Mandeville –

to help them in their tragic task. Voluntary helpers were present in scores, and the main road was lined for a long distance with all manner of vehicles.

'Doctors and nurses worked desperately to clean and bandage wounds and splint fractures, some on the banks of the track, some inside the coaches, while the injured tossed and moaned, cried, or lay unconscious. Many less seriously hurt wandered about, their clothes bloody, or sat at the fires, staring into space…A girl of about seven was led away, weeping, and saying over and over again: "Ah got no mammy now. Mammy gone dead, gone dead."

'Besides the magnificent work done by the medical personnel of all ranks, on the spot and in the hospitals, Mr C.C. Michelin, of Monymusk, Commissioner of the St John Ambulance Brigade, organised ambulances, and stretcher bearers, of whom – and of stretchers – there were never enough. The police have done a splendid job throughout, remaining on their feet and without food for many hours to help with rescue work, maintain order, guide the congested traffic, and record the injured and dead. The only policeman actually on the train was Corporal S.C. Davis, of Montego Bay station, who was in the leading coach, but escaped with minor injuries. He too stayed on duty for hours until relieved. Police rigged hand ropes up the steep bank north of the track so that laden stretchers could be hauled up and loaded into vehicles. Very good work was also done by the staff, cadets, and students of the West Indian Training College in Mandeville, who were early on the scene and helped with first aid, rescue, and relief work all through the night.

'Tribute was paid by police officers and other witnesses to the contribution of Alumina Jamaica Ltd., which supplied many vehicles and willing hands and the heavy equipment with which to drive a road to the site and clear away the wreckage from the lines. Outstanding was Mr Roger Hardy, vehicle superintendent at Kirkvine works, who took charge of the wrecking operations and worked through the night and far into the day, worn and dirty but indefatigable, to get the wreckage lifted and into the gully.

'The cause of the wreck is obscure and must no doubt await for elucidation on an official inquiry, but several travellers remarked on the speed with which the train appeared to be travelling just prior to the accident. One, who was in the fourth coach from last, said that the excursionists were singing happily, but that the train was swaying from side to side and then, suddenly, came the crash. Another, a member of the Gleaner's staff {*The Daily Gleaner*, the newspaper from which this report is taken} mentioned that the train gave "a great sway" about five minutes before the disaster and was moving at speed. A teenage girl, who was also in the fourth coach, pointed out the body of her aunt lying in the cutting below and went on to speak of "trouble with a wheel of the engine" on the way to Montego Bay, where something was done to it, she said, and also on the way back. She said the wheel "gashed fire" and seemed to be "slipping." Just before the crash, she continued, "the train was moving fast. When the first shock came, someone cried, "Wheel slip off." Then, she continued, "Everything come off."'

By midday the next day 159 bodies had been recovered from the wreckage, three injured had died in Mandeville Public Hospital and nine more at Spaldings.

It was quickly announced that a commission of inquiry would be set up to investigate the accident. [786]

Unfortunately not everyone involved in the disaster had noble motives. Glendale

---

[786] *The Daily Gleaner*, Kingston, Jamaica, 3rd September 1957

Russell, 17, a newspaper office help who had been a passenger on the train, related afterwards that on the outward journey a number of pickpockets 'and other undesirables' had joined the train and had "cursed, fought, stole, and generally made themselves a nuisance. After the derailment these individuals had made themselves busy looting the dead and the injured, taking the watch of an injured man who told them, "Have mercy, wait until me dead, no."

The driver of the leading engine of the train, Garnish Lurch, was said to have been in tears as he spoke to a reporter, telling him, "I don't know how it happened, I don't know how it happened." 'Lurch said, "Coming down hill, I, as is customary, made the test application which should have slowed up the train. To my surprise I felt no change and I made another application which should have brought the train to a stop. Instead the air pressure fell to zero and the train picked up speed. I then shouted to the observer Theophilus James, who was beside me, 'We dead now, we dead now.' I prepared myself for death and made three blasts of the whistle to warn those behind that something was wrong."

'The driver said somehow he was able to find enough pressure to take him around the "S" bend and from the jaws of death. He then told of the horror that struck him when he felt the pull of the coaches as they tore away from the engine and were dashed this way and that. He said when he saw the shattered coaches "a terrible feeling" came over him. He felt "absolutely lifeless." As he sat in the engine cab with his observer and a few others he had a vacant look and the mental agony which he was going through was obvious.

'Lurch, asked for his views on the possible cause of the disaster said he thought that either somebody had tampered with the angle cock in one of the coaches or the air hose had broken.' [787]

The following day the newspaper wrote, 'The driver of the train, Garnish Lurch, according to the interview published in yesterday's Gleaner, does not know how the train got away. Now, let us compare it with the story I got from one of the survivors of the tragedy – one of the few who escaped almost completely unhurt. (His name and address, unless required by the Commission, will of course remain undisclosed).

'"The driver stop the train at Appleton station to see the stationmaster. But them tell him that the stationmaster was gone to Kendal. Him g'wan the train all right till him done climb the hill after you pass Appleton. Then the train start race. I go down through the coach them to him and I ask him why he driving so fas'. And him continue drivin' the same way. Bwoy, all me do is go back down to my seat." This survivor goes on to say that the train never slowed down till after the crash. Here is what another survivor has to say: "The train was travellin' very fas', and the coach was rocking very fas'.

'These are the facts without any varnish or polishing. All the survivors I met or spoke with said that the train was going at too merry a clip – so much so that it was a bit too much for their comfort and peace of mind. One of them – he says he knows quite a bit about calculating speeds – said that the train was travelling at between sixty and seventy miles an hour. Most of them remember that it was raining. Not heavily, just enough to make things a bit slippery all round and all along.

'There is more to it than this though. Somebody interfered with the air brake

---

[787] Ibid, 4th September 1957

mechanism in the seventh coach. This is no mere rumour. This statement was made by the woman over whose head the statement was made. I have her name and address.' [788]

News of the catastrophe continued to fill *The Daily Gleaner* for some days as the commission of inquiry went on with its investigation. Almost a week later the newspaper found at least one happier story to write about in relation to the accident. 'Barbara is a little girl of three years who was travelling with her mother on last Sunday's fateful excursion, while her father remained in Kingston. Her mother was injured in the crash and afterwards taken to Spaldings Hospital. A friend of her father, who was on the train, shielded Barbara with his body, but she suffered head injuries notwithstanding and was one of those warded at the Mandeville Maternity Home. Her saviour lost sight of her and went on to Kingston, where he met the frantic father and told him that his child was alive. The father came to Mandeville to find her but sought her in vain at the Mandeville Public Hospital. Someone who had heard of a little girl being saved from the wreck and taken to a children's home at Kendal, offered to take him there to see if she was his lost daughter. But before leaving the father "had a feeling" to look in the Maternity Home and there saw and recognised his daughter. Little Barbara, who up to then had been the quietest of patients, cried {so} piteously when he attempted to leave her in the nurse's care that later on he was allowed to take her away. No doubt she is now as happy with her Daddy as her injuries will allow.' [789]

By 12th September, with the death of Mr Cyril Ismay at Kingston Public Hospital, the number of deaths in the accident had risen to 181. [790] Another 700 were more or less injured.

At the beginning of the investigation into the accident held by the Commission of Inquiry, chaired by the Chief Justice, the Hon. Colin MacGregor, evidence was given by engineer Arthur Sloly, who stated that the two leading coaches of the train, Nos. 514 and 500, had derailed as a result of centrifugal course on taking the curve, which he attributed to the speed at which the train had been travelling.

The engineer was followed by Permanent Way Superintendent Altamont Donegal, who spoke of a conversation that he had had with Driver Lurch after the derailment. "Lurch said that he left Greenvale with the train in the normal way. He applied the brake when he was approaching the level crossing at Dunkeld between the 57 and 57¼ mile posts and at the beginning of the short climb. When he turned over the hill as was customary, he made a small application of the train brakes – a 10 lb application – and waited to feel the brakes on the train. He said he felt nothing. He said he noticed that the train was going faster. He made another application of about 20 lbs of air that would ordinarily stop the train, but this did not affect the train at all. He said he noticed that the air indicator was dropping. He started blowing the whistle to signal the train crew that, if any of them had any valves down, they should turn them up. He said he then applied the emergency brake and when that did not work, he told his companion in the cab what had happened, adding the remark 'We must pray!' He said the train developed fast and when it went around the curve he noticed the air coming up back to indicate that air was in the pipe then. He said he then stopped the engine."

Air Brake Inspector Sydney Macfarlane described removing a conductor's valve from

---

788 Ibid, 5th September 1957
789 Ibid, 7th September 1957
790 Ibid, 12th September 1957

coach No. 403 and an angle cock from coach No. 502 on 10th September under instruction from Brigadier C.A. Langley, the British inspecting officer who acted as technical adviser to the commission. He stated that the valve had been closed and the angle cock had been jammed. These conditions, he felt, must have existed before the accident in which case the jamming of the cock and the closing of the valve would have "isolated coaches 7 to 12 from the driver's control," so that he would only have had control over the leading six coaches.

Father Charles Eberle, Rector of St Anne's Roman Catholic Church, was the final principal witness that day. He said that there had been between 1,600 and 1,800 persons on the train and that shortly before the crash the train had been travelling at between 40 and 50 miles an hour. He denied having been assaulted but said that he had been abused by a Rastafarian at Montego Bay and had had filthy insults hurled at him after the accident. He said that at first tickets had sold slowly but then the sales had speeded up and when he had realised how many passengers there were likely to be he had asked the railway people for a second train, even if it meant paying £125 or £150 more, rather than have anything happen, but had been told that no second train was available. When he had arrived at Kingston station at about 4.35 a.m. on Sunday morning there had been "a great crowd" of people there. Later there was an announcement that no more tickets would be sold, when, he was informed, some 200 or 300 people had rushed from the waiting room to board the train. It was easy for joy-riders to evade the ticket collectors and board the train. Tickets had been sold to members of the congregation and their friends but there might have been 200 or 300 outsiders as well who had 'crashed' the train. The train had left Kingston at 5.20 a.m., and had arrived at Montego Bay late, so he had made arrangements for the return journey to be put back an hour.

On the outward journey he had gone from coach to coach telling people that the train was crowded {which must have been obvious to them}, to be careful and to pray for a safe and successful journey. The return journey had been further delayed because the engine had gone to get oil and they had finally left Montego Bay at 6.20 p.m. rather than the planned 6 p.m. Regarding the behaviour of the passengers on the return journey he recalled that two men right in front of him had had their pockets picked. He thought that the returning train had gone rather fast between Montego Bay and Greenvale, around the curves between Raheen and Balaclava, but then very slowly up the hill at Greenvale. After leaving Greenvale the speed of the train had picked up to the extent that he had begun to get tense and the youngsters had teased him for his nervousness over the speed. He could also remember one application of the brakes but he was not sure exactly where. Going around a corner there had been a jerk but this had not bothered him as he had had confidence in Driver Rhone, whom he had thought was driving. Another priest, Father Mallette, had spoken to him of the dangers of the overcrowding, the wooden coaches, people walking up and down the train and collecting on the platforms, and had predicted that "only a calamity would correct the situation." Then suddenly there had been a lurch, the sound of breaking glass, and the train had crashed.

Roy Walker, Senior Clerk, Jamaican Government Railways, gave evidence as to the selling of the tickets for the excursion and of the journey. He said that there had been five clerks besides himself collecting tickets on the train, named Craig, Estoffery, Estick, Patterson and Shaw, Brakesmen Burnett, Davidson and Lawson had also assisted with the ticket collecting, but not Guard Lewis. Davidson was one of those who had been killed. Some passengers had not had tickets, some had bought them from the ticket collectors and others had moved about the train to avoid the collectors. He agreed that the train had been crowded. According to his records 1,443 tickets had been sold. The passengers had been in 'a holiday mood' with much disorderliness and indecent language.

Driver Leopold Rhone told the inquiry that he had driven the train on the outward trip. There had been two locomotives, No. 90 piloting No. 85. There had been no trouble on the way to Montego Bay other than a 'hot-box' (overheated axle box), which had been dealt with at Maggotty. There had also been two stops due to the communication cord being pulled, but that was 'a normal thing,' and a matter to be dealt with by the guard. He had never had any problem controlling diesel engines on falling gradients. He was regarded as a 'moderate' driver. Reluctantly, he agreed that diesel hauled trains had become uncoupled from their coaches five or six times, he said he could not remember when.

Brakesman Norman Lawson said that the behaviour of some of the passengers had been 'unspeakable' and 'very, very bad.' Men asked for their tickets would say that their wives or sisters or girlfriends had them and if you let them go for them you would not see them again. As to his job as brakesman he explained that on some sections of the line the air valves had to be operated, taking off the speed of the train. There was an air valve on the platform of each coach. The valves were very easy to operate, you just lifted them with your finger when the train was going down a bank and released them at the bottom. After leaving Montego Bay he had had to lecture several men about not playing with the valve and had had to tell them off again later on. He added, "Big, big men play with it like it was a toy, Sir." On the return journey the behaviour of the passengers had been worse than on the way out. There had been four fights near Caladupa and coach lights had been turned off. After Greenvale there were two steep slopes. He had turned on the valves in the fourth and fifth coaches and the train had come down that one all right. Speed had been moderate at the start of the second slope but then had accelerated and "I had never experienced such speed in a train that size." The speed kept increasing to the "S" curve at the bottom of the grade. "My coach came right through the first half of the curve all right, but going into the second half, I felt when it was over. After I got up the first thing I did was to give thanks to God for my still being alive." [791]

The following day Acting Traffic Manager Lenval Fidler said that Father Eberle told him that he had sold 980 tickets and he offered the priest a further 100. He had not anticipated further tickets being sold on the train, though he knew that such things happened. He said that 1,146 passengers had been accounted for before the train had left Kingston and another 20 were picked up at Spanish Town. He reckoned on 100 passengers per coach, though there were only seats for 80. Passengers were not permitted to ride on the platforms; he had left it to the guard to prevent them from doing so. Grilled as to why he had not taken steps to prevent the train becoming overcrowded he said that he had spoken to Father Eberle. He had not spoken to the police or restricted the sale of tickets. He had not expected the train to be overcrowded, nor that some people would board it without tickets. He had been at home on the Sunday morning.

Steam Section Supervisor Rupert Gibson stated that he was in charge of coach maintenance. Six of the coaches in the train had been 71 years old, the others were 31 years old, having been rebuilt from the 71-year-old ones. All had a seating capacity of 80. The railway had four steel coaches.

Brakesman Frank Burnett said that the communication cord had been pulled after

---

[791] Ibid, 13th September 1957

Maggotty because someone had been left behind. Leaving Greenvale he had been riding in the first coach and when the speed of the train had increased he had turned up the valves on the first two coaches. The speed had not dropped and he had released them again "because you have to at a certain point." He had moved to the rear locomotive before the derailment. He did not recall seeing any signs of braking such as sparks.

Driver Lurch next gave evidence to the inquiry. He said that he had worked a train to Montego Bay on the Saturday afternoon and had stayed there overnight. The return trip to Greenvale had been normal. There he and Guard Lewis had made a brake test and he signed the usual certificate. Brake tests were always made at Greenvale, which was the summit of the line. Describing his journey from Greenvale he said, "Shortly after leaving Kendal {sic, this is clearly a misprint for Greenvale}, I have to run down a grade. I had no difficulty in controlling the train down that grade. I made my applications and got the desired results. I made no more use of my brakes again until we reached the Dunkeld crossing. On reaching that crossing, Sir, I reduced my speed to negotiate it. My speed was reduced from about 35 miles an hour to about 20 miles an hour for the crossing. Then I started to throttle the engine to take my upgrade from the crossing to a place called Balcarras. I turned over the grade and after turning over I made a test application. My speed was then about 15 miles an hour. I waited on the results of my test application to feel if the train was under control. I did not feel it; I did not feel the brakes resetting on the train. I went to make a heavier reduction on the pressure. I did not feel the brakes responding. I felt what seemed the surging of the train into the engine and observed that the speedometer was registering 30 miles per hour. I realised that I was in difficulty Sir and I gave about four blasts on the whistle…I went on to make an emergency application. I felt the violent surge of the train into the engine. At that time the engine was then into the cutting. I felt a slight check on the engine. I observed that the train line pressure was down to zero. The engine brakes were holding hard. I kicked off the independent brake valve to give the engine freedom in the curve, then reapplied the independent brake valve. The engine came to a stop." Driver Lurch said that he was aware that the speed limit was 20 miles an hour around the curves. He described in some detail the carrying out of the brake test at Greenvale. He denied being worried about arriving late back at Kingston.

The afternoon session of the inquiry that day began with the evidence of Guard Joseph Lewis, who stated that there had been 1,500 passengers from Kingston, with another 50 joining at Greenwich Town and another 100 joining at Spanish Town. Many passengers had been standing on the platforms. They had been very rough and rude though he had not noticed any tampering on the outward journey. After they had left Greenvale a passenger complained to him that the lights had been turned off in the third coach. He had gone there and a woman told him that a man had interfered with a guard valve. He had dealt with it, then as the train was at the top of the grade he turned up the remaining valves in the second and third coaches, finding that in the first coach it had already been done {by Brakesman Burnett}. Standing on the platform of the first coach some men had then tried to push him off, saying that they did not want any lights in the coach, and he had moved to the engine. There had been ganja smoking and gambling on the platform of the first coach.[792]

Little seems to have occurred on the next day of the inquiry, other than the production of a brake certificate signed by Drive Rhone rather than Driver Lurch and the latter

[792] Ibid, 14th September 1957

insisting that he had signed one, plus some further questioning of Guard Smith and Air Brake Inspector Macfarlane. [793]

The next day of the inquiry (Monday 16th), Drive Rhone was recalled and admitted signing brake certificates in advance for convenience. Guard Lewis was then recalled and gave more evidence regarding the brake certificates and the matter ended with Chairman MacGregor accusing Drivers Lurch and Rhone and Guard Lewis having conspired to cover up the fact that no brake test had been made at Greenvale on the return journey.

Mechanical Engineer Arthur Powney then gave evidence as to the routine examination of the engines and as to his examination of the engine and track after the accident. [794] He then faced further extensive questioning on the subject of the braking of the train and Driver Lurch's actions. He testified that there were two possible causes of the train's excessive speed. One was negligence or a lack of judgement on the part of the driver in making too small and/or too late a brake application. The other was that the brakes had been tampered with in some way so that the driver had not had as much brake power as he anticipated. Brigadier Langley then suggested that Driver Lurch might have exhausted his air by making a series of small brake applications without allowing time for the air to be recharged in between. Regarding the operation of the retaining valves Powney produced a book of instructions on the use and operation of brakes, issued to drivers and brakesmen, in which it stated that brakesmen must pass over all the vehicles of the train and turn the handles of the pressure retaining valves to the horizontal position upon all the vehicles upon receiving whistles signals from the driver at or near the top of a grade. Powney said that if some irresponsible person had interfered with an angle cock at Greenvale Driver Lurch should have immediately noticed it when he went to apply the brakes.

Stationmaster Arthur Brown at Montego Bay was then questioned. He spoke of being given a brake certificate, signed by Guard Smith and Driver Rhone, and agreed that the return excursion had been very crowded, with passengers riding on the platforms, which he was aware was unauthorised. He admitted that the brake certificate which he had been handed at 5.40 p.m. was for the incoming train and he had not been given another certificate for the brake test which was supposedly done before the train departed in the evening, after refuelling and with engine No. 87 substituted for No. 85.

On Tuesday 17th Fitter Hermon Balmer testified that he had examined the brakes on the coaches of the excursion train in conjunction with Fitter Peters though he (Balmer) was a yardman and not a regular examiner, and they had adjusted the length of the brake pistons to six inches. Airbrake Fitter Peters and Artisan Stanley Laidlaw also gave evidence as to the brake testing.

Next Police Corporals Sidney Davis and Eply McGregor gave evidence. Corporal Davis said that he had thought it wise for a policeman to be travelling on the train. There had been people riding on every platform. On the return journey the behaviour had been worst in the middle of the train, where there had been some fighting between Montego Bay and Anchovy over seats and also some drunkenness. The bad behaviour had not continued after Greenvale as a proportion of the passengers had alighted at previous stations. Corporal McGregor estimated that there had been 2,000 people on the train. He said that at Maggotty on the outward journey he had spoken to Driver Rhone about the slow progress of the train and had been told that you had to be careful as anything could

793 Ibid, 15th September 1957
794 Ibid, 17th September 1957

happen if you tried to go too quickly. On the return journey there had seemed to be organised gangs on each platform who had made it difficult for him to pass, though he had not noticed them interfering with the valves.

Carriage and Waggon Examiner Roydon Russell stated that he had helped with the testing of the brakes at Montego Bay before the return journey. He had seen the brake certificate but could not read it as he did not have his glasses. He had ridden back on the excursion with Mr McFarland and Examiner Stewart as far as Catadupa. It was then shown that the certificate which Russell had signed referred in fact only to engine No. 87, which had worked the train to Montego Bay in the morning, and not to the test which he had done with a different driver. He agreed that that was the case.

Observer Theophilus James said that his duty had been to observe signals given on his side of the track and any obstacles on the line. Sand had had to be applied at various points on the return journey due to slippery rails but the supply of sand had not been exhausted. He saw Driver Lurch make a brake test at Greenvale. Referring to the last miles of the return journey leading up to the derailment he said, "After we got to the summit we turned down the hill, Dunkeld Hill. When the two engines and about nine coaches, that is about three-quarters of the train, turned down the hill the driver made an application of his brakes. After making the application he said to me, 'James, something is wrong in the back. The train is not holding.' A little from that he made another application and found that he did not get any result. He blew four whistles to call the attention of the train men. After blowing those whistles I, of myself, discovered that there was a heavy impact as if the weight of the whole train rebounded on the engine. It started to develop faster. The driver, while operating with both his hands on the brakes made a full release on the independent valve, a direct release, and made a sudden emergency that should bring the train to a dead stop. After making that emergency the train developed very fast. The driver said to me, 'James, we gone now,' and he started to pray Sir. I, of myself, held on to the diesel and called on God to have mercy on me. We come down to the curve and saw that the engines were able to make that curve but still fretting about another curve we have. But both engines were able to make it. We got near to the quarter mile post when we found that the engine stopped. I turned my eyes behind and I saw no coach behind us. I said to him, 'Lord have mercy, no coach is behind the engine.'"[795]

The Kingston stationmaster, Vincent Ferguson, was asked about the capacity of the excursion train, the fact that there were passengers riding on its platforms and its overcrowding. He stated that if a train were overcrowded he would report the matter to the Traffic Manager by telephone, though he had never in fact done so. He was also questioned about his responsibilities regarding brake certificates.

Booking Clerk Alfred Powell at Kingston told the inquiry that 773 adult tickets and 207 seven children's tickets had been sold for the excursion previous to Sunday 1st and another 140 adult tickets and 26 children's tickets on that day, when he had been told by the stationmaster not to sell any more as he should stop at 1,100. The station had been very crowded that morning.

Daily Gleaner employee Keith Brown described speaking to Driver Lurch and Observer James after the accident.

Passenger Richard Lumsden maintained that there had been over 100 pickpockets riding on the excursion, while conceding that he had not actually seen any pickpocketing

---

[795] Ibid, 18th September 1957

in his coach.

Acting General Manager Stanley Magnus described the working of the railway and stated that every employee should have one, though at the moment not all stationmasters, drivers and guards had. The rule book dated from 1933, was now out of print and had not been issued for 20 years, but retiring employees were meant to hand in their rule books to be given to new employees. A copy of the rule book given to the Commission at the inquiry lacked amendment slips which had not been added by its last owner. There was also a book entitled 'Rules and Instructions for the Operation of Hand and Automatic Air Brakes.' It dated from January 1921 and had not, he thought, been revised. It was supposed to be issued to all guards, fitters, drivers, examiners, brakesmen, shed foremen, waggon examiners and stationmasters but he did not think that the department had sufficient copies for everyone to have one. In addition there was a working timetable for the Kingston – Montego Bay route with instructions regarding features of the line. He was not sure if any instructions had been issued regarding the loading of passenger trains pulled by diesel locomotives. He considered it the stationmaster's duty at ordinary stations to see that there was no overcrowding, though at Kingston that duty would fall on the Acting Traffic Manager, as the stationmaster's superior. In his experience he had never known severe overcrowding of a train with passengers standing on the platforms, though passengers stood on the platforms of trains which were not overcrowded. He had not issued any orders regarding their prosecution. He blamed Acting Traffic Manager Fidler for the overcrowding inasmuch as the latter had given Father Eberle only 100 extra tickets but had been told by the priest that there were 300 more intending passengers. The traffic manager should have known that the remaining 200 people would try to take the train if they could. He thought that the traffic manager should have contacted the police in advance to warn them of likely overcrowding. There had been no extra coaches available to attach to the train. It was forbidden for passengers to interfere with the conductor's emergency valve. It was the Traffic Manager's duty to see that brake certificates were properly filled out.

Asked by Brigadier Langley if he had any recommendations to suggest, the acting general manager replied:

(1) Brake certificates should be issued in duplicate or triplicate and each set numbered

(2) New all-steel carriages should be provided

(3) There should be fencing at Kingston and other stations to keep out intruders

(4) There should be rigid enforcement of the ban on passengers riding on platforms but this would require an increase in accommodation as it would be unfair to prosecute passengers with tickets for riding on the platforms when there was no room inside the coaches.

Driver Lurch was then recalled and said that he could not remember having driven any other 12-coach trains; it had been a pretty heavy load. He was again questioned about his handling of the train and about the brake certificate.

Passenger Ferdinand Madden said that he had heard six people complain on the return journey that they had been pickpocketed, he had stopped a fight and had been threatened by a male passenger carrying a knife. [796]

---

[796] Ibid, 19th September 1957

Passenger Ruix Moore told the inquiry that he had ridden on the platform with about 20 other people on the outward journey after being repeatedly 'bounced' (forced from the seat in which he was sat) inside. He had seen a passenger interfering with the conductor's valve; the culprit had been told to desist by the guard. Other passengers gave evidence, then Clerk Victor Craig, who denied having seen anyone interfering with the brake valves.

Clerk Edward Escoffery said that the behaviour of the passengers had been terrible. He had found passengers without tickets on the return journey. He had not seen anyone interfering with an angle cock. He had moved to the rear engine at Greenvale. After Grove Park (between Greenvale and Kendal) he had realised that the engine was going much too fast. Clerk Estick had remarked to him that a man like Driver Lurch should not be going that fast at that point of the road. On reaching a curve the engine had given a hard pull and rocked around the curve. When the two engines were around the corner he had seen that the coaches had overturned. He had then gone to the driver, who asked him what had happened. "You have no train, it is over in the gully," he had told the driver. Driver Lurch said that he did not know what had gone wrong, but he had had no brake. He had told the driver that it was no use doing any more talking, he had best go to Kendal for assistance. Clerk Escoffery said that there had been two people in the cab of No. 2 engine; one had been Mr McFarlane, the other was probably Constable Roberts.

Acting Traffic Manager Fidler gave further evidence and stated that the General Manager, Assistant Manager and Traffic Manager of the railway were all aware that there had been overcrowding on all passenger trains for some time. He mentioned specific examples and said that he had sent reports of these instances of overcrowding to the General Manager. He maintained that he had done all that he could to limit overcrowding on the train involved in the accident by limiting the number of tickets sold. Regarding passengers riding on the platforms Fidler said that guards had told him that their lives would be in jeopardy if they tried to stop passengers doing so and thus there had been two Special Constables riding on the train. He denied hearing Father Eberle say there were likely to be another 300 people wanting to travel or hearing the father offering to pay £125 or £150 for a second train.

Clerk Lawson Estick spoke of seeing people forcing their way through the gate onto the platform at Kingston. Whilst checking tickets he saw many people evade him by moving forward along the train, alighting when it stopped at a station and reboarding it towards the rear in a section which he had already checked. The behaviour of the passengers "had been nothing to boast about." There were gangs of pickpockets and he had been threatened. On the return journey people had been huddled on the platforms "like sardines," some of them asleep. Leaving Greenvale on the return journey Mr McFarland and Special Constable had been in the cab of the second engine and he, Clerk Craig, Clerk Escoffery and perhaps a guard had been on its platform. He had not witnessed any tampering with the valves.

Special Constable Edgar Roberts said that he had become ill on the return journey and Air Brake Inspector McFarlane had allowed him to ride in the engine cab. He had not touched anything there.

Passengers Rupert Cover and Wilfred Thompson also gave evidence. [797]

The next day Clerk Craig and Acting Manager Magnus gave further evidence, then District Constable William Wiles, then several passengers. One of the latter, carpenter

---

[797] Ibid, 20th September 1957

Leslie Price, spoke of seeing a boy fiddling with a valve for about 15 minutes. [798]

The inquiry dragged on. Guard Joseph Smith admitted on Tuesday 24th September that the air brake certificate purportedly signed by him at Greenvale on the day of the accident had in fact been signed a week later in Air Brake Inspector McFarlane's office. He insisted, though, that the test had been carried out. [799]

The following day, Wednesday 25th September, Driver Lurch was again questioned and admitted that he sometimes signed blank brake certificates for convenience.

Air Brake Inspector McFarlane was then grilled on the keeping or not keeping of records of air brake inspections, then on the question of the bogus air brake certificate. He admitted that in conniving with Guard Smith and Driver Lurch to prepare a false air brake certificate "I have made an ass of myself. Having done a foolish thing I must admit that I must have been crazy when I did what I did. What I did was a very stupid and dangerous thing to have done."

Guard Lewis reiterated that brake tests had been done at Montego Bay and Greenvale on the return journey. Brakesman Frank Burnett admitted that he did not really know the whistle signals for a brake test. The taking of evidence for the inquiry was then finally completed. [800]

The conclusion of the inquiry was that the cocks of the brake valves were loose enough to be easily turned by a passenger or they could have come into contact with a coupler on a curve and been accidentally closed. The guard had been riding on the engine contrary to the regulations and the brakesmen had failed to take action when the speed had become excessive.

On Sunday 22nd June 1958 a Mexico City – Veracruz night electric passenger train of the Ferrocarril Mexicano was stopped at Botas station, Mexico, at or near the summit of the line, due to a power failure. Power was then restored without warning, whereupon the train, consisting of two engines and seven coaches, started off down a steep gradient. Some reports stated that neither of the locomotives had been manned, another that Driver Porfirio Vega had tried desperately to stop it and had been injured in the subsequent accident. [801] Driver Vega or other crew members had reportedly tried to put down sand to help brake the train and some of its passengers had jumped from it during its three-mile run. [802] Newspaper reports on the accident varied in several details but they agreed that the train had run downhill for three miles to Maltrata station, where it had left the tracks. Both engines, three passenger cars, a mail car and a luggage van had become derailed, had overturned, had demolished the rear wall of the station, and had come to rest against a freight warehouse. The Pullman sleeping car and an empty dining car had remained on the line.

Different accounts stated that 10, 20 or 40 people had died. A Red Cross official estimated that 80 people had been injured. Ambulances took the injured to hospital in Orizaba. Among the victims were train passengers and also villagers and fruit peddlers

---

[798] Ibid, 21st September 1957
[799] Ibid, 25th September 1957
[800] Ibid, 26th September 1957
[801] *Valley Morning Sun*, Harlingen, Texas, United States, 23rd June 1958
[802] *The Bryan Daily Eagle*, Bryan, Texas, United States, 23rd June 1958

who had been waiting on the station platform at Maltrata to catch the train to Orizaba. [803]

The line was expected to be blocked for two days.

The Drachenfelsbahn rack railway in West Germany suffered a serious accident on Sunday 14th September 1958. The line was worked under the Riggenbach system and electric power had partly replaced steam operation since 1953, but steam trains were still run at peak times, and it was one which was involved in the accident. The steam locomotive left the track on a left-hand curve directly beneath a concrete post supporting the overhead line equipment and the first of three overcrowded coaches became derailed on the following right-hand curve and crashed into a tree, whereupon the second coach collided with the wreckage of the first. Eighteen people died and 112 were injured.

The railway climbed 1,520 metres with a gradient of 1 in 7 in its middle section, steepening to 1 in 5 near the top, while it was 1 in 8 at the point of the derailment. The carriages were not coupled to the engine, which was always placed below them, whether ascending or descending the ramp. The maximum speed was limited to 10.8 kilometres an hour. The maximum load of the three carriages was meant to be 150 passengers; at the time of the derailment the train was slightly overloaded, with 160 passengers, but this was not the cause of the accident.

On the evening of the accident the train left the summit station, Drachenfels, at 18:45 and its speed immediately became too high, so that when it reached the steepest section after 55 metres it was already travelling at 12 kilometres an hour instead of the usual 8. The driver noticed the increased speed and immediately made a full brake application of the Riggenbach counter-pressure brake, without effect. He then told his fireman to apply the hand brake. Thirty-two metres further on, with the train now travelling at 16 kilometres an hour, the engine reared up on the mountain side of the line and disconnected itself from the rack. As soon as this happened all the brake power of the engine was lost. According to the driver's later statement he had blown the engine whistle before he jumped off, though three conductors on the train did not remember hearing it. The fireman jumped off as well, somewhat later and at the wrong moment, so that he was crushed against a road overpass.

The central buffer of the engine must have become entangled with that of the leading carriage, since this did not detach itself from the engine, even though the conductor had applied the hand brake on the platform at the lower end of the carriage. It was estimated that the carriages reached a speed of 40 to 50 kilometres an hour before they became derailed. The engine slid along for 150 metres and then became derailed, overturning after another 50 metres. Although with the application of the hand brake the second car of the train was a little distance from the first, it still ran into it with considerable force. The third car, better braked, collided with the second but caused minimal damage.

The major cause of the accident was stated to be the instability of the steam engine when both brakes were applied at the same time. Experienced drivers were nevertheless able to operate steam locomotives safely when they were in use all the time, but since the electrification of the line steam locomotives were only in use occasionally, and the driving expertise had been lost. The engine was making its last trip of the day when the accident occurred; the boiler pressure was less than half what it normally was and its fire

---

[803] *The Daily Oklahoman*, Oklahoma City, Oklahoma, United States, 23rd June 1958

was low. A detailed explanation of what then happened is given in the German book from which this account is taken but it is unfortunately beyond the understanding of the author. Suffice it to say that it would appear that because of the low steam pressure the Riggenbach counter-pressure brake did not work when it was first applied, then when the hand brake was applied both brakes suddenly came into operation, and the combined pressure of 27 tonnes was sufficient to throw the engine from the rack.

A court case followed the accident; the driver was acquitted for lack of evidence, the manager of the railway was sentenced to 7 months' imprisonment and a supervisor to 9 months, both sentences being suspended. Following an appeal there was a retrial, as a result of which the manager was acquitted for lack of evidence and the supervisor was likewise acquitted. [804]

[804] *Katastrophen der deutsche Bahnen Teil II* by H.J. Ritzau, pgs. 127-9 (in German)

CHAPTER 8

BRIDGE FAILURES

Some of the most serious railway accidents have resulted from the failure of bridges. Bridges may fail for a variety of reasons. The inherent design of the structure may be faulty, the materials used in its construction may be poor and inferior, the materials may deteriorate over time, or its foundations may be washed by a stream or river which has become exceptionally swollen following abnormally heavy rainfall. With improvements in bridge engineering failures due to faulty design had greatly diminished by the middle of the twentieth century, but accidents, including major disasters, continued to occur from causes such as bridge foundations being washed away. A new danger, mainly though not entirely affecting road underbridges, would be so-called 'bridge bashing' by vehicles which were too high to pass beneath the structure.

A severe storm on the evening of Sunday 16th June 1946 led to floodwaters which greatly swelled a tributary of the Killbuck River and weakened the supports of a bridge carrying the Erie Railroad at a location variously described as two miles south of Friendsville and near Creston, Ohio, United States. When eastbound freight train No. 76, consisting of three diesel locomotives and 96 cars reached the bridge at 9.45 p.m., travelling at 30 or 35 miles an hour, Engineer J. Shank had just relieved Engineer E.W. Flickinger at the controls. Just as the leading engine was clearing the structure the bridge gave way and the three locomotives and 23 cars piled up in a tangled, splintered, mass. Engineer Flickinger was thrown from the engines and killed. [805] [806]

Three railwaymen had a lucky escape at Ottumwa, Iowa, United States, on the morning of Saturday 5th April 1947. They were riding on a Chicago Milwaukee St Paul and Pacific Railroad switch engine which was pulling 15 cars across a 200-foot wooden piling bridge over the Des Moines when the structure, weakened by floodwater, collapsed. The engine was completely submerged but Engineer Kurt Adcock, Fireman Billy Wells and Switchman Cleve Carnahan were able to escape from the cab and were pulled from the water below the bridge. [807]

---

[805] *Medina County Gazette*, Medina, Ohio, United States, 18th June 1946
[806] *Walla-Walla Union Bulletin*, Walla-Walla, Washington, Union States, 19th June 1946
[807] *Creston News Advertiser*, Creston, Ohio, United States, 5th April 1947

Mail train No. 1355 from Soria to Calatayud was crossing a bridge over the Ribota River between Clarés de Ribota and Villaroya de la Sierra, Spain, at about 18:00 on Saturday 30th August 1947. A severe storm which had lasted for several hours had greatly swollen the river and the floodwaters had undermined the abutments of the bridge, which was ten metres long and about three metres high. As often seems to have been the case the engine of the train managed to negotiate the structure but its tender, the leading mail van, and the following third-class coach, in which were about 90 passengers, fell into the river.

By about 02:00 the following morning 30 injured had been transported to an emergency hospital set up at Villaroya de la Sierra in military trucks of the 45th Regiment of Calatayud. Relief trains were sent from Madrid and Zaragoza. One of the injured passengers, trapped in the wreckage by one of his feet, was stoical and smoked cigarettes incessantly while he was waiting to be released.

Two days after the accident 13 complete bodies and one of the limbs of a 14th victim had been recovered from the river. The driver of the train was among the fatalities; he had been crushed underneath the tender. About 60 people were injured. The number of fatalities could have been higher as the *Mundo Obrero* newspaper spoke of at least 18 dead.[808] [809]

Engineer Quentin Briggs, Fireman John Faulkner, and Brakeman William Nunn lost their lives at 11.35 p.m. on Monday 12th September 1949 when the engine and seven cars of a St Louis-San Francisco Railway freight train plunged from a bridge over a creek weakened by floodwater near Seventy-Six, Missouri, United States.[810]

The Interstate Commerce Commission (ICC) investigated the accident. It described the structure as Bridge 94.3, 87 miles south of S.E. Junction and just over a mile north of Seventy-Six. Constructed in 1926, the bridge was an 11-span creosoted pile trestle, 143 feet long and 18 feet high. The catchment area of the stream crossed by the bridge was 1.56 square miles and the bridge was about seven times larger than was necessary to cope with the normal volume of the stream, as a precaution against backwater from the nearby Mississippi River.

The train involved was Extra No. 4107 South, consisting of 2-8-2 steam locomotive No. 4107, an auxiliary water car, 55 cars and a caboose, in charge of Conductor Luther King. It had left St Marys, the last open office, at 10.12 p.m., was stopped at mileposts 91.5 and 91.65 because of landslides, and was crossing the bridge at 15 miles an hour when the structure collapsed.

Three hours before the accident the train dispatcher had been advised of heavy rain between Chaffee and S.E. Junction and had issued order No. 128 advising train crews to be alert at all locations because of heavy rain. It was not addressed to Extra No. 4107 because from his information that train was already south of the area with the heaviest rainfall. However, Extra No. 4107 crossed train Extra No. 4123 North at St Mary's and the front brakeman of the latter train boarded the engine of the southbound train and advised its crew to proceed with caution as the northbound train had encountered heavy

---

[808] *La Vanguardia*, Barcelona, Spain, 31st August 1947
[809] Webpage: Federación Castellano Manchega de Amigos del Ferrocarril (in Spanish)
[810] *The Southern Illinoisian Daily Free Press*, Carbondale, Illinois, United States, 13th September 1949

rain. Extra No. 4107 did so and after encountering the second landslide its conductor sent the flagman back to Menfro, 7¼ miles north of Seventy-Six, to notify the section foreman and operator there of the situation.

The ICC determined that there had been heavy rainfall at Seventy-Six from 3.30 p.m. up to the time of the accident. At Chester, Illinois, 15 miles distant, 6 inches of rain had fallen in the 24 hours up to 7 a.m. on 13th September. It noted that previously the backwater from the Mississippi River had been sufficient to retard the velocity of the stream when it was running high and prevent scouring, but on the day of the accident the Mississippi River had been at a low stage, there had been very little backwater, the velocity had not been retarded and the stream bed had been washed away to a maximum depth of 15 feet. The bridge piles had thus been sufficiently weakened for the bridge to collapse under the weight of the train. Signal No. 929, in the rear of the bridge, had apparently been displaying a *Proceed* aspect.

According to the rules, the ICC continued, in stormy weather the section foreman must patrol his section if he thought it necessary. The Seventy-Six section foreman had only been in charge for seven days and said that he had not thought the storm severe enough to warrant patrolling. One of his sectionmen, with 17 years' experience, had on his own initiative patrolled the line from milepost 99 to Seventy-Six, about four miles. Arriving there about midnight he had called up the section foreman and the two of them, with another sectionman, had continued northwards to bridge 94.3, where they had found the wrecked train. The section foreman did not think that he would have discovered the weakened state of the bridge had he patrolled the line earlier. [811]

One of two spans of a bridge over the Turkey River just east of Fort Atkinson, Iowa, United States, collapsed beneath an eastbound Chicago Milwaukee St Paul and Pacific Railroad freight train from Sioux Falls, South Dakota, to Chicago, Illinois, at 1 a.m. on Thursday 29th September 1949. The diesel engine made it across the span but 16 cars were derailed and eight cars fell into the river or were left hanging from the bridge. The train was carrying alfalfa, grain, agricultural machinery, and meat. No one was hurt. The line was expected to be blocked for several days and meanwhile trains were diverted via Austin, Minnesota, and Mason City, Iowa. [812] [813]

Torrential rains led to the failure of a bridge in Italy late on Monday 10th October 1949. A Milano – Lecce passenger train found the line blocked near San Vito and was forced to return the way it had come, but when it was near Ortona at 23:00 a temporary bridge over the River Moro gave way beneath it and its rear vehicle fell into the water with the collapsing bridge. Sixteen passengers, who had probably been travelling at the front of the affected coach, were able to escape from it but three railwaymen were less fortunate. Conductors Alfredo Bassano and Nicola Pavoni and Platelayer Carmine di Florio fell into the river and were drowned, their bodies being recovered later. It was possible that other victims might have been carried away by the raging waters. The line was expected to be out of use for at least a fortnight.

The bridge which had failed was a wartime structure built by the Allies, which

---

[811] Interstate Commerce Commission report
[812] *The Telegraph-Herald*, Dubuque, Iowa, United States, 29th and 30th September 1949
[813] *The Democrat and Leader*, Davenport, Iowa, United States, 30th September 1949

remained in use in 1949 and was only subject to annual inspection. Local politicians condemned the Fellovie dello Stato for not having replaced it sooner, especially in a location which was known to be prone to flooding. [814]

Nine passengers died and 20 were injured when the engine, three coaches and a goods waggon of a mixed train fell from a bridge into a river at Muroan, Japan, on Wednesday 2nd August 1950. It was believed that the foundations of the bridge had been weakened by two days of heavy rain swelling the river. [815]

Early on Thursday 31st May 1951 an Illinois Central Railroad freight train met with an accident a mile from Woodbine, Iowa, United States, as it was crossing a bridge over a creek which was so small that it lacked a name. The creek entered the Boyer River just beyond the bridge.

Eastbound second-class freight train No. 74 bound for Chicago was in charge of Conductor Z.M. Decker with Engineer Edward (Eddy) Holland and Fireman Cecil Lyerly. It consisted of 2-8-2 steam locomotive No. 1569, an auxiliary water car, 64 cars and a caboose. Leaving Council Bluffs at 4.45 a.m., an hour and 45 minutes late, it passed Logan, just under 7 miles west of the point of the accident and the last open office, at 5.42 a.m. an hour and 57 minutes late, and was travelling at 40 miles an hour when it reached Bridge WA 476.6 at about 5.55 a.m. The bridge gave way beneath the engine, which fell to the bank of the stream below, and the tender, water car and leading 14 cars of the train piled up on top of or next to it. The 15th car was derailed. Engineer Holland, Fireman Lyerly and Front Brakeman M.L. Jenison were all killed. Conductor Decker and Flagman Sigart Hanson, riding in the caboose, were unhurt. They said that the brakes had not been applied prior to the accident. [816]

Bridge WA 476.76 was a five-span open-deck creosoted pile trestle, 83 feet in length, consisting of a 27-foot central span with two 12-foot approach spans on either side, with a normal maximum height of 17 feet. It dated from 1920. The drainage area of the stream was about 2.58 square miles and the bridge was about two or three times larger than needed to cope with the normal runoff. The watercourse was usually dry.

The Interstate Commerce Commission (ICC) report on the accident stated that heavy rain had fallen in the Woodbine area from about 1.30 a.m. At 2.30 a.m. the dispatcher had been informed that though there was heavy rain at Dow City, 17½ miles east of Woodbine, there was no heavy rain at Denison, 27¼ miles to the east, nor at Logan, 8 miles to the west, while a section foreman at Dow City reported that despite the heavy rain at that point, there was no damage to the track in his area. Information from a U.S. Weather Observation Station 9 miles west of Woodbine was that there had been one and a half inches of rain there between 1 a.m. and 5 a.m. on the morning of the accident. Subsequent to the accident, when the waters had subsided, a large tree was found downstream which had evidently been carried down the creek and had displaced the I-beams of the centre spans of Bridge WA 476.6, without dislodging the track structure. A Chicago and North Western Railroad bridge 100 feet south of Bridge WA 476.6 had escaped damage but a county highway bridge 580 feet further south again had also been

---

814 *L'Unita*, Roma, Italy, 14th October 1949

815 *Lubbock Evening Journal*, Lubbock, Texas, United States, 2nd August 1950

816 *Boone News Republican*, Boone, Iowa, United States, 31st May 1951

carried away.

According to the rules the section foremen were to patrol their tracks whenever they thought it necessary, such as during severe weather, but the section foreman concerned, who had seven years' experience at Woodbine, stated that he had not considered the storm severe enough to require such an inspection, adding that the bridge had never previously been damaged by high water.

The bridge had been examined by a bridge inspector on 6th July 1950, by the section foreman on three days before the accident and by a track supervisor from a track motor car the next day. Eastbound freight train No. 76 had crossed the bridge about nine and a half hours earlier without its crew noticing any problems.

The ICC concluded simply that the accident had been caused by the collapse of the bridge. [817]

Several Australian newspapers reported that seven people had been killed and 24 injured in the derailment of an express train on Sunday 26th August 1951 220 miles west of Gauhatis (Guwahati?), Assam, India. [818] A single Australian newspaper published a couple of weeks later bore a photograph of the wreck of an Assam Link express at an unstated location where a culvert, weakened by floodwater, had collapsed beneath the weight of the train. [819] It is assumed that this was the same accident.

A bridge accident occurred in Southern Italy on the morning of Saturday 17th November 1951. Railcar No. 341 was crowded with workers bound for the cement factory in Vibo Marina, with teachers en route to various schools in the area and with a party of carabinieri on their way to an exercise at a shooting range, also in Vibo Marina. At about 06:00 the train had passed through Pizzo station, had negotiated a tunnel, and was crossing the 20-metre-high Ciliberto bridge, 2¼ kilometres from Pizzo, when the accident occurred. The railcar had reached the last arch of the bridge when the structure gave way. The train allegedly hung suspended over the void for several minutes, to the terror of its passengers, before falling.

The alarm was immediately raised after the accident and a rescue operation was launched. Seven bodies were recovered from the wreckage and an eighth victim died after being rescued. Two women, who were identified as a mother and her daughter, were found embracing one another; the mother was dead, her daughter had survived, though badly hurt. In all 20 people were seriously injured and were taken to hospital in Vibo Valencia, while various officials went to the scene of the accident. The cause of the bridge failure was not immediately apparent; an investigation was opened. [820]

Chicago Great Western Railway time freight train No. 82 from Omaha, Nebraska, to Chicago, Illinois, was involved in what was described as 'a spectacular accident' on the morning of Thursday 27th March 1952. It consisted of an engine, 87 cars and a caboose. The engine and leading 14 cars had safely crossed a 150-feet-long bridge over a creek at

---

[817] Interstate Commerce Commission report

[818] *The Daily Examiner*, Grafton, New South Wales, Australia, 27th August 1951

[819] *The Northern Star*, Lismore, New South Wales, Australia, 11th September 1951

[820] *La Stampa*, Torino, Italy, 18th November 1951

Halbur, Iowa, United States, when a rail reportedly came loose, derailing the train, and tearing out the bridge, so that the next 33 cars, loaded with lumber, canned fish, hides, processed feed, wheat, and canned goods, fell to the stream below. The remaining 40 cars and caboose stayed on the track. No one was hurt but it was anticipated that the line would be blocked for at least a week. [821]

Prolonged heavy rain in the southern part of the American state of Minnesota led to flooding in early June 1953. On the evening of Monday 8th a Chicago St Paul Minneapolis and Omaha Railway freight train, consisting of a diesel locomotive, 51 cars and a caboose, in charge of Conductor C.L. Johnson, with Engineer H.L. Jordan and Fireman H.C. Langworth, came to grief at a trestle bridge over the swollen Spring Creek, three miles south of Madelia, Minnesota. The engine cleared the structure but the leading 13 cars, which included two loaded with material to repair a washout west of Butterfield, fell into the stream when the trestle collapsed. The remainder of the train stopped short of the bridge. No one was hurt but a spokesman for the company predicted that the line would be "closed indefinitely." The track had been damaged by floodwaters at numerous other points. Cloudbursts on the previous day had reportedly deposited 10 inches of rain. [822]

Thirteen railwaymen lost their lives in a bridge disaster in south-western Spain on Monday 2nd August 1954. The train involved had started out from Huelva, bound for Zafra. It had run as a mixed train as far as Fregenal de la Sierra, where the only passenger coach had been detached and it had continued as a pure freight train. Before it reached the next station at Los Jarales disaster struck at a location given in an English newspaper as Petruegano, when a 25-metre-high bridge collapsed beneath it. A Spanish newspaper states that the accident occurred at about three in the morning but as the whole distance from Huelva to the accident site is only about 50 kilometres this seems unlikely, unless for some reason the train made a prolonged stop en route. Conductor Simón García Medina, Driver Antonio Zamora López, Fireman José Rodríguez Merchon, Brakeman Nicholás Fernández and eight other railwaymen, including a pointsman, lost their lives. The one survivor was a drover travelling with some cattle, named Antonio Ribero Moreno, who only suffered a slight leg injury. A road worker who witnessed the accident reported it to Fregenal station, from where a relief train was despatched. [823] [824]

During the monsoon rainy season in India, which in the area of the next accident that we are to consider lasted from roughly June to October, special monsoon patrols inspected the track nightly, passing over the same point three times in each direction. Unfortunately this precaution failed to avert a major catastrophe on the night of Monday 27th September 1954 at Bridge 393 at mileage 174/5, between Jangaon and Raghunathpalli in Hyderabad state.

The train involved in the accident was No. 319 Down, the Hyderabad – Kazipet Express, which consisted of an engine and seven bogies (carriages). When it departed from its last stop, Alir (now Alair) it was estimated that it was carrying 180 civilian

---

[821] *Council Bluffs Nonpareil*, Council Bluffs, Iowa, United States, 27th March 1952
[822] *The Evening Tribune*, Albert Lea, Minnesota, United States, 9th June 1953
[823] *La Vanguardia*, Barcelona, Spain, 3rd August 1954
[824] *The Times*, London, England, 3rd August 1954

passengers, 120 military personnel and 19 railwaymen, or 319 people in all, coincidentally the same figure as the number of the train. It had left Hyderabad at 19:45, passed through Jangaon at 22:45, and the accident must have occurred a short time later, as the bridge was only 2¼ miles further on.

The bridge involved in the accident consisted of four 20ft girder spans, with a maximum height of 11 feet above the river bed. At the time of the accident the water level was reportedly two feet below rail level.

The average annual rainfall at the location of the accident was 27 inches. In the 24 hours preceding the accident about six inches of rain had fallen. As a result of this heavy rainfall two bunds (earth embankments designed to hold back water and prevent flooding) had burst, the water thus released flooding the track, and washing away the foundations of the bridge piers. Three bridge piers and all four girder spans were washed away. The line had also been breached at three other points.

When the train reached the bridge the engine and tender made it across the structure, though the rear of the tender was derailed. The leading third-class bogie was smashed, the second third-class bogie overturned, the next two bogies, which were second-, inter- and third-class composites, fell into the water and were swept away, as was the next bogie, a third-class, luggage and brake composite. The sixth vehicle, another second-class, inter- and third-class composite, was thrown onto the river bank and the rear vehicle, a third-class and ladies' coach, was derailed and damaged. Only 37 minutes before the accident No. 101 Down passenger train had crossed the bridge without incident and without anything abnormal being noticed.

Patrolman Venkati brought the first news of the accident to Jangaon station at 00:10 on Tuesday 28th. The stationmaster there telephoned the Divisional Offices at Secunderabad and a breakdown train was sent from that station but due to breaches in the line elsewhere could not reach the accident site until 08:05. Meanwhile permanent way and other inspectors had arrived from Kazipet with medical equipment by trolley at 04:41. The Chief Minister of Hyderabad and other officials arrived at 13:00 and the General Manager appeared at about 14:45, having flown from Bombay.

Due to the various breaches in the track the injured were taken on trollies and R.T.D. (Nizam State Rail & Road Transport Department) buses to the nearest point where they could be taken on coaches and brake vans to the civil hospital an Jangaon, the railway hospital at Lallaguda and the government hospital at Warangal. Twelve doctors and the same number of nurses went to the scene. There were 22 seriously injured and 50 slightly injured. One of the passengers on the train was Mr M.R. Krishna, M.P., who was initially reported to have lost his life but was in fact uninjured, though trapped for some time in a derailed coach. Food was sent to the site from Jangaon at 03:40 and enquiry offices for relatives were opened at Secunderabad and later also at Hyderabad, Kacheguda and Kazipet stations. Uninjured passengers were forwarded to Secunderabad station on R.T.D. buses, where attempts were made to record their names, but many of them left the station before this could be done.

Early on it was reported that 63 bodies had been recovered from the river bed and flooded fields, including six railwaymen on duty, three military personnel and two senior Hyderabad Government officials. The search for further bodies continued and by the 30th the number of known dead had risen to 137. The final figure quoted in the press was 139 but the official report on the disaster gives a slightly lower figure of 136. [825] [826]

The report on the accident attributed it to the scouring of the bridge foundations. Among its recommendations were:

(a) Bridge No. 393 at Mile 174/4-5 should be relocated at a better site and an adequate waterway should be provided.

(b) The stability of all bridges should be investigated in the light of the maximum possible flood and action where wanting should be taken.

(c) A list of all danger points should be made out, a predetermined limit of safety in each case should be clearly laid down, full-time watchmen at all such points should be posted during the monsoon and detailed instructions as to the action the watchmen should take when danger is apprehended should be specified.

A new girder bridge was subsequently provided with eight 40ft girder spans. [827]

Two railwaymen lost their lives when a Norfolk Southern Railway freight train plunged through a trestle on Saturday 26th March 1955 near Midland, North Carolina, United States. Three other railwaymen had a lucky escape.

The train involved in the accident was northbound second-class freight train No. 64, consisting of diesel-electric units Nos. 1610 and 1510, coupled in multiple-unit control, 21 cars and a caboose. It was in charge of Conductor Fred Styres with Engineer Bill Andrews and Fireman Guy Williams, and had left Charlotte, North Carolina, at 3.43 a.m., 43 minutes late.

The trestle involved in the accident, otherwise Bridge 366.9, crossed the Rocky River just under a mile north of Midland station. It consisted of a 10-bent open-deck framed wooden trestle, 123 feet in length, a three-span plate-deck steel girder bridge with concrete piers across the river itself, 301 feet long, and a five-bent open-deck wooden trestle, 53 feet in length. The wooden portions of the trestle consisted of untreated fir.

As the train was crossing the trestle at 4.45 a.m., travelling at about 35 miles an hour, the structure collapsed and the second diesel unit and the first eight cars of the train fell into the river. The unit and the first six cars were destroyed and the next three cars were badly damaged. Engineer Andrews, Fireman Williams, and Head Brakeman Joe Williams, riding in the first unit, escaped unhurt, but Conductor Styres and Flagman John Belvin, in the second unit, were killed. [828]

Engineer Andrews told the Interstate Commerce Commission (ICC) inquiry that as his engine had rounded the curve south of the bridge he had seen a small flickering blaze and as he got closer he had seen that it was the trestle itself which was on fire. The fire had seemed small and the trestle had appeared undamaged, but he had applied the brakes anyway with the intention of stopping and extinguishing the blaze. The fireman and head brakeman agreed that the fire had only seemed small, though the latter said that the fire had flared up immediately after the derailment. The engineer thought that the speed of the train had dropped to 15 miles an hour when the trestle had given way.

The ICC could not determine the origin of the blaze. The fuel tank of both units had

[825] https://rsdebate.nic.in/bitstream/123456789/584722/2/ID_07_29091954_28_p3898_p3901_1.pdf
[826] *The Times*, London, England, 29th September, and 1st October 1954
[827] https://ia903408.us.archive.org/24/items/dli.ministry.20007/1262.pdf
[828] *The Daily Independent*, Kannapolis, North Carolina, United States, 27th March 1955

been punctured during the accident and about 750 gallons of diesel fuel had leaked and added to the intensity of the fire.

The section foreman had crossed the bridge at 4 p.m. the previous day and had seen nothing amiss. A southbound freight train, No. 63, had passed over the bridge 5¾ hours earlier. Its crew had likewise noticed nothing wrong with the bridge.

The ICC concluded that the accident had resulted from the weakening of the trestle by fire. [829]

A bridge spanning the Ibrahimya Canal near Maghagha, Egypt, collapsed under a Cairo – Luxor express early on Friday 23rd March 1956. Eight people died, all of them Egyptians, and including Bishop Anba Thomas, secretary of the Holy Synod of Egypt's Coptic Christian Church, and a Coptic priest who was with him. On the other hand some foreign tourists who had been on their way to visit the antiquities of the Upper Nile Valley escaped injury. The failure of the bridge was blamed on the weakening of its supports by the swollen waters of the canal. [830] [831]

Early on Wednesday 4th April 1956 a 50-foot-long bridge over the Duchesne River at Manseau, Quebec, Canada, collapsed beneath an 85-car Canadian National Railways freight train from Quebec City to Montreal. Forty-one cars were derailed. Twenty box cars, mostly loaded with paper bales, fell into the river itself and the other 21 were strewn along the side of the track on either side of the bridge. No one was hurt. Passenger and freight trains were diverted via Victoriaville. The cause of the failure of the bridge was not immediately apparent. [832]

Train No. 565 Down of the Central Railway was passing over a 20-foot-long single-span bridge (Bridge 229) between Jadcherla and Mahbubnagar stations, India, on the metre gauge Secunderabad – Dronachallam line at 00:39 on Sunday 2nd September 1956 when the bridge gave way with disastrous results. Heavy monsoon rains had swollen the Pochani Nalla rivulet crossed by the bridge and the waters of the stream had risen 15 feet. The engine of the train just cleared the bridge but its tender and the following two bogies (coaches) 'were wrenched away and crashed into the swirling stream below.' The third coach was left hanging precariously over the eastern bank of the stream, its front end smashed. The rear six coaches of the train, two of which were for Bangalore, escaped damage.

The leading bogie, crowded with third-class passengers for Dronachallam, was stated to have been 'smashed to smithereens.' It had a seating capacity of 70 and had been crowded. Not a single passenger in it was thought to have escaped death. The second coach included a brake van and postal sorting compartment and contained only 14 people, including five postal staff and two policemen. The postal workers all escaped with minor injuries but only one person escaped from the brake van. The third coach, for upper-class passengers, only had two occupants, who escaped, though one had to cling to

---

829 Interstate Commerce Commission report

830 *The Times*, London, England, 24th March 1956

831 *The Progress-Index*, Petersburg-Colonial Heights, Virginia, United States, 23rd March 1956

832 *The Lethbridge Herald*, Lethbridge, Alberta, Canada, 4th April 1956

his seat for over three hours before he was rescued. The driver and two firemen also had a narrow escape.

The first news of the accident was received at Mahbubnagar station, five miles away, by a passenger who cycled there on a bicycle he chanced to have with him on the train, while a second passenger walked back to Jadcherla. News of the accident reached Secunderabad about 3 a.m. and a rescue operation was immediately set in progress.

'Mr K. Swaminathan, an engine fitter of the I.A.F. {Indian Air Force} who was travelling in the composite First and Second Class coach which sat on the brink of disaster, told Pressmen at Kurnool that when he opened the door after the crash he saw swift currents of the stream almost washing the steps of his compartment. The door of the compartment just ahead was ajar, and the couple he had met at Hyderabad was missing. In the ice vendor's compartment the vendor sat dazed little realising that he had {had} a close call from death.

'Mr Swaminathan added that within an hour of the accident a car with a doctor arrived on the scene from Mahbubnagar. The doctor started rescue operations immediately with the help of volunteers from among the passengers. With the dawn came more help and succour from the villages around. Waterpots, books, papers and clothes strewn all over the area told their own tales, he said.

'Merchants of Mahbubnagar brought food to the survivors of the midnight disaster, another passenger told Pressmen.

'A man who was brought out of a compartment collapsed as he was laid on a bench of a safe compartment.

'A ticket examiner who was travelling in a rear bogie and was one of the first to get out and go to the rescue of those affected, said that the accident took place at 12.40 a.m. There was "a terrific jolt" and his coach came to a sudden stop. As he got out immediately, he said, he saw the front portion of the train sinking. It was pitch dark and rain was heavy and nothing much could be done with the aid of his torch light.'

'By morning the water level in the stream fell to its normal knee depth. Twenty-four bodies were picked up from the site. Forty-six bodies were recovered later from places downstream before 10 a.m. Further search continued. Beddings, clothes, sarees, and other articles of personal wear were seen scattered on either side of the stream for a mile. Some of the bodies had been carried down a mile by the swollen waters. The rescue party soon collected the bodies of the dead and arranged to send them by train to Hyderabad for identification. All the mail bags were reported to be safe. The mail sorters, who had a very narrow escape, managed to save the mails. The mail bags were intact as the rear portion of the second carriage where these were kept had not disintegrated.

'On the left bank of the stream lay the bodies of the dead passengers and a few of these had been mutilated beyond recognition. The garments, shoes, and other belongings of the victims were stuck up to trees and bushes on both banks to form weird patterns. Of about 100 persons believed to have been travelling in the two first carriages six were known to have escaped. The rest were feared to have lost their lives. According to a ticket examiner who checked the first coach at Jadcherla, there were not more than half a dozen women in it. Among the bodies recovered were a child.

'The Chief Minister of Hyderabad, Mr B. Ramakrishna Rao, went to the scene of the accident by car as soon as he learnt of the tragedy this morning. He was one of the first to reach the spot and look after the rescue operations. Dr G.S. Melkote, Public Works

Minister, also was there. Mr K. Chandramouli, Minister for Religious Endowments, Andhra, whose train was held up in Mahbubnagar because of the line block was also on the spot this morning.' [833]

By Tuesday 4th September the number of dead in the disaster was put at 114 and there were about 30 injured, several seriously. The final number of fatalities was 121, with 37 injured, though the possibility of other fatalities could not be ruled out.

On Monday the Railway Minister, Mr Lal Bahadur Shastri, went to the scene of the accident and was reportedly deeply moved by what he saw. He stated that he proposed to appoint a committee of experts to examine every railway bridge in Hyderabad state and suggest steps to strengthen them where necessary. Meanwhile questions were asked in both Houses of Parliament about the safety of railway bridges after a statement from the Deputy Railway Minister, Mr O.V. Alagesan, that the bridge had given way due to a sudden rush of water from a breached tank nearby. Opposition members charged the Railway Administration with "criminal neglect" and "dereliction of duty."

Prime Minister Jawaharlal Nehru, speaking in the Lok Sabha (lower house of parliament) said that 'every possible step to be taken not only to find out what had happened in the present case, but to ensure that such a thing did not happen again. He would have no objection to a judicial inquiry into the accident, as demanded by some members, but he doubted if it would be helpful in this highly technical matter.'

Mr Alagesan stated that the bridge had been examined by the Divisional and Civil Engineers and the Permanent Way Inspector on 2nd August and declared safe. The previous train over the bridge had been No. 305 Down goods, which had crossed without incident at 20:30. After the accident a relief train had left Secunderabad at 04:05 and had reached the scene at 06:49, with surgeons and nurses, and had been followed by a second relief train at 12:30 with more medical staff and Osmania University medical students.

Allegations were made by Opposition members that the bridge had not been repaired since 1920, that it had been constructed of lime mortar, which was less strong than concrete and that the Railway Administration did not know how to keep bridges in a safe condition. [834]

The accident was the subject of a very lengthy inquiry, running to 134 pages. It heard that a mobile patrolling system at night had been introduced on the Mahbubnagar – Jadcherla section during the monsoon season, with a single watchman concentrating on bridges considered vulnerable which were within a mile of one another. Watchman Bheemadu told the inquiry that he had been at Bridge 229 at Mile 66/15-16, then had made his way to Bridge 233 at Mile 67/9-10, rather more than half a mile away, arriving there about 23:30, both bridges being under his supervision. At the latter bridge he seen a danger signal from the patrolman on the Mahbubnagar side and had placed detonators on the line at Mile 67/10-11. After 15 or 20 minutes he had met the other patrolman who informed him that there was serious damage to the embankment at Mile 68/5-12. He (Bheemadu) had then made his way back towards Bridge 229 showing a red light but within about 1,200 feet of it he had heard the crash of the passenger train coming from Jadcherla.

Driver Mohammed Yousef stated that it had been a dark night and he had had no warning of the bridge having failed. He had been travelling at 20 to 25 miles an hour and had gone down with his engine, the water coming up to his neck. Firemen Narasimhah

---

[833] *The Indian Express*, Madras, India, 3rd September 1956

[834] Ibid, 4th September 1956

and Satiah confirmed the account of their driver.

Guard Prakash Rao put the speed of the train at 30 miles an hour. The lights of the train had gone out, he had relit them, sent an urgent memo via a gateman, then assisted in the rescue work.

Gangmate Jalal Sab stated that he was in charge of the gang whose length included the bridge. There had been no problems at the bridge before it had been breached on 1st August 1956 (when Watchman Bheemadu had been able to stop a Down goods train). His gang had been active in the restoration work after that breach. His instructions were to protect the line whenever there was heavy rainfall.

Watchman Adivigadu Balliah said that there had been thunder with heavy rain at about 23:00 and fearing a heavy rush of water between Miles 68/7 and 68/10 he had gone in that direction, though his beat ended at Mile 67/8. Seeing a red flag exhibited by the watchman at Mile 68/9 he had returned towards Jadcherla and met Watchman Bheemadu and they had both gone in that direction.

Permanent Way Inspector Seshachalem stated that there had been no previous breaches Bridge 229 before that of 1st August. He had inspected the bridge between 01:00 and 03:00 the previous morning after a weather warning telegram.

Much more evidence was given at the inquiry, far too much to be repeated here.

The Government Inspector of Railways, C.R. Sule, concluded that:

(a) The engine and rolling stock of the train had been in good condition and had not contributed to the accident.

(b) The general maintenance of the track had been excellent and it had not contributed to the accident.

(c) After the breach of 1st August the bridge had been examined by the Divisional Engineer, Assistant Engineer and Permanent Way Inspector and no defects or scouring had been found.

(d) A sudden cloudburst in the catchment area of Pochani Nalla had caused erosion and subsidence of the Jadcherla approach of the bridge.

As to the responsibility for the disaster the inspector found:

(a) The bridge had been designed according to the practice of H.R.H. The Nizam's State Railway at the time that it was constructed.

(b) The water level at the bridge had never previously risen above 2 ft 11 in of the floor level so that there had been no reason to suspect the adequacy of the waterway.

(c) The rainfall in the catchment area had been abnormally heavy and sudden.

(d) The vulnerability of the bridge had been removed by the breach of Pochani Kunta a month earlier.

(e) All the precautions that human ingenuity could think of had been taken but had been outwitted by nature.

(f) No one, therefore, could be held responsible for the accident.

Finally the inspector ruled that the relief measures had been prompt and adequate under the circumstances. [835]

A logging train plunged from a bridge into a deep ravine at an unstated location in northern Formosa on Friday 28th June 1957. Chen Chin-tung, 54, the deputy director of the Forestry Administration, was killed, as was a second man, while 21 other workers were injured. [836]

Also on Friday 28th June 1957, a Pennsylvania Railroad work train of about 25 cars was heading west towards Terre Haute because the track beyond there had been washed away. It was proceeding cautiously for fear of encountering further washouts a mile east of Reedsville, Indiana, United States, and crossing a bridge over the swollen Big Walnut Creek, when the structure gave way. The engine and one car managed to get over the bridge before it collapsed; the next 18 cars fell into the stream. The rear cars and the caboose came to a stand on the approach to the bridge. The engineer, fireman and a brakeman at the front of the train and the conductor and a flagman at the rear of it all escaped injury. The accident happened at 11:40 a.m. At 5.30 p.m. the water in the creek was still rising and clearance of the wreckage and repair of the bridge had to wait until the waters had receded. Another bridge failed at about the same time on the New York Chicago and St Louis Railroad at Shelbyville, Indiana, this time without a train being involved. [837]

A week after the Reedsville accident a bridge failure near Edenton, North Carolina, United States, on Friday 5th July 1957, resulted in the death of two railwaymen and the injury of three others. The bridge in this case was an open-deck trestle of just over 5 miles in length across Albemarle Sound. The trestle consisted of 2,131 cypress and pine full-pile bents and dated from 1910.

The Norfolk Southern Railway train involved in the accident was southbound second-class freight No. 64. It consisted of diesel units Nos. 1605 and 1610, coupled in multiple, 76 cars and a caboose, in charge of Conductor S.R. Bray with Engineer W.R. Munden and Fireman Bert Rankin. It had left Marsden, the last open office, at 9.20 p.m. the previous evening, 20 minutes late, and was crossing the trestle at less than 10 miles an hour when the two diesel units and the leading car of the train fell through it. Conductor Bray and Engineer Munden were drowned; Fireman Rankin floated and swam to the north shore of the sound, Brakeman C.E. Jones crawled painfully back along the trestle with a broken arm and Road Foreman J.G. Lowery was picked up from the water by a fishing boat as he held onto debris. The engineer and fireman had been in the cab of the first unit, and the other three men had been in the cab of the second.

The trestle included a draw span and the accident was witnessed by Bridge Tender Joshua Griffin, half a mile to the south, who said, "It happened like the snap of your fingers." He added that the trestle had been swaying and vibrating beneath the bridge

---

[835] Ministry of Railways: Report of the Commission of Inquiry to Ascertain the Causes of the Accident to 565 Down Passenger Train between Jadcherla and Mahbubnagar on the Central Railway on the night of 1st/2nd September 1956.

[836] *The Abilene Reporter-News*, Abilene, Texas, 30th June 1957

[837] *The Brazil Daily Times*, Brazil, Indiana, United States, 29th June 1957

before it failed. [838]

The surviving members of the train crew told the Interstate Commerce Commission (ICC) that they had noticed no brake application before the trestle failed.

Examination of the diesel units by the ICC showed nothing wrong with them which could have caused them to derail. They were fitted with speed-recording devices but these were water damaged and unreadable. There was no indication that the train had become derailed before the trestle had collapsed.

Examination of the trestle, on the other hand, disclosed multiple defects. A large number of ties (sleepers) were in various states of decay. A number of bridge stringers were decayed. The tops of the piles supporting the frames of a considerable number of frame-type bents had deteriorated to the extent that the sills were not bearing on them. Some piles were missing or had moved from their proper location, other piles had deteriorated to the extent that the tops of them were not in contact with the sills, The sills of two bents had failed because the top of the outside batter-pile was not in contact with them, and so on.

Two regularly assigned bridge crews inspected the bridge. Each crew inspected a portion of the bridge each day. Bridge patrolmen inspected the bridge from a velocipede car after every train. A bridge inspector, often accompanied by an assistant chief engineer, made weekly inspections by train, track motor car or on foot; the last had been on 3rd July. The last inspection by boat had been in January.

Since 1949 there had been a programme of replacing the original full-pile and frame-type bents with full-pile treated pine bents and steel stringers. The ICC noted that at the 1957 rate of replacement it would take another 19 years to replace the lot.

The ICC concluded that the accident had resulted from the failure of the trestle. It recommended:

(a) That the railroad company immediately arrange a thorough and systemic inspection by competent engineers to determine its present overall condition and its ability to withstand the traffic loads using it.

(b) That any member of the structure found during the inspection to have deteriorated to the extent that it could not perform its intended function, as well as those deteriorated structures already identified in the ICC report, should be immediately replaced to forestall another collapse.

(c) That the company should immediately initiate an adequate maintenance programme, including frequent periodic inspections of the entire structure by competent personnel, to ensure that the trestle was maintained at all times in a safe and suitable condition to carry the traffic passing over it. [839]

A bridge over the Jumarella stream in Southern Italy, on the Ferrovie Jonica portion of the Fellovie dello Stato, collapsed beneath freight train No. 7395 at 01:20 on Saturday 28th October 1957. All of the short train, which consisted only of an engine, a luggage van, and three waggons, fell from the bridge, which was situated between kilometres 130 and 131, near Taurio station. Driver Luigi Momaco, 60, from Crotone depot, was trapped in the wreckage of his locomotive and killed. Fireman Francesco Pupa, 30, was trapped

838 *The Columbus Daily Telegram*, Columbus, Nebraska, 5th July 1957

839 Interstate Commerce Commission report

briefly under water but managed to struggle free; he suffered a broken left leg. Brakeman Traverso, 34, also survived, but sustained various injuries. Assistance was sent from Sibari infirmary and the two injured railwaymen were taken to Corigliano hospital. A breakdown train was despatched from Reggio Calabria. The line was of course blocked and passengers were transhipped between Sibari and Reggio Calabria.

Heavy rain had swollen the stream and the floodwater had eroded the left approach embankment of the bridge over a distance of 20 metres, causing the structure to fail.[840]

A bridge over Medway Creek, Queensland, Australia, near Bogantungan, was washed away by the swollen waters of the stream in 1956. Just after 2.30 a.m. on Friday 26th February 1960 the bridge failed again, this time as *The Midlander*, an air-conditioned express running between Brisbane and Rockhampton, was crossing it. The train consisted of two C17 class locomotives, Nos. 826 and 712, a generator car, three sleeper cars, a dining car, three coaches and two vans, ten vehicles in all. It was in charge of Guard Kerrod Jaques with Driver Sambo Dean and Fireman Leonard Wilson on the pilot engine and Driver George Krause and Fireman Neville Helmuth on the train engine. The railwaymen on the train also included Conductor Samuel Hedges.

Driver Dean told the subsequent inquest, "I had a good headlight on the engine. I noticed that the creek had roughly about 8 feet of water in it…The bridge was quite normal. The rails and transoms looked perfect as far as I could see in the big headlight."

Unfortunately the bridge was far from normal. The first (pilot) engine just managed to get across it, then lurched to the right, then to the left and overturned onto the side of the line. The second (train) engine, ended up in the creek, as did the generator car and following three coaches. The dining car ended up hanging off the bridge and the remainder of the train stayed on the line.

Driver Krause, Fireman Helmuth, Conductor Hedges, and four passengers, two of them children, lost their lives and 43 people were injured.

Driver Dean related what happened immediately after the accident. "I was sort of stunned and felt my way out over the tender. All the lights went out and the cab was in darkness. I saw Fireman Wilson round the front of the engine. He asked me how I felt. I said, 'No good, the bridge gave way.' He said, 'I heard it crack.' Wilson sang out to the second crew, 'Are you there George? Are you there Neville?' There was no answer…I heard women and children screaming. Everything was in darkness."

Fireman Wilson made his way to Bogantungan station and there aroused Stationmaster Brian McGann, who in turn immediately aroused all the local railwaymen and villagers and then set off to the bridge on a quadricycle with stretchers and first-aid equipment, arriving there about 3 a.m.[841]

Prominent in the rescue work were three passengers, all of whom were also railwaymen: John Bennett, a workshop fitter of North Rockhampton, travelling with his wife and family; Lawrence Murray, a fettler from Pioneer, also travelling with his wife and family; and Alan Streeter, a fireman from Gladstone. The fettler was the last witness to give evidence at the inquiry into the accident. A newspaper wrote, 'With his wife, three children and an infant

---

[840] *La Stampa*, Torino, Italy, 29th October 1957

[841] https://judithsalecich.com/hear-the-children-cry-bogantungan-and-the-1960-medway-creek-rail-accident/

granddaughter he was travelling in the last carriage of the Midlander when it crashed. He was among the first down to the wrecked carriages and waded into the rushing stream to carry the children to the bank as they were handed to him.

'Then he swam across the creek to the Bogantungan side and helped hand up tools to men working on the wrecked carriages. He also helped four injured men and a woman to the creek bank. After that he went through the darkened carriages searching for other people to help to safety. He continued to work on the wreckage until daylight.

'Murray said he had taken off his shoes and could not find them afterwards. He said he had a sliver of glass in his foot early. Eventually he received treatment for his injuries in Emerald. Telling about the loss of his shoes Murray said, "That is why I am barefooted now." The Chairman of the Board of Inquiry, Mr V.F. Hall, warmly complimenting Murray on his actions, said the Board would do something about getting him more shoes.' [842]

The Aborigine fettler died suddenly of a heart attack not long after having attended the inquiry. He was posthumously awarded the Queen's Commendation for Brave Conduct. Bennett and Streeter were also honoured with the award.

Most of the passengers from the train were taken to Borgantungan where they were cared for by the inhabitants. Relief trains with Dr Frederick Bennett of Alpha and Dr Charles Whitchurch of Emerald left Alpha and Emerald stations respectively at about 4 a.m. and 4.30 a.m. and arrived at the scene about the same time. Twenty-four passengers plus rescuers Murray and Streeter were treated at Emerald Hospital. Most of the passengers eventually continued by train to Rockhampton.

The Queensland Railways conducted a very extensive inquiry into the accident and there was an inquest on those who had died. At the conclusion to the latter Coroner C.H. Smith delineated the cause of the accident. 'A tree was uprooted above the bridge and was carried downstream by flood waters, resulted from a heavy downpour in the catchment area of Medway Creek. The tree apparently struck the upstream pile in pier five, breaking the mortise and dislodging the pile from its bed. The weight of the leading engine caused this pile to give away, and the second engine, power car and three leading coaches plunged into Medway Creek.

'Although the line was not run that night, it is evident that the weakening of the bridge would not have been discovered even if this had been carried out. There is no evidence of any mechanical defect in the engine and coaches, and the running crew cannot be blamed in any way.' [843]

An accident near Logan, New Mexico, United States, which cost five lives on Saturday 9th July 1960 received little attention in the press and was apparently not the subject of an Interstate Commerce Commission report. The Chicago Rock Island and Pacific Railroad train was en route from Amarillo, Texas, to Tucumcari, New Mexico. It consisted of an engine, 41 cars and a caboose in charge of an unnamed conductor with Engineer R.H. House and Fireman B.J. Beavers. The accident happened at a wooden trestle over Ravuelto Creek, a normally dry arroyo which had been swollen by recent

[842] *The Canberra Times*, Canberra, Australian Capital Territory, Australia, 12th March 1960

[843] https://judithsalecich.com/hear-the-children-cry-bogantungan-and-the-1960-medway-creek-rail-accident/

heavy rains, fatally weakening the trestle. The engine just about made it across the structure but was derailed on the far side, while the following 17 cars also left the rails, eight of them plunging 30 feet to the creek bed below, catching fire and setting fire in turn to the trestle.

The five who died were hobos or transients who had been riding on the train. Three of them were quickly identified as Herman Goff, of Sherman Gap, Kentucky; Sammy Ross Miller, 17, of Waurika, Oklahoma; and Lesley Delmar Yancy, 51, of Long Beach, California. Three of the surviving hobos were credited with saving the life of Engineer House by swimming across the creek and pulling him from the wrecked engine. The engineer then lay beside the track for four and a half hours with back injuries before rescuers were able to reach the scene, the approach by rail being delayed by a soft road bed and by road by thick mud. Fireman Beavers and one of the surviving hobos were also injured and taken to hospital. [844] [845]

The other two hobos who lost their lives were later identified by their fingerprints as Harold Goochey, 57, of Wilmington, California, and Walter Earl Stonestreet, 53, of Akron, Ohio. [846]

---

[844] *The Galveston Daily News*, Galveston, Texas, United States, 10th July 1960
[845] *Waterloo Sunday Courier*, Waterloo, Iowa, United States, 10th July 1960
[846] *Amarillo Daily News*, Amarillo, Texas, United States, 16th July 1960

CHAPTER 9

OPENING BRIDGE ACCIDENTS

Accidents involving opening bridges appear to have occurred in only a handful of countries, with the majority occurring in the United States. It seems that most countries managed to a large degree to avoid building such bridges, which from a railway point of view are a constant source of expense and worry. Robert Shaw states that the early railroads on the eastern seaboard of the United States were forced by opponents of the new form of transport to build such bridges over even insignificant waterways where they were never really needed. [847]

The first opening bridge accident during the period covered by this book appears to have occurred on Saturday 22nd February 1947 on the New York Central Railroad at Charlotte, New York, United States. The accident occurred at 4.23 a.m. during a blizzard. The train involved was a switching (shunting) movement, consisting of steam engine No. 1599 and 38 cars. The drawbridge involved crossed the Genessee River close to its outlet into Lake Ontario. Bridge 0-151 was a steel, through-lattice truss swing bridge, 308 feet long, situated 1,826 feet east of Charlotte station. The draw span was powered by a steam engine and normally stood open for railroad traffic but had been opened for river traffic shortly before the accident. Protection for westbound movements over the bridge was provided by a rectangular stop sign, 428 feet to the east of it, 2 feet wide and 4 feet long, mounted upon a 6-foot mast and worded 'DRAWBRIDGE 400 FEET STOP' in white letters on a red background. In addition a lighted red lantern was placed in the centre of the track immediately east of the span. Average movements over the bridge in the 30 days preceding the accident was 10.4 trains, 26.4 engines and only 3.8 ships.

The switching movement that led to the accident had begun 1,500 feet east of the bridge, which had been opened for the passage of a car ferry. The engine passed the stop sign with its headlight shining brightly. Its fireman was busy firing and unaware that the bridge was open. A brakeman on the tenth car and a conductor and two brakemen at the rear of the train gave stop signals with white lanterns when the shunt move passed the stop sign and they saw that the bridge was open but the engine did not pull up and travelling at only three miles an hour it fell into the river from the open span. The bridge tender had also given stop signals with an electric flashlight. The engineer was killed in

---

[847] *A History of Railroad Accidents, Safety Precautions and Operating Practices* by Robert B. Shaw, pg. 187

the accident and the fireman was injured. [848]

Like the previous accident, one at Clarksville, Tennessee, United States, on the Louisville and Nashville Railroad involved a low-speed switching movement. The opening bridge over the Clarksville River, Bridge No. 19, was situated just south of Clarksville station. From the north the 270 feet through steel truss draw span was preceded by a 21-foot-long trestle and a 59-foot-long steel deck girder span. South of the opening span were a 213-foot steel truss span and a 2,195-foot-long trestle. The bridge was powered by a gasoline engine and operated by a bridge tender from a control room on the draw span.

A warning sign was provided 5,044 feet north of the draw span. It was rectangular, 17 inches wide and 28 inches long, mounted on a 6-foot mast and worded 'DRAWBRIDGE ONE MILE' in black letters on a yellow background. In contrast to Charlotte there also was signal protection here, provided by signals Nos. 1771 and 1781, situated 5.054 and 191 feet north of the drawbridge. The colourlight signals indicated *Yellow* (*Approach) – Prepare to stop at next signal.* Trains exceeding medium speed must at once reduce to that speed, and *Red (Stop) – Stop*, respectively. The signals and bridge were interlocked such that signal No. 1781 must display *Stop* before the drawbridge could be opened and the draw span must be closed before the signals could be cleared. If signal No. 1781 was showing *Proceed* and was then placed to *Stop* two minutes must elapse before the draw span could be unlocked and it then took another four minutes to open the span.

On the day of the accident southbound second-class freight train No. 123 arrived at Clarksville at 10.45 a.m. and stopped 226 feet north of signal No. 1781. A quarter of an hour earlier the bridge tender had gone to the station and found from the operator that train second No. 123 had left Guthrie, just over 14 miles to the north, at 10.15 a.m. He had then returned to his control cabin, placed his signals at *Stop* and opened the draw span, not for a ship but for a government inspection.

At about 10.48 a.m. the engine and five cars were detached from the front of the freight train and moved southward. The conductor and front brakeman were in the third car and the flagman and swing brakeman were in the fifth car, The first two thought that signal No. 1781 had been showing *Proceed* when their train had first stopped but they had not noticed it afterwards. The engine and cut of cars moved southwards at about 4 miles an hour past signal No. 1781, which the Interstate Commerce Commission determined must, despite the statements of the conductor and front brakeman, have been displaying *Stop* for at least a minute before the arrival of train No. 123 at Clarksville, and the engine and first two cars fell into the river at about 10.50 a.m. Engineer John Black and Fireman M.E. (Red) Carter were drowned in water between 20 and 30 feet deep. [849] [850]

Southbound Chesapeake and Ohio Railway through freight train No. 45 was approaching the drawbridge over the St Joseph River between Benton Harbor and St Joseph, Michigan, United States, at about noon on Monday 20th June 1949, when its big *Berkshire* class 2-8-4 steam locomotive, No. 1230, became derailed, together with its tender and the leading car. No one was hurt but the line was blocked until early the

---

848 Interstate Commerce Commission report

849 Ibid.

850 *Wilmington Morning Star*, Wilmington, North Carolina, United States, 14th June 1947

following morning, a wrecking train having to be brought from Grand Rapids. The drawbridge had doubtless been open but the derail had done its job and prevented the engine from going into the river. [851]

Various local newspapers reported on an accident at Mermentau, Louisiana, United States, at about 11.45 a.m. on Sunday 15th October 1950 but the accident was not investigated by the Interstate Commerce Commission and its exact circumstances are unclear. It appears that the bridge over the Mermentau River had been raised to allow a barge to pass, and had been lowered again, but the second section of eastbound Southern Pacific Railroad freight train No. 252 had started across it prematurely, before the derails had been properly closed. The locomotive, tender and two box cars managed to cross the bridge but six cars, including two gondolas loaded with coal and a tank car, fell into the river, while 400 feet of track was torn up and 800 feet more was damaged. No one was hurt but it was clearly quite an expensive accident. Repairs were expected to take several days and in the meantime detours were likely to add two or three hours to train journeys. [852] [853]

An accident on an opening bridge at Maassluis, the Netherlands, on Thursday 15th November 1956, was not as serious as it might have been. A three-car boat train from Utrecht via Amsterdam to Hoek van Holland, travelling at an estimated 90 kilometres an hour, had almost crossed the bridge when its rear car began to jump about violently, giving the passengers a good shaking. The car then left the rails altogether, struck a signal box and then a metal post, probably a support for the overhead line equipment, which prevented it falling into a lineside ditch. The Vlaardingen fire brigade were soon on the scene and set up floodlights to aid the rescue work. Fortunately the casualties were limited to two English passengers who were slightly hurt but there was much damage to rails, sleepers, and the overhead line equipment while one of the cars of the train, presumably the last one, was detached from its undercarriage. The line was blocked and passengers from the boat train and three following ones were forwarded to Hoek van Holland by bus, where the night ferry to Harwich awaited them. [854]

Examination of the scene by experts from the Nederlandse Spoorwegen disclosed that the derailment had been caused by the bridge not having been properly closed. It transpired that work had been carried out on the structure in an effort to reduce the amount of noise generated when trains crossed it, doubtless after complaints from local residents. The bridge tender had been instructed to check that the bridge was properly secured after the completion of the work and stated that he had done so before he cleared his signals for the train. [855]

A 'misunderstanding of signals' was reportedly to blame for what was said to have been the first serious accident for half a century to occur at the Illinois Central Railroad opening bridge over the Ouachita River at Monroe, Louisiana, United States. The train involved belonged to the Tremont and Gulf Railroad but was being operated over the

---

[851] *St Joseph Herald Press*, St Joseph, Michigan, United States, 21st June 1949
[852] *The Austin American*, Austin, Texas, United States, 15th November 1956
[853] *Lubbock Evening Journal*, Lubbock, Texas, United States, 17th November 1956
[854] *Nieuwe Haarlemsche Courant*, Haarlem, the Netherlands, 15th October 1956
[855] *De Telegraaf*, Amsterdam, the Netherlands, 17th October 1956

Illinois Central because flooding had damaged its own tracks. Arriving at Monroe the train had stopped, its engine had run around it, and begun setting back the cars into a siding of the Olin Mathieson Paper Plant near West Monroe.

Meanwhile, after the arrival of the train Bridge Tender Malcom Slayton in the cabin had opened the bridge for a barge. Signal protection was provided on either side of the bridge, possibly flashing light signals, as the local newspaper states that they had been flashing for a full two minutes before the setting-back movement was begun. The 41-car train of paper and pulp wood was stopped fairly quickly, but not before the rear box car loaded with paper had fallen into the river and the flatcar loaded with pulp wood next to it had been left hanging off the bridge.

Further damage could have been caused had the loaded box car been carried by the swollen river against the adjoining road bridge, but this was averted by a police officer. 'The quick thinking of Officer M.L. Spangler, of the Monroe Police Department, averted possible serious damage to the Old Traffic Bridge. Spangler was inside the bridge house when the railroad car backed off into the river. His quick actions in opening the traffic bridge allowed the loaded car to pass just as the bridge swung open. Police Chief J.P. Kelly stated at the scene, "Spangler deserves to be commended. He was in a bad spot up in that house. If he hadn't reacted as he did, there's no way of knowing just how badly the bridge might have been damaged." The bridge has been under tremendous pressure from the flooded river.'

The box car sank on the far side of the road bridge after having been pushed towards the west bank of the river.

The accident had happened just before noon and the bridge was reopened just after 6 p.m. A passenger train passed over it at 7.30 p.m. [856]

The other drawbridge accidents described in this chapter were minor affairs compared to that which occurred on the Central Railroad of New Jersey on the drawbridge over Newark Bay, New Jersey, United States, on the morning of Monday 15th September 1958. A passenger train fell from the bridge, causing the death of 44 passengers and four railwaymen and injuries to another 46 passengers and two railwaymen. It was easily the most serious drawbridge accident to occur anywhere in the world since that at Atlantic City, New Jersey, not too far away, back in October 1906 (described in *World Railway Disasters 1901-1915*).

Between Elizabethport and West 8th Street, Bayonne, the four tracks of the Central Railroad of New Jersey (CNJ) crossed Newark Bay by the Newark Bay Bridge, which was almost a mile and a half long and included four double-track lift spans over the two shipping channels of the waterway. The spans were designated as northwest, northeast, southwest and southeast. The northwest and southwest spans were 310 feet in length and the northeast and southeast spans were 210 feet long. Tracks Nos. 1 and 3 used the southwest and southeast spans and tracks Nos. 2 and 4 used the northwest and northeast spans. The lift spans were of the through truss vertical lift type. When lowered they had clearances over high water of 35 feet and could be raised to a height of 135 feet. They were operated electrically by DY interlocking. At the time of the accident, the northwest and southwest spans were raised to a height of 108 feet above water level. The

[856] *Monroe Morning World*, Monroe, Louisiana, United States, 31st May 1958

approaches to the spans were protected by derails. That of track No. 1, on which the train involved in the accident was running, was situated 490 feet to the rear of the southwest span. The spans were also protected by automatic and semi-automatic colourlight signals. Those applicable to the accident and their indications were:

Signal L48, controlling movements off the Perth Amboy branch to track No. 1, which was displaying red over green over red (*Medium clear*), indicating *Proceed, medium speed through interlocking limits.*

Signal 82, displaying yellow over staggered red (*Approach*), indicating *Prepare to stop at next signal*, trains exceeding medium speed must at once reduce to that speed.

Signal R26, displaying red over red over red (*Stop*) indicating *Stop*.

Before the southwest span could be opened signal R26 had to be displaying *Stop* with signal 82 displaying *Approach* and the derail open.

On an average day 140 trains crossed the bridge and the spans were opened 28 times for ships.

The train involved in the accident was eastbound first-class passenger No. 3314, consisting of road switcher diesel-electric units Nos. 1532 and 1526, coupled in multiple-unit control, four coaches and a passenger baggage car, in charge of Conductor John Kelly with Engineer Lloyd Wilburn and Fireman Peter Andrew, all of whom lost their lives. The other railwayman killed was Front Brakeman Theodore Dayton.

Starting from Bay Head Junction, on the New York and Long Branch line, at 7.28 a.m., on time, the train had passed Woodbridge Junction, 12¼ miles from the bridge, at 8.44 a.m., stopped at Elizabethport station, 1¾ miles from the bridge, and left there at 8.57 a.m., still on time. It had passed signal L48, displaying *Approach*, had passed onto track No. 1, and passed FH interlocking station at 8.59 a.m. Proceeding onto Newark Bay Bridge it passed signal 82, displaying *Approach*, passed signal R26, displaying *Stop*, and was still moving at 42 miles an hour when it derailed on the derail protecting the southwestern span, which was raised. The derail was not designed to stop a train going at such a speed and although it did its job the derailed train continued along the ties (sleepers) until it reached the bridge opening, and the two diesel units and first two cars fell into the water, which was 35 to 40 feet deep, and were submerged. The third car was left hanging off the bridge with its front end in the water; two hours after the accident it, too, fell into the water and was totally submerged. The two rear cars remained on the bridge. The leading car was fortunately empty and many of the occupants in the third car were able to jump from it into the water to be picked up by small boats. Almost all of the passengers in the second car drowned.[857]

A newspaper reporter wrote, 'Survivors told of swimming around in the sunken cars, looking for a way out of the watery trap. One was Rafael A. Leon, of Holmdel, N.J., who was on the train with his wife. "Suddenly we were in the water," Leon said. "I started to pray I would drown quickly. My wife was beneath me in the water. I swam down and tried to find her but she had slipped down. I went back to the surface of the water inside the coach. There was a window open."' Mrs Teresa Leon was listed among the dead.

'Small boats from all over busy New York Harbor converged beneath the bridge. There were Coast Guard vessels, police and fireboats, and small private vessels.'

'Edward McCarthy, who owns the Elco Marina Boating Dock in Bayonne, was at the

---

[857] Interstate Commerce Commission report

dockside when he saw steam gushing from the water beneath the bridge. McCarthy immediately put out in a small boat. In three trips, he brought back three bodies and nine survivors.

'Thomas Messina, a brewery truck driver also at the scene, said the first two passenger coaches appeared to be completely submerged. Messina said an oil tanker had been approaching to pass under the bridge, but that it had passed through when the train fell into the water. The tanker immediately put down its launches and began taking aboard survivors.

'Four tugs of the Moran Company, usually engaged in towing ocean liners to piers, were on the scene. Two New York City fireboats and three police department launches sped to the scene. Every available ambulance in Bayonne and Jersey City was summoned.' [858]

At least one of the survivors of the accident, G.H. Planitz, of Fair Haven, New Jersey, was doubly lucky, or unlucky, depending on how one looks at it, for he had already been in the Woodbridge disaster of February 1951 (see Other Derailments chapter). Describing the new disaster he said, "There was a jamming, screaming sound of brakes and I was thrown forward, head over heels. It has taken me all this time to get over it {the Woodbridge disaster}. And now this."

Bridge Tender Patrick Corcoran at DY interlocking told the Interstate Commerce Commission (ICC) inquiry that he had received a signal from an approaching vessel, the barge *Sand Captain*, that it needed the draw spans raised for its passage. Signal R26 already indicated *Stop*, signal 82 indicated *Approach*, and the derail on track No. 1 was open, these being their normal positions when no train was approaching. The bridge tender manipulated the levers of the operating machine and the bridge control apparatus, then applied power, raising the spans of the west channel to 108 ft. at 8.55 a.m. He then looked out of the window to see where train No. 3314 was, as he knew that it was due to leave Elizabethport at 8.57 a.m. and he had to record any delays to trains. To his horror he saw that the engine and leading car of the train had already passed signal R26 and there was nothing that he could do to stop it. He told a newspaper reporter, "I heard a rumble and knew it was going into the water. I couldn't believe my eyes. I've been working on the railroad 35 years and I've never seen such a mess." [859]

The signals were examined an hour after the accident and found to be working normally, while exhaustive tests of DY interlocking showed nothing which could have caused the accident. The operator of FH interlocking had also seen signal 82 at *Approach* (signal R26 was not visible from his post). Examination of the track showed no evidence of braking before the train had reached the derail, though there were indications that the brakes had been applied from the rear of the train after it had derailed and the brakes of the rear two cars of the train were found to be fully applied. The speed recorder of engine No. 1532 indicated a speed of 42.5 miles an hour at the moment it had struck the derail. There was no indication of any problems with the brakes of the two diesel units and no indication that they had been applied before the accident.

Post-mortem examinations found that Engineer Wilburn had died of asphyxia due to drowning and Fireman Andrews had died of multiple fractures. Previous examinations had shown no medical problems with regards to the engineer; the fireman had moderately

---

[858] *The Daily Courier*, Connellsville, Pennsylvania, United States, 15th September 1958

[859] *Chicago Daily Tribune*, Chicago, Illinois, United States, 16th September 1958

raised blood pressure (hypertension). Engineer Wilburn had waved to the operator at FH interlocking.

Automatic cab signals were provided on locomotive No. 1532 for operation over the New York and Long Branch line, 12 miles further west. The ICC noted that had this system been in use over the bridge the brakes would have been applied automatically when signal R26 was passed at *Danger* and the accident would probably have been averted.

The ICC recommended that the automatic cab signal system be extended over the bridge or that some alternative form of automatic train stop system be installed. [860]

Why Engineer Wilburn had failed to stop must remain a mystery.

A two-car westbound Lake Street Elevated electric train of the Chicago Transit Authority (CTA) was approaching the lifting bridge over the Chicago River in Chicago, Illinois, United States, on Tuesday, 2nd June 1959 as the structure was about to be raised. The protecting signals should have warned the motorman of the train but it failed to stop until its leading wheels were derailed on the derail protecting the bridge. There were about 20 people on the train. Conductor Stanley Low, 51, and one passenger, Mrs Mamie Burgard, 54, were slightly hurt. An investigation was opened by the CTA to determine the circumstances of the accident. [861]

"I opened the bridge and all at once they were going by. They shouldn't have done that." Those were the words of Bridge Tender Gabe Clark, describing an accident at Brazos, California, United States, on Sunday 12th July 1959. He had opened the swing bridge over the Napa River for a 24-foot vessel (variously described as a fishing boat, a pleasure boat, and a small power cruiser) when the train, which consisted of three 186-ton diesel locomotives coupled together and a caboose, was approaching the structure. Seeing the danger and realising that they could not stop in time Engineer E.E. Panelli, Fireman Berthal Farmer, and Conductor M.N. McCoy jumped from the train before it reached the river. Brakeman R.E. Adams hesitated and ended up in the shallow water near the river bank. Brakeman J.H. Traughber decided not to jump from the caboose. The engines fell into water 35 feet deep; the caboose remained hanging off the bridge. Two of the railwaymen, Farmer and Traughber, were slightly hurt while Mr R.D. and Mrs Hazel Pittsley on the boat suffered bruises. Mr Pittsley said, "I happened to look up, and here's a locomotive falling on my boat." The impact of the engines drove the boat under water briefly before it popped back to the surface. Three feet of its bow were shattered. [862] [863] [864]

Donald Miller, Assistant Superintendent of the Southern Pacific Railroad, Western Division, told reporters that the explanation for the accident would not be obtained until the three locomotives, valued at $500,000 in total, had been lifted from the river and their speed recorders had been analysed. [865]

---

[860] Interstate Commerce Commission report

[861] *Freeport Journal-Standard*, Freeport, Illinois, United States, 2nd June 1959

[862] *The Modesto Bee*, Modesto, California, United States, 13th July 1959

[863] *The Herald Journal*, Logan, Utah, United States, 13th July 1959

[864] *San Mateo Times*, San Mateo, California, United States, 13th July 1959

[865] *Amarillo Globe-Times*, Amarillo, Texas, United States, 13th July 1959

CHAPTER 10

TRAIN ACCIDENTS INVOLVING BRIDGES

Bridges could feature in train accidents in one of two ways. Either the bridge itself could collapse for a number of reasons such as faulty design, faulty construction or because its foundations had been damaged; or a train could fall from a bridge, the structure itself remaining intact. Bridge failures have been dealt with in the previous chapter; this chapter will concern itself solely with accidents where the bridges itself was not at fault. It may seem perverse to list such accidents separately instead of in the chapters dealing with primary causes of the accidents, the procedure adopted elsewhere in this book, but the fortuitous presence of a bridge has such a potential for transforming what might have been only a minor accident into a major disaster that it has been decided to treat such accidents separately, in the same way as accidents occurring in tunnels have been given a chapter of their own.

Four coaches of a northbound Canton – Hankow passenger train were derailed as it was crossing the Shaochi Bridge in Hunan province, 60 miles south of Hengyang, China, on Thursday 30th May 1946. According to the Canton news agency about 90 passengers were killed and 60 injured. [866] [867]

A Taubenbach – Unterwellenborn ore train travelling at over 60 kilometres an hour became derailed at Gräfenthal, Germany, on the Probstzella – Neuhaus line on Wednesday 23rd October 1946. A luggage van and 13 waggons fell from a viaduct, partly onto some houses below. Only the engine and one waggon remained on the track. The Erfurt head office of the railway initially suspected a failure of the viaduct but the accident was then put down to a failure of the driver's automatic brake valve. [868]

A major catastrophe occurred on Japanese National Railways on Tuesday 25th February 1947.* A Down passenger train on the Hachiko line got out of control on a 20‰ (1 in 50) falling gradient between Higashi Hanno and Komagawa, Japan, ran away, became derailed and fell 5 metres from a bridge. No less than 184 people died and 495

[866] *Reno Evening Gazette*, Reno, Nevada, United States, 3rd June 1946
[867] *The Singapore Free Press*, Singapore, 4th June 1946
[868] *Eisenbahnunfälle bei der Deutschen Bahn* by Erich Preuß, pgs. 175-6 (in German)

were injured. The last four cars tumbled down an embankment into a wheat field unbeknown to the driver, who remained unaware that there had been an accident until he reached the next station. The coupling between the second and third cars gave way and four wooden coaches, each of which had held about 200 passengers, ended up as bloodstained, splintered wreckage over a distance of 200 metres, while the bodies of men, women and children lay all around. It was believed that most of the passengers had been on their way from Tokyo into the country to buy food, most of them having canvas bags or large scarf-like handkerchiefs in which to carry their purchases. [869] The accident was blamed on the overcrowding of the train, the use of rolling stock built with inferior wartime materials, and insufficient driver training. [870]

*(A Japanese source states that the accident occurred in February 1946 but reports of it occur in American newspapers of February 1947.)

The engine and ten carriages of a 15-coach passenger train from Kukong to Canton reportedly plunged from a weakened bridge into the North River near Yingtak, China, on Thursday 10th July 1947. The railway authorities stated that 21 people had died and 56 had been injured but there were fears that the true death toll was considerably higher with many bodies not recovered. It was estimated that there had been 200 passengers on the train; the authorities could not give a precise number. [871] [872] A later statement was that only two carriages had become derailed as the train was crossing the bridge, one of which had gone off the bridge but remained above the surface of the water and the other had remained on the bridge. The 20 who died had drowned. [873]

A family of six had a lucky escape when a Baltimore and Ohio Railroad freight train was derailed on a 50-foot trestle near Chester, Pennsylvania, United States, on Sunday 21st March 1948. A wheel of the 21st car of the 45-car train bound for Philadelphia became derailed due to a broken wheel, dragging the remainder of the train after it, and some of the cars fell from the bridge into Ridley Creek, others falling down the 20-foot embankment, one of which crashed against a house owned by the Maisch family. 'Five-year-old Anthony Maisch said he was standing in the kitchen of his house dressing for church when he looked up and saw "wood flying through the air right at me." He screamed to his father, James, and the two fled through the front door moments before the steel pipe and copper ingot crashed into the kitchen. Mrs Maisch, who was upstairs with three other members of the family, was taken to hospital and treated for shock.'

A number of Western Union telegraph poles were snapped off by the derailed cars, disrupting communications in the area, and the trestle was badly damaged. The railroad said that reconstruction was likely to be a long and difficult task. No one was hurt. [874]

In 1949 mixed trains operated by the Witwatersrand Native Labour Association ran twice weekly from Johannesburg in South Africa to Lourenco Marques in Portuguese

---

[869] *Waterloo Daily Courier*, Waterloo, Iowa, United States, 25th February 1947
[870] Webpage – Japanese Railway Safety and the Technology of the Day by Masio Saito
[871] The Luddington Daily News, Luddington, Michigan, United States, 16th July 1947
[872] *The Straits Times*, Singapore, 14th July 1947
[873] Ibid, 16th July 1947
[874] *Pottstown Mercury*, Pottstown, Pennsylvania, United States, 22nd March 1948

East Africa carrying 'native' miners back to their homes and families, including, a modern writer notes, men who were dying from silicosis, tuberculosis, and pneumonia. There were equivalent trains in the opposite direction, carrying 'natives' in locked third-class carriages and cattle waggons from Portuguese East Africa. The same writer adds that the trains ran at night so that people along the line would not see them and that the Up trains ran quickly, to get the cheap labour to the mines, whilst the Down train dawdled, in the hope that the miners might spend their small wages before they reached home. [875]

One of the Down trains, No. 305, which had left Booysens, Johannesburg, the day before, was involved in a catastrophe at a viaduct over the Elands River, a couple of miles on the Portuguese East Africa side of Waterval Boven (now Emgwenya), in the Transvaal, South Africa. The Elands viaduct or bridge was curved and consisted of five 50-foot spans with a maximum height of 60 feet. As the train was crossing the bridge just after midnight on Tuesday 15th November 1949 the two engines became derailed when they neared the far end of the bridge. Behind them six coaches fell from the bridge into the icy river, another coach was left hanging from the bridge and six other coaches were overturned. The brake van at the rear of the train remained on the line. Of the 500 miners on the train 62 were killed and another 117 were injured, many seriously. Driver Welsh Hilly Green on the pilot engine was killed and Driver P.J. van Rooyen on the train engine and both firemen were seriously injured. Guard Swart, riding in the rear van, climbed down to the track and seeing what had happened ran back along the line towards Waterval station. After about a mile he found a lineside telephone from which he was able to ring the station, where the siren was sounded and most of the townspeople made their way to the accident scene. Some of the most seriously injured were evacuated by South African Air Force Dakota aircraft to Pretoria. The dead miners were buried at the scene in a mass grave.

An inquiry into the accident blamed it on Drivers Green and van Rooyen operating the train at excessive speed. [876] [877]

First reports stated that 20 people had lost their lives and 45 were injured when the Delhi-bound Kashmir mail train left the rails on a bridge near Gurdaspur, India, in the northern Punjab, and the engine and three coaches plunged down an embankment. [878] The revised death toll was reduced to 12. In Parliament subsequently the accident was blamed on 'an error of judgement' without a further explanation of what that meant. [879]

Twenty-nine Korean civilians who had 'hitchhiked' a ride on a United Nations freight train, operated by the Third Transportation Military Railway Service and bound for the

---

[875] https://www.timeslive.co.za/sunday-times/books/news/2019-09-22-charles-van-onselen-tracks-the-dark-history-of-south-africas-mine-worker-trains/

[876] https://www.geocaching.com/geocache/GC797YG_1949-train-crash?guid=2d0a4b69-6299-46ba-8700-f0c5704c6964

[877] *The Times*, London, England, 16th November 1949

[878] *The Times*, London, England, 4th September 1950

[879] https://indianrailways.gov.in/railwayboard/uploads/directorate/finance_budget/Previous%20Budget%20Speeches/1951-52.pdf

port of Pusan, lost their lives when the train was derailed and fell from a bridge near Iwon, South Korea, 23 miles south of Taejon, late on Friday 2nd January 1953. Forty-one other riders were injured. The train consisted of an engine and 31 cars, 27 of which were empty. The cause of the derailment was not immediately clear. [880] [881]

The morning of Saturday 15th August 1953 saw a collision between two local trains, one steam hauled and the other electric, on a viaduct at Irk Valley Junction, not far from Manchester Victoria station, England, on British Railways, London Midland Region. As a result of the collision part of the electric train broke through the parapet of the viaduct and fell 40 feet to the bank of the shallow river below. Nine passengers and the motorman of the electric train were killed and 58 passengers were taken to hospital, of whom 22 were admitted.

The trains involved in the accident were the 7.20 a.m. stopping train from Bury to Manchester Victoria and the 7.36 a.m. stopping train from Manchester Victoria to Bacup. The Bury train was a five-coach electric service consisting of a motor coach, a trailer, a motor coach, a trailer, and a motor coach, in charge of Guard John Collinge with Motorman Alfred Hardman. The Bacup stopping train consisted of a Class 4P 2-6-4 tank engine and four coaches in charge of Guard Robert Roberts with Driver Fred Heap and an unnamed fireman.

The Bury train had started on time. Guard Collinge told the Ministry of Transport inquiry into the accident that his train had last stopped at Woodlands Road Halt at 7.38 a.m. It had then restarted and passed Queen's Road signal box at normal speed, 36 to 40 miles an hour, for an apparently clear run into Manchester Victoria. He had not been able to see any signals until his train passed them and had had no warning of the collision.

The Bacup train had likewise left Manchester on time. It was brought to a stand at the branch home signal at Irk Valley Junction, where its route to Bacup crossed the Up line from Bury by a flat junction. The signal then cleared, Driver Heap opened the regulator, and as his engine was passing over the junction at about five miles an hour it was struck by the electric train, which neither the driver nor his fireman had seen coming.

Much of the evidence given at the inquiry into the accident was that of the various signalmen involved, beginning with that of Signalman William Upton at Woodlands Road. The latter initially stated that the times for the signalling of the Bury electric train were those recorded in his train register book but later admitted that the times recorded at the next box at Queen's Road were the correct ones. He said that he had not, as he should have done, sent the *Train Approaching* signal for the train. He also admitted, having previously denied it at the internal inquiry, that he had received a query from Queen's Road about 7.38 a.m. regarding the whereabouts of the electric train, to which he had replied that it would reach him shortly.

Signalman Raymond Davenport at Queen's Road box stated that he had accepted the electric train at 7.21 a.m., had received the *Train Entering Section* signal for it at 7.30 a.m. and had offered it to Irk Valley Junction at that time, when it had been accepted and he had cleared his signals, except the Outer Distant, which he did not usually clear until the Irk Valley Outer Distant had been cleared. At about 7.39 a.m. the Irk Valley signalman had phoned him and asked where the electric train was, so he had said he

---

[880] *Council Bluffs Nonpareil*, Council Bluffs, Iowa, United States, 3rd January 1953

[881] *Defiance Crescent-News*, Defiance, Ohio, United States, 3rd January 1953

would ask Woodlands Road. He had duly done so and had been told that it was just leaving there. Then he saw the train approaching and so told the Irk Valley signalman, to whom he had sent the *Train Entering Section* signal at 7.40 a.m.

It may be here noted that Woodlands Road box was only about 5 furlongs (1,100 yards) from the previous box at Crumpsall; Queen's Road box was about the same distance from Woodlands Road and Irk Valley box was only about 2 furlongs (440 yards) from Queen's Road.

Signalman Davenport said that when the train was offered to him at 7.21 a.m. he did not know which train it was. If he were able to he just accepted trains as they were offered. He did not know what trains ran on Saturdays and no extract from the Working Timetable had been made and exhibited in the box. He said the electric train was frequently checked at Queens Road because a branch train was crossing the junction at Irk Valley but it was usually accepted before it was stopped at his starter. On the day of the accident the train had been travelling at normal speed. Although the booked time for the train between Woodlands Road and Irk Valley was only two minutes he said he had not been concerned when it had not appeared in nine minutes as this was not unusual. Neither he nor Signalman Upton could recall seeing the motorman of the electric train.

Relief Signalman Arthur Clayton at Irk Valley Junction testified that he had only begun working there on the previous Monday, having learnt the box, which he had found easy. He said that as soon as he had been offered the Bury train by Queen's Road box at 7.29 a.m. he had accepted it and offered it forward to Newtown No. 1 box but it had been refused by them. At 7.33 a.m. he had accepted the Bacup train from Newtown No. 1, offered it forward to Smedley Viaduct on the branch and had it accepted. At 7.37 a.m. he had received the *Train Entering Section* signal from Newtown No. 1. (The Bacup train was booked to pass Irk Valley Junction at 7.39 a.m., crossing the Up main line on which the Bury train was approaching on its way to Smedley Viaduct; the Bury train was booked past the junction a minute later.)

When the engine of the Bacup train had come into view he had still not received the *Train Entering Section* signal for the Bury train so he had telephoned Queen's Road where Signalman Davenport had said that he would enquire of Woodlands Road. It was at this point that Signalman Clayton made a fatal mistake. Assuming that the Bury train was still some distance away he had reversed the junction points and cleared the Down Main to Branch Home signal for the Bacup train to pass onto the branch to Smedley Viaduct, across the path to be taken by the Bury train, in defiance of the block regulations. Going back to the telephone he had been told that the Bury train was passing there. By then the Bacup train was starting away over the crossing and there was nothing he could do, though the electric train should have stopped at his Up Main Home signal, a mere 19 yards short of the fouling point of the junction. Immediately after speaking to Signalman Davenport again he said he had seen the electric train coming and had realised that it was travelling at its usual speed and was not going to stop. The collision had quickly followed.

Signalman Clayton pointed out that this was his first Saturday at Irk Valley and said that he had not known the timetable. He had been surprised at the delay in receiving the *Train Entering Section* signal from Queen's Road and during the week had refused trains from Queen's Road so as to cross trains to the branch. He said that he was aware that Regulation 4 did not permit him to foul the junction after having given *Line Clear* to Queen's Road but pleaded that he had thought that the electric train was a mile away.

Signalmen Smith and Robinson at Newtown No. 1 box confirmed that they had refused the electric train when it had been offered to them by Irk Valley because it was so much ahead of its booked arrival time.

Porter H. Knight at Woodlands Road Halt said that he had received a greeting from Motorman Harding, whom he knew well. The motorman had looked well and had been smiling.

Signalman H.B. Burstow at Smedley Viaduct said he had witnessed the collision 30 seconds after he had received the *Train Entering Section* signal for the Bacup train. He could not see the Up Main Home signal.

Ganger J.W. Page testified that he had heard the noise of the collision and immediately afterwards had seen that the Irk Valley Up Main and Up Branch Home signals were at Danger.

Assistant Stationmaster Thomas Bevan said he had arrived at Irk Valley Junction at 7.56 a.m. He had confirmed that all the signals were at *Danger* and said that Signalman Clayton had denied clearing his signals for the Bury train.

As part of his investigation the Ministry of Transport inspecting officer, Colonel D. McMullen, arranged for an inspection of the train registers at Woodlands Road, Queen's Road, and Irk Valley for the fortnight preceding the accident. Various irregularities were found:

(a) The Woodlands Road signalman had regularly failed to send the *Train Approaching* signal to Queen's Road. This had not been reported by the Queen's Road signalman.

(b) Signalman Upton at Woodlands Road had adopted a practice known as "knock for knock," asking *Line Clear* for an Up train immediately *Line Clear* had been given for a Down train, irrespective of whether he had been offered an Up train by Crumpsall. As a result there had been intervals of up to 18 minutes between the *Line Clear* and *Train Entering Section* signals.

(c) *Train Entering Section* signals had been sent from Woodlands Road to Queen's Road long before the signal had been received from Crumpsall.

(d) The times of the *Line Clear* and *Train Entering Section* signals were often incorrectly entered in the Woodlands Road train register.

(e) Signalman Davenport had repeatedly switched out Queen's Road box, supposedly to go to the lavatory or fetch coal, without exchanging any bell signals. He had telephoned the adjacent boxes when switching out but not when switching in. On one occasion he had switched out for 47 minutes, writing the times of the trains which had passed while the box was closed in the train register, recording 6 trains though only 5 had passed.

(f) The switching out of Queen's Road box had not been recorded in the train registers at Woodlands Road or at Irk Valley before Signalman Clayton had started working there.

Four cases of serious irregular block working by Signalman Clayton at other boxes he had worked were also found.

The inspecting officer then interviewed supervisory staff.

District Signalman's Inspector W.E. Farnell stated that he had trained Signalman Clayton at Ilex Hall Carriage box in 1950 and had later recommended him for the post of

Relief Signalman Class 2. He considered Clayton to be capable, keen and intelligent. He said that when visiting a signal box he checked the train register for omissions, failures of trainmen to carry out Rule 55 and incorrect signing on and off. Only when investigating some specific incident did he compare the train registers of different boxes.

Assistant District Signalman's Inspector J. Washington said that in three months he had visited Woodlands Road, Queen's Road and Irk Valley boxes each twice to introduce an inspector and pass out new signalmen. He said that it was unusual to visit a box "unless you happen to be passing or have something definite to investigate." Half of his time was taken up with new entrants and he had no time for routine inspections. Asked what he did when checking the book he said that he asked the signalman if everything was alright, looked at the last two pages and checked if Rule 55 had been carried out. He was unaware that "knock for knock" was practised at Woodlands Road or that the *Train Approaching* signal was not sent. He did not think it was usual to record the *Train Approaching* signal. The inspecting officer noted that Woodlands Road was the only one of 21 boxes in the District which did not record it. Mr Washington said that the times in the train registers of adjacent boxes were only compared in special train registers used for the three-monthly train census. He considered Signalman Upton to be "average." Signalman Davenport he thought particularly intelligent and good on the Rules and Regulations. Signalman Clayton's knowledge of the Rules and Regulations was satisfactory. He was not aware that Signalman Davenport was switching out irregularly, and there was no need for him to do so to go to the lavatory, which was attached to the box.

District Signalman's Inspector A.W. Kelly said he had only been in the district since May. He had only visited the three boxes for a specific purpose. There were 64 boxes in his district and he was concentrating on the more important boxes first. He would only check the train registers of adjoining boxes in connection with a specific incident. It was he who had discovered the infringements of the block regulations by Signalman Clayton and the irregular working at Woodlands Road and Queen's Road after the accident, which he considered should have been found by the Crumpsall stationmaster. He described the irregular switching out of Queen's Road as a dangerous practice.

Crumpsall Stationmaster J.J. Johnson said that he had visited Woodlands Road box three times weekly. He was unaware that the *Train Approaching* signal should have been sent there, though it was in the signal box local instructions.

Three out of four other stationmasters who had relieved at Crumpsall admitted that they had not read the local instructions at Woodlands Road, the fourth one had and was unaware that they had not been carried out.

Concluding his report, Colonel McMullen stated that two things had caused the accident. One was Motorman Hardman overrunning the Irk Valley Junction signals, the other was the committing of a serious signalling irregularity by Signalman Clayton.

He had no doubt that the Irk Valley Junction Up Main Inner and Outer distants and the Up Main Home had been at *Caution* and *Danger* when the Bury train had passed them. The interlocking with the junction points prevented them having been cleared and there was no defect in them which could have caused a false clear. Furthermore Ganger Page had seen the home signal at *Danger* immediately after the collision. Calculations showed that the Up Main Inner Distant and Home could not have been cleared and then returned to *Danger* early enough for Motorman Hardman to have seen them at *Clear*. The Outer Distant could have been seen at *Clear* by the motorman but Signalman Davenport had seen the Outer Distant at *Caution* and Signalman Clayton strongly denied having cleared the signals.

Why Motorman Hardman had failed to observe the signals could of course not be known but in the 110 times that the Bury train had run since the end of March it had been checked at Queen's Road on all but 9 of them. It was thus rare for the Irk Valley Distant signals to be clear but the train had not once been stopped previously at the Irk Valley Home signal. When Hardman had found the Queen's Road off he had therefore doubtless assumed that he had a clear road through Irk Valley Junction as usual. This did not explain, of course, why he had not braked upon seeing the Home signal at *Danger*. There was no evidence of him having suffered from any medical condition or of any problem with the brakes of the train and it seemed likely that he had just relaxed his attention. He had clearly realised the danger at the last moment as he had attempted to vacate the cab immediately before the collision. Hardman had been 64 years old and a respected local figure, an alderman and a former Mayor of Bury.

Signalman Clayton was little, if any, less responsible for the accident than the motorman, as the Block Regulations specifically forbade the crossing over of the steam train while the electric train was approaching, unless there was an Outer Home signal, which there was not at Irk Valley Junction. He should either have held the Bacup train at the Down Main Home signal until the Bury train had passed or else have waited until the Bury train had come to a stand at the Up Main Home signal before letting the Bacup train across in front of it. Undoubtedly he had been mislaid as to the position of the electric train but that was no excuse.

The various irregularities committed by the other signalmen, while having no direct bearing on the accident, the inspecting officer continued, indicated slipshod methods of working due to ineffective supervision. Signalman Upton was an unsatisfactory witness and probably deliberately untruthful. Signalman Davenport's irregular switching out was unpardonable and his explanation that the closing of the box was connected to visits to the lavatory was difficult to believe. In the inspecting officer's view 'neither Upton nor Davenport was the type of man who was suitable for the respected and responsible position of a signalman.'

Turning to remarks and recommendations, the inspecting officer noted that the outside interests of drivers were now (in 1953) more varied than previously but it was vital that while they were at work they did not relax their concentration on the observance of signals. It was unlikely that Automatic Train Control of the Warning type would have prevented the accident.

Likewise block controls and track circuits at Irk Valley Junction would not have prevented the accident. The safety of trains depended on signalmen obeying the Block Regulations.

The narrow margin between the passing of the Bury and Bacup trains was undesirable.

Signalman Clayton had no railway background. His initial training had been adequate but his promotion to relief signalman after only 15 months and before he had worked any signal box controlling a junction was premature. Relief signalmen should also be tested on their knowledge of all the boxes they were sent to, as was the practice in some regions. His earlier breaches of the Block Regulations had only been discovered after the accident. Had they been found earlier he would have been relieved of his post or relegated to an unimportant box. It was only when train registers were checked with care and compared between boxes that such irregularities could be uncovered.

Commenting on the various irregularities at Woodlands Road and Queen's Road, Colonel McMullen wrote, 'The above irregularities cumulatively indicate a very unsatisfactory state of affairs. The Rules and Regulations have been designed as a result of experience over many years to prevent accidents, and they are there to be obeyed and not disregarded. Signalmen who get into the habit of displaying contempt for the Regulations or who think they can do things in a better or easier way are certain sooner or later to make a serious mistake which may cause an accident, more usually when someone else makes a mistake at the same time.

'These facts recall the necessity for effective supervision at all levels to ensure that train working is in accordance with the Rules and Regulations. I am sorry to say that it was lacking as far as these boxes were concerned. Indiscipline as disclosed in this and in some other accidents in recent years will invariably result from lax supervision.'

In the final paragraph of his report the inspecting officer wrote, 'It is satisfactory to note that immediately after this accident the Railway Executive Headquarters inaugurated a campaign against irregular working in signal boxes, and I am sure that this will have good results. I do, however, consider that arrangements should be made for a closer scrutiny of Train Registers than I am informed has been possible in the past, and I recommend that this should be applied particularly at junctions and to signalmen whose experience is limited.' [882]

A passenger on the Bury train, Mr Frant Latham, of Prestwich, gave evidence at the inquest on the victims of the disaster. He said that just before the accident the train had been travelling normally, then he had got the impression that the brakes had been applied. "A few seconds later glass started flying round and the coach tilted violently. I was flung on to the roof when the coach turned over. I grabbed hold of a strap and managed to hold on. Other passengers were flung all over the place. When the coach came to rest I looked round to see what had happened. It was too steep for me to go to the front end to get out, and I could not get out of the bottom end because of the wreckage, so I had to stay in until help came. Some people were trapped by the broken seating."

The three signalmen were given a hard time at the inquest by Coroner Jessel Rycroft but at the conclusion of the evidence he told the jury that he doubted whether any jury at the Assizes would convict of manslaughter any of the persons who had been mentioned as being involved in this case. After an absence of 15 minutes the jury returned a verdict of Accidental Death, adding a rider that: "We find negligence in the observance of the regulations on the part of the signalling staff, both active and supervisory, and to a lesser degree by the driver of the electric train." [883]

Sixteen people were killed, 14 were seriously injured and 28 were slightly hurt when No. 4 Up passenger train bound for New Delhi was derailed in the Punjab as it was crossing a canal bridge near Bhatinda, India, at 05:00 on Monday 4th January 1954. The first four of the six coaches of the train were derailed and there was reportedly some telescoping. The accident closely followed a collision between passenger and freight trains near Howrah in which a number of railwaymen had been injured and the collision near Rauxal in Bihar in which ten lives had been lost. [884] [885] The accident was blamed on

---

[882] Ministry of Transport report

[883] *The Times*, London, England, 27th and 28th August 1953

[884] *Wisconsin Rapids Daily Tribune*, Wisconsin Rapids, Wisconsin, United States, 4th January 1954

the erosion of the embankment by a torrent of water following the failure of a nearby embankment of a canal distributary. It remained unclear whether the railway or the irrigation authorities were to blame for the disaster. [886]

A heavily loaded logging train of the Insular Lumbar Company, which was also carrying about 130 passengers, consisting of employees of the company, their wives, and children, who were riding on empty flat waggons, was descending a steep incline on the northern tip of Negros Island, between Fabrica and Sagay, in the Philippines, on Thursday 2nd September 1954, when it became divided. Driver Pablo Villarete, 46, attempted to keep his engine and the seven waggons which were still attached to it ahead of the 23 following waggons, which were loaded with logs. He failed, and the rear waggons caught up with the front of the train on a bridge. The bridge was wrecked in the collision and five waggons fell from it. Chains holding the logs snapped and they hurled off the waggons, crushing many passengers and sweeping others into a 100ft-deep rocky gorge. At least 82 passengers died and over 50 were injured, many seriously. The train had been taking logs from mountain forests to sawmills on the coast.

The following day the Philippines News Service said that rescuers were still digging bodies out from beneath the logs, while a correspondent for *The Manila Times* at Bacolod stated that the death toll might reach 100 or more. 'Families of the victims picked their way through the piled up bodies, their screams and moans marking the identification of a loved one. Some bodies, too mangled for immediate identification, were placed to one side and covered by banana leaves.' [887] [888]

A two-car Japanese electric train on the 3ft 6in (1,067mm) gauge Lida line, carrying about 50 passengers, met with an accident on the night of Thursday 20th January 1955. Between Chiyo and Tenryūkyō stations it ran into a large boulder which had rolled down the mountainside, was derailed and reportedly fell 100 feet from a bridge over the Tenryū River. At least two and possibly five people died and about 30 were injured. [889] [890]

A Ferrocarril Nacional de México passenger train left Guadalajara on the evening of Saturday 2nd April 1955 for Manzanillo. Like all the other trains on the line it was packed with holidaymakers heading for the Pacific beaches for the Holy Week weekend, was strengthened with extra carriages, and was packed with passengers, some standing in the aisles or the vestibules. Its journey was interrupted for 12 hours at Ciudad Guzman, a division point, awaiting the repair of a bridge, and it was still en route to its destination at 20:45 on Sunday evening when it became derailed just after it had left a tunnel and part of it fell from a bridge near the town of Alsaba, Mexico, in Colima state. Three cars of the 15 in the train tumbled into the so-called "Canyon of Death."

An early report in the *El Informador* newspaper in Guadalajara spoke of 300 dead but as was often the case this was a grossly exaggerated figure. A later, more accurate report,

---

885 *Bluefield Daily Telegraph*, Bluefield, West Virginia, United States, 5th January 1954

886 https://archive.org/stream/dli.csl.1063/1063_djvu.txt

887 *Wisconsin State Journal*, Madison, Wisconsin, United States, 3rd September 1954

888 *The Morning Bulletin*, Rockhampton, Queensland, Australia, 4th September 1954

889 *The Brownsville Herald*, Brownsville, Texas, United States, 21st January 1955

890 *The Abilene Reporter-News*, Abilene, Texas, United States, 22nd January 1955

stated that nine passengers had been killed outright, four had been fatally injured and four others were seriously hurt. The Colima state governor, Jesús Gonzáles Lugo, added that another 60 passengers had been slightly hurt. Railway officials said that the casualties were less than they might otherwise have been because the train had been composed of steel coaches. Three relief trains were sent from Colima and Guadalajara with doctors, nurses, and breakdown crews, and the dead and injured were brought to Colima early on Monday morning. Meanwhile, 'Frantic relatives besieged offices of railroad and airport depots in an effort to reach the scene.' [891]

The accident began with the derailment of the rear truck (bogie) of the diesel locomotive pulling the train within the tunnel itself and before the driver could stop the train it had emerged onto a narrow mountain ledge. A dozen coaches jumped the track. Most of them remained upright on the right-of-way but the collapse of a bridge led to one first-class and two second-class coaches falling into the canyon, rolling, and sliding 200 feet down its side. [892]

Eastbound Erie Railroad freight train No. 98 from Meadville, Pennsylvania, to Buffalo, New York, consisted on Wednesday 12th October 1955 of four diesel units and 120 cars. As it was crossing a three-span steel trestle at 12.40 a.m. one of its cars, possibly the ninth, became derailed. Other cars were dragged after it and about 30 fell from the trestle into Cattaraugus creek on the border between Cattaraugus and Erie counties. The front of the train continued, pulling some of the derailed waggons, until it reached a 20-foot wooden trestle a little further on, which collapsed in turn. In all about 60 cars were derailed in the accident at Gowanda, New York, United States. Two of the waggons were tank cars, one containing propane and the other lubricating oil, but there was no fire. Other derailed cars had carried lumber, foodstuffs, washing machines, machinery, and other general merchandise. No one was hurt. The cause of the accident was not clear but it was believed that the derailment of the train had preceded and caused the failure of the first trestle and not the other way around.

Sergeant John Callinan of Gowanda Police Department commented, "It will take several days to clear the wreckage. I've never seen anything like it in my life." It was estimated that it would in fact take two weeks to clear the wreckage and repair the damage. The cost of the accident was put at $500,000. [893] [894]

On the evening of Wednesday 2nd May 1956 Canadian Pacific Railway slow freight train Extra No. 5186 had to drop off 14 cars in the yard at Galt, Ontario, Canada. It was hauled by steam engines Nos. 5607 and 5126 and in charge of Conductor Gerry Rivard with Engineer William Hunter and Fireman A.H. Holden on the lead engine and Engineer Tim Ready and Fireman Jack Easby on the second engine.

Travelling in the same direction as train No. 5186 was through freight train No. 903, consisting of engines Nos. 2205 and 5420, 40 cars and a caboose, bound for Windsor, and in charge of Conductor Jack Cowring with Engineer William Palmer and Fireman Henry Blocka on the lead engine and Engineer George Nutkins and Fireman William

---

[891] *The New Mexican*, Santa Fe, New Mexico, United States, 4th April 1955

[892] *The Daily Chronicle*, Centralia-Chehallis, Washington, United States, 4th April 1955

[893] *Dunkirk Evening Observer*, Dunkirk, New York, United States, 12th and 13th October 1955

[894] *The Oneonta Star*, Oneonta, New York, United States, 13th October 1955

Hopkins on the second engine. This train was not booked to stop at Galt.

The connections to the yard having been reset for the main line, the automatic block signals were standing at clear for train No. 903. When this train passed Killean, 5 miles short of Galt, a block indicator in the office at Galt showed that the train was approaching, and there was a second block indicator at the points leading from the yard back to the main line. The cars having been dropped off, the Fireman Easby of train No. 5186 reset the points for the main line and their train moved out of the yard as train No. 903 was close at hand. According to the rules after having reset the points the crew should have waited three minutes in case an approaching train had already passed the last protecting signal, but this apparently was commonly often not done in order to save time. Operator James Stewart in the Galt office was aware of the approach of train No. 903 and attempted to warn the crew of the freight train to stay in the yard, but his warning was apparently misunderstood.

Train No. 903 struck the rear of the cut of cars on a bridge over Water Street, which collapsed, and the two engines fell into the road below. Engineer Palmer and Front Brakeman Tommy Watson of train No. 903 were killed, Engineer Nutkins was critically injured but survived and Firemen Blocka and Hopkins were less badly hurt.

At the close of the inquest on the two railwaymen the jury returned a verdict that no individual could be held responsible for the accident but it had been due to a 'combination of negligence on the part of the crew of the "extra" train 5186, and a lack of proper communication with the station dispatcher.' The jury decided that 'the traffic blocking signals are not visible for all areas of the Galt railway yards. We therefore recommend a closer system of communication between the train crews, especially in connection with "extra" train crews and during unusual circumstances of the moving extra trains. We also recommend the use of a red flag by the dispatcher in circumstances similar to those that occurred at this accident, instead of taking a chance of someone not hearing, or misunderstanding an emergency order. We also recommend the installation of a better and closer blocking signal system within the Galt yard limits.' [895]

Engineer Hunter and Brakeman R. Gates (who should have reset the switch rather than Fireman Easby) were dismissed by the railway, though the engineer at any rate was able to rejoin the railway after two years, and later became Chairman of the Brotherhood of Locomotive Engineers. [896]

Having suffered catastrophic bridge accidents between Jangaon and Raghunathpalli in September 1954 and between Jadcherla and Mahbubnagar in September 1956, India suffered a third railway catastrophe involving a bridge on Friday 23rd November 1956. The train involved this time was No. 683, the Madras – Trichinopoly Express, and the accident occurred between Ariyalur and Kallagram stations, Tamil Nadu, on the Villupuram – Trichinopoly Chord Line of the Southern Railway at Bridge No. 252 at Mile 170/10-14 at 05:29.

Train No. 683 consisted of steam engine Y.P. 2869 and 12 bogies (coaches). The flooded Marudayar River, which the bridge crossed, had washed away the approach embankment of the bridge, causing a breach of about 70 feet just beyond the Kallagam abutment of the bridge. The engine became derailed when it reached the breach, capsized,

[895] *Tragedy in Galt* by Paul Langan

[896] http://www.trainweb.org/oldtimetrains/CPR_London/galt_wreck.htm

and sank into loose earth of the embankment. The tender crashed into the engine and the leading six carriages piled up on top, were telescoped and in the words of the official report 'some of them were smashed to smithereens.' The seventh carriage ended up hanging off the bridge with its leading end submerged in the river. The eight coach was derailed and the four rear carriages remained on the bridge undamaged. Of the approximately 368 passengers on the train about 151 lost their lives and 114 were injured. The driver and both firemen also died.

The bridge concerned dated from 1928 and consisted of eight 62ft 3in girder spans each resting on two cast-iron screw piles of 3ft diameter. The abutments consisted of U-shaped reinforced concrete boxes resting on cast-iron piles. Problems with erosion had been found after the bridge was finished and stone revetments were provided on the north (Ariyalur) side of the bridge.

During the North East monsoon season (beginning of October to mid-January) three patrolmen monitored the Ariyalur – Kallagram stretch each night, each carrying a sheet which he exchanged with his colleague at the end of his length. The sheets were signed by the stationmasters at the start and end of each beat.

Very heavy rainfall had been experienced in the area on 20th November and on the night of 22nd to 23rd, but no weather warning had been issued on the morning of the accident.

Relieving Assistant Stationmaster Vincent Raj was on duty at Ariyalur station on the morning of the disaster. Train No. 683 arrived at 05:18 and left again at 05:21. It should have been reported back from Kallagram about 05:40. When the assistant stationmaster had heard nothing from there by 05:50 he tried telephoning his colleague there but could get no reply. Attempts to contact his Control by telephone and telegraph also failed. He then instructed a pointsman to make his way to Kallagram to try and find what was wrong as his relief, Assistant Stationmaster Vasudevan, arrived. At about 06:25 a villager appeared and told him that a train had fallen into the river and shortly afterwards two passengers from the train came running into the station and said the same thing, and that seven bogies had gone into the river. Passenger train No. 699 was then at the station and its engine and one bogie were detached and proceeded to the scene, with the permanent way inspector, men and first-aid equipment.

The bridge was in the section covered by Patrolman R. Inassimuthu. The latter told the inquiry into the disaster that he had started the second beat of the night at 00:30 at Ariyalur, reached the end of his beat at 03:00 and had arrived back at Ariyalur after 05:00. At 21:45 the water had been 5 feet below the level of the girder at Bridge No. 252. When he had returned at 04:00 the water had only been 2 feet below the girder, but there had been no rain and he had assumed that the level would then fall. He had arrived back at Ariyalur at about 05:20 while the ill-fated train was in the station and had spoken to the assistant stationmaster, then had gone to see Permanent Way Inspector T.V. Soundararajan and Assistant Permanent Way Inspector V.N. Jagannathan, informing all three, the inquiry into the disaster concluded, about the water level at the bridge.

Assistant Stationmaster Raj had recorded the patrolman as having returned at 05:00, which was the time that he should have returned rather than 05:20, the time that he apparently actually returned. If the patrolman did not return at the correct time the stationmaster was meant to caution the train, which he had not.

High Court Judge H.J. Bose, who conducted the inquiry into the disaster, was critical in his report of the railway engineers for not being more concerned about risks to the bridge and its approaches, of Patrolman Inassimuthu, of Relieving Assistant Stationmaster Raj, of Permanent Way Inspector Soundararajan and of Assistant Permanent Way Inspector Jagannathan. He did not consider that the driver of the train could have realised the danger in the darkness, nor could Guard K.S. Vaidyanathaswami, who stated that he had known nothing of the accident until he had felt a violent shock and jolting as it had happened. The judge did not believe, however, that any individual could be held responsible for the disaster.

As to the cause of the disaster, the judge concluded, 'With regard to the cause of the accident, my finding is that {an} unprecedented and unforeseen flood in the River Marudyar and in the areas around the embankments of the river caused erosion or scour under and around the southern abutment of the bridge, No. 252, and its approach, and when the express train No. 603 reached the spot of the accident, seventy feet of the approach to the bridge beyond the southern embankment, collapsed or had already collapsed, due to this scour or erosion, and this resulted in the accident.

Judge Bose went on to list 15 suggestions to prevent a recurrence of the disaster:

(1) The foundations of abutments of bridges resting on screw piles should be strengthened and construction of revetments or pitching of the river banks or margins on either side of the abutments to a sufficient length should be at once undertaken.

(2) The rules in chapters X, XI, XIV and XVII of the Indian Railway Way & Works Manual should be rigidly observed and mere nominal compliance should be avoided. Engineers and Inspectors should be instructed to explain the rules to the staff before and during the monsoon each year.

(3) On sections where railway quarters are not provided for patrolmen at least two patrollers should be appointed for each beat from experienced gangmen specially trained for the purpose.

(4) The length of beat of patrollers should not exceed three miles.

(5) Night patrolmen should have as part of their equipment a strong electric torch to enable them to observe the level of water and the strength of current flowing through the bridges or along the abutments.

(6) Patrollers should be trained to take soundings and probings near abutments and piers to determine the depth of scour in order to decide whether trains should be slowed down or stopped and piloted over the danger point, and the selection of men for this work should be from people who can usefully function in this manner.

(7) Stationary watchmen should be posted at the bridge during floods or heavy rainfall whether there is imminent danger to the bridge or not, until the floodwater subsides and the water reaches the normal level.

(8) In case of doubt the patrolman should not allow trains to pass over bridges when the river is flowing at high velocity.

(9) Signboards should be put up at either end of the bridges at adequate distances from them.

(10) The speed of trains should be restricted to 5 to 10 miles an hour over bridges whenever there is a rise in the water level or an increase in its flow during the monsoon period.

(11) Good glass should be fitted to the YP engines to make it possible for the drivers to see through the glass at a distance, especially during rains.

(12) Daily water readings should be taken at all important bridges during the monsoon and the register should be available for perusal by the superior Engineering staff at any time. The staff should be properly trained for the purpose.

(13) The rules in chapters X and XI of the Way and Works Manual should be amplified to include an appendix on river training, control of big and small rivers and indications to be looked for indicating future trouble clearly described and explained. The Railway Assistant Engineers and District Engineers should undergo a short course on this and also visit river model research stations to acquaint themselves with the behaviour of rivers and their response to training works.

(14) The High Flood Level mark should be prominently displayed on every bridge and a danger level mark below the High Flood Level should be fixed by the Engineering authority and marked in red. When the level is exceeded, the patrolmen should be prepared to stop trains and pilot them if necessary.

(15) Passenger trains should carry portable telephone instruments so that they can communicate with the nearest railway station in the event of trains being halted between stations by patrolmen. [897]

Railway and Transport Minister Lal Bahadur Shastri had tendered his resignation to Prime Minister Nehru after the Jadcherla disaster and had been persuaded to withdraw it. After the Ariyalur accident he tendered it again and this time it was reluctantly accepted. [898]

Virginian Railway fast freight train No. 72 from Norfolk to Roanoke was crossing a 150ft-high steel trestle over Cub Creek near Phenix, Virginia, United States, on Monday 2nd December 1957 when Engineer B.E. Lovern realised that it had become derailed. He managed to bring the two diesel units and four cars off the 720-foot-long trestle before the following cars began to topple from the structure. Three cars and the caboose at the rear of the train also remained on the line but 44 cars went into the creek. No one was hurt. A broken rail was blamed for derailing the train. [899]

If they had not already heard of it on the radio or perhaps seen it on the television, readers of a national newspaper on the morning of Thursday 5th December 1957 would have found headlines stating: '52 KILLED IN LONDON RAIL DISASTER' and, 'VIADUCT FALLS ON TO A CROWDED TRAIN.' They would have learnt that the disaster had happened during the previous evening rush hour on the British Railways, Southern Region, near the station at St John's, Lewisham, England.

At least one newspaper report included a brief official statement giving an outline of the disaster:

The 4.56 p.m. steam train from Cannon Street to Ramsgate crashed into the rear of the 5.18 p.m. Charing Cross to Hayes, Kent, electric train on the down through line between St John's and Park's Bridge, under the Nunhead fly-over. The fly-over collapsed onto the

---

897 https://archive.org/details/in.gazette.e.1957.358/mode/2up

898 https://theprint.in/pageturner/excerpt/lal-bahadur-shastri-sent-nehru-his-resignation-set-gold-standard-for-politicians/304341/

899 *Raleigh Register*, Beckley, West Virginia, United States, 4th December 1957

rear coach of the 4.56 Cannon Street train. The 5.22 p.m. Holborn Viaduct to Dartford train had been derailed on the fly-over. [900]

The Hayes train consisted of two four-coach units, with a motor saloon second brake and a trailer compartment second, and a trailer saloon second in between, plus a two-coach unit consisting of a motor saloon second brake, and a driving trailer semi-saloon second; 10 coaches in all. It was in charge of Guard Robert Reynolds with Motorman J.B. Skilton.

The Ramsgate train consisted of 4-6-2 Battle of Britain Class engine No. 34066 *Spitfire* and 11 coaches, including a buffet car. It was in charge of Guard E.W. Humphries with Driver William Trew and Fireman Cyril Hoare.

The line between London Charing Cross and London Cannon Street via London Bridge to St Johns and Parks Bridge Junction was equipped with multi-aspect colour light signalling and continuous track circuiting. The only signal boxes between London Bridge and Park's Bridge Junction were those at North Kent East Junction and St Johns. There were no manual block working, trains were described on Walker's train describers. The line was extremely busy, with an average of 990 trains in 24 hours on the four tracks. During the evening rush hour 44 Down and 37 Up trains passed St Johns between 5.1 p.m. and 6.01 p.m. To cope with the intense traffic the signals were placed close together and could display one of four aspects:

Green – *Clear. Proceed.*

Double Yellow – *Preliminary Caution. Prepare to find next signal showing Caution* (or in exceptional cases with very short headways Preliminary Caution)

Yellow – *Caution. Prepare to stop at next signal*

Red – *Danger. Stop.*

The four signals directly involved in the accident were automatic signal No. A 42 at New Cross and signals Nos. L 16, L 17, and L 18, the latter three all controlled from St John's box.

It had become foggy on the afternoon of the day of the accident and although no special fog service had been introduced the trains were all running considerably behind time. The first of the five Down trains through St John's before the accident was the 5.05 p.m. steam service from London Cannon Street to Hastings. This train was 47 minutes late past St John's. It received green signals to A 42 at New Cross, which was at double yellow, and was almost stopped at signal L 17. Signal L 18 and the Parks Bridge signals were yellow and it was stopped for six or seven minutes at Hither Green awaiting *Line Clear*.

The 5.16 p.m. electric train, London Cannon Street to Orpington (35 minutes late past St John's) was checked after New Cross by the Hastings train and was stopped at Hither Green waiting for it to clear the section ahead.

The 5.25 p.m. diesel-electric train, London Charing Cross to Hastings (30½ minutes late past St Johns) received double yellows and yellows after North Kent East Junction until it was stopped, unnecessarily, at signals M10/12 at Parks Bridge Junction at 6.09 p.m.

The 5.18 p.m. electric train, London Charing Cross to Hayes (37 minutes late past St John's), received clear signals up to A 82, which was showing double yellow. The fog was much thicker at signal L 16, which was yellow, and thereafter the signals changed from red to yellow as the train approached them, St John's box being passed at 6.11 p.m. The train stopped at Parks Bridge signals M 5/8 at about 6.12 p.m. and was still

[900] *The Times*, London, England, 5th November 1957

stationary when it was run into by the Ramsgate train.

The 4.56 p.m. steam train, London Cannon Street to Ramsgate, did not start until 6.08 p.m. and was 76 minutes late past St John's. With the Hayes train stopped at signals M5/8 at Parks Bridge signal L 18 in its rear would have been red and signals L 17 and L16 at yellow and double yellow respectively, but the brakes on the Ramsgate train were not applied until Fireman Hoare saw signal L 18 at red at short range and shouted a warning to his driver. The brakes had barely begun to take effect when the steam engine ran into the rear of the electric train only 138 yards in advance of the signal. As a result of the collision the engine of the Ramsgate train embedded itself in the rear of the last coach of the Hayes train, killing Guard Reynolds, while the front of the tenth coach and the rear of the ninth coach of the Hayes train were crushed together. The ninth coach in turn overrode and destroyed the body of the eighth and it was in this last coach that most of the casualties in the electric train occurred.

What made the accident considerably worse was that it occurred directly beneath a lattice girder bridge carrying the double-track Nunhead – Lewisham line over the St John's – Park Bridge line. The derailed tender of the steam locomotive struck the stanchion between the Down Through and Up Local lines which supported the girder bridge, and the bridge collapsed onto the leading coaches of the Ramsgate express, the first two of which were wrecked and the third partly so. An even worse disaster was averted when the motorman of the 5.22 p.m. electric train from Holborn Viaduct to Dartford on the upper line managed to stop it just short of the failed bridge.

As a result of the rear collision and subsequent bridge collapse no less than 89 passengers lost their lives, in addition to Guard Reynolds. Thirty-seven passengers were believed to have been in the electric train and 49 in the steam train; the location of the other three passengers was not known. Another 109 people were sufficiently seriously injured to be detained in hospital and another 67 sustained slight injuries or shock. Driver Trew was not hurt but suffered severely from shock. A lineside resident called the ambulance service at 6.22 p.m. and another resident called the police about a minute later. A passing ambulance was at the scene at 6.25 p.m. and the first ambulance to be called out arrived at 6.29 p.m. and the fire service at about the same time. As the extent of the disaster became apparent more and more services were called out and the St John's Ambulance Brigade, Women's Voluntary Service and Salvation Army also attended, while lineside residents provided what help they could. The last of the injured was removed at 10.30 p.m. [901]

'The early injured were taken to four local hospitals, the Miller Hospital, St Alfege's, St John's and Lewisham. Some of the dead were taken to Lewisham mortuary. Police loudspeakers twice repeated an urgent instruction for doctors from certain hospitals to return to their wards to care for the many injured.

'The last of the injured recovered from under the fallen bridge was a young man, who was not seriously hurt. He came out carrying his brief case and told firemen, "I'm lucky to be alive."

'While casualties were still being brought from the stretch of track 200 yards south of St John's station the three nearest hospitals reported that they could not take any more cases. Emergency arrangements were made to receive the injured in hospitals as far away as Bromley. At the scene of the crash doctors, nurses, and other rescue workers fought

[901] Ministry of Transport report

relentlessly to reach those still trapped, ignoring the peril of the damaged viaduct, which might have fallen on them at any moment. Later, work on the coaches trapped under the viaduct was halted until daylight. Firemen believed that if they continued cutting, there would be danger of the viaduct collapsing further. Rescue workers were standing by all night.'

Mr L. Ladbrook, a porter who was on the platform at St John's four minutes after the accident, said he heard a crash and though the train had gone past the signals and jerked its brakes on. "I could not see properly because of the thick fog," he said, "but I walked quickly down the track and discovered what had happened. I helped people out of the coaches and on to the safe side of the track."

Mr A. Dexter, aged 65, who was stationmaster at St John's until he retired recently, said that he was one of the first on the scene. His house overlooks the railway line. "I heard the crash and went upstairs to see if I could see anything, but the fog was too dense," he said. "I went down my garden and climbed over the fence. I could hear cries and screams, but it was impossible to see what had happened because of the fog. I directed firemen and ambulance men where they could get up on to the track from the road below."

David Wilmot, aged 18, was travelling home to West Wickham in one of the middle carriages of the electric train. He told your Correspondent that after leaving London Bridge the train had progressed in stops and starts, never travelling at much above 20 m.p.h. The already crowded train was boarded by more passengers at New Cross and later passed through St John's, stopping about 200 yards out of the station. "Suddenly there was a terrific impact and all the lights on the left-hand side of the carriage went out," he said. "I got down on to the track and found that the train had been lifted off of the lines and that one track separated us from the edge of the embankment. People from the houses below the embankment scrambled up on to the track to give assistance, some carrying tea. I could see a number of people stretched out on the ground. Some of the injured were taken down the embankment to the houses below."

'Mrs Mary Raffell, of Forest Hill, who had been doing some Christmas shopping, said she was in a crowded compartment of the electric train. "The compartment toppled on to its side," she said, "and we were one struggling mass of tangled arms and legs. It seemed we had been there for an age before the silence was broken by one man who said, 'Let's have cigarettes all round.' When the rescue workers arrived with the arc lamps I could see that both his legs were badly torn and bleeding."

'Mr C. Morris, of Bermondsey, said he was cycling along Thurston Road when the crash occurred on the line above him. He paid tribute to firemen, who, he said, "worked like Trojans, many with their coats off." Mr Morris said that families in Thurston Road "turned out to a man" to provide blankets, tea, and hot water for the injured.'

'Because of the dislocation of train services and the delay in bus services due to the fog, hundreds of people were queueing at bus stops in the locality three hours after the accident. Many hundreds more were walking home through the streets.' [902]

The main lines between St John's and Parks Bridge could finally be reopened at 5 a.m. on Thursday 12th December after about 7½ days. The Nunhead Loop line was reopened using a temporary bridge at 6 a.m. on Monday 13th January 1958.

---

[902] *The Times*, London, England, 5th November 1957

The Ministry of Transport was opened by Lieut. Colonel G.R.S. Wilson, who died before the report was finished, and the investigation was completed by Brigadier C.A. Langley,

The first witnesses to give evidence at the inquiry were the two St John's signalmen. Signalman L.R.V. Presslee had been in charge of Down traffic. He remembered dealing with the Orpington train and then a Hastings train and was sure that he had described both as "Main Line – Main Electric" to Parks Bridge (although the 5.25 p.m. Hastings train was actually diesel hauled it was described as an electric train on the Walker's describers, which dated from pre-diesel haulage days). He added that he had "sent the clock round twice" to draw attention to the fact that he was sending a second identical description. He had then dealt with the Hayes train, clearing his signals for it, and replacing them after it had passed, by which time the Ramsgate express had been described to him from North Kent East Junction box as "Main Line – Main Steam Passenger." As soon as the Hayes train had cleared track circuit No. 52 he had cleared signals L 16 and L 17 to allow the Ramsgate train forward as far as signal L 18, and he saw the indicators of the signals change to double yellow and yellow respectively, while the indicator for signal L 18 was showing red. Almost immediately the Ramsgate train had passed his box and he heard the noise of the collision a few seconds later, before he had had time to send the *Train Running Away on Right Line* signal. He had sent the *Obstruction Danger* signal to North Kent East Junction at 6.20 p.m. and his mate had sent it to Parks Bridge Junction and Blackheath at the same time.

Signalman T.E. Messer, in charge of Up traffic, confirmed Signalman Presslee's evidence and added that the fog had been thick but fluctuating all afternoon and it was common in foggy conditions for Down trains to be stopped at Parks Bridge Junction.

Signalman S.G. Beckett was in charge of Down traffic at Parks Bridge Junction. He testified that after one description of a "Main Line – Main Electric" (for the 5.16 p.m. Orpington train) the next description which he had seen on the Walkers train describer had been for a "Mid Kent Loop – Main Electric" which he had assumed was the 5.18 p.m. Hayes train. The Hayes train had to cross the junction at his box onto the mid Kent line but this could not be done immediately because of a preceding mid Kent train and a train from Bromley North on the Up Through line. It was therefore held at signal M10/12. He said that either the second "Main Line – Main Electric" description was not sent from St John's or else he had missed it. A telephone message from the driver of 5.25 p.m. diesel electric train did not dispel his illusion, for he thought that the driver was ringing from signal M5/8 (the omnibus telephone circuit did not indicate which signal post telephone the call was being made from). He did not recall any telephone call from the motorman of the Hayes train, though his recollection of events was confused.

Signalman T.F. Cunningham, in charge of Up traffic at Parks Bridge, said that after the accident Signalman Beckett had told him, "The Ramsgate train has run into the back of a diesel." On receipt of the *Obstruction Danger* signal from St John's Signalman Cunningham had replaced the Up signals and been able to stop two Up trains.

In charge of the engine of the 5.05 p.m. Hastings train was Driver F.J. Frewin with Fireman A.J. Emery.

Driver Frewin said that he had noticed on an Up journey as a passenger that the fog was thickest between New Cross and St John's. Driving the Hastings train he had seen signal A 42 at a range of between 30 and 50 yards, had shut off steam and applied the brakes. Knowing that the following signals on the right-hand side of the line would be hard to see from his (the left-hand) side of the cab he had asked his fireman to look for them. Fireman Emery had told him that signal L 16 was yellow, and when they were close to it that signal L 16 was red. That signal had then changed to yellow and subsequent signals to Hither Green had also been yellow. Driver Frewin said that the right-hand position of signals A42, L16, L17 and L18 was not a problem in clear weather but in a fog he would either have to cross to the other side of the footplate to see them or rely on his fireman, which he was well happy to do in the case of Fireman Emery.

Fireman Emery testified that he had had a good fire before starting and had put on no more coal after leaving London Bridge but had concentrated on looking out for signals, which he knew his driver would not be able to see. It had been relatively clear to start but had got foggy after New Cross. The fog had been very thick at St John's and he had not been able to see signal L17, 'a great blazing colourlight' in the words of the inspecting officer, until the last moment. Driver Frewin had not crossed over to his side of the footplate to see the signals but had relied upon his doing so.

Driver C.W. Bishop of the 5.25 p.m. Hastings train was of course in a much better position to see signals from the cab of a diesel engine without the obstruction of a boiler. He confirmed that the visibility had not been too bad until after New Cross. Signal A 42 he had seen from about 200 yards, signal L16 from about 50 yards and signals L17 and L18 from about 30 yards. Signal M5 had cleared from red to yellow as he had approached it and he had stopped at signals M10/12 at red. After three minutes he had telephoned the Parks Bridge signalman and told him, "I am the 5.25 p.m. Charing Cross to Hastings train standing at signal M10." The signalman had, he thought, replied, "All right."

Motorman Skilton said that he had had Driver J.A. Crane and Fireman D.T. Nash with him in the cab. They were riding home after duty and were helping him observe the signals. His statement as to the visibility was described by the inspecting officer as 'not very definite.' He said that the fog had been especially bad at St John's where his view of the signals had been less than 10 yards. He had had a double yellow at signal A 42, a succession of single yellows and then been stopped at signals M5/8 at about 6.12 p.m. After about a minute he had telephoned Parks Bridge box and had identified himself. Signalman Beckett had replied, "All right when you get the aspect." After a few more minutes he had felt a jolt and realised that his train had been run into from behind, though he had not appreciated the severity of the collision at first.

Driver Crane confirmed Motorman Skilton's account of the signal aspects. Crane said that after the accident he had helped Driver Trew to a nearby house. Trew had been clearly suffering severely from shock and he "could get nothing out of him."

Fireman Nash described the fog as having been patchy and not too bad as far as New Cross. After signal A42 they had been "creeping from signal to signal." At St John's the fog had become very dense and signal L18 had "just loomed out of the fog." The motorman had got onto the telephone almost at once.

Driver Trew had been taken to hospital after the accident, being allowed home the next day. Two days later he was interviewed at home but gave incoherent answers and appeared to be on the point of breaking down. He was questioned again by Colonel Wilson over a month later, on 10th January 1958, when he appeared to have somewhat recovered. He stated that he had come on duty at Ramsgate at 12.30 p.m. to work the 12.55 p.m. express to London Charing Cross. It had been cold and foggy to Tonbridge and from Orpington onwards. Their arrival in London had been at 4.10 p.m., 23 minutes late. After drinking tea in the staff canteen he and Fireman Hoare had gone by train to

London Cannon Street to work the 4.56 express back to Ramsgate. A long, very cold wait on the platform had followed before the empty stock for the Ramsgate train had finally arrived at 5.55 p.m. Driver F. Jeffrey, who had brought the empty stock in, had told him that with all the signal delays getting to Cannon Street the water in engine No. 34066 must be getting low, so he (Driver Trew) had decided to stop for water at Sevenoaks rather than waiting until the booked stop at Tonbridge.

The 4.56 p.m. Ramsgate express had finally started at 6.08 p.m. Driver Trew said that visibility had not been very good even between London Bridge and New Cross, with visibility down to 30 to 40 yards. Signal A 42 he had seen at green and signal L16 at double yellow. "I just saw it over the boiler, just at the bottom, just the glimmer of a double yellow." He alleged that he had seen signal L17 at yellow but could not say how far away it had been. His answers about what had happened after that were described by the inspecting officer as "contradictory." He said that he had asked Fireman Hoare about signal L18 and had been told that it was red, when he had applied the brakes. As to his speed, he had not been crawling through St John's but had not been going as fast as 20 or 30 miles an hour. He could not explain why he had not asked Fireman Hoare to look out for the signals earlier when they were on the fireman's side. It was not because he did not trust Hoare, who, he said, was a good fireman. There had been no problems with his engine. He had been in good health, though anxious about losing time with the extra stop for water at Sevenoaks. He at first denied having failed to see signals L16 and L17, but later contradicted himself, admitting that he had not seeing them and implying that he had not been worried about failing to see them because he had not been expecting to be stopped at St John's. He had not asked his fireman to look out for the signals.

Fireman Hoare was badly hurt in the accident and was interviewed in January 1958, after Driver Trew. He said that the fog had been patchy, the visibility "not good and not bad" to New Cross. Having seen the New Cross starter (signal A42) at green he had occupied himself with the firing, preparatory to the climb to Knockholt summit, until asked by his driver to look out for signal L18. Asked why he had not looked out for signals L16 and L17 to help his driver he replied that he had had to do his work (firing) and that he had done his duty by looking out for signals A42 and L18. Confronted with Fireman Emery's account of what he had done, Fireman Hoare still maintained that he had done his duty, adding that with signal A42 at green he had not expected the next two signals to be anything but green as well, that it was unusual to be stopped at St John's and that it was a good place to fire. He also said that he had not expected the fog to be so thick at St John's and that if he had realised how bad it was, or if signal A42 had been anything but green, he would have looked out for signals L16 and L17.

Guard Humphries stated that he had joined the Ramsgate train at Cannon Street at 6 p.m. He had not gone forward to the engine but had been given the driver's name and an inspector had advised him of the extra stop at Sevenoaks. Visibility had he said been "nil." The train had been going slower than usual, perhaps 30 miles an hour. He had felt no brake application before the collision.

Several passengers on the Ramsgate train also said that they had not noticed any brake application before the accident.

A motorist, Mr Fieldgate, who had been driving a car in the area of the accident at the time that it happened also described a very thick bank of fog, which had resulted in him driving through a set of traffic lights without seeing them.

The signalling was of course carefully examined after the accident and nothing was found to have caused any false clear aspects.

Beginning the conclusions to his report Brigadier Langley pointed out that although the Hastings train had been unnecessarily stopped at Parks Bridge Junction, and thus the Hayes train had been held up behind it, due, probably, to an error on the part of Signalman Beckett in missing a train description, the latter signalman was in no way responsible for the collision as the Hayes train should have been fully protected by signals.

Driver Trew had given contradictory evidence as to whether or not he had seen signals L16 and L17. It was clear that visibility as far as New Cross, over which stretch the line ran mostly on viaduct, had not been too bad for observing signals, but that the visibility had been much less in the cuttings and short tunnel beyond. In such conditions the signals on the right-hand side of the line would not have been visible to the driver, on the left-hand side of a *Battle of Britain* Class steam engine before they were obscured by its boiler. Driver Frewin had not been able to see them even from the smaller *Schools* class engine that he had been driving and had relied upon his fireman. It was clear, therefore, that Driver Trew could not have seen signals L16 and L17. The speed of the Ramsgate train had probably been at least 30 miles an hour at the moment of collision. Various witnesses did not recall any braking but it was likely that Driver Trew, taken by surprise, had applied the brakes at the last moment on learning that signal L18 was red, but they had not had time to take effect at the rear of the train in the short distance.

It was difficult to understand how a man of Driver Trew's knowledge and experience could have made no attempt to observe signals L16 and L17 or else to have asked his fireman to do so. It could only be concluded that he had not realised how severely restricted the visibility was and that having seen that signal A42 was green he had unjustifiably assumed that he would not be stopped at signal L18, especially as he had never been stopped there before. He must therefore be held solely responsible for the accident. Fireman Hoare's statement that he would have looked out for signals L16 and L17 had he realised the density of the fog or had been asked to do so by his driver was accepted. Given the different attitude of Fireman Emery, Fireman Hoare was perhaps lucky to avoid more criticism from the inspecting officer.

Turning to remarks and recommendations, the inspecting officer predictably brought up the subject of Automatic Train Control (ATC). He noted that after extensive trials the British Railways Board had approved the British Railways system of Automatic Warning Control (AWS) in November 1956 and planned to install it on all main lines. As far as the Southern Region was concerned the main lines from London Waterloo to Exeter and Bournemouth, which had mostly semaphore signals, would be equipped first, then the Kent and South Coast lines, including that through New Cross. Accidents such as that at Harrow and Wealdstone (see Sidelong, Head-on, Double and Fouling Collisions chapter) had shown that drivers could overrun even colourlight signals in fog, so they would be included in the AWS scheme, but semaphore signals would be given priority.

The use of train stops, as employed on London Transport, had been suggested but their use on main lines would pose difficulties, be very expensive and add little in terms of safety to AWS alone.

Little would be gained by re-siting the signals between New Cross and St John's on the left-hand side of the line and in any case the problem would disappear with the substitution of electric and diesel for steam traction. The signals on the line were closely spaced but there was still ample distance between a double yellow and a red signal to stop a steam- or diesel-hauled train and electric trains had more powerful brakes.

The *Battle of Britain* Class engines were no longer running on the Hastings line, with its restricted loading gauge, and it would be useful if they could be fitted with wider windscreens to improve visibility until they were rebuilt with wider cabs.

The replacement of the Walker's train describers and omnibus telephones with more modern equipment was being reviewed.

It was a stroke of great misfortune that the accident had occurred beneath a heavy overbridge. The supports of the bridge had been quite adequate for normal purposes and he knew of no other case of such a thing happening. The action of Motorman D.S. Corke of the Holborn Viaduct – Dartford train in stopping it and preventing an even worse accident was commendable. [903]

The inquest on the victims of the disaster at the end of December returned a verdict of 'Accidental Death,' despite which towards the end of February 1958 Driver Trew was charged with manslaughter. He stood trial at the end of April 1958 but at the end of the trial, after the jury had been out four hours, they were unable to agree. Mr Justice Jones dismissed the jury and ordered a new trial. In the end, after a question in the House of Commons from the Brixton Labour M.P., Mr Marcus Lipton, and a petition signed by 7,000 people the Crown decided to drop the case. [904]

Mr Christmas Humphreys, the prosecuting counsel, explaining the Crown's decision, stated, 'It now seemed, however, that the cumulative effect of those threefold proceedings {the inquest, the driver's appearance at the magistrate's court and the first trial – to which might be added the railway's internal inquiry and the Ministry of Transport inquiry} had had a grave effect upon his mind and to some extent upon his physical health. In those circumstances it might be felt that further proceedings would prove a serious danger to his mental condition.' Driver Trew's ordeal was over. [905]

A wooden trestle over the Sipsey River near Tuscaloosa, Alabama, United States, collapsed beneath a Gulf Mobile and Ohio Railroad freight train on Wednesday 15th January 1958 and 34 cars in the middle of the 122-car train fell into the river along with the trestle. No one was hurt. The accident was blamed on the shifting of a load of steel pipes on a car. [906]

A disastrous derailment near Meldrim, Georgia, United States, on the afternoon of Sunday 28th June 1959 resulted in the deaths of 23 people and the injury of seven others. All the dead and all but one of those injured were locals, many families with children, who had been taking a respite from the summer heat along the east bank of the Ogeechee River, a popular spot near Savannah for swimming, boating, fishing, and picnicking.

The train involved in the accident was Seaboard Air Line Railroad eastbound second-class freight train No. 82 from Montgomery, Alabama, to Savannah, consisting of diesel electric units 1509, 1751 and 1502, coupled in multiple-unit control; 123 cars and a caboose, in charge of Conductor Ira Hines with Engineer G.R. Parsons. It had left

---

903 Ministry of Transport report

904 *The Times*, London, England, 1st January, 19th February, 8th March, 22nd, 23rd, 24th and 26th April and 7th May 1958

905 Ibid, 8th May 1958

906 *Tucson Daily Citizen*, Tucson, Arizona, United States, 16th January 1958

Americus at 7 a.m., 4 hours and 15 minutes late, stopped at Vidalia at 12.05 p.m., left there at 1.30 p.m., 4 hours and 55 minutes late, stopped at Collins, 42½ miles from Meldrim, where the 21st car was detached because of an overheated journal, and reached the trestle where the disaster occurred at about 3.40 p.m.

The trestle was 2,337 feet long and spanned a slough (swamp) and the Ogeechee River. It had 188 pile bents, each containing five or six piles. Three hundred and fifty feet of the trestle spanned the river itself.

As the train was crossing the trestle and its front was near its east end it became derailed, the 107th to 122nd cars leaving the rails and the 109th to 119th cars falling from the trestle to the east bank of the river, where there was a picnicking area, and into the river itself. It so happened that the 113th and 114th cars were tank waggons containing liquified petroleum gas (LPG) and in their fall to the ground the coupler of the 114th car penetrated the tank head of the 113th car. The gas escaped from the damaged waggon, blanketed the surrounding area, and was ignited from an unknown source. The resulting flash fire spread over the area, killing and injuring the people on and in the river below, as well as destroying the derailed cars and a considerable portion of the trestle. [907]

Conductor Hines told his story to a reporter. "The train was moving along between 45 and 50 miles an hour. I was looking out of the cab of the caboose. All of a sudden, I saw the cars begin to pile up in front. It looked like they were piling up to the sky. I told Kimble {William Kimble, the flagman}, 'You'd better say your prayers now because we're gone.' Just as we stopped, gas began to shoot out in every direction. It was like water coming out of a fire hydrant under tremendous pressure. One stream was shooting up the other river and one was shooting down.

"I ran out on the platform of the caboose and began hollering to the people swimming to get out of there, that the gas was going to explode any minute. Some of them started to swim away, but most just looked at me and talked among themselves.

"My flagman heard me telling them that the gas was going to explode and stepped out onto the trestle. I told him. 'Son, you better get back in here right away because the gas is going to explode.' Then there was a tremendous explosion and flames were everywhere. Fire was everywhere. There was one tremendous sheet of flame. I jumped back into the car and pulled the door shut and slammed down all the windows. I thought I was going to be roasted alive. Then I saw an opening in the flames and started out on the trestle to get to it and climb down and get away from there.

"I began to yell to Kimble and he didn't answer. I kept yelling. Just as I got to the opening in the flames I heard him answer from down in the water. I don't know how I got my clothes off and down to the water, but I did and got him and hauled him to the shore. He would not let me take him out of the water. He was in terrible pain. I left Kimble in a mudhole on the edge and climbed back on the trestle and got on the phone and called Parsons {the engineer}. I told him to send everything he could – ambulances, doctors, everything.

"It must have been 15 minutes after the wreck that I called Parsons. Then I saw two men way down the track behind us and began waving to them with a white handkerchief. They finally came down and I got them to take Kimble back with them to a hospital. Then a man who had been swimming and had swum over to the other side when the

[907] Interstate Commerce Commission report

explosion happened swam back and came through the swamp and called me. He said there were people everywhere screaming and dying. I couldn't see anything for all of the smoke but I could hear people screaming. I called Parsons and said to get some helicopters because I didn't think they could get any ambulances in here." [908]

Another newspaper wrote, 'Survivors said the gas spread like a ground fog over the river, then flashed into flame with a tremendous explosion. This ignited a second car loaded with 10,000 gallons of butane. The impact of the double explosion wrecked several cars on the banks. Within seconds the area became a screaming mass of humanity. Sobbing fathers searched for their families. Injured persons yelled with pain as they struggled to escape the fire in woods along the bank.

'George Hodges, Jr, of near Meldrim, said, "The train made a funny bump, bump, sound when it came across the trestle and I stopped swimming and looked at it. Then some of the cars fell into the water and we began to get scared. When I saw the gas coming out of a freight car I grabbed my wife and ran. We heard the explosion but we didn't look back until we got back in the woods. When we looked back we saw a great sheet of fire sweeping over the water and into the woods. It burned everything it touched. When it first went off it sounded just like a bomb. We went back and tried to help some of the people who weren't killed. They were all burned, charred black. It was horrible."

'L.M. Nicholls, 33, of West Savannah, saw the accident in time to save his wife and three daughters. "When I saw the gas leak out it looked like a white, ground fog," he said, "I thought at first it might be poison and I got my wife and kids out of there. Then the explosion came and the whole river seemed to be burning."

'Robert W. Morris, of Meldrim, one of the first on the scene, said, "Some of the people never made it out of the water. And some that did got as far as 500 yards away before the fire overtook them. I never saw so much fire. Everything was on fire."' [909]

Engineer Parsons and Conductor Hines told the Interstate Commerce Commission (ICC) inquiry that they had noticed no problems with the train other than with the car which had been taken out at Collins. Engineer Parsons stated that he had not noticed anything wrong with the track prior to the derailment. Conductor Hines described the derailment, giving similar details to those he had given to the reporter, as detailed above.

Examination of the track on the trestle showed no indications of dragging equipment or of any obstruction on the line. Marks on the rails, other parts of the track structure and on the wheels of the 107th car of the train indicated that as it had been travelling over the trestle the wheels on the south side of the rear truck of its 107th car had dropped inside the south rail at a point 117 feet west of the east end of the trestle, and the pressure on the south rail had forced it outwards and broken it. The fact that the wheels on one side of the car had dropped inside the gauge while the wheels on the opposite side had remained on the rail, indicated that either the wheel and axle assembly of the car or the track had been out of gauge. As there were no problems with the vehicles of the train it was apparent that the track must have spread as the train was passing over it.

The ICC noted that at 3 p.m. on the day of the accident the temperature had been 98° Fahrenheit, a higher temperature than experienced during the previous 11 days. Examination of the track west of the point of derailment showed that the rails had moved longitudinally eastwards about 17 inches. The rails on the trestle were not anchored to

[908] *The Monroe News-Star*, Monroe, Louisiana, United States, 29th June 1959

[909] *Hattiesburg American*, Hattiesburg, Mississippi, United States, 29th June 1959

prevent movement though they were on each side of it. The anchors on the westward side of the bridge had moved, those on the eastward side had not. It was believed that the expansion of the rails due to the heat and their eastward movement on the trestle, which was restricted at its east end, had resulted in an excessive compression of the rails, causing the south rail to deflect beneath the 107th car. No guard rails were provided on the trestle 'as a matter of policy' according to an official of the railroad, though no reason could be advanced for such a policy. Had they existed, the ICC commented, the derailed cars would probably not have fallen from the trestle. [910]

Fourteen cars of an Illinois Central Railroad freight train piled up in Catfish Creek, two miles south of Dubuque, Iowa, United States, early on Saturday 16th April 1960 when the bridge over the creek collapsed beneath the train. The engine and one car made it over the bridge and no one was hurt. A 15-car was left hanging from the bridge, 40 feet above the ground. The cars had been loaded with lumber, meat, and lime according to railway officials. The bridge foundations had apparently been weakened by floodwater in January and March. [911]

Thirteen or fourteen people died, including men of the armed forces, and about 35 were injured when a passenger train was derailed in north-western Pakistan on the evening of Sunday 3rd July 1960. Three coaches left the track and two of them fell from a bridge into a dry river bed in the Abbottabad district between Taxila and Haveli, about 30 miles west of Rawalpindi. [912] [913] [914] [915]

---

[910] Interstate Commerce Commission report

[911] *The Cedar Rapids Gazette*, Cedar Rapids, Iowa, United States, 16th April 1960

[912] *The Times*, London, England, 5th July 1960

[913] *The Canberra Times*, Canberra, Australia Capital Territory, Australia, 6th July 1960

[914] *Pacific Stars and Stripes*, Tokyo, Japan, 6th July 1960

[915] *Star News*, Pasadena, California, United States, 5th July 1960

CHAPTER 11

TUNNEL ACCIDENTS

In his classic work on British railway accidents, *Red for Danger*, L.T.C. Rolt remarks on the acute trepidation felt by Victorian travellers when they were passing through a tunnel, and their oft-repeated pleas to the railway authorities that all tunnels should be artificially illuminated, pleas which, probably for practical reasons if nothing else, fell upon deaf ears. He goes on to point out that these fears were by no means unjustified, especially in the period before the introduction of the absolute block system, when train were merely spaced by the time interval system. If a train should chance to break down in the open there was a good chance that the crew of a following train might be able to see it in time to pull up, even if there was insufficient time for the guard of the first train to go back and protect its rear. In the smoke- and steam-filled darkness of a tunnel there was no such hope. Furthermore, should a serious collision occur in a tunnel the confined space would be likely to result in the wreckage being hideously compressed, and should a fire break out in a long tunnel provided with ventilating shafts, the latter could act as chimneys. In addition the rescue work would be hampered by the darkness and the difficulty of working in a confined space where the use of breakdown cranes was impossible, and the only resort was to drag the wreckage bodily out of the tunnel.

A fire and explosion occurred in the motor waggon of a passenger train on the Arosa – Chur line shortly after the train had started from Arosa, Switzerland, on Friday 1st March 1946. The brakes came on automatically and the train came to a stand in Arosa Tunnel. The passengers were able to escape the blaze in time, though some of them suffered minor injuries in disembarking from the train. The luggage of two passengers was destroyed or damaged and one third-class compartment of the motor coach was wrecked. [916]

A serious misunderstanding led to the death of two railwaymen and injuries to four others on Saturday 27th April 1946. The accident occurred in the 1,000-yard Cherry Creek Tunnel of the Canadian Pacific Railway, 12 miles west of Kamloops, British Columbia, Canada. A party of 17 men were making their way eastwards through the single-line tunnel in charge of Foreman B. Fredbrek, riding on three hand cars which were being kept 20 feet apart. The foreman had been under the mistaken impression that no westbound trains were due for several hours. When the leading trolley was about 40

[916] *Journal de Genève*, Genève, Switzerland, 1st March 1946

feet from the east portal of the tunnel it was struck by the engine of a westbound freight train, the headlight of which had only become visible at the last moment. Railwaymen Joe Bodner and James Orr were killed and four others were hurt, as already related. Coroner J.S. Burris ordered an inquest to determine what had gone wrong. [917]

Fifteen people died and 68 were injured in a rear collision in Azuchi Tunnel, Japan, on the Tokaido line, on Friday 26th July 1946. [918]

On Wednesday 4th December 1946 the eastern extension of the Central line of the London Passenger Transport Board (LPTB) was opened as far as Stratford, England. The continuation of the line in a tunnel to Leyton was not yet open for public traffic but as trains could not cross over to the opposite running line at Stratford they continued to a temporary crossover at Draper's Fields, just short of Leyton. The line was signalled in accordance with the usual LPTB practice with red and green colourlight signals, train stops and continuous track circuiting. There were three signals between Stratford station and the tunnel; A 489 18 yards from the end of the platform, LDX.491, 587 yards from the platform, and LD.1, 1224 yards from the platform and there was telephone communication between some of the signals and a shunter's cabin at the crossover. There was no telephone provided at signal A 489 and motormen were permitted to pass it at *Danger* and proceed at caution (the Stop and Proceed rule).

Seven trains from Stratford to Leyton were directly or indirectly involved in the accident. In order these were Nos. 21, 10, 25, 14, 5, 31 and 18. Just after 10 a.m. the electro-pneumatic train stop at signal LD.1 failed so that the trip cock remained in position to stop train No. 21, despite the signal being *Green*. The fault was indicated to the motorman by the *Red* aspect remaining lit as well; he telephoned the shunter at Draper's Fields and was instructed to proceed. Trains Nos. 10 and 25 were likewise delayed at signal LD.1 and with train No. 10 delayed in front train No. 25 was also stopped at signal LDX. 491. Train No. 14 waited at Stratford until signal A 489 cleared and with train No. 25 in front it was stopped three minutes at signal LDX.491. The lineman had been at Draper's Fields when train No. 21 arrived there and its motorman reported the fault. He returned to Stratford on that train, thence back on train No. 14 to signal LD.1, where he was quickly able to rectify the fault.

The next train, No. 5, waited at signal A 489 about two minutes, then passed it at *Danger* after its guard had been told of the signal failure by the station staff. It proceeded cautiously and came to a stand behind train No. 14, still stopped at signal LDX. 491, then stopped at that signal, telephoned Draper's Fields, waited as instructed until the signal cleared and had almost reached signal LD.1 when the current failed.

Train No. 31 arrived at Stratford at 10.26 a.m. Signal A 489 was at *Danger*. Guard Evans heard from the station staff that there had been a signal failure and he advised Motorman Porter by telephone. The guard verified that the electric and oil tail lamps were alight and after about five minutes Motorman Porter set off. His train was tripped at Signal A 489, he reset the trip cock and set off down the 1 in 30 gradient at 10 miles an hour. He had no difficulty in stopping behind train No. 5. When that train started forward he had not immediately followed and before he could do so his train was run into from behind.

---

917 *The Lethbridge Herald*, Lethbridge, Alberta, Canada, 29th April 1946

918 http://www.ikjeld.com/japannews/00000147.php

Train No. 18 arrived at Stratford shortly after 10.30 a.m. Signal A.489 was at *Danger*, Guard Bosher was likewise advised by the station staff of the signal failure, and he advised Motorman George Winter over the telephone. Signal Engineers Executive Assistant Henry Hadaway chanced to be at Stratford station and finding that there was a problem with the signalling he had decided to investigate, taking with him Signal Department Assistants Henry Wheat and H.S. Maddams. The three of them boarded train No. 18 and after a couple of minutes Mr Hadaway told the motorman that he could pass signal A.489 in accordance with Rule No. 55 (g) (the Stop and Proceed rule). Mr Hadaway said that he had soon thought that the train was going too fast and he had reminded Motorman Winter that they might find a previous train at any point. Motorman Winter did not recall this remark later but Assistant Wheat confirmed the comment. Because of the curvature of the line Messrs Hadaway and Wheat had both seen the rear of train No. 31 a moment before Motorman Winter could and both had exclaimed, "There she is," or words to that effect. Neither had thought that their train was going too fast or that it would be unable to stop in time.

Motorman Winter said that he had seen the train immediately after the warning and had been surprised that he had not been able to pull up; he had no complaint about the brakes.

The rear collision was violent enough to trap the four men in the wreckage of the train and it took rescuers four hours to release them, working by the light of arc lights. Assistant Maddams was killed and the other three men were seriously injured.

Tests carried out after the accident showed that if a train was allowed to run unchecked down the steep gradient from Stratford and the brakes were first applied at the point where train No. 31 could first be seen by the motorman it attained a speed of 25 miles an hour and stopped 23 feet beyond the collision point.

The Ministry of Transport inspecting officer, Lieut. Colonel E. Woodhouse, concluded that the accident had been due to Motorman Winter's failure to drive with sufficient caution, possibly because he had not appreciated the steepness of the gradient, or the speed which he had attained, or had not appreciated that there might be two trains in the rear of the next signal. At any rate his judgement had been seriously at fault.

The inspecting officer noted previous accidents on the LPTB due to lack of caution when driving under the Stop and Proceed rule, at Northwood in 1945 and at Harrow-on-the-Hill in 1947, but concluded that there was no reasonable alternative on the underground lines where there were many automatic signals and few signalmen. He could only suggest that motormen should be continually reminded of the risks that they incurred if they failed to drive with sufficient caution. [919]

An electric train stalled and caught fire in the long Ikomayama Tunnel, Japan, on the Osaka – Nara line of Kinki Nippon Railways, on Wednesday 16th April 1947. Twenty people died and about 70 were hurt. [920] [921] Some reports put the number of dead at 35. The fire was blamed on an overheated motor. [922]

---

[919] Ministry of Transport report

[920] *Atlantic News Telegraph*, Atlantic, Iowa, United States, 16th April 1947

[921] http://www.ikjeld.com/japannews/00000147.php

[922] *The Straits Times*, Singapore, 17th April 1947

The partial failure of a tunnel at New Lexington, Ohio, United States, caused the derailment of a New York Central Railroad empty waggon train from Monongahela, Pennsylvania, on Friday 25th April 1947. Fireman Don C. Mellor and Head Brakeman Henry R. Maley were slightly injured. The engine and 25 cars of the train left the track inside and outside the tunnel. [923]

Seven people lost their lives on the morning of Saturday 7th June 1947 when passenger train No. 602 from Vyšehrad collided with a light engine coming from track No. 28 just inside of the second Vinohrady Tunnel, near Praha Vršovice station, Czechoslovakia. Seven other people were seriously hurt and 49 were slightly hurt. Both lines through the tunnel were blocked until 19:45. The main cause of the accident was stated to have been that a safety device had been disabled. [924]

Seven passengers died and others were injured according to initial reports when the walls (roof?) of a tunnel fell onto a train on the evening of Tuesday 10th February 1948 between Malatya and Fevzipasha, Turkey. Four of the coaches of the train were destroyed by fire. [925] [926]

A serious accident occurred in Italy on Ascension Day 1948, which was Thursday 6th May. Freight train No. 5706 was passing through a 400-metre-long tunnel between Mornago Cimbro and Ternate, in Lombardy, when for some reason it became divided and its rear four waggons became separated. The train was stopped and its crew had just finished recoupling the detached waggons when it was run into by passenger train No. 1564, running at full speed, at about 15:25. The last waggon of the freight train was smashed and the leading coach of the passenger train piled up against the tender of the engine. It was in this carriage that most of the deaths and injuries occurred. At first there was a panic among the remaining passengers on the train, then once they had recovered themselves they went forward to help their less fortunate fellow travellers. Outside help was provided by the Varese police commissioner and the Savora-Corona colonel of carabinieri, who arrived with a strong force of men. The injured were conveyed to hospitals at Gallarate and Vergiate. In all, six people died and about 20 were injured. Those who lost their lives were Agostino Reggiori, 63, of Leggiuno San Giano; Augusto Albertari, 41; Gianni Rissone, Piera Castiglioni, and Maria Chiesa, 49; all of Milano; and a person tentatively identified as Ernesto Barella. The injured came from Milano, Varese, Gallarate and Porto Valtravaglia.

An investigation quickly found that the collision had resulted from the negligence of 25-year-old Relief Stationmaster Luigi Passera at Mornago, who had permitted the passenger train to proceed before he had been advised by Ternate that the freight train had arrived there. He immediately admitted his guilt and was arrested. [927]

[923] *The Marion Star*, Marion, Ohio, United States, 26th April 1947

[924] https://www.koridory.cz/vinohradske-tunely-poprve/

[925] *Journal de Genève*, Genève, Switzerland, 11th February 1948

[926] *Nebraska State Journal*, Lincoln, Nebraska, United States, 12th February 1948

[927] *La Stampa*, Torino, Italy, 7th May 1948

An unusual accident happened in Kaiser Wilhelm Tunnel, Germany, on Monday 22nd November 1948, involving an international express, train D 21 from Paris Est via Luxembourg to Koblenz Hbf. There were 700 passengers on the train, including about 200 members of the French military and administration of the Allied occupied zone. The train engine had been changed at Trier and the booked replacement locomotive having been found to have a broken spring, a replacement Class 50 freight engine was substituted. In addition the booked driver from Trier, rushing to prepare the freight engine, managed to injure himself so Driver August Vochtel was substituted. The lights of the engine also failed so the driver and fireman were forced to resort to carbide lamps to study timetables and suchlike. The train left Trier at 19:07, 25 minutes late.

Kaiser Wilhelm Tunnel was 4,205 metres long and special regulations applied to trains passing through it, for example the firebox door was not to be opened within the bore. The train entered the tunnel at 20:29, travelling at about 70 kilometres an hour. When the train was about three-quarters through the tunnel numerous spurts of flame erupted on the driver's side of the cab, setting the cab on fire. It was supposed afterwards that a defective carbide lamp had ignited an explosive mixture of coal dust and air. The driver's clothes were set alight and much heat was generated. The fireman escaped the fire but the engine controls were too hot for him to touch and the train was now running uncontrolled down a falling gradient towards Cochem station. Driver Vochtel made his way to the side of the engine and attempted to apply the brake from there but like his fireman was unable to do so, because of the heat, and he suffered severe burns to his face and arm. He then made his way along the side of the locomotive to its front where he was able to open the air brake pipe with his foot and stop the train 300 metres from Cochem station. The conductor of the train was unaware that there was a problem until his train stopped in mid-section, when he made his way to the engine and found the badly injured driver. He reported what had happened to the Cochem *fahrdienstleiter* via a lineside telephone and took the driver to the station on foot. Later a passenger on the train, who was a qualified driver, extinguished the fire with the water hose used for wetting the coals and brought the train to Cochem, from where it continued with another Class 50 engine, arriving in Koblenz 73 minutes late. No passengers were hurt.

Driver Vochtel was honoured by both French military governor Marie Pierre Kœnig and Federal President Theodore Heuss. Unfortunately, he was too badly hurt to ever drive again. After the accident the regulations were amended such that an assisting engine was always to be placed at the rear of the train, carbide and kerosene lamps were forbidden to be used in the tunnel and it was decreed that engine cabs must be thoroughly swept clean of coal dust and the coal in the tender wetted before entering the bore. [928]

Six railwaymen, described in the Interstate Commerce Commission report as maintenance-of-way employees, lost their lives when the track motor car on which they were riding was hit by a Norfolk and Western Railway passenger train in Kimball Tunnel, West Virginia, United States, at 7.14 a.m. on Monday 24th April 1950.

Kimball Tunnel was on the line between Williamson and Bluefield, just west of Kimball station, and was 697 feet long. Two tracks ran through the tunnel and the line was electrified and was used by both electric and steam trains. Trains running with the current of the traffic were controlled by timetable, train orders and an automatic block

[928] https://de.wikipedia.org/wiki/Eisenbahnunfall_im_Kaiser-Wilhelm-Tunnel_(1948)

signal system; those running against the current of traffic were controlled by train orders.

The passenger train was eastbound first-class train No. 24 and consisted of steam engine No. 137, two baggage express cars, a mail storage car, a passenger baggage car, and a coach. At Williamson it had received train order No. 4, which read:

Effective 701 AM April 24 Account Laying Rail on Eastward Track Eastward Trains will be Detoured over Westward Track from Big Four Cross Over to Eckman Cross Over. Eastward Trains Approach Big Four Cross Over and Westward Trains Approach Eckman Cross Over Prepared to Stop and Proceed Only On Instructions From Official on Ground.

Big Four Crossover and Eckman Crossover were rather more than a mile west and 3¾ miles east of the site of the accident respectively.

Train No. 24 left Williamson at 5.20 a.m. and passed Tug, about 6½ west of Kimball and the last open office, at 6.59 a.m., 14 minutes late. At Big Four Crossover the assistant trainmaster had given its train crew instructions which read:

I will protect you over westward track Big Four Crossover to Eckman Crossover. Block clear.

The train crossed over to the westward track at 7.12 a.m. and was moving at 35 miles an hour when it struck the track motor car at about 7.14 a.m., about 156 feet from the east portal of the tunnel. [929]

Track motor car No. 195 was occupied by Foreman Henry Houchins, John Arthur, T.J. Bowles, Leonard Cochran, Walter Lambert, and Henry Rickard. It had left Kimball at 7.12 a.m. and was travelling at 10 miles an hour when it was struck by the passenger train and demolished. Five of the men were killed outright and T.J. Bowles was fatally injured and died later. [930]

The ICC noted that the brakes of the passenger train had been tested and were working properly. Its engine had whistled four times between the crossover and the tunnel and its bell had been ringing but its headlight had not been lit. Because of the curvature of the tunnel the view of Engineer Arthur Reynolds had been restricted to about 200 feet and his fireman's view was much less. When he had seen the track motor car he applied his brakes in emergency immediately before the collision. Engineer Reynolds said that the men on the track motor car had apparently not seen his train at all.

According to the rules of the company the operators of track motor cars must, where practicable, inform themselves of the location of all trains from either the nearest telegraph operator or train dispatcher. Here it was usual to communicate with the telegraph operator. The carpenter force had arrived at Eckman on 18th April and had used the track motor car daily between Tuesday 18th and Friday 21st April. A man who had been with the crew on those days but was not with them on the day of the accident, the next day that the track motor car was run, said that before leaving Kimball each morning Foreman Houchins had obtained a line-up of train movements by telephone which he had repeated from memory to the operator of the motor car; they had not been written down. About 7.05 a.m. the agent at Kimball saw the foreman on the telephone but did not know to whom he was speaking. A signal maintainer and a section foreman both heard the foreman speaking to the someone they said was the operator at Eckman, who gave a line-

---

[929] Interstate Commerce Commission report

[930] *Bluefield Daily Telegraph*, Bluefield, West Virginia, United States, 25th and 26th April 1950

up of westbound trains but did not mention that eastbound trains were being diverted over the westbound track. No employee at Eckman would, however, admit to issuing a line-up to the foreman on the day of the accident, or on the previous week.

Neither the assistant trainmaster at Big Four Crossover nor the assistant road foreman of engines at Eckman Crossover was aware of the carpentry force or the movements of the track motor car.

The ICC commented that, 'The operators of track motor cars are given line-ups when they request them. A line-up neither confers rights over trains or other track motor-cars nor relieves the operator of responsibility for collision if the movement of a train is omitted from the line-up or if the line-up is otherwise erroneous. A line-up does not confer authority for a track motor-car to occupy the main track. Train crews and the operators of other track motor-cars are not informed when a track motor-car is occupying the main track. Trains may be created at any time after a line-up has been issued without the issuance of a superseding line-up. A line-up of train movements is issued as information only and track motor-car operators are required to provide the same degree of protection when they have a line-up as when they do not have it. Since January 1, 1944, the Commission has investigated twenty-eight collisions, including the instant case, in which track motor-cars were involved. These accidents resulted in the death of 58 persons and the injury of 77 persons and were caused by failure to provide adequate protection for the movement of track motor-cars.'

The ICC recommended that the Norfolk and Western Company provide adequate protection for the movement of its track motor cars. [931]

Just after midnight on the morning of Sunday 3rd December 1950 fire broke out in the mail coach which was leading vehicle of a Gijón – Madrid train and spread to the following third-class coach. The train stopped partially inside Tunnel No. 13, between Pajares and Busdongo, Spain. Five passengers died and a further 16 were injured attempting to escape the blaze. The cause of the fire was not stated. [932] [933]

The driver and firemen were killed and 30 passengers were injured when passenger train No. 1564 from Gallarate to Luino ran into fallen rocks and mud in a tunnel near Mornago Cimbro, Italy, on the Fellovie dello Stato, on a line between Milano and Luino on the Swiss frontier. [934] The train had left Gallarte at 18:19 on Wednesday 21st November 1951 and the accident happened after it had travelled about ten kilometres. The train was packed with workmen on their way home. The engine was wrecked by the impact with the obstruction and Driver Pierino Zonca and Fireman Giordano Torri were fatally crushed and scalded by steam from the damaged boiler of the engine. The first of the six coaches of the train was also crushed and panicking passengers made for its doors and windows.

Mornago station raised the alarm and two Italian Red Cross ambulances were despatched from Gallarte shortly afterwards, plus others from Sommalombardo and Varese. The Gallarte and Varese fire brigades attended and breakdown trains were sent

[931] Interstate Commerce Commission report
[932] Webpage: Federación Castellano Manchega de Amigos del Ferrocarril (in Spanish)
[933] *The Times*, London, England, 4th December 1950
[934] *The Railway Gazette*, London, England, 14th December 1951

from Gallarte and Milano. Drs. Vincente and Colomba, the Varese prefect and commissioner respectively, also went to the scene. The injured were taken to hospitals in Sommalombardo and Gallarte. It was anticipated that the line would remain blocked for several days. Heavy rain had preceded the failure of the tunnel roof. [935] It is not clear if this was the same tunnel as that which featured in the 1948 accident (see above).

An unusual accident happened in England on British Railways, North Eastern Region, on Monday 28th January 1952. It was a bitterly cold day as a train from Middlesborough to Scarborough arrived at Grinkle Tunnel, just south of the station of the same line on the coastal branch between Middlesborough and Whitby. Icicles eight feet long and a foot thick were hanging down from the roof of the tunnel and they smashed the windows of the cab of the steam engine hauling the train. The driver and fireman suffered cuts to their faces from the broken glass. [936]

At 6.56 p.m. on the evening of Wednesday 8th April 1953 there was a rear collision in a tunnel on the Central line of London Transport at Stratford, England, very close to the site of the December 1946 accident described earlier in this chapter. This time the results were much more serious; nine passengers were killed outright, three more were fatally injured, four passengers and the motorman of the colliding train were seriously injured and 41 passengers were slightly hurt or suffered from shock. It was the worst accident in the history of the London Underground up to that time, and would remain so until the Moorgate disaster 22 years later.

The signalling on the line was unaltered from what had existed in 1946. The line was fully track circuited with two-aspect colourlight signals equipped with train stops. As in 1946, the Stop and Proceed rule was in operation, whereby a train stopped at a red signal after waiting one minute could proceed with caution to the next signal at such a speed such as to enable the motorman to stop short of any obstruction.

The events which led up to the disaster had begun three quarters of an hour earlier when a heavy hook attached to the loose side chain of an eastbound train had struck and damaged the train stop at signal No. A 491. The result of this was that both this signal and the signal in its rear, No. A 489, remained at *Danger*. Stop and Proceed working was instituted at 6.12 p.m. and two porters were positioned at Stratford station to advise the motormen and guards of eastbound trains to pass signal No. A 489 at *Danger* in accordance with the rule. It was not known at Stratford at the time that signal A 491 was also permanently at *Danger*. Train No. 71, the 5.42 p.m. service from Greenford to Hainault, was the 12th train to be thus cautioned at Stratford. It proceeded cautiously past signal No. A 489, entered the tunnel, and drew up behind train No. 39, which was stopped at signal No. A 491. Train No. 39 then moved off and train No. 71 drew forward to signal No. A 491. As it was doing so it was run into from behind by train No. 59, the 5.37 p.m. service from Ealing Broadway to Epping and pushed forward three feet.

The damage resulting from the accident exceeded that caused in any previous accident to tube stock. The front of the leading car of train No. 59 was lifted until it struck the tunnel roof and the second car of the train telescoped the leading car for a distance of six feet. All the deaths and serious injuries occurred in the leading two cars of train No. 59.

---

935 *La Stampa*, Torino, Italy, 22nd November 1951

936 *Birmingham Daily Post*, Birmingham, England, 29th January 1952

The emergency services were alerted immediately after the accident. The West Ham fire brigade arrived at 7.15 p.m., closely followed by doctors and nurses from Queen Mary's Hospital and Whipps Cross Hospital, the St John's Ambulance Brigade and Metropolitan, London Transport and British Railways police. The rescue operation was very difficult in the confines of the tunnel. The first stretcher case was removed at 8.20 p.m. and it was 2.15 a.m. the next morning before the last two seriously injured casualties could be released from the crushed coaches.[937]

A newspaper account of the accident wrote, 'An official who went into the tunnel shortly after the crash said that the front coach of the second train appeared to have been completely telescoped. He said, "The sides of the coach had burst out against the wall of the tunnel, and passengers inside had apparently been thrown out of the windows and crushed."

'The first 30 people trapped in the train were brought out of the tunnel after about two hours. Most of them were able to walk, but many were assisted by volunteer rescuers. Lack of space was a great handicap to rescuers. In the first hour and a half only three serious stretcher cases were brought out.

'Immediately after the crash a general call went out for all available railway policemen. They were drafted to the area to control traffic and help in rescue operations. Many were brought from their homes after having gone off duty. Doctors from several hospitals went to the scene. Nurses were summoned from all parts of London to help in the ambulances and first aid stations.'[938]

The first witness to be interviewed at the Ministry of Transport inquiry was Motorman R.H. Goodin of train No. 39, who stated that when he had arrived at Stratford a porter had told him that he could pass signal No. 489 at *Danger*. Not being satisfied he had demanded to speak to 'the stationmaster' (correctly the station inspector). The latter official had told him the same thing and after resetting the trip cock, which had operated as he passed signal No. 589, he proceeded cautiously into the tunnel. At a distance of 12 to 15 yards he had seen the tail lamps of a previous train and by making a full brake application and releasing the "dead man's handle" he had been able to pull up within a foot of it. There had been no problem with visibility. The preceding train had then moved on and he had drawn up to signal No. 491, waited the prescribed time, then proceeded.

Guard S.J. Bass of train No. 39 said that a porter at Stratford had advised him that his train could pass signal No. 489 at *Danger* after waiting three minutes. He had spoken on the telephone to Driver Goodin and the train had then proceeded, going very slowly. At signal No. 491 he had heard another train coming, had checked that his tail lamps were lit and the other train (No. 71) had stopped 5 or 6 yards away. His own train had then proceeded until it was stopped by the current being cut off (by the collision).

Motorman A.T. Bryan of train No. 71 stated that upon arrival at Stratford a porter had told him simply that signal No. A 489 was remaining at red and he could pass it after waiting one minute. The porter had no flags and said nothing else. He had spoken on the telephone to his guard, then proceeded, resetting the trip cock, and running at less than walking pace with continual brake applications. His first sight of train No. 39 had been a refection "through the haze" and he estimated that he had seen the tail lamps from 50 feet away, the oil lamp before the electric one, then the whole of the rear of the train, which

[937] Ministry of Transport report
[938] *The Times*, London, England, 9th April 1953

he had stopped 3 or 4 yards from. His guard had spoken to him on the telephone and he had asked him to check the tail lamps. As he had moved away from the telephone there had been a loud crash, a blue flash and his cab had been filled with smoke, dirt, and dust. His train had been pushed forward and he had been thrown backwards. He said that he had at once clipped his telephone to the tunnel wires to cut the current and told the substation attendant not to restore it. Then he had gone back and had been surprised to find his guard was still alive and had seen that there was nothing that he could do to free the trapped motorman of the colliding train.

Guard H.E. Dell of train No. 71 testified that the tail lamps of his train had been burning brightly at Bethnal Green and Mile End. At Stratford he had been given the same information as his motorman and his train had then gone forward with extreme caution. When it stopped he had been just about to check his tail lamps when there was a terrific impact. He had seen the motorman of the colliding train pinned in the wreckage of his cab, had tried to go forward but a door of his train had been jammed, had tried to go back but could not get past the colliding train and had gone forward again, had met Motorman Bryan, and between them they freed the jammed door. The injured motorman of train No. 59 had told him that there had been much haze and that there had only been one tail light on his train.

Motorman J.S. Besley of train No. 59 was badly hurt in the accident and it was four months before he could give evidence. He stated that at Stratford a porter had told him, "the signal has failed – you are to pass it at *Danger*." There had been no mention of trains being ahead, of proceeding at caution or of tripping past the signal. As the porter had told him that, another porter was advising his guard he had not spoken to the latter on the telephone. After a normal stop at Stratford he had started off, been tripped at signal No. 489, had reset the trip arm and proceeded. According to his account he had proceeded cautiously at 5 to 8 miles an hour, controlling the speed with the brakes. Suddenly his train had run into a thick bank of dust 6 or 7 feet deep. As it had emerged from the dust he had seen the rear of train No. 71 only about 2 yards ahead. He had released the "dead man's handle" and applied the brakes at about the same moment as the collision had taken place. He insisted that his speed had not been any higher than he had stated.

Questioned closely about the dust Besley said, "It was more like a cloud, as the lights caught it was going round and round as though it was being blown by air pressure – as if my train was pushing it one way and the air from the other train the other way. It obscured my view of the train in front; I could see no tail lamp of any sort. It was a whitish dust as it was going round in the lights – a greyish, whitish colour." He added that it had not been fog, smoke, or condensation on the windscreen. He had not experienced such a thing before, nor heard of any other motormen having done so. He had not known that signal A 491 was also at *Danger*. "I was expecting to find a signal at *Danger* but definitely not a train in that position. Really I was looking for anything ahead. Actually I never stopped looking. I was going cautiously in case anything was in front of me whatever it was." The brakes of his train had worked satisfactorily. He could not remember when he had last applied the 'Stop and Proceed' rule, though he had done so several times previously.

Guard P.J. Newling of train No. 59 stated that a porter at Stratford had told him that signal A 489 had failed and that his train could pass it after waiting 2 to 3 minutes. After 2½ minutes a porter at the front of the train had raised his arm, he (the guard) had given his motorman the start signal, and the train had set off, then stopped; he was unsure if this was because it had been tripped. Then the train had proceeded, at, he thought, 8 miles an hour and certainly slower than usual, though he had been occupied attending to a boy left in his charge. Then there had been a heavy impact which he had realised must be a collision. He had protected the train with a short-circuiting bar and seeing that it was serious had gone back to Stratford to report the collision.

Several passengers gave evidence. Mr R. Mackenzie-Smith, a civil engineer, who had boarded train No. 59 at Holborn, said that he was quite familiar with the "tripping" of trains and did not think that train had been tripped after leaving Stratford. He said that when the train had set off from Stratford, "I became nervous again and thought to myself – 'we are going down here very fast – in fact, I thought we were going down at a normal rate; it was certainly not a slow speed." He said the speed had "called attention to itself." Then there had been an "almighty crash."

Mr. H.F. Hodges, a regular traveller on the line, thought that train No. 59 had left Stratford at normal speed after a slow journey from Liverpool Street. "It made me look at my watch as I had an engagement that night and I was trying to estimate my time for getting home. I thought that whatever the trouble there was, we had cleared it and we were back more or less on to normal running again."

Mr J. Roberts, a regular passenger who had joined train No. 59 at Mile End, thought the train had gone down the tunnel from Stratford at 8 to 10 miles an hour, adding that he drove a car a lot and considered himself a good judge of speed. The train had certainly been travelling more slowly than usual. He had helped release Motorman Besley, who had kept repeating that "they had no business to let him out of Stratford" and "that he could not see for dust."

Mr R.H. Judge, travelling from Mile End in the same car as Mr Roberts, put the speed of train No. 59 after leaving Stratford at 5 to 10 miles an hour, and definitely slower than usual. He had helped release the motorman and confirmed his statements to Mr Roberts.

Other passengers gave similar estimates of the speed of train No. 59 after leaving Stratford and some recalled the motorman speaking of dust in the tunnel.

Mr J.W. Higgs, a passenger on train No. 71, who had been in its rear car, recalled seeing train No. 59 coming down behind them. "It occurred to me that at the speed I thought it was travelling it would never stop before it hit us, and of course the next thing was the crash."

Station Inspector E.G. Brown at Stratford stated that he was in charge of the station. Station Foreman J. Rowe had advised him that there was a signal failure on the eastbound line and that porters had been instructed to tell motormen and guards to wait four minutes and then proceed.

Station Foreman Rowe stated that at 6.14 a.m. he had realised that signal A 489 was not clearing and had told Porter A.E. Wiseman to inform control. The reply had been that all trains should be cautioned. He (Rowe) had then told Porter A. Bolingbroke to instruct all motormen and Porter K.C. Clowser to likewise instruct all guards to wait four minutes and then proceed with caution. He thought that the Stop and Proceed rule required trains to wait two minutes but had increased the time to four minutes as an extra precaution.

Porters Wiseman, Bolingbroke and Clowser confirmed what Foreman Rowe had said. Porter Wiseman said that all control had told him was to, "Advise all trains."

The Ministry of Transport inspecting officer, Colonel D. McMullen, concluded that the cause of the disaster had been the failure of Motorman Besley to comply with Rule

55(g) (the Stop and Proceed rule) and that he had not exercised even reasonable caution. The rule included a warning that trains might be found ahead of a signal at *Danger* and used capital letters to emphasise the need for caution. Considering the witness statements and the damage caused, the speed of train No. 59 at the moment of collision had probably been 20 miles an hour. There was no evidence of a brake application before the collision.

The inspecting officer dismissed the statement of Motorman Besley that he had suddenly run into a thick cloud of dust, as none had been seen by the motormen of preceding trains and it was hard to see what could have caused such a phenomenon. The motorman had been seriously injured in the accident, including a brief spell of unconsciousness. The probable explanation for his statements as to a cloud of dust, made amongst others to his rescuers, was that there had been a quantity of dust caused by the collision itself and in his confused state after the collision the motorman had convinced himself that the dust had been there before the collision and had caused it. He probably had no real recollection of events just before the accident due to retrograde amnesia. His statements that the Stratford staff should not have let him down and that he had not been expecting to find a train where it was, were also revealing. It seemed likely, Colonel McMullen wrote, that Besley had allowed his train to coast down the gradient after leaving Stratford with little or no brake application so that its speed had reached 20 miles an hour. Given the gradient and curvature of the line 'such a speed was *very much too high. It should not have exceeded 5 m.p.h.*' (italics in report).

Motorman Besley was 43 years old. He had worked as a guard since 1946 but for only a short time as a motorman. He said that he had been fit and well, had not been tired, and had no worries at home. His eyesight had been good when last tested.

Turning to recommendations, the inspecting officer noted that this was the first accident on a tube line resulting in a passenger fatality since the first tube railway had opened in 1890. There had, however, been a previous accident resulting in the death of a railway servant in almost exactly the same place and in near identical circumstances in 1946. Both of the Stratford accidents and several other, less serious ones had resulted from a failure of the human element whilst trains were being operated under the Stop and Proceed rule.

It would be unreasonable to abolish the Stop and Proceed rule as this would completely dislocate the service and cause hardship to many thousands of Londoners.

One possible solution was the provision of governors to control the speed of trains but this was found to be impracticable, firstly because the lowest safe speed under any circumstances, say 5 miles an hour, would be too low on straight open stretches and would cause delays, and secondly because of the difficulty in maintaining equipment which would only be occasionally used on all of the Underground trains.

Another suggestion was the provision of speedometers but these did not accurately register the low speeds involved and in any case the prime duty of a motorman operating under Stop and Proceed rules was to keep a constant lookout ahead, without being distracted by having to monitor a speedometer.

Other possible solutions were being considered.

The Rules needed clarifying as some of the instructions given to motormen by the porters were irregular, especially an instruction from Porter Bolingbroke who due to a misunderstanding had told motormen to isolate the trip cock.

The inspecting officer also made recommendations regarding avoiding delays in the detraining of passengers after an accident and the establishment of a means for motormen to contact control directly. [939]

The Simplon – Orient express coming from Paris was derailed in Mesnay-Arbois Tunnel, in the French Jura mountains, on the morning of Saturday 16th July 1955. Seven carriages left the track due to a broken rail; and four of them overturned, fouling the opposite running line, and were struck by a light engine. Twenty people were slightly injured, 'including an Englishwoman.' The line was completely blocked and international trains were diverted via Geneva. Traffic was re-established with single-line working at 20:00 on Sunday evening. [940] [941]

One passenger died and 35 were injured when a short-circuit caused a fire to break out on a three-car electric train in a tunnel near Wakayama, Japan, on Monday 7th May 1956. Another 150 passengers bound for Osaka escaped injury. [942]

Two eight-car subway passenger trains were involved in a rear collision in Brooklyn, New York, United States, on the evening of Friday 19th April 1957. Both were en route from Brooklyn to Queens with several hundred passengers. The first train stalled in the tunnel when its brakes jammed. Motorman Victor O'Britis had climbed down from his cab and released the brakes and was just getting back into the cab when the second train ran into it. He said, "There was a crash. Some people were knocked to the floor and some of the injured got panicky and began screaming. I said to them, 'Take it easy.' I had to calm them down. They used the emergency exits themselves and a lot of them walked out."

Motorman William Hoke, in the colliding train, was pinned in his cab for an hour and a half before firemen could release him. He was badly hurt, with a crushed right leg as well as suspected internal injuries. The impact knocked out the lights and shattered windows in the second train but after a moment of confusion the conductor switched on the emergency lighting. Conductor David Rotunda of the front train was knocked unconscious for some minutes and was taken to hospital along with the motorman. Sixty-three passengers were also slightly hurt.

The exact location of the accident was between Bedford-Nostrad Avenues and Myrtle-Willoughby Avenues stations.

New York City Transit Authority General Manager Thomas McLernon commented, "After checking at the scene of the accident, we feel that it was due to man failure." Assistant District Attorney Elliot Golden said that Motorman Hoke explained that he had seen a signal showing two flashing red lights and a single yellow which meant "proceed at your own caution," presumably under some version of the Stop and Proceed rule. [943]

Two subway trains of the New York City Transportation Authority collided in the tunnel between Eighth Avenue and 59th Street stations on the Sea Beach branch of the

---

[939] Ministry of Transport report

[940] *Journal de Genève*, Genève, Switzerland, 18th July 1955

[941] *The Times*, London, England, 18th July 1955

[942] *The Daily Chronicle*, Centralia-Chehalia, Washington, United States, 7th May 1956

[943] *The High Point Enterprise*, High Point, North Carolina, United States, 20th April 1957

former Brooklyn Manhattan Transit Corporation (BMT) on Saturday 21st November 1959. A six-car work train was stopped in the tunnel about 1,000 feet short of 59th Street station when a three-car local passenger train travelling at about 20 miles an hour ran into its rear. About 40 passengers were injured, the most serious case being an unidentified man who suffered a possible broken leg. [944]

Train No. 532, the *Freccia dei due mari* (Arrow of the two seas) *direttisimo* electric train from Firenze (Florence) to Bologna was travelling at 110 kilometres an hour through the 7,113-metre-long Monte Adone Tunnel in Italy when a fierce fire broke out due to an electrical fault in the pantograph of the rear coach and the train came to a stand two kilometres within the tunnel at 08:45 on Sunday 7th February 1960. The rear coach of the train was soon well alight, causing a panic amongst the passengers. Fortunately one of those travelling on the train was a paratrooper named Gætano Argento, who took charge of the situation and shepherded the passengers towards the front of the train. Some of the passengers left their possessions behind in the rush to escape and these were subsequently destroyed by the fire. Conductor Ettore Cardinali then realised that damage to the pantograph caused by the fire made it impossible for the leading three cars of the train to continue the journey and there was no option but for the passengers to make their way out of the tunnel on foot. It was only 2 kilometres to the Monzuno-Vado (Firenze) end of the tunnel and over 5 kilometres to the Pianoro (Bologna) end but the choking smoke from the burning train was making its way towards Monzuno-Vado so he instructed the train crew to see that the passengers went towards Pianoro. He, himself, then ran for over a kilometre to an emergency telephone in the tunnel to raise the alarm.

The 160 passengers thus set off on a long and stressful journey. Their way was lit for the first few hundred metres by the light of the fire but then was in darkness. At the head of the single file of travellers was Dante Cia, a horse dealer, who described the journey to a reporter. "The women screamed that a train would come along from the opposite direction. More than the darkness they feared being caught by another train. The children then began to scream as well because their mothers were. We made our way slowly, step by step, by the light of a few candles, cigar holders and torches made from newspapers." Having gone some distance the passengers met Conductor Cardinali returning from the lineside telephone, who assured them that he had telephoned Pianoro and no trains would come from there. The horse dealer continued, "During the long trek some of the women fainted from exhaustion and others had to carry them. Some had left their suitcases and clothes behind in their haste to escape and some of the women removed their shoes, being no longer able to walk in them, and continued barefoot or wrapped their feet in paper and suchlike. As we went on in almost total darkness the column split into several sections. Unable to see we had to grope our way along by feeling the tunnel wall with our left hands. When, after two hours, we finally reached the exit, the fingertips of many of our hands were bleeding."

As soon as news of the fire reached Pianoro from the conductor's telephone call a message was sent to Bologna Central station and a rescue train was despatched from there as quickly as possible, reaching Pianoro at 10:00. The Red Cross, Army, firefighters and carabinieri were also alerted. Two fire engines had already been loaded onto waggons when the train from Bologna arrived and these waggons were attached to the

944 *Kenosha Evening News*, Kenosha, Wisconsin, United States, 21st November 1959

rescue train which set off into the tunnel at 10:10, complete with a powerful headlight to pick up anyone making their way back through the tunnel. The train went very slowly and after about a kilometre met the passengers coming back. Doctors and nurses alighted from the train to treat the passengers but the latter preferred to continue on foot to the exit, where they were assisted by carabinieri who took them to Pianoro station. Arriving at the burning train the firefighters were able to extinguish the fire using large quantities of water; they were forced to use gas masks because of the smoke. The intensity of the fire melted the metal of the car directly involved such that its frame collapsed onto the track. The remaining three cars were removed by the rescue train at about noon but it took eight hours to remove the last car. Single line working was resumed at 17:30 and both lines were back in operation at 21:20.

Sixty-six passengers were slightly injured in the fire.[945]

International rail traffic between France and Italy was interrupted for several hours on Tuesday 8th November 1960 after a bull escaped from a waggon at Modane, France, and ran into the tunnel. Here it met an incoming train from Italy and derailed its rear coach, being itself injured, though not killed. No one on the train was hurt.[946]

945 *La Stampa*, Torino, Italy, 8th February 1960

946 *The Capital Times*, Madison, Wisconsin, United States, 9th November 1960

CHAPTER 12

LEVEL CROSSING COLLISIONS

Wherever roads and railways intersect they must do so either by means of a bridge or on the level. Bridges have the advantage of allowing road and rail traffic to operate entirely without interference from one another, but they are much more expensive to provide, especially in flat countryside, due to the need to provide extensive earthworks for approach embankments or cuttings as well as the cost of providing the span itself. Where underbridges are provided there is often also the need to build retaining walls and there can be problems with drainage.[947] The provision of level crossings, especially if, as in many parts of the world, they are unstaffed, is the much cheaper alternative, but where they exist, they introduce an additional source of danger to both road and rail users. Robert B. Shaw in his outstanding book on American railway accidents *A History of Railroad Accidents, Safety Precautions and Operating Practices* writes that he has excluded ordinary grade (i.e. level) crossing collisions from consideration on the grounds that they are not proper train accidents, inasmuch trains usually suffer much less harm from such accidents than the road vehicle involved. This series of volumes, on the other hand, will attempt to include, amongst others, all such accidents which resulted in five or more fatalities, even if, as was often the case, none of those who died were train passengers or railwaymen. As in previous volumes, accidents involving school buses have also been included. The First World War provided a great stimulus for the development of the internal combustion engine and the number of motor vehicles in use, and consequently level crossing collisions involving them, greatly increased following the war. The surge in the ownership of motor vehicles continued in many richer countries after the Second World War and the removal of the most dangerous level crossings, or at least the provision of more sophisticated crossing protection, did not keep pace, so there was a substantial increase in serious level crossing accidents. This chapter is therefore a long one.

Early on Thursday 17th January 1946 a Chicago Rock Island and Pacific Railroad Chicago – Des Moines passenger train collided with a car on a level crossing at Rockdale, Illinois, United States. Six people in the car were killed. They were Donald Emerick, 32; his wife Esther, also 32; his brother Calvin, 22; Ernest Chisholm, his wife Mary, 26, and Miss Lavaughan Voight, 17. Rose Dulevich, 15, was critically hurt with

---

947 *Backtrack*, Volume 24, pg. 88.

two broken legs and a broken arm. All were from Joliet, Illinois.

Crossing Flagman Thomas Harbus attempted to halt the car by waving a red flag but it failed to stop and he had to leap to one side to avoid being run down. He said the vehicle was demolished by the train and its occupants were thrown clear of it. [948] [949]

Six Mexicans lost their lives when the automobile in which they were riding was struck by a westbound New York Chicago and St Louis Railroad passenger train on a grade crossing with State Highway No. 500 a mile from Payne, Ohio, United States, on the afternoon of Monday 21st January 1946. Those who died were Louie Martinez, 27; his wife Apalonia, 25; their children, Adela, 8, Louie Jr., 7, and Rogelo, 3; and Mrs Hermalinda Romos, 38. Mr Recarda Romos, also 38, husband of Hermalinda and the driver of the car, suffered a fractured shoulder, cuts and bruises but his condition was not regarded as serious. What sort of protection, if any, was provided at the crossing was not stated in the report. [950]

Goods train G 6921 collided with a Soviet Army lorry on a level crossing at Triptis, Germany, on the Saalfeld – Gera line, on Friday 1st February 1946. Seven people died and the same number were injured. The crossing keeper and his assistant had failed to lower the makeshift crossing barriers in time. [951]

A crowded car driven by Charles Palmer and carrying a number of people home from a movie matinee was struck by a Southern Pacific Railroad troop train on an unobstructed and presumably unprotected crossing two miles north of Salinas, California, United States on Saturday 9th February 1946. The car driver, his son (?) Richard, 11; Private First Class Hugh Smith, 23; his wife Katherine Smith, 20, their daughter Barbara Ann, 3 weeks; Smith's brother Hayward, 16; and Charles Wright, 10, were all killed. Dixie Palmer, 11, and Patricia Wright, 7, presumably sisters of Richard and Charles respectively, were injured. Trainmen said that the car had been dragged more than 200 yards. [952]

A Calais – Basel express on the last lap of its journey from Strasbourg collided with a Salta bus loaded with workmen and on its way from Colmar to Kayersburg on a level crossing at Ingersheim, France, on Monday 11th February 1946, cutting it in two. Seven workmen were killed outright, three more died soon afterwards and by the evening there were 12 dead and 28 injured. The final death toll was 15. Some of the train passengers were also hurt. The crossing barrier had not been lowered and the crossing keeper was taken into custody. Already an hour late leaving Strasbourg, the train was three hours late arriving in Basle. [953] [954]

---

948 *The Evening Democrat*, Fort Madison, Iowa, United States, 17th January 1946

949 *The Daily Register*, Harrisburg, Illinois, United States, 17th January 1946

950 *The Daily Times Bulletin*, Van Wert, Ohio, United States, 22nd January 1946

951 *Deutsche Eisenbahn-Katastrophen* by Hans Joachim Ritzau, Jürgen Hōrstel and Thomas Wolski, pg. 212 (in German)

952 *The Modesto Bee*, Modesto, California, United States, 11th February 1946

953 *Gazette de Lausanne*, Lausanne, Switzerland, 12th February 1946

954 https://fr.wikipedia.org/wiki/Liste_des_accidents_ferroviaires_en_France_au_XXe_si%C3%A8cle

Southbound Grand Trunk Western Railroad passenger train No. 54 collided with a bus filled with singing children on their way to Sunday School on Sunday 17th February 1946, on the outskirts of Detroit, Michigan, United States. The Rev. James Patton, pastor of the Carmel Baptist Church, and the driver of the bus, was killed, as were Joyce Jetton, 9, and Richard Scott, 7. Church deacon Homer Haggerd, 39, and children Margaret McCarthy, 10; Connie Scott, 7, sister of Richard; Andrew Samov, 7, and Nancy Paul, 6, the only other passengers on the bus, were all injured. Pastor Patton was driving the bus in the absence of the regular driver, and was ahead of schedule, a chance which meant that some of the children missed the bus and thus avoided likely death or injury.

Engineer H.J. Carlson stated that he sounded the whistle when he had seen the bus approaching the crossing and it had stopped, then it had started up again and had moved directly into the path of his engine. The bus was demolished and carried 100 feet down the line.

The same train had been involved in a disastrous collision with a bus in Detroit back in 1942. [955]

A lorry carrying about 20 men to work was struck by a train on a level crossing at Veracruz, Mexico, on Wednesday, 5th June 1946, killing six of them. [956]

A bus carrying Soviet soldiers and their families on a sightseeing tour to Praha (Prague) was struck by an express train bound for the capital on a level crossing near Rynholec, Czechoslovakia, in West Bohemia, on Friday 19th July 1946. Two soldiers, 12 of their wives and one other person, perhaps the bus driver, were killed. [957] [958]

Balmuckety Crossing, Scotland, on the Kirriemuir branch of the London Midland and Scottish Railway, was the scene of a serious accident involving a bus carrying a party of Forfar day trippers who were returning to the town from a visit to Glen Clova, in the Cairngorm Mountains, on Thursday 25th July 1946. [959]

The crossing with the A 926 was protected by four hand-worked wooden gates, two on each side of the road, and distant signals, which were interlocked with the gates so that they could only be cleared when the gates had been locked across the road. The road fell at 1 in 20 to a point close to the crossing. Ten passenger and four goods trains in each direction passed over the crossing daily. A census in 1935 showed 344 motor vehicles, including motorcycles, and 89 bicycles, hand carts and horse-drawn vehicles using the crossing daily; the figures for 1946 would probably be rather higher. The branch was closed between 6 p.m. and 7 a.m. weekdays and all day on Sundays.

The train involved in the accident was the 5.08 p.m. Up passenger train from Kirriemuir to Forfar. It consisted of 0-4-4 tank steam engine No. 15190, running bunker first, and three bogie coaches, in charge of Guard John Small with Driver Stewart and an unnamed fireman.

---

955 *The Evening Star*, Washington, D.C., United States, 18th February 1946

956 *Reno Evening Gazette*, Reno, Nevada, United States, 6th June 1946

957 *The Singapore Free Press*, Singapore, 20th July 1946

958 *The Sandusky Register Star-News*, Sandusky, Ohio, United States, 19th July 1946

959 https://www.thecourier.co.uk/fp/past-times/2399773/kirriemuir-rail-tragedy/

The bus was a four-wheeled Commer dating from 1933 with seating for 20 persons, owned by Robert Davidson of Forfar. It was being driven by James Petrie, who had been employed by Davidson as a bus driver since his demobilisation in March and had spent six years driving tank transporters in the Army during the war.

Crossing Keeper Andrew Pullar had received the usual warning signal that the train was on its way, actuated by a treadle 1,542 yards from the crossing, had closed the gates across the road and cleared his Up Distant signal for the train. Standing at his ground frame he had seen the bus approaching but had thought no more of it until the bus had suddenly crashed through the crossing gates and stopped on the line. Driver Stewart had likewise seen the bus and had assumed that it would pull up until, when his engine was close to the crossing, he had seen that it was not stopping. He had at once applied the brakes. Guard Small had also gone for his brake when he had seen the bus on the crossing but had found that his driver had already destroyed the vacuum. The train had been travelling at 30 miles an hour, probably reduced to 15 miles an hour when it struck the bus.

As a result of the collision the bus was crushed between the engine and a stout sleeper fence and demolished. Negligible damage was caused to the train except to its brake pipes. First aid was at once rendered by the railwaymen using equipment from the train and messages were telephoned to Kirriemuir and Forfar. The accident happened at 5.11 p.m. and all the injured had been removed by 6 p.m. Seven bus passengers were killed outright, three more died in hospital and another 14 were admitted there. Only two bus passengers escaped unhurt. No one on the train was injured.

Colonel A.C. Trench investigated the accident on behalf of the Ministry of Transport and found that none of the railwaymen involved was at fault. He did make two observations which were not directly related to the accident as the bus driver was a local man who was familiar with the crossing:

(1) The crossing gates were painted with creosote or a similar preservative and hence were dark in colour and not as visible as they might be. They should be painted white when the wartime shortage of paint was over.

(2) There had been no advance warning notices of the crossing. These were desirable, though the crossing was easily visible and the gates were seldom closed across the road after dark. They had since been provided.

The circumstances of the bus driver losing control of his vehicle were investigated by a separate inquiry. This found that the premises at Forfar were too small for the bus to enter and it had to be parked outside, where all repairs and maintenance were done. There was no pit for inspecting the underside of the vehicle and 'the facilities for inspection and repair are virtually non-existent.' Owner Robert Davidson said that the brakes had been examined and adjusted by him in April.

Driver Petrie stated that he had had no problem with the brakes of the bus until he approached the level crossing. Seeing the closed gates he had depressed the brake pedal to the floor but to no effect. He had then tried the hand brake, still without any effect. Finally he had changed down from top to third to second gear, bringing the speed of the bus down from 20 to 5 miles an hour, when it had crashed through the gates and stopped on the crossing. He had adjusted the hand brake a week before the accident. Davidson said that he had replaced the foot brake rod the previous November, using a coach bolt.

It was pointed out that there were four more passengers on the return journey than there had been seats (there had been five more on the outward journey). It was answered that three passengers had been children who had shared seats with adults, while an unsecured household chair had provided a further seat.

The inquiry concluded that since the accident was apparently not due to an error of judgement on the part of the bus driver there must have been a failure of the brakes. The coach bolt used by Davidson to connect the footbrake rod to the pedal was not suitable for such a purpose. It was likely that the footbrake rod had become detached from the pedal on the approach to the crossing. The handbrake might have failed due to the extra stress place on it by Driver Petrie in the emergency or its rod might have fractured.

In conclusion the inquiry found that: 'The maintenance of the vehicle has been neglected and it was not in a fit mechanical condition to operate on the roads for the carriage of passengers. This opinion is based on the following:-

(a) Worn out brake linings

(b) The presence of oil and grease on the brake drums indicating faulty oil seals in the rear axles

(c) The evidence of unworkmanlike repairs, for example the use of a broken clevis to attach the offside front brake rod to the cross shaft arm

(d) The probability that the vehicle had been operating with a broken front brake cable.'[960]

At 12.50 a.m. on the early morning of Tuesday 13th August 1946 a car containing five young men from Valparaiso was driven into the side of a slow-moving Elgin Joliet and Eastern Railway freight train in Hobart, Indiana, United States. Junior Dalke, 20; Gordon La Tour, 19; and Donald Pierce, 20, were killed outright. William Beckwith and John Hofferth, both 19, were extricated alive from the car, which was wedged under a gondola waggon, but Hofferth died upon reaching the Methodist Hospital in Gary and Beckwith died there an hour later.

The five men, who had all recently been released from the Services, were proceeding northwards along Highway No. 130 when they arrived at the crossing, which was reportedly well lit. The initial rescue attempts were made by Engineer Eugene Gallaher and Fireman Joseph Peart, who said that they had only felt a slight shock when the automobile hit their train. There were no witnesses to the accident. Skid marks indicated that Dalke, who had been driving, had not seen the train until the car was within 10 feet of the train.[961 962]

On the afternoon of Tuesday 29th August 1946 at about 16:00 a passenger train from Buitenzorg collided with an Army lorry on the level crossing at Tandjoeng-West Halt, Dutch East Indies. The road over the crossing was part of a short cut between Pasar Minggoe and Depok. As a result of the impact the lorry was thrown against a pole carrying high-voltage electric wires, one of which fell onto the roof of the train. The train, which was carrying workmen home to celebrate the end of Ramadan, was seriously overcrowded, and as a result some of the passengers were forced to ride on the roofs of the coaches, something which is not unknown in Third-World countries. Nine of them

---

[960] Ministry of Transport report

[961] *The Vidette-Messenger*, Valparaiso, Indiana, United States, 13th August 1946

[962] *The Tipton Daily Tribune*, Tipton, Indiana, 13th August 1946

were fatally electrocuted by the live wire and a number of others were hurt. The two people in the lorry were also seriously injured. After the accident the train took the injured and the bodies of the dead on to Manggarai station, which was reached at about 18:00. From there 14 ambulances took the victims to the military hospital. The crossing was an unguarded one. [963]

Five local people died when the car in which they were riding was struck by a New York Central Railroad passenger train on a level crossing with Bolander Avenue, Dayton, Ohio, United States, early on Tuesday 3rd September 1946. They were Mary Belcher, 18; Robert Brown, 21; Cora Fritz, 36; Virginia Johnson, 20; and Duanna Knaff, 34. The driver of the car, Robert Evans, was seriously injured and another passenger, Arthur Fritz, 41, escaped unhurt. [964]

A bus of the Ancona-based Reni company providing a service between Ancona and Firenze left the latter place at 09:00 on Tuesday 15th October 1946, well filled with passengers. At 13:00 it reached Remaggio level crossing, Italy, between Ponticino and Laterina stations. The bus driver found the crossing barriers raised and had started to cross when his vehicle was struck by an express railcar from Roma to Firenze. The force of the impact threw the engine of the bus 15 metres and the main body of the bus was carried 50 metres and demolished. The railcar was derailed. Rescuers retrieved the bodies of nine dead bus passengers from the wreckage. Eight injured were taken to hospitals at Alrezzo, Montevarche and Firenze. Three of the injured later succumbed. Among the injured were illustrious Franciscan glottologist Father Eusebio Tintori and Prince Carlo Isamburg of Rapallo. The crossing keeper was arrested. [965]

An accident at Sheldon Avenue Crossing, Lancaster, New York, United States, just before midnight on Wednesday 30th October 1946, resulted from the familiar scenario of two trains approaching the crossing from opposite directions, one shortly after the other. Brothers Leo and Frederick Weber aged 21 and 18 respectively, their 15-year-old sister Coletta; George Lanthier, 20, and John Stiegler, 18, all from Lancaster, had been to the movies and were returning home in two cars. Arriving at the crossing the two vehicles stopped to await the passage of an eastbound New York Central Railroad train. As soon as this had cleared the crossing the two cars were restarted and both were hit by train No. 67, the westbound *Commodore Vanderbilt* luxury passenger train, en route from New York to Chicago and travelling at 80 miles an hour. All five lost their lives, their bodies and wreckage being strewn for half a mile along the track. The line was blocked for two hours. According to the police visibility was good at the time of the accident and the flashing lights at the crossing were working properly. [966] [967]

Only four days after the Lancaster accident, on Sunday 3rd November 1946, five people died in the same way at Sartell, Minnesota, United States. Mrs Charles Blaske Sr.,

---

[963] *Het dagblad*, Batavia, Dutch East Indies, 29th August 1946

[964] *The Coshocton Tribune*, Coshocton, Ohio, United States, 3rd September 1946

[965] *La Stampa*, Torino, Italy, 16th October 1946

[966] *Dunkirk Evening Observer*, Dunkirk, New York, United States, 31st October 1946

[967] *Warren Times-Mirror*, Warren, Pennsylvania, United States, 31st October 1946

47; her sons Lawrence, 27, and Harvey, 19; her daughter Beatrice, 21; and Viola Poganske, 21, fiancée of Lawrence, all of Popple Creek, near Sauk Rapids, had attended a birthday party and were returning to St Cloud to pick up Mr Poganske. Their car stopped at the crossing for a freight train, then proceeded and was struck by Northern Pacific Railway passenger train No. 8 from Fargo to Minneapolis. The car was carried several hundred yards down the line and the victims were badly mangled, delaying identification. [968]

Mr William Wetherell Jr., 43, and his wife Marie, 38, had planned to spend Armistice Day, 11th November 1946, with their family and William's father, William Wetherell Sr., 75. They had driven to Shumway to pick up the older man and had almost reached their home at Illiopolis, Illinois, United States, when their car was struck by an Illinois Terminal Railroad three-car electric interurban passenger train, en route from Springfield to Decatur and travelling at about 55 miles an hour. All three were killed, as were four of William Jr's children, Eugene, 21; Donald, 16; Billy, 14; and Melba, 12. Two other children, Leonard, and Marvin, were away on military service.

Statements of the moments leading up the accident, given to the police by Motorman C.L. Musick, varied according to different newspaper accounts. According to one he told them that the car had stopped suddenly on the crossing as if it had stalled. Another version of his statement was that the car, approaching the crossing along Route 36, had slowed down as it neared the railroad but had then suddenly spurted across in front of the train. Motorman Musick said that he had braked about 100 feet from the crossing but had nevertheless hit the car and carried it about 900 feet. Mr Wetherell Jr. had been an electrician at the former Illiopolis ordnance plant. [969] [970]

An automobile, apparently travelling at high speed, was driven into a coal car which was the third vehicle from the end of a moving Grand Trunk Western Railroad freight train at Sherman Boulevard level crossing, Muskegon, Michigan, United States, at 2.05 a.m. on the morning of Tuesday 19th November 1946. The vehicle was demolished and five of its occupants were killed instantly. They were Benjamin Senecal, 25, a former Marine; his 23-year-old wife of four months, Helen; Miss Armene Senecal, sister of Benjamin, 23; Kenneth Jorgesen, 26; and Lawrence Matthews, 24. Another sister of Benjamin and Armene, Miss Marian Senecal, 25, was critically injured. Kenneth Jorgesen was her fiancé and was on terminal leave from the Navy, having fought in the Pacific during the war.

The Senecal couple and sisters all worked at the Continental Motors factory. Lawrence Matthews was apparently driving the car, which was taking the party home from an evening out.

Engineer Clarence Harrington and Conductor Leo Reterstaff declared that the crew of the train had been unaware of the collision until they stopped at a car ferry dock a mile distant, where they had found part of the wreckage of the automobile on the coal car. They had then returned to the crossing and found more remains of the vehicle.

---

[968] *The Evening Tribune*, Albert Lea, Minnesota, United States, 4th November 1946
[969] *The Tipton Daily Tribune*, Tipton, Indiana, United States, 11th November 1946
[970] *Dixon Evening Telegraph*, Dixon, Illinois, United States, 11th November 1946

The crossing was regarded as dangerous because of the high volume of both road and rail traffic over it. Coroner Wood decided that there was no need for an inquest. [971] [972] [973]

A brother and sister on their way home from school lost their lives when the school van in which they were riding was struck by a westbound Canadian Pacific Railway train at about 5.30 p.m. on Thursday 12th December 1946 at Pilot Mound, Manitoba, Canada. They were Margaret Pouffe, 15, and her brother Albert, 12, of Deleu. Their brother Wilfred, 10, suffered a broken arm and possible internal injuries while another brother, Edward, 7, escaped unhurt. [974]

The first of several post-war disastrous level crossing accidents in the United States involving a school bus occurred on the morning of Wednesday 18th December 1948 near Old Town, South Carolina. Eleven bus passengers (children) died in the accident, as did the bus driver, and another ten children were injured.

The train involved in the accident was Southern Railway eastbound first-class passenger train No. 18, consisting of engine No. 1361, two express cars, a baggage express car, a baggage mail car, a passenger baggage car, a coach, and a Pullman sleeping car, in charge of Conductor O.C. Crouch.

The bus was a 1935 Chevrolet with a composite body of metal and plastic wood lining, owned and operated by Richard Sanders, 41, who held a South Carolina special chauffeur's licence. The windscreen and side windows of the vehicle gave its driver an unrestricted view ahead and to the sides. It was carrying 21 passengers at the time of the accident.

The crossing was with what was described as an 'unimproved road' and the crossing itself was surfaced with a mixture of cinders, stone dust, and soil. The road crossed the line at right-angles. An advance sign 250 feet from the crossing consisted of vertical and horizontal lines intersecting at right-angles and the letters "R.R." in black on a white background. The road descended for a short distance and then climbed slightly to the crossing. Adjoining the crossing there were the standard crossbucks with signs marked "RAILROAD CROSSING", again in black letters on a white background. The usual whistle boards for trains were provided in advance of the crossing. The regulations required school buses to stop dead before attempting to cross the line. The average number of trains over the crossing was only 3.7 daily. During a 24-hour period beginning at 4 p.m. on 20th December 1946 the crossing had been used by 26 cars, 2 lorries and 4 trains.

The train had left Greenwood, the last open office, 27¾ miles short of the crossing, at 7.05 a.m., an hour and 35 minutes late. As it neared the crossing it was travelling at 45 miles an hour (the maximum authorised speed was 50 miles an hour). It had whistled on the approach to the crossing. The fireman was the first to see the bus and he shouted a warning to his engineer, who applied the brakes in emergency, but the speed of the train was not appreciably reduced before the collision, which occurred at about 8.10 a.m. The impact demolished the bus, while the locomotive was only slightly damaged and the train was not derailed. No one on the train was hurt.

---

[971] *The Ludington Daily News*, Ludington, Michigan, United States, 19th November 1946
[972] *South Haven Daily Tribune*, South Haven, Michigan, United States, 19th November 1946
[973] *Canaba Daily Press*, Escanaba, Michigan, United States, 20th November 1946
[974] *The Winnipeg Tribune*, Winnipeg, Manitoba, Canada, 13th December 1946

The Interstate Commerce Commission (ICC) stated that the bus driver was experienced and familiar with the route. He had picked up all of his pupils and was on the last leg of his run to the Consolidated High School at Silverstreet, 12 miles from Newberry. Newspaper accounts stated that the weather was foggy, with a low cloud ceiling and poor visibility, but the ICC report, on the other hand, stated that visibility had not been impaired by the weather conditions and the bus driver could have seen the train for a considerable distance. According to one of the surviving pupils Driver Sanders had not stopped the bus before entering the crossing, as he should have done. [975] It was theorised that he had probably not been expecting to meet a train, given that there were so few of them and that the one involved was running an hour and a half late.

Newspapers accounts added some details not to be found in the ICC report. 'Policeman John Livingston said the accident happened where a narrow, winding, unpaved road crosses the single track of the Southern. There are banks on each side of the road, but they are fairly low.

'W.J. Martin of the South Carolina Highway Patrol, returned to town from Silverstreet with reports that "bodies were scattered up and down those tracks for a thousand yards. It was the most awful sight I've ever seen."

'The children were in their last week of school before the Christmas holidays and were making plans to exchange gifts with classmates as the bus wound along the country road to Silverstreet. [976]

'Two girls and a boy were thrown clear of the bus. It was not immediately determined whether they suffered internal injuries, but the girls walked with O.C. Crouch, conductor of the train, to the home of R.E. Hanna, where Crouch put in a telephone call for ambulances.

'Four children living two miles from the crossing escaped being in the accident by missing the bus. Because of the bad weather they waited in their homes, planning to flag the bus, but it went by before they had time to stop it.' [977]

Another accident involving a train and a school van, in this case a horse-drawn vehicle, occurred near Bield, Manitoba, Canada, at 8.50 a.m. on the morning of Thursday 19th December 1946. The van was described as a canvas-sided waggon with a wooden top and was being drawn by two horses. The morning was foggy and as the van was negotiating a level crossing two and a half miles east of the consolidated school at Bield, which was its destination, it was struck by Canadian National Railways passenger train No. 9. Although the train was only moving at 10 miles an hour the van (described in one report as a sled) was dragged about 30 feet. Stella Titanich, aged 8, was killed almost instantly, and Elaine Sawicki, aged 6, died on the train taking her to hospital. Joe Horecy, aged 13, was seriously injured with a fractured skull but Billy Titanich, aged 7, brother of Stella, was only slightly hurt. The driver of the van, George Gunchowski, aged 67, had presumably failed to see the train due to the fog, as the view from the crossing was unobstructed in clear weather and there was a window in the canvas of the van on its right side, the direction from which the train came. He was fatally hurt in the accident and died in the King Edward Hospital at Kamsack the same evening. The horses escaped,

---

975 Interstate Commerce Commission report

976 The *Anniston Star*, Anniston, Alabama, United States, 18th December 1946

977 *The Robesonian*, Lumberton, North Carolina, United States, 18th December 1946

breaking loose, and running away in their fright. [978]

A westbound Pennsylvania Railroad New York – St Louis passenger train, *The Spirit of St Louis*, was running about five hours late when it collided with a car on Bloomington Street Crossing, Greencastle, Indiana, United States, shortly before 2 p.m. on Friday 17th January 1947. Mrs Charles Poe, her children Michael, 4½ and Terrill Gay, 6 months; her sister, Miss Martha Kennedy, 19; and a friend, Thomas Rainey, 19, all died. Mr Clarence Grimes, following the Poe automobile in his own vehicle, saw the passenger train nearing the crossing and expected the car in front of him to pull up. Instead, to his horror, it continued directly into the path of the train. Mr Grimes was struck by a piece of wreckage but not hurt. The car was reportedly carried over a mile down the line. Warning lights were provided at the crossing but it was thought that the attention of Mrs Poe had been distracted by an engine which was taking water on a side track adjoining the crossing and that she had thus failed to notice the passenger train. The party had been on their way home from Cloverdale, where Mrs Poe had formerly resided. [979]

A party of young people returning from a dance were the victims of a level crossing accident at 3.55 a.m. on Sunday 2nd February 1947 at Aplington, Iowa, United States. One member of the party, Bernice Ackerman, had already been dropped off at her home just before the accident. The remainder of the party were travelling along Highway 20, intending next to drop off Ernestine Cuvelier, 17, when they reached a crossing with the Illinois Central Railroad. Approaching the crossing at the same moment was train No. 11, *The Hawkeye Limited*, en route from Fort Dodge to Chicago, running an hour late and travelling, according to Engineer Claude Byeray, at its usual speed of 70 miles an hour.

The car was carried about 100 feet when it was struck by the train and its occupants were all killed. In addition to Miss Cuvelier they were Dorothy Huisman, aged about 21; Private Howard Sessler, 19, home on leave from the army; his brother Kenneth, 23; and Richard Strahorn, 23. The victims were identified by a passenger on the train, Robert McGreevy, who had intended to ride home with the party but had missed them.

Visibility at the crossing was partly obscured by a pile of cement blocks on one side and stockyards on the other. A boxcar was also standing on a side track near the crossing. Three women had died in an accident at the same crossing the previous summer. [980]

Five prominent citizens of Nashville, Tennessee, United States, lost their lives when the car in which they were riding was struck by a Nashville Chattanooga and St Louis Railway passenger train in a suburb of the city on the evening of Sunday 16th February 1947. The victims were Pat Estes, 73, retired general counsel and a founder of the Life and Casualty Insurance company; Johnson Bransford, 74, president and treasurer of the Bransford Realty company; his wife; Mrs Bransford's sister Mrs Julia Dake and Mrs Susan Warner.

Fireman C.W. Smith saw the car moving onto the crossing a moment before his engineer. Engineer L.F. Smith related, "I set the brakes immediately and opened the

---

978 *Winnipeg Free Press*, Winnipeg, Manitoba, Canada, 20th December 1946
979 *The Daily Banner*, Greencastle, Indiana, United States, 18th January 1947
980 *Mason City Globe-Gazette*, Mason City, Iowa, United States, 3rd February 1947

whistle. That was all I could do. Then we hit the car."

The train, inbound from Memphis, was travelling at about 60 miles an hour when it struck the car and it continued for half a mile before it could be pulled up. Witnesses said that the car appeared to hang in the air for a moment before it was thrown across an embankment, knocked down a signal tower and finally came to rest on the highway. Police described it as the most disastrous wreck of its kind for some years. [981]

About 30 section workers had finished their day's labours at Andry Siding, Indiana, United States, and had boarded a bus driven by Glenn Morris which would take them back to Michigan City on the afternoon of Monday 17th February 1947. The bus had barely started on its journey when it was struck by Chicago South Shore and South Bend Railroad passenger train No. 26, from South Shore to Chicago, on a level crossing. The bus was cut in two, its front half being carried down the line and its rear half left at the crossing. Bodies were scattered along the track for 200 yards. The bus driver and 12 of the workers were killed and 14 other workers were injured. The train did not entirely escape either, as windows in its leading car were smashed. Its motorman and one passenger were slightly hurt.

John Steinhagen, a truck driver employed by the railroad, said that he had been working with the section gang a quarter mile down the track from the crossing. The bus had been parked on the north side of the crossing and his vehicle on the south side. As he had been walking to his truck, he stated afterwards, he turned around and had waved to the bus driver to indicate that a train was coming. Turning around again he had heard a crash and had seen several bodies flying through the air.

Motorman W.E. Willard commented that he had seen the bus stop at the crossing, which had an unobstructed view in both directions, but then the bus had slowly moved forward into the path of his train. His train had been travelling at 65 to 70 miles an hour. [982] [983]

A party travelling from Clio, Michigan, to celebrate a 61st wedding anniversary met their deaths on a level crossing at Cardington, Ohio, United States, when the car in which they were riding was struck by a New York Central Railroad train on Sunday 2nd March 1947. Those who died at the scene were Mrs Ethel Peterson, 55; her daughters Mrs Wilma Pashby, 24, and Miss Bonnie Peterson, 19; Mrs Pashby's baby, born prematurely after the accident; and Lawrence Pashby, brother-in-law of Mrs Pashby, 25. Another occupant of the car, Miss Norine Conen, 16, died later of her injuries in Marion City Hospital, while Donald Pashby, 22, husband of Mrs Pashby, was seriously injured. The party had been en route to Waldo, Ohio, to the home of Mrs Peterson's parents, Mr and Mrs Elmer Kramer. Mrs Peterson's husband Donald and other relatives arrived at the scene of the tragedy ten minutes later in a second car. [984]

In early March 1947 businessmen at West Frankfort, Illinois, United States were

---

[981] *The Decatur Daily*, Decatur, Alabama, United States, 17th February 1947

[982] *The Evening News*, Jeffersonville, Indiana, United States, 18th February 1947

[983] *The Dubois County Daily Herald*, Jasper, Indiana, United States, 18th February 1947

[984] *The Evening Independent*, Massillon, Ohio, United States, 3rd March 1947

petitioning the Interstate Commerce Commission demanding the elimination of two adjoining level crossings, either by the relocation of the road or the construction of an over or underbridge. Less than a fortnight after their meeting nine young people would lose their lives on one of the level crossings concerned, where Route 110 crossed the line of the Chicago Burlington and Quincy Railroad.

At 2.30 a.m. on Sunday 16th March 1947 a car containing the victims of the accident was returning from a nightclub at Zeigler when it was struck by a switch engine. Those who died were Golda Arnett, 19, her sister Wilma, 20; Samuel Coffel, 21; John Gaddis, 22; Ronald Harris, 26; Howard Niblett; Mike Puskar, 23; Thelma Vineyard, 18; and Peggy Williams, 18. All but Wilma Arnett were killed outright; she died of her injuries en route to hospital a few hours later. John Gaddis had arrived at Zeigler by other means but had begged a lift home with the party, while Thelma Vineyard would not have been in the car had she not had a row with her boyfriend, who left the nightclub a few minutes later and arrived at the crossing to find the accident. He told a reporter, "Thelma's body was the first I saw when I got to the wreck. She had asked me to take her out Saturday night but I thought I'd teach her a lesson. I'm the one who's sorry now."

'Leroy Sloan, West Frankfort, said that he arrived at the crossing a couple of minutes after the crash and was flagged to a stop by one of the engine crew who asked him to help get the bodies out of the wreckage. He said that he parked his automobile and went down the track to find the crumpled automobile wedged under the coal tender, bodies mingling with the debris. "It's a sight I never want to see again as long as I live," Mr Sloan said, "I got home after 4 o'clock and couldn't sleep at all."'

Engineer C.R. Guthrie said that the car had been travelling fast and its driver had apparently not seen the train at all until it was too late.

Fireman F.J. King added that it appeared to him that the car driver had first attempted to stop and then, realising that he would not be able to, accelerated in an attempt to get across the line in front of the train. He put the speed of the car at about 60 miles an hour, while the engine was only moving at 12 or 15 miles an hour.

Engineer Guthrie agreed as to the speed of the engine and said that after the collision he and the fireman had run to a telephone 200 feet away and they had called their office for assistance. Then they had flagged down cars. He remembered hearing a terrific noise but not feeling any impact. Part of the automobile was entangled in the engine and it was dragged about 50 feet.

State Highway Patrolman Peter Carnahgi stated that there was no evidence that any of the young people had been drinking at the nightclub.

Road and railway crossed one another at an angle. At the time of the accident the only protection at the crossing was standard warning notices, though as a result of the aforementioned meeting with local businessmen it had been agreed to add flashing warning lights. There had been previous fatal accidents at the crossing in 1937, 1942 and 1944. [985]

A bride, groom and four members of a wedding party were killed only two hours after their wedding when the car they were riding in was hit by a Pennsylvania Railroad Holiday Special en route to Chicago on a level crossing at Munster, Indiana, United States, on Friday 30th April 1947. They were Mrs Juanita Graham, 17, the bride; her husband Chelso,

---

[985] *The Daily Independent*, Murphysboro, Illinois, United States, 17th March 1947

20; the bride's parents, Mr and Mrs Raymond Brock; her sister Greta, 15, And her brother Kenneth, 5, all from Highland, Indiana. They had been on their way to a photographic studio to have their wedding photographs made. The car was carried 400 yards down the line before the train could stop. All but the groom were killed outright.

Other members of the wedding party who were following witnessed the collision. They said that a low building and a brickyard obscured the view of the line at the crossing.

Fireman P.T. Parker said that he had seen the automobile coming onto the crossing from his side of the engine of the train but only when it was too late to do anything about it.[986]

Only ten miles south of West Frankfort, the scene of the March tragedy, there was a level crossing collision at Herrin, Illinois, United States, on Sunday 24th August 1947 in which five young people died. A Chicago Burlington and Quincy Railroad freight train was again involved. The northbound train struck the car at about 5 p.m. Engineer J.W. Malleck could not see the car from his side of the engine before he hit it but Fireman R.J. Alexander and Brakeman Robert Carpenter testified at the inquest on the victims that as the car had neared the crossing it had become clear that it was not going to stop. They had then both shouted to their engineer but he had been unable to hear them above the sound of the engine whistle and bell.

Those who lost their lives were Clyde Cobb, 20, of Johnston City; Paul Hale, 16, of Herrin; Joseph Miller, 21, of Johnston City (the driver of the car); and Dante Underwood, 14, and Dale Yantz, 21, both of Herrin.

As a result of the collision the cowcatcher of the engine was knocked off and its braking system was damaged. The train, pushing the wreckage of the car, came to a halt half a mile from the crossing. All the occupants of the automobile were dead when police removed them from the wreckage 30 minutes after the collision.[987] [988]

The same day as the Herrin accident, Sunday 24th August 1947, ten people died in what police termed 'the worst auto accident in Colorado's history.' Those who lost their lives were nine members of the same family and a relative. The Colorado and Southern Railway local train from Denver to Colorado Springs and Pueblo travelled three-quarters of a mile after striking the car at Littleton, Colorado, United States.

Those who died were Harold Rathburn, 40; his wife Nellie, also 40; their children Amarylis Ann, Bonna Beth, Harold Kent, Loyce, Marjo, Modine and William Albert, aged between four and 16, all from Vesper, Kansas; and relative Earl Strange, 42, from Lincoln, Nebraska.

Mrs Bartlett Bankston, of Denver, an aunt of one of the victims, heard news of the accident on the radio and was able to identify them. She said that the party had been touring Colorado.

The train crew testified that the car had first appeared to slow down for the crossing, then had speeded up, so presumably it was yet another case of the car driver thinking that

---

[986] *The Daily Register*, Harrisburg, Illinois, United States, 31st May 1947

[987] *The Daily Register*, Harrisburg, Illinois, United States, 25th August 1947

[988] *The Daily Independent*, Murphysboro, Illinois, United States, 26th August 1947

he could outrace the train.

County Coroner Ivan Joss stated that it was clear from the initial investigation that no blame attached to Engineer C.E. Roper and that he did not intend to hold an inquest into the accident. [989] [990] [991]

Although the war with Germany had ended in May 1945 the last prisoners of war (POWs) were not repatriated from Great Britain until November 1948 for several reasons, including the wish to repatriate all the Italian POWs first, the need for labour for agriculture and building reconstruction and the collapse of Germany's economy. [992] Thus it was that ten German POWs lost their lives in a level crossing accident in a foreign country on Wednesday 17th September 1947.

The accident happened on the London and North Eastern Railway at Burton Agnes, England, on the Hull – Scarborough line. The train involved was the 5.55 a.m. Down passenger train from Hull Paragon to Bridlington consisting of 4-4-0 tender steam locomotive No. 2772, a four-wheeled van and five bogie coaches, with Driver Charles Stephenson, Fireman Kenneth Robinson, and an unnamed guard. It was not booked to stop at Burton Agnes and was travelling at about 55 miles an hour when it reached the crossing at 6.42 a.m.

The Army lorry was a four-wheeled Bedford three-tonner with a canvas-covered body. It had Lockheed hydraulic and vacuum servo brakes and had had a routine service at the Royal Army Service Corps workshops at York the day before the accident. A test drive after the service found no problems with the brakes, nor was there anything found to be wrong with them when the wreckage of the lorry was examined after the accident.

The level crossing at Burton Agnes station was placed between its staggered platforms. Four wooden crossing gates were provided, worked by a wheel in the adjoining signal box. The gates were dirty white in colour, each had a square red target and one gate on each side had a lamp. Two hundred yards from the crossing, around a corner, a sign read, "Gated Level Crossing." Around the corner the crossing gates were easily visible from a distance of 105 yards. The road approaching the crossing was more or less level.

The work of harvesting in the area having been completed, 50 prisoners from No. 250 POW Camp were being moved to Billingham to do work there, [993] taking the 7.09 a.m. train from Burton Agnes, and some of them were already at the station. The lorry had already take the POWs' baggage to the station and was taking the second batch of 25 prisoners there when the accident occurred. On this trip there were three people in the cab of the lorry, Staff Sergeant James Wadey, acting as interpreter, who was driving, Sergeant Roland Cramer, in charge of the escort party, and Hans Graf, a prisoner. Warrant Officer Lee told the inquiry into the accident that Wadey had no authority to drive the lorry as he only had a motorcycle licence; Graf was the authorised driver. Lee said that he had never heard of Wadey driving the lorry previously but a prisoner said that Wadey had done so occasionally.

---

[989] *Joplin News Herald*, Joplin, Missouri, United States, 25th August 1947

[990] Mobberly Monitor Index, Mobberly, Missouri, United States, 25th August 1947

[991] *Neosho Daily Democrat*, Neosho, Missouri, United States, 25th August 1947

[992] Wikipedia – German prisoners of war in the United Kingdom

[993] *The Times*, London, England, 18th September 1947

According to Private Adams, prisoners Hoermann and Reichenbach waiting at the station and prisoners Schlupper and Jungblut who were riding on the lorry, Graf had driven the lorry to the station and back on the first trip and to the station on the second, when Wadey had taken over the driving. Schlupper and Jungblut said that Wadey had driven unusually fast to the station on the third trip, so that the prisoners in the back of the lorry had been thrown about. Adams and Reichenbach saw the lorry arrive at the station, at an usual speed, hardly slowing down and striking the gates at 15 to 20 miles an hour according to Adams. Adams, Schlupper, Hoermann and Reichenbach all spoke of the engine of the lorry racing at the last moment, as if the driver was changing down to a lower gear. Porter Campbell John at the station and Signalman Miles Gray in his box said much the same thing.

Signalman Gray stated that he had closed the gates at 6.35 a.m. for an Up train and they had not subsequently been opened.

Driver Stephenson and Fireman Robinson said that their train had already been alongside the Down platform when the lorry had broken through the gates, far too late for them to do anything.

A motorcyclist, Stanley MacKenzie, also witnessed the engine of the lorry racing as it broke through the gates. He said there had been no skidding and no brake application.

The train struck the bonnet of the lorry and dragged it along the line, forcing it against the signal box. 'Wreckage was strewn along the line for 200 yards as the lorry was smashed to pieces, and the brickwork of the signal cabin was badly damaged and fencing along the side of the station torn down.

'Police and doctors were sent at once from Bridlington, with ambulances. In the meantime other German prisoners waiting on the platform and the British escort carried the injured into the station waiting room. The seriously injured were taken to hospitals at Bridlington and Driffield, and others with lesser injuries to the camp hospital.' Seven of the prisoners and the two British soldiers were killed outright, ten prisoners were seriously injured, three of whom died in hospital, and the other nine were slightly hurt. The train was not derailed and only suffered superficial damage; no one on it was hurt.

Lieut. Colonel E Woodhouse, who conducted the Ministry of Transport inquiry, stated that there could be no criticism of the railway arrangements at the crossing. The accident, which could have been even worse had the train been derailed, was entirely due to the careless handling of the lorry by the unauthorised and apparently inexperienced Staff Sergeant Wadey. He noted that the arrangement of the control pedals in the lorry was such that an inexperienced driver such as Wadey might easily let his foot slip from the brake to the accelerator, which would account for the racing of the engine described by the witnesses, especially if the lorry had also been declutched. Wadey was also several inches taller than Graf and his uncomfortable position in the driving seat might also have made his mistake more likely. [994] [995]

Railbus N 9153 collided with a lorry on a level crossing between Marktredwitz and Schirnding, Germany, on Saturday 20th September 1947. Five people died and one was

[994] Ministry of Transport report
[995] *The Burton Agnes Disaster* by Richard M. Jones

injured, either on the train or the lorry or both. The barriers had been left open. [996]

Five men died and another was seriously injured when their car was struck by a train on a level crossing near Port Germain, South Australia, on Friday 10th October 1947. The men were James Baker, a painting contractor and five of his employees and they had been returning to their homes in Whyalla. The body of one victim was carried 200 yards. [997]

Police blamed the lack of any form of protection at a level crossing on the outskirts of Muncie, Indiana, United States, for a collision which caused the death of six people on a level crossing with the New York Chicago and St Louis Railroad early on Sunday 12th October 1947. The occupants of the car had attended a football match and dance at Ball State Teachers College in Muncie and were on their way home to Anderson, Indiana, at the time of the accident.

The victims were Mr Joseph Jackson, 43; his 17-year-old son Richard; James Huff, 17; Florene Saunders, 19; Beatrice Thomas, 22; and Marilyn Wilson, 18.

The light sedan was struck at 1.18 a.m. by a St Louis – Cleveland passenger train, which was travelling at about 65 miles an hour, and carried about 100 feet. Police theorised that the elder Mr Jackson, who had been driving the automobile, had been unable to see the train because his vision had been obscured by several coal cars standing on a siding. [998] [999]

A Trafikförvaltingen Göteborg-Dalarna-Gävle passenger train from Göteborg to Kil collided with a bus on a level crossing at Edsvalla, Sweden at about 08:00 on Tuesday 14th October 1947. There was a dense fog at the time with visibility reduced to four or five metres. When the bus arrived at the open level crossing its driver asked his passengers to help him see if it was safe to cross but as the front wheels of the bus got onto the crossing the driver noticed the faint glow of a train's headlights. Unable to judge in the fog exactly how far away the train was, the bus driver nevertheless attempted to cross in front of it. He only made it halfway across when the bus was struck by the train, which was travelling at 90 kilometres an hour, and cut in two. Fourteen people died and another seven were injured. After the accident the train driver alighted and gave what assistance he could before taking his train on to its destination.

There was subsequently a court case in which the train driver, rather than the bus driver, was the accused, it being alleged that he had failed to whistle for the crossing. Witnesses testified that he had, and he was acquitted. [1000]

Sunday 4th January 1948 saw a tragedy at Archbold, Ohio, United States. A considerable quantity of snow had fallen on Ohio and farmer Jesse Wyse arranged to take a party of children out for a ride on a large bobsled pulled by a tractor. Originally the trip had been planned for the Saturday but it was postponed until the next day. The farmer's

---

[996] *Katastrophen der deutschen Bahnen Teil II* by H.J. Ritzau pg. 240 (in German)
[997] *The Canberra Times*, Canberra, Australia Capital Territory, Australia, 11th October 1947
[998] *The Vidette-Messenger*, Valparaiso, Indiana, United States, 13th October 1947
[999] *The Washington Herald*, Washington, Indiana, United States, 13th October 1947
[1000] Wikipedia – Järnvägsolyckan I Edsvolla

son, John, aged 13, wanted to take the sled out himself but his father decided that he was too young and inexperienced to do so.

As the tractor and sled reached a level crossing with the New York Central Railroad at 1.30 p.m. the farmer failed to notice the approaching train. He related afterwards, "I started up the tractor again and just as it was clearing the tracks I saw the train bearing down on the sled. It was horrible." Explaining the accident to another reporter the farmer said, "I pulled up to the tracks and made my safety stop. A car was following me. It had some boys in it. There were some snowballs thrown like kids will do. I thought I'd better be moving on before some of the kid's glasses were broken. I looked back to see if the children were all right. When I turned around again there was the train. I stiffened up but gave it the gas. I didn't make it." [1001]

Approaching the crossing and travelling reportedly at 80 miles an hour was the *Chicagoan* express passenger train. Seeing the tractor and bobsled Engineer D.O. Scherer frantically sounded his whistle and applied his brakes but could not avoid the collision. Conductor R.L. Snook said that two of the children had jumped from the sled before it was struck – one of them was John Wyse, who escaped unhurt.

Ten children lost their lives in the accident, and they belonged to only three families, all neighbours of Farmer Wyse. They were four children of Mr and Mrs Lawrence Grieser – twins Gayle and Charlene, 13; Patricia, 8 and Sharon, 6; four children of Mr and Mrs Raymond Grieser – Donald 13; Marlin, 11; Arlene, 10, and Janet, 4; and two daughters of Mr and Mrs Lyle Osborne – Jean, 12, and Marjorie, 16.

Shirley Grieser, daughter of Lawrence, suffered a broken leg. She described the accident from Wauseon hospital. "We didn't see the train because we were having so much fun. All of a sudden we looked up – and there the train was, big as everything and whistling like mad. I was so scared I didn't know what to do, so I just sat there. Janice Wyse {Farmer Wyse's daughter, aged 7} and I were sitting up front in the bobsled. The next thing I knew we were flying right through the air." Janice Wyse was in the same hospital with two broken legs. [1002]

At the inquest on the victims Engineer Scherer stated that he had begun whistling earlier than was required whilst Farmer Wyse admitted that he had not looked to see if a train was coming before he had entered the crossing. He said that he had not heard the whistle until the last moment. A verdict of 'Accidental Death' was returned by Coroner Milton Warner. [1003]

A Société nationale des chemins de fer vicinaux tram which had left Namur at 08:10 for Andenne was in collision with a freight train bound for Dinant on a level crossing at Jambe, Belgium, on Thursday 4th March 1948. The train also struck a lorry. The tram and the lorry were stopped at the lowered crossing barrier. A passenger train had passed, whereupon the crossing keeper had raised the barrier despite the freight train approaching from the opposite direction, possibly because it was closer than he had realised. The tram began to cross and was in the middle of the track when the engine of the freight train emerged out of the fog. The motor unit at the front of the tram escaped damage but the following passenger car was smashed to pieces, while the lorry was thrown violently

---

[1001] *The Evening Tribune*, Marysville, Ohio, United States, 6th January 1948
[1002] *Vincennes Sun-Commercial*, Vincennes, Indiana, United States, 5th January 1948
[1003] *The Marion Star*, Marion, Ohio, United States, 15th January 1948

against the crossing keeper's cabin. Barrier, tramcar, and lorry were transformed into a pile of scrap metal. There were pieces of broken wood, bundles of yeast (from the load of the lorry?) and shreds of bloodstained clothing everywhere, while the body of one victim was hanging from the tender of the engine.

Assistance was immediately on hand and a representative of the parquet (public prosecutor) was soon on the scene while Pères Oblats (Catholic priests) gave what religious comfort they could to the dying. The road was closed with traffic being diverted over the nearby Velaine level crossing. A dozen injured were transported to the Clinique Ste Elisabeth and the bodies of the dead were taken to the Parc Reine Astrid. Governor Gruslin arrived at the site at 10:00 and conferred with the police and Namur officials of the Société nationale des chemins de fer belges (the railway company).

Nine people lost their lives in the accident; among them was M. Boinon, the conductor of the tram. All the casualties were on the tram; the lorry driver had escaped death or injury by jumping from his moving vehicle when he saw that the train was close at hand.

The driver of the freight train apparently did not see the tram on the line until the last moment. The sound of the collision was heard all over Jambre and news of the accident caused consternation throughout the area. [1004]

A Ferrocarril Nacional de México passenger train bound for Veracruz collided with a bus near San Lucas Xolox, Mexico, on Tuesday 30th March 1948 at a level crossing with the main highway to Veracruz. The train driver stated that he had begun whistling a kilometre short of the crossing but the bus had kept going, implying that either the bus driver was trying to beat the train or else that he was fatally inattentive. The bus driver and 14 or 15 bus passengers died and 18 more of the occupants of the bus were seriously injured. [1005] [1006]

A goods train collided with a bus on a level crossing at Mount Isa, Queensland, Australia, on Friday 2nd April 1948. Nine bus passengers were injured and were taken by ambulance to Mount Isa hospital where four of them died shortly after admission. They were Mrs Euphemia Meares, 21, a recent immigrant from Scotland; her daughter Norma Regina, 3; Victoria Herbert, 46, and Owen Stapleton, 2. Owen's mother, Mrs Elizabeth Stapleton, 22, became the fifth fatality when she died in hospital of her injuries a couple of days later.

The bus driver, Charles Reid, escaped with abrasions and shock and was not admitted to hospital. He told police that he had seen the train when it was 100 yards away. He said that he had been travelling slowly because of the rough road and accelerated when he had seen the train but had not been able to get the heavy vehicle clear in time. The train caught the last three feet of the bus, dragging it 20 feet. Crushed against the gateposts the bus was described as having had its framework splintered to matchwood. [1007] [1008]

[1004] *Vie Mosane*, Andenne, Belgium 5th to 11th March 1948

[1005] *The Montana Standard*, Butte Anaconda, Montana, United States, 31st March 1948

[1006] *The Bakersfield Californian*, Bakersfield, California, United States, 31st March 1948

[1007] *The Courier Mail*, Brisbane, Australia, 3rd April 1948

[1008] *The Northern Miner*, Charters Towers, Queensland, Australia, 6th April 1948

The bus driver was subsequently charged with unlawfully killing the five victims of the accident but after hearing the evidence of ten witnesses the Stipendiary Magistrate, Mr O'Connor, dismissed the charge, ruling that though there had been some negligence it was not sufficient to warrant committing Reid to trial. [1009]

The General Manager of Martin's Light Railways stated that six people had died and seven others had been seriously injured in a collision between Arrah and Sasaram, India, involving a passenger train on the narrow (2 ft 6 in) railway and a loaded lorry on Wednesday 14th April 1948. The brief report did not state if the casualties had been on the train or the lorry. [1010]

Nine people died and another 45 were injured when an express train collided with an American government lorry on a level crossing east of Patras, Greece, on Saturday 24th April 1948. The engine and two carriages of the train were derailed and overturned and the dead and injured, who included 16 Greek soldiers, were all on the train. The uninjured lorry driver was arrested. [1011]

A new car owned by Vern Brown, 32, was on its way from a grocery store he had recently purchased at Scotts, Michigan, to a farm he also owned near Bloomingdale, Michigan, on Tuesday 8th June 1948. Accompanying Mr Brown were his wife Marie, 24, who was eight months pregnant, their children Verna, 9, and Bonnie Lou, 10 months; and three children of some Scotts' neighbours, Mr and Mrs William Denton – Tommy, 10; James, 9; and Jean, 3. The Denton children were presumably coming along for the ride. As the car was approaching the Kilgore Road Crossing with the Pennsylvania Railroad, on the outskirts of Kalamazoo, Michigan, United States, a northbound train, consisting of an engine and a caboose, was also approaching the crossing. Engineer W.R. Bets, observing the Browns' car, sounded his whistle several times but the car failed to stop. Skid marks suggested that Mr Brown had belatedly seen the train and applied his brakes, but by the time they had begun to take effect his automobile was only 50 feet from the crossing. Although it was cloudy at the time of the accident it was not raining and the road was dry. The car was caught broadside by the engine and smashed to pieces. Five of its occupants were killed outright. Mrs Brown was still living when she arrived at the hospital but she died three minutes later. Doctors found that her unborn child had no heartbeat so they did not attempt a Caesarean delivery from its dead mother. James Denton, the only survivor of the accident, was said to be in a serious condition in Bronson Hospital. [1012] [1013]

The *Ocean Limited* express of the Canadian National Railways collided with an automobile on a level crossing just west of Sayabec, Quebec, Canada, on Sunday 12th September 1948. Six people in the car lost their lives, five of them members of the same family. [1014]

[1009] *Townsville Daily Bulletin*, Townsville, Queensland, Australia, 14th April 1948
[1010] *The Indian Express*, Madras, India, 17th April 1948
[1011] *The Sunday Star*, Washington, D.C., United States, 25th April 1948
[1012] *South Haven Daily Tribune*, South Haven, Michigan, United States, 8th June 1948
[1013] *The News-Palladium*, Benton Harbor, Michigan, United States, 8th June 1948
[1014] *The Brandon Daily Sun*, Brandon, Manitoba, Canada, 13th September 1948

Two families from Des Moines, Iowa, who were neighbours, the Isermans and the Kauzlarichs, set off together in one car on a fishing trip on Sunday 19th September 1948, a trip which was to end in tragedy at Avon Lake, Iowa, United States, 6½ miles south of Des Moines. Shortly after 12 noon the car carrying the party arrived at a level crossing with the Chicago Burlington and Quincy Railroad at the same time as an Ottumwa – Des Moines motor passenger train, travelling at about 30 miles an hour. Engineer C.C. Coffin said later that he had pulled the whistle cord until it had broken and applied the brakes but had been unable to avoid hitting the car. He thought that its driver, Anton Kauzlarich, 29, had failed to notice his train.

Mrs Lorene Kauzlarich, 29, the Kauzlarichs' son David, 7; Mr Milton Iserman, 36, his wife Sandra, 34, and their daughter Dorothy, 11, all lost their lives. Anton Kauzlarich and Doris Iserman, 9, were injured; the former was suffering mainly from shock but the Isermans' daughter was said to be unconscious and in a critical condition in the Mercy Hospital at Des Moines after the accident. [1015] [1016]

Gettysburg Avenue Crossing, on the west edge of Dayton, Ohio, United States, was the scene of a tragedy which could apparently easily have been avoided on the evening of Thursday 30th September 1948. Mr Ollie Patrick, 51, his wife Carille and their children Mary, 16; Shirley Belle, 10; Michael, 5; and Roma Jean, 3, were returning home from a visit to Mrs Patrick's mother, Mrs Mabel Rose. When they reached the intersection with the Pennsylvania Railroad their car stalled on the unprotected and unlit crossing.

Engineer R.C. Essert of the Columbus – Chicago train saw the stationary car from a quarter mile back and sounded his whistle. He related later, "A little girl got out, looked at the train, then got back in. All could have gotten out." The car was wrecked and carried almost a quarter mile. Mrs Patrick, Mary, and Michael were killed outright and Mr Patrick and Shirley Belle died in hospital early the next morning. Roma Jean was critically injured. There would have been time for all the occupants of the car to have got out but as often happens in such situations they doubtless panicked, while Mr Patrick would have been anxious to try and save his car. [1017] [1018]

County Auditor Carl Bauer stated that the crossing was "one of the worst in the district," and that county officials had offered to share with the railroad company the cost of installing flashing lights there, though that would not, of course, have prevented the accident just described. [1019]

A 1936 automobile from Alberta province carrying 12 people struck by a Canadian Pacific Railway freight train on Erickson Crossing, two and a half miles south-east of Creston, British Columbia, Canada, at about 1.30 p.m. on Sunday 17th October 1948. All the occupants of the car were thrown from it, some as far as 60 feet, and five of them were killed. Those who died were Nicholas Gorgiehuck, 33; Larry Morozoff, 15; Michael Nesplah, 23; Peter Olynyk, 23; and William Olynyk, 50. The other seven people in the

---

1015 *The Telegraph-Herald*, Dubuque, Iowa, 20th September 1948

1016 *The Charleston Daily Mail*, Charleston, West Virginia, United States, 20th September 1948

1017 *The Evening Tribune*, Marysville, Ohio, United States, 1st October 1948

1018 *The Middletown Journal*, Middletown, Ohio, United States, 1st October 1948

1019 *The Lima News*, Lima, Ohio, United States, 1st October 1948 car were all injured and were treated by doctors brought to the scene by private cars before being taken to hospital. Driver Alex Sedar, in charge of the automobile, which had been heading northwards towards Creston, could give no clear account of what had happened, but the view of trains at the crossing was obscured by a high bank which masked the view of trains from road users. The car had almost cleared the crossing when its rear was hit by the train and it was not badly damaged. Engineer Cope immediately stopped the train and the train crew went to give first aid to the victims. No one on the train was hurt. [1020]

Another major level crossing accident on Sunday 17th October 1948 happened in the United States. Mr Albert Freel, 22, was driving a car containing his father, Matthew, 69; his brother Lewis, 19; and two nephews, George, and Wayne Broda, aged 11 and 9 respectively. Following in a pickup truck were Mr Matthew Freel's wife, his daughter and his son-in-law, Mr and Mrs Paul Broda. The party were on their way from South Bend to visit a daughter of the older Mr Freel, Mrs Loren VanDeusen, who lived at Dunn's Bridge. Three miles east of Kouts, Indiana, the vehicles had to negotiate a level crossing with the Pennsylvania Railroad and it was here that tragedy struck. As the first car, a 1937 Ford coach, reached the crossing at 2.24 p.m. it was struck by eastbound train No. 208. Engineer Frank Messersmith estimated the speed of his diesel-hauled train at 70 miles an hour and the car was carried almost a mile down the line, wedged beneath the locomotive, as far as Grasmere Crossing. The bodies of four of the victims remained in the wreckage, the fifth was found alongside the line.

Mr Broda, who was driving the pickup truck, was following some little distance behind and did not witness the actual collision, but as he turned off Indiana Highway 8 onto the road leading to the crossing he saw the engine of the train going by with a car attached to its front, which he at once recognised as his brother-in-law's, and he made for Grasmere Crossing.

A pair of crossbucks was the sole protection provided at the crossing, while a thicket of trees obscured the vision of road users.

Deputy Coroner Herman Kosanke soon arrived at the scene. Surviving family members were understandably too upset to give statements but he interviewed the train crew – Engineer Messersmith, Fireman Carvel Parker, and Conductor C.C. Newhard. The train was delayed for an hour and a half whilst the wreckage of the car was cut free from the front of the locomotive. Speaking of the dangers of such unprotected crossings the coroner remarked that, "There are too many dangerous crossings of that type almost every mile between the east and west boundaries of Porter County." [1021]

Six people died and three more were seriously injured in a collision between a station waggon and a Chicago Burlington and Quincy Railroad coal train on Highway 60 southeast of Des Moines, Iowa, United States, on the evening of Saturday 6th November 1948. Those who died were Harry Bollard, 36, the owner of the vehicle and presumably its driver; his wife Grace, 30; their children Marilyn, 12; Elaine, 9; and Bob, 4; and Bob Miller, 3. Seriously hurt were Bud and Bea Miller, aged 28 and 24 respectively, the parents of Bob, and their daughter Joanne, 4. The Bullards were from Dows, Iowa, and

---

[1020] *The Lethbridge Herald*, Lethbridge, Alberta, Canada, 18th October 1948

[1021] *The Vidette-Messenger*, Valparaiso, Indiana, United States, 18th October 1948 the Millers were from Bradford, Iowa. Three of the victims were killed outright and the other three passed away shortly afterwards at the scene, the last dying in the arms of a woman passer-by who had stopped to help. Five ambulances took the victims to Broadlawns General Hospital in Des Moines.

Engineer E.S. McIntosh told highway patrolmen, "The car was going pretty fast. It tried to make it across the crossing but just didn't get across in time."

The vehicle was knocked down a 25ft embankment and landed on its roof. [1022]

The accident happened only two miles from the scene of the Avon Lake tragedy of 19th September (see above).

A Manistee and Northeastern Railroad freight train collided with a school bus on Cherry Bend Crossing, near Greilickville, Michigan, United States, on the afternoon of Tuesday 16th November 1948. The bus was being driven by Fred Smith, aged 36, and was bringing seven young children home from first grade and kindergarten classes at Norris school in Greilickville. Five of the children and the driver escaped with small cuts but Jean Couturier and Barbara Crawford, both aged six, were thrown through the door of the bus as it was spun round. Jean was fatally injured and died after admission to Munson hospital whilst Barbara had to undergo amputation of a badly crushed leg and was reportedly critically hurt. What sort of protection, if any, was provided at the crossing was not stated. [1023]

By the beginning of December 1948 the civil war between the Chinese Nationalists and the Communist forces was reaching its final stages, with city after city falling in quick succession to the Communists. By May of the following year well over a million people had fled China via Shanghai. At least two ships which were grossly overcrowded with refugees sank during their mass flight, leading to the reported loss of over 9,000 lives. On a much smaller scale, on about Saturday 4th December 1948 a bus carrying refugees was struck by a train at Wusih, China, 90 miles west of Shanghai, presumably on a level crossing, with the loss of eight lives. [1024]

Railbus N 8034 collided with a lorry on an unprotected level crossing between Kronach and Nordhalben, Germany, on Tuesday 14th December 1948. Five people died and 24 were injured on the train, lorry or both. [1025]

Five people died in a level crossing accident at Cavan Crossing, South Australia, Australia, about 10 miles from Adelaide, on Sunday 19th December 1948. Those who lost their lives were Mrs Edith Wyatt, 54, of Penfield, South Australia; her daughters Emily, 23; Dulcie, 18; and Thelma, 16; and Hurtle Rosewarne, 53, a relative of Payneham. Mrs Wyatt's husband, George, 61, a tanner and the driver of the tourer car, suffered slight injuries and shock and was admitted to the Royal Adelaide Hospital. The family had been on their way to a Christmas reunion and police later had to inform 50 relatives. The

---

[1022] *Waterloo Sunday Courier*, Waterloo, Iowa, United States, 7th November 1948
[1023] *The News-Palladium*, Benton Harbor, Michigan, United States, 17th November 1948
[1024] *Clovis News-Journal*, Clovis, New Mexico, United States, 6th December 1948
[1025] *Katastrophen der deutschen Bahnen Teil II* by H.J. Ritzau pg. 241 (in German)

engine, running tender or tank first, carried the car along the track and smashed it to pieces. It was thought that Mr Wyatt might have been distracted by a herd of cattle which were being driven along the road in front of him. [1026]

Argentina suffered a disastrous level crossing collision on Sunday 19th December 1948. A Ferrocarril Central Argentino train collided with a bus carrying 45 picnickers. The bus was proceeding along Roque Sáenz Peña street towards Barracas station in the San Isidro suburb of Buenos Aires. As the bus came down a hill to the crossing the barriers were lowered for a train and the bus driver applied his brakes. These failed to work and the bus crashed through the crossing barrier and onto the line just as an electric train was approaching at speed. At least 41 of the 55 passengers on the bus were killed outright or fatally injured and most of the others were hurt. The train driver, Enrique Carlos Loyet, was also injured.

It further emerged that a little earlier the bus had almost come into collision with a car. The bus driver, Francisco Erasmo Venti, had then climbed out of his cab and declared that he would not drive the vehicle any further as its brakes were defective. Upon hearing this a taxi driver, Fernando Barriti, who was one of the passengers on the bus, had jumped up and seizing the steering wheel had said to Venti, "I'll show you how to drive." He had started the bus moving again and the next thing was it had plunged forward and demolished the crossing barrier and come onto the line immediately in front of a crowded train from Buenos Aires Retiro to Tigre. [1027]

By 23rd December the number of fatalities in the accident had risen to 44. [1028]

A diesel-powered Chicago Burlington and Quincy Railroad passenger train No. 6, bound for Chicago and travelling at 80 miles an hour collided with a sedan just west of the station at Wiggins, Colorado, United States at 8.45 p.m. on Friday 31st December 1948, killing all of the people in the sedan. The victims were three brothers, Everett, Francis, and Kenneth Sharp, aged 18, 22, and 19 respectively; Bruce Harris, 22, and a man tentatively identified at Patrick O'Leary. O'Leary was the owner and probably the driver of the car. All were from Fort Morgan, Colorado. The car was carried some 300ft to the station and all the victims were dead when rescuers reached the scene. [1029]

Recklessness on the part of the driver of a Jeep appears to have been the cause of a level crossing accident near the Maiko Kan Hotel, 23 miles from Nagoya, Japan, on Sunday 23rd January 1949 which resulted in the death of five Americans of the US Fifth Air Force and one Japanese man. A lorry was already stopped at the crossing waiting for the train to pass and it was overtaken by the jeep, which was then struck by an electric train and carried 175 feet down the track. [1030] [1031]

Those who lost their lives were later identified as Master Sergeant James Gordon, Mrs Gordon, Lieutenant Ernest Harper, Lieutenant Jack Kaufman, James Hollon Jr., aged

---

[1026] *The Northern Miner*, Charlton Towers, Queensland, Australia, 20th December 1948
[1027] *El Orden*, Santa Fe, Argentina, 19th December 1948
[1028] Ibid, 23rd December 1948
[1029] *The Sunday News and Tribune*, Jefferson City, Missouri, United States, 2nd January 1949
[1030] *Cumberland Evening Times*, Cumberland, Maryland, United States, 24th January 1949
[1031] *Chester Times*, Chester, Pennsylvania, United States, 24th January 1949

two, son of Staff Sergeant and Mrs Hollon; and Kiyomatsu Nakamura. Mrs Hollon survived the accident, though she was seriously hurt. [1032]

Another level crossing collision involving a school bus occurred in the United States on Tuesday 25th January 1949 at Lansing's or Brown's Crossing, near Gansevoort, New York. The bus was taking pupils home from South Glen Falls Central School and four had already been dropped off before it reached the level crossing with the Delaware and Hudson Railway. The bus was a converted station waggon which was being driven by William Derocher Jr., 26. The train involved was the 3.58 p.m. local service from Fort Edward to Albany, in charge of Conductor James Toomey with Engineer Charles Mercier and Fireman Leo Oulette.

Driver Derocher explained after the accident that he had stopped the bus about 30 feet from the crossing and had looked both north and south along the line, though an embankment partly obscured his view. He had then restarted and on reaching the crossing the bus had slipped on the icy road surface and then stalled. Seeing the train bearing down upon him he and one of the three remaining students, Mary Bahr, 15, had leapt to safety. Derocher said that he had tried to pull another pupil, Elizabeth Vance, 16, with him, but had failed. The train, moving at about 50 miles an hour, had then struck the left front fender (bumper) of the bus, and had "whipped it away from me." The station waggon was dragged about 150ft and ended up in a ditch, surrounded by wreckage, school books and fluttering pages of note paper. Elizabeth Vance and Mary Bahr's brother, Lyle, 11, were killed outright. Derocher said that after the accident he had sent Mary Bahr for help and she had been taken to the house of Mrs Irving Peters, who had telephoned the State Police. [1033]

Sunday 6th February 1949 saw two major level crossing accidents in the United States. The slightly more serious in terms of casualties was at Tulsa, Oklahoma. Seven people died in this accident: Mrs Kenneth Tinsley, 21; her daughter Claudia Sue Aery, 3; Mrs Claude Litterell, 55; her daughter Manry Lue, 15; Mrs Helen Litterell, 20; Mrs Wanda Litterell, 14, and Jimmy Eugene Litterell, 10 days, son of Wanda, who succumbed later to his injuries. In addition Mr Kenneth Tinsley, the driver of the vehicle, which was described as a 1937 motor coach; Johnny Wayne Litterell, 2, son of Helen; and Bobby Lee Aery, 1, son of Mrs Tinsley, were badly hurt with fractured skulls.

Train and car were both reportedly travelling at 35 miles an hour. Engineer W.F. Brooks could not see the train from his side of the locomotive but Fireman W.E. Merger could and shouted to him to "put on the brakes," but it was too late. An ambulance taking the injured to hospital was involved in another accident, injuring its two-man crew. [1034]

The other accident on Sunday 6th February 1949 occurred at Onawa, Iowa, United States. The automobile, a 1930 Chevrolet, was carrying farmer and war veteran Clarence Dahl, aged about 25, his wife Kathleen, 23; their sons Richard, 4, and Dennis, 2; and Clarence's sister Phyllis Dahl, 18, all from nearby Whiting, Iowa. They all lost their lives

---

[1032] *Walla Walla Union Bulletin*, Walla Walla, Washington, United States, 27th January 1949

[1033] *The Troy Record*, Troy, New York, United States, 26th January 1949

[1034] *Miami Daily News-Record*, Miami, Oklahoma, United States, 7th February 1949 in the evening collision. Phyllis Dahl was apparently driving the car.

Engineer Leonard Erickson of the Chicago and North Western Railway train stated that he had been travelling at about 40 miles an hour and had had no chance of avoiding the accident. Station Agent C.F. Freyer said that the train had been due to stop at his station, four blocks from the crossing, and that it had only been travelling at a low speed, so the engineer's estimate of his speed was perhaps on the high side. The car was dragged two blocks and demolished.

There were no other witnesses to the accident but visibility at the unprotected crossing was said to be very poor and there was also the possibility that the windscreen of the car had been frosted up, further restricting the driver's view. [1035] [1036]

The train was delayed an hour and 25 minutes.

A taxi carrying four men, believed to have been lumberjacks on their way to the Kokotow Lumber Company camp next to Larder Lake, was struck by a train on a level crossing eight miles east of Kirkland Lake, Ontario, Canada, on the evening of Wednesday 16th February 1949. The four passengers were all killed, as was the driver of the taxi, Emile St Louis, who was the owner of Walter's taxi company. It was theorised that the taxi driver had been momentarily blinded by a snow squall which had swept the area at the time and had thus failed to see the train. [1037]

A school bus carrying about 50 black children to the Unity Negro School at Foreman, Arkansas, United States, had almost reached its destination when it was struck by a St Louis-San Francisco Railway train, described as a two-unit motor, and en route from Hope, Arkansas, to Ardmore, Oklahoma. The accident happened at about 8.45 a.m. on Wednesday 16th February 1949. Eight or nine children were seriously hurt and although none of them were killed outright it was feared that several might die of their injuries. They were taken to hospitals at Texarkana and Ashdown.

The bus driver, Paul Roberts, who like the remainder of the pupils had escaped harm, explained that a fog and some boxcars stabled on a siding had prevented him from seeing the passenger train. [1038]

Five Chinese soldiers died and others were seriously injured when the lorry in which they were riding was struck by a train at Lou Far Kio, China, a suburb of Canton (now Guangzhou) on Sunday 13th March 1949. [1039]

A train running from Arequipa to Juliaca, Peru, collided with a lorry carrying 30 people on a level crossing situated on a curve in the line on Tuesday 19th April 1949. Fourteen of the passengers on the lorry were killed. [1040]

[1035] *The Democrat and Leader*, Davenport, Iowa, United States, 7th February 1949
[1036] *Carroll Daily Times Herald*, Carroll, Iowa, United States, 7th February 1949
[1037] *Winnipeg Free Press*, Winnipeg, Manitoba, Canada, 17th February 1949
[1038] *The Sedalia Democrat*, Sedalia, Missouri, United States, 16th February 1949
[1039] *Indian Daily Mail*, Singapore, 14th March 1949
[1040] *The Montana Standard*, Butte-Anaconda, Montana, United States, 20th April 1949

A two-and-a-half-ton farm truck carrying 11 people back to their homes in Hogan's Creek after they had attended an evening church service was struck by fast Tennessee Central Railway freight train No. 81 on Sunday 24th April 1949. The accident occurred on the Gordonsville – Lancaster road about eight miles south of Carthage, Tennessee, United States. The vehicle was cut in two, dragged 100 yards down the line, and burst into flames. Ten of its occupants died, nine of them members of the same family, and the remaining person was badly hurt.

The victims were Jess Bennett, 50, the driver of the truck; his wife Mattie Bell, 45; their sons Douglas, 12; U Lee, 10, and Melvin Earl, 8; the Bennetts' daughter, Mrs Paulie Bennett Dickens, 24; their son-in-law Paulie Dickens, 27; the Dickens' daughter Catherine, 1; Jess Bennett's brother W.E. (Coonie) Bennett, 49; and Miss Linnie Gibbs, 17, who died shortly after arrival in hospital at nearby Lebanon. Ruth Robinson, the only survivor of the accident, was taken to Lebanon hospital with a broken pelvis. Her condition was described as "fair."

J.T. Waddell, general manager of the railway, ordered an immediate investigation of the tragedy and State Safety Commissioner Sam Neal hurried to the scene. The latter said that it appeared that five people had been sat in the cab of the truck and the other six had been on chairs set up at the back of the vehicle.

John Overstreet, another son-in-law of the Bennetts, told the Commissioner that he had stopped his car at the crossing to drop off some passengers when the truck had come up behind him. He continued, "I knew the train was coming. I tried to wave the truck down, but my father-in-law pulled around me and went on in front of the train." Jess Bennett was reportedly blind in one eye.

The conductor of the train said that none of the train crew had seen the truck until it had loomed up in the lights of the diesel engine.

All available ambulances from Carthage and Gordonsville went to the scene. It took rescuers over an hour to extricate all of the bodies pinned in the cab of the truck and work went on into the early hours to remove the wreckage from the locomotive with crowbars and acetylene torches. [1041]

As a westbound Canadian Pacific Railway freight train was approaching a private level crossing near Tompkins, Saskatchewan, Canada, on Sunday 1st May 1949, Engineer C. Carlson and Conductor E. Toole glimpsed a car also nearing the crossing. The vehicle slowed up and stopped at the crossing, then started up again and tried to cross the line directly in front of their train. Struck by the engine the car was carried a mile down the line. Frank McBee and Mrs Thelma Hughes, both 25 and both from Tompkins; and sisters Ellen and Jean Kewley, 22 and 16 respectively, of Porcupine Plains, were all killed outright. Howard Kamroski, 27, also of Tompkins, died later in Gull Lake hospital. The sixth occupants of the car, Helen Fetis, 16, of Porcupine Plains, was badly hurt, but was given a "fighting chance" to pull through by Gull Lake doctors. The party had been returning from a dance to the home of Mrs Hughes. [1042]

A family of five were wiped out when their car was in collision with a northbound

---

[1041] *Kingsport Times*, Kingsport, Tennessee, United States, 25th April 1949

[1042] *The Lethbridge Herald*, Lethbridge, Alberta, Canada, 2nd May 1949

Chicago Rock Island and Pacific Railroad passenger train near Northwood, Iowa, United States, on the morning of Sunday 22nd May 1949. Mr Marvin Holstad, 27; his wife, 26; and their children Marilyn, 4; Montie Lou, 3; and Donavan, 18 months, had been on their way to Deer Creek Lutheran church, six miles further on. Also killed was Diane Leidal, 4, daughter of farmer Milford Leidal and niece of Mrs Holstad, who had been living with the family since the death of her mother two years earlier.

The main part of the 1934 Chevrolet was dragged 162ft by the train. The bodies of the adults were found close to it and those of the children were scattered along the line.

In charge of the locomotive of the train was Engineer George Duncan, with Fireman Obert Arendits.

The crossing was unprotected but the weather was fine and clear and there was good visibility along the line. Possibly the attention of the car driver was distracted by one of the small children. [1043]

A major disaster occurred at Schlebusch, Germany, on the Köln – Opladen line, on Monday 8th August 1949 when express D 99 collided with a bus carrying 80 children. The children had come from a Catholic holiday camp and were only a kilometre and a half from the bus station where their parents were waiting for them. The bus was pulling two trailers. Its driver attempted to negotiate the crossing when the barriers were already half lowered, while in addition the crossing keeper had only lowered them at the last moment, when the train was close at hand. The train struck the rear of the bus, killing 18 of the children and injuring 12 more. Both the bus driver and the crossing keeper were arrested. [1044] [1045]

Five firemen were killed instantly and two others were seriously injured when their fire engine was in collision with a three-coach electric train on a level crossing in Nagoya, Japan, on Tuesday 16th August 1949. They had been engaged in a training exercise. The train was little damaged and no one on it was hurt. [1046]

A prominent local farmer who had started out as an immigrant from Mexico was killed together with five members of his family in a level crossing accident at Ogallala, Nebraska, United States, on the evening of Tuesday 16th August 1949. Ascension Jiminez, aged about 60; his wife, aged about 57; their daughters Dora, 21, and Margie, 12; and grandsons Augustine Jr., 3, and Carl, 18 months, all lost their lives. Another daughter, Annie, aged 13 or 15, was critically injured. The family were returning home after going to a movie.

Patrolman William Affleck said that Jiminez had waited at the crossing for a westbound Union Pacific Railroad freight train to pass, had then assumed that it was safe to cross and his car had been struck by the eastbound *San Francisco Overland* [1047] passenger express. The 1948 automobile was struck broadside, hurled 90ft, and

---

1043 *Waterloo Daily Courier*, Waterloo, Iowa, United States, 23rd May 1949
1044 *De Locomotief*, Semarang, Dutch East Indies, 10th August 1949
1045 *Katastrophen der deutschen Bahnen Teil II* by H.J. Ritzau pg. 241 (in German)
1046 *Indian Daily Mail*, Singapore, 18th August 1949
1047 Wikipedia – *Overland Limited* (UP train)

demolished. Its occupants, except Jiminez, were thrown from it. A steam pipe on the locomotive was broken. Yellow flashing lights were provided at the crossing and were working normally. [1048] [1049]

A lorry laden with gelignite collided with a train about two miles from N'dola, Northern Rhodesia (now Zambia) on either Sunday 21st or Monday 22nd August 1949. The gelignite exploded as a result of the collision, killing 11 Africans and two Europeans, and seriously injuring 17 more Africans, who were taken to hospital. Those who died comprised the engine driver, three people on the train, seven men on the lorry, the crossing keeper and one other person, perhaps a passer-by. All that remained of the lorry were a few scraps of twisted iron, while the engine and three waggons of the train were blown to pieces and a portion of the track was ripped up. Pieces of wreckage were found over a mile away. [1050]

Probably on Sunday 21st August 1949 a taxi tried to beat a Canadian National Railways train over a level crossing near Desbiens, Quebec, Canada. It failed and was demolished, killing Henri Caouette, 30; his 23-year-old wife; their six-year-old daughter, Francine; Albert Fortin, 30; and Norman Larouche, 26. [1051]

A serious accident was narrowly averted at Genoa, Illinois, United States, when a train collided with a school bus on Tuesday 30th August 1949. Genoa-Kingston school superintendent Clarence Louderback and bus driver Maylon Davis were taking the bus on a trial run to check its schedule. Eight children, four of whom were the superintendent's, were coming along for the ride. The bus was stopped short of a level crossing with the Chicago Milwaukee St Paul and Pacific Railroad, then restarted, but as it got onto the track it stalled. At the same moment the crossing alarms began to sound. "Here comes the Flyer," the bus driver cried. "We're in trouble. Let's get out of here!" The superintendent picked up two of his sons, whilst the driver shepherded the other six pupils through the bus doors. The group had got about 10 or 15 feet away when the streamliner, travelling at about 80 miles an hour, rammed and demolished the 40-seater bus, which was thrown 75 feet. The Chicago – Omaha train was stopped and its crew ran back to the crossing, but when they found that no one was hurt the train proceeded on its journey. [1052] [1053]

A bus carrying servicemen from the March Air Force base, girls from a local service club which helped out with parties and entertainments for the troops, and a chaperone, was returning from Corona Del Mar and Long Beach on the evening of Sunday 2nd October 1949 when it was involved in a disastrous level crossing collision at Ontario, California, United States. The bus was travelling along Highway 60, which paralleled the Union Pacific Railroad tracks, when it was turned left into Archibald Street and directly into the path of the oncoming eastbound *Pony Express* passenger train from Los Angeles

---

[1048] *The Telegraph*, Sidney, Nebraska, United States, 19th August 1949

[1049] *Beatrice Daily Sun*, Beatrice, Nebraska, United States, 17th August 1949

[1050] *The Townsville Daily Bulletin*, Townsville, Queensland, Australia, 22nd August 1949

[1051] *Medicine Hat Daily News*, Medicine Hat, Alberta, Canada, 23rd August 1949

[1052] *The Register-News*, Mt Vernon, Illinois, United States, 31st August 1949

[1053] *Jacksonville Daily Journal*, Jacksonville, Illinois, United States, 31st August 1949 to Kansas City. In the ensuing collision 11 servicemen, five girls and Mrs Ruby McLaughlin, the chaperone, were killed. Three servicemen and two girls, twin sisters Marquita and Omega Pearce, 18, survived the accident with injuries, while several other girls had already been dropped off at their homes. The accident happened at about 9 p.m.

Engineer Al Hall said later that he had just begun picking up speed when he had seen the bus. "I saw it start creeping slowly toward the tracks and I began to blow the whistle. When I saw that it wasn't going to stop I slammed on the air brakes, but it was too late."

Fireman P.K. Sweeney, speaking of the stretch of track leading to the crossing, added, "It is a very fast area. A high ball piece of road." He estimated the speed of the train at the moment of the collision as 73 miles an hour; his engineer put it at 71 miles an hour.

One of the surviving servicemen, Sergeant Peter Grisolia, provided a partial explanation for the accident. He stated that the driver of the bus, tentatively identified as Corporal Gordon Crimin, had been unsure of which way to go and his passengers had been shouting directions to him. "People would yell out 'turn left' or 'turn right.' Just before we were hit someone said, 'turn left,' and I closed my eyes because I was sleepy." Then he was thrown from the bus.

It took the train a mile to stop. Acetylene torches had to be used to cut the wreckage of the bus from the front of the locomotive and it was three hours before the train could continue, either back to Los Angeles or onwards towards Kansas City, according to different accounts. No one on the train was hurt and many passengers were reportedly hardly aware of the crash. [1054] [1055]

On Friday 28th October 1949 *direttissimo* No. 82 from Siracusa to Messina collided with a bus on a level crossing at Mili, Italy, just before 18:00. Finding the crossing barriers raised the bus driver naturally assumed that it was safe to cross, not realising that the train was close at hand. The train hit the bus and crushed it against a wall, causing a panic among its passengers, and women screamed in terror. The train pulled up a few hundred metres down the line and train staff ran back to give assistance. Five people died and about 20 were injured, five very seriously. Those who lost their lives were passengers Annunziata Camiata, 42, and Santa De Luca, 35, both of Messina; Caterinaz Famularo, 25, of Lipari; and Tullio Ferigno, 32, of Villa Estense. The fifth fatality was the bus conductor, Giovanni Zagami, 51, of Messina.

Crossing Keeper Salvatore Saievo, 27, was arrested. [1056]

The exertions of a crossing watchman early on Saturday 5th November 1949 were not enough to stop a car which was struck by a New York Central Railroad passenger train at Dayton Pike Crossing, Sharonville, Ohio, United States. Travelling at 45 or 50 miles an hour, the train demolished the automobile; five of the occupants of the car were killed and the sixth was critically hurt.

Crossing Watchman George Ringeisen, 52, who had been standing on the yellow line in the middle of the highway as the train approached, related afterwards, "I saw the car coming. It slowed down for a moment, then seemed to speed up. I held the flag up and

---

[1054] *The Daily Review*, Hayward, California, United States, 3rd October 1949
[1055] *The Independent*, Long Beach, California, United States, 4th October 1949
[1056] *La Stampa*, Torino, Italy, 29th October 1949

hollered. The crossing bell was ringing. I yelled, 'Hey! Hey!' but they just weren't paying any attention. The engine seemed to hit just the rear of the auto, and the car just rolled down the track end over end. I was all quivery. I haven't quit shaking yet."

Identification of the victims was difficult due to conflicting names on draft registration cards, fishing licences, identification cards and tags and other paperwork found on the bodies. [1057] The dead were provisionally identified as Robert Former, 23; Otha Glandon, 20; Walter Palmer, 20; and James Raby, 22, all of Detroit, Michigan; and William Malone Jr., of Louden, Tennessee. [1058]

Early on the morning of Friday 18th November 1949 what was described as a Jeep panel truck was involved in a level crossing collision with a westbound Missouri Pacific Railroad train on Oak Grove Crossing, 3½ miles north of Sedalia, Missouri, United States. The vehicle was carrying workmen of the National Tree Company of Missouri, who were going to work on a contract job for the Rural Electrical Administration. Three men were riding in the cab of the truck and the other five were sat on a rear platform. The truck was demolished in the collision and carried 175ft down the track.

Five workmen were killed outright: Harvey Bird, Mart Reed (the driver), T.J. Reese Jr., Marvin Schumate and Joe Townsend. Of the remainder, John Walterscheid was fatally injured and died in shortly in hospital, Eugene McCoy was described as critically injured and Frank Kabler was said to be seriously hurt.

Two more workmen were following the truck in a car but were not close enough to be able say what had happened, and there were no other witnesses to the accident apart from those on the train. The coroner intended to question the train crew, Conductor Sherman Grazier, Engineer Harry Salmon, and Fireman W.D. Phillips, upon their return from Kansas City in the evening. The accident happened at 7.45 a.m.

A warning bell was provided at the crossing which was still working when patrolmen and ambulances arrived at the scene. [1059]

A truck carrying 16 Latin American cotton pickers on their way from Lubbock to the fields where they were to labour for the day was struck by a Panhandle and Santa Fe Railway freight train just before 9 a.m. on Wednesday 23rd November 1949, killing or fatally injuring eight of them and injuring the remainder. The accident happened near Shallowater, Texas, United States, and the truck was bound for the Werner Teggeman farm, only a mile and a half further on.

The train was in charge of Conductor M.C. Hodge with Engineer William Rust and Fireman W. Hollingsworth. The train crew said that they had been moving at 40 miles an hour. Fireman Hollingsworth saw the truck turn off the main highway onto the side road leading to the crossing. Seeing that the truck was not stopping he had tried to shout to its occupants. He said, "They were that close to the locomotive. After we hit, the train travelled seven telephone poles before it could be stopped."

Wayland Penney, 15, witnessed the collision from a point 120 yards away. He told a highway patrolman that he had seen the truck turn off the main road onto the north-south

---

[1057] *Pharos Tribune*, Logansport, Indiana, United States, 5th November 1949
[1058] *The Coshocton Tribune*, Coshocton, Ohio, United States, 5th November 1949
[1059] *Moberly Monitor-Index*, Moberly, Missouri, United States, 18th November 1949

dirt road and said, like the fireman, that the truck had apparently made no effort to stop short of the crossing. Other, more distant witnesses, spoke of seeing the truck turning over completely in mid-air and of bodies being thrown from it.

The crossing was protected by the usual crossbucks, a sign reading "Look Out for the Cars," and also, according to one account, by a wig-wag swinging disc signal.

Those who lost their lives in the collision were: A.C. Quirino, 45, the owner of the truck and boss of the crew; his sons Francisco, 15, and Arthur, 9; Nicholas Padron, 87, who died in Lubbock Memorial Hospital the same evening; his grandson Fred, 14; Angolo Gonzales, 13; Raymond Hernandez, 19; and Joe Moreno, 16. Isaac Quirino, 18, another son of A.C., was only slightly hurt. He had been the driver of the truck and doubtless in view of the fact that he was facing possible manslaughter charges he refused to speak to newspaper reporters except through his attorney, Winston Brummett. The latter quoted Isaac as saying, "If the train ever whistled I never heard it." Other survivors could not recall any details of the accident. Not apparently having yet been informed of the deaths of their friends and relatives they were given evasive answers concerning their fellow travellers' conditions. [1060]

Seven people died and 'many' were hurt in a collision between a train and a bus, presumably on a level crossing, at an unstated location about 100 kilometres south of Quito, Ecuador, on Monday 12th December 1949. [1061]

A bus carrying 26 teenage Catholic students and two nuns, Sisters Mary Lenore, and Mary Vinventia, on its way to the Jackson Home for the Aged where they were to perform a musical program, was in collision with the westbound *Missouri River Eagle* streamliner of the Missouri Pacific Railroad on Saturday 4th February 1950. The train was en route from St Louis to Omaha and was probably travelling at about 79 miles an hour. The accident happened on a level crossing near Kansas City, Missouri, United States. Sister Lenore was the only person badly hurt; nine students suffered slight injuries when the train caught the rear of the bus.

Sergeant A.H. Bell of the Missouri Highway Patrol told reporters that the driver of the bus, Clarence Miller, aged 53, would be charged with careless and reckless driving in connection with the accident. [1062] [1063]

Miller was subsequently found guilty of careless driving and was fined $105 in the magistrates' court at Independence, Missouri. [1064]

A level crossing at Richmond, Indiana, United States, was one of those American ones which was provided with a crossing watchman but lacked gates. A diesel-powered Pennsylvania Railroad passenger train from Cincinnati to Chicago consisting of an engine and 17 cars was approaching North 15th Street Crossing on Sunday 26th February 1950 at the same time as an automobile driven by Woodrow Dadisman, 24, was also

---

[1060] *Lubbock Morning Avalanche*, Lubbock, Texas, United States, 24th November 1949
[1061] *Nebraska State Journal*, Lincoln, Nebraska, United States, 13th December 1949
[1062] *San Antonio Sunday Light*, San Antonio, Texas, United States, 5th February 1950
[1063] *The Sunday News and Tribune*, Jefferson City, Missouri, United States, 5th February 1950
[1064] *The Sedalia Democrat*, Sedalia, Missouri, United States, 24th February 1950

nearing the crossing, travelling at speed. Crossing Watchman Willie Thomas signalled frantically to the car driver with his lamp and a whistle but it did not stop. Engineer Ray Essart said that he had seen the lights of the automobile but could not halt his train either, and train and car collided. Wreckage of the car and bodies were strewn along the track for two blocks.

Orville Howard, 46, of Indianapolis; his sister-in-law, Mrs Dolly May Johnson, 41; Mr Dadisman, her son; and his wife Carmello, 30, all of Richmond; and Robert King, 28, a son-in-law of Mrs Johnson, who lived near Paris, Ohio, were all killed outright. Another passenger in the car, Edgar Mapheny, 22, of Richmond, was fatally injured and died later in Reid Memorial Hospital, Richmond. [1065] [1066]

A school bus had only picked up four children when it turned off Highway 6 onto Jasper County trunk road H and descended a slope to Turner's Crossing, near Grinnell, Iowa, United States, on the morning of Monday 27th February 1950. The bus driver, Dennis Gaines, 22, told police afterwards that the road had been covered with snow and ice and his bus had skidded on the slippery surface so that he had been unable to pull up short of the track. Seeing a train and realising that he could not stop, he had tried to get his bus across in front of it. The train was the *Des Moines Rocket* of the Chicago Rock Island and Pacific Railroad, which had just started from Des Moines for Chicago.

Travelling at probably 50 miles an hour the train carried the bus 50 feet along the line before the vehicle rolled off onto one side in a heap of twisted metal. 'Galoshes, school books, lunch packages and bits of apparel were scattered along the right of way.' The door of the bus was thrown into an adjoining pasture. The accident happened at 8.10 a.m. A passing motorist took one of the children to hospital and the others and the bus driver were taken to Grinnell station on the train.

One of the pupils, Jerry Hay, 8, died shortly after the collision, and his sister, Shelvey, 12, was fatally hurt and died later in hospital. Another pupil, Shirley Taylor, 15, was reportedly critically injured and the fourth child, Joyce Weidendorf, 14, and the bus driver were rather less badly hurt. [1067] [1068] [1069]

Only a day after the Grinnell accident, at 7.55 a.m. on Tuesday 28th February 1950, two children again died in a level crossing accident involving a school bus, this time at Sharp's Siding, near Monticello, Maine, United States. Temperatures were near zero Fahrenheit and there were three or four feet of snow on the ground as the bus taking the pupils to Monticello Grammar School was struck by the diesel-hauled Bangor and Aroostook Railroad freight train. There were 23 children on the bus. Pamela Davis, 8, and Bonnie Rae Flewelling, 9, were killed and ten other pupils were injured. Private ambulances took most of the injured to Houlton hospital. [1070]

'Leo C. Good, a member of Maine's governor's council, was among the first at the scene. He took one child to hospital after assisting in removing others from the broken

---

[1065] *Rushville Republican*, Rushville, Indiana, United States, 27th February 1950
[1066] *The Tipton Daily Tribune*, Tipton, Indiana, United States, 27th February 1950
[1067] *The Evening Democrat*, Fort Madison, Iowa, United States, 27th February 1950
[1068] *Creston News Advertiser*, Creston, Iowa, United States, 27th February 1950
[1069] *The Muscatine Journal*, Muscatine, Iowa, United States, 27th February 1950
[1070] *The North Adams Transcript*, North Adams, Massachusetts, United States, 28th February 1950

bus wreckage. "The bus was pinned up against the train," said Good. "The badly frightened children were crying. We took them out of the wreck as quickly as possible and took them off to hospitals." As the bus was pushed along the railroad tracks it struck a utility pole and broke it off.' [1071]

The bus driver, Weldon Parent, 32, told reporters after the accident that he had never known a train to pass at the hour that the accident had happened and he had neither seen nor heard it before it struck the rear of his bus. A railroad spokesman stated that Engineer W.O. Milbury had whistled for the crossing and applied his brakes when he had seen that the bus was not stopping. The bus driver, who was reportedly shaken up by the accident and was taken to hospital, was charged with two counts of manslaughter by negligence. [1072]

Appearing before the court in Houlton at the beginning of March, Parent testified that he had only been moving at about four miles an hour as his bus had reached the crossing but acknowledged that he had not come to a complete stop. He repeated that he had neither seen nor heard the train. A lorry driver, Eugene Amnoit, who had been nearby at the time, also denied having heard the train whistle.

Fireman Kenneth Mayo said that he had noticed that the doors of the school bus were frosted over. He had called to his engineer to sound the whistle and ring the bell and the emergency brakes of the train had locked about 15 to 20 feet from the crossing.

Two of the schoolchildren, James Foster, and Merle Flewelling, told the court that they had been seated towards the front of the bus at the only window which was not frosted over. When the bus driver had asked them if there was a train coming they had replied that they could not see or hear one.

Municipal Judge Harry Baulch found probable cause and committed the bus driver to appear before the Grand Jury, allowing him bail to the sum of $6,000. [1073] Despite this Parent seems to have continued to enjoy the support of the children's parents, who a month later were asking when he would be able to return to duty. [1074]

It has not been possible to locate any report of a trial, so possibly the case was thrown out by the Grand Jury.

Near Bergheim, West Germany, on the line to Rommerskirchen, passenger train P 2518 collided with a lorry carrying workmen on an unprotected level crossing on Wednesday 8th March 1950. Five people died and 21 were injured. [1075]

Five young people died when the car in which they were travelling collided with a Spokane International Railroad train on a level crossing three miles south of Rathdrum, Idaho, United States, early on Saturday 8th April 1950. It seemed likely that the driver of the car had failed to see the train at all and his vehicle was driven into the side of the locomotive at high speed. The car was demolished, its engine being thrown 20 feet. Although the train was not derailed the battery box of its locomotive was badly damaged

---

[1071] *Fitchburg Sentinel*, Fitchburg, Massachusetts, United States, 28th February 1950

[1072] *Portland Press Herald*, Portland, Maine, United States, 1st March 1950

[1073] Ibid, 7th March 1950

[1074] Ibid, 4th April 1950

[1075] *Katastrophen der deutschen Bahnen Teil II* by H.J. Ritzau pg. 241 (in German)

and it came to a halt a quarter mile from the crossing.

LeRoy Banks, 21, of Rathdrum; Betty Lou Buckenberger, 16, of Trentwood, Washington; Betty Griggs, 17, of Coeur d'Alene; and Allen Manweiler, 21, of Rathdrum, were all killed instantly, while Edward Dutro, 21, the apparent driver of the car, died on the way to hospital. The sixth occupant of the car, Kay Olson, 17, of Post Falls, Idaho, who had been sat in the back, was thrown from it, and escaped with cuts and severe shock. She was taken to hospital as a precaution.

The three-car train had been en route from Spokane to Yahk, British Columbia. The accident happened at 1.14 a.m. [1076] [1077]

A taxi which served as a school bus for the rural pupils of a school at Campus, Illinois, United States, was struck by the *Banner Blue* streamliner of the Wabash Railroad only a short distance from the school on the afternoon of Thursday 27th April 1950. Running about 20 minutes late, the St Louis – Chicago train was travelling at about 75 miles an hour when it struck the taxi on an unprotected level crossing. The vehicle was carried about a mile before the train could be pulled up, with the bodies of all but one of its occupants still inside it. It took over two hours to remove the victims from the wreckage.

Killed in the accident were George Frederick Johnson, 8; his sister Judith Kay, 9; Alice Mae McArthur, 8; her sister Ethel, 13; Barbara Seabert, 8; Mary Helen Ziehr, 13; and the school bus driver, Leo Kanne, 52. Kanne's wife, Helen, was the principal of the two-room Campus consolidated school. [1078] She ordered all classes suspended until after the funeral of the victims and the inquest. The dead represented almost a quarter of the complement of the school. There should have been another pupil in the taxi but a nine-year-old girl who usually travelled on the school bus was picked up from the school that day by her mother to take her to a birthday party. [1079]

Deputy Sheriff F.J. Fitzpatrick commented after the tragedy that it would renew a 12-year fight by the town of Camps for the provision of gates or at least flashing lights at the crossing, adding that three people had lost their lives in a previous accident at the crossing about three years previously. [1080]

At the inquest on the victims Engineer Michael Flynn and Fireman George Klinger stated that their train had been going at 75 miles an hour when it hit the taxi, which neither of them had seen until a moment before the collision. The engineer added that the maximum authorised speed at that point was 78 miles an hour.

The inquest jury returned a verdict of 'Accidental Death' and made three recommendations to Governor Adlai Stevenson and the Illinois Transport Commission:

(a) That additional warning devices should be installed at the crossing.

(b) That until such warning devices had been provided the speed of trains through the town should be limited to 45 miles an hour.

(c) That there should be a more thorough examination of engineers.

[1076] *Salt Lake Telegram*, Salt Lake City, Utah, United States, 8th April 1950

[1077] *Idaho Sunday Journal*, Pocatello, Idaho, United States, 9th April 1950

[1078] *Daily Journal-Gazette*, Mattoon, Illinois, United States, 28th April 1950

[1079] *Edwardsville Intelligencer*, Edwardsville, Illinois, United States, 28th April 1950

[1080] *Moline Daily Dispatch*, Moline, Illinois, United States, 28th April 1950

The commerce commission had already set a day of 17th May for an examination of the case and had summoned the railroad, Livingston County and the Campus authorities to appear at Springfield on that date and show cause why warning signals should not be installed at the crossing. [1081]

On Sunday 21st May 1950 a party of 13 people had been fishing near Sturgeon Falls, Ontario, Canada. Just after midnight on Monday 22nd the lorry in which they were returning to their homes in nearby River Valley ran into the side of a Canadian Pacific Railway transcontinental passenger train in the village of Verner, striking the train between the third and fourth cars of its sleeper section. Mrs Eddie Cayen, 39; her son Herve, 19; Mrs Edouard Forget, 26; her daughter Denise, 5, and son Robert, 3; and Julien Cormier, 25, were all killed outright. Mr Eddie Cayen, his sons Hector and Roger; Mrs Julien Cormier and her son Gerald Guay; Mr Edouard Forget and Donat St Martin were all injured.

The train crew remained unaware of the accident until they stopped at Sturgeon Falls, 11 miles further east, when they found pieces of cloth and body parts wedged under the platform between the Pullman cars. Police said that the cab and every part of the body of the truck had been filled up. [1082]

Hector Cayen and Gerald Guay, both aged 4, died of their injuries in hospital later in the day. [1083]

Seven children being taken to a convent school had a lucky escape on the morning of Monday 29th May 1950 when the bus in which they were riding was struck by a train on Brown Street Crossing, Queenstown, Tasmania, Australia. The bus was halfway over the crossing when its driver, William Peter Pennyston, saw the train only 10ft away. He accelerated and swung the bus to one side but the engine of the train caught the rear left corner of the vehicle and tore out its bodywork.

The pupils were thrown out of their seats but were not hurt. They had also been sat immediately behind the driver and had reportedly distracted him, so that he had failed to see the train earlier. [1084]

The circumstances of a horrific accident between Guayaquil and Salinas, Ecuador, on Thursday 22nd June 1950 are somewhat unclear. According to some reports 22 railway workmen died and 13 were injured when a train collided with a handcar. [1085] Other reports stated that a train had collided with a bus (on a level crossing?) touching off an explosion and fire which had killed 18 and injured 11 more. According to these accounts there had been about 35 people aboard the bus. Most of them had been trapped when the petrol tanks exploded after the collision and flames had swept into the vehicle. It was feared that several of the injured would die. The charred bodies of the victims could not

---

1081 *Dixon Evening Telegraph*, Dixon, Illinois, United States, 4th May 1950

1082 *The Lethbridge Herald*, Lethbridge, Alberta, Canada, 22nd May 1950

1083 *The Brandon Daily Sun*, Brandon, Manitoba, Canada, 23rd May 1950

1084 *The Examiner*, Launceston, Tasmania, Australia, 30th May 1950

1085 *The Ogden Standard Examiner*, Ogden City, Utah, United States, 23rd June 1950 immediately be identified. [1086]

A disastrous level crossing collision occurred near Saint-Amour in the Jura department of France on Monday 3rd July 1950. The train involved was an express from Paris to Aix-le-Bains and was carrying 500 children from the Douai mining district bound for a holiday camp at Annecy. The motor coach was carrying a party of holidaymakers, composed of veterans of the war and their families, and was en route from Foissiat to Besançon Saint-Fouet, near Besançon.

The train was running under clear signals and travelling at 90 kilometres an hour when it struck the bus, which was carried over 500 metres and demolished. Of the 26 occupants of the bus 21 were killed outright and four more were fatally injured. The sole survivor, a 70-year-old man, was taken to hospital in Lons-le-Saulnier, apparently only slightly hurt.

By the time that he saw the bus Driver Penet of the express was completely unable to avoid the collision. When his train hit the coach his locomotive shook and swayed dangerously but fortunately was not derailed. The accident happened at 04:17.

The crossing was protected by full barriers operated by a crossing keeper. The initial investigation indicated that the crossing barriers had been lowered when the coach had arrived at the crossing. The coach driver had then sounded his horn for the barriers to be raised and the crossing keeper, who had been chatting to a passer-by, had without thinking opened them in the face of the oncoming train. [1087]

The gatekeeper, who had himself suffered facial injuries from flying glass, was subsequently charged with homicide through negligence by the local *juge d'instruction* (examining magistrate) and was committed to prison. [1088]

Five members of one family were killed in a level crossing collision near Algoma, Iowa, United States, on Tuesday 4th July 1950. Their train was struck by a southbound Chicago and North Western Railway. Those who lost their lives were Dr William Howard, 74, a retired physician; his 70-year-old wife; their son-in-law, Chris Tofft, 45; their daughter Lillian, 44; and a son of the Toffts, William Arthur, 20. The car was registered to Dr Howard but it was not clear who had been driving it.

Engineer Fred Watson said that his train had been travelling at about 25 miles an hour and he estimated that the car had been travelling at 45 miles an hour. He related that he had whistled and had applied the brakes when it had become apparent that the car was not going to stop. Rain had made the rails slippery and the car was dragged 479ft down the track before he could pull up the train. The crossing lay on a straight stretch of road and was unobstructed, though it was also unprotected by automatic signals. It was believed that the family had been visiting relatives at Cherokee, Iowa, and had been on their way home at the time of the accident. [1089]

Probably because the victims were all black (African Americans) a level crossing

[1086] *The Evening Star*, Washington, D.C., United States, 23rd June 1950
[1087] *Gazette de Lausanne*, Lausanne, Switzerland, 4th July 1950
[1088] Ibid, 5th July 1950
[1089] *The Cedar Rapids Gazette*, Cedar Rapids, Iowa, United States, 5th July 1950

collision which killed five people on Sunday 9th July 1950 only received brief coverage in the press. It happened two miles south of Denmark, South Carolina, United States, when a car which according to an eyewitness had stalled on the track was hit by a northbound Seaboard Air Line Railroad passenger train and demolished. Three men were killed outright and another man and his 13-year-old daughter died shortly afterwards. [1090]

Five members of a family were killed when the truck taking them to a Sunday service was struck by a train a short distance from the United Church at Uno Park, Ontario, Canada, six miles north of New Liskeard, on Sunday 3rd September 1950. Market gardener Delbert Peters, 54; his wife Mary, 38; and their children Patsy, 9; Mary Ann, 6; and Harry, 4, were all killed. Six other Peters children had been left at home. The northbound train consisted of a pair of two Ontario Northland Railway diesel engines and a van bound for Engleheart. [1091]

A man named G. Mantha, who operated a store near the accident scene, saw it happen. He said, "It all happened very quickly. The train ripped into it and the air was filled with flying bodies." There was a good view from the crossing in both directions and other eyewitnesses had heard the train whistling but Mr Peters, who had been driving, had the truck windows rolled up and he presumably neither saw the train nor heard its whistling. [1092]

Nine soccer players on their way home from a match were killed when the truck in which they were riding was struck by a train on a northern Natal level crossing near Vryheid, South Africa, on the evening of Sunday 10th September 1950. [1093]

Shortly after noon on Wednesday 27th September 1950 a freight train from Chiquinquirá to Bogotá, Colombia, collided with a bus coming from Zipaquirá on a level crossing at El Mortiño station, on the Norte line.

Bus No. 30113 of the Ubaté fleet had a seating capacity of 37 passengers but was carrying rather more than 40 at the time of the accident. Driver Manuel Rozo stated that his bus had had a normal run until it reached kilometre 57 of the Norte line from Zipaquirá to Nemocón, that is, El Mortiño. The barriers at the level crossing here were lowered, but his bus, somewhat overloaded and travelling at high speed, was unable to stop. It smashed through the crossing barrier and came to a stand on the line. A few moments later it was struck by the train, which was hauled by engine No. 15, with Driver Pedro Nel Rivera.

The collision occurred at 12:10. The bus was dragged some metres along the track and completely demolished. Its occupants were thrown in various directions and several were killed outright. Others were very badly hurt and passed away whilst they were being taken to Zipaquirá hospital while others again suffered possibly life-threatening injuries. At the time of the report it was believed that there were six dead passengers and another 20 who were more or less seriously injured. The remaining passengers escaped with

[1090] *The Bee*, Danville, Virginia, United States, 10th July 1950

[1091] *The Winnipeg Tribune*, Winnipeg, Manitoba, Canada, 4th September 1950

[1092] *Winnipeg Free Press*, Winnipeg, Manitoba, Canada, 4th September 1950

[1093] *The Brandon Daily Sun*, Brandon, Manitoba, Canada, 11th September 1950 slight injuries. [1094]

A drizzling rainstorm, darkness and a road approach to an elevated crossing lined by trees were factors in a crossing collision on a country road at 11 p.m. on Thursday 16th November 1950. A fast Canadian Pacific Railway freight train collided with a car three miles north of Beeton, Ontario, Canada. The car ran into the side of the train, was carried over 200 feet down the line, and then rolled down a 30-foot embankment. Motorists who arrived at the scene shortly afterwards said that there was no sign of the accident at the crossing itself but that wreckage was strewn along the right-of-way from a short distance beyond it. Provincial police from Alliston and Bradford put up road blocks to deter spectators and all available local ambulances and hearses were sent to the scene. Killed in the collision were Edwin Hewitt, Raymond Pether, George Morrison, Mrs Inez Rushton, and her daughters Marion, 10, and Joan 7. Hewitt was probably driving the car; he was the foreman of a construction company and the other men worked under him. There had been several previous accidents at the crossing. [1095]

A few days later a drawing of a train and a car at a level crossing appeared in a local newspaper, headed, 'YOU CAN STOP FASTER THAN THE BIG FELLOW!' Beneath the drawing the text read: 'ANOTHER TRAGIC ACCIDENT LAST WEEK SNUFFED OUT THE LIVES OF SIX ALLISTON DISTRICT RESIDENTS. IT COULD HAVE BEEN YOU! A modern passenger train rolling along at 70 miles an hour is a 1500-ton projector of tremendous weight and velocity…Give the big fellow the right of way!

Remember…140 people were killed and 549 injured in 443 railway crossing accidents last year.' [1096]

Six people lost their lives when a Southern Railway freight train collided with a car on a level crossing at Belle Mina, Alabama, United States, on Friday 17th November 1950. Those who died were all members of one family. They were William Bragaw, 27, of Southport, North Carolina, a political science instructor; his wife Louise, 30; their children Louise, 5, and Margaret, 4; Mr Bragaw's mother, Mrs Helen Bragaw, 65; and Mrs Bragaw's mother, Mrs Eleanor Niernsee, 70. The family were returning home after a visit to Mrs Niernsee's son in Memphis, Tennessee. According to Brakeman T.R. Martin of the train, the new Plymouth Tudor car was struck at 1.5 p.m.

There were several witnesses to the accident, who informed highway patrolmen that the warning lights and bell at the crossing had been working properly and that they had heard the engine of the 48-car train whistling well before it had reached the crossing. Mrs Eugene Stewart, the wife of a filling station attendant, and Mr Harry McCurdy, filling station attendant, agreed that the car had flashed by at between 50 and 60 miles an hour. They believed that had the car simply maintained that speed it would probably have just got across the line in front of the train, but its driver, whom it was assumed was Mr Bragaw, had jammed on his brakes at the last moment. There were 30ft skid marks leading up to the crossing. Striking the right side of the train the car was carried down the line till it hit an empty coal tender on an adjacent track. There had been no other vehicles at the crossing at the time.

[1094] *El Tiempo*, Bogotá, Colombia, 28th September 1950

[1095] *The Lethbridge Herald*, Lethbridge, Alberta, Canada, 17th November 1950

[1096] *The Georgetown Herald*, Georgetown, Ontario, Canada, 22nd November 1950

Harry McCurdy said that he had just finished repairing a flat tyre when Mrs Stewart had cried something like, "He'll never make it!" Looking up, he had seen the speeding car and the train nearing the crossing.

Farmer Tommie Steele and his helper, Hosey Ellison, were showered by broken glass from the wrecked car. They had been in a lorry running along a road which paralleled the railway, going in the opposite direction. Steele said that he had heard the train whistling and the crossing bell ringing but had taken no particular notice of them. Then suddenly he had seen the car drive into the path of the train and simultaneously had heard the crash of the collision. At that point he had shut his eyes because of the flying glass.

Mrs Lance Pickett, who lived near the crossing, said, "I was washing when I heard the train blowing hard. I knew something was wrong. I dashed to the front of the house." Although a clump of trees partly hid the line she could see the railwayman on the right-hand side of the engine cab leaning out and looking down. At the same time she heard the sound of ripping metal as the car was dragged along. "It's a noise you aren't likely to forget," she added. Alberta Parker, another witness, spoke of the seeing the car crumpled up against the front of the locomotive. She and Mrs Picket agreed that there had been no sign of life in any of the victims directly after the accident. [1097]

Twelve people died and six were badly enough hurt to be taken to hospital when a train bound for Tokyo collided with a bus on a level crossing at Urawa, Japan, on Monday 18th December 1950. 'Faulty operation of a crossing gate' was blamed for the accident. The brief reports on the disaster did not make it clear whether this phrase meant human error or an equipment failure. [1098]

A woman and four of her five children lost their lives in a level crossing accident at Grandview, Michigan, United States, at 5.55 p.m. on the evening of Wednesday 20th December 1950. The car in which the family was riding had turned off the main road, M-21, onto a side street, when it reached a crossing with the Chesapeake and Ohio Railway line and it was struck by a freight train. Engineer John Reed said that the car had appeared at first to stop but it had then lurched in front of the train. He had already been in a collision with a truck earlier in the day at Michigan City in which the truck driver had been injured. The car was carried 2,600 yards down the line. There was no protection at the crossing, not even warning lights.

Mrs Kathryn Smith, 30; her sons Boyd, 12; Larry, 10; and Donald, 4; and her daughter Sandra, 7, were all killed. Another son, Leroy, 9, suffered fractures of both legs, head lacerations and possible concussion. He told attendants at St Mary's Hospital, Grand Rapids, that his mother had been shopping for Christmas presents and they had been on their way home to Jenison, Michigan, at the time of the accident. His father, Varian Smith, collapsed when informed of the accident and it was several hours before he had recovered sufficiently to visit his surviving son in hospital.

Kent County Prosecutor Roger McMahon announced an investigation into the accident, which would include a determination into whether the train had been going too

---

[1097] *The Decatur Daily*, Decatur, Alabama, United States, 19th November 1950

[1098] *Sarasota Herald-Tribune*, Sarasota, Florida, United States, 18th December 1950 fast. [1099] [1100]

A mistake by a crossing keeper led to a disastrous accident near Podivin, Czechoslovakia, on the evening of Thursday 21st December 1950. A large bus of the ČSAD (the state bus company) had stopped at the lowered barrier of the crossing on the Břeclav – Brünn line. A passenger train went over the crossing, whereupon the crossing keeper raised the barriers and the bus started over the line, only for it to be struck by a Bratislava – Brünn express travelling in the opposite direction. The bus was cut in two. No less than 34 bus passengers were killed and 56 were injured. [1101] The crossing keeper, as might be expected, was arrested. [1102]

Boxing Day, Wednesday 26th December 1950, saw a serious level crossing collision in Argentina. Ferrocarril General San Martin train No. 10 from Rufino, in Santa Fe Province, was on its way to Buenos Aires Retiro, where it was due to arrive at 16:10. It was in charge of Conductor Alberto Gonzáles with Driver Justo A. Muñoz and Fireman Ideal Singuida. The road vehicle involved was a bus, or possibly a lorry performing the function of a bus, carrying 14 passengers. A party of Jews living in Buenos Aires had travelled on the previous Saturday to the Julio Levi Jewish Colony, situated between Agote station on the General San Martin line and the nearby Gowland station on the Ferrocarril Domingo Faustino Sarmiento line. After lunch on the day of the accident the party decided to return to the Argentinian capital by train and left the Julio Levi Colony at 14:15 for Goldney station, one stop closer than Agate to Buenos Aires Retiro, on the bus or lorry. The accident happened on a level crossing between Agate and Goldney stations.

Nine bus passengers died in the collision and the other six were hurt. Various vehicles took them to hospitals in Mercedes and Luján. [1103]

Valencia – Barcelona mail train No. 782 collided with a lorry carrying textiles, parcels, and miscellaneous goods travelling from Barcelona to Valencia, on a level crossing at Vilallops, Spain, five kilometres from Tortosa, at 04:00 on Tuesday 23rd January 1951. The lorry was dragged about 300 metres and the engine, leading brake van, a mail van and three third-class coaches of the train were derailed. The sudden stoppage of the train resulted in the telescoping of two of the third-class carriages, one embedding itself about halfway into the other, and it was in this part of the train that the deaths and injuries occurred.

The mayor of Tortosa chanced to be a passenger on the train and was unhurt. He at once commandeered a car and proceeded thence to organise a rescue operation. The Tortosa Red Cross, the Civil Guard, doctors, firemen, and a large number of locals were soon on the scene and worked to release the injured for transfer to the hospital and clinics, a task made more difficult by the pre-morning darkness. Buses, lorries, and private cars helped to transport the casualties. Red Nacional de los Ferrocarriles

[1099] *Traverse City Record-Eagle*, Traverse City, Michigan, United States, 21st December 1950
[1100] *The News-Palladium*, Benton Harbor, Michigan, United States, 21st December 1950
[1101] Wikipedia – Liste von Eisenbahnunfällen in Tschechien
[1102] *The Newcastle Sun*, Newcastle, New South Wales, Australia, 23rd December 1950
[1103] *Clarin*, Buenos Aires, Argentina, 26th December 1950

Españoles (RENFE) also sent a rescue train to the scene with men and equipment. [1104]

Fourteen people died in the accident; six men, four women and four children; and about 30 were injured, a number seriously. The three occupants of the lorry were amongst those who were hurt. Uninjured and slightly injured passengers were evacuated by train or remained temporarily in local residences.

Driver Felipe Lagos told reporters that as his train had come around the corner he had spotted the headlights of the lorry on the level crossing. He had applied his brakes in emergency but had been unable to avoid the collision. His engine had then begun to ride very roughly, which he attributed to pieces of cloth from the load of the lorry having got under its wheels, and it had then left the rails altogether, taking the leading vehicles of the train with it. [1105]

A second report on the disaster stated that Crossing Keeper Justo Fernández had been arrested, implying that it was through a mistake on his part that the lorry had been permitted to enter the crossing when the train had been close at hand. [1106]

A car driver returning from an evening church service with his family apparently failed to notice the ringing of the warning at a level crossing at Sheridan, Indiana, United States, and his car was struck by the Chicago – Indianapolis *Hoosier* streamliner of the Chicago Indianapolis and Louisville (Monon) Railroad on Thursday 25th January 1951. Edmond Snow, 33; his wife Jewell, 34; and their children Robert, 15; Iris Joan, 11; Iva Jean, 9; and James Lee, 7, were all killed when the 1940 automobile in which they were riding was struck and carried 200 feet along the track. Iris Joan was the only one still alive when she was removed from the wreckage; she died in the ambulance taking her to an Indianapolis hospital. [1107] [1108]

Engineer Leo Goodman said that the car driver "drove right onto the crossing and I never had chance to stop."

The accident occurred at 9.54 p.m. The train was running 24 minutes late.

The warning bell was the only protection provided at the crossing; it had neither barriers nor warning lights. [1109]

Friday 9th February 1951 was a bitterly cold day in Coniston, Ontario, Canada, with a lowest temperature of -47° Fahrenheit (-44 Celsius). There was also a thick fog. At 8 a.m. Canadian Pacific Railway train No. 7, due in nearby Sudbury at 7 a.m. but running rather more than an hour late, had stopped at Coniston station, and was just picking up speed again as it approached a level crossing. Also approaching the crossing was a Nickel Belt Coach Lines, carrying mainly men returning home after finishing their night shift at the Coniston plant of the International Nickel Company, together with five women going to work in Sudbury, seven miles away. Engineer D.F. Dowdall was driving the train and Edouard Carriere was in charge of the bus.

1104 *El Adelanto*, Salamanca, Spain, 24th January 1951

1105 *La Vanguardia*, Barcelona, Spain, 24th January 1951

1106 Ibid, 25th January 1951

1107 The Rushville Evening Daily Republican, Rushville, Indiana, United States, 26th January 1951

1108 *The Tipton Daily Tribune*, Tipton, Indiana, United States, 26th January 1951

1109 *The Daily Plainsman*, Huron, South Dakota, United States, 26th January 1951

Dino Santi of Coniston said that he had heard the train whistle as it neared the crossing with highway No. 17.

Driver Carriere said that he had stopped his bus as usual at the crossing and even opened the door to look down the tracks. He told reporters, "I didn't see or hear the train. You couldn't see a dammed thing. The frost fog was as thick as molasses. I didn't hear any train whistle so I closed the door and moved on. Then it came – a terrible crash. The train caught the rear corner of the bus and swung around…into a snowbank."

Adrian Trepanese, 49, one of the workmen on the bus, told a reporter, "All at once there was a big crash. I can just remember bodies flying past me. A girl sitting in the front seat – I don't know her name – flew directly over my head and landed at the rear of the bus.. I was sitting just behind the rear door of the bus. I think it stopped at the crossing but I am not sure. The bus was crowded and people were standing in the aisles. The bodies just piled up at the rear of the bus. Two men pulled me out through the side."

Seven men died at the scene and 18 men and the five women were injured. One of the injured men died as he was being taken to hospital. Both the dead and the injured were taken to Sudbury on the train. Most of the injured went to St Joseph's Hospital, where 12 doctors worked on them, though three were taken to Sudbury General Hospital. Trepanese was among the injured and was said to have been scarcely able to speak afterwards.

'A first-aid team of seven men headed by the nickel company's safety engineer, E. O'Brien rushed to the scene after the resounding crash which echoed a quarter of a mile away. They were joined by a dozen first-aid men from Coniston.'

There had been at least 49 people on the bus including the driver, more passengers than usual because some of the workmen had decided to catch the bus rather than walk because of the severe cold.

Enoch Cooper, who had been very seriously injured in the accident, died in hospital four days later, increasing the number of dead to nine. [1110]

Driver Carriere, 21, was subsequently charged with manslaughter but was acquitted at his trial.

It would appear that there were no warning devices at the crossing, not even a bell or flashing lights to warn drivers of approaching trains. [1111] [1112]

A Bond's Tours coach, on a round trip from Adelaide, South Australia, via Melbourne, Victoria, to Sydney, New South Wales, and back, was on its return journey when it was in collision with a goods train on Dimboola Road Crossing, on the outskirts of Horsham, Victoria, Australia, on Saturday 22nd February 1951.

The Natimuk – Horsham train caught the centre of the coach and carried it 50 yards along the line. The coach was badly damaged, the side which was struck being forced in, trapping passengers in their seats before they could escape. Rescuers worked for an hour and a half to free them, watched by a large crowd of bystanders. First on the scene were the members of two cricket teams who had been playing nearby, who tried to free the victims

---

[1110] *The Lethbridge Herald*, Lethbridge, Alberta, Canada, 13th February 1951

[1111] *Medicine Hat News*, Medicine Hat, Alberta, United States, 9th February 1951

[1112] https://www.sudbury.com/local-news/68-years-ago-today-remembering-the-deadly-coniston-bus-crash-1235367 with borrowed axes and hacksaws but could make little progress on the metalwork of the coach until oxy-acetylene equipment was available to cut away portions of the coach roof and seats. Meanwhile nurses crawled through the wreckage to administer morphia to trapped passengers. Three women who had jumped from the coach at the moment of the impact reportedly fell beneath the wheels of the train and had limbs amputated. Prominent in the rescue work until he collapsed from exhaustion was the coach driver, Reginald Forbes, who had only suffered minor injuries in the accident. As each casualty was released they were handed over to a team of four doctors and eight nurses for treatment before being rushed by ambulance and Jeep to the Wimmera Base Hospital.

The accident happened at 3.43 p.m. and the last victim was released at 5.20 p.m. Of the 22 passengers on the bus three were killed outright, six died on the way to hospital or shortly after arrival there and two more succumbed the following morning. The train crew of three was unhurt.

The only outside witness to the accident was 82-year-old Fred Holland, who lived in a house only 20 yards from the crossing; he had been stood outside trimming a hedge at the time and heard the shrill whistle of the train's engine. He said, "It whistled twice more, and then out of the corner of my eye I saw the coach. I knew they could not miss." Then he had heard the noise of the crash followed by the screams of passengers. "I ran over to the wreckage but couldn't do much because of my age." [1113]

At the start of the inquest on the victims of the accident in July Victorian Railways Locomotive Superintendent George Burnell testified that the train would have been travelling at about 15 miles an hour when it struck the bus. It had run 60 yards after the emergency brake application.

Salesman John Stewart, a passenger on the bus, said that he had heard a crash and had found himself in the wreckage. He went on, "I felt the bus swerve. I remember hearing a scream and then a terrific noise and the bus was rolling over." The passengers had not been singing, and he had heard no whistle or other warning of a train. The bus had been "just cruising slowly."

Another passenger, Mrs Florence Rigby, limped into the witness box and was allowed to sit down to give her evidence. She recalled, "A woman sitting beside me, a few seconds before the smash, said, 'Oh, my God, look out!' I then saw a railway engine level with the bus. The train hit the bus and I did not know anything for a few moments."

Fireman James Doherty testified that as the train had neared the crossing his driver had sounded the whistle three times – a quarter of a mile, 300 yards and 150 yards from the crossing. He continued, "When the front of the engine was about 40 yards from the crossing I saw a bus about 60 yards from it, travelling about 25 miles per hour. It appeared to slow down but then picked up speed as if to try to get across. I called out, 'Red light!' and Driver Roberts immediately put on the Westinghouse brakes. We hit the bus somewhere in the middle and carried it along for about 45 yards." He added that the crossing "was not such a dangerous one." [1114]

Driver William Roberts confirmed his fireman's statement that he had whistled three times for the crossing. He explained that from his side of the engine he had not been able to see the bus but as soon as his fireman called out he had applied his brakes in emergency.

[1113] *The Age*, Melbourne, Victoria, Australia, 26th February 1951

[1114] *The Herald*, Melbourne, Victoria, Australia, 24th July 1951

A statement from the bus driver was read out. "I looked to my right and could see the line was clear. I looked to the left and the sun struck the windscreen and caused a momentary blur. As the bus was on the crossing I could see something causing a shadow across the bus. I looked to the left and could see a train bearing down upon us and realised that it was that which was blotting out the sun. I accelerated to swing the bus to the right, and then the train struck us." [1115]

At the close of the inquest the Coroner, Mr. P.R. Biggin, returned a verdict of "Misadventure" on the victims. Commenting on the absence of any protection at the crossing he stated that all were aware of the impossibility of stopping heavy trains at all road crossings. However, it was not justice to give the railways paramount right of way. The road user was entitled to a clear view of the line. "When engineering difficulties make it impossible to give the oncoming driver this view, I feel that some warning signal should be provided," he said. "In this case the intersection was on a main highway on the outskirts of a city with a large population and there was not a clear view of the line. It is not sufficient, in my opinion, to say that this places an extra burden on the road user. Human life is at stake." [1116]

The line over the crossing was only used by a dozen trains a week.

Bad weather was doubtless a factor in a minor level crossing collision involving a school bus at Eau Claire, Wisconsin, United States, on Friday 2nd March 1951. During a blizzard a bus of the Eau Claire Transportation Company was struck by a Minneapolis St Paul and Sault Ste Marie (Soo Line) Railroad switch engine. The bus very likely stalled on the crossing, as some reports spoke of its quick-thinking driver, Gordon Cole, getting the 25 children, who had been returning from school, off of the vehicle just in time. The engine struck the bus near its rear, the pupils escaping via the side and front doors. Four children were treated in hospital for minor injuries and sent home. [1117] [1118]

The level crossing accidents involving school buses continued with a considerably more serious one on the afternoon of Tuesday 13th March 1951, at Elam Crossing, Virginia, United States, about 14 miles west of Farmville. Crowded with 44 black schoolchildren, the bus was struck by a Norfolk and Western Railway train as it crossed the main line between Roanoke and Norfolk. The train caught the rear of the vehicle, shearing off about a quarter of it and breaking it into three or four pieces. Five of the children, who had been returning home from the elementary and high schools, were killed, and 10 more were hurt, one seriously. The injured were brought back to Farmville, to the Southside community hospital there.

The bus driver, John Arthur Robinson, told police afterwards that he had not seen the train and had been in the process of crossing the line when he had felt a blow. [1119] [1120]

Five people died on Thursday 12th April 1951 when the car in which they were riding

---

[1115] *Warwick Daily News*, Warwick, Queensland, Australia, 25th July 1951
[1116] *The Horsham Times*, Horsham, Victoria, Australia, 27th July 1951
[1117] *The Winona Republican-Herald*, Winona, Minnesota, United States, 3rd March 1951
[1118] *Racine Journal-Times*, Racine, Wisconsin, United States, 3rd March 1951
[1119] *The Charleston Gazette*, Charleston, West Virginia, United States, 14th March 1951
[1120] *The Oil City Derrick*, Oil City, Pennsylvania, United States, 14th March 1951

was struck by the northbound *Texas Special* passenger train of the Missouri Kansas Texas Railroad of Texas on a level crossing at Lorena, Texas, United States, 12 miles south of Waco. Those who died were Mrs James Hay, 34, of Lorena; her 11-month-old daughter Rita Alice; her mother, Mrs Ethel McNeil, of Waco; and Mr and Mrs Arthur Denning, also of Waco. Mrs Hay was Mrs Denning's niece.

Mr R.A. Yates, of Lorena, who witnessed the accident, said that the car had almost cleared the track when the train hit it. He speculated that the strong wind that had been blowing at the time had prevented Mr Denning, the driver of the car, from hearing the whistle of the train. [1121]

The high speed of a car was blamed for a level crossing collision on Highway 133 near Payne, Minnesota, United States, at about 10 a.m. on Friday 4th May 1951. The fast-moving automobile struck the baggage car of a Duluth Missabe and Iron Range Railway passenger train 70 feet behind the front of the engine. The driver of the car, Mr Edward Maloney, 32, of Los Alamos, New Mexico; his sister-in-law, Mrs Marcel Duvivier, 31; and Mrs Duvivier's daughters, Ann Marie, 4, and Marie Therese, 18 months, were all killed instantly. Mrs Duvivier had been seven months pregnant and her baby was born alive prematurely as a result of the accident. It died in the ambulance taking it to hospital.

Mrs Esther Maloney the wife of the driver and sister of Mrs Duvivier, survived the accident with slight injuries, a fact which she attributed to having been sat in the back seat of the car surrounded by piles of clothing and baggage, which had shielded her from the impact. She was treated at Hibbing hospital and released. She told investigating officers, "I don't remember much about it. I heard a train whistle. Then Mrs Duvivier cried out, 'There's a train!' I was knocked out almost as soon as the words cleared her lips."

There was an unobstructed view along the line at the crossing for 200ft in either direction.

Mr Maloney had been reportedly on his way to Oklahoma to report for military service and was taking his sister-in-law and her children to their home in St Paul, Minnesota. [1122]

After consideration the Coroner, Dr John Ekblad, decided that there was no need to conduct an inquest. [1123]

A car carrying seven people on their way to a Saturday night dance was struck by a Chesapeake and Ohio Railway freight train on a level crossing on the outskirts of Blenheim, Ontario, Canada, on Saturday 12th May 1951. The car was dragged 150 yards along the track before the train could stop. Four of the occupants of the car were killed outright, two died later in Chatham Hospital, 10 miles away, and the seventh was critically injured.

Those who died were Mary Doreen Como, 18; Kenneth Crawford, 18; Marjorie Graham, 14; William Hughes, 19; William Merritt, 22; and Rose Marie Therrien, 17. Neil Merritt, 19, was said to be "only fair" in hospital.

It was believed that the party had been en route to a dance at Rondeau Park on Lake

---

[1121] *Valley Morning Star*, Harlingen, Texas, United States, 13th April 1951
[1122] *Austin Daily Herald*, Austin, Minnesota, United States, 4th May 1951
[1123] *The Winona Republican-Herald*, Winona, Minnesota, United States, 5th May 1951

Erie. Two other cars had already been stopped at the crossing when the victim's car overtook them and drove into the path of the train. Police said that it was not clear who had been driving but that the accident had happened in ideal weather conditions. The following day no decision had been made regarding an inquest. [1124]

On Saturday 2nd June 1951 a convoy of five lorries and a motor coach took a party of foundry workers from Sarnico on an excursion to Lake Isco. They arrived at their destination safely but disaster struck at 17:15 on the return journey via Louvre when the convoy reached an unprotected level crossing with the Brescia – Isco – Edolo line at Stanghe, Italy, near Pisogne. The first four lorries passed over the crossing safely but the fifth lorry was struck by a train, a two-car multiple unit, travelling reportedly a full speed. The violent impact smashed the lorry and carried the remains of it, with the bodies of its occupants, 200 metres along the track. The first car of the train was derailed and the driver's cab was smashed but Motorman Ernesto Ferrari and the train passengers were unhurt. Of the 37 workers on the coach seven were killed outright, six died en route to hospital or soon after arrival there, and another 20 were injured, two of whom were thought to be dying.

The line was soon cleared after the accident and normal train services resumed four hours later. A good many people went to the accident scene, as did Prefect Aria [1125] [1126] (in Italy the prefect is the national government's representative in each province).

The Estrada de Ferro Central do Brasil suffered a large number of accidents in the early 1950s, including a horrific level crossing collision at Nova Iguaçu, Brazil, on Thursday 7th June 1951, which was not in any sense the railway's fault.

The circumstances of the disaster are a little unclear, but it would appear that at about 04:15 a Standard Oil of Brazil road tanker, loaded with 18,000 litres of petrol, was approaching the level crossing when it broke down, or possibly stalled, on the hill leading to the crossing. The driver of the tanker, Orlando Madeira, then asked the driver of a following lorry to give his vehicle a push. The latter agreed to do so but in the darkness neither of them realised the proximity of the Dr Tibau level crossing, close to Nova Iguaçu station, the barriers of which had already been lowered for a train. As a result the petrol tanker was pushed through the barrier onto the level crossing, where a second or so later, at about 04:20, it was struck by electric passenger train UM-12, which had left Queimados at 04:03 for Rio de Janeiro Dom Pedro II. There was immediately a violent explosion and the leading car of the train was enveloped in flames. The train ran for about another 200 metres before it stopped. Motorman J. Santana apparently attempted to unlock the train doors from his cab but failed, perhaps because of the damage caused by the collision with the tanker. Like the remainder of the front car his driving cab was on fire and in desperation he then jumped through a window, but was burnt to death outside the train, nevertheless. The train conductor likewise jumped through a window and suffered a similar fate. Of the passengers in the leading car, most of whom had been asleep, only four escaped from it alive; two hours after the disaster firemen had recovered the bodies of 51 victims, including that of Madeira, who had fled from the cab of his

---

1124 *The Lethbridge Herald*, Lethbridge, Alberta, Canada, 14th May 1951

1125 *La Stampa*, Torino, Italy, 3rd June 1951

1126 *The Times*, London, England, 4th June 1951

tanker but perished from burns on the road nearby. The driver of the lorry which had been pushing him fled and survived.

As soon as he learnt of the disaster Lieutenant Colonel Souza Gomez, the director of the company, and his assistant, Engineer Niconar Pereira, took immediate measures for the assistance of the injured, as did Stationmaster João Martins Senna, in charge of Dom Pedro II station, and his assistants Sebasto Vicira and Claudionor Gomez. In addition to those who were killed outright, 47 more passengers were injured, of whom 37 were taken to the Carridade Hospital in Nova Iguaçu. Three of the injured succumbed in the days following the disaster. The line was blocked for a considerable time as a result of the accident.

While accepting that the accident had arisen from the negligence of persons unconnected with the railway and admitting the serious financial position of the railway company, a newspaper urged the speedy replacement of level crossings by over or underbridges, especially given the increased speed and frequency of trains following the replacement of steam by electric traction.[1127] [1128]

Sixteen people died and 11 were injured when a goods train collided with a bus on a level crossing between Sedusar and Bhagat Ki Kothi stations, India, on the Jodhpur State Railway. The accident happened on the evening of Friday 9th September 1951.[1129]

A serious level crossing accident occurred on Tuesday 10th June 1951 on the Moselbahn, a small railway operated at this date by Vereinigte Kleinbahn AG, on an open level crossing between Ruwer West and Kenn, West Germany. A so-called Bahnbus, owned by Deutsche Reichbahn, had been hired by the Cäcilia mail choir, together with a trailer, to convey 75 of its members from Zewen to Wittlich, where their late conductor was interred. The party then went to Bremm, spent some hours there and set off home at 21:00 via Schweich. At about 22:46 the bus and trailer reached the level crossing on federal highway 49 at the same time as a steam-hauled train from Trier. The train was only moving at 15 kilometres an hour, its driver whistled for the crossing, and the warning lights at the crossing were operating. However, visibility at the crossing for vehicles going towards Ruwer was poor, the noise of the bus's engine and its passengers probably drowned out the sound of the train's whistle, and perhaps most to the point, the bus driver had been drinking. Ignoring the warning lights he tried to cross in front of the train, the rear of the bus was torn open by the engine and its trailer was overturned and smashed. Six bus passengers were killed outright, nine died later of their injuries, about 20 were seriously hurt and 13 were slightly hurt. The bus driver escaped injury but was sentenced to nine months' imprisonment for drunk driving and causing the accident.[1130]

Ten people died when an Oran – Algiers express train collided with a horse-drawn mail van on a level crossing near Les Attafs, Algeria, on Wednesday 13th June 1951. The

---

1127 *Jornal do Brasil*, Rio de Janeiro, Brazil, 8th June 1951

1128 *O Estado de São Paulo*, São Paulo, Brazil, 8th June 1951

1129 *The Indian Express*, Madras, India, 10th November 1951

1130 https://de.wikipedia.org/wiki/Eisenbahnunfall_von_Kenn

victims were all passengers in the mail van. [1131]

Passenger train P 1745 collided with a bus on the line between Waldshut and Immendingen, West Germany (via Oberlauchringen and Hintschingen) on Tuesday 19th June 1951. The crossing was protected by warning lights which the bus driver failed to observe. Seven people died and 14 were hurt. [1132]

A lorry carrying eight workmen of the Benaventa Electra company was in collision with a coal train on a level crossing between Pobladura del Valle and Benavente stations, Spain, on Friday 24th August 1951. Workmen Ángel Abril, Nicanor Barros, Diego Blas Cara, Eufemio Oliva, Pablo San Francisco and Victoriano Sendin were killed, together with a seventh man who could not immediately be identified. The only survivor, Santos Zanseno, was badly hurt. As soon as news of the accident was received a relief train was sent to the scene from Zamora. [1133]

A car collided with the tender of the engine of a Canadian National Railways freight train of 41 cars at a level crossing at McGivney Junction, New Brunswick, Canada, on Friday 7th September 1951. August Lapeigne, the driver of the car; Antoine Nowlan, Melaine Saulnier, and her daughter Odecia, aged seven, were all killed outright and the fifth person in the car, a woman who could not immediately be identified, died of her injuries en route to hospital at Fredericton. There were no witnesses to the accident. [1134]

A lorry loaded with pilgrims who were returning to Mexico City from a religious festival at the Catholic shrine at San Juan de los Lagos, north-east of Guadalajara, was struck by a freight train at Irapuato, Mexico, about 260 kilometres from the capital. Nineteen people lost their lives in the accident on Wednesday 21st November 1951. [1135]

Frank Miles, aged 51, was killed when the school bus that he was driving was struck by a Chicago Milwaukee St Paul and Pacific Railroad freight train on a level crossing two miles east of Odin, Iowa, United States, early on the morning of Friday 30th November 1951. The bus was empty, either because he had not yet picked up any pupils or because he had already dropped them off at school. Fog was probably a contributary factor. [1136]

Eight bus passengers were killed and seven were injured when a country bus was struck by an express train on a level crossing at Mahalla el Cobra, Egypt, outside of Cairo on Wednesday 5th December 1951. [1137]

1131 *The Times*, London, England, 14th June 1951
1132 *Katastrophen der deutschen Bahnen Teil II* by H.J. Ritzau pg. 245 (in German)
1133 *La Vanguardia*, Barcelona, Spain, 25th August 1951
1134 *Winnipeg Free Press*, Winnipeg, Manitoba, Canada, 7th September 1951
1135 *El Paso Herald-Post*, El Paso, Texas, United States, 22nd November 1951
1136 *Fort Madison Evening Democrat*, Fort Madison, Iowa, United States, 1st December 1951
1137 *Singapore Standard*, Singapore, 6th December 1951

An Oran – Algiers express collided with a lorry on a level crossing near Oran, Algeria, on Tuesday 11th December 1951. Five passengers were killed and three injured. The brief report did not indicate whether they were on the train or the lorry. [1138]

The engineer of the Grand Trunk Western Railroad train which struck a car on a level crossing at Fenton, Michigan, United States, and the driver of a car which was following that which was involved in the collision, agreed that the driver of the vehicle had been attempting to beat the train over the crossing. Engineer Thomas Kruse of the Detroit – Durand train said later, "I saw the car coming, but I had no idea it was going to try to beat my train." The motorist, Karl Wykoff, said, "The driver of the other car appeared to see the train. All of a sudden the other car speeded up as if trying to beat the train to the crossing."

Among the seven who lost their lives in the accident on Sunday 13th January 1952 were three married couples. The victims were Leroy Bexell, 23, the owner of the car, and his wife Sheila, 21, of Fenton; William Chapman, 24, and his wife Nancy, 16, of Flint; Frank Reitano, 29, and his wife Donna, 21, of Fenton; and Kenneth Tipplar, 26, also of Fenton. The Bexells had only been married on 22nd December and the Chapmans on 10th November. The Reitanos had three children. Tippler was survived by his widow and one child.

According to the police the occupants of the car had stopped briefly at a tavern but it was not believed that they had been intoxicated. The automobile was badly wrecked and the bodies were thrown some distance from it, so that it was not possible to determine who had been driving. [1139]

A major disaster occurred near Wystepy, Poland, on Saturday 19th January 1952. At 04:00 that morning a 'bus', in reality an American army surplus lorry, set off from Kamieniec to take workmen to Kielce and Skarźyskow. Its driver was very tired, having had little sleep. Passengers boarded the bus, it set off, other passengers got on en route and it arrived at Kielce, where some of the workmen alighted, and it then continued its journey towards Skarźyskow. This part of the journey included the level crossing, where the barriers had been lowered for a freight train at about 07:00. The lorry driver had probably dozed off at this point. He failed to stop before his vehicle collided with the crossing barrier, when he instinctively swung it to the left, but its right side struck the penultimate waggon of train, it was spun around and its left-hand side then hit the rear waggon of the train. Seventeen of the 35 or 40 bus passengers died and others were injured. The lorry driver survived and fled into the forest but was later apprehended by the Citizens Militia. [1140]

Six people from one family died when a Southern Pacific Railroad train hit their car on a level crossing north of Fresno, California, United States, on the evening of Wednesday 23rd January 1952. The victims were identified as Juilo Sabala Sr., 52; his wife Victoria, 27; their sons Julio Jr., 9, and David, 5; and relatives Raymond Agular, 19

[1138] *The Railway Gazette*, London, England, 21st December 1951

[1139] *The Ludington Daily News*, Ludington, Michigan, United States, 14th January 1952

[1140] https://echodnia-eu.translate.goog/swietokrzyskie/te-straszna-katastrofe-przez-lata-ukrywaly-wladze-dzis-odslaniamy-jej-tajemnice/ar/8410492?_x_tr_sl=pl&_x_tr_tl=en&_x_tr_hl=en&_x_tr_pto=sc

and his sister Emily, 17. [1141]

A southbound St Louis-San Francisco Railway freight train, in charge of Conductor R.R. Wilson with Engineer Vinson and Fireman Jackson, collided with a Ford station wagon on a crossing south of the station at Exeter, Missouri, United States, on Sunday 9th March 1952. The vehicle was carried a distance of seven telegraph poles along the track. Three of its six occupants were thrown clear of it and the other three were caught under the wheels of the diesel engine. All were killed outright or fatally hurt. They were Mr O.J. McQueen and his wife, both aged 59, who lived on a farm nearby; their niece Mrs Corrine Tarvin, 31; and Mrs Tarvin's children, Barbara Beasley, 14; Jerry Beasley, 10; and John Tarvin, 5. Mrs Tarvin's husband, Kelly, was a soldier serving in Germany; she and her children had been planning to join him in three weeks.

Visibility at the crossing was somewhat restricted by a box car stabled on a siding and it was presumed that this had prevented the driver of the station wagon (variously stated to have been Mr McQueen and Mrs Tarvin) from seeing the train. The accident happened on State Highway 44 just after 3 p.m. Barbara Beasley was taken from the wreckage alive but died a few minutes after having been taken by ambulance to the Berry County hospital in Cassville. The train, bound for Fort Smith, Arkansas, was delayed three hours. [1142] [1143]

The unprotected Stange Road Crossing with the New York Central Railroad near Graytown, Ohio, United States, was the scene of a fatal collision at 8.05 p.m. on Wednesday 16th April 1952. A car owned by Mr K. (Gerry) Wangran, 33, and containing him, his wife Eileen, 34; their daughter Janet, 3; Mr John R. (Rudy) Danschroeder, 69; and his wife Antonia, 64, was involved. All were killed instantly. Engineer F.J. Smith of westbound passenger train No. 43 stated that he had not seen the car until it had passed directly in front of his engine and it was presumed that Mr Wangran, whom it was supposed was driving the automobile, had failed to see the train either. The car was dragged about a mile, the two men remaining in it but the two women and the child being thrown out. The Danschroeders came from Graytown and the Wangrans from nearby Genoa. Other than the engineer there were no witnesses to the accident. [1144]

On the afternoon of Thursday 17th April 1952 the *Hummingbird* streamliner of the Louisville and Nashville Railroad collided with a fire engine on a level crossing at Pascagoula, Mississippi, United States. Volunteer Fireman Joseph Williams was killed outright, Assistant Fire Chief H. Kline Thornton died in hospital about an hour after the accident and Fireman Dewitt Monroe and Fire Chief Jim Hudson were injured, the former seriously. The train was in charge of Conductor E.E. Burnett with Engineer Oscar Hood.

School bus driver John Coe, sitting in his vehicle, witnessed the accident. He said that he had heard the fire truck approaching, its siren sounding, and at the same time saw the train approaching from the west, apparently at low speed, with the bell of its locomotive ringing. The driver of the fire engine apparently realised too late that there was not time

---

[1141] *Panama City News Herald*, Panama City, Florida, United States, 24th January 1952
[1142] *Joplin Globe*, Joplin, Missouri, United States, 11th March 1952
[1143] *Neosho Daily Democrat*, Neosho, Missouri, United States, 10th March 1952
[1144] *Ottawa County News*, Port Clinton, Ohio, United States, 18th April 1952

for him to get across and attempted to swing to the right parallel to the line but the train caught the rear of the vehicle. The fire truck scraped along the side of the locomotive before coming to rest about 50 yards from the crossing, while its occupants were thrown out on either side of the track.

Another witness, Emory Warwick, insurance man, who had crossed from north to south shortly before, said that the crossing lights and bells had been working and the train had seemed to be stationary in Pascagoula station, 500 yards to the west.

The fire engine was virtually demolished. It had been bought by the city of Pascagoula for $16,000 only a month before. The train, in contrast, was little damaged, and continued eastwards after about 45 minutes. [1145]

An Atlanta and West Point Railroad train bound for Montgomery, Alabama and New Orleans, Louisiana, collided with a school bus at a level crossing on the Central of Georgia and Atlanta and West Point Joint track just outside of East Point, Georgia, United States, on the morning of Friday 9th May 1952.

Car salesman A.T. Pittman witnessed the accident and was one of the first to begin the rescue work. He said that there had been two school buses. The first had got over the four tracks at the crossing safely. The second had halted at the crossing momentarily, then had started over the crossing directly in front of the engine of the train, which had been running at 20 or 25 miles an hour. This bus, too, had almost made it but the train had struck its left rear end and pushed it 75ft down the track. He said, "We started getting them out as soon as we got the emergency door open. It was locked. They were piled up on the floor and there were pocketbooks and lunches scattered all over. There was a lot of screaming and the first five or six who got out of the bus under their own power appeared to be stunned."

Also early on the scene was Motorcycle Patrolman L.V. Roberts, who said that he had found the children stunned and sprawled in the aisles of the bus, across seats and slumped under them. He related, "They were bleeding, some from the nose, many from cuts, while others were unconscious. They were bleeding too." He added that people who had rushed from nearby stores and residences had begun rescue work and a few of the casualties had been laid out alongside the track.

There had been 35 high school girls on the bus (the boys presumably had been on the other bus) of whom 17 were hurt, one possibly fatally. The injured were transported in ambulances and cars to Grady Hospital in Atlanta. [1146]

What sort of protection was provided at the crossing was not stated. [1147]

Australia suffered another bad level crossing accident involving a bus on the evening of Sunday 1st June 1952. This time the location was Boronia, in the state of Victoria. At 8.36 p.m. a bus carrying 32 Churches of Christ students from Caulfield, Coburg and Essendon churches was struck by an electric train coming from Fern Tree Gully.

Mr John Sutton, an eyewitness to the accident, spoke of the bus being torn in two and its driver's seat being carried yards up the track. He continued, "I saw the driver lurch from the

---

[1145] *The Daily Herald*, Biloxi, Mississippi, United States, 18th April 1952
[1146] *Thomasville Times-Enterprise*, Thomasville, Georgia, United States, 9th and 10th May 1952
[1147] *The News and Views*, Jacksonville, North Carolina, United States, 9th May 1952

wreckage and collapse at the side of the railway line. People were lying all over the road in the darkness. I asked one young man if he would like a drink of brandy. 'No, give it to the girls.'" Later, Mr Sutton said, he heard that the young man had been taken straight to the operating theatre at the Alfred hospital in a last-minute attempt to save his eyesight.

A local shopkeeper, Mr Jim Ewen, was quickly on the scene. He said that he had heard a tremendous impact and that the flash of light from the collision had "lit the road like lightning." Rushing to the scene he had seen dazed people from the train standing by the wreckage and injured lying on the ground. He had straight away gone and rung for police and ambulances.

'Nine ambulances raced to Boronia, together with three carloads of police patrolmen. Working in the light of torches and car headlights, about 50 people from the town and from the train tore at the wreckage to free trapped passengers who were still living and could be heard moaning.

'One injured woman was crying for her child, and the rescuers were able to tell her that the child was alive.

'Townspeople praised Dr John Merry, the first doctor on the scene, who performed amputations on the spot with the aid of a young constable. Two trainee nurses from the train worked with rescuers, together with an air force man and a soldier, whom bystanders said did splendid work leading the rescuers.'

Neither the bus driver nor the train driver was much hurt but both suffered from shock. The bus driver, Donald Mountjoy, 40, was somehow thrown from his vehicle. He was too shocked to immediately be interviewed by the police.

The crossing was widely regarded as dangerous. Five people had died there in an October 1944 accident (described in the previous volume of this series) and in total 19 people had reportedly died and 34 had been injured in accidents at it. A number of roads converged at it and the main road approach from one side was down a steep slope. There had been protests raised in Parliament and made to the Railway Commissioners, but the latter maintained that the warning devices at the crossing were adequate.

An initial police estimate was that 13 people had died and that eight had been injured. In fact only seven were killed more or less outright, though two more would subsequently die of their injuries. Eighteen more were injured.[1148]

The inquest on the victims of the accident was originally to be held before a jury of nine, but after protests from Mr J. Starke, the counsel for the relatives of the victims, that the jury was likely to be partisan because it included railwaymen and representatives of the bus company, the Coroner, Mr J.M. Burke, elected to discharge it and continue the inquest alone.[1149]

The second day of the inquest proper heard evidence from Raymond Lumsden, a motor mechanic who lived near the crossing and who described hearing two blasts of the train whistle and then the noise of the crash. He described the crossing as dangerous. Another resident, baker James Marshall, said that he had only heard one whistle but agreed that there could have been two. Boronia Assistant Stationmaster William Jones testified that there had definitely been two blasts. The other assistant stationmaster, Allan Canterford, stated that he considered the crossing to be "perfectly safe."

[1148] *The Age*, Melbourne, Victoria, Australia, 2nd June 1952
[1149] *The Herald*, Melbourne, Victoria, Australia, 7th August 1952

Mrs Betty Drayton, a passenger on the bus, stated that there had been no singing. On a previous journey the bus driver had stopped at the crossing and opened the bus door to listen for a train but she could not recall if he had done so on the evening of the accident. Other passengers agreed that there had been no singing and the radio had not been on, that the bus had slowed down but they could not remember it stopping, and that they had not heard any train whistle. [1150]

Bus driver Donald Mountjoy testified on the third day of the inquest. He said that he had been warned before he had left the Church of Christ at Boronia that he must turn sharp left over the crossing and that it was a bad one. He was not familiar with it. He only had a vague recollection of the journey to the crossing. "I think I started the bus and waited until the passengers got on. Then I reversed the bus for a short distance. I can vaguely remember travelling along the road. My next recollection was an explosion. Then a brief time after that there was a smaller explosion." Then there was a clear recollection of an ambulance. He could not remember if he had stopped at the crossing, if he had seen the train, or if the wireless had been playing.

Earlier that day Assistant Chief Mechanical Engineer Edgar Brownhill had given evidence as to the stopping distances of trains travelling at various speeds and had agreed that the whistles of electric and diesel-hauled trains were less effective than those of steam engines. He was also questioned about the efficiency of the marker lights of the train in warning road users at crossings. [1151]

On the final day of the inquest Motorman Ronald Kellet gave evidence. He said that his train had been travelling at 40 miles an hour as it approached the crossing and he had whistled for five seconds at the quarter mile post. "Almost immediately I noticed a bus coming towards the crossing from the right-hand side. The bus was about 30 yards from the crossing. When I first saw it I was about 300 yards from the crossing. The bus was travelling very slowly – about 5 to 10 m.p.h. I flicked my headlights from dim to bright to dim to bright. The bus seemed to be gradually slowing. I thought he was going to stop, he was going so slowly. I blew a series of short whistles, about four or five. The bus was getting very close, and I could see he was not going to stop. I applied the brakes in emergency. I stepped from the cab and waited for the smash. It was only a matter of seconds." He was grilled by Mr Burke but stressed that he had done all that he could and added that he considered the crossing to be perfectly safe, with a good view. [1152]

Returning his verdict the Coroner exonerated the motorman and found that the bus driver had also complied with the regulations. He was satisfied that the bus passengers had not been singing and that the radio had not been on but he considered that just the noise of the engine of the bus combined with the passengers' talking would have been sufficient to have masked sound of the train whistle. He also believed that the bus driver had stopped before crossing the line but that something had blocked the driver's view of the train, possibly a passenger or a lineside straining post.

Coroner Burke returned a verdict simply that the deceased passengers had died from injuries caused when the bus in which they had been riding came into collision with a train. [1153] He added remarks criticising the lack of more sophisticated protection devices

[1150] Ibid, 21st August 1952
[1151] Ibid, 22nd August 1952
[1152] Ibid, 25th August 1952
[1153] Ibid, 1st September 1952

at the crossing, stating, "I do not think the Commissions have discharged their duty to make this crossing safe for the travelling public. Previous happenings at the crossing should have made them alert to the responsibility they owe to the public to make the crossing safe." [1154]

The Reverend A. Dicke, 55, pastor of Trinity Lutheran Church, Long Prairie, Minnesota; his wife, also 55; their son Martin, 18; their daughter Mrs Dorothy Minke, about 28; and Edwin Kroll, 18, son of Mr and Mrs William Kroll, farmers living near Long Prairie and a chum of their son, were all killed when the car in which they were riding was hit by a westbound Northern Pacific Railway express passenger train on a level crossing at Elk River, Minnesota, United States on Friday 6th June 1952. The engineer of the train said that it had been travelling at 30 miles an hour and that he had only seen the car for a split second before he had hit it and so he had no chance to avoid the collision.

Mr Ralph Williams was close to the crossing when the accident occurred. He described what happened. "The car drove out on the crossing while the warning bells were ringing. The car hesitated for a second on the tracks, then it looked like the driver tried to speed up to get off the tracks. But the train hit the car with an awful crash – like an explosion. I didn't hear any screams."

Another witness was Deputy Sheriff Charles Nelson, who related, "The bells and flashing light signals were working at the time. I turned to my wife and said, 'There's a car going to be hit!' but before I finished the train smashed into the car."

'The club coupe was rammed into one of the railroad signal posts. The post and its concrete base was flipped into the wrecked car. Two of the victims were carried about 100 feet down the tracks and hurled into a food elevator. Church candles, the pages of a Bible, and graduation congratulation cards were scattered about the wreckage area. Dr Tesch said the five were evidently instantly killed.'

Martin Dicke had graduated only the previous night from Concordia College High School, St Paul.

There had been previous accidents at the crossing, though none recently. [1155]

A car containing six people was struck by a northbound Missouri Pacific Railroad passenger train as it was going over Plum Street Crossing, Monroe, Louisiana, on the evening of Thursday 3rd July 1952. All the occupants of the automobile were thrown from it and killed or fatally hurt as it was carried about 100 feet down the track. The victims were Mr and Mrs Buford Whittington of Monroe, both 43; Mr Clayton Weems, 45; his wife Lilburne, and the Weems' daughters Cheryl Ann, 6, and Sally Kaye, 1. The Weems family were from New Orleans. Two other Weems children, Carol, 14, and Niel, 12, were visiting Monroe with their parents but were not in the car. Mr Whittington worked as a mechanic in Monroe and Mr Weems for the New Orleans federal alcohol and tobacco tax board. Mr Whittington and Mrs Weems were brother and sister. The train was bound for McGee and was in charge of Conductor J.J. Irving with Engineer Coleman Rush.

Mrs W.C. Bell of Monroe, who had just left a market, witnessed the accident. She said she had stopped her car at the crossing when signalled to do so by a flagman. The

---

1154 *The Age*, Melbourne, Victoria, Australia, 2nd September 1952

1155 *Austin Daily Herald*, Austin, Minnesota, United States, 7th June 1952

Whittington car had been on the opposite side of the track. "Suddenly," she recounted, "the train smashed into the car and I could see the car plunge into the ditch with an impact that sent the people flying into the air." It had been raining hard at the time of the accident and it was supposed that with the reduced visibility Mr Whittington, who had been driving the car, had failed to see the flagman signalling to him. [1156] [1157]

Recklessness on the part of a car driver in trying to beat the train was blamed for an accident near Central Islip, New York, United States, at 4.20 p.m. on Sunday 20th July 1952 which resulted in seven deaths. The car was struck by a diesel-hauled Long Island Railroad Ronkonkoma – New York train on a crossing with Lowell Avenue, equipped only with a stop sign. The vehicle was dragged almost 1,000ft by the train and fell down a 20ft embankment. Three of its occupants were thrown out and the other four were described as having been 'burnt to a crisp' when the petrol tank of the car exploded and it caught fire. Hopefully they were all killed in the actual collision. Three firemen also suffered burns fighting the blaze and it took them two hours to extricate the bodies from the wreckage, using acetylene torches.

Assistant Stationmaster John Schmidt had already stopped his car at the crossing for the train and witnessed the accident. He said that the car had overtaken him on the left-hand side and driven onto the crossing in front of the train.

The victims of the accident were wholesale butcher Paul Lamber, 47 (the driver) and his wife Eileen, 37, of Islip Terrace; Mrs Frances Hunter, a practical nurse at Meadowbrook Hospital, 44, of Queens; Mrs Catherine Dempsey, 44, of Cambridge, Massachusetts; Mrs Margaret Harris, 39, of Cohasset, Massachusetts; Mrs Harris' daughter Sheila, 12; and Harold O'Donnell, 9, a nephew of Mrs Harris who lived with her. Mrs Hunter, Mrs Dempsey, and Mrs Harris were all sisters of Mrs Lamber. Those who lost their lives were all dressed in bathing suits and according to one report on their way to the Lake Ronkonkoma resort. Another report stated that they had come from Mrs Hunter's home and were on their way to the Lamber home.

The train was held up over an hour by the accident and its passengers, none of whom was hurt, were eventually transferred to a following train. The line was only single at this point and in the opposite direction a New York – Greenport train was delayed 40 minutes. [1158] [1159] [1160]

The Touseant family had visited relatives, then gone shopping on Monday 4th August 1952, and they were on their way home when their car was in collision with an eastbound Canadian Pacific Railway extra freight train travelling at 20 miles an hour at Petawawa, Ontario, Canada. The car was thrown ten feet, demolishing several fence posts as it rolled over, and its occupants and the groceries that they had purchased were scattered along the right of way. Glass, bits of bread and smears of red paint decorated the front of the locomotive after the collision. Mrs Evelyn Touseant, 25, and her son Edward, 2, were killed outright, her sons Nelson, 5, and Leonard, 3, died shortly after an army ambulance

---

[1156] *The Monroe News-Star*, Monroe, Louisiana, United States, 4th July 1952
[1157] *The Daily Herald*, Biloxi, Mississippi, United States, 4th July 1952
[1158] *The Kingston Daily Freeman*, Kingston, New York, United States, 21st July 1952
[1159] *The Norwich Sun*, Norwich, New York, United States, 21st July 1952
[1160] *Adirondack Daily Enterprise*, Saranac Lake, New York, United States, 21st July 1952

from nearby Camp Petawawa had brought them to hospital in Pembroke and Mr Touseant, the driver, died six hours after the accident. A daughter, Idella, 7, survived with serious injuries.

The crossing was unguarded. According to witnesses the Touseant car had overtaken another car stopped at the crossing, though police were unable to trace the other car afterwards. Visibility at the crossing might have been affected to some degree by the weather, as a light drizzle had been falling at the time. The Pembroke coroner, Dr. J.C. Bradley, had not decided at the time of the newspaper report whether to hold an inquest. [1161]

Flames from a burning petrol tank alerted neighbours at Kirkwood, New York, United States, on the evening of Sunday 17th August 1952 and they called firemen and police. When the emergency services arrived they found the wreckage of a car which had been struck by a westbound Erie Railroad passenger train on a private level crossing and pushed 220ft down the track. The six occupants of the car, consisting of two families of three, had all been thrown out of the vehicle and killed. The petrol tank had been knocked clear of the automobile it had formed part of, which was not burnt.

The victims were identified as hospital attendant Adelbert Lee Houghton Sr., 19; his wife Florence, 21; and their two-month-old son Adelbert Jr., of Binghamton; plus Russell Tompkins, 2; his wife Hazel, 22; and their daughter Victoria, 2½, of Langdon, near Kirkwood. Russell Tompkins was Mrs Houghton's brother.

There were no witnesses to the accident but the families were apparently on their way to the Tompkins' home. Given that it was a private level crossing it is doubtful that it would have had any protection and the car driver had very likely failed to observe the train. [1162]

A woman and her four children lost their lives when the car in which they were riding was struck by a diesel-hauled passenger train on Cootamundra Road Crossing, three quarters of a mile from Temora, New South Wales, Australia, late on the afternoon of Thursday 21st August 1952.

Those who died were Mrs Clara Muir Lang, 46; her children Hilda Meale, 9, and Jean Muir Lang, 18 months; and her grandchildren, Robert, and Phillip Murray, aged 4 and 2 respectively. Mrs Lang was driving the 16 miles home to Dirnaseer from Temora District Hospital, where Mrs Eileen Murray, her daughter, had given birth to a baby girl the previous day. Mrs Murray was informed of the tragedy the same evening, it being thought better not to withhold the news from her.

The 1936 sedan had been about halfway over the crossing when it was struck at about 5.15 p.m. and carried about 300 yards along the line. As the car was dragged along it tore out heavy "Cattle Stop" posts at the crossing. It was wrecked with pieces of it strewn along the line. The children were thrown from the vehicle while Mrs Lang was trapped underneath the engine, the front wheels of which were lifted from the rails. The train was packed with schoolchildren returning from Sydney, Goulburn, and Bowral, who were going home for the holidays, and who had changed from the Riverina Express at

[1161] *The Lethbridge Herald*, Lethbridge, Alberta, Canada, 5th August 1952

[1162] *The Kingston Daily Freeman*, Kingston, New York, United States, 18th August 1952

Cootamundra. None of them was hurt, but they screamed and wept when they saw what had happened. A steam engine brought their train on to Temora. [1163]

Although it lacked gates the crossing concerned was considered to be the safest of the five in the Temora area, as there was good visibility from it along the line in each direction. No eyewitnesses to the accident could be found. The driver of the train, J. Cormack, had run down a pensioner at another Temora level crossing only six weeks previously. [1164]

A train collided with a crowded bus near the station at Rengo, Chile, on the evening of Thursday 28th August 1952. The bus caught fire after being struck, presumably from a ruptured fuel tank, and 15 bus passengers died and 25 others were injured, some seriously. [1165]

The island of Jamaica suffered a major level crossing disaster on Sunday 31st August 1952, involving a train and a 'bus'. Two miles west of May Pen, at Jacob's Hut Crossing, the main Kingston – Mandeville highway crossed the Kingston – Montego Bay main line of the Jamaica Government Railway on the level. At one time the crossing had been equipped with lifting barriers but by the time of the accident these had been removed, probably to spare the expense of a crossing keeper, and an open crossing had been substituted. There were, however, warning signs, which read, 'RAILWAY CROSSING', 'LOOK OUT FOR THE TRAINS', 'WARNING', 'SLOW DOWN', 'STOP! LOOK! LISTEN!'

The train involved in the accident was the newly introduced diesel-hauled, steel and aluminium *Cornwall Special* express, on its return journey from Montego Bay to Kingston. The 'bus' (in reality a motor truck) had made a regular journey from Carron Hall to Spanish Town on Saturday and the next day a special trip was arranged from Spanish Town to Vernam Field, a World War Two American air base slightly less than 35 miles from Kingston, which was now used to stage motor rallies. The truck set off from Spanish Town with about 30 passengers and the driver and owner of the truck, named Leopold Tucker. Among the passengers were the bus driver's father, James Tucker, whose wife had tried to dissuade him from making the trip. He had gone anyway and was amongst those killed. The bus driver's daughter, Jean, whose 14th birthday fell on the following day, was also a passenger; like her father she survived the accident, though both were seriously injured. In all 19 bus passengers died in the crash and five more passengers and driver Tucker were injured. Several other passengers escaped by jumping from the truck immediately before the impact. No one on the train was hurt.

Arnold Robertson, the train driver, told reporters afterwards that he had seen a car and the truck approaching the crossing from the Old Harbor direction. "The car crossed the lines. Just then the truck, which was proceeding behind the car, reached the crossing. Apparently the driver saw the trolley and seemed to check his speed. I continued, at the same time blowing my whistle. Apparently the truck did not check at all but had hesitated to make another gear to beat the trolley. He could not make it and the trolley hit the truck

---

1163 *The Daily Advertiser*, Wagga Wagga, New South Wales, Australia, 22nd August 1952

1164 *Newcastle Morning Herald and Miner's Advocate*, Newcastle, New South Wales, Australia, 22nd August 1952

1165 *The Mercury*, Hobart, Tasmania, Australia, 30th August 1952 on the broadside. I saw the driver of the truck leap from the cab just at the impact."

Mr J.B. Lieba, who had also been on his way to Vernam Field, said, "It was unfortunate that it should have happened. I was going in the same direction as the car which was in front of the truck. It all happened in a flash. The motor car got through and the truck tried to get through also. The diesel on its way in was blowing its horn all along. I stopped when I heard it. If the truck driver had also stopped nothing would have happened. It is all unfortunate."

Mr Wellesley White of Jacob's Hut told a reporter, "I was standing just at the crossing. The diesel was coming in and blowing all the way. I saw the motor car get through, and the truck tried to get through also, but I believe his timing was bad. The diesel hit it right between the cab and body. Some of the people on the truck seem to have realised what was going to happen. They jumped off."

Mr Wilbert Reid, also of Jacob's Hut, related, "I was riding my bicycle and when I got to the crossing I heard the diesel blowing. I stopped. A motor car was in front of the truck and it got through. The truck tried to get through also, but the diesel came down on it and hit it right in the middle. I saw some people jumping from the truck. We have never had this here before. It is a pity." [1166]

The inquest on the victims began in mid-November. General Manager John Atkinson explained to the court on its second day that a sign instructing the drivers of trains to reduce speed to three miles an hour for level crossings did not apply to scheduled services. Train drivers were expected to sound one long and two short whistles for crossings but there was nothing in the regulations requiring that they reduce speed. If the drivers of express trains were to reduce speed for every crossing they would not be able to maintain schedules.

Mechanical Engineer Peter Martin gave evidence as to braking distances. A two-coach diesel train such as that involved in the accident, approaching Jacob's Hut Crossing at 35 miles an hour, would be able to stop in 200 feet, though it would take the driver two or three seconds in addition to react and apply the brakes.

Dr Dudley Jacobs of Lionel Town hospital stated that the bus driver had suffered severe concussion and his condition had deteriorated upon learning of his father's death. He was not yet in a fit state to give evidence. [1167]

At the resumed inquest at the end of the November evidence was given by Mr Gabriel DeCordova, a merchant. He testified that had been driving towards Vernam Field at about 9.45 a.m. There had been a stream of traffic. He had been following a car, a truck, and a pick-up truck, in that order. He had seen the diesel approaching, about 90 or 100 yards from the crossing. It had been whistling and was travelling at 25 or 30 miles an hour. The car had crossed the line, with a very small margin. The truck had followed it and the accident had occurred. The speed of the truck at the crossing he put at 25 miles an hour. He agreed that in a stream of traffic the driver would be concentrating on the vehicle in front. Neither the truck nor the diesel had appeared to reduce speed. He did not think that the driver of the truck had been in a worse position than he had been to hear the whistle of the diesel and he believed that the truck could have stopped, provided that he had good brakes.

1166 *The Daily Gleaner*, Kingston, Jamaica, 1st September 1952

1167 Ibid, 18th November 1952

Mr Robert Marley, representing the train driver, stated that the latter was anxious to tell the jury his side of the story, but he (Mr Marley) had strongly advised the train driver against it as there might be further proceedings.

Mr Eric Duelle, the manager of the Texaco Super Gas Station at St Andrew and the driver of the car preceding the car immediately in front of the truck, said that he had crossed because it had been safe for him to do so, unless he had stalled his vehicle. He agreed that the car which had crossed behind him had got across "by the skin of the teeth."

The Attorney General stated that it not been possible to trace the driver of the latter car, a Buick.

This concluded the inquest, and the jury duly returned a verdict of 'Accidental Death', finding no one criminally responsible. [1168]

A lorry loaded with dynamite was in collision with a shunting engine on a level crossing near the Orange Free State station of Welkom, South Africa, on Friday 5th September 1952. The force of the impact was sufficient to detonate the explosives in the truck. The shunting locomotive was blown to pieces, the station buildings, nearby houses were damaged, and it was feared, according to a report in a London, England, newspaper, that three Europeans and four Africans had been killed, while several other people were seriously injured. Parts of the train were found nearly two miles away. [1169] Other reports spoke of 12 or even 25 fatalities. [1170] [1171]

A serious level crossing collision occurred on the outskirts of Valencia, Spain, on the evening of Tuesday 16th September 1952. The gatekeeper of a level crossing at Mislata had placed a chain across the road and tramway prior to the arrival of a steam train proceeding from Liria to Valencia. In the darkness and with the poor lighting at the crossing the motorman of a tram on the Torrente line failed to see that the crossing was closed, smashed through the chain, and entered the crossing, where the tram was stuck by the train. Five tram passengers were killed or fatally hurt and another 26 were injured, several seriously. All of them were workmen and poor people who had been returning from Valencia to their homes. The sudden braking of the train derailed its locomotive but no one on the train was hurt. [1172]

At about 10:45 p.m. on Tuesday 16th September 1952 a Missouri Pacific Railroad freight train in charge of Conductor Howard Handley with Engineer J.C. Barron collided with a car on a level crossing just outside of Austin, Texas, United States. The car, a 1939 Chevrolet, was thrown 125ft and overturned. All eight people in the car were killed. They were Thomas Fort, 32; his wife Ruby Inez; the couple's children Shirley Jean, 12; Brenda Joyce, 9; and twins Bobby Gale and Thomas Dale, 3, of Eden; Jack Miller, 22, a soldier stationed at Fort Hood; and Ben Smith, aged about 15, of San Angelo, the soldier's half-

---

1168 Ibid, 29th November 1952

1169 *The Times*, London, England, 6th September 1952

1170 *The Cedar Rapids Gazette*, Cedar Rapids, Iowa, United States, 5th September 1952

1171 *Pacific Stars and Stripes*, Tokyo, Japan 6th September 1952

1172 *La Vanguardia*, Barcelona, Spain, 17th September 1952

brother. [1173]

There were no witnesses to the accident other than the train crew. The crossing was protected by warning lights and bells which were operating properly afterwards. [1174] It was the worst single car/train accident in the history of the state.

Collinsburg, Pennsylvania, United States, was the scene of a bad level crossing collision involving a school bus on the morning of Wednesday 17th September 1952. Four children died in the accident and 39 children and the bus driver were injured.

The train involved in the accident was eastbound Pittsburgh and Lake Erie Railroad freight train No. 9402, consisting of a 2-8-4 steam locomotive, 64 cars and a caboose, with Engineer Frank Goldsboro. It had passed Douglas, just under 3 miles west of Collinsburg and the last open office, at 7 a.m., and was travelling at 35 miles an hour, the maximum authorised speed, when it struck the school bus at 7.05 a.m.

The school bus was owned by Doernte Bus Lines and operated under contract with the Rostraver Township School Board. It had a 1947 GMC chassis surmounted by a body built by the Hicks School Bus Body Company. There was a seating capacity for 60 passengers. It was carrying 48 passengers at the time of the accident and was being driven by Edward Steiner, aged 22.

The level crossing with German Street, Collinsburg, had no gates (barriers), only standard crossbucks accompanied by signs reading "RAILROAD CROSSING" in black letters on a white background. For northbound vehicles (the direction in which the bus was travelling) the road fell at gradients of 1 in 9, 1 in 25 and 1 in 17 to the crossing. German Street met the Collinsburg – West Newton road on the far side of the crossing. All school buses were required to stop dead before crossing the railway. Trains were required to whistle on the approach to crossings, have their headlights lit by day as well as night and the engine bell ringing. On average 17 trains daily used the crossing. Over a 24-hour period from 12.01 a.m. on 24th September 1952 the crossing was used by 939 cars, 91 lorries, and 21 other vehicles.

Engineer Goldsboro told the Interstate Commerce Commission (ICC) inquiry that he had whistled for the crossing, the bell of his engine had been ringing and its headlight had been lit. As his engine had passed over the level crossing preceding that at German Street he had run into a bank of fog which had extended to within 200 feet of German Street crossing. When he had emerged from the fog he had seen the bus entering the crossing. He had at once closed the throttle and applied the brakes in emergency but speed had not been reduced before his train struck the bus. As a result of the collision the body of the bus was torn from its chassis, carried about 60 feet and demolished. The chassis was carried 85 feet and badly damaged. Slight damage was caused to the engine. The train was not derailed.

The bus was on its regular run between Collinsburg Grade School and Rostraver Junior-Senior High School in Pricedale, Pennsylvania. It stopped to pick up some students 60 feet from the crossing and again at the crossing itself. Driver Steiner said that he had opened a door of the bus and looked and listened for a train but neither he nor the student behind him had seen one or heard a whistle. It was hazy with fog in the distance

---

[1173] *Brownwood Bulletin*, Brownwood, Texas, United States, 17th September 1952

[1174] *The Taylor Daily Press*, Taylor, Texas, United States, 17th September 1952 but there was no fog at the crossing. Satisfied that nothing was coming he had started over the crossing and only then had he seen the train and heard its whistle. The collision had occurred immediately afterwards. [1175]

Steiner suffered a broken leg in the accident but afterwards, after sending one of the uninjured pupils to a nearby house for help, he had 'worked frantically with the other men to get the dying and injured children from the smashed vehicle.'

A newspaper report on the accident continued, '"I didn't have any idea I was hurt because my leg didn't bother me at all until after all the kids were out of the bus and I was in a car. Then it started to stiffen up and I knew something was wrong," Steiner said.

'A driver of the Doernte Bus Lines-operated school bus since last February, Steiner first was taken to the West Newton police station by State Police, then returned to Collinsburg, before being admitted to Charleroi-Monessen {hospital} for treatment.

'Steiner, who has been driving trucks and other large motor vehicles since he was 16, said his bus was in good mechanical condition, although he was a few minutes behind schedule when the accident happened.

'"I have never seen a train on that track before," he said.

'Steiner, visibly shocked by the tragedy that enveloped him a few minutes after 8 a.m. {sic – the ICC report says 7 a.m.} yesterday, made barely audible replies to the questions put to him by the reporter. His eyes were sunk into his deeply tanned face and his hands moved jerkily as he talked. They were only two of the many indications that he was still suffering a personal hell nearly 12 hours after the accident.

'Separated from his wife, Steiner lives with his mother, Mrs Clara M. Steiner, near Smithton.' [1176]

The ICC noted that the visibility of a road user approaching the crossing was obscured by a building, trees, and undergrowth. It carried out a test with a similar bus and locomotive. The bus was stopped at the point where the school bus had, with its windows closed and door open. An occupant of the bus could hear the engine whistling for the preceding crossing but could not hear its bell until it came into view. There was no fog and the engine became visible from the bus driver's seat when it was 348 feet from the crossing, and the bus to the engineer when it was 296 feet from the crossing. Moving at 35 miles an hour a train would cover these distances in 6.8 and 5.8 seconds respectively. Starting in second gear it took the bus 10.5 seconds to clear the crossing. With no automatic warning devices the only protection provided for a road user against a train moving at 35 miles an hour was the sound of its whistle, which the bus driver had not heard on the day of the accident. In other words, the crossing was inherently dangerous, or, in the words of the ICC, 'the conditions present constitute a definite hazard.' There had been two previous collisions with motor vehicles, not resulting in fatalities since 1st January 1942.

The route of the school bus was changed after the accident so that it no longer used the crossing. [1177]

A diesel railcar reportedly travelling at 120 kilometres an hour on the Paris – Brittany

---

[1175] Interstate Commerce Commission report

[1176] *The Monessen Daily Independent*, Monessen, Pennsylvania, United States, 18th September 1952

[1177] Interstate Commerce Commission report

line collided with an oil tanker on a level crossing at Basse-Indre, France, near Nantes, on the evening of Tuesday 4th November 1952. Passengers attempting to escape from the wrecked railcar were blinded by oil which had spurted from the damaged tanker but there does not seem to have been a fire. Four adults and a child lost their lives and possibly a sixth person, while several others were injured. According to different reports the tanker either broke down on the crossing or alternatively smashed through the crossing barrier directly in front of the train. [1178] [1179]

Ten 'European' (white) schoolchildren between the ages of seven and 14 were killed when their school bus was struck by a goods train on a level crossing near Springs, South Africa, a city in the Transvaal. The bus driver was also killed and almost all the other children on the bus were hurt, five of them critically. The accident happened on Thursday 13th November 1952. [1180]

An attempt by a car driver to beat a train over a level crossing was believed to have been the cause of five deaths at Saint Vincent de Paul, Quebec, Canada, on the evening of Sunday 16th November 1952. The rear of the car driven by Adelard Manville, 32, was struck by a Canadian Pacific Railway passenger train and sheared off, while the remainder of the vehicle was strewn half a mile along the track along with the bodies of the victims. Those who lost their lives were Leo Preston, 25; Jean Baptiste Preston, 22; Roland Fortin, 32; his unnamed wife, 30; and the Fortins' daughter Gisele, 8; all from nearby Pointe aux Trembles. Mr Manville and his wife were only slightly hurt. [1181]

Jean-Paul Giraud, 30; his wife, 29; his brother Jean-Marie, 22; his sister Solange, 17; and his son Girard, 4, were all killed when the car in which they were travelling was hit by a train on a level crossing at St Perpetue, Quebec, Canada, on Christmas Eve, Wednesday 24th December 1952. [1182]

Towards the end of 1952 the Clayton family had moved to Crowell, Texas, United States, from Lawton, Oklahoma. On Boxing day, Friday 26th December 1952, the parents went to visit Mr Clayton's parents at Dublin, Texas, while their three young daughters went to see their maternal grandparents, Mrs Clayton's parents, Mr and Mrs Charlie Williams, at nearby Harbin. It was at Harbin that tragedy struck that day when the car containing Mr and Mrs Williams and their grandchildren was struck by a Gulf Colorado and Santa Fe Railway train. Mr Williams, 67; Mrs Williams, 72; Christie Clayton, 6 and Pamela, 4, were killed instantly. Janna Doyle was thrown clear of the wreckage of the car and died three hours later in Dublin hospital. The others were dragged about 400 yards, trapped in what remained of the car, before the fast freight train could pull up. The engineer of the train was J.T. Harrell. [1183] [1184]

[1178] *The Times*, London, England, 5th November 1952
[1179] *Journal de Genève*, Genève, Switzerland, 5th November 1952
[1180] *The Times*, London, England, 14th November 1952
[1181] *The Medicine Hat News*, Medicine Hat, Alberta, Canada, 17th November 1952
[1182] Ibid, 27th December 1952
[1183] *Wichita Daily Times*, Wichita, Texas, United States, 27th December 1952
[1184] *Lubbock Avalanche-Journal*, Lubbock, Texas, United States, 28th December 1952

A bus carrying 20 Mexican farm workers (*braceros*) was struck by a Southern Pacific Railroad mixed train en route from Niland to Calexico on a level crossing at Brawley, California, United States, early on Thursday 19th February 1953. Six of the *braceros* were killed outright and the other 14 were all injured, three so badly that they were not expected to live. The injured were taken to the small Pioneer Memorial Hospital at Brawley, which reported that supplies of blood plasma there, at Calexico, and at El Centro were exhausted and requested further supplies from more distant places.

Jacob Bynum, who witnessed the accident, said, "The collision sent a streak of fire 50 feet along the railroad tracks. The bus was carried 150 yards and crumpled into twisted debris. I don't see how anyone could escape alive."

A Brawley produce-packing firm, the Arena Imperial Company, owned the bus. [1185]

Engineer Barton Brown stated that he had sounded the whistle of his locomotive for the crossing and its bell had been ringing. [1186]

Seven people from Tampa lost their lives when the car in which they were travelling was in collision with an Atlantic Coast Line Railroad train on a level crossing near Rubonia, between Palmetto and Gillette, Florida, United States, at about 5.30 p.m. on Sunday 22nd March 1953. Travelling northwards at high speed on Highway 41 the car had overtaken several other vehicles which had already stopped for the train and hit the right side of the engine. It then bounced against the light signals at the crossing, flew about 75ft and burst into flames. All the occupants of the vehicle were killed in the initial collision; their bodies were partially burnt in the subsequent fire. The engine of the car was torn loose and ended up 80 yards away in the yard of a novelty shop. After the accident motorists, pedestrians and members of the train crew rushed to the car and extinguished the fire.

The victims of the accident were named as Mrs Julia Bearden, 68; her son Wilbur, 34; his wife Ruth Irene, also 34; their children Terry, 12; Kay, 10; and Gerald, 8; and Mrs Bearden's brother, Horace Morris, 12.

The train was about 30 minutes late and was travelling at 59 miles an hour, within the permitted speed limit. The accident caused a major traffic jam on the highway, with cars lined up back to Bradenton and Palmetto. It was not possible to determine who had been driving the car. There was no indication of braking, indicating that the car driver had not seen the train at all. [1187]

A muddy road making progress slow appears to have contributed to a level crossing collision at New Cambria, Kansas, United States, on Sunday 12th April 1953. The driver of the car reportedly accelerated in order to achieve a higher speed on the muddy surface and then, when the car reached a dry piece of road on the approach to the crossing, he was unable to stop, and the car lurched into the path of an Atchison Topeka and Santa Fe Railway freight train. The car was carried 500 yards by the train and its occupants were all killed. They were Meredith Seagrove, 45; his son Phillip, 22; John Lantz, 38; his wife

---

[1185] *Albuquerque Journal*, Albuquerque, New Mexico, United States, 20th February 1953
[1186] *The Post-Tribune*, Jefferson City, Missouri, United States, 19th February 1953
[1187] *Sarasota Herald-Tribune*, Sarasota, Florida, United States, 23rd March 1953

Mamie, also 38; and their children Francis, 13, and Eddy, 3. The party had only just left the Seagrove home, with the elder Seagrove driving. [1188] [1189]

A mother, her three daughters and three grandchildren died in a level crossing collision with a Michigan Central Railroad passenger train on the outskirts of Chicago on the evening of Wednesday 3rd June 1953. The accident happened at Burnham, Illinois, United States, at a crossing provided with warning lights but no barriers.

Engineer Claire Phillips stated that his train, the *Canadian Niagara* bound from Chicago to Detroit, had been travelling at 40 miles an hour at the time of the collision and the car, he estimated at 60 miles an hour. Thrown about 150ft by the impact, the automobile damaged the signalling system. The *Canadian Niagara* was delayed about an hour and the westbound *Twilight Limited* streamliner, from Detroit to Chicago, lost about 90 minutes.

Those who lost their lives were Mrs Catherine Zawacki, 60; Mrs Genevieve Kovack, 26; Mrs Irene Skala, 22; and Miss Patricia Zawacki, 18 (her daughters); and Barbara Davis, 11; Jaqueline Kovack, 5; and Arelene Zawacki, 7 (her grandchildren). Mrs Skala's son David, 4, and Miss Virginia Warren, 17, the daughter of a neighbour, were injured and were taken St Margaret's Hospital in Hammond. They were expected to survive. The party had been returning home from a picnic. No one on the train was hurt. [1190] [1191]

Six members of one family and a relative of theirs lost their lives when the car in which they were travelling was struck by Dominion Atlantic Railway passenger train on a rain-shrouded level crossing at Clifton, Nova Scotia, Canada, not far from Truro, on Wednesday 26th August 1953. The victims were Mr and Mrs Murray Burgess, their Esther, Geraldine, Heather, and Roger, all of nearby Beaverbrook; and their niece Margaret Taylor, aged about 16, of nearby Harmony. Five sons of the Burgesses survived. The Truro to Kentville train carried the car about 150 feet before the vehicle fell down an embankment. [1192] [1193]

A car driver failed to observe local goods train Ng 9905 on a crossing protected only by warning and 'Slow' signs on the line between Varel and Neuenburg, West Germany, on Saturday 5th September 1953. The train hit the car and six people, probably all its occupants, were killed. [1194]

Seven women and girls on their way to a family reunion died in a level crossing accident near Alafia, Florida, United States, on Monday 7th September 1953 when the car in which they were riding moved into the path of a Seaboard Air Line Railroad train. The party had bought a cake for the occasion only shortly before.

1188 *The Daily Tribune*, Great Bend, Kansas, United States, 13th April 1953

1189 *The Leavenworth Times*, Leavenworth, Kansas, United States, 13th April 1953

1190 *The Terre Haute Tribune*, Terre Haute, Indiana, United States, 4th June 1953

1191 *McKenney Daily Courier-Gazette*, McKinney, Texas, United States, 4th June 1953

1192 *Walla-Walla Union-Bulletin*, Walla-Walla, Washington, United States, 27th August 1953

1193 *The Times-Recorder*, Zanesville, Ohio, United States, 27th August 1953

1194 *Katastrophen der deutschen Bahnen Teil II* by H.J. Ritzau pg. 245 (in German)

Engineer J.F. Gleason of Seaboard train No. 321 heading for Boca Grande said later that as he had approached he had seen the women's car stopped at the crossing with its front bumper foul of the track. At this point he had sounded the whistle of his engine when, instead of moving back a little so that it was completely clear of the crossing, the car had jerked forward into the path of his train, and the collision ensued. The train demolished the car and carried it a quarter of a mile down the line before it could stop. Rescuers found six of the bodies still in the wreckage of the car and all were dead. One victim, probably the driver, was thrown clear and was still alive but died in hospital an hour and a half later. Highway Patrolman B.B. Smith theorised that the driver of the car had got confused upon hearing the train whistle and instead of putting the vehicle into reverse had inadvertently put it into forward gear.

Those who lost their lives were Mrs Alene Turner, 48; her mother, Mrs Lila Hunter, 68; Mrs Turner's daughters, twins Katie and Mattie, 18, and Linda, 11, all of Durant; plus Miss Pearla Mae Lastinger, 19 (the apparent driver of the car) and her sister Joyce, 14, of Lithia. The party were bound for a Lastinger family reunion at a location on the Lithia – Bloomingdale road a few miles away. [1195]

A car containing a family of six was struck by a Pennsylvania Railroad freight train at 4.55 p.m. on Sunday 13th September 1953 near Mahaffey, Pennsylvania, United States, killing all of them outright. The victims were Mr Calvin Small, 28; his wife Betty, also 28; their children Minnie Regina, 6, and John, 2; together with Mrs Small's brothers Phillip and Lynn Stumpf, aged 19 and 15 respectively.

The diesel engine of the train carried the car 350 yards before it could be stopped. The train was only a short one, consisting of two diesel units, four loaded coal cars, a loaded box car, an empty box car and a caboose. It was travelling at 30 miles an hour when it hit the car. How fast the car was travelling was not known but its speed was believed to have been fairly high.

Fireman Glen Bannon told police, "I saw he (the driver of the car) wasn't going to make it and I hollered to the engineer to throw on the brakes. We were almost on the crossing when I saw the car."

Engineer H.E. Weaver said that the train had been en route from Punxsutawney to Cresson. Its bell had been ringing on the approach to the crossing.

On the morning of the accident the Stumpf brothers had visited the home of their mother, Mrs Minnie Stumpf, at Mahaffey. They had then driven a mile up the hill to the home of the Small family, had picked them up, and had taken them for a ride through McGees Mills, and over the new McGees Mills – Mahaffey road. They had then visited another sister, Mrs Stella Peace, who lived near the crossing, and were returning to Mrs Stumpf's home when the accident occurred. When the car left Mrs Pearce's the three men had been sat in the front and the women and children in the rear. Phillip Stumpf, the owner of the car, had been driving.

There was a short rise to the crossing and no clear view of the line in the direction from which the train had been coming until within 15 yards of the line because of lineside trees. The line was straight for 150 yards in the direction from which the train was coming and for 500 yards in the other direction.

[1195] *The Charleston Gazette*, Charleston, West Virginia, United States, 8th September 1953

The car, a 1939 sedan, was demolished, its right side being completely caved in. Damage to the diesel was estimated at $500. [1196]

The inquest into the accident was told that Phillip Stumpf had been driving the car while his license was suspended, also that the registration plate of the vehicle was issued to a Dodge but the vehicle involved was a Ford.

Fireman Bannon testified that the freight train had had no schedule and therefore could not be late. He said that his engineer had been pulling the whistle cord when he (the fireman) had seen the car and shouted to the engineer to stop, and the brakes had been applied in emergency. A terminal brake test had been made at Punxsutawney. After leaving there four cars had been picked up at Hillman and three left at McGees Mills. He was sure that the train's brakes had been working properly and that its bell and whistle had been sounded.

Engineer Weaver stated that he had been "on the blind side" and so had not seen the car before his fireman had shouted. His train had 100 pounds brake pressure and sand had begun falling immediately the brakes had been applied. Questioned by the coroner as to why there had been little sand on the rails for the last 500 feet to the crossing he said, "We were slowing down for the last 300 feet. We couldn't have done any more than was done that day." He added that the engine had no speedometer and no paper tape for recording its speed.

Conductor T.E. Bredley, Brakeman Frank Hall and Second Brakeman J.T. Goulden also gave evidence, though they could not add much.

Mrs Daisy Anderson and Mrs Rose Pennington, who lived near the crossing, both testified as to having heard the train whistle.

After an absence of 20 minutes the inquest jury returned a verdict exonerating the trainmen from blame for the accident. Their verdict read in part, "We find this accident was not caused by negligence on the part of any or all of the train crew because we feel that everything possible was done by the train crew under the circumstances. We therefore adjudge it an unavoidable accident." [1197]

Just before 17:00 on the afternoon of Friday 9th October 1953 a major disaster occurred in Chile, when a passenger train on the narrow-gauge branch from Linares to Colbún collided with a minibus on a level crossing between Arrayanes and Yerba Buenas. As a result of the collision the boiler of the 60cm-gauge locomotive was badly damaged and the minibus passengers, who included women and children, were scalded by escaping steam and boiling water. Twenty-two people died and another 25 were badly hurt. The minibus was wrecked and two carriages of the train overturned. [1198] [1199]

All but one of a family from Oakland County, Michigan, lost their lives when the car in which they were riding was in collision with a Grand Trunk Western Railroad passenger train on a level crossing three miles east of Durand, Michigan, United States, on Sunday 29th November 1953. Those who lost their lives were Roy Dalton, 40; his wife

---

[1196] *The Progress*, Clearfield, Pennsylvania, United States, 14th September 1953

[1197] Ibid, 7th October 1953

[1198] https://amigosdeltren.cl/tren-chico-linares-colbun

[1199] *Ironwood Daily Globe*, Ironwood, Michigan, United States, 10th October 1953

Ruth, 38; and their daughters Judy, 14; Diana, 6; and Sherry, 18 months. Their son Douglas, 10, was taken to Durand hospital, reportedly with critical injuries. He was conscious and told police that no one in the car had seen the train. It had been snowing heavily at the time of the accident. Engineer Leo Dunton stated that his train had been travelling at about 50 miles an hour and that because of the blizzard he likewise had not seen the automobile. The train dragged the car about 100 feet and all six of its occupants were thrown out. [1200]

A Dallas, Texas, family of five and their black maid lost their lives when the car in which they were riding was struck by a Missouri Kansas Texas Railroad of Texas streamliner, the *Texas Special*, on a level crossing near Royse City, Texas, United States, on the evening of Sunday 13th December 1953. The train did not come out unscathed either; its two diesel locomotives and ten cars were derailed. About 30 passengers suffered minor injuries and a shaking while three passengers and the engineer required medical attention.

Those who died were Victor Saufley, aged about 44, a Dallas real estate agent and retired Army officer and Southern Methodist University football player; his wife Harriet, 35; their sons Bill, 5, and twins Jim and John, aged about 2; together with the maid, identified only as Esther and aged about 28.

Engineer M.B. Hudgens, who sustained a back injury in the accident, said that the station wagon had been driven slowly onto the crossing as his train had borne down upon it. When he realised that the vehicle was not stopping he had applied his brakes but it had been too late to avoid the collision.

There were neither barriers nor warning lights at the crossing, only warning bells, which were still clanging hours after the accident. Mr Saufley was deaf and normally wore a hearing aid, which could not be found after the accident. It was theorised that because of his deafness and the fact that perhaps he had not been wearing his hearing aid, he had failed to hear the ringing of the warning bell. He had presumably also failed to see the train. [1201]

On Monday 14th December 1953, 11 Dupo Community High School students were injured, none seriously, when the school bus in which they were riding was struck by a Missouri Pacific Railroad engine pulling a single car on a level crossing at Dupo, Illinois, United States. The children had been returning from a basketball game with Lebanon High School. [1202]

On about Saturday 19th December 1953 a Canadian National Railways freight train collided with a taxi on a level crossing at St Jean Port Joli, Quebec, Canada. Five young men were killed and two were seriously injured. [1203]

A fateful attempt by a car driver to beat a train at a level crossing resulted in the death

---

[1200] *The Holland Evening Sentinel*, Holland Michigan, United States, 30th November 1953
[1201] *The Mexia Daily News*, Mexia, Texas, United States, 14th December 1953
[1202] *Dixon Evening Telegraph*, Dixon, Illinois, United States, 15th December 1953
[1203] *Brandon Daily Sun*, Brandon, Manitoba, Canada, 21st December 1953

of him and six members of his family at the Palatine Road Crossing, near Wheeling, Illinois, United States, on the evening of Sunday 20th December 1953.

Fireman E.B. Smiley of the southbound Wisconsin Central Railway freight train said that he had seen the car 300 feet away and that it had apparently speeded up as it neared the crossing. He thought that its driver, realising that he was not going to be able to stop in time, had accelerated in a vain attempt to get across ahead of the train. The fireman added that he had signalled to Engineer J.P. Johnson to apply the brakes, which had been done, but they had struck the car anyway. The vehicle was caught broadside and its wreckage was strewn along the track for 150 yards.

The driver of the car, Fred Galle, 37; his wife Violet, 33; and their children Audrey, 12; Evelyn, 10; John, 8; and Carol Ann, 1, were killed outright. Another child, Elaine, 6, was fatally injured and died later in the Resurrection Hospital, Des Plaines (some reports state that Elaine was killed outright and that it was Audrey who died in hospital). [1204] [1205]

The first bad level crossing accident of 1954 happened at 7.30 p.m. on Tuesday 4th January at Snyder, Texas, United States. A car containing a family of five was in collision with a Panhandle and Santa Fe Railway freight train consisting of a three unit diesel locomotive and 73 cars in charge of Conductor J.L. Reed with Engineer Tom Bickers and Fireman Weker McDonald. The westbound train was travelling at between 30 and 55 miles an hour. The car, a 1951 Hudson sedan, was carried about 50 feet by the impact and the bodies of all its occupants but one were thrown from it in this distance. All had been killed instantly.

The victims were Mr Lemye Miller, 40; his wife Georgia, 42; and their children Gary Edmond, 15; Linda Kay, 11; and Brenda Joe, 10. It was believed that Mrs Miller had been driving.

Following the accident Engineer Bickers ran to the home of Mr and Mrs Billy Thompson to call an ambulance. "The train crewman was so badly shocked after having had to run through brush and climb over two fences to get to the house, that I called the ambulance for him," Mr Thompson said.

Both road and railway climbed to the crossing, which was unprotected, though there was an unrestricted view along the line. The car was southbound

Damage to the engine was estimated at only $100, while there was no damage to the track. [1206]

On the night of Sunday 10th February 1954 an express train bound for Taipei collided with a military lorry on a level crossing near Zhongli, Formosa, despite the best efforts of the train driver to avert the collision. The lorry had apparently stalled on the crossing. Nine people aboard the lorry were killed and 30 were seriously hurt. [1207]

In a similar accident to that at Snyder, Texas, less than a fortnight earlier, Harlan Vail, 35, a farmer and truck driver for a wholesale company; his wife Ruth, 34; and their

[1204] *The Brazil Daily Times*, Brazil, Indiana, United States, 21st December 1953
[1205] *The Racine Journal-Times*, Racine, Wisconsin, United States, 21st December 1953
[1206] *The Snyder Daily News*, Snyder, Texas, United States, 6th January 1954
[1207] *Pacific Stars and Stripes*, Tokyo, Japan, 13th January 1954

children Sandra Ann, 14; Jerry, 11; and Michael, 17 months, died in an accident at Foley's Crossing, two miles from Milroy, Indiana, United States, on the evening of Sunday 17th January 1954. Their car, a 1952 Plymouth, was struck by a southbound New York Central Railroad freight train at 10.25 p.m. Four of the occupants were killed outright; little Michael was taken to the home of Mrs Carrie Ralston, who lived near the crossing, but succumbed seven minutes after arriving there.

Engineer George Gimm said that he had applied the brakes of his train when he had seen the car bearing down on the crossing but had not been able to avoid the collision. The car, which was travelling west, struck the cowcatcher of the engine and was thrown, badly battered, into a lineside fence. Mr Vail was driving the car. The train, in charge of Conductor John Bahler, was travelling at about 40 miles an hour.

The family was returning to their farm north-west of Sandusky after a visit to the Willard Patterson family near Clarksburg. Mrs Patterson was a sister of Mr Vail.

As at Snyder the crossing was unprotected but there was good visibility along the line. It was theorised that fogging of the car's windscreen from the breaths of its passengers in the cold night air might have been a factor. [1208]

Eight people died when a Canadian National Railways passenger train collided with an overcrowded car on a level crossing at Napanee, Ontario, Canada, on the evening of Saturday 23rd January 1954. The party of nine people in the car had been returning from Napanee to Newburg, seven miles to the east.

Those who lost their lives were Harry Lewis, 56; his wife Bessie, 51; their son Joseph, 12; and Mrs Lewis' brother, Clifford Thompson, 39, all of Newburg; twins Sandra and Sharon McQuaig, 11, and Max Fox, 56, of Napanee; and Borden Cox, 32, of Marlbank. The only survivor of the accident was Donna McQuaig, 10, sister of the twins, who was taken to hospital with a broken leg and other injuries but was said to be in a favourable condition.

The wreckage of the car and the bodies of the victims were carried 250 feet along the double-track main line by the train. [1209]

On the evening of Sunday 31st January 1954 a passenger train travelling from Pusan to Seoul collided with an Army lorry on a level crossing near Suwon, South Korea, with disastrous results. The engine of the train was derailed as a result of the collision, the leading coach of the train piled on top of it, the second coach overturned and the third coach slammed into it. First reports spoke of 40 dead and 100 injured. [1210]

An updated report stated that American Army crews had recovered the bodies of 57 victims from the wreckage. The report continued, 'At least six other bodies were believed to be in the twisted wreckage of three wooden passenger cars which had been telescoped into the space of one. One man, pinioned to the wall as if he had been crucified, could be seen from the railroad siding. Stray arms and legs littered the frozen rice paddy where the dead had been placed in lines under straw mats. Three legs, each with a different shoe, protruded from under one mat. Bundles of rice and snacks for the trip were piled neatly

---

[1208] *Rushville Republican*, Rushville, Indiana, United States, 18th January 1954
[1209] *The Medicine Hat News*, Medicine Hat, Alberta, Canada, 25th January 1954
[1210] *The Frederick Post*, Frederick, Maryland, United States, 1st February 1954

beside the corpses. A small black prayer-book lay in the snow next to a food-filled basket.' [1211]

All of the victims were Koreans. President Syngman Rhee sent his deepest condolences to the families of the victims.

As is often the case the reports from different sources varied in some details, for example some reports implied that the train had come from Seoul rather than was going to the capital and that the vehicle with which the train had collided had been a fish truck rather than an Army lorry. [1212]

Passenger train P.J. 3 of the Companhia Paulista de Estradas de Ferro collided with a bus on a level crossing 3 kilometres from Adamantina, Brazil, on Friday 12th February 1954. An initial report stated that 30 bus passengers had been killed in the accident. The train, which was on time, was not derailed and no one on it was hurt. There was good visibility at the crossing. [1213]

A second report confirmed that 30 people had died in the accident, which had occurred between Adamantina and Lucelia, and that a dozen more had been injured. It explained that the two places were about 6 kilometres apart. A circular bus route ran between them, with buses running at 30 minute intervals. The buses were often crowded, especially towards the end of the week when there was a large influx of rural workers and a natural movement of people going to spend Sundays with their families. It was a bus coming from Lucelia towards Adamantina which was in collision with the train. There were 42 people aboard the vehicle. The collision was a violent one, the bus was dragged 150 metres along the line and its petrol tank exploded, setting it on fire. Five people escaped with critical injuries, five more were seriously injured and two people, one of them the bus driver, only sustained slight injuries. The other 30 passengers lost their lives. Five of their bodies were so badly burnt as to be unidentifiable and they were buried in a common grave at Lucelia.

There were several versions as to how the accident had happened. According to one account the bus driver, although aware of the oncoming train, had tried to cross in front of it. In the process of doing so he had changed gear and stalled the bus. Seeing the imminent danger he then abandoned it and fled. Another version accused the bus driver of great recklessness, alleging that he had seen the train close at hand but had kept going, thinking that by accelerating he still had time to get the bus across in front of it. His judgement of the speed and distance of the train had been faulty and he had failed to make it across. There was a curve in the line on the approach to the crossing which prevented a train driver from seeing an obstruction on the crossing early enough to be able to stop the train.

Pedro Lopes, the bus driver, was taken into custody by the police. [1214]

The *Bluebonnet* passenger train of the Missouri Kansas Texas Railroad of Texas collided with a car on North Sayle Street Crossing, Greenville, Texas, United States, at

---

1211 *Waterloo Daily Courier*, Waterloo, Iowa, United States, 1st February 1954

1212 *The Daily Independent*, Kannapolis, North Carolina, United States, 1st February 1954

1213 *O Estrada de São Paulo*, São Paulo, Brazil, 13th February 1954

1214 Ibid, 16th February 1954

about 8.35 a.m. on Tuesday 16th February 1954. Five people died in the accident, all killed instantly, and three more were injured, two critically. All but one of the victims were children. Lewis Gipson, custodian of the National Exchange Bank, was driving the children to the Booker T. Washington Grade School, as he did every day.

Those who lost their lives in the accident were Mr Gipson, 52; his daughter Sandra Jane, 8; Robbie Ann Batie, 8; her brother Rodney Gale, 6; and Janet Hardaway, 7. James Oliver Gipson, 7, the son of the car driver; Dermond Batie Jr., 7; and John Wayne Ward, also 7, were injured, the first two critically.

The crossing was unprotected except for the standard crossbucks. A witness said that he had tried to warn Mr Gipson of the approaching train but Mr Gipson had just made a friendly wave back, clearly not realising the danger. The car was hurled through the air and landed in a ditch 50 feet away. The train was over an hour late. [1215]

The day after the Greenville accident, Wednesday 17th February 1954, six members of the Sprinkle family lost their lives in an accident on Quincy Avenue Crossing, Sedalia, Missouri, United States. The accident occurred at 5.43 p.m. when the car was struck by train No. 14 or 15 (both numbers are given in the same report), the eastbound *Texas Eagle* passenger train of the Missouri Pacific Railroad. The automobile was a 1948 Kaiser sedan. It was carried 150 feet along the right-of-way and demolished.

Those who died were Mr William Sprinkle, 43, the driver of the vehicle; his wife Frances Dillard, 35; their children Lyla Jo, 15; Carol Lee, 9; and Richard Leslie 'Dickie', 2; plus Mrs Sprinkle's father, Leslie Albert Dillard, 71. The other person in the car William 'Butch' Sprinkle, 11, was critically injured.

Charles Hurt, 15, a playmate of 'Butch' said he had talked to him a few minutes before the accident at a grocery store and 'Butch' had said that he was going for a ride with his folks. "He left the grocery store and went home and a few minutes later it happened. I heard a crash like an explosion and ran out and saw the train had hit something. I didn't see the accident though," Hurt explained. The Sprinkles' residence was only a short distance north of the crossing.

Conductor R.H. Hammons, Engineer H.E. French, and Fireman George Paul declined to discuss the accident with a local newspaper reporter.

Again the crossing lacked warning devices. The only reason which could be advanced for the accident was that the occupants of the car had been preoccupied discussing where they should drive to and had forgotten about the crossing. [1216]

Twenty-two people lost their lives and four more were seriously injured when a train collided with a bus at Coronel Suarez, Argentina, on Monday 29th March 1954. All those who died had been on the bus. It has not proved possible to find any reference to this accident in the Argentinian press. [1217]

Another school bus accident occurred in the United States on Friday 2nd April 1954, this time at Yazoo City, Mississippi. The school bus was struck by a northbound Illinois

[1215] *The Greenville Banner*, Greenville, Texas, United States, 16th February 1954
[1216] *The Sedalia Democrat*, Sedalia, Missouri, United States, 18th February 1954
[1217] *The Daily Independent*, Kannapolis, North Carolina, United States, 30th March 1954

Central Railroad freight train consisting of 96 cars loaded with bananas. Engineer A. Cutrer said that he had not seen the bus before his engine struck it. Fortunately his train was only moving at about 12 or 15 miles an hour so the results of the accident were not as serious as they otherwise might have been. There were 30 students on the bus, of whom 23 were injured, Twelve of them were sent to hospital but only one was seriously hurt and no one was killed. Twelve more students who would normally have been on the bus had gone on a field trip to New Orleans.

An injured girl who had been sat just behind the bus driver, J.H. Moore, 50, said that she had seen the train and shouted to the driver to stop. He had then jammed on the brakes but the bus had rolled onto the crossing. Moore explained after the accident, in a shaking voice, that he had been unable to stop because his brakes had failed. The train was carried about 25 feet as screaming students fought to get out of it. Witnesses said that the train whistle had been sounded and its bell had been ringing. The crossing itself lacked any form of protection. [1218] [1219]

On about Thursday 15th April 1954 a train collided with an intercity bus on an unprotected crossing in the Quilmes suburb of Buenos Aires, Argentina. The bus was demolished and eight people were killed and five others were seriously injured. [1220]

A car containing six young people returning from a dance at Trenton, Illinois, United States, was in collision with a Baltimore and Ohio Railroad passenger train near there early on Sunday 16th May 1954. A man and a woman in another car told authorities that they had already pulled up at the crossing for the train when the car containing the teenagers had approached at high speed, swerved around their vehicle, and crashed into the side of the train. Five of the occupants of the car were killed and the sixth was seriously injured.

Eleanor Clark, 16, of St Jacob; and Marlin Goff, 19; Horace Hooks, 19; Raymond King, 22; and Kenneth Ray Stewart, 18, all of Edwardsville, lost their lives. Donna Reed, 18, of Highland, was critically hurt and was taken to St Joseph's Hospital, Highland.

The accident happened at 1.23 a.m. on an 'oiled road' at a crossing with County Line Road two miles west of Trenton. The northbound train was en route from St Louis to New York. The car was owned by Raymond King's mother but it was not known who had been driving it. Like many others, the crossing was protected only by crossbucks. [1221]

At about 14:00 on Tuesday 25th May 1954 a bus carrying 30 wedding guests from the Achtkarspelen area was struck by a diesel passenger train on a level crossing at Zandbulten, the Netherlands, pushed 50 metres down the line and almost cut in two. Initial reports stated that three bus passengers had been killed outright, a fourth was fatally and a fifth seriously injured. About 15 other bus passengers were less badly hurt. Those who died included two sisters and a sister-in-law of the bridegroom. Neither the train nor the bus driver was hurt.

[1218] *Hattiesburg American*, Hattiesburg, Mississippi, United States, 2nd April 1954
[1219] *The Daily Herald*, Biloxi, Mississippi, United States, 2nd April 1954
[1220] *The Lovington Leader*, Lovington, New Mexico, United States, 15th April 1954
[1221] *Alton Evening Telegraph*, Alton, Illinois, United States, 17th May 1954

The Nederlandse Spoorwegen train involved was No. 3649, the 13:25 service from Leeuwarden to Groningen. It was only slightly damaged.

The wedding guests were bound for Blije, near Ferwerd, and had hired a bus from the Veenstra garage at Quatre-Bras, Belgium.

The line was blocked and the 13:24 passenger train from Groningen to Leeuwarden was terminated at Buitenpost, from where passengers continued their journeys by bus. The line was reopened in the evening.

The bus had stopped just before the crossing to pick up more passengers.

A number of ambulances and vehicles of the GGD (Gemeentelijke Gezondhiedsdienst – the Municipal Health Service) from Groningen went to the scene, though several slightly injured victims were still awaiting evacuation at 15:45. [1222] The casualties were taken to two hospitals in Leeuwarden and the University Hospital at Groningen.

The crossing was unprotected but there was good visibility in either direction and the Leeuwarden public prosecutor decided that the accident had clearly resulted from inattention on the part of the bus driver. [1223]

Two more bus passengers died of their injuries subsequently, increasing the total number of fatalities to six. The bus driver, 39-year-old Pieter V., was sentenced to a term of three months' imprisonment in August for having caused the accident through negligence. [1224]

Very heavy rain was a likely factor in a level crossing collision on Monday 31st May 1954 (Memorial Day) at Trevor, Wisconsin, United States. As was often the case a car load of young people was involved. A 96-car Chicago – Milwaukee Wisconsin Central Railroad train, hauled by three diesel units and travelling at 45 miles an hour, was already passing over the crossing when the car, travelling at an estimated 50 to 60 miles an hour, smashed into it at the point between the first and second units. The car was crumpled into a steel ball and carried 75 feet along the line. Damage was caused to the diesel units which it had struck, part of the car being embedded in the leading unit, but the train was not derailed.

Mary Ann Czarny, 19; Delores Jefferson, 17; George Kostoff, 4 (a nephew of Delores); and Joel Seres, 19, whose father owned the automobile, were all dead when rescuers reached the scene. Some hope was initially held out for Casper DeStefano, 23, but he died while been transported to Waukegan hospital.

Charles Dexter, 15, and George Terrill, 16, witnessed the accident from the gas station owned by Dexter's father. They saw the car coming fast along County Trunk Road C and said that it had made no attempt to stop. There were no skid marks on the road leading to the crossing, confirming that the brakes of the car had not been applied. The crossing was protected by wig-wag signals which were operating correctly. It appeared that the driver of the car (it could not be determined who this had been) had failed to see either the crossing signals or the train in the driving rain. It was not known where the victims had been or where they were going, nor did the father of Joel Seres know that his son had taken the car. [1225]

[1222] *Nieuwsblad van het Noorden*, Groningen, the Netherlands, 25th May 1954

[1223] *Algemeen Handelsblad*, Amsterdam, the Netherlands, 26th May 1954

[1224] *Friese Koerier*, Heerenveen, the Netherlands, 4th August 1954

[1225] *Kenosha Evening News*, Kenosha, Wisconsin, United States, 1st June 1954

A car containing six young people returning from a pilgrimage was struck by a diesel railcar on a level crossing in Zamora province at Coreses, Spain, on Tuesday 21st June 1954. All six were killed. [1226]

A taxi containing no less than 11 passengers (four of them children) in addition to its driver was struck by a train on a level crossing in Rabat, French Morocco, on Tuesday 29th June 1954 and ten people were killed, including all four children. [1227]

Five people died and 18 were hurt in a level crossing collision involving a bus and a freight train near Coronel, Chile, on the night of Friday 9th July 1954. The bus reportedly 'went up in flames' following the collision, presumably due to the rupture of its fuel tank. [1228]

On Saturday 24th July 1954 a bus carrying 29 employees on a works outing approached a level crossing at Abenheim on the Worms – Gundheim line in West Germany. The crossing was an unprotected one, but there was a clear view along the line. Despite this, the bus driver failed to see that passenger train P 3912 was close at hand. His passengers did see the train, and when the bus was still 60 metres from the crossing they shouted, "Der Zug kommt!" ("There's a train coming!") Although he still had time to pull up short of the line the bus driver instead accelerated to get across in front of the train. It was a fatal mistake. The train driver applied his brakes but the bus was caught and carried 50 metres down the line before the train could be stopped. The tachograph of the bus showed a speed of 69 kilometres an hour. Only one of the people on the bus escaped with slight injuries, 25 were killed outright and six seriously hurt. [1229]

A car crammed with ten members of two Central City families was struck by a train on a level crossing at Coal City, Illinois, United States, on the evening of Friday 13th August 1954, killing five of its occupants, all from one family. The car was driven past the wig-wag warning signal protecting the crossing directly into the path of an Atchison Topeka and Santa Fe Railway train and was thrown against the wig-wag signal on the opposite side of the crossing. The passengers in the car, who included several small children, were thrown out. Four people were killed outright and a fifth died en route to hospital.

The victims who were killed were Mrs Clara Mae Barnes, 33; and her children Irene, 12; Edna, 11; Wilma, 9; and Howard Jr., 4. Mr Howard Barnes Sr., 39, the driver of the car, was only slightly hurt. Also injured and taken to St Joseph's hospital in Joliet were another Barnes son, Johnny, 13; Doris Shane, 19; and her son Jimmy, 19 months. The other occupant of the car, Earl Shane, was shaken up but did not need to go to hospital. [1230]

[1226] Wikipedia - Anexo: Accidentes ferroviarios en España

[1227] *The Billings Gazette*, Billings, Montana, United States, 30th June 1954

[1228] *The Montana Standard*, Butte-Anaconda, United States, 11th July 1954

[1229] *Katastrophen der deutschen Bahnen Teil II* by H.J. Ritzau, pg. 246 (in German)

[1230] *Waukesha Daily Freeman*, Waukesha, Wisconsin, United States, 14th August 1954

A prominent Westminster, Maryland woman, the wife of a road-building contractor, was one of those who lost their lives when the westbound *Metropolitan Special* of the Baltimore and Ohio Railroad, en route from Baltimore to St Louis, Missouri, struck the car in which she was riding at a little-used crossing at Weverton, Maryland, United States. The accident happened at 10.13 a.m. on Saturday 21st August 1954.

The woman, Mrs John Bennett, aged about 64, was on her way to a summer house at Potomac Shores, accompanied by her niece, Mrs Travis Meredith, 41, and three of Mrs Travis's sons – John, 8; Donald, 7; and Michael, 5. All died in the accident.

A spokesperson for the railroad company stated afterwards that the train would have been travelling at about 40 miles an hour. Its crew had sounded the whistle when they had seen the car approaching the tracks and had taken all the usual precautions. According to the engine crew the car had first stopped at the crossing, then proceeded across in front of their train, suggesting that perhaps the car driver had underestimated the speed of the train, or else had simply failed to see it at all. The track was straight at that point and there was no obstruction to a motorist's view, though it had been raining at the time.

Following the collision the car was ground beneath the front wheels of the diesel locomotive and dragged about a fifth of a mile. 'The late model Buick sedan looked like giant hands had picked it up and swished it flat,' a reporter wrote. It took a breakdown crew from the nearby Brunswick yard an hour and a half to remove the wreckage of the vehicle from beneath the engine. [1231]

Five members of what was described as a 'migratory cotton picking family' lost their lives when the flat-bottomed truck in which they were riding was struck by a Texas and New Orleans Railroad freight train just after 7 a.m. on Monday 6th September 1954. The accident happened on Westbrook Gin Crossing, near Hearne, Texas, United States. Those who died were Mr Frank Serablo, 36; his wife Marie, 41; and their children Albert, 17; Genoviva, 14; and Domingo, 3. Another son was serving with the US Army in Korea. All those who died had been riding in the cab of the vehicle except for Genoviva, who had been riding on the bed of the truck. Also riding on the truck bed had been Matilda Arrendondo, aged about 16, who was quite badly hurt, and Henry Pena, who saw the danger in time to jump off and was unhurt.

There was an unimpaired view for about half a mile from the crossing and police theorised that Mr Serablo had panicked when he had seen the southbound train approaching and stalled his truck. The vehicle was carried about 350 yards down the track, then caught fire and was burnt out, the bodies of the occupants of its cab being charred almost beyond recognition. It was believed that the victims had been heading for the cottonfields at the time of the accident. They had been in Hearne for four weeks, having come from Tahoka, Texas. [1232]

On the evening of Wednesday 15th September 1954 an express bound for New Delhi, India, collided with a lorry loaded with students at a level crossing 80 miles north of there, killing 10 of them and injuring 18 others. The students had been on their way to

---

[1231] *Cumberland Sunday Times*, Cumberland, Maryland, United States, 22nd August 1954
[1232] *The Hearne Democrat*, Hearne, Texas, United States, 10th September 1954

carry out volunteer work on a canal. [1233]

Five members of the same family lost their lives when the car in which they were riding was struck by a Pennsylvania Railroad train on the North College Street Crossing, Newcomerstown, Ohio, United States, at about 10.40 p.m. on Saturday 16th October 1954. They were Mrs Helen Mullins Duty, 27; three of her children by a former marriage – Edward, Robert, and Marylena Mullins, aged 8, 7, and 6 respectively; and a baby daughter of her second husband, Loretta Marie Duty, three weeks. Mr Raymond Duty, 19, the driver of the car, was very seriously hurt and another child, Karen Mullins, 3, was only slightly hurt. The family came from Tazewell County, Virginia.

The car, a 1946 Oldsmobile sedan travelling south, was struck by a northbound freight train, dragged about 150 feet, and caught fire.

Newspaper reports did not indicate what sort of crossing protection was provided. [1234]

A car carrying eight sophomores (second-year high school students) returning home from a football game was struck by the northbound *Winnipeg Limited* passenger train of the Great Northern Railway at St Cloud, Minnesota, United States, at about 10.15 p.m. on Friday 15th October 1954. The car was described as having been "crumpled like an accordion," by the impact. Shirley Hinkenmeyer, Janice Johnson, Nancy Minars and Sharon Roder, all students of the Cathedral High School, St Cloud; and William Ehli of Duluth and Larry Stoltman, of Ardoch, North Dakota, both students at St John's High School, Collegeville, were all killed. James Johnson, brother of Janice, and Ralph Mahowald, of Grand Forks, North Dakota, were injured. After the accident they were said to be semi-conscious but it was thought that both would survive. The eight students, who were all 15 or 16, had been returning from a game between Cathedral and De La Salle, Minneapolis.

Engineer A.B. Leander of the train, which had been slowing up for the St Cloud stop, said that he had seen the car and applied his brakes. He had whistled for the crossing and the oscillating light on his diesel locomotive had been operating. The crossing was illuminated by an amber street light but had neither barriers nor warning lights. Janice Johnson had been driving the car; her father was a Great Northern Railway conductor. [1235] [1236]

On Monday 18th October 1954 a train collided with a lorry near Pegu, Burma, killing 11 people. There had reportedly been 50 people riding in the truck, bound for a nearby Buddhist pagoda. [1237]

A ballast train collided with a bus on a level crossing at Rajabhatkhawa, West Bengal, India, on about Sunday 31st October 1954, killing two women and their three daughters.

[1233] *Youngstown Vindicator*, Youngstown, Ohio, United States, 16th September 1954

[1234] *The Coshocton Tribune*, Coshocton, Ohio, United States, 17th, 18th and 23rd October 1954

[1235] *Fergus Falls Daily Journal*, Fergus Falls, Minnesota, United States, 16th October 1954

[1236] *The Sunday Tribune*, Albert Lea, Minnesota, United States, 17th October 1954

[1237] *The Evening Independent*. Massillon, Ohio, United States, 18th October 1954

Forty-five others were hurt, 13 seriously. [1238]

On Saturday 6th November 1954 local goods train Ng 9715 collided with a car on an unprotected level crossing between Stadthagen-Leese and Stolzenau, West Germany. Five people died. [1239]

Ten people died and 25 more were injured when a bus carrying workmen to the port was in collision with a train at the level crossing with the Haifa – Acre road in Haifa, Israel, on the morning of Monday 22nd November 1954. The victims were mostly recent immigrants from North Africa. [1240]

A major disaster occurred in Mexico late on the night of Saturday 11th December 1954 when a train collided with a bus on the highway south of Tierra Blanca. Fire broke out after the collision at an unprotected crossing and 29 people were killed more or less outright, while seven others were reportedly in a grave condition. The bus was destroyed by the impact and fire. [1241]

Spain suffered a serious level crossing accident on Monday 13th December 1954 when a Costa Azul company bus operating a passenger service between Alicante and Cartagena was in collision with a shunting engine on a level crossing between San Javier and Los Alcázares. Bienvenido Góméz Navarro, the bus driver; Antonio Bas, the bus conductor; and passengers Señor Cuesta and his wife, Francisco Doménech Grau and Cecilio Martinez were killed and another 15 bus passengers were seriously hurt. [1242]

The level crossing accidents involving buses continued with the third serious one in a week at Draper, South Australia, Australia, at 4.50 p.m. on Friday 17th December 1954. An MTT (Municipal Tramways Trust) bus, travelling from Port Adelaide to Osbourne, was struck by an express train at the level crossing with the Old Outer Harbor Road, adjoining Draper station. The bus was carried 150 yards, ripping up the double-deck planking of the station platform and damaging buildings on it, before the train came to a stop. About 250 workmen who had been passengers on the train then rushed to nearby homes for axes, crowbars, hacksaws, and anything else which could be used to free passengers trapped in the wreckage of the vehicle, while later oxy-acetylene torches became available.

Bus driver Rupert Pash and passengers Charles Flaxman, O'Rita Okkes and Christopher Rynason were killed outright and 11 other passengers were admitted to the Royal Adelaide Hospital with various injuries. One of the injured passengers, Mr Graham Le Cere, spoke to a newspaper reporter from his hospital bed. "I was seated about the middle of the bus. I can remember the warning lights were flashing and then the bus moved forward. I don't remember anything after that until I regained consciousness.

[1238] *Indian Daily Mail*, Singapore, 1st November 1954

[1239] *Katastrophen der deutschen Bahnen Teil II* by H.J. Ritzau pg. 247 (in German)

[1240] *The Australian Jewish News*, Melbourne, Victoria, Australia, 26th November 1954

[1241] *Kenosha Evening News*, Kenosha, Wisconsin, United States, 13th December 1954

[1242] *La Vanguardia*, Barcelona, Spain, 14th December 1954

Men, women, and children were screaming. My legs were pinned and I could not move. I lay in the wreckage for about 10 minutes until somebody freed me."

Several people who had not been on the bus witnessed the accident. Mr Benton, of Broken Hill, related, "I had just crossed the line on my way back to the {Broken Hill Proprietary holiday} camp when the bells began to ring. It was then that I saw the bus coming straight for the crossing. He was not speeding but seemed to be going too fast to stop at the crossing. I turned and watched and suddenly realised that he would not get across the line in time."

James Niemann, 14, said that he had heard the train coming and had rushed out of the camp to get the engine number. "I heard the train give a long whistle, but the bus kept on going. He did not seem to slow down. The bells were ringing and the lights working all the time till the bus hit them."

It was suggested that the smoke from a train which had just passed in the opposite direction might have obscured the vision of the bus driver. [1243]

The death of bus passenger Mrs Laura Stevenson two days after the accident reportedly increased the number of fatalities in the accident to six. [1244]

Five people died and two others were critically injured when a local train collided with a lorry, presumably on a level crossing, near Debrecen, Hungary, on the night of Tuesday 28th December 1954. The brief announcement from Radio Budapest did not indicate if the casualties had belonged to the lorry, to the train, or to both. [1245]

The driver of a Petroleo Mexicanos (Pemex, the Mexican state-owned petroleum company) lorry made the fateful decision to try and beat a train over a level crossing two miles west of Valadeces, Mexico, just after noon on Wednesday 19th January 1955. As a result eight of the 12 people on the lorry, including its driver, Francisco Garcia, were killed and three others were injured, one critically. The 12th person was thrown from the lorry and escaped injury. Those who died were oilfield workers from Pemex Seismological Station No. 3 and their foreman, Gabriel Barocio.

Road and railway ran alongside one another for some distance. The lorry managed to get somewhat in advance of the train, which was travelling in the same direction, and its driver then attempted to cross the line in front of the engine when "something happened"; perhaps the lorry driver panicked and stalled his vehicle. [1246]

An icy and slippery road surface led to a minor level crossing collision involving a school bus on the afternoon of Friday 4th February 1955 near Ostrander, Ohio, United States. A total of 28 Scioto Valley children were being taken home on the bus when its driver, Bill Stults, 31, found that he was unable to stop when he reached the crossing at about 3.45 p.m. and his vehicle slid into the side of the engine of a New York Central Railroad freight train. Two ambulances and a passing motorist took ten of the children to the Jane Case hospital in Delaware but only three were detained. Those injured included

---

[1243] *The Advertiser*, Adelaide, South Australia, 18th December 1954
[1244] *Sydney Morning Herald*, Sydney, New South Wales, Australia, 20th December 1954
[1245] The *Corpus Christi Times*, Corpus Christi, Texas, United States, 29th December 1954
[1246] *The Brownsville Herald*, Brownsville, Texas, United States, 20th January 1955

two of the bus driver's own children – Connie, 10, and Mickey, 5 – who escaped with cuts and bruises.

Although the Stover Road Crossing was unprotected even by warning lights this was not really a factor in the accident as there was a good view of approaching trains. More significant was the fact that the road descended to the crossing and was covered with glare (black) ice.

The eastbound local freight train from Springfield to Delaware was in charge of Conductor C.R. Geer with Engineer E.L. Brittenham. It consisted only of a switch engine, four cars and a caboose and stopped very quickly after hitting the bus. The front of the bus was badly damaged and it was thrown into a ditch. [1247]

A minibus from the Chilean city of Iquique to the Victoria saltpetre office, 115 kilometres to the south east, and which went by the name of El Relámpago (The Lightning) had only just begun its journey on the night of Saturday 19th February when the trip ended in disaster. Not far from its starting point it was struck by train No. 3523 on a level crossing. The impact threw the minibus into a ditch and the train was badly derailed. At least one of the badly injured victims was apparently evacuated by air ambulance. Sixteen bus passengers were killed and 15 passengers and its driver, Pedro Valenzuela, were injured. The train driver, Gregorio Herrera, was not hurt but he was arrested and lodged in jail after the accident pending an investigation into who was responsible for it. [1248] [1249]

Southbound train No. 17 from Seoul collided with a bus on a blind level crossing two miles outside of Pusan, South Korea, at 7.25 p.m. on Tuesday 22nd February 1955. The impact reportedly rolled the bus over 15 or 20 times and of the 35 bus passengers 9 were killed and 11 injured. Among those who died was a mother with a baby strapped to her back. American ambulances rushed the injured to the Swedish hospital in Pusan. Attached to the train was a single coach for American troops. Only two soldiers were riding in it, Sergeant Curtis McDowell, and Corporal Thomas Putnam, and they assisted in the rescue work. The train was not derailed and no one on it was hurt. [1250]

A freight train collided with a bus on a level crossing in Johannesburg, South Africa, on Saturday 12th March 1955, killing eight passengers and injuring 27. All were described as 'Africans.' 'The bus was ripped open and hurled down the track, leaving a trail of dead and injured.' [1251]

The rear of a crowded bus which was crossing the line was struck by a train on a level crossing near Metan, Argentina, on Friday 29th April 1955. The bus was thrown against a barrier and four bus passengers were killed outright and a fifth was fatally injured.

1247 *The Evening Journal-Tribune*, Marysville, Ohio, United States, 5th February 1955
1248 https://chile-catastrofes-tragedias.blogspot.com/2020/08/blog-post_24.html
1249 *The Independent Record*, Helena, Montana, United States, 21st February 1954
1250 *Pacific Stars and Stripes*, Tokyo, Japan, 24th February 1955
1251 *Council Bluffs Nonpareil*, Council Bluffs, Iowa, United States, 13th March 1955

Another 14 bus passengers were injured, eight seriously. [1252]

A car was struck by express railcar E 938 at Muhlheim, West Germany, on the Hanau – Offenbach line, on Sunday 8th May 1955. Five people lost their lives. The barriers had not been lowered. [1253]

Six members of one family, who had recently moved from Floresville to work on a farm and ranch near Howth, Texas, United States, lost their lives when the truck in which they were riding was struck by the *Sunbeam* passenger train of the Southern Pacific Railroad on a crossing at Howth on Sunday 22nd May 1955. They were Juan Gonzales, 55; his wife Vaceliva, 43; and their adult children Paul, 22; Neachure, 21; Consuella, 20, and Urwado, 18. A seventh family member, Bernardo Gonzales, was critically injured and was taken to hospital. The accident happened at about 9.20 a.m. in clear weather when the truck pulled in front of the train and was demolished. The engine of the train was demolished but it was able to proceed to Dallas after an hour's delay. No one on the train was hurt. Mr R.E. Hamer, who lived near the crossing, heard the crash, and looked up to see debris flying in all directions. [1254 1255]

Another of the Sunday level crossing tragedies in the United States occurred at 4.19 p.m. on 19th June 1955 two miles north of Lepsic, Ohio. A car containing six people, five of them residents of Lima, was returning from a visit to Toledo. Mrs Helen Reinecke, 20; her children David Lee, 2, and Randy, 4 months; Miss Mary Martha Goss, 18, and cousins Reuben and Robert Alsept, 21 and 20 respectively, were all killed. The train was the northbound *Cincinnatian* passenger train of the Baltimore and Ohio Railroad, running between Baltimore, Maryland, and Cincinnati, Ohio. Torn apart by the impact, the car was carried 300 feet along the track, bodies and wreckage being distributed along the way. The victims were all pronounced dead at the scene. It was believed that Robert Alsept, of Toledo, whose father owned the car, had been driving it.

The crossing, with Route 65, had an unobstructed view in each direction of half a mile. Engineer Charles Matthise said that he had seen the car slow down as if it was going to stop short of the crossing, then it had started across in front of his train, suggesting that the driver thought he could beat the train.

By the following day Coroner J.R. Echelbarger had already returned a verdict of 'Accidental Death' on the victims. [1256]

A mother and five of her ten children were killed in an accident at what was described as an unauthorised level crossing near Bardwell Crossing, Wisconsin, United States, at 8.28 a.m. on Tuesday 5th July 1955. Mrs Harriet Smith, 53, of Beloit and her children Larry, 12; Keith, 6; Jackie, 4; Debbie, 1; and 'Skippy', 3 months, all lost their lives. Beatrice Smith, 21, Harriet Smith's sister-in-law and the driver of the car and Harriet's

---

[1252] *Bluefield Daily Telegraph*, Bluefield, West Virginia, United States, 30th April 1955
[1253] *Katastrophen der deutschen Bahnen Teil II* by H.J. Ritzau, pg. 247 (in German)
[1254] *The Taylor Daily Press*, Taylor, Texas, United States, 22nd May 1955
[1255] *Brownwood Bulletin*, Brownwood, Texas, United States, 22nd May 1955
[1256] *The Lima News*, Lima, Ohio, United States, 20th June 1955

children Leland, 12; Betty Lou, 9; Susie, 7; and Cheryl, 2, were all badly hurt.

The family had been staying at the home of Mr Eddie Smith's parents, Mr and Mrs Harold Smith, at the George Joslin farm, only 270 feet from the crossing. His parents had left for a vacation in northern Wisconsin only 15 minutes before the accident. The accident victims had just set off for nearby Allen's Grove for a pound of coffee. Mr Smith and the couple's 16-year-old son witnessed the accident from the farm yard of the elder Smith's property and the distraught husband, after identifying the bodies of his wife and children, was hoping to borrow a car to start after his parents and inform them of the tragedy.

Diesel-hauled Chicago Milwaukee St Paul and Pacific Railroad passenger train No. 22, from Madison to Chicago, which struck the car, was in charge of Conductor Charles Wright with Engineer Robert Roberts. The 12-car train, which did not whistle for the crossing, was travelling at an estimated 60 miles an hour when it struck the car. The vehicle was cut in two, half of it being carried 104 feet down the line and its motor further.

Signs on the gravel road adjoining the crossing read "Private Property, No Thoroughfare, Keep Out." Only 300 yards further on was an overpass (overbridge) over the line which offered a safe alternative to the crossing. However, the overpass had recently been out of use because its piers were disintegrating and in need of repair, necessitating a detour over the crossing. By the day of the accident the overpass was back in use but Beatrice Smith perhaps was not aware of this. Because of the overbridge visibility at the crossing was limited to a mere 30 feet. [1257] [1258] [1259]

On about Sunday 31st July 1955 a Canadian Pacific Railway passenger train from Ottawa to Montreal collided with a taxi on a level crossing at Talon station, Quebec, Canada, about five miles from Hull. The taxi driver, Alexandre Lauzier, and his four passengers, two men and two women, were all killed. [1260]

On Sunday 8th August 1955 a Paris – Warsaw international express collided with a car on a level crossing at Aulnoye, France, killing eight occupants of the vehicle, among them two three-year-old girls. The crossing was protected by barriers which had already been lowered for the approaching train but the crossing keeper, perhaps because he had been intimidated by the waiting motorists, had then lifted them again prior to the arrival of the train to let a few vehicles across. The last car to attempt to cross was the one struck by the train. [1261] [1262]

A school bus taking children home from an elementary school was struck by a train with disastrous results at Spring City, Tennessee, United States, at about 3 p.m. on Monday 22nd August 1955. The seating capacity of the bus was 36 but there were 49 pupils on it at the time of the accident, of whom 11 were killed and the remaining 38 were injured, as was the driver.

[1257] *Janesville Daily Gazette*, Janesville, Wisconsin, United States, 5th July 1955
[1258] *The Daily Telegram*, Eau Claire, Wisconsin, United States, 5th July 1955
[1259] *Las Vegas Daily Optic*, Las Vegas, Nevada, United States, 5th July 1955
[1260] *The Lethbridge Herald*, Lethbridge, Alberta, United States, 1st August 1955
[1261] *Overijsselsch Dagblad*, Zwolle, the Netherlands, 9th August 1955
[1262] *The Times*, London, England, 9th August 1955

The train involved in the accident was southbound second-class Cincinnati New Orleans and Texas Pacific Railway freight train No. 51, consisting of diesel units 4146, 4428, 4355 and 4159, coupled in multiple, 103 cars and a caboose, with Engineer Paul Giles.

The bus was a 1948 Chevrolet. Its driver, Raymond Moore, aged 38 or 44 according to different accounts, was described as a part-time farmer. He had only recently bought the vehicle and had only been operating it was a school bus since the start of the autumn term, a week earlier, under an agreement with the Rhea County School Board. He had no previous experience as a bus driver.

The crossing was with Picadily Avenue, close to its intersection with US Highway 27. There were no barriers at the crossing but it was provided with flashing warning lights. Standard crossbucks were also provided, with signs marked 'RAILROAD CROSSING'. In addition above the warning lights were signs reading '2 TRACKS' and below the lights signs read 'STOP ON RED SIGNAL'. The regulations required trains to whistle two long, one short and one long approaching the crossing.

Engineer Giles told the Interstate Commerce Commission (ICC), which investigated the accident, that he had watched the bus approaching the crossing from the east travelling at a speed which made it appear that it was stopping for the train, but as his train neared the crossing it was clear that the bus was not giving way. He had then applied the brakes in emergency but the speed of the train had not been materially reduced before the collision. The flashing lights at the crossing had been working. [1263]

Driver Moore, who was only slightly hurt, said that the traffic lights at the intersection with US Highway 27 had been green and he alleged that the warning lights at the crossing had not been operating. He told a newspaper reporter, "I stopped at the crossing, but didn't see the train. When I started across the tracks I saw it and heard the engineer blowing his whistle. I prayed I would make it but knew I was too late." The train struck the rear of the yellow bus, demolishing it and carrying it about 30 feet down the track.

A newspaper wrote, 'Police Chief Raymond Powers said he heard the crash, looked up and saw "a puff of smoke. When the smoke cleared, people began picking up bodies of the children and loading them into cars and ambulances," he said.

'Bill Barton said he and other bystanders saw the crash was imminent and began shouting to the bus driver, "Go on, go on."

'Paul Giles, engineer of the train, estimated its speed at 45 miles an hour at the time of the accident.

'"This is one of those things you read about, but don't believe will happen," said Denton Smith, principal of the Spring City school. "It is the worst thing that ever happened here."

'The impact of the collision scattered torn, bloody bodies along the right-of-way for several hundred feet, Mrs A.W. Hardaway reported.

'R.E. Cox, the station railroad agent, said he saw the crash, covered his face with his hands, and cried, "Oh, my God."

'{illegible word} said the blinker at the crossing was working right after the accident, indicating it apparently was working before Moore drove onto the tracks.

'"The bus looked awful," Mrs Hardaway said. "The bodies of the children were scattered over the ground, all covered with blood and mangled.

[1263] Interstate Commerce Commission report

'Scattered {illegible word} school books with their pages fluttering lay along the tracks. The body of one boy, Jimmy Hackler, had a notebook beside it, Mrs Hardaway said. Inscribed on the back of the book in red crayon was the word "love."

'"The children's bodies were tossed into high weeds along the track and broken just like some giant hand had picked them up and crushed them all together," Mrs Hardaway said.

'All five doctors in this small town of 1,725 population dropped their work and rushed to the scene where the painful screams of the injured emerged from the battered vehicle. Other doctors were called in from outside the city and ambulances came from the surrounding area to assist in taking the injured to nearby cities. When citizens heard of their plight, they began lining up at Rockwood's Chamberlain Memorial hospital to offer themselves as blood donors.

'The railroad tracks run through the center of town, paralleling U.S. 68, and only last month the council considered passing an ordinance prohibiting trains from travelling more than 30 miles an hour through town. No action was taken on the proposal.' [1264]

Despite the statements of the bus driver a witness told the ICC that the bus had been moving at 15 miles an hour as it crossed highway 27 and had not stopped before it crossed the tracks. Another witness put the speed of the bus at the crossing at 4 to 6 miles an hour. Several witnesses had heard the train whistling. Perhaps most significantly two of the surviving bus passengers told the ICC that they had seen the warning lights flashing and had heard the train approaching. They had warned Driver Moore that a train was coming, to which he had replied that he intended to cross in front of it. They said that the bus had not stopped at the crossing.

The ICC noted that the crossing lights on the side from which the bus had approached had been demolished in the accident but no fault was found with them and the lights on the far side of the crossing were still operating after the accident. Cars stabled on a spur close to the crossing obscured the view of approaching trains to some degree.

The ICC ruled that 'the accident was caused by a school bus occupying as rail-highway grade crossing immediately in front of an approaching train.' [1265]

Driver Moore was charged with manslaughter and released on bond. [1266]

Six people lost their lives when a Canadian National Railways train struck a car on an unmarked level crossing at Laval-sur-le-Lac, Quebec, Canada, on the evening of Wednesday 31st August 1955. The vehicle was carried several hundred feet. Armand Beausoleil, of Minnow Lake, near Sudbury, Ontario, the owner and presumably the driver of the car, was critically injured and was taken to the Sacre Cœur Hospital in Cartierville. Several of those who died were children. [1267] [1268] [1269]

Briefly reported in a large number of American newspapers and elsewhere was a

---

[1264] *Lubbock Morning Avalanche*, Lubbock, Texas, United States, 23rd August 1955
[1265] Interstate Commerce Commission report
[1266] *The Daily Telegram*, Eau Claire, Wisconsin, United States, 23rd August 1955
[1267] *The Daily News*, St Johns, Newfoundland, Canada, 6th September 1955
[1268] *The Lethbridge Herald*, Lethbridge, Alberta, Canada, 7th September 1955
[1269] *The Evening Journal-Tribune*, Marysville, Ohio, United States, 1st September 1955

collision between a freight train and a bus carrying workmen on a level crossing at Polpaico station, Chile, about 40 kilometres north of Santiago. Ten people died and 40 were hurt in the accident on the morning of Friday 16th September 1955. [1270]

A disastrous level crossing accident near Gomez Palacio, Mexico, on Friday 23rd September 1955 was caused by what can only be described as gross stupidity on the part of two lorry drivers. Each lorry was carrying four and a half tons of dynamite, used for mining operations, despite which the two lorry drivers were apparently racing one another after dark and without lights, and were so intent on their race that they failed to notice a train going over a level crossing. Travelling at full speed both lorries collided with a mixed local train from Torreon to Chihuahua and there were two terrific explosions. The engine, mail car, two passenger cars and freight car of the train were destroyed and its engine crew and most of its passengers were killed. The boiler of the engine was blown 300 metres and the motor of one of the lorries was found 700 metres away, while no trace could be found of the two lorry drivers. A gaping hole was left in the ground and houses near the crossing were levelled, while the shock of the blasts caused the balcony of a cinema in Gomez Palacio to collapse, causing a panic among the cinema goers, though none of them was hurt. Many windows of buildings in the town were broken. As well as the people on the train the inhabitants of houses near the crossing were killed or injured and the scene was described as being like a battlefield, with bodies and parts lying all around. Chickens, pigs, and burros kept in pens alongside the houses were also killed.

Because the bodies were so mangled an accurate estimate of the number of victims was impossible but it was thought at first that at least 65 people had died, including 12 children, and 100 more were injured, a number fatally. Ambulances were despatched from Torreon, 12 kilometres away, and many of the injured were taken to houses in Gomez Palacio as the little hospital there was quickly filled up.

The accident occurred at 19:15. The authorities stated that the leading lorry had crashed into the train just as the latter had reached the crossing and its load had exploded. The second lorry, a few metres behind, had followed it into collision a moment or two later and there was a second explosion.

Railway authorities said that it would be several days before the line could be reopened, while the road as well as the railway was blocked by the accident and hundreds of cars were left stranded on each side of the crossing while highway crews worked to fill in the big crater and make the road passable again.[1271] [1272]

By Sunday night 20 bodies had been identified and 14 other people known to have been on the train were "assumed dead." The fragments of other bodies recovered appeared to come from about 20 persons.

A newspaper found an eyewitness to the accident, a train passenger who had survived. 'Young Rodolfo Astorga of Torreon, one of the few survivors of the blast, told authorities Sunday he saw the racing trucks trying to beat the train to the crossing. He said the first truck disappeared from view through the window and suddenly "there was a blinding

[1270] *Kenosha Evening News*, Kenosha, Wisconsin, United States, 16th September 1955

[1271] *Las Vegas Daily Optic*, Las Vegas, Nevada, United States, 24th September 1955

[1272] *The Progress*-Index, Petersburg, Virginia, United States, 24th September 1955 flash and a terrific blast of hot air shook me. The whole earth shook." Astorga said he found himself sitting dazed on the ground several yards from the wreckage. "There was smoke everywhere," he said, "and the air for 500 yards around was filled with a rain of debris and flying objects." Astorga said he vaguely remembered a second explosion but he had no recollection of being blown from the train. He was badly burned about the face and arms by the giant flame that shot through the train when the 48,000 sticks of dynamite exploded.

'Medical brigades still combed the surrounding territory Sunday night for remains of blasted bodies.' [1273]

A runaway lorry caused a serious level crossing collision in West Newton, Pennsylvania, United States, on Tuesday 4th October 1955, resulting in the death of the lorry driver and four people in West Newton and the injury of two other persons.

The train involved in the accident was Baltimore and Ohio Railroad freight train No. 829 West, which consisted of diesel-electric units 829-A, 833-X and 829-0, coupled in multiple, 70 cars and a caboose with Engineer William McManus.

The road vehicle involved was a tractor-semi trailer (articulated lorry) owned by Sophie Lines Inc. of North Carolina and loaded with lumber. Its driver was Claude Brandt, 31, of Warrenton, North Carolina. Damage to the tractor unit of the lorry in the accident was such that little information could be gained as to its mechanical condition before the accident. No condition such as could have caused a brake failure was found.

The crossing, with Main Street, West Newton, was protected by standard railway signs and manned by a crossing watchman 24 hours a day. Commercial builds flanked both the road and the railway at the crossing. In the direction from which the lorry approached the crossing there was a short rise at 1 in 20 half a mile from the crossing, near the start of which there was a sign warning road users of the steep hill coming into West Newton, which read, "DANGER HILL - All trucks shift to low gear." Beyond the rise the road descended into West Newton at a gradient of more than 1 in 10 and there were a series of further signs warning drivers to stay in low gear and reading, "SLOW", "59.00 FINE: TRUCKS must stay in low gear", "WARNING - Speed trap in operation", "TRAFFIC SIGNAL AHEAD" and "NO PASSING ZONE". The speed of trains was restricted to 30 miles an hour over the crossing.

Engineer McManus told the Interstate Commerce Commission (ICC) that the headlight of the front unit had been lit, its bell had been ringing and he had whistled as he neared the crossing. As his unit passed over the crossing he had seen the lorry coming down the hill, 450 or 500 feet away, he could not estimate its speed. His train was travelling at 30 miles an hour when the lorry had struck its second car. The second and third cars of the train were derailed and demolished a hardware store, a fruit market, and a beauty parlor. [1274] In addition to Driver Brandt, the owner of the fruit store, Jacob Casale, 50; Sheryl Rose, 6, the daughter of a West Newton restaurant proprietor; Sheryl's grandmother, Mrs Dora Whiteman, aged about 70, of Industry, Pennsylvania; and Mrs Martha McConnell, 70, of West Newton, were killed. [1275]

[1273] *Lubbock Morning Avalanche*, Lubbock, Texas, United States, 26th September 1955
[1274] Interstate Commerce Commission report
[1275] *Chester Times*, Chester Pennsylvania, United States, 5th October 1955

It developed that as the lorry had approached the short rise east of West Newton it had had to stop behind a bus and two cars at a bus stop. It had then needed a push from another lorry to restart up the rise. At this point the driver of the Sophie lorry had inquired of the other driver the way to Elizabeth, showing that he was unfamiliar with the road. Starting down the hill he had applied the brakes, but the speed of his lorry had increased to 10 or 12 miles an hour. Rounding a curve he had overtaken a road grader travelling at 5 miles an hour, sounding his horn as he did so. It was unclear whether he had overtaken the grader by choice or whether his lorry was already out of control. An eastbound articulated lorry was ascending the hill and to avoid a collision with it Driver Brandt released his brakes, changed to a higher gear, and increased his speed. At the same time the road grader pulled over as far as it could to the right and the Sophie lorry narrowly missed both it and the ascending lorry. The speed of the Sophie lorry had now increased to 40 miles an hour and although Driver Brandt was attempting to brake as well as sounding his horn continuously its speed continued to increase as it neared the crossing at the bottom of the hill. The runaway lorry held the centre of the road until it struck the freight train.

The ICC investigation found much to criticise in the operation of the lorry. It determined that the Sophie company had no authority to transport property for hire in interstate commerce, despite which it frequently leased three lorries to lumber companies for interstate runs. The weight of the load of the lorry had not been recorded and it could have been 'well-seasoned' lumber weighing over 43,000 pounds, a 'dry shipping' weighing 45,733 pounds or 'green' lumber weighing 60,529 pounds. In any case, the gross weight of the lorry must have been at least 56,000 pounds, the weight of the load might have exceeded the Pennsylvania state limit of 50,000 pounds and it was an excessive weight for such a light tractor to handle, especially on a road with steep hills.

Driver Brandt was qualified to drive in interstate commerce. It was not known if he was experienced in driving over a route with steep hills and he evidently was unfamiliar with the road through West Newton. He had previous convictions for assault and 'careless and reckless driving.' It could not be established if the brakes of the lorry had failed but the gross weight of the lorry had been too much for its light tractor to handle safely, as was shown by the fact that it had not been able to restart up a 1 in 20 rise. In addition, Driver Brant had not kept to a low speed and gear as he began to descend the hill, nor had he steered into a ditch when he felt the lorry beginning to run away but had increased its speed and shifted to a higher gear, beyond the possibility of regaining control.

The ICC concluded, 'This excessive loading of light equipment, and the use of an unqualified or irresponsible driver, are symptoms of the basic lack of responsibility with which the operation was conducted. Sophie Lines Inc. was unlawfully transporting lumber in interstate commerce for hire, although its president was aware of the legal requirements and of his company's lack of authority from this Commission to perform such service. Required records relative to driver qualification, hours of service of drivers, and maintenance of vehicles, were not being kept. Such irregular practices invariably breed unsafe conditions on the highway.'

The ICC noted that it would seek additional facilities to enable it to strictly enforce compliance with its regulations by all carriers subject to them. [1276]

[1276] Interstate Commerce Commission report

At about 7.15 a.m. on Thursday 27th October 1955 a panel truck carrying a party of Mexican labourers was struck by a Southern Pacific Railroad train on Sonavi Road Crossing, near Chualar, California, United States. Eight men were killed outright or died on the way to Salinas Valley Memorial Hospital and six more were admitted to the hospital with injuries. A southbound local freight train knocked the truck about 50 feet. The truck had just turned off US Highway 101 onto the private road. [1277] It was thought that labour contractor Gregorio Guerrero, the owner and driver of the truck, had been blinded by the early morning sun and had thus failed to notice the train. He was amongst those who were killed. [1278] [1279]

An impromptu farewell party for a teenager going into service ended in the death of him and his five friends at about 10.16 p.m. on the evening of Thursday 10th November 1955. About to go into the military the following Monday was Robert Luedtke, 18, of New Auburn, Minnesota. The others who lost their lives were Roger Frauendiest, 17, also of New Auburn; and Milton Brelje, 16; Margaret Fuller, 15; Kari Olson, 16, and her brother Maurice, 14, the last four all of Plato, Minnesota, United States. Luedtke had graduated the previous spring from Glencoe High School, where the others were still pupils. The party had only just picked up Brelje and Miss Fuller, who were half-brother and half-sister, from the home of their grandparents in Plato where they lived, and the accident happened at the crossing at Plato station.

The train was a Chicago Milwaukee St Paul and Pacific Railroad local passenger service from Minneapolis to Aberdeen, South Dakota. Engineer Frank Webster said his train had been travelling at 70 to 75 miles an hour. Fireman Glen Rogers related that he had seen the lights of the car, which had been northbound, but had assumed that it would stop at the crossing. The wreckage of the car and the bodies of the victims were scattered along the track over a distance of 200 yards. Sheriff Leon Odegaard of McCloud County commented that the car was so badly smashed that, "I doubt that a junk yard would want it."

Ted Kingsley, who owned a restaurant in Plato, said that everyone in the small community was stunned. "To say that we are all shocked would be putting it mildly, he commented, adding that, "practically everyone here is related." [1280]

Saturday 12th November 1955 saw a rather unusual accident in Hong Kong, close to the Chinese border, when a passenger train was in collision on a level crossing with a British Comet tank. The tank was pushed about 100 yards along the track before the brand new diesel-electric locomotive tumbled down an embankment. Fortunately the train was almost empty as some of its carriages were telescoped. Three passengers were injured. An N.C.O. of the 7th Hussars in the tank and the guard of the train were the principal casualties in the accident; both were killed. [1281]

Nine members of a railroad section crew were killed when the truck in which they were riding was struck by a New York Central Railroad passenger train on South

---

1277 *The Bakersfield Californian*, Bakersfield, California, United States, 27th September 1955

1278 *Nevada State Journal*, Reno, United States, 28th October 1955

1279 *Lubbock Morning Avalanche*, Lubbock, Texas, United States, 28th September 1955

1280 *The Winona Daily News*, Winona, Minnesota, United States, 11th November 1955

1281 *The Times*, London, England, 14th November 1955

Newport Road Crossing, Michigan, United States, on Friday 18th November 1955. The lorry had just pulled onto the line and the mutilated bodies of the victims were scattered along the track for three-quarters of a mile.

Engineer Fred Foshag of the Indianapolis to Detroit train said afterwards that he had been "leaning on the whistle," when he had seen the lorry approaching the crossing. He continued, "There wasn't a chance of stopping the train in time. The bodies seemed to fly every way. Some of the men were so badly mutilated you could hardly tell they were human. Bodies were scattered by the impact."

The driver of the lorry, J.C. Martin, and two men riding with him in the cab of the vehicle survived the accident, though they were injured and taken to the Mercy Hospital in Monroe, Michigan. Asked about the accident Martin replied, "I don't know what happened," and said that he had not realised that a train was approaching. There was no protection at the crossing, not even warning lights or bells.

New York Central officials stated that the work truck had left section headquarters at Monroe, a few miles to the south, and they had been on their way to do some track repairs at Rockford, Michigan.

The train was not derailed but its engine was damaged and disabled and its passengers were forwarded on another train. [1282] [1283]

On the afternoon of Sunday 20th November 1955 a party of Illinois relatives was on its way from Orangeville to a roller skating rink at Polo, a few miles to the south, travelling in two cars. They were not far from their destination when they reached a level crossing with the Chicago Milwaukee St Paul and Pacific Railroad on the South Freeport road two miles south of Forreston, Illinois, United States. The leading car passed over the crossing safely. Its driver, Leslie Becke, said later, "I stopped my car and looked back to see if Alvin's car had crossed. As I did, I saw the train hit the car." The second vehicle had been struck by the *Challenger* streamliner passenger express of the Union Pacific Railroad, running over Milwaukee tracks under trackage rights (equivalent to British running powers). The streamliner was travelling, according to Engineer W.G. Lewis, at about 75 miles an hour. No one on the train was hurt and it continued its journey after a short delay.

As a result of the collision the car was wrapped around the front of the locomotive of the train and carried about three-quarters of a mile. Its six occupants were all killed. They were Alvin Wagner, the driver, and his wife June, both 20; Wayne Zettle, 25; his wife Joanne, 24; their daughter Becky, 3; and Marilyn Zettle, 12, a half-sister of Wayne. Leslie Becke was accompanied in the front car by his wife, who was Wayne Zettle's sister, and Earl Zettle, a half-brother of Wayne. All but Becky were killed outright, she was thought to be still alive when rescuers reached her but was dead on arrival at Freeport Hospital.

Railroad officials stated that visibility at the crossing had been good and that the warning device (a wig-wag signal perhaps) there had been working properly. [1284]

A 35-year-old Colorado man and five children died in an accident on the afternoon of

[1282] *South Haven Daily Tribune*, South Haven, Michigan, United States, 18th November 1955
[1283] *Syracuse Herald Journal*, Syracuse, New York, United States, 18th November 1955
[1284] *Freeport Journal-Standard*, Freeport, Illinois, United States, 21st November 1955

Saturday 26th November 1955 half a mile from Matheson, Colorado, United States, when the car in which they were riding was in collision with the eastbound *Rocky Mountain Rocket* of the Chicago Rock Island and Pacific Railroad, en route from Colorado Springs to Chicago. The victims were Steve Puzick, his daughters Junie, 11; Marilyn, 9; Suzanne, 4; and Cynthia, 3; and a probable cousin, Eli White, 9. The automobile was carried 200 feet and all those in it were dead when rescuers reached the scene. It was assumed that Puzick had failed to see the train as he was driving the vehicle southwards along a country lane. [1285]

The engineer of the train was named as G. Garnhart. [1286]

Very restricted visibility due to fog was clearly a factor in a level crossing collision at 7.58 a.m. on the morning of Tuesday 29th November 1955. A car containing six students was driven into the side of the northbound *Chief*, a crack passenger train of the Atchison Topeka and Santa Fe Railway, from Los Angeles to Chicago, at a crossing eight miles south of Hanford, California, United States. The students had been on their way to attend classes at College of Sequoia in nearby Visalia. The train had been travelling at 80 miles an hour when the car had slammed into its fourth coach, scattering bodies and wreckage along the track for several hundred yards. Patrolmen said skid marks 110 feet long showed that the driver of the car had only seen the train in the last few seconds.

The dead students were named as Loreto Anthony Fabbri, 19, the owner and presumed driver of the car; John Earl Champness, 20; Marlene Rae Faubion, 19; William Eugene Valentine, 22; and Jerald Dale Villaborgi, 20, all of Lemoore; and Frances Joann Quist, 18, of Stratford. All died at the scene except Valentine, who passed away after arrival at Hanford Hospital.

Mrs B.C. Hampshire of Hanford was following the students in another car and witnessed the accident.

The crossing was unprotected by warning signals. [1287] [1288]

Only a few days later there was yet another bad level crossing collision in the United States involving a crack passenger train. This time it was the *Crescent* of the Southern Railway, which at dusk on Sunday 4th December 1955 collided with a car on Waddell Road Crossing, Paris, South Carolina, just east of Greenville. Engineer G.F. London said that the northbound train from Los Angeles to New York had been travelling at 75 miles an hour when it had struck the vehicle, carrying it half a mile. He had watched the car approach the crossing, going at about 10 miles an hour, and it had just continued onto the track without stopping.

James Brooks, 37; his wife Louise, 26; Holland Benny Sutton, 29; his wife Rosa Lee, 23; and their daughter Wanda, 3, all of Greenville, lost their lives. A sixth fatality, a woman, could not immediately be identified. The seventh passenger in the car, Benny Sutton, 10 months, survived the accident with only minor injuries.

The train had only just left Greenville and was going at its usual speed. As usual in

---

1285 *Wichita Daily Times*, Wichita Falls, Texas, United States, 27th November 1955

1286 *The Greely Daily Tribune*, Greely, Colorado, United States, 28th November 1955

1287 *The Modesto Bee*, Modesto, California, United States, 29th November 1955

1288 *Oakland Tribune*, Oakland, California, United States, 29th November 1955 this sort of accident no one on the train was hurt. The car was owned by Mr and Mrs Brooks and Mrs Brooks had been driving it. [1289]

A minor school bus collision at Huntingtonville, New York, United States, on the morning of Wednesday 14th December 1955 was apparently the fault of the bus driver. Engineer Clinton Mack of the New York Central Railroad train, which consisted only of an engine and one car, said that as his train had approached the train, travelling at 40 miles an hour, the bus had come up to the crossing and stopped. Then, as the train had got closer, the bus started up again and had been driven onto the crossing into the path of his train. Struck by the engine the bus had been knocked down a 12-foot embankment. Of the 16 children who were passengers on it five were taken to hospital, as was the driver, but only two children were detained, one with a broken leg and the other suffering from shock.

The bus driver, Frank Williams, 45, told police that he had stopped at the crossing but had not seen the train. There was a clear view along the line for 500 feet. Investigations into the accident were continuing at the time of the report. [1290]

Five teenagers died at 7.05 p.m. on the evening of Saturday 17th December 1955 when the pickup truck in which they were riding was struck by Chicago Burlington and Quincy Railroad passenger train No. 42 from Billings, Montana, to Lincoln Nebraska, on a crossing just west of Gillette, Wyoming, United States. Struck broadside, the truck was knocked into the air and over the locomotive of the train. It was demolished but remained largely in one piece, although small pieces were distributed along the lineside for over 150 yards. No one on the train was hurt, though many passengers felt the shock of the impact, and the train made its normal stop at Gillette.

Those who died were Daryl Johnson, 16; Glenna Phillips, 15; and brothers Dale and Gary Snylor, aged 16 and 15 respectively, all of Gillette; and Robert Seifert, about 18, of Scottsbluff, Nebraska. It was not known who had been driving, though the truck was registered to Robert's father George Seifert. It was believed that the teenagers had been returning to Gillette from the residence where Robert Seifert had been staying. [1291]

Just before noon on Friday 23rd December 1955 a family of five on a Christmas visit were killed when their car was struck by a northbound Gulf Colorado and Santa Fe Railway freight train on a level crossing at View, Texas, United States on US Highway 277. They were Mr and Mrs Elda Adron Starr and their children Roberta, 8; Everett Lynn, 7; and Chole Ann, 5. They had been on their way from Levelland to Jacksboro to visit their parents and grandparents for Christmas. A telegram found in the wreckage of the car read, "To Roberta, Lynn, and Chloe Ann Starr, 114 Wilson, Levelland, Tex. Christmas Eve is almost here, and with my sleigh and eight reindeer, I'll start out over the snow and ice, to bring you something really nice, to your grandmother's at Jacksboro, Tex. Santa Claus."

The car was twisted around the front of the diesel engine, apart from the motor, which was thrown to the east of the track. It was carried three-quarters of a mile down the track

---

[1289] *The Derrick*, Oil City, Pennsylvania, United States, 5th December 1955

[1290] *The Post-Standard*, Syracuse, New York, United States, 15th December 1955

[1291] *The Billings Gazette*, Billings, Montana, United States, 18th December 1955 before the train stopped and the car caught fire.

Roy Willis and Smokey Owens from Sweetwater were waiting in a truck for the train to pass on the opposite side of the crossing. 'Willis said he saw the other car coming east and wondered when it was going to stop. He said the driver never applied the brakes and made no attempt to stop. He thought the man never saw the train. Both the witnesses and Santa Fe officials said the red blinking lights were in operation, as well as the bells at the crossing. Santa Fe officials said the train had just blown its whistle seconds before the crash.'

The train was in charge of Conductor E.K. Dixon with Engineer Ed Land, Fireman J.K. Carroll, and Brakeman M.O. Hill. The accident happened at 11.25 a.m. [1292]

The *Maple Leaf*, a Chicago, Illinois, to Montreal, Quebec, Canada, limited express of the Grand Trunk Western Railroad, collided with a station waggon on an unguarded level crossing near New Carlisle, Indiana, United States, on Saturday 7th January 1956. The car involved in the accident was the third of a procession of vehicles carrying members of the New Carlisle High School 'B' basketball team to a competition at nearby Walkerton. The coach of the team, Loyal Marker, who had been driving the first car, advised the sheriff's office of the accident.

Killed in the accident were Mrs Louis Berzal, 37, the wife of the manager of a South Bend Exterminator Company; her daughters Elizabeth, 14; Barbara Ann, 10; and Josephine, 8; Thomas McCord Baird, 16, and Roger Lindsey, 17, members of the basketball team; and a young girl, Martha Kleine. Louis Berzal Jr., son of Mrs Berzal and baseball team manager; and Janet Paul Ruther, 15, were badly injured.

Engineer J.H. Hansom said that his train had been travelling at 79 miles an hour as it had approached the crossing. Visibility had been good and he had been sounding his whistle repeatedly but the procession of cars had all been blowing their horns. The station waggon had failed to give way, had been struck and carried 100 feet down the track, and demolished a telegraph pole.

Despite the tragedy after taking a vote the remaining team members decided to go ahead with their tournament and they won their first game. [1293]

Eight people died when a train collided with a vehicle variously described as a truck and a station wagon being used as a bus at a level crossing 15 miles north-east of Monterrey, Mexico, on the night of Tuesday 17th January 1956. The truck driver, Aurelio Pompa Vela, had apparently dozed off. He was the only survivor of the accident, though according to most reports he was critically injured.

The next morning two people died and 20 were injured when a city bus collided with a shunting engine on the outskirts of Mexico City. The bus driver was arrested for negligence. [1294] [1295]

Tragedy followed an act of kindness on the afternoon of Tuesday 31st January 1956.

[1292] *The Abilene Reporter-News*, Abilene, Texas, United States, 23rd and 24th December 1955

[1293] *The News Palladium*, Benton Harbor, Michigan, United States, 9th January 1956

[1294] *The Galveston News*, Galveston, Texas, United States, 19th January 1956

[1295] *Albuquerque Journal*, Albuquerque, New Mexico, United States, 20th January 1956

Mrs Emily D'Ascenzo, 26, had gone with her young children, Joseph, 4; and Donna, 2; plus her mother, Mrs Esther Benedetto, 57; to visit the home of Mr and Mrs Serafini, Mrs Edith Serafini being ill. At the end of the visit Mr Joseph Serafini offered to drive the visitors home, and it was on this return journey that the tragedy occurred. When the car in which the five were riding reached the level crossing at Secane Station, Pennsylvania, United States, it was struck by the 2.50 p.m. four-car commuter passenger train from Philadelphia and carried 300 feet down the line. The accident happened at 3.18 p.m. Four of the five occupants of the car were killed outright. Donna was lifted out of the wreckage still alive but died in the Fitzgerald Mercy Hospital early the same evening. Mr Joseph D'Ascenzo, who had spent the day job hunting, came home to find that he had lost his whole family. The crossing was protected by warning lights which were working at the time of the collision.

The sequel to the accident was the arrest of the engineer of the train, Myron Adams, on a manslaughter charge; quite why is not clear. He was released on $2,000 bail. Asked as to the speed of his train he refused to reply on the advice of his attorney. [1296]

The inquest on the victims of the accident in early March was described by Coroner Joseph Tercha as the longest in his memory, lasting for eight and a half hours. Evidence was given that the blinker lights at the crossing operated for such a long period and so frequently because of shunting operations in the vicinity of the crossing that people tended to become careless and ignore them, as perhaps had Joseph Serafini on the day of the accident. At the conclusion of the inquest the jury exonerated Engineer Adams, who was released from bail, instead finding the Pennsylvania Railroad company guilty of criminal negligence. The jury recommended the provision of gates and a watchman at the crossing with Bishops Avenue. [1297]

The southbound *Silver Meteor* streamliner of the Seaboard Air Line Railroad from New York to Miami collided with a car on the outskirts of Ocala, Florida, United States, on Sunday 4th March 1956, knocking the vehicle into a concrete signal standard. Mrs R.S. Floyd, the driver of the car, her children Cecilia, 7; Carolyn, 5; and Edward, 1; plus Richard Smith, 12, all of Ocala, were killed. Morris Floyd, 9, was taken to Monroe Memorial Hospital in Ocala, critically hurt, while Douglas Floyd, 11, managed to jump from the automobile before it was hit and escaped unhurt. The warning lights at the crossing were reportedly working properly.

Engineer G.W. Haven, whose train was delayed half an hour by the collision, was unwise enough to state that his train had been travelling at 40 miles an hour at the time. [1298] A city ordinance set a maximum speed of 20 miles an hour through Ocala, so a week after the accident the engineer found himself in court, where he was fined $35 for exceeding the speed limit through the town. [1299]

Six residents of Auburndale were on their way to Winter Haven, Florida, United States, on the evening of Tuesday 20th March 1956 to begin their night shift at the citrus canning plant there. They had almost reached their destination when their car was struck

---

[1296] *Chester Times*, Chester, Pennsylvania, United States, 1st February 1956
[1297] Ibid, 8th March 1956
[1298] *Fort Pierce News-Tribune*, Fort Pierce, Florida, United States, 4th March 1956
[1299] *Winchester Evening Star*, Winchester, Virginia, United States, 10th March 1956

by a Seaboard Air Line Railroad train on a level crossing and all were killed. The impact carried the vehicle 200 yards down the track, its occupants trapped inside.

The victims were James Ruben Moore, 20, the car driver; his wife Reba Jean, 18; his mother Mrs Jessie Moore, 55; Mattie Shirey, 64; his daughter-in-law, Mrs Sarah Shirey, 18; and Rachel Elizabeth Pelham 48.

Engineer J.E. Haddock said his train had been travelling at 50 miles an hour and the car at 20 miles an hour. He did not believe that the car driver could have seen his train. Citrus groves lined the country road on either side and could have obscured the view of the train from the car driver. [1300] [1301]

Eight people died when a narrow-gauge train collided with a packed taxi at Tanta, Egypt, in the Nile delta, on Saturday 28th April 1956. The train was derailed as a result of the collision and fell into an irrigation canal. The brief report did not indicate if the deaths were in the train or the taxi. [1302]

After exactly two months free of what might be described as major level crossing accidents the United States suffered another bad one on Sunday 20th May 1956. A car containing eight members of two families was struck by passenger train No. 2 of the New York Central Railroad, the eastbound *Pacemaker* from Chicago to New York. The accident happened at the unprotected Murray Ridge Road Crossing, not far from Elyria, Ohio, where four running lines were traversed. Wreckage of the automobile, a new Oldsmobile, was reportedly welded to the front of the diesel engine of the train and carried a mile down the track. All eight people were killed outright.

The victims were Mr Millard Martin, 37; his wife Eugenia, 33; their children David, 15; Donald, 14; Charleen, 9; and Timothy, 3; together with Mr Warren Edwards, 46; and his wife Mary, 45. All were from Akron, Ohio. They had been attending an evangelists' meeting at the Junior Chamber of Commerce Hall in Elyria. The meeting had finished at 9.15 p.m. and the families indicated that they would return to Akron via the Ohio Turnpike, joining it at the Rt. 57 interchange. When the accident happened at 9.30 p.m. their car was only a few hundred feet from the turnpike but two miles from the interchange so it was theorised that they had got lost.

Conductor Clyde Coffman stated that the whistle of the engine of his train had been blowing and its bell had been ringing as the southbound car had driven onto the crossing directly in front of the train. The engineer of the train was W.G. Karchner and the fireman L.D. Krall. Immediately after the accident the train crew had lit a fusee and thrown it into the path of westbound mail train No. 129, fearful that it might run into the wreckage of the car and be derailed. The latter train was stopped close to the scene by Engineer M.N. Sly, having just scraped the wreckage.

As might be expected the tragedy received extensive coverage in local newspapers and the danger of this and many other similar level crossings was commented upon:

'As after every railroad crossing fatality, there were assertations on what should have been done, what should be done now.

[1300] *Fort Pierce News-Tribune*, Fort Pierce, Florida, United States, 21st March 1956

[1301] *The Daily Independent*, Kannapolis, North Carolina, United States, 21st March 1956

[1302] *The Modesto Bee*, Modesto, California, United States, 30th April 1956

'Elyria Township residents remembered that Maplewood School has one bus which, loaded with youngsters, crosses the track twice daily. It was recalled too, that township trustees, three weeks ago, asked the New York Central to put up warning signals – a continuation of a several-year campaign for automatic controls.

'The spectators could see for themselves, or were all too familiar with, the conditions at the crossing with the heavily travelled black top. They could see that, going southward, as the Akron families were, the motorist has a very limited view of the tracks to the right while he is some distance back from the crossing. There is a cornfield along the right. It slopes upward gently, until, at the crossing, it is a bank high enough to obscure the motorist's view of any train.

'For the southward motorist, the view is slightly better at the left, but even so a row of small trees and brush partially obscure the view. For the northbound motorist, the view to the left is good and the view to the right is improving slightly – thanks to grading work under way for construction of a Northern Ohio Growers packing house. That view has been blocked by a high bank, but part of the bank has been cut away now.

'Maybe the catastrophe will result in action, installation of flashers or some controls at the Murray Ridge crossing. Maybe it won't.

'In any case what will happen at the other some 120 crossings in the county which, like Murray Ridge, have no automatic controls; which, in many cases offer track views just as obscured as Murray Ridge?

'Where will the crossing accident blight strike next, and whose family circle will be broken, or obliterated?'

An editorial in the same newspaper carried headlines such as 'HOW MANY MORE MUST BE KILLED?', 'IT COULD HAVE BEEN YOUR FAMILY', (unprotected crossings with poor visibility) 'MUST NOT BE TOLERATED ANY LONGER', and, again, 'HOW MANY MORE MUST BE KILLED BEFORE WE ACT?????' [1303]

An accident occurred in the town of Young, Uruguay, on Monday 21st May 1956. A passenger train from Montevideo, hauled by Class R steam locomotive No. 112, approached a level crossing near the station there at the same time as bus No. 225 of the ONDA bus company. The bus driver, failing to notice that the train, moving at speed, was close at hand, entered the crossing directly in front of it. It seems likely that there were no barriers at the crossing. The train driver was completely unable to avoid a collision and the bus was wrecked, while the engine of the train suffered slight damage to its front. The number of casualties among the occupants of the bus is unclear, different accounts speak of 30 deaths or of only two or three. [1304]

On about Tuesday 22nd May 1956 a Gonda-bound train collided with a bullock-drawn(?) cart between Biswan and Sarayan stations, India, on the Burhwal Junction – Sitapur Junction section of the North Eastern Railway. Eight people died and others were injured. [1305]

1303 *The Chronicle-Telegram*, Elyria, Ohio, United States, 21st May 1956
1304 http://www.inforio.com.uy/noticia.php?id=9418
1305 *Indian Daily Mail*, Singapore, 23rd May 1956

No. 563 Down train collided with a bus on an unmanned level crossing near Varvala, India, on the Rajkot Junction – Okha line in Saurashtra state, at about 3.30 p.m. on Thursday 24th May 1956. Eleven bus passengers lost their lives and 25 others were injured, four seriously. No one on the train was hurt. [1306] [1307]

Five occupants of a car were killed outright and a sixth was fatally injured when it was struck by a southbound Pennsylvania Railroad freight train from Richmond, Indiana, to Cincinnati, Ohio, on Williamsdale Crossing, near Hamilton, Ohio, United States, at 1.29 a.m. on Thursday 28th June 1956.

Those who lost their lives were Miss Opal Brogan, 16; Mrs Leona Collier, about 33; Mrs Oma Blackwell Jones, 28; James Neace, 30; Claude Raleigh, 38; and Jerry Stidham, 30. Mrs Collier and Mrs Jones were sisters and Miss Brogan was their niece.

Registered to Claude Hamilton, the car was travelling eastwards on the Hamilton – Trenton road. It was struck on its left-hand side by the diesel engine of the train, pushed about 200 feet along the track, knocked to the right-hand side of the single line, then caught by the waggons of the train and dragged another 50 feet. The car, a 1952 Ford, was demolished and parts of it were scattered along the right-of-way. The train was delayed an hour while the wreckage of the car was removed from under it.

Conductor P.H. Conley and Engineer D.J. Calloway said afterwards that car driver had made no attempt to stop at the crossing. Wooden crossbucks were the only protection at the crossing.

Jackson Bosch, the Butler County prosecutor, was preparing a report to submit to the State Public Utilities Commission in the hope of getting flasher signals installed at the crossing with a road which had heavy traffic. He stated, "We consider it a dangerous crossing even though some time ago the Pennsylvania Railroad cleared weeds and other obstacles from the railroad in order to provide a better view of the track." Much needed flashers had been installed at the nearby Bobmyer Road Crossing with the Pennsylvania line after much litigation between the county commissioners, the railroad company, and the utilities commission. "We waited for years to get lights at Bobmyer Road," Bosch added, "and we hope the similar devices will be installed at the Williamsdale Crossing. The accident this morning is certainly an example of the need for such signals." [1308]

Engine 90 843 B, running light, collided with a car on a level crossing at Emmerthal, West Germany, on Sunday 8th July 1956, killing five people. The crossing keeper had mistakenly raised the barriers too soon. [1309]

A widow, her male companion and four of the woman's seven children were killed in a level crossing collision at Crocker, Indiana, United States, at 9.40 p.m. on Saturday 4th August 1956. The car in which they were riding was struck by an Elgin Joliet and Eastern Railway freight train and carried about 750 feet.

The victims were Mrs Valeria Shupe, 32; Elgie Rose, 33; and Mrs Shupe's children,

1306 *The Daily News*, Saint Johns, Newfoundland, Canada, 26th May 1956
1307 *Indian Daily Mail*, Singapore, 30th May 1956
1308 *Hamilton Journal*, Hamilton, Ohio, United States, 28th June 1956
1309 *Katastrophen der deutschen Bahnen Teil II* by H.J. Ritzau, pg. 247 (in German)

Roberta Jean, 11; Valeria Rae, 9; Penny Jo, 7; and David Marion, 2. Two other Shupe children who had been in the car, Jessie, 8, and Rickie, 3, were injured, while a seventh child, Gale, 12, was not in the car but was visiting neighbours at the time. Mrs Shupe owned the car but Mr Rose was driving it at the time of the accident. Skid marks 103 feet long indicated that he must have seen the train at the last moment. What sort of protection was provided at the crossing on Indiana 149 was not stated; very likely there was none. [1310]

The Levine family of Chicago and their guest, David Strinel, aged 12, had just finished a week's holiday at the Forest Beach YMCA camp, a mile south of New Buffalo, Michigan, United States, on Sunday 19th August 1956 and were on their way home when tragedy struck. Their car, a 1956 Buick, was still on the private road leading from the camp when it was in collision with a New York Central Railroad passenger train at 4.10 p.m. The car was wrecked and its occupants were all killed or fatally injured. Apart from the Strinel boy they were Mr Peter Levine, 49; his wife Mollie, aged about 43; and their children Paula, 19; and Norman, 14. Four of the victims were killed outright and Mr Levine died after admission to the Doctor's Hospital in Michigan City.

Engineer George Tickfer told police that his train had been travelling at about 80 miles an hour along a straight stretch of track. He said that the Levine car had emerged from woodland travelling at about five or six miles an hour, when he had sounded his whistle, although the crossing was not one where he was obliged to do so. The car, however, had kept going until it reached the track when it had suddenly stopped. The road climbed to the crossing and it was believed that Mr Levine had been looking in the opposite direction and had failed to see the train. Being a private road crossing it was not provided with any warning devices.

'Broken tennis rackets, phonograph records and clothing items were strewn for two-tenths of a mile along the railroad tracks.' The car itself was smashed into three pieces and its engine was found about 300 feet from the point of impact.

Mr Levine operated a small manufacturing business in Chicago along with his brother. [1311]

A school bus stalled on a level crossing at Columbia, South Columbia, United States, on Wednesday 17th September 1956 and was struck by the Seaboard Air Line Railroad *Palmland* passenger train, running between Miami and New York. Fortunately the results were not as serious as they might have been. The driver, Jack Terry, explained afterwards that he had sent his 13-year-old brother out of the bus to check if there were any trains approaching, as per the school bus driving regulations. His brother had indicated that there was indeed a train coming and that he needed to go back a little. He had then attempted to do so but in his haste he had put the bus into low gear instead of reverse and it had moved forward, then he had stalled the vehicle. At that point he had ordered the children to disembark and nine of them had had time to do go. Jack Terry himself and his sister Margie, Georgie Dial and two sisters, Betty Lee, and Mary Gates, were the only ones on the bus when the train hit it. Margie Terry and Georgie Dial escaped unhurt, Terry and the Gates sisters were injured; the two sisters were detained in hospital for further examinations.

[1310] *The Vidette Messenger*, Valparaiso, Indiana, United States, 6th August 1956

[1311] *The News-Palladium*, Benton Harbor, Michigan, United States, 20th August 1956

The train was in charge of Conductor E.W. Southern with Engineer J.R. Lucas. The engineer said that he had seen the stalled bus, had braked, and reversed his engine, but had struck the bus and dragged it about 25 feet.

Jack Terry was only 16 years old, which seems very young for someone in such a responsible position, though it was apparently permitted, he had been trained under the auspices of the Highway Department and officials claimed that the training programme had an usually high safety record since it had begun. [1312]

On the evening of Friday 5th October 1956 a car driven by Mrs Marie Benda, 33, of Buckingham, Iowa, and also containing her daughter Marie, 7; her sister, Mrs Violet Dungy, 25, of La Porte City, Iowa; and Mrs Dungy's children, Richard, 2; and Debra, 1, turned off US 218 onto a gravel track leading to the Kenneth Eastman farm near Washburn, Iowa, United States, where the women intended to purchase some chickens.

According to Engineer C.D. Shimerda and other members of the train crew the car appeared to hesitate for a moment and was then driven onto the track in front of an 18-car Chicago Rock Island and Pacific Railroad freight train. The 1948 model automobile was struck broadside and almost cut in two. Four of its occupants were killed outright and the Benda child died en route to a Waterloo hospital. The husbands and fathers of the victims arrived at the scene as the clearance work was in progress and both almost broke down.

It was believed that the setting sun had blinded Mrs Benda and prevented her from seeing the approaching train. [1313]

Five 'youths,' as they were termed at the time, were killed when the car in which they were riding was driven into the side of a moving southbound St Louis-San Francisco Railway freight train on the evening of Thursday 18th October 1956 at Linden, Alabama, United States. Sylvester Alston, 16; Levin Besteder Jr., 16; M.C. Robertson, 16, believed the driver of the car; his brother John Lee Robertson, 14; and Lorenza Scott, 17, were all killed. Lieutenant Johnston, 17, the sixth occupant of the vehicle, was taken to Demopolis Hospital critically injured. Sheriff T. Wilmer Shields commented, viewing the results of the accident, that, "It looked like a slaughter pen." The victims had been returning from a football game in Linden.

The accident was clearly due to the car being driven at excessive speed, so that its driver was unable to pull up when he reached the level crossing and discovered that the freight train was passing over it. The speed of the automobile was such that the heavily loaded flat car which it struck midway along the train was partly derailed and bumped along the line for about a mile, damaging cross ties (sleepers) and a switch (set of points). The automobile meanwhile was thrown against a small frame building which had previously served as a waiting room. Skid marks were left over a distance of 258 feet leading up to the crossing. [1314]

Eight children of Mr Richard and Mrs Ruth Hammes died in a level crossing collision at 8.50 a.m. on Thursday 25th October 1956 near Sigourney, Iowa, United States. They

[1312] *Florence Morning News*, Florence, South Carolina, United States, 18th September 1956
[1313] *The Boone News-Republican*, Boone, Iowa, United States, 6th October 1956
[1314] *The Monroe News-Star*, Monroe, Louisiana, United States, 19th October 1956

were twins Donnie and Ronnie, 9; Karen, 8; Linda, 7; Gary, 6; Rosemary, 2; and twins Ricky and Vicky, 18 months. All but Ronnie were killed outright; he succumbed to his injuries at noon in Keokuk County Hospital. Mrs Hammes, 38, was herself seriously injured. She had been taking the older children to a grade school when her car was struck by a westbound Chicago Rock Island and Pacific Railroad train consisting of a diesel engine and a caboose, which dragged the four-door automobile 684 feet and split it open. School books and lunch pails were strewn along the line.

Visibility at the crossing was restricted by a cutting and vegetation. Deputy Sheriff Keith Bryant described the crossing as a bad one, "close to being blind."

Mrs Paul Webb, who lived near the crossing, had just returned from taking her own children to the same school, Plank township No. 9. She witnessed the accident. "It was terrible. I heard the train honk. The honk drew my attention and that's why I saw her. My husband was feeding the hogs and the train stopped near where he was. He was the first to the scene, but he just won't talk."

Mrs Hammes herself told a highway patrolman that as her car had reached the crossing one of her boys had cried, "Here comes a train!" She said that she had then instinctively hit the brakes, despite which the train had hit the right rear door of the car.

Engineer George West put the speed of his train at 30 miles an hour.

Richard Hammes had said goodbye to his family only a few minutes earlier. He had then gone to a cornfield with his tractor but had returned to his house when his tractor had developed a problem and had still been there when a neighbour, Mrs Jack Sellers, had arrived, advised him what had happened and had taken him to the scene. There he stood over the bodies of his children with no outward display of emotion. "He was shocked beyond any reaction," the deputy sheriff explained. [1315]

Mrs Hammes survived the accident but a further tragedy was to follow. A son, Myron, was born in 1957 but in April 1960, shortly before she was due to give birth to a tenth child, Mrs Hammes died of a heart attack, aged 41. Mr Hammes said afterwards, in a faltering voice, "The doc said her heart was just too heavy with grieving about the children. It had to give. She kept the grief all to herself. She never talked about what happened." [1316]

At least 20 people lost their lives and 24 were hurt in a collision between a bus and a freight train on Saturday 24th November 1956 at Mixcoac, Mexico, on the Cuernavaca highway south of Mexico City. A number of those injured suffered burns and some were critically hurt. It was raining hard when the bus, carrying 50 passengers, crossed the track. According to survivor Raul Munoz the bus driver had been changing into second gear when the vehicle had been struck by the train and carried about 30 feet, into the side of another engine standing on an adjoining line. Crushed between the two locomotives the fuel tank of the bus exploded and it was quickly enveloped in flames. Fire engines and 18 ambulances rushed to the scene of the disaster. [1317]

Six occupants of a crowded car were killed and the other four were badly hurt when it

---

[1315] *The Cedar Rapids Gazette*, Cedar Rapids, Iowa, United States, 25th October 1956
[1316] https://toprad.com/hammes/getperson.php?personID=I157&tree=tree1
[1317] *Albuquerque Journal*, Albuquerque, New Mexico, United States, 25th November 1956

was struck by a Canadian Pacific Railway train on a level crossing at Belleville, Ontario, Canada, on Sunday 2nd December 1956. Killed in the accident were Mr Harry Hill, his wife, their daughters Audrey, aged about 14, and Marion, aged about 12, their son Robert, aged about 8, and Archie Maracle, aged about 50. All were from the Tyendinage Indian Reserve, about 18 miles east of Belleville. Injured were Mr Don Robertson, his wife Thelma, and their son, aged about 7, and daughter, aged about 5. All were severely injured and it was feared that Mrs Robertson in particular might succumb to her injuries. [1318]

At about 1.10 a.m. on Monday 17th December 1956 a car, reportedly travelling at between 70 and 80 miles an hour, was in collision with the *Golden State Limited* passenger train of the Southern Pacific Railroad, moving at about the same speed, at the level crossing with 35th Avenue on the outskirts of Phoenix, Arizona, United States. The engineer of the Los Angeles to Chicago train, Carl DePriest, said later that he had felt a "jar" as the car had struck the train between its leading and second engines, while the passengers knew nothing of the accident until the train stopped at the Union station, three miles further on. No one on the train was hurt but the results were disastrous as far as the car was concerned, 12 of its 13 occupants being killed and the 13th critically hurt. The car was completely demolished; part of it was jammed beneath the train and wreckage was scattered over a wide area.

Eleven of those who lost their lives, all Phoenix residents, were identified as Juan Alires Sr., 59; Juan Alires Jr., 21; Victor Alires, 28; Sofia Alires, 28; Clara Alires, 21; Steven Alires, 10; David Alires, 8; Bobby Alires, 3; Mike Alires, eight months; Elvira Mary Cline, Leroy Johnson, and John Massey, the last three all 36. Crucila Alires, 18 months, survived the collision but sustained a fractured skull. Crucila and her six brothers and sisters, who were not in the car, were orphaned by the accident and a group of local radio stations made an appeal for financial aid for the orphans. A California family offered to foster some or all of them.

There were two witnesses to the accident. Ray Gromek, an employee at a nearby metal plant, said that he had seen the car, a 1938 four-door Chevrolet sedan, approaching the crossing at high speed. He said, 'The driver ran red traffic lights and used his horn to get other cars off the road. The car was packed and it looked like they were in a hurry."

Walter Vandergriff had just driven over the crossing on his way to work. He related, "I had seen the headlight of the approaching train. Then I saw the sedan headed south and travelling pretty fast. I looked into my rear view mirror to see if the car was going to make it. It didn't." [1319]

Engineer DePriest stated that he had sounded his horn and flashed the headlight of his engine as his train had neared the crossing, which was protected only by crossbucks.

The car was apparently on its way to the hospital, Leroy Johnson having suffered an injury to his side.

A possible explanation for the accident was that John Massey, the owner and presumed driver of the automobile, was somewhat intoxicated, with an alcohol level 8 points above the level at which a person was legally declared to be under the influence of alcohol.

[1318] *Winnipeg Free Press*, Winnipeg, Manitoba, Canada, 3rd December 1956

[1319] *The Galveston Daily News*, Galveston, Texas, United States, 18th December 1956

Following the disaster the Phoenix City Council discussed speed limits for trains within the city limits and authorised City Attorney William Eliot to request the State Corporation Commission to provide signals at certain level crossings within the city limits. [1320]

Early on Thursday 3rd January 1957 a Detroit Toledo and Ironton Railroad switch engine pulling two cars collided with a Bath Township school bus on North Street Crossing, Lima, Ohio, United States. No one was killed but at least 20 of the 55 pupils on the bus were hurt, though only four of them were admitted to the Memorial Hospital. The collision, at about 8.10 a.m., pushed the bus southwards about 100 feet until it struck a box car on a siding. The box car, in turn, was pushed to the High Street Crossing, where it narrowly missed vehicles using the crossing. The fuel tank of the bus was punctured and firemen were busy spraying water on it to prevent a possible fire and or explosion.

Engineer O.G. Smith told police that the bus had stopped short of the crossing only to then start across in front of his engine.

Driving the bus was Oliver Hager, 47, who was also a teacher at Bath school and was among those injured. Interviewed by police he confessed that he had neither heard nor seen the train. He said that the children had been quiet and had not distracted him.

The local newspaper added that, 'School authorities who answered "at least 30" telephone calls from frantic parents after the crash, said all those aboard the bus were examined by school doctors and nurses after they were taken there by police, private autos, and another school bus. Several were taken home by their parents, a school spokesman said. Worried parents also rushed to the Memorial Hospital emergency room.

'Mrs John A. Pierce, whose home at 954 E. North Street is adjacent to the tracks, said she was in the kitchen when she heard the crash. "When I looked out the trainmen were trying to pry open the front bus door. They tried for several minutes and finally had to get the kids out through a side door. Mrs Pierce took about 20 youngsters into her home. "They were all crying and very scared," she said. "I thought I had all of them from the bus in my house. But every time I looked outside more were coming from the bus." [1321]

A mother driving her daughter and three classmates back to school for afternoon lessons on Friday 8th February 1957 took her car along a road which paralleled the Norfolk and Western Railway main line, slowed almost to a stop, then turned her vehicle to the left to cross the railway. As she did so her vehicle was struck by the eastbound *Powhatan Arrow* passenger train, en route from Cincinnati, Ohio, to Norfolk, Virginia. The impact knocked the 1955 automobile 220 feet and it ended up on a spur track, clear of the two running lines. All five occupants of the car were killed in the accident, which happened only 300 yards from the High School at Fort Gay, West Virginia, United States.

Those who died were Mrs Beatrice Higgins, 37; her daughter Deloris, 18, a junior and majorette in the high school band; Lois Ann Ratcliff, 16; her sister Dorothy Gail, 12; and Kenneth David Carroll, 15.

Travelling in a special car on the train were company president R.L. Smith and 25 other officials. It took the train a quarter mile to stop, after which it was backed up to the crossing but proceeded after a short delay. It was in charge of Conductor Roy Sampson

---

1320 *Tucson Daily Citizen*, Tucson, Arizona, United States, 18th and 20th December 1956

1321 *The Lima News*, Lima, Ohio, United States, 3rd January 1957 and Engineer R.T. Chapman. Engineer Chapman said that his train had been 14 minutes late and travelling at 60 miles an hour at the time of the collision. Witnesses testified that the warning lights at the crossing had been working normally. Less than two hours later the same train collided with a truck loaded with cinder block on a level crossing at Chattaroy, West Virginia, injuring its driver. [1322]

Another school bus accident occurred at Bancroft, Maine, United States, on the afternoon of Friday 1st March 1957. The accident happened during a snowstorm on a blind level crossing. A 28-car Maine Central Railroad freight train carried the bus 750 feet from the crossing and the diesel engine and 16 cars of the train were derailed, though they stayed upright. The train included seven mail cars from the *Queen Elizabeth* liner from Halifax, Nova Scotia, Canada. Ten children had already left the bus and its only remaining passenger, seven-year-old Faylene Hannington, could have alighted as well, but she chose to stay on the bus to the end of its route "just for the ride," with the bus driver, Ardie McKenney, 67. Both died in the accident. Faylene's father had escaped injury in a collision at the same crossing the previous autumn and the bus driver's brother had been killed in a level crossing accident at Lincoln, Maine, some years earlier. [1323]

A Pennsylvania Railroad train bound for the Standard Steel Works Division of the Baldwin-Lima-Hamilton Corporation at Burnham collided with the rear of a school bus on Knepp Crossing, near Lewistown, Pennsylvania, United States, on Thursday 5th April 1957. Engineer R.R. Hoffman said that his train had been travelling at 15 miles an hour. The bus had stopped short of the crossing, then started across directly in front of him. Thanks to the relatively low speed of the train the bus did not overturn and the consequences of the accident were less serious than they might have been. Only one pupil, Patsy Henry, aged 9, was detained in Lewistown Hospital, suffering from a fractured and dislocated right leg. Six others of the 55 pupils on the bus were treated for minor injuries and bruises and sent home.

The bus had been taking the children home from Burnham Elementary School. Its driver, Henry Wilson, 35, said that he had not noticed the train. [1324] [1325]

On Tuesday 30th April 1957 the *Powhatan Arrow* passenger train of the Norfolk and Western Railway was again involved in a level crossing collision, this time with a school bus. This accident occurred just outside of Waverly, Virginia, United States, at about 8 a.m. As chance would have it there were no passengers on the bus, which had only just started its journey to pick up pupils for the Waverly High School, but the bus driver, James Wilson Cornett, 49, was killed. In addition to driving the school bus Cornett had worked as a janitor for the State Highway Department. Lohring Staples, the principal of Waverly High School, described Cornett as "one of the best drivers we had."

There were no automatic signals at the crossing.

The *Powhatan Arrow* had been on its way from Norfolk, Virginia, to Cincinnati, Ohio

---

1322 *Bluefield Daily Telegraph*, Bluefield, West Virginia, United States, 9th February 1957
1323 *Biddeford Daily Journal*, Biddeford, Maine, United States, 2nd March 1957
1324 *The Daily News*, Huntingdon and North Union, Pennsylvania, United States, 5th April 1957
1325 *The Titusville Herald*, Titusville, United States, 5th March 1957

and consisted of a diesel engine and five coaches. It was delayed about 15 minutes by the accident.[1326]

A minister and four other persons on a sight-seeing tour between sessions of a church conference lost their lives when the car in which they were riding was struck by a Pennsylvania Railroad passenger train on a level crossing in Austin, Indiana, United States, on Saturday 4th May 1957.

Zeb Ernest Burge, 33, was a minister of the Church of God at Mitchell, Indiana. The other victims were his wife Mildred, their three-year-old daughter Diana; Mrs Ella Strange, 60; and Miss Betty Taylor, 23. The Burges had told friends that they were going for a ride around the small town.

The northbound train from Louisville to Chicago was in charge of Conductor H.W. Higgins with Engineer Clarence Thornley. It carried the crushed remains of the automobile half a mile along the track with the bodies of the victims inside. It took an hour for the wreckage to be removed, with the aid of cutting equipment, and in all the train was delayed an hour and a half. In the opposite direction the *South Wind* passenger train from Chicago, Illinois, to Miami, Florida, was held up for an hour.

Police stated that the flashers and bell signal at the crossing had been operating normally.[1327]

In the worst accident of its type for some time a light truck carrying 17 farm workers to their labours in onion and sugar beet fields was struck by an Atchison Topeka and Santa Fe Railway freight train on a level crossing just west of Vroman Siding, Colorado, United States, at 6.05 a.m. on Tuesday 11th June 1957. Twelve of the workers were killed and the other five were injured, three of them seriously. Jack Lee of La Junta radio station KBNZ told a reporter that, "bodies, arms, legs and heads were strewn along the right of way." A number of the victims were caught under the train and dragged along the line before it could be stopped. The front part of the quarter-ton truck was carried 25 feet down the track while its rear part was sheared off. Those who died included labour contractor Sylvester Licano, who had been driving the truck. The victims all lived a few miles away in Rocky Ford.

The engineer of the 56-car eastbound train was W.S. Inskeep. He said that his train had been moving at about 40 miles an hour. He had seen the truck turn south from US Highway 50, which was on a track leading from the main road to the David Smith farm. The truck had first stopped at the crossing, then had started across when his train had been too close for him to stop.

Two of the survivors, Joe Lucero, 23, and Betty Licano, 13 (the daughter of the labour contractor), stated that they thought that box cars stabled on Vroman Siding had obscured the view, so that the truck driver had failed to see the train. Jack Lee, on the other hand, thought that the truck driver had seen the approaching train but had either thought that it was stationary or else had misjudged its speed. Joe Lucero's wife Rose, 20, and Betty Licano's sister, Erma, 17, were amongst the other injured survivors.

An indirect minor casualty of the accident was lorry driver Charles Jones, who came

---

[1326] *The Progress*-Index, Petersburg-Colonial Heights, Virginia, United States, 30th April 1957
[1327] *The Terre Haute Tribune*-Star, Terre Haute, Indiana, United States, 5th May 1957

across the scene soon after the accident. Stopping his lorry he rushed to the scene with his first aid kit, which, of course, proved wholly inadequate. Overcome by what he had seen he returned to his lorry after other rescuers had arrived, then fainted and fell out of his cab onto the road. He was treated at Rocky Ford hospital for cuts and bruises. [1328] [1329]

An automobile was apparently driven into the side of a Canadian National Railways train near Matapédia, Quebec, Canada, on Dominion Day (now Canada Day), 1st July 1957. The road vehicle was trapped beneath a car of the train for several hours and six passengers in it were killed. The victims were Lionel Cormier, 32; his wife Aline, 30; Albert Benwill, 24; his 20-year-old wife and two unidentified children. A seventh person was critically hurt and was taken to hospital in Campbelltown, New Brunswick. All were from New Richmond, Quebec. [1330]

Eight 'youths' out on a Sunday afternoon joy ride were killed when their car apparently stalled on Clifty Crossing, near Somerset, Kentucky, United States, on 21st July 1957. The unnamed crew of the Southern Railway freight train told police that the car, moving at between 35 and 45 miles an hour, "didn't seem to hesitate," before it was driven onto the crossing. The train, travelling at about 50 miles an hour, took roughly half a mile to stop. Three of the occupants were thrown clear of the car while the other five were trapped within it. All were killed instantly. The automobile caught fire after the collision.

Those who died were siblings Josephine, Frances, and Charles Allen, aged 15, 14, and 13 respectively; Betty Bray, 15; James Calhoun, 12; Roland Davis, 17; James Richards, 16; and Charles Simmonds, 16, the driver of the 1941 "jalopy." They were all from Somerset.

Sheriff Gilmore Phelps said that a 15-year-old boy who had ridden in the vehicle earlier in the day had told him that its brakes had been bad, while two witnesses had reported that the car had had a flat tire as it had approached the crossing. The youngsters, all friends, had gone swimming earlier in the day and had been just "messing about," driving around visiting various people. [1331] [1332]

Just after 6 p.m. on Thursday 1st August 1957 five men on their way home from work were killed when the station wagon in which they were riding was hit by a Cleveland Cincinnati Chicago and St Louis Railway (Big Four) passenger train on Lake Road Crossing, near Fortville, Indiana, United States. The sixth occupant of the car was critically injured. Westbound train No. 407, in charge of Conductor C.J. Sleche and Engineer Bill Bixler, was travelling at 80 miles an hour and the station wagon was going in a south-westerly direction. The vehicle was cut in two by the impact and its occupants were thrown out.

Those who lost their lives were Gordon Ankrom, 41, of Muncie; Joe Padgett Heuer, 31, also of Muncie; Marian Lawhorn, 27, of Pennville; Floyd Leiter, 48, of Eaton; and Gene Richardson, 32, also of Eaton. The other person in the car, Harry Howell, 31, was

---

[1328] *The Greeley Daily Tribune*, Greeley, Colorado, United States, 11th June 1957
[1329] The *Amarillo Daily News*, Amarillo, Texas, United States, 12th June 1957
[1330] *Winnipeg Free Press*, Winnipeg, Manitoba, Canada, 2nd July 1957
[1331] *Middlesboro Daily News*, Middlesboro, Kentucky, United States, 22nd July 1957
[1332] *Washington C.H. Record-Herald*, Washington Court House, Ohio, United States, 22nd July 1957

critically hurt and was taken to the Hancock County Hospital in Greenfield. It was believed that Richardson had been driving the car, which was driven in front of the train.

The crossing at which the accident happened was on a road leading to State Road Indiana 67. It was unguarded, though there was a clear view along the line in both directions. [1333] [1334]

Late on Thursday 22nd August 1957 a diesel-hauled passenger train collided with a bus en route from the border town of Juarez to Mexico City on a level crossing at Iraputo, Mexico. The bus was reportedly cut in two, eight bus passengers were killed and another 16 were badly enough hurt to be admitted to hospital. Those who lost their lives included two Americans, named as Samuel Alvardo Lascari of Los Angeles and Elizabeth Mary Sahlman (address unknown). The other dead were three Mexican women, two children and a man. None of the train passengers was hurt. Police blamed the accident on the carelessness of the bus driver. [1335]

Two men and three children died when the panel truck in which they were travelling was struck by Canadian Pacific Railway passenger train No. 634 on a level crossing with a country road off of Highway 2 near Tecumseh, Ontario, Canada, on Friday 20th September 1957. The train on the Windsor – London line was probably travelling at about 70 miles an hour and the van was broken into three pieces by the impact and carried 170 feet. Those who lost their lives were Pietro Vozza, 33, Paul Drefko, 30, and three children, believed relations of Drefko. All were from the Windsor area.

No one witnessed the accident but nearby residents Mrs Annabelle Gauthler and her neighbour Jack Oates heard the noise of it and rushed to the scene. They found a sixth victim, a little girl, alive. She was unconscious at first but soon regained consciousness and was crying when taken away to hospital in an ambulance. [1336] [1337]

Five cotton pickers died on Monday 14th October 1957 when their stalled vehicle was struck by a Louisville and Nashville Railroad freight train at Jackson, Tennessee, United States. The accident happened when the automobile stalled on a level crossing with a private track. Desperate efforts made by the car driver to get the vehicle by ramming it into gear and using the starter to turn the motor were unsuccessful.

Killed in the accident were Wardell Clendennin, 46; Yolanda Coleman, 10; Mrs Rosie Gault, 71; Mrs Maurine Hendrix, 58; and Mrs Addie Jones, 41. The other five occupants of the car survived; Tommy Howell, 18; Frank Jones, 41, husband of Mrs Addie Jones; and Mrs William Hubert Smith, 39, the driver of the car, were slightly hurt; while Clyde Coleman, 13, brother of Yolanda; and David Howell, 9, brother of Tommy, escaped unhurt. The latter two had managed to scramble through the car windows before the collision.

One of the first police officers to reach the scene, Chief Deputy Sheriff Alfred Evans, found that his mother-in-law, Mrs Hendrix, was one of the victims.

1333 *The Anderson Herald*, Anderson, Indiana, United States, 2nd August 1957

1334 *The Evening News*, Jeffersonville, Indiana, United States, 2nd August 1957

1335 *The Pampa Daily News*, Pampa, Texas, United States, 23rd August 1957

1336 *Saint John's Daily News*, Saint John's, Newfoundland, Canada, 21st September 1957

1337 *Brandon Daily Sun*, Brandon, Manitoba, Canada, 21st September 1957

A few minutes after the accident another freight train collided with a pickup truck 15 miles down the line, killing its driver, Walter Chism, 70. [1338]

The same day as the Jackson accident, Monday 14th October 1957, a family from Shidler, Oklahoma, had gone to visit a relative who was home on leave from Navy "boot" training and lived near El Dorado, Kansas. They had completed their visit and were on their way home when tragedy struck at Haverhill, Kansas, United States, about midway between Leon and Augusta and only a few miles from El Dorado. Their 1957 station wagon was struck by a St Louis-San Francisco Railway train running on the branch line from Wichita, Kansas, to Joplin, Missouri. Engineer Vernon Hammond stated that his train had been running at about 45 miles an hour at the time. He went on, "I saw the car about 150 feet from the crossing and it came right up on the crossing incline in front of my engine."

The occupants of the station wagon were all thrown from the vehicle and all were killed. They were Charles Carrigan, 80; his grandson, A.P. Busby, 28; Mrs Busby, 27; and their children Stephen, 4; Michael, 3; and Rickie, three months.

It was raining heavily at the time of the accident and was thought likely that with the poor visibility the driver of the station wagon, presumably Mr Busby, had failed to see the train at all. [1339]

A collision on County Line Road Crossing, New Carlisle, Indiana, United States, claimed six lives on the night of Wednesday 6th November 1957. The car was being driven by Mrs Jean Vanderberg, 39, and her passengers were Mrs Pauline Barker, 52; Mrs Aleon Spencer, 50; Mrs Spencer's son Michael, 11; Mrs Irene Williams, 45; and Mrs Williams' daughter Mary Ann, 17. All were locals from the New Carlisle – Hudson Lake area. Arriving at the crossing Mrs Vanderberg had stopped her car and waited for a westbound freight train to pass but then had assumed that it was safe to cross. The car was driven onto the crossing with the double-track line and reportedly stalled. A few seconds later it was struck by an eastbound New York Central Railroad passenger train travelling at rather more than 75 miles an hour. The automobile was dragged about 200 feet and demolished; its occupants were all thrown out. The train, which was in charge of Conductor Harry Roy, with Engineer August Jelka, continued five blocks (about 500 yards) to New Carlisle depot before it could be stopped. It was then backed up to the crossing but nothing could be done for the victims. 'Flashing signals' were provided at the crossing. The party had been on their way to a 'social gathering.' [1340] [1341] [1342]

Five people died and three were injured, one seriously, when their car was in collision with a Canadian National Railways on a level crossing at Jacola, Quebec, Canada, on Saturday 16th November 1957. The accident happened close to the mining town of Val d'Or. [1343] [1344]

[1338] *Kingsport Times*, Kingsport, Tennessee, United States, 15th October 1957
[1339] *Arkansas City Daily Traveler*, Arkansas City, Kansas, United States, 15th October 1957
[1340] *Logansport Pharos Tribune*, Logansport, Indiana, United States, 7th November 1957
[1341] *The Tipton Daily Tribune*, Tipton, Indiana, United States, 7th November 1957
[1342] *Marshall Evening Chronicle*, Marshall, Michigan, United States, 7th November 1957
[1343] *Syracuse Herald Tribune*, Syracuse, New York United States, 17th November 1957

Fellovie dello Stato Class ETR 200 electric multiple-unit express train R 127, which had left Milano for Roma at 17:45 on Monday 9th December 1957 met with disaster at Codogno, Italy, at 18:19 when it collided with a lorry on a level crossing while travelling at 130 kilometres an hour. The impact with the lorry derailed the train, which broke in two, and the front of it collided with the concrete pillar of an overbridge then came to a stand in the station. The remainder of the train continued undamaged along the line. First reports indicated that 15 people had died, comprising 14 train passengers and the lorry driver, and over 30 people had been hurt, including the train driver and the lorry-driver's mate.

Driver Edoardo Ascè of train R 127 stated afterwards that he had had clear signals as his train approached Codogno, where it was not booked to stop. Suddenly, through the fog, he had seen the lorry stopped on the level crossing and had jammed on his brakes, but it had been too late.

As the lorry driver, Enrico Bigoli, had arrived at the crossing, the nearside crossing had already been descending but he had entered, nevertheless. Possibly, due to the fog, he had not seen the barrier coming down until the last moment. Having moved onto the crossing he found the barrier on the far side down and stopped his lorry on the line in the hope, perhaps, that it would be raised again to let him proceed. When this did not happen he then sent his mate, Franco Dall'Olio, to the station to notify the staff in the hope of getting someone to stop the oncoming train. Meanwhile he remained in the cab of his lorry, which was loaded with feed and bound for Cremona. He was still sat behind the steering wheel when the train struck the lorry, smashing it and carrying it about 300 metres along the line, where his badly mangled body was found. His mate, Dall'Olio, was found wandering aimlessly about the place after the collision, calling out his driver's name. He was taken to hospital in a state of severe shock and had also suffered some slight chest injuries from flying pieces of wreckage.

Driver Ascè and Assistant Driver Luigi Luvazza escaped with their lives, though they were hit by a shower of broken glass from the windscreen of the train. Lavati said afterwards that he and Driver Ascè had initially hoped to pull up their train after it had struck the lorry, while as his driver had been braking he (Lavati) had turned to the passengers in the leading car and shouted, "Relax, everybody calm down!" Then there had been a powerful shock, an alarming noise, and screams, as the leading coach had struck the concrete pillar and was broken in two.

Among the passengers who lost their lives in the disaster was a young French dancer, Ginette Michèle Raymonde David, who bled to death after a piece of sheet metal had severed an artery. In her handbag was found the novel that she had been reading: *L'Immortalite* (Immortality). More fortunate was Police Commander General Scavia, who was uninjured and who began to organise the rescue operation immediately after the accident. By this time it was dark, and a blanket of fog hid the station lights and those of the houses in Codogno, as moans and cries came from the wreckage. A mother with a baby in her arms called out in German for her husband to be saved; after two hours' effort with cutting equipment he was released from the wreckage; one of his legs was broken above the knee. Another passenger was the daughter of Signor Agnesina, the inspector general of public safety, she was unscathed. There were also a number of members of parliament on the train; Deputies Riccardo Lombardio, Alcide Malagugini,

[1344] *The Lethbridge Herald*, Lethbridge, Alberta, Canada, 16th November 1958

Lelio Basso, Giancarlo Mattcotti and Albertto Cavaliere all escaped injury, only Deputy Pezzi suffered some bruises.

As soon as news of the disaster reached Milano, a rescue operation was begun from there and firemen, doctors, nurses, blood for transfusions and a breakdown train were despatched from there. The Lodi State Prosecutor was also advised and immediately opened an investigation. Crossing Keeper Zefferinio Grisi was taken into custody for interrogation, though it seemed clear early on that he had not been responsible for the accident. The crossing barriers were operated from a cabin some distance from the crossing. They did not descend simultaneously; one came down a short time before the other. (There were presumably four barriers, each closing across half of the road.) They were lowered about five minutes before the arrival of the train. There had been no problem with their operation and the accident was clearly due to the actions of the lorry driver. About 40 workmen in an adjoining factory had seen the lorry come to a stand on the crossing and the train approaching while the lorry driver remained in his cab. "Jump out, run, flee!" they had shouted, but Bigoli, paralysed with fright, had just sat there. Had he simply restarted it he could have driven his lorry through the lowered barrier, saving both himself and the train.

At 23:00 the police arrested a looter (a jackal, as the newspaper called him) who had pocketed the wallet of one of the dead or injured passengers.

By about midnight sufficient wreckage had been cleared, after frantic activity, to allow traffic to resume on one line, while a crowd of onlookers had assembled outside the station and telephone calls were being received from all over the country, enquiring about relatives or friends who might have been on the train.

Signor Angelini, the Minister of Transport, issued a statement to ANSA news agency: 'Tomorrow we will begin a rigorous inquiry into the cause of the accident. From the information that I have so far received, however, I do not believe that the railway has any responsibility for the disaster, which it appears was due to the lorry entering the level crossing as the barriers were descending and then remaining stationary on the crossing, unable to clear the tracks. I have communicated news of the accident to the President of the Republic and he is greatly affected by it. I shall leave at 00:20 to reach the scene of the disaster as soon as possible, so as to personally assess the situation.' [1345]

Friday 13th December 1957 proved to be a lucky day for a bus load of schoolchildren in Winnipeg, Manitoba, Canada. The driver of a Greater Winnipeg Transit Commission school bus turned from Archibald Street into Cottonwood Road and across the Canadian Pacific Railway line at about 8.08 a.m. As he did so he failed to see an approaching train, which sheared off the rear end of the bus. There had been 12 pupils on the bus at the time. Those sat towards the rear saw what was about to happen and rushed towards the front of the vehicle with the result that only four of them, all teenagers, were slightly hurt. There were two further level crossing collisions in the Winnipeg area over the next 48 hours, leading to calls for improvements in level crossing protection, for example the Cottonwood Road crossing had neither barriers nor even wig-wag signals. [1346]

[1345] *La Stampa*, Torino, Italy, 10th December 1957

[1346] *Winnipeg Free Press*, Winnipeg, Manitoba, California, 13th and 16th December 1957

Six members of a New York state family lost their lives when the car in which they were riding was struck by a Rutland Railroad freight train on a level crossing a mile south of the station at Knapp's, New York, United States, at 10.20 p.m. on Saturday 11th January 1958. The 1952 two-door Ford automobile was driven onto the crossing directly in front of westbound freight train No. 201, which consisted of three diesel units and 99 cars.

Killed in the accident were farmer Howard Michaud Sr., 32; his wife Betty, 25; and their children Nancy, 6; Linda, 5; Caroline, 2; and Howard Jr., 11 months. A fifth child, Wayne, 3, was taken to Potsdam Hospital, critically hurt with head injuries. The impact carried the car 1,000 feet down the line and Mrs Michaud was thrown out; the other victims remained in the car.

Conductor William Mandigo, Engineer Floyd Crellar and Fireman Hubert Paquette told state police that the bell of the front engine had been ringing and its horn had been sounded as the train approached the crossing but the car had just kept going. Protection at the crossing was limited to two warning signs: a yellow disc 300 feet from the line, supplemented by a crossbuck at the crossing itself. [1347]

Mrs Cora Jeffers, 50; her sons Eddie, 20 (the driver of the car); Herman, 15; and Homer, 13; plus Linda Logan, 12; were killed when their automobile was in collision, on Friday 31st January 1958, with a Pennsylvania Railroad Cincinnati to New York passenger train near South Charleston, Ohio, United States. The car was carried 450 feet down the line before the train could stop and was demolished. No one on the train was hurt. Its engineer was Eldon Goodyear of Cincinnati. Warning bells at the crossing with the Jamestown Pike were reportedly ringing at the time of the accident. [1348] [1349]

A pickup truck in which a Navajo couple and their four young children were riding was struck by the westbound *Chief* passenger train of the Atchison Topeka and Santa Fe Railway, en route from Chicago to Los Angeles, close to the couple's home at Powell Crossing. Arizona, United States. Dale Nelson, 26, a section hand on the railroad; his wife Elsie, 23; and their children Tinson, 7; Helen, 6; David, 3; and Sarah, 1; were all killed in the accident, which happened at 5.08 p.m. on Thursday 6th March 1958.

According to one witness the train had been travelling at between 90 and 100 miles an hour, though something in excess of 75 miles an hour, as stated by another witness, seems more likely. At any rate the train, running down a gradient at speed, struck the pickup with great force and carried it nearly a mile. No one on the train was hurt and it continued its journey after about an hour.

It was reportedly unclear how the driver of the pickup had failed to see the train, though it was twilight and visibility was reportedly poor. [1350] [1351] [1352]

A freight train collided with a bus at an unstated location north of Lima, Peru, on

[1347] *Courier Freeman*, Potsdam, New York, United States, 16th January 1958
[1348] *The Chronicle-Telegram*, Elyria, Ohio, United States, 1st February 1958
[1349] *The Daily Reporter*, Dover, Ohio, United States, 1st February 1958
[1350] *The Arizona Republic*, Phoenix, Arizona, United States, 7th March 1958
[1351] *Tucson Daily Citizen*, Tucson, Arizona, United States, 7th March 1958
[1352] *The Yuma Daily Sun*, Yuma, Arizona, United States, 7th March 1958

Saturday night, 15th March 1958. Six people, presumably all on the bus, were killed and 36 were injured. [1353]

A family of five, going on a Sunday visit to relatives, were killed when their car was hit by a Chicago Burlington and Quincy Railroad passenger and mail train on a gravel road crossing at Ontario, [1354] near Henderson, Illinois, United States, just north of Galesburg, on 16th March 1958. Mr William Edgar Barry, 23, a potato chip salesman; his wife Sylvia, 22; and their children Dina, 3; David, 2; and Donald, 1, lived near Oneida, a few miles from the crossing.

Engineer F.C. Healy stated his train had been travelling at 50 miles an hour. The Barry car had been moving "pretty fast," he said. It had slowed down as it neared the crossing but had then picked up speed again and had been struck by his engine.

The car was carried about a quarter mile, Mr Barry and his daughter being thrown from it, the others trapped inside. It took rescuers several hours to remove the bodies of Mrs Barry and the two boys from the wreckage. [1355] [1356] Given that the crossing was on a rural road it seems likely that it lacked any sort of protection other than, perhaps, a pair of crossbucks.

Albert Witte, a school bus driver, acted quickly when his vehicle stalled on a snow-covered level crossing with the Long Island Railroad at Syosset, New York, United States, on Thursday 20th March 1958. Opening the front emergency door, he shouted, "Outside!" and reportedly threw some of the slower reacting students off the bus 'into a swirling snowstorm.' When all of his passengers were outside he herded them to a position of safety. Shortly afterwards a New York bound train arrived. Its engineer applied the brakes but it just nudged the corner of the bus, and the vehicle was not so badly damaged that it could not be later driven away. [1357] [1358]

Near blizzard weather conditions, with very little visibility, and a seriously overcrowded car, were factors in a bad level crossing collision at Eastray, about 20 miles west of Sherbrooke, Quebec, Canada, on Easter Sunday, 6th April 1958. A car travelling at an estimated 50 miles an hour collided with a Canadian Pacific Railway train, was wedged under the train, and dragged about 300 feet. No one on the train was hurt but all 11 occupants of the car were killed. They were Eliodor Larose, 38, of Waterloo, Quebec, his wife, also 38, their children Ronald, 14; Ginette, 12; Jean-Pierre, 9; Alain, 5 and Andre, 2; and Jean Claude Roy, 27, of Joliette, Quebec, his wife, 23, and their children Marcel, 4, and Micheline, 3. The two families were related. [1359] [1360]

An Italian Alfa Romeo car, registration number Re 35484, which was about cross the

---

[1353] *The Titusville Herald*, Titusville, Pennsylvania, United States, 17th March 1958
[1354] *The Cumberland News*, Cumberland, Maryland, United States, 17th March 1958
[1355] *Logansport Pharos Tribune*, Logansport, Indiana, United States, 17th March 1958
[1356] *Centralia Sentinel*, Centralia, Illinois, United States, 17th March 1958
[1357] *Adirondack Daily Enterprise*, Saranac Lake, New York, United States, 21st March 1958
[1358] *The Bridgeport Telegram*, Bridgeport, Connecticut, United States, 21st March 1958
[1359] *Medicine Hat News*, Medicine Hat, Alberta, Canada, 7th April 1958
[1360] *The Daily Jeffersonian*, Cambridge, Ohio, 7th April 1958

border back to its home country, was involved in a level crossing accident at Menton-Garavan station, France, on the evening of Saturday 26th April 1958, when it was struck by a railcar on the Vintimille-Nice line. The driver of the car was thrown 50 metres onto the opposite running line and almost decapitated. Most of the vehicle was embedded under the engine of the train, though pieces of it were scattered along the right of way. The work of removing the wreckage lasted until 03:00 the following morning, when it could be determined that it contained the bodies of four further victims. All those who died were Italians. The crossing barriers had not been lowered.

The crossing keeper, M. Piombo, suffered a nervous breakdown after the accident and had to be taken to hospital. He was not the regular crossing keeper but a replacement employee.

There were conflicting accounts as to whether the warning bell had sounded at the crossing before the arrival of the train. Some witnesses stated that it had rung before the train reached the crossing, others stated that the bell – which should have given the crossing keeper a minute and a half's warning of the approach of a train, so that he had sufficient time to lower the barriers – had only sounded after the accident. The newspaper report on the accident concluded that a failure of the warning bell as a cause of the collision could not be ruled out. [1361]

A family from Springfield, Ohio, United States, was almost wiped out when the station wagon in which they were travelling was hit by a New York Central Railroad passenger train about a mile west of their home on Sunday 4th May 1958. Charles Dempsey, 38; his wife Doris, 35; and their children Ronald, 9; Ray, 7; and Linda, 5; were all killed. The sole survivor, an unnamed 11-year-old boy, was admitted to the Miami Valley Hospital, Dayton, critically injured. Engineer Elmer Hast stated that he had seen the Dempsey vehicle approaching the crossing with the country road but had assumed that it would stop. It had not and the impact dragged it 1,200 feet down the line. [1362] [1363] [1364]

A tram carrying workers who were returning from a Mass in memory of Eva Perón, wife of deposed president Juan Perón, was struck by a train at La Plata, Argentina, the capital of Buenos Aires Province. Six of the workers were killed and 50 were injured in the accident on the evening of Wednesday 7th May 1958. [1365]

A school bus taking 90 children from an elementary school on an outing was struck by a steam train on a level crossing at Kameoka, Japan on Tuesday 10th June 1958. Four of the children were killed and 80 other people were injured, 30 of them, comprising 26 children and four adults, seriously. Witnesses said that the weather had been clear and the train had whistled several times but these warnings had been ignored by the bus driver. [1366]

[1361] *Journal de Genève*, Genève, Switzerland, 28th April 1958
[1362] *The Newark Advocate*, Newark, Ohio, United States, 5th May 1958
[1363] *The Steubenville Herald-Star*, Steubenville, Ohio, United States, 5th May 1958
[1364] *The Evening Journal-Tribune*, Marysville, Ohio, United States, 5th May 1958
[1365] *The Salt Lake Tribune*, Salt Lake City, Utah, United States, 9th May 1958
[1366] *The Dominion News*, Morgantown, West Virginia, United States, 12th June 1958

A car was struck by a train on a level crossing between Hanau and Eberbach, West Germany, on Tuesday 10th June 1958. Five people died. The car driver had failed to observe or disregarded the warning lights. [1367]

Diesel railbus 19 387, running light, collided with a tractor and trailer on an unprotected crossing between Heide and Neumünster, West Germany, on Sunday 27th July 1958. The impact derailed the railcar and five people died and two were injured. [1368]

A rash assumption by a car driver was apparently the cause of a level crossing collision with a New York Central Railroad passenger train from Montreal, Quebec, to Malone, New York, on the evening of Tuesday 29th July 1958. The train, which consisted of a diesel engine and six coaches, had left Montreal at 5.35 p.m. and the accident happened at 7.30 p.m. at Saint Stanislas de Kostka, Quebec, Canada. Mrs Evelyn Millette, 40, the car driver; her children, Rejean, 5 and Suzanne, 5; Mrs E. Labbe and her son Francois, 5, were all killed. Richard Millette, 5, was seriously injured, while his brother Jacques, 7; his sisters Armelle and Catherine, both 3; Michel Leduc, 5, and Michel Bibault, 4, were all unhurt. All were from the Montreal suburb of Verdun. Richard told ambulance men that his mother had thought that the train would stop at a station near the crossing but it did not. [1369] [1370] [1371]

A bus travelling from San Isidro to Once was struck by a train from Buenos Aires Retiro to Tigre on a level crossing in the Buenos Aires suburb of Vincente Lopez, Argentina, on Sunday 3rd August 1958. The bus was demolished and six of its passengers were killed and 15 injured. [1372] [1373]

All five members of Robert M. Harrison's family lost their lives when their car was struck by train No. 21, the *Chippewa-Hiawatha* streamliner passenger service of the Chicago Milwaukee St Paul and Pacific Railroad, on the afternoon of Sunday 7th September 1958. In addition to Mr Harrison, 37, they were his wife Kathryn, 32; and their children Robert J., 3; JoAnn, 1; and John Mark, 3 weeks. Mr Harrison was general manager of the Paper Box and Speciality Co. in Sheboygan, Wisconsin.

The accident happened about 300 yards from the family home at a level crossing with County Trunk Road F, two miles south of Waldo, Wisconsin, United States. The Harrisons had just set off on their journey. Mr Harrison, who had been driving, apparently failed to see or hear the northbound train, en route to Plymouth and Green Bay. Travelling at 50 miles an hour the train hit the car, a new two-door Chevrolet, broadside, killing its occupants instantly, while the vehicle was carried several hundred yards along the track. The car was demolished though the train locomotive was only slightly damaged.

[1367] *Katastrophen der deutschen Bahnen Teil II* by H.J. Ritzau, pg. 249 (in German)
[1368] Ibid.
[1369] *The Daily News*, St John's, Newfoundland, Canada, 30th July 1958
[1370] *Winnipeg Free Press*, Winnipeg, Manitoba, Canada, 30th July 1958
[1371] *Adirondack Daily Enterprise*, Saranac Lake, New York, United States, 30th July 1958
[1372] *Clarin*, Buenos Aires, Argentina, 4th August 1958
[1373] *The Idaho Free Press*, Nampa, Idaho, United States, 5th August 1958

The train was in charge of Conductor Cyril Morgan with Engineer B.R. Berman and Fireman Fred Huettner. Engineer Berman stated that he had applied the brakes when he had seen the car was not stopping but the distance had been too short to stop.

Farmer Ransom Meyer and his 15-year-old son Ralph witnessed the accident. The farmer said that he watched the train going through "about six days out of seven," and that he had not seen the car approach the crossing but "saw a flash" as the train hit the car.

A dozen men were unable to free the wreckage of the automobile from the engine and eventually the railwaymen had to chain the wreckage to the track and then reverse the train until the remains of the car were pulled clear of the locomotive. The collision occurred at 4.32 p.m. and it was 5.26 p.m. when the train could finally continue its journey.

There were no automatic signals at the crossing and there was no clear view for a motorist until his or her vehicle was within 20 feet of the line. [1374] [1375]

Guard dogs, returning to their kennels still muzzled and trailing their leads, were the first warning to a dog handler at Marlborough Farm Camp, an Army Ordnance depot near Kineton, England, that something was wrong. What he discovered had happened on Tuesday 23rd September 1958, was that a lorry carrying soldiers and their guard dogs to their posts on the camp had been in collision with a train consisting of an engine and a number of waggons loaded with ammunition. The camp had its own internal railway system. It was raining heavily when the collision happened at 6.45 p.m.

Six soldiers died in the accident and eight were taken to hospital, though only three were detained. Six guard dogs also died. The War Office named the men who had been killed as: Lance-corporal Brian Duffy and Privates Norman Holt, Donald McIver, John Murphy, Ronald Ridings and Roy Alexander Vandermark. Both the lorry driver and the train driver were among those injured. [1376]

A Colombo – Chilaw train collided with a bus carrying office workers on an unprotected level crossing near Nattandlya, about 35 miles north of Colombo, Ceylon, on Tuesday 21st October 1958. Nine bus passengers were killed and 30 were seriously injured. The engine of the train jumped the rails as a result of the collision, dragging the bus with it. [1377] [1378]

Five bus passengers lost their lives and 32 were injured when the vehicle was struck by a passenger train at Puebla, Mexico, on Tuesday 28th October 1958. [1379]

Nine lives were lost when a train collided with a taxi on a level crossing at an unstated location on the line between Dakar and Saint Louis, Senegal, on or about Wednesday 29th October 1958. [1380]

1374 *The Sheboygan Press*, Sheboygan, Wisconsin, United States, 8th September 1958
1375 *Wisconsin State Journal*, Madison, Wisconsin, United States, 8th September 1958
1376 *The Times*, London, England, 24th and 25th August 1958
1377 *The Canberra Times*, Canberra, Australian Capital Territory, Australia, 23rd October 1958
1378 *The Singapore Free Press*, Singapore, 22nd October 1958
1379 *Independent*, Long Beach, California, United States, 29th October 1958
1380 *The Times*, London, England, 30th October 1958

The Pennsylvania Railroad's westbound *Spirit of St Louis* passenger train, en route from New York to St Louis, Missouri, was involved in an accident at about 11.20 p.m. on Friday 14th November 1958, when it collided with a car on Clow Crossing, just south of Coshocton, Ohio, United States. The car, a new Chevrolet convertible, was carried about 1,300 feet along the right-of-way and 'reduced to a mass of twisted steel.' Its five occupants were killed. They were Carl Ames, 22, of Fresno, the driver; Bonnie Grace, 29, of Route 5; and Phyllis Holder, 16; Josephine Stiteler, 18; and Mrs Margaret Stiteler, 35, all of Coshocton. A sixth person, Jerry Clough, 23, of Coshocton, was removed to the Memorial Hospital suffering from severe shock. It was at first thought that he had also been a passenger in the car but later surmised that in view of his lack of injuries he must have been an eyewitness to the tragedy, as he had been found wandering along the line in a dazed condition. He had been seen in the company of the victims shortly before they had left the nearby Slim's Inn tavern. The train was delayed about an hour while the wreckage of the car was removed. Its conductor was W.P. Johnson. None of the newspaper reports on the accident indicated what sort of protection, if any, was provided at the crossing. [1381] [1382]

On the evening of Thursday 4th December 1958 a school bus carrying 11 toddlers home from the Small Fry Nursery and Kindergarten in Blue Island was struck by an Indiana Harbor Belt Railroad freight train on Western Avenue Crossing, Dixmoor, Illinois, United States, in the suburbs of Chicago. The bus was carried 100 feet and one child, Diane Lisik, 3, was killed. All of the other ten children on the bus were taken to hospital with various injuries, as were the bus driver, David Berry, 31, and Fireman John Matthews, 36. The latter was credited with preventing further injury or death by scooping up a child who was lying on the track and then proceeding to put down flares to stop an approaching eastbound freight train. Diane's brother Eugene, 4, was another of the children on the bus.

The accident came only four days after the disastrous fire at Our Lady of the Angels School in Chicago in which 92 children and three nuns had died.

There were no gates at the crossing but it was provided with flashing light signals, which witnesses said were working at the time of the accident, and the bus driver was subsequently charged with manslaughter. At the same time the owner of the school, Peter Kooyenga, was charged with having improper license plates on the bus and failure to have the vehicle properly insured. [1383] [1384]

A hearing held in late January 1959 regarding providing increased safety at the crossing was told that the crossing alarms were activated by a treadle 4,500 feet from the crossing, designed for a train travelling at 60 miles an hour, so that a slow freight train, travelling at 5 miles an hour, could take as long as 13 minutes to reach the crossing, which naturally led to impatient motorists disregarding the warnings. In a traffic survey during the hour between 4.15 p.m. and 5.15.pm on 7th January five trains blocked the crossing for 27½ minutes while a total of 1,136 road vehicles used the crossing during the

---

1381 *The Coshocton Tribune*, Coshocton, Ohio, United States, 15th November 1958
1382 *The Daily Jeffersonian*, Cambridge, Ohio, United States, 15th November 1958
1383 *Dixon Evening Telegraph*, Dixon, Illinois, United States, 5th December 1958
1384 *Bristol Daily Courier*, Bristol, Pennsylvania, United States, 5th December 1958

same period.

Suggestions for improving safety at the crossing included permitting trains to whistle – something was usually banned in urban areas, posting policemen to catch drivers who disobeyed the warning lights, and the provision of a crossing keeper. A grade separation, the otherwise ideal solution, would cost from two to three million dollars and was thus ruled out on the grounds of cost. [1385]

The Illinois Commerce Commission duly ordered the provision of modern automatic flashing signals, short arm gates, warning bells and selective speed controls (to prevent long delays between the activation of the warning controls and the arrival of the trains). In total these improvements were estimated to cost $43,800, the sum to be shared between the railroad company and the state. Pending the installation of these safeguards trains were permitted to whistle at the crossing. [1386]

Shortly before 9 p.m. on Saturday 6th December 1958 a northbound Missouri Kansas Texas Railroad of Texas passenger train collided with a car in downtown Trenton, Texas, United States, killing James Thomas Chesher, 36, and his children, Judy Carolyn, 10; Thomas Darrell, 7; Connie Diane, 4; and Patricia Anne, 2. The car was rolled 128 feet by the diesel-hauled train before being pushed off to the left, with its driver inside, the children having been thrown out. Mr Chesher had been taking his children on a visit to his parents, who resided at a farm a mile outside of Trenton. Mrs Chesher, who was recovering from an operation, had stayed at their home in Grand Prairie. Lewis Owens, a close family friend, was given the unenviable task of telling her what had happened.

The train was in charge of Conductor H.G. Hoskins with Engineer W.M. Kelly and Fireman J.W. Kennedy. Engineer Kelly said afterwards, "We were within 50 feet of the crossing when the car drove onto the track. It got squarely in the middle of the track when we hit. I couldn't see anything as we swept by. We were so close to the crossing that I couldn't tell how fast the car was moving." He said that he had automatically applied the air brakes when he caught sight of the car but the train had run some distance beyond the crossing before it could be stopped. The train was delayed 40 minutes. Protection at the crossing consisted of a warning bell, which Mr Chesher had presumably failed to hear. [1387] [1388]

A school bus driven by 24-year-old Albert Slacht had picked up ten pupils bound for St Ignatius School and was about to pick up five more from the highway outside of their homes at Essex, Ontario, Canada, as it approached an unmarked level crossing with the New York Central Railroad on the morning of Friday 12th December 1958. Driver Slacht said afterwards, "It was foggy and my windows were frosted over. When I saw the train coming I put my brakes on but it seemed as though they wouldn't hold and I started to slide. When I realised what was happening I gunned her and the motor cut out on me."

At that moment *The Detroiter* express passenger train was approaching the crossing, travelling reportedly at 80 miles an hour. Its engineer told police afterwards that its headlight had been on and its whistle blowing. When he saw the bus he had applied his

---

1385 *The Harvey Tribune*, Harvey, Illinois, United States, 27th January 1959
1386 Ibid, 19th February 1959
1387 *The Bonham Daily Favorite*, Bonham, Texas, United States, 8th December 1958
1388 *The Bryan Daily Eagle*, Bryan, Texas, United States, 8th December 1958

brakes in emergency but it took the train three-quarters of a mile to stop.

The bus was almost over the crossing when the train hit its rear, killing David Purvis, 11, and Paul Bosse, 8. The bus was thrown down an embankment and overturned. Driver Slacht, who suffered multiple cuts and bruises, carried the eight surviving children up the steep slope to the roadway and most of them were able to walk to the house of a Miss Margaret Seymour, 300 yards away, where they were treated by a doctor. Most had multiple bruises and three had broken arms or legs. [1389]

The United States had another level crossing accident involving a school bus just before 9 a.m. on Monday 15th December 1958, on the Williamson Road Crossing, Bridgeport, Michigan. The bus was full, with 60 passengers, and only just over a mile from its destination. It stopped at the crossing, then moved onto it and promptly stalled. One newspaper report stated that bus driver Russell Knisley had left his seat and run down the line in an attempt to flag down an approaching slow-moving Chesapeake and Ohio Railway freight train. According to another newspaper it was service station operator Joseph Bonk who had seized a red rug and run towards the train. Perhaps both men did. Engineer Charles McKendree [1390] succeeded in slowing but not stopping the train, which clipped the front of the bus, spinning it around and damaging it somewhat, but it did not overturn. Many of the passengers had time to evacuate the vehicle but others froze in their seats; five were badly enough hurt to be taken to Saginaw Hospital, one with a suspected fractured skull, and 16 others suffered cuts and bruises.

Pupil Pat Thayer told a reporter, "It was awful. Some of the kids started screaming."

School superintendent Thomas White commented, "We were mighty lucky." He referred to the Essex accident in Canada three days earlier (see above) and added, "The best we know there were only cuts and bruises."

The crossing was protected by warning lights. A second school bus had come up behind the stalled one but did not dare give the front bus a push lest it became trapped on the crossing in turn. [1391] [1392]

There were no witnesses to an accident on a level crossing four miles west of Lincoln, Nebraska, United States, late on Tuesday 30th December 1958, and Chicago Burlington and Quincy Railroad Engineer J.H. Doerr did not realise that his train had been hit until he neared Friend, about 35 miles west of Lincoln. How he realised was not stated, possibly by the rough running of his train, which was stopped at Friend, where a damaged freight car, 12th from the end of the train, was detached before the train continued to Hastings. This vehicle was clearly the point in the train where it had been struck by the automobile. Meanwhile other motorists had become aware of the accident by discovering wreckage and the bodies of the victims at the crossing at about 12.15 a.m.

Those who died were tentatively identified from papers found on their bodies as Sylvester Ivory, Barbara Keller, Philip Silva, John Warren, and Karen Wiedeman.

The train had left the yards at Lincoln at 10.45 p.m. so the collision must have

---

[1389] *The Lethbridge Herald*, Lethbridge, Alberta, Canada, 12th December 1958

[1390] *Marshall Evening Chronicle*, Marshall, Michigan, United States, 16th December 1958

[1391] *The Herald-Press*, St Joseph, Michigan, United States, 15th December 1958

[1392] *The Journal-Tribune*, Marysville, Ohio, United States, 15th December 1958 occurred at about 10.55 p.m. Mrs Louise Kleine, who lived only 50 yards from the crossing, said, "I was asleep when the passing train woke me up. I heard a loud noise, looked out the window, but saw the train and nothing else."

The crossing was protected only by a "Railroad Crossing" sign, while a rise in the road just west of it restricted visibility for the drivers of vehicles proceeding towards Lincoln, as the car concerned had been. [1393]

The bus driver and six of his passengers lost their lives near Osaka, Japan, on Saturday 3rd January 1959 when the vehicle was struck on a level crossing by the Osaka – Kyoto Hankyu express. Fourteen train passengers were injured. [1394]

As a school bus driven by Walter Kronz arrived at a level crossing at Cudahy, Wisconsin, United States, on the morning of Thursday 22nd January 1959 the warning lights were flashing. He applied the brakes but the bus skidded onto the crossing on the slippery snow- and ice-covered road and then stalled. Unable to restart it he shepherded his four passengers, two boys and two girls, all aged about 11, off of the vehicle, and they stood and watched as a northbound Chicago and North Western Railway passenger train from Chicago struck the front of it. The bus was pushed off the line and its engine and two front wheels were demolished. The train was not derailed and no one was hurt. [1395]

The immunity which the United States had enjoyed at the beginning of 1959 from really serious level crossing accidents ended on Wednesday 18th March when a new station wagon carrying eight Girl Scouts, all aged 11, and two of their mothers was struck on Factory Road Crossing, Alpha, Ohio, 3½ miles from Xenia, by a Pennsylvania Railroad freight train. All ten of the occupants of the car were killed or fatally injured. The victims were Patricia Lapinski, Connie LaPrise, Cindy Moorman, Ann North, Mrs Jeanette Randall, 39, and her daughter Paulette; Linda Ward, Mrs Lucille White, 44, the driver of the station wagon, and her daughter Sharon; and Anna Wilvert. Six were killed outright, three were dead on arrival at Greene County Memorial Hospital, and Anna Wilvert died there about two hours after the accident.

The girls and the two women, described as their advisers, were returning from the Greene County Library, where they had spent an hour and a half studying for their Scout merit badges in reading. Their car had reached the county road crossing, situated 100 feet north of Ohio Route 35, at 4.45 p.m. East of the crossing, which was protected only by crossbucks, the railway was on a long curve, making the crossing a danger point according to witnesses. As well as the Pennsylvania line the road crossed a spur track of the Baltimore and Ohio Railroad, with a gap between the two lines capable of accommodating one or two vehicles.

Engineer Bill Murray of the Columbus, Ohio, to Richmond, Indiana, train stated that the station wagon had crossed the Baltimore and Ohio tracks, "seemed to pause on the tracks as if going to stop, then continued across the track."

1393 *The Lincoln Star*, Lincoln, Nebraska, United States, 31st December 1958

1394 *The Racine Times-Journal Sunday Bulletin*, Racine, Wisconsin, United States, 4th January 1959

1395 *The Sheboygan Press*, Sheboygan, Wisconsin, United States, 22nd January 1959

Fireman Joe Stvartak said, "I laid on the whistle half a mile down the track. But you can't stop a thousand tons of train on a dime. I threw on all the emergencies. It was terrible. There was a rending crash. Then I saw the kids flying through the air."

Conductor Bill Jacobs declared that the headlight of the leading engine had been on.

Volunteer fireman Kenneth Ward had heard the emergency call on the fire radio at his home and had rushed to the scene with other firemen in ambulances. Not realising that his only child had gone to the scout meeting he was one of the first rescuers to reach the scene, to find that his daughter was one of the victims. "I saw a school book," he said, "it drew me to the spot. I lifted the cover of the blanket. I felt her pulse. I tucked the blanket back around Linda Ellen."

The Rev. Alvin Klotz of the Beaver Creek Church of the Brethren was having his car repaired at a service station a mile from the crossing when a man he did not know had appeared and had shouted, "There's been an accident, call the emergency squad!" Fearing that another man whom he had been waiting for might have been involved, the minister had hurried to the scene. He said, "I couldn't believe it. There were two or three bodies close to the car but the others were scattered in the field. The car was lying on its side. I checked for life, but there didn't seem to be any. I picked up a wallet lying near one of the bodies and glanced at it. It belonged to Anna Wivert, a sixth grader. I realized Anna was still alive, but she was not conscious. I started to take my coat off to cover her and some of the others. But there were so many I couldn't do it. It was the most frustrating experience I have ever known, as I went from person to person – most of them unrecognizable – knowing they were probably all dead. Curiously, even though members of four families of my church were killed, it was only when I found Anna's billfold that I recognized some of my flock."

County School Superintendent Vaughn Lewis ordered school buses to take a different route rather than that which went over the crossing, and the flags at the Beaver Township school which the girls had attended flew at half-mast. He complained that school officials had pushed demands for the installation of warning lights at more level crossings with the Ohio Public Utilities Commission and after three or four years they had finally got a second set of lights installed. As to the Factory Road crossing, "The biggest tragedy was that there was no flasher light there. That's a bad crossing and we (the school board) fought for flashers and didn't get it. This is inexcusable in a populated area." He added that five crossings in his district were used by 30 school buses a day. Two of these crossings had warning lights, the other three only had crossbucks and signs reading, "Stop, Look, Listen."

Ohio Republican representative Richard Metcalf attempted to make political capital out of the accident, claiming that his bill to install warning devices at all Ohio level crossings, which he had introduced in February, had been bottled up in the House Representative Committee for a month and a half. This, he alleged, was because the Democrats had stifled it as it would involve an extra $1 on the auto licence fee to fund the installation. Democrat representative Martin Gavin responded that this was a "cheap political trick," and that the bill would be sent to a committee for a hearing the following week. The bill proposed that the most dangerous crossings would be dealt with first, with the remainder following as the money became available. The State would meet 85% of the costs and the railroad companies the remaining 15%. [1396] [1397] [1398]

[1396] *The Fairborn Daily Herald*, Fairborn, Ohio, United States, 19th March 1959

Six 'Africans' died and two more were hurt when a crowded electric train was in collision with a three-ton lorry on an unguarded level crossing at Kliptown, South Africa, near Johannesburg, on Sunday 5th April 1959. The brief report on the accident did not state whether the victims had been on the train or the lorry. [1399]

Eight people died and 19 were injured when the lorry on which they were riding was struck by a train on an unguarded level crossing at Hajnowka, in north-east Poland, on Sunday 10th May 1959. The only person on the lorry to escape injury was its driver. [1400]

The same day as the accident in Poland, Sunday 10th May 1959, another major level crossing collision occurred at a location 13 kilometres north of Mexico City, Mexico. A passenger train which had left Mexico City only a short time before was in collision on the Laredo highway at about 21:30 with a crowded second-class bus carrying passengers returning to the capital. Fifteen or sixteen bus passengers and its driver died and another 29 bus passengers were hurt and were brought into Mexico City hospitals using every available ambulance. Most of the victims were women and children who had been on Mother's Day visits to San Juan Teotihuacán and surrounding villages. The bus was completely destroyed and pieces of it were scattered along both sides of the line. [1401] [1402]

A car carrying eight teenagers who were out for a ride after the annual junior-senior prom at Shelbyville High School was struck by southbound Illinois Central Railroad passenger train No. 3 from Chicago to Louisville, Kentucky, the *Louisiane*, on a level crossing with a dirt track two miles from Mattoon, Illinois, United States. The car had just turned off US Highway 45. The accident happened at 12.05 a.m. on Saturday 16th May 1959. The train was travelling at about 75 miles an hour when it struck the car broadside. Six of the occupants of the car were killed outright, a seventh was fatally injured and the eighth critically hurt.

Those who died were Jerry Hill, 18; Judy Keen, 18; Irene Moon, 16, Lorraine Pfeiffer, 17, Lyle Pfeiffer, 18; and Dorothy Thomas, 18, all of Shelbyville; and Jerry Hayes of Woodlawn. Miss Thomas succumbed from her injuries in the Memorial Hospital at 8.40 a.m. Ivan Moon, 22, of Shelbyville, who had been the driver of the car, was confined to the same hospital, gravely injured. Hayes and Moon were students at the University of Illinois, the others had attended Shelbyville High School.

The train was in charge of Conductor Fred Maneely with Engineer William Ward. It wrecked the car, parts of which were scattered for three-quarters of a mile along the line, though the front portion of the vehicle remained in one piece.

Mattoon police were notified of the accident at 12.14 a.m. Identification of the victims was difficult and the police initially called on the parents of other students who had

---

1397 *The Chronicle-Telegram*, Elyria, Ohio, United States, 19th March 1959

1398 *Wisconsin State Journal*, Madison, Wisconsin, United States, 19th March 1959

1399 *The Times*, London, England, 6th April 1959

1400 *Charleston Daily Mail*, Charleston, West Virginia, United States, 11th May 1959

1401 *The Corpus Christi Times*, Corpus Christi, Texas, United States, 11th May 1959

1402 *El Paso Herald-Post*, El Paso, Texas, United States, 11th May 1959 attended the prom in an attempt to identify them, in some cases before the students had arrived home, causing great distress to the parents. Eventually someone thought of identifying the victims from the initials on the class rings they were wearing.

It was theorised that due to the cold weather the windows of the car might have been steamed up, obscuring the view of the driver and passengers. Protection at the crossing consisted of a pair of crossbucks marked "Railroad Crossing" beneath which a smaller sign indicated "2 Tracks". One of the crossbucks was knocked over in the collision. [1403] [1404]

The Brownfield family of Elliott, Indiana, were returning from a Memorial Day visit to relatives on the evening of Sunday 31st May 1959 when their car was struck by Pennsylvania Railroad train on a level crossing where the visibility was obscured by vegetation at Crown Point, Indiana, United States. Mrs Louise Brownfield, 28; and children Doyle, 10; Rebba, 4; Rickie, 3; and Barbara, six months, were all killed. Mr Robert Brownfield, 30, probably the driver of the car, and his daughter Faylene, 9, were injured, the father seriously and the child critically. [1405]

Also on Sunday 31st May 1959 a train bound for Llay-Llay in the Andes collided with a lorry carrying 41 football players and their supporters on a level crossing near the station at Chagres, Chile. Eleven of the people on the lorry were killed and another 20 were hurt. Everyone on the train escaped. [1406]

On Saturday 6th June 1959 a passenger train collided with a lorry on an open level crossing at Kretschau, East Germany, on the Zeitz – Molau line, due to inattention on the part of the lorry driver. Six people died. [1407]

Another level crossing accident wiping out a family occurred at Helena, Arkansas, United States, on Friday 12th June 1959. The victims were Orange Jones Sr., 67; his wife Willie, 37; and their children Orange Jr. (from a previous marriage), 7; Helen Thomas, 6 (from a previous marriage); Lena, 6; Mary Ellen, 4; and George, 9 months. A two-car diesel Missouri Pacific Railroad passenger train caught the vehicle broadside and dragged it 400 yards. Orange Jones Sr., the driver of the automobile, was thrown from the vehicle but the other victims were trapped within it and it took almost two hours to release their bodies after the train had been backed to the crossing.

Engineer J.Q. Hastings told police that the Jones' car had overtaken another vehicle before entering the crossing. The warning lights provided at the rural crossing were said to have been working normally. [1408] [1409]

(The Christian names of the car driver and his son are given in some newspapers as Aaron rather than Orange. It is not known which is correct.)

[1403] *Daily Journal-Gazette*, Mattoon, Illinois, United States, 16th May 1959
[1404] *Mt Vernon Register-News*, Mt Vernon, Illinois, United States, 16th May 1959
[1405] *Logansport Pharos Tribune*, Logansport, Indiana, United States, 1st June 1959
[1406] *The Lowell Sun*, Lowell, Massachusetts, United States, 1st June 1959
[1407] *Katastrophen der deutschen Bahnen Teil II* by H.J. Ritzau, pg. 289 (in German)
[1408] *Daily Capital News*, Jefferson City, Missouri, United States, 13th June 1959
[1409] *The Holland Evening Sentinel*, Holland, Michigan, United States, 13th June 1959

A disastrous level crossing collision occurred in West Germany on Saturday 20th June 1959 at Lauffen am Neckar. The accident occurred at a level crossing with both the double-track standard-gauge Frankenbahn and the single-track narrow-gauge Zabergäubahn. Ironically, the bus involved in the accident was one of a number put on by the Deutsche Bundesbahn in place of a Zabergäubahn train, the railway company being keen to close the narrow-gauge line. However, although the narrow-gauge trains were slow, the buses were slower still, and in addition they lacked the capacity of the trains and so were often overcrowded. Thus, the bus involved in the accident, which had a capacity of 59 passengers, was actually carrying 71.

The crossing was protected by barriers, worked by the crossing keeper at *Streckenposten* 47 by means of a winch. The keeper was advised by telephone by a *fahrdienstleiter* at Kirchheim (Neckar) of the approach of trains so that he could lower the barriers in time, an operation which took 12 to 15 seconds. On the day of the accident the keeper was advised by the *fahrdienstleiter* of the approach of train E 867 from Tübingen via Stuttgart to Würzburg, which was often worked by a Class 38 steam locomotive, but on the day of the accident was being worked by electric locomotive No. 01 1094, which had better acceleration than a steam engine. The travel time from Kirchheim (Neckar) to the crossing was only two minutes. For unexplained reasons the crossing keeper was slow in lowering the barriers. When he began to do so a car was approaching the crossing from one side and the railway bus from the other. He apparently hesitated even then, halfway through lowering the barriers, which the car driver took as an invitation to cross, whereupon the bus driver decided to cross as well. Meanwhile the train was close at hand and its driver, seeing that the crossing barriers were still not down, began to whistle and to apply his brakes in emergency. Despite this the train was still travelling at 80 kilometres an hour when it struck the bus, which was carried 400 metres down the line, though the car escaped. The accident happened at about 17:30.

The bus driver and 36 bus passengers were killed outright or died shortly afterwards and eight more bus passengers died in the following days. All the other bus passengers except one were seriously injured.

Both the crossing keeper and the car driver subsequently found themselves in court. During their trial various safety deficiencies came to light, for example neither at the crossing nor at Kirchheim station were the clocks accurate, there were problems in turning off the current to the overhead line equipment, and the regulations did not make it clear at what point the crossing keeper should close the barriers. Initially the crossing keeper and the car driver were both acquitted but subsequently the verdict on the crossing keeper was reversed by the Federal Court of Justice; he was found guilty and received a suspended sentence of nine months' imprisonment for involuntary manslaughter.

Rather than being closed the Zabergäubahn was subsequently converted to standard gauge while the level crossing was replaced by an underpass. [1410]

Two families travelling from their homes to a park on Lake Ontario for a swim lost their lives when the car in which they were riding was struck by a Canadian National Railways passenger train near Winona, Ontario, Canada, on the evening of Tuesday 23rd

[1410] Wikipedia – Busunfall von Lauffen

June 1959. The crossing, on a country road, was unprotected.

Those who lost their lives were Walter Ukrainetz, his wife Mary, their children George, 9, and Anna, 5, of Hamilton, and Mike Papish, 40, and his son Dennis, 6, from Stoney Creek.[1411]

Five members of a Glastonbury, Connecticut, family on holiday in Alabama, where they had formerly lived and where they had relatives, lost their lives in a level crossing collision on Thursday 30th July 1959 at Columbiana, Alabama, United States. The station wagon in which they were riding was driven into the side of the locomotive of a Southern Railway freight train bound for Rome, Georgia.

Otis Zuiderhoek, his wife Dixie, and their children Roy, 13; Deborah, 7; and Russell, 3, all succumbed. The parents and the two younger children were killed outright and Roy died in hospital two hours later. Two other children, Rebecca, 9, and her brother Vann, 1, survived the accident. Rebecca was seriously injured, her baby brother rather less so. The victims were all thrown from the car, one as far as 68 feet.

Engineer Mark Lyde said that his train had been travelling at about 45 miles an hour. He had glimpsed the station wagon immediately before the collision.

The crossing with Highway 25 only boasted a stop sign; there were no warning lights or bells, let alone barriers, though there was a relatively unobstructed view of the line. Highway Patrolman Billy Bradshaw pointed out that there were no skid marks to indicate that the driver of the station wagon had tried to stop. He admitted that there was a slight rise in the road leading to the crossing but commented, "I paced off about 200 feet back from the accident in which the train should have been in full view."[1412] [1413]

A major school bus accident occurred in the United States on Thursday 10th September 1959 when a bus stalled on what was variously described as the Mountain Lake Crossing and the Lake Lynn Crossing on State Route 135 in Maryland. The bus driver and owner, Leroy Campbell, 49, told authorities that his bus had stopped as soon as it entered the crossing, for some reason he could not explain, and at that moment the crossing lights and bells had begun to operate. He had then tried to get the children off the bus but only a few had been able to do so before the bus had been hit by the eastbound *Diplomat* passenger train of the Baltimore and Ohio Railroad, en route from St Louis, Missouri, to New York. As a result of the collision the bus was almost demolished and seven children were killed and 11 others seriously injured. Eight more children and the bus driver were treated for minor injuries and discharged.

The children who lost their lives were named as sisters Janet and Nancy Deems, aged 12 and 14 respectively, Merle B. Harvey Jr., 11, and his sister Nancy Lee, 12; Richard Hinkle, 11; Lee Hoffman, 11; and Shirley Lee, 12.

Engineer Otto Droege of the diesel-hauled train gave a long account of the accident to a newspaper reporter. He began, "I never felt so helpless. I knew we were going to hit that bus and couldn't do a thing about it. If something will come out of this horrible

---

[1411] *Winnipeg Free Press*, Winnipeg, Manitoba, Canada, 24th June 1959

[1412] *The Decatur Daily*, Decatur, Alabama, United States, 31st July 1959

[1413] *The Daily Independent*, Kannapolis, North Carolina, United States, 31st July 1959 accident which will benefit us in the future maybe it won't be as bad as it seems now. If that bus driver had carried a red flare which could have been tossed on the ground I would have seen it a quarter of a mile further from the crossing. Maybe I still couldn't have stopped in time but I know the engine would not have struck the bus with the impact it did. It is just a suggestion, but I think those school buses should be compelled to carry danger flares in case of emergency."

He continued, "The rolling, pitching motion of the train gives the impression that other objects are moving. It was some seconds after I saw the tail end of the bus on the tracks before I realized it was stopped. I had one hand on the whistle board and the other near the brake when I knew the bus wasn't moving. I immediately closed the engine throttle and threw on the emergency brakes, but we just couldn't stop in time. It was some moments after rounding the curve before I could see the crossing. When I first saw the bus I thought it was crossing the tracks and as it was some distance away it did not immediately occur to me that there was trouble. We just braced ourselves and hung on before the train hit. My immediate concern was for the approximately 400 passengers on board the eight cars behind our three diesel units. If the train had left the tracks it could really been a mess. I'll remember this for a long time. I just hope nothing like it ever happens again."

Fireman A.L. Holland apparently did not realise that the bus had stalled on the crossing until the last moment. [1414]

Mountain Lake Station Agent Clinton Shrout witnessed the accident. "There wasn't anything I could do. I saw the school bus stalled on the crossing. I saw the train coming. There wasn't anything I could do but watch. The engineer on the locomotive applied his air brakes maybe 300 to 400 feet before he hit the bus but he couldn't stop. He was going too fast. The train plowed into the rear part of the bus. It turned it over and threw it maybe 35 to 50 feet. The air was full of sound – the grinding of the wheels against the track. The hiss of air and the sound of the metal. It all seemed to be in slow motion, the bus turning over as it moved through the air, the train stopping after it moved about its own length.

"And there lay the bus, knocked off its wheels, the side ripped open. I could hear the children screaming. I called state police and the hospital for ambulances. Then I ran down the track as fast as I could. There were about six children lying very still scattered on the ground. Maybe eight or ten others were conscious but crying. Some were cut and bleeding.

"I tried to do what I could but what could I? I started to carry some of the children out of the bus, laying them on the ground outside. I carried three out before help started to come. Police, deputies, and maybe seven or eight ambulances. There are a lot of people in this area. They came from homes and stores, crowding around. There must have been a lot of parents of those kids in the crowd. Pretty soon mothers began calling their children's names. Many were hysterical. I'll never forget it."

In another piece of the extensive press coverage of the accident a newspaper reporter wrote, 'Mrs Phyllis Paugh, who resides in a second storey apartment atop Helbig's Grocery in Loch Lynn, was almost an eye-witness. Mrs Paugh, who was getting her two children ready to go to school, heard the crash and ran to the window. "It was the most horrible scene I have ever viewed. Children were crying, running around, walking in a daze, blood streaming from them, their clothing in tatters. I ran to the bus and gave first

---

[1414] *The Cumberland News*, Cumberland, Maryland, United States, 11th September 1959 aid to as many children as I could. In a matter of seconds, the place was jammed with cars, ambulances, doctors, and nurses."

'Gerald Minnich, operator of the Minnich Funeral Home in Oakland, said the accident was "the worst I have encountered in my 10 years in this business. By the time I got to the scene most of the bodies had been removed from the bus," he said. "Everyone pitched in to give a hand in the rescue work. There were milkmen and bakery truck drivers who stopped to help…everything was orderly, but it certainly was a terrible thing." Minnich, who said four of the dead were taken to his funeral parlor, described the bus after the crash as "not entirely demolished." He said the side was caved in and the roof was buckled and partially torn away.

'One resident at the Mountain Park Hotel, which overlooks the accident, said she didn't know where all the nurses came from in such a brief time after the mishap. She said Route 560 in Loch Lynn and Route 135 toward Mountain Lake Park was black with cars just minutes after the crash.

'A Cumberland Evening Times reporter at the scene said after the bus was removed, only bits of children's clothing, parts of shoes and books along the right-of-way served as a grim reminder of the tragedy.'

An account from 14-year-old Ray Dixon, one of the surviving pupils who had been on the bus, was also published. He said, "Everybody knew the train was gonna hit us. And everybody wanted to get out. It seemed everybody rushed to the doors of the bus all at once. One girl – she got stuck in the door. I didn't know what to do. Finally somebody pushed her out. I ran across the road. Then I saw it…I couldn't believe it. It was just like a dream. It just smashed it right in the middle. At first I didn't know about my sisters. Then I saw Gladys…she was out of the bus too – she was crying terrible. I guess I was crying too. Neither of us knew what happened to Frances. We were crying because we were scared, too.

"The first time I saw Frances she was in an ambulance. We weren't the only ones crying. That's all we heard when it hit. Later I saw some blood on the floor. I guess some of the kids got hit just as they were getting out. People started to come from everywhere. Anybody that asked me anything, I told them, 'There's nothing good to see here.' They took me in an ambulance to the hospital, then we were sent home.

"I knew those kids who were killed. My sister has a broken back and a cut eye or something. She's pretty bad. I don't know what happened. I think the gear went wrong or something." [1415]

The accident was duly investigated by the Interstate Commerce Commission (ICC) who noted that train No. 4, the one involved, consisted of diesel units Nos. 4507, 1432 and 1415 coupled in multiple control, a mail car, a passenger baggage car, two coaches, two sleeping cars, a dining car, and a sleeping car. It had left Terra Alta, West Virginia, at 7.05 a.m., had passed Mountain Lake Park station at 7.24 a.m., five minutes late, and had been moving at 45 miles an hour when the collision had occurred immediately afterwards.

The school bus was described as a 1951 GMC with a Wayne school bus body. It was 22 feet in length and had seating for 36 passengers. The bus body was painted school bus chrome (yellow) with the words "SCHOOL BUS" in black at its front and rear ends and

[1415] *Cumberland Evening Times*, Cumberland, Maryland, United States, 10th September 1959

"GARRETT COUNTY SCHOOLS" in white on its sides.

The crossing was protected by advance warning signs and at the crossing itself crossbucks and signs reading "3 TRACKS" and "STOP ON RED SIGNAL." In addition there were flashing light signals and warning bells. The bells began to sound and the red light signals to flash alternately when an eastbound train occupied track No. 2, 2,250 feet west of the crossing. There was also a whistle board for eastbound trains 1,683 feet west of the crossing.

Maryland State motor laws decreed that 'The driver {of a school bus} must be sure there is no danger from approaching trains or cars before attempting to drive the bus across the track. The door must be open while the vehicle has stopped, so that the driver may hear any possible noise of the locomotive or whistle,' and, 'The driver of any school bus before crossing at grade any track or tracks of a railroad shall come to a full stop within fifty (50) feet but not less than 10 feet from the nearest rail of such railroad and while so stopped shall listen and look in both directions along such track for any approaching train, and for signals indicating the approach of a train and shall not proceed until he can do so safely.'

The maximum speed of trains at that point was limited to 50 miles an hour.

The ICC report noted that the bus driver had stopped short of the crossing as required. He had been unable to see or hear any approaching trains and the crossing lights and bells had not been operating. He had then moved forward onto the crossing when the engine of his vehicle had stalled. After unsuccessful attempts to restart the engine or to move the bus off the crossing using the starter he had seen train No. 4 approaching. At this point he had shouted a warning to the children, who moved towards the front door, and he had assisted them in alighting until shortly before the collision, when he had been on his way to the emergency door at the rear of the vehicle to assist a child who had fallen onto the track. About ten children had alighted before the bus was struck.

The engineer of the train stated that he had whistled at the whistle board and intermittently until he reached the crossing. Although the weather was clear and the visibility was good he had not recognised the vehicle on the crossing as a school bus until he was in the vicinity of signal 28L (1,291 feet from the crossing). He had not been sure if the bus was stopped or moving slowly but had applied the brakes in emergency, reducing the speed of the train to 45 miles an hour at the crossing.

Several witnesses agreed that the crossing warning devices had not begun to operate before the bus had stalled.

Inspection of the school bus showed that its engine was undamaged and could be started using the starter switch.

The bus driver had acted as such for eight years and had driven buses over the crossing concerned for two years. There was no regulation requiring him to drill children in the emergency evacuation of the bus and he had not done so.

The ICC noted that the bus driver had made no attempt to evacuate the children in the short interval between the stalling of the bus and his becoming aware of the train's approach.

In conclusion, the ICC started simply that the accident had been due to the bus stalling on the crossing. [1416]

[1416] Interstate Commerce Commission

At the beginning of October Thomas Pullen Jr., the Maryland State Schools Superintendent, proposed a long series of recommendations to prevent a repetition of the accident, the simplest of which was to reroute buses where possible so as to avoid any of the 480 level crossings in the state, which he said were currently used by school buses 1,649 times daily. Other recommendations included:

(1) Reducing the maximum age of school bus drivers from 65 to 60

(2) Requiring surprise spot checks of buses for mechanical faults

(3) Requiring bus evacuation drills

(4) Retiring all buses from service after ten years' use

(5) Expanding the driver training programme

(6) Requiring trains to signal at least one minute before reaching a crossing

(7) Regulating the speed of trains

(8) Requiring unscheduled trains to reduce their speed to 5 miles an hour at unprotected crossings. [1417]

The morning after the Mountain Lake crash, Friday 11th September 1959, another crossing tragedy occurred in Minnesota, when a mother and her six children were killed at Waseca, Minnesota, United States. Mrs Irene Zimmerman, 39, was taking her children to school from the family farm, four miles away. Her car, a late-model station wagon, was less than a block from the Sacred Heart Catholic School when it was hit by a Minneapolis and St Louis Railroad freight train, thrown 100 feet, and rammed against a depot dock and a freight car standing on a siding. Five of the Zimmerman children – Michael, 10; Constance, 8; Barbara, 7; James Jr., 5; and Janice, 2 – died at the scene. Kathryn, 13, died soon after reaching a Waseca hospital and Mrs Zimmerman succumbed three hours after admission. As she lay dying she had moaned, "Help me, help me." Told by a priest of the accident, Mr James Zimmerman Sr. had to be hospitalised, suffering from shock. A further casualty of the accident was Robert Szyska, aged about 12, a school patrol boy, who was struck in the back or leg by a piece of flying wreckage and also had to be admitted to hospital for surgery.

Several classmates of the Zimmerman children witnessed the accident. The school principal, a nun, said, "The students were quite upset. For a while they milled around, almost in shock. After about two hours, they started to calm down."

Dean Fleming heard the crash from his nearby store. He related, "It only sounded like a small thud. It sounded like somebody dropped something on the floor. When I went outside, the train had thrown the station wagon up against a loading platform and box cars." [1418]

The 64-car freight train, driven by Engineer Glen Fairbanks, was only moving slowly. The crossing was a blind one, with a view along the line impeded by lineside buildings. It was, however, equipped with warning lights which reportedly had begun operating 35 seconds before the collision, while Engineer Fairbanks had been sounding his whistle. A newspaper wrote, 'A motorist driving directly behind Mrs Zimmerman said, "If it hadn't been her it would have been me." He said he had planned to do just as Mrs Zimmerman

[1417] *Cumberland Evening Times*, Cumberland, Maryland, United States, 2nd October 1959

[1418] *Fergus Falls Daily Journal*, Fergus Falls, Minnesota, United States, 11th September 1959 did, slow down for a moment and then proceed,' thus implying that motorists did not take much notice of the warning lights. [1419]

Nine people died when a train collided with a bus in Morelos state, Mexico, on or about Monday 28th September 1959. [1420]

A lorry carrying Post Office and communication workers was struck by an express train on an unprotected level crossing at a location variously given as Dabrowka and Jezioki Podlaskie, Poland, a village about 16 kilometres south of Warszawa, on Friday 2nd October 1959. Nineteen of the workers, who had been going to work, were killed and about 15 were injured, nine seriously. [1421] [1422] Four of those injured died in hospital in the following days. [1423]

A high road-side bank on the approach to Millville Road Crossing, three miles north-west of Lapeer, Michigan, United States, was blamed for an accident there on Sunday 4th October 1959 which killed five people. It was supposed that Lynn Morris, 38, of Otisville, had failed to see the approaching *Beeliner* passenger train of the New York Central Railroad. The blacktop road on which he had been driving was wet with afternoon rain and skid marks indicated that his vehicle had skidded 105 feet after he applied the brakes. (A blacktop road was similar to an asphalted one but had a larger proportion of crushed stone to its bitumen content.)

Mr Morris and his children Marsha, 9; Michael, 3; George, 2; and Joy, 7 months; were killed or fatally injured. His wife Betty, 35; and their eldest children, Earl, 16; and Lynda, 11, were critically hurt. The car was completely demolished, its engine being thrown out and found 35 feet away from the body of it, and all its occupants except Lynda were thrown from it; she was pinned inside it. The victims were scattered along the line over a distance of 356 feet. Identification of the victims was delayed for five hours, until the arrival of Morris' brother, James.

The only protection at the crossing was a pair of crossbucks. [1424] [1425]

Five residents of Alto Pass village lost their lives early on Wednesday 7th October 1959 when the car in which they were riding was hit by an Illinois Central Railroad freight train on an unguarded level crossing on Green Street Crossing, Anna, Illinois, United States. The car was carried over a quarter mile, with bodies and wreckage strewn along the roadbed. Driving the car was Lolo Minton, 57, a gardener at the Anna Street Hospital, who was taking four other people to work. His passengers were Miss Barbara Loann Brown, 24; Donald Leland "Pete" Marks, 18; his sister Phyllis Marks, 26; and Mrs Iva Myers, 46. Four of the occupants of the car were killed more or less outright and

---

[1419] *Kingsport News*, Kingsport, Tennessee, United States, 12th September 1959

[1420] *The Times*, London, England, 29th September 1959

[1421] *Kenosha Evening News*, Kenosha, Wisconsin, United States, 2nd October 1959

[1422] *The Times*, London, England, 3rd October 1959

[1423] Ibid, 6th October 1959

[1424] *The Holland Evening Sentinel*, Holland, Michigan, United States, 5th October 1959

[1425] *The Daily Oklahoman*, Oklahoma City, Oklahoma, United States, 5th October 1959

Donald Marks died on the way to the Union County Hospital.

A newspaper reported, 'Jimmy Karraker, operator of a service station at 101 W. Davie St., said he first saw the car pinned to the front of the engine as it approached the Main street crossing. He said the train was whistling for the crossing even though the car had been hit a block earlier. "There was a lot of blue smoke and sparks. I don't see how the car kept from catching fire," Karraker said. He said the train was going "fairly fast" but stopped quickly after passing the Main street crossing.'

The train was in charge of Conductor E. Payne, with Engineer Aud Spain.

There were crossbucks at the crossing but no warning lights or bells. The accident happened at about 6.37 a.m. [1426]

In September 1959 Japan was devastated by Typhoon Vera, which killed over 5,000 people and caused great devastation. [1427] In the aftermath of the disaster, on the night of Sunday 11th October, a vehicle bringing relief supplies to Nagoya was struck by an electric train near the station at Osato, Japan, about 32 kilometres north of Nagoya. The crowded six-coach train was derailed and seven of its passengers were killed; all of them had been riding in the leading coach. About 100 other train passengers and the other four people of the six on the truck, including its driver, were injured, the latter seriously. [1428]

The motorman of the train said that he had been travelling at about 95 kilometres an hour and had applied his brakes in emergency when it had become evident that the truck was not stopping. Witnesses said there had been a blinding shower of sparks at the moment of the collision and then the leading coach of the train had "appeared to hurdle over the truck." The truck was reportedly 'smashed into an almost unrecognizable hulk,' when it was squeezed between the train and the station platform. The train continued for another 300 metres before the leading derailed first coach fell down a low embankment into a rice paddy a metre below the line. The next four coaches were also derailed but remained upright and the rear coach, uncoupled in the shock of the collision, remained on the track.

The driver of the pickup truck had apparently disregarded the warning lights protecting the crossing. [1429]

A road tanker carrying gasoline (petrol) which was speeding along a gravel road through the prairies ran into the side of the Canadian Pacific Railway *Dayliner*, a two-car diesel unit passenger train en route from Lethbridge to Calgary, on Saturday 24th October 1959. Engineer W.F. Warner estimated that the tanker was travelling at 60 miles an hour while his train was going at about 70 miles an hour. He sounded his horn to warn the lorry driver and until the last moment thought that his train had just cleared the crossing when there was a jolt as the rear of his train was struck. The accident occurred two miles from Parkland, Alberta, Canada.

Farmer Ellis Oviatt witnessed the accident from his farmhouse a quarter mile away. He said that at the moment of the impact there was an explosion, then a huge wall of flame. The train was derailed but ran 200 yards further before it stopped. The railway

---

1426 *Southern Illinoisian*, Carbondale, Illinois, United States, 7th October 1959

1427 Wikipedia – Typhoon Vera

1428 *The Montana Standard*, Butte-Anaconda, Montana, United States, 12th October 1959

1429 *Pacific Stars and Stripes*, Washington, D.C., United States, 12th October 1959 right of way was set ablaze and the five ton truck, whose tanks were probably almost full, was torn into three pieces. The lid of one of the tanks was found 70 yards away. The rear coach of the train was engulfed in flame. Five passengers at its rear, three women, a girl, and a man, were unable to escape and were burnt to death. They were Mr and Mrs Thomas Benson, 72 and 67, Mrs Anne Patterson, 82, and Sherry Bilton, 15, all of Lethbridge, and Mrs Arthur Murdoch, about 50, of Calgary. Sixteen other passengers were taken to hospital in Claresholm, though only one of them, Mrs Margaret Bilton, Sherry's mother, was badly hurt. The tanker driver and part owner of the tanker company, Arthur Boyd, 27, of Lethbridge, was also killed. [1430] [1431]

Farmer Oviatt, who was also a local councillor, suggested that the lack of visibility of the silver-coloured train, known locally as the "Gray Ghost," was part of the problem at the unprotected level crossing. "It slips up so fast you often don't notice it, especially against stubble wheat fields at this time of the year. The dayliners should be painted a bright red or yellow so they can be easily seen. The approach to crossings like this should not be at right angles. The road should run parallel with the tracks a short way before turning across so drivers would have to slow down and would also give more attention to the crossing." [1432]

A freight train collided with a bus at Avellaneda, Argentina, on the night of Wednesday 29th October 1959. Eleven people died and about 50 were hurt. [1433] [1434]

A school bus carrying about 32 passengers, the St Anthony Junior Catholic Seminary first- and second-year high school football team, who were playing rival St Girard School on the other side of San Antonio, Texas, United States, stalled on a level crossing at Sherman and Austin Streets at the end of a short rise. It was then hit by a Southern Pacific Railroad train at about 6.50 p.m. on Thursday 12th November 1959. Father Paul Mabry, the coach of the team, said, "We were going east, crossing the tracks, when the bus stalled. We saw the train bearing down as we came up the steep hill but we didn't see any light flash. When we stalled the light began to flash and the train came on with the lights flashing and bells ringing. We were trying to evacuate the bus when the train hit us. I was one of the first to see the train coming. I opened the front door and told the boys to get out and take it easy."

Most of the passengers had time to scramble out of the bus before the train arrived, knocking the bus more than 50 feet, and spinning it around. The train hit the left rear side of the bus and those who were injured had been trying to get out of the rear door, which had jammed. Six boys, all aged 15, suffered fairly minor head injuries and cuts due to glass from broken windows while a seventh, Richard Hrynyshn, of Elmira, New York, was badly hurt, sustaining a possible fractured skull and internal injuries.

The bus had been driven by Brother Edward Krusinski, who stated that its motor had given trouble previously.

Engineer Bill Fowles of the southbound train said that he had slammed on the brakes

---

[1430] Victoria Daily Colonist, Victoria, British Columbia, Canada, 25th October 1959
[1431] *Winnipeg Free Press*, Winnipeg, Manitoba, Canada, 26th October 1959
[1432] *Winnipeg Tribune*, Winnipeg, Manitoba, Canada, 26th October 1959
[1433] *Pacific Stars and Stripes*, Washington, D.C., United States, 31st October 1959
[1434] *Clarin*, Buenos Aires, Argentina, 29th October 1959

but had not been able to pull up his train in the distance available.

Despite the accident the St Anthony B team went on to win their game against St Girards B, 22-6. [1435] [1436] [1437]

Eight people lost their lives on the evening of Saturday 14th November 1959 when the pickup truck in which they were riding was struck by a northbound Chicago Rock Island and Pacific Railroad passenger train on a level crossing on a rural road near Houston, Texas, United States. Their bodies were badly mutilated and identification was difficult but they were believed to be Doyle Patton, 31, a maintenance engineer for an air conditioning company; his wife Juanita, 29; their children Janice Diane, 12; Don LaVerne, 10; Sandra Sue, 6; and Randy Dale, 5; plus two of Mrs Patton's nieces, Rena, and Lynn Bryant, aged 2 and 18 months respectively.

The impact carried the truck half a mile down the track and pieces of it were scattered along the right-of-way. The train was little damaged; it had been running through near-freezing mist at about 80 miles an hour when the accident happened shortly after 6 p.m. at an unprotected crossing.

The same evening four men died when their car was struck by a Panhandle and Santa Fe freight train at Sudan, Texas. [1438]

What was described in some reports as a 'speeding bus' which was on its way from Rosario to Cordoba was in collision with a passenger train on a level crossing near Cañada de Gómez, Argentina, late on Thursday 17th December 1959. Eleven people were killed outright and 25 injured, two fatally. Most of the casualties were said to have been on the bus. [1439] [1440]

The apparent rashness and stupidity of a motorist cost the lives of himself, his wife and son and five other people at Fowlersville, Michigan, United States, at 5.44 p.m. on Friday 18th December 1959. Albert Simpson Sr., 29, attempted to race the *Pere Marquette* streamliner passenger train of the Chesapeake and Ohio Railway, travelling at 65 miles an hour, and lost. The train struck the car broadside, dragged it a quarter mile along the line and 'strewed mangled bodies along the track for 150 feet.'

Floyd Stewart witnessed the accident from his parked car 40 feet from the line. He related, "I had just left my office and was ready to go home. I heard the train coming and I was waiting for it to come through. I didn't realise what had happened until the train had cleared the crossing. I didn't see the bodies fly out of the car, but I saw them scattered along the tracks."

Victor Epley, also on his way home from work, saw what was going to happen and stood in the middle of the road trying to stop Simpson's car. He said that he had shouted and waved his arms but Simpson had merely slowed his car, then had lurched forward

---

[1435] *San Antonio Express*, San Antonio, Texas, United States, 13th November 1959
[1436] *The Taylor Daily Press*, Taylor, Texas, United States, 13th November 1959
[1437] *The Daily News-Texan*, Arlington, Texas, United States, 13th November 1959
[1438] *Big Spring Daily Herald*, Big Spring, Texas, United States, 15th November 1959
[1439] *The Holland Evening Sentinel*, Holland, Michigan, United States, 19th December 1959
[1440] *Lawrence Daily-Journal World*, Lawrence, Kansas, United States, 18th December 1959

into the path of the train.

Engineer Grover Lamport of the Grand Rapids to Detroit train said that he had sounded his whistle as he approached the crossing. He was clearly overcome by the accident and said simply, "It was awful."

Nineteen-year-old John Hubbard also witnessed the accident and Victor Epley's vain efforts to prevent it. He confirmed that the flashing warning lights at the crossing had been operating and that the car driver had ignored both them and Epley's waving and shouting.

In addition to Simpson his common law wife Mrs Mildred Cameron, 26; their son Albert Simpson Jr., three months; Mrs Simpson's daughter Dawn, 7; Clyde Ballard, 22; his wife Bertha, 24; their son Keith, seven months; and George Copeman, 40, lost their lives. [1441] [1442]

The vast majority of level crossing collisions are the fault of the road user but there are always exceptions. The first really serious level crossing accident of 1960 was the fault of a railway employee rather than the road user. It happened near Opole, Poland, in Lower Silesia, at 20:57 on Wednesday 6th January. A Polish State bus was struck passenger train No. 4623 from Bytom to Kamiemiec and carried about 200 metres down the line before the train could be stopped. Fifteen or 16 people were killed and another 39 were injured. The train driver, Stanislaw Behounek, said that the signals had been clear for him and he had only caught sight of the bus at the last moment by the light of the street lamps. His conductor Jan Szymaszek, said that the impact had been so violent he had thought that his train had been derailed,

The crossing was equipped with barriers, which had been lowered for the passage of a freight train and the bus and several cars had duly come to a stand at them. As soon as the freight train had passed the crossing keeper raised the barriers, unaware or perhaps forgetting that the passenger train was approaching on the opposite running line. The second train and the bus reached the crossing at the same moment. [1443] [1444] [1445]

As tractor driver Harold Kampmeier, pulling a hay waggon along County Highway PP, which linked Highway 21 and the village of Oakdale, Wisconsin, United States, reached a level crossing with the Chicago and North Western Railway on the early evening of Sunday 10th January 1960, he heard the whistle blasts of an approaching train and stopped. Shortly afterwards his tractor and waggon were overtaken by a car travelling at at least 50 miles an hour, which entered the crossing to be struck by the train, which itself was travelling at 70 miles an hour. The car was demolished and its remains were carried half a mile along the track.

Engineer H.G. Wilcox related that he had seen the car at the last moment and had one hand on the brake and the other on the whistle when his engine hit the car. It had been

---

[1441] *South Haven Daily Tribune*, South Haven, Michigan, United States, 19th December 1959

[1442] *The News-Palladium*, Benton Harbor, Michigan, United States, 19th December 1959

[1443] *The Times*, London, England, 8th January 1960

[1444] *Jefferson City Post-Tribune*, Jefferson City, Missouri, United States, 7th January 1960

[1445] https://opole-naszemiasto-pl.translate.goog/katastrofa-kolejowa-w-opolu-6-stycznia-1960-autobus-wjechal/ar/c15-7502353?_x_tr_sl=pl&_x_tr_tl=en&_x_tr_hl=en&_x_tr_pto=sc

impossible to stop.

The six occupants of the car were all killed. They were George Emrich, 43; his children Duane, 15; Patrick, 11; and Sandra, 5; his son-in-law James Lyon, 29, of Round Lake, Illinois; and John Monarch, 75, a neighbour of the Emrich family. [1446]

In another bad school bus level crossing accident in the United States a train collided with a bus carrying 12 children as it was using a crossing with a dirt road at Fackler, Alabama, just after 7 a.m. on the morning of Monday 18th January 1960. Four of the children were killed and the other eight and the bus driver, Charlie Beavers, 50, were injured.

In a hospital interview Beavers said, "The brakes gave out…I never saw the train but it wouldn't have made any difference if I had. I must have been knocked unconscious but I remember hearing the children holler and I remember getting out of the bus. I started pulling them out of the water." He said that the brakes of the bus had failed several times recently. He had reported the problem and had assumed that the brakes had been repaired. On the other hand Jackson County School Superintendent E.G. Thomas Sr. stated that the 1951 bus had been checked the previous Monday and it had been in good condition when it had left the shop.

Engineer C.E. Edwards, of the Southern Railway freight train which had struck the bus, said he had thought that the bus would stop. The train, travelling at 55 miles an hour, cut the bus in two and its rear portion was dragged three-quarters of a mile. Some of the children were thrown into a ditch.

Two of the children who died, Leonard Beavers, 12, and his brother Foster, were nephews of the bus driver, while a nephew and a niece of his were among the injured. The other two children who died were Frances Corbitt, 10, and her sister Mary Grace, 8.

The bus was taking the children from Fackler to the high school at nearby Stenson. It had been near the start of its journey and would later have had 45 passengers. The train was on its regular return run from Stevenson to Sheffield and consisted of two engines, a single car, and a caboose.

Alabama Governor John Patterson called the accident shocking and ordered an investigation. [1447] [1448] [1449]

Farm to Market Road 1512 intersected the line of the Panhandle and Santa Fe Railway on the outskirts of Amarillo, Texas, United States, at Washington Street Crossing, which was provided not only with warning lights and gates but also automatic half barriers. These measures failed, however, to prevent a serious collision at 6.24 p.m. on Tuesday 19th January 1960. A 1957 model four-door sedan driven by Dr Robert Lee Brown of Canyon was struck by passenger train No. 2, the *San Francisco Chief,* en route from Kansa City to Chicago.

Dr Brown, 37, and his sons Robert Lee Jr., 9, and David, 6, were killed instantly. His

---

1446 *Appleton Post Crescent*, Appleton-Neenah-Menasha, Wisconsin, United States, 11th January 1960

1447 *The Decatur Daily*, Decatur, Alabama, United States, 19th January 1960

1448 *Sarasota Herald-Tribune*, Sarasota, Florida, United States, 18th January 1960

1449 The *Anniston Star*, Anniston, Alabama, United States, 18th January 1960 wife, Mrs Dorothy Brown, 34, died in the Northwest Texas Hospital at about 7.30 p.m. and their daughter Sharon, 11, died at 12.40 a.m. the following morning after undergoing surgery for a head injury.

The train had been in charge of Conductor Floyd Ridner with Engineer W.W. Wheeler and Fireman W.J. Westcoat. Engineer Wheeler said afterwards that his train had been moving at 79 miles an hour at the time of the accident. He said that the Browns' automobile had first stopped at the lowered barrier, then had pulled out around it into the path of his train. The impact threw the car 171 feet along the snow- and ice-covered track; it was said to have exploded upon being struck. Damage was also caused to the locomotive of the train, which was delayed about three hours.

Dr Brown had been on his way to attend a veterinarians' meeting at the Herring Hotel in Amarillo. Why he had chosen to go around the half barriers could of course not be determined; possibly he had feared being late for his meeting.

The road described a double bend as it passed over the railway. There were already plans for a bridge across the line in place of the level crossing; the bridge would also eliminate the double bend in the highway. A contract for its construction was expected to be let by the state highways department during the year. [1450]

Three Erin District High School pupils lost their lives when a Canadian Pacific Railway freight train collided with a school bus on a level crossing on the outskirts of Hillsburgh, Ontario, Canada, on the afternoon of Thursday 28th January 1960. The small bus was carried about 385 feet down the line and its petrol tank exploded following the impact, spraying the front and inside of the vehicle with fuel, and it was engulfed in flames. The pupils kicked out the emergency doors of the bus, through which 25 of them were able to escape, but the bodies of Karen Awrey, 14; Donna Jackson, 15; and Gudrun Koch, 15, were found in the wreckage. The bus driver, Calvin Leitch, 29, was fatally hurt and died later in Guelph Hospital.

Trainmen and locals rushed to free pupils from the bus and some of the injured were taken to hospitals in private cars. With telephone lines still down after a recent ice storm it was some hours before many of the parents learnt of the accident. At the time of the accident the bus had been on the last stage of its homeward journey. [1451]

An inquest on Karen Awrey, as representative of the victims of the accident, was held at Hillsburgh in mid-February. Surviving pupils stated that the bus driver rarely stopped at crossings though he usually slowed down. There were also allegations that the train had been travelling faster than usual. Engineer Frank Baker (surname illegible in newspaper) estimated his speed at 20 miles an hour, though he had not looked at his speedometer.

The jury returned a verdict that the accident had been due to "limited sight and unusual road conditions" and made a whole series of recommendations:

(a) That the drivers of all school buses should be required to stop at all railway crossings which were not protected by gates or flagmen, and be required to open the front door, select a proper gear, and proceed in that gear across the railway.

[1450] *Amarillo Globe-Times*, Amarillo, Texas, United States, 20th January 1960

[1451] *Acton Free Press*, Acton, Ontario, Canada, 4th February 1960

(b) That the character, qualifications, and capabilities of all school bus operators should be more thoroughly supervised and that there should be more supervision between the local school board and the bus company.

(c) That because a thorough inspection of school buses was important an inspector highly qualified with mechanical experience should be used.

(d) That some consideration should be given to the use of the monitor system, whereby senior students would help maintain discipline in the bus, allowing drivers to devote their full attention to driving.

(e) That the proper department of the government should issue a booklet or circular with rules and regulations for the use of school bus drivers.

(f) That modern signals be installed at all crossings where sight lines are limited.

(g) That a better agreement should be arranged between road and railway patrols to eliminate the accumulation of snow and ice on the railway adjacent to the level crossing.

(h) That sanding done at the proper time, and thoroughly done, was a very necessary procedure and the operation should not be curtailed in the low assessment areas within the snow belt; more provincial grants should be made available for such conditions.

(i) That consideration should be given to the erection of black and yellow reflector signs which designated the number of tracks and the direction of the railway in relation to the road or highway.[1452] [1453]

A wigwag was installed at the crossing in late October.[1454]

Some mystery surrounded a level crossing collision involving a school bus at Centralia, Illinois, United States, on Monday 8th February 1960. The 46-seat bus had only just begun picking up its passengers when the accident happened at 7.50 a.m. at the Beech Street crossing. It was struck by a Southern Railway train from East St Louis, Illinois, to Princeton, Indiana, the train striking the rear of the vehicle, which was knocked about 22 feet but remained upright. Eight pupils of Centralia High School were badly enough hurt to be admitted to hospital. The most seriously injured was Billy Joe Bryant, 15, who was thrown through the emergency door at the back of the bus and sustained head injuries and a broken arm. He succumbed to his injuries the following day. A girl pupil was transferred to a St Louis hospital for special treatment. Several other students of the 15 on the bus were treated and released. The students remained calm after the accident, unlike some of their parents who rushed to the scene and became hysterical when they saw what had happened.

The crossing was protected by warning lights and a bell. The bus driver, Otto Beck, 54, said that he had stopped on the approach to the crossing and had opened the door of the bus, but he had not seen or heard any train, though he admitted that his view had been obstructed by some freight cars standing on a siding. In addition, he alleged that the crossing bell and warning lights had not been operating, and in this assertion he was backed up by the testimony of a lorry driver who had been behind him, and who said that he would have moved his lorry into the path of the train had the school bus not been in front of him. Two other witnesses reportedly agreed that the crossing alarms had remained inert. Beck

---

[1452] Ibid, 18th February 1960

[1453] *Georgetown Herald*, Georgetown, Ontario, Canada, 25th February 1960

[1454] *Acton Free Press*, Acton, Ontario, Canada, 27th October 1960 said that as he had pulled onto the crossing one of the students had shouted, "Look out, stop, here comes a train!" but it had then been too late to do anything.

The train consisted of a locomotive and 11 cars of fruit and its speed at the point of the accident had been limited to 20 miles an hour. Engineer Willard Veith agreed that the bus had first come to a stand at the crossing before moving forward onto it in front of his engine.

An inspection of the crossing controls by the local police on the day after the accident found them to be working correctly, despite the allegations that they had been out of order at the time of the accident. [1455] [1456]

Mr John Denneus, 45; his wife Agatha, 40; and their children Donna, 16, Phillip, 9, and Eileen, 7, lost their lives when the station waggon in which they were returning home from church was struck by an eastbound diesel-hauled Canadian National Railways freight train on a level crossing within sight of their home at Tofield, Alberta, Canada, on Sunday 14th February 1960. Two other children survived. John Michael, 12, sustained a broken collarbone and a broken wrist and a two-week-old baby girl was critically hurt. The vehicle was carried several hundred feet along the line. Mr Denneus and the children were killed outright and Mrs Denneus died very shortly after admission to hospital in Edmonton. There was no immediate explanation for the accident, which happened on a sunny day. [1457] The crossing was presumably an unprotected one.

In the vast majority of level crossing accidents the casualties were principally of wholly road users, but there were exceptions, such as the disaster at Rosedale, near Bakersfield, California, United States, on Tuesday 1st March 1960.

Atchison Topeka and Santa Fe Railway passenger train No. 2, *The San Francisco Chief*, had left San Francisco at 11.15 a.m. with 75 passengers. It consisted of four diesel units and 11 cars, with Engineer L.A. Snyder and Fireman A. H. Braley. At 5.10 p.m. it reached Rosedale, where the line was crossed on the level by the four-lane Hageman Road. At the same time an oil rig, owned by the Oglesby Brothers Trucking Co and consisting of a tractor and two trailers loaded with 6,465 gallons of crude oil, reached the crossing, failed to stop, and drove onto the crossing. The leading diesel unit of the train, which was travelling at about 75 miles an hour, struck the second of the two trailers, which was wrapped around the engine, and a tremendous explosion followed which could be heard four miles away. The forward tanker was thrown 100 yards, damaged and leaked oil but did not catch fire. The leading three diesel units were set ablaze and the force of the impact derailed nine of the cars of the train, which 'jacknifed into one another like a row of huge steel dominoes knocked ashore. Flames from the initial explosion swept back along the train, but only forward cars in the tangled jumble caught fire.' The first three of the diesel unts were charred to blackened rubble – reduced, one newspaper suggested, to hardly more than half their original size. The last two cars remained on the track.

Engineer Snyder and Fireman Braley were cremated in the cab of the engine. The

---

[1455] *Mt Vernon Register-News*, Mt Vernon, Illinois, United States, 8th and 9th February 1960.

[1456] *Edwardsville Intelligencer*, Edwardsville, Illinois, United States, 8th February 1960

[1457] *The Medicine Hat News*, Medicine Hat, Alberta, Canada, 15th February 1960 lorry driver, John Garrett, fared no better. An eyewitness saw him jump from the cab of his lorry and run but the flames caught up with him. In desperation he jumped into an irrigation ditch where he was 'burned to a crisp.' In total 14 people reportedly lost their lives in the accident. Another 68 people, some badly burned, were taken to hospital in ambulances from Bakersfield, Delano, Shafter, and Wasco. Some of the casualties were evacuated by helicopter. Fire crews fought the blaze for two hours. [1458]

A surviving passenger said, "The train started to twist and jerk and seats flew through the air. Most of us were thrown around like rag dolls...Children inside were crying. The passengers broke a window and carried children out to me. They were very calm about it...I looked out of the window and saw a huge fireball at the crossing." [1459]

The Interstate Commerce Commission stated that the cause of the disaster was 'undetermined.' Witnesses stated that Driver Garrett had paused momentarily when he had reached the crossing but had not come to a full stop before entering. At the last moment Garrett had apparently attempted to turn his rig parallel to the track before fleeing the cab. [1460]

On the grounds that the lorry had failed to stop at the crossing in conformance with California law the Santa Fe company sued the trucking company for $2 million in damages. [1461]

In 1957 Mr John Whittaker and his wife were cited by the Catholic Transcript for their care of foster children, having looked after 16 children in 14 years. They continued to do so and on the morning of Tuesday 8th March 1960 Mrs Whittaker was taking seven children to St Paul's School in Kensington, Connecticut. The 1955 station wagon had just left the Whittakers' home and was proceeding along Norton Lane when it shortly arrived at a level crossing with the New York New Haven and Hartford Railroad near Berlin, Connecticut, United States. As the station wagon reached the unprotected crossing it was struck by passenger train No. 97, en route from Springfield, Massachusetts, to New York; the accident happening at 8.33 a.m. Mrs Whittaker had waited at the crossing for a freight train to pass and after it had done so her station wagon had crept forward into the path of the passenger train on the opposite track. A flat tyre had possibly contributed to the accident. The vehicle was hit broadside and thrown down an embankment into a pond called Silver Lake, which was ice covered and five feet deep. Three of the children were thrown from the car, one remaining on the embankment and the bodies of the other two skidding over the ice. The other five occupants were trapped in the station wagon. Parts of the vehicle were scattered along the line for a fifth of a mile and scattered school books and briefcases marked the location of the tragedy. Mr Whittaker witnessed the accident from their home, only 100 yards away, and ran to the scene, but found that he could do nothing. All eight occupants of the car lost their lives, some it was thought from drowning. No one on the train was hurt.

Apart from Mrs Whittaker, the victims were named as Byron Battle, 6; Joseph Cormler, 14; Mary Ann DeCruze, 12; Doris Gomez, 6; Patricia Heanure, 14; Kenneth

[1458] *The Billings Gazette*, Billings, Montana, United States, 2nd March 1960
[1459] *Waterloo Daily Courier*, Waterloo, Iowa, United States, 2nd March 1960
[1460] *Denton Record-Chronicle*, Denton, Texas, United States, 2nd March 1960
[1461] *Star News*, Pasadena, California, United States, 4th March 1960

Mirage, 6; and a seven-year-old girl who first name was Shelley. [1462]

A school bus being driven by Thomas Mattingly, 31, was struck by a Louisville and Nashville Railroad train at Loretto, Kentucky, United States, near Lebanon, at about 8.30 a.m. on Friday 18th March 1960. There were 36 children on the bus, aged between 7 and 13, on their way to Catholic grade and high schools in Loretto. Seven of them were hurt, according to some reports seriously, though it seems more likely that only one child was badly hurt and none were critically injured. The train was a switch engine making a short run to pick up some cars from a distillery at Gethsemane. Its engineer, C.E. Stanley, said that the bus had been in motion but had not quite cleared the crossing when his engine hit it. Most of the children scrambled out of the big hole made in the back of the bus after the collision. [1463] The crossing, with Kentucky 52, had neither gates (barriers) nor warning lights. [1464]

On Tuesday 22nd March 1960 a diesel railcar travelling from Maffra to Traralgon collided with a school bus carrying children from three schools in the Traralgon area at an unprotected level crossing three-quarters of a mile south of Traralgon station, Victoria, Australia. The bus was struck on its left-hand side, spun completely around, and overturned. Bus driver William Todd and children Margaret Hayes, 12, of Princes Highway, Traralgon; John Leslie, 11, of Willung South; Jane and John Warburton, aged 12 and 10 respectively, of Traralgon South; and Marie Wimbridge, 14, of Cowwarr, were killed. Nine other children were admitted to Traralgon hospital and the remaining 14 were treated for minor injuries and allowed to go home. The bus was owned by the Latrobe Valley Bus Line. Visibility at the crossing was good. [1465]

The very same day as the Traralgon accident, Tuesday 22nd March 1960, there was yet another level crossing collision involving a school bus in the United States, and the second serious one of the year in Alabama. It occurred near Coy and involved a St Louis-San Francisco Railway through freight train from Pensacola, Florida, to Memphis, Tennessee and St Louis, Missouri, and a school bus carrying 25 pupils to a consolidated Negro grammar school in Coy and another school in Camden. [1466] The bus was being driven by David Jackson, a 20-year-old student. The latter said afterwards that the brakes of the 10-year-old vehicle had failed. He had frantically changed gears in an attempt to stop the bus and when this failed had stepped on the accelerator in an attempt to beat the train over the crossing. He related, "There was a jarring crash and the bus came apart. There was no panic. Most of the children didn't even scream."

State highway patrolman Sergeant C.E. Canterbury said that the train had almost ripped the bus into two pieces and that the wreckage and the bodies of the children had been strewn for 1,500 feet along the tracks. The bus driver escaped with an injured thumb and some of the older students who had only sustained slight injuries picked up the bodies of the dead and critically injured children and carried them to a nearby house,

[1462] The News, Frederick, Maryland, United States, 8th March 1960
[1463] The Hammond Times, Hammond, Indiana, United States, 18th March 1960
[1464] *Sandusky Register*, Sandusky, Ohio, United States, 18th March 1960
[1465] *The Canberra Times*, Canberra, Australian Capital Territory, Australia, 23rd March 1960
[1466] *Laurel Leader-Call*, Laurel, Mississippi, United States, 22nd March 1960

where they lay on a big bed awaiting the arrival of ambulances. There were no telephones within 15 miles of Coy, a place described as 'a tiny unincorporated farming town in the Alabama bottomlands, 17 miles south of Camden.' To get help residents drove over rural back roads to Camden. The injured were taken to the Wilcox Memorial Hospital there, those critically hurt then being transferred to Selma, 45 miles away. In total six children aged between 10 and 18 died and 15 were hurt, three critically. [1467]

The following month at Camden David Jackson pleaded guilty to second-degree manslaughter in connection with the accident, was given a one-year suspended sentence and had his driving licence revoked. [1468]

A party of five people returning from a wedding at Layton lost their lives at 2.15 p.m. on the afternoon of Saturday 23rd April 1960 in a collision on West Point Road Crossing, two miles from Clearfield, Utah, United States. The victims were Mrs Ida Martinez Bolastig, 45, of West Point; her adopted son Walter Martinez Bolastig, 11; and her nephews Teddy Martinez, 14; and twins Julian and Richard Martinez, 13. Their car was struck by a Denver and Rio Grande Western Railroad extra freight train travelling at about 50 miles an hour, was split open along its entire right-hand side, damaged a piece of lineside signalling equipment, and was knocked onto an island of ground between the track and a canal. The train stopped a quarter mile down the line. Mrs Bolastig survived the accident for a few minutes but died either before the ambulance arrived or on the way to hospital. The others were probably killed outright.

The crossing was protected by warning lights and bells which were working at the time of the accident. The engineer of the northbound train was W.S. Buckley. Witnesses said that another car had crossed in front of the train immediately before the Bolastig one. [1469] Farmers in nearby fields witnessed the accident and heard the train whistle; they believed that the Bolastig car had tried to beat the train. [1470]

Five middle-aged residents of Pampa, Texas, who were returning from a meeting of Jehovah's Witnesses in Fort Worth, were killed when their automobile was struck by a 78-car Chicago Rock Island and Pacific Railroad train on a level crossing on State Highway 70 on the north edge of Jericho, Texas, United States. The accident happened at 1.05 a.m. on Monday 27th June 1960. The driver of another car had heard the whistle of the approaching train and stopped at the crossing to wait for it. The victim's car pulled into the left-hand lane of the road to overtake the waiting vehicle, pulled back into the right-hand lane and immediately collided with the locomotive of the train. The impact demolished the car and several of its occupants were thrown out of it.

The victims were identified as Mr Daudie Mize and his wife, Mr Tom Hammonds and his wife and Mrs Ethel York. Four of the victims were killed more or less outright and Mrs Hammonds died in Groom Hospital after having been rushed there by a motorist who had come across the wreck. Mr and Mrs Hammonds were both blind. Mr Mize was the owner and apparent driver of the car.

No one on the train was hurt and the train was only slightly damaged. After a short

---

1467 *The Cumberland News*, Cumberland, Maryland, United States, 23rd March 1960
1468 *The Journal-Tribune*, Marysville, Ohio, United States, 21st April 1960
1469 *The Salt Lake Tribune*, Salt Lake City, Utah, United States, 24th April 1960
1470 *The Sunday Herald*, Provo, Utah, United States, 24th April 1960

delay it continued its journey towards Amarillo. [1471] [1472]

A holiday coach owned by Smiths Tours of Wigan, in the north of England, had set off on 2nd July 1960 on a 14-day tour of the Swiss and Italian lakes, and was returning from Italy to Interlaken, near the end of the tour, when it was involved in a level crossing accident near Meiringen, Switzerland, at about 17:00 on Sunday 10th July 1960. The coach was struck on its right-hand side by a Chemins de fer fédéraus suisses (CFF) narrow-gauge train running on the Brünig line, dragged some distance down the line and ended up in a field. Of the about 20 passengers on the bus five were killed and 13 were hurt and admitted to Meiringen Hospital. It was difficult at first to identify those who had lost their lives but they were later named as a Mr Herbert Fisher, Mrs Olive Fisher, Mrs Olive Halliwell, Mrs Phoebe Twist and Miss Annie Wolfendale. The following day Dr Rolf Stadler at the hospital stated that all those injured were likely to recover, though four days after the accident it was stated that one of the injured had still failed to regain consciousness.

There were no barriers at the crossing, but it was protected by flashing lights and bells. It appeared that the bus driver-courier, Mr Walter Rushton, of Wigan, who suffered severe shock in the accident, had failed to notice the warnings. [1473] [1474]

Six people died and 13 were seriously injured when a train collided with a bus on Friday 30th September 1960 on a level crossing about 50 kilometres south of Bangkok, Thailand. [1475]

"I was blinded by the sun and couldn't hear the train because of the shouting, laughing children," school bus driver Edward Hallowell, 40, told police after a collision on Bancroft Crossing, Grant City, New York, United States, on Monday 7th November 1960. The vehicle had been struck by a Staten Island Rapid Transit passenger train from Tottenville to St George. None of the eight passengers on the two-car electric was hurt. [1476]

The train hit the rear of the bus, knocking it over. 'Sobbing hysterical children, many bloodied from injuries, crawled from the overturned bus, assisted by police and other rescuers. As word of the crash spread, terrified parents converged on the scene, then hurried to hospitals or checked door to door, searching for their children,' a newspaper reported.

There had been 51 children on the bus, aged between 5 and 8. One child, Caroline Tarr, 7, was killed and 44 others were injured, of whom 18 were admitted to the Richmond Memorial Hospital, three of them in a critical condition.

The crossing was protected by warning lights and bells, which were operating normally. The bus had stopped at the crossing, then proceeded.

It had already been decided that crossings on the line should be done away with. The newspaper report continued, 'A Richmond County (Staten Island) grand jury was investigating the island's unguarded rail crossings last December when a high school girl

[1471] *Amarillo Globe-Times*, Amarillo, Texas, United States, 27th June 1960
[1472] *Amarillo Daily News*, Amarillo, Texas, United States, 28th and 29th June 1960
[1473] *The Times*, London, England, 11th and 12th July 1960
[1474] *Journal de Genève*, Genève, Switzerland, 11th and 15th July 1960
[1475] *The Singapore Free Press*, Singapore, 30th September 1960
[1476] *The Republican Courier*, Findlay, Ohio, United States, 8th November 1960

was killed by a train near the scene of yesterday's accident. The jury filed a report urging that all such crossings be eliminated immediately. But the Transit System, operated by the Baltimore and Ohio Railroad, had been granted an extension of time to erect protective gates, police said. Dist. Attorney John M. Bralsted said the bus accident would be referred to a new grand jury, which is being convened now.'

The children were being driven home at the end of their lessons at Public Schools 38 and 41. [1477]

Six people died and eight were injured, two critically, in an accident which occurred at 07:25 on the morning of Tuesday 15th November 1960 when a local bus carrying a dozen people from the village of Cineto, including five schoolchildren, was struck by the 06:33 *direttissimo* (express) from Roma Termini to Pescara at a level crossing with the Via Triburtina at Mandela, Italy.

The crossing was protected by barriers worked by a crossing keeper. The train was running a little late, possibly having been delayed by a fog, and the bus driver was accustomed to finding the crossing open when he reached it. The train approached the crossing around a curve, hidden by cuttings and high vegetation, and travelling at 70 kilometres an hour, so it was unlikely that those on the bus had much if any warning of what was about to happen. Carried a few metres by the force of the impact, the whole front of the bus was smashed. It only took the train a relatively short distance to pull up and its passengers then disembarked and ran back to the crossing to render what help they could. Someone ran to a telephone to call for ambulances, the fire brigade and the police and the rescue services were soon on the scene. It took oxy-acetylene torches to release the bodies of some of the victims from the twisted wreckage of the bus.

Crossing Keeper Antonina Valle, aged 50, who had operated the crossing for 21 years, had to be taken to hospital suffering from severe shock after the accident and could not immediately be interviewed, as to how it had happened. Likewise the bus driver, Alberto Civili, who somehow escaped with slight injuries despite the damage to the front of his bus, was too shocked to be able to give an explanation of the tragedy. It was theorised, however, that the crossing keeper, anxious to prevent unnecessary delay to road users, had delayed lowering the crossing barriers until the last moment. As she had then begun to do so the bus had appeared and tried to cross the line as the barriers were descending. There were no skid marks, indicating that the bus driver had not made any attempt to stop, and his vehicle had smashed through the crossing barriers to reach crossing at the same moment as the train.

Four bus passengers were killed outright. They were Francesco Latini, 12, and his cousin Quinti Latini, 13, both students at the San Cosimato vocational school, near Mandela; Elisa Colasanti, 30, a janitor at a Roma elementary school; and Antidio Pucci, 56, going on a hunting trip. Carolina Cinti, 54, died shortly before arrival at Tivoli Hospital and Bruno Meloni, 13, another schoolboy, died in the casualty department of the hospital. Nineteen-year-old Giovanni Latini, a mechanic and probably a relative of the other Latinis, was critically hurt and doctors feared for his survival. The two train drivers, Sante Cerrone, 51, and Guerrino de Santis, 24, were slightly cut by broken glass and two train passengers, workman Attilo Sammartino, 32, and Naval Foreman Quirino Maiolino, 41, were slightly injured by being thrown to the floor.

[1477] *The Lowell Sun*, Lowell, Massachusetts, United States, 8th November 1960

The number of passengers on the bus would have been greater had not some dozen schoolboys chosen to walk the five kilometres from Cinetro to the Via Tributina in the fine weather and catch another bus into Tivoli. [1478]

The cause of a level crossing collision at Toltec, Arizona, United States, on Monday 21st November 1960, which resulted in the loss of seven lives, remained a mystery. Russell Martin and Ralph Jewel were automatic cotton machine operators on the McCarthy – Hildebrand Ranch south of Eloy and a short distance from the scene of the accident. They had been paid on the day before the accident and their wives had been into Eloy to shop and were on their way to bring lunch to the men at work when the accident occurred at about 12.35 p.m. at a crossing near Highway 84. The automobile was struck by the westbound *Blue Streak Merchandise* fast freight train of the Southern Pacific Railroad from East St Louis, Illinois, to Los Angeles, California. It was pushed 900 feet along the line, knocked down some signalling apparatus and was wrecked. Two of its occupants were thrown out along the line, the others remained in the wreckage.

Mrs Annabelle Martin, 30; her children Elliott, 5, and Barbara Ann, 3; Mrs Katherine Jewel, 26; her children Leroy, 3, and Rusty, 15 months; and Mrs Martin's brother-in-law, Benjamin Martin, 37, were all killed. Mrs Martin, who was pregnant, had been driving the car when it left the store 12 miles from the crossing.

Engineer H.L. Kennedy said that he had whistled for the crossing and that the car had appeared to "drift" up onto the tracks. The car had approached the crossing slowly, as if it were going to stop, but when he had been 150 feet from the crossing he realised that the car was still moving forward. He had then applied the brakes in emergency and the train had stopped half a mile beyond the crossing. Another witness thought that the automobile had stopped before rolling ahead onto the line.

The speed limit over this section of the line was 60 miles an hour and Head Brakeman Russell Randall reported that the train had been travelling at 58 miles an hour.

A railwayman ran back towards the crossing and found Barbara Martin to be still breathing, but she was beyond help and soon died. Another passer-by said, "I went up and down the tracks and saw the bodies lying all around…There was nothing anyone could do."

The crossing was protected solely by a pair of crossbucks but there was a good view up and down the line. [1479]

An inquest on the victims of the accident at Eloy found simply that they had died as a result of a collision between their car and a freight train and did not assign responsibility for it. [1480]

A passenger train from Guadalajara had almost arrived in Mexico City, Mexico, when it collided with a heavily loaded bus, on the night of Sunday 27th November 1960, killing six people and injuring 15. It was alleged that the bus driver had attempted to beat the train over the crossing. [1481]

1478 *La Stampa*, Torino, Italy, 16th November 1960

1479 *Tucson Daily Citizen*, Tucson, Arizona, United States, 22nd November 1960

1480 *Arizona Daily Sun*, Flagstaff, Arizona, United States, 29th November 1960

1481 *The Eau Claire Leader*, Eau Claire, Wisconsin, United States, 30th November 1960

A school bus was involved in a major disaster at Lamont, Alberta, Canada, on the morning of Tuesday 29th November 1960. The bus was carrying 41 pupils from the village of Chipman and neighbouring farms to the high school in Lamont and was within sight of the school when it had to cross the Canadian National Railways line at an unprotected crossing, approached by a slight slope in the roadway. It was travelling along Highway 15 parallel to the track before it made a right-hand turn onto a side road to cross the line, stopped momentarily, and then moved onto the line in the path of the train. It seemed likely that the early-morning sun would have been in the eyes of driver Frank Budley as he looked towards the approaching westbound through freight train, which consisted of two diesel engines, four loaded boxcars and 23 empty cars, en route from Vermilion to Edmonton and was travelling at an estimated 50 miles an hour. There was a clear view along the line in each direction from the crossing. The bus was struck in its middle and carried a quarter mile along the track. The rear end of the vehicle was thrown into a field, the main part of it was wrapped around the engine of the train and a tractor and chain had to be used to pull the wreckage apart. The accident occurred at 9 a.m. in near-zero weather. Classes were due to start at 9.15 a.m. After the accident the school was closed.[1482] [1483]

First warning of the disaster was probably given by Lamont agent T.A. Peters, who saw the train approaching the station with the wreckage of the yellow bus attached to its engine and rang the hospital for ambulances.

Driving behind the bus as it neared the crossing was teacher Lois Stefura. She said, "I thought, 'If he doesn't stop, the train will hit the bus.' But I saw he stopped, and I thought everything was all right. When I looked again, the bus was pulling out onto the tracks. The train horn was blowing so loudly, I thought everyone must hear it. The next instant, everything seemed to fly up, books and bodies. I thought everyone must be dead. Then I saw a boy move." Many of her pupils had been on the bus. She took an injured boy and girl to hospital.

Keith Tompkins, 12, the youngest pupil on the bus, spoke to a reporter from his hospital bed. "I heard a bang and the next thing I knew I was sitting in a ditch." His brothers Robert, 15, and Tom, 13, had been sat at the rear of the bus. "I found Tom in the wreckage; he didn't know where he was. Robert was walking up the tracks, looking a little dazed."[1484]

All the injured were taken to Lamont's new 91-bed Archer Memorial Hospital, but, 'The town's two ambulances were not enough. Farm trucks and cars started pulling up to the site, as workers from the municipality, farmers and passing motorists rushed to help. Men used doors from grain bins as stretchers as they pulled injured teenagers from the twisted wreckage and drove them to hospital in the backs of pickup trucks. Acetylene torches were used to free some of the injured. Some victims had their clothes ripped off in the crash; others were injured so badly they were unrecognizable to people who had known them all their lives. A grim reminder of the crash was left strewn along the railway right-of-way. Books, lunch baskets, clothing and other articles were scattered along the embankment.'

The hospital was reportedly taxed to capacity. Four doctors, blood and medical

---

[1482] *The Lethbridge Herald*, Lethbridge, Alberta, Canada, 29th November 1960
[1483] *Winnipeg Free Press*, Winnipeg, Manitoba, Canada, 29th November 1960
[1484] *Edmonton Journal*, Edmonton, Alberta, Canada, 29th November 1960

supplies were rushed from Edmonton, 45 miles away.

'Throughout the morning a tragic procession moved to and from the hospital. Parents who heard reports of the crash and to the scene to check on the condition of their children. Many returned in tears. Only relatives were permitted into the hospital by the RCMP who guarded the door.

'Women of the United Church began to serve coffee and sandwiches to the frantic parents. Their headquarters was a community hall a few steps from the hospital where news could be brought quickly. It was a sad scene in the Hall. Men and women, some of them accompanied by younger children, just sat there, jumping to their feet every time the door opened.

'Back in Chipman there was another gathering of parents, also in a community hall, but this one was different. These mothers and fathers knew their children were dead. They were trying to make themselves believe it, and to plan what to do next. Members of the Homemakers Club, a local women's organisation, made tea and sandwiches for them. Some of these people are Roman Catholic, some are Greek Catholic, some are Protestant. Their backgrounds are different, their names are varied. Their common grief has smoothed away their differences.

'No real plans have been made yet, but the parents spoke of a funeral service where the children who had died together could be buried together. It might be in the high school auditorium, some thought.

'No family has lost more than one child, though perhaps Mrs Eleanore Tompkins came closest. Her three boys were on the bus – the only lads not of high school age – Keith, 6, Tommy, 7, and Robert, 8, were all injured and in hospital, but all are expected to recover. {It will be noted that the ages of the Tompkins boys stated in this newspaper are different from those given in another newspaper the previous day; it is not known which are correct}. Keith had his shoes and socks torn off by the impact and he was found wandering in his bare feet.

'It was a nightmare and many keeping the long vigil kept muttering that they couldn't believe it was real. But the lights burning late in Chipman's stricken homes showed that it did happen. Under the cold, clear sky Chipman mourned its youthful dead.'

Fifteen pupils were killed outright in the collision, a sixteenth died in hospital later the same day and a seventeenth, Bernadette Hayduck, 17, of Chipman, died of head injuries the following day. Twenty-three other pupils were in hospital, nine seriously hurt. Only one escaped any injury. Bus driver Budley was also seriously hurt. All but two of those who died were girls.

'John Winnick, husky 16-year-old Grade 10 student, came through a school bus-train crash with only skimmed knees. He was the only one who did not require hospital treatment. Winnick, a six-footer with square shoulders, told his story before heading back to his farm home to help his father with the night's chores. "Sat in my usual place, three seats from the rear, and over the right wheels. I sat in the centre of the seat with a friend on each side. Donald Dzioba was on one side and Peter Lipkewich on the other. The kids were laughing and talking loudly but we were trying to figure a math problem. We had an exam to write. The windows were covered with frost and the windows make a lot of noise rattling. I couldn't see out. I don't remember the bus stopping and I can't remember the driver opening the door when he came to the crossing. I heard a loud bang. I must have been knocked out because the next thing I knew I was crawling out of the tail section of the bus. I wanted to find my notebook. I saw the train standing on the track. My cousin, Robert Antoniuk, was in the snow crying for help. He kept telling me his leg was sore and he was cold. I tried to lift him up but couldn't. I gave him my gloves and walked back 50 feet to the end of the bus. Others were crawling in the back section of the bus. There were a lot of bodies, mostly girls, around the rear section of the bus. I found two girls face down in the snow and turned them over. I helped those we could help. I asked where the other part of the bus was. A girl told me to look on the other side of the track. I ran over but saw nothing. Then I realized it must have been on the front of the train. There were still bodies in the snow when I got on the bus carrying the injured to hospital. Then I put my hands to my face and found I had lost my glasses. I looked for my notebook and glasses and found only my notebook." Winnick, who lives near Chipman, 10 miles south east of here, said he believes the blow of the crash was cushioned by his two friends. Both were injured.' [1485] He added, "I still don't believe it happened. Seems like I'll go back tomorrow and see them all just like I always did."'

The day after the accident the Manitoba government announced that it would begin testing all school bus drivers in the province to make sure that they were up to provincial driving standards. Manitoba Motor Vehicles Branch officials feared that many school bus drivers might be unable to meet the standards required of Winnipeg motorists. All 18 school bus drivers in the Melita area had failed a test the previous year, though they subsequently passed it after receiving special training.

Meanwhile the department of education was preparing to conduct safety tests on all vehicles used as school buses in the province. Farmers in some areas were reportedly hauling children in on feeder routes to the bus in the back of their pickup trucks, sometimes heated with stoves. [1486]

The inquest on the victims of the accident was held in Edmonton at the beginning of February 1961. The principal witness should have been the driver of the school bus, Frank Budley, but the court was told by Dr Morley Young, who had treated him, that he had no memory of the disaster. He had been critically injured, was unconscious for a time and doubtless was suffering from post-traumatic amnesia. [1487]

Fireman Morris Gonnet was the first person to give evidence on the second day of the inquest. He said that he had first seen the bus when the train was a mile from the crossing and had told his engineer, "Hang on to the whistle, Bill, there's a school bus coming." He had kept sight of the bus until it was within 50 feet of the tracks, when he had yelled to his engineer to apply the brakes in emergency. He added, "All the time that I was observing it, it did not stop." The fireman had been sat on the left side of the cab, the side from which the bus came. He said that the bus had been "moving at a reasonable speed and I thought it was going to stop. The windows of the bus were badly frosted over, but I didn't notice the windows of the door."

Sat behind the fireman on a high seat was Headend Trainman Herbert Duneer, who testified that he had first noticed the school bus as it turned off the highway. He said, "In my opinion it would stop, but it did not stop at all." He thought that the bus windows were all frosted up.

Engineer Donald William McIvor stated, "As we were coming into Lamont, I blew

---

[1485] *The Medicine Hat News*, Medicine Hat, Alberta, Canada, 30th November 1960
[1486] *Winnipeg Tribune*, Winnipeg, Manitoba, Canada, 30th November 1960
[1487] *The Medicine Hat News*, Medicine Hat, Alberta, Canada, 1st February 1961

the whistle one mile from Lamont, and again at the whistle marker a quarter mile from Lamont. While I was blowing the second whistle, the fireman said, 'Keep blowing, there's a school bus coming.' I kept blowing and about 150 feet from the crossing I saw the bus coming over the track. I put the train on emergency just as the fireman yelled, 'Goose her!' The train was moving at 50 miles an hour, 51 at the most, when I applied the brakes, and at least 45 when it struck the bus. If I had seen the bus at the quarter mile marker I couldn't have stopped and it might have been worse; the train might have struck the bus more in the centre and killed them all."

Driver Budney did then give evidence. He stated that he had driven the school bus in total for four months in three periods, including for a month before the accident. He admitted that he had never held the Class A licence required for school bus drivers in Alberta and that he had failed a written examination for a Class B licence in 1959. Gregory Moshanski, the owner of the school bus, said that Budney had told him before he had begun driving the school bus that he had a Class A licence. Budney said he could not remember being asked if he possessed such a licence. As to the accident itself, he could recall nothing between going to bed the day before and waking up in a hospital bed a month later. Asked if the sun had ever interfered with his vision at the crossing, he replied, "Yes, sometimes the glare from the tracks interfered with my vision, but it never prevented me from seeing if a train was coming. He said that he had opened the door of the bus whenever the windows were steamed up and that he always stopped 25 or 30 feet from the tracks and crossed the line in first gear. Before he had started driving the bus he had an interview with the school board and had been given a letter of recommendation signed by Marshall Moshanski, secretary of the Lamont school division. The letter was based on good recommendations given by two previous employees of the bus driver.

School bus inspector Kenneth Parson testified that the bus had been "mechanically in good condition" when he had examined it in April 1960.

The Coroner, Dr M.M. Cantor, pointed out that the whistle boards at the crossing should have been re-sited further away from it to compensate for the higher speed of diesel-hauled trains. [1488]

Full-time school bus driver Stanley Chernecki, who used the crossing twice daily, gave evidence that a bus had to be close to the crossing before there was a good view of a train. Because the road went up an incline to the crossing, "If we are too far back you don't see far enough. I have practised with two buses and if you stood too far back you could see to the east one quarter of a mile. If you pull further up – I won't say the exact distance – you can see half a mile. Closer still and you can see all the way." He added, "I accompanied Budney on my longest route, which has 40 stops and nine railway tracks to see how he would get through a route like that. I was completely satisfied with Budney and with the distance he stopped from railway tracks."

There was a difference of opinion as to whether the school bus had stopped before crossing on the day of the accident. Two witnesses, Fireman Gonnet and Trainman Dunheer, said that it had not, three that it had, and one was not sure. School caretaker Terry Lucas, who had been 400 yards from the crossing, said, "As far as I'm concerned the bus was stopped before it moved onto the tracks. I've never yet seen one come over the crossing without stopping." [1489]

[1488] Ibid, 2nd February 1961

[1489] Ibid, 3rd February 1961

At the close of the inquest the jury returned a verdict that the students had died from shock due to multiple injuries after bus driver Frank Budney had proceeded onto the railway crossing when it was not safe to do so. They decided that the train had approached the crossing at an unsafe speed and noted that the bus driver had not had a proper licence. They made five recommendations:

(a) That the front of diesel locomotives should be painted in the most easily visible colour.

(b) That locomotives should be fitted with flashing headlights and a distinctive auxiliary horn or whistle for use in emergency only.

(c) That the operators of school buses should be required to stop at all crossings.

(d) That while stopped the entry door should be opened to give an unobstructed view and allow any warning sound to be clearly audible.

(e) That the regulations covering school bus drivers should be rigidly enforced. [1490]

Driver Budney was subsequently charged with manslaughter but at the conclusion of his trial in October 1961 he was acquitted. At the trial Mr Justice J.V.H. Mulvain said that he was satisfied that the bus driver had stopped short of the crossing. The bus driver "may have exercised a degree of care that was less than it should have been in his failure to arise from the seat of his bus and look both ways up and down the track" which he stated was negligence of a type but could not be criminal negligence. He suggested that legislators should give serious thought to forming a set of clearly defined regulations to control the operation of school buses and that bus drivers should be subject to "proper methods of licensing with examination and teaching." [1491]

What was described as a small bus taking a group of young Mitsubishi Co. workers to an athletics meeting was struck by an electric passenger train on a level crossing in Yokohama, Japan, at noon on Friday 2nd December 1960. Eight bus passengers were killed and ten were injured. Two crossing keepers, who had allegedly failed to lower the crossing barriers prior to the arrival of the train, were arrested. [1492] [1493]

On the morning of Wednesday 7th December 1960 Mr Robert Wrenn, 29, made a delivery at the Koppers Co. wood preservation plant on Koppers Road, near Nashua, New Hampshire, United States. With him in the three-quarter-ton panel truck that he was driving were his wife Patricia, 28, and their children Christopher, 19 months, and Jennifer, one month. Leaving the plant Mr Wrenn continued southwards along Koppers Road, then turned right onto Hill's Ferry Road close to a level crossing with the Boston and Maine Railroad main line from Montreal, Quebec, Canada to Boston, Massachusetts. His truck reached the crossing at about 8.45 a.m. at the same moment as a Buddliner diesel multiple-unit passenger train, the 7.15 a.m. service from Lincoln and Concord to Boston. The truck was travelling at 15 or 25 miles an hour and the train at 70 miles an hour.

Engineer Rudolph Aubut stated afterwards that he had first seen the truck when his train was about 850 feet from the crossing. His view of it had then been blocked by a

---

[1490] Ibid, 4th February 1961

[1491] Ibid, 12th October 1961

[1492] *Pacific Stars and Stripes*, Tokyo, Japan, 3rd December 1960

[1493] *Lubbock Avalanche-Journal*, Lubbock, Texas, United States, 2nd December 1960

Koppers Co. switch engine and boxcar standing on a siding to the east of the main tracks. He had given long and short blasts on his whistle as per the regulations and the Koppers engine had also been whistling. The Wrenns' truck then reappeared and he saw it make a right turn onto Hill's Ferry Road and expected it to stop short of the crossing. When his train had been about 50 feet from the crossing he had seen that the truck was not stopping and applied his brakes in emergency but the collision had occurred almost at once.

Being carried on the Wrenn truck were about 10 bottles of liquified petroleum gas which exploded upon impact. The truck was almost destroyed and the train, which continued for about a third of a mile before it could be pulled up, was enveloped in flames, causing its passengers and crew to jump from it as soon as it stopped. A passenger who spoke to reporters, Mrs Daniel Macdonald, of Manchester, who had been travelling with her son, said, "Flames seemed to fill the entire train. My head aches and I am shocked." [1494]

The Wrenn family were all killed and their bodies were found by the side of the track. Walter Kobzlk, chief clerk at Koppers, related, "Bodies were strewn all over the track and roadway." Most of the 70 employees of Koppers raced to the scene. Some of them had thought at first that the explosion had been a sonic boom. [1495] There had been about 50 passengers on the train, of whom 28 were taken to St Joseph's and Memorial Hospitals in Nashua. One of the passengers, John Wilder, of Manchester, died of his injuries on the way. Conductor Vaughan Talbert suffered serious burns and became the sixth fatality in the accident when he died in St Joseph's Hospital just after 1 p.m. [1496]

The protection at the crossing consisted of advance warning discs consisting of the letters 'RR' and a diagonal cross plus standard crossbucks at the crossing itself, beneath which were signs reading 'Railroad Crossing' and '3 Tracks.' The line was used by 14 passenger trains and the odd freight train on weekdays, while the road was used by 'a considerable amount of traffic.' Mr Wrenn had been very familiar with the crossing, having used it perhaps 200 times in the previous five years.

Mrs Talbert subsequently sued the railroad company on the grounds that the protection provided at the crossing was inadequate. It emerged that there had been discussions between the Boston and Maine and Koppers regarding installing flashing warning lights at the crossing. Koppers had agreed to pay half the cost of the installation but the railroad company had refused to pay the other half and so nothing had been done. Mrs Talbert was awarded substantial damages. [1497]

There was a minor school bus accident at New Brockton, Alabama, United States, on Friday 9th December 1960. About 50 high school pupils were returning to Brundridge, 40 miles away, and had not gone far when the rear of their bus was hit by a slow-moving Atlantic Coast Line Railroad freight train. The pupils had been taking part with other Negro bandsmen in a homecoming parade in New Brockton. The impact was sufficient to overturn the bus. Thirty-five students of the Pike County Training School were treated for minor injuries, mostly cuts and bruises, at a hospital in nearby Enterprise. One girl

[1494] *Newport Daily News*, Newport, Rhode Island, United States, 7th December 1960
[1495] *The Lowell Sun*, Lowell, Massachusetts, United States, 7th December 1960
[1496] *Nashua Telegraph*, Nashua, New Hampshire, United States, 7th December 1960
[1497] https://law.justia.com/cases/federal/appellate-courts/F2/360/286/390870/

complained of a back injury but all were discharged after treatment. [1498]

A train collided with a bus on an unprotected level crossing at Maniwa, Japan, on the Kishin line of Japanese National Railways, on Monday 12th December 1960. Ten people died and about 50 were injured. [1499] [1500]

A Southern Railway freight train collided with the rear of a school bus on a level crossing near Gastonburg, Alabama, United States, on Monday 12th December 1960, injuring three children and two adults. The children, aged 11, 8, and 6, were taken to hospital, as was Mrs Sarah Thomas, 65, a teacher, who was the most seriously injured of the five.

The serious level crossing accidents at the end of 1960 continued with another in Italy on Wednesday 14th December. It happened in the south of the country, at a crossing between Noicàttaro and Triggiano. At 13:38 that day a Ferrovie del Sud Est company train collided with a small Fiat Belvedere car. The vehicle was dragged 500 metres and the bodies of its passengers were thrown out along the line. Of the six people in the vehicle five were killed outright and the sixth was critically hurt.

The car was being driven by Giacomo Suglia, 56, and was carrying his father-in-law, Michelangelo Boccazzi; his mother-in-law, Maria Boccazzi (née Cellamare), 66; his wife Rosa Suglia, 31, who was seven months pregnant; Rosa Sciannameo, 6; and her brother Pasquale, 11. The newspaper report did not state the relationship of the last two to the others but they were all said to have been from the same family. The boy Pasquale was the only survivor of the accident. He was rushed to the Onarmo Hospital in Triggiano where he was operated upon.

The party had been to Bari to collect the body of Oronozo Cellamare, Maria Boccazzi's brother, who had been a lay confrere of the Preziosissimo Sangue Catholic Church in Bari. Returning to Noicàttaro they had attended a Mass in the church there, followed by a funeral and the internment of the body in Noicàttaro Cemetery. They had been coming from the cemetery when the accident had occurred.

Rapido AD 63 111 from Bari to Putignano was travelling at 70 kilometres an hour when the collision occurred. Fortunately the train was not derailed and its passengers escaped with a few bruises and a considerable fright. As the accident blocked the line for many hours they were forced to disembark and make their way along the track to a point where they could be collected, presumably by a replacement road service. News of the accident quickly spread and among the first people to arrive were two Augustine fathers, who gave absolution to the victims, whose bodies lay among prickly pear trees and rows of vines.

Warning lights and bells protected the crossing. Driving the railcar had been Michel Dentamaro, 48, who was arrested (?) (*fermato* in Italian) after the accident, though it was expected that he would be released in the evening. The authorities opened an immediate investigation into the tragedy. It was clear that the crossing signals had been operating correctly. It transpired, however, that Michelangelo Boccazzi had held an all-night vigil over the body of his brother-in-law and it was thought probable that due to his tired state

---

[1498] *The Decatur Daily*, Decatur, Alabama, United States, 10th December 1960
[1499] http://www.ikjeld.com/japannews/00000147.php
[1500] Wikipedia – List of level crossing crashes

he had either suffered a microsleep or had at any rate been too slow to react when he had seen the crossing signals warning him of the train. [1501]

A station wagon ferrying nine children to school was in collision with a Great Northern Railway passenger train from St Paul, Minnesota, United States, to Winnipeg running on Canadian National Railways tracks on a level crossing at St Jean Baptiste, Manitoba, Canada, at 8.50 a.m. on Wednesday 14th December 1960. The station wagon ran into the second car of the train. The crossing was protected by flashing lights and a bell but there was an early-morning fog. In addition the warning lights were coated with frost and the station wagon was being driven into the glare of the rising sun, as at Lamont. Farmer Alphonse Beaudette, the driver of the vehicle, and seven-year-old Marcel Rojotte were slightly hurt and were taken to hospital. Marcel's sister Lucille was also very slightly hurt, the other children were unhurt. Mrs Beaudette explained that her husband usually drove the Rojotte children, Jeanette, Gilbert, Leo, Lucille, and Marcel, and another neighbour's daughter, Lucille Lallier, to school together with their own children, Diane, Leonne, and Raymond. [1502] [1503]

Five young men decided to go for a short drive during a coffee break from their work at Joyner's Silver and Electroplating Inc. works in the Brooklyn Park suburb of Minneapolis, Minnesota, United States, at about 9.30 p.m. on the evening of Monday 19th December 1960. They had only travelled a few yards when their car was struck by passenger train No. 7 of the Great Northern Railway, the *Winnipeg Limited*, en route from St Paul, Minnesota, to Winnipeg, Canada. The car was carried half a mile down the track and was described as having been 'crushed almost beyond recognition.' The five men were killed. The train was little damaged and no one on it was hurt.

Those who lost their lives were named as Roger Bronson, 20; Larry Frederick Danner, 19; Ambrose Hoppe, 22; Gary Frank Lambert, 19; and Paul Arthur Scharber, 18. Four of the men were from Anoka, Minnesota; Scharber was from Rogers, Minnesota.

Engineer Harold Pearo stated that his train had been travelling at 65 miles an hour. He said that the automobile had been moving "very fast." His engine had struck it broadside. There was a clear view along the track at the crossing, which was protected by crossbucks. It was theorised that the windscreen of the car might have been fogged or frosted over. There had been no recorded previous accidents at the crossing.

A police official described the five as "real sensible kids" and an official at the plant called them, "The kind of fellows you become fond of because they assume responsibility." [1504]

Six people died and 16 were injured in a level crossing accident on the evening of Christmas Day, Sunday 25th December 1960, involving a bus on the Nova Iguaçu – Praço Mauá route and freight train No. 3.303 of the Rede Ferroviária Federal near the Manguinhos Refinery in Rio de Janeiro, Brazil. The bus, registration number 10-77-85,

---

[1501] *La Stampa*, Torino, Italy, 15th December 1960
[1502] *Winnipeg Free Press*, Winnipeg, Manitoba, Canada, 14th December 1960
[1503] *Brandon Daily Sun*, Brandon, Manitoba, Canada, 15th December 1960
[1504] *Fergus Falls Daily Journal*, Fergus Falls, Minnesota, United States, 20th December 1960

was reportedly racing another bus working the Caxias – Praço Mauá route and disregarded the warning signals at the crossing. Struck broadside by the engine of the train, it was wrecked. Two passengers who had been sat at the point where the train struck the bus were killed outright and four more succumbed during the rescue operations. The accident was blamed on the bus driver, whose name was not immediately known, and who apparently escaped injury.

According to the statements of surviving bus passengers, their driver had been operating his vehicle dangerously since he left the Presidente Dutra highway, racing other buses. On the straight piece of road near the refinery the bus driver had been trying to overtake another bus as they approached the level crossing, the warning signals of which were already operating. Whistling continuously, the train had reached the crossing at the same time and hit the left side of the bus. The bus was pushed along for over 30 metres until it demolished a lamppost. A number of passengers had jumped from the bus before or as it was struck.

The clearest picture of what had happened was given by Crossing Keeper Jacob de Almeida, who told his story to a newspaper reporter. He related how he had seen the two buses racing towards the crossing at high speed, despite the fact that its warning lights were already flashing and the freight train was very close at hand. He went on, "I tried to stop the buses by waving my flag but they took no notice of me. The Caxias bus, whose registration I didn't get, managed to get across safely, and the Nova Iguaçu bus, which was just behind it, was struck by the train. I saw lots of people jumping from the second bus and ran to a telephone to summon help. The train driver was running under clear signals and could not pull up his heavy train in time when he saw the danger." [1505]

[1505] *Folha de S. Paulo*, São Paulo, Brazil, 27th December 1960

CHAPTER 13

RAIL/RAIL CROSSING COLLISIONS

At locations where railways and roads intersected the idea of having them cross on the level had obvious advantages in terms of economics over the cost of building an over or underbridge, especially if the level crossing were unmanned and the construction of a bridge would require the building of substantial approach embankments or possibly approach cuttings leading to an underbridge. With road users being limited in the early days to pedestrians, cyclists, and horse-drawn vehicles the dangers to a train from a collision on such a level crossing were also likely to be minimal. For railways crossing other railways on the level, though, the situation was very different. A collision with another train on a crossing was likely to be far more serious than a collision with, say, a horse and cart, and furthermore a train driver approaching a level crossing and seeing another train nearing or on it could not pull up in anything like the distance that a road user could. It is no surprise therefore that in most countries, where the railway system was either built directly by the state or under state supervision, that rail/rail crossings were not provided. It was only in countries where the railways were built by competing private companies without state interference that such crossings existed, most especially in the United States, where lines were often built as cheaply as possible and the economy of not needing to construct a bridge was doubtless an important consideration. Thus, the United States had many rail/rail level crossings, Great Britain at one time had about 20 or so and its colonies, such as Ireland and Canada, had a handful. There were also a number in Argentina, where the railways were constructed by British companies, such as Circunvalacion Crossing, Argentina, where the broad-gauge Buenos Aires Great Southern and the metre-gauge Province of Buenos Aires crossed, [1506] and one or two in other countries, such as at Istravoz (later Hilal), Turkey, where the Oriental Railway crossed the Smyrna – Cassaba Railway, and at Warthausen, Germany, where a major disaster had occurred in 1944. [1507] As has been pointed out, the American companies also seemed to have a perverse proclivity for building rail/rail crossings where they could easily have been avoided, perhaps because in the early days train speeds were low, trains infrequent and night operation unusual so that the dangers were not fully appreciated. [1508] By 1946 almost every one of the 15 or so remaining rail/rail crossings in Great Britain

[1506] Facebook – Signalboxes and Signalling page

[1507] Wikipedia – Hilal railway

[1508] *A History of Railroad Accidents, Safety Precautions and Operating Practices* by Robert B. Shaw, pg. 99

was fully signalled and there were hardly any further collisions on them. In the United States, on the other hand, where many crossings remained with little or no signal protection, collisions on them continued to occur regularly and about 50 of them occurred in the period covered by this volume.

On the last day of 1946, Tuesday 31st December, there was a collision at Ashland Junction, Wisconsin, United States, where the Chicago St Paul Minneapolis and Omaha Railway and the Northern Pacific Railway, both single track, intersected. The Omaha line ran southwest to northeast and the Northern Pacific line southeast to northwest. Trains of the Duluth South Shore and Atlantic Railway were operated by trackage rights (running powers) over the Northern Pacific line and it was a train of the Duluth company which was involved in the accident. The crossing was only protected by stop signs. There was very little traffic over it, an average of less than 6 trains daily on the Northern Pacific line and less than 4 daily on the Omaha line.

Eastbound second-class Duluth freight train No. 632 consisted of engine No. 1050, 25 cars and a caboose. It had left East End at 3 a.m., 2 hours and 45 minutes late, stopped about 80 feet short of its stop sign, then proceeded, and travelling at about 5 miles an hour had entered the crossing when it was struck by the Omaha train.

Westbound second-class Omaha freight train No. 75 consisted of engines Nos. 361 and 240, 35 cars and a caboose and had left Hayward, 53¾ miles from Ashland Junction, at 5 a.m., an hour and a half late. Passing the stop sign 107 feet from the crossing it had collided with the Duluth train while also moving at about 5 miles an hour.

All three engines were damaged in the collision, especially No. 240 in the Omaha train. The first two cars of the Duluth train and the first three cars of the Omaha train were derailed and damaged. The main victim of the accident was Engineer Ralph Carlson on engine No. 240, who was trapped in his cab and killed. [1509] The other railwaymen all jumped, though in doing so the engineer, fireman and front brakeman of the Duluth train were injured.

The engineer of the Duluth train told the Interstate Commerce Commission (ICC) inquiry into the accident that he had first seen the Omaha train when he had been about 300 feet from the crossing, but it appeared to be under control. About 100 feet from the crossing he had realised that the Omaha train was not stopping and applied his brakes in emergency but had not been able to pull up in time.

The Omaha train had been travelling at 35 miles an hour as it neared Ashland Junction. The engineer of engine No. 361 stated that a mile from the crossing he had made a 20-pound brake pipe application, followed by a 10-pound reduction 2,000 feet further on. He had believed that he had his train under control until 800 feet from the crossing it had become apparent that the speed of the train was still too high. He had then made an emergency brake application and his fireman warned him that there was a train approaching on the other line.

No problems were found with the brakes of the Omaha train.

The ICC noted that lineside vegetation and an overhead bridge restricted the views of the engineers of trains on both lines at the crossing. Nothing was said about the lack of

[1509] *The La Crosse Tribune*, La Crosse, Wisconsin, United States, 1st January 1947 any protection at the crossing other than stop signs. [1510]

A destructive collision at Higbee, Missouri, United States, at 10.50 a.m. on Tuesday 13th May 1947, also involved a passenger train, probably the *Mark Twain Zephyr*, despite which it does not appear to have formed the subject of an Interstate Commerce Commission inquiry.

The main line of the Chicago Burlington and Quincy Railroad main line was crossed at Highbee by a line of the Missouri-Kansas-Texas Railroad (MKT). The protection provided at the crossing, according to a newspaper report, consisted only of a colourlight signal at the crossing itself, though this seems rather doubtful.

The passenger train had left Kansas City at 8 a.m. and was due in St Louis at 1.25 pm. It was in charge of Conductor R.A. Markcell with Engineer Ed Donohew and Fireman George McConnell. The engineer told a reporter, "We were going from 20 to 25 miles an hour when I noticed the red light of the signal. I set my brakes and waited. There wasn't anything else I could do."

A northbound MKT freight train was on the crossing. 'The Zephyr hit the side of a box car near its rear end and ploughed about half way through as the box car moved on to the north. Then the locomotive wedged between the box car and a gondola filled with rock. The locomotive stopped a few feet past the track intersection. The left bottom part of the locomotive front was embedded in a bank of earth on the north side of the Zephyr's track. The front wheels of the locomotive were derailed, but the rest of the Zephyr remained on the tracks. Two freight cars were derailed and one was turned almost completely over. The wheels of the freight car were scooted from under it, and a signal at the crossing was smashed down to the ground.'

Fred Harper, who worked at Hicken and Sons Garage, near the crossing, said, "It sounded like thunder when the train hit. Orlin Hicken, Lestel Hicken, Roy Hicken and I ran over to the crossing right away and helped get the engineer out of the cab. Blood was flowing from the engineer's head when we got him out. The engineer was not unconscious but appeared to be dazed and shaken up." Engineer Donohew, who suffered cuts to the top and right side of his head, was the only casualty in the accident.

Passenger Frank Chrisman, of Loveland, Ohio, put the speed of the train at 60 miles an hour before it began to slow down for the curve approaching the crossing. He said, "The jolt was so slight that none of us knew that there had been an accident. When the brakes were applied I just thought we were stopping at a station."

The train had been running 14 minutes late at the time of the accident, which it was estimated would delay it about five hours. A wreck train and a replacement engine had to be sent from Mexico, Missouri. [1511]

The rail/rail crossing at Winston Tower, about a mile and a quarter south of the station at Litchfield, Illinois, United States, was the site of a collision on the morning of Wednesday 6th August 1947. Here the Chicago Burlington and Quincy Railroad (CBQ) and Illinois Central Railroad (IC) lines crossed on the level at an angle of 30 degrees and there was signal protection, with lower quadrant semaphores on the CBQ line and four-aspect

---

[1510] Interstate Commerce Commission report

[1511] *Moberly Monitor-Index*, Moberly, Missouri, United States, 13th May 1947 colourlight signals on the IC line. Audible and also in the case of the IC line visual annunciators announced the approach of trains to the tower operator, in the case of CBQ trains when they were a mile and a quarter south of the tower and in the case of IC trains when they were three miles south of the tower. The signals were interlocked with one another and a time release prevented the route being set for an IC train for two minutes after the route had been set a CBQ train and the latter train had entered the approach circuit.

The signals and their indications at the time of the accident were:

40A (CBQ line) (3,130 feet from the crossing) 60 degrees (*Proceed Clear – Distant)*

40 (CBQ line) (639 feet from the crossing) 60 degrees (*Proceed Clear – Home*)

D2392 (IC line) (10,507 feet from the crossing) Yellow (*Proceed prepared to stop at next signal. Trains exceeding medium speed must at once reduce to that speed – Approach*)

AL (IC line) (405 feet from the crossing) Red over Red (*Stop – Stop*)

On the morning of the accident northbound CBQ freight train No. Extra 6106 North consisted of engine No. 6106, 70 cars and a caboose, in charge of Conductor C.J. Wright with Engineer Paul Brown. Having passed signals 40A and 40 at *Proceed* it was moving over the crossing at 15 miles an hour when its first car was struck by the IC train.

The IC train was northbound first-class passenger train No. 22, otherwise, the *Green Diamond.* It consisted of diesel multiple units Nos. 4004 and 4003, coupled in multiple-unit control, a mail-express car, four coaches, a café-lounge car, and a parlor-observation car. It had left Mount Olive, the last open office, just under seven miles south of the crossing, at 10.08 a.m., passed signal D2392 at yellow, passed signal AL at red, and was still moving at 65 miles an hour when it was derailed on a derail 350 feet south of the crossing. It continued in line with the track and was still moving at 30 miles an hour when it hit the freight train. The engine and first eight cars of the freight train and the two diesel units and first five cars of the passenger train were derailed and damaged. Twenty-nine passengers, eight dining car employees, two coach attendants, a mail clerk, the fireman and a baggageman, all on the passenger train, were injured.

Engineer Bert Christian of the IC train told the Interstate Commerce Commission (ICC) inquiry that about five and half miles from the crossing a warning signal had sounded in the cab of his engine indicating a problem in the engine compartment. His fireman had gone back to investigate. He (the engineer) had been distracted and failed to see signal D2392. A mile and three-quarters south of the crossing he had made a light brake application, reducing the speed, he thought, to 75 miles an hour. Some distance south of signal AL he had seen that it was at *Stop* and had then applied his brakes in emergency.

The ICC remarked that in the previous six years it had investigated six accidents where the fireman had not been in the control compartment of the engine and these had resulted in 22 deaths and 122 injuries. The IC rules required train crews to communicate signal indications to one another when practicable but at the same time the fireman was required to check the performance of his engines. It made no recommendations [1512]

A newspaper report stated that by chance the IC train had struck the only "soft car" in the CBQ train – a box car loaded with pulp wood. [1513]

[1512] Interstate Commerce Commission report

[1513] *The Illinoian-Star*, Beardstown, Illinois, United States, 8th August 1947

Chicago Great Western Railway (CGW) and Illinois Central Railroad (IC) freight trains collided at the crossing of the two lines about a mile from Waverly, Iowa, United States, early on the afternoon of Thursday 26th May 1949. One railwayman lost his life in the accident and two were injured. Both lines were single track and neither was worked in the block system. The lines crossed at an angle of 36 degrees, the CGW line running east and west and the IC line running southeast and northwest. An automatic interlocking was provided at the crossing.

The eastbound signals and their indications at the time of the accident were:

Fixed Approach (CGW line) (3,041 feet from crossing) Horizontal (*Approach Home Signal with Caution*)

2 (CGW line) (518 feet from crossing) Diagonal over Horizontal (*Proceed*)

Fixed Approach (IC line) (2,992 feet from crossing) Diagonal (*Proceed preparing to stop at next signal*)

4 (IC line) (490 feet from crossing) Horizontal over Horizontal (*Stop*)

Each stop signal was provided with an approach circuit and both normally indicated stop. The first train to enter its approach circuit automatically cleared its stop signal and locked the opposing stop signal until the train had passed through the interlocking or a predetermined time interval had elapsed. The operating rules of both companies limited the speed of their trains to 20 miles an hour over the crossing.

Eastbound CGW Freight train No. 82 consisted of diesel-electric unit No. 111A, 43 cars and a caboose. It had passed Allision, the last open office, 18¾ miles from the crossing, at 12.18 p.m., 2 hours and 36 minutes late, passed the Waverly approach signal, passed signal No. 2, indicating *Proceed*, and was going over the crossing at 33 miles an hour when its 29th car was struck by the IC train. [1514]

Eastbound IC freight train No. 1586 East consisted of engine No. 1586, 35 cars and a caboose. It had passed Plainfield, the last open office, 7¾ miles from the crossing, at 12.51 p.m., passed the Waverly approach signal, passed signal No. 4, indicating *Stop*, and moving at 5 miles an hour struck the CGW train. As a result of the collision the engine of the IC train was derailed and damaged and the 29th to 38th cars of the CGW train were derailed, many of them being badly damaged. Engineer E.C. Bradfield was killed and Fireman Arthur Drew, Conductor B.T. Crummer and Front Brakeman A.M. Walsh, were injured. All had been on the IC train. [1515]

The enginemen of the CGW train told the Interstate Commerce Commission (ICC) that they had observed signal No. 2 change to *Proceed* and the speed of their train had then been increased. When their engine had been some distance past the crossing the fireman and front brakeman had seen that the IC train had passed signal No. 4. The speed of the CGW train had then been further increased but the train could not clear the crossing in time.

Fireman Drew of the IC stated that the first brake application had been made 200 feet short of the approach signal when the train had been travelling at about 30 miles an hour. A second brake application was made 400 feet beyond the signal, when speed was down to 20 miles an hour, and the throttle was placed in drifting position. Because of the curvature of the line he was the first to observe the *Stop* signal and he had called its

[1514] Interstate Commerce Commission report

[1515] *The Oelwein Daily Register*, Oelwein, Iowa, United States, 26th May 1949 indication to the engineer and front brakeman. The speed of the train was then 15 miles an hour and it had fallen to 7 miles an hour 500 feet short of signal No. 4, where the engineer had made an emergency brake application. About 100 feet from the signal the engineer had reversed the engine and he (the fireman) had jumped.

Front Brakeman Welsh put the speed of the train at 30 miles an hour, said that the first brake application had been 900 feet short of the approach signal and this had reduced the speed to 20 miles an hour. He thought the fireman had shouted a warning 500 feet beyond the approach signal and that the engineer had applied the brakes in emergency about 1,200 feet short of signal No. 4.

Conductor Crummer put the speed of the train at 25 miles an hour before the brakes were applied 200 feet from the approach signal. Speed had been reduced, he thought, to 20 miles an hour at the approach signal and there had then been another brake application. He thought the emergency brake application had been made when his caboose was 1,200 feet beyond the approach signal and the engine would have been within 200 feet of the crossing. The speed at that moment, he said, was 18 miles an hour.

Flagman J.C. Hecklsmiller in the caboose put the speed at 30 miles an hour at the approach signal, with a brake application 300 feet from it. He put the speed at 15 miles an hour when the brakes were applied in emergency.

The ICC noted that the IC braking charts indicated that the stopping distances for freight trains travelling at 25, 30, 35 and 40 miles an hour were 2,500, 3,300, 4,200 and 6,400 feet respectively. The IC signals protecting the crossing were 2,502 feet apart {the ICC states 2,492 feet which does not agree with the given distances}. Estimates of the train's speed varied between different witnesses but it seemed likely that it had been more than 20 miles an hour passing the approach signal. The interlocking was so arranged that an IC train travelling at the maximum authorised speed of 40 miles an hour would be about 2,300 feet from signal No. 4 before that signal indicated *Proceed*, provided no CGW train was in the vicinity. A train travelling at the maximum authorised speed would therefore not be able to pull up short of the crossing.

After the accident the IC reduced the maximum authorised speed at the approach signal to 20 miles an hour but the ICC felt that this was still not sufficient and that 'the carrier should immediately make any further changes necessary to provide adequate stopping distance for movements through the interlocking.' [1516]

A minor rail/rail crossing accident occurred in England on Friday 26th August 1949 on the Burnham branch of the former Somerset and Dorset Joint Railway, now part of British Railways, Southern Region. The trains involved were the 8.05 a.m. mixed train from Glastonbury to either Highbridge or Burnham-on-Sea and one of the Eclipse Peat Company, the latter consisting a light petrol driven locomotive pulling two peat trucks. The 2ft-gauge line of the peat company connected its factory to some peat beds via a crossing of the standard-gauge line near Ashcott station. On the day of the accident the engine of the peat train stalled on the crossing. After unsuccessful attempts to restart his engine or uncouple the engine from the trucks the peat company driver ran along the line to warn the passenger train, which he could hear approaching, but the morning was foggy and he was not seen. The crew of the mixed train leapt to safety and their engine, 0-6-0

[1516] Interstate Commerce Commission report

Class 3F No. 3260, was derailed and fell into the water and mud of the adjacent South Drain, which at this point was 10 feet wide and some feet deep. It took eight days to recover the engine after the drain had been diverted around it and even then it could not be recovered whole but had to be cut up in situ, its parts destined for the scrap yard. [1517]

Early on Friday 21st October 1949 Atlantic Coast Line Railroad (ACL) and Seaboard Air Line Railroad (SAL) freight trains, both northbound and both loaded with phosphate ore, collided on a crossing at Owensboro, Florida, United States. The crossing was protected by signals operated from an adjoining tower. Operator W.C. Crosby said that the engine and leading cars of the ACL train had cleared the crossing when the SAL line train had overrun the signal protecting the crossing and gone into the side of the ACL train. Six or eight cars were piled up beneath an overbridge carrying Highway No. 301, which was damaged, other cars were derailed, and the tracks were ripped up. The operator said that he had made a hasty exit from his post when he had seen that a collision was about to occur. The crew of the SAL engine jumped for their lives while a brakeman who was trapped in the wreckage escaped with a few scratches. [1518]

In 1950 the Minneapolis St Paul and Sault Ste Marie Railroad (Soo) line from Harvey to Enderlin was crossed by the Northern Pacific Railway branch from Casselton to Marion at Lucca, North Dakota, United States. The Soo line ran from southeast to northwest and the Northern Pacific line from northeast to southwest; they crossed one another at an angle of 70 degrees. There was a manual block system for following trains on the Soo line and no block system on the Marion branch. The crossing was protected only by stop signs, 200 feet from the crossing for westbound Northern Pacific trains and 400 feet from the crossing for eastbound Soo trains. Trains on both lines were required to approach the crossing prepared to stop unless the crossing was clear.

On Wednesday 16th August 1950 westbound Northern Pacific second-class mixed train No. 137 consisted of diesel-electric unit No. 803, eight cars, two cabooses and an all-steel combination mail-baggage-passenger car. It had left Alice, the last open office, 8¾ miles from the crossing, at 9.58 a.m., 26 minutes late, and stopped 50 feet short of the stop sign at Lucca. It then proceeded and was passing over the crossing at 7 miles an hour when its rear car was struck by the Soo train.

Eastbound Soo second-class freight train No. 90 consisted that day of steam engine No. 1027, diesel electric units Nos. 210A and 210B, coupled in multiple unit control, 103 cars and a caboose. It had left Rogers, 35½ miles from Lucca, at 9.22 a.m., 2 hours and 15 minutes late, passed Fingal, 5¼ miles from Lucca, at 10.21 a.m., 2 hours and 11 minutes late, passed the stop sign at Lucca and was still moving at 28 miles an hour when it struck the Northern Pacific train. The three engines and nine cars of the Soo train were derailed and wrecked, eight of the cars being destroyed, while the rear car of the Northern Pacific train was considerably damaged and passenger Gunnar Larson, of Superior, Wisconsin, riding in it, was killed. Engineers Marvin Fredericks and Ed Hawkins on the first and second engines of the Soo train, their firemen, Kenneth Wallner and Lloyd Carlson, and Conductor Roy Carr, Mail Clerk Donald Ness, and passenger Mrs Rosie

[1517] *The Somerset and Dorset Railway* by Robin Athill, pgs. 129-30

[1518] *Panama City News-Herald*, Panama City, Florida, United States, 21st October 1949

Percel, all on the Northern Pacific train, were all injured or suffered from shock. [1519]

The trainmen of the Northern Pacific train told the Interstate Commerce Commission inquiry that as they had approached the crossing they had seen the Soo train about half a mile away. After their engine had passed over the crossing their view had been obscured by the Lucca station buildings. When the engine had passed the station buildings they had heard a series of short whistles from the engine of the Soo train, which had then been 1,500 feet from the crossing, and had seen Lucca Station Agent F.A. Fish giving them "high ball" signals to move quickly ahead.

The inquiry was told that the Soo train had left Harvey at 4.45 a.m., had stopped at Carrington, Bordulac, Kensal, Wimbledon, where five cars had been detached, and Rogers, where helper (assisting) engine No. 1027 had been attached. After leaving Rogers the brakes had been effective in reducing the speed of the train for a restriction at Valley City. On the falling gradient about a mile from Lucca the speed of the train had been 51 miles an hour. Engineer Fredericks on engine No. 1027, who had been in control of the brakes, said that a 10-pound brake application had been made at this point, followed by a second similar reduction 4,000 feet from the crossing. About 2,500 feet from the crossing he had become apprehensive about stopping and made an emergency brake application, to no effect. At a point 1,500 feet from the crossing he had seen the Northern Pacific train and had begun whistling.

After the accident it was found that the independent brake valve handle of engine No. 1027 was being held in the full release position by a wire loop around the brake valve handle, preventing the engine brakes from being applied. In addition the brakes of one car of the train were cut out, two were inoperative, and the piston travel on two cars and the caboose was excessive. The speed recorder on the first diesel unit showed that the speed of the train had dropped from 51 miles an hour a mile from the crossing to 28 miles an hour at the point of collision.

The ICC concluded that the accident had resulted from failure to control the speed of the Soo train approaching the crossing. [1520]

Rail/rail crossing collisions continued to occur regularly in the United States during this period; there were no less than 23 of them between 1950 and 1958, though after the aforementioned one at Ashcott in 1949 there were none anywhere else in the world, at least before 1961, to the author's knowledge. The next two accidents, at Galatea, Ohio, on 8th October 1950 and at Penford, Michigan, on 15th March 1951 only resulted in injuries and damage but one, at Walkerton, Indiana, on Thursday 7th February 1952, was rather more serious.

At Walkerton the double track of the Baltimore and Ohio Railroad intersected with the single track of the New York Central Railroad, the crossing being protected by fixed signals controlled from an adjoining tower (the New York Chicago and St Louis Railroad also crossed the Baltimore and Ohio on the level at this point but this crossing was not involved in the accident). The collision occurred between New York Central freight train Extra 2800 West, consisting of engine No. 2800, 65 cars and a caboose; and Baltimore and Ohio track motor car No. 1143 propelling two trailers. The track motor car and trailers, in charge of an extra maintenance gang, had been engaged in carrying coal from

[1519] *The Winona Republican-Herald*, Winona, Minnesota, United States, 17th August 1950
[1520] Interstate Commerce Commission report

a car in the freight yard at Walkerton to the tower and station, both of which were on the opposite side of the crossing with the New York Central line to the yard. The section foreman had arranged over the telephone for the Walkerton operator to provide protection against east and westbound Baltimore and Ohio trains for movements to and from the yard but had not arranged for protection against New York Central trains at the crossing. The New York Central train approached the crossing travelling at 39 miles an hour under clear signals, with its crew unable to see the track motor car until the last moment and having no knowledge of it. A view of the New York Central train was likewise obscured from the operator of the track motor car by the tower and a car on a siding and by the time he was able to see it, it was too late to stop. [1521] The track motor car was demolished in the collision and employees Sam Blair and Samuel Brennam were killed, while Joe Jankowiac and Marion Scarbaugh were injured, the former critically so. A fifth, unnamed employee, jumped clear just in time. [1522]

The Interstate Commerce Commission (ICC) noted that the operator (driver) of the track motor car had had no knowledge of the arrangement made by the section foreman with the operator (signalman), had not been examined on the rules regarding the operation of track motor cars and had not been qualified to operate them. It concluded that the accident had been due to the failure of the Baltimore and Ohio company to provide adequate protection for the movement of track motor cars. [1523]

Two passenger trains were involved in a side collision at Carrington, North Dakota, United States, on Christmas Eve, Wednesday 24th December 1952. The companies involved were the Northern Pacific Railway and the Minneapolis St Paul and Sault Ste Marie Railroad (Soo Line). The crossing of the two lines was protected only by stop signs. The operating rules of both companies stated that all trains must come to a complete stand at the crossing and only proceed if the line was clear. Less than four Northern Pacific trains and less than eight Soo Line trains used the crossing daily. [1524]

Westbound Northern Pacific first-class passenger train No. 157, the Jamestown – Carrington local, consisted of 4-6-2 steam locomotive No. 2154, a mail express car and a coach, with Engineer Vernon Campbell and Fireman Tom Olson. It had left Carrington, just under a mile east of the crossing and the last open office, at 1.09 p.m., 2 hours and 42 minutes late, had stopped at the crossing, then had proceeded at 7 miles an hour and the collision had occurred at 1.13 p.m.

Eastbound Soo Line first-class passenger train No. 12, the Minot – Enderline local, consisted of 4-6-2 steam locomotive No. 706, a baggage car, a mail express car and a coach, with Engineer Steve George and Fireman Ralph Mhyre. [1525] It had left Cathay, 14¼ miles west of the crossing and the last open office, at 12.55 p.m., had stopped at the stop sign of the crossing, then had proceeded at 12 miles an hour into collision with the other train.

The trains met engine to engine and both locomotives were derailed. Engineer George was killed and Engineer Campbell, both firemen and both conductors were injured.

[1521] Ibid.

[1522] *Rushville Republican*, Rushville, Indiana, United States, 8th February 1952

[1523] Interstate Commerce Commission report

[1524] Ibid.

[1525] *The Bismark Tribune*, Bismark, North Dakota, United States, 26th December 1952

Engineer Campbell told the Interstate Commerce Commission (ICC) inquiry that after he had stopped at the stop signal his fireman told him that the Soo Line train was approaching but was slowing down. He had then proceeded until his fireman had shouted a warning, when he had applied the brakes in emergency and the trains met one another.

Fireman Mhyre stated that he had told his engineer that the Soo Line train had stopped, glanced at the steam gauge and water glass, and had then seen that the Soo Line train was moving forward again.

Fireman Olson said that he had told his engineer that there was nothing coming from his side. When his engineer had restarted after stopping he had assumed that the engineer had seen that it was also clear on his side.

The ICC concluded that the accident had resulted from the failure to properly control the speed of both trains at the crossing. The accident was therefore unusual inasmuch as both rather than just one train were at fault.[1526]

An accident similar to that at Walkerton, Indiana, in 1952, occurred at Eureka, Texas, United States, on Friday 20th November 1953. The single lines of the Missouri Kansas Texas Railroad of Texas (MKTT) and the Texas and New Orleans Railroad crossed one another here. As at Walkerton the crossing was protected by fixed signals worked from an adjoining tower and a track motor car was involved, its wheels insulated as at Walkerton so that it did not operate track circuits and place automatic signals at danger.

The eastbound Texas and New Orleans train was a switching (shunting) movement consisting of diesel electric unit No. 30 and eight cars, with Engineer Frank Ebelt. It was proceeding over the crossing under clear signals at 12 miles an hour when it hit the track motor car.

The northbound MKTT train consisted of track motor car No. 100163 and a trailer loaded with bridge timbers and other material, accompanied by a maintenance of way foreman and six workmen. [1527]

As a result of the collision Maintenance of Way Foreman Joe Shubert was killed outright and workmen Joe Hall, L.J. Moon and A. Wright were injured, one of them, who had been driving the motor car, fatally. The other three workmen, June Ewing, Albert Kasper, and W.F. Zimmerman had had time to jump to safety and were unhurt. [1528]

The surviving workmen told the Interstate Commerce Commission (ICC) inquiry that they had been on their way to a bridge three miles north of Eureka. Before they had started the foreman obtained a line-up of train movements from the Eureka operator. They had then entered the main track from a siding. Usually the foreman told two of them to look out for trains each way at the crossing but he had not done so on this occasion. On the main track their speed had increased to 10 miles an hour, then had been reduced to 4 miles an hour for Kansas Street level crossing and it had been maintained at that level until they reached the Texas and New Orleans crossing. None of the men had been aware of the switching movement before they reached the crossing, then several of them had shouted warnings and all had attempted to jump off.

Engineer Ebelt stated that he had sounded his horn approaching the crossing. Neither

---

[1526] Interstate Commerce Commission report

[1527] Ibid.

[1528] *Lubbock Avalanche Journal*, Lubbock, Texas, United States, 22nd November 1953 he nor the conductor had seen the track motor car until the last moment. The engineer told a newspaper reporter, "I didn't see the motor car until it shot out in front of me from behind the tower."

The Eureka operator stated that it was not unusual for motor cars to be operated over the crossing without asking his permission, though according to the rules they were supposed to do so. One employee usually walked on ahead of the motor car to check that it was safe to cross.

The ICC concluded that the accident had been caused by failure to operate a track motor car in accordance with the rules governing movements over railroad crossings.[1529]

The most serious rail/rail crossing accident to occur during the period covered by this volume occurred on Saturday 29th June 1957 at Guthrie, Kentucky, United States. Three passengers, two dining car employees and an engineer lost their lives and 12 passengers, eight dining car employees and a railwayman were injured. Unusually only one company, the Louisville and Nashville Railroad, was involved, the lines being part of the Evansville and Louisville Divisions of the company. The rail/rail crossing dated from the openings of the different railroads as independent companies which were later taken over by the Louisville and Nashville.

Movements over the crossing were governed by interlocking signals controlled from Guthrie station. Southbound semi-automatic searchlight signals 2171 and 3R on the Evansville Division were 1¼ miles and 162 feet north of the crossing, both were showing green over red, *Proceed*, indications at the time of the collision. On the Louisville Division southbound searchlight signals 1621 and 1L were respectively 1¼ miles and 200 feet north of the crossing. At the time of the collision signal 1621 was displaying a yellow, *Approach*, aspect, meaning 'prepare to stop at next signal', and signal 1L was displaying a red over red, *Stop*, indication.

Southbound Evansville Division first-class passenger train No. 11 from Chicago, Illinois, to Miami, Florida, *The Dixieland*, consisted of diesel-electric units Nos. 604 and 612 in multiple control, a dormitory baggage car, two sleeping cars, a dining car, a tavern car, a sleeping car, and six coaches. It had left Evansville, Indiana, 109½ miles from Guthrie, at 2.18 p.m., three minutes late, passed Henderson, Kentucky at 3.8 p.m., five minutes late, passed signals 2171 and 3R at *Proceed* and was moving over the crossing at 35 miles an hour when its dining car was struck by train No. 121 at 4.45 p.m.

Southbound Louisville Division second-class freight train No. 121 consisted of diesel-electric units Nos. 806, 703 and 812, in multiple-unit control, 30 cars and a caboose, with Engineer J.T. Raspberry (who was killed), Fireman J.T. Smith and Conductor A. M. Lashlee. It had left Strawberry, Kentucky, 158 miles from Guthrie, at 12.43 p.m., three hours and 43 minutes late, passed Russellville, 20¼ miles from Guthrie and the last open office, at 4.23 p.m., passed signal 1621 at yellow, passed signal 1L at red over red and was still moving at 6 or 8 miles an hour when it ran into the passenger train. The dining car of the passenger train was destroyed, the third and fifth cars were badly damaged, the first, second and sixth cars were damaged and the first seven cars were wholly or partly derailed. The first two diesel units of the freight train were derailed.[1530]

[1529] Interstate Commerce Commission report

[1530] Ibid.

Ervin Brownlee, a machinist of Chicago, who had been travelling as a passenger to Jacksonville, Florida, for a vacation, spoke to a newspaper reporter. He said, "I had just gone in the diner to eat. I sat down, looked up, and saw this train coming. It never stopped. It hit the side of the diner I was on and threw me out in the middle of the floor. I was only five or six feet from where the train hit. My nose was banged up and I lost my glasses. The crash was terrific. I saw steam coming out of the car. I thought the car was on fire so I jumped out the big hole that was torn in the side. I held this woman in front of me and helped her out of the hole. I think she probably had a fractured skull. She was bleeding from the side of the head but still walking."

Dining Car Steward Joe Stibuerk told the reporter, "I had just left the dining car and gone back to get some people. I got them – a woman and child who had asked me to call them. I looked out. Suddenly I saw this train coming. I said, 'Jump into the rooms.' They stood there dumfounded, but I pushed them in. Then we hit, and it felt like the whole world was shaking loose. I came out and saw the way was blocked. My first thought was of the diner where I worked. I went back and came around. Chairs were thrown everywhere. The car was ripped open." [1531]

Railwayman C.T. Swallows of Guthrie, possibly the operator there though not identified as such, said, "Everything is a mess up here and the sheriff's having trouble keeping the crowd away so the ambulances can get to the injured. I imagine both trains were going at a pretty good clip. The Dixieland doesn't even stop at Guthrie."

In the cab of the leading unit of the freight train with the engineer and fireman was the unidentified front brakeman, who said, "I was on the walkie-talkie, talking to A.M. Lashlee, of Paris, the conductor. Smith was beside me and Raspberry was at the wheel. I looked up. I saw we couldn't make it. I {jumped and} landed on both feet, stumbled, but kept up. Smith was right behind me. Smith said, 'I'm going to jump. Jump quick!' Raspberry didn't have time." [1532]

The engineer of the passenger train told the Interstate Commerce Commission (ICC) inquiry that as he neared the crossing he had seen the Guthrie station staff moving away from the platform, indicating that something was wrong. He had applied the brakes in emergency and the collision had occurred. His fireman said that he saw the freight train after it had passed signal 1L.

Fireman Smith stated that he had been promoted to engineer in February 1951 and had been operating the locomotive. There had been no problem with the brakes. Signal 1621 had been passed at 55 miles an hour. It had displayed *Prepare to stop at next signal* and he had applied the brakes, reducing the speed to 25 miles an hour, then released them. His train had had work to do at Guthrie, and it was the practice to stop it north of a highway crossing 1,563 feet north of the rail/rail crossing, or to allow it to come forward as far as signal 1L if the train was short enough to be able to stop at that signal with its rear clear of the level crossing. Conductor Lashlee had instructed him via the head brakeman to stop at signal IL. Fireman Smith stated that 2,000 feet short of signal IL he had seen it showing *Stop* and made an 8-pound brake application, but the brakes had not appeared to function properly. About 1,200 feet from the crossing he had made an emergency brake application, closed the throttle, and applied sand. His engineer had shouted a warning and he (the fireman) and the front brakeman had alighted from the engine.

[1531] *Florence Morning News*, Florence, South Carolina, United States, 30th June 1957

[1532] *Sunday Bulletin*, Racine, Wisconsin, United States, 30th June 1957

Conductor Lashlee though that the brakes had been applied in emergency 500 feet from the crossing, the head brakeman thought 600 feet and the flagman thought 800 feet. The ICC concluded that the train had at any rate, been closer to the crossing than the fireman had stated when the emergency brake application had been made. [1533]

Post-accident investigation disclosed that the brakes of the ninth car had been cut out and the brake cylinder travel of seven cars and the caboose was excessive, so that the brakes of only three-quarters of the cars of the freight train had been in proper condition.

The ICC concluded simply that the accident had been caused by failure to operate the Louisville Division train in accordance with signal indications.

The last rail/rail crossing accident before the end of 1960 was a fairly minor one at Rushville, Indiana, United States, on Sunday 12th January 1958. Here the New York Chicago and St Louis Railroad (Nickel Plate) and the Pennsylvania Railroad intersected on the level. There was another level crossing between the Nickel Plate and the Baltimore and Ohio Railroad a little further north, while New York Central Railroad trains operated over the Nickel Plate crossing and it was a train of the New York Central that was involved in the accident. The Interstate Commerce Commission (ICC) report states that the crossing was protected by fixed signals, which was true, but these were only stop boards, such as might be found at an ordinary road junction. All trains were, however, required to stop before entering the crossing. Less than two trains of either company used the crossing daily.

The Pennsylvania Railroad train involved was northbound freight train Extra No. 9142 North, consisting of diesel-electric unit No. 9142, a caboose and eight cars, the caboose being immediately behind the engine. It had left Columbus at 5.20 p.m., stopped short of the crossing, then moved forward onto it and come to a stand with its fourth car over the crossing at about 11.43 p.m. It was still stopped over the crossing at about 11.52 p.m. when it was hit by the New York Central train.

The New York Central train was northbound second-class freight train No. 78, consisting of diesel multiple units Nos. 1811 and 1810 coupled in multiple, 62 cars and a caboose. It had left Jeffersonville at 7 p.m. It had left Greensburg, 19 miles south of Rushville and the last open office, at 11.05 p.m., an hour and 5 minutes late and was still moving at 18 miles an hour when it had reached the crossing and struck the Pennsylvania train. The fourth car of the Pennsylvania train was destroyed, the fifth car was considerably damaged and the fourth, fifth and sixth cars were derailed, as was the leading unit of the New York Central train. The engineer, fireman and brakeman of that train were injured.

The crew of the Pennsylvania train told the ICC inquiry that after their train had stopped on the crossing the conductor had gone to the station for information regarding switching operations. After he had returned but before these could begin to switch they had seen the New York Central train coming and there had been no time to move their train off the crossing.

The engine crew of the New York Central train had understood that they were to stop short of the crossing. The fireman, who was not a qualified engineer, had been operating the train. He stated that a mile from the crossing he had made a service brake application,

1533 Interstate Commerce Commission report

reducing the speed to 32 miles an hour. The fireman maintained that he had then kept the brakes applied but the tachograph of the engine indicated that its speed had in fact then increased to 34 miles an hour. The engineer stated that 1,100 feet from the crossing he had become concerned at the speed of his train and instructed the fireman to apply the brakes in emergency. The latter had done so but the train had not been pulled up short of the crossing. The engineer and the front brakeman had jumped from the engine just before the collision. The brakes of the New York Central train were found to be in order after the collision.

The ICC concluded that the accident had been caused by failure properly to control the speed of the New York Central train. [1534]

[1534] Ibid.

CHAPTER 14

FIRES

Fires often broke out on trains following derailments or collisions. While wooden coaches were still in use in passenger trains an overturned stove used for heating or spilt engine coals from a steam locomotive might set them ablaze. Later on the substitution of gas for oil for train lighting proved to be another hazard, and a number of accidents were made much more severe by fires started by leaking or damaged gas cylinders until its dangers led to its replacement by electricity. This chapter will confine itself, however, to what might be termed primary fires, that is, accidents where there was no precipitating derailment or collision but merely a fire. Serious primary fires were considerably rarer than what might be termed secondary fires, but they did occur from time to time from a variety of causes, while more minor fires were fairly common.

For the fire in Ikomayama Tunnel, Japan, 16th April 1947, see Tunnel Accidents chapter above.

The leading car of a multiple-unit passenger train caught fire at Gravanches, France, a marshalling yard north of Clermont-Ferrand and between that station and the station at Gerzat at 13:30 on Tuesday 3rd June 1947. The train was full and the fire quickly became an inferno. An attempt by the passengers to flee to evacuate the coach was held up for precious seconds because a man with a large suitcase accidentally blocked the exit. Twenty-two passengers were taken to Clermont-Ferrand Hospital suffering from burns and examination of the remains of the damaged coach revealed that it contained five charred bodies, one of them a child. By the following day a sixth passenger had died of his or her injuries. [1535] [1536]

A can of petrol was placed on the luggage rack of a passenger train in Germany on Saturday 23rd August 1947. It was leaking and the petrol fell onto a package of films. The discarded cigarette end of a smoker then set the package alight and a fierce fire rapidly developed. A ticket collector stopped the train between Velten and Kremmen and the panicking passengers made a rush for the door at the end of the coach, their only means

[1535] *The Times*, London, England, 5th June 1947
[1536] *Gazette de Lausanne*, Lausanne, Switzerland, 4th June 1947

of escape. In the panic a returning prisoner of war fell, other passengers stumbled over him and the doorway became blocked. Twenty-four passengers died and 35 others were injured, many seriously. [1537] One coach was burnt out and another was damaged. [1538]

As a Baltimore and Ohio Railroad mail train of 10 cars from Toledo to Cincinnati was passing through Trenton, Ohio, United States, early on Tuesday 11th November 1947, a crew member noticed smoke coming from the penultimate vehicle. The train was stopped momentarily at New Miami for the alarm to be raised and then proceeded the short distance to Hamilton, where it was brought to a stand at 8.40 a.m. at the level crossing with Fourth and Vine Streets for the attention of Company No. 2 of the Hamilton Fire Department. Flames burst out as soon as the door of the affected car was opened but after a while the firemen were able to get the blaze under control. Firemen and members of the train crew then threw out the mail bags onto the road, from where they were loaded onto a lorry and taken to the railway depot to be loaded into another mail car. The train was delayed an hour. Only one car was involved. The fire was apparently not completely put out because roof of the car was reportedly still ablaze when the train arrived at the Union Terminal in Cincinnati and station employees there had to deal with it.

Fifty of the 1,050 mail bags in the car were destroyed and about 25 were damaged. The loss was estimated at $13,000, made up of $8,000 of damage to the contents of the mailbags and $5,000 worth of damage to the mail car itself. [1539] The damaged mail was destined for Tennessee, West Virginia, Virginia, and other Southern states. It was to be sorted and classified at Hamilton, with deliverable items being forwarded to their destinations and the remainder being sent to the dead letter office. [1540]

Another destructive fire occurred on train No. 55 of the New York Central Railroad at 6 a.m. on Thursday 18th December 1947. The ten-car New York – Watertown – Massena passenger and mail train was between Utica and Altmar, New York, United States, when the fire was discovered. Conductor J.E. Jeffery arranged for the train to be divided and the engine and five vehicles, which included the burning baggage car, were taken forward to Altmar, where the village volunteer fire department dealt with the blaze. Eleven passengers in the snack car immediately in the rear of the baggage car had to be moved when it began to fill with smoke, while when the fire from the sealed baggage car began to spread to the postal car in front of it postal employees were forced to throw out sacks of mail.

Despite the best efforts of the firemen the wooden baggage car was destroyed. It had been attached to the train at Utica. No one was hurt but an estimated 2,500 parcels, probably mostly Christmas gifts for communities north and south of Watertown, were destroyed. No letter mail was lost. The cause of the fire was unknown but postal and railroad authorities were investigating. [1541] [1542]

A disastrous fire broke out in a third-class carriage of a passenger train on the Xindian

[1537] *The Times*, London, England, 25th August 1947
[1538] *Miami Daily News Record*, Miami, Oklahoma, United States, 24th August 1947
[1539] *Hamilton Journal* and Daily News, Hamilton, Ohio, United States, 11th November 1947
[1540] *Cincinnati Times-Star*, Cincinnati, Ohio, United States, 11th November 1947
[1541] *Courier and Freeman*, Potsdam, New York, United States, 24th December 1947
[1542] *Syracuse Herald-Journal*, Syracuse, New York, United States, 18th December 1947

River Bridge, Formosa, between Wanhua and Banqiao stations, on Friday 28th May 1948. An early report from Taipei stated that nine bodies had already been recovered and 54 injured had been taken to hospital. Other passengers had died when they jumped from the burning train into a river. [1543] [1544]

A second report indicated that the accident was considerably worse than had been first indicated. It said that 40 bodies had been recovered from the burnt-out coaches and another 30 bodies had been fished from the river, which the train had been crossing at the time of the accident.

Another report, issued about five days after the event, gave even higher figures, stating that there were 175 people dead or missing. It added that surviving passengers had reported a deafening explosion and that petrol, camphor oil and celluloid picture film were believed to have been involved in the fire. [1545]

A Swiss newspaper stated that the train had been crossing the 300-metre-long bridge between Taipei and Chiay and that the fire, which had spread to four coaches, had originated in the accidental ignition of petrol which was being carried clandestinely by one of the passengers. [1546] Reports from Formosa stated that the source of the fire was unknown, so perhaps the petrol-carrying passenger was just a theory.

On the night of Friday 18th June 1948 a passenger train caught fire between Saint Étienne and Villars, France. The fire reportedly had begun with an explosion in the leading passenger coach. Three people burnt to death and 15 others were injured. [1547] Another source stated that six passengers had died and that a bag of hunting powder had exploded. [1548]

On Friday 6th August 1948 a fire was discovered in a steel box car attached to a westbound Pennsylvania Railroad express in the vicinity of Tyrone, Pennsylvania, United States. Upon the arrival of the train at nearby Altoona trainmen and station staff tackled the fire with water hoses and put it out. Several thousand dollars' worth of second-, third- and fourth-class mail, including newspapers, magazines, books, and parcel post, mostly consigned from eastern cities to Chicago, was destroyed. Postal inspectors were investigating the cause of the blaze. [1549]

That common cause of train fires in the days when steam locomotion was general, a spark from the engine, was believed to have led to a fire in a "W" waggon of a goods train bound for Winton on the night of Tuesday 17th August 1948. The train was two miles on the Barcaldine side of Dartmouth, Queensland, Australia, when the fire was discovered. The train was then divided immediately behind the burning waggon and the engine proceeded to Dartmouth with the front portion of the train, where the burning truck was

---

[1543] *Morning Tribune*, Singapore, 31st May 1948

[1544] Wikipedia – List of Rail Accidents (1940-1949)

[1545] *The Singapore Free Press*, Singapore, 31st May and 3rd June 1948

[1546] *Gazette de Lausanne*, Lausanne, Switzerland, 31st May 1948

[1547] *European Stars and Stripes*, Pfungstadt, Germany, 20th June 1948

[1548] https://www.leprogres.fr/loire-42-edition-saint-etienne-couronne/2019/08/13/villars-le-passage-a-niveau-de-la-gare-entre-trains-et-drames

[1549] *The Daily News*, Huntingdon and Mount Union, Pennsylvania, United States, 6th August 1948 shunted into a siding. The engine then returned for the remainder of the train. This included a tank waggon fully loaded with petrol and care had to be taken in passing the burning waggon stabled in the siding lest the heat of the fire should cause an explosion.

It was necessary to let the fire burn itself out and the contents of the waggon – 328 bags of chaff from Darling Downs – were lost. The intensity of the blaze was such that the sleepers of the siding were burnt and its rails were buckled so that they had to be renewed. [1550]

A spark from the engine was thought to have been the cause of another train fire near Picton Junction, Western Australia, at about 3.45 p.m. on Saturday 4th September 1948. The blaze, in two waggons loaded with wool and flax, was noticed by the train driver half a mile outside of Picton. To prevent the fire spreading the two affected waggons were detached. Bunbury Fire Brigade were called and soon extinguished the fire but both waggons, all of the flax and most of the wool was destroyed. Damage to trucks and contents was estimated at close to £2,000. [1551]

In contrast to the two previous Australian accidents, a fire which broke out in the sleeping car of a westbound New York Central Railroad passenger train at Fonda, New York, United States, on Saturday 15th January 1949 was believed to have been electrical in origin. Seven passengers suffered from smoke inhalation and one passenger sustained cuts to his right hand when he broke a window to let out the fumes. The cause of the fire was thought to have been a short circuit in an electric motor attached to the ventilation system, with burning insulation causing the smoke and fumes. [1552]

Twelve miles from The Pas, Manitoba, Canada, the baggage car of the twice-weekly mixed Canadian National Railways train from La Pas to Churchill caught fire near Tremaudan and was completely destroyed at noon on Thursday 20th January 1949. No one was hurt and the track was cleared within a few hours. The cause of the fire was undetermined. [1553]

A crowded electric workers' train caught fire at Matsuzaka, Japan, on the morning of Tuesday 8th March 1949. Eight people died and several others were injured. The fire, caused by a short circuit, completely destroyed the train. [1554]

The 7.30 p.m. Up express passenger train from Edinburgh Waverley to London King's Cross met with an accident at 8.40 p.m. on Thursday 23rd June 1949 when it caught fire at the entrance to Penmanshiel Tunnel, between Cockburnspath and Grantshouse, Scotland, on British Railways, Scottish Region. The train was in charge of Front Guard W.G. Prime and Rear Guard W.C. Miller, with Driver H. Brown and Fireman E.M. Mcleod. It consisted of Type A-3 4-6-2 steam locomotive No. 60035, a

---

[1550] *The Longreach Leader*, Queensland, Australia, 20th August 1948

[1551] *The Blackwood Times*, Greenbushes, Western Australia, 10th September 1948

[1552] *The Post-Standard*, Syracuse, New York, United States, 15th January 1949

[1553] *Winnipeg Free Press*, Winnipeg, Manitoba, Canada, 21st January 1949

[1554] *The Daily Gleaner*, Kingston, Jamaica, 9th March 1949 mail van, a brake third, two third-class, a first- and third-class composite, a luggage brake van, a first- and third-class composite sleeper, a first- and third-class composite, a third-class, a first- and third-class brake composite, a buffet car, and a third-class brake composite coach.

The train left Edinburgh on time, passing Innerwick at 8.27 p.m. and Cockburnspath at 8.31 p.m. Two and a half miles beyond there, fire broke out suddenly in the tenth vehicle of the train, brake composite No. 1148, and spread with such ferocity that the whole coach was ablaze within a matter of seconds. A strong brake application had been made almost at once from the burning brake composite and the train stopped within 30 seconds, partly within the tunnel. Most of the passengers in the coach were able to flee to the guard's compartment at the rear but three were trapped and had to break windows and jump down to the track, in the process of which one passenger, an American woman, was seriously hurt, while a second passenger suffered severely from smoke inhalation and burns. Seven other passengers were less badly hurt and all the personal possessions of those in the coach where the fire first broke out were lost. An ambulance arrived from Eyemouth at 9.45 p.m. but buses and private cars had already taken the injured to hospital. The Eyemouth fire brigade arrived at 9.47 p.m., closely followed by the Duns fire brigade at 9.50 p.m. and the fire was extinguished at 10.30 p.m.

Soon after the train had stopped the fire had spread to the ninth coach, No. 1498. The train was divided behind the tenth coach and behind the eighth coach and the front portion was drawn forward to Grantshouse, arriving there a few minutes later. The ninth and tenth coaches had to be left to burn themselves out. The passengers from the rear four coaches made their way to Grantshouse on foot and the train eventually left Grantshouse at 10.30 p.m., arriving at Newcastle an hour and 45 minutes late. The lines were blocked until 3.10 a.m. the following morning and trains were diverted via Carlisle or Kelso.

The accident was the subject of a comprehensive Ministry of Transport report.

Carriage and Waggon Examiner N. Ireland and Train Lighting Examiner F. Whyte stated that they had found no problems with the running gear or electrical equipment of the train vehicles when they had checked them at Edinburgh prior to the departure of the train.

Signalman G. Masterton at Innerwick was sure that there had been nothing wrong with the train when it had passed his cabin.

Signalman J. Hunter at Cockburnspath was equally confident that he had seen no smoke or signs of fire when the train had passed him. He added that the Grantshouse signalman had telephoned him at 8.46 p.m. to say that the train had not yet reached there, only 4½ miles from Cockburnspath, after 15 minutes. At 9 p.m. the Grantshouse signalman had rung again to tell him that the front of the train had now reached him, and that the rear portion of it was on fire at Penmanshiel Tunnel.

The testimony of various passengers on the train followed in the report. Mr J. J. Wilson, a Railway Traffic Agent at Newcastle, spoke of the smoke becoming very thick in a few seconds so that it was almost impossible to breathe, of assisting passengers from the train and then of helping the (rear) guard to divide the train.

Mr W.A. Fletcher spoke of seeing black smoke and flames in the corridor, pulling the communication cord, leaving his compartment, passing two or three compartments, and then pulling the cord again as the train did not seem to be slowing down.

Mr P. Silverman of New York said he had been travelling with his wife and a friend, Mr G. Hirsh. He related that his attention had first been attracted by seeing a boy running along the corridor who was evidently alarmed. At the same moment Mr Fletcher had shouted, 'Fire!' pulled the communication cord and left the compartment. The fire had immediately developed to such a degree that it had been impossible for the three of them to follow Mr Fletcher, nor due to the spread of fire could they reach the communication cord in the compartment. After Mr Hirsh had failed to break the coach window with a beer bottle he (Mr Silverman) had succeeded in kicking it out with his foot and the two of them made the hole big enough for them to jump through by kicking more glass out. The three of them had then jumped from the compartment through the broken window, his wife first, then he, and finally Mr Hirsh.

Mrs D. Silverman and Mr Hirsch corroborated Mr Silverman's account. Mr Hirsch put the period between the first alarm and his jumping out of the window at only 15 to 30 seconds.

Mr Whitworth, a Doncaster engine driver, said that by chance he had gone out to the corridor to look for the location of the previous year's severe floods when he had seen a sudden burst of flame spurting from the floor level. Startled, he had shouted, 'Fire!' had told his wife to run to the front of the train, and had gone to the guard's compartment, warned the guard and applied the vacuum brake. He estimated that only 10 seconds had elapsed between his seeing the fire and applying the brake.

A boy, Master G. Flintham, spoke of his attention being drawn to the fire by Mrs Whitworth and of his running forward through the train until he had met the ticket collector, to whom he reported the fire.

Mrs Davison described seeing a flicker of light and being curious as to its nature going to the corridor to investigate. There she had seen flames "like a huge tongue." Warning her friends in the compartment she had made her way to the guard's department just as someone (Driver Whitworth) was applying the brakes. She had been helped from the train as soon as it stopped.

Mrs Reavely said that she had been speaking with Mrs Davison when the latter had left the compartment. In what she thought could not have been more than two seconds she had seen strong flames coming along the roof of the corridor and had feared that they were trapped, but her husband had told her to "duck under the flames and run." In the confusion of their flight their son had not followed them and her husband had had to go back and fetch him.

Mr I. Sutherland spoke of going back with Mr Reavely for the Reavelys' child and of the flames chasing the three of them along the corridor.

Mrs M. Owen described having experienced some delay in escaping to the guard's compartment due to trying to open a door the wrong way. Due to the delay she suffered badly from both burns and smoke inhalation and was a long time in hospital.

Mr L. Orrick said he had heard two or three men shouting, "Pull the cord, a fire has broken out," about ten miles after passing Dunbar, but the train had already been stopping. He had picked up his son and told his wife to follow him but had they lost one another in the thick smoke. He had eventually found her at the tunnel entrance.

Mr R.H. Lodge, a Parcels Van Driver at South Shields, related he had heard men shouting, had seen black smoke, had pulled the communication cord, and helped people out of the train.

Rear Guard Miller stated that his first warning of the blaze had been a man and a woman rushing along the corridor shouting that the train was on fire. He had at once seized a fire extinguisher. Going to the door of his compartment he had seen flames and dense black smoke, then the smoke had become too thick to see anything and passengers running back along the corridor had pushed him back into his compartment. The train had quickly stopped and he had helped the passengers disembark, had taken part in the division of the train behind the brake composite he had been riding in and in front of the burning ninth coach, then had gone back to Cockburnspath Up Intermediate Block Signal (IBS) Home signal to protect the rear of his train, though without his flags or detonators, which he had not been able to locate in the smoke.

Restaurant Car Attendant J. Allen said a passenger in the buffet had told him of seeing a body on the line (Mrs Silverman). He had informed the conductor and at the same time the train had been pulling up. Several passengers had then run back into the buffet car and dense smoke had come through the vestibule to it. After the train had come to a stand he had helped passengers disembark and the rear guard to divide the train. Restaurant Car Attendant R.H. Down and Conductor T. Carruthers corroborated his statement. The latter added that there had been intense heat and a 'smell resembling celluloid'.

Sleeping Car Attendant W. Fyvie spoke of the train stopping, his hearing a noise and then finding Ticket Collector Haig assisting passengers from coach No. 1498, the rear end of which was beginning to burn. He had assisted the ticket collector, then helped with the division of the train.

Front Guard Prime stated that he had been riding in the luggage van sixth in the train, which had stopped with his van in the tunnel. Hearing children crying and screaming and seeing men jumping from the train he had gone back to see what was wrong and had met Sleeping Car Attendant Fyvie, who had told him that there was a "terrific fire" at the rear of the train. He had at once decided what must be done and instructed the ticket collector to help the rear guard to divide the train while he would go forward to tell the driver what to do.

Ticket Collector Haig stated that he had been in the eighth coach when an agitated boy had run up to him and told him that the train was on fire. Calming the boy (Master Flintham), he had gone to investigate but a rush of people coming forward along the corridor had blocked his progress. Going to an offside window and looking out he said, "I could not believe my eyesight to see the flames that were coming out." Helping the passengers disembark he had then occupied himself removing passengers' luggage from coach No. 1198, which had not at that point been on fire, though the heat had been "terrific" and the smoke blinding. By the time that he had gone to all seven compartments to remove baggage the rear of the coach had been well alight.

Driver Brown said the train had been running normally until the brakes had been applied from the rear of the train approaching Penmanshiel Tunnel and the train had stopped. His fireman had gone back to investigate, had met the front guard, and had found what was required. After the train had been divided he had run with the forward part to Grantshouse, watching for oncoming traffic, and at Grantshouse his fireman had at once gone to the signal box. His train had run easily up the bank from Cockburnspath and he did not think that any hot cinders from his engine could have flown back as far as the ninth or tenth coaches.

Fireman Mcleod confirmed his driver's evidence and added that the only Down train that they had passed in the vicinity of the accident had been stopping at Innerwick, making it unlikely that it would have generated any sparks which could have lodged in the vehicles of their train.

Signalman R. Thomson at Grantshouse said that the 7.30 p.m. train had been described to him from Cockburnspath at 8.30 p.m., followed by another at 8.46 p.m., at which time the 7.30 p.m. train had not reached his berth track circuit. Conferring with Signalman Hunter at Cockburnspath they had decided that the first train must have stalled on the bank, but at 8.49 p.m. there had been a telephone call from the Up IBS Home signal and Rear Guard Miller had advised him as to what had really happened, adding that they needed a doctor, the fire brigade and perhaps an ambulance. Very shortly the front portion of the train had arrived and its driver and fireman had come to his box and confirmed the situation, whereupon he had telephoned Control and informed them. The Grantshouse stationmaster had then proceeded to the tunnel by car and returning had also telephoned for the fire brigade and the police.

Grantshouse Stationmaster J.T. Moffat stated that his first concern had been to protect the Down line. Having ensured that this had been done he had gone to the scene by car and from there had attempted unsuccessfully to telephone Signalman Thomson (the connection had been damaged by the fire). Having assured himself that arrangements were being made to speedily remove to Dunbar the seriously injured passenger (Mrs Silverman) he had returned to Grantshouse to call the fire brigade and the police.

The Ministry of Transport inspecting officer, Colonel R.J. Walker, noted that in the alarm and excitement caused by such a fire it was always difficult for witnesses to judge time in seconds, but a close examination of the evidence showed that the different statements fitted together remarkably well, and it was clear that the fire had spread 'with a rapidity and fierceness which it is difficult to conceive.' The three Americans had been trapped in their compartment within 15 or 20 seconds of the fire breaking out 15 feet away and the time from the discovery of the fire to the stopping of the train had been only about 75 seconds and possibly as little as a minute.

The two coaches that had been destroyed had been built by the London and North Eastern Railway in 1947. Their bogies, underframes and panelling had been of steel, other parts had been of teak. They had been painted with the usual oil-based paints, except for the corridor panelling, which had been sprayed with a clear cellulose lacquer. The electrical equipment was of the normal type.

Initially it was suspected that the fire had resulted from an electrical fault, but two months after the accident the evidence of Driver Whitworth, that the fire had originated at floor level, where there was no wiring, came to light, indicating that this could not be the case. Tests had already been made on the paints used in the coaches, with negative results. Further tests were then made on the cellulose lacquer used in the corridors of the coaches. An initial experiment, involving the lighting of a small piece of paper in the corridor, produced strong flames 2 feet high in 10 seconds and within 25 seconds the whole corridor was ablaze. A similar attempt with a cigarette end did not work but the inspecting officer had no doubt that under different conditions of draught and position a cigarette end, cigar end or the smouldering dottle from a pipe would have been sufficient to start such a fire. Further tests at the Scientific Research Department of the Railway Executive indicated that it was the clear cellulose lacquer which had burnt so freely.

A panel from a similar coach was then sent to the Fire Research Department of the Department of Scientific and Industrial Research, who reported that the flammability of the cellulose lacquer was, by a very large margin, the worst in the records of the Joint Fire Research Organisation.

Although an electrical fire had been ruled out the inspecting officer was critical of the standard of workmanship of the coach wiring, which he considered might have led to an electrical fire in other circumstances.

In conclusion, Colonel Walker stated that the fire had clearly originated in the floor of the brake composite coach, probably from a discarded cigarette end, cigar end or pipe dottle and had spread very rapidly due to the flammability of the cellulose lacquer, and doubtless accelerated by the draughts. The burning lacquer had also given off noxious and asphyxiating nitrous gases. He ruled out the possibility of the fire having been caused by an electrical fault or a cinder from the engine.

Turning to remarks and recommendations, he noted that there had only been two serious train fires in the preceding 20 years (at Micheldever in 1936 and Westborough in 1941, both dealt with in the previous volume of this series). The experience of train fires was therefore limited and the possibility of one was perhaps not sufficiently considered when designing coaches. The great flammability of clear cellulose lacquer was no new discovery.

Although the injury and damage sustained in the accident were bad enough, it could have been considerably worse had not it occurred close to the brake in the guard's compartment and not been seen by an experienced engine driver who had taken immediate action (these were prophetic words in view of the Beattock train fire the following year, see below).

All other coaches known to have been sprayed with cellulose lacquer had been withdrawn and other rolling stock was being checked. The question was how such a dangerous substance had come to be used in the first place and what could be done to prevent such a thing happening again.

It emerged that the coaches had been built under contract and the contractors had been given specifications as to the materials to be used, which included ordinary oil paints which were not specially flammable. At some point there had been a deviation from these specifications; it was not possible to determine when or why, but clearly no thought had been given to the fire risk.

The inspecting officer went on to point out that even a so-called all-steel coach, such as in the new British Railways standard design, had wooden interior furnishings, while the upholstery of all-steel and composite steel and wood coaches, such as those involved in the accident, was the same. A strictly all-steel coach was undesirable from the point of view of passenger comfort and amenity.

More attention should be given to the standard of workmanship of coach electrical wiring.

The inspecting officer ended his report by commending the actions of the train crew in the fire.

There were two more fires on passenger trains in Great Britain on Saturday 2nd July 1949. The 10.48 a.m. Cardiff – Bournemouth train was stopped at Poole, England, by a passenger pulling the communication cord. The leading carriage was found to be on fire and was shunted into a siding and left to burn itself out. About 40 holidaymakers reportedly jumped from the train. Probably rather earlier in the day a coach of the 12.24 a.m. Carstairs to Birmingham train had caught fire at an unstated location and had to be detached from the train. 'Delay was caused all along the line' and 'some trains were held up for 140 minutes.' [1555]

At about 3 p.m. on the afternoon of Saturday 17th September 1949 a goods train from Julia Creek was about 7 miles from Cloncurry, Queensland, Australia when Guard L. Honeycombe noticed a fire in an open waggon. He stopped his train and investigation found that a tarpaulin covering the waggon was well ablaze. With the help of the fireman of the train dirt and sand was thrown onto the fire, which was eventually put out, though by this time the truck and some of its contents were badly charred. The waggon had been carrying a 'new silky oak upright refrigerator cabinet and electrical contact gear for the Cloncurry Bowling club-house.'

The fire was again blamed on a hot spark from the engine having landed on the tarpaulin. The alertness and prompt action of the guard was credited with saving the other contents of the waggon. [1556]

Fire broke out in the rear coach of a three-car electric train of the Keihan Electric Railway on the Kyodo – Osaka main line near Kōrien station, Japan, on Tuesday 27th September 1949. According to different accounts 116 or 140 passengers were injured, 40 seriously, mostly suffering burns. The driver reportedly stopped the train after passers-by had frantically signalled to him, whereupon passengers with their clothes ablaze jumped out of doors and windows. [1557] The rear of the last car was said to have been destroyed. [1558] The fire was blamed on 'the carelessness of an inspector.' [1559]

There was another train fire in Queensland, Australia, on Tuesday 1st November 1949, this time involving a Mackay – Rockhampton mixed train. As the train was nearing The Caves, 15 miles short of Rockhampton, at 7 a.m., the rear of a waggon containing a refrigeration unit was seen to be on fire. By the time the train had stopped the fire had spread to the following waggon which contained 24 bales of wool. The two waggons were detached and pushed into a siding at The Caves.

Rockhampton Fire Brigade were alerted and were soon on the scene. They began by using water from the engine tender, which was then supplemented by running 1,000 feet of hose to a stand pipe on the Yaamba pipe line, by which the blaze was soon brought under control.

Unfortunately the wool waggon and its contents were destroyed. The compartment holding the refrigeration unit was burnt out and the machine considerably damaged, while the wall dividing the refrigeration unit from the cooling chamber, which contained fish, was damaged. When the fish, 17,000lbs of mackerel from Townsville for the Brisbane market, was later retrieved, it was found to be smoke damaged and was dumped. The damage was estimated at £4,000. [1560]

[1555] *The Times*, London, England, 4th July 1949

[1556] Cloncurry Advocate, Cloncurry, Queensland, Australia, 23rd September 1949

[1557] *The Singapore Free Press*, Singapore, 28th September 1949

[1558] *Long Beach Press Telegram*, Long Beach, California, United States, 27th September 1949

[1559] *Pacific Stars and Stripes*, Tokyo, Japan, 28th September 1949

[1560] *The Daily Mackay*, Mackay, Queensland, Australia, 2nd November 1949

Fire broke out in the engine of a nine-car Chicago Milwaukee St Paul and Pacific Railroad passenger train as it was climbing the steep Beverly incline on the night of Sunday 29th January 1950. Fumes drove Engineer Eddie Maxwell and Fireman Liberty from the electric locomotive, though they were able to cut the power from the overhead line equipment before they vacated the cab. They then attempted to extinguish the fire by throwing snow upon the burning engine but matters were complicated when the air in the brakes began to leak off and the train started to roll backwards down the hill, leaving the engineer and fireman behind.

The runaway train passed through the station at Kittitas, Washington, United States, at about 30 miles an hour but about a mile beyond Kittitas the momentum of the runaway decreased, presumably when it reached the bottom of the bank, and a brakeman was able to halt it using the manual brakes. It had covered seven miles. Operator G. J. Bohnen at Kittitas stated that no one had been hurt and that the train had eventually continued its journey to Spokane, 4 hours and 47 minutes late. [1561]

A special goods train which included a consignment of 35 bales of wool from Barcaldine to Bulimba in an FG waggon caught fire on Saturday 4th February 1950. Driver H.G. Daniel became aware that the third waggon from his engine was ablaze near Talarena, Queensland, Australia, and by then it was well alight. Before the waggon could be detached the fire had spread to an adjoining waggon loaded with empty fruit barrels. Both waggons and their contents were destroyed, leaving only a mass of twisted iron fittings and the waggon wheels. The train crew were, however, able to prevent the spread of the fire to other waggons loaded with wool in the train.

After a breakdown gang had cleared the wreckage of the destroyed waggons it was discovered that the fire had also caused considerable damage to the road bed. Twenty-three sleepers and five lengths of rail had to be replaced and the line was blocked for nine hours.

This was stated to have been the second fire in which wool had been destroyed on the Barcaldine – Alpha line in the previous few months and police under the direction of Inspector Sproul of Longreach were investigating. A suggestion that the fires had been deliberately set in order to conceal thefts of wool was ruled out. [1562]

The 11 a.m. Down express from Birmingham New Street to Glasgow Central was involved in a serious fire near Beattock Summit, Scotland, on British Railways, Scottish Region, on the late afternoon of Thursday 8th June 1950, in which five passengers lost their lives.

The train consisted of Class 6P 4-6-0 steam locomotive, a third-class brake composite, a first- and third-class composite, a first-class coach, a first-class dining car, five third-class and a third-class brake composite coach, in charge of Passenger Guard George Johnston with Driver T. Hannah and Fireman H. McKerracher. It had started on time but reached Carlisle 15 minutes late at 3.56 p.m. Three minutes was regained by quick station work and it left there at 4.02 p.m., 12 minutes late, and passed Beattock North box at 4.46 p.m. only 10 minutes late. All appeared normal as the train passed Greskine box at 4.53

[1561] *Walla Walla Union-Bulletin*, Walla Walla, Washington, United States, 30th January 1950

[1562] *The Longreach Leader*, Longreach, Queensland, Australia, 10th February 1950 p.m. Three minutes later the communication cord was pulled in coach No. 4851, the first- and third-class composite marshalled second from the engine. Travelling up the bank at only 30 miles an hour the train was quickly stopped, when it was seen that the rear end of coach 4851 and the gangway at the end of coach No. 1073, the first-class coach placed third in the train, were burning fiercely. A ganger who chanced to be at the lineside where the train stopped entered coach No. 4851 and rescued a woman passenger, then bravely re-entered the burning coach to search for further passengers but could find none, before he was forced to withdraw due to the heat, smoke, and flames. It was not known for certain at this time whether there was anyone left on the train but after the fire had been extinguished the fire brigade found the bodies of two women, a man and two children. They were later identified as Mrs Campbell, of Hillhead, Glasgow; her children Mirabel, 4, and John, 2; Mrs Chaffin, of Oxford; and Mr Jacobs, of Glasgow.

Signalman T. Nichol at Greskine was the first person to give evidence at the subsequent Ministry of Transport inquiry. He stated that he had watched the train pass and he was certain that there had been no sign of any smoke or fire and there had been no sparks coming from the engine. At 5.07 p.m. he had noticed that the train had apparently still not reached the Harthope Intermediate Block Section (IBS) 4,000 yards further on. Thinking that possibly the track circuit indicator in his box had failed he had telephoned his colleague Beattock Summit box, 3,500 yards further on, to ask if the train had reached there. Told that it had not he had sent the *Train unusually long time in Section* signal. A minute later a telephone call from Harthope IBS Home signal had informed him of the fire and he had sent the *Obstruction Danger* signal to Beattock Summit. At 5.30 p.m. a further telephone call from Harthope IBS had asked him to allow the following Down *Royal Scot* express into the section to pull up to the rear of the Birmingham train, for fear that the coaches of the latter should begin to run back down the incline, and after obtaining permission from Control he had done so.

Driver Hannah said that his engine had been steaming well, the load of his train had been within the limits for one engine and he had not needed to stop for a banker at Beattock station (which slightly confusingly was situated 10 miles short of Beattock Summit). Only a few sparks had been coming from his engine. Passing over Harthope Viaduct he had noticed that the vacuum was falling and had realised that the communication cord must have been pulled. Looking back he had seen dense smoke coming, it seemed, from the rear of the second coach and the front of the third and had made a full brake application, stopping the train in about 14 seconds. As his train had stopped the smoke had been replaced by fierce flames. Initially he had gone back with a bucket of water to fight the fire, after arranging with another driver travelling as a passenger for him to protect the Up line. Then seeing the intensity of the fire he had realised that he needed to divide his train and had done so, aided by his fireman, between the first and second coaches and the third and fourth coaches.

Fireman McKerracher agreed with his driver's evidence and added that he had looked back on passing Beattock North and Greskine boxes and seen no sign of fire.

Miss Goodwin, a passenger, said that she had been sat in a compartment with Mrs Chaffin. A quarter hour before the stoppage of the train she had thought that she smelt smoke and felt heat coming from beneath her seat but looking and seeing nothing had dismissed the idea. As the train had passed Greskine she had gone to the lavatory at the end of the coach. She was only there two or three minutes and as she opened the lavatory door to return to her compartment the train was stopping. The corridor was now smoky and hot and she must have briefly collapsed here as it was where she still had been when she had been helped from the train by the ganger. Outside she had walked along the side of the carriage to find that her compartment blazing furiously.

Passenger Miss M.R. Tomkinson had gone to the dining car with her mother after leaving Carlisle. When they had returned at 4.45 p.m. their compartment had seemed rather warm. About ten minutes later they had seen dirty black smoke going past the window, then smelt burning. Going to the corridor they had seen smoke and flames pouring from the windows of the coach ahead. Grabbing their handbags they had made for the rear of the coach but had not been able to find their way through the dense smoke or find the communication cord. On the point of collapsing they had, between them, managed to lower a window and were about to jump from a side door but a dining car attendant had shouted for them to wait until the train had stopped, which they had done before alighting.

Mrs H. Tomkinson confirmed her daughter's evidence and added that her daughter had cried, "I am almost done." She described the smoke as having been "peculiarly stupefying," so that they had both been on the point of collapse and her daughter had had difficulty in opening the coach window. Outside they had seen that the coach ahead was burning fiercely. Later a man (Ganger Adam Moffat), black with smoke and clearly shocked, asked them if they had seen a woman from the burning coaches, saying that he had rescued one but feared that there was another.

Restaurant Car Attendant William Ritchie described the earlier positions of the five passengers who had died and his later smelling smoke. Leading Restaurant Car Attendant Albert Bastien had then told him to go and tell the guard that the train was on fire.

Restaurant Car Attendant W. Russell told of smelling burning, having been thrown back by heat and smoke and of his having warned two women (Mrs and Miss Tomlinson) not to jump before the train had stopped lest they should fall beneath its wheels.

Leading Restaurant Car Attendant Bastien told of his attention being drawn to the burning by Attendant Russell, and of opening a vestibule door to be confronted by "thick smoke and very intense heat."

Ticket Collector Smith described noticing an unpleasant small and seeing black smoke. Leading Attendant Bastien had then told him that the train was on fire and he would not be able to get through to the next coach. Nevertheless, he had tried to do so, but thick choking smoke had forced him to retreat.

Travelling Carriage Cleaner A. Levy stated that she had been cleaning a lavatory compartment in the fifth coach when she had smelt smoke and looking out had seen thick dark green smoke coming from near the front of the train. Running back to the guard's van she had reached it just in front of Attendant Ritchie. There she heard a noise, "like a small explosion of some kind," followed by a burst of flame which, "recalled to my mind a flame which would be caused by petrol or paraffin thrown on a fire."

Ganger Moffat said that at 4.50 p.m. he had been checking expansion joints in the track and had stopped to exchange a few words with Signalman A.J.B. Millar, who lived in a lineside cottage and had been about to do some gardening. The two of them had heard the Birmingham train approaching and had seen smoke and flames coming from the second coach. They had held up their hands to stop the train but it was already pulling up hard. Signalman Millar had gone for some buckets of water and he (Moffat) had gone towards the train. He had helped two women from the rear of the first coach (Mrs and

Miss Tomkinson) then found a woman in a dazed condition at the end of the second coach (Miss Goodwin). Despite the dense smoke he had re-entered the second coach and searched the lavatory and first three compartments, which had all been empty. As he had reached the fourth compartment a burst of flame filled it; he had glimpsed a woman sat there. He had again tried to re-enter the compartment but it had been impossible; the terrific heat had forced him down to his knees. There had been great difficulty in breathing and he feared losing consciousness, but on his hands and knees he had managed to crawl out of the coach. Like Mrs Levy, he spoke of the flames resembling "petrol thrown on a fire." Afterwards he had helped the engine crew divide the train, had scotched the rear coaches and had gone through the train searching for the woman (Mrs Chaffin) whom Miss Goodwin told him had been in the compartment with her. Earlier, before he had entered the train, he had told a girl in the cottages to go to a nearby farm which had a telephone to call the fire brigade, a doctor, and the police. Later he had helped the fire brigade retrieve the five bodies. According to a newspaper reporter he commented later, "The poor people must have been trapped like rats. It was just impossible to get near."

Signalman Millar described attempting to fight the fire with two buckets of water and others handed to him by people from the cottages to little effect. Later he had helped to scotch the standing train and had given other help.

Lengthman A. McGaw had just returned home from work when his wife, Mrs Eileen McGaw, looked out of the window, saw the stopped train, and exclaimed to her husband, "The Scot is on fire!" A sleeper fence hid all but the top of the train and the fire had not appeared at that moment to be too serious but the lengthman and his wife had knocked on neighbour's doors and then rushed to the train with buckets of water. [1563] There they had realised that due to the magnitude of the blaze these would have little effect. Concerned about protecting the Up line the lengthman said he had then collected detonators and a flag from a platelayer's hut and had run towards Harthope IBS signals. It was he who had phoned Greskine box from the Down IBS home signal, then Beattock Summit box from the Up IBS home. At this point he had been joined by Driver Byars, who had been a passenger on the train, had handed him the detonators and flag and returned to the train.

Passenger Guard Johnston said that he had been in the van at the rear of the train when the brakes had come on. Looking out of a window he had seen a little smoke, then Travelling Cleaner Levy and Attendant Ritchie had appeared and told him of the fire. By the time that he had gone forward towards the front of the train the driver and fireman had already begun dividing it and told him that the Up line was protected, so ensuring that the rear coaches were scotched he had gone back with detonators to protect the rear of his train. Later, the *Royal Scot* had come up to within a yard of the rear coach of the Birmingham train.

Beattock Stationmaster G.M. Scott stated that he had been advised of the fire at 5.15 p.m. and reached the scene at 5.30 p.m. Moffat fire brigade had just arrived and were laying out hoses (they had been called from the farm at 5.10 p.m. and were followed by crews from Abington who arrived at 5.40 p.m. and Lanark who arrived at 5.55 p.m.). Ganger Moffat had told him that he feared that there was someone still in the burning coaches and they had spoken to Miss Goodwin. A fireman had found the first body at 6 p.m.

[1563] *The Times*, London, England, 9th June 1950

Shunter J. Cochrane, travelling as a passenger in the leading coach, stated that he had looked out passing Greskine and had seen nothing wrong. When the brakes had been applied he looked out again and had then seen black smoke coming from the second coach. He had opened the door to that coach but had been so choked by smoke that he could not continue. He instead helped passengers out of the leading coach, including his own wife and child.

The Ministry of Transport inquiry was again conducted by Colonel R.J. Walker. Coming to the conclusions of his inquiry, he noted that unfortunately there were no living witnesses as to what had happened in coach No. 4851 after Miss Goodwin had gone to the lavatory. From the evidence, though, it was possible to give an account of what had happened in what, he said, was a most unusual accident. He then summarised the sequence of events:

Twenty minutes before the train had stopped an attendant had noticed a peculiar smell and Miss Goodwin noticed a smell and unusual heat, but neither could find anything wrong.

Three minutes before the stoppage the train had passed Greskine box and neither the signalman nor the fireman had seen anything wrong. Miss Goodwin had left Mrs Chaffin and gone to the lavatory, seeing Mrs Campbell and her children in the next compartment.

Twenty seconds before the train had stopped the communication cord had been pulled in the second coach. Two women in a compartment of the third coach had seen smoke and flames coming from the second coach, their corridor had filled with smoke and they had struggled to escape. Two restaurant car attendants had noted thick smoke, and intense heat and smoke had been seen by the driver, fireman, ganger, and signalman.

A minute after the accident Ganger Moffat had entered the second coach, searched three compartments but been completely unable to do any more. He had glimpsed the body of a woman in a fourth compartment who was evidently already dead.

An hour after the train had stopped firemen found five bodies, all in one compartment.

There were a number of questions to be answered:

(a) Why had the five victims made no effort to escape or protect themselves?

(b) In view of (a) who had pulled the communication cord?

(c) Why had Mrs Chaffin moved to the compartment ahead and why were her handbag, hat and shopping bag found on the floor at the end of the corridor by Ganger Moffat?

(d) Where had Mr Jacobs come from and why was he in the compartment?

It was clear, the inspecting officer continued, that the victims must have been killed instantly and the other significant point was that the sixth compartment of the coach had been empty and closed after Carlisle, while the others had been open.

The sudden death of the victims could have been caused by direct burning, carbon monoxide poisoning or heat blast with contributory asphyxiation.

The quiet and relaxed position of the bodies found by the firemen ruled out direct burning.

The gradual accumulation of carbon monoxide could be ruled out by the presence of Mr Jacobs and the pulling of the communication cord. A very high concentration of carbon monoxide sufficient in itself to cause immediate unconsciousness was very unlikely.

Heat blast, with a single breath of extremely hot air, probably accompanied by a high level of carbon monoxide, causing acute pulmonary oedema, seemed the likely explanation. Such cases had been common during wartime bombing and sometimes occurred in house and factory fires. The hot gases must have been generated elsewhere in the train and entered the compartment almost explosively.

Colonel Walker theorised what had probably happened. Before the train had reached Carlisle a lighted match or cigarette end had been carelessly discarded by a passenger in the sixth (smoking) compartment. It had found its way into the space beneath the seat where there was a collection of dust and small pieces of paper. There, smouldering had continued for the next 50 minutes, forming carbon monoxide and other gases, and considerably raising the temperature of the space, but generating little or no smoke. Eventually the fire had burnt through the wooden partition between the sixth and seventh compartments, probably a minute or so after Miss Goodwin had gone to the lavatory. Mixed with more air the gases had cooled and produced smoke and fumes. Mr Jacob, sitting in the third coach, had seen the smoke, and had gone to investigate. He had warned Mrs Chaffin of the fire, then gone to warn Mrs Campbell and her children. Mrs Chaffin had thrown her possessions along the corridor so as not to be encumbered with them while helping Mrs Campbell with her young children and Mr Jacob, entering Mrs Campbell's compartment, had pulled the communication cord. None of them had realised the imminent danger that they were in. At that moment the hole in the partition had enlarged to the point where the conditions of temperature and the gas/air mixture were such as to cause what was in effect a slow explosion in the corridor and perhaps also in the compartment. The heat thus generated had killed the five people instantly and set the whole coach on fire.

A coincidence of several factors had caused this unusual accident. These were an unnoticed slow smouldering fire, a completely closed and empty compartment, a long non-stop run, and few people passing along the corridor who had not noticed anything wrong, probably because the fire was concealed by the upholstery.

The inspecting officer remarked that he had examined a number of coaches similar to No. 4851. All had been well cleaned but most still had a collection of items beneath the seats, consisting of bits of paper, cigarette cartons, toffee and sweet wrappings, general dust and fluff, fibres, hairs and so on (the same thing would be found 37 years later under the escalators after the disastrous King's Cross Underground fire). The location was almost impossible to clean, sited as it was behind a heater and beneath a non-removable seat. The design was not satisfactory and removable seats, though they had disadvantages, were the obvious answer. Otherwise special cleaning measures should be adopted as it was difficult to completely prevent the accumulation of rubbish. The wooden partitions should also be replaced by ones of asbestos board (there was no appreciation of the dangers of asbestos in 1950). Wire mesh grilles fitted under the seats would reduce the fire risk but would not prevent the entry of fluff or lighted matches and were not a substitute for removable seats.

Colonel Walker concluded his remarks by praising the actions of Mr Jacob and Mrs Chaffin, who had given their lives in trying to help Mrs Campbell and her children; of Ganger Moffat, whose actions had also almost cost him his life (he was later awarded the British Empire Medal for his bravery); and the conduct of Driver Hannah and Fireman McKerracher.

The report continued with general remarks regarding fire precautions in passenger coaches. It was noted that over the years coach design had passed through half a dozen stages, from all-wooden coaches to all-steel coaches which included partitions of metal or plastic. All-steel construction, widely used in Europe and the United States, was not considered ideal for Great Britain. Weight, strength, safety, and maintenance and manufacturing costs all had to be taken into account in coach design and fire risk alone could not be considered a sufficient justification for introducing all-steel coaches. Even so, coaches with wooden interiors had featured in serious fires at Micheldever in 1936 and Westborough in 1941 and it was not wholly clear why these fires had spread so quickly. At Penmanshiel Tunnel in 1949 the rapid spread had been due to highly flammable lacquer. Here again, as at Beattock, the coach had had a wooden interior. If such coaches were to be continued with, the fire risk must be kept to a minimum by:

(a) Careful attention to detail in design

(b) The provision of a rapid means of escape in emergency

(c) Adequate fire-fighting preparations

As regards to detail of design, mention was made of the highly flammable lacquer which had caused the Penmanshiel Tunnel fire. It was still not possible to determine exactly how many coaches were involved but 'active and vigorous special measures' were being taken and it was hoped that the task of finding all the coaches so treated would be completed in another 12 months. The Railway Executive had given an assurance that the lacquer would not be used again. The Beattock coaches were finished with C.R. Matt Lacquer, which also contained nitrocellulose, though in a smaller proportion. It was recommended that the use of this lacquer should also be discontinued as quickly as possible.

Originally British coaches had a door to each compartment. From about 1923 the number of doors had been reduced, and from about 1930 corridor coaches had end doors only, though from about 1939 further doors had again been provided to ease the operation of crowded trains. In fires the corridors were likely to become blocked and for coaches with wooden interiors a form of emergency exit from each compartment was desirable. Ventilator openings were only for the young and thin and smashing the glass of the main windows would probably be too much for a woman to manage, quite apart from the risk of cuts from broken glass. Windows might perhaps be redesigned so that the lower portion could be lowered to sill level. Sleeping cars posed a special problem in view of the fact that passengers who were asleep would be unaware of a fire, the amount of flammable bedding and so on (prophetic words again, as shown 28 years later by the Taunton train fire). This was especially true for third-class cars carrying 28 passengers, which needed an intermediate door.

Fire extinguishers on British Railways trains were generally carried in kitchen, dining and sleeping cars and in guards' vans, though on the Great Western and Southern Railways they had been provided in every carriage. They were mostly of the soda-acid pattern, with some of the water-CO_2 type or of the CO_2, gas, or foam type where electric power was used. Extinguishers were only of use if employed at the start of a fire and the four previous serious fires of the preceding 15 years had shown how quickly things could develop. Extinguishers should therefore be provided in every coach. The training of railwaymen also needed to be improved as at Beattock two of them had not been able to get an extinguisher to work because they did not know how to use it. [1564]

The afternoon express from Hobart to Launceston on Thursday 17th August 1950 included girls from two basketball teams, from St Mary's College, Hobart and St Mary's Convent School, Hobart, some of whom described their experiences of it catching fire near Conara, Tasmania, Australia. Gwen Martin, of Launceston, told a newspaper reporter that smoke had drifted into their compartment. She said, "A railwayman came along then and asked us to alight. The train was separated into three units and we had removed our luggage and were standing in the paddocks for some minutes when the front of the compartment we had left burst into flames. The fire spread from the engine through the compartment and very soon it was a mass of flame and burning inside fiercely. It was too hot to approach closer than about 10 yards. The glass fell out of the windows and the paint sizzled."

Pat Begent of Launceston said that there had been no panic among the passengers, who had been going to leave their luggage but were told to take it with them.

Barbara Kingsley, Gwen Tomas, and Mary Kelly of Hobart agreed that it had been an adventure once it was known that all the passengers and their luggage were safe. They had been in the compartment behind the unit that caught fire and had felt some anxiety regarding their luggage, which included evening frocks, until they had found that it was safe.[1565]

The train was a diesel multiple unit consisting of articulated motor and passenger car, a passenger car and a rear guard's van. The fire had broken out in the motor unit and remained undetected until it had damaged the brake pipe, causing the brakes to apply automatically. The train crew had managed to detach the articulated unit and had attempted to tackle the fire with extinguishers which proved ineffective in controlling it. As passengers from the affected portion of the train waited in adjoining fields flames rising to a height of about 30 feet could be seen and the smoke was visible two miles away. [1566]

A fire in a mail car of the *Black Diamond* express of the Lehigh Valley Railroad resulted in the loss of thousands of dollars of letters and parcels post on Friday 13th October 1950. The blaze broke out in a sealed mail car which had been loaded in Philadelphia, Pennsylvania, and was destined for Buffalo, New York. As the train pulled into Geneva, New York, United States, at 6.55 p.m. passengers waiting on the platform could see that it was on fire. Railwaymen who tried to access the car, which was marshalled immediately behind the diesel locomotive, pending the arrival of the fire brigade, were frustrated by its locked doors. Firemen contained the blaze to the rear of the car and once it was out postal officials loaded the charred and sodden remains, damaged by both fire and water, into trucks. An eastbound freight train on an adjoining line was stopped by the use of flares. The damaged vehicle was placed in a siding under guard pending an investigation by postal authorities, local police, and railroad detectives, hopefully working together. The train was delayed an hour. [1567]

[1564] Ministry of Transport report

[1565] *The Examiner*, Launceston, Tasmania, Australia, 18th August 1950

[1566] *The Advocate*, Burnie, Tasmania, Australia, 18th August 1950

[1567] *Syracuse Herald-Journal*, Syracuse, New York, United States, 14th October 1950

Fire broke out on a crowded six-car Chicago North Shore and Milwaukee Railroad commuter train bound for Chicago as it reached the city limits and switched onto the tracks of the Chicago Transit Authority (CTA) on the evening of Friday 22nd December 1950. There were 70 passengers in a coach which filled with smoke following the outbreak of the fire and they panicked, smashing all the coach windows before the train, which had been travelling at 50 miles an hour, stopped at Argyle station after travelling a mile and a half. A passenger on the station said that she had heard cries of, "Let me out, let me out! Then the front door of the car opened and several persons were knocked down in the rush to get out," she added.

CTA trainman James Hillquist said that the panic had been among the civilians, while sailors on the train had helped to calm them. Eleven people were injured, including four sailors with severe burns on their hands and faces, and two Catholic priests.

The fire had reportedly originated in an electrical short-circuit switch box as the power had been changed from the overhead line equipment to the third rail of the CTA elevated line. It was out by the time that the fire brigade had arrived.

Eight people were still in hospital the following day. [1568]

A catastrophic train fire occurred at Sakuragichō, Japan, on Thursday 24th April 1951. The line was electrified on the overhead system and a maintenance crew were renewing electrical insulators when at 1.38 p.m. a wire was accidentally cut, allowing the contact to hang down. Four minutes later five-car MoHa 63 series train 1271B arrived from Yokohama. Fifty metres short of Sakuragichō the train changed tracks but the pantograph of the leading car of the train collided with the hanging contact wire. The motorman lowered the pantograph, which fell sideways and struck the wooden carriage. Sparks set fire to the roof, which consisted of wooden planks over steel bars, the whole roof being covered with canvas painted over with tar and sprinkled with sand.

The fire spread to the interior of the carriage and burned so fiercely that most of the passengers in the leading carriage were dead within eight minutes. Escape to the second carriage was impossible because the electrically operated doors could not be opened with the current having been cut off. Eight doors in the coach could have been released by using safety handles to release the compressed air which held them shut but the method of operating the safety handles had been kept secret by the authorities lest "someone turn them in fun." Not one passenger succeeded in opening a single door. Nor was escape through the train windows possible because as a cost-cutting measure a single large pane had been replaced in the wartime 63 series by three small panes which were too small for adults to climb through. One report did speak of passengers scrambling out of windows only to fall to their death from the elevated railway top the roadway 20 feet below, possibly they were small children pushed through the windows by their parents. [1569] The estimated 150 passengers in the leading carriage were therefore trapped. As a result 106 people died and 92 were injured.

An investigation into the catastrophe recommended improved fireproofing of carriages and the provision of through corridors between them. Also the marking of the

---

[1568] *Moline Daily Despatch*, Moline, Illinois, United States, 23rd December 1950

[1569] *The Times*, London, England, 25th and 26th April 1951 safety handles which released the doors and which were situated beneath the seats with red signs. Hideo Shima, the director of the rolling stock department, resigned. [1570]

Four people were burnt to death and ten more were injured when a coach of the Calcutta to Bombay mail train caught fire from an unknown cause at Bilaspur, India, on Friday 11th May 1951. [1571] The location of the accident was also given as near Nagpur. [1572]

A goods train from the West to Brisbane early on Sunday 10th June 1951 included a box waggon containing 8 tons of flour, consigned from Dalby to Ipswich. As the train was passing through Lockyer, Queensland, Australia, it was seen to be on fire and it was stopped at Laidley, about 18 miles further on, at about 1.30 a.m. The waggon loaded with the flour, which was the one on fire, was detached and the guard of the train extinguished the blaze, as he believed, in the undercarriage of the vehicle, with buckets of water. The waggon was left in the siding at Laidley and when Night Officer V. Francis left duty at 4.30 a.m. all seemed to be in order. At 5.15 a.m., however, Mr. M. Davenport, a milkman, noticed smoke and flames belching from the top of the waggon and raised the alarm. This time the local fire brigade was called and they put out the fire using water from railway tanks. 'The fire is said to have started as a result of friction in the undercarriage.' It was estimated that about 3 tons of the flour had been destroyed either directly from the fire or as a result of water damage. [1573]

The third serious train fire in Great Britain in just over two years occurred a mile north of Huntingdon, England, on Saturday 14th July 1951 on the East Coast Main Line of British Railways, Eastern Region.

The train involved was the 3.45 p.m. Down express from London King's Cross to Leeds Central, *The West Riding*. On the day of the accident the train consisted of Class A3 4-6-2 steam locomotive No. 60058, *Blair Athol*, twin articulated brake and third-class coaches, twin articulated first-class coaches, a twin articulated restaurant car and third-class coach, a first- and third-class composite, a third-class, a brake third-class composite, a brake first and third-class composite, a first- and third-class composite, two third -class and a brake third-class composite coach, in charge of Guard J.W. Nunns with Driver A. Cartwright and an unnamed fireman.

The train left King's Cross three minutes late at 3.48 p.m. At 4.30 p.m. a woman passenger noticed a wisp of smoke at the rear end of the second coach, No. 1737, and told a pantry boy, who notified the restaurant car conductor. The latter came and had a look, saw some smoke but no fire, and went back and notified the guard. The guard in turn saw the smoke and decided to write a note to throw out at Huntingdon, 10 minutes' journey down the line. He went back to his van to write out the note, stopping at the restaurant car to find a potato to attach the message to, but he could not find one. Meanwhile the fire was becoming more serious, until finally a passenger pulled the communication cord, after some difficulty finding it, while a soldier travelling as a passenger went back to

[1570] Wikipedia – Sakuragichō train fire
[1571] *Sunday Times*, Perth, Western Australia, 13th May 1951
[1572] *The Kingston Daily Freeman*, Kingston, New York, United States, 12th May 1951
[1573] *The Queensland Times*, Ipswich, Queensland, Australia, 11th June 1951

fetch the guard again. The guard threw out his message, unattached to anything, as the train was passing Huntingdon No. 1 box, and just as the soldier called him forward again. This time the guard did realise the seriousness of the situation and went to apply the brake in his van but the driver, having realised that the communication cord had been pulled, was already stopping the train. By now the fire had spread rapidly and with it burning at the rear of the second coach the only exit for about 65 passengers in the twin set was a single door at the front of the leading coach. The centre corridor of the open-plan coaches became blocked by passengers and those at the rear of the second coach found themselves trapped between the advancing fire behind them and the blockage of people in front of them. Their only escape was to smash the windows and jump from the train.[1574] 'One woman, who was in the restaurant car at the time, ran back to her compartment, where she found her six-month-old baby sleeping in a portable cot which was burning. She grabbed the child, handed him to a sailor on the track, and then jumped to safety herself.' [1575]

The train stopped 300 yards from a road but several people in the vicinity had witnessed the fire and called the fire brigade and police. The train stopped at about 5 p.m., the first fire engine arrived at 5.08 p.m. and the first ambulance at 5.12 p.m. The signalman in Huntingdon No. 2 box also saw the fire and blocked the lines, while the driver and fireman put down detonators. It proved impossible because of the heat to uncouple the train behind the second coach and instead it had to be divided behind the second set, with the result that all four leading coaches were burnt down to their underframes. Twenty-two passengers and the guard suffered cuts and burns, nine of whom were detained in Huntingdon County Hospital.

Colonel R.J. Walker, the inspecting officer, found himself again investigating a train fire for the Ministry of Transport.

Signalman S.L. Luff in Paxton box, five miles south of Huntingdon, told the inquiry that the train had passed him at 4.49 p.m., travelling at about 20 miles an hour. He had seen nothing wrong with it.

Signalman W.G. Pateman in Huntingdon No. 1 box said that the train had passed him travelling at 50 to 55 miles an hour. Everything had seemed normal. He had not seen the guard throw out any message (the note was found beneath a platform coping the following day).

Signalman D.L. Munro in Huntingdon No. 2 box thought that the train had passed him travelling at about 40 miles an hour. He could see nothing amiss with it, but a couple of minutes later saw that it had stopped down the line and that smoke was coming from it. He had telephoned Stukeley box to stop Up trains, then Huntingdon No. 1 box to call the fire brigade and an ambulance.

Guard Nunns stated he had been such for 14 months, usually working on local trains. He related his actions as detailed above. Asked why he had not at once stopped the train, or at least done so as it was passing Huntingdon, he explained that he had thought his first duty was to get a message out to have the train stopped and examined at Peterborough (17 miles further on).

Driver Cartwright put his speed on passing Huntingdon No. 2 box at 20 miles an hour. He said that he had noticed the vacuum drop 5 inches so had made a full brake

---

[1574] Ministry of Transport report

[1575] *The Times*, London, England, 16th July 1951

application. After his train had stopped he had got down from the cab of his engine and had seen passengers and suitcases pouring out of doors and windows on both sides of the leading two coaches. There had been very black smoke and then, after windows had been broken, flames as well.

Salvation Army Officer Boardman, a passenger, described seeing smoke coming from between an arm rest and the side of the coach half an hour before the train had stopped. The guard had been called, had lifted a seat cushion, reassured the passengers that it was probably a hot axle box and said that the train would be examined at Peterborough. Five minutes later the smoke had intensified, with acrid fumes, and the passengers, knowing that Peterborough was still half an hour's journey away, had not been happy with the guard's response. Two passengers had moved elsewhere in the coach while Mr Boardman had decided to investigate the matter further. He had found an area which was distinctly hot, had seen that some sponge rubber had half melted, and decided that it was time to pull the communication cord. After some difficulty he had found the alarm cord and pulled it, had asked a soldier to get the guard back again and advised the passengers in the coach to move forward. By the time that the guard had returned a small flame had appeared as well and uttering an exclamation, the guard had returned to his van to apply the brakes. He had then come back a third time with a fire extinguisher but had been unable to get it to work. As the train had slowed up and stopped the flames had suddenly spread with great speed up the sides to the roof of the coach. With the crowd of passengers it had been impossible for them all to escape through the single door at the end of the coach away from the fire and with some difficulty windows had been smashed, through which he and other passengers had got out. Some of the women passengers, frightened of the drop to the track, had had to be dragged out of the train. Mr Boardman stressed that the flames had spread very rapidly, enveloping both coaches of the articulated set, while the fumes blowing across the line had choked some of the passengers even after they had escaped from the train.

Other passengers gave similar accounts, agreeing that:

(a) There had been between 20 minutes and half an hour between the first appearance of smoke and the stopping of the train.

(b) Only Mr Boardman, of several passengers who had tried, had been able to find the communication cord.

(c) In general it was thought that there had not exactly been a panic, though there had been much excitement and rushing about.

(d) The soldiers and sailors among the passengers had done good work. Had it not been for their efforts, done at personal risk, some passengers would probably have suffocated or burnt to death.

The inspecting officer concluded that there could be little doubt that the fire had begun in the hole in the coach floor through which the air duct to the toilet was taken. A piece of live coal had probably lodged there soon after the train had left London King's Cross. The wood of the coach floor had smouldered for some time until the fire had spread to the sponge rubber floor covering, which had melted, giving off fumes of isoprene. Finally the sponge rubber had burst into flame and the fire had come into contact with the leather cloth covering the side of the coach and spread rapidly.

Other sources of fire, such as a dropped match or cigarette end, or an electrical fault, were, Colonel Walker ruled, extremely unlikely. The passengers in the coach had not been smoking and in any case a dropped match would not have smouldered for so long. None of the electric wires in the floor of the coach had been live at the time.

Turning to remarks, the inspecting officer noted that this was the third serious fire on a passenger train in a little more than two years. All three fires, and that at Westborough in 1941 (dealt with in the 1931-45 volume of this series) had been characterised by extremely rapid spread of the fire and the difficulty passengers had had in escaping from the burning coach. The Huntingdon fire again emphasised the need for proper and careful consideration of the fire hazard in designing coaches, the need for providing adequate means of escape in case of emergency, plus the need for the elementary training of train crews in fire fighting.

Several points about the Huntingdon fire stood out clearly:

(1) Had there been asbestos packing between the edges of the holes in the coach floor and the sides of the air duct the fire would have been prevented altogether.

(2) The dangerous and extremely rapid spread of the flames within the coaches would not have occurred had not their interiors been covered with highly combustible materials.

(3) The injuries to the passengers and the grave dangers to which they had been exposed might have been avoided if:

(a) The fire extinguisher used by the guard had worked.

(b) The communication cord had been readily available in the coach, and certainly if:

(c) The restaurant car conductor and the guard had stopped the train promptly when they were first told of the fire.

(d) There had been sufficient doors or opening windows in the coach for passengers to get out of it quickly.

Dealing with these points in turn, the inspecting officer wrote:

(a) The asbestos packing around the air vent hole had probably been missing for some time. Staff should be reminded of the importance of correcting such issues.

(b) The Penmanshiel and Huntingdon fires had been strikingly similar in their behaviour and the materials which had caused them. Their most dangerous feature, even more than the flames, had been the thick black smoke and fumes, which had blinded and choked the passengers. It was well known that fumes, rather than flames, often were the cause of death in fires. Materials containing nitrocellulose had caused the initial rapid spread of the fire in both cases; at Penmanshiel the clear cellulose lacquer had contained 68% nitrocellulose, at Huntingdon the coated cloth had contained 28% nitrocellulose. At Huntingdon the sponge rubber flooring had also been a factor. Nitrocellulose was clearly not suitable for use in coaches. Coaches containing nitrocellulose lacquer were being tested and dealt with as quickly as possible and their elimination was well in hand. The use of nitrocellulose cloth had been abandoned in 1949 but many coaches containing it remained. However, provided that the cloth did not extend to floor level the fire risk was less and metal or other skirtings were being fitted to 1,200 coaches, though the sooner that nitrocellulose cloth was removed from all coaches the better.

(c) The failure of two of the six fire extinguishers used at Huntingdon was disturbing. Those which had failed had both been of the soda acid type. As they had been destroyed in the fire the reason for their failure could not be determined. It appeared that the guard had attempted to use his extinguisher correctly. Fire extinguishers were now being provided in every coach, of the water rather than the soda acid type.

(d) The lack of a communication cord throughout the entire length of a double articulated coach was 'a particularly unsatisfactory feature of this accident' the inspecting officer stated. Nor was there any notice to tell passengers where the cords were, so it was not surprising that only Mr Boardman had found one. Notices were now being provided, together with access to the cords within the body of each articulated set.

(e) Had the restaurant car conductor or guard taken reasonable precautions the fire would have been easily quelled. The conductor had taken no steps to investigate where the fire had been coming from. The guard's actions, though well-intentioned, had been particularly ill-judged. He had rashly assumed that there was a hot-box and that it would be sufficient to examine the train half an hour later at Peterborough. Both assumptions had been wrong. It had not been a hot-box, and if it had been, he should not have delayed dealing with it for half an hour. Having viewed the fire he had walked the length of the train to his van to write a note, leaving no one to watch the fire and wasting valuable time. When he had returned the second time he had acted correctly in stopping the train and trying to fight the blaze with a fire extinguisher. The fact that the extinguisher had not worked was not his fault and he had sustained burns in fighting the fire. He had then helped passengers to escape and carried out his other duties as a guard. After the Beattock fire the Railway Executive had been asked to consider issuing a standard fire fighting procedure to staff but had thought it impracticable due to the need of staff to protect the train, then had adopted the suggestions of the inspecting officer only in part. Colonel Walker urged the institution of some form of fire fighting procedure.

(f) The general design of the articulated coaches, with only an exit door at each end, a door between the two coaches of the set, inward opening doors and narrow corridors, prevented an adequate means of escape. The need for a more adequate means of escape, with a centre door in sleeping cars, had been stressed in the report on the 1950 Beattock fire. Like the recommendation for the training of staff in fire fighting the recommendations for additional doors in corridor coaches and sleeping cars had not been accepted by the Railway Executive. The executive pointed out that any number of improvements could be made, but a balance had to be made between the risk of fire and the cost of reducing that risk, also that the risk would be reduced and centre doors made unnecessary by:

(i) The removal of nitrocellulose from all coaches containing it.

(ii) The protection of surfaces under seats around heaters with asbestos millboards.

(iii) The fitting of wire mesh grills beneath seats, and the making of all seats removable, to reduce the accumulation of rubbish and make it easier to clean beneath them.

(iv) The closure of the gaps between seat backs and partitions.

(v) The lower portions of compartment partitions to be of steel.

(vi) The provision of fire extinguishers in each coach, including dining, kitchen, and sleeping cars.

(vii) The provision of additional fuses in the main circuits of electrical wiring.

(viii) The lowering of the compartment windows in sleeping cars so that passengers could escape through them.

In addition, centre doors in open coaches would reduce sleeping capacity. In the new all-steel sleeping cars the only combustible material would be bedding and carpets. A centre door in sleeping cars would render the control of passengers at stations difficult, increase the risks of theft and of passengers mistaking the door at night and falling out of the train.

The inspecting officer was not impressed by the Railway Executive's objections. In the final part of his report he stated:

'To sum up, I do not consider that the reasons put forward by the Railway Executive, either in regard to risk, incidence, cost, or other disadvantages, are sufficient to justify their reluctance to carry out the few remaining improvements. Their unwillingness seems the more strange in view of the relatively small cost involved, particularly when the importance of the improvements has been confirmed so strongly by the experience at Huntingdon. I must therefore say that I do not think that the present position is satisfactory, or that the matter should be allowed to stand where it is.

'It is not in the nature of things that fire can he absolutely prevented, but it can be very largely so, and its effects reduced, by the adoption of suitable measures. The wide extent of the precautions which are taken in public buildings, however fire resisting their main structures may be, is well known. The incidence of fire in such buildings is probably very small indeed, yet experience has shown the necessity of the protection, and much of it is obligatory by law. A railway train, carrying six or seven hundred people at high speeds is, also in the nature of things, a considerably greater hazard than a solidly constructed building; it surely demands, therefore, at least as high a standard of protection in so far as it is possible to provide it. The first object in any fire is to save life, and the first and obvious way to do this is by providing adequate means of getting away from the fire. This is as vital in trains as it is anywhere else; certainly the lack of means of escape should not be allowed to be the direct cause of deaths or injuries, as it was at Westborough and Huntingdon.

'I do not wish to give the impression, and it would he entirely wrong to suggest, that the Railway authorities do not give constant thought and care to safety; indeed, it is their vigilance in this respect over many years which has resulted in the very high general standard which exists today. Nevertheless, consideration of the details of the recent series of fires, the design and equipment of the coaches concerned, and the lack of staff training, all seem to indicate that of late years the awareness of fire risk has become less acute than, perhaps, it should be. Neither the Penmanshiel nor the Huntingdon fires should have been allowed to occur for, if it was not already known, a simple test when the coaches were designed would have shown how highly combustible the offending materials were. Furthermore, after the Penmanshiel and Beattock fires had so clearly pointed out the dangers, it is difficult to understand why the particular coaches at Huntingdon were allowed to remain in service without modification; with their inflammable interiors, inadequacy of communication chains, lack of fire extinguishers and, above all, lack of sufficient exits, they were a very poor fire risk indeed.

'I therefore think that, even more important than the detailed improvements which remain to be done, there is need for a new outlook on the subject of fire in passenger trains. It seems to me that the problem should now be unreservedly recognised and squarely faced, and that it should be tackled with the object of deciding how much can be done in all directions to improve the situation and to remove unnecessary risk.'

An appendix to the report described a further train fire which had occurred on British Railways, London Midland Region, at Fordhouses, on Friday 14th March 1952.

Passengers had observed smoke and alerted the guard, who had been close at hand. He had attacked the fire with an extinguisher, it had apparently been extinguished, and the train had continued a quarter mile to the station. As it had stopped there, the compartment had burst into flames. Despite the use of all the fire extinguishers on the train and the attendance of the fire brigade the coach had been burnt out. No one was hurt. A discarded match was believed to have caused the fire. Although it had been built as recently as September 1950 the coach had contained nitrocellulose. [1576]

A 10-ton open waggon loaded with 67 carboys of hydrochloric acid caught fire when the train of which it formed a part was running between Deri Junction and Fochriw, Wales, on British Railways, Western Region, on Friday 17th August 1951. The train was divided and its engine proceeded to Fochriw with the front portion, including the waggon which was ablaze. There, railwaymen and firemen extinguished the fire and only seven carboys of acid were lost. [1577]

A complex accident occurred in Buenos Aires province, Argentina, on Wednesday 29th August 1951. A passenger train was stopped between Suipacha and M.J. Garcia stations after a fire had broken out in a dining car. While the fire was being fought a following train arrived and a rear collision ensued in which two people were killed and 16 were injured. Meanwhile the fire spread to involve five sleeping cars before it was extinguished. [1578]

Sparks and cinders emitted by steam engines have often been blamed for causing lineside fires but on Tuesday 4th September 1951 the opposite thing happened in Australia. A goods train was stopped by signals outside of Innisfail, Queensland, at a point where a grass fire was in progress close to the line. Several people witnessed the fire and called the fire brigade, who were therefore on hand when the train pulled into the station at 12.20 p.m., the train crew still being unaware that the blaze had spread to their train. A southbound mail train was standing on the track between the fire brigade and the goods train and apparently could not be moved so the firemen had to excavate a hole beneath the track through which they could feed the fire hose.

By the time that the fire had been extinguished one waggon had been extensively damaged. It had been loaded with a consignment of porcelain and earthenware goods packed in dry grass and straw, destined mostly for Wyper Brothers of Cairns. The consignment was valued at £1,500 and damage was estimated at £500. Mr K.E. Glasheen, manager of the company, said that the fire was ‘a big knockback’ as the goods had been on order for seven months and were in short supply. [1579]

When the Paris – Strasbourg express arrived at Pagny-sur-Moselle, France, either late on Sunday 2nd December or early on Monday 3rd December 1951, the passengers in a *Wagons-Lit* carriage were awakened by a burning smell. One of them was Mr Henry

---

[1576] Ministry of Transport report

[1577] *The Railway Gazette*, London, England, 24th August 1951

[1578] *Dixon Evening Telegraph*, Dixon, Illinois, United States, 30th August 1951

[1579] *Maryborough Chronicle*, Maryborough, Queensland, Australia, 5th September 1951

Hopkinson, the British Secretary of State for Overseas Trade, who was on his way to Strasbourg to take over from Mr John Foster, Under-Secretary for Commonwealth Relations, as leader of the British delegation.

A British newspaper outlined what happened next. 'When he {Mr Hopkinson} opened the door of his compartment he found the coach filled with smoke and having tried the window in vain without further delay he made for the door at the end of the corridor. The other passengers did the same or were able to get out through the windows, as in the meantime the communication cord had been pulled. Mr Hopkinson arrived at Strasbourg in a railwayman's overcoat over his pyjamas and was able to borrow clothes in order to carry out the day's business.' [1580]

Mr Hopkinson lost all his personal possessions in the fire and suffered slight burns to his feet. A number of the other passengers likewise suffered slight burns, though none was seriously hurt. The cause of the fire was not stated. [1581]

Mystery surrounded the origin of a fire at Beerwah, Queensland, Australia, on Monday 3rd December 1951. The box waggon, loaded with refrigerators, washing machines and other goods burst into flames when the waggon was opened at 11.45 a.m. The flames spread to a second waggon loaded with chaff. The box waggon and the chaff in the second waggon were destroyed but the second waggon was possibly saved. Rails were buckled by the heat and three must be replaced, blocking the North Coast main line for five hours. The Gympie mail train was delayed at Landsborough and three northbound goods trains were stopped at Glasshouse Mountains for several hours. [1582] [1583]

At 04:00 on Thursday 20th December 1951 a Société nationale des chemins de fer français employee noticed that the rear of a *Wagons-Lit* carriage was on fire when the train was near Montbard, France, in the Côte d'Or. The train was stopped and the passengers were hurriedly asked to alight. The heat prevented the burning coach from being detached from the vehicle behind it and that soon caught fire in turn, so that both had to be removed to a siding. No one was hurt. [1584]

A fire on a New York City Transit Authority train at 8.40 a.m. on Wednesday 2nd August 1952 caused something of a panic amongst its passengers. Matters were not helped by the fact that the fire occurred as the Brooklyn-bound train, the incoming *Sea Beach Express*, was on the Manhattan Bridge across the East River at the time, about midway between its centre and the Brooklyn end, and smoke began to fill some of the vehicles of the eight-car train.

A newspaper reported, 'Passengers screamed. Clothing was torn as men and women struggled to get off the train. One passenger said, "Everybody was fighting. Everybody was scared." A passenger said that one woman, "almost fell between the ties into the river but I grabbed her." He was Max Fleischer, 48, who escaped through a window with his wife,

---

[1580] *The Times*, London, England, 4th December 1951
[1581] *Journal de Genève*, Genève, Switzerland, 4th December 1951
[1582] *The Courier Mail*, Brisbane, Queensland, Australia, 4th December 1951
[1583] *The Morning Bulletin*, Rockhampton, Queensland, Australia, 4th December 1951
[1584] *Gazette de Lausanne*, Lausanne, Switzerland, 21st December 1951

Rita. He and his wife were taken to Long Island hospital suffering from lacerations.'

In all, seven people were taken to hospital suffering from injuries caused by breaking windows and perhaps smoke inhalation. Another person was hospitalised suffering from severe hysteria.

The fire was thought to have been caused by a grounded third rail contact. [1585]

As dawn was breaking on Friday 5th December 1952 and a Roma – Milano express travelling at 120 kilometres an hour was in the vicinity of Parma, Italy, three of its coaches were found to be on fire. Its passengers raised the alarm and evacuated to another part of the train. The train then continued to the next station where firemen were quickly able to deal with the fire. The remainder of the train was not affected. [1586]

A van on a northbound mail train from Brisbane was discovered to be on fire at 11.30 p.m. on Wednesday 17th December 1952 when the train was three miles south of Palmwoods, Queensland, Australia. The train continued to Palmwoods as quickly as possible with flames reportedly shooting 30 feet into the air from the van, and on arriving there the burning vehicle was detached and shunted into a loop line.

According to Mr Faragher, the Deputy Director of Posts and Telegraphs, 74 bags of Sydney, Melbourne and Brisbane mail destined for the area between Gladstone and Rockhampton and west of Rockhampton as far as Longreach were destroyed, the loss amounting to £4,000. Only 15 of 230 bags were undamaged and passengers' luggage was also lost. Fortunately 141 bags of Rockhampton mail had been transferred to the Townsville mail at the last moment to relieve congestion.

The cause of the fire was unknown and police were investigating. [1587] [1588]

Seven people died and five had to be taken to hospital, four of them Republic of Korea soldiers, as the result of a fire which was discovered when a northbound train from Taegu arrived in the station at Chikjisa, South Korea, on Christmas night, 25th December 1952. The fire occurred in a boxcar at the rear of the train and it was theorised that the people riding in the car, in order to keep warm, had collected firewood and built a fire, which had accidentally got out of control. [1589]

The carelessness of a passenger and the fact that he was travelling with hazardous goods led to a serious fire on a train in India on Thursday 26th February 1953. After No. 73 Up passenger train on the Eastern Railway had left Gaya a compartment of one of its coaches was seen to be on fire and the train was stopped at 14:55. Despite the best efforts of railwaymen and fellow passengers five passengers were burnt to death, nine others suffered burns and seven more were injured in jumping from the moving train to escape the fire. Damage to property was estimated at 40,000 rupees. It transpired that among the luggage of a passenger was a tin of petrol and a match carelessly discarded after lighting

---

[1585] *The Daily Herald*, Biloxi, Mississippi, United States, 2nd April 1952
[1586] *Journal de Genève*, Genève, Switzerland, 6th December 1952
[1587] *Maryborough Chronicle*, Maryborough, Queensland, Australia, 19th December 1952
[1588] *Brisbane Telegraph*, Brisbane, Queensland, Australia, 18th December 1952
[1589] *Pacific Stars and Stripes*, Tokyo, Japan, 29th December 1952

a cigarette, it was believed by the same person, set the petrol alight. He was one of those who died. [1590]

At about 13:30 on Monday 13th July 1953 a fire broke out on an electric train from Tagliacozzo to Roma near Tivoli, Italy, not far from the capital. A short-circuit in the overhead line equipment had set fire to a coach. The frightened passengers fought to escape to the corridor, which was already occupied by other panicking passengers. One woman, Maria Torre, 51, tore herself from her husband's grasp and threw herself from the train in her fright. An ambulance took the poor woman to a polyclinic with a suspected fractured skull. Other passengers were taken to hospital for treatment of cuts and bruises but none was seriously hurt. According to a report in an American newspaper the train stalled in a tunnel but no tunnel is mentioned in an Italian newspaper account. [1591]

On Saturday 24th October 1953 a mail van in a train from Dunkerque to Paris was seen to be ablaze as the train ran into Corbie, France. The fire was a serious one and over 100 of the 350 mail bags in the van were destroyed, resulting in the loss of thousands of letters from Great Britain. The Post Office announced that second-class mail to France, Italy, Portugal, Spain, Gibraltar, and Malta (including mail for Naval personnel) might have been lost. First-class and registered mail was not affected. No one was hurt and the cause of the fire was unclear. [1592]

Shortly after Canadian National Railways westbound ten-car transcontinental passenger train No. 1 from Montreal had left Pembroke Junction, Ontario, Canada, early on Wednesday 30th December 1953 fire broke out in its rear Pullman sleeping car, at 2.05 a.m. The train had only covered about a mile when its crew realised that it was on fire, it was brought to a stand and the rear sleeping car was detached.

The passengers in the coach were unable to escape to the coach ahead, presumably because of the location of the fire, and instead they made their way to the rear, where a woman was able to smash a window with her travelling case and eighteen of them climbed through the window and jumped down to the track in their night clothes. None of them was severely hurt, though they suffered cuts and bruises and probably some degree of exposure given the sub-zero temperatures outside. The passengers, all bound for northern Ontario, also lost all their possessions.

It took firemen over two hours to extinguish the blaze, when they found the body of a man, badly burnt, lying in the rear vestibule of the coach. It was thought that he had suffocated while he was waiting his turn to climb through the window. He was later identified by his brother as Dr H.A. Collins, of Ottawa, who had been travelling to Haileybury, Ontario.

The cause of the fire was not immediately clear but the police held a man, Charles Faucault, of Noranda, Quebec, for questioning, and he was later 'charged' in the wording of a newspaper report, with being mentally ill, implying that arson might have been

1590 https://archive.org/stream/dli.csl.1063/1063_djvu.txt
1591 *La Stampa*, Torino, Italy, 14th July 1953
1592 *The Times*, London, England, 26th October 1953

involved. [1593] [1594]

On Tuesday 15th June 1954 another New York City Transit Authority subway train caught fire. A contact shoe had become jammed onto the third rail, generating acrid smoke which filled the cars. As on the Manhattan Bridge rather less than two years earlier the fire caused a panic among the passengers. This second accident happened near 86th Street station on the Lexington line. The southbound train, which was en route to Brooklyn Bridge, was about 150 yards north of the station when the fire broke out at 8.01 a.m.

When the cars began to fill with smoke the frightened passengers began smashing windows. About 20 pushed through the doors and climbed to an emergency exit, only to find that their path was blocked by a sidewalk grating. The group, who included many high school students, then began to scream, and shout and the noise attracted a man in a nearby church, who was able to prise the grating open with a crowbar and release the passengers. Meanwhile transit patrolman Andrew Kenders, on duty on the station platform, smelt the smoke and saw the train stopped north of the station. He made his way to the train and assisted a large number of passengers along the catwalk to the station platform. "There was yelling and screaming," he said. About 20 people were attended for various injuries. [1595]

Quick action by Guard J. Newly, in charge of a mixed train from Charleville, saved two crates of pigs and probably prevented major fire damage on the night of Tuesday 31st August 1954. Three miles short of Amby, Queensland, Australia, he noticed that an A.L.G. box waggon was on fire (it has not been possible to determine what the initials A.L.G. stood for). The train was stopped, the box waggon was detached from the carriages behind and the vehicles in front and shunted into a siding. Railwaymen were able to rescue the crate of pigs and some empty beer kegs before the intensity of the fire forced them to stop. The other contents of the van: wools, skins, and hides, were destroyed. The Amby fettling gang stood by to clear the line once the van had burnt itself out. Superintendent Buchanan stated that the cause of the fire was unknown but an inquiry would be held. [1596]

In what was something of a freak accident five miles north of Seoul, South Korea, on Thursday 4th November 1954, three Korean students were burnt to death and two passengers and the driver were injured. Sparks from the steam engine hauling the train had ignited petrol flowing from a broken pipeline adjoining the track and the fire had spread to the coaches of the train, three of which were destroyed. Most of the passengers on the train were students on their way to an engineering class, many of whom had jumped from the train and rolled on the ground to extinguish their burning clothing. [1597]

Fire broke out on a crowded express train in the station at Pusan, South Korea, at

[1593] *The Brandon Daily Sun*, Brandon, Manitoba, Canada, 30th December 1953
[1594] *The Lethbridge Herald*, Lethbridge, Alberta, Canada, 30th December 1953 and 2nd and 4th January 1954
[1595] *Record Herald*, Waynesboro, Pennsylvania, United States, 15th June 1954
[1596] *The Western Star*, Roma, Queensland, Australia, 3rd September 1954
[1597] *Pacific Stars and Stripes*, Tokyo, Japan, 6th November 1954

about 6.15 p.m. on Wednesday 2nd March 1955 as the train was about to leave for Seoul. Frantic passengers found themselves trapped behind a jammed door in the burning coach. Thirty-nine passengers lost their lives and their bodies were burnt beyond recognition. Twenty-eight more were injured, 19 seriously. 'There were no American casualties.' One coach was totally destroyed and three others were badly damaged.

The fire broke out when a candle carried by a vendor accidentally came into contact with a 15-gallon can containing a rubber solution mixed with petrol. The contents exploded and flaming jets of petrol shot through the coach so that the fire spread very rapidly. [1598]

Korean and American firefighters were able to get the blaze under control within about 15 minutes. [1599]

The Swiss first-class coach of a Basle – Paris express was found to be on fire as the train arrived at Thomery station, France, about 70 kilometres from Paris, on Tuesday 24th May 1955. Alerted by the conductor its passengers and those of the adjoining couchette coach exited the train with their luggage. The two burning vehicles were then shunted into a siding where the fire took hold and they were burnt out. The train was able to resume its journey after an hour and other traffic was not affected. The fire was blamed on an electrical short-circuit. [1600] [1601]

During the evening rush hour of Tuesday 18th September 1956 fire broke out on a seven-car elevated electric express train of the Chicago Transit Authority. Flames spread through several of the wooden cars of the train, which dated from the 1900s, and almost 100 passengers leapt from the train at Morse Avenue, Illinois, United States, many of them risking electrocution by skirting the live third rail in their desperation to escape. The current was quickly shut off on all four tracks to prevent any fatalities. Five people were injured, the most seriously hurt being Conductor Bernard Devenby, who suffered burns to his face and hands as he made his way through the cars calming passengers as the train came to a halt. Four of the train cars were almost totally destroyed and the accident resulted in delays to thousands of homegoing workers. A newspaper wrote that the fire 'was accompanied by a mysterious blast believed to be an electrical flash under the antiquated wooden coaches from the fuse box beneath the third and fourth cars, from where the fire apparently started.' [1602]

A hundred women and children escaped serious injury, if not worse, when fire broke out in five coaches of a Ceylon Government Railway Colombo – Jaffa night mail train between Palal and Kodikaman, Ceylon early on Tuesday 25th September 1956. Railwaymen were able to detach the burning vehicles from the train. So fierce was the fire that telegraph and telephone wires melted and the five coaches were completely burnt out. [1603]

1598 *McKinney Daily Courier-Gazette*, McKinney, Texas, United States, 3rd March 1955
1599 *Fairbanks Daily News-Miner*, Fairbanks, Alaska, United States, 3rd March 1955
1600 *The Times*, London, England, 25th May 1955
1601 *Journal de Genève*, Genève, Switzerland, 25th May 1955
1602 *Lubbock Evening Journal*, Lubbock, Texas, United States, 19th September 1956
1603 *Indian Daily Mail*, Singapore, 1st October 1956

Less than two months after the Morse Avenue fire, on Friday 7th December 1956, a more serious fire occurred on another Chicago Transit Authority train, this time a four-car one bound for downtown Chicago from Evanston. By chance Russell Lane, an Associated Press reporter, was a passenger on the train. He had been sat in the second car and said that the first car, where the blaze had begun, "went up like a torch." Some of the passengers in the forward part of the train had fought their way to its rear while others had leapt through closed windows to the elevated right of way. The train had been scheduled to stop at Howard Street station but continued through it, though it slowed down and came to a halt just south of the station. Meanwhile, passengers from the front car were streaming through the second amid shouts of "fire." As often happens in such situations the rush of passengers in the narrow passage between the cars resulted in a jam and at this point some of them jumped through windows instead. Lane himself was one of those who made their way to the rear of the train before descending to the track. The front car of the train was destroyed and the second badly damaged.

About 40 passengers were taken to various hospitals on the North Side suffering from smoke inhalation and cuts. The main casualty, however, was Motorman Daniel Cullen, 47, whose badly burnt body was found in what remained of his cab, where the fire had begun. [1604]

The passengers in the Swiss coach of an international train which had left Milano at 06:18 for Domodossola on Tuesday 7th January 1958 had some moments of panic at about 07:00. While the train was travelling at high speed they became aware that an electric radiator under a seat had set fire to some wooden panelling. Frightened, they raised the alarm and unfortunately opened windows to let out the smoke. The train staff were not able to put the fire out, which grew in magnitude. However, when the train reached the station at Gallarte, Italy, firemen were able to extinguish the blaze in a few minutes. The passengers vacated the affected coach unharmed and complete with their luggage and after 30 minutes' delay the train continued its journey, minus the badly damaged coach, which was left at Gallarte. [1605]

As a southbound Southern Pacific Railroad freight train was approaching Heather Siding, Oregon, United States, on the night of Thursday 25th June 1959, Conductor P.W. Conley saw smoke and fire coming from the 29th car back from the engine. He notified his engineer and the train was put into the siding, where the burning box car exploded. Later the bodies of two men, believed to have been transients (hobos) were found in the remains of the vehicle. None of the train crew was hurt and the train was not derailed but the line was blocked for about four hours.

The box car involved had been carrying water melons packed in straw. It was theorised that the two men had been using canned heat (a device containing alcohol, water, and a petroleum gel, used for outdoor cooking) to prepare food, and had accidentally set the straw alight. They had then been overcome before they could escape and the intense heat generated had transformed the moisture in the melons into steam, causing the explosion. Earl Morton, another transient from Texas, said that he had seen

[1604] *Moline Daily Despatch*, Moline, Illinois, United States, 7th December 1956

[1605] *Journal de Genève*, Genève, Switzerland, 8th January 1958 the two transients board the train at Oakridge, Oregon, and the remains of what appeared to be four tins of canned heat were discovered among the wreckage. [1606]

A crowded six-car electric local train from Avila to Madrid caught fire at kilometre 16 of the Madrid – Hendaye line, a short distance from the station at El Plantío, Spain, on the evening of Sunday 12th July 1959. Two of the cars were reportedly of wooden construction and the fire spread quickly, causing some of the passengers to jump from windows to escape. Six passengers were injured, none seriously. A kilometre of track and overhead line equipment was damaged and both lines were blocked for a time; later single-line working was introduced. The cause of the fire was believed to either have been a discarded cigarette end which had fallen into the space between a window pane and the wooden framing of the carriage, or else an electrical short circuit. [1607]

Fire broke out on the 5.07 p.m. British Railways Eastern Region Down electric passenger train from London Liverpool Street to Southend on Thursday 13th August 1959. The train stopped out of course at Wick Lane Crossing, just after leaving Wickford, England, where an ambulance picked up three women passengers suffering from the effects of inhaling fumes and took them to hospital. A following train pushed the affected one four miles to Rayleigh station, where the passengers were transferred to another train and firemen dealt with the fire. [1608] An investigation found that 'a blockage in the pneumatic controls for the raising and lowering of the pantographs of a two-set 1500 volts D.C. electric train approaching Wickford had caused the pantographs of one set to make only intermittent contact with the overhead wire with the result that at times the eight coaches were being driven by only two motors. Consequently the resistance of one motor became very hot and burnt a hole through the heat shield above it, starting a smouldering fire in the coach floor. The protective asbestos prevented any fire in the compartment, but fumes generated when the fire extinguishing fluid came into contact with the hot resistance causing some distress to passengers. Arrangements have been made to improve the fire barrier between the resistances and the coach floors.' [1609]

The postal van of train No. 4881 from Winterthour to Schaffhouse, Switzerland, was found to be on fire when it arrived at its destination at 06:01 on Friday 23rd October 1959. The fire had begun in a cupboard where two parcels had been placed. A heater had set fire to the parcels and the fire had then spread to the body of the van itself. As soon as the blaze was discovered it was dealt with using fire extinguishers. Other than the two parcels none of the mail was harmed, and no one was hurt, but 25,000 francs worth of damage was done to the mail van. [1610]

Mail vans seem to have had rather a disposition to catch fire and on Friday 18th December 1959 it was the baggage and mail car of the eastbound *Super Continental*

1606 *Tri-City Herald*, Pasco, Washington, United States, 26th June 1959
1607 *La Vanguardia*, Barcelona, Spain, 14th July 1959
1608 *The Times*, London, England, 14th August 1959
1609 RAILWAY ACCIDENTS: Report to the Minister of Transport on the Accidents which occurred on the Railways of Great Britain during the year 1959
1610 *Journal de Genève*, Genève, Switzerland, 24th October 1959

passenger train of the Canadian National Railways which caught fire at a point east of Winnipeg. The train proceeded as far as Minaki, Ontario, Canada, where the fire was extinguished.

There were 49 mail bags in the van at the time. These were returned to Winnipeg where they were sorted, dried out, and forwarded again. According to a Post Office official none of the mail was so badly damaged as to be undeliverable. The mail consisted largely of Christmas cards and first-class business mail weighing more than eight ounces. The remainder of the mail went by air.

A note attached to mail delivered in Ottawa was headed "Winnipeg Damaged Mail" and read, "A severe fire recently occurred among a number of mail bags on a train leaving Winnipeg Dec. 18. This article of mail matter was damaged in the fire. The inconvenience and delay which this unfortunate circumstance has caused you is very much regretted by the post office department."

No one was hurt in the fire, the cause of which was not stated.[1611]

A motor unit of shuttle train No. 17, which had left Romanshorn at 20:41 on Tuesday 14th June 1960, caught fire between Sulgen and Amriswil, Switzerland. The train was able to continue as far as Amriswil, from where it had to be towed by a steam locomotive after the fire had been extinguished. No one was hurt but the motor unit was seriously damaged. There was also serious delay caused to a number of other services.[1612]

[1611] *Winnipeg Free Press*, Winnipeg, Manitoba, Canada, 23rd December 1959
[1612] *Gazette de Lausanne*, Switzerland, 16th June 1960

CHAPTER 15

CHEMICAL EXPLOSIONS

Railway accidents involving explosions were of two types; those which might be defined as chemical explosion and locomotive boiler explosions. Accidental chemical explosions, such as those due to mishaps with blasting powder or munitions, were rare, whereas in the early days in particular locomotive boiler explosions were much more common, though by the mid-twentieth century they were becoming much rarer, while chemical explosions were perhaps becoming a little more common. In wartime the latter often involved the accidental explosion of munitions. This chapter will deal with chemical explosions and also those involving electrical equipment, even though these are not strictly chemical in origin.

Eighteen waggons of a Soviet ammunition train were destroyed in a mysterious explosion near Wolmirstedt, Germany, on Thursday 12th May 1949. The train had been on its way from Stendal to Madgeburg and had been shunted near the town, in the Soviet zone, to allow a passenger train to pass. That train had just gone by when the explosion occurred, ripping up 500 metres of track and damaging several lineside houses. West Berlin newspapers stated that several railwaymen and Soviet guards had been killed but this was denied by officials. Damage to the track was such that the line was blocked for several days. Officials and police were investigating the cause of the blast. [1613] [1614]

A westbound Chicago Great Western Railway train of 134 cars from Chicago to Oelwein, in charge of Conductor Al Basler, was wrecked between Aurora and Stanley, Iowa, United States, at 1.15 a.m. on Thursday 25th May 1950. Twenty-two cars of the train left the track and it appears that the precipitating cause of the derailment was the explosion of a gasoline (petrol) tank car which had formed part of the train. Following the derailment the gasoline caught fire and the blaze spread to other cars of the train. Further complicating the accident was the leakage of another car loaded with muriatic (hydrochloric) acid. This at first made it impossible for section workers and wreck crews to linger near the wreckage as the acid fumes were burning their eyes and making breathing difficult. Later in the morning a breeze drove away the flames and the rescue work could continue.

[1613] *Indian Daily Mail*, Singapore, 17th May 1949
[1614] *The Billings Gazette*, Billings, Montana, United States, 16th May 1949

The Oelwein fire department had been called at 1.45 a.m. and the fire truck had to make its way through the Kyle Ingamel farmyard to reach the accident site, only for the firemen to find when they got there that the fire was intense and they reportedly could do nothing to save the burning cars. According to another newspaper account the firemen could not get closer than 600 yards to the fire anyway because their vehicle had become stuck in a field. The fire having eventually burnt itself out, as it seemed, the firemen left, only to be called out again at 2 p.m. to deal with a further outbreak. This time a smaller fire truck was taken to the scene loaded onto a railway car, doubtless to ensure that it did not get bogged down, as had apparently happened earlier.

Cars of farm implements, sheet steel and starch were involved and the damage caused was estimated at several thousand dollars. Some hundreds of feet of both tracks were ripped up in the derailment. [1615] [1616]

When an Egyptian Mersa Matruh – Alexandria passenger train was about six miles from its destination on Monday 21st August 1950 an explosion occurred which wrecked a coach, killing five people and injuring seven more. It was believed that a sack containing explosives had accidently fallen from the train and theorised that desert Bedouins might have salvaged the explosive from Second World War battlefields around Mersa Matruh and have been smuggling it into Alexandria. [1617]

Twenty-three people died and over 200 were injured when an ammunition train standing alongside a passenger train at Yongchong, South Korea, exploded on Thursday 28th September 1950. News of the accident was only reported the following week when it was also announced that nineteen members of the American military police had been recommended for the soldier's medal for heroically rescuing Korean civilians and military personnel while the ammunition was still exploding and burning. [1618]

A Soviet ammunition train exploded near Wittenberg, East Germany, on Tuesday 5th August 1952, killing an unknown number of German railwaymen and Soviet soldiers. The authorities, rightly or wrongly, blamed the explosion on an act of sabotage. [1619]

The Second World War in Europe had ended in May 1945 with the occupation of Germany by the Allies. The Cold War between the Western Powers and the Soviet Union had followed, with the division of the occupied zones of Germany into East and West Germany in 1949. Thus it was that the Americans still had a strong military presence in Germany in late 1952. Ammunition for United States army and air force bases in West Germany was shipped from the United States to the La Rochelle – Bordeaux region of south-western France, from where it was transported by rail to Germany. This was the background to the explosion of a munitions train at Châlons-sur-Marne, France, on Thursday 26th November 1952.

1615 *The Oelwein Daily Register*, Oelwein, Iowa, United States, 25th May 1950
1616 *The Telegraph-Herald*, Dubuque, Iowa, United States, 25th May 1950
1617 *The Norwich Sun*, Norwich, New York, United States, 22nd August 1950
1618 *Lincoln Evening Journal*, Lincoln, Nebraska, United States, 4th October 1950
1619 *Defiance Crescent-News*, Defiance, Ohio, United States, 5th August 1952

One waggon of an ammunition train initially exploded at 09:30. This in turn set off explosions in two further waggons of the 12 which made up the train and eventually 25 waggons on adjoining tracks in the marshalling yard were also involved in the fire, while 20 lengths of rail were ripped up. The accident could have been much worse but for the courage of a French railwayman, Shunter Georges Girard, who had uncoupled the nine remaining waggons of the ammunition train so that they could be moved out of harm's way.[1620]

French police evacuated people from an area within a radius of 600 metres. The main line between Paris and Strasbourg was blocked and trains were diverted. Initial reports stated that no one had been hurt.

An investigation was opened to determine whether the explosion had been accidental or the result of an act of sabotage.[1621] [1622] [1623]

Just as passengers were alighting from a Canadian Pacific Railway train from Edmonton on its arrival in the station at Calgary, Alberta, Canada, at 7.30 p.m. on the evening of Sunday 26th April 1953 there was a powerful explosion. Mr Charles Blake and his wife chanced to be the first ones to alight from the train and it was they who thus caught the full force of the blast. Mrs Blake was killed outright and Mr Blake was critically injured and died in hospital five days later. The Blakes had been thrown 10 or 12 feet as they had walked past a refrigerator car. The car itself was almost destroyed, windows were smashed within 100 feet and platform trucks were overturned. Three railwaymen – Stanley Eby, Ernest Edwards, and James McKenna – suffered burns to their faces and bodies but were able to walk to a waiting ambulance.[1624] [1625]

The cause of the blast was initially a mystery. After an investigation General Superintendent C.E. Mayne stated that the explosion had been due to the ignition from a railwaymen's kerosene lantern of an accumulation of propane gas. The gas had escaped from a tank being shipped as empty from Fairview to Calgary due to it having a defective valve. The tank had originally been used as a fuel tank on a tractor and was being returned by its owner due to the defective valve. There had been no markings on the shipment to indicate its dangerous nature and hence no railwayman had been aware of the hazard. The explosion had occurred as soon as railwaymen opened the door of the refrigerator car to unload it.[1626] [1627]

A violent explosion occurred in a third-class bogie (coach) of No. 313 Katihar – Kanpur Express at 20:33 on Wednesday 31st March 1954 in Jagatbela station, India. Twenty-eight passengers were killed outright and another 16 people were seriously injured, including the guard. Three of the injured died on the way to hospital. The lights of the train had gone off and then back on again twice of their own accord, suggesting, a

---

1620 *Albuquerque Journal*, Albuquerque, New Mexico, United States, 27th November 1952
1621 *Journal de Genève*, Genève, Switzerland, 27th November 1952
1622 *Tucson Daily Citizen*, Tucson, Arizona, United States, 26th November 1952
1623 *Ironwood Daily Globe*, Ironwood, Michigan, United States, 26th November 1952
1624 *The Medicine Hat News*, Medicine Hat, Alberta, Canada, 27th April 1953
1625 *The Brandon Daily Sun*, Brandon, Manitoba, Canada, 27th April 1953
1626 *The Lethbridge Herald*, Lethbridge, Alberta, Canada, 4th May 1953
1627 *The Medicine Hat News*, Medicine Hat, Alberta, Canada, 16th May 1953

newspaper wrote, that a leak of electricity had set off a stick of gelignite. There had been a terrific flash and the entire train and the station platform had been shaken. The entire roof of an 85-seat third-class composite carriage, which included railway patrolling police and women's compartments, was blown off.

The newspaper report continued, 'Eye-witnesses said there was considerable confusion for 20 minutes after the explosion. Groans and cries of the injured and dying were heard from a third-class bogey. Hearing the noise of the explosion, a small crowd collected at the scene of the accident and helped in rescuing the people trapped in the wreckage. According to the eye-witnesses, limbs, and bodies of some of the passengers were mutilated beyond recognition.' An ambulance train was sent from Gorakhpur at 21:30 and returned there with the injured at 23:10.

The District Manager, Gorakhpur, stated that there had been three packages in the compartment where the explosion had occurred. One of the packages had contained 380 detonators, the second about 300 sticks of gelignite and the third a quantity of fuse wire, the three packages all being properly sealed and kept separate. Two constables of the civil armed police had been taking the packages to Agra and had been accompanied in the same compartment by three men of the railway protection police. All, of course, were killed instantly. The remainder of the coach, placed fourth from the engine, was occupied by ordinary passengers.

Reference to the accident was made in Parliament, where it was explained that the explosives were a part of about 1,500 gelignite sticks which had been intercepted by the police, and they were being taken for testing at the time of the accident. The matter was being investigated. [1628]

A bottle of chemicals being carried by a passenger exploded on a train near Taejon, South Korea, on Thursday 22nd April 1954. This caused a panic which led to a stampede in which two passengers died and 80 were injured, 27 seriously. [1629]

An explosion in the kitchen of a restaurant car of the 10.40 a.m. Liverpool to Plymouth express on the evening of Sunday 11th September 1955 occurred while the train was stationary at Bristol Temple Meads station, England. It was followed by a fire which damaged tables, seats and equipment in the car and injured three men who fought it. The three kitchen staff, chef Edward Maher, and attendants Frederick Martin, and Kenneth Sweet, all suffered burns and were taken to hospital, the first two being admitted. The car was detached from the train and shunted into a siding, where the fire was extinguished, and the blaze was confined to the one vehicle. [1630] The train continued its journey 23 minutes late and arrived at Plymouth North Road only 13 minutes late. [1631] A leakage of gas used for cooking was the cause of the explosion. A defective gas pipe had been poorly repaired two days earlier and again one day earlier but had continued to leak, until the gas had accumulated sufficiently to cause the explosion. [1632]

1628 *The Indian Express*, Madras, India, 2nd April 1954

1629 *The Winona Republican-Herald*, Winona, Minnesota, United States, 23rd April 1954

1630 *The Times*, London, England, 12th September 1955

1631 *The Singapore Free Press*, Singapore, 20th September 1955

1632 Report to the Minister of Transport and Civil Aviation upon the Accidents which have occurred upon the Railways of Great Britain during the year 1955

Six people were seriously burned when blank ammunition for use in toy pistols exploded in a compartment of a Katowice – Warsaw train in the station at Czestochowa, Poland, on or about Tuesday 21st May 1957. [1633]

A series of electrical explosions occurred at 8.40 a.m. on Friday 2nd August 1957 on a Broad Street six-car northbound subway train at Erie Avenue station, Philadelphia, Pennsylvania, United States. The train was very full, with standing passengers, as it was towards the end of the morning rush hour. There was something of a panic amongst the passengers as they rushed to escape and 81 were injured by glass from broken windows or suffered shock, though none was seriously hurt. A newspaper wrote, 'Some of the passengers panicked. A few took off their shoes to smash windows and climb out onto the station platform. Women screamed. Shattered glass and blood covered the platform.' Three of the cars of the train had already been past the platform as the train had been pulling out of the station when the explosion occurred but the motorman opened the doors of the rear three cars. The fire alarm was sounded and fire engines joined a stream of ambulances heading for the station. There was a large traffic jam on the surface while the subway service was suspended. Injured passengers were taken to the Temple University Hospital, the Einstein Medical Center and the Women's Medical College hospital. Other passengers left the station and purchased sedatives to calm their nerves at nearby drug stores (chemists). The damaged train was towed away to a depot and normal working was resumed within an hour. [1634]

Motorman Edward Caswell said that the blast had derailed the leading car of his train. He added that he had been temporarily blinded by the flash of the explosion but had soon recovered, when he had opened the rear train doors, by which time many of the passengers had been screaming and hysterical. [1635]

A spokesperson for the Pennsylvania Transportation Company explained that a motor on the leading car of the train had jammed, causing a short circuit and the explosion of a circuit breaker, with several very loud reports, electrical flashes, and thick smoke. [1636]

Five people were killed outright and 12 were injured when a wagonload of crackers exploded at 17:30 on Wednesday 21st August 1957 at Katpadi Junction, India. [1637]

An explosion in the station at Malinitz, Austria, on Saturday 22nd December 1957, occurred in the mail van of the Balkan Express and was followed by a fire. No one was hurt but a large number of Christmas parcels bound for Yugoslavia were destroyed or damaged. Police were investigating the possibility that a bomb was involved. [1638]

A powerful explosion occurred in a marshalling yard at Niagara Falls, New York,

[1633] *The Times*, London, England, 22nd May 1957
[1634] *The Titusville Herald*, Titusville, Pennsylvania, United States, 3rd August 1957
[1635] *The Gettysburg Times*, Gettysburg, Pennsylvania, United States, 2nd August 1957
[1636] *Cumberland Evening Times*, Cumberland, Maryland, United States, 2nd August 1957
[1637] Wikipedia – List of railway accidents and incidents in India.
[1638] *Panama City News*, Panama City, Florida, United States, 24th December 1957

United States, at 9.32 a.m. on Wednesday 22nd January 1958. [1639] The explosion left a crater more than 40 feet deep and 150 feet across. Over 60 people were injured, though no one was killed, and property damage was estimated at over a million dollars. A line of tank cars was shattered and one of them appeared to have been at the centre of the blast, but a chemist for E.I. Du Pont de Nemours & Co, the company which owned the contents of the waggon, stated that it had been loaded with dimethylformamide, a non-volatile solvent which should not therefore have exploded. Railroad superintendent Norbert Fitzgerald stated that an explosion of the solvent had been ruled out as a cause.

Some of those who were injured were pupils at Hyde Park Elementary School, which was closed for the day while broken glass was cleaned up and windows were replaced. Pupils at a second school were treated for injuries caused by broken glass by a medical unit from Niagara Falls Air Force Base. Only two people, Engineer John Geeran and Brakeman Richard Billson, both Niagara Junction Railway employees, were badly hurt. Thirty-three people were treated at the Memorial Hospital and 27 at St Mary's Hospital. The explosion had fortunately happened in a sparsely populated area. The windows of houses two miles distant were broken and dishes were rattled five miles away, as if by an earthquake. The operator of a nearby service station said, "We could feel it before we heard it. There was a surge of pressure in the air and then the ground rocked." Ordnance disposal experts from a nearby anti-aircraft artillery unit examined the area with Geiger counters and were unable to detect any sign of radioactivity but the Niagara Country sheriff's department ordered the area evacuated anyway for fear of harmful fumes from the blast. Other damage caused by the explosion included damage to utility poles and the roofs of homes, falling plaster, collapsed house porches and so on, while power to the city of 100,000 was cut off briefly. City Manager Edward Potthoff Jr. said that damage to the city incinerator plant, half a mile from the site of the explosion, might total $100,000. Five employees of the plant were among those injured. [1640]

As for the cause of the explosion, Public Safety Director Charles Gorman stated, "I'm not satisfied it was a completely surface explosion; in my book it could well have come from underneath." He added that investigators were studying the possibility that material in subsurface pipes might have caused or contributed to the blast.

By the following day Superintendent Fitzgerald had changed his mind as to the cause of the explosion, stating that the explosion of an underground pipeline had now been ruled out, and it had been concluded that the dimethylformamide in the tank car had exploded after all, though it was not clear why it had done so. The car had been on its way from the Commercial Solvents Corporation at Peoria, Illinois, to the International Minerals and Chemical Corporation at Niagara Falls. He added that the crater had been filled in using 3,000 cubic yards of dirt and the yard should be back in full operation by Friday 24th. [1641]

According to the Interstate Commerce Commission report on the June 1958 explosion at Mount Pulaski, Illinois (see below), the tank car had actually contained nitromethane rather than dimethylformamide.

On Tuesday 18th March 1958 a 64-car New York Central Railroad freight train on the branch line between Kankakee and South Bend included three cars loaded with propane,

---

[1639] *Dunkirk Evening Observer*, Dunkirk, New York, United States, 22nd January 1958

[1640] *The Bradford Era*, Bradford, Pennsylvania, United States, 23rd January 1958

[1641] Ibid, 24th January 1958

placed 14th, 15th and 16th in the train. Half a mile west of North Judson, Indiana, United States, one of these cars exploded, set the second on fire and caused the third to explode in turn. The shocks derailed the diesel locomotive and 18 leading cars of the train and another ten cars at the end of the were also thrown off the line. Police kept spectators a mile from the scene because of the intense heat and the danger of further explosions; a level crossing with Indiana 10 close to the scene was blocked off. Railwaymen estimated it would take 48 hours to clear the line. The cause of the explosion was not determined. No one was hurt. [1642] [1643]

The explosion of a railroad tank car during switching (shunting) operations at Mount Pulaski, Illinois, United States, on Sunday 1st June 1958 killed two railwaymen, injured four others, and left 1,000 people in the town temporarily homeless. The accident formed the subject of a full Interstate Commerce Commission (ICC) investigation.

Illinois Central Railroad northbound second-class freight train No. 292 consisted of road switcher diesel-electric units Nos. 9188 and 9136, coupled in multiple, 31 cars and a caboose. It had left Mattoon, Illinois, at 9.05 a.m., 3 hours and 5 minutes late, and arrived at Mount Pulaski at 1.05 p.m., 3 hours and 15 minutes late, when switching operations began.

Following train No. 292 was northbound second-class freight train No. 68, which consisted of road switch diesel-electric units 9174 and 9019, 78 cars and a caboose. The 36th vehicle in the train was a tank car, GATX 29633. This train had left East St Louis, Illinois, at 9 a.m., 3 hours late and arrived at Mount Pulaski at 2.35 p.m., 4 hours late. Switching operations were then performed involving the cars of these two trains and at 3.10 p.m. tank car GATX 29633 exploded.

A crater approximately 100 feet deep and 36 feet in diameter was caused by the explosion. Tank car GATX 29633, and 17 other cars were destroyed and 12 more cars were heavily damaged by the explosion and further damaged by the fire which had followed it. Conductor Robert Glover and Swing Brakeman Jack Griffin of train No. 292 were killed and the engineer, fireman, front brakeman, and flagman of that train were injured. [1644]

The shockwave shattered windows and tore off the doors and porches of buildings in the town of 1,500 people. Officials estimated that two-thirds of the 400 homes in the town were damaged and not immediately habitable. Damage was estimated by town and Civil Defence officials at $1,000,000. State police ordered the evacuation of the town for fear of further explosions from other tank cars, while National Guardsmen were ordered into the town to prevent looting and help with the clean-up. Once the fire had been extinguished about 500 residents whose homes were not too badly damaged were allowed to return. Other residents, fearing a second blast, had driven to Lincoln, ten miles away. For those who could not return to their badly damaged homes food and shelter were provided at the Central High School and a 200-bed emergency Civil Defense hospital unit. Other evacuees went to local farms and nearby communities. A resident commented, "It looks like a war-ravaged city, a ghost town." [1645]

[1642] *The Terre Haute Star*, Terre Haute, Indiana, United States, 19th March 1958
[1643] *The Brazil Daily Times*, Brazil, Indiana, United States, 18th March 1958
[1644] Interstate Commerce Commission report
[1645] *The Daily Gazette*, Sterling-Rock Falls, Illinois, United States, 2nd June 1958

The ICC report only included the testimony of only one witness, the engineer of train No. 292, who said that he had seen a white vapour coming from the cut of cars which included tank car GATX 29633 just before the explosion.

The tank car concerned had been loaded with about 10,000 gallons of nitromethane on 28th May at the Commercial Solvents Corporation plant at Sterlington, Louisiana, on the Missouri Pacific Railroad. It had passed onto the Alton and Southern Railroad and then the Illinois Central Railroad at East St Louis and was bound for Peoria, Illinois.

The ICC noted that tests done on nitromethane in 1952 by the Bureau of Explosives of the Association of American Railroads had determined that it was not a dangerous substance and hence it was not regulated by the Commission. After the January 1958 Niagara Falls explosion further tests were carried out and it was decided that if certain contaminants were present the nitromethane could become very explosive, a fact which had not previously been recognised.

The ICC concluded that the accident had resulted from the explosion of a tank car loaded with nitromethane. Exactly why the explosion had occurred could not be determined but it was clear that under certain conditions the substance could become a dangerous explosive. On 10th September the ICC issued an order prohibiting the bulk transportation of nitromethane by either rail or road. [1646]

Further research suggested that the Mount Pulaski accident had resulted from adiabatic compression, with small, entrained air bubbles becoming compressed and superheating with a rapid rise in pressure. It was thought that a railwayman had rapidly closed a valve, creating a 'hammer-lock' pressure surge. [1647]

[1646] Interstate Commerce Commission report

[1647] Wikipedia – Nitromethane

CHAPTER 16

BOILER EXPLOSIONS

The railways of the mid-twentieth century were still powered largely using steam engines, though electric traction was making inroads, mostly on suburban and underground railway systems and since the end of the 1920s gasoline- or diesel-powered locomotives had been in use to some degree. By the period dealt with in this book there were no full boiler explosions in Great Britain, as opposed to crown sheet failures, but they continued to occur in the United States until the replacement of steam by diesel traction, while a number have come to light in Germany and Poland, plus a couple elsewhere.

The boiler of Missouri Pacific Railroad engine No. 1712 exploded near Everest, Kansas, United States, on Tuesday 15th January 1946. Fireman Glen Mosley was killed outright and Engineer Fred Maisenbacker died of his injuries two days later. [1648] The Interstate Commerce Commission attributed the explosion to the failure of the crown sheet due to low water. It noted that the feed water pump of the engine had been erratic in operation, 'apparently due to the feed water by the exhaust steam from stoker which was discharged into the leg of the water tank about 30 inches from the tank outlet and by steam discharged into the feed pipe from the antifreeze pipe which interfered with the prime of the pump.' Problems with the feed pipe not working properly had been reported 26 times since the beginning of November of the previous year. [1649]

The crown sheet of the locomotive of an International-Great Northern Railroad hospital train collapsed about a mile west of Thrall, Texas, United States, at about 10.55 a.m. on Friday 15th February 1946. Conductor C.E. Keevert, riding in a caboose at the rear of the train, realised that there was a problem when he saw clouds of smoke billowing back over the train. Looking out of a window he saw a man lying beside the track and immediately applied the air brakes, stopping his train in roughly 400 yards. At first he imagined that his train had collided with a car at a level crossing and did not realise until he ran back to investigate that the man that he had seen lying in the ditch was Fireman R.H. Donaldson.

Soldiers on the train saw Engineer H.J. Owens crawl out of his cab and start towards

[1648] *Moberly Monitor-Index*, Moberly, Missouri, United States, 18th January 1946
[1649] Interstate Commerce Commission Thirty-fifth Annual Report of the Director, Bureau of Locomotive Inspection

the rear, his face blackened by soot. After covering about a car's length he collapsed and was dead when help reached him. He had not been scalded but had been killed by the force of the blast.

The main force of the explosion had apparently vented itself on the fireman's side and Fireman Donaldson was thrown from the cab. He was taken to hospital in a critical condition suffering from severe shock, scalds to his face, a broken shoulder and leg bruises.

A farmer whose home was 500 yards from the tracks was just about to have a shave when he heard the noise of the explosion and felt his house shake. Looking out of a window all that he could see at first was steam and smoke, then the air cleared, he could see the red crosses on the side of the train and he realised that something must have happened.

The train had been carrying 350 wounded soldiers from various parts of the South Pacific who had only been back in the state for three days and were on their way to hospitals in the east for further treatment. [1650]

The boiler of a Canadian Pacific Railway engine exploded near Rumford, Ontario, Canada, at 2.30 a.m. on Thursday 18th April 1946. Engineer Cleve Eaton and Head Trainman Angus McKinnon were killed and William Balec, who presumably was the fireman, was taken to hospital suffering from severe burns. The engine was at the head of a freight train from North Bay to Sudbury; it and two cars were derailed. [1651]

Shortly after midnight on Friday 26th July 1946, as train No. 88, a Cairo – Shellal express, was running between Etlidem and El-Roda, Egypt, and moving at about 75 miles an hour, the boiler of its engine, 2-6-0 locomotive No. 631, exploded. As a result, the leading brake van, a sleeping car, a Pullman car, two first-class coaches and another sleeping car were derailed. The rails at the point of the blast were badly twisted and both running lines were blocked. The driver and fireman were killed instantly and a pointsman attached to the Mechanical Department was seriously injured, while the guard and a brakeman were slightly hurt, as were two child passengers.

The Down main line was cleared at 5.40 a.m. on Saturday 27th and single-line working was opened up between Etlidem and El-Roda. A shortage of water in the boiler was blamed for the accident. [1652]

Germany suffered several boiler explosions in the immediate post-war period. The first in 1946 occurred at Rastatt on Wednesday 25th September 1946 when engine No. 44 1368 was the victim. Less than a fortnight later, on Monday 7th October 1946, the boiler of engine No. 44 1514 gave way at Nürnberg. Both explosions were blamed on low water. [1653]

Poland likewise suffered a number of boiler explosions in the first years after the war, the first of which occurred on Monday 14th October 1946, between Charsznica and

[1650] *The Taylor Daily Press*, Taylor, Texas, United States, 15th February 1946
[1651] *The Lethbridge Herald*, Lethbridge, Alberta, Canada, 18th April 1946
[1652] *The Railway Gazette*, London, 30th August 1946
[1653] https://de.wikipedia.org/wiki/Kesselzerknall#cite_note-4

Wolbrom, to engine No. Tr203-58. [1654]

Late on the afternoon of Monday 11th November 1946 an eastbound Southern Pacific Railroad freight train No. 3175 was proceeding up the seven-mile incline east of Gila Bend, Arizona. 4-10-2 steam locomotive No. 5037 was being assisted in front by Alco diesel switch locomotive, No. 1333. At Bosque Siding, Arizona, United States, the train crossed westbound a freight train, Extra No. 5008. As the eastbound train cleared the siding the westbound train set off and Conductor C.W. Birk, looking back at the departing eastbound train, saw the boiler of engine No. 5037 explode.

The force of the explosion lifted the 80-ton boiler of engine No. 5037 over the top of the diesel and it landed 265 feet up the track, breaking a rail and depressing the roadbed 30 inches. Bouncing back into the area it finally stopped 390 feet ahead of the site of the explosion and 17 feet to the right of the track. Engineer Harry Hall and Fireman Walter Glisson on the steam locomotive were killed instantly. The explosion broke the connection between the diesel and steam engines and the diesel ran forward until it fell into the depression made by the detached boiler. Engineer John Rhodes and Fireman Frank Bogulas on the diesel were severely injured and died in hospital the next day. A fifth railwayman, Brakeman Theodore Holaway, was seriously injured. Conductor C.W. Adams in the caboose was unhurt. None of the cars behind the engines was derailed. The line was cleared at 2.30 a.m. the next day, previous to which one passenger train, the *Argonaut*, had been diverted via the Phoenix line. [1655]

A coroner's jury at Gila Bend returned a verdict of 'Accidental death' on the four men. [1656]

The Interstate Commerce Commission investigation determined that locomotive 5037 had been serviced at Yuma only two and a half hours before the accident and that no faults had been found with its injector, feedwater heater, or water glasses. A boiler wash had been carried out four days earlier and its fusible plugs had been renewed 12 days earlier. The water level in the boiler had fallen to 6¾ inches below the bottom of the water glasses and the low water had resulted in the softening of the top three inches of the crown sheet, the "burning" of which extended almost 19 rows of staybolts back, almost half the length of the firebox. Tested on a sister engine the feedwater heater and injector worked correctly. A possible clue to how the water had been allowed to get too low was that the reflective surface on the inside of the water glass on the engineer's side, the only one to have survived the explosion, had worn dark, perhaps contributing to its being misread. The fusible plugs had melted but it was theorised that the noise of the hard-working engine might have masked the sound of the steam escaping from them. [1657]

As a coal train was coming to a halt at the *Einfahrsignal* (home signal) at Haldensleben, Germany, on Monday 18th November 1946 the boiler of its locomotive, No. 50 582, exploded. Torn from the frame of the engine the boiler was thrown 15

[1654] https://www-polskieparowozy-pl.translate.goog/index2.php?wykaz=wybuchy&_x_tr_sch=http&_x_tr_sl=pl&_x_tr_tl=en&_x_tr_hl=en&_x_tr_pto=sc

[1655] *Tucson Daily Citizen*, Tucson, Arizona, United States, 12th and 13th November 1946

[1656] Ibid, 14th November 1946

[1657] https://www.trainorders.com/discussion/read.php?10,4696263

metres. The driver, fireman and a third railwaymen riding on the engine were killed. The cause was low water, either due to inattention on the part of the engine crew or a faulty water gauge, it could not be determined which. [1658]

The *Empire Builder*, the crack express passenger train of the Great Northern Railway, was travelling at 65 miles an hour when the boiler of its locomotive exploded at Devil's Lake, North Dakota, United States, on the evening of Thursday 9th January 1947. Engineer Reginald Bennett, Fireman Wilbert Hoff and Roadmaster E.A. Zinther, who had boarded the train at Grand Forks, were all killed. A smoking car, two sleeping cars and the front wheels of a dining car were derailed and two passengers were bruised and one was shaken. Division Superintendent J.P. Cameron described it as "only a happy miracle" that scores of passengers on the transcontinental train had not been injured. The other seven cars of the train remained on the track.

The cause of the explosion was not known and an investigation was opened.

The engine had been completely demolished according to a Devil's Lake banker who visited the scene. He said that the scalded body of the fireman was recovered near the engine cab and the bodies of the other two men had been found among the twisted remains of the boiler, adding, "I never saw more complete destruction. Scared, but unhurt, passengers were milling around while workers were playing flashlights on the wreckage. Lack of light added to the confusion as workers searched the wreckage in the darkness for possible other injured. Nearby telephone lines were hanging limp and several {had been} apparently cut by flying debris." [1659]

The circumstances surrounding the explosion of the boiler of locomotive No. 52 2753 at Sontra, Germany, on Monday 27th January 1947, were a little unusual. The engine was to be used to assist a train of supplies destined for Czechoslovakia. There was a long delay before the train was ready to depart, during which time its driver and fireman fell asleep. When they awoke they failed to notice that during their rest the water level in the boiler had dropped to a critically low level. The explosion which followed killed them both for their negligence as well as another driver in the van adjoining the engine. The boiler was thrown 50 metres. [1660]

Engine No. 52 6992 exploded at Ingolstadt, Germany, on Monday 3rd February 1947 due to low water. [1661]

As a 65-car Central of Georgia Railway freight train from Atlanta to Macon was passing through Bolingbroke, Georgia, United States at 40 miles an hour, early on Sunday 1st June 1947, the boiler of its engine exploded, killing Engineer C.S. Simson and Fireman J.H. Cannon. The boiler was thrown 18 feet and pieces of the engine were

---

[1658] *Deutsche Eisenbahn-Katastrophen* by Hans Joachim Ritzau, Jürgen Hōrstel and Thomas Wolski, pg. 212 (in German)
[1659] *Kenosha Evening News*, Kenosha, Wisconsin, United States, 10th January 1947
[1660] https://de.wikipedia.org/wiki/Kesselzerknall#cite_note-4
[1661] Ibid.

scattered about. An automatic brake application stopped the train. [1662] The Interstate Commerce Commission found that the explosion was due to overheating of the crown sheet as a result of low water. Water had reportedly been taken at Barnesville, 20 miles back, but had run low by Bolingbroke. It was theorised that the small amount of remaining water had run to the front of the boiler as the train was descending a grade, exposing the crown sheet, then, when the train had reached a level stretch the water had run back again onto the overheated crown sheet, triggering the explosion. [1663]

An eastbound 55-car Chicago and North Western Railway freight train was wrecked three and a half miles south of Blencoe, Iowa, United States, on the evening of Tuesday 17th June 1947 when the boiler of its engine exploded. Engineer Harry Elkins and Fireman Warren Smitts were killed and Brakeman Dan Nice was injured and taken to hospital in Onawa. The bodies of the engineer and fireman were blown 200 yards in different directions by the force of the explosion. Conductor John Beckwith and Brakeman Tony Kunnen, riding in the caboose at the rear of the train, were unhurt. Several carloads of livestock were thrown into a ditch. Some of the cattle in them were killed, others were released and roamed about the countryside. Carloads of meat at the rear of the train remained on the track. [1664]

Engine No. 98 896 suffered a boiler explosion at Lennep, Germany, on Tuesday 7th October 1947 due to low water. [1665]

On the evening of Tuesday 7th October 1947 a Chicago and North Western Railway was stabled in the yard at Lake Preston, South Dakota, United States. The four-man crew of the train elected to spend the night in Huron, 40 miles further west, and only night watchman Boyd Cavanaugh was in the yard when the boiler of the oil-fired engine exploded at 12.15 a.m. the following morning. The 20-ton boiler was hurled 175 feet and the night watchman was blown 50 feet. He was very badly hurt and was rushed to Huron hospital where he died of his injuries shortly after admission.

Railroad officials pronounced that the cause of the explosion might never be known. Damage was estimated at $100,000. Within 24 hours the wreckage had been cleared and work had begun to repair the damaged tracks. [1666] [1667]

At 9.15 a.m. on Wednesday 15th October 1947 a driver and fireman began to prepare 0-6-0 goods engine No. 1572 at Hither Green, England, on the Southern Railway, one of a class dating from 1900, to work a train across London to Old Oak Common. Three-quarters of an hour later, while both men were in the engine cab, the top of the firebox gave way with such force as to lift the engine from the rails. Somehow neither man was

---

[1662] *The Butler Herald*, Butler, Georgia, United States, 5th June 1947

[1663] http://www.mymcr.net/news/70-years-ago-train-engine-explodes-in-bolingbroke/article_27c9a858-462f-11e7-9890-db367ac12f2e.html

[1664] *The Daily Times Herald*, Carroll, Iowa, United States, 18th June 1947

[1665] https://de.wikipedia.org/wiki/Kesselzerknall#cite_note-4

[1666] *Atlantic News-Telegraph*, Atlantic, Iowa, United States, 8th October 1947

[1667] https://www.kingsburyjournal.com/stories/1947-steam-locomotive-explodes-kills-night-watchman,5135?

killed but both were burnt and scalded, the fireman very seriously. The 112 one-inch mild-steel stays which had secured the firebox roof were found after the accident to be 'shockingly corroded.'

The engine had been taken into Ashford works for general repairs in July 1944. It had been examined by the boiler shop shoreman who had written in a notebook 'roof stays bad.' His note had been overlooked and nothing had been done to the roof stays when the other repairs were undertaken. The engine was then returned to service. In addition, a boiler inspector had examined the engine no less than nine times between 1944 and the day of the accident without detecting the corroded stays. [1668]

Four locomotive boiler explosions occurred in Poland towards the end of 1947. On Saturday 22nd November that of engine Ty2-361 gave way at Szumirad station. Only three days later, on Tuesday 25th November, engine No. Ty2-1050 was shunting at Michalin Siding, on the Częstochowa – Kaatowice line, when its boiler exploded. On Wednesday 10th December (or possibly Sunday 12th October, the dates are unclear) engine number Tr203-163 exploded at Bukowo Czluchowskie station. Finally, on Wednesday 17th December, the boiler of engine No. Ty2-947 exploded while it was shunting at Jaworzyna Śląska station. The last-named accident was unusual. All the other boiler explosions in Poland between 1946 and 1960 were attributed to low water but that to engine No. Ty2-947 was attributed to a stick of explosive which had accidentally got into the coal supply and not been noticed. [1669]

A Chicago Milwaukee St Paul and Pacific Railroad mixed train consisting of an engine, two combination passenger and baggage cars, five refrigerator cars and a cattle car had left Portage at about 7 a.m. on Sunday 11th January 1948 for Madison. It left Poynette at 7.20 a.m. and ten minutes later the boiler of its locomotive exploded 2½ miles south of Arlington, Wisconsin, United States. At the moment the boiler gave way it had been ascending an incline at about 15 miles an hour. The blast was heard six miles away. The boiler landed 40 feet from the remainder of what was described as an 'old-type locomotive' which was reduced to a skeleton. The rails and sleepers in the vicinity of the explosion were badly damaged while the two coaches were only slightly damaged and the freight cars escaped harm.

Fireman Raymond Hovey was killed instantly and Engineer Walter Woodrow, who was blown from the cab of the engine, suffered severe burns and cuts, and was taken to St Saviour's Hospital, Portage, where his condition was described as "serious." There had only been six passengers on the train, none of whom was hurt. Conductor Arch McPherson, Baggageman Vern Clemmons and Brakeman Jesse Taylor also escaped injury. A taxi took the passengers on to Madison and buses replaced trains for the remainder of the day. Traffic resumed at 10.30 p.m. after section crews had worked all day to clear the line. In the early afternoon a driving snowstorm began which hindered the rescue work and snowed up the cars of local residents who had driven out to see the

---

[1668] *Locomotive Boiler Explosions* by C.H. Hewison, pgs. 130-2

[1669] https://www-polskieparowozy-pl.translate.goog/index2.php?wykaz=wybuchy&_x_tr_sch=http&_x_tr_sl=pl&_x_tr_tl=en&_x_tr_hl=en&_x_tr_pto=sc

wreck. Some of their cars had to be pulled out of the snow by a breakdown truck. [1670]

The 9.25 p.m. Up express passenger train from Glasgow Central to London Euston on Sunday 7th March 1948 consisted of 4-6-2 Coronation Class steam locomotive No. 46224 *Princess Alexandra* and 11 coaches, in charge of Guard Colin Sargent with Driver John Wetton and Fireman John Wilson. It had almost reached the station at Lamington, Scotland, about 40 miles from Glasgow, when the crown sheet of its firebox gave way. When the train stopped for no reason Guard Sargent went forward to the engine where he found Driver Wetton lying in the four-foot of the Down line. He helped the driver to the side of the line, could not find the fireman, and proceeded to Lamington signal box, where he found Wilson. With the assistance of the fireman of a Down train which the Lamington signalman had stopped because the Up train had been a long time in the section, the guard then helped the driver to the station. Both driver and fireman were badly burnt and scalded and Driver Wetton died shortly afterwards.

The cause of the failure was the subject of a comprehensive investigation by Employment Inspector J.L.M. Moore. He found that both boiler water gauges on the engine had been defective. They were of a type, the design of which had much to criticise, but which had nevertheless been adopted by the London Midland and Scottish Railway (by 1948 here British Railways, Scottish Region).

The right gauge glass had been renewed at Polmadie by Fitter John Daglish and wrongly assembled so that it caused the water in the glass to rise to a much higher level than it was in the boiler. The left gauge glass had a different and less serious defect which made it slow to fill. Driver Routledge discovered these problems when the engine was next in use on 6th March. Becoming suspicious of the right-hand gauge constantly being full when the left gauge was only half full, he told his fireman to rely on the left-hand one and on arrival at Carlisle wrote a report: "Both gauge frames showing false water."

Fitter Maxwell at Carlisle found the same problems but decided that the boiler must be full and the left-hand gauge was faulty. He told the crew of the next train that the left-hand gauge was not to be relied on. The train was worked to Glasgow Central and thence to Polmadie shed, where Driver Runcie reported, "Left gauge glass column to examine, right gauge glass showing false water." Leading Fitter H. Hillhouse gave the repair car to Fitter W. Begg and later both examined the engine together, but Hillhouse's examination of the gauges was, Mr Moore commented, 'unsatisfactory and inconclusive.' At 5.30 p.m. Driver Clark Robertson found that the right-hand gauge glass was full but there were only 1½ inches of water in the left-hand glass. He informed Hillhouse who told Fitter William Matheson to renew the left-hand glass. The latter had some difficulty and concluded that the handle and crank of the left-hand gauge were on the wrong squares (they were not). Matheson told Driver Wetton, Fireman Wilson and Hillhouse of the supposed problem with the left-hand gauge and that the gauge was otherwise working normally. It was now nearly time for the engine to be got ready to work the 9.25 p.m. train, so instead of investigating further Hillhouse released the engine, the driver and fireman being told to rely on the right hand gauge.

Fireman Wilson, when he had finally recovered enough to give evidence, said that he had not been surprised that the water level shown by the right-hand gauge had not varied much with a light load and a free-steaming engine. On the journey he and his driver had

1670 The *Wisconsin State Journal*, Madison, Wisconsin, 12th January 1948 heard the escape of steam due to the failure of the front fusible plug, but they had not realised its significance.

Mr Moore put most of the blame for the accident on Leading Fitter Hillhouse, whom he considered to have been gravely negligent. He made three recommendations:

(a) Consideration should be given to the advantages and disadvantages of coupled water gauge cocks.

(b) If coupled, assembly should be carried out in the works.

(c) A stop should be incorporated in the design of each cock, to limit the travel of the plug. [1671]

A more recent writer wonders if the leading fitter had not been under considerable pressure from his superiors not to withdraw an engine from service if it could possibly be avoided. [1672]

On Wednesday 14th April 1948 the boiler of locomotive No. Tr203-397 exploded between Warsaw Gołąbki and Józefinów, Poland. [1673]

The boiler of Clinchfield Railroad Class L-3 Mallet 2-8-8-2 steam locomotive No. 740, pulling a 70-car southbound coal train from Elkin City, Kentucky, to Spartanburg, South Carolina, exploded at 1.14 a.m. on Wednesday 21st April 1948 at Milepost 146, a location given in newspaper reports as Caney Bottom, North Carolina, United States. The boiler was carried 360 feet, ending up lying at right-angles across the track. The body of Engineer Ed Davenport was found 250 feet from the wrecked locomotive. The severely injured fireman, Rex Phillips, was found lying on top of the coal tender and he expired about an hour after the explosion. Before he succumbed he told rescuers that they had had plenty of water and he could not understand how the boiler could have given way. [1674] [1675] [1676]

Chesapeake and Ohio Railway locomotive No. 3020 suffered a boiler explosion as it has hauling a train of 156 coal cars bound for Columbus, Ohio, on the evening of Wednesday 12th May 1948. The explosion occurred at Charleston Pike, just south of Chillicothe, Ohio, United States. The engine cab was sprayed with boiling water. Fireman T.R. Lauter and Brakeman W.C. Bush died of their injuries a few hours after the accident and Engineer W.R. Terry died in hospital the following day, all having been badly scalded. Railroad officials stated that the front end of the locomotive had been ripped off and parts of it scattered over half a mile.

Dick Vincent, a farmer living near the scene, reportedly stopped a southbound train 600 yards from the wreckage by placing a flare on the track. (Where did he get a flare

---

[1671] Ministry of Transport report

[1672] *Locomotive Boiler Explosions* by C.H. Hewison, pgs. 132-7

[1673] https://www-polskieparowozy-pl.translate.goog/index2.php?wykaz=wybuchy&_x_tr_sch=http&_x_tr_sl=pl&_x_tr_tl=en&_x_tr_hl=en&_x_tr_pto=sc

[1674] *Hattiesburg American*, Hattiesburg, Mississippi, United States, 24th April 1948

[1675] *Kingsport Times*, Kingsport, Tennessee, United States, 22nd April 1948

[1676] *When Steam Ran the Clinchfield* by James A. Goforth, pg. 98

from?) [1677] [1678]

The boiler of engine No. Tr203-197 exploded on the Czempiń – Iłówiec line in Poland on Tuesday 25th May 1948. [1679]

The boiler of Union Pacific Railroad Class 9000 4-12-2 locomotive No. 9018, pulling a freight train of about 100 cars, exploded near Upland, Kansas, United States, at about 5.25 p.m. on Wednesday 20th October 1948. Engineer Elmer Schreck and Fireman John Delich were thrown almost 300 yards from the track and killed outright. Head Brakeman David Blevins was terribly injured and died in the Randall Hospital in nearby Marysville the next morning. Like the engineer and fireman he had been blown from the cab. The boiler of the locomotive was blown between 100 and 150 yards and all but the very front of the engine was demolished. The line was blocked until early the following morning. None of the cars of the train was derailed and the engine, minus its boiler, was towed back to Marysville.

Chief Clerk G. Haber stated that he was unable to explain the accident to the train, which had left Marysville shortly after 5 p.m. The Union Pacific company and the Interstate Commerce Commission were to conduct a joint investigation into the accident. It was theorised that the fact that the train had been ascending a bank at the time of the explosion might have been a factor in water getting low in the boiler. [1680] [1681]

Freight train No. 94 of the Charleston and Western Carolina Railway left Spartanburg, South Carolina, for Augusta, Georgia, at 2.50 a.m. on Thursday 18th November 1948, with locomotives No. 812 and 853. At Laurens both engines took on coal and water, the load of the train was reduced and it continued with engine No. 863 alone at 6.15 a.m. Water was taken at Maxwell Yard and McCormick and the train left McCormick at 9.27 a.m. with 25 loaded cars, 11 empty cars and a caboose. As the train was negotiating a 6-degree left-hand curve and ascending a 0.84 (1 in 120) gradient the boiler of the engine failed 1,620 feet west of the station at Clark's Hill, South Carolina, United States. The engineer was killed outright, the fireman was fatally hurt and died a few minutes after reaching hospital, and the head brakeman was badly hurt, having either jumped or been thrown from the engine and fallen down a 60-foot embankment. The accident was officially attributed to the overheating due to low water in the boiler leading to rupture of the crown sheet. This was disputed by the surviving head brakeman who attributed the explosion to a build-up of sediment in the crown sheet. [1682]

Westbound St Louis-San Francisco Railway (Frisco) passenger train No. 3, the *Will Rogers Special*, en route from St Louis, Missouri, to Oklahoma City, Oklahoma, was a

---

[1677] *Mansfield News-Journal*, Mansfield, Ohio, United States, 13th May 1948
[1678] *The Republican Courier*, Findlay, Ohio, United States, 14th May 1948
[1679] https://www-polskieparowozy-pl.translate.goog/index2.php?wykaz=wybuchy&_x_tr_sch=http&_x_tr_sl=pl&_x_tr_tl=en&_x_tr_hl=en&_x_tr_pto=sc
[1680] *Beatrice Daily Sun*, Beatrice, Nebraska, United States, 21st October 1948
[1681] *Atchison Daily Globe*, Atkinson, Kansas, United States, 21st October 1948
[1682] https://www.flickr.com/photos/michaeljy/3514234654

mile and a quarter short of Niangua, Missouri, United States, and passing through a cutting, when the boiler of its engine exploded at about 4.20 a.m. on Saturday 26th February 1949. For once none of the engine crew was killed, though Engineer Frank Fulton and Fireman Lewis Breewsowitz were badly hurt. Express Messenger B.H. Willoughby was slightly hurt.

Following the explosion the engine rolled over onto its right-hand side up against the side of the cutting and four mail and baggage cars were piled up behind it crosswise over the track. Another baggage car was derailed but did not overturn. Lyle Jones, a Frisco company materials inspector, who had been a passenger on the train, was one of the first to reach the injured engineer. He said that Fulton had been wandering about in a dazed condition, clutching the stump of his right arm, which had been traumatically amputated by the blast. With the help of other trainmen and passengers Jones had improvised a tourniquet to stem the bleeding. Fireman Breewsowitz was badly burnt. Both men were taken to hospital in Springfield. The five passenger cars of the train, two Pullmans and three chair cars, were drawn back to Conway, from where the passengers were forwarded to Springfield on buses. [1683]

Chief Mechanical Officer W.B. Berry suggested that the explosion had resulted from the usual culprit, low water, but this was disputed by Fireman Breewsowitz who insisted that the water level had not been low and speculated that a thorough inspection would reveal a defect in the boiler. [1684]

At 8.05 p.m. on Wednesday 6th April 1949 the firebox of British Railways, Southern Region 0-6-2 engine No. 2557 (ex-London Brighton and South Coast Railway) burst inwards, blowing two enginemen out of the cab at Bevois Park, Southampton, England. It was found that 137 of the 207 copper stays which should have supported the firebox had given way because they were badly corroded. Until three weeks before the accident the engine had been stationed at Horsham, where the water was taken from a borehole. This water contained much sodium carbonate and rapidly corroded boilers. An inspector had examined the boiler only a month earlier and had noted some corrosion but had decided that it was fit for another six months' service. It had been six years since the boiler had had a general repair. [1685]

A Class B-5 locomotive was being steamed up in the yard of the Pecos Valley Southern Railway at Pecos, Texas, United States, when its boiler exploded early on Saturday 23rd July 1949. The smokebox door of the engine was blown about 1,000 feet but most of the damage was in the firebox and cab area. Engine Watchman Francisco Reyes, who had been preparing the engine, was critically hurt, and later died. The manager of the short line, W.C. Tutley, said that the explosion could have had a number of causes. The boiler had been rebuilt in January. [1686] The Interstate Commerce Commission attributed the explosion to a crown sheet failure due to low water. [1687]

1683 *Blytheville Courier*, Blytheville, Arkansas, United States, 26th February 1949
1684 *The Sunday Star*, Washington, D.C., United States, 27th February 1949
1685 *Locomotive Boiler Explosions* by C.H. Hewison, pgs. 137-8
1686 *El Paso Herald-Post*, El Paso, Texas, United States, 23rd July 1949
1687 Interstate Commerce Commission, Annual Report of the Director, Bureau of Locomotive Inspection

Engine No. Tr203-476 suffered a boiler explosion between Ozimek and Opole, Poland, on Monday 19th December 1949. [1688]

On the morning of Friday 24th February 1950, Atlantic Coast Line Railroad local freight train known as the Williamston Turn left South Rocky Mount, North Carolina, at 9.30 a.m. Williamston was reached without incident and at 3.50 p.m. southbound extra freight train No. 411 left Williamston for South Rocky Mount, with engine No. 411, 28 cars and a caboose. This train left Tarboro, North Carolina, United States, at 5.57 p.m. and at about 6.09 p.m., two miles south of Tarboro, the boiler of the engine exploded as the train was travelling up an incline at about 20 miles an hour. Separated from the frame of the engine by the force of the explosion the boiler was thrown 150 feet, while the lower part of the smoke box and the front end of the boiler were stove in and the engine cab was thrown 170 feet forward of the boiler. Engineer C.E. Boyette, Fireman Bud Tillman, and Brakeman J.H. Carson were killed instantly. The body of the engineer was found 425 feet from the site of the explosion, that of the fireman was on top of the tender and of the brakeman in the wreckage of the cab. [1689]

Conductor Howard Gunn and Flagman Hubert Speight in the caboose were unhurt. The conductor told newspaper reporters afterwards that he and the flagman had been attending to records when the explosion occurred. They did not hear the blast but he had been thrown beneath the water cooler by what he had thought was an emergency stop, with the flagman on top of him. The flagman had then gone back to protect the rear of the train while he (Gunn) had gone forward, when he discovered that the engine boiler had given way. [1690]

The explosion was investigated by the Interstate Commerce Commission who produced a detailed report on it, which included a comprehensive account of their post-accident examination of the remains of the locomotive and the evidence of various witnesses. Their report concluded that the accident had resulted from the overheating of the crown sheet due to low water, but they could not determine why this had happened. [1691]

The boiler of a locomotive on the Hunan – Kwangsi Railway exploded near Canton, China, on Thursday 29th June 1950, killing three drivers and two fireman. Reporting the accident, the *Ta Kung Pao* newspaper stated that the explosion 'was caused by the fact that there was no water in the boiler.' [1692]

The Norfolk and Western Railway suffered a locomotive boiler explosion near Eastwood, Ohio, United States, at about 7.30 p.m. on Wednesday 23rd August 1950. Eastbound freight train third No. 84 had left the yard at Clare, Ohio, at 6.45 p.m. bound for Portsmouth, in charge of Conductor W.R. Newman with Engineer Carl Clark and Fireman R.D. Aldred. It consisted of Class Y engine, of the 2-8-8-2 Mallet type, No.

1688 https://www-polskieparowozy-pl.translate.goog/index2.php?wykaz=wybuchy&_x_tr_sch=http&_x_tr_sl=pl&_x_tr_tl=en&_x_tr_hl=en&_x_tr_pto=sc

1689 Interstate Commerce Commission report

1690 *The Evening Telegram*, Rocky Mount, North Carolina, 25th February 1950

1691 Interstate Commerce Commission report

1692 *Pacific Stars and Stripes*, Tokyo, Japan, 5th July 1950

2114, 39 loaded cars, 7 empty cars and a caboose and was moving at 20 miles an hour up a 1.1 per cent (I in 90) gradient when the boiler exploded without warning, killing the engineer and fireman, whose bodies were blown several hundred feet.

Head Brakeman Paul Keyes was thrown from a cupola atop the coal tender but escaped with a sprained ankle. He said the cupola door had suddenly flown open, and added, "Boy, it got hot in there. The next thing I remember I was standing on the track about 10 feet in front of what was left of the engine and boiler and about 60 feet from my cupola." The blast was heard in Williamsburg, a mile or so away, and telegraph lines along the right-of-way were knocked down. Three cars of the train loaded with hogs were derailed. Some of the animals were killed and 'the others squealed in a stampeding uproar.' The line was well blocked, with rails ripped up and a big hole dug in the ground by the blast. [1693] Conductor Newman and Flagman Goodlow Fleming were unhurt. [1694]

The Interstate Commerce Commission carried out a detailed investigation of the accident but could not come to any firm conclusions other than that the explosion had been caused by overheating of the crown sheet due to low water. The locomotive was provided with a low-water alarm but Head Brakeman Keyes did not remember hearing this sound before the explosion. Eight different problems, such as repeated failures of the water pump and the injector not taking up water, were, however, noted by trainmen as needing attention in a slightly over three-week period before the accident, which seems excessive. [1695]

Southbound Florida East Coast Railway freight train extra No. 815 left Fort Pierce, Florida, at 7.25 a.m. on Wednesday 4th April 1951 for Miami. When it left West Palm Beach at 9.15 a.m. it consisted of engine No. 815, 13 loaded, 24 empty cars and a caboose. All went well until the train, travelling at 45 miles an hour, was rather less than a mile north of Lake Worth, Florida, United States, when the engine boiler exploded. The engineer and fireman were injured, the former was sent home after treatment while the latter was admitted to Fort Worth Hospital. An automobile standing close to the track was overturned and 19 lineside buildings were damaged; two people in the buildings and one who had been at the lineside suffered shock.

The Interstate Commerce Commission (ICC) investigated the explosion. It described the engine as an oil-burning 4-8-2 dating from 1926. Its report gave the usual description of the locomotive and its condition after the accident, followed by the evidence of various witnesses.

The engineer stated that he had noted a leak in the steam pipe to the water glass and had tried to fix it by tightening a nut with a wrench, without success. He had not considered the leak bad enough to affect the water level and the water glass had been more than half full just before the explosion.

The fireman agreed that the water level in the glass had been more than half full just before the blast and said that he had had no difficulty in maintaining the water level with the feed water pump.

The conductor and brakeman agreed that the driver and fireman had not appeared

---

[1693] *The Evening Independent*, Massillon, Ohio, United States, 24th August 1950

[1694] *The Evening Tribune*, Marysville, Ohio, United States, 24th August 1950

[1695] Interstate Commerce Commission report

concerned about the operation of the engine before the accident.

The ICC conducted tests which this time clearly showed the cause of the explosion. Its report stated, 'The crack in the copper pipe which extended from the top of the gauge glass to the valve at the water column and was located at the valve union caused a differential in pressure which resulted in maintenance of a water level indication materially higher than that which existed in the boiler. Tests made by our inspectors subsequent to the accident…showed that a safe water level would be indicated by the gauge glass when the level of the water in the boiler had receded below the danger point. The leak at the top gauge glass valve could also affect the column-mounted gauge cock indication of water level. It thus appears that the boiler was being operated with a falsely indicated and apparently safe water level while the quantity of water in the boiler deceased until the crown sheet overheated with disastrous consequences. Had the boiler been equipped with a second water glass the difference in reading of water level shown by the two gauge glasses would have indicated the necessity for immediate action to remedy a defective and potentially dangerous situation, and this accident might have been avoided.'

The ICC noted that the necessity for at least two devices providing independent indications of the water level in the boiler at all times had been stressed in numerous annual reports of the Bureau of Locomotive Inspection since 1920. It was also unhappy that various boiler controls had been 'meddled with' by unknown persons after the accident, before its inspectors had had a chance to examine them. [1696]

Steam was raised in the boiler of locomotive No. 95 6679 in the *Reichsbahnausbesserungswerk* (Repair Shop) at Meiningen, East Germany, on Friday 4th May 1951 despite the fact that the pressure gauge was known to be defective. A disastrous explosion followed this rash act, killing nine workshop employees and a passer-by in an adjoining street, the boiler landing in the garden of a nearby hospital. [1697] [1698]

The boiler of the engine of through goods train Dg 19136 from Hamburg to Mōnchengladbach exploded on Tuesday on Tuesday 28th August 1951 as the train was between Tostedt and Konigsmoor, West Germany, on the Hamburg – Bremen main line. The train was derailed and fouled the opposite running line in the path of goods train G 6507, which was derailed in turn. In all, 24 waggons left the track including two tank waggons carrying oil, and the wreckage caught fire. Three railwaymen lost their lives and the material loss was considerable. [1699]

Engine No. 42 1893 was pulling a freight train between Klotten and Colchem, West Germany, on the Koblenz – Trier line, when its boiler exploded on Friday 12th October 1951 when its boiler gave way. The locomotive and eight heavily laden coke waggons were derailed and badly damaged and the line was blocked. The driver and fireman were badly hurt and the fireman died in hospital. The Moselle road, which paralleled the railway at this point, was also blocked by large quantities of spilt coke. Low water was given as the cause of the explosion. This would be the last accident of its type before the

[1696] Interstate Commerce Commission report
[1697] Wikipedia – Meiningen Steam Locomotive Works
[1698] https://de.wikipedia.org/wiki/Kesselzerknall#cite_note-4
[1699] *Katastrophen der deutschen Bahnen Teil II* by J.R. Ritzau, pg. 122 (in German)

end of steam traction in West Germany. [1700] [1701]

The new *El Marplatense* air-conditioned express on the General Roca line was on its return journey from Mar del Plata to Buenos Aires and was between the stations of Coronel Vidal and General Pirán, Argentina, on Monday 21st January 1952 when it met with an accident. It was running at 120 kilometres an hour when the roof of its firebox collapsed and its cab was filled with high-pressure steam, water, fuel oil and flames. An inspector riding on the footplate was thrown off and killed and the driver and fireman suffered severe burns. The fireman died of his injuries within 24 hours. None of the train passengers was hurt and the train later continued its journey with a replacement locomotive. [1702]

Locomotive 52 1515 suffered a boiler explosion at Döbeln, East Germany, on Saturday 24th May 1951 due to low water. [1703]

The boiler of a locomotive pulling a freight train over the mountainous stretch of line between Concepción and Santiago exploded near Santiago, Chile, on Sunday 1st June 1952. The driver and fireman were killed, as were a woman and her two young daughters in a lineside house which was struck by the remains of the boiler. [1704]

The boiler of a Chicago Rock Island and Pacific Railroad exploded near Okarche, Oklahoma, United States, on the evening of Friday 18th July 1952. The engineer, fireman and head brakeman were killed.

Freight train Extra No. 5026 consisted at the time of the explosion of engine No. 5026, 42 loaded, two empty cars and a caboose. It had left Caldwell, Kansas, at 7.30 p.m. for El Reno, Oklahoma, and was travelling at about 15 miles an hour and about half a mile north of Okarche when the boiler had given way at 10.55 p.m. The force of the explosion tore the boiler from the frame of the locomotive and hurled it upward and forward 215, gouging a hole 18 inches deep in the road bed when it fell to earth again. [1705] Witnesses said that a huge ball of flame had spread over the right-of-way on either side. Flaming oil (the locomotive was oil-fired) was spattered over a 100-yard-diameter area but no major fires were started. Houses a mile distant were rocked by the blast. At least one telegraph pole was demolished and water from the damaged tender washed away a lineside fence. None of the cars of the train was derailed or harmed, the main force of the explosion having been directed upwards. Station Agent Jack Eischen at Okarche witnessed the explosion. Engineer Roy Tomlin, Fireman Hutch Spivy and Head Brakeman Jack Gorman lost their lives. [1706]

The Interstate Commerce Commission (ICC) issued its usual detailed report. As in the Eastwood explosion (see above) there was a long list of reports of defects with the engine

---

1700 *The Railway Gazette*, London, England, 9th November 1951
1701 https://de.wikipedia.org/wiki/Kesselzerknall#cite_ref-RITZAU_12_17-1
1702 *The Railway Gazette*, London, England, 15th January 1952
1703 https://de.wikipedia.org/wiki/Kesselzerknall#cite_ref-RITZAU_12_17-1
1704 *The Daily Herald*, Biloxi, Mississippi, United States, 2nd June 1952
1705 Interstate Commerce Commission report
1706 *The Daily Oklahoman*, Oklahoma City, Oklahoma, United States, 20th July 1952

in the month before the accident. Also as at Eastwood the locomotive had been equipped with a low-water alarm, which had not prevented the accident.

The train crew who had worked the engine from Herington to Caldwell testified that they had noted no problems with it.

A machinist at Caldwell said that he had seen the engineer test the right water glass and the fireman test the left water glass. The low-water alarm had not been tested there.

The conductor and rear brakeman stated that the engineer and fireman had appeared normal and they had noticed nothing unusual before the blast. Neither of them had heard the low-water alarm at any time.

Various lineside residents said that they had been awakened by the explosion. None of them remembered hearing the low water alarm.

The ICC concluded that the accident had been caused by an overheated crown sheet due to low water.[1707]

The engineer and fireman in charge of the engine of a local freight train were seriously injured when its boiler exploded at McIntyre, Georgia, United States, at 1.20 p.m. on Friday 29th August 1952. Engine No. 629 of the Central of Georgia Railway had started work that day at Gordon, Georgia, at 7.30 a.m., and was doing its normal work of switching cars at four kaolin mines at and between Gordon and McIntyre. It arrived at McIntyre with three loaded and four empty cars, which were placed on a siding. The engine was then moved onto a house track and was stationary there when it exploded shortly afterwards. The engineer and fireman jumped from the engine as it blew up which did not save them from being badly hurt. The engineer was taken to his home in Gordon and the fireman to Hilledgeville Hospital and later transferred to Savannah Hospital.

The Interstate Commerce Commission (ICC) noted that engine No. 629 was of the 2-8-2 type and dated from 1916. As at Lake Worth the previous year (see above) the ICC concluded that the explosion had resulted from an overheated crown sheet due to low water and was able to pinpoint the cause. Its report read, 'The leak at the top of the right glass of the water gauge caused a reduction in pressure on the surface of the water in the water gauge which resulted in the maintenance of a water level indication materially above that existing in the boiler. Tests made by our inspector subsequent to the accident showed that a safe water level would be indicated by the gauge glass when the level of water in the boiler had receded to a danger point. It thus appears that the boiler was being operated with a falsely indicated and apparently safe water level while the quantity of water in the boiler decreased until the crown sheet became exposed and overheated with disastrous consequences.' Again there was no second gauge glass. The report continued, 'Had the boiler been equipped with a second water glass the difference in level shown by the two gauge glasses would have indicated {the} necessity for immediate action to remedy a defective and potentially dangerous situation and this accident could have been avoided. The installation of a second gauge glass to improve safety of locomotive operation has been emphasized and recommended in numerous annual reports of the Bureau of Locomotive Inspection, accident reports, and on various other occasions.'[1708]

[1707] Interstate Commerce Commission report

[1708] Ibid.

An unusually serious boiler explosion in terms of casualties occurred on Wednesday 17th September 1952 Oryu-Dong, South Korea. A three-coach passenger train was carrying between 400 and 500 passengers, most if not all schoolchildren en route from the port of Inchon to schools in the Seoul area. By chance the explosion happened as the train was crossing a bridge and the force of the blast threw the leading coach against a bridge pier, ripping it open and throwing its occupants 40 feet to the river below. The other two coaches were also blown off the track and one, probably the second, overturned and was broken open. Major Edwin Mercer estimated that there were 20 dead, mostly aged between eight and 14, and about 200 more injured, of whom all but 50 had been treated and released from a nearby military hospital. [1709]

A reporter described the scene in somewhat grisly detail. 'US army doctors sped to the scene in 25 ambulances and performed 10 emergency amputations to free passengers trapped in the twisted wood and steel. Koreans and US soldiers watched in horror. The screams of an 18-year-old girl silenced the crowd as a blood-splattered American doctor worked for 20 minutes to amputate her right arm. She was the last living passenger to be extricated from the wreckage. Still inside the twisted mass hanging over the trestle were five bodies. Army engineers began removing with blowtorches and hacksaws. Only two bodies were visible. Directly underneath the bridge were the head and legs of a schoolboy dangling over the water and rescue workers could see the bloody arms of a little girl pinned in the wreckage. Mothers who hurried from nearby Inchon went from stretcher to stretcher. School books and jackets were piled up neatly beside the tracks. British troops from a nearby warehouse were first on the scene. "Pieces of jet black stuff flew through the air into our compound," said Sgt. T.S. Nixon of Midlands, England. "I'd rather see men wounded in action than kids hurt this way. You can get at them on the battlefield." American doctors from the 121st evacuation hospital said they had to make several "dissections" to separate injured children from the dead.' [1710]

The Korean War was in progress at the time of the accident but Lt. Col. Clinton C. Langford of the Third Transportation Military Railway Service, which was operating the train, discounted sabotage as a cause of the disaster. [1711]

Several days later an investigation confirmed that it was an ordinary boiler explosion. The engine crew had permitted the water level to fall below the level of the crown sheet and cold water belatedly added to the boiler had caused the weakened crown sheet to fail. [1712]

The Denver and Rio Grande Western Railroad suffered a serious boiler explosion at about 11:10 a.m. on Sunday 19th October 1952 at Louviers, Colorado, United States, south of Denver.

Southbound freight train No. 67-S consisted of engine No. 3703, 31 loaded and 64 empty cars and a caboose. It left the Burnham yards in Denver at 10.25 a.m. for Pueblo, Colorado, and proceeded without incident until it was approaching Louviers, 19½ miles out, at 11:10 a.m., when the boiler gave way. Torn from its frame, the boiler was thrown 260 feet and parts of the engine were scattered in various directions. Lineside fences were

---

[1709] *Pacific Stars and Stripes*, Tokyo, Japan, 18th September 1952

[1710] *The Altoona Mirror*, Altoona, Pennsylvania, United States, 17th September 1952

[1711] *Pacific Stars and Stripes*, Tokyo, Japan, 18th September 1952

[1712] Ibid, 21st September 1952

blown down and fires started in adjoining stubble fields. None of the cars of the train was derailed or damaged. [1713] Engineer R.R. Malloy, Fireman John Taylor, Head Brakeman Sidney Messeraul and a fourth man named David Clint were killed instantly. Clint was a railway enthusiast, a member of the Railroad Fans Club. Newspaper reports explained that enthusiasts were allowed to accompany trains at times provided that this had been cleared by head office, though it was not immediately known if Clint had been authorised to accompany the train on this occasion. A crushed Speed Graphic camera which had clearly belonged to Clint was found among the wreckage. [1714] According to the Interstate Commerce Commission (ICC) report he was a trespasser.

The ICC noted that No. 3703 was a Class L-105 simple articulated 4-6-6-4 Baldwin Works locomotive constructed in April 1938. Its report described in considerable detail the construction of the engine and the findings of an inspection of the wreckage. There was no low-water alarm or fusible plugs.

The engineer and fireman who had worked the engine on Friday 17th on its last trip before the fatal Sunday run testified that they had had no particular problems with it. Various engine house employees stated that the locomotive had been in good condition and working normally on the morning of the accident. The conductor and rear brakeman of the train stated that the engine crew had appeared normal and in good spirits.

The ICC concluded that the explosion had resulted from an overheated crown sheet due to low water. It did not come to any conclusion as to how this might have happened. [1715] The possibility that the engineer and fireman might have been distracted by the presence on the engine of the railway enthusiast was not mentioned.

The 8.20 p.m. British Railways, Western Region, vacuum-fitted freight train from Birkenhead to London on Monday 22nd December 1952 was climbing Gresford bank (nearly 4 miles of 1 in 80) just north of Wrexham and was near Wheatsheaf Junction signal box, England, when the crown sheet of its engine, 4-6-0 steam engine No. 6859, *Yiewsley Grange*, failed. Fireman C.A. Cook sustained severe burns and shock, Driver W. Brazier escaped injury but was shaken and Guard R.W.H. Knight was thrown about in his van by the sudden stoppage and badly bruised. The signalman heard the report of the explosion and protected the line, while local residents cared for the fireman and guard until the arrival of an ambulance. The explosion was due to low water, the enginemen having been misled by a faulty gauge glass. Near United Collieries box, half a mile north of Wheatsheaf Junction, a fusible plug had failed and the sound of escaping steam had been heard by the enginemen, but as the water had been near the top of the gauge glass they had assumed that a tube had failed.

The gauge glass tube had been changed at Birkenhead on 18th December by Apprentice Fitter Michael Crean, supervised by Fitter William Fenton. In fitting the new tube Crean had apparently tightened the tube with a spanner a little too hard and broken it. Neither he, Fenton, nor Fitter Robert Tyrer, who had examined the engine when it was in steam, had realised anything was wrong. The break of the glass had occurred inside a rubber packing washer, which over the next two days had almost obstructed the tube, reducing the pressure of steam on the water at the top of the glass and allowing the water

---

[1713] Interstate Commerce Commission report

[1714] *The Salt Lake Tribune*, Salt Lake City, Utah, United States, 20th October 1952

[1715] Interstate Commerce Commission report

level to rise above that in the boiler, giving a false reading. In accordance with former Great Western Railway practice there was only one steam gauge.

At Chester on 20th December, Fireman Cyril Morgan had noted that the gauge glass was only filling slowly after the drain cock had been used. This was mentioned to Driver R.A. Taylor when he took over the engine at Oxley Sidings and the engine was examined at Oxley by Fitter Charles Knowles, who concluded that it was only a minor problem. Driver E. Langford noticed the same problem when he worked the train back to Birkenhead and he wrote on a repair card, "the water gauge frame wants changing." Fitter Edwin W.R. Ash also failed to detect the problem but omitted to hand in the repair card when he left duty, so no one else knew there had been a problem. Driver William Robertson and Fireman Dennis Partington worked the Birkenhead to London train with the engine to Chester on 22nd December, where they handed over to Driver Brazier and Fireman Cook.

Employment Inspector J.L.M. Moore noted that the type of fracture which had occurred was most unusual. Tests were therefore carried out to try and reproduce the breaking of the gauge glass and they succeeded in doing so. It was then determined that this was because not only had there been excessive tightening but the gauge glass had been slightly out of alignment.

Mr Moore did not blame either Crean or Fenton for the breakage of the gauge glass. He considered, on the other hand, that Fitter Knowles had been seriously at fault for having returned the engine to traffic when he knew that there was a problem with its single-gauge glass which he had been unable to locate.

Fitter Ash was at fault to a lesser degree for not having made a much more thorough examination of the gauge frame before deciding that there was no need to change it as the driver had requested.

The footplate staff were also to blame to some degree for not having carried out the instructions to, "Operate both test cocks to ensure that they are in working order."

Mr Moore made four recommendations:

(a) Staff should be reminded of the importance of opening test cocks on locomotives which only had one water gauge.

(b) Drivers, firemen and perhaps fitters should be made aware of the dangerous significance of the slow return of water to the glass after draining.

(c) Drivers should be reminded of the importance of giving as much information as possible when completing repair cards.

(d) The attention of fitters should be drawn to the possibility of a gauge glass snapping inside the rubber packing with little or no subsequent leakage of water. [1716] [1717]

The next locomotive boiler explosion in the United States occurred at Hinton, West Virginia, at 5.25 p.m. on Tuesday 9th June 1953 on the Chesapeake and Ohio Railway.

Involved in the ambulance was eastbound freight train Extra 1642 East, which consisted at the time of the accident of engine No. 1642, 123 loaded and two empty cars. It had left Handley, West Virginia, at 1.20 p.m., had taken coal and water at Thurmond at

[1716] Ministry of Transport report

[1717] *Locomotive Boiler Explosions* by C.H. Hewison, pgs. 138-40

2.55 pm., and water at Quinnimont at 4.12 p.m. where some cars were picked up. It was travelling at about 20 miles an hour on level track when the boiler of its locomotive exploded, killing Engineer W.H. Anderson, Fireman Joe Sullivan, and Head Brakeman O.L. Richmond. The force of the explosion carried the boiler forward 440 feet while the engine cab was blown back and to the right, ending up on the bank of the New River, while parts of the engine fell into the river. The tender was derailed, as were nine cars, which piled up in a heap behind the tender. [1718]

Operator H.S. Gwinn in nearby CW Cabin witnessed the explosion. He told a newspaper reporter that the blast had sounded like an earthquake. The blast threw the westbound track about three feet out of alignment, seared vegetation within a radius of over 150 feet and threw debris across Route 20. Hot parts of the firebox started small blazes in the woods above the railway. Mr and Mrs W.L. Graham, of Hinton, who had been planning to drive down Route 20, also witnessed the accident. Mr Graham said that the explosion had roared out and the entire river valley had seemed to be filled with smoke, fire, and flying debris. Hundreds of cars immediately rushed to the scene, the newspaper added, and city and state police had found themselves unable to keep the traffic moving. The line was expected to be blocked for 10 or 12 hours. [1719]

The Interstate Commerce Commission (ICC) described No. 1642 as a Class H-8 Alleghany 2-6-6-6 locomotive, which had been built in December 1944. The usual comprehensive details of the engine and its condition after the accident were given in the report, including the information that it had been fitted with a Nathan Type B low water alarm.

Reports on the condition of the engine in the days leading up to the accident included several entries that the water pump had stopped working.

The engineer and fireman who had operated the engine on its previous trip testified that they had had no problems with it.

A machinist at Handley stated that he had tested the water pump and found nothing wrong with it.

The foreman at Thurmond said that when the train had arrived there on the afternoon of the accident its engineer had asked him to examine the cold water governor, which was stuck. He (the foreman) had done so and had found it to be free.

A machinist helper who had been about 160 feet from the line at a point one and a half miles from CW Cabin stated that the engine had been working at medium throttle as it had passed him and the low-water alarm whistle had been sounding.

Operator Gwinn said that he had had a telephone inquiry as to where the freight was. Looking down the line he had seen it 600 feet away and the engine had appeared to be working at medium throttle; then, looking again after he had got up, he had seen the boiler give way. He described the engine as appearing to disintegrate before it was obscured by smoke and steam. As soon as pieces of debris had stopped falling he telephoned the dispatcher, informed him that the engine of the freight train had exploded and was wrecked, and requested an ambulance.

The ICC reached its usual conclusion that the explosion had resulted from an

[1718] Interstate Commerce Commission report

[1719] Beckley Post-Herald, Beckley, West Virginia, United States, 11th June 1953 overheated crown sheet due to low water. [1720]

The boiler of locomotive No. Tr203-135 gave way between Tychowo and Grzmiąca, Poland, on Saturday 10th July 1954. [1721]

The use of steam locomotives in the United States, other than on branch and minor lines operated by enthusiasts, was rapidly coming to an end by the late 1950s (the last steam locomotives in ordinary service were those on the Grand Trunk Railroad which were replaced by diesels in 1961). [1722]

The very last American boiler explosion involving a locomotive in day-to-day operation occurred near Wytheville, Virginia, United States, on Monday 12th December 1955. The train involved was eastbound Norfolk and Western Railway freight No. 52 from Bristol to Roanoke. It was about two miles short of Wytheville when the engine, Class Y-6 No. 2153, blew up just before 5 a.m. Fireman Harry Hawthorne, who had only qualified as such the previous week, was killed almost instantly. Engineer William Shumate and Student Brakeman M.B. Shorter were critically injured, Fireman H.S. Sledd was rather less badly hurt. The blast derailed the engine but the boiler did not detach from the frame. The tender and the remainder of the train remained on the track.

A local newspaper reported the reactions of local residents. 'People leaving near the scene of the explosion described it as, "something like a terrific blast of thunder, though with a hissing to it." Or "It sounded like another world war had started and someone had dropped an A-bomb on us."

'Mrs Robert Cassell, whose home is on the south side of the tracks at the scene of the explosion, said that she heard the train come to a stop. She went out on the front coach to turn on the light to let the men on the train know she was up in case they needed to call for assistance, Mrs Cassell said.

'"Something told me to go back inside the house," she added. "Just as I stepped inside the door there was an explosion." She said the blast shook the house and sprayed it with coal, steam, and dirt. A piece of flying metal took a chunk out of the side of the house. Mrs Cassell said she and Minnie Groseclose, a woman living with her, rushed outside, and found Shumate lying about 20 yards from the engine, wounded in the neck and head, and badly burned.' [1723]

'Mrs Paul Walters, who lived 40 yards from the scene of the accident, said she had been up at 4 a.m. feeding the baby, and had just gone back to bed when the explosion occurred. The house was sprayed with debris, windows were broken, and the latch was knocked off the door. {Mr} Walters said he saw a piece of steel go by the window and later found a large steel pipe that had gone past the house. Mrs Walters phoned Wythe County lifesaving crew and tried to get doctors. Walters took blankets to the scene of the accident. He found Shorter lying in a pool of boiling water and steam. Shorter remains in very critical condition with burns over two-thirds of his body and a skull fracture.

[1720] Interstate Commerce Commission report

[1721] https://www-polskieparowozy-pl.translate.goog/index2.php?wykaz=wybuchy&_x_tr_sch=http&_x_tr_sl=pl&_x_tr_tl=en&_x_tr_hl=en&_x_tr_pto=sc

[1722] https://worldwiderails.com/when-did-diesel-trains-replace-steam/

[1723] *Bluefield Daily Telegraph*, Bluefield, West Virginia, United States, 13th December 1955

'Hawthorn's body was found on the top of a nearby embankment. He had been blown 25 or 30 feet in the air by the force of the explosion.

'The mishap is holding up two N&W passenger trains – the eastbound *Birmingham Special* and the westbound *Pelican*.'

It has unfortunately not proved possible to trace an Interstate Commerce Commission report on the explosion. According to one newspaper there was a rumour that the train had stopped at a new water tank shortly before the explosion but the tank had been defective and unknown to the crew no water had been taken.

Another newspaper wrote that the locomotive had been giving trouble for several miles because it had not been producing enough steam to haul its load, so it had been decided to put the train into a siding at Wytheville and await a replacement engine. [1724] Possibly the combination of a very inexperienced fireman and a student brakeman was relevant.

An Austerity 2-8-0 steam locomotive which had been supplied to Hong Kong at the end of the Second World War had proved a constant source of trouble. This culminated at 3.30 p.m. on Friday 27th April 1956 in its boiler exploding in the Kowloon – Canton Railway workshops at Hong Kong. Eleven men were scalded, two of whom died the same night, three more the next day and a sixth later. Following the accident the remaining engines were subject to a detailed examination, resulting in six being withdrawn for repairs, two of them receiving completely new boilers. The replacement of all the steam locomotives with diesel engines was meanwhile accelerated. [1725] [1726]

Engine No. 03 1046, hauling express train D 78, the Balt – Orient Express from Berlin and Dresden to Varna, Bulgaria and Budapest, Hungary, [1727] suffered a boiler explosion of an unusual kind on Friday 10th October 1958 at Wunstorf, East Germany. It was not the usual case of the roof of the firebox giving way but rather the boiler split along its length, a crack having formed and propagated due to metal fatigue. [1728]

Portugal suffered a boiler explosion on Saturday 5th December 1959. It happened on the Norte line at about 06:00 to freight train No. 63, which had left Entroncamento at 04:25 for Vila Nova de Gaia, in charge of Conductor Alfredo Pereira Anastácio with Driver João Lopez Dourado and Fireman Manuel Piedade de Freitas on engine No. 863. The train was ascending an incline between the stations of Albergaria-dos-Doze and Caxarias, and approaching a tunnel, when its boiler gave way at Kilometre 146.200. The violent explosion destroyed the engine (which was wrongly described as a diesel in the newspaper report) and injured a number of pine trees in the woods bordering the line.

Immediately after the explosion the conductor and brakesman went to the remains of the locomotive. They found that the fireman had been killed outright. The driver was still

---

[1724] *The News Journal*, Radford, Virginia, United States, 12th December 1955

[1725] https://industrialhistoryhk.org/the-kowloon-canton-railway-british-section-part-5-post-war-years-1945-to-1978/

[1726] *The Singapore Free Press*, Singapore, 30th April 1956

[1727] http://www.eurailtracks.co.uk/page90.html

[1728] https://de.wikipedia.org/wiki/Kesselzerknall#cite_ref-RITZAU_12_17-1

alive, though barely recognisable, and he died of his injuries on the way to hospital.

The Porto – Lisbõa line was blocked until 11:20 and passengers were transhipped. The remains of the engine were taken back to Entroncamento and the train continued its journey with a fresh one. [1729] The locomotive was a 2-8-2 (*Mikado*) type, built by the American Locomotive Company in 1945. [1730]

[1729] *Diario de Lisbõa*, Lisbõa, 5th December 1959

[1730] Wikipedia – List of Portuguese locomotives and railcars

CHAPTER 17

VANDALISM

'Accidents' due to vandalism and sabotage have always been a problem for the railways. They are obviously very closely related, inasmuch as both represent deliberate attempts to cause trains to crash, as opposed to accidents caused through non-deliberate carelessness, negligence or purely accidentally. Here they will, however, be treated separately, as they are slightly different. This separation is perhaps a little artificial, and if the perpetrator of a given act is not caught or his or her motives not sufficiently obvious, it may be difficult to decide which it was, but it has nevertheless been felt that the distinction is worth making.

Vandalism is usually, though not invariably, carried out by either children or young adults and its results are most commonly, though not always, less serious than those following acts of sabotage. The motives for train accidents resulting from vandalism are usually either simple curiosity, in the case of younger children – 'I wanted to see what would happen' – or in the case of older children and teenagers, a reflection of general anti-social behaviour; so that such individuals who attempt to derail trains also carry out other acts of 'mindless vandalism,' and perhaps also commit burglaries or carry out arson attacks.

Sabotage on the other hand is usually undertaken by adults, who employ methods more likely to cause death and injury, so that while a vandal might for example just place an object upon the track or jam a stone into a set of points, a saboteur would be more likely to employ methods such as blowing up a section of track using explosives.

An 'accident' due to vandalism which had serious consequences occurred on the Pennsylvania Railroad at Walton, Indiana, United States, on Monday 27th January 1947. The train involved was westbound first-class passenger train No. 207, *The Union*, from Cincinnati to Chicago, consisting of 4-6-2 steam engine No. 5377, a baggage car, two coaches, a sleeping car, two coaches, a dining car, and a sleeping car, in charge of Conductor Allen Terhane, with Engineer N.J. Lamartine and Fireman C.A. Wisler. It had left Kokomo, Indiana, slightly under 12¾ miles east of Walton, at 5.27 p.m., two minutes late, and was travelling at about 60 when the engine and first six vehicles became derailed. The fireman and three passengers were killed and the engineer, conductor, Brakeman George Calkin and 33 passengers were injured. [1731]

It quickly became clear that the derailment had been caused by the engine striking a

---

[1731] Interstate Commerce Commission report

400 lb roll of wire fencing. Before the accident there had been nine rolls of the fencing at a section shanty south of Walton. After the derailment only eight rolls remained. The ninth roll had been struck by the engine, carried several hundred feet down the track as Engineer Lamartine was braking his engine, then got caught in a switch (set of facing points) at a spur leading to Farmer's grain elevator. It was here that the initial derailment had occurred. [1732]

State Police Lieutenant John Fisher commented, "This couldn't have been an accident. It was either sabotage or larceny. Someone either placed the rail on the track or it caught there while they were trying to steal it." [1733]

The culprits were soon tracked down. On Thursday 30th Sheriff Harold Smith announced that Lysle Graves, 11, and Jack Sprinkle, 12, had confessed to derailing the train after Sprinkle had suggested, "Let's wreck a train tonight." 'The two boys had pushed a roll of fence wire off a pile near the tracks, thrust an iron bar through, and pushed the roll onto the tracks. A moment before they had placed a wooden plank, six feet long and eight inches wide, on the tracks. They said they first tried to put railroad ties on the track but found they were too heavy.' The two told the sheriff that they had rolled the wire onto the track "to see what would happen," a typical motive for vandals who place obstructions on the line. Prior to the derailment the pair had committed two burglaries. [1734]

The boys appeared before Circuit Judge Clifford Wild at the beginning of March. He stated simply that they had been "sent away" and would not elaborate, saying that under State Law juvenile cases were "sacred." [1735]

A little more typical of 'accidents' due to vandalism than the Walton accident, inasmuch as it was a lot less serious, was the derailment of a freight train at Westwood, California, United States, on or about Thursday 20th March 1947. A similarity with the Walton case was that one of the perpetrators had also committed at least one act of burglary. Two youths had first broken into a business establishment and stolen various items, including keys to railroad switch boxes. One of the two, with a third youth, had then used a stolen key to unlock a switch leading to the Fruit Growers Supply Co siding at Westwood and had reversed the points. Almost at once a freight train had appeared, travelling at moderate speed. Three box cars were derailed and one railwayman was slightly hurt. The two railway vandals were aged 11 and 12. One was already under supervision for a previous burglary. [1736]

'Hoodlums or maniacs' were blamed by a newspaper for an attempt to derail a train on the Up main line near Blackheath, New South Wales, Australia, early on the morning of Monday 23rd February 1948. Just before 1 a.m. a goods train carrying perishables from Bathurst to Darling Harbor, travelling at 35 miles an hour, struck a pile of blue metal (crushed aggregate rock) which had been piled up on both rails. Two bogies of the engine were derailed and ploughed along the track for 550 yards. The engine crew were not hurt.

[1732] *The Kokomo Tribune*, Kokomo, Indiana, United States, 29th January 1947
[1733] Ibid, 28th January 1947
[1734] *The Tipton Daily Tribune*, Tipton, Indiana, United States, 30th January 1947
[1735] *The Evening News*, Jeffersonville, Indiana, United States, 7th March 1947
[1736] *Reno Evening Gazette*, Reno, Nevada, United States, 25th March 1947

Fettlers and a breakdown crew were able to rerail the engine after about four hours. It was obvious that a deliberate attempt to derail the train was involved, the obstructions having been placed on the line at a late hour, under cover of darkness. It was pointed out that the first train to arrive at the obstruction could have been the Mudgee Mail, travelling at 50 miles an hour, which could have plunged down a 15 foot embankment with great loss of life.

Eighteen months earlier a passenger train had struck a sleeper placed across the line at the same place but had just pushed it aside.

Then at about 9 p.m. on Thursday 15th April 1948 a member of the public drew the attention of the Blackheath stationmaster to a piece of crushed concrete lying on the Down line beneath an overbridge. The stationmaster wired Mt Victoria and was told that Driver Dews of *The Fish* passenger train from Sydney to Lithgow had just reported striking an obstruction at the same point. He had examined his engine at Mt Victoria and found that it had suffered some damage and there were pieces of concrete beneath it. The police theorised that the concrete had come from a disused step taken from the overbridge and placed on the line. The possibility of the step having accidentally fallen from the bridge was discounted. [1737]

On Tuesday 24th February 1948 eastbound Minneapolis St Paul and Sault Ste Marie Railroad (Soo) first-class passenger train No. 62 consisted of 4-6-2 steam locomotive 716, a mail-express car and two coaches. It left Centuria, Wisconsin, the last open office before the scene of the accident and just over 6 miles therefrom, at 3.10 p.m., 6 minutes late, and was moving at about 40 miles an hour when it became derailed at Milltown, Wisconsin, United States, at about 3.12 p.m. Engineer Bruce Staunchfield and Fireman John Corrigan were badly hurt and were taken to Centuria Hospital. [1738] The conductor, a baggageman, 15 passengers and two mail clerks were also injured. The fireman had been trapped in the wreckage and it took two hours to release him.

Engineer Staunchfield told the Interstate Commerce Commission (ICC) that he had applied the brakes preparatory to stopping at Milltown station and immediately afterwards he had felt the front of his engine drop and applied the brakes in emergency. [1739]

Investigation disclosed no defects in the locomotive, the vehicles of the train or the track. However, at the first mark of derailment a crushed stone was found on a rail and it was found that four small boys had been playing near the line. The ICC report states that the boys were aged about 11, a newspaper report gives their ages as 5 and 6. Stones had been placed on the rails by two of the boys, one or two of which had been big and hard enough to lift a leading wheel of the engine and cause the derailment. The boys confessed that they had put stones on the rails and left them there when they had been called back to school, two blocks away, after the dinner recess, by the ringing of the school bell. A Soo official said the boys would probably be taken to the juvenile court, "if for no other reason than to impress them with the seriousness of what they have done." [1740]

[1737] *Lithgow Mercury*, Lithgow, New South Wales, Australia, 23rd and 26th February and 16th April 1948

[1738] The *Wisconsin State Journal*, Madison, Wisconsin, United States, 25th February 1948

[1739] Interstate Commerce Commission report

[1740] *The La Crosse Tribune*, La Crosse, Wisconsin, United States, 28th February 1948

The points to a siding at an unstated location on the single-line branch from Hatfield to St Albans, England, on British Railways, Eastern Region, were kept padlocked when not in use and had apparently been left locked on the evening of Thursday 20th May 1948. Later that night, however, someone, it was thought probably children, had tampered with them, and when a goods train arrived early next morning, Friday 21st May, it was diverted into a siding. Its driver and fireman jumped clear as it demolished the buffers at the end of the siding, went through an iron fence, crossed a footpath, and smashed another fence at the back of a house before stopping. Waggons were derailed and the branch was blocked. [1741]

In a typical act of vandalism two boys enter the yard of the Pennsylvania Railroad at Reedsville, Pennsylvania, United States on Wednesday 8th December 1948 and opened a switch. Later, the engine that was to work a stone train struck the open switch and was derailed. No one was hurt but it cost the railroad $500 to re-rail the locomotive and repair the damage. The culprits were subsequently apprehended 'and will be taken to Mifflin County Juvenile Court later.' Meanwhile Pennsylvania Railroad police officers warned parents and teachers to keep children off railroad property, pointing out the danger to the children themselves as well as to railroad personnel and property, as evidenced by the Reedsville derailment. [1742]

Not far away from the scene of the incident of the previous May, children were suspected of being responsible for the derailment of a train from Watford Junction near the British Railways, London Midland Region station at St Albans Abbey, England, on Monday 25th April 1949. They had rolled a tree trunk about a yard long and weighing two hundredweight down a steep embankment onto the line. No one was hurt but the line was blocked for some time. Possibly the same vandals were involved in both incidents. [1743]

On about Saturday 29th October 1949 a nine-year-old boy placed an iron screw on a rail "just to see what would happen," near Pöchlarn. Austria. A passenger train was derailed and its fireman was seriously injured. [1744]

New Year's Day, 1st January 1950, saw the derailment of a New York New Haven and Hartford Railroad passenger train at Springfield, Connecticut, United States.

Westbound first-class passenger train No. N35 was a two-car diesel multiple unit. It left New Canaan at 5.39 p.m. and was travelling at about 25 miles an hour when it struck an open switch at Springfield, leading to auxiliary track No. 6. The leading car was derailed and separated from the second car, which continued along the auxiliary track and was diverted onto another auxiliary track, No. 8. At the end of the latter track it collided with the buffers and struck a building before stopping. Both cars were badly damaged

---

[1741] *The Times*, London, England, 22nd May 1948

[1742] *The Daily News*, Huntingdon and Mount Union, Pennsylvania, United States, 13th December 1948

[1743] *The Times*, London, England, 26th April 1949

[1744] *Malaya Tribune*, Singapore, 4th November 1949

and three railwaymen and 22 passengers, one of them an off-duty railwayman, were injured. It was dark and cloudy when the accident happened at 5.48 p.m.

The switch to track No. 6 had last been used at about 2.30 a.m. on 31st December. The front brakeman of the freight train involved insisted that afterwards he had lined the switch for the main track, placed the hasp of the switch lock in the keeper and locked the switch lock. Between 3 a.m. on 31st December and the time of the accident 21 westbound passenger trains had passed the switch safely. It developed that on the afternoon of New Year's Day two 10-year-old boys had passed the switch and found the switch lock open. The hasp was in the keeper but was not closed. At about 3.15 p.m. one of the boys had returned to the switch, removed the lock from the keeper and experimented with pulling the lever. When he had finished and attempted to return the lever to its normal position he had found that he was unable to move it far enough for it to drop into the slot in the casting. Unaware of the danger, he had then simply left the switch open and gone away, not telling anyone what he had done until after the accident. [1745]

An eight-year-old boy residing in a trailer camp told investigators that he had placed a pair of two-by-four pieces of timber on some bricks upon the rails because he wanted to hear the bricks crack and see the timber break and fly off under the wheels. Instead, a 59-car Baltimore and Ohio Railroad extra freight train in charge of Conductor Herbert Sammons, with Engineer M.J. Reed, was derailed on the evening of Saturday 4th February 1950 at New Haven, West Virginia, United States. The pony trucks of the engine jumped the track at the obstruction and the train continued in this state until it reached a switch, where the remainder of the engine and nine cars left the rails. Damage was estimated at $7,500 to the cars and $2,000 to the track, plus labour costs. The freight had been travelling at 18 or 20 miles an hour according to its engineer.

The boy added that although he had been told by his parents not to play on the tracks he had first placed obstructions on the line on the Saturday afternoon before the passage of passenger train No. 73 from Wheeling to Kenova. When this train had brushed them aside he had replaced the obstructions and then gone home. He denied getting the idea from movies or books.

'Investigators did not say Sunday whether any charges would be lodged.' [1746]

Two 12-year-old boys were taken into custody pending an appearance before District Judge John Chambers after admitting an act of vandalism which had derailed a Chicago Great Western Railway track motor car near Durango, Iowa, United States, on Monday 20th March 1950. A local boy from Durango and his visiting cousin from Minneapolis, Minnesota, admitted placing spikes, bolts and angle irons along a stretch of track. Signal Maintainer Elmer Larson was thrown from the car but escaped serious injury. Moderate damage was caused to the track motor car.

Two freight train derailments had recently occurred nearby but it was decided that there was no evidence to link the youths to these earlier accidents, which it was believed had been due to broken rails. [1747] [1748] [1749]

[1745] Interstate Commerce Commission report

[1746] *The Athens Messenger*, Athens, Ohio, United States, 6th February 1950

[1747] *Mason City Globe-Gazette*, Mason City, Iowa, United States, 23rd March 1950

Serious derailments due to vandalism continued in the United States with one at Holland, Missouri, on the St Louis-San Francisco Railway on Sunday 23rd July 1950. Northbound first-class passenger train No. 806 from Memphis, Tennessee to St Louis, Missouri, consisted of 4-6-4 steam locomotive No. 1067, two baggage cars, a mail car, two coaches, a buffet-lounge car, and a sleeping car. It left Blytheville, 9¾ miles south of Holland and the last open office, at 12.55 a.m., eight minutes late, and was travelling at 57 miles an hour when it became derailed at a switch at Holland at 1.09 a.m., the engine, tender and leading five cars leaving the track. Engineer Ira Woods was killed instantly and Fireman Fred Thurman, the unnamed conductor, 18 passengers, a mail clerk and a steward were injured, the fireman seriously. [1750] [1751]

Examination of the engine and train after the derailment failed to reveal anything which could have caused it. There was no indication of dragging equipment or of an obstruction on the line. On Thursday 20th the crew of a northbound train had reported that two of the switch lamp lenses at the switch where the derailment had occurred were broken and the target was bent, though the switch was properly closed and locked. The section foreman replaced the lenses and straightened the target the next day. Shortly after the accident a 15-year-old sharecropper's son, William Godsey, who lived at a nearby farm, was taken into custody. After two days of questioning he confessed to sawing through the shackle of the switch lock with a hacksaw and lining the switch for the siding, breaking the lenses of the switch lamp, and bending the target to indicate the switch was set for the main line, after a freight train had passed on Saturday night. He blamed his actions on having watched a movie thriller featuring the 19th-century Dalton gang of bank robbers and train wreckers. He was subsequently charged with first-degree murder and 'criminal intent to kill or injure passengers on a train by tampering with a switch.' Godsey denied wanting to wreck the train and said that he had just wanted to turn it into a siding. [1752] [1753]

Because he had pleaded guilty to derailing the train the initial charge of first-degree murder was reduced to one of second-degree murder and in early December, at Caruthersville, Missouri, Godsey was sentenced to 15 years' imprisonment. [1754]

A Chicago Great Western Railway passenger train from Minneapolis to Omaha, consisting of an engine and six coaches, in charge of Conductor W.J. Barth with Engineer Frank Boggs and Fireman Orrin O'Bele, was approaching Austin, Minnesota, United States, at 11.42 p.m. on Sunday 29th October 1950. It was about half an hour late. As it neared a switch at about 40 miles an hour the switch lamp, which should have indicated to the engineer how the switch was set, was extinguished. Engineer Boggs thus had no warning that the points were set for the storage track until they were illuminated by the headlight of his engine at close range. According to Brakeman W.P. Whitlock the

---

[1748] *The Telegraph-Herald*, Dubuque, Iowa, United States, 22nd March 1950
[1749] *The Oelwein Daily Register*, Oelwein, Iowa, United States, 23rd March 1950
[1750] Interstate Commerce Commission report
[1751] *Moberly Monitor-Index*, Moberly, Missouri, United States, 24th July 1950
[1752] *Daily Capital News*, Jefferson City, Missouri, United States, 26th July 1950
[1753] Ibid, 25th July 1950
[1754] *Blytheville Courier*, Blytheville, Arkansas, United States, 11th December 1950

engineer had immediately applied the brakes in emergency as the train was diverted into the storage track but it left the track at a derail about 200 feet from the switch. The engine was uncoupled from its train and overturned and four coaches ended up in a ditch, fortunately upright. Seven railwaymen were injured; the six passengers on the train were unhurt. Conductor Barth ran to a nearby tower after the accident to report it.

Sheriff Albert Reinart said that he believed that youngsters had tampered with the switch. The switch lamp and the lamp of the home signal had been shot out the previous week and had been replaced. [1755]

Two boys aged 7 and 10 piled rocks on the tracks at Berkeley, California, United States, on Sunday 13th July 1952 "out of curiosity." The leading bogies of a Key System interurban train were derailed. No one was hurt but the service was disrupted for several hours. '"We just wanted to see the train bust the rocks. We didn't mean anything," the tearful lads told their parents, after police took them home.' [1756] [1757]

The derailment of a freight train as it was rounding a curve in the Benning Yard, D.C., United States, on Monday 17th November 1952, cost the Pennsylvania Railroad $3,500. Two boys aged 8 and 10 were blamed for four cars leaving the rails. The younger boy said, "The train was going crooked and I wanted to see it go straight – I pulled the switch." The boys were handed over to their parents pending a likely appearance in the juvenile court. [1758]

The engine and first coaches of the 3.50 p.m. train from Port Adelaide on the Hendon line were derailed about 150 yards from the Albert Park Crossing, South Australia, when they struck a four-foot length of two-inch piping on Wednesday 26th August 1953. Inquiries led the police to the home of a six-year-old boy who denied having placed the piping on the line. The following day he was questioned again at his school and he eventually admitted having committed the act. The derailment was said to be likely to cost the railway several hundred pounds in penalty payments to the crew of the breakdown train sent to the scene and in interrupted services. The line was cleared at midnight and trains were running normally the next day. [1759]

On Tuesday 23rd August 1955 an eight-car Long Island Railroad electric train from New York with 400 passengers struck a 10-foot steel I-beam laid across the westbound track at Kew Gardens, New York, United States. The beam was dragged some hundreds of feet until a short-circuit occurred. No one was hurt.

David Alex, 15, was taken into custody and later freed on $20,000 bail. A second boy, William Hannigan, 18, was arrested as an accessory to the vandalism. Asked why he had placed the beam on the track Alex said simply, "I wanted to wreck the train." [1760] [1761]

[1755] *Austin Daily Herald*, Austin, Minnesota, United States, 30th October 1950
[1756] *Southern Illinoisian*, Carbondale, Illinois, United States, 14th July 1952
[1757] *San Mateo Times*, San Mateo, California, United States, 14th July 1952
[1758] *The Republican-Courier*, Findlay, Ohio, United States, 20th November 1952
[1759] *The News*, Adelaide, South Australia, 27th August 1953
[1760] *The Oneonta Star*, Oneonta, New York, United States, 27th August 1955

The eastbound *Knickerbocker* passenger train of the New York Central Railroad had a narrow escape on the evening of Monday 29th August 1955 near the Otter Creek bridge at Markells, about 5 miles north of Terre Haute, Indiana, United States. It struck a two-by-eight-inch plank which had been laid across the track and crushed it without being derailed. Bailing wire had been used, according to a newspaper report, to short circuit the absolute block system. Had it left the track and fouled the opposite running line the *Knickerbocker* might have been run into shortly afterwards by a westbound freight train. The plank had been taken from the bridge. [1762] [1763]

Residents living near the scene saw two teenage boys and a brown dog in the neighbourhood. They described the animal and police were able to follow it to the home of one of the boys, who were aged 12 and 15. The boys admitted the act, saying that they had "wanted to see the splinters." The younger boy was released to his parents. The older boy was already on probation for vehicle theft. He was also released on a $500 bond paid by his parents. [1764]

Canadian National Railways (CNR) train No. 3, the westbound *Super Continental*, was over five hours late when it was derailed at Nourse, Manitoba, Canada, at about 4 p.m. on Sunday 28th July 1957. Its two diesel engines and a mail car passed over a switch safely but the remainder of the train, composed of 16 passenger, express and mail cars. was thrown off the line, eight of the cars being jack-knifed and one overturned. 'The cars crumpled and twisted and finally lay strewn from the tracks across a ditch and into track-side bushland.' The train had been travelling at about 60 miles an hour. No one was killed but 27 people were taken to hospital by ambulance, 24 of whom were admitted and two of whom were in a critical condition. Another 59 were treated in CNR clinics at Nourse itself and in Winnipeg. There had been about 400 passengers on the train, many of whom had been immigrants riding in the first three cars and were among the injured.

CNR Vice President J.R. McMillan announced that, "Evidence has been obtained that a switchlock was removed from its position and the switch handle raised by juveniles immediately prior to the approach of the train, causing the derailment. Investigation is proceeding."

Working in a field about 100 yards from the wreck was farmer Rudolph Riter, who also worked for the railway company and whose home was close to the railway at that point. He related, "There wasn't any particular noise. There just seemed to be a big cloud of dust and I could see the cars going the wrong way." He ran to the tracks and said that by the time he reached them people were pouring out of all the exits which were not blocked by twisted rails or other cars. When he learnt that the train's telephone was not working he and the conductor ran across a field to a telephone to summon help. Seven ambulances took the injured to hospital, the police having to control the heavy traffic on the adjoining Highway 15 so that the ambulances could get through. Ten buses brought the uninjured passengers to Winnipeg, from where those going further continued on a

---

[1761] *The Norwich Sun*, Norwich, New York, United States, 26th August 1955

[1762] *The Kokomo Tribune*, Kokomo, Indiana, United States, 30th August 1955

[1763] *Pharos Tribune*, Logansport, Indiana, United States, 30th August 1955

[1764] *The Kokomo Tribune*, Kokomo, Indiana, United States, 31st August 1955 special train. [1765]

The accident received extensive coverage in the newspapers, including the experiences of various passengers. 'First it was a car accident. Now a train wreck. And today 45-year-old Lloyd White of Sioux Lookout, Ont., is wondering what next. He is one of 24 persons in hospital following Sunday's 20-car CNR transcontinental passenger train at Nourse, Man. He still bears a leg cast from a serious automobile crash in Ontario last April.

'Mr White, who suffered broken ankles and ribs in the head-on collision, was on his way to Winnipeg to see his wife. She is seeing her parents here after being discharged from hospital last Thursday. Mr White bears facial cuts and bruises from Sunday's train smash up – but, for a few moments, he wondered if he would leave the crippled train alive. He had been riding with his cast-encased left foot resting under a foot bar. When the train hit a shaky switch, it lurched and his foot was trapped under the bar. By this time the day coach was on its side and people were thrusting their way around him to safety. Finally a male passenger stopped beside him, turned the seat to free Mr White's leg and kicked out the train window – helping Mr White outside.

'Mr White said he first felt the train shudder. Then, "my coach took to the air and started turning over. People were piled into one another."

'One woman was screaming for her baby. Another, as she was hurtled from her seat, missed by inches being hit by a section of rail which pierced the bottom of the coach. The rail tore through the floor, lanced upwards and veered just before reaching the roof. "The train was a big one and we were late. I guess it was making up for lost time – we were travelling very fast."

'Mrs Torchia, who is in St Boniface hospital with deep cuts about the face and neck, was combing her hair in the ladies room of her coach when the train was derailed. She said the scene was bedlam, and she frantically tried to open the door, then finally succeeded in getting into the corridor. "I was travelling with my mother," she said, "and I wanted to see if she was all right. Thank heaven she was." Mrs Torchia said suitcases were thrown everywhere in the coach, seats were twisted and people badly cut were lying in the aisle.

'Fred Pritchard, who was working in the train kitchen, was working in the train kitchen, is improving in St Boniface hospital from leg burns. He was standing near a large container of boiling water, and the water poured over his legs. Fred has been working for the railroad for the summer months, and it was his fifth trip as a cook in the dining car. He graduated from Grade 12 in Gordon Bell high school in June and hopes to accept a scholarship from North Dakota state college in Fargo in the fall. He is to try out with the Winnipeg Rods football club.

'Mr Craig, in Winnipeg General hospital, was unable to see visitors Monday morning, but hospital officials said his condition appears satisfactory. Although he has cuts and bruises, the remainder of his injuries have not yet been determined and he will be given further examination.

'Mrs Schwitz, also in Winnipeg General, is believed to have suffered a neck injury in the wreck but pending studies of X-rays Monday it was not determined how serious the injury may be. She also suffered cuts and bruises.

[1765] *The Lethbridge Herald*, Lethbridge, Alberta, Canada, 29th July 1957

'St Boniface hospital reported Gilbert Cook, of Gypsumville, to be in good condition. He received back and head injuries in the accident but was much improved Monday.

'At Winnipeg General hospital Monday the switchboard was jammed with telephone calls from relatives and friends inquiring into the condition of those injured in the derailment. One harassed operator said it was one of the busiest mornings in a long time.' [1766]

Not long after the accident it emerged that it was the two sons of the previously mentioned Mr Riter, Bobby, and Ralph, aged 6 or 7 and 7 or 9 according to different reports, who had caused the derailment by interfering with a switch. And had admitted what they had done to the police. Their mother, Mrs Riter, was not convinced they were responsible though. She said that they had been out picking raspberries on the afternoon of the accident. '"You know how cops scare children," she said. "They might say anything." She said she thought it peculiar that the children could open the switch "because my husband said it takes a lot of strength to move it." There have been "lots of lies" written about the accident, she said.

'One report said the boys used a crowbar. "This is not true," the mother said. "Nobody saw them and it is still doubtful in my mind they did it. However, I don't want to commit myself. My husband is an employee of the railway and I don't like to say too much." Nevertheless, she said, the boys "have been given a talking to and they realize what they've done. I don't think it will happen again."

Mrs Riter added that she didn't think her sons were to blame, "if the lock was left open. After all, kids are kids." A CNR spokesman said that whether or not the switch had been unlocked was being investigated.

Vice President McMillan said it might be several days before a decision was taken as to what action would be taken against the boys. Juvenile authorities were being consulted, he stated. [1767]

A six-year-old boy with a fascination for trains "wanted to see a train go on two tracks," so he threw a switch beneath a stationary Baltimore and Ohio Railroad freight train at Winchester, Virginia, United States, on Saturday 10th August 1957. When the train moved off again six of its 51 cars were derailed in the middle of town, damaging four automobiles, striking a grocery warehouse, knocking out the town's traffic lights system and so on. No one was hurt but damage was estimated at over $35,000. Police said that no official action would be taken against the boy because of his age. [1768]

At least 18 people died and 100 were injured in the derailment of the engine and eight of the 11 coaches of a passenger train near Taoyuan, Formosa,* on Monday 9th December 1957. Police said that there were probably more bodies buried beneath the first two cars of the train which according to witnesses were buried like matchboxes. The driver and his assistant were among the dead while about 40 more passengers were seriously injured. Railway officials said that the accident might have been worse had the train not been slowing to stop at a station about a kilometre further on.

[1766] *Winnipeg Free Press*, Winnipeg, Manitoba, Canada, 29th July 1957
[1767] *The Medicine Hat News*, Medicine Hat, Alberta, Canada, 1st August 1957
[1768] *The Bee*, Danville, Virginia, United States, 12th August 1957

(* The country is referred to in contemporary newspapers as Formosa rather than as Taiwan so that name has been used here.)

Over 500 firemen, police and soldiers took part in the rescue operation, which was still in progress late on the night of the accident. About 150 metres of track was destroyed and it was thought that it would be at least a day before traffic on the main west coast line could be resumed. The cause of the derailment was not immediately clear. [1769]

By the following day the number of dead had increased to 19 and it had become obvious that the derailment had been due to unidentified children placing stones on the rails. A warning was issued to parents reminding them of the dangers of their children doing so. [1770]

A further report stated that a witness, Pai Hung-Fei, had been found who said that he had seen two five-year-old boys pile stones onto the rails and then stand aside to watch what would happen. The boys, Wang Yun-Chi and Lao Yao, were said to have confessed to police but the mother of Yun-Chi, like Mrs Riter at Nourse (see above), charged that her son had made the confession only after repeated coaxing and offers of sweets, saying, "He is so young he doesn't know what he is talking about." The boys were in any case too young to be held responsible for their actions under Taiwanese law. [1771]

Baltimore and Ohio Railroad trains ran into obstructions placed on the line in Athens, Ohio, United States, on two occasions in January 1958. On Saturday 18th a train ran into a 'highway-type' barricade near Mulberry Street Crossing, then on Thursday 23rd two cinderblocks obstructed the line in roughly the same location. In both cases the trains pushed the obstructions aside, though on the second occasion the train was delayed for half an hour for slight repairs. A mail and a freight train were involved. A sharp curve at this point gave engineers little warning of the obstructions.

Three teenagers, all freshmen at Ohio University and 'all sons of well-to-do Cleveland area parents,' were subsequently arraigned by the Federal Bureau of Investigation (FBI) to appear before a grand jury, charged with attempting to derail the trains. John Hamm, 18; Gerald Hinkley, 18; and James Onder, 19; told US Commissioner Willard Walter that it had been a college prank, with no malice intended and no intent to derail the trains, but they were told that they faced a maximum penalty of $10,000 fines and/or 20 years' imprisonment before they were released on bail of $2,000 paid by their fathers. University dean Mural Hunkins recommended that the university disciplinary board should take no action and leave the matter in the hands of the FBI. Students had previously placed dummies and snowmen on the track near the university campus. [1772]

A Chicago and Eastern Illinois Railroad freight train ran into an obstruction on the track and was almost derailed on Thursday 12th May 1960 near Goreville, Illinois, United States. It suffered some damage. A man in an adjoining field saw a youth on the track around the time of the incident, though he did not know what he was doing. Sheriff Glenn

---

[1769] *San Mateo Times*, San Mateo, California, United States, 9th December 1957

[1770] *The Southwest Times*, Pulaski, Virginia, United States, 10th December 1957

[1771] *The Daily Courier*, Connellsville, Pennsylvania, United States, 13th December 1957

[1772] *East Liverpool Review*, East Liverpool, Ohio, United States, 27th January 1958

Dallas took a description of the youth from the witness and questioned teenager Lloyd Shelton, who admitted placing a bridge timber across the rails, saying that he was "just trying to scare somebody." [1773]

State Attorney William Dyer filed a petition to have Shelton declared a delinquent child. [1774]

Head Brakeman Francis Blackwell suffered a broken leg when the engine and one car of a 25-car Union Pacific Railroad freight train travelling at about 12 miles an hour were derailed on Monday 30th May 1960 at Lakewood, California, United States. The brakeman and Engineer Herman Jensen told police that they had seen a group of boys tampering with a switch immediately before the derailment. It was found that the lock of a switch had been sawn through and the points had been thrown under the train. The lock had clearly only recently been tampered with as the edges of the metal were still bright and particles of metal filings were lying around. The vandals do not seem to have been caught, at least not in the following few days. [1775]

On Sunday 5th June 1960 an attempt was made near Askham-in-Furness, England, on British Railways, London Midland Region, to wreck a troop train carrying Territorial soldiers from the London area but a police spokesman attributed the act to vandalism by irresponsible youths rather than sabotage. The train had been en route from London St Pancras to Millom, where the men were to attend a Civil Defence training school. Driver T. Parkin felt a bump as his train passed over a board (a gradient sign). He stopped the train and the train crew removed the board and four piles of pipes from the line. Other trains were delayed while the line was searched for further obstructions. There had almost been a derailment for it was stated that "the engine wheels must have left the track momentarily before jumping back on the rails again." A British Railways spokesman agreed with the police view that it was vandalism, stating, "There have been a number of outbreaks of vandalism in the north recently. This seems to be the work of young people." [1776]

[1773] *Southern Illinoisian*, Carbondale, Illinois, United States, 15th May 1960
[1774] Ibid, 17th May 1960
[1775] *Independent*, Long Beach, California, United States, 31st May 1960
[1776] *The Times*, London, England, 6th June 1960

CHAPTER 18

SABOTAGE

Four distinct motives for sabotaging trains may be distinguished (after Robert B. Shaw):

(a) As an aid to intended robbery, either of the train passengers or to steal, for example bullion being carried by the train.

(b) To satisfy a personal grudge against the railway company, such as by an ex-employee who feels that he has been wrongly dismissed.

(c) To hinder the operation of the railway, as for example during a strike.

(d) As an act of guerrilla warfare or terrorism, often against an occupying enemy power. Sabotage against the railways has been a feature of many conflicts from the American Civil War onwards.

The period covered by this volume saw numerous acts of sabotage in various countries such as Palestine, Algeria, Burma, Malaya, the Dutch East Indies and so on, often in connection with post-Second World War independence movements, most of the sabotage, therefore, belonging to the fourth of the categories outlined above. In a departure from the method used in other chapters of this book, acts of sabotage have been to some degree grouped together according to the country in which they occurred, rather than being listed in a strictly chronological order.

Civil war had raged in China since August 1927 between the Chinese Nationalists and the Chinese Communist Party, though the hostilities between them were largely put on hold between 1937 and 1945 as both fought the Japanese. At the end of the Second World War, Russia had declared war on Japan and with the Japanese surrender in August 1945 the Russians occupied Manchuria. The Nationalist Chinese forces then took over Manchuria from the Russians and the civil war resumed, with the Chinese Communist forces continuing with the guerrilla war tactics that they had employed against the Japanese.[1777] A particular target was the Peking (or Peiping and now Beijing) – Mukden (now Shenyang) Railway.[1778]

At dusk on possibly Sunday 6th January 1946 an estimated 50 bandits derailed a Canton – Hankow passenger train between Yuentam and Canton, China. The passengers were then lined up and robbed, the bandits reportedly escaping with several million US

1777 Wikipedia – Chinese Civil War

1778 *The Times*, London, England, 13th April 1946

dollars' worth of loot. [1779]

On the night of Monday 8th April 1946 a passenger train on the railway was mined near Wanchiatun, China, killing over 300 people according to a report from Tientsin. After the derailment the Communists attacked the survivors and Wanchiatun station. [1780]

Eight people were killed and a number were wounded when an express train consisting of a diesel engine and two coaches was mined near Canton, China, on the line to Kowloon, Hong Kong, on Friday 10th May 1946. Two large explosions threw the train from the track and it caught fire. The driver and his assistant were among the dead and some passengers were trapped in the wreckage. About ten bandits then swooped down from the hills, held up and systematically robbed the passengers, stripping the wounded. One man, probably a Chinese official, was forced to hand over the equivalent of $70,000,000 in Chinese currency. [1781] A passenger said that the bandits had carried two large baskets, presumably to hold the possessions taken from the passengers. The train involved was an extra fare working catering principally for businessmen travelling between Kowloon and Canton. [1782]

Speaking to the *Kuomintang New Times* newspaper, a government official who had been on the train said that the attackers had definitely been Communists as their leader had shown passengers a card identifying him as belonging to the Communist army. The bandits had been equipped with Japanese weapons.

As the attack had occurred in an isolated location it was almost 12 hours before help arrived. Later Chinese troops scoured the hills in an unsuccessful hunt for the attackers.

On Friday 9th August 1946 a Chinese Nationalist coal train from Kuyeh to the port of Tanku, east of Tientsin, was mined by Chinese Communists, who derailed the engine and eight of the 25 waggons of the train at a point about 5 kilometres north of Lutai, China, and some 100 kilometres north-east of Tientsin. Following the derailment the attackers at once opened fire on the train. Four American marines escorting the train returned fire and held off the attackers until they were later rescued. Four waggons and about ten metres of track were destroyed. [1783]

In what was regarded as another setback for the Nationalists, the Communists derailed a passenger train on their "lifeline" to Manchuria, the Tientsin – Mukden line, at a point about 30 kilometres north of Shanhaikuan, China, 'where the Great Wall of China meets the sea,' on Tuesday 20th August 1946. No casualties were mentioned. [1784]

On about Tuesday 27th August 1946 three attacks were made on trains between Tientsin and Chinwangtao, an important coal port, and two trains were derailed. The explosion of a mine derailed an armoured train near Hanku, China, precipitating a 20-minute clash between the Nationalist guards aboard the train and the Communist attackers. Meanwhile, 6 kilometres from Lutai, China, a mine derailed a coal train. [1785]

The final removal of the American marines who had been guarding the Peking – Mukden railway since the surrender of the Japanese was announced in December 1946.

[1779] *Southern German Stars and Stripes*, Altdorf, Germany, 8th January 1946
[1780] *Helena Independent Record*, Helena, Montana, United States, 11th April 1946
[1781] *The Straits Times*, Singapore, 12th May 1946
[1782] Ibid, 14th May 1946
[1783] *The Daily Mail*, Hagerstown, Maryland, United States, 10th August 1946
[1784] *Lebanon Daily News*, Lebanon, Pennsylvania, United States, 20th August 1946
[1785] *The Galveston News*, Galveston., Texas, United States, 29th August 1946

The newspaper report noted that, 'While the marines guarded the line, service between Peiping and Chinwangtao seldom was interrupted, but since the Government took over Chinese Communists have attacked trains, destroyed rails or blown up bridges almost daily.' On Friday 6th December 1946 'Communist demolition units,' as they were termed, wrecked a passenger train near Chating, China, and a freight train near Peitang, China. [1786]

Following the end of the First World War, Palestine, which had formerly been part of the Ottoman Empire, was administered by Great Britain in what was referred to as the British Mandate. Between 1936 and 1939 the Great Palestinian Revolt took place, fuelled by Arab resentment against both British occupation and Zionist (Jewish) settlement. The revolt was supressed by the British towards the end of 1939. Also in 1939 a White Paper had proposed restricting Jewish immigration to Palestine. This led to the formation of Jewish terrorist groups aimed at forcing the British out of Palestine and in early 1944, with the Axis retreat from the Mediterranean and Russian advances in Eastern Europe indicating that Nazi Germany was clearly going lose the Second World War, a Jewish campaign of violence against the British presence in Palestine began. [1787]

There were no attacks on the railway during the Second World War, but on the morning of Saturday 12th January 1946 a payroll train from Haifa struck a landmine at Hadera, Palestine, and the engine and seven coaches were derailed. A gang of 70 Jews, some of them women, who had been concealed in bushes on either side of the line, then opened up with automatic fire, slightly injuring a British policeman and an Arab temporary constable and causing extensive damage to the train. The railwaymen were unarmed and were forced to hand over the safe containing the staff's payroll, though £15,000 of its contents had already been paid out. The gang then withdrew as they were fired on by a police escort, who had extricated themselves from an overturned coach.

Following the attack police dogs were used to trail the attackers but the scent was lost when they reached a main road, where the gang had clearly had a vehicle waiting. Pits in which the gang had hidden arms and explosives were found in neighbouring woods. The next day 17 suspects were arrested by police in the Yemenite Jew quarter of Hadera. [1788]

Armed bandits attacked three trains in Palestine on the night of Monday 10th June 1946. The Jerusalem – Jaffa train was held up between Sarafand Junction and Tel Aviv. Its engine was detached and run over a mine and the coaches were set on fire. A Jaffa – Lydda train was held up by Jewish terrorists near Aquir (20 miles from Tel Aviv). Its passengers were ordered off at pistol point and it was then set on fire. A third train was derailed at an unstated location. [1789] Women armed with Tommy guns were among the attackers. [1790]

Masked Jewish extremists blew up an oil train on the line between Tel Aviv and Haifa on the night of Monday 23rd September 1946, causing a fire which was visible 40 miles away. The train, running south from Haifa, was mined near Hadera, Palestine. When the driver and conductor got down to inspect the damage they were confronted with the

---

1786 *Walla Walla Union-Bulletin*, Walla Walla, Washington, United States, 6th December 1946
1787 Wikipedia – Jewish Insurgency in Mandatory Palestine
1788 *The Times*, London, England, 14th January 1946
1789 *The Times*, London, England, 11th June 1947
1790 *Malaya Tribune*, Singapore, 11th June 1946

terrorists who ran out of the woods alongside the line, waving submachine guns. The attackers then set off further explosives which threw the engine and ten tank cars off the track and set fire to the petrol being transported. More than 5,000 British troops rushed to the area, cordoned off the entire district and screened the entire population of Hadera and nearby Benyamina. The attack was possibly linked to the sailing earlier in the day of the British transport ship *Ocean Vigour*, which was carrying nearly 400 intending Jewish immigrants to a detention camp on the island of Cyprus. [1791]

Another goods train was derailed north of Haifa, Palestine, probably on Saturday 2nd November 1946 and its engine was derailed. [1792]

Mines exploded under a goods train near Ras El Ain, Palestine, north of Tel Aviv, late on Sunday 3rd November 1946. The engine and six waggons were derailed and the firemen and two other members of the train crew were injured. [1793]

In another attack at Ras El Ain, Palestine on the night of Tuesday 5th November 1946 a train was blown up and caught fire. [1794] The driver and fireman were injured.

A train from Haifa carrying troops going on leave to Egypt ran over mines near Lydda, Palestine, early on Thursday 7th November 1946. Three carriages in the middle of the train were damaged and three soldiers were slightly hurt. The mines had been detonated from an adjoining orange grove, from which a wire was found leading to the track. [1795]

A wooden trolley powered by a Ford engine and carrying six railway policemen – four Arab and two British – struck a contact mine laid on the track at Beit Safafa, Palestine, about four miles outside of Jerusalem, before dawn on Wednesday 13th November 1946. The policemen had been patrolling the line between Jerusalem and Jaffa at the same time for the previous four days. As already stated the mine was set off by contact rather than by an observer. Three of the men were blown wide of the track on one or the other side of it. One went through the telegraph wires, breaking them, and another landed in the middle of an olive grove, leaving bits of clothing on the olive trees. Five of the six were killed outright and the other succumbed after an hour. The two British were a constable acting as a sergeant, who was in charge of the party, and a young man who had only been out from England five weeks and was making his first patrol.

When Arabs from a nearby village arrived at the scene after hearing the sound of the explosion they found that the trolley was on fire and the track was torn up. Within an hour police patrol dogs had arrived and they traced the scent as far as Rehavis, a smart Jewish suburb of Jerusalem. There the scent was lost, the attackers having doubtless escaped in cars. Trains on the Jerusalem – Lydda line were suspended following the attack.

Also on Wednesday 13th November 1946 a goods train was derailed near Jaffa, Palestine. Its Arab fireman was badly hurt and died the next day. [1796]

In an effort to counter the terrorist attacks the British started using armoured rail cars. On the morning of Friday 15th November two of these vehicles, travelling together, detonated a mine at Binamina, Palestine. A terrorist hiding in sand dunes 100 yards away

---

1791 *Nevada State Journal*, Reno, Nevada, United States, 24th September 1946
1792 The Mirror, Perth, Western Australia, 2nd November 1946
1793 *The Times*, London, England, 5th November 1947
1794 Ibid, 7th November 1947
1795 Ibid, 8th November 1947
1796 Ibid, 15th November 1946

detonated the explosive. Both cars were derailed and one soldier was injured. [1797]

A goods train was derailed by a mine near Rehoboth, Palestine, on Monday 18th November 1946. [1798]

An attempt to wreck No. 76 Down passenger train was made on the evening of Tuesday 30th April 1946 between Jullunder and Amritsar, India. 'Machinery' was wired to the track. The driver noticed the obstruction and was able to reduce speed so not much damage was done. The engine and two bogies were damaged. [1799]

A special train carrying Indian nationalist leader Mahatma Gandhi and his party from Bombay to Poona ran into two or three boulders on the line near Neral, India, 68 miles from Poona, at about 2 a.m. on Sunday 30th June 1946.. The train was travelling at 35 to 40 miles an hour and by the time that the driver became aware of the obstruction it was too late to pull up. The cowcatcher of his engine struck the boulders and threw them to one side. There was no derailment, though the engine was damaged. No one was hurt, the passengers escaping with a violent shaking. On arrival at Poona, at 9 a.m. instead of 2.30 a.m., Gandhi profusely thanked the driver and guard of the train. They replied, "We must be thankful that you were safe and received no injuries."

A police investigation into the incident was opened and it was noted that a passenger train had passed the spot half an hour earlier without noticing any obstruction. There was also no high ground near the scene of the accident from which the boulder could have fallen, while the boulders were heavy enough to require two or three persons to lift them.

At a prayer meeting in Poona later Gandhi made a statement. "I should have never reached here today but for the grace of God. I did not know anything about the accident, but I woke up at 4 a.m. and found that the train had stopped about two miles from Neral. I was told that big boulders were kept on the track and the engine had dashed against them. Had it not been for the vigilance and skill of the driver, I do not know what would have happened. I have never done any harm to anyone and not even in my dreams have I ever thought of injuring anyone. I do not understand why anyone should wish to do injury to me. But I feel God did not wish me to die. And now that he has saved me, I hope for 125 years to serve my people." [1800] His wish was, alas, not to be, for in January 1948, aged 78, he was assassinated by a Hindu fanatic.

A goods train was derailed at Rhimyarkhan, in Bhawalpur State, India, on Wednesday 3rd July 1946, one or more rails having been removed. As a result, the North Western Railway made arrangements to provide armed police and military guards on passenger trains. A pilot engine was also to precede trains between Sukkur and Bhawalpur. The police and guards were to be drawn from the Sukkur Police, Sind Police Rifles and Baluch Regiment.

It was noted that the derailment of the goods train had very likely saved a large number of lives, as the *Karachi Mail* and the *Sind Express*, proceeding in opposite directions and which were both held up by the blockage of the line, might have been derailed instead. Dacoits were believed to have planned the two passenger expresses so that they could rob the passengers, as they had done four years previously near Derolal.

[1797] Ibid, 16th November 1946
[1798] *The Sun*, Sydney, New South Wales, Australia, 18th November 1946
[1799] *The Railway Gazette*, London, 31st May 1946
[1800] *The Indian Express*, Madras, India, 1st July 1946

The fireman of the goods train was injured. [1801]

Several people were injured when a passenger train from Mandalay jumped the rails at midnight on Saturday 12th October 1946 near Pegu, Burma, 50 miles from Rangoon. Two pairs of rails were found to be missing and the accident was attributed to sabotage. [1802]

Four people died and 47 were injured, 15 seriously, in the derailment of the northbound mail train from Colombo to Jaffna in the jungle six miles from Anuradhapura, Ceylon, in the early hours of Friday 18th October 1946. The train was travelling at 45 miles an hour at the time. Although the engine remained on the track and was undamaged, behind it one passenger coach was completely smashed, three were badly telescoped and five were derailed. Two goods waggons, two coal bogeys and a water tank wagon were thrown off the rails and piled up in a heap in the boggy ground by the trackside, reduced to matchwood.

The passengers were naturally dazed after the accident and in the pandemonium which followed some of them were robbed of their possessions, one losing 3,500 rupees and all his belongings. 'Further thefts were, however, prevented.' A mobile X-ray van was rushed from Colombo and Ceylon Government Railway Acting General Manager J.E.S. Bodger set off for the scene with a breakdown train. As a result of the derailment all long-distance night mail trains from Colombo had to be suspended on Friday, including to Talaimannar which connected with the Indo-Ceylon Express.

The derailment was the result of sabotage, fishplates having been removed and rails moved out of position. Strike supporters were suspected to be responsible as 4,000 railwaymen were out on strike, as were 6,000 harbour workers and 2,000 factory and power station workers, the strikes all being for pay rises. [1803]

Since 1st October 1945 the Toledo Peoria and Western Railroad had been involved in a bitter strike over pay, rules and working conditions. Several violent incidents had occurred, and in February 1946 two union pickets had been shot dead and three others wounded. The working of the railroad was resumed in late November 1946 following a court injunction restraining "all parties" from interfering with the operation of the line. [1804] On Tuesday 26th November 1946 the engine, tender and caboose of a small work train were derailed and overturned into a shallow ditch alongside the line near Secor, Illinois, United States. Two railwaymen were slightly hurt. A rail had been sprung several inches out of line and an angle iron was found in the weeds nearby. Railroad officials promised a thorough investigation. [1805]

A civil war raged in Greece between late 1944 and October 1949 between the Royalist forces and the Communists, who wished to seize power. It ended with the defeat of the Communists. Presumably in connection with this conflict Guerrillas mined a train near

[1801] *The Railway Gazette*, London, 16th August 1946
[1802] *Morning Tribune*, Singapore, 15th October 1946
[1803] *The Indian Express*, Madras, India, 20th October 1946
[1804] *Ironwood Daily Globe*, Ironwood, Michigan, United States, 26th November 1946
[1805] *The Lima News*, Lima, Ohio, United States, 28th November 1946

the station at Gravia, Greece, 145 kilometres north of Athens on Sunday 9th February 1947. One man was injured and the engine was damaged. [1806]

Guerrilla bands derailed a Salonika – Istanbul express on Tuesday 2nd December 1947 in the Greek Macedonian province of Kavalla. [1807]

The Mandalay Mail was derailed at a small bridge near Nyanglon, Burma, at 1 a.m. on Sunday 23rd February 1947. Nyanglon was a small halt between Pyinmama and Yamtein on the Rangoon – Mandalay main line. Ten people died, including the engine driver, who succumbed to his wounds in Pyinmama hospital, and both firemen, while 11 were injured. The loss to the railway was estimated at 50,000 rupees. According to the railway about 70 feet of track had been removed, throwing the engine down an embankment. The Chief Railway Commissioner of Burma Railways, J.C.B. Wakeford, announced that it had been definitely established that the accident was due to the deliberate removal of fishplates but that full details were lacking as the telephone lines between Rangoon and Pegu had been cut. [1808]

Twenty-eight passengers, including six railwaymen not on duty (?) were killed and 14 people were injured when a passenger train was derailed near Thazi Junction, Burma, on Friday 7th March 1947. Those who lost their lives included seven British soldiers and an Indian non-commissioned officer. A communique from the Chief Commissioner of the Burma Railways stated that the derailment had been due to the "deliberate removal of fishplates and dogspikes," adding that, "simultaneously with the derailment of the train, there was a concerted attack from the neighbouring hillsides by dacoits, armed with various weapons, which resulted in the death of 28 passengers and injuries to 14."

The Burma Command of the South East Asia land forces stated that large forces of British, Indian and Burma army infantry, supported by armoured cars, had gone into an all-out drive to smash the "dacoit dictatorship" in Central Burma. [1809]

Another train was derailed near the station at Dabein, Burma, 40 miles north of Rangoon, on Monday 10th March 1947. After the derailment armed bandits attacked the passengers and ransacked the carriages. [1810]

Two trains were derailed by mines north and south of Jerusalem on the night of Thursday 13th March 1947. On the Jerusalem – Haifa line an oil train was derailed and subjected to small arms fire near Petah Tiqva, Palestine, about 53 miles north of Jerusalem and outside the martial law area. The attack reportedly occurred as the train was climbing an incline, derailing 19 waggons, and overturning seven whilst the remaining vehicles ran back down the bank. Almost 50 yards of track were uprooted. No casualties were reported.

The other attack was at Beit Safafa, Palestine, just south of Jerusalem on the Jaffa line. In this attack a double-headed freight train was derailed by an electrically detonated mine.

[1806] *Malaya Tribune*, Singapore, 11th February 1947
[1807] *Joplin Globe*, Joplin, Missouri, United States, 5th December 1947
[1808] *The Indian Express*, Madras, India, 25th February 1947
[1809] Ibid, 9th March 1947
[1810] *The Straits Times*, Singapore, 12th March 1947

An Arab driver was killed and a fireman and a British officer were injured. [1811] [1812] [1813]

The Cairo – Haifa train was mined early on the morning of Tuesday 22nd April 1947 near Rehoboth, Palestine. The train was regular working which ran through the night but did not enter Palestine until after dawn, night running having been stopped for fear of terrorist attacks. Patrols were also made along the line but it was not known if one had been made that day. The train always carried a mixture of civilians and military, the latter usually riding in the front portion of the train and the civilians in the rear portion. Four coaches were piled up in a heap by the blast, and victims were thrown onto the line. The saboteurs then opened fire on the guards of the train with small arms and grenades before making their escape in a black car. Five British soldiers and three civilians died in the attack and four civilians, a Polish soldier a Polish Auxiliary Territorial Service girl were seriously injured and 21 soldiers and 14 civilians were slightly hurt. [1814]

Mines were placed at different points on railway lines in Palestine on the night of Wednesday 14th May 1947 and early on Thursday 15th May 1947 two trains were derailed. Near Rehoboth the engine, tender and four waggons of a goods train left the rails and the Arab driver was injured and taken to hospital. Between Haifa and Jaffa a workmen's train was derailed and one workman was hurt. Other mines were laid between Jaffa and Lydda and near Jerusalem, some of which failed to explode. In the most serious incident, which did not involve a train, two British officers were killed and two officers and a sapper were wounded when a bomb disposal squad attempted to deal with a mine laid on the Haifa – Acre line and it exploded without warning. [1815]

Two separate attacks were made on trains, it was presumed by the Jewish underground, on Wednesday 4th June 1947. In the first attack, south of Haifa, Palestine, four tank waggons and an escort van carrying several Arab guards were derailed. In the other attack on the Tel Aviv – Lydda line three-quarters of the waggons of a freight train were derailed by an electrically operated mine and a fireman was injured. [1816]

Three goods trains were derailed at different locations in Palestine on Sunday 20th July 1947, though no serious damage resulted from any of the attacks. Four more mines at various locations were discovered before they could be set off by trains. Patrolling of the lines to detect mines continued but it was regarded as inevitable that some mines, especially those of the contact type, would not be detected. It was noted that the attacks coincided with the arrival of *Exodus 1947*, an illegal immigrant ship, from which 4,500 Jewish refugees had been transhipped and returned to France. [1817] A few days later, however, it was stated that the attacks against the railway had again become general and they no longer merely coincided with the arrival of ships carrying attempted illegal immigrants. Thus in the 19 days between 21st July and 8th August there were 21 cases of sabotage directed at trains, track, and bridges, despite intensive daylight patrols by both Engineering Branch staff and the Army. On 24th July mines found beneath a bridge at kilometre 38, near Benyamina, had to be detonated by the army and the bridge was destroyed. The line was closed for three days for repairs. Then on 7th August a goods train was blown up and 12

[1811] *The Republican-Courier*, Findlay, Ohio, United States, 14th March 1947
[1812] *The Lowell Sun*, Lowell, Massachusetts, United States, 14th March 1947
[1813] *Waterloo Daily Courier*, Waterloo, Iowa, United States, 14th March 1947
[1814] *The Times*, London, England, 23rd April 1947
[1815] Ibid, 16th May 1947
[1816] *Reno Evening Gazette*, Reno, Nevada, United States, 4th June 1947
[1817] *The Railway Gazette*, London, 5th September 1947

empty bogie oil tanks were derailed and piled up. No serious injuries to either passengers or staff occurred as a result of any of the attacks during this period.

A British railway journal commented that, 'The unhappy record of the Palestine Railways is thought to be without parallel in the history of transport. During the Arab riots of 1936-39, sabotage of railway installations was serious and intensive, but the methods now adopted by dissident Jewish groups greatly exceed, in severity and efficiency, anything previously experienced. Considerable credit is due to both management and staff of all races for the devotion to duty and patience displayed, often at great personal risk.' [1818]

In spite of the intensive patrols of the railway and after several previous failed attempts, at 10.15 a.m. on Saturday 9th August 1947 terrorists succeeded in blowing up the main Cairo – Haifa passenger train just north of Lydda, Palestine. A contact mine exploded under the locomotive, which was derailed and ploughed along the line for about 100 yards before overturning. The five leading coaches of the train, which were full of troops, were badly derailed but did not overturn. The first three coaches were Egyptian State Railways steel stock, the other two were Palestinian Railways first- and second-class coaches. The remainder of the train, which included dining and sleeping cars, remained intact. The Jewish driver of the train was trapped in the cab of his engine and died before he could be released; the fireman was also trapped but survived. The train passengers escaped with bruises.

Salvage operations began immediately and the passengers were forwarded by road. The track had been cleared and repaired by 7 p.m. the same day. [1819]

Two Jewish youths who had carried out the sabotage were caught by mounted Arab police shortly after the attack, following a gun battle. A third saboteur, who had been armed with a submachine gun, managed to escape. The Irgun Zvai Leumi, a right-wing Jewish underground organisation, announced that the attack was the opening of a campaign to cut every railway line in Palestine. [1820]

The engine, tender and two coaches of the *Egypt Express* were derailed by a mine near Benyamina, Palestine, 20 miles south of Haifa, on Tuesday 30th September 1947. No one was hurt. [1821]

French Indochina had been occupied by the Japanese during the Second World War. When the war ended the French attempted to reassert colonial control and refused to grant their pre-war colonies independence. As a result, a war of liberation began in December 1946, which lasted until the final defeat of the French in July 1954.

What were referred to as Annamite saboteurs derailed a train near Hải Dượng, Vietnam, French Indochina, on Sunday 6th April 1947. A soldier was killed and three were wounded. [1822]

On Wednesday 28th May 1947 train carrying about 200 military personnel and some civilians was mined and looted by an armed band near Xuân Lộc, Vietnam, French

[1818] Ibid, 12th September 1947
[1819] Ibid, 19th September 1947
[1820] *The Telegraph-Herald*, Dubuque, Iowa, United States, 10th August 1947
[1821] *The Corpus Christi Times*, Corpus Christi, Texas, United States, 30th September 1947
[1822] *Malaya Tribune*, Singapore, 7th April 1947

Indochina, 100 kilometres north of Saigon. The explosion killed the engine driver and five passengers were reportedly injured. [1823]

On Monday 14th July 1947 the engine of what was variously described as a passenger train and a troop train from Saigon to Phanthiet, was blown up by a mine at Biên Hòa, Vietnam, French Indochina, east of Saigon. The attackers killed the small military escort of the train and then murdered the passengers. The Vietnamese 'bandits' were eventually put to flight by the arrival of French reinforcements. Forty people died and a number of others were seriously injured. Among those killed was a French Army officer, Captain Saint Remy [1824] [1825]

A brief report from a news agency quoting a French Army communique stated that a passenger train had been derailed by sabotage between Saigon and Mỹ Tho, Vietnam, French Indochina, on about Thursday 6th November 1947. Seventeen civilians were killed and 15 wounded. [1826]

The Surma express was derailed on the Bengal and Assam Railway between Kamalsagar and Nayanpur, India, on Saturday 17th May 1947. Twenty-two people were killed outright, four people, including the driver and fireman, died in the relief train taking them to Comilla and by the next day the number of dead had risen to 31. A relief train arrived at Comilla early the following morning with 45 injured, including ten women, and about 100 uninjured passengers. The injured were admitted to Sardar hospital. At the same time the District Magistrate and the Superintendent of Police made their way to the accident scene. About half an hour before the derailment the Surma Up Mail had passed the spot without incident. The bodies of some of the victims were barely recognisable.

Mr Dhirendra Nath Dutta, Congress Party leader in the Bengal Legislative Assembly, interviewed at Comilla on his return from the accident scene, said that he had been told by a senior railway official that the accident had been caused by the removal of fishplates by miscreants. Mr Dutta commented, "Such a terrible train disaster never took place within living memory on this section. During the last 17 months, two more accidents took place in the District due to the removal of fishplates by miscreants. The police should make thorough and immediate investigation to find out the culprits." [1827] There seemed no doubt that the accident was the result of sabotage as relief workers found screwdrivers and implements near the scene.

The number of dead may have increased further for a report from Calcutta published in the American press speaks of 36 dead. The engine and nine of the 15 coaches of the train on the metre-gauge line were derailed and rolled down an embankment. Three relief trains in all reportedly went to the accident. Injured passengers complained that their possessions had been looted immediately after the accident. [1828]

[1823] *Morning Tribune*, Singapore, 3rd June 1947
[1824] *The Times*, London, England, 17th July 1947
[1825] *Malaya Tribune*, Singapore, 17th July 1947
[1826] *Newcastle Morning Herald and Miners' Advocate*, Newcastle, New South Wales, Australia, 10th November 1947
[1827] *The Bombay Chronicle*, Bombay, India, 17th May 1947
[1828] *Pottstown Mercury*, Pottstown, Pennsylvania, United States, 19th May 1947

In mid-August 1947 British India was divided into predominantly Hindu India and predominantly Muslim East and West Pakistan. Immediately there was a massive migration of Muslims east and west from India to the two Pakistans and of Hindus from the two Pakistans to India. The migrations were accompanied by a terrible and unpredicted outbreak of sectarian violence in which Hindus and Muslims who had co-existed for centuries attacked one another, with massacres, arson, rapes, and mass abductions. By 1948 15 million people had been uprooted and between one and two million had died. [1829]

A few days before the granting of independence to Pakistan two coaches of a special train carrying personnel of the new Pakistan government were derailed by an explosion between New Delhi and Bhatinda, India, on Saturday 9th August 1947. Four passengers were killed and 20 injured. Three relief trains were sent from Lahore. [1830]

On Saturday 13th September, in what was presumably an act of sabotage, a special train carrying Muslim refugees from India and Pakistan was derailed between Amritsar and Cheherta, on the Indian side of the new frontier. A bomb was thrown at the train after the derailment. Nineteen people died and others were wounded. [1831]

Nine waggons of a coal train bound for Tientsin, China, on the Peiping (Peking, now Beijing) – Mukden railway were derailed by a mine on the night of Saturday 17th May 1947. [1832]

A 13-car train bound for Mukden ran over a mine laid by Communist guerrillas and was derailed at Lutai, China, on or about Friday 30th May 1947. Officials stated that about 100 passengers had been killed and many others injured. The American consular office in Tientsin stated that no Americans or other foreigners were known to have been on the train. [1833] [1834] The explosion hurled the locomotive and three cars from the track and two of them were upended. After the derailment the Communists reportedly looted the train while the survivors were screaming in pain. [1835]

On about Thursday 12th June 1947 a fully loaded refugee train from Manchuria struck a mine in a tunnel 5 kilometres east of Yehpaishou, China, 190 kilometres north-east of Chengteh. Further casualties were caused when a following train ran into the rear of the derailed train in the darkness of the tunnel. Traffic on the line was suspended. How many died and were injured in the accident was not known. [1836]

A train from Shanghai hit a mine near Tangfang, China, about 110 kilometres east of Tientsin, on the morning of Wednesday 23rd July 1947. Twenty-seven passengers were killed, 12 were seriously injured and many others were less badly hurt. [1837] The train guards exchanged fire with the attackers. [1838]

[1829] Wikipedia – Partition of India
[1830] *The Straits Times*, Singapore, 11th August 1947
[1831] *The Times*, London, England, 15th September 1947
[1832] *Malaya Tribune*, Singapore, 20th May 1947
[1833] *The Daily Independent*, Murphysboro, Illinois, United States, 31st May 1947
[1834] *Creston News Advertiser*, Creston, Iowa, United States, 31st May 1947
[1835] *The Racine Journal-Times*, Racine, Wisconsin, United States, 31st May 1947
[1836] *Malaya Tribune*, Singapore, 14th June 1947
[1837] *The Times*, London, England, 24th July 1947
[1838] *Malaya Tribune*, Singapore, 24th July 1947

About Friday 25th July a mixed train had been derailed by a land mine south of Tangshan, China, 145 kilometres north-east of Tientsin. In this attack five coaches were overturned; one person, a student, was killed and seven were hurt. [1839] A southbound express was held overnight at Tangshan while the line was cleared. [1840]

Three explosions in the vicinity of Chating, China, 70 kilometres east of Tientsin on the Peiping – Mukden railway, indicated the presence of Communists in the area. An armoured train was rushed to the scene at 10.40 a.m. on Monday 28th July 1947 and exploded a mine on the way. [1841]

On or about Wednesday 13th August 1947 a Peking – Mukden express was derailed by Communist sabotage at an unstated location north of Tientsin, China. No one was killed but about 30 people were injured. [1842]

A northbound freight train from Tientsin to Mukden ran over two land mines as it crossed a bridge south of Tangshan, China, about 110 kilometres north of Tientsin on or about Monday 18th August 1947. The engine of the train plunged into the river but no casualties were reported. [1843]

On the night of Sunday 28th September 1947 the Communists raided and destroyed five small stations between Chinshi and Shanhaikuan and destroyed a Nationalist armoured train at a point 105 kilometres north-east of Shanhaikuan, China. The following day, Monday 29th September 1947, they wrecked a freight train with land mines near Janku, China, 50 kilometres north-east of Tientsin. The attackers then fired on the train but withdrew on the arrival of Nationalist troops. [1844]

Two attacks on trains were made by the Communists on or about Thursday 30th October 1947. In one a train was mined about 50 kilometres south-west of Shanhaikuan, China, and then attacked. In the other case a train was mined 25 kilometres east of Tientsin, China. [1845]

The train crew and a large number of workmen were reportedly killed when a train exploded a land mine as it was crossing a bridge near Pengpu, China, 145 kilometres north-west of Nanking, on Sunday 19th October 1947. [1846]

An empty goods train ran over a mine and its engine and six waggons were derailed near Yangthun, China, 20 kilometres west of Tientsin on the line from Peking on the night of Saturday 27th December 1947. The line was blocked for a considerable time. [1847]

On Tuesday 30th December 1947 the Communists blew up a coal train on the Peking – Chengteh railway '20 miles south of the Great Wall.' [1848]

Between 1948 and 1960 Malaya was in the grip of what was variously known as the Malaya Emergency and the Anti-British National Liberation War, a conflict between

---

[1839] *Corpus Christi Times*, Corpus Christi, Texas, 24th July 1947
[1840] *The Singapore Free Press*, Singapore, 28th July 1947
[1841] *The Straits Times*, Singapore, 29th July 1947
[1842] *The Evening Tribune*, Marysville, Ohio, United States, 13th August 1947
[1843] *Sarasota Herald-Tribune*, Sarasota, Florida, United States, 18th August 1947
[1844] *The Amarillo Daily News*, Amarillo, Texas, United States, 1st October 1947
[1845] *The Evening Star*, Washington, D.C., United States, 31st October 1947
[1846] *The Singapore Free Press*, Singapore, 20th October 1947
[1847] *Morning Tribune*, Singapore, 29th December 1947
[1848] Ibid, 30th December 1947

military forces of the British Empire and Communist fighters of the Malayan National Liberation Army, who were fighting for independence and a socialist economy. [1849] The conflict was marked by a number of instances of railway sabotage, the first of which probably preceded the announcement of the emergency by a year.

The 8 p.m. mail train from Singapore to Kuala Lumpur with 218 first- and second-class passengers was derailed at 10.30 p.m. on the night of Tuesday 24th June 1947 between Sedenak and Kulai, Malaya. Three passengers were killed and 16 more were wounded. Several Civil and Military ambulances and medical personnel were rushed to the scene from Kluang and Johore Bahru and emergency first aid tents were erected to render immediate aid. The most seriously injured were rushed to hospitals in Kluang and Johore Bahru. It was found that unknown persons had sawn the heads from the bolts of fishplates and the weight of the train had dislodged the rails. 'Bandits,' it was assumed, had intended to derail, and loot the waggons of a goods train but the mail train had unexpectedly preceded it. [1850] [1851] [1852] The 15-coach train had been travelling up an incline at 15 to 20 miles an hour at the time of the derailment. The tender of the engine fell into a culvert and the driver and fireman had a lucky escape, the fireman suffering a cut leg. The leading luggage van and following four second-class carriages left the track but only the third vehicle in the train was wrecked. After leaving the track the engine fell down an embankment and was half buried in the soft earth on its left side with the coaches alongside it. The derailment occurred in an isolated area with neither houses nor roads nearby. The Kuala Lumpur-bound third-class mail train had passed the spot safely a short time earlier.

A newspaper reporter described the accident in detail. 'Tribute was paid to the excellent work by the military personnel who were either returning to duty or on their way to rest camps on leave. One signaller picked up a rifle and went on a short patrol, then walked all the way to Sedenak, one and a half hour's walk away, to report the disaster to the Rail Transport Officer. Mr Ong Eng Kuang, a medical student on his way to Penang on holiday, said that he and his friends had spoken just before the accident of a possible derailment. "I would say we were derailment-conscious," he said. "I looked at my watch just before 11 o'clock. I remember the train rumbling and baggage falling on my head. My reaction was one of annoyance. Only then did I realise that I was sitting under a carriage and there was baggage all around me. There were two Selangor footballers sitting in our carriage and one man whom I understood had travelled to Singapore for the Sing Tao match as a supporter. This man was killed and one of the Selangor footballers was injured." The Selangor footballer who was injured was Shi Kam Hoi, who was by the carriage and was picked up by two soldiers and later taken to Johore Bahru hospital with an injury on his head and bruises. Military personnel numbering about 40 were in the second coach in the train. Signalman Vincent McEwan said he was knocked out of his bunk. The door was jammed and he smashed through the windows. "There was no panic," he said. "I saw the engine crew who had been thrown clear all filthy and muddy but uninjured."' [1853]

[1849] Wikipedia – Malayan Emergency

[1850] *The Times*, London, England, 26th June 1947

[1851] *Malaya Tribune*, Singapore, 25th June 1947

[1852] *Indian Daily Mail*, Singapore, 26th June 1947

[1853] *Straits Budget*, Singapore, 3rd July 1947

Like Germany, Austria had been occupied at the end of the Second World War. At 10.30 p.m. on the evening of Tuesday 12th August 1947, in what was apparently an isolated incident, a troop train was derailed in Carinthia, in the British zone, by a bomb which had been placed on the rails. A few passengers were slightly hurt. A second bomb which had failed to explode was later found and a third bomb exploded the next day at an army office. The reason for the attacks was unknown. [1854]

Railwaymen in French West Africa went on strike on 10th October 1947 for improved conditions. Near the beginning of the strike, on Friday 24th October 1947, a train was derailed by sabotage near Koulikoro, Senegal. The strike dragged on until the end of February 1948 but eventually ended with a settlement which was favourable to the railwaymen. [1855]

Severe post-war inflation combined with continued rationing and difficult living conditions led to a series of strikes in France from April 1947. In December of that year there were a series of train derailments due to sabotage in continuation of the strike action.

On the morning of Monday 1st December 1947 a freight train was derailed between Rennes and Le Mans, France, on the Paris – Pontoise line after the rails had been unbolted. Also on Monday morning a Lyon – Bourg passenger train was derailed by sabotage between Mionmay and Saint André-de-Courcy, France. A number of passengers were slightly hurt. [1856]

On Tuesday 2nd December 1947 four coaches of a passenger train were derailed at an unstated location on the Paris Metro. No one was hurt.

The sabotage culminated early on Wednesday 3rd December 1947 in the derailment near Arras, France of the night Paris – Lille express. Activists of the Pas-de-Calais branch of the Confédération Générale du Travail trade union had unbolted two lengths of rail, taking care not to disturb the track circuits and thus place the signals to danger. Twenty people died (some reports say 16) and between 50 and 60 were hurt. The saboteurs had believed that the train was carrying members of the Compagnies Républicanes de Sécurité (a police reserve force) to support scab workers from Arras. [1857] [1858]

The last act of sabotage of the strike was the derailment of a Geneva – Lyon express passenger train on Tuesday 9th December at the entrance to Rossillon Tunnel, France, near Virleu-le-Grand. The rails had been unbolted over a length of 30 metres. The engine remained on the rails and the coaches, although off the track, were not badly damaged. A following express was stopped in time to avoid running into the obstruction. Only three people were hurt but the line through the tunnel was blocked. [1859] [1860]

[1854] *The Times*, London, England, 14th August 1947

[1855] https://nvdatabase.swarthmore.edu/content/french-west-african-railway-workers-strike-greater-benefits-1947-1948

[1856] *Journal de Genève*, Genève, Switzerland, 2nd December 1947

[1857] Wikipedia – 1947 Strikes in France

[1858] *The Times*, London, England, 4th December 1947

[1859] *Journal de Genève*, Genève, Switzerland, 8th December 1947

[1860] *The Times*, London, England, 10th December 1947

On about Sunday 18th January 1948, guerrillas blew up a train on the eastern Thessalian railway in Greece, derailing 20 waggons. The UK Special Balkans Commission stated that the guerrillas had received aid from Albania. Two bridges were also destroyed but no one was hurt. [1861]

On about Tuesday 17th February 1948 a passenger train struck a landmine in a tunnel near Xanthe, Greece. The engine and four coaches were derailed but passenger injuries were reportedly slight. [1862]

The engine of a 'guard' train preceding that carrying King Paul of Greece, General James van Fleet, head of the American military mission, and a group of American and British officers, who were on their way to Tripoli, Greece, was derailed in the central Peloponnesus region, near Parthenion station, about 17 kilometres short of Tripoli, and five following vehicles went down a 15-metre embankment. Four soldiers of the military guard of the train were slightly hurt in the accident on Thursday 8th April 1948. The train carrying the king and general was following about 100 metres behind and was stopped in time. The authorities stated that there was no definite evidence of sabotage and possibly it was an accident but three weeks earlier about 45 kilograms of dynamite had been discovered on the track ahead of General van Fleet's train at Kilkis. [1863] [1864] [1865]

On about Wednesday 19th July 1948, guerrillas blew up a train between Comotini (now spelt Komotini) and Xanthe, Greece, west of Iasmos in Thrace, killing 14 women and wounding six more. [1866]

Guerrillas mined a train and then attacked it on the line between Serres and Salonika, Greece, on Sunday 1st August 1948. A soldier escort was killed and two others were wounded. On about the same date a youth convicted of guerrilla activities in the Athens area was sentenced to death by a court martial. [1867]

After the defeat of Japan in 1945, Korea, which had previously been a Japanese colony, was divided into Soviet and United States occupation zones pending the establishment of an independent state. The occupying forces were unable to reach an agreement and in early 1948 the United Nations was invited to the country to try and solve the problem. There was pressure in Korea, though, for the UN commission to leave the country and for the Soviet and United States occupations to end. In connection with the efforts in this direction a 12-car train was derailed about 30 miles east of Seoul on Friday 6th February 1948. No casualties were reported. The track was damaged at five other locations, damage was caused to about 50 locomotives and telegraph lines were cut. [1868]

The Yeosu-Suncheon Rebellion occurred in South Korea in October 1948, led by leftish soldiers opposed to the regime of Syngman Rhee. On Wednesday 27th October 1948 a government train consisting of an engine, a boxcar loaded with ammunition and a guards van was derailed by the rebels between Suncheon and Koksong, a rail having been

---

[1861] *The Straits Times*, Singapore, 19th January 1948
[1862] *San Antonio Express*, San Antonio, Texas, United States, 18th February 1948
[1863] *Het Dagblad*, Batavia, Dutch East Indies, 9th April 1948
[1864] *The Holland Evening Sentinel*, Holland Michigan, United States, 8th April 1948
[1865] *Altoona Mirror*, Altoona, Pennsylvania, United States, 8th April 1948
[1866] *Press-Telegram*, Long Beach, California, United States, 20th July 1948
[1867] *The Salt Lake Tribune*, Salt Lake City, Utah, United States, 2nd August 1948
[1868] *Kenosha Evening News*, Kenosha, Wisconsin, United States, 7th February 1948

removed. The rebellion had been brutally supressed by the end of November. [1869] [1870]

In the most serious act of sabotage against a train committed in Palestine so far, on the morning of Sunday 29th February 1948 a Cairo – Haifa train which was carrying troops who were returning from leave or on duty visits was mined just north of Rehoboth. An initial report stated that 28 soldiers had been killed and another 35 injured and taken to hospital. Three coaches were blown up, each by a separate mine placed so that it exploded under that vehicle. A mine which should have exploded under a fourth coach failed to detonate; it was found to contain 100 pounds of ammonal.

The Stern Gang claimed responsibility for the attack, claiming that it was in retaliation for the explosion in Ben Yehuda Street, Jerusalem, a week earlier, in which three Army lorries preceded by an armoured car, driven by Arab irregulars and two British deserters, had exploded, killing about 55 Jewish civilians.

A British newspaper wrote, 'Tonight as darkness fell rescuers were still digging in the wreckage of the train, and it was feared that the final death toll might be heavier. Some of the wounded were not expected to recover. There were no reports of civilian casualties. An Army hockey team was among the military passengers. Later Army reports said there were 110 passengers in the four coaches.

'The train was the normal daily passenger express, to which four military coaches had been attached. It had left Rehoboth station only a few minutes before the explosion occurred as the train was going up an incline. It is understood that the mines were detonated electrically through a cable running from the tracks into an orange grove nearby, where the terrorists lay hidden and pressed a plunger as the train passed. More than 100 yards of the track were ripped up by the force of the blast.

'A Jewish eyewitness said that the shattered coaches were flung on both sides of the railway lines, and the rescue workers were guided by the cries of the injured buried and trapped under the wreckage. Among the first to arrive on the scene, this eyewitness said, were Jewish Boy Scouts, who gave first aid to the injured as they were carried out of the wreckage. A Jewish ambulance helped to transport the dead, dying, and injured to the military hospital at Beer Yaacov, about five miles away.' [1871]

As might be expected the magnitude of the sabotage provoked a strong reaction from the British government, as was reported in the press the following day. 'All the bitterness, injustice and anger felt by the Government over the loss of 27 more British lives – 19 soldiers and eight members of the R.A.F. – in the attack on the train at Rehoboth yesterday, and over events which during the better part of two years have led up to such outrages, is expressed in an official statement issued today. It is addressed again to the Jewish Agency and to the Jewish community generally because of their collective responsibility for Jewish deeds, and its points are made with the directness allowed by the knowledge that the curtain soon will fall on Britain's stay in Palestine.

'The Government first is careful to say that its remarks do not imply any change in its declared policy to do all it can to maintain law and order impartially; but in the body of the statement it acknowledges indirectly that British soldiers are human, and that their

---

[1869] Wikipedia – Yeosu – Suncheon rebellion
[1870] *Bluefield Daily Telegraph*, Bluefield, West Virginia, United States, 28th October 1948
[1871] *The Times*, London, England, 1st March 1948

own feelings may play now upon the way they carry out their orders to give Jews protection. The relevant passage reads:- "In this neglect of its responsibilities (to bring the terrorists to justice) the Jewish Agency has attempted to excuse itself by resort to calculated innuendoes, falsehoods and propaganda directed against British members of the security forces, who are, in fact, every day protecting Jewish property and saving hundreds of Jewish lives, even at the risk to their own. This propaganda had already discredited its authors, and not only among those accredited with the facts. It has also the serious consequences that it much henceforward be clearly more difficult for British troops to look upon members of the Jewish community as persons entitled, as they are, to protection."

'Nobody outside Jewish circles, the statement continues, believes the Agency's repetition that a British Army convoy was responsible for the Ben Yehuda Street outrage, and there can be no other purpose in repeating it than to stir up racial hatred.' [1872]

The Cairo – Haifa train was mined again on Wednesday 31st March 1948, this time with even more serious consequences. Its engine and four coaches were derailed near Benyamina, Palestine. Again the mine was electrically detonated from a nearby orange grove. Forty people were killed and 60 injured. As in the previous attack the train was a mixed civilian and military one and there were more than 30 British soldiers on it, though none was killed or injured. Those who died were mostly Arabs. Following the attack a curfew was imposed on the coastal highway between Benyamina and Zikhron Yaacov, four miles further north, until further notice, breaking the stream of Jewish traffic between Haifa and Tel Aviv. Trains between Haifa and Egypt were cancelled for the remainder of the week. [1873]

Not far from the scene of the attack on the last day of March, three more attacks on the railway were made on the night of Monday 5th April between Zikhron Yaacov and Athlit, Palestine. A goods train was derailed and the engine of a breakdown train was attacked in turn (derailed?). The motives of the saboteurs were unclear, as the newspaper report noted. 'When the mail train was mined it was at first thought that it was another attempt on the lives of British troops; but last night's attacks on non-passenger trains could not have such an object, and the possibility arises that Jews in this area are trying to break communications, to interfere with the Army as it withdraws northwards.' [1874]

A British troop and ammunition train en route from Lydda to Egypt was attacked by armed Jews hiding in an orange grove near Rehoboth, Palestine, on Monday 12th April 1948 and a large quantity of arms was looted. There were no casualties. [1875]

Three British soldiers were killed and one was wounded when two trolleys were derailed near Lydda, Palestine, on Thursday 15th April 1948. Their bodies were found stripped of clothes. [1876]

Three more British soldiers died in a derailment due to sabotage on the afternoon of Sunday 25th March 1948 near Ramleh, Palestine, south of Lydda. [1877]

[1872] Ibid, 2nd March 1948
[1873] Ibid, 1st April 1948
[1874] Ibid, 7th April 1948
[1875] Ibid, 13th April 1948
[1876] Ibid, 16th April 1948
[1877] *The West Australian*, Perth, Western Australia, 27th April 1948

In Vietnam, French Indochina, a Hanoi – Haiphong train was blown up by a mine on Wednesday 5th May 1948 at an unstated location. One passenger was injured and three waggons were damaged. The French Press Agency blamed Indochinese nationalists. There had been a previous attack on a train five days earlier. [1878]

A passenger train from Saigon to Phan Thiết was mined and derailed between Biên Hòa and Trảng Bom, Vietnam, French Indochina, on the evening of Tuesday 17th August 1948. Twelve people died. A newspaper commented, 'Observers here connected the incident with a fresh outbreak of activity by the Viet Minh (Indo Chinese autonomy movement) coinciding with the third anniversary of the first Viet Namese government's formation.' [1879]

A Reading Railroad Allentown to Philadelphia local train was derailed at Valley Forge, Pennsylvania, United States, at 7.55 p.m. on Sunday 9th May 1948, killing Engineer Buch Demy and Fireman Howard Jackson [1880] and injuring 66 passengers. The train consisted of a 4-6-0 steam engine, a baggage car and four coaches. It had left Perkiomen Junction, the last open office, 2¼ miles short of the point of derailment, at 7.48 p.m., had stopped at Valley Forge station, left there at 7.52 p.m., passed automatic signal No. 151 A, showing *Proceed*, and was moving at about 30 miles an hour when it left the track. The engine, tender and all but the last coach were derailed, the engine was badly damaged and the derailed coaches were considerably damaged.

The conductor and flagman told the Interstate Commerce Commission that the train had been running normally before the derailment. Examination of the track showed that a rail had been displaced and that rail holding spikes, two pairs of joint bars, joint bar bolts, lock washers and nuts had been removed, apparently with a wrench and a claw bar which had been stolen from a section toolhouse rather more than a mile away. Care had been taken not to disturb the rail bond wires so as not to place the signals at danger. [1881]

An inquest on the two railwaymen who had lost their lives returned a verdict that the men had died due to the 'deliberate removal of bolts and spikes from the tracks by a person or persons unknown.'

Suspicion fell on two railroad workers who had been discharged from the Bridgeport labor camp, five miles from the scene of the derailment, four days before it had happened. Investigations were in progress. [1882]

Fifteen people were killed more or less outright and 80 injured, 12 seriously, when train No. 9 Up, the *Dehra Dun Express* of the East Indian Railway, was derailed at 1.30 a.m. on Saturday 15th May 1948 between Chota Ambona and Prodhan Khanta, India, nine miles from Dhanbad, on the Grand Chord line of the company. The train had left Howrah at 7.20 p.m. the evening before. The engine and leading bogie capsized, the second and third bogies overturned and the fourth, fifth and sixth were derailed.

The Divisional Medical Officers of Asansol and Dhanbad and doctors from Dhanbad

---

1878 *Malaya Tribune*, Singapore, 6th May 1948

1879 *The Lethbridge Herald*, Lethbridge, Alberta, Canada, 18th August 1948

1880 *Chester Times*, Chester, Pennsylvania, United States, 10th May 1948

1881 Interstate Commerce Commission report

1882 *Somerset Daily American*, Somerset, Pennsylvania, United States, 15th May 1948

Civil Hospital went to the scene, followed by other officers from Asansol and Dhanbad and Mr V. Nilakantan, the General Manager of the company from Calcutta, accompanied by other senior officials. Mr S.D. Bamjee, the Government Inspector of Railways, announced that he would open an inquiry into the disaster.

Two of the passengers who were admitted to Dhanbad hospital died shortly afterwards and several others were said to be in a precarious condition.

'Mr P.N. Sidhwa, a passenger from Calcutta, who was proceeding to Musoorie with his family, said that about 1.30 a.m. he suddenly felt that his carriage, the fourth bogie of the train, was going off the rails. Later he discovered that the first three bogies had overturned and that his compartment was in a slanting position. It was with great difficulty that he could extricate himself and his family.

'The accident took place near a culvert with an embankment 15 feet high. One third and one inter-class bogie were reduced to matchwood. Relief operations are still going on. The General Manager of the E.I. Railway, the Chief Operating Superintendent and other railway officials along with the Deputy Commissioner and Superintendent of Police are supervising the operations.'[1883]

By the following day the number of dead had risen to 30, 28 of whom were still unidentified, and there were now said to be 101 injured. Lord Louis Mountbatten, at that time the Governor General of India, expressed his sympathy with the injured and the families of the dead.[1884]

It was believed that the derailment had resulted from an act of sabotage.

A fast passenger train from Batavia to Bandung was derailed by an armed gang three kilometres east of Tjikampek, Dutch East Indies, on Wednesday 23rd June 1948. Saboteurs then fired on the train for ten minutes both from a four-metre-high embankment on the right-hand side of the line and from the valley on the left-hand side. Five people died and eight were hurt according to an initial report. Shortly after the derailment a Chinese passenger photographed the derailed train, unaware, presumably, at the moment that his wife and granddaughter were among those who had died. The train had left Tjikampek at 13:45 and the derailment occurred near the small village of Tjiboengoer. It would have been more serious than it was but for the prompt action of the driver but even so, four carriages left the track. At the time of the accident the train was climbing an incline and not moving at more than 30 kilometres an hour.

Major van Giessenburg, who was travelling on the train, urged the civilian passengers to take cover in a lineside ditch. There had been a good many soldiers on the train but because they were travelling on leave most of them were unarmed. Even so, they were able to clear the area around while a patrol made its way to Tjikampek for assistance. It was likely that the attackers, on finding how many soldiers there were on the train, had fled the scene. When the situation was more or less under control an attempt was made to treat the victims. A telephone handset was found on the train with which it was possible to communicate with both Tijkampek and Poerwakarta. Soldiers were sent from both places. They scoured the whole area and two firefights occurred in which one soldier was killed.

It was discovered that the rails had been loosened at two points and in addition,

---

1883 *The Indian Express*, Madras, India, 16th May 1948

1884 Ibid, 17th May 1948

fishplates connecting the rails had been removed. Judging by the volume of the gunfire there must have been about 20 attackers. Automatic weapons were apparently not used.

A Dutch statement noted that, "This is the climax of a series of a series of incidents throughout West Java. It is an indication of growing unrest and lawlessness. We do not believe the Republican Government are responsible but we think that some Republican elements have taken matters into their own hands." [1885]

A second sabotage attempt on the Tijkampek – Cheribon line was discovered on Thursday morning at Tandjoengrasa, six kilometres east of Tjikampek, where the rails had been loosened over a distance of 100 metres. The sabotage was discovered because the military telephone had been cut.

It was hoped that the blockage would be cleared by Friday night so that passenger and goods trains could run normally on Saturday. Meanwhile, passengers were having to change trains at the accident point and goods traffic was suspended. [1886]

On the night of Thursday 5th August 1948 an act of sabotage was carried out on the Buitenzorg – Soekaboemi (– Bandoeng) line. The first passenger train out of Buitenzorg for Soekaboemi was derailed at Pondok Leungsir, Dutch East Indies, the screws of several rails having been loosened. The engine and tender crossed the damaged section of track safely but some of the following coaches were derailed. No one was hurt and there was only minor damage to the rolling stock. The passengers were forwarded to Soekaboemi in MLD lorries. (MLD was an abbreviation for Marineluchtvaartdienst – the Dutch naval aviation branch.) Military and civil police patrols immediately searched the area for the attackers. [1887]

At about 22:20 on Saturday evening, 21st August 1948, a goods train was derailed by sabotage on the Poenwakarta – Bandoeng line. The accident happened at Bedoel Bridge, about 5 kilometres short of Poerwakarta, Dutch East Indies. Following the derailment the bridge the two engines, a passenger coach and a tank waggon fell from the bridge into the ravine, which was about 20 metres deep. The saboteurs opened fire on the train after it had left the rails. Five members of the train crew and accompanying guard were known to have been killed and another 13 were missing. The track had been damaged over a distance of about 50 metres. The remaining vehicles of the train were badly damaged and came to a stand in a criss-cross fashion across the line. One of the pillars of the bridge was damaged by the falling engines.

The saboteurs had reached the railway via Goenoeng Semboeng kampong (village) where they had ordered the villagers to remain indoors. After the attack the perpetrators disappeared in the direction of Tegelharendoeng. They left a note in which the Dutch military were threatened. A similar notice was written on a pole. Both drivers and both firemen were killed. The passenger coach, in which were riding the conductor and the police guard, was completely smashed. Initially it was not possible to determine how many victims were trapped in the wreckage. There were eight members of the police guard of the train, two of whom were seriously hurt and six of whom were missing.

The military detachment at Goenoeng Poetri was informed at 23:30 that a train had been derailed at Bedoel by a gang of 20 or 30 men. A telephone message from Poerwakarta stated that the goods train which should have reached there some time

---

[1885] *The Straits Times*, Singapore, 25th June 1948

[1886] *Nieuwe Courant*, Soerabaja, Dutch East Indies, 25th, 26th and 28th June 1948

[1887] Ibid, 9th August 1948

earlier had not arrived. A patrol was immediately despatched from Goenoeng Poetri. Another patrol was already out in the area of the attack. A relief train was sent from Poerwakarta and returned at 05:00 on Sunday morning with two bodies and seven injured. At that time rescuers were still busy trying to free those who were trapped in the wreckage in the ravine, though they had little hope of finding anyone alive. A large number of workmen came from Bandoeng on the Sunday morning express to the accident site and immediately began clearance and repair work. Damage to engines, rolling stock, track and bridge was estimated at three million guilders. 'Military countermeasures have been taken,' it was reported. [1888]

At the end of September it was announced that two of the attackers, Oewasta bin Asman and Darja bin Sarta, had been sentenced to death, and two others had been sentenced to 20 years' imprisonment for their part in the affair. [1889]

An initial report stated that six people, including one soldier, had been killed and many wounded in the sabotage of a Batavia – Bandoeng express, again at Bedoel Bridge, Dutch East Indies, just south of Poerwakarta, at 16:50 on Friday 22nd October 1948. The track had been damaged over a length of 20 metres. Immediately after the derailment a gang of about 50 attackers opened fire on the train with automatic weapons and rifles from a point to the west of the line, wounding a number of people, including the driver and fireman. There were a number of military on the train and the attackers' fire was at once answered. (According to the state railway company shots were fired at the train from both sides of the line.)

The two engines and a luggage van were derailed, overturned, and stopped in a criss-cross position across the track. The three following coaches, one of them a new sleeping car which had just arrived from the Netherlands, left the rails but remained upright. The six dead comprised one soldier, one Chinese and four Indonesians. It was not known if they were killed in the initial derailment or as a result of the shooting which followed. Their bodies were taken to Poerwakarta and would probably be forwarded to Batavia the next day. The line was blocked and until it could be cleared passengers had to make their way between trains on either side of the obstruction. Those who were on the train that was attacked were taken to Bandoeng on the rescue train that was sent from there. Goods traffic was interrupted. The sabotage occurred only a short distance from the scene of the August attack, when ten people died (see above).

A later report stated that there were eight dead; the six previously mentioned plus two unknown. The number of wounded was 23, including seven Dutch soldiers, most of whom were hurt during the post-derailment gun attack.

It also emerged that the midday Batavia – Bandoeng express had been fired upon with Lee Enfield rifles in the same area on Thursday 21st wounding both firemen and a soldier, who were taken to hospital.

On the evening of Christmas Day, Saturday 25th December 1948, a local train from Poerwakarta to Tjikampek, Dutch East Indies, was derailed due to sabotage at an unstated location, the engine and one coach leaving the rails. No one was hurt. The Batavia – Bandoeng main line was blocked at that point and passengers had to make their way between trains on either side of the obstruction. It was hoped to clear the line

---

[1888] *Het Dagblad*, Batavia, Dutch East Indies, 23rd August 1948

[1889] Ibid, 30th September 1948

during Sunday. [1890]

A Canton – Kowloon express carrying 1,000 passengers was derailed at Lilong, China, just over the border from Hong Kong, on Friday 3rd September 1948. Twenty-five people were killed and over 100 others injured, 30 seriously. Of the 11 carriages of the train, one fell 50 feet into a stream, two were telescoped and two more overturned. Two special trains brought the injured to Kowloon, where the authorities set up a first aid station, and a fleet of ambulances carried the seriously injured to hospital. According to one account the derailment was the result of track subsidence following heavy rain. [1891] Another account, however, stated that there were indications of sabotage to the track. [1892]

A Communist mine derailed a Pukow-bound train near Pengpu, China, 145 kilometres north-west of Nanking, on the night of Tuesday 14th September 1948. A Reuter report stated only that a locomotive had been destroyed and the engine crew killed [1893] but according to an Associated Press report railway officials stated that both the engine and the leading coach of the train had been completely destroyed and at least 100 people had been killed. [1894] The truth, perhaps, lies somewhere between.

A Kowloon – Canton express was derailed and attacked 20 kilometres south of Canton, China, by 200 'bandits' on the night of Thursday 4th November 1948. One report stated that a guard had been killed and the passengers searched, some of them four times. In all, four people were reported to have lost their lives and 100 to have been wounded. Bolts had been removed from the track to derail the train. The engine, one first-class and two second-class carriages left the rails. [1895] [1896] [1897]

About 200 Communist guerrillas attacked a government passenger train travelling from Kalgan to Peiping, China, on Saturday 20th November 1948. The train ran over a mine and was derailed, then the guerrillas robbed the passengers and took about 100 hostage before they were driven off by arriving government forces. Five passengers died and 60 of the 'Reds' were reportedly killed. A few of the hostages were released a little later. [1898]

A train was attacked on the main line between Singapore and Kuala Lumpur on Wednesday 15th September 1948 at a point between Sedenak and Layang Layang in Johore, Malaya. The locomotive of the southbound night mail train from Kuala Lumpur was preceded by a pilot train consisting of an engine pushing a "crash" waggon (a heavily weighted truck) and pulling an armoured car as a safety measure. The crash waggon was derailed by a mine as the train was passing over a small bridge. No one was hurt but two lengths of track were torn up. [1899]

The Singapore chief of Malayan Railways, Mr Morgan Griffiths, spoke to a

---

[1890] *De Locomotief*, Semarang, Dutch East Indies, 27th December 1948
[1891] *Malaya Tribune*, Singapore, 4th September 1948
[1892] *Salt Lake Telegram*, Salt Lake City, Utah, United States, 4th September 1948
[1893] *The Straits Times*, Singapore, 16th September 1948
[1894] *Lubbock Morning Avalanche*, Lubbock, Texas, United States, 15th September 1948
[1895] *The Singapore Free Press*, Singapore, 6th November 1948
[1896] *The Daily Gleaner*, Kingston, Jamaica, 6th November 1948
[1897] *The Straits Times*, Singapore, 7th November 1948
[1898] *Walla Walla Union-Bulletin*, Walla Walla, Washington, United States, 22nd November 1948
[1899] *The Times*, London, England, 16th September 1948

newspaper reporter. "The mail train is safe in Layang Layang station, heavily guarded. We have no reports of anyone being injured. The crash car did the job for which it is intended. Being forty-odd feet long, its wheels pass over any trap some time before the engine. By eight o'clock we had the line open for light traffic, but the mail train is not likely to get into Singapore before one or two o'clock." [1900]

The track north of Tapah, Malaya, was blown up on the night of Tuesday 19th October 1948 and the patrol train preceding the southbound night mail train from Kuala Lumpur was derailed. No one was hurt. Telephone lines were also cut. The northbound train was held at Tapah pending repairs to the line. [1901]

On Monday 29th November 1948 the pilot train preceding the northbound Kuala Lumpur – Penang night mail was derailed less than an hour after starting, 18 miles out, just north of Rawang, Malaya, fishplates having been removed from the track. Following too close behind the pilot train, the main train collided with it when it was derailed, damaging the engine. Driver Stephen Joseph, Fireman Arokiasamy and Fireman Kamil of the pilot train were slightly hurt. The passengers on the mail train were taken back to Kuala Lumpur, arriving there about 4 a.m. Three hours later they were taken by bus to Serendah, where they were transhipped with passengers from the southbound train. It was expected that the line would be cleared about noon. About 100 yards of track was torn up. [1902] Driver Joseph said, "My mates and I crawled out somehow and ran to warn the police inside the armoured turret at the rear. We shouted to them to get out as the mail train would certainly smash into us." [1903]

French coal miners went on strike in late 1948 in response to cuts in pay and worsening conditions. Several trains were derailed in connection with the strike. On the morning of Tuesday 2nd November a train carrying miners (presumably scabs) from Vieux Condé to Abscon was derailed 400 metres from the station at Hérin, France, rails having been unbolted. Four coaches left the rails but no one was hurt. [1904] On the same day in southern France, near Marseille, a train bound for the Fuveau mine was derailed in a similar manner between Aubagne and Labarque. [1905]

A third train was derailed in the Marseille area on Saturday 6th November 1948, the engine and two coaches leaving the track because a rail had been loosened. Again there were no casualties. Interference with the track near Douai in northern France at about the same time was discovered by a track walker before a train arrived. [1906] There seem to have been no more derailments. The strike was brutally supressed by the French government.

A Soviet general and 17 other people reportedly died when a Berlin – Moscow express was derailed about 10 kilometres west of Łowicz, Poland, on Christmas Eve, Friday 24th December 1948. An explosion occurred beneath the second coach of the train immediately after the derailment and it was in this coach that the general, who could not

---

[1900] *The Singapore Free Press*, Singapore, 15th September 1948
[1901] Ibid, 20th October 1948
[1902] Ibid, 30th November 1948
[1903] *The Straits Times*, Singapore, 1st December 1948
[1904] *Gazette de Lausanne*, Lausanne, Switzerland, 3rd November 1948
[1905] *The Bakersfield California*, Bakersfield, California, United States, 2nd November 1948
[1906] *Creston News Advertiser*, Creston, Iowa, United States, 6th November 1948

be identified by Western sources, had been riding. A number of other Soviet and Eastern European officials were riding in the same carriage, en route to attend the Pan Slavic celebrations early in the New Year, and some were amongst the others killed. Ten people were also injured. Immediately after the event tight security measures were imposed by the Polish government, who admitted in a Press report that a three-coach train had been derailed but said that there had been no deaths or injuries, while private reports, circulating in Warsaw, and denied by the Polish government, spoke of 200 dead. The figure of 18 dead came from Western sources. Polish partisans were said to have sabotaged the train. [1907]

On the morning of Sunday 2nd January 1949, when a passenger train from Hanoi to Haiphong was 25 kilometres from Hanoi, Vietnam, French Indochina, it ran over a mine. One passenger was killed and seven were injured. The train resumed its journey at 14:00. On Sunday night Vietminh guerrillas laid another mine in the same place and another act of sabotage was only averted because of the alertness of the military escort of a second Hanoi to Haiphong train. The military said that the guerrillas were hidden in the vegetation adjoining the line ready to explode the mine. A grenade and an aircraft bomb were placed on the track. A Vietminh prisoner who was captured was found to be carrying two hand grenades. Twenty minutes later a train coming from the opposite direction exploded another mine anyway which had been laid some hundred metres away. Its engine was slightly damaged. [1908]

A train exploded a mine about 15 kilometres south of Saigon, Vietnam, French Indochina, on Friday 28th January 1949. Two people died and ten were injured. [1909]

Vietminh forces blew up a train bound for Nha Trang 225 kilometres north-east of Saigon, Vietnam, French Indochina, on Sunday 11th September 1949, then opened fire on it, but were repulsed by its armed guard. [1910] In other incidents on about the same date a goods train was derailed 30 kilometres north of Saigon and a train from Lang Son to Cao Băng was ambushed. Five people died, 15 were wounded and 11 were described as ‘missing’ after the latter attack. [1911]

Communist forces blew up a train near Serres, Greece, on about Wednesday 26th January 1949. No other details were given. [1912]

Greek terrorists derailed a train between Adrianople (now Edirne), Turkey, and Pythio, Greece, on the night of Sunday 4th September 1949. The train was running at full speed when it exploded two time bombs laid on the track. Five coaches of the northbound train from Istanbul left the track and 20 people were seriously hurt. [1913] [1914]

[1907] *The Daily News*, Huntingdon and North Union, Pennsylvania, United States, 30th December 1948

[1908] *De Locomotief*, Semarang, Dutch East Indies, 5th January 1949

[1909] *The Straits Times*, Singapore, 30th January 1949

[1910] *De Vrije Pers*, Dutch East Indies, 13th September 1949

[1911] *Le Courrier Australien*, Sydney, New South Wales, Australia, 16th September 1949

[1912] *The Times*, London, England, 28th January 1949

[1913] *Het Dagblad*, Batavia, Dutch East Indies, 6th September 1949

[1914] *The Singapore Free Press*, Singapore, 6th September 1949

A Tegal – Poerwakarta passenger train was derailed by sabotage near Boemiajoe, Dutch East Indies, on Sunday 23rd January 1949. The engine was overturned and the coaches derailed and wrecked. The military guard of the train drove off the attackers after a firefight but seven passengers lost their lives. [1915] It was believed that the losses amongst the attackers were greater and some of them were taken prisoner. [1916]

The train from Malang to Soerabaja ran over a mine at a bridge between Sengon and Lawang, Dutch East Indies, on Monday 21st February 1949. As the line had been damaged at this point the week before, the train was only travelling at 5 kilometres an hour, and the carriages did not fall into the ravine, though the engine, a luggage van and two coaches were seriously damaged. The exact number of victims was not immediately known, but it was thought that there were three dead and two wounded. The passengers walked back to Lawang from where they were returned to Malang. [1917]

A Staatsspoorwegen (strictly Staatsspoor en Tramwagen in Nederlansche-Indië) work train was derailed between Serang and Rangkastbitoeng, Dutch East Indies, on Wednesday 2nd March 1949. Four Indonesian railwaymen were killed and ten were wounded when the train fell into a ravine. The accident resulted from the sabotage of a viaduct, which had only been repaired the same morning, on the outward journey of the train. Some shots were fired at the train after it had been wrecked. Radio-communication remained intact after the accident so assistance could be quickly summoned. [1918]

At about 15:00 on Saturday 2nd April 1949 a slow train from Djember to Probolinggo was derailed between Klakah and Probolinggo, Dutch East Indies, because the rails had been loosened. One of the Indonesian passengers was killed and four were wounded. [1919]

On Monday 25th April 1949 as the last train of the day from Garoet to Tjibali was approaching Tjimurah, Dutch East Indies, it was derailed through sabotage and then fired on. One brakesman was wounded in the shooting. The military escort of the train fired back at the attackers. The engine and one vehicle were derailed. A train reached the scene of the attack on Monday evening to begin the clearance work. [1920]

The first train from Cheribon to Tjikampek was derailed on Sunday morning, 8th May 1949, near the first stop at Tjangkring, Dutch East Indies, due to a rail having been loosened. The engine, luggage van and two coaches overturned and two firemen were killed. It was anticipated that the line would be blocked for two days. [1921]

On Tuesday 10th May 1949 the first train from Mageland to Moentilan was derailed 6 kilometres south of Mageland, Dutch East Indies, the track having been interfered with. The engine buried itself a metre deep in the soil and some coaches were derailed. Immediately after the derailment a band of terrorists opened fire on the train and six passengers and a railwayman were wounded. [1922]

On the morning of Thursday 26th May 1949 the mixed train to Malang and Djember ran over a mine near Bangil, Dutch East Indies. The mine exploded under a third-class

---

[1915] *De Locomotief*, Semarang, Dutch East Indies, 2nd February 1949

[1916] Ibid, 24th January 1949

[1917] *Het Dagblad*, Batavia, Dutch East Indies, 23rd February 1949

[1918] *Het Niewsblad voor Sumatra*, Kesawan, Dutch East Indies, 4th March 1949

[1919] *Het Dagblad*, Batavia, Dutch East Indies, 5th April 1949

[1920] Ibid, 26th April 1949

[1921] *Nieuwe courant*, Soerabaja, Dutch East Indies, 10th May 1949

[1922] *Het Dagblad*, Batavia, Dutch East Indies, 17th May 1949 carriage towards the rear of the train. The train continued about 100 metres along the line, when the coach behind that which had exploded the mine, fell into hole created by the blast. Only one coach was derailed. Three passengers died, seven were seriously and four slightly wounded. [1923]

The train from Setjang to Magelang on the morning of Monday 6th June 1949 was derailed due to sabotage at Pomagelang, Dutch East Indies. A soldier and a railwayman were killed outright and two railwaymen were fatally wounded. The line was reopened on Wednesday. [1924] [1925]

Also on Monday 6th June 1949 (Whit Monday) a passenger train from Buitenzorg to Sukabumi ran into an obstruction loaded with explosives between Seladjambe and Parungkuda, Dutch East Indies. The resulting blast threw the sand waggon being propelled by the engine off the rails and the rear coach of train was also derailed. A band of about 100 terrorists then fired upon the train from both sides with Bren guns and other automatic weapons. While the train escort returned fire the driver uncoupled his engine and proceeded to Parungkuda for help, accompanied by one of the escorts, and returned half an hour later with police and soldiers. One passenger was wounded, two Chinese passengers and a student at the police school were kidnapped and a police agent was missing. [1926]

A train from Soekoredjo to Wonokerto, Dutch East Indies, ran over a mine at about 08:00 on Sunday 19th June 1949 and was badly damaged. Three soldiers and two railwaymen were killed, four soldiers and four railwaymen were seriously wounded and two railwaymen were slightly wounded. [1927]

Also on the morning of Sunday 19th June 1949 a train from Tjibatu to Garut was fired upon by a band of terrorists Tjimurah and Wanaradja, Dutch East Indies. The attack happened at a bridge where the rails had been loosened and the engine of the train was derailed. The terrorists then opened up heavier fire on the engine, which the armed guard of the train returned. The two firemen, the conductor and a passenger were wounded, one of the firemen seriously. The line was expected to be blocked three days. [1928]

The first train from Magelang to Moentilan on Wednesday 22nd June 1949 was derailed at Mertjojoedan, Dutch East Indies, the track having been interfered with. Three locomotives and some waggons were derailed. No one was hurt. [1929]

On the morning of Sunday 3rd July 1949 the train from Ngawen to Blora, Dutch East Indies, ran over two mines. Five people died and several were injured. [1930]

A train from Malang to Soerabaja ran over a mine near Soekoredjo, Dutch East Indies, on Saturday 9th July 1949. Some waggons were derailed and two brakemen were wounded. The line was reopened on Monday. [1931]

On Tuesday 19th July 1949 a train ran over a mine between Bangil and Pasoeraeon,

---

[1923] Ibid, 28th May 1949

[1924] *De Vrije Pers*, Dutch East Indies, 10th June 1949

[1925] *De Locomotief*, Semarang, Dutch East Indies, 10th Jun 1949

[1926] *Het Niewsblad voor Sumatra*, Kesawan, Dutch East Indies, 9th June 1949

[1927] *De Vrije Pers*, Dutch East Indies, 20th June 1949

[1928] *Nieuwe courant*, Soerabaja, Dutch East Indies, 21st June 1949

[1929] *Het Dagblad*, Batavia, Dutch East Indies, 30th June 1949

[1930] *De Locomotief*, Semarang, Dutch East Indies, 6th July 1949

[1931] *De Vrije Pers*, Dutch East Indies, 12th July 1949

Dutch East Indies. Two brakemen were injured. [1932]

A train from Blitar to Garoem, Dutch East Indies, ran over a mine on about Tuesday 2nd August 1949. Two people were seriously and four slightly wounded. [1933]

The train which left Malang at 13:00 for Soerabaja ran over a mine near Wonokerto, Dutch East Indies and was derailed on Monday 8th August 1949. Some of the coaches were badly damaged and 12 people were badly hurt, four of whom later succumbed to their wounds. Others were slightly hurt. [1934]

On Wednesday 24th August 1949 a work train on its outward journey from Koetoardjo to Keboemen ran over a mine between Wonosari and Koetowinangoen, Dutch East Indies and four of its vehicles were badly damaged. Ten people died, including two soldiers and a police agent, while 20 more were seriously wounded. [1935]

At about 10:00 on Saturday 8th October 1949 an express left Soerabaja at 07:30 for Toeloengagoeng. It ran over a mine near Modjokerto, Dutch East Indies, and was derailed. Eight Chinese passengers in a second-class coach eighth from the rear of the long train were killed. Two other passengers were seriously hurt and a number were slightly wounded, while the material damage was considerable. A second mine was exploded prematurely beneath the third vehicle of the train, which was fortunately empty. Efforts to clear the line were begun as soon as possible and it was reopened at midday on Sunday. [1936] [1937]

A goods train to Madioen was 6 kilometres on the Kertosono side of Modjokerto, Dutch East Indies, on Tuesday 8th November 1949 when it ran over several mines. Twenty-five waggons were derailed and piled up in a heap of wreckage. Five people – brakemen and unauthorised passengers – lost their lives, four others were seriously injured and four were slightly injured. Soldiers took the wounded to the civilian hospital at Modjokerto. It was expected to take two days to clear the line. [1938]

At 13:00 on Monday 19th December 1949 a train was derailed four kilometres from Tjibatoe, Dutch East Indies, the track having been damaged. The engine, luggage van and a coach were derailed and overturned, and the train was then fired on. The armed guard of the train and 40 men of the TNI (Tentara Nasional Indonesia – the Indonesian Army), who chanced to be passengers on the train, returned fire and the attackers fled. A brakeman was killed and six passengers were wounded. [1939] Two firemen were missing. A work train left Tjibatoe at 16:20 for the scene. [1940]

Indonesia achieved independence from the Netherlands eight days later.

An Egyptian troop train derailed near Rafa, Egypt, on about Tuesday 4th January 1949 by Israelis during the final phase of Israel's War of Independence, reportedly with heavy

---

[1932] *Nieuwe courant*, Soerabaja, Dutch East Indies, 20th July 1949

[1933] *Het Dagblad*, Batavia, Dutch East Indies, 3rd August 1949

[1934] *Nieuwe courant*, Soerabaja, Dutch East Indies, 10th August 1949

[1935] *De Locomotief*, Semarang, Dutch East Indies, 30th August 1949

[1936] *De Vrije Pers*, Dutch East Indies, 10th October 1949

[1937] *Nieuwe courant*, Soerabaja, Dutch East Indies, 11th October 1949

[1938] Ibid, 9th November 1949

[1939] *De Vrije Pers*, Dutch East Indies, 21st December 1949

[1940] *De Locomotief*, Semarang, Dutch East Indies, 21st December 1949 casualties.[1941] The train was stopped by 'demolitions' and then subjected to a 'withering fire' killing or injuring most of the troops.[1942]

Express train No. 803 from Barcelona to Madrid was derailed at Km. 536.8 between Els Guiamets and Mora de Nova, Spain, at 22:15 on Saturday 12th February 1949. According to Francoist sources the derailment was an act of sabotage carried out by terrorists who had crossed the frontier from France. A Barcelona newspaper railed at the cowardice of the Communist attackers, who had not accepted that they had been defeated in the Spanish Civil War, now finished ten years earlier. Exiled Republican sources strongly denied, however, that there had been any sabotage and blamed the derailment on the poor state of the track, alleging that the Francoists were blaming the derailment on an act of sabotage for political purposes.

The train consisted of a steam engine and tender, a mail van, a baggage van, a second-class coach, a couchette coach (?), a first-class coach, three sleeping cars and a restaurant car. The derailment occurred on a curve and at a bridge, at the end of descent. Three vehicles in the middle of the train – the couchette car, the first-class coach, and a first-class sleeping car – fell down the embankment into the River Amat. The remaining vehicles of the train, although derailed, were little damaged. The couchette car, which was of steel construction and had only entered service a short time before, fell 15 metres. It was the most badly damaged part of the train, despite its metal construction, and the location of most of the fatalities. Conductor Francisco Pino Felipe, Assistant Conductor Félix Maria García Argüello and 29 passengers in total were killed and about 80 passengers were injured. Forty-nine passengers had needed admission to Reus hospital, seven of whom died in the days after the accident.

The track had reportedly only been renovated in the days before the accident, with new rails and sleepers. Interference with the rail fastenings was said to be the cause the derailment.[1943] [1944]

Sabotage was suspected in the derailment of No. 18 Down Punjab express from Saharanpur to Calcutta near Burna, India, in the early hours of Sunday 10th April 1949. Ten people died, six of whom were members of the Shah family, who had only boarded the train a few minutes earlier. Another 51 were injured, eight seriously. Three coaches capsized, one of which went down an embankment and was 'reduced to matchwood.' Medical aid was rushed to the scene and most of the injured were taken to hospital.

A passenger related his story to a newspaper reporter. 'Somnath Behere, a clerk of the Banaras Municipality who was travelling in the bogie which was reduced to pieces, told the A.P.I. (Associated Press of India) that the train was running behind the scheduled time. It had reached Fyzabad one hour late. The driver of the train, he said, appeared to have made up 45 minutes between Fyzabad and Sheopur in a run of 118 miles. The train was running quite fast after it left Jaunpore. As it crossed the Burna bridge, he heard a terrific crashing sound and lights went off. All the persons travelling in the bogie were thrown in different directions, Behere said. He was thrown off the compartment about 12 feet down the

---

[1941] *The Railways of Palestine and Israel* by Paul Cotterell pg. 84
[1942] *The Mexia Daily News*, Mexia, Texas, United States, 6th January 1949
[1943] *La Vanguardia*, Barcelona, Spain, 15th February 1949
[1944] Webpage: Federación Castellano Manchega de Amigos del Ferrocarril (in Spanish)

embankment. Four bodies were found lying under the 'debris' there, while two or three had been crushed under the wheels. He, however, escaped with minor injuries.

'Another passenger, who escaped with five other members of his family, including a boy of five, said that just after the train had passed over the Burna bridge, he heard a grating sound. His bogie tilted. The luggage kept on the upper berth toppled over him and he heard a big crash. He himself was sandwiched between two planks and was later saved by other passengers.

'Members of the rescue party described how a six-year-old boy was found trying to extricate his father from the 'debris' of the ill-fated train. The victim's palm was pressed hard under a bogie. The rescue party freed the man and found he had lost all the fingers of one hand.'

Another passenger said that the majority of those riding on the train had had a narrow escape. Those who had been fortunate included a party of 45 pilgrims returning from Ayodhya after attending the Rama Navami '(a Hindu festival celebrating the birth of the deity Rama) who had been travelling on one of the coaches which were derailed. Only the engine and a single coach remained on the line; they were later brought to Banaras Cantonment station. The other eight bogies, either capsized or at least derailed, were described as presenting a grim sight. The passengers in the capsized coaches were pulled out through the windows by other passengers. As the line was completely blocked trains between Benares and Fyzabad were diverted via another route. The uninjured passengers from the train were forwarded in a replacement train which left Banaras Cantonment station at 10.30 a.m.

An East Indian Railway Public Relations officer stated that a Government Inspector would examine the site on Monday 11th and begin an inquiry the following day. Senior railway and police officials who went to the scene shortly after it had occurred reportedly found that six bolts joining two rails had been removed and the district magistrate stated that sabotage was suspected. [1945]

The Bombay-bound mail train from Calcutta was derailed on the night of Wednesday 12th October 1949 near Kharagpur, India, on the main line of the Bengal – Nagpur Railway 80 miles west of Howrah Junction. Five passengers were killed outright and a sixth was fatally injured. Another five were seriously hurt and 24 were slightly hurt. A Government inspector opened an inquiry into what railway officials described as an act of sabotage, as fishplates and bolts had been removed from a length of rail. [1946] [1947]

On about Wednesday 20th April 1949 gunmen believed to be members of the White Band People's Volunteer Organisation (a Burmese paramilitary force fighting for independence from British rule and also against the Communists) [1948] attacked an armoured train a mile south of Insein, Burma, and made off with arms and ammunition. In a separate attack unknown rebels attacked a passenger train in the Shwebo district of Burma, killing a civilian. [1949]

[1945] *The Indian Express*, Madras, India, 11th April 1949

[1946] *The Taylor Daily Press*, Taylor, Texas, United States, 13th October 1949

[1947] *The Daily Gleaner*, Kingston, Jamaica, 14th October 1949

[1948] Wikipedia – Burma Independence Army

[1949] *Indian Daily Mail*, Singapore, 24th April 1949

In 1949 Japan was still under American occupation following its defeat in the Second World War. During the first two years of the occupation (1945-7) the emphasis was on demilitarization but from 1947 an economic crisis and fears of a Communist-led revolution led to a shift in emphasis. Matters were exacerbated by the Japanese Communist Party increasing the number of its seats in the Diet (Parliament) from four to 35 in the January 1949 election. To tackle the economic problems American financial adviser Joseph Dodge advised a policy of austerity, with reductions in public expenditure and a substantial cut in the number of government employees, the so-called Dodge Line. His proposals were accepted and in one of the first reductions on 4th July 1949 it was announced that more than 30,000 railwaymen were to be laid off. The following day Sadanori Shimoyama, the President of the Japanese National Railways (JNR), met his death in unexplained circumstances. His body was discovered on the railway and it was not clear whether he had committed suicide or had been murdered. Two further mysterious events followed later in the year, both being acts of sabotage directed against the railway.

On Friday 15th July 1949 a Class 63 multiple-unit train, unmanned and with its operating handle tied down, ran away into Mitaka station, where it demolished the buffer stops and part of the station beyond. Railwaymen and onlookers began the rescue work but the US military police soon intervened and barred members of the public. Six people were killed and 20 injured. There had been a further wave of dismissals of about 63,000 JNR employees a few days earlier and the sabotage was immediately blamed on members of the National Rail Workers Union. Ten union members were arrested and charged with causing death through sabotage. Nine of them were members of the Japanese Communist Party but it was the tenth person, Conductor Kaeisuke Takeuchi, who allegedly confessed to the sabotage. As a result the other nine defendants were acquitted, while Takeuchi was sentenced to life imprisonment by the Tokyo District Court, even though he later recanted his confession. Evidence from a fellow worker that the conductor had been with him in a public bath at the time of the accident was not allowed to be presented at his trial.

In 1955 the Tokyo High Court overruled the sentence of the lower court and Takeuchi was instead sentenced to death by hanging. The Supreme Court upheld the death sentence. In the following year he filed an appeal. He was still in prison awaiting execution in January 1967, when, aged 45, he died of a brain tumour. He had continued to maintain his innocence throughout and in 2010 a lawyer claimed that his confession, "was driven by a sense of hopelessness amid harsh, prolonged and coercive interrogations by prosecutors." In 2019 the Tokyo High Court refused a request for a retrial, to the great disappointment of his son. [1950] [1951]

On Wednesday 17th August 1949 in Japan a passenger train was derailed between Kanayagawa and Matsukawa stations and close to the latter. The nuts and bolts of rail joints had been loosened and spikes had been removed. The CT11 Class steam engine pulling the train overturned. Three railwaymen died and eight passengers were injured. Ten railway union members and ten members of a local Toshiba factory were subsequently arrested and following the same coercive police tactics that had featured in the Mitaka incident all 20 pleaded guilty. At the trial five of the accused were sentenced to death, a number reduced to four after a second trial. A staunch defence of the accused

[1950] https://www.tokyoweekender.com/2018/05/the-three-big-rail-mysteries-that-defined-japans-summer-of-1949/

[1951] https://www.japanpowered.com/history/the-japanese-national-railways-incidents-enduring-mysteries-from-post-war-japan was then launched by novelist Kazuo Hirutsu, who led a citizens' support group and wrote two publications, *Matsukawa Saiban* (The Matsukawa Trial) and *Izumi e no Michi* (The Road to Spring). Numerous literary figures such as Nobel prize winner Yasunari Kawabata joined the fight. The forced confessions extorted by the police and the burial of evidence by the prosecution were revealed, and as a result the convictions were all overturned by the Sendai High Court in 1961, a verdict confirmed two years later. The real saboteurs were never apprehended. [1952] [1953]

Between 30 and 50 armed terrorists stopped a Malayan Collieries private railway near Batu Arang, Malaya, on Tuesday 25th January 1949. They drove off the crew and passengers, then damaged the track and derailed the train, the engine and a number of waggons ending up down an embankment three miles from the colliery offices. The train had been carrying timber and about 50 passengers – woodcutters who were coming into Batu Arang for the Chinese New Year holidays. A collieries official said, "The derailment will not affect coal production in any way. The train was used solely for carrying timber. The permanent way will be repaired quickly enough, but it will be a bigger job to lift the engine and the derailed waggons. We have sent out breakdown trains to the area and hope to obtain mechanical assistance from Malayan Railways." [1954]

Both the pilot train and the main train were derailed by sabotage between Banir and Temoh stations, Malaya, early on Wednesday 2nd February 1949, one landing on top of the other. Fishplates had been removed at the 141-mile bridge. The crash waggon, engine and pilot box-waggon were derailed, fell down the embankment and was fired on and the engine of the Kuala Lumpur – Prai train was derailed in turn and fell on the pilot engine. The leading two steel waggons of the pilot train were also derailed, the following third-class coach stopped on the bridge. The drivers of the two trains were slightly hurt. The passengers of north and southbound trains were transhipped with seven hours' delay. [1955]

An armoured train from Mentakab to Kuala Krau, Malaya, was derailed soon after starting its journey at the 84th milepost at 10.30 p.m. on Wednesday 31st August 1949. The leading three coaches left the track and the first fireman was slightly hurt. Four fishplates and several nuts and bolts had been removed from the track. The morning Gemas – Jerantut passenger train was terminated at Mentakab. [1956] [1957]

The police station at Kuala Krau, Malaya, was attacked at 4.40 a.m. on Sunday 11th September 1949 by between 200 and 300 terrorists and several people were killed. A police party was sent from Mentakab to investigate in an armoured train, which was derailed at about 7 a.m. and attacked a quarter mile south of Kuala Krau station. Six terrorists were killed in this attack; no one on the train was hurt. [1958]

A train was travelling between Jerantut and Mentakab, Malaya, on Sunday 2nd

---

[1952] https://www.tokyoweekender.com/2018/05/the-three-big-rail-mysteries-that-defined-japans-summer-of-1949/

[1953] https://www.japanpowered.com/history/the-japanese-national-railways-incidents-enduring-mysteries-from-post-war-japan

[1954] *The Straits Times*, Singapore, 27th January 1949

[1955] *The Straits Times*, Singapore, 3rd February 1949

[1956] *Indian Daily Mail*, Singapore, 2nd September 1949

[1957] *The Straits Times*, Singapore, 2nd September 1949

[1958] *Indian Daily Mail*, Singapore, 12th September 1949

October 1949 when its driver felt it lurch and stopped it. It was then surrounded by 'bandits' who opened fire on it, killing a Malay woman and a policeman and wounding several passengers in a fierce gun battle. The people on the train were then ordered to disembark and the attackers drove the train away a short distance, probably onto a section of track which had been ripped up, so that it was derailed, and they set fire to the coaches before they withdrew. [1959]

A Chinese woman passenger and Guard Shariff were killed and two Tamil men and a Tamil boy passenger were wounded when a train was derailed by bandits at the 38th mile between Ayer Hitam and Kemayan, Malaya, in Negri Sembilan state, at 10.15 a.m. on Saturday 10th December 1949. The ticket inspectors were also injured. A mine had been detonated by a piece of string leading into the undergrowth, sending the engine and four coaches down a 15-foot embankment. Three hundred feet of rail were torn up and twisted. After the derailment the attackers attempted to close in on the train but were driven off by the escort riding the armoured car which formed part of the train and was not derailed. The passenger was the first to lose her life as a result of sabotage in Malaya since the Emergency had begun. The injured were taken to Kual Pilah hospital. A relief train despatched from Gemas at 1 p.m. with a fully equipped jungle squad was also fired on by the attackers who made two more unsuccessful attempts to rush the wrecked train. Security forces later pursued the attackers into the jungle. [1960] [1961]

What was said to have been the third derailment of a train between Canton and Hankow, China, in 12 days, resulted in serious injury to four passengers on Saturday 27th August 1949. Fishplates had been removed from the rails. [1962]

Twenty-four Chinese passengers died and 30 were hurt in the derailment of a Canton – Kowloon passenger train at Namchung, China, on Friday 30th September 1949. There had been about 1,000 passengers in the 18-coach train. Most of the deaths occurred in a single wooden coach which was telescoped. Some of the injured were taken to hospitals in Hong Kong. The cause of the derailment was not immediately clear but railway officials asked the Nationalist Government to investigate the possibility of sabotage by Communist guerrillas, who were known to be active in the area. [1963] [1964]

Two explosive devices tore up the track and derailed a train between Arau and Padang Besar, Malaya, close to the frontier with Thailand, on Sunday 1st January 1950. Driver G.F. Williams, a Eurasian, was shot and killed by the Communist attackers, one fireman was seriously burned, the other wounded, and a mail attendant was stated to be missing. There were no passengers on the train at the time. The leading engine overturned on its right-hand side and coaches were derailed. The 'bandits' set fire to the train after the attack.

Driver N.S. Nathan of the rear engine told a reporter, "We had just passed Bukit Ketri and were going up a slight slope when there was a sudden burst of gunfire from the right side of the embankment. The gunfire was followed by a terrific explosion. The bandits

[1959] *The Townsville Daily Bulletin*, Townsville, Queensland, Australia, 5th October 1949
[1960] *The Straits Times*, Singapore, 11th December 1949
[1961] Ibid, 12th December 1949
[1962] Ibid, 29th August 1949
[1963] *The Malaya Tribune*, Singapore, 30th September 1949
[1964] *The Straits Times*, Singapore, 1st October 1949

fired again and I saw them come down. There were about 20 of them, all Chinese dressed in jungle green. They ordered me in Malay to stay with the firemen. They ordered a ticket collector, a guard and brakeman to "Go away." The three then walked towards the direction of Padang Besar, five miles away. Next the bandits collected some rubber and kerosene and set fire to the coaches and engine. After this they left."

Two charges were believed to have been laid on the track, over 40 metres of which was damaged.

The train had left Alor Star at 4.40 a.m. for Padang Besar and was attacked at the 93rd milepost.[1965] [1966] It had been on its way to inaugurate a new service.[1967]

The Singapore – Kuala Lumpur night mail was ambushed at 11.45 p.m. on Wednesday 4th January 1950 by 12 terrorists armed with automatic weapons three miles south of Labis, Malaya. A Malay woman and a soldier were wounded and were rushed to Segamat hospital. Security forces on the train engaged the terrorists.[1968]

An attempt to blow up a goods train between Chemor and Tanjong Rambutan on Tuesday 10th January 1950 failed, but the Singapore – Kuala Lumpur train was derailed, presumably by sabotage, at 12.15 a.m. on Wednesday 12th January between Labis and Tenang, Malaya. No serious damage was caused and no one was hurt. The train returned to Labis with the assistance of another engine.[1969]

Shots were fired at the night mail train from Singapore to Kuala Lumpur between Sedanak and Layang, Malaya, at 9.45 p.m. on Tuesday 24th January 1950. Two Chinese men and a Chinese woman were seriously wounded and one of the men later died.[1970]

Three people were hurt when a train was derailed by sabotage between Kemayan and Bahau, Malaya, in Negeri Sembilan State, on Sunday 5th February 1950. A rail had been removed and two coaches ran off the track.[1971] Five Gurkhas travelling on the train reportedly beat off the attackers.[1972]

A passenger train from Gemas to Triang, was derailed by sabotage near Sungei Lui, Malaya, in Pahang state on Wednesday 1st March 1950. The attackers had loosened the rail by pulling it away with a long wire as the train approached, derailing the engine, a crash waggon, and a goods waggon. A 40-minute gun battle followed between the Communist attackers and the police and Gurkha troops on the train in which one of the 'bandits' was killed and three were wounded. One Gurkha was also wounded in the gun battle and the fireman, a Chinese passenger and an auxiliary constable were wounded in the initial derailment.[1973] [1974]

A northbound mail train from Singapore was attacked in Malaya at 9.25 p.m. on Friday 3rd March 1950 three miles south of Layang Layang, a Chinese woman being

1965 *Newcastle Morning Herald and Miners' Advocate*, Newcastle, New South Wales, Australia, 2nd January 1950
1966 *The Straits Times*, Singapore, 3rd January 1950
1967 Ibid, 5th January 1950
1968 *Malaya Tribune*, Singapore, 5th January 1950
1969 *Indian Daily Mail*, Singapore, 12th January 1950
1970 *Malaya Tribune*, Singapore, 25th January 1950
1971 *The Times*, London, England, 7th February 1950
1972 *The Straits Times*, Singapore, 9th March 1950
1973 *The Times*, London, England, 2nd March 1950
1974 *The Straits Times*, Singapore, 2nd March 1950

killed and two Malay soldiers, a Chinese woman, an Indian woman, and an English sailor being wounded. In other attacks terrorists fired on the night mail train from Kuala Lumpur to Penang and an unsuccessful attempt was made to derail a mixed train north of Kuala Lumpur. [1975] [1976]

A southbound Jerantut – Gemas passenger train was derailed by a mine on the afternoon of Monday 6th March 1950, between Mentakab and Gemas, Malaya, at about milepost 67, derailing several carriages. A police constable was killed and a sergeant, a Scots Guard guardsman and a ticket collector were wounded. Five persons seen running away from the track were fired on by the soldiers on the train. [1977] [1978]

A southbound train on the Mentakab – Bahau line was derailed by sabotage at 4.35 p.m. on Monday 27th March 1950. Fishplates and dog spikes had been removed from the track and the engine and five coaches capsized between Kemayan and Ayer Hitam, Malaya. Another waggon was derailed and a bridge was damaged in the derailment. The attackers then opened fire on the train from a high embankment but were driven off by the armed escort of the train. A relief train sent from Triang took the passengers back to Kemayan. [1979]

An armoured train from Triang was derailed on the evening of Saturday 1st April 1950 at the 53 milepost between Triang and Mengkuang, Malaya. Fishplates had been removed from the track, derailing the engine and two crash waggons. No one was hurt. [1980]

The Singapore – Kuala Lumpur night mail was fired on between Kulai and Sendanak, Malaya, at 9.45 p.m. on Monday 3rd April 1950 by terrorists armed with Bren guns and automatic weapons. Three civilians and a soldier were taken to Kluang civilian and military hospitals. A Tamil woman was critically hurt. [1981]

The pilot engine of the Kuala Lumpur – Ipoh night mail was derailed between 2 a.m. and 3 a.m. on Wednesday 5th April 1950 between Kampar and Timoh, Malaya, about 30 miles short of Ipoh, fishplates and 30 feet of rail having been removed from the rails. According to one account the mail train was terminated at Kampar and passengers and luggage were forwarded to Ipoh by bus. [1982] Another account stated that the Up and Down mail trains were held until dawn, then worked to the scene of the derailment and their passengers transhipped. The engine and crash waggon went down an embankment but the turret vehicle was only derailed. One of the Indian firemen was hurt when he was thrown into a ravine and he was lucky that the engine did not fall onto him, while the Chinese driver and the other Indian fireman were trapped briefly in the engine cab but were not hurt. The line was blocked until midday. [1983] The crew of the pilot engine fired their Verey pistol and were thus able to warn the mail train. Communist pamphlets were left at the scene. [1984] [1985]

[1975] *The West Australian*, Perth, Western Australia, 6th March 1950
[1976] *Malaya Tribune*, Singapore, 4th March 1950
[1977] *The Times*, London, England, 7th March 1950
[1978] *Indian Daily Mail*, Singapore, 8th March 1950
[1979] *The Straits Times*, Singapore, 29th March 1950
[1980] *Malaya Tribune*, Singapore, 4th March 1950
[1981] Ibid, 4th April 1950
[1982] Ibid, 5th April 1950
[1983] *The Singapore Free Press*, Singapore, 5th April 1950
[1984] *The Straits Times*, Singapore, 6th April 1950
[1985] *Malaya Tribune*, Singapore, 6th April 1950

In another attack on the Kuala Lumpur night mail on Thursday 6th April 1950 between Senai and Kulai, Malaya, a Chinese man named as Chong Keng Sew, aged 55, was killed and three others were injured. One of those injured was a Singapore footballer, 25-year-old Tang Tuck Choon, the Tiger Sports Association goalkeeper, who was injured in the left arm, left shoulder and thigh by machine gun bullets. Other footballers said that the advice of the team manager, who had suggested that they lie in their bunks throughout the journey in case of an attack, had prevented further casualties. The train had been packed with holidaymakers going north to spend the Easter holidays. All the attacks had happened in the 20 miles between Senai and Layang Layang. [1986]

The engine and tender of the pilot engine preceding the northbound mail train from Singapore detonated a charge under the track 62 miles north of Johore Bahru, near Chamek station, Malaya, 8 miles from Kluang, on the night of Saturday 8th April 1950. Rails and fishplates had also been removed. The locomotive and the crash waggon overturned and fell down a 10-foot embankment but Driver Poulu and Second Fireman Ramakrishna escaped with slight injuries. The carriage carrying the pillbox was derailed and slightly damaged. A crater 2½ feet deep and 5 feet wide was caused by the explosion and almost 300 feet of track was damaged. The mail train returned to Kluang to await the clearance of the line. [1987] Rather than immediately fill the crater, a breakdown gang constructed a bypass line around it and the northbound and southbound trains resumed their journeys at 1.30 p.m. on Sunday. [1988] [1989]

On Saturday 15th April 1950 about 50 armed and uniformed 'bandits' at Bukit Kuin stole railway labourers' crowbars and spanners just before dusk. Three hours later, at 8.40 p.m., an Up passenger train was derailed between Mentakab and Bukit Kuin, Malaya, and three crash waggons were wrecked. In a separate accident on the same day a Gemas – Bahau passenger train was derailed near a bridge six miles outside Gemas, Malaya, damaging the structure. A dead buffalo, which had apparently been struck by the train, was found at the scene, and this accident was therefore not attributed to sabotage. No one was hurt in either accident. [1990]

Four waggons of an Up train from Gemas, Malaya, were derailed by sabotage on Thursday, 20th April 1950 at an unclear location (illegible in newspaper). No one was hurt. [1991]

A pilot engine preceding the night mail from Singapore exploded a charge as it was crossing a bridge a quarter mile from the station at Bekoh, Malaya, on Thursday 27th April 1950. Driver George Hay said, "The engine was lifted up in the air. Luckily for me it fell back on the line. I put my lever forward and went full steam ahead. The bandits had opened fire by that time." Reporting what had happened at the station, he raced back with a Verey pistol. "I was just in time to fire the pistol to warn the northbound mail train," he added. A seven-foot-square hole was made in the centre of the bridge by the explosion. [1992] [1993]

[1986] *The Straits Times*, Singapore, 8th April 1950
[1987] *Malaya Tribune*, Singapore, 10th April 1950
[1988] Ibid, 10th April 1950
[1989] *Indian Daily Mail*, Singapore, 11th April 1950
[1990] *The Straits Times*, Singapore, 17th April 1950
[1991] *The Singapore Free Press*, Singapore, 21st April 1950
[1992] *The Straits Times*, Singapore, 29th April 1950

On Sunday 30th April 1950 'bandits' fired on a northbound mail train with automatic weapons and rifles a mile and a half south of Layang Layang, Malaya. Two British soldiers and two civilians were wounded. [1994]

The pilot engine preceding the Kuala Lumpur to Singapore night mail was derailed and capsized on the night of Wednesday 3rd May 1950 between Ayer Kuning and Batang Malaka, Malaya, because the rails had been tampered with. No one was hurt. [1995] Trains were held at the stations on either side as the area was considered too dangerous for passengers to be transhipped between trains. [1996]

An armoured train was derailed near Triang, Malaya, on Monday 22nd May 1950 by an explosion on the line, wounding a police constable and a soldier. [1997]

The pilot train of the Kuala Lumpur night mail was derailed by sabotage between Bidor and Tapah Road, Malaya, at about 1.45 a.m. on Friday 26th May 1950. No one was hurt but the line was temporarily blocked. [1998]

The patrol train preceding the night mail from Kuala Lumpur to Penang was derailed by sabotage at 1.45 a.m. on Monday 29th May 1950 near Temor, Malaya, and the fireman was wounded. [1999] The passengers and contents of the night mail trains were transhipped. [2000]

The Singapore bound night mail train was fired at by terrorists between Labis and Bekok, Malaya, on Tuesday 6th June 1950, injuring five passengers, including the wife of a British army officer. An Indian woman was fatally hurt. [2001] [2002]

The night mail from Singapore to Kuala Lumpur was fired on by terrorists near Segamat, Malaya, at about 1.30 am on Sunday 25th June 1950, slightly wounding one passenger. [2003] At 3 a.m. the same morning the corresponding Kuala Lumpur – Singapore night mail was fired at near Paloh, Malaya. No one was hurt. [2004]

The Kuala Lumpur – Singapore mail train was fired on by terrorists four miles from Senai. Malaya, on the evening of Thursday 20th July 1950. A Chinese woman passenger was killed and four Malay passengers were wounded. [2005]

Early on Monday 24th July 1950 the pilot train preceding the Penang – Kuala Lumpur was derailed about three miles from Ipoh, Malaya. About 40 yards of rail had been blown up. No one was hurt. [2006] Passengers, parcels, and mails were transhipped by road between Chemor and Ipoh and the morning train from Ipoh to Penang was cancelled. [2007]

[1993] *Indian Daily Mail*, Singapore, 1st May 1950
[1994] *The Singapore Free Press*, Singapore, 21st April 1950
[1995] Ibid, 4th May 1950
[1996] *The Straits Times*, Singapore, 5th May 1950
[1997] Ibid, 23rd May 1950
[1998] *Indian Daily Mail*, Singapore, 27th May 1950
[1999] *The Singapore Free Press*, Singapore, 29th May 1950
[2000] *Malaya Tribune*, Singapore, 30th May 1950
[2001] *The Straits Times*, Singapore, 7th June 1950
[2002] *Indian Daily Mail*, Singapore, 8th June 1950
[2003] *The Straits Times*, Singapore, 26th June 1950
[2004] *Malaya Tribune*, Singapore, 27th June 1950
[2005] *Singapore Standard*, Singapore, 21st July 1950
[2006] *The Singapore Free Press*, Singapore, 24th July 1950
[2007] *Singapore Standard*, Singapore, 25th July 1950

The same day a mixed train was derailed in Pahang, Malaya. No one was hurt. [2008]

The Singapore – Kuala Lumpur night mail was attacked at 9.45 p.m. on the night of Wednesday 2nd August 1950 in central Johore, Malaya, and the three coaches caught fire and had to be detached. The train was fired on from both sides of the line and a guard and a Chinese passenger were wounded in the heavy gunfire. [2009] The ignition of a quantity of celluloid film was responsible for the fire. [2010] As a result of the fire 75 bags of letters and 57 bags of parcels from Singapore were destroyed. [2011] The Penang – Kuala Lumpur mail train was also fired on at an unstated location and a Gurkha warrant officer was killed and two Malay passengers and a ticket collector were wounded. [2012]

The Singapore night mail was again attacked in Johore province, Malaya, on Friday 4th August, a Chinese civilian, Lim Tim Jay, a waiter on the train, being killed and two Europeans and an Indian were wounded. Meanwhile two waggons of a southbound goods train were derailed in Wellesley Province, Malaya, by the removal of fishplates. [2013]

Four crash waggons of a Down train were derailed in Negri Sembilan Province, Malaya, on Saturday 5th August 1950, presumably as a result of sabotage. [2014]

The patrol train preceding the Kuala Lumpur – Singapore night mail was derailed by sabotage just after midnight on Tuesday 8th August 1950 in Negri Sembilan state, Malaya. No one was hurt but the line was blocked and passengers were transhipped between the two night trains. [2015]

The Up train from Gemas to Mentakab, Malaya, was derailed by sabotage on Tuesday 15th August 1950. There were no casualties. [2016]

The pilot train preceding the night mail from Kuala Lumpur to Prai was derailed by sabotage in Perak state, Malaya, on Thursday 31st August 1950. Passengers and mail were transhipped between the Up and Down trains. [2017]

The pilot train preceding the night mail from Kuala Lumpur to Prai was again derailed by sabotage in Perak state, Malaya, on the night of Saturday 16th September 1950 and passengers and mail were transhipped between the Up and Down trains. [2018]

The pilot train preceding the Penang – Kuala Lumpur night train was derailed by sabotage in Wellesley state, Malaya, on the night of Tuesday 26th September 1950. The engine and crash waggon overturned and the driver and fireman were slightly injured. [2019]

The pilot train preceding the Kuala Lumpur – Singapore night mail was derailed in the early hours of Thursday 28th September 1950 between Tampin and Gemas, Malaya. No one was hurt. Passengers and mail were transhipped. [2020]

[2008] *The Times*, London, 25th July 1950
[2009] *Malaya Tribune*, Singapore, 3rd August 1950
[2010] *The Singapore Free Press*, Singapore, 3rd August 1950
[2011] *Singapore Standard*, Singapore, 4th August 1950
[2012] *The Straits Times*, Singapore, 4th August 1950
[2013] *Singapore Standard*, Singapore, 6th August 1950
[2014] *Indian Daily Mail*, Singapore, 7th August 1950
[2015] *The Straits Times*, Singapore, 9th August 1950
[2016] Ibid, 17th August 1950
[2017] *Indian Daily Mail*, Singapore, 2nd September 1950
[2018] *Singapore Standard*, Singapore, 18th September 1950
[2019] *Indian Daily Mail*, Singapore, 28th September 1950
[2020] *Singapore Standard*, Singapore, 29th September 1950

On Saturday 30th September 1950 a trolley carrying seven labourers, five Indians and two Malays, ran over a charge placed on the track in Perlis state, Malaya. All were injured, four of the Indians seriously. The same day a goods train was derailed by sabotage in Negri Sembilan state, Malaya. [2021]

The Kuala Lumpur – Singapore was fired on by 'bandits' on Monday 2nd October 1950 in Johore state, Malaya. Two Chinese and a Malay were wounded, one of the Chinese, aged 60, fatally. [2022]

The pilot train preceding the Kuala Lumpur – Penang night mail was derailed due to sabotage on Thursday 5th October 1950 in Perak state, Malaya. No one was hurt but the line was blocked and passengers and mails were transhipped. [2023]

On the night of Saturday 14th October 1950 the pilot train of the Kuala Lumpur to Penang night mail was derailed by sabotage in Selangor state, Malaya, and the same night another pilot train was derailed in Perak state. Passengers and mail were transhipped. [2024]

The day train from Kuala Lumpur to Singapore was derailed by sabotage on Wednesday 18th October 1950 near Senai, Malaya. Its engine was blasted off the track and overturned and Driver Albert Thyriah was pinned beneath it and killed. The two firemen, two other railwaymen and three passengers were slightly hurt. Two coaches and a goods waggon were derailed. The blast caused a crater 18 inches deep and five feet in diameter and 300 feet of track were torn up. At the time of the accident, 8.30 p.m., the train was running four hours late. Day trains were not usually preceded by pilot trains but given its lateness, which meant that darkness had fallen, Driver Thyriah could have requested one. It was believed that the mine had been intended for the pilot train of the northbound night mail from Singapore, which had been waiting at Johore Bahru station to cross the southbound train. After the accident the pilot train was rushed to the scene with police and troops and returned with the wounded to Johore Bahru. As rescue work was in progress late in the night news was received that the pilot train for the southbound mail train from Kuala Lumpur had been derailed at Gemas. [2025] [2026] Dahlan, a Malay passenger studying Arabic, on the train, spoke to a reporter. "It was an act of God which saved us from sudden death or injury. My brother Ismail and myself are coming to Singapore to obtain a visa for him for Egypt. He has been awarded a scholarship to Egypt by the Government. It was about 8.10 last night. I was dozing on the lower berth of the first passenger coach, which is the third from the engine. Suddenly I was thrown onto the floor and everything went black. Babies and small children started to scream and they created a terrific row. After a while all became quiet. Everybody was flat on his or her belly. They were in this position for more than one hour, while the soldiers who accompanied the train searched the surroundings. After a while I got out of the train and surveyed the damage. The engine and the first five coaches including mine were derailed." About 80 tired-looking passengers finally arrived in Singapore at 9.30 a.m. the next morning after having spent 12 hours in pitch darkness. [2027]

[2021] Ibid, 3rd October 1950

[2022] *The Straits Times*, Singapore, 4th October 1950

[2023] *Malaya Tribune*, Singapore, 7th October 1950

[2024] *Singapore Standard*, Singapore, 16th October 1950

[2025] *The Straits Times*, Singapore, 19th October 1950

[2026] *The Times*, London, 20th October 1950

[2027] *Malaya Tribune*, Singapore, 19th October 1950

The pilot train preceding the Kuala Lumpur – Penang night mail train was derailed in Perak State on Friday 20th October 1950. The line was blocked and passengers were transhipped. [2028]

The pilot engine preceding the night mail train from Kuala Lumpur was derailed at Batang Malaka, Negri Sembian, Malaya, on Friday 27th October 1950, presumably by sabotage. [2029]

After a passenger train had been derailed in Pahang State, Malaya, on Saturday 28th October 1950 security forces and bandits fought a fierce gun battle. [2030]

The pilot train preceding the night mail from Kuala Lumpur to Singapore was derailed on Wednesday 1st November 1950 between Tampin and Kluang, Malaya. The line was blocked and passengers, parcels and mails were transhipped between trains. [2031]

The Singapore – Kuala Lumpur night mail was derailed early on Wednesday 8th November 1950 at Layang Layang, Malaya, the engine, three coaches, a luggage van and an armoured car all leaving the rails in dense jungle. It was described as a miracle that no one was badly hurt or killed. The train ploughed up 150 feet of track and the engine turned over several times before coming to rest on its side. Driver K.P. Singham, 45, climbed out of a side window with a few scratches to his arms and shoulders and in an interview thanked God for his escape. He said that he remembered "loud explosions" and then rolling around several times in the cab of his locomotive. When all was quiet he had looked for his fireman and then got out through a window, hearing some shots being fired in the darkness. The passenger coaches did not overturn but ended up standing a right-angles to one another across the track. The armoured car did overturn and stopped resting on the roof of the baggage car.

Some railwaymen who had been in the vicinity were locked into a hut by the attackers. The police and military found a thick roll of steel wire leading from the track into the jungle, indicating that the mine(s) had been set off electrically by the 'bandits' after the pilot engine had passed. The location where the sabotage took place could only be reached by rail. Two hours after the attack breakdown gangs were already at work with the aid of floodlights. [2032]

The northbound night mail train from Singapore to Kuala Lumpur was derailed at about 10 p.m. on Monday 20th November 1950, in between Layang Layang and Sedenak, Malaya, again after the pilot train had passed the spot. Immediately after the passing of the pilot train a length of rail had been removed. The engine and one coach overturned and the driver and fireman suffered slight burns. [2033]

The Singapore – Kuala Lumpur night mail was derailed on Wednesday 29th November 1950, 12 miles from Johore and then attacked, but the attackers were repulsed by eight young Gurkha soldiers who jumped from the derailed coaches and counter-attacked, 'screaming, kukris drawn, Sten guns firing.' One Gurkha was slightly hurt and three people on the train were wounded, one fatally. The events were reported in detail in a newspaper. 'Train casualties were three – one fatal. Mohammed Bhi, 28-year-old Malay

[2028] *Singapore Standard*, Singapore, 22nd October 1950
[2029] *Malaya Tribune*, Singapore, 28th October 1950
[2030] *The Straits Times*, Singapore, 30th October 1950
[2031] *Malaya Tribune*, Singapore, 3rd November 1950
[2032] Ibid, 10th November 1950
[2033] *The Singapore Free Press*, Singapore, 21st November 1950

saloon attendant, had his legs shot off and died on the way to the Johore Bahru hospital. His companion, another attendant, Chelliah, an Indian, who only a few weeks ago carried a splinter on his neck and survived, was shot on the thighs. A European soldier, lying in his bunk, escaped with a slight graze across his back. "Just lucky," he grinned. "Not much to write about, eh?" The Gurkha, who returned with his companions ten minutes later with blood streaming from his neck, calmly smoked a cigarette and spoke in perfect Malay, "Diah, lari, lari" (they run, run). A medical officer bandaged his neck and sent him off to hospital in a car with the other injured men. Describing the attack, a young RASAC lieutenant told *The Standard* that the train was passing through two hillocks just two miles away from Kempas station when they were thrown back by an explosion. "The engine and two coaches had passed the danger spot and I was in the third coach about to go to bed," he said. "Immediately after, there was another explosion and four coaches were thrown off the rails. Just as suddenly, there was a Bren gun firing from all sides. We knew the devils were at it again – and we crouched beneath the derailed coaches to return fire as best we could. It was at this time that the Gurkhas jumped out of the coaches yelling like mad into the darkness. Ten minutes later the firing stopped. We knew they had that done their work," he said.' Two land mines, placed on the track a few yards apart, had apparently been detonated electrically from the hillside. [2034]

Probably on Thursday 1st December 1950 two crash waggons of an armoured train were derailed in Negri Sembilan state, Malaya, and four crash waggons and the engine of a mixed train were derailed in Pahang state. Malaya. In the latter attack the train was then fired on but no one was hurt. [2035]

In the early hours of Sunday 10th December 1950 an armoured train was derailed, presumably due to sabotage, between Kerdau and Mentakab, Malaya, two crash waggons, an armoured coach and the engine and tender leaving the track. [2036]

The pilot train of the Penang – Kuala Lumpur night mail was derailed in Kampar, Perak State, Malaya, on the morning of Monday 11th December 1950. [2037]

A mixed train was derailed by sabotage near Mentakab, Pahang state, Malaya, on Monday 18th December 1950. The engine and a crash waggon left the rails. [2038]

The engine and some waggons of the second portion of the Singapore – Kuala Lumpur night mail were derailed between Sedenak and Layang Layang, Malaya, on Thursday 21st December 1950. No one was hurt. The second portion followed ten minutes after the first portion, which was itself preceded by the pilot train. [2039]

The pilot train of the Kuala Lumpur – Penang night mail was derailed between Batang Kali and Rasa, Selangor state, Malaya, on Sunday 31st December 1950. [2040]

The overnight express from Strasbourg to Paris Est was derailed near Châlons-sur-Marne, France, at 04:40 on Sunday 15th January 1950. Five passengers were injured, a woman seriously. Sabotage was suspected, a rail having been detached from its sleepers.

[2034] *Singapore Standard*, Singapore, 30th November 1950
[2035] Ibid, 2nd December 1950
[2036] Ibid, 11th December 1950
[2037] *The Singapore Free Press*, Singapore, 11th December 1950
[2038] *Singapore Standard*, Singapore, 20th December 1950
[2039] Ibid, 22nd December 1950
[2040] *Indian Daily Mail*, Singapore, 2nd January 1950

It pierced the floor of a carriage and caused the injuries. A worse accident was prevented because the driver, seeing that something was wrong with the track, had reduced speed. Another derailment, of a Paris – Briançon fast train, at Livron station, on Monday 17th, was also thought to have been due to sabotage. There had also been attempts at sabotaging signalling installations. [2041] [2042]

Six passengers, four men and two women, were killed and 93 were injured, 45 seriously, when the Calcutta Mail which from Madras was derailed at about 11.30 p.m. on Friday 3rd March 1950 two miles short of Sulurpet, India, on the Madras and Southern Mahratta Railway.

An official communique from the company stated, "The General Manager of the M and S M Railway regrets to report that at about 23-30 hours on the night of Mar. 3, No. 1 Down Madras-Calcutta Mail, which left Madras on Mar. 3, met with a serious accident at mile 48, two miles south of Sulurpet station. As a result of this accident the engine and first four coaches of the mail are damaged and six more have derailed. Information at present available indicates that six persons have been killed, eight seriously injured and 85 have sustained minor injuries. A breakdown special and medical specials are already at the site of the accident and every possible assistance is being rendered. Local medical assistance from Sulerpetta has also been availed of. Senior operating engineers and medical officers who left Madras by special train very early this morning are already at the scene of the accident. An enquiry is being ordered into the accident. So far as is known it is suspected to be due to sabotage by the removal of a pair of fishplates."

Those who lost their lives were all upper-class passengers, and five of them had been travelling in the bogie next to the engine, which was smashed to smithereens. The mail van, on the other hand, was undamaged, and the mail bags intact.

One of the passengers who very luckily escaped with slight injuries was Mr R.B. Ramakrishna Raju, Chairman of the Legislative Council, who gave a long statement to a newspaper reporter. He said, "I do not know how it happened. I was sleeping in the first bogey behind the engine. Suddenly I heard a terrific crash and our carriage was blown to pieces. Our carriage was thrown a hundred feet from the track. The chassis and superstructure got separated. Fortunately the roof flew off, otherwise it would have fallen on my head. I found myself in the midst of splinters, and as my limbs were whole I climbed out of the carriage. Ours was the only carriage that was completely smashed; others were slightly damaged. There were six passengers in our compartment. There was a representative of Messrs Kutty and Rao and his sister-in-law, Mrs Rathnamma. Mrs Rathnamma got crushed under the chassis. It was not possible to extricate her and Mr Jayant was calling out to people to liberate her from the chassis. Many people tried to lift her from the debris but could not do so.

"Mr Kala Venkato Rao, who was also in the train, was not injured. Mr K Lingaraju, M.L.A., and Mr Pyda Venkatanarayana, who were accompanying Mr Rao, were also not injured. Mr Kala Venkato Rao, who came out after the crash, stopped a passing lorry, and arranged for the removal of the injured persons to the Sulurpet Hospital. The accident occurred two miles before Sulurpet. Local doctors who arrived on the scene of the crash rendered medical aid to the injured. Some ladies rendered first aid.

[2041] *The Times*, London, England, 17th January 1950

[2042] *Gazette de Lausanne*, Lausanne, Switzerland, 18th January 1950

"The engine of the train had flown off the track. Telegraph wires were cut, and I do not know whether they were cut by design or by the falling of the carriages on the telegraph posts. I was unable to contact Madras, and in fact nobody was able to do so. Mr Venkato Rao went to the railway station and informed Vijayarada and Guntakal and returned to Madras. The District Medical Officer of Nellore came to the place at 5 a.m. today. Later the District Collector, Mr Malayappan, the Deputy Inspector-General of Police, Mr V.R. Rajarathnam, and others arrived. Meanwhile the Chief Secretary and the General Hospital at Madras were informed. Three ambulances came and I am told one was punctured on the way. These ambulances removed the serious cases to the General Hospital. Meanwhile Mr Kala Venkato Rao, who had returned at 5.30 a.m., contacted Mr Gopala Reddi by phone. A car was sent for me and I came away.

"My injuries are not severe. It was a very miraculous escape. Mardini Narasimha, President of the East Godavari District Board, who also travelled by the train, sustained a bleeding wound and his collar bone is also suspected to have been broken.

"The local doctors and ladies were very helpful. The District Medical Officer, Nellore, and other doctors attended to the comforts of the injured."

A newspaper reporter also went to the scene of the accident. 'The *Indian Express* reporter who rushed to the spot this morning found the scene of accident humming with activity, thousands of workers straining every nerve to clear the debris and attending on the injured. The railway relief train, which arrived with medical supplies and doctors, started work very early in the morning.

'The engine of the ill-fated train, a recently imported heavy American type, had been thrown off the rails and it had fallen on the sandy ground to the east of the railway track. It is stated that the crew of the engine, realising that the engine was flying off the rails, jumped clear. They are safe, the fireman alone having received slight injuries on his legs and arms. About 40 injured persons were brought to Madras by train about 12.30 p.m. today.'

Speaking in the Legislative Assembly after the accident Mr Gopala Reddi stated, "It is a clear case of sabotage by the removal of fishplates according to an official message," and he read out a message from the Deputy Inspector-General of Police. "The Calcutta Mail was derailed at the 48th mile between Tada and Sulurpet at 11-25 hours last night. Out of eleven carriages the engine and first bogie capsized. The first bogie turned over completely and was smashed. Four other bogies also capsized. All others except the last are off the track. As far as is known, five died and 41 were injured, five of them seriously. The Superintendent, Railway Police, travelling in the first bogie, was injured. Mr R.B. Ramakrishna Raju was also injured. Both not serious. Clear case of sabotage by removal of fishplates. All injured were removed to Sularpet Hospital. All relief arrangements were made. The relief arrangements were made. The D.I.G. (Northern Range) camping at Tada is on the spot from 1 a.m. District Magistrate, Nellore, and District Superintendent of Police have arrived. Investigation is proceeding. Message sent from scene of occurrence." [2043]

The following day it was reported that a gangmen's tool box had been found nearby with the lock open, this presumably being the source of the tools used to loosen the fishplates. It was also reported that the only railwayman who had been injured was Guard

---

[2043] *The Indian Express*, Madras, India, 5th March 1950

Malan, who was slightly hurt. [2044]

Other apparent attempts at sabotage were reported about the beginning of March between Ottapalam and Mankara, where the driver of a goods train saw a boulder on the line in time to stop his train, and on 4th March near Ihaksar on the East Indian Railway, where the driver of the Dehra Dun Saharanpur express pulled up after seeing that fishplates had been removed from the track. [2045]

The inquiry into the derailment began on 6th March, with Driver J.W. Rodriques being examined for over four hours. [2046]

The report of the Government Inspector of Railways concluded that the derailment was indeed due to sabotage, by the removal of a right-hand rail and at least two fishplates on the inside of a two-degree curve, by some unknown person or persons, and that no blame was attached to the driver and fireman, who had been vigilant and cautious. [2047]

An initial report stated that 50 people were thought to have died and another 75 to have been injured in the derailment of No. 10 Down train of the Oudh and Tirhut State Railway, the *Kumaon Express*, at 1.40 a.m. on Wednesday 12th April 1950, between Bitroi and Kachla stations and half a mile from Bitroi, Uttar Pradesh, India. The engine and five leading bogies (coaches; three third-class and two upper-class) fell from a bridge onto the dry bed of a stream. Those who had lost their lives included the fireman and the 'whalasi' (meaning of this word unknown), while the driver had suffered burns all over his body. A further report stated that 40 bodies had been recovered. An ambulance train with medical officers, doctors and nurses had left Izatnagar for the scene while villagers from neighbouring areas had collected in their hundreds to help with the rescue work. It was almost immediately clear that an act of sabotage had been committed. A newspaper wrote, 'Preliminary investigation revealed that fishplates on both sides of the bridge and a number of dog spikes from the rails of the track were found removed. A rail on the right side of the bridge was offset.' [2048]

As is often the case the initial casualty figures proved to be rather too high, for on 15th April it was reported from Lucknow that with the death of three further victims in hospital the number of dead had reached 37. [2049]

Train No. 5, the Up *Punjab Mail*, of the East Indian Railway, met with a disaster early on Sunday 7th May 1950 between Jasidih and Sumultala, Bihar, India. A communique from the Divisional Superintendent of the company at Asansol stated: "The engine and three leading bogies, including an upper-class carriage, derailed, and capsized down the embankment. The fourth bogie, an R.M.S. {Rail Mail Service} carriage derailed all wheels and the following two bogies derailed partially and seventh (last) bogie was on rail. Sixty dead bodies have been recovered up to 11-25 hours and about 70 injured persons have been removed to hospital. Operations to recover the dead and the injured from under the debris are still going on."

One of the passengers on the train, who had a narrow escape, was a Bihar state minister, Abdul Qaiyum Ansari, who gave a detailed account of his experiences to a

---

2044 Ibid, 6th March 1950
2045 Ibid, 7th March 1950
2046 Ibid, 8th March 1950
2047 Ibid, 14th May 1950
2048 Ibid, 13th April 1950
2049 Ibid, 16th April 1950

newspaper reporter. 'Mr Ansari said the number of women killed also must be "considerable" as the compartment behind the engine was a ladies' compartment, which had been smashed to smithereens. Asked about the cause of the accident, Mr Ansari said, "I found fishplates removed at two places and the bolts and nuts thrown down the embankment. The rails had been brought closer together after the removal of the fishplates at one place." Mr Ansari said he had himself seen about one hundred dead bodies, while others were still lying under the wreckage. He told a P.T.I. {Press Trust of India} correspondent: "I cannot describe the horrors of the disaster." Two bogies of the train were smashed to pulp, while two others and the engine were badly damaged. The driver was thrown clear and had a miraculous escape. The fireman and a cleaner were killed. The relief train, Mr Ansari said, arrived six hours after the accident took place.

'Mr Gouri Sankar Dalmia and Mr M. Yasin, two Bihar M.L.A.s {Members of the Legislative Assembly} arrived earlier with a number of doctors and volunteers and rendered first aid. Mr Ansari said he had found a number of persons "indulging in looting the victims of the disaster." Mr Ansari said when he asked the people to stop looting they took up a stubborn attitude. "I then ordered my personal guard to shoot if they did not stop. Then only they stopped," he said. Mr Ansari, who had boarded the train at Madhupur, said that immediately after the train had left Jasidih he heard a rumbling noise and just as he was about to enquire about it of his co-passengers, he felt that he was rolling down from his berth. After recovering from the shock of his fall Mr Ansari with three of his companions broke open the window and got out and saw screaming men, women, and children around. A high police official told the PTI that at the same spot where the derailment occurred, fishplates had been found removed a month ago when a Military Special was passing the station. Timely detection of the removal of fishplates had saved the train from derailment.' [2050]

By the following day the number of dead had risen to either 71 or 81 (both figures are quoted in consecutive paragraphs of a newspaper report). It was noted that the mail train had been preceded by a train carrying foodgrains and that possibly this might have been the real target of the saboteurs but it had been shunted at a wayside station for the Punjab Mail to pass. Mr A.K.P Sinha, Deputy Inspector-General of Police, C.I.D. {Criminal Investigation Department} visited the scene and later spoke to the Press. He stated, "It is a clear case of sabotage. But it is difficult to say who has done it. There are various clues and theories about it, but they are at present under investigation." 'The fishplates, he said, had been removed, and two ends of one rail length had been disconnected and "pushed slightly inwards" in such a manner that it left no doubt in his mind that it was the handywork of a person or persons. Mr Sinha declined to answer a question if he suspected Communists for this act of sabotage.

'All the wreckage that could be removed by human agency had been cleared, Mr Sinha said, and the rest would have to be removed by crane. Giving an account of how the accident took place, Mr Sinha said that two pairs of fishplates on either end of a joint on the line were removed. The bolts had been taken off and the rails shifted from their places. The engine had crushed one of the loose ends of the rail almost flat and ran not less than 30 feet, cutting the sleepers and tearing them from their moorings. Asked whether it was a fact that only one bogie remained intact, Mr Sinha said that people out of excitement seemed to have said so. He himself saw three bogies standing on the track.

[2050] Ibid, 8th May 1950

They were the last three of a total of seven bogies. The bogie in which Mr A.Q. Ansari, Bihar Minister, was travelling had jumped off the rails and was lying on its side. The other three were reduced to matchwood. The engine and one of the bogies were absolutely upside down. Describing how the three bogies had escaped disaster, Mr Sinha said that the cowcatcher of the engine had gone into the mud immediately after coming off of the track. This ought to have reduced the force behind the engine and the three bogies must have acted as buffers to the rest of the bogies. In reply to questions, Mr Sinha said that all the plates and screws removed from the permanent way were found near the place of the accident. In addition an empty bottle of country liquor was also found. In the course of rescue operations volunteers dug out of the ladies' compartment which had been smashed to smithereens a woman who had remained unscathed. The entire compartment had been wrecked but she was alive, not even receiving a scratch. Ornaments and valuables were found strewn at the site of the accident. They were collected under the supervision of the police. From the body of a Marwari passenger currency notes worth Rs 13,000 and a draft for Rs 6,000 were recovered. Eight more injured passengers, mostly belonging to Gorakhpur, were brought down here {Patna} by the Up Janata Express that arrived here at 7 p.m., 12 hours behind schedule. Of the eight injured, the condition of four was reported to be serious. They were all removed to the Patna General Hospital. One of the injured persons told U.P.I. {United Press International} that although the accident took place at 2.30 a.m. no relief was in sight till 6 a.m., when a batch of local people arrived and rendered first aid. From the time of the accident till sunrise, there was indiscriminate looting. Ornaments on the persons of injured women were also snatched away. About a dozen persons in military uniform were arrested at the instance of Mr Abdul Qaiyum Ansari, Minister for Rehabilitation, on charges of looting and intimidation, he said.' [2051]

Mr N. Gopalaswami Ayyangar, the Transport Minister, in a broadcast on the evening of 8th May, appealed to the public to develop a social conscience which would make impossible the planning and execution of sabotage and bring saboteurs to book. He stated that the latest figures for casualties was 82 killed and 100 injured. [2052]

The Minister of State for Railways, Mr K. Santahanam, gave an assurance on 9th May that the victims of the disaster would be compensated even if it was proved to have been an act of sabotage. Questioned as to the positioning of the ladies' carriage next to the engine, he replied that the marshalling of vehicles in a train was dependent on various considerations. As to the apparent delay in the arrival of relief at the accident site, he stated, "I have personally gone into the question of the time lag and propose to take steps to reduce it to the absolute minimum. In this particular case the accident took place rather in the middle of a ten mile block section which made it unusually difficult to reach the place quickly." He added that 83 bodies had been recovered from the wreckage by 3 p.m. the day before, six injured had died in Deoghar hospital, three had died in Patna hospital and three bodies remained in the wreckage, bringing the number of fatalities to 95. [2053]

India suffered another major accident in Bihar State early on Sunday 13th August 1950, again due to sabotage. A Press note from the Public Relations Officer in Calcutta of the railway involved explained what had happened: "The 7 Up *Toofan Express* which

---

[2051] Ibid, 9th May 1950
[2052] Ibid, 10th May 1950
[2053] Ibid, 11th May 1950

left Howrah on Saturday Aug. 12 met with an accident today shortly after midnight between Durgauti and Karmanasa, 24 miles below Moghalsarai on the Grand Chord of the East Indian Railway. The ill-fated train collided at 12.35 a.m. on Sunday with some waggons of a Down goods train which had derailed as few minutes earlier at mile 394 from Howrah. The engine and the three leading bogies of the *Toofan Express* derailed and went down the embankment and another was hanging between the track and the embankment. According to information received so far (1 p.m. Sunday) in Calcutta 17 persons were killed and 79 persons are reported to have received injuries, of whom 27 seriously. The injured were being sent into hospitals at Banaras and Gaya. A relief train with medical aid was sent from Moghalsarai. The General Manager and other high officials of the railway have left for the site of the accident. The cause of the accident is under investigation."

The engine and leading coaches of the train fell 20 feet down the embankment into floodwater four feet deep and two bogies and a luggage van telescoped one another. The telescoped bogies were described as having been 'reduced to matchwood' and the bodies of a number of victims were sandwiched between them. The engine and eight waggons of the goods train had fallen down the embankment on the opposite side. Apart from the engine and leading luggage van two third-class, one inter-class and a first- and second-class composite bogie of the express were involved in the accident, which had occurred near a small bridge. Five undamaged bogies were taken to Gaya with a number of the injured. The driver of the express survived the accident. Among the first rescuers to reach the scene were district officials from Benares, who had to wade through flooded paddy fields to reach the railway and render aid to the victims. They were followed by doctors and police officials who arrived at 5.30 a.m. in drenching rain. Later armed pickets were posted around the wreckage. Over 50 victims were admitted to the railway hospital at Moghalsarai and another 20 were brought to the Government hospital at Banaras. By the evening the number of dead was stated to be 23 and of injured almost 200, of whom 35 were serious.

That the initial derailment of the goods train had been due to sabotage was indicated by the fact that a fishplate, four bolts and some screws were found to have been removed at the small bridge where the derailment had happened. These were later recovered by police and railway officials from the nullah (stream bed) beneath the bridge. [2054]

More dead bodies had been recovered by the next day but the exact number was not at that point known. A Punjabi child, aged about three, whose mother and father were both killed, had been taken to Moghalsrarai hospital. As is often the case some people had lucky escapes, for example a family of four riding in the inter-class bogie third from the engine, which was badly damaged, all escaped unscathed.

A detailed account of the fate of other passengers on the train was given by a newspaper. 'Mrs Charles, wife of a stenographer employed in a Kanpur firm, died while her husband and son, a railway employee, were reciting prayers at her side. She lost a leg and hand and also bore severe injuries on her body. Her husband was travelling in the inter-class while her son had left the second-class compartment at Gomo station to chat with the guard of the train and later travelled with him. Her daughter, who was also severely injured, later died in the Government hospital at Benares.

'A servant of Mr M.H. Rau, Assistant Superintendent of the E.I.R. Printing Press, Howrah, said that he was sleeping on the luggage rack of the servants' compartment

[2054] Ibid, 14th August 1950

attached to the mixed first and second-class bogie and was violently thrown and sandwiched between two bogies. He was unconscious when the police extricated him with great difficulty, several hours after the accident.

'Mr Lakshmi Chand, stated to be a mill owner of Jaipur, was among the injured who lost his wife and two children in the accident.

'Owing to the darkness and the rain the injured had to suffer pain and agony for some hours till morning when the rescue and relief parties reached the site of the accident. A crane was transhipped to the site of the accident by noon and salvage operation continued till the evening. A high official stated that it was not possible to ascertain the correct number of dead till the two bogies were cleared and the inundated paddy fields near the site of the accident were thoroughly combed. So far six children, seven girls and five boys, have been identified {children presumably refers to babies and toddlers, and girls and boys to older children}.

'Some passengers of the ill-fated *Toofan Express* boarded the No. 5 Up *Punjab Mail* at Moghalsarai. A passenger who had earlier travelled by the *Toofan Express* bound for Delhi – from Gaya, gave an eyewitness account of the disaster to a P.T.I. correspondent who was travelling by the *Punjab Mail*. He said: "There was a rattling of brakes and swaying of bogies at 00.24 a.m. and the train came to a sudden stop between Durgauti and Karmanasa railway stations. It was pitch dark with thundering clouds," he added. The train was running at a speed of about 45 miles an hour. The passenger, who was travelling in the fifth bogie which escaped destruction, said that the engine of the *Toofah Express* fell about 50 yards away from the line. He added that a moving sight at the place was the sobbing of a two-and-a-half year old girl, who was the only survivor of a family travelling by the train.

'Eight injured persons arrived at Allahabad from Kalka Mail, which arrived six hours behind schedule. These passengers were brought to Moghul Sarai from the scene of the accident by a relief train. Hundreds of people had assembled at Allahabad railway station, anxiously awaiting the arrival of the train to inquire about the welfare of their friends and relatives, travelling by the ill-fated train. Senior railway officials were also present to look after the injured. Two women with marks of injuries on their faces, one of them carrying in her arms an injured baby were the focus of attention as they alighted. She burst into tears when a P.T.I correspondent asked her about the tragedy. A railway official travelling by the same train described the scene of the disaster as "ghastly." "It is beyond human imagination to describe the agony of the passengers who were injured," he said.

'All trains coming from Calcutta are delayed indefinitely. Mr. R.C. Sood, Government Inspector of Railways, will, it is learnt, hold an inquiry into the accident on August 15, says a New Delhi message.' [2055]

On the evening of Monday 14th the Railway Minister, Mr N. Gopalaswami Ayyangar told Parliament that his information was that 20 persons had died at the scene, 30 had been seriously injured and 80 were slightly injured. One of the injured had succumbed in the Civil hospital, Banaras.

As to the cause of the derailment he stated, "A rail joint on the left-hand side of the Down track almost at the centre of the girder bridge was found to have been tampered

---

[2055] Ibid, 15th August 1950

with. Fishbolts, fishplates and removed coach screws had also been found. The rail at joint was displaced. The local rail officers, the District Magistrate, and the police, are stated to be of the opinion that the displacement of the rail was an act of sabotage." He added that the site of the accident was one patrolled by the Railway Protection Police, who had passed the bridge half an hour before midnight. The work of introducing anti-sabotage joints had begun, but they were not yet installed in that area.

The driver of the goods train, realising that the opposing line was obstructed, had gone forward exhibiting a red light, which had been seen by the driver of the express, who had applied his brakes, but the distance had been too short to stop. The goods train had derailed at 12.22 a.m. and the express had arrived at the bridge at 12.25 a.m., thus only three minutes later. Following the goods train had been No. 4 Down Bombay – Calcutta Mail. When this train had reached Karmanasa at 12.45 a.m. the goods train had still not cleared the section ahead and the engine of train No. 4 had been sent forward to examine the line. The guard of either the express or the goods train had gone back towards Karmanasa, had met the engine at 12.50 a.m. and had returned with it to Karmanasa and information of the accident had been given to control.

The minister was asked about pilot engines preceding express trains and about placing two or three goods waggons between the engine and the first passenger coach of the train. He replied that neither was always feasible. Mr Ballram Baghat complained that every time there was a serious accident the government stated that it was a case of sabotage and the accident could have been due to track defects. Mr Ayyangar replied that the disaster would be investigated by a fully autonomous board of inquiry, the results of which the House should await.

Despite the statement of the minister that there were only 21 known fatalities, it was stated later in the same newspaper account that with the death of another victim in Gaya hospital the official number of dead had risen to 24. In addition there were an unknown number of bodies possibly still in the third-class women's coach which still lay in 17 feet of water below the embankment.

Mr A.K.P Sinha of the Bihar Criminal Investigation Department stated that, "I am sure to the extent of 14 annas in the rupee that it is a case of sabotage. I have doubt only to the extent of the remaining two annas. This doubt will also be cleared after an actual demonstration of the accident is held today." [2056]

Confusion appears to have reigned over the number of dead, for the following day it was stated that two more bodies had been recovered from the scene, and in one paragraph that the number of dead was 25, and in a subsequent paragraph that it was 24! [2057]

Anti-Government insurgents blew up a train about 100 miles north of Mandalay, Burma, on about Tuesday 28th March 1950, killing nine passengers. [2058] The train had been coming from the town of Sagaing. Dynamite placed on the track blew up the engine and derailed the seven leading coaches. The attackers then fired on the train from behind bushes. Afterwards they set it ablaze. A relief train was sent to the scene and all other train services were suspended. [2059]

[2056] Ibid, 17th August 1950
[2057] Ibid, 18th August 1950
[2058] *Malaya Tribune*, Singapore, 29th March 1950
[2059] *Indian Daily Mail*, Singapore, 29th March 1950

Communist rebels derailed a Prome – Rangoon passenger and mail train on a bridge near Rangoon, Burma, on or about Wednesday 26th July 1950. They fired on people trying to flee the overturned coaches and captured several guards. The Government announcement of the attack did not state the number of casualties. [2060]

Nationalist guerrillas derailed a passenger train near Canton, China, on Sunday 16th April 1950, killing the driver, fireman and several passengers. [2061]

Anti-Red (Chinese Nationalist) guerrillas derailed a Communist troop train near the Honan-Hubei border, China, on Monday 17th July 1950. The explosion, which could be heard five kilometres away, caused heavy casualties to General Lin Piao's Fourth Field Army from Hankow. [2062]

A train carrying 2,000 soldiers of the Communist Fourth Field Army to a northern destination had just left Kukong, China, on Sunday 22nd October 1950, and was entering a valley when a mine exploded beneath it. Three coaches crammed with soldiers, ammunition, food, signal equipment and oil were overturned. A 40-man guard was immediately thrown around the wrecked coaches and rescue work was begun. As this was in progress Chinese Nationalists suddenly appeared on the higher ground overlooking the line, poured fire onto the train and reportedly killed all those guarding the train, then seized as much ammunition, arms, and food as they could carry, and withdrew, leaving a rear guard to fend off any pursuers. [2063]

'Rebels' attacked a train near Nha Trang, Vietnam, French Indochina, on Saturday 22nd April 1950. The armed guard of the train, with aircraft assistance, beat off the attackers after an hour. Six people died and eight were wounded. [2064]

On about Wednesday 25th October 1950 'a strong force of guerrillas' derailed and then attacked a train at My Chank, Vietnam, French Indochina. No further details were given. [2065]

The derailment of the leading wheels of a two-unit diesel engine hauling a 59-car New York Central Railroad freight train at the Collinwood Yard in Cleveland, Ohio, United States, on Friday 12th May 1950, was clearly related to a nationwide strike of engineers and firemen over pay and conditions. A piece of wood was wedged in a switch to keep it open. The strike was not settled until May 1952. [2066]

Radio Pyongyang announced that their forces had derailed an American military train between Fousan and Ulsan, South Korea, on Monday 21st August 1950. [2067]

[2060] *The Singapore Free Press*, Singapore, 27th July 1950

[2061] *The Pampa Daily News*, Pampa, Texas, United States, 20th April 1950

[2062] *The Evening Star*, Washington, D.C., United States, 24th July 1950

[2063] *The Singapore Free Press*, Singapore, 30th October 1950

[2064] *The Times*, London, England, 24th April 1950

[2065] *Chester Times*, Chester, Pennsylvania, United States, 26th October 1950

[2066] The *Sandusky Register Star News*, Sandusky, Ohio, United States, 12th May 1950

[2067] *Ce Soir*, Paris, France, 27th August 1950

The engine and three coaches of a Dhaka – Chittagong express were derailed as the train approached a bridge between Daulakandi and Bhairabazar, East Bengal, East Pakistan on Saturday 9th September 1950. The engine and carriages fell into 10 feet of water. Fishplates and bolts were recovered by police from the river, which it appeared saboteurs had removed from the track 40 feet from the bridge. Initially it was reported that 50 people had been killed and the same number injured. [2068] An official statement a couple of days later indicated that 37 people had died and 70 had been injured. [2069]

A diesel passenger train bound for Keren ploughed into a stone barricade erected by bandits across the track at a point 29 miles from Asmara, Ethiopia, on Friday 6th October 1950. Bandits in hiding then fired on the train and threw grenades at it. Stones in the barricade pierced the oil sump of the train forcing it to stop further along the line. A British soldier of the Royal Berkshire Regiment was killed in the attack. Troop reinforcements were sent to the location by road and rail to hunt for the bandits. [2070]

An armoured train was dynamited by the Communists between Rangoon and Toungoo, Burma, on Saturday 6th January 1951. The engine and one waggon were destroyed. [2071]

On Monday 8th January 1951 another armoured train piloting a Rangoon – Prome express was wrecked when a 50-yard bridge 57 miles north west of Rangoon was dynamited and it fell into the river. The number of casualties was not immediately known. [2072]

Communist rebels blew up an armoured train 13 miles north of Rangoon, Burma, on or about 20th February 1951. Six of the attackers were killed by police as they withdrew. [2073]

The crew of a ballast train were injured when it was derailed by a mine which had been laid on the track between Sagaing and Shwes, Burma, on Sunday 25th March 1951. The saboteurs withdrew after firing shots at the train. In a separate attack the same day the engine and two coaches of a passenger train from Pyinmana bound for Rangoon were derailed; the rails and sleepers had been removed. No one was hurt in this attack. [2074]

A passenger train descending a long incline to Mandalay was dynamited by rebels at a point 12 miles south of Maymyo, Burma, on Sunday 22nd April 1951. Men and women dived for cover as the attackers fired on the derailed coaches. The train crew, one of whom was injured, were captured by the rebels who also seized the telegraph equipment from a nearby station. [2075]

Mines laid by the Communists derailed a passenger train near Mandalay, Burma, killed eight and wounded ten on Monday 9th July 1951. After the derailment the attackers looted valuables from the passengers. Train services between Mandalay and Bamyo, 40

[2068] *The Times*, London, England, 11th September 1950

[2069] *The Bradford Era*, Bradford, Pennsylvania, United States, 13th September 1950

[2070] *The Times*, London, England, 7th October 1950

[2071] *The Straits Times*, Singapore, 8th January 1951

[2072] Ibid, 10th January 1951

[2073] *The Singapore Free Press*, Singapore, 21st February 1951

[2074] *Singapore Standard*, Singapore, 26th March 1951

[2075] *The Daily Oklahoman*, Oklahoma City, Oklahoma, United States, 23rd April 1951 miles away, were stopped after the attack. [2076]

A train from Mentakab to Gemas was derailed between Kemayan and Ayer Itam, Malaya, on Sunday 21st January 1951. No one was hurt. [2077]

A train was derailed near Kluang, Malaya, on Sunday 25th February 1951, blocking the line between Gemas and Kluang. Whether the derailment resulted from sabotage is unclear. [2078]

The engine and five waggons of the night goods train from Singapore to Kuala Lumpur were derailed by sabotage between Paloh and Bekok in Johore State, Malaya on Tuesday 27th March 1951. Four of the waggons were empty and the fifth contained a police escort, only one of whom was slightly hurt. [2079]

The night mail train from Kuala Lumpur to Penang was derailed by sabotage in Perak State, Malaya, on Wednesday 28th March 1951. Passengers, mails, and parcels were transhipped. [2080]

The night mail train from Singapore to Kuala Lumpur was derailed near Kluang, Malaya, in the early hours of Wednesday 11th April 1951. [2081]

The 5.30 a.m. Down goods train from Prai to Kuala Lumpur was derailed between Simpang Ampat and Sungei Bakap in Wellesley State, Malaya, on the morning of Thursday 12th April 1951, the engine and 13 waggons leaving the track. A bracket had been removed from the track and sabotage was suspected. As the train had been moving at low speed the remainder of its 37 waggons remained on the rails. The night mail from Kuala Lumpur to Penang was terminated at Nebong Tebal and six buses took the passengers the 22 miles to Prai. [2082]

The pilot train was derailed early on Monday 16th April 1951 between Tampin and Tebong, Malaya, on the line between Kuala Lumpur and Singapore. No one was hurt but the line was blocked. [2083]

A mixed train was derailed by sabotage in Mentakab, Malaya, on Monday 30th April 1951. No one was hurt. [2084]

The 8.30 p.m. Up train from Johore Bahru to Kuala Lumpur was derailed by sabotage on Wednesday 9th May 1951 between Senai and Kulai, Malaya. Some military personnel and goods were being carried by the train. An explosion threw four goods waggons off the track and damaged about 120 yards of the permanent way. No one was hurt. The 9.20 p.m. train from Johore Bahru was stopped at Senai and returned to Singapore and the next day's day mail was cancelled between Gemas and Singapore. [2085]

The Kuala Lumpur to Penang day mail was derailed between Alor Pongsu and Bagan Serai, Malaya, on Thursday 17th May 1951. No one was hurt. The cause of the derailment

---

2076 *The Straits Times*, Singapore, 10th July 1951
2077 *The Singapore Free Press*, Singapore, 22nd January 1951
2078 *Singapore Standard*, Singapore, 27th February 1951
2079 *The Singapore Free Press*, Singapore, 28th March 1951
2080 *Singapore Standard*, Singapore, 30th March 1951
2081 *The Singapore Free Press*, Singapore, 11th April 1951
2082 *The Straits Times*, Singapore, 13th April 1951
2083 Ibid, 16th April 1951
2084 *Singapore Standard*, Singapore, 1st May 1951
2085 Ibis, 11th May 1951

was unknown.[2086]

A goods train was derailed between Kulai and Sedenak, Malaya, on Tuesday 5th June 1951, four lengths of rail having been removed. No one was hurt, though the engine overturned and two waggons left the track. Day mail trains between Gemas and Singapore did not run as a result.[2087]

The passenger train from Kemayan to Gemas, Malaya, was derailed by sabotage on Thursday 7th June 1951 and the driver and two firemen were slightly hurt.

Three trains were derailed due to sabotage in Malaya on Friday 8th June 1951. No one was hurt in the derailment of the day mail train from Gemas in the southern part of Negri Sembilan State in the afternoon. About five miles north of Gemas, the removal of three lengths of rail derailed the engine and seven or eight coaches of the day mail from Singapore to Kuala Lumpur. None of the 20 passengers or the engine crew was hurt. Finally, three crash waggons of an armoured train from Bahau, approaching the site of the latter derailment from the south, were also derailed.[2088] [2089]

The engine and two coaches of a passenger train were derailed at Kulai, Malaya, early on Tuesday 12th June 1951.[2090] Later the same day the railway was dynamited and 40 'bandits' opened fire on a railway Jeep and trailer carrying a party of telecoms workers with a police escort near Kluang, Johore, Malaya.[2091] The Jeep blew up and caught fire and Police Constable Ghazami bin Yup, Extra P.C.s Abdul Thani bin Mohamed and Abdul Waha bin Ali; Linesmen Salat and Rahem and Junior Technical Assistant A. Kathirasu were killed, while P.C. Mahedi, E.P.C. Rabas, Special P.C. Hassam, Linesmen Velusamy and Ramasamy and Shanmugam, the driver of the Jeep, were wounded. The wounded were taken to hospital.[2092]

The Singapore – Kuala Lumpur train was again derailed, presumably by sabotage, early on Friday 15th June 1951, at Kluang, Johore, Malaya. No one was hurt.[2093] On the evening of the same day a goods train was derailed by sabotage between Ipoh and Tanjong Malim, Malaya.[2094]

The engine and four crash waggons of the Down train from Kuala Krau to Gemas were derailed by terrorists in the Mentakab area of Pahang State, Malaya, on Tuesday 19th June 1951. Shots were then fired at the train from the side of the line for 15 minutes. No one was hurt.[2095]

The mail van and seven passenger coaches, but not the engine of a train, were derailed on the morning of Sunday 24th June 1951 at Ayer Kuning, Malaya, near Gemas. No one was hurt and the cause of the derailment was being investigated.[2096]

'Bandits' derailed a goods train between Rangam and Layang Layang, Malaya, at 10

[2086] Ibid, 19th May 1951
[2087] Ibid, 7th June 1951
[2088] Ibid, 9th June 1951
[2089] *The Straits Times*, Singapore, 10th June 1951
[2090] *Singapore Standard*, Singapore, 13th June 1951
[2091] *The Times*, London, England, 14th June 1951
[2092] *Singapore Standard*, Singapore, 14th June 1951
[2093] *The Straits Times*, Singapore, 16th June 1951
[2094] Ibid, 17th June 1951
[2095] *The Singapore Free Press*, Singapore, 20th June 1951
[2096] *Singapore Standard*, Singapore, 25th June 1951

p.m. on Wednesday 27th June 1951. The engine and two waggons were derailed, blocking the line. [2097] The engine was sent 20 or 30 feet down a ravine and the three engine crew were injured. [2098]

The Singapore – Kuala Lumpur night mixed train was derailed between Kulai and Gemas, Malaya, on the night of Friday 29th June 1951, several fishplates having been removed from the track. [2099]

The engine and four crash waggons of a passenger train were derailed at Bahau, Malaya, in Negri Sembilan state, on Thursday 2nd August 1951. No one was hurt. Sabotage was confirmed as the cause of the accident. [2100]

Two crash waggons attached to a pilot engine were derailed by sabotage north of Sedenak, Malaya, on the night of Monday 27th August 1951. The train from Kuala Lumpur was thus delayed 6½ hours. [2101]

An Up goods train was derailed between Tapah Road and Temoh, Malaya, in Perak State on the night of Wednesday 24th October 1951. Only the engine and leading waggon left the track. The day mails between Ipoh and Tapah Road were cancelled. [2102]

A pilot engine and two goods waggons of a Down goods train was derailed by sabotage near Johore Bahru, Malaya, on Sunday 28th October 1951. The day mails between Gemas and Singapore were cancelled on Monday. [2103] [2104]

The engine and two crash waggons of the Up pilot train from Kuala Lumpur were derailed just after midnight on Wednesday 31st October 1951 between Trolak and Slim River, Malaya. No one was hurt. [2105] Another train appears to have been derailed at about the same time in Selangor State and Royal Air Force aircraft provided protection while the rescue operation was in progress. [2106]

The Down train from Prai to Kuala Lumpur was derailed by sabotage on the night of Sunday 4th November 1951 between Tanjong Malim and Kuala Lumpur, Malaya. The engine remained on the track but six waggons ran off and some went down an embankment. The day and night mail trains were cancelled between Kuala Lumpur and Tanjong Malim as a result. [2107]

Two crash waggons, the engine, tender, and two armoured vehicles of the Triang armoured train were derailed by sabotage near Triang, Malaya, on the afternoon of Tuesday 6th November 1951. [2108]

The Penang to Kuala Lumpur night mail was derailed by sabotage at 2 a.m. on Sunday 11th November 1951 a mile and a half from Temoh, Malaya. The saboteurs let the pilot train go through and removed two sections of rail in the seven-minute interval between it

---

[2097] *The Singapore Free Press*, Singapore, 28th June 1950
[2098] *Singapore Standard*, Singapore, 29th June 1951
[2099] *The Straits Times*, Singapore, 1st July 1951
[2100] *Singapore Standard*, Singapore, 4th August 1951
[2101] *The Straits Times*, Singapore, 29th August 1951
[2102] *Singapore Standard*, Singapore, 26th October 1951
[2103] Ibid, 30th October 1951
[2104] *The Singapore Free Press*, Singapore, 30th October 1950
[2105] Ibid, 31st October 1950
[2106] *Singapore Standard*, Singapore, 1st November 1951
[2107] Ibid, 6th November 1951
[2108] Ibid, 8th November 1951

and the main train. The engine, three coaches and two goods waggons were wrecked and the coaches telescoped one another. A Malay police lance corporal and an Indian special constable were killed and the driver, fireman and five passengers were injured. The train had been packed with pilgrims returning home after visiting Mecca. Following the derailment heavily guarded gangs worked all day to clear the line, but they had not finished doing so by Sunday evening. [2109]

Among the 300 passengers on the day mail from Singapore to Kuala Lumpur on Wednesday 14th November 1951 were the ruler of Negri Sembilan State, the Yang di Pertuan Besar, his consort, two grandchildren and other members of his party, who were returning from a trip to the United Kingdom. Riding in a special carriage at the rear of the train, the Royal party escaped when the train was derailed by sabotage and fired on about four miles from Tampin, Malaya. They were conveyed from the scene by the police and military. Less fortunate were Driver K. Vethivelu, another railwayman known as a scavenger (?) and two child passengers, a Malay boy and girl who were all killed. The second driver was slightly hurt and the two firemen were badly hurt. Initial reports stated that the engine and three coaches were derailed.

Father Abraham, a Syrian Christian priest, said, "I suddenly heard two shots being fired and there was a crash. It was my first experience of being involved in an ambush though I have been travelling up and down the country during the past few years in various types of vehicles."

A railway official stated that the engine and three coaches had telescoped and four coaches were derailed. Apart from some children who had begun screaming, the passengers had lain flat and remained calm until the arrival of the relief train. The official said that seven passengers had been wounded, four seriously. [2110]

Tungu Mustapha, a cousin of Yang di Pertuan, told a reporter, "I was seated with the rest of the Sultan's party in the special coach at the end of the train when three miles out of Tampin and one mile before Tebong I heard four heavy bursts of fire followed immediately by an explosion. This was at about four p.m. Everyone in the carriage fell onto the floor of the carriage and crouched there. Then the carriage was thrown violently against the next carriage and the train jolted several times before coming to a stop. We found out later that the engine and several of the leading coaches had jumped the rails and were lying on their sides. There were dreadful moments when women and children were screaming and the sound of crockery and glasses being smashed rent the air."

Twenty minutes before the derailment, 10 to 15 'bandits' armed with Sten guns and rifles had tied up 10 track labourers to rubber trees. They had taken the "keys" from the *mandor* (Malay for foreman) and removed about ten yards of rail and fishplates. They released the labourers after the derailment. [2111]

The pilot train of the night mail from Prai to Kuala Lumpur was derailed by sabotage between Malim Nawar and Kampar, Malaya, on the night of Tuesday 20th November 1951. The driver fired his Verey pistol and the mail train was able to pull up in time. Railway officials boarded the pilot train of the corresponding Kuala Lumpur to Prai night mail to get to the scene and that train was derailed in turn between Temoh and Kampar, not far from the scene of the attack on the 'pilgrim train' 11 days earlier. No one was hurt

---

2109 *The Straits Times*, Singapore, 12th November 1951
2110 *Singapore Standard*, Singapore, 15th November 1951
2111 *The Straits Times*, Singapore, 15th November 1951

in either derailment. [2112]

The driver and fireman of a northbound goods train from Kuala Lumpur were slightly hurt when their engine was derailed by sabotage and fell down an embankment between Bidor and Tapah Road, Malaya, on the night of Sunday 2nd December 1951. Several hours' delay were caused and passengers and mails on the day and night mail trains had to be transhipped. [2113]

Two carriages of the southbound mail train were derailed by 'bandits' in the Tanjong Malim area of Selangor State, Malaya, on Tuesday 11th December 1951. [2114]

A northbound goods train was derailed on the afternoon of Saturday 15th December 1951 in the Tanjong Malim area, Malaya. No one was hurt.

The engine and tool van of the Singapore – Kuala Lumpur train were derailed on the morning of Sunday 16th December 1951 between Tenang and Labis in Johore state, Malaya. Again no one was hurt. [2115]

On Saturday 29th December 1951 the East Coast mixed train from Kuala Krau in Pahang state to Gemas in Negri Sembilan state was derailed by sabotage in Negri Sembilan State, Malaya. One wheel of the engine and two crash waggons left the rails. The fireman twisted his ankle. [2116]

Nationalist Chinese sources stated that their guerrillas had derailed a Chinese Communist troop train carrying reinforcements to Korea on Tuesday 12th June 1951 at a location 220 kilometres north of Hankow, China, adding that more than 1,000 'Reds' had fled the wrecked train and joined the guerrillas and another 200 had been killed or injured. [2117]

Anti-Communist guerrillas blew up the engine and two vans of a train in China near the border with Hong Kong on Monday 24th December 1951. [2118]

Vietminh guerrillas laid a mine which derailed the engine and seven waggons of a goods train probably on Monday 17th September 1951. Ten people were injured in the derailment and subsequent machine gun attack about 40 kilometres north of Saigon, Vietnam, French Indochina. The guerrillas maintained an all-night attack on the derailed train and the French sent in troops and fighter planes, the latter dropping napalm on the guerrillas. The guerrillas were eventually driven off, leaving a reported 100 dead, while the army lost eight dead and ten injured. [2119]

On Sunday 30th December 1951 a train between Haiphong and Hanoi was derailed about 15 kilometres from Haiphong, Vietnam, French Indochina, when it ran over a mine. The engine and several carriages were derailed and two people were hurt. [2120]

2112 Ibid, 22nd November 1951
2113 *Singapore Standard*, Singapore, 4th December 1951
2114 *The Straits Times*, Singapore, 13th December 1951
2115 Ibid, 17th December 1951
2116 *The Straits Times*, Singapore, 31st December 1951
2117 *The Daily Herald*, Biloxi, Mississippi, United States, 22nd June 1951
2118 *The Singapore Free Press*, Singapore, 27th December 1951
2119 Ibid, 18th September 1951
2120 Ibid, 31st December 1951

The engine and five waggons of a large military train were derailed at 6.30 a.m. on Saturday 15th December 1951 between Adabya Port and Suez, Egypt. The next day they were still embedded in sand and water but salvage work was in progress. Sabotage was suspected, especially as news of the accident had appeared the same day in Cairo newspapers, which normally went to press at 2 a.m., four and a half hours before it had occurred. [2121]

The night mail train from Penang to Kuala Lumpur was derailed in the early hours of Friday 11th January 1952 near Tapah, Malaya, in Perak State. [2122]

A pilot train was derailed in Perak State, Malaya, on the night of Wednesday 16th January 1952, delaying the arrival of the Up and Down Kuala Lumpur – Penang night mail trains. [2123]

Early on the morning of Thursday 17th January 1952 the pilot train preceding the Kuala Lumpur – Penang night mail was derailed by sabotage between Temoh and Kampar, Malaya, in Perak State. The mail train was following too close behind and ran into the wreckage before it could be stopped. No one was hurt. [2124]

The night mail trains between Kuala Lumpur and Singapore were both attacked on the night of Saturday 26th January 1952. The pilot train of the northbound mail was derailed in Johore State, Malaya, four wheels of the leading crash waggon leaving the track. No one was hurt. In Malacca State the southbound train was fired upon and its escort fired back. A Malay soldier jumping for cover was injured. [2125]

What was stated to be the worst bandit attack so far on a train in Malaya occurred on Monday 3rd March 1952 when the Up mixed train from Gemas to Mentakab was derailed and fired upon in the Bahau district, three miles from Rompin. One civilian was killed and four railwaymen were missing, believed dead, in the initial derailment, and two civilians lost their lives in a half-hour shoot-out between the attackers, who numbered about 40, and the police escort of the train. Fifteen passengers were injured. Those who died included a Malay relief clerk and an Indian labourer and telecoms clerk. The train was crossing a bridge at Batu Belang when the mine exploded under a refreshment coach, which was blown in half. The engine and leading vehicle travelled some distance beyond the bridge, pulling half of the wrecked coach. Behind the rear half of refreshment coach, which toppled over, was the armoured waggon, the roof of which was ripped off when it ran into the broken end of the bridge. This saved the police escort in it from being trapped and enabled them to open fire on the attackers. The escort was credited with preventing the 'bandits' from looting the train. After ten minutes Driver S. Govindan crept back, uncoupled the front of his train from the wreckage and proceeded to Rompin station for reinforcements. Troops and police went to the scene to find that the attackers had fled. Meanwhile the wounded had been treated by two soldiers of the Royal Electrical and Mechanical Engineers and a Gurkha medical officer was on the relief train from Rompin.

[2121] *The Times*, London, England, 17th December 1951
[2122] *Indian Daily Mail*, Singapore, 12th January 1952
[2123] Ibid, 17th January 1952
[2124] Ibid, 18th January 1952
[2125] Ibid, 29th January 1952

The wounded were taken to Seremban and Kuala Pilah Hospitals. [2126]

A pilot train was derailed due to sabotage between Gemas and Paloh, Malaya, early on Saturday 15th March 1952, the wheels of one of the crash waggons leaving the rails. No one was hurt. [2127] [2128]

The pilot train of the Prai – Kuala Lumpur night mail was derailed between Malim Nawar and Kampar, Malaya, at 1.45 a.m. on Thursday 1st May 1952. A fishplate had been removed and the crash waggon, engine and tender but not the 'pill box' left the rails. No one was hurt and the crew of the pilot train had time to warn the following night mail. [2129]

The engine of a pilot train of the southbound night mail were derailed and went down an embankment between Malim Nawar and Kota Bahru at 1 a.m. on Thursday 8th May 1952, bolts and fishplates having been removed from the track. [2130] The engine landed in a marsh 15 feet below the track. The three crash waggons and 'pill box' were also derailed. Driver "Fordy" Kelaart and the two firemen had a narrow escape and were able to stop the mail train. [2131]

The southbound passenger train from Penang was derailed by terrorists near Temoh, Malaya, in Perak State at 2 a.m. on Sunday 11th May 1952. The two rear coaches were derailed but no one was hurt. The train resumed its journey after several hours and arrived in Kuala Lumpur at 11 a.m.[2132] In what was apparently a new approach the terrorists had let the pilot train pass and then felled a tree across the line. Seeing the obstruction the driver had reversed the train but the terrorists had cut a section of the line and the rear of the train was then derailed. Some shots were fired at the train, soldiers on board fired back and the terrorists withdrew. [2133]

Early on Monday 12th May 1952 terrorists derailed a goods train in south Perak State, Malaya, delaying the night mail from Penang to Kuala Lumpur for the second night in succession. [2134]

At about 2 a.m. on Thursday 15th May 1952 the pilot train from Kuala Lumpur to Ipoh was derailed between Kampar and Malim Nawar, Malaya, several lengths of rail having been removed. Trains were run to Kampar and Ipoh and passengers were transhipped. [2135]

The pilot train preceding the northbound express was derailed by sabotage in the Slim River area of Perak State, Malaya, early on Sunday 1st June 1952. [2136]

Ten armed, uniformed terrorists derailed a goods train from Kuala Lumpur to Ipoh between Bidor and Tapah Road, Malaya, in Perak State, on Monday 30th June 1952. They then held up the crew of three, while they looted tools from the train. [2137]

Three people, the driver, fireman, and a police wireless operator, were hurt, one

---

2126 *The Straits Times*, Singapore, 4th March 1952
2127 *Sunday Standard*, Singapore, 16th March 1952
2128 *The Straits Times*, Singapore, 17th March 1952
2129 Ibid, 2nd May 1952
2130 *Singapore Standard*, Singapore, 9th May 1952
2131 *The Straits Times*, Singapore, 9th May 1952
2132 *Singapore Standard*, Singapore, 12th May 1952
2133 *The Straits Times*, Singapore, 12th May 1952
2134 Ibid, 13th May 1952
2135 *Singapore Standard*, Singapore, 16th May 1952
2136 *The Straits Times*, Singapore, 2nd June 1952
2137 *Singapore Standard*, Singapore, 1st July 1952

seriously, when the engine and two coaches were derailed by sabotage in the Kuala Krai district of Kelantan State, Malaya, on Friday 1st August 1952. [2138]

A train was derailed by sabotage between Gemas and Paloh, Malaya, in Johore State, on Sunday 3rd August 1952, disrupting train services. [2139]

A mine exploded beneath a train as it was crossing a bridge near Phnom Penh, Cambodia, French Indochina, on Wednesday 16th January 1952. The bridge collapsed and the engine and six carriages fell into the water. Twenty passengers were reportedly injured. [2140]

Thirteen kilometres from Saigon, Vietnam, French Indochina, a passenger train to Mỹ Tho ran over a mine on Saturday 26th January 1952. Two people died and seven were injured. [2141]

On about Thursday 6th March 1952 a train on the Hanoi – Haiphong line was blown up by a mine near Haiphong, Vietnam, French Indochina. [2142]

Eighty-five people reportedly died and 23 were injured when Communist saboteurs derailed a passenger train near Kompong Lat, Cambodia, French Indochina, 30 miles north west of Phnom Penh, on Friday 7th March 1952. Fifteen coaches of the train fell from a bridge into a ravine. Those who died were all Cambodians or Vietnamese. Communist rebels belonging to the Issarak tribe were blamed. A large section of track on a high bridge between Phnom Penh and Battambeng had been torn up and several mines were also laid on the track. [2143] A further official report, however, stated that the true number of dead was only 8, not 85, with 26 injured, and that the higher figure was a result of a transmission error. [2144]

A French armoured train was derailed by a mine and attacked by Vietminh guerrillas either late on Sunday 23rd or early on Monday 24th March 1952 about 100 kilometres north-east of Saigon on the line to Nha Trang, Vietnam, French Indochina. The troops fought off the attack and a second armoured train was sent to rescue them. Six soldiers were killed and another six injured according to one report. [2145] [2146] [2147]

The Vietminh mined a maintenance train near Saigon, Vietnam, French Indochina, on Monday 5th May 1952. [2148]

Seven civilian and five soldiers were killed and 29 people were wounded when a convoy of trains exploded mines between Trang Bom and An Loc, Vietnam, French Indochina, on Tuesday 10th June 1952. It was estimated that the Vietminh had lost 20 dead, including two officers, and suffered about 60 wounded in the fighting which had followed the attack; most of their wounded they took with them as they withdrew. The

---

[2138] Ibid, 2nd August 1952
[2139] *The Straits Times*, Singapore, 4th August 1952
[2140] *Singapore Standard*, Singapore, 17th January 1952
[2141] *The Straits Times*, Singapore, 26th January 1952
[2142] Ibid, 7th March 1952
[2143] *Singapore Standard*, Singapore, 11th March 1952
[2144] *The Straits Times*, Singapore, 12th March 1952
[2145] *The Albuquerque Tribune*, Albuquerque, New Mexico, United States, 24th March 1952
[2146] *The Eau Claire Leader*, Eau Claire, Wisconsin, United States, 25th March 1952
[2147] *The Straits Times*, Singapore, 25th March 1952
[2148] Ibid, 7th May 1952

convoy consisted of five trains, the first of which was an armoured one and the second a mixed train. These were allowed to pass before the mines were exploded remotely under the third and fourth trains. Four carriages were destroyed and four more were derailed. The terrorists then attacked and the passengers fled into the jungle. Troops on the train opened fire on the attackers and were supported by the front armoured train and a second armoured train, the last of the five trains in the convoy. Half an hour after the attack further support came from fighter aircraft and local defence forces, whereupon the Vietminh withdrew.

A newspaper correspondent, writing from Paris, France, commented that, 'The convoy was maintaining the service between the capital and the rubber plantations of north-eastern Cochin-China. The attack shows that however weakened the rebels may have been through their recent defeat in the region to the north-west of Saigon, they are still able to wage operations of some scale. Observers conclude that they are as anxious to win the battle for rubber in the south as they are to win that for rice in Tongking.' [2149]

A train was derailed by a mine near Dian station, Vietnam, French Indochina, 24 kilometres north-east of Saigon, on Friday 13th June 1952. [2150]

In their third attack within a week the Vietminh mined a French supply train on Sunday 15th June 1952 about 65 kilometres west of Saigon, Vietnam, French Indochina, killing two soldiers and wounding several others. A co-ordinated attack on supply columns was theorised. [2151]

Vietminh mines derailed a train from Saigon bound for Nha Trang, Vietnam, French Indochina, on Thursday 18th September 1952. The attackers then fired on the train with machine guns. The sleeping car, restaurant car and five other vehicles were derailed and three passengers were hurt. [2152]

On about Thursday 4th December 1952 a mine exploded beneath a passenger train at a location about 20 kilometres north of Saigon, Vietnam, French Indochina. One soldier of those escorting the train died of his wounds en route to hospital and three others were seriously hurt. [2153]

A train on the Saigon – Loc Ninh line, Vietnam, French Indochina, was derailed by the explosion of a mine on about Monday 29th December 1952 and about 100 Vietminh rebels then opened what was described as a withering fire on the train. Twenty-three Vietnamese civilians, 12 soldiers and an officer were killed, four civilians and two soldiers were missing and eight civilians and 25 soldiers were wounded. [2154]

A 15-waggon freight train was derailed on the night of Monday 21st January 1952 between Sousse and Sfax, Tunisia. No one was hurt. The French authorities stated that saboteurs had loosened the rails. The line was expected to be blocked for some time. [2155]

Two further freight trains were derailed in Tunisia on Friday 25th January 1952. Early in the morning a slow train was thrown off the line at Depienne, about 30 kilometres west

[2149] *The Times*, London, England, 12th June 1952
[2150] *Singapore Standard*, Singapore, 14th June 1952
[2151] *The Straits Times*, Singapore, 16th June 1952
[2152] Ibid, 19th September 1952
[2153] Ibid, 5th December 1952
[2154] Ibid, 2nd January 1953
[2155] *Beatrice Daily Sun*, Beatrice, Nebraska, 22nd January 1952

of Tunis, and its driver was killed. The other derailment was at Djedeida, 25 kilometres west of Tunis. It was believed that the derailments might have been carried out with the aim of delaying French reinforcements in the protectorate. [2156]

An empty coaching stock train was derailed between Sousse and Sbeitla, Tunisia, by Destour Nationalists on Saturday 16th February 1952. Six railwaymen were hurt. [2157] [2158]

Six railwaymen were seriously injured on Sunday 2nd March 1952 when a train consisting of 36 waggons of ore from Djerissa Mine to Tunis was derailed by sabotage between Gafour and Sidi Ayed stations, Tunisia. The driver suffered a fractured skull and later died, the fireman's arm was crushed and the other four men, three Europeans and a Tunisian, suffered severe internal bruising. The train fell 25 feet into a ravine. A length of rail had been unbolted. Four railwaymen who had been dismissed by the railway company earlier were arrested. [2159] [2160] [2161]

A bomb exploded on a crowded passenger train at Gabes, Tunisia, on Thursday 13th March 1952, killing three Tunisians and two Europeans and wounding 17 people. [2162]

A British military train was blown up by a mine near Kantara, Egypt, on the afternoon of Monday 28th January 1952. Four British other ranks were slightly hurt and 20 yards of track were destroyed. Soldiers riding on the train as an escort fired on Egyptians seen near the line, killing one and it was thought wounding three others. [2163]

A southbound British War Department oil train from Port Said to Ismailia was derailed by two powerful mines roughly halfway between the two places at El Kap, Egypt, six miles north of Kantara, on Friday 15th February 1952. The lines left craters 5 feet deep with diameters of about 70 feet. The engine tender was derailed and six tankers were "completely wrecked." Some soldiers suffered scratches. Egyptian terrorists were suspected. [2164]

A train from Aranya Prades on the border with Cambodia, French Indochina, to Bangkok, Thailand, was derailed on about 26th April 1952, killing several passengers. At the end of the month the Siamese Railway Organisation stated that it was certain that derailment had been due to sabotage. [2165]

Saboteurs tore up the track and derailed a passenger train of the Régie des chemins de fer Abidjan – Niger on about Sunday 1st June 1952 near Banfora, French West Africa. Seven 'natives' were killed and six injured. [2166]

2156 *The Racine Journal-Times*, Racine, Wisconsin, United States, 25th January 1952
2157 *The Singapore Free Press*, Singapore, 18th January 1952
2158 *The Straits Times*, Singapore, 19th January 1952
2159 *The Times*, London, England, 3rd March 1952
2160 *The Singapore Free Press*, Singapore, 3rd March 1952
2161 *Indian Daily Mail*, Singapore, 15th March 1952
2162 *The Straits Times*, Singapore, 14th March 1952
2163 *The Times*, London, England, 29th January 1952
2164 Ibid, 16th February 1952
2165 *The Straits Times*, Singapore, 30th April 1952
2166 *El-Paso Herald Post*, El Paso, Texas, United States, 2nd June 1952

On about Thursday 19th June 1952 Karen rebels derailed a train 80 miles north of Rangoon, Burma, killing seven people and injuring 11 others. After the derailment about 50 armed rebels looted the passengers' belongings. [2167]

Communist rebels mined an army patrol train near Thazi Junction, Burma, on or about Wednesday 9th July 1952, then attacked it, killing one soldier and wounding nine in a 90-minute gun battle. [2168]

On or about Friday 11th July over 100 rebels attacked a passenger train near Moulmein, Burma, killing one person and injuring seven. The engine was damaged but the train crew managed to drive the train another half a mile and government patrols saved the 300 passengers. [2169]

On or about Sunday 24th August 1952 Communists derailed a passenger train near Pegu, Burma, dynamiting the track. [2170]

On about Monday 10th November 1952 rebels derailed and looted at passenger train at Htongyi, Burma, 31 miles east of Rangoon. Eight people were reportedly killed and 20 injured while the rebels seized three army patrols as hostages. Armoured cars were sent to the scene. [2171]

Forty-six people were reportedly killed, one of them an American soldier, when 'bandits' attacked a Korean National Railways train near Sagari, South Korea, west of Pusan, at 2.44 p.m. on Tuesday 24th June 1952. A mine derailed the engine of the train and eight coaches caught fire or were set on fire by the attackers. Apart from the one American the dead, all Koreans, comprised 21 South Korean soldiers, 13 civilians, 7 policemen, and the driver, conductor, baggageman and a train fire marshal. Two American soldiers, a Korean soldier and a Korean civilian were wounded. Another 80 passengers were taken hostage. [2172]

A further report stated that 75 bodies had been found at the scene of the derailment and a further 16 of those taken hostage had been killed by the guerrillas during the police pursuit of them. Another 80 persons had been wounded; this report added. [2173]

Darul Islam guerrillas derailed and plundered an express train on the line from Bandung to Yogjakarta on Saturday 26th July 1952. A guerrilla band of between 30 and 40 had ambushed the train in mountainous territory south-east of Bandung, between Malangbong and Warungbandrek, Indonesia, damaging the track. The locomotive – a large, newly arrived Krupp engine, luggage van and two third-class carriages overturned and were wrecked. Initial reports spoke of eight dead and a large number of wounded. The train lacked a troop escort. [2174] [2175] (Darul Islam was a terrorist group, which having fought the Dutch, continued to fight the Indonesian government after independence at the

[2167] *The Straits Times*, Singapore, 21st June 1952
[2168] Ibid, 10th July 1952
[2169] Ibid, 12th Jul 1952
[2170] Ibid, 25th August 1952
[2171] Ibid, 11th November 1952
[2172] *Pacific Stars and Stripes*, Tokyo, Japan, 25th June 1952
[2173] Ibid, 30th June 1952
[2174] *Indian Daily Mail*, Singapore, 28th July 1952
[2175] *Indische Courant boor Nederland*, Amsterdam, the Netherlands, 31st July 1952

end of 1949 in the hope of establishing an Islamic state.)

Part of a goods train from Jerusalem to Haifa was derailed, a viaduct was blown up and a length of track was destroyed by an electric mine of military pattern a mile or so north of Qalqilya, Israel, on Thursday 2nd February 1953. Small arms fire was heard after the derailment. No one was hurt. 'A spokesman of the Israeli Army stated that the attack had been planned and carried out by regular Arab forces. A serious view is taken of the incident here {Tel Aviv}, and it is pointed out that, quite apart from the damage, one effect will be to destroy faith in the railway as a secure means of transport for passengers and goods.' [2176]

A freight train was derailed between Tulkarm and Qalqilya, Israel, close to the border with Jordan, on Wednesday 21st October 1953, 13 of its 52 waggons being thrown from the track when it exploded a land mine. No one was hurt. There were angry accusations of sabotage from the Israeli side. An urgent meeting of the armistice commission formed after the 1948 Arab – Israeli war was requested. [2177] [2178]

A goods train was derailed by sabotage in south Perak State, Malaya, on Sunday 15th February 1953. No one was hurt but a number of trains were cancelled while the line was repaired. [2179]

The Up Nainital Express of the North Eastern Railway was derailed at 9.40 p.m. on Wednesday 29th April 1953 near Baheri, India, 34 miles from Bareilly. The engine, dining car and a third-class bogie overturned, five bogies were derailed and only three remained on the track. The driver, one fireman and a third person, presumably a passenger, were killed and ten people were injured. Medical aid was rushed to the scene while the entire area was cordoned off by the police. The cause of the accident was not known, but the Regional Mechanical Engineer and the Regional Transport Superintendent of the railway, together with the Bareilly Superintendent of Police were examining pieces of loose bolts and nuts at the site of the derailment and a railway officer at Izatnagar stated that tampering with the track was suspected. [2180]

Communist guerrillas mined a French supply train just east of Quang Tri, Vietnam, French Indochina, about 480 kilometres north-east of Saigon, on Friday 8th May 1953, detonating the charge by remote control. After waiting a short time they then attacked the train but were driven off by French artillery from a nearby garrison and fighter aircraft, losing six men. Four vehicles of the train were derailed and 10 Franco-Vietnam soldiers and nine civilians were killed and six soldiers and nine civilians were wounded. [2181]

Land mines laid on a viaduct at the Col des Nuages (Pass of the Clouds), Vietnam, French Indochina, on the line between Hué and Tourane, exploded beneath a passenger

---

2176 *The Times*, London, England, 4th February 1953
2177 *The Times*, London, England, 23rd October 1953
2178 *The Straits Times*, Singapore, 24th October 1953
2179 Ibid, 16th February 1953
2180 *The Indian Express*, Madras, India, 1st May 1953
2181 *Wisconsin State Journal*, Madison, Wisconsin, United States, 10th May 1953

train late on Monday 22nd June 1953. The engine and 18 vehicles fell about 15 metres into a ravine, an 8-metre span of the viaduct giving way. At least 35 people died, possibly as many as 100, and 20 or more were injured. Those who died included at least six soldiers and a number of 'fairly important' civilians. For what it was worth the train had included an armoured vehicle. [2182] [2183] [2184]

Rebels dynamited the track and derailed a passenger train 50 miles east of Rangoon, Burma, on or about Wednesday 20th May 1953. Five were killed, four of them soldiers, and many injured. After the derailment the rebels looted the passengers. [2185]

At least four people were killed and about 40 were taken hostage after 100 Karen rebels mined a passenger train on the line between Moulmein and Thaton, Burma, on Wednesday 28th October 1953. The passengers were also looted of their possessions at gunpoint. [2186] [2187]

A Mandalay – Maymyo passenger train ran over six mines laid by Communist rebels, killing 15 and injuring 23, on Tuesday 17th November 1963. The attack took place near Zibingyi, Burma. Shots were fired at the train after it was derailed but the attackers were driven off by the military police escort after an hour's gun battle. [2188]

A strong force of Karen rebels who were aided by Communist Nationalist guerrillas mined that track between Pegu and Thaton, Burma, on Sunday 22nd November 1953. The armoured train which was preceding a Rangoon – Moulmein passenger train ran over land mines and three trucks were derailed. The passenger train was then set back about half a mile but in the meantime rebels had laid more mines behind the train which it ran over and was derailed in turn. A gun battle then took place between the rebels and the train's escort, during which the passengers were able to escape. They were followed by the escort when their ammunition began to run out. Two police and one civilian were killed and several were wounded. Once the escort had withdrawn the rebels looted the train, set fire to its carriages, and blew up the track. [2189]

Eleven people died and over 30 were wounded when a passenger train was derailed due to sabotage and overturned between Jogjakarta and Bandung, Indonesia, on Sunday 21st June 1953. [2190] The well-armed attackers fired on the derailed carriages. In response to the attack on the following Friday 4,000 railwaymen marched through the centre of Jakarta to the presidential palace demanding better protection for railway workers taking trains through the bandit infested area of West Java. [2191]

The West Berlin *Telegraf* newspaper reported that a Russian military supply train had

---

[2182] *The Times*, London, England, 24th and 25th June 1953
[2183] *The Modesto Bee*, Modesto, California, United States, 24th June 1953
[2184] *Lubbock Morning-Avalanche*, Lubbock, Texas, United States, 24th June 1953
[2185] *The Straits Times*, Singapore, 21st May 1953
[2186] Ibid, 29th October 1953
[2187] *Singapore Standard*, Singapore, 30th October 1953
[2188] *The Times*, London, England, 20th November 1953
[2189] Ibid, 25th November 1953
[2190] *Singapore Standard*, Singapore, 23rd June 1953
[2191] *The Straits Times*, Singapore, 27th June 1953

been blown up by anti-Communist partisans between Toruń and Poznań, Poland, on about 10th July 1953 or possibly a few days earlier. It added that "many" Russian soldiers had been killed. [2192]

A Calais – Basle express was derailed at Berlaimont, France, near Aulnoye, at 22:30 on 22nd August 1953. Normally the train would have been travelling at 100 kilometres an hour at that point but because of a strike it had stopped to pick up passengers at Berlaimont and so was only moving at 40 kilometres an hour. The engine came to rest across both tracks; its tender and three coaches were derailed. The driver and some of the 200 passengers were slightly hurt. Sabotage in connection with the strike was the cause of the derailment, a rail and four sleepers having been tampered with. The police were investigating. [2193]

French Morocco was said to have been on the edge of a religious and political war at the end of August 1953 before Sultan Sidi Mohammed Ben Youssef was deposed and exiled to Corsica. The sultan had been demanding Moroccan independence from the French. He was replaced by the pro-French Sultan Mohammed Ben Aarafa, installed by the French as a puppet ruler. [2194] Three days after the departure of the former sultan, on the evening of Sunday 23rd August 1953, a Casablanca – Algiers express travelling at 60 miles an hour was derailed by sabotage about 30 kilometres north-east of Rabat, French Morocco. A section of track had been unbolted after the passing of the previous train an hour earlier. All but two of the carriages were derailed and the Moroccan fireman and according to some reports one passenger were killed and about seven passengers were injured. Rescue vehicles rushed to the scene and French and American ambulances took half a dozen European passengers to hospital [2195] [2196] [2197]

A Casablanca – Algiers express was again targeted on the evening of Saturday 7th November 1953. Just after it had left Rabat, French Morocco, and as it was crossing a bridge over the Bou Regreg River, a bomb exploded on it. As the guard was making his way to the point from which the explosion had come a second, more powerful blast, occurred, followed by a third. All the explosions had originated in the lavatories. Four Frenchmen and three Moroccans died. [2198] [2199]

Probably on the night of Saturday 25th September a Casablanca – Oued Zem freight train was derailed by sabotage near Berrechid, French Morocco. One person was slightly hurt.[2200]

Communist saboteurs derailed a train near Saigon, Vietnam, on Tuesday 19th January

---

[2192] *Statesville Daily Record*, Statesville, North Carolina, United States, 11th July 1953
[2193] *The Railway Gazette*, London, England, 28th August and 11th September 1953
[2194] Wikipedia – French protectorate in Morocco
[2195] *The Tipton Daily Tribune*, Tipton, Indiana, United States, 24th August 1953
[2196] *Lubbock Evening Journal*, Lubbock, Texas, United States, 24th August 1953
[2197] *Nevada State Journal*, Reno, Nevada, United States, 25th August 1953
[2198] *The Times*, London, England, 9th November 1953
[2199] *The Straits Times*, Singapore, 9th November 1953
[2200] *Journal de Genève*, Genève, Switzerland, 27th September 1954

1954. No one was hurt. [2201]

On two successive days, probably Friday 29th and Saturday 30th January 1954, the Vietminh derailed by mines the pilot trains of convoys about 70 kilometres north-east of Saigon, Vietnam.. [2202]

A train carrying troops and supplies between Haiphong and Hanoi, Vietnam, was mined by the Vietminh early on Sunday 31st January 1954. It was announced that there were 'losses in men and material.' [2203]

By early April 1954 the French colonies in South East Asia were disintegrating. Laos had declared its independence on 22nd October 1953, quickly followed by Cambodia on 9th November 1953. By April 1954 the French were fighting what was to prove a decisive battle at Diên Biên Phủ, which would end in their defeat and the partition of Vietnam into North and South at the end of July.

On about Friday 2nd April the Vietminh mined a train near Hải Dượng, Vietnam, about halfway between Hanoi and the port of Haiphong. [2204]

A Communist radio source stated that on Monday 12th April 1954 the Vietminh had mined a 22-waggon train proceeding towards Hanoi from Haiphong at a unstated location between the two places, killing or wounding 182 French soldiers. [2205]

Two trains were blown up using radio-controlled mines between Hanoi and Haiphong, Vietnam, on Wednesday 12th May 1954. 'Many' civilians were injured according to the French high command. [2206]

Another radio-controlled mine blew up a French supply train near Hanoi, Vietnam, on the line to Haiphong, on the night of Monday 31st May 1954. No one was hurt. [2207]

About 500 Vietminh guerrillas mined a train busing a remotely controlled device near Kraing Skea, Cambodia, on either Sunday 11th or Monday 12th April 1954, causing its coaches to pile up and overturn. After the train had stopped they emerged from hiding and opened fire on its passengers with pistols, rifles, sub-machine guns and grenades, reportedly dousing some of the victims in petrol and burning them to death, though some passengers, including children, priests, and elderly persons, escaped, and fled through the jungle to report the attack. Other fleeing passengers, according to a French Press Agency report, were caught and put to death with knives and hatchets.[2208] The train of 38 vehicles had been carrying mostly poor people to the capital, Phnom Penh, 80 kilometres from the scene of the attack, to celebrate the Cambodian New Year. About 100 people were killed. It took two days to identify the bodies of all the victims. [2209] [2210] Thirty yellow-robed Buddhist monks, women, and children were among the victims. [2211]

2201 *The Monessen Daily Independent*, Monessen, Pennsylvania, United States, 20th January 1954
2202 *The Advertiser*, Adelaide, South Australia, 1st February 1954
2203 *European Stars and Stripes*, Pfungstadt, West Germany, 1st February 1954
2204 *Bluefield Daily Telegraph*, Bluefield, West Virginia, United States, 3rd April 1954
2205 *The Fairborn Daily*, Fairborn, Ohio, United States, 16th April 1954
2206 *Panama City News*, Panama City, Florida, United States, 14th May 1954
2207 *The Lowell Sun*, Lowell, Massachusetts, United States, 1st June 1954
2208 *The Straits Times*, Singapore, 18th April 1954
2209 *Fergus Falls Daily Journal*, Fergus Falls, Minnesota, United States, 14th April 1954
2210 *The Greeley Daily Tribune*, Greeley, Colorado, United States, 17th April 1954
2211 *Delphos Daily Herald*, Delphos, Ohio, United States, 17th April 1954

Hundreds of passengers travelling to visit friends and relatives for the Chinese New Year escaped injury on Monday 31st January 1954 when the pilot train preceding the Singapore to Kuala Lumpur night mail exploded a mine seven miles north of Kluang, Malaya, in Johore state at 11.45 p.m.. The crash tender in front of the engine was thrown off the track and overturned. No one was hurt and there was time to stop the main train. The southbound mail to Singapore was stopped at Labis. Work on clearing the track began in the morning. The northbound train returned to Kluang where the passengers spent the night on the train before being taken to Kluang on buses and lorries in the morning, arriving in Kuala Lumpur at about 7 p.m., almost 12 hours late. The explosion caused a crater four feet wide and three feet deep and damaged the engine, an armoured waggon and two ballast waggons. [2212]

The armoured Wickham trolley which was piloting the night mail train from Singapore to Kuala Lumpur was blown up early on the morning of Tuesday 23rd March 1954 near Ayer Kuning South, Malaya, six miles from Gemas. The explosion made a five-foot crater and damaged 30 yards of track. The trolley driver and a special constable were slightly hurt. Fire was opened up on the terrorists but was not returned. The night mail train itself was delayed about seven hours. [2213]

Mau Mau rebels removed a length of rail four miles north of Nairobi, Kenya, either late on Wednesday 3rd or early on Thursday 4th February 1954, derailing a goods train. The same thing happened again at a point on the Nairobi – Mombasa line on Sunday 7th February 1954. There had been a previous derailment due to sabotage four months previously. [2214]

On Thursday 18th February 1954 about 100 rebels, it was believed composed of Communists and Karens, derailed a passenger train 40 miles north of Moulmein, Burma, then opened fire on it and robbed the passengers before they came under fire from the military escort train. The engine and four coaches of the passenger train were thrown off the track and wrecked. Three people were killed and several hurt, probably in the initial derailment. Seven Burmese soldiers were wounded and many rebels were believed to have been killed in the shootout between the escort troops and the rebels. An attempt to blow up the escort train failed. Two other trains were attacked elsewhere in Burma, with casualties and damage. [2215]

On Wednesday 11th August 1954 terrorist chose a 15-metre-high bridge over the Tji Tiis ravine between Leles and Nagrek, Indonesia, as a target, ripping up the track on the structure. At about midday a mixed train from Tjibatu to Bandung arrived at the bridge. It was preceded by two road/rail Jeeps, manned by the military. The leading Jeep fell from the bridge into the ravine. The second Jeep, following a short distance behind, stopped on the bridge and the terrorists immediately opened fire upon them. The soldiers on the Jeep returned fire but the main train was close behind it and although its driver braked hard it hit

---

2212 *The Straits Times*, Singapore, 3rd February 1954
2213 Ibid, 24th March 1954
2214 The West Australia, Perth, Western Australia, 8th February 1954
2215 *The Times*, London, England, 20th February 1954

the second trolley and pushed it into the ravine as well. The mixed train was just able to stop before it, too, fell from the bridge. About midnight, after its passengers had left it, the mixed train also fell into the ravine. Four people, three of them soldiers, were believed to have died while at least 12 were injured. The *Eengaagse* (Daily) express which followed shortly after the mixed train was stopped at Leles and returned to Tjibatu. [2216]

A mixed train from Casablanca to Oued Zem, French Morocco, was derailed by sabotage early on Thursday 21st October 1954. No one was hurt. [2217]

A train was derailed between Mexico City and Vera Cruz, Mexico, on about Friday 29th October 1954, using a homemade bomb. One person was killed and two injured. The perpetrator was supposedly a 78-year-old man who had also bombed two trams, two cinemas and a timber yard. No explanation was given for his acts. [2218]

According to the Tatao News Agency, which was operated by the Chinese Nationalist Ministry of the Interior, two ammunition cars attached to a Chinese Communist troop train exploded 95 miles south of Shanghai, China, on Saturday 20th November 1954. The news agency attributed the explosion to the mining of the track by Nationalist guerrillas. [2219]

Fellagha terrorists derailed what was variously described as an armoured and a freight train in southwestern Tunisia on Thursday 25th November 1954. They then fired on the train and swarmed up the derailed vehicles. There were reportedly 20 soldiers on it and an officer and a private were wounded. Troops from a nearby post then appeared and the attackers withdrew. [2220] [2221]

Terrorists in Malaya ripped up about 34 feet of track of the East Coast line, resulting in the derailment of a southbound goods train at about 7 p.m. on Tuesday 4th January 1955. The engine, tender and 17 of the 33 waggons of the train fell down one or the other side of a steep embankment and the fireman hurt his hand. A reporter described the scene. 'The length of track where the terrorists struck is ideal for their purposes. The nearest town is Ayer Hitam, about five miles south. On each side of the track there are tall trees. The countryside is desolate, wooded, and swampy. The engine and its tender sprawled on its side on the grassy embankment. Waggons loaded with logs lurched down each side of the track. A giant crane was at work clearing the line. Repair gangs which had been rushed to the scene were preparing to replace the track. Down below, hidden in the deep forest, 30 policemen were guarding the area while the repair work was going on.' The train had been en route from Gemas to Kuala Lipis with timber, palm oil, rubber, and general merchandise. The derailed waggons had been loaded with timber or

---

[2216] *De nieuwsgier*, 13th August 1954
[2217] *The Examiner*, Launceston, Tasmania, 23rd October 1954
[2218] *The Kokomo Tribune*, Kokomo, Indiana, 30th October 1954
[2219] *Pacific Stars and Stripes*, Tokyo, Japan, 9th December 1954
[2220] Ibid, 27th November 1954
[2221] *The Lethbridge Herald*, Lethbridge, Alberta, Canada, 25th November 1954

were empty. Acting General Manager G.S. Warren promised, "We will have the line cleared by midnight." [2222]

A Wickham trolley preceding the southbound mail train exploded a mine buried beneath the track between Layang Layang and Sedenak, Malaya, early on Friday 19th February 1955. No one was hurt and the trolley was apparently not derailed but its headlamps were damaged and a portion of the track was torn up. The train was delayed two hours while the damage was repaired. [2223]

A Wickham trolley piloting the night mail train from Johore Bahru to Kuala Lumpur was again derailed by sabotage on the night of Tuesday 7th June 1955 three miles north of Sedenak, Malaya, it being blown from the rails as it passed through a cutting at 9.45 p.m. A police corporal was seriously injured and was admitted to the British Military Hospital at Kluang. Three constables who had sustained bruises were treated as outpatients. The officer in charge of the guard of the night mail hearing the noise of the explosion and seeing a red Verey light being exhibited from the trolley stopped the mail train. Troops then guarded the line while it was being repaired. The night mail eventually reached Kuala Lumpur at 3.30 p.m. Through coaches from Singapore and Kuala Lumpur for Kelantan left Gemas on the *Golden Blowpipe* seven and a half hours late and the train was expected to reach Kota Bahru nine hours late. [2224]

On about Monday 21st February 1955 what was referred to as a strong band of Karen rebels mined a mail and passenger train 128 miles south of Rangoon, Burma, blowing up two coaches. The rebels then looted the train, taking arms, cash, and passengers' valuables, plus about 40 hostages, the latter including several girls, who survivors said had obviously been picked for their looks. Ten hostages, including two women, were released a short time later. [2225]

At least 20 persons reportedly died when a passenger train was mined on Friday 25th March 1955 between Sagaihn and Monyma, Burma, about 30 miles west of Mandalay. The insurgents, this time reportedly Communists, then looted the train and reportedly shot several passengers who resisted them. About 40 people were wounded. [2226]

A band of 300 rebels mined an armoured train 90 miles north of Rangoon, Burma, on Saturday 22nd October 1955. They then opened fire on the train from both sides. Eleven people died, seven of them police. [2227]

A train was derailed by sabotage at an unstated location between Casablanca and Oued Zem, French Morocco, on Monday 28th February 1955. No casualties were mentioned. [2228]

The morning train from Langsa (Atjeh Province) to Besitang (East Sumatra) was

---

[2222] *The Straits Times*, Singapore, 6th January 1955
[2223] Ibid, 19th February 1955
[2224] Ibid, 9th June 1955
[2225] *Pacific Stars and Stripes*, Tokyo, Japan, 23rd February 1955
[2226] Ibid, 29th March 1955
[2227] *Austin Daily Press*, Austin, Minnesota, 22nd October 1955
[2228] *Pacific Stars and Stripes*, Tokyo, Japan, 2nd March 1955

derailed in a Darul Islam terrorist attack near Medan Ara, Indonesia, on Monday 30th May 1955, the rails on a bridge across a river having been interfered with. The engine and coaches fell into the river and the driver and fireman were killed. Casualties among the passengers were not immediately known. [2229]

Six people died when a train was derailed by sabotage at Myslowice, Poland, on Sunday 6th March 1955. [2230]

A Canadian National Railways Sydney to Halifax passenger train was derailed on Friday night, 10th June 1955, near West River, Nova Scotia, Canada, by a damaged set of points. The company's Atlantic headquarters stated that, "preliminary investigation" indicated that part of the switch had been ripped away "by a person or persons unknown." Fireman Don Taylor was injured when the engine and seven coaches jumped the rails. None of the passengers was hurt but considerable damage was done to the track.

The same night the Wolverine express of the New York Central Railroad, which twice crossed into Canada and ran non-stop between Windsor, Ontario, and St Thomas, Ontario, collided with a steel splice laid on the track near Jeanette, Ontario, Canada, 10 miles west of Chatham. The steel bar was carried 500 feet along the line before it buried itself in the sand. The train was not derailed but the diesel locomotive and carriages suffered extensive damage. One passenger was slightly hurt by flying glass. The splice had been taken from a lineside pile of permanent way supplies. Whether this incident was related to the Nova Scotia derailment or they were just coincidental is unclear. [2231]

A rash of five derailments due to sabotage occurred at different points on the East Coast Railway of the colony of British Guiana on Tuesday 22nd, Wednesday 23rd, Thursday 24th and Friday 25th November 1955, probably related to a growing independence movement in the country (whose constitution had been suspended by the British Conservative Party in 1953). One person died and five were injured in the Thursday derailment of a goods train 15 miles from Georgetown. Independence was achieved in 1966. [2232] [2233] [2234]

A train was derailed by sabotage in Algeria on Monday 19th December 1955. [2235]

Saboteurs unbolted a rail on the line between Algiers and Constantine, Algeria, on the night of Wednesday 18th January 1956, and part of a French troop train carrying 300 men was derailed. Two railwaymen and one soldier were wounded, none of them seriously. [2236]

2229 *Algemeen Indisch dagblad: de Preangerbode*, Bandoeng, Indonesia, 3rd June 1955

2230 Author's own notes, source mislaid

2231 *The Lethbridge Herald*, Lethbridge, Alberta, Canada, 11th June 1955

2232 *The Times*, London, England, 26th November 1955

2233 *Corpus Christi Caller-Times*, Corpus Christi, Texas, United States, 27th November 1955

2234 *The Winnipeg Tribune*, Winnipeg, Manitoba, Canada, 28th November 1955

2235 *The Times*, London, England, 21st December 1955

2236 *The Bridgeport Telegram*, Bridgeport, Connecticut, United States, 19th January 1956

A passenger train between Batna and Biskra, Algeria, on Sunday 29th April 1956 chanced to be carrying a British newspaper correspondent. It was derailed 'in a valley amid rocky mountains near the famous El Kantara gorge.' Two lengths of rail had been unbolted and the train tore up 40 metres of wooden sleepers before stopping 'precariously tilted above the valley slope.' Had it been going faster it would probably have fallen down the slope. No one was hurt. The military escort of the train at once left it and searched for the rebels, while reinforcements were brought up. 'After two hours' wait the passengers, stoical and good-humoured about the affair after 18 months' rebellion in this Aurès area, were taken on to Biskra by goods train. Senior French officers and veiled, grubby Arab women clutching tiny children all suffered alike.' The correspondent added that, 'The rebels usually prefer to act by night, but today they must have tampered with the line in broad daylight, for the crew of a military engine that inspected it at 8 a.m. met with no trouble.' The sabotage must have been carried out either by local shepherds obeying rebel orders or by one of a band of 40 fellagha believed to live in the Metlili massif nearby. [2237]

A train was derailed at an unstated location in western Algeria on about Sunday 6th May 1956. [2238]

Rebels set fire to the station at El Hachin, Algeria, 125 kilometres south of Constantine, Algeria, on or about Wednesday 16th May 1956 and derailed a train near there. [2239]

Rebels machine-gunned a passenger train from Tunis to Algiers near Souk Ahras, Algeria, killing two Muslims, on Saturday 16th June 1956. [2240] On probably on the same day they derailed a train in the Constantine area, overturning 11 waggons and causing considerable damage, though no one was hurt. [2241]

Nationalist rebels blew up a French military supply train with a land mine near Biskra, Algeria, on Sunday 15th July 1956. The fire which followed the blast destroyed 14 waggons of ammunition and vehicles which had been destined for troops in the southern portions of Constantine and Bône Departments. Officials gave no indication of casualties; unofficial reports spoke of eight dead and nine wounded. [2242] Reconnaissance planes and helicopters searched for the saboteurs, who for the first time rather than loosening the rails or felling a tree across the track had employed an anti-tank mine. [2243]

The engine pulling the Algiers – Constantine express was derailed between Bouira and Setif, Algeria, on Wednesday 25th July 1956, the rebels having loosened a rail and moved it out of position. They then opened fire on the train but the military escort drove them off. One rebel died, no one on the train was hurt. [2244]

A train, probably only carrying freight, was mined just north of Ain Sefra, Algeria, 400 kilometres south of Oran, on Monday 30th July 1956. As usual, the rebels then fired on the train and five of its armed escort were wounded. [2245]

[2237] *The Times*, London, England, 2nd May 1956
[2238] *Pacific Stars and Stripes*, Tokyo, Japan, 7th May 1956
[2239] *Oxnard Press-Courier*, Oxnard, California, United States, 17th May 1956
[2240] *The Times*, London, England, 18th June 1956
[2241] *Pacific Stars and Stripes*, Tokyo, Japan, 18th June 1956
[2242] *The Albuquerque Tribune*, Albuquerque, New Mexico, United States, 16th July 1956
[2243] *The Mexia Daily News*, Mexia, Texas, Texas, United States, 18th July 1956
[2244] *European Stars and Stripes*, Pfungstadt, West Germany, 26th July 1956
[2245] *The Albuquerque Tribune*, Albuquerque, New Mexico, United States, 30th July 1956

Rebels derailed and fired on an ore train south of Constantine, Algeria, on Sunday 5th August 1956. No one was hurt and the attackers were reportedly routed by French reinforcements. [2246]

An Oujda – Oran train was ambushed by rebels near Lamoriciere, Algeria, on Monday 10th September 1956. Machine gun fire pierced the fuel tank of the diesel-hauled (?) train and forced it to a stop. The military guard reportedly routed the rebels. A French soldier was killed. [2247]

Terrorists derailed a petrol train north of Constantine, Algeria, on the night of Wednesday 14th November 1956. A fire followed the derailment, involving an estimated 230,000 litres of petrol. One terrorist was killed. [2248]

Rebels machine-gunned an Oran – Colomb-Bécher train in Algeria on Tuesday 27th November 1956, killing the driver and a second railwayman and wounding 15 others. [2249] [2250]

Eight people died and seven others were seriously wounded when a train ran over a mine about 25 kilometres north of Colomb-Bécher, Algeria, on the night of Tuesday 11th December 1956. [2251]

A train was derailed between Bougie and Algiers, Algeria, on about Saturday 15th December 1956. [2252]

Train No. 41 Down Bombay – Calcutta mail train was derailed at mile 308/18-21 between Malkapur and Biswa Bridge stations, Bombay State, India, on the Bhusaval – Nagpur section of the Central Railway, at about 4.17 a.m. on Tuesday 24th January 1956. The engine and leading six coaches of the train left the rails, the rear six coaches were not derailed. There were almost 1,500 passengers on the train of whom two were seriously injured, one was slightly injured and 38 suffered trivial injuries. Railway officials said that loose fishplates had been found near the scene and the rails had been loosened, indicating that sabotage was to blame. [2253] [2254] [2255] The preliminary report of the Government Inspector of Railways came to the same conclusion, stating that the cause of the derailment was 'the removal, by some person or persons unknown, of three pairs of fishplates and all steel keys from two consecutive left-hand rails and the shifting of one of these rails by a few inches towards the right.' [2256]

About 30 insurgents mined and attacked an armoured army escort train near Thrawaw, Burma, about 100 miles from Rangoon, on Wednesday 25th January 1956, according to

[2246] *Pacific Stars and Stripes*, Tokyo, Japan, 8th August 1956
[2247] *Albuquerque Journal*, Albuquerque, New Mexico, United States, 12th September 1956
[2248] *Wichita Falls Times*, Wichita Falls, Texas, United States, 15th November 1956
[2249] *The Times*, London, England, 29th November 1956
[2250] *The Daily Chronicle*, Centralia-Chehalis, Washington United States, 28th November 1956
[2251] *Pacific Stars and Stripes*, Tokyo, Japan, 13th December 1956
[2252] *The Times*, London, England, 18th December 1956
[2253] Ibid, 26th January 1956
[2254] *Nevada State Journal*, Reno, Nevada, United States, 25th January 1956
[2255] *The Bakersfield Californian*, Bakersfield, California, United States, 24th January 1956
[2256] India Parliament Lok Sabha Debates 1956 (Google Books)

reports. [2257]

On Friday 20th July 1956 insurgents, variously identified as Karen rebels and Communists, derailed the Rangoon – Prome mail train between Palon and Thametchaung, Burma, [2258] some 50 miles north of Rangoon. Having allowed the pilot train with its armed escort to pass the spot some minutes earlier they then detonated which blew the engine, tender, two waggons and four coaches off of the track. About 100 attackers then emerged from the jungle, fired a hail of bullets at the train and threw grenades at it, some of them operating from tree tops. As might be expected the passengers panicked and ran in all directions or took cover beneath the overturned coaches. The attackers then systematically robbed the passengers of cash, jewellery, and other possessions before setting fire to the coaches. Becoming aware of the attack the pilot train reversed to the scene, whereupon the attackers retreated back into the jungle, taking some passengers with them as hostages. The attack had lasted half an hour. A relief train was sent from Rangoon and returned there with the bodies of a police sub-inspector and a constable who had been killed in the exchange of gunfire. Others who lost their lives included three Indians travelling in an upper-class compartment, who had burnt to death. At least 35 people died and about 75 more were injured. The driver and two firemen were critically hurt. Some reports spoke of 80 or even 100 dead. [2259] [2260]

Communist rebels mined a train near Kambalu, Burma, 100 miles north of Mandalay, on Thursday 6th September 1956. They then took a railway guard and six policemen hostage. [2261]

A train ran over four landmines, was derailed, and then attacked by about 200 Communist insurgents who opened fire on it near Maymyo, Burma, on Wednesday 26th September 1956. A fireman was killed and 11 police escorts were seriously wounded. [2262]

Six Portuguese passengers were seriously injured when a time bomb exploded in a second-class carriage of a train at an unstated location in Goa on about Monday 13th February 1956. At the time Goa was still a Portuguese colony and the attack was blamed on Indian terrorists. [2263]

‘Bandits’ derailed a goods train between Kalay and Colen, Goa, on about Sunday 12th August 1956. The driver and a passenger were wounded and one person (an attacker?) was killed. [2264]

In another attack an ore train was derailed ‘on the Mormugao railway,’ in Goa on Friday 17th August 1956. Rail spikes had been pulled out. No one was hurt. [2265]

Goa was annexed by India by force late in late 1961 after Portugal had refused to grant it independence.

[2257] *The Singapore Free Press*, Singapore, 27th January 1956
[2258] *New Castle News*, New Castle, Pennsylvania, United States, 21st July 1956
[2259] *Winnipeg Free Press*, Winnipeg, Manitoba, Canada, 21st July 1956
[2260] The *Sunday Standard*, Madras, India, 22nd July 1956
[2261] *The Straits Times*, Singapore, 8th September 1956
[2262] Ibid, 28th September 1956
[2263] *The Times*, London, England, 15th February 1956
[2264] Ibid, 15th August 1956
[2265] *Nevada State Journal*, Reno, Nevada, United States, 18th August 1956

In what was described as 'the biggest derailment in recent years,' a train was dynamited by 'Communist terrorists' on Wednesday 11th April 1956 between Bannir and Tapah Road stations, in Perak State, Malaya. Driver S.R. Kuppusamy, 37, and Firemen A. Gangan, 26, and P. Raman, 26, were injured and taken to hospital. The firemen were discharged next morning but the driver was more badly hurt, being unconscious when admitted, and he was detained. The engine, tender and eight of the 48 waggons left the track and the engine and one waggon went down a 10-foot embankment. The night mail from Prai, due in Kuala Lumpur at 7 a.m., was not expected to arrive until 12 noon. Large numbers of passengers had to be transhipped between Tapah Road and Kampar using eight buses and four lorries.[2266] It was believed that not more than five perpetrators had taken part in the sabotage; they had used dynamite to blow up about 150 feet of track and damage was estimated at over $100,000. By the following day the driver had recovered sufficiently to be able to speak to a newspaper reporter. He said, "I was driving the train from Ipoh when I heard a loud explosion shortly before 10.15 p.m. This was a mile before we reached Tapah Road. I pulled the brakes and the engine and some of the waggons went off the rails. I lost consciousness after my chest struck the regulator. The next thing I knew was waking up in the Tapah hospital very early this morning." Malays living nearby heard the sound of the explosion but fearing gunfire kept to their houses. Breakdown gangs were said to be working feverishly to repair the damage and replace rails, using 20- and 40-ton cranes, but the line was not repaired until noon on Friday. [2267] [2268]

An armoured Wickham trolley piloting a passenger train was derailed by the explosion of an electrically detonated mine between Gemas and Rompin, Malaya, in Negri Sembilan state on Saturday 3rd November 1956. A Malay corporal was slightly wounded in the ensuing engagement and there was some delay to the train, though repairs were immediately begun. [2269]

Ukrainian rebels derailed a Soviet Army troop train near Shepetovka, Ukraine, USSR, on the main line between Lvov and Kiev, on Sunday 20th May 1956, according to reports reaching Wien, Austria. An unknown number of troops were killed. Soviet troops and members of the MVD, the Soviet secret police, sealed off a large area and seized a group of heavily armed Ukrainians. [2270]

On Monday 2nd July 1956, the eve of the second anniversary of the overthrow of its democratically elected government, the engine of a passenger train 'owned by U.S. interests,' was derailed by sabotage in Guatemala, possibly near Ayutla. [2271]

A DSM (Deli Spoorweg Maatschappij) train from Kisaran to Belawan was derailed on the night of Wednesday 19th September 1956 near Teluk Mengkudu, Sumatra, Indonesia. The engine and 39 waggons carrying rubber and other products were derailed. Conductor A. Latief was killed and two other railwaymen, presumably the driver and fireman, were

---

[2266] Ibid, 12th April 1956
[2267] *The Straits Times*, Singapore, 14th April 1956
[2268] *Straits Budget*, Singapore, 19th April 1956
[2269] *The Straits Times*, Singapore, 5th November 1956
[2270] *Fairbanks Daily News-Miner*, Fairbanks, Alaska, United States, 9th June 1956
[2271] *The Post Standard*, Syracuse, New York, United States, 3rd July 1956

injured. The track had been destroyed over a length of 45 metres. The driver of the train only saw that the track was damaged at the last moment and was thus unable to pull up in time. Darul Islam insurgents were believed to be to blame. [2272]

A train from Tel Aviv to Beersheba, Israel, was derailed by a mine early on Sunday 11th November 1956. Sabotage by Palestinian Fedayeen guerrillas was presumed. No casualties were reported. [2273]

Nineteen people died and 18 were injured in a collision between two express trains at 09:40 on Sunday 16th December 1956 between Taguasco and Jatibonico, Cuba. The collision was apparently a sidelong one, the engine of a train bound for Santiago de Cuba colliding with the leading coach of a train bound for Habana. It was in this coach that most of the casualties occurred. The injured were treated at Jatibonico, Sancti Spiritus and Ciego De Avela. It was alleged that anti-government saboteurs had diverted the Havana-bound train into the path of the other one. A revolutionary movement against the corrupt regime of Fulgencio Batista had begun in July 1953, led by Fidel Castro, and martial law and censorship had been imposed in Camaguey Province, where the disaster had occurred, the previous week. Collisions due to sabotage, as opposed to derailments, are unusual though, and possibly it was in reality just an accident. [2274]

The attacks on trains in Algeria in 1957 began already on the night of Wednesday 2nd January when a train was derailed between Oran and Relizane. When the attack began the train crew attempted to reach the lineside telegraph wires to send a message but they were driven back by the advancing rebels, who machine-gunned the train and entered some of the coaches. A four-hour battle with the train's escort followed before a relief column, attracted by flares, arrived, and the rebels retreated with ten hostages. Six Europeans (presumably French) were killed. [2275] [2276]

An Algiers – Oran passenger train was derailed by saboteurs just before midnight on Monday 7th January 1957 near Miliana, Algeria. There were no casualties. [2277]

Three European crewmen died and ten Europeans and Muslims were wounded when an ore train ran over a mine between Beni-Saf and Sidi-Safa, Algeria, on Friday 1st February 1957. A second derailment due to sabotage happened on the same day, this time due to a rail having been unbolted between Algiers and Bougie. No one was hurt this time. [2278]

Three people were wounded on about Sunday 17th February 1957 when a train was derailed by sabotage near Soukahras, Algeria. [2279]

Algerian rebels exploded a mine beneath an Oujda – Oran train on Tuesday 2nd April

---

[2272] *Java-bode*, Jakarta, Indonesia, 21st September 1956

[2273] *The Times*, London, England, 12th November 1956

[2274] *The Arizona Republic*, Phoenix, Arizona, United States, 17th December 1956

[2275] *European Stars and Stripes*, Pfungstadt, West Germany, 4th January 1957

[2276] *The Times*, London, England, 4th January 1957

[2277] *The Canberra Times*, Canberra, Australian Capital Territory, Australia, 9th January 1957

[2278] *European Stars and Stripes*, Pfungstadt, West Germany, 2nd February 1957

[2279] *Pacific Stars and Stripes*, Tokyo, Japan, 19th February 1957

1957. Seven people were wounded and several coaches were reportedly destroyed. [2280]

A Casablanca – Algiers passenger train was blown up by a mine near Turenne, western Algeria, on Thursday 18th April 1957. Two coaches fell into a ravine and several people were injured and were evacuated by helicopter. [2281]

Three people died in the derailment of a train between Batna and Niska, Algeria, on about 4th June 1957. [2282]

On about Sunday 30th June 1957 the rebels derailed a freight train near Bou Medfa, Algeria, using an old artillery shell as a mine. No one was hurt. [2283]

A remote-controlled mine exploded beneath a train carrying 150 children on the first leg of a trip to summer camps in France on about Friday 19th July 1957. The attack occurred about 30 kilometres north of Djelfra, Algeria. The engine and tender were derailed but no one was hurt. [2284]

Rebels derailed a train near Aumale, Algeria, on about Sunday 28th July 1957. [2285]

A train from Oran to Orleansville, Algeria, was fired on by rebels on Wednesday 30th October 1957. An eighteen-month-old baby was killed and a French soldier was gravely wounded. [2286]

On about Sunday 3rd November 1957 rebels derailed 13 waggons of a freight train near Orleansville, Algeria. [2287]

Two engines and 20 waggons of a heavy freight train were derailed by rebels on Wednesday 4th December 1957 between Bône and St Charles, Algeria. Four French soldiers were killed and three were injured. [2288]

The engine and 17 waggons of an Algiers – Oran train were derailed by sabotage near Perrégaux, (now Mohammadia) Algeria on about Saturday 28th December 1957

Also on Saturday 28th December 1957 a freight train was derailed between Bône and Constantine, Algeria. Eighteen of the 30 waggons of the train were derailed and some were telescoped. The body of an Algerian railwayman and five injured were found crushed in the van which normally carried the armed guard of the train. [2289]

Communist rebels derailed the Mandalay – Rangoon mail train on Monday 4th March 1957 near Pyinmana, Burma, about 220 miles north of Rangoon. They had removed plates and bolts over a stretch of half a mile, causing the engine to leave the rails and some the following 16 coaches to telescope one another. Thirty people reportedly died and 40 were injured. [2290] [2291]

2280 *Walla Walla Union-Bulletin*, Walla Walla, Washington, United States, 2nd April 1957
2281 *The Canberra Times*, Canberra, Australian Capital Territory, Australia, 20th April 1957
2282 *Press Telegram*, Long Beach, California, United States, 5th June 1957
2283 *Pacific Stars and Stripes*, Tokyo, Japan, 1st July 1957
2284 Ibid, 22nd July 1957
2285 *The Charleston Gazette*, Charleston, West Virginia, United States, 30th July 1957
2286 *The Emporia Daily Gazette*, Emporia, Kansas, United States, 31st October 1957
2287 *The Bakersfield Californian*, Bakersfield, California, United States, 4th November 1957
2288 *Sarasota Herald-Tribune*, Sarasota, Florida, United States, 6th December 1957
2289 *The Amarillo Daily News*, Amarillo, Texas, United States, 28th December 1957
2290 *The Anniston Star*, Anniston, Alabama, United States, 5th March 1957
2291 *Pacific Stars and Stripes*, Tokyo, Japan, 5th March 1957

About 100 Communist rebels dynamited and mined a passenger train north of Mandalay, Burma, on Monday 28th October 1957. Ten people were reportedly killed, including a police escort, and several passengers were seriously injured. [2292]

Automatic fire was directed at an Israeli goods train travelling near Tulkarm, close to the border with Jordan and about 20 miles north-east of Tel Aviv, on Friday 29th March 1957. No one was hurt. [2293]

A freight train struck a mine and was derailed near Jerusalem, Israel, again close to the Jordan border and on the line from Tel Aviv, on Friday 26th April 1957. No one was hurt. Police with dogs set out to try and track the saboteurs. [2294]

On about Saturday 1st June 1957 a 29-waggon freight train was derailed by a mine near Bejucal, Cuba, about 30 kilometres south of Habana. [2295]

At an unknown date in October 1957 a passenger train was derailed near Zalaszentivan, Hungary, apparently by an explosion, 30 people being killed and 100 injured. Unfortunately it has not proved possible to obtain any further information about this event, which was presumably an act of sabotage, perhaps directed against the occupying Soviet forces, after the abortive uprising of the previous year. [2296]

Train No. 1 Down Bombay – Calcutta Mail was derailed at mile 97/4 between Padli and Asvali stations, Bombay State, India, on the Igatpuri – Bhusaval double-line section of the Central Railway at 22:45 on Saturday 23rd November 1957. [2297] The engine and leading coach were derailed, the next four coaches were derailed and capsized and the next two coaches were also derailed. The rear four coaches remained on the track. There were estimated to have been over 1,000 passengers on the train and initial reports spoke of 50 dead, later corrected to nine dead and 46 injured. Relief trains with food and medical supplies were rushed to the scene in the morning. Sabotage was suspected as the track had been interfered with in two places, causing it to go out of alignment. It was supposed that the saboteurs had expected a goods train, which should have preceded the express. [2298] [2299]

A goods train was derailed near Lusaka, Northern Rhodesia (now Zambia) on Wednesday 27th November 1957, bolts having been unscrewed from the rails. Ten waggons left the track and over 50 cattle died. The perpetrator, Benjamin Chirwa, 30, an African, was quickly apprehended, pleaded 'Guilty' and was sentenced to 20 years'

2292 *The Singapore Free Press*, Singapore, 30th October 1957
2293 *The Times*, London, England, 30th March 1957
2294 *The Bee*, Danville, Virginia, United States, 26th April 1957
2295 *The Sunday News*, Sarasota, Florida, United States, 2nd June 1957
2296 *Singapore Standard*, Singapore, 8th November 1957
2297 https://www.advocatekhoj.com/library/lawreports/commissionsofinquiryact/1b.php?Title=Commissions%20of%20Inquiry%20Act,%201952&STitle=Ministry%20of%20Railways
2298 *Pacific Stars and Stripes*, Tokyo, Japan, 25th November 1957
2299 *The Times*, London, England, 25th November 1957

imprisonment. His motive was not stated. [2300]

El Tucumano, a luxury express train in Argentina, was derailed by sabotage at an unnamed location on Monday 30th December 1957. It was only travelling slowly when five coaches left the rails and fell into muddy ground. No one was hurt. Examination showed that spikes had been removed from the track. It was believed that Peronist terrorists were responsible. [2301] The Peronist Party had been outlawed after a coup had overthrown President Juan Perón in 1955.

The French ran their first ever oil tanker train from Touggort, Algeria, to the port of Phillipeville, 360 kilometres to the north, on Wednesday 8th January 1958, the oil coming from a newly opened field in the Sahara Desert. Elaborate security was arranged, with fortified posts at 10-kilometre intervals and patrols by soldiers and aircraft. Despite this, a goods train was derailed by a mine on the same line at Conde Smendou, about 20 waggons leaving the track. No one was hurt but the line was blocked and it thus took the oil train three days to complete its journey. [2302] [2303]

A freight train was derailed near Berrouaghia, Algeria. On about Sunday 12th January 1958. No one was hurt but seven waggons were damaged and 50 metres of track was torn up. [2304]

An oil train from Touggort ran into a pile of rocks placed on the line near Phillipville (now Skikda), Algeria, on Tuesday 14th January 1958. The locomotive was damaged. [2305]

Rebels blew up a railcar between El Kantara and Biskra, Algeria, on the line from the Sahara oilfield, on Sunday 26th January 1958, killing its driver and six soldiers of the escort party and wounding three others. [2306]

An Algiers – Oran train was derailed by a mine near Perrégaux, Algeria, on the night of Friday 14th February 1958. The driver and two passengers were injured. [2307]

Saboteurs unbolted the rails on the line between Algiers and Oran, Algeria, on Friday 4th April 1958. A freight train was derailed, its driver was injured and the line was blocked. [2308]

An Algiers – Oran passenger train was derailed by sabotage near Relizane, Algeria, on about Tuesday 24th June 1958 and two passengers were injured. [2309]

A passenger train was derailed by a mine near Orleansville, Algeria, on Monday 11th August 1958. Three soldiers of its military escort were killed and many people were wounded. [2310]

2300 Ibid, 5th December 1957

2301 Ibid, 1st January 1958

2302 *Chicago Daily Tribune*, Chicago, Illinois, United States, 10th January 1958

2303 *The Times*, London, England, 10th January 1958

2304 *The Racine Journal-Times*, Racine, Wisconsin, United States, 13th January 1958

2305 *Winnipeg Free Press*, Winnipeg, Manitoba, Canada, 16th January 1958

2306 *The Times*, London, England, 28th January 1958

2307 *Sunday Standard*, Singapore, 16th February 1958

2308 *Chicago Daily Tribune*, Chicago, Illinois, United States, 5th April 1958

2309 Ibid, 26th June 1958

2310 *The Racine Journal-Times*, Racine, Wisconsin, United States, 11th August 1958

During a two-day visit by French President Charles de Gaulle rebels blew up a military train between El Kantara and El Outaya, Algeria, on Thursday 28th August 1958, killing two soldiers and wounding 15 soldiers and five railwaymen. [2311]

Six people died when a passenger train exploded a mine 30 kilometres from Blida, Algeria, on Monday 10th November 1958. Three of the dead were French soldiers. Six other passengers were slightly hurt. A second-class carriage was destroyed and several others were damaged. [2312]

An engine and 15 waggons of a freight train were derailed by a mine near Medea, Algeria, south of Algiers, on Saturday 15th November 1958. [2313]

On Tuesday 9th December 1958 a mine derailed an Oran – Algiers express near Alfreville, Algeria, killing its two drivers. [2314]

A train ran over three mines and was derailed between Algiers and Constantine, Algeria, on about Monday 29th December 1958. [2315]

An eastbound Canadian Pacific Railway passenger train on their Kettle Valley branch was derailed by a bomb which had blown out two ties (sleepers) and left a hole 18 inches deep beneath one rail near Grand Forks, British Columbia, Canada, on Wednesday 15th January 1958. The engine, baggage car and a day coach were derailed but no one was hurt. The Royal Canadian Mounted Police sent for a dog to follow tracks leading from the derailment location. Freedomites, a splinter group from the Doukhobor religious sect, were thought to have been responsible for the derailment. [2316]

Fidel Castro's rebels derailed a train loaded with sugar at Guantanamo, Cuba, on about Thursday 20th February 1958. Another train was diverted into a row of waggons. [2317]

The defeat of dictator Fulgencio Batista was in its final days when rebels led by Che Guevara blew up a Government train loaded with ammunition and weapons in Santa Clara Province, Cuba, on about Tuesday 30th December 1958. [2318]

Communist rebels reportedly mined a Mandalay – Myitkyina express on Thursday 20th March 1958 near Htangon, Burma, 108 miles north of Mandalay, then fired upon the wrecked train. Fifteen people died and 85 were seriously hurt. [2319] [2320]

The long-standing conflict in Ceylon between the majority Sinhalese population and the minority Tamils led to an outbreak of violence in 1958 following the passing of

---

[2311] *Tucson Daily Citizen*, Tucson, Arizona, United States, 29th August 1958
[2312] *Pacific Stars and Stripes*, Tokyo, Japan, 11th November 1958
[2313] *San Antonio Express*, San Antonio, Texas, United States, 17th November 1958
[2314] Ibid, 11th December 1958
[2315] *Sunday Standard*, Singapore, 29th December 1959
[2316] *The Medicine Hat News*, Medicine Hat, Alberta, Canada, 15th January 1958
[2317] *Lubbock Evening Journal*, Lubbock, Texas, United States, 21st January 1958
[2318] *Wisconsin State Journal*, Madison, Wisconsin, United States, 31st December 1958
[2319] *The Albuquerque Tribune*, Albuquerque, New Mexico, United States, 21st March 1958
[2320] *The Times*, London, England, 22nd March 1958

legislation making Sinhala the only official language of the country. In connection with this outbreak the night mail train was derailed in Batticaloa. It was not clear whether the saboteurs were Sinhalese or Tamils but most of the passengers were Sinhalese. Three of them, Victor Fernando, Police Sergeant Appuhamy and Constable Pararajasingham – all Sinhalese, were killed, probably in a post-derailment attack. [2321] [2322]

The derailment of a passenger train near Tepic, Mexico, on Tuesday 5th August 1958 was related to a wildcat railway strike which had then been in progress for several days. No one was hurt. [2323]

A passenger train was derailed 9 miles south of Lusaka, Northern Rhodesia (now Zambia) on Monday 29th September 1958. The 300 African passengers on the train escaped serious injury. An investigation indicated sabotage as the cause and Mr D.C. Hall, Acting Chief Secretary of the protectorate, declared a state of emergency, with no person permitted to come within 100 feet of the track on either side. [2324]

An Algiers – Orleansville passenger train ran over a mine near Ruperre station, Algeria, not far from Blida, on Friday 2nd January 1959. The engine, tender, and possibly the leading carriage were derailed and a few passengers were hurt. [2325] [2326] [2327]

Eighteen people were hurt when a train was derailed near Oran, Algeria, on Monday 19th January 1959. [2328]

In what was stated to have been their first successful attack for more than a year on an oil train from the Hassi Messaoud field, rebels derailed a train from Touggourt to Phillipeville on Friday 23rd January 1959 south of Constantine, Algeria. Nine tank waggons caught fire, an estimated 590,000 litres of oil were lost and the line from Constantine to Algiers and Biskra was blocked. [2329]

Another oil train was derailed by a mine in eastern Algeria on Wednesday 11th February 1959. [2330]

A Bougie – Algiers passenger train was mined and machine-gunned near Akbou, Algeria, on about Saturday 28th February 1959. No one was hurt. [2331]

On Sunday morning, 1st March 1959, a freight train from Blida to Djelfa was derailed near Brazza in southern Algeria and two people were injured. [2332]

Rebels derailed a passenger train at El Kantara, Algeria, on the line between the Sahara oil field and the Mediterranean coast on Thursday 26th March 1959. Two people

---

[2321] *Pacific Stars and Stripes*, Tokyo, Japan, 25th May 1958

[2322] Wikipedia – 1958 anti-Tamil pogrom

[2323] *Miami Daily News-Record*, Miami, Oklahoma, United States, 6th August 1958

[2324] *The Times*, London, England, 1st October 1958

[2325] *Pacific Stars and Stripes*, Tokyo, Japan, 4th January 1959

[2326] *The Daily News*, St Johns, Newfoundland, Canada, 3rd January 1959

[2327] *The Brandon Daily Sun*, Brandon, Manitoba, Canada, 5th January 1959

[2328] *The Times*, London, England, 22nd January 1959

[2329] Ibid, 26th January 1959

[2330] *The Singapore Free Press*, Singapore, 12th February 1959

[2331] *The Straits Times*, Singapore, 3rd March 1959

[2332] *Pacific Stars and Stripes*, Tokyo, Japan, 3rd March 1959 were killed and one hurt. [2333]

A mine derailed a train at dawn between Algiers and Orleansville, Algeria, on Friday 17th April 1959. The engine and eight vehicles were derailed but no casualties were reported. [2334]

Rebels derailed a train 40 kilometres south of Algiers. Algeria, on Sunday 1st November 1959. [2335]

A mine exploded beneath a train from Algiers to Tizi Ouzou, Algeria, on Monday 7th December 1959. Two people died and nine were injured, three seriously. [2336]

A mine exploded beneath an ore train near Bône, Algeria, on about Sunday 20th December 1959, killing one railwayman and wounding another. [2337]

A train from Bougie to Algiers, Algeria, ran over a mine and was derailed, probably on Christmas Eve, Thursday 24th December 1959. Ten people were wounded. [2338]

On the evening of Tuesday 14th April 1959 a train was dynamited near the entrance to a tunnel at Puerto de Pajares, Spain, by guerrillas led by Cuban Antonio Rodriguez, a Spanish Civil War veteran. [2339]

A disastrous runaway accident near Tasikmalaja, Indonesia, early on Thursday 28th May 1959 was at first blamed on rebel sabotage but the initial results of an investigation into it attributed it rather to the fault of its train crew, specifically its conductor. The accident in West Java involved a Bandjar – Bandong express carrying an estimated 500 passengers. Several carriages ran away down an incline, covering a distance of about 3 kilometres and reaching a reported speed of about 95 kilometres an hour before they became derailed and fell into a ravine. At least 140 passengers were said to have been killed and another 125 seriously injured. The location of the accident was so remote that it was hours before officials learnt of it and almost a day before any substantial outside help arrived. The train consisted of a converted armoured car with a military guard, an engine, two wooden coaches, two steel coaches and a conductor's van.

The events leading up to the catastrophe began with the stalling of the train on a steep incline. A second locomotive was sent from the nearest station but even with two engines it proved impossible to restart the train from a dead stand. The investigation determined that at this point the conductor, on his own initiative, decided to divide the train. The coaches were uncoupled and as soon as this had been done they began to run back down the hill, gaining speed as they did so. [2340]

The derailment occurred in 'thick jungle country, infested by Darul Islam rebels who have been fighting the government for years in trying to force the establishment of a theocratic Moslem state.' Two planeloads of doctors and medicines were flown to the

---

2333 *San Antonio Express*, San Antonio, Texas, United States, 27th March 1959
2334 *Kingsport News*, Kingsport, Tennessee, United States, 18th April 1959
2335 *Tri-City Herald*, Pasco, Washington, United States, 2nd November 1959
2336 *The Winnipeg Tribune*, Winnipeg, Manitoba, Canada, 8th December 1959
2337 *The Arizona Republic*, Phoenix, Arizona, United States, 21st December 1959
2338 *Elmira Advertiser*, Elmira, New York, United States, 26th December 1959
2339 *The Daily Colonist*, Victoria, British Columbia, Canada, 22nd May 1959
2340 *The Farmington Daily Times*, Farmington, New Mexico, United States, 31st May 1959

scene and the Health Ministry alerted all medical staff in the region to stand by. [2341]

Whether the runaway was the result of an accident or sabotage remains unclear, for a week into June Indonesian Communications Sukardan told Parliament that the accident was due to sabotage, an unknown person, and not the conductor, having disconnected the coaches from the engine. [2342]

No. 3 Up Howrah – Madras mail was derailed between Chinna Ganjam and Uppugundur, India, on the Southern Railway at 23:53 on Sunday 23rd August 1959. The results were not serious; nine people were slightly hurt. The train consisted of engine No. WP 7493 and 12 bogie coaches, in charge of Guard S.T. Wilson with Driver Alfred Sundaraj and First Fireman E. Adhinarayana. The train was only travelling at 25 miles an hour at the time which probably reduced the effects of the accident; the engine and leading four coaches left the rails. The accident was the subject of a detailed investigation which found that it was due to the removal of fishplates and fish-bolts from two rail joints and the keys from one rail. Two suspected saboteurs were arrested close to the scene and brought to trial but were acquitted. [2343]

The Jaffna – Colombo night mail was derailed by sabotage on the night of Sunday 27th December 1959 as it passed through Vavuniya, Ceylon. Thirty passengers were injured. The area had been the scene of serious rioting between Sinhalese and Tamil groups the previous year. Mr B.D. Rampala, general manager of the government railway, said that several attempts had already been made in the days before the incident to derail the mail train and an investigation had been opened to try and determine if these were linked to the intercommunal rivalry. The government had ordered a state of emergency in an effort to halt the bloody fighting between the Sinhalese majority and the Tamil minority. [2344] [2345]

A train from Algiers to Tizi Ouzou, Algeria, ran over a landmine close to its destination early on Monday 4th January 1960. It was badly damaged but reportedly only a few people were slightly hurt. [2346]

A mine derailed the engine and eight waggons of a freight train near Batna, Algeria, on about Sunday 10th January 1960. [2347]

A Bougie to Algiers, Algeria, train was blown up by a mine early on Thursday 18th August 1960. The driver was wounded and four waggons placed at the front of the train were damaged. [2348]

About 200 Karen rebels mined the Moulmein – Rangoon mail train near the station at

---

2341 Ibid.

2342 *The Daily Telegram*, Eau Claire, Wisconsin, United States, 8th June 1959

2343 Government of India Ministry of Transport and Communications: Report on Derailment of No. 3 Up Howrah – Madras Mail between Chinna Ganjam and Uppugundur stations on 20th August 1959

2344 *Appleton Post-Crescent*, Appleton, Wisconsin, United States, 28th December 1959

2345 *Pacific Stars and Stripes*, Tokyo, Japan, 30th December 1959

2346 *The Straits Times*, Singapore, 5th January 1950

2347 Ibid, 12th January 1960

2348 *The Times*, London, England, 19th August 1960

Donwun, Burma, on Sunday 10th January 1960 and then opened fire on it. At least 10 passengers and three policemen (part of the train's escort?) were killed and 16 wounded. [2349] [2350]

A passenger train on the Rangoon – Mandalay line ran over two mines near Kyauktaga, Burma, 105 miles north of Rangoon, on Saturday 18th June 1960, and six coaches were blown off the track. One passenger was killed and three were described as 'missing'. 'The area, formerly a stronghold of Communist insurgents, had been thought cleared by the army,' the newspaper report added. [2351]

The Paris Lyon – Marseille express which had left the French capital at 20:00 the evening before was derailed at 01:30 on Wednesday 9th March 1960 near Sérézin-du-Rhône station, not far from Vienne, while it was travelling at 130 kilometres an hour. The engine was detached from its train during the derailment, six coaches and three luggage vans left the track and one of the coaches overturned. Twelve people were injured, nine of them passengers and three postal sorters. Only four of the injured were badly enough hurt to need to go to hospital. One of them was Mrs Clifton Brown, a Norfolk county alderman, who was on her way with six other Britons to attend a congress of municipalities at Cannes. She was not that seriously hurt and her hospital stay was expected to be brief. It was theorised that the derailment might have been the result of an act of sabotage by the Front de liberation nationale (FLN) which was fighting for Algerian independence. [2352]

A paytrain of the Rede Ferroviária Federal was dynamited and derailed on the Estrada de Ferro Central do Brasil line at Kilometre 72 between Japeri and Botais, Brazil, at 08:22 on Tuesday 14th June 1960. The train was carrying 27 million cruzeiros, the wages of several thousand employees. The money had arrived in Japeri as part of train R-1, and the employees there having been paid, it then continued in a new train, which consisted of steam locomotive No. 1147 and a single vehicle, the paycar. The train was in charge of Brakeman Pedro José da Silva, with Driver Venceslau José de Castro and an unidentified fireman, while there were about eight officials riding in the paycar. After the train had been derailed, it was attacked by masked and well-armed robbers and a gunfight ensued between the robbers and the armed guard of the train, as a result of which one of the railwaymen, Francelino Pauline Correia, was killed, and three others were injured. The shooting ended when the paycar staff ran out of ammunition and the robbers then boarded it, subjugated the railwaymen and led them to a nearby ravine. One of the gang wanted to kill the captives but he was dissuaded by the others. The robbers then fled with their loot in a car, apart from a crate containing 2 million cruzeiros, which they accidentally dropped as they made their escape. [2353]

Large boulders placed on the track derailed a Djibouti – Addis Ababa express just after midnight on Thursday 11th August 1960 between Addis Ababa and Diredawa,

---

[2349] Ibid, 12th January 1960

[2350] *Moberly Monitor-Index*, Moberly, Missouri, United States, 11th January 1960

[2351] *The Pharos-Tribune and Logansport Press*, Logansport, Indiana, United States, 19th June 1960

[2352] *The Times*, London, England, 10th March 1960

[2353] *Jornal do Brasil*, Rio de Janeiro, Brazil, 15th June 1960

Ethiopia. The bandits, who according to the train crew numbered about 300, and were believed to be Somali tribesmen, then attacked the train. The passengers were robbed of all their money and possessions and stripped of their clothes, while three goods waggons attached to their train were stripped of their contents. Three passengers were killed and a number of Ethiopians and a French Roman Catholic nun were wounded. One of the passengers who died was M. Maurice Malavays, the Franco-Ethiopian Railway Company's stores and equipment chief. Ethiopian troops and police scoured the area for the attackers.[2354] [2355]

On Sunday 6th November 1960 a bomb exploded on a New York City Transit Authority subway train on the former Independent line at 125th Street, New York, United States. The dynamite bomb had been hidden under a straw-covered seat at the rear of the fifth car of a six-car train. The passenger sat on the seat directly over the bomb, 15-year-old Sandra Breland, was killed. The bomb was believed to have been the work of the so-called Holiday Bomber, responsible for a series of bombings in the previous months, though before he had employed low-grade powder and placed bombs where they were less likely to be lethal.[2356]

[2354] *The Times*, London, England, 15th August 1960

[2355] *Pacific Stars and Stripes*, Tokyo, Japan, 12th August 1960

[2356] *The Post Standard*, Syracuse, New York, United States, 8th November 1960

CHAPTER 19

MISCELLANEOUS ACCIDENTS

The vast majority of railway accidents fall into one of a small number of categories such as derailments, collisions, bridge failures and so on. A few, though, do not easily fall into any obvious group, hence this chapter on miscellaneous accidents.

What was referred to in newspapers as a 'freak wreck' occurred on the Pennsylvania Railroad near Warrior Ridge, Pennsylvania, United States, at 1.41 a.m. on Thursday 1st May 1947. *The American* express, en route from New York to St Louis, Missouri, was struck by the displaced steel plates of a freight train in an accident reminiscent of that at Beighton, England, in February 1942, described in the previous volume of this series. As a result of the accident four passengers were killed more or less outright, one died later, and 53 were injured.

There were four tracks at the scene of the accident, which from south to north were No. 1, eastward passenger; No. 2, eastward freight; No. 3, westward freight; and No. 4, westward passenger. The freight train was on track No. 3 and the passenger train was on track No. 4. Automatic approach-lit position light signals were provided.

Westbound passenger train No. 67, *The American,* consisted of 4-4-4-4 steam locomotive No. 5512, a mail car, a baggage car, a passenger baggage car, two coaches, two sleeping cars, a dining car, a lounge car and six sleeping cars, in charge of Conductor J.V. Givler with Engineer R.B. Pee and Fireman R.A. Zeigler. It had passed Hunt, the last open office, 4¼ miles from Warrior Ridge, at 1.36 a.m., 35 minutes late, had passed signal No. 2067, showing *Proceed*, and was moving at 57 miles an hour on track No. 4 when it was struck by the displaced plate.

Westbound freight train Extra No. 6744 West consisted of engine No. 6744, 65 cars and a caboose in charge of Conductor W.M. Ryan with Engineer G.R. Cone and Fireman E.H. Benson. It had passed Hunt at 1.20 a.m., stopped at signal No. 2067 {the signals for different tracks had the same numbers} showing *Stop then proceed*, had restarted and was moving at 5 miles an hour when the accident occurred.

Neither train was derailed. The displaced plate, from the seventeenth car of the freight train, scraped the side of the engine of the passenger train, dented its first, third and fifth cars and sheared through the side of the fourth car, in which the casualties occurred. The displaced plate then came into contact with plates on the cars in front of it in the freight train, and a plate from the sixteenth car fell off and fouling track No. 2 was struck by eastbound freight train Extra No. 6893 East, the engine of which was derailed. [2357]

A newspaper described the aftermath of the accident. 'Ambulances from the Blair Memorial hospital and from Mount Union, along with emergency ambulances from the Temple funeral home in Petersburg as well as those of Huntingdon funeral homes, were utilized to remove the injured from the scene of the wreck to the hospital in Huntingdon. Notified of the wreck, the hospital staff was organized on an emergency basis by Administrator Leon I. Houck, with additional physicians and nurses called to the institution. As the wreck victims arrived at the hospital their apparent injuries were quickly determined and they were admitted to beds, many of them necessarily in the second floor corridor because of lack of space.

'The American was travelling through the Juniata river gap when it overtook and passed the freight on the adjacent track. Witnesses said there was a brilliant flash suddenly as the sliding steel ripped into the express locomotive. "It seemed like a long metal arm reached into our coach," a passenger said. Rescue crews were forced to use acetylene torches to free the injured and recover bodies of the dead. The steel scraped a mail car, an express car and a combination passenger and baggage car after the striking the locomotive. Then it slashed into the train's first passenger coach, gouging out the entire left side of car 4344. Emergency brakes halted the express as the dislodged steel tore out the lavatory in the forward part of the second passenger coach, car 4199.

'Most of the dead and injured were riding in car 4344. Those who escaped injury huddled in the darkness awaiting a relief train. Many went to the nearby Community hall in Petersburg. Most of the passengers appeared stunned, but one couple calmly changed the diaper on their laughing baby. Most of the passengers in the coach were asleep when the crash occurred. They were awakened first by the brilliant flash of fire from the grinding steel, then tossed from their seats.

'Emergency crews quickly cleared the wreckage from three of the four tracks of the right-of-way – the main east-west line of the Pennsylvania railroad. It was five hours before normal traffic was restored.' [2358]

A passenger described his experiences. 'Seaman Ernest Lubin, 19, of Pittsburgh, a passenger on the Pennsylvania railroad's American express when it was wrecked near Huntingdon, yesterday, said that almost all of the train's passengers were asleep when it was struck by the load of steel freight. Lubin, who was coming home on a short leave, said that he was in the fifth coach and had fallen asleep in his seat. "Suddenly I heard the shattering of glass and saw giant sparks flying from the coach ahead," Lubin said. "The train grated to a stop. I jumped up and started forward. As I reached the door to the fourth car, a man opened it and stumbled out. His face was a mass of blood." He brushed past me and went on. Broken glass was everywhere and there were bodies lying on the floor. The steel had ripped out all the partitions between the windows on one side – it was just one giant, jagged, window. In front of me on the floor was a baby. Its mother was dead beside it. I think the baby had a broken leg and a broken back. Three soldiers had been sitting, one behind the other, on the side that was hit. The first one was dead, the second unhurt, and the face of the third cut to pieces. It seemed that almost everyone in the car was cut by flying glass. I went back and finally managed to find a doctor and two nurses. Later, some of us service men and some of the railroad workers made beds on the track

---

[2357] Interstate Commerce Commission report

[2358] *Altoona Mirror*, Altoona, Pennsylvania, United States, 1st May 1947 out of blankets and sheets. The injured were carried there until ambulances could pick them up. Those of us who weren't injured were put on a special train."' [2359]

The Interstate Commerce Commission (ICC) inquiry into the accident noted that the shipment of steel plates had been loaded onto three gondola cars at the plant of the Lukens Steel Company at Coatesville, Pennsylvania, destined for East Pittsburgh, in the same state, via the Reading and Pennsylvania railroads, and had become part of Extra 6744 West at Enola Yard, Harrisburg.

The crew of train Extra 6744 West told the inquiry that they had been unaware that there had been anything wrong until they had heard a call to stop all trains on the train phone from someone on the passenger train, when they had duly stopped.

The first warning that the crew of train No. 67 had of the accident was hearing an unusual noise at the left side of the engine. The engineer had applied the brakes, there had been a stop signal sounded by the brakeman from the rear of the third car, and the engineer had then applied the brakes in emergency.

The engine of Extra No. 6893 East had struck the obstruction on track No. 2 before there had been time to protect the adjoining running lines.

The crew of train Extra 6744 West stated that their train had been running normally up to the moment of the accident.

The loading foreman at the Lukens Steel Company stated that he had inspected the seventeenth car after it had been loaded and he was sure that it had been loaded in conformity with specifications. Inspectors at the Reading in Coatesville, at the Enola Yard, and by the train crew during the journey had found no problems with the loading.

The foreman of the Pennsylvania repair shop at Altoona, who examined the car after the accident, stated that in his opinion the bottom edges of the top plates had not been placed in contact with the bearing pieces on the car flooring when the car had been loaded at the point of origin. He thought that because this was the case, the weight of the top plates at the west end of the car had been sufficient during the movement of the car to force the bottom plate upward and that, since the overlapping ends of the anchor brackets were not secured, the two anchor brackets at the west end had been displaced by the upward movement of the plate.

After the accident the Association of American Railroads issued additional instructions pertaining to the diagonal loading of flat, dished or flange plates. These instructions now required a 3/4-inch rod extending diagonally across the top of the lading and secured to each anchor bracket and to the opposite side of the car near the car flooring.

The ICC concluded that that the accident had been caused by operating a freight train in which the loading on one of the cars was not adequately secured. [2360]

An American Army Air Force plane, both of the engines of which had failed, crashed into the rear of a goods train crossing a bridge near Hachioji, Japan, on Tuesday 15th July 1947. One of the waggons of the train was knocked off the bridge into the river. A railwayman was killed and six were injured. The two aircrew were unhurt. [2361]

[2359] Ibid, 2nd May 1947
[2360] Interstate Commerce Commission report
[2361] *The Straits Times*, Singapore, 16th July 1947

Having agreed in November 1946 by the Lingaddadjati Agreement to the establishment of an Indonesian republic on 1st January 1949, on 20th July 1947 the Dutch launched a military offensive, Operatie Product, aimed at destroying the republic, claiming violations of the agreement. At the beginning of the offensive Dutch aircraft strafed a Soerakarta (now Surakarta) – Sraden train in central Java. [2362] A little later, possibly on 21st July, another train was machine-gunned near Lawang in East Java. [2363]

Situated as it is on the Pacific Ring of Fire, Japan has always been liable to earthquakes. At 17:14 on Monday 28th June 1948 the Fukui earthquake, magnitude 6.8, occurred, killing over 3,700 people. All the coaches of one passenger train and three coaches of another were overturned between Morita and Kitafukui. The Nagoya Railway stated that it believed that the casualties would be heavy. [2364]

The Salta – Antofagasta railway, Trasandino del Norte or Huaytiquina line, a 1,000mm railway linking Argentina and Chile, only opened in late February 1948. On 28th September 1948 it was the location of a serious accident near Socompa, Argentina.

The usual customs formalities having been completed at Socompa, passenger train No. 232 set off for Salta, Argentina, hauled by engine No. 859, with Driver Laurentino Cinari and Fireman Mario Samaniego. Riding in the train were a second train crew consisting of Driver Juan León Reyes and Fireman Cristóbal Gómez. The train had only covered ten kilometres when without warning it left the track at a location known as Salar del Diablo. The engine, a tank waggon, two vans and a second-class carriage overturned. The two firemen and passengers Fabio Agudo, Lizardo Mamani and Pastora Nolasco, the last three all from San Antonio de los Cobres, lost their lives. Fireman Samaniego was trapped in the engine cab and scalded by escaping steam from two broken pipes but before he lost consciousness he managed to turn off the valves supplying oil to the boiler. Seven other people were injured.

When news of the accident reached Salta a motor car and then a special train were sent from there and subsequently doctors and nurses were flown from Buenos Aires with blood plasma and penicillin.

The line ran on an embankment at the site of the accident. It was theorised that the derailment had possibly resulted from track subsidence due to an earthquake tremor. Another severe earthquake had affected the country only a month earlier. [2365]

Fifteen Hong Kong passengers, including women and children, were killed when a fast train from Canton to Hong Kong was machine-gunned by two single-engine Nationalist aircraft on Wednesday 22nd February 1950, about 30 kilometres north of Shumchun (now Chenzhen), China. Between 15 and 20 passengers were injured. Nine of them were rushed across the British border for treatment. [2366]

2362 *The Independent-Record*, Helena, Montana, United States, 21st July 1947

2363 *The Queensland Times*, Ipswich, Queensland, Australia, 22nd July 1947

2364 *The Chronicle Telegram*, Elyria, Ohio, United States, 28th June 1948

2365 https://www.eltribuno.com/salta/nota/2019-11-30-23-8-0-el-terremoto-de-1948-y-el-primer-accidente-ferroviario-en-el-ramal-c-14

2366 *The Straits Times*, Singapore, 24th February 1950

The Korean War had begun on Sunday 25th June 1950 when the north invaded the south. On Friday 30th June American Superfortress aircraft attacked various North Korean targets including two locomotives and a number of waggons, either pulling trains or in goods yards. [2367]

On Sunday 2nd July American fighters attacked a train between Seoul and Suwon, South Korea. [2368]

Fighter bombers of the American Fifth Air Force attacked locomotives and waggons in South Korea on Wednesday 12th July 1950. [2369]

Unites States aircraft attacked a train near Kumchon, South Korea, on Wednesday 26th July 1950. [2370]

American aircraft blew up an ammunition train at Wonju, South Korea, on Monday 8th January 1951. [2371]

A major disaster occurred near Montenegro, Columbia, on Friday 5th October 1951, when a passenger train was struck by a landslide. The train to which the accident happened consisted of steam engine No. 53, a luggage van, two first-class and four third-class coaches, in charge of Conductor Bernardo Zafra with Driver Rodolfo Contreras. The train was between Carmelita station and Montenegro, and about a kilometre from the latter place, at the Puente del Robre, when a large rockfall fell onto it from the hillside overlooking the railway just after 19:00. The main body of the rockfall struck the third of the four crowded third-class coaches, which was crushed and destroyed, while the second and fourth third-class coaches were also badly damaged. An initial report stated that 26 people had lost their lives and another 18 had been injured, several very seriously. They would almost all have been peasants from the surrounding area, on their way to the Saturday markets in different towns.

As the accident happened close to the town of Montenegro, help was quickly on hand from townspeople, assisted by the Montenegro fire brigade in charge of Señor Hernando Ocampo and firemen from nearby Armenia. Later the administrator of the Ferrocarril del Pacifico left Cali with further assistance and the Red Cross sent plasma and medical supplies from Bogotá by air to Armenia. There were allegations in the press, however, that one of their correspondents had not been made welcome at the scene by railway officials, who had also not permitted firemen from Quimbaya and Pereira to reach the scene by rail. The rescue operation continued for some time and there were fears that more bodies might be found in the wreckage or in the adjoining Rio Roble, but none were.

The dead and injured were taken to the San Vicente hospital in Montenegro, where distressing scenes such as are common in such circumstances were witnessed; an old man clutched the bodies of his dead grandchildren, aged seven and nine; a worker at the hospital

---

2367 *The Times*, London, England, 2nd July 1950

2368 *The Bakersfield Californian*, Bakersfield, California, United States, 3rd July 1950

2369 *The Times*, London, England, 13th July 1950

2370 Author's notes, source mislaid.

2371 *Southern Illinoisian*, Carbondale, Illinois, United States, 8th January 1951 came across the disfigured body of his own father; Julio Garcia, who had gone to the hospital in case he could identify someone from the town, was horrified to discover the bodies of his own wife and small son, whom he had not even realised had been passengers on the train. Others were luckier; Conductor Zafra, who had been busy collecting tickets, had just left the ill-fated third coach; so had Jesús Urrea, who just before the accident had gone from the third coach to join a friend in the second one. It was also fortunate that the train came to a stand just short of the bridge over the Rio Roble.

By the following day the number of dead had risen to 28, though some of the less badly hurt had been discharged from hospital. At that time a number of the bodies, which had not been found with any identification papers, had still not been identified. [2372]

At 10.23 on Tuesday 4th March 1952 Japan suffered the Tokachi earthquake, with a magnitude of 8.1, occurred off the island of Hokkaidō, accompanied by a tsunami. It was not an exceptionally severe earthquake in terms of casualties but at least 28 people died and many houses were damaged or destroyed. [2373] As far as the railway was concerned five or six trains were reportedly derailed, including a passenger and a freight train derailed in a tunnel at Nemuro with unknown casualties [2374] and a freight train between Tomewaka and Rikubetsu. [2375] The location of the other derailments was not stated. [2376]

A Caminhos de Ferro Portugueses electric train en route from the seaside report of Cascais to Lisbon was struck by a landslide near Caxias, Portugal, on Monday 31st March 1952. Train No. 10 had left Cascais at 11:00 and stopped at Caxias station at 11:30. It then restarted, had covered less than a kilometre alongside the estuary of the Tagus River, and was rounding a curve below the Gibalte lighthouse when its penultimate third-class car was struck by a landslide composed of soil and large boulders. There had been persistent heavy rain in the days leading up to the accident, which had caused flooding elsewhere.

The coach that the landslide struck was demolished and a number of its passengers were killed and others injured. The first help, from the train crew and uninjured passengers, was soon supplemented by assistance from outside, from firemen, police, soldiers, and locals. The landslide had damaged the overhead line equipment but a rescue train arrived at 12:50, drawn by a steam locomotive with a third-class coach carrying a large number of railway workers. The injured were taken to the San José Hospital in Lisbon. Numerous spectators also appeared at the scene and the police were busy keeping them at bay. Early on there were known to be seven dead and about 50 injured. Señor José Frederico Ulrich, the Minister of Public Works, Dr. Albano Ribeiro, Secretary to the Minister of Communications, Engineer Nazaré Oliveira and various other company and government officials went to the scene of the accident.

By the following day the bodies of three more victims had been discovered in the wreckage and it was feared that there might be others. The rain and wind continued, and the work of clearing the line was slow and difficult. There were further landslides and work had to be stopped until the foundations of the lighthouse overlooking the railway

---

[2372] *El Tiempo*, Bogotá, Colombia, 6th and 7th October 1951

[2373] Wikipedia – 1952 Tokachi earthquake

[2374] *Pacific Stars and Stripes*, Tokyo, Japan, 4th March 1952

[2375] *Humboldt Standard*, Eureka, California, United States, 4th March 1952

[2376] *The Railway Gazette*, London, England, 7th March 1952 had been shored up. It transpired that Señor Fernando da Silva Marreiros, the keeper, had heard noises beneath his lighthouse and had begun to see cracks in it. He had telephoned the General Director of Lighthouses, and an engineer and another man had been sent to investigate, but the accident had happened before they arrived. The following day it was decided that the lighthouse was a factor in causing the landslide, and the keeper and his family were ordered out of it, while its lamp was removed to another location nearby.

The day after the accident a newspaper reported the measures which were taken in the aftermath of the disaster:

(1) The Batalhao de Sapadores (fire brigade) had installed two pumps which were working continuously to drain excess water.

(2) The lighthouse and the adjoining lighthouse keeper's residence were being shored up. The movement of the lighthouse was being checked in two dimensions. In a day it had apparently moved 36 millimetres towards the river and at the same time a plumb line was being used to check if it was also leaning over (as per the Leaning Tower of Pisa). To support it, holes were being drilled through a five-metre-thick wall and steel cables were being passed through them.

(3) Checks were being made to determine if the lighthouse and the residence had shifted from their foundations.

(4) The geology of the terrain at the accident site was being examined.

Two days after the disaster, on 2nd April, it having been determined that the lighthouse was built upon loam limestone, it was agreed at a meeting of the Minister of Public Works and Engineers Oliveira and Arantes that the lighthouse must be demolished and a new, lighter one, be built on a more solid rock foundation. [2377]

Collisions with animals on the line do not usually have serious consequences, except to the beast involved, but on Wednesday 6th August 1952 the headlight of the engine of a passenger train had failed as it ran into the station at Drownino, USSR, at night. Thus, in the darkness, its driver failed to see a horse standing on the line. The collision with the animal was sufficient to derail the locomotive and the coaches of the train piled up into it. No less than 109 people lost their lives and another 211 were injured. [2378]

'Allied' Sabre aircraft attacked a freight train at an unstated location in North Korea on Friday 11th April 1952. [2379]

On Wednesday 24th September 1952 a supply train moving down the east coast of North Korea was attacked by American aircraft at a location given in the newspaper report as Chongjin (actually renamed Kimchæk in 1951). As the train continued southwards it was then shelled by the British (United Nations) destroyer *Charity*, which poured almost 2,400 shells into it. [2380]

[2377] *Diario de Lisbôa*, Lisbon, Portugal, 31st March and 1st, 2nd and 3rd April 1952

[2378] https://de.wikipedia.org/wiki/Liste_schwerer_Unf%C3%A4lle_im_Schienenverkehr#1941%E2%80%931950

[2379] *The Doylestown Intelligencer*, Doylestown, Pennsylvania, United States, 11th April 1952

[2380] *The New Mexican*, Santa Fe, New Mexico, United States, 25th September 1952

The Korean War dragged on into 1953 and on Wednesday 5th March 1953 American light B-26 night bombers, despite overcast conditions, destroyed a train of 13 boxcars 'near Chongjin' (Kimchæk), North Korea. [2381] The war ended with an armistice on 27th July 1953.

On Saturday 13th December 1952 a massive boulder rolled down a mountainside and smashed into a coach of a São Paulo – Rio de Janeiro passenger train in the vicinity of Barra Mansa, Brazil, killing six passengers and injuring 15. There had been heavy rain the previous night.[2382]

What up to the present time (October 2022) is by far the most serious railway accident in New Zealand occurred on Christmas Eve, Thursday 24th December 1953, at Tangiwai. The accident was also, as far as the author is aware, unique in its cause, which was the destruction of a bridge by a lahar, a mixture of water and pyroclastic debris flowing from a volcano.

The train involved in the disaster was the 3 p.m. express from Wellington to Auckland, consisting of Class Ka locomotive No. 949, five second-class coaches, four first-class coaches, a van, and a mail van, in charge of Guard William Inglis with Driver Charles Parker and Fireman Lance Redman.

Shortly before reaching the fateful bridge the express had passed through Tangiwai station. Here it should have crossed a southbound goods train, but the latter train was late so the crossing was moved northwards to Karioi. Had the goods train been on time it, rather than the express, would probably have met with disaster. Station Agent Raymond Hall at Tangiwai thought the express had been running a little slower than usual.

Postal Clerk Cyril Ellis was driving along the main road adjoining the railway, on his way to visit his parents, when he found the road bridge over the Whangehu River covered in floodwater, although it was not raining. He stopped his vehicle to investigate, then saw the light of the approaching express and anticipating danger ran towards it waving his torch. He said later, "I did not know what train, but I realised that with the river in that state it had no chance of crossing the rail bridge. I jumped the railway fence and ran up the embankment. I ran down the middle of the track brandishing my torch and shouting at the top of my voice, in an endeavour of the engine driver or fireman. I ran away from the bridge to try and give the crew more room to pull the train up, but it was no use – they just didn't see the torch. I jumped clear and yelled to the driver when the locomotive roared past me about 200 yards from the bridge, but he had no chance of hearing me above the roar of both river and engine. As the train passed I realised for the first time that it was the express. I didn't know the bridge was damaged, I had merely realised that with the flood they could not get across."

In fact, it seemed that the train crew had either seen Ellis's torch or the damaged bridge as the brakes of the engine were applied in emergency, but the train could not be stopped in the short distance available. The engine, tender, and five second-class coaches fell into the gap left by the collapsed bridge, and the leading first-class coach teetered on the edge of the gap and then fell into the river as well. The 198-foot-long bridge consisted

[2381] *The Daily Dispatch*, Moline, Illinois, United States, 5th March 1953

[2382] *The Daily Ardmoreite*, Ardmore, Oklahoma, United States, 14th December 1952 of seven plate girder spans on concrete piers; piers 2, 3, 4 and 5, (numbered from the Auckland end) had been damaged or carried away by the lahar.

Guard Inglis did not know what had happened when his train suddenly stopped until he met Ellis, who told him, "Half your train is in the river." The guard and Ellis entered the leading first-class carriage and the guard went to its front end and looked out the door, when he exclaimed, "Good God, we have lost five carriages and the engine." He urged the passengers in the coach to walk back through it but at that moment the coupling gave way and the coach plunged into the river. Describing this moment, Ellis said later, "She rolled over a number of times and then came to a rest on her side, with the water flowing through. It was wicked. I had a couple of women hanging on to me, but I managed to knock out a window with an elbow and my feet." With a passenger, John Holman, Ellis succeeded in rescuing all the passengers from the leading first-class coach except a girl trapped in her seat with her head under water, who drowned. Ellis and Holman were later awarded the George Medal for their bravery, while Guard Inglis and Arthur Bell, a passing motorist who had saved 15 passengers while his wife drove to the New Zealand Forestry Service in Karioi to raise the alarm, received the British Empire Medal.

Of the 285 people on the train 151 lost their lives, comprising Driver Parker, Fireman Redman, 148 second-class passengers and one first-class passenger. A number of the bodies of those who died were washed down the river into the ocean and never recovered.

A board of inquiry investigated the catastrophe and found that the bridge had been fatally damaged by a lahar from Mount Ruapehu, a dormant volcano. It found that:

'The relative positions of the 44 ft girder (span 3) and the first car (A. 1907) dictate that the girder was removed before the passage of the first car. This girder must either have been removed before the locomotive arrived on the span, or during the passage of the locomotive and tender over the span. If this girder was removed during the passage of the locomotive, there is a strong inference that some part of the heavy locomotive would have hit girder No. 2. There is, however, no evidence of extensive damage to girder No. 2, and it may therefore be inferred that girder No. 3 was not removed during the passage of the locomotive across it. Span 3 was therefore carried away before the locomotive entered upon it and there is a strong inference that Pier No. 4 was carried away before the train passed over it. This would cause span 4 to be carried away. There is also a strong inference that Pier No. 4 was removed by the lahar and that this caused the accident. The bridge had been designed for foreseeable flooding and for previous lahars that had different characteristics to the 1953 lahar, the forces of which had been unpredictable.'

There had been lahars in the previous 93 years in 1861, 1889, 1895, 1925. The 1861 and 1925 floods had been cold lahars due to the release of water from lake water below the glacier; the 1889 and 1895 lahars had been associated with eruptions in the crater lake of Mount Ruapehu. The 1953 cold lahar had been due to the sudden release of a huge amount of water from the Mount Ruapehu crater lake through an outlet cave below the Whangaehu Glacier, which had been channelled into the Whangaehu River, carrying with it large amounts of volcanic ash from a 1945 eruption of the volcano and blocks of ice from the glacier. The lahar had proceeded down the mountainside as a huge wave, carrying massive quantities of sand, silt, and boulders. The bridge had been struck between 10.10 p.m. and 10.15 p.m. and the train had reached the bridge at 10.21 p.m.

The board of inquiry found that no one was directly or indirectly to blame for the disaster. It made a number of observations and recommendations:

(a) A bridge division should be created within the Chief Engineers portion of the Railway Department to deal with the design, erection, and maintenance of railway bridges

(b) Battery-operated floodlights which could be plugged in, in each carriage, should be provided for emergency use

(c) It was not feasible to place luggage and mail vans next to the engine

(d) Emergency exits in the roofs of carriages were problematical and would be dangerous on electrified lines

(e) Tangiwai station could be repositioned nearer the bridge, staff could be stationed at the bridge or a tunnel could be constructed beneath the river

(f) Different kinds of lahar warning devices should be considered

(g) Research into the thermal activity of Mount Ruapehu should be conducted. [2383] [2384]

Guatemala was subject to a coup d'état in the summer of 1954, whereby the democratically elected President Jacobo Árbenz was ousted by dictator Carlos Castillo Armas, who had American support. During the coup, on Friday 25th June 1954, a passenger train was strafed by an aircraft near Zacapa, Guatemala, killing seven people. [2385]

Typhoon Harriet, which struck Japan on Thursday 27th September 1956, was blamed for causing two shipwrecks and a train crash. Heavy rain associated with the typhoon was responsible for a landslide which struck the coach of a train near Nagoya and hurled it 24 feet down a slope into a flooded river. Only three of the estimated 20 passengers in the coach could be rescued, the remaining 17 were presumed to have drowned. [2386]

On the evening of Monday 20th January 1958 a passenger complained to railway staff that the 5.28 p.m. passenger train from London Cannon Street and the 5.30 p.m. passenger train from London Cannon Street to Orpington, running on parallel lines, had scraped one another, on the curve near London Bridge, England, on British Railways, Southern Region. An investigation showed that this was true. Slightly wider carriages were now being used, an official explained. The tracks had to be realigned to solve the problem. [2387]

On Monday 10th August 1959 the Milano – Amsterdam express was struck by lightning near Monza, Italy. The train was slightly damaged and was held up for 30 minutes. [2388]

[2383] *Tragedy on the Track* by Geoff Conly and Graham Stewart, pgs. 1-62

[2384] Wikipedia – Tangiwai disaster

[2385] *Bluefield Daily Telegraph*, Bluefield, West Virginia, 26th June 1954

[2386] *The Montana Standard*, Butte-Anaconda, Montana, United States, 28th September 1956

[2387] *The Times*, London, England, 21st January 1958

[2388] Ibid, 11th August 1959

INDEX

TABLE OF MAJOR RAILWAY ACCIDENTS 1946-1960

Date	Location	Railway Company	Description	Passenger Fatalities
1 Jan 1946	Lichfield (Trent Valley), England	London Midland and Scottish	Train diverted onto wrong line, collision	20
5 Jan 1946	Browney signal box, England	London and North Eastern	Double collision	10
21 Jan 1946	Japan	Fujisanroku-Dentetsu	Collision	26
26 Jan 1946	Ravenstein, The Netherlands	Nederlandse Spoorwegen	Head-on collision, single line	5
28 Jan 1946	Tsurumaki, Japan	Odakyu	Runaway train, derailment	30
4 Mar 1946	Bhagauli, India	East Indian	Head-on collision	46
14 Mar 1946	Mashike, Japan	Japanese National	Derailment	17
20 Mar 1946	B Laranjeiras and Riachuelo, (Near Aracaju), Brazil	Viação Férrea Federal Leste Brasileiro	Derailment	185
8 Apr 1946	Wanchiatun, China		Derailment, sabotage	300
18 Apr 1946	Eisleben, Germany	Deutsche Reichsbahn	Rear collision in fog	23
25 Apr 1946	Naperville, IL, USA	Chicago Burlington and Quincy	Rear collision	44
10 May 1946	Near Canton, China		Derailment, sabotage, fire	6
30 May 1946	Shaochi Bridge, China		Derailment, train fell from bridge	90
19 Jun 1946	Ponte Cagnano, Italy		Train diverted onto wrong line, collision	19
26 Jul 1946	Azuchi Tunnel, Japan	Japanese National	Rear collision in collision	15
28 Jul 1946	Bhatni Junction, India	Oudh Tirhut State	Rear collision	23
12 Aug 1946	Kokura, Japan		Rear collision	7
13 Aug 1946	Near Kinosho, Japan	Onomichi	Runaway train, derailment	37

29 Aug 1946	Tandjoeng-West Halt, Dutch East Indies		Collision with bus on open level crossing	9
9 Sep 1946	B Talsahar and Bramhanbaria, India	Bengal and Assam	Derailment	7
18 Sep 1946	B Reichenbach and Linde block post, Germany		Rear collision	6
22 Sep 1946	Near Héricourt, France	Société nationale des chemins de fer français	Rear collision	7
23 Sep 1946	B Dighwara and Barra Gopal, India	Oudh Tirhut State	Derailment	29
26 Sep 1946	Near Oro Grande, CA, USA	Atchison Topeka and Santa Fe	Derailment, excessive speed	5
28 Sep 1946	Łódź Kaliska, Poland		Rear collision?	21
28 Sep 1946	B Veynes and Montmaur, France		Head-on collision, single line	7
4 Oct 1946	Ongole, Madras, India	Madras and Southern Mahratta	Train diverted onto wrong line, collision	36
4 Oct 1946	Szczecin Turzyn, Poland		Rear collision?	7
31 Oct 1946	La Mata, Mexico		Collision	11
1 Nov 1946	Tröglitz, Germany	Deutsche Reichsbahn	Train diverted onto wrong line; buffer stop collision	30
12 Nov 1946	Yeongdeungro, Korea		Collision	41
12 Nov 1946	Revigny, France	Société nationale des chemins de fer français	Rear collision in fog	32
13 Nov 1946	Beit Safafa, Palestine	Palestine	Train mined by terrorists	6
13 Dec 1946	Near Guthrie, OH, USA	Pennsylvania	Double collision	15
15 Dec 1946	Santa Fe, Near Minturno, Italy		Collision	5

17 Dec 1946	Cinco Casas, Spain	Red Nacional de los Ferrocarriles Españoles	Collision with shunting goods	21
19 Dec 1946	Near Taguchi, Japan	Japanese National	Derailment, train ran into snowdrift?	11
2 Jan 1947	Gidea Park, England	London and North Eastern	Rear collision in fog	7
17 Jan 1947	Lerdo, CA, USA	Southern Pacific	Derailment, broken rail	8
8 Feb 1947	Segarcea (Near Craiova) Romania		Train diverted onto wrong line in blizzard, collision	27
18 Feb 1947	Near Gallitzin, PA, USA	Pennsylvania	Derailment, excessive speed	21
25 Feb 1947	B Higashi Hanno and Kompagawa, Japan	Japanese National	Derailment, excessive speed, train fell from bridge	184
7 Mar 1947	Near Thazi Junction, Burma		Derailment, sabotage	28
16 Apr 1947	Ikomayama Tunnel, Japan	Kinki Nippon	Fire in tunnel	28
22 Apr 1947	Near Tabata, Japan		Rear collision	5
22 Apr 1947	Near Rehoboth, Palestine	Palestine	Train mined and fired on by terrorists	8
1 May 1947	Near Warrior Ridge, PA, USA	Pennsylvania	Collision with load projecting from freight train on adjoining line (derailment, obstruction on line)	5
5 May 1947	Camp Mountain, QD, Australia		Runaway train, derailment	14
5 May 1947	Husby, Germany		Rear collision	5
10 May 1947	Emperador, Spain	Red Nacional de los Ferrocarriles Españoles	Train diverted onto wrong line, head-on collision	12
17 May 1947	B Kamalsnagar and Nayanpur, India	Bengal and Assam	Derailment, sabotage	34
21 May 1947	Bukit Tengah, Malaya	Malayan	Derailment, fire	11
30 May 1947	Lutai, China		Derailment, sabotage	100

1 Jun 1947	Welver, Germany		Derailment, track distorted by the heat	7
3 June 1947	Gravanches, France	Société nationale des chemins de fer français	Fire	6
7 Jun 1947	Vinohrady Tunnel, Czechoslovakia		Collision in tunnel	7
12 Jun 1947	Near Yehpaishou, China		Derailment in tunnel, sabotage, rear collision	?
15 Jul 1947	Near La Cruz, Argentina	Nordeste Argentino	Collision with cow, derailment	16
1 Jul 1947	B Kudamatsu and Hikari, Japan	Japanese National	Derailment	15
10 Jul 1947	Near Yingtak, China		Derailment, part of train fell from bridge	20
14 July 1947	Biên Hòa, Vietnam, French Indochina		Derailment, sabotage	?
20 Jul 1947	B Soerakarta and Sraden, Dutch East Indies		Train attacked by Dutch aircraft	?
20 Jul 1947	Lawang, Dutch East Indies		Train machine-gunned by Dutch aircraft	?
21 Jul 1947	Near Polesworth, England	London Midland and Scottish	Derailment, track defects	5
23 Jul 1947	Near Tangfang, China		Train mined by Communist guerrillas	27
26 Jul 1947	B Biberbrücke and Einsiedeln, Switzerland	Chemin-de-Fer du Sud-Est de la Suisse	Head-on collision, single line	7
29 Jul 1947	Near Pematangsiantar, Sumatra, Dutch East Indies		Head-on collision, single line	400
5 Aug 1947	Near Kuty, Czechoslovakia		Train diverted onto wrong line? derailment	16
9 Aug 1947	Near Doncaster, England		Rear collision	18

16 Aug 1947	Vila Franca de Xira, Portugal		Derailment, broken axle	17
23 Aug 1947	B Velten and Kremmen, Germany		Fire	24
27 Aug 1947	Kolaghat, India	Bengal-Nagpur	Rear collision	23
30 Aug 1947	B Clarés de Ribota and Villarroya de la Sierra, Spain	Red Nacional de los Ferrocarriles Españoles	Bridge collapsed under train	15
1 Sep 1947	Dugald, MB, Canada	Canadian National	Head-on collision, fire	27
11 Sep 1947	Karlburg block post, Germany		Collision in fog	8
13 Sep 1947	B Amritsar and Cheherta, India		Derailment, sabotage	19
30 Sep 1947	B Temanggoeng and Kranggan, Dutch East Indies		Runaway coaches, derailment	25
9 Oct 1947	Jullundur, India		Derailment, obstruction on line	9
10 Oct 1947	Near Karnal, India		Collision	32
12 Oct 1947	Senohraby, Czechoslovakia		Runaway train, collision	6
18 Oct 1947	Tamaree, QD, Australia		Train diverted onto wrong line, collision	6
19 Oct 1947	Near Pengpu, China		Train exploded mine on bridge	?
24 Oct 1947	Near South Croydon, England	Southern	Rear collision in fog	31
26 Oct 1947	Goswick England	London and North Eastern	Derailment, excessive speed through crossover	27
6 Nov 1947	Saigon and Mỹ Tho, Vietnam, French Indochina		Derailment, sabotage	17
24 Nov 1947	Shambhu, India	Eastern Punjab	Train diverted onto wrong line; buffer stop collision, sabotage?	171

3 Dec 1947	Near Arras, France	Société nationale des chemins de fer français	Derailment, sabotage	20
8 Dec 1947	Near Kanchrapara, India	East Indian	Rear collision	7
12 Dec 1947	Flensburg, Germany		Rear collision	5
12 Dec 1947	Pontmort, France	Société nationale des chemins de fer français	Head-on collision, single line, fire	17
22 Dec 1947	Near Neuwied, Germany		Sidelong collision in fog	42
1 Jan 1948	Near Syracuse (Otterville), MO, USA	Missouri Pacific	Rear collision	12
5 Jan 1948	Moriyama, Japan	Nagoya	Derailment, excessive speed?	35
14 Jan 1948	Near Wykes, QC, Canada	Canadian National	Rear collision	8
10 Feb 1948	B Malatya and Fevzipasha, Turkey		Collision with rubble which had fallen from tunnel roof? Tunnel roof fell onto train? fire	7
17 Feb 1948	B Thumeries and Moncheaux, France	Compagnie Générale de Voies Ferées d'Intérêt	Head-on collision, single line	22
22 Feb 1948	B Burghalden and Wädenswil, Switzerland	Chemin de fer du Sud Est de la Suisse	Runaway train, buffer stop collision	21
25 Feb 1948	B Seddon and Blind River, New Zealand		Derailment, excessive speed	6
29 Feb 1948	Near Rehoboth, Palestine	Palestine	Train mined by Jewish terrorists	27
31 Mar 1948	Near Kawachi-Hanazono, Japan	Kinki-Nippon	Rear collision	49
31 Mar 1948	Near Benymina, Palestine	Palestine	Train mined by Jewish terrorists	40
14 Apr 1948	B Arrah and Sasaram, Bihar, India	Martin's Light	Collision with lorry on open? Level crossing	?

15 Apr 1948	Jungshahi, Pakistan		Head-on collision?	15
17 Apr 1948	Winsford, England	British Railways, London Midland Region	Rear collision	24
24 Apr 1948	Near Patras, Greece		Collision with lorry on open? Level crossing, derailment	9
30 Apr 1948	B Sultangange and Gangania, India	East Indian	Head-on collision, single line?	7
7 May 1948	B Mornago Cimbro and Ternate, Italy	Fellovie dello Stato	Rear collision in tunnel	6
15 May 1948	B Chota Ambona and Prodhan Khanta, India	East Indian	Derailment, sabotage?	30
18 May 1948	Wath Road Junction, England	British Railways, London Midland Region	Derailment, track distorted by the heat	7
25 May 1948	Rákospalota, Hungary		Collision	8
28 May 1948	Xindian River Bridge, Formosa		Fire, explosion	175
18 Jun 1948	B Saint Etienne and Villars, France		Fire	6
23 Jun 1948	Tjiboengoer, Dutch East Indies		Derailment, sabotage	5
28 Jun 1948	B Morita and Kitafukui, Japan		Derailment, track damaged by earthquake	?
28 Jun 1948	B Morita and Kitafukui, Japan		Derailment, track damaged by earthquake	?
9 Jul 1948	Uroševac, Yugoslavia,		Collision	?
19 Jul 1948	B Comotini and Xanthe, Greece		Derailment, sabotage	14
1 Aug 1948	B San Francisco and Asnazú, Colombia		Derailment	6

6 Aug 1948	Near Namyangju, South Korea		Derailment, track subsidence	9
17 Aug 1948	B Biên Hòa and Trảng Bom, Vietnam, French Indochina		Derailment, sabotage	12
21 Aug 1948	Bedoel Bridge, Dutch East Indies		Train derailed and fired on by terrorists, part of train fell from bridge	?
3 Sep 1948	Lilong, China		Derailment, sabotage? Track subsidence?	25
14 Sep 1948	Napnin, South Korea		Rear collision	38
14 Sep 1948	B Puranpur and Shahgarh, India	Oudh – Tirhut State	Head-on collision, single line?	28
19 Sep 1948	Gangajagar, East Pakistan	Eastern Bengal	Rear collision	9
22 Sep 1948	B Jamraniwah and Lodhran, West Pakistan		Head-on collision, single line?	?
10 Oct 1948	Anapoima, Colombia		Derailment, broken axle	20
22 Oct 1948	Bedoel Bridge, Dutch East Indies		Train derailed and fired on by terrorists	?
28 Oct 1948	Near Samsun, Turkey		Collision	13
29 Oct 1948	B Irmak and Ankara, Turkey		Derailment, broken axle	49
15 Nov 1948	B Furkating and Kamarbandha Ali, Assam, India	Bengal and Assam	Derailment	12
20 Nov 1948	B Kalgan and Peiping, China		Derailment, sabotage	5
30 Nov 1948	B Heaton Norris and Stockport, England	British Railways, London Midland Region	Rear collision in fog	5
2 Dec 1948	B Drusenheim and Herrlisheim, France		Head-on collision during temporary single line working	8
3 Dec 1948	B Falkensee and Spandau, Germany		Collision	17

5 Dec 1948	Bissendorf, Germany		Collision in fog	6
24 Dec 1948	Near Łowicz, Poland		Derailment, sabotage?	18
c4 Jan 1949	Near Rafa, Egypt		Derailment, sabotage	?
14 Jan 1949	Shimo Goto, Japan		Derailment?	7
23 Jan 1949	Near Boemiajoe, Dutch East Indies		Derailment, sabotage	7
26 Jan 1949	Near Serres, Greece		Train blown up by guerrillas	?
13 Feb 1949	B Els Guiamets and Mora la Nova, Spain	Red Nacional de los Ferrocarriles Españoles	Derailment, track defects? sabotage?	29
18 Feb 1949	Pont d'Atelier, France	Société nationale des chemins de fer français	Collision with light engine	42
4 Mar 1949	Near Eisenach, Germany		Collision	9
8 Mar 1949	Matsuzaka, Japan		Fire	8
13 Mar 1949	B Pipraich and Bodarwar, India	Oudh Tirhut State	Head-on collision, single line?	19
6 Apr 1949	Near Romeas, Cambodia		Collision	20
10 Apr 1949	Near Burna, India		Derailment, sabotage	10
23 Apr 1949	Chuchow, China		Derailment	11
28 Apr 1949	Near New Canada Junction, South Africa		Rear collision	66
18 Jun 1949	Mahinathpur, Bihar, India		Derailment, broken axle	6
22 Jun 1949	Bucay, Ecuador		Derailment	7
3 Jul 1949	B Ngawen and Blora, Dutch East Indies		Train ran over mines	5

4 Jul 1949	Emberménil, France	Société nationale des chemins de fer français	Derailment, track distorted by the heat	7
15 Jul 1949	Mitaka, Japan		Runaway train, Buffer stop collision, sabotage	6
5 Aug 1949	Near Pula, Yugoslavia		Derailment	14
15 Aug 1949	Near Plavno, Yugoslavia		Head-on collision, single line, stationmaster intoxicated	21
24 Aug 1949	B Wonosari and Koetowinangoen, Dutch East Indies		Train ran over mine	10
c11 Sep 1949	B Lang Son and Cao Băng, Vietnam, French Indochina		Derailment, sabotage	5
30 Sep 1949	Namchung, China		Derailment, sabotage?	24
8 Oct 1949	Near Modjokerto, Dutch East Indies		Derailment, sabotage	8
11 Oct 1949	Buenos Aires Presidente Peron, Argentina	General Mitre	Sidelong collision	18
12 Oct 1949	Near Kharagpur, India		Derailment, sabotage	6
22 Oct 1949	Nowy Dwór Mazowiecki, Poland		Derailment, excessive speed?	200
25 Oct 1949	Barbentane, France	Société nationale des chemins de fer français	Derailment, broken axle	11
2 Nov 1949	Near Dairut, Egypt		Collision, fire	6
7 Nov 1949	Las Matas and Las Rosas, Spain	Red Nacional de los Ferrocarriles Españoles	Rear collision	14
15 Nov 1949	Near Waterval Boven, South Africa		Derailment, excessive speed, part of train fell from bridge	62
7 Dec 1949	Hagar Nawatteth, Egypt		Rear collision	7

30 Dec 1949	B Empalme and Pilar, Argentina		Derailment, track washed away	8
24 Jan 1950	Near Dar-es-Salaam, Tanganyika	East African	Derailment	11
26 Jan 1950	B Moncey and Rigney, France	Société nationale des chemins de fer français	Head-on collision, single line	9
29 Jan 1950	Sirhind, Punjab, India		Derailment, defective vehicle, collision	65
8 Feb 1950	B Abidjan and Dioulasso, Côte d'Ivoire	Chemins de fer de la Côte d'Ivoire	Derailment	7
11 Feb 1950	Near Lisle-sur-Tarn, France	Société nationale des chemins de fer français	Head-on collision, single line	18
17 Feb 1950	Near Rockville Center, NY, USA	Long Island	Head-on collision	31
22 Feb 1950	20 miles north of Shumchun, China		Train machine-gunned by Nationalist aircraft	15
3 Mar 1950	Near Sulurpet, India	Madras and Southern Mahratta	Derailment, sabotage	6
c28 Mar 1950	100 miles north of Mandalay, Burma		Derailment, sabotage, train fired on by terrorists	9
30 Mar 1950	Bobadilla, Spain	Red Nacional de los Ferrocarriles Españoles	Derailment on points	8
6 Apr 1950	Near Villayana, Spain	Red Nacional de los Ferrocarriles Españoles	Derailment, broken axle	18
6 Apr 1950	B Tanguá and Rio Bonito, Brazil	Estrada de Ferro Leopoldina	Derailment, track subsidence	110
7 Apr 1950	Ripanj, Yugoslavia		Derailment, points moved under train	5
8 Apr 1950	Near General Macarthur, The Philippines		Derailment	14
12 Apr 1950	Near Bitroi, India	Oudh and Tirhut State	Derailment, sabotage, part of train fell from viaduct	35

22 Apr 1950	Near Nha Trang, Vietnam, French Indochina		Train attacked by terrorists	6
7 May 1950	B Jasidih and Sumultala, Bihar, India	East Indian	Derailment, sabotage, train fell from bridge	93
26 May 1950	B Zilina and Košice, Czechoslovakia		Collision	12
8 Jun 1950	Near Beattock Summit, Scotland	British Railways, Scottish Region	Fire	5
6 Jul 1950	Near Monica, IL, USA	Atchison Topeka and Santa Fe	Derailment, defective vehicle, collision	9
12 Jul 1950	B Hartenstein and Fährbrüke, East Germany		Head-on collision during temporary single-line working	18
17 Jul 1950	'Near the Honan-Hupeh Border, China		Derailment, sabotage	?
c26 Jul 1950	Near Rangoon, Burma		Derailment, sabotage	?
30 Jul 1950	B Abidjan and Dioulasso, Côte d'Ivoire	Chemins de fer de la Côte d'Ivoire	Derailment?	10
2 Aug 1950	Muroan, Japan		Bridge collapsed under train	9
6 Aug 1950	Near Castelvertrano, Sicily, Italy		Derailment, excessive speed?	6
12 Aug 1950	B Acari and Pavuna, Brazil	Estrada de Ferro Central do Brasil	Head-on collision, single line	6
13 Aug 1950	B Durgauti and Karmanasa, India	East Indian	Derailment, sabotage, collision	25
21 Aug 1950	B Fousan and Ulsan, South Korea		Derailment, sabotage	?
21 Aug 1950	Near Alexandria, Egypt		Explosion	5
27 Aug 1950	Penmaenmawr, Wales	British Railways, London Midland Region	Collision with light engine	5

27 Aug 1950	Havinnes, Belgium		Derailment	5
2 Sep 1950	Hales Corners, WI, USA	Milwaukee Rapid Transit and Speedrail	Head-on collision, single line	10
3 Sep 1950	Near Gurdaspur, Punjab, India	Eastern Punjab	Derailment, sabotage? Bridge collapsed under train?	11
6 Sep 1950	Near Pantojo, São Paulo, Brazil	Sorocabana	Derailment	36
9 Sep 1950	B Daulakandi and Blairab Bazar, East Pakistan		Derailment, sabotage, part of train fell from bridge	37
11 Sep 1950	Coshocton, OH, USA	Pennsylvania	Rear collision	33
17 Sep 1950	Sarpenzele, Hungary		Collision, Explosion, fire?	100
25 Sep 1950	Lexoville, France	Société nationale des chemins de fer français	Derailment, excessive speed through crossover	6
28 Sep 1950	Yongchong, South Korea		Explosion	?
1 Oct 1950	Svitavy, Moravia, Czechoslovakia		Head-on collision	7
22 Oct 1950	Near Kukong, China		Train derailed and fired on by terrorists	?
28 Oct 1950	B Godegård and Mariedamm, Sweden	Statens Järnvägar	Head-on collision, single line	7
15 Nov 1950	B Hjuksebø and Holtsås, Norway	Norges Statsbaner	Runaway waggons, collision	13
21 Nov 1950	Near Canoe River, BC, Canada	Canadian National	Head-on collision, single line, fire	17
22 Nov 1950	Jamaica (Richmond Hill), NY, USA	Long Island	Rear collision	78
3 Dec 1950	Tunnel No. 13, B Pajares and Busdongo, Spain	Red Nacional de los Ferrocarriles Españoles	Fire in tunnel	5
6 Dec 1950	Cecebre, Spain	Red Nacional de los Ferrocarriles Españoles	Head-on collision	5

23 Dec 1950	Engano, Brazil	Estrada de Ferro Leopoldina	Derailment, excessive speed?	15
22 Jan 1951	B Kinstaby and Marmaverken, Sweden	Statens Järnväger	Head-on collision, single line	9
23 Jan 1951	Vilalops, Spain	Red Nacional de los Ferrocarriles Españoles	Collision with lorry on chained? Level crossing, derailment	14
6 Feb 1951	Woodbridge, NJ, USA	Pennsylvania	Derailment, excessive speed	84
16 Mar 1951	Bridge Junction, Doncaster, England	British Railways, Eastern Region	Derailment, excessive speed and track defects	14
18 Mar 1951	Near Thea, Argentina	General Belgrano	Derailment, excessive speed?	6
24 Apr 1951	Sakuragicho (Yokohama), Japan	Japanese National	Fire	106
11 May 1951	Darshanga, India		Collision	9
18 May 1951	Near Bryn Mawr, PA, USA	Pennsylvania	Rear collision	7
7 Jun 1951	Nova Iguaçu, Brazil	Estrada de Ferro Central do Brasil	Collision with petrol tanker on FLB level crossing, explosion, fire	51
12 Jun 1951	220 km north of Hankow, China		Derailment, sabotage	?
17 Jun 1951	Ghotki, Pakistan	Pakistan Western	Rear collision	40
23 Jun 1951	Hipórito Yrigoyen, Argentina	Ferrocarriles Argentinos	Rear collision in fog	6
9 Jul 1951	Near Mandalay, Burma		Derailment, sabotage	8
22 Jul 1951	B Tatranská Lomnica and Studený Potok, Czechoslovakia	Československe Státní Dráhy	Runaway coach, collision, runaway vehicles	19
5 Aug 1951	Near Ford, England	British Railways, Southern Region	Rear collision	8

10 Aug 1951	Near Lettsworth, LA, USA	Texas and Pacific	Head-on collision, single line, fire	6
18 Aug 1951	Székesfehérvár, Hungary		Rear collision?	150
21 Aug 1951	Yugoslavia		Derailment, excessive speed?	10
24 Aug 1951	Sanry-sur-Nied, France	Société nationale des chemins de fer français	Rear collision	23
26 Aug 1951	'About 220 from Gauhati,' Assam, India		Derailment, track washed away	11
3 Sep 1951	'11 Miles from Delhi,' India	Shahdara – Saharanpur Light	Derailment, track defects?	10
19 Sep 1951	B Antonio Carlos and João Ayres, Brazil	Estrada de Ferro Central do Brasil	Rear collision in fog	21
21 Sep 1951	Weedon, England	British Railways, London Midland Region	Derailment, engine defect	14
25 Sep 1951	Langenwang, Austria		Collision with shunting goods	22
5 Oct 1951	Near Montenegro, Colombia		Train struck by landslide	28
6 Nov 1951	Near Vinkovci, Yugoslavia		Rear collision	17
8 Nov 1951	Walpertskirchen, West Germany		Train diverted onto wrong line, head-on collision	16
12 Nov 1951	Wyuta (Evanston), WY, USA	Union Pacific	Rear collision in blizzard	11
17 Nov 1951	B Pizzo and Vibo Marina, Italy		Bridge collapsed under train	8
25 Nov 1951	Woodstock, AL, USA	Southern	Head-on collision, single line in fog	15
11 Dec 1951	Near Oran, Algeria		Collision with lorry on open? Level crossing	?
17 Dec 1951	Near Piquet Carneiro, Brazil		Derailment, excessive speed?	53

4 Mar 1952	Near Anchieta, Brazil	Estrada de Ferro Central do Brasil	Derailment, broken rail, collision	119
7 Mar 1952	Near Kompong Lat, Cambodia, French Indochina		Derailment, sabotage, train fell from bridge	8
13 Mar 1952	Gabes, Tunisia		Bomb exploded on train	5
30 Mar 1952	Near Stvolinky, Czechoslovakia		Head-on collision, single line	12
31 Mar 1952	Near Caxias, Portugal	Caminhos de Ferro Portugueses	Train truck by landslide	10
c26 Apr 1952	B Aranya Prades and Bangkok, Thailand		Derailment, sabotage	?
7 May 1952	Berala, NSW, Australia		Rear collision in fog	10
17 May 1952	B Ulukiska and Adana, Turkey		Derailment? Collision?	32
18 May 1952	B Palana and Bikaner, India	Northern	Head-on collision, single line	44
20 May 1952	B Padrón and La Esclavitud, Spain	Red Nacional de los Ferrocarriles Españoles	Runaway waggons, collision, fire	8
2 Jun 1952	Near Banfora, Upper Volta, French West Africa	Régie des chemins de fer Abidjan-Niger	Derailment, sabotage	7
10 Jun 1952	B Trang Bom and An Loc, Vietnam, French Indochina		Derailment, sabotage	12
20 Jun 1952	'80 Miles North of Rangoon,' Burma		Derailment, sabotage	7
24 Jun 1952	Near Sagari, South Korea		Derailment, sabotage, fire	70
9 Jul 1952	B Boczow and Rzepin, Poland		Derailment, faulty train makeup	160
26 Jul 1952	B Malangbong and Warungbandrek, Indonesia		Derailment, sabotage	8

6 Aug 1952	Drownino. USSR		Collision with horse, derailment	109
24 Aug 1952	Near Pegu, Burma		Derailment, sabotage	?
26 Aug 1952	Suchdol Nad Odrou, Czechoslovakia		Derailment	12
17 Sep 1952	Oryu-Dong, South Korea		Boiler explosion, derailment	20
18 Sep 1952	Near Hamburg Hbf, West Germany		Derailment, broken rail	8
7 Oct 1952	Seadong, South Korea		Rear collision	5
8 Oct 1952	Harrow and Wealdstone, England	British Railways, London Midland Region	Double collision	108
13 Oct 1952	Jhatpat, West Pakistan		Collision with shunting goods	7
15 Oct 1952	B Ogunshile and Illugun, Nigeria		Runaway breakdown crane, collision	50
20 Oct 1952	Heavitree, South Africa	South African Railways	Derailment, excessive speed?	23
30 Oct 1952	B Furstenstein and Nammering, West Germany		Runaway waggon, collision	7
4 Nov 1952	Basse-Indre, France		Collision with lorry on FLB level crossing	5
10 Nov 1952	Htongyi, Burma		Derailment, sabotage	8
13 Dec 1952	Barra Mansa, Brazil		Train struck by rockfall	6
21 Dec 1952	Dobrinje, Yugoslavia		Collision	13
23 Dec 1952	Near Verona, Italy		Derailment, inadequate protection of PW work, fire	6
25 Dec 1952	Chikjisa, South Korea		Fire	7
c29 Dec 1952	B Saigon and Lon Ninh, Vietnam, French Indochina		Train derailed and fired on by Vietminh guerrillas	36

2 Jan 1953	Near Iwon, South Korea		Derailment, train fell from bridge	29
2 Jan 1953	Barraganetal, Ecuador		Runaway waggon, collision	10
15 Feb 1953	Benevento, Italy	Fellovie dello Stato	Derailment, excessive speed	23
15 Feb 1953	Punta Piedras, Argentina		Derailment, excessive speed, motorman intoxicated	36
26 Feb 1953	Gaya, India		Fire	5
23 Mar 1953	Clemente Falcão, Brazil	Estrada de Ferro Central do Brasil	Double collision	13
27 Mar 1953	Near Conneaut, OH, USA	New York Central	Derailment, track defects (faulty loading), double collision	17
8 Apr 1953	Near Stratford, England	London Transport	Rear collision in tunnel	12
8 May 1953	Near Quang Tri, Vietnam, French Indochina		Derailment, sabotage	19
20 May 1953	'50 Miles East of Rangoon,' Burma		Derailment, sabotage	5
13 Jun 1953	B Battulapuram and Mulucalacheruvu, India	Southern	Head-on collision, single line	67
16 Jun 1953	B Katihar West and Semapur, Bihar, India	North Eastern	Head-on collision, single line	6
21 June 1953	B Jogjakarta and Bandung, Indonesia		Derailment, sabotage	11
22 Jun 1953	Col des Nuages, Vietnam, French Indochina		Train mined while crossing viaduct	35
c10 Jul 1953	B Toruń and Poznań, Poland		Derailment, sabotage	?
25 Jul 1953	B Monistrol de Monserrat and Montserrat Monastery, Spain	Cremallera de Montserrat	Runaway train, collisions (rack railway)	8
9 Aug 1953	Samsong, South Korea		Rear collision?	15
14 Aug	Buenos Aires		Collision in fog	5

1953	Presidente Perón, Argentina			
15 Aug 1953	Irk Valley Junction, England	British Railways, London Midland Region	Sidelong collision, part of train fell from viaduct	9
16 Oct 1953	Sérézin-du-Rhône, France	Société nationale des chemins de fer français	Rear collision in fog	12
7 Nov 1953	Near Rabat, French Morocco		Bombs exploded on train	7
17 Nov 1953	Near Zibingyi, Burma		Derailment, sabotage	15
19 Dec 1953	Sydenham, NSW, Australia	New South Wales Government	Rear collision	5
24 Dec 1953	Tangiwai (Waiouru), New Zealand	New Zealand Railways	Bridge collapsed under train	149
24 Dec 1953	Sakvice, Czechoslovakia		Rear collision	133
2 Jan 1954	Near Raxaul, Bihar, India		Collision	8
4 Jan 1954	Near Bhatinda, Punjab, India		Derailment, part of train fell from bridge	15
20 Jan 1954	Pozanti, Turkey		Collision	18
21 Jan 1954	B Jhampir and Braudabad, West Pakistan	North Western	Derailment, collision, fire	60
26 Jan 1954	Sątopy Samulewo, Poland		Head-on collision	18
31 Jan 1954	Near Suwon, South Korea	Korean National Railways	Collision with lorry on open? Level crossing	63
31 Jan 1954	B Hanoi and Haiphong, Vietnam		Derailment, sabotage	?
24 Feb 1954	Swebodzice, Poland		Collision	10
28 Feb 1954	B Mi-Sahage and Luchon, France	Société des chemins de fer et hôtel de montagne aux	Runaway train, derailments (rack railway)	6

		Pyrénées		
30 Mar 1954	B Kottbus and Guben, East Germany		Collision	?
31 Mar 1954	Jagatbela, Uttar Pradesh, India		Explosion	31
3 Apr 1954	Pomucho, Mexico		Derailment	11
11 Apr 1954	Near Kraing Skea, Cambodia		Derailment, sabotage, fire	?
12 Apr 1954	B Haiphong and Hanoi, Vietnam		Derailment, sabotage	?
25 Jun 1954	Near Zacapa, Guatemala		Train machine-gunned by aircraft	7
3 Jul 1954	Châteaubourg, France		Head-on collision	35
3 Aug 1954	Near Humbria, Portuguese West Africa		Derailment?	24
2 Sep 1954	B Fabrica and Sagay, the Philippines	Insular Lumber	Collision between portions of divided train, part of train fell from bridge	82
6 Sep 1954	Rörvik, Sweden		Derailment, excessive speed	6
9 Sep 1954	Engenheiro Clemente Falcåo, Brazil	Estrada de Ferro Central do Brasil	Collision in fog	12
13 Sep 1954	Sabola and Pereiro, Portugal	Caminhos de Ferro Portugueses	Derailment, track and engine defect, high speed	27
27 Sep 1954	B Jangaon and Raghunathpalli, Hyderabad, India	Central	Bridge collapsed under train	136
13 Oct 1954	Elst, the Netherlands	Nederlandse Spoorwegen	Train diverted onto wrong line, collision	6
24 Oct 1954	Fürstenwalde, East Germany		Train diverted onto wrong line, double collision	?
20 Nov 1954	'150 Km south of Shanghai,' China		Derailment, sabotage? Explosion	?
2 Dec 1954	Wilsele, Belgium	Société nationale des chemins de fer belges	Derailment, excessive speed	19

17 Dec 1954	B Le Kroub and Rahmoun, Algeria		Head-on collision, single line? Fire	15
18 Dec 1954	Dortmund Hbf, West Germany		Train diverted onto wrong line, collision	15
9 Jan 1955	Near São Cristãvel (Aracaju), Brazil	Viação Férrea Federal Leste Brasileiro	Derailment, defective vehicle? Excessive speed?	9
22 Jan 1955	Meknès, French Morocco		Derailment, broken axle	7
23 Jan 1955	Sutton Coldfield, England	British Railways, London Midland Region	Derailment, excessive speed	15
21 Feb 1955	'128 Miles South of Rangoon.' Burma		Derailment, sabotage	
26 Feb 1955	Near Plauen, East Germany		Derailment, engine defect?	70
1 Mar 1955	B Itapevi and Amador, Brazil	Estrada de Ferro Sorocabana	Derailment, high speed with SWB vehicles?	11
2 Mar 1955	Pusan, South Korea		Fire	39
14 Mar 1955	Myslowice, Poland		Derailment, sabotage	6
25 Mar 1955	B Sagaihn and Monywa, Burma		Derailment, sabotage	20
3 Apr 1955	Near Alsaba, Mexico	Ferrocarril Nacional de México	Derailment, bridge collapsed under train	13
26 Apr 1955	Deodoro, Brazil	Estrada de Ferro Central do Brasil	Sidelong collision	7
12 May 1955	B Tjiamis and Manondjaja, West Java, Indonesia		Derailment, excessive speed?	36
22 May 1955	Sidi Yahia du Gharb, French Morocco		Derailment	50
30 May 1955	Near Medan Ara, Indonesia		Train fell from bridge, sabotage	?

16 Jun 1955	Vicario Geral, Brazil	Estrada de Ferro Leopoldina	Double collision	6
17 Jul 1955	San Bernando, Chile	Empresa de los Ferrocarriles de Estado	Rear collision in fog	45
24 Jul 1955	B Rawicz and Wroclaw, Poland		Collision	9
16 Aug 1955	'South of Tehran,' Iran		Collision	12
23 Aug 1955	Ciudadela, Argentina		Rear collision in fog	21
31 Aug 1955	Near Managua, Nicaragua		Collision	6
11 Sep 1955	B St Martin-Sittich and Mautbrücken, Austria		Head-on collision, single line	6
23 Sep 1955	Near Gomez Palacio, Mexico		Collision with lorries on open level crossings, explosions	?
20 Nov 1955	Milton, England	British Railways, Western Region	Derailment, excessive speed	11
2 Dec 1955	Barnes, England	British Railways, Southern Region	Rear collision	11
24 Dec 1955	B São Silvestre and Guararema, Brazil	Estrada de Ferro Central do Brasil	Runaway waggons, collision, fire	15
7 Jan 1956	B Katihar West and Semapur, Bihar, India	North Eastern	Head-on collision, single line	11
13 Jan 1956	Near Ställdalen, Sweden	Statens Järnväger	Runaway train, collision	20
22 Jan 1956	Redondo Junction (Los Angeles), CA, USA	Atchison Topeka and Santa Fe	Derailment, excessive speed	30
14 Feb 1956	Near Cerrillos, Chile		Rear collision	23
25 Feb 1956	Bornitz, East Germany		Sidelong collision in fog	48

28 Feb 1956	Swampscott Junction, MA, USA	Boston and Maine	Rear collision in blizzard	11
22 Mar 1956	Near Maghagha, Egypt		Bridge collapsed under train	8
28 Mar 1956	B Akkavare and Arvidsjaur, Sweden	Statens Järnväger	Head-on collision, single line	15
28 Apr 1956	Tanta, Egypt		Collision with taxi on open level crossing, derailment	?
19 May 1956	B Hadmatiya and Jamwanehali, Saurastra, India	Western	Derailment	7
20 May 1956	Near Shepetovka, Ukraine, USSR		Derailment, sabotage	?
14 Jun 1956	Fismes, France	Société nationale des chemins de fer français	Derailment, defective vehicle? Track defects?	11
23 Jun 1956	Santa Catalina, Spain	Red Nacional de los Ferrocarriles Españoles	Rear collision	9
30 Jun 1956	Near Oaxaca, Mexico	Ferrocarril Nacional de México	Derailment, excessive speed	15
15 Jul 1956	Near Biskra, Algeria		Derailment, sabotage, fire	8
20 Jul 1956	B Palon and Thametchaung, Burma		Train derailed and fired on by terrorists	35
1 Aug 1956	Barra do Pirai, Brazil	Estrada de Ferro Central do Brasil	Rear collision	17
7 Aug 1956	Přerov, Moravia, Czechoslovakia		Rear collision?	9
2 Sep 1956	B Jadcherla and Mahbubnager, India	Central	Bridge collapsed under train	119
5 Sep 1956	Robinson (Springer), NM, USA	Atchison Topeka and Santa Fe	Train diverted onto wrong line, collision	15
6 Sep 1956	Near Kambalu, Burma		Derailment, sabotage	?

27 Sep 1956	Near Nagoya, Japan		Derailment, train struck by landslide	17
13 Oct 1956	B Weaversdown Halt and Liss Forest Road, England	Longmoor Military	Head-on collision, single line	6
15 Oct 1956	Rokken, Japan	Japanese National	Buffer stop collision at end of trap siding?, collision	38
5 Nov 1956	Wilson, Chicago, IL, USA	Chicago Transit Authority	Rear collision	8
23 Nov 1956	B Ariyalur and Kallagam, Tamil Nadu, India	Southern	Derailment, track washed away, train fell from bridge	151
1 Dec 1956	Wallumbilla, QD, Australia	Queensland Railways	Head-on collision	5
11 Dec 1956	Colomb-Bécher, Algeria		Derailment, sabotage	8
16 Dec 1956	Taguasco and Jatibonico, Cuba		Train diverted onto wrong line, collision, sabotage?	19
21 Dec 1956	Ludwigshafen, West Germany		Collision with crane	5
28 Dec 1956	Near Quereño, Spain	Red Nacional de los Ferrocarriles Españoles	Derailment, broken rail	5
2 Jan 1957	B Oran and Relizane, Algeria		Train derailed and machine-gunned by terrorists	6
25 Feb 1957	Alta Córdoba, Argentina		Derailment, excessive speed?	9
4 Mar 1957	Near Pyinmana, Burma		Derailment, sabotage	30
15 Mar 1957	B Kuurila and Iittala, Finland		Head-on collision, single line in blizzard	24
7 Apr 1957	Magno and Turiaçu, Brazil	Rede Ferroviária Federal	Rear collision in fog	18
8 Apr 1957	Woodstock, South Africa	South African Railways	Collision	20

28 Apr 1957	B Brumov and Bylnice, Czechoslovakia		Head-on collision, single line	10
2 Jun 1957	B Sewri and Vadala Road, India	Central	Rear collision	23
5 Jun 1957	Near Cortazar, Mexico		Collision	20
18 Jun 1957	Boisleux-au-Mont, France	Société nationale des chemins de fer français	Derailment, broken axle? Track distorted by the heat? Track subsidence?	11
20 Jun 1957	B Indaiaçu and Poço D'Antas, Brazil	Rede Ferroviária Federal	Derailment	?
19 Jul 1957	Bollène, France	Société nationale des chemins de fer français	Derailment, excessive speed	30
4 Aug 1957	Villaverde, Spain	Red Nacional de los Ferrocarriles Españoles	Collision with light engine	17
12 Aug 1957	Near Woensel, the Netherlands	Nederlandse Spoorwegen	Sidelong collision	5
20 Aug 1957	Near Bahia Blanca, Argentina		Derailment	6
1 Sep 1957	Near Kendal, Jamaica	Jamaica Government	Runaway train, derailment	186
7 Sep 1957	Nozieres-Brignon, France	Société nationale des chemins de fer français	Derailment, excessive speed over track under repair	27
11 Sep 1957	Vrbátky, Czechoslovakia		Train diverted onto wrong line, collision	15
29 Sep 1957	Gambar, West Pakistan	Pakistan	Train diverted onto wrong line, collision, fire	250
29 Sep 1957	B Lalupon and Odo-Oba, Nigeria		Derailment, track washed away	68
1 Oct 1957	Stuttgart, Bad Canstatt, West Germany		Rear collision in fog	10
20 Oct 1957	B Ispartakule and Yarimburgaz Thrace, Turkey	Türkiye Cumhuriyeti Devlet Demiryollan	Head-on collision, single line, fire	90

23 Oct 1957	Near Fortin Uno, Argentina		Derailment, broken axle	12
28 Oct 1957	'North of Mandalay,' Burma		Derailment, earthquake	10
Oct 1957	Near Zalaszentivan, Hungary		Derailment, sabotage?	30
16 Nov 1957	Near Chantonnay, France	Société nationale des chemins de fer français	Head-on collision, single line, fire	29
23 Nov 1957	B Padli and Asvali, India	Central	Derailment, sabotage	9
4 Dec 1957	St Johns, Lewisham, England	British Railways, Southern Region	Rear collision in fog, overbridge collapsed onto train	89
5 Dec 1957	Near Zagreb, Yugoslavia		Collision, fire	5
9 Dec 1957	Near Taoyuan, Formosa		Derailment, vandalism	17
9 Dec 1957	Codogno, Italy	Fellovie dello Stato	Collision with lorry on FLB level crossing, derailment	14
31 Dec 1957	Otwock, Poland		Rear collision	6
1 Jan 1958	Mohri, India	Northern	Head-on collision in fog	32
10 Jan 1958	Palermo, Argentina		Derailment, collision, fire	6
26 Jan 1958	B El Kantara and Biskra, Algeria		Derailment, sabotage	6
30 Jan 1958	Dagenham East, England	British Railways, Eastern Region	Rear collision in fog	10
26 Feb 1958	Sonanpur, India		Sidelong collision	5
7 Mar 1958	Near Paciência, Brazil	Rede Ferroviária Federal	Rear collision	54

20 Mar 1958	Near Htangon, Burma		Derailment, sabotage	15
8 May 1958	B Mangueira and Triagem, Brazil	Rede Ferroviária Federal	Head-on collision, single line	112
15 May 1958	Leipzig Hbf, East Germany	Deutsche Bundesbahn	Train diverted onto wrong line, head-on collision	53
21 May 1958	Chamraj, Karnataka, India	Western	Derailment	30
22 Jun 1958	B Botas and Maltrata, Mexico	Ferrocarril Mexicano	Runaway train, derailment	10
26 Jun 1958	Near New Canada Junction, South Africa		Derailment	6
18 Jul 1958	Vilvoorde, Belgium	Société nationale des chemins de fer belges	Derailment, track distorted by the heat	5
5 Sep 1958	B Grycksbo and Sågmyra, Sweden	Statens Järnvägar	Head-on collision, single line	7
14 Sep 1958	Near Drachenfels, West Germany	Drachenfelsbahn	Runaway train, derailment (rack railway)	17
15 Sep 1958	Newark Bay Bridge, NJ, USA	Central of New Jersey	Derailment on derail, train fell from open drawbridge	44
24 Sep 1958	B Nettersheim & Gronrechtsmühle Block Post, West Germany		Head-on collision, single line	7
8 Nov 1958	Palermo, Argentina	Empresa Ferrocarriles del Estado Argentino	Rear collision, fire	22
10 Nov 1958	'30 km from Blida,' Algeria		Derailment, sabotage	6
14 Nov 1958	Near Cairo, Egypt		Collision	7
19 Nov 1958	Laksham, East Pakistan		Head-on collision	16
13 Dec 1958	Near Diogo Siqueira, Brazil	Rede Ferroviária Federal	Derailment, excessive speed?	20
5 Mar 1959	Xometla, Mexico		Rear collision	11

28 May 1959	Near Tasikmalaja, Java, Indonesia		Runaway coaches, derailment	140
2 Jun 1959	Minino, USSR		Rear collision, fire	56
5 Jun 1959	B Engenheiro Goulart and Engenheiro Trindade, Brazil	Rede Ferroviária Federal	Head-on collision, single line	55
16 Jun 1959	B Vieira Fazanda and Maria da Graça, Brazil	Rede Ferroviária Federal	Rear collision	5
6 Jul 1959	Near Sivas, Turkey		Head-on collision, single line?	10
24 Jul 1959	Near Chittagong, East Pakistan		Collision	5
24 Aug 1959	Morés, Spain	Red Nacional de los Ferrocarriles Españoles	Train diverted onto wrong line, head-on collision	5
27 Aug 1959	Near Banja Luka, Yugoslavia		Derailment	14
11 Oct 1959	Near Osato Japan		Collision with pickup truck on open level crossing, derailment	7
24 Oct 1959	Near Parkland, AB, Canada	Canadian Pacific	Collision with petrol tanker on open level crossing, derailment, explosion, fire	5
24 Dec 1959	Near Rotifunk, Sierra Leone		Derailment	13
5 Jan 1960	Near Monza, Italy	Fellovie dello Stato	Derailment, excessive speed	14
6 Jan 1960	Near Tlanalapan, Mexico		Collision	7
10 Jan 1960	Near Donwun, Burma		Train derailed by Karen rebels	13
21 Jan 1960	Near Settle, England	British Railways, London Midland Region	Derailment, engine defect, collision	5
1 Feb 1960	Near Devon, Transvaal, South Africa		Rear collision	15
7 Feb 1960	Near Sewell, Chile		Derailment, excessive speed?	33
1 Mar 1960	Rosedale (Bakersfield), CA, USA	Atchison Topeka and Santa Fe	Collision with oil tanker on open level crossing, derailment, fire	11

3 Jul 1960	B Taxila and Haveli, West Pakistan		Derailment, part of train fell from bridge	14
9 Jul 1960	Near Logan, NM, USA	Chicago Rock Island and Pacific	Trestle collapsed under train, fire	5
20 Jul 1960	'North East of Saigon,' South Vietnam		Collision between portions of divided train	13
28 July 1960	Near Camaguey, Cuba		Rear collision	7
25 Aug 1960	Martin Brod, Bosnia, Yugoslavia		Collision	6
14 Nov 1960	Near Stéblová, Czechoslovakia		Head-on collision, single line in fog, fire	116

ABOUT THE AUTHOR

The author has been interested in railways for many years, ever since he was invited up to a signal box at the age of 14 back in 1966. Not long afterwards he 'cabbed' the engine of a goods train which was shunted out of the way of a faster train and its driver showed him a copy of *Red for Danger*, the late L.T.C. Rolt's book about British railway accidents. This was the first railway book that he ever bought, several months saved up pocket money going towards a paperback copy of it. Later he came across a copy of *A History of Railroad Accidents, Safety Precautions and Operating Practices* by Robert B. Shaw, dealing with American railroad accidents, and his interest in the subject grew. When the local signal boxes were all done away with in a modernisation scheme in the mid-1980s he began a more detailed study of world railway accidents. There was no internet back then of course and many hours were spent in the local reference library slogging through bound volumes of *The Times*, writing out longhand in large notebooks details of every accident found. Later it was possible to view the paper on microfilm and print pages, which made things a little easier, and with the internet the amount of information available vastly increased. Some years ago, the author contributed to *Railway Disasters of the World* by P.W.B. Semmens (which deals only with 20th-century accidents) and more recently the idea came of writing his own book or books on world accidents, starting at the beginning. This book is the sixth book of the series, with previous volumes covering different periods from 1811 to 1945. Further volumes will hopefully follow.

Made in the USA
Middletown, DE
05 November 2026

36587687R00451